T0142049

# Communications in Computer and Information Science 1469

More information about this series at http://www.springer.com/series/7899

Qinglong Han · Sean McLoone ·
Chen Peng · Baolin Zhang (Eds.)

# Intelligent Equipment, Robots, and Vehicles

7th International Conference on Life System Modeling and Simulation, LSMS 2021
and 7th International Conference on Intelligent Computing for Sustainable Energy and Environment, ICSEE 2021
Hangzhou, China, October 30 – November 1, 2021
Proceedings, Part III

Springer

*Editors*
Qinglong Han
Swinburne University of Technology
Melbourne, VIC, Australia

Chen Peng
Shanghai University
Shanghai, China

Sean McLoone
Queen's University Belfast
Belfast, UK

Baolin Zhang
Qingdao University of Science
and Technology
Qingdao, China

ISSN 1865-0929 ISSN 1865-0937 (electronic)
Communications in Computer and Information Science
ISBN 978-981-16-7212-5 ISBN 978-981-16-7213-2 (eBook)
https://doi.org/10.1007/978-981-16-7213-2



This Springer imprint is published by the registered company Springer Nature Singapore Pte Ltd.
The registered company address is: 152 Beach Road, #21-01/04 Gateway East, Singapore 189721, Singapore

# Preface

This book series constitutes the proceedings of the 2021 International Conference on Life System Modeling and Simulation (LSMS 2021) and the 2021 International Conference on Intelligent Computing for Sustainable Energy and Environment (ICSEE 2021), which were held during October 30 – November 1, 2021, in Hangzhou, China. The LSMS and ICSEE international conference series aim to bring together international researchers and practitioners in the fields of advanced methods for life system modeling and simulation, advanced intelligent computing theory and methodologies, and engineering applications for achieving net zero across all sectors to meet the global climate change challenge. These events are built upon the success of previous LSMS conferences held in Shanghai, Wuxi, and Nanjing in 2004, 2007, 2010, 2014, and 2017, and ICSEE conferences held in Wuxi, Shanghai, Nanjing, and Chongqing in 2010, 2014, 2017, and 2018, respectively, and are based on large-scale UK-China collaboration projects on sustainable energy. Due to the COVID-19 pandemic situation, the themed workshops as part of these two conferences were organized online in 2020.

At LSMS 2021 and ICSEE 2021, technical exchanges within the research community took the form of keynote speeches and panel discussions, as well as oral and poster presentations. The LSMS 2021 and ICSEE 2021 conferences received over 430 submissions from authors in 11 countries and regions. All papers went through a rigorous peer review procedure and each paper received at least three review reports. Based on the review reports, the Program Committee finally selected 159 high-quality papers for presentation at LSMS 2021 and ICSEE 2021. These papers cover 18 topics and are included in three volumes of CCIS proceedings published by Springer. This volume of CCIS includes 79 papers covering eight relevant topics.

The organizers of LSMS 2021 and ICSEE 2021 would like to acknowledge the enormous contribution of the Program Committee and the referees for their efforts in reviewing and soliciting the papers, and the Publication Committee for their editorial work. We would also like to thank the editorial team from Springer for their support and guidance. Particular thanks go to all the authors, without their high-quality submissions and presentations the conferences would not have been successful.

Finally, we would like to express our gratitude to our sponsors and organizers, listed on the following pages.

October 2021

Minrui Fei
Kang Li
Qinglong Han

# Organization

## Honorary Chairs

| | |
|---|---|
| Wang, XiaoFan | Shanghai University, China |
| Umezu, Mitsuo | Waseda University, Japan |

## General Chairs

| | |
|---|---|
| Fei, Minrui | Shanghai University, China |
| Li, Kang | University of Leeds, UK |
| Han, Qing-Long | Swinburne University of Technology, Australia |

## International Program Committee

### Chairs

| | |
|---|---|
| Ma, Shiwei | China Simulation Federation, China |
| Coombs, Tim | University of Cambridge, UK |
| Peng, Chen | Shanghai University, China |
| Chen, Luonan | University of Tokyo, Japan |
| Sun, Jian | China Jiliang University, China |
| McLoone, Sean | Queen's University Belfast, UK |
| Tian, Yuchu | Queensland University of Technology, Australia |
| He, Jinghan | Beijing Jiaotong University |
| Zhang, Baolin | Qingdao University of Science and Technology, China |

### Local Chairs

| | |
|---|---|
| Aleksandar Rakić | University of Belgrade, Serbia |
| Athanasopoulos, Nikolaos | Queen's University Belfast, UK |
| Cheng, Long | Institute of Automation, Chinese Academy of Sciences, China |
| Dehghan, Shahab | Imperial College London, UK |
| Ding, Jingliang | Northeastern University, China |
| Ding, Ke | Jiangxi University of Finance and Economics, China |
| Duan, Lunbo | Southeast University, China |
| Fang, Qing | Yamagata University, Japan |
| Feng, Wei | Shenzhen Institute of Advanced Technology, Chinese Academy of Sciences, China |
| Fridman, Emilia | Tel Aviv University, Israel |
| Gao, Shangce | University of Toyama, Japan |
| Ge, Xiao-Hua | Swinburne University of Technology, Australia |
| Gupta M. M. | University of Saskatchewan, Canada |

**Members**

| | |
|---|---|
| Huang, Deqing | Southwest Jiaotong University, China |
| Jahromi, Amir Abiri | University of Leeds, UK |
| Jiang, Lin | University of Liverpool, UK |
| Jiang, Ming | Anhui Polytechnic University, China |
| Kong, Jiangxu | China Jiliang University, China |
| Li, MingLi | China Jiliang University, China |
| Li, Chuanfeng | Luoyang Institute of Science and Technology, China |
| Li, Chuanjiang | Harbin Institute of Technology, China |
| Li, Donghai | Tsinghua University, China |
| Li, Tongtao | Henan University of Technology, China |
| Li, Xiang | University of Leeds, UK |
| Li, Xiaoou | CINVESTAV-IPN, Mexico |
| Li, Xin | Shanghai University, China |
| Li, Zukui | University of Alberta, Canada |
| Liu Jinfeng | University of Alberta, Canada |
| Liu, Kailong | University of Warwick, UK |
| Liu, Mandan | East China University of Science and Technology, China |
| Liu, Tingzhang | Shanghai University, China |
| Liu, Xueyi | China Jiliang University, China |
| Liu, Yang | Harbin Institute of Technology, China |
| Long, Teng | University of Cambridge, UK |
| Luo, Minxia | China Jiliang University, China |
| Ma, Hongjun | Northeastern University, China |
| Ma, Yue | Beijing Institute of Technology, China |
| Menhas, Muhammad Ilyas | Mirpur University of Science and Technology, Pakistan |
| Naeem, Wasif | Queen's University Belfast, UK |
| Nie, Shengdong | University of Shanghai for Science and Technology, China |
| Niu, Qun | Shanghai University, China |
| Pan, Hui | Shanghai University of Electric Power, China |
| Qian, Hong | Shanghai University of Electric Power, China |
| Ren, Xiaoqiang | Shanghai University, China |
| Rong, Qiguo | Peking University, China |
| Song, Shiji | Tsinghua University, China |
| Song, Yang | Shanghai University, China |
| Sun, Qin | Shanghai University, China |
| Sun, Xin | Shanghai University, China |
| Sun, Zhiqiang | East China University of Science and Technology, China |
| Teng, Fei | Imperial College London, UK |
| Teng, Huaqiang | Shanghai Instrument Research Institute, China |
| Tian, Zhongbei | University of Birmingham, UK |
| Tu, Xiaowei | Shanghai University, China |
| Wang, Binrui | China Jiliang University, China |
| Wang, Qin | China Jiliang University, China |

| | |
|---|---|
| Wang, Liangyong | Northeast University, China |
| Wang, Ling | Shanghai University, China |
| Wang, Yan | Jiangnan University, China |
| Wang, Yanxia | Beijing University of Technology, China |
| Wang, Yikang | China Jiliang University, China |
| Wang, Yulong | Shanghai University, China |
| Wei, Dong | China Jiliang University, China |
| Wei, Li | China Jiliang University, China |
| Wei, Lisheng | Anhui Polytechnic University, China |
| Wu, Fei | Nanjing University of Posts and Telecommunications, China |
| Wu, Jianguo | Nantong University, China |
| Wu, Jiao | China Jiliang University, China |
| Wu, Peng | University of Jinan, China |
| Xu, Peng | China Jiliang University, China |
| Xu Suan | China Jiliang University, China |
| Xu, Tao | University of Jinan, China |
| Xu, Xiandong | Cardiff University, UK |
| Yan, Huaicheng | East China University of Science and Technology, China |
| Yang, Aolei | Shanghai University, China |
| Yang, Banghua | Shanghai University, China |
| Yang, Wenqiang | Henan Normal University, China |
| Yang, Zhile | Shenzhen Institute of Advanced Technology, Chinese Academy of Sciences, China |
| Ye, Dan | Northeastern University, China |
| You, Keyou | Tsinghua University, China |
| Yu, Ansheng | Shanghai Shuguang Hospital, China |
| Zan, Peng | Shanghai University, China |
| Zeng, Xiaojun | University of Manchester, UK |
| Zhang, Chen | Coventry University, UK |
| Zhang, Dawei | Shandong University, China |
| Zhang Xiao-Yu | Beijing Forestry University |
| Zhang, Huifeng | Nanjing Post and Communication University, China |
| Zhang, Kun | Nantong University, China |
| Zhang, Li | University of Leeds, UK |
| Zhang, Lidong | Northeast Electric Power University, China |
| Zhang, Long | University of Manchester, UK |
| Zhang, Yanhui | Shenzhen Institute of Advanced Technology, Chinese Academy of Sciences, China |
| Zhao, Chengye | China Jiliang University, China |
| Zhao, Jianwei | China Jiliang University, China |
| Zhao, Wanqing | Manchester Metropolitan University, UK |
| Zhao, Xingang | Shenyang Institute of Automation Chinese Academy of Sciences, China |
| Zheng, Min | Shanghai University, China |

| | |
|---|---|
| Zhou, Bowen | Northeast University, China |
| Zhou, Huiyu | University of Leicester, UK |
| Zhou, Peng | Shanghai University, China |
| Zhou, Wenju | Ludong University, China |
| Zhou, Zhenghua | China Jiliang University, China |
| Zhu, Jianhong | Nantong University, China |

## Organization Committee

### Chairs

| | |
|---|---|
| Qian, Lijuan | China Jiliang University, China |
| Li, Ni | Beihang University, China |
| Li, Xin | Shanghai University, China |
| Sadegh, Azizi | University of Leeds, UK |
| Zhang, Xian-Ming | Swinburne University of Technology, Australia |
| Trautmann, Toralf | Centre for Applied Research and Technology, Germany |

### Members

| | |
|---|---|
| Chen, Zhi | China Jiliang University, China |
| Du, Dajun | Shanghai University, China |
| Song, Yang | Shanghai University, China |
| Sun, Xin | Shanghai University, China |
| Sun, Qing | Shanghai University, China |
| Wang, Yulong | Shanghai University, China |
| Zheng, Min | Shanghai University, China |
| Zhou, Peng | Shanghai University, China |
| Zhang, Kun | Shanghai University, China |

### Special Session Chairs

| | |
|---|---|
| Wang, Ling | Shanghai University, China |
| Meng, Fanlin | University of Essex, UK |
| Chen, Wanmi | Shanghai University, China |
| Li, Ruijiao | Fudan University, China |
| Yang, Zhile | SIAT, Chinese Academy of Sciences, China |

### Publication Chairs

| | |
|---|---|
| Niu, Qun | Shanghai University, China |
| Zhou, Huiyu | University of Leicester, UK |

### Publicity Chair

| | |
|---|---|
| Yang, Erfu | University of Strathclyde, UK |

### Registration Chairs

| | |
|---|---|
| Song, Yang | Shanghai University, China |
| Liu, Kailong | University of Warwick, UK |

### Secretary-generals

| | |
|---|---|
| Sun, Xin | Shanghai University, China |
| Gan, Shaojun | Beijing University of Technology, China |

## Sponsors

China Simulation Federation (CSF)
China Instrument and Control Society (CIS)
IEEE Systems, Man and Cybernetics Society Technical Committee on Systems Biology
IEEE CC Ireland Chapter

## Cooperating Organizations

Shanghai University, China
University of Leeds, UK
China Jiliang University, China
Swinburne University of Technology, Australia
Life System Modeling and Simulation Technical Committee of the CSF, China
Embedded Instrument and System Technical Committee of the China Instrument and Control Society, China

## Co-sponsors

Shanghai Association for System Simulation, China
Shanghai Instrument and Control Society, China
Zhejiang Association of Automation (ZJAA), China
Shanghai Association of Automation, China

## Supporting Organizations

Queen's University Belfast, UK
Nanjing University of Posts and Telecommunications, China
University of Jinan, China
University of Essex, UK
Queensland University of Technology, Australia
Central South University, China
Tsinghua University, China
Peking University, China
University of Hull, UK

Beijing Jiaotong University, China
Nantong University, China
Shenzhen Institute of Advanced Technology, Chinese Academy of Sciences, China
Shanghai Key Laboratory of Power Station Automation Technology, China
Complex Networked System Intelligent Measurement and Control Base, Ministry of Education, China
UK China University Consortium on Engineering Education and Research
Anhui Key Laboratory of Electric Drive and Control, China

# Contents – Part III

**Fuzzy, Neural, and Fuzzy-Neuro Hybrids**

## Intelligent Modelling, Monitoring, and Control



## Intelligent Manufacturing, Autonomous Systems, Intelligent Robotic Systems

# Intelligent Robots and Simulation

# Robotic Manpower Feedback Study on Lower Limb Rehabilitation

Wei Sun, Zhiyuan Guo, Haining Peng, Dong Zhang, and Li Li(✉)

School of Mechatronics Engineering and Automation,
Shanghai University, Shanghai 200072, China

**Abstract.** The application of robots in the field of lower extremity rehabilitation is becoming more and more extensive. Research on the real-time feedback of force and torque of lower extremity rehabilitation robots is very important to reduce the error during feedback. At present, multidimensional force sensors are mostly used for feedback, but the structure of multidimensional force sensors is complicated. This paper proposes to attach the resistance strain gauge to the structure of the lower limb rehabilitation robot to simplify the sensing structure and make the structure itself a sensing system. The traditional bridge used in the force sensor is redesigned, and the structure is analyzed with ANSYS. The strain changes with the length of the structure when the force is applied. Finally, MATLAB is used to simulate the output of the bridge, and an effective signal output is obtained.

**Keywords:** Lower limb rehabilitation robot · Bridge simulation · Strain gauge · ANSYS analysis

## 1 Introduction

With the development of national economy, the improvement of living standards, the aging population continues to increase, people pay more and more attention to for your health, the pursuit of more and more high level of medical care, an aging population, resulting in a decline in motor function and cerebral stroke, spinal cord injury, brain injury such as illness or accident resulting in a decline in exercise capacity and lower limb rehabilitation robot force feedback is the premise of safe rehabilitation, for rehabilitation robot detection is mainly rely on the force sensor to measure the force and moment, the force feedback in the robot involves multiple force and moment measurement, The multidimensional force sensor has complex structure and low precision.

In this paper, the resistance strain gage is applied to the lower limb rehabilitation structure. In the field of rehabilitation machine, the resistance strain gage is more maneuverable, and it is widely used in automatic testing and control technology. The resistance strain gauge in the resistance strain sensor has the strain effect of metal, that is, mechanical deformation is generated under the action of external force, so that the resistance value changes accordingly.


Q. Han et al. (Eds.): LSMS 2021/ICSEE 2021, CCIS 1469, pp. 3–10, 2021.
https://doi.org/10.1007/978-981-16-7213-2_1

## 2 Lower Limb Force Model

Lower limb rehabilitation robot is an auxiliary rehabilitation tool designed according to the walking of normal people. The force feedback should be analyzed according to the force of human walking. The human lower limb mainly has three joints, and the force on the thigh is shown in the figure during normal walking (Fig. 1).

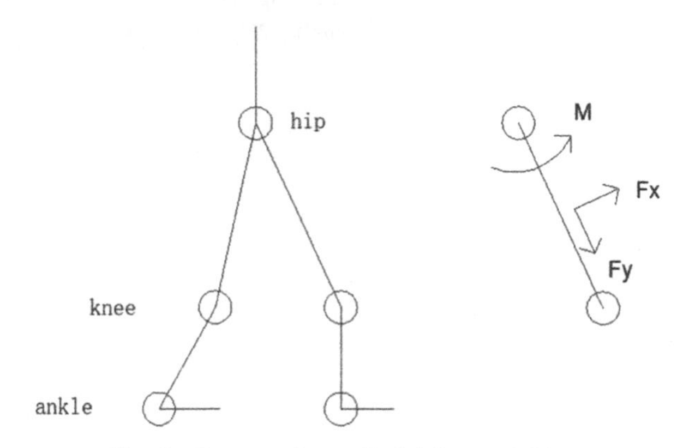


**Fig. 1.** Stress on lower limb joints and thighs

The structures of human lower limbs are connected by hip joint, knee joint and ankle joint. The movement of each joint is shown in the Table 1:

**Table 1.** Movement of lower limb joints

| Joint | Movement |
|---|---|
| Hip | Flexion and extension, adduction/abduction, pronation/pronation |
| Knee | Flexion and extension |
| Ankle | Pronation/pronation, dorsiflexion/plantarflexion, varus/valgus |

The most prominent motor dysfunction in stroke patients was the loss or limitation of walking ability, most patients are characterized by walking at a slower pace, asymmetry of time and space, gait, inefficient, family and social activity limitations on foot, safety walk is to decide whether or not a patient to continue the social activities and the key of the occupational activities, so for the force and moment in the direction of the detection is particularly important, the structure of mechanical legs are modified, make itself a sensing system with feedback ability, greatly enhance the support capability of safety walk.

# 3 Measurement Principle and Error Compensation

## 3.1 Stress Deformation Model

Dressed in patients with lower limb rehabilitation institutions, patients generally divided into no independent movement ability and has a certain capacity for independent movement of patients, no voluntary movement ability of mobile users under the mechanical legs, for patients with a certain sport ability, mechanical legs can provide a booster effect, when patients in power under the action of movement, once appear, resistance force, the lower mechanical legs sensing system detected, stop motion, to avoid additional damage on the patients, the force deformation caused by the structure of the micro structure of the strain gauge produces change, resistance strain gauge is converts the change of the strain to relative changes.

Only by converting the change in resistance into a change in voltage or current can the measurement be made with a circuit measuring instrument (Fig. 2).

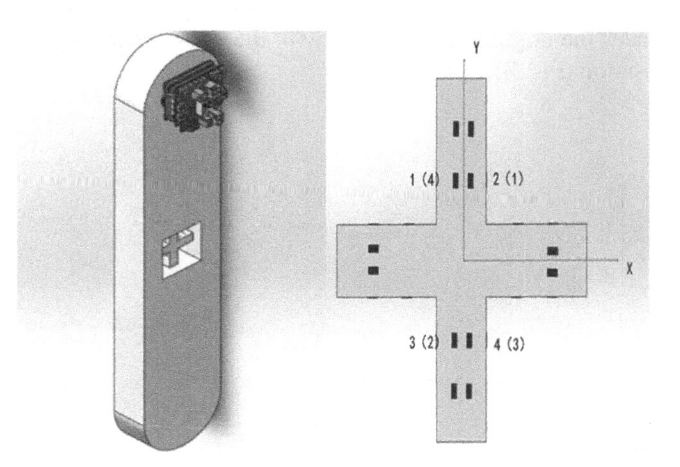


**Fig. 2.** Mechanical leg structure and cloth mode

A physical phenomenon in which the resistance value of a resistance wire changes with deformation.

Formula principle:

$$R = \rho \frac{L}{A}. \tag{1}$$

R is wire resistance; $\rho$ is the resistivity of the wire; L is the length of the wire;

A is the cross-sectional area of the wire. In order to obtain the relationship between deformation and resistance change, the above equation is processed to obtain.

$$\frac{\Delta R}{R} = K_0 \varepsilon. \tag{2}$$

$K_0$ is the sensitivity coefficient of a single wire, and its physical significance is: the ratio of the resistance change rate to the strain when the wire changes per unit length.

### 3.2 Temperature Compensation Circuit in Lower Limb Structure

The bridge composed of strain gauge is assembled in the structure of the mechanical leg. The micro-deformation of the structure causes strain, and the resistance value in the process of use is greatly affected by temperature. The resistance change caused by the change of ambient temperature is basically the same order of magnitude as the resistance change caused by the actual strain, thus resulting in a large measurement error.

The resistance value of the strain gage arranged in the structure of the lower limb rehabilitation robot is greatly affected by the ambient temperature (including the temperature of the structure) due to its material characteristics. There are two main factors that cause the change of resistance due to the change of ambient temperature: one is the resistance temperature coefficient of the resistance line of the strain gage; the other is that the linear expansion coefficient of resistance wire material is different from that of specimen material.

The temperature compensation methods of resistance strain gage mainly include bridge compensation and self-compensation of strain gage. The circuit introduced in this paper replaces the original compensator with the same strain gage on the basis of bridge compensation (Fig. 3).

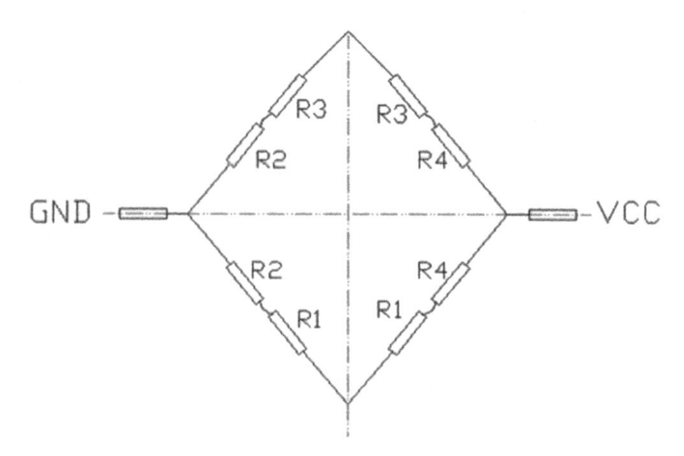


**Fig. 3.** Circuit

When the strain gauge is pasted on the surface of an elastic element, only the action of load and temperature is considered, the output strain value can be expressed as follows:

$$\varepsilon = \varepsilon_{\sigma} + \varepsilon_{t}. \tag{3}$$

$\varepsilon$ is the total strain value of the resistance strain gauge;$\varepsilon_{\sigma}$ is the strain value generated by the action of load; $\varepsilon_{t}$ is the strain value caused by temperature action, which is generally called false strain.

Using strain gauge and compensator is tie-in, the strain gauge on the specimen of measurement point 1 and 4, the same characteristics of strain gauge 1 and 2, 3 and 4 on the adjacent bridge arm, 2 and 3 on the relative bridge arm, at the same temperature field of the same characteristics of strain gauge, when specimens subjected to pressure and temperature changes, when resistance strain gauge 1, using the principle of "adjacent is presupposed, relative addition" can use the bridge circuit compensation method for temperature compensation.

When the specimen is subjected to pressure and temperature changes, the resistance change rate of strain gage $R_1$:

$$\frac{\Delta R_1}{R_1} = (\frac{\Delta R_1}{R_1})_\varepsilon + (\frac{\Delta R_1}{R_1})_t. \tag{4}$$

Where, $(\frac{\Delta R_1}{R_1})_\varepsilon$ is the resistance change rate of resistance strain gage $R_1$ caused by strain. $(\frac{\Delta R_1}{R_1})_t$ is the resistance change rate of strain gage $R_1$ due to temperature.

It can be seen from the expression that the resistance change rate caused by temperature is subtracted, which eliminates the influence caused by temperature change and achieves the purpose of temperature compensation.

### 3.3 The Initial Imbalance is Set to Zero

Under ideal conditions, when the input is zero, the output is also zero. The circuit layout of the lower limb rehabilitation robot is relatively complex. When the input is zero, the output is often not zero, which causes the imbalance of the bridge (Fig. 4).

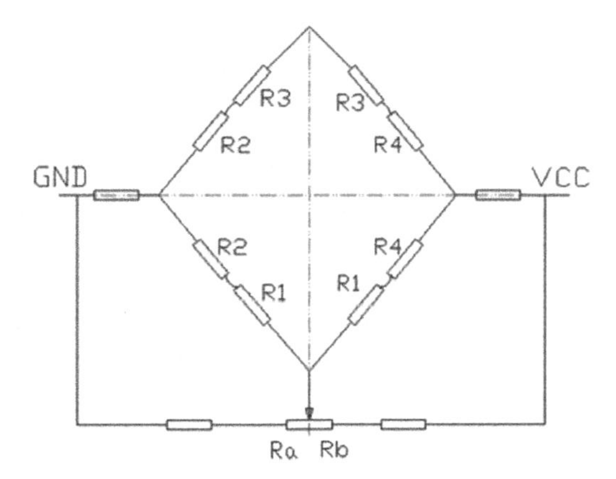


**Fig. 4.** Zeroing circuit

Practical application, the bridge balance should satisfy the conditions of the initial voltage output is zero, each bridge arm resistance sensor has certain internal work in addition to the strain gauge and sensor there will be some compensation also became the bridge unbalanced factors, in order for the circuit input to be zero and the output to be zero, so the circuit should be reset to zero, and taking line resistance into account, balance it with a variable resistance adjustment.

The left and right components of the sliding rheostat are regarded as $R_a$ and $R_b$. From the perspective of resistance voltage division, as long as the ratio of the circuit resistance on the left and right sides is equal, the balance can be satisfied. Therefore, to realize the circuit zero adjustment, the resistance distribution of the sliding rheostat needs to meet the following conditions.

$$\frac{R_2 + R_3}{R_3 + R_4} = \frac{\frac{1}{R_a} + \frac{1}{R_1 + R_2}}{\frac{1}{R_b} + \frac{1}{R_1 + R_4}}. \tag{5}$$

In practice, the circuit balance can be realized by adjusting the sliding rheostat according to the resistance value of the strain gauge.

## 4 Simulation Result

Firstly, ANSYS is used to model the fabric structure of the lower limb rehabilitation structure. A 200N force is added in the direction of the X axis, and a section of the X axis is selected for strain analysis (Fig. 5).

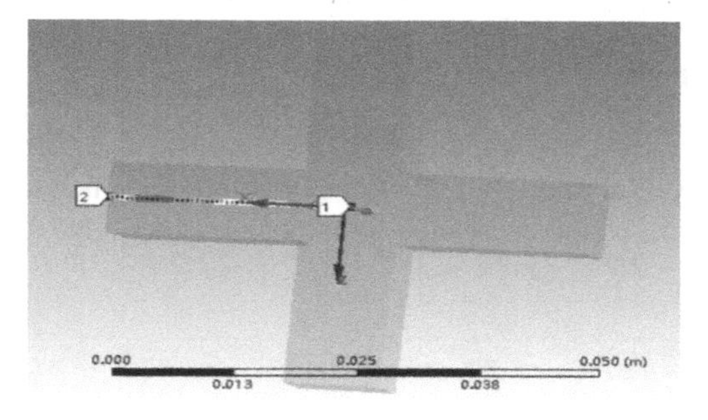


**Fig. 5.** Ansys model

By creating model of Ansys finite element analysis for Stress curve, the curve can be seen that the distribution of Stress in structure is linear, according to $\varepsilon = \delta / E$, E is the modulus of elasticity, ε is the strain, and δ is the stress, the result is can be seen that the strain and the approximate linear relationship between distance. After selecting the parameters of the strain gage, the arrangement of the strain gage can be carried out on this basis (Fig. 6).

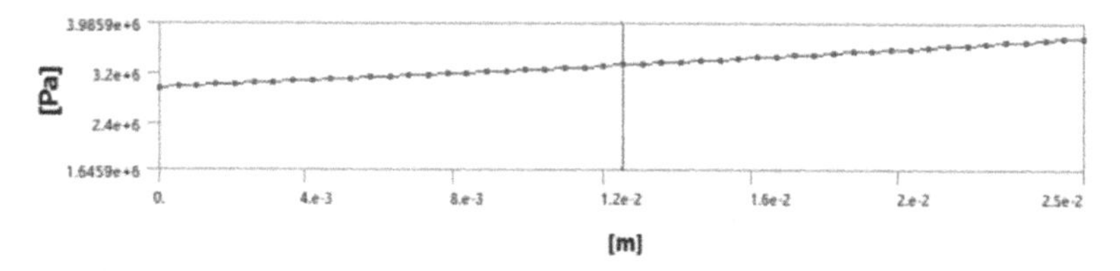


**Fig. 6.** The stress curve

When applied to the exoskeleton of lower limbs, the force received can be divided into $F_x, F_y, F_z, M_x, M_y, M_z$ in space. The strain gauge resistance will produce corresponding deformation according to the relationship between the force and the deformation.

The type of strain gauge is selected, the initial resistance value and sensitivity coefficient can be determined, and the theoretical rate of change can be obtained.

When subjected to external force, the position of the patch will be deformed, and the strain gauge will have corresponding deformation. When stretching, the resistance value will increase, and when compressing, the resistance value will decrease. $R_1 = R_2 = R_3 = R_4 = R$.

$$R_1 + R_4 = 2R - 2\Delta R_{Fx} + 2\Delta R_{Mz}.$$

$$R_1 + R_2 = 2R + 2\Delta R_{Fx} - 2\Delta R_{Mz}.$$

$$R_2 + R_3 = 2R - 2\Delta R_{Fx} - 2\Delta R_{Mz}.$$

$$R_3 + R_4 = 2R + 2\Delta R_{Fx} + 2\Delta R_{Mz}.$$

$$U_{Fx} = \left(\frac{R_1 + R_2}{R_1 + R_4 + R_1 + R_2} - \frac{R_2 + R_3}{R_2 + R_3 + R_3 + R_4}\right)U = \frac{\Delta R_{Fx}}{R}U. \quad (6)$$

By the same logic, we can figure out $U_{Fy} = \frac{\Delta R_{Fy}}{R}U$, $U_{Fz} = \frac{\Delta R_{Fz}}{R}U$, $U_{Mx} = \frac{\Delta R_{Mx}}{R}U$, $U_{My} = \frac{\Delta R_{My}}{R}U$, $U_{Mz} = \frac{\Delta R_{Mz}}{R}U$.

Through MATLAB simulation, it is assumed that all directions after loading force and moment produced by deformation, according to the mechanical deformation after type can be derived, under the condition of known resistance change rate and constant voltage power supply, can get the output voltage, the reference strain gauge on the market, assuming that the initial resistance value is 350 Ω, resistivity K = 2.16, the initial voltage is 5 V (Fig. 7).

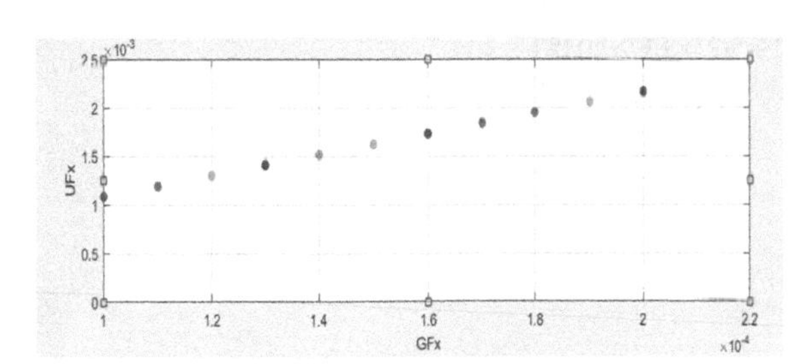


**Fig. 7.** The simulation output

Assuming that the force in the X direction does not affect the output in other directions, when the deformation in the X direction changes from 0.0001 to 0.0002, the output voltage increases from 0.0011 to 0.0022, the voltage changes significantly, and the effective output is obtained. However, the output is too small, which requires signal amplification to detect.

## 5 Conclusion

Deformation resistance strain gauge is used in lower limb rehabilitation robot, which not only measures the lower limb robot motion process of force and moment, but also facilitates the application of been robot in the future better to combine with the mechanical structure, improving the alignment of human and machine, this paper put forward the solution of decreasing the temperature error and zeroing circuit, finally the simulation gets the effective output, but in this structure will exist between multiple output coupling, still need to decouple the results, the simulation results of the output signal is small and need to be further enlarged.

## References

1. Wang, W., Qin, L., Yuan, X., et al.: Bionic control of exoskeleton robot based on motion intention for rehabilitation training. J. Adv. Robot. **12**, 590–601 (2019)
2. Zhang, X., Zhang, Y., Jiao, Z.: Research progress of rehabilitation robots. J. Med. Health Equip. **41**(04), 97–102 (2020)
3. Zhang, W., Zhang, W., Ding, X., Sun, L.: Optimization of the rotational asymmetric parallel mechanism for hip rehabilitation with force transmission factors. J. Mech. Robot. **12** (2020)
4. Martinez, B.L., Durrough, C., Goldfarb, M.: A velocityfield-based controller for assisting leg movement during walking with a bilateral hip and knee lower limb exoskeleton. J. IEEE Trans. Robot. **35**(2), 307–316 (2019)
5. Luna, C.O., Rahman, M.H., Saad, M., Archambault, P., Zhu, W.-H.: Virtual decomposition control of an exoskeleton robot arm. J. Robotica **34**(7), 1587–1609 (2016)
6. Manna, S.K., Dubey, V.N.: Comparative study of actuation systems for portable upper limb exoskeletons. Med. Eng. Phys. **60**, 1–13 (2018)
7. Liu, J., Li, M., Qin, L., et al.: Active design method for the static characteristics of a piezoelectric six-axis force/torque sensor. J. Sens. **14**(1), 659–671 (2014)
8. Zhang, X., Elnady, A.M., Randhawa, B.K., et al.: Combining mental training and physical training with goal-oriented protocols in stroke rehabilitation: a feasibility case study. J. Front. Hum. Neurosci. **12**, 125–137 (2018)

# Object Detection of Basketball Robot Based on MobileNet-SSD

Chen Chen(✉), Wanmi Chen, and Sike Zhou

School of Mechatronic Engineering and Automation, Shanghai University, Shanghai 200072, China
cehncehn@shu.edu.cn

**Abstract.** Object detection is one of the research hotspots in the field of computer vision. In this paper, we use the lightweight network MobileNet combined with Single Shot Multibox Detector (SSD) to realize the object detection of the robot. SSD combined with MobileNet can effectively compress the size of the network model and improve the detection rate. The method does automatic extraction on the image features first, and add different size feature maps after the basic network, and then do convolution filtering on the multi dimension feature maps to get the object coordinate value and the object category. In the experiment, compared with the original vision method based on OpenCV, the MobileNet-SSD algorithm was less affected by illumination conditions in the object recognition process, and achieved the rapid and accurate recognition of the basketball robot on the ball.

**Keywords:** Basketball robot · MobileNet-SSD · Neural network · Object detection

## 1 Introduction

In the past few years, the field of computer vision for object detection has been extensively studied. Many researchers have developed different algorithms and models to study this problem. Object detection algorithms based on deep learning can be roughly divided into two categories. One is the two-stage method, which divides the whole process into two parts, generating candidate boxes and identifying objects in the boxes, mainly including Region-Convolutional Neural Networks (R-CNN), Mask-CNN, etc.; the other is the one-stage method, which unifies the whole process and directly gives detection results, mainly including Single Shot Multibox Detector (SSD) and You Only Look Once (YOLO) series [1].

The two-stage model is named for its two-stage processing of images, also known as region-based approach. In [2], an R-CNN method is proposed: Convolutional Neural Networks (CNN) can be used to locate and segment objects based on region. In case of shortage of training samples, the pre-trained model on additional data can achieve good results through Fine-Tuning. In [3], He proposed a mask-CNN network. In this paper, the links involving the size changes of feature maps in the network were not rounded, but the pixels at non-integer positions were filled by bilinear difference values. In this way,


Q. Han et al. (Eds.): LSMS 2021/ICSEE 2021, CCIS 1469, pp. 11–22, 2021.
https://doi.org/10.1007/978-981-16-7213-2_2

there is no position error when the downstream feature map is mapped to the upstream, which not only improves the effect of target detection, but also makes the algorithm satisfy the accuracy requirements of semantic segmentation task.

The one-stage model has no region detection process, and the prediction results are obtained directly from the images, also known as the region-free method. In [4], Joseph proposed the YOLO network, which represented the detection task as a unified, end-to-end regression problem, and obtained the position and classification by processing the image only once. The disadvantage is that the grid is rough, which limits the detection of small objects. However, the subsequent YOLOV2 [5] and YOLOV3 [6] have improved the original network and achieved better detection effect. SSD [7] was proposed by Liu. Compared with Yolo, it contains multi-scale feature map, which can improve the detection accuracy of small objects. In addition, SSD has more anchor boxes, each grid point generates boxes of different sizes and aspect ratios, and the category prediction probability is based on box prediction (YOLO is based on the grid).

In this paper, considering the time of basketball robot competition and the uncomplexity of target recognition, we choose the SSD algorithm in the one-stage class with faster target recognition speed and slightly lower accuracy In order to improve the accuracy and speed of SSD algorithm in ball detection, MobileNet [8] is used to replace VGG-16 network as feature extraction network, which reduces the model volume, significantly reduces the amount of calculation and improves the detection speed.

## 2 Related Work

Basketball robot competition is one of the most important events of the international robot live competition and China robot competition initiated by Taiwan. Basketball robot technology combines artificial intelligence, automatic control, sensor technology, image processing, wireless communication and new materials and other cutting-edge technologies. The structure diagram of the basketball robot is shown in Fig. 1.

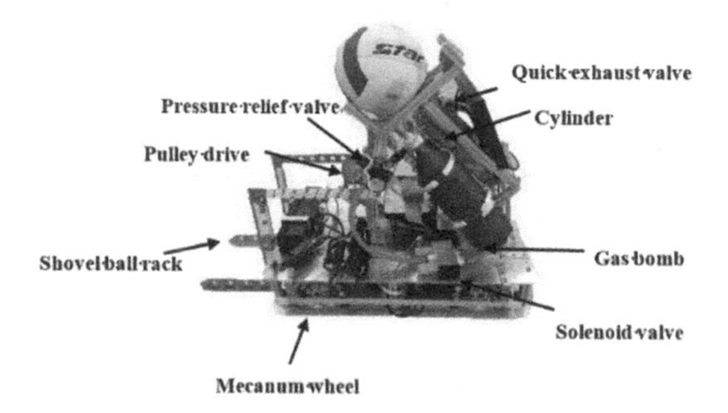


**Fig. 1.** Structure diagram of basketball robot

The basketball robot is a fully autonomous mobile robot with the capabilities of independent image processing and intelligent decision-making. The robot designed by us is composed of decision-making subsystem, perception subsystem, control subsystem and communication subsystem. The functions of each subsystem are as follows:

The decision-making subsystem processes the data collected by the perception subsystem, and then transforms the data into decision instructions to complete the corresponding action flow of the basketball robot. The perception subsystem collects the data of each sensor, and then transmits it to the decision-making subsystem through the communication subsystem. The sensors used include: Kinect sensor and laser radar for vision, and odometer and gyroscope for pose estimation; The control subsystem converts the decision instructions sent by the decision-making subsystem into various control signals, such as motion control, pitch control, ball-picking control; The communication subsystem completes the data transmission between the upper computer and the lower computer as well as between the controller and the sensor.

Hardware system design: this paper focuses on the target detection task of the robot, whose hardware is usually composed of visual sensors. In order to improve the speed, accuracy and robustness of target detection, we adopt the combination of Kinect sensor [9] and laser radar sensor (for positioning).

Software system design: the original vision scheme based on OpenCV was abandoned because of the influence of illumination, the poor robustness of the algorithm and a lot of parameter tuning needed. Instead, the deep learning algorithm MobileNet-SSD convolutional neural network was used to recognize the target in the framework of TensorFlow.

## 3 MobileNet Network Structure

### 3.1 Depthwise Separable Convolution

MobileNet network is one of the lightweight convolutional networks. The biggest difference between MobileNet network and convolutional neural network is that the former uses standard convolution, as shown in Fig. 2, while the latter uses depthwise separable convolution.

In the Fig. 2, M represents the number of input channels, N represents the number of output channels, and $D_K \times D_K$ represents the size of the convolution kernel. Assume that the size of the input image is $D_F \times D_F$, then the input is $D_F \times D_F \times M$, and the output is $D_F \times D_F \times N$. Therefore, the amount of calculation of the standard convolution is as follows:

$$D_F \cdot D_F \cdot D_K \cdot D_K \cdot M \cdot N. \quad (1)$$

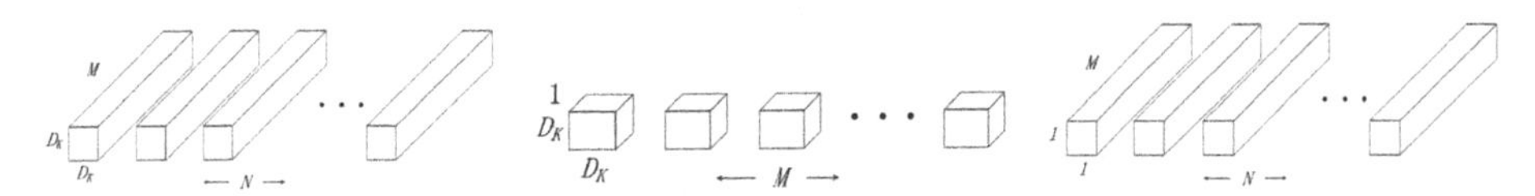

**Fig. 2.** Standard convolution **Fig. 3.** Schematic diagram of channel by channel convolution **Fig. 4.** Schematic diagram of point-by-point convolution

In the MobileNet network structure, the channel-by-channel convolution and point-by-point convolution are obtained by standard convolution decomposition. The size of the convolution kernel is $D_K \times D_K$ and $1 \times 1$ respectively and the channel-by-channel convolution is shown in Fig. 3.

When the feature map is input to the channel-by-channel convolution layer, the convolution operation will produce a single output, and the amount of calculation will be compressed for the first time. The amount of calculation of the channel-by-channel convolution is as follows:

$$D_K \cdot D_K \cdot M \cdot D_F \cdot D_F. \tag{2}$$

The schematic diagram of point-by-point convolution is shown in Fig. 4.

The output of the channel by channel convolution layer is used as the input of the point by point convolution layer. After the convolution operation, the output of the depth feature will be obtained, and the amount of calculation is compressed for the second time. The amount of calculation of point by point convolution is as follows:

$$\mathrm{M} \cdot \mathrm{N} \cdot D_F \cdot D_F. \tag{3}$$

After the standard convolution is decomposed into channel by channel convolution and point by point convolution, the amount of calculation is as follows:

$$D_K \cdot D_K \cdot M \cdot D_F \cdot D_F + M \cdot \mathrm{N} \cdot D_F \cdot D_F. \tag{4}$$

Finally, the ratio of the amount of calculation after the standard convolution decomposition to that before the decomposition is as follows:

$$\frac{D_K \cdot D_K \cdot M \cdot D_F \cdot D_F + M \cdot \mathrm{N} \cdot D_F \cdot D_F}{D_F \cdot D_F \cdot D_K \cdot D_K \cdot M \cdot N} = \frac{1}{N} + \frac{1}{D_K^2}. \tag{5}$$

MobileNet usually uses the convolution kernel of $3 \times 3$. Through Formula 5, it can be calculated that the amount of calculation of standard convolution is 8~9 times of the combination of channel-by-channel convolution and point-by-point convolution, and the amount of the corresponding parameter is also 8~9 times.

### 3.2 Network structure and Advantage

In the MobileNet network structure, there are a total of 27 convolution layers, including a standard convolution layer, 13 channel-by-channel convolution layers and 13 point-by-point convolution layers. To compare the performance of convolution neural network using standard convolution and depthwise separable convolution, reference [10] used standard convolution and depthwise separable convolution to train and test on Imagenet dataset respectively. In terms of accuracy, standard convolution is 1.1% higher than depthwise separable convolution, but in terms of the amount of calculation and parameter, the former is 8~9 times that of the latter. It shows that the depthwise separable convolution can ensure the accuracy and reduce the amount of calculation and parameters. Accordingly, it can reduce the difficulty of network model training, reduce the training time, and reduce the performance requirements of hardware equipment.

## 4 Mobilenet-SSD Neural Network

SSD model recognition method is based on feedforward convolution network, which is divided into two parts: basic network and multi-scale feature map detection. The basic network mainly extracts features from the image with data size of 300 × 300, and then adds feature maps of different sizes to detect the feature maps of different sizes, so as to obtain the detection effect of different sizes of targets.

### 4.1 Network Structure

The basic network carries the function of image feature extraction, and its ability to extract features affects the quality of multi-size feature map detection. The common basic networks include VGG [11], Googlenet [12] and ResNet [13]. MobileNet-SSD, the network framework used in this paper, is based on the SSD network framework, using lightweight convolutional neural network MobileNet instead of the original basic network VGG-16, and removing the full connection layer and softmax layer in MobileNet network, and adding 8 convolutional layers to complete the feature extraction of images. The visualization model of some MobileNet-SSD models is shown in the Fig. 5.

As can be seen from Fig. 6, the 8 additional convolution layers are Conv14_1, Conv14_2, Conv15_1, Conv15_2, Conv16_1, Conv16_2, Conv17_1, Conv17_2 respectively In the process of feature extraction, the method used in MobileNet-SSD network model is similar to SSD network model. The idea of feature pyramid is adopted to obtain the feature information of 6 convolutional layers for multi-scale and multi-target object detection.

The six convolution layers of MoblieNet-SSD network model used for target detection are conv11, conv13, conv14_ 2, Conv15_ 2, Conv16_ 2 and Conv17_ 2. Among them, the size of the output feature map of each layer is 19 × 19, 10 × 10, 5 × 5, 3 × 3, 2 × 2, 1 × 1. The network structure of MobileNet-SSD is shown in Fig. 6. The size of the input image is normalized to 300 × 300, and the target classification and candidate box regression are performed after 6 layers of convolution and pooling operations.

### 4.2 Multi-scale Detection

The target features of the image are extracted through the basic network, and several feature maps are added after the basic network. These feature maps are of different sizes. A convolution filter is used for these feature maps to obtain the coordinate value and the category of the target. For example, if there is a feature map with p channels and size of m × n, convolution operation can be performed with convolution kernel of 3 × 3 × p to obtain the confidence of a default box for each target category and the coordinate offset value of the default box. Multiple convolution kernels are used to convolute m × n feature map for many times to obtain the confidence and coordinate offset values of each category of the feature map in different default boxes.

In MobileNet-SSD network, six feature maps (Conv11, Conv13, Conv14_2, Conv15_2, Conv16_2, Conv17_2) of different scales are selected, and 3 × 3 convolution operation is used to perform object classification prediction and bounding box regression on the default box, which becomes an end-to-end frame on the whole. As

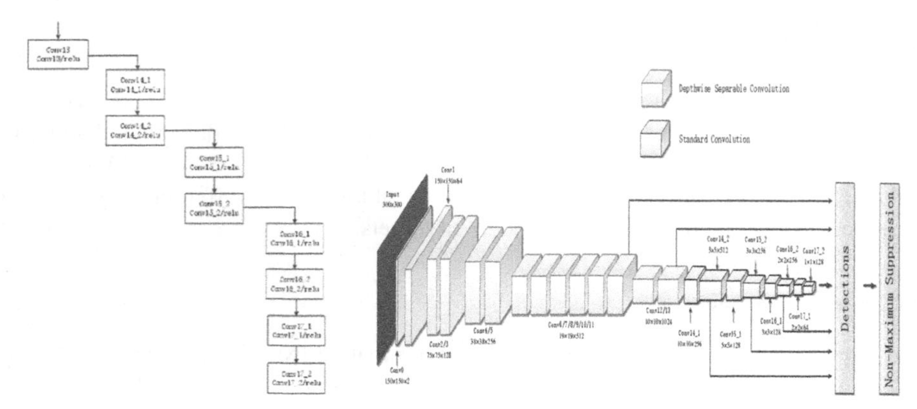


**Fig. 5.** Visualization of MobileNet-SSD partial network model

**Fig. 6.** MobileNet-SSD network structure model

shown in Fig. 7, there are two feature maps with size of 8 × 8 and size of 4 × 4, which have the same proportion of default box. The default box range of the former target object is significantly smaller than that of the latter. The former selects the smaller object in the whole picture, while the latter selects the larger object. According to this principle, 6 feature maps with different sizes are selected in the SSD framework for classification and regression, so that the model can well learn the features of different sizes of target objects in the image, and ensure that the model has a strong sensitivity to the size of the target object.

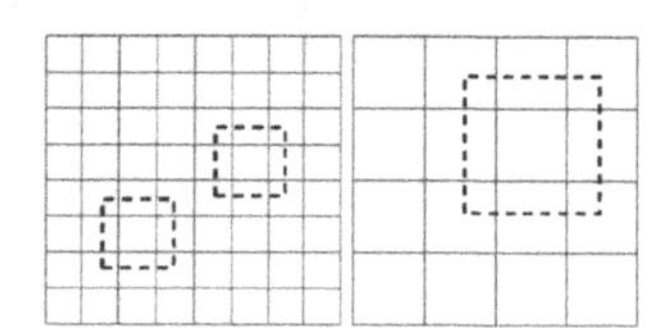

**Fig. 7.** Multi-scale detection

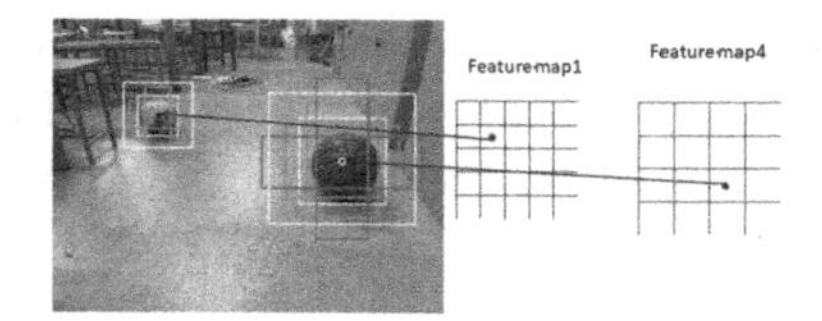


**Fig. 8.** Default box generated by feature maps

## 4.3 Generation of Default Box

The default box is generated by feature map. In convolution network, the size of feature map is reduced by network depth. The feature maps of different size are obtained for target detection under different sizes. Assuming that m feature maps are used for prediction, the proportion of default box of each feature map is calculated as follows:

$$S_k = S_{min} + \frac{S_{max} - S_{min}}{m - 1}(k - 1), k \in [1, m]. \tag{6}$$

Where $S_{min} = 0.2$, $S_{max} = 0.9$, according to the difference of m, the default box scale of different feature maps is calculated. Then different aspect ratios are applied to the default box to increase the number of default frames, which are represented as

$a_r \in \left\{1, 2, \frac{1}{2}, 3, \frac{1}{3}\right\}$. The width and height of the added default box are calculated as follows:

$$w_k^a = S_k\sqrt{a_r}. \tag{7}$$

$$h_k^a = \frac{S_k}{\sqrt{a_r}}. \tag{8}$$

When the aspect ratio is 1, an additional default box is added, and its scale calculation formula is as follows:

$$S_k^{'} = \sqrt{S_k S_{k+1}}. \tag{9}$$

The center of each default box is set to $\left(\frac{i+0.5}{|f_k|}, \frac{j+0.5}{|f_k|}\right)$.

Where, $f_k$ is the size of the $k_{th}$ feature map, i, j $\in \left[0, |f_k|\right]$ represents the horizontal and vertical positions of the feature map.

In this paper, for the feature layers of conv11, conv13 and conv17_2, we ignore the ratio of $\frac{1}{3}$ and 3, so only four default boxes will be generated; for the feature maps corresponding to the other three convolution layers, six default boxes will still be generated. The specific implementation effect is shown in Fig. 8.

## 4.4 Loss Function

The loss function of MobileNet-SSD algorithm is mainly divided into two parts. The first part is the loss function of confidence between the predicted value of the default box and the target category (conf). The second part is the location regression loss (loc) between the prediction box after the adjustment of the default box and the real box The total loss function is the weighted sum of the location loss function and the confidence loss function.

$$\mathrm{L}(x, c, l, g) = \frac{1}{N}\left(L_{conf}(x, c) + \alpha L_{loc}(x, l, g)\right). \tag{10}$$

Where, N is the number matched to the default box, N = 0 represents the loss is 0. α is a parameter used to adjust the ratio between the loss of confidence (conf) and the loss of location (loc), and the default value is 1. c is the predicted value of category confidence, 1 is the bounding box predicted by the model, and g is the truth box of the sample.

For the confidence loss in the first part, in order to improve the classification effect of Softmax, we can adopt the cross entropy cost function as the confidence loss function, which is defined as:

$$\hat{c}_i^p = \frac{\exp\left(c_i^p\right)}{\sum_p \exp\left(c_i^p\right)}. \tag{11}$$

$$\mathrm{L}_{conf}(x, c) = -\sum\nolimits_{i \in Pos}^{N} x_{ij}^p \log\left(\hat{c}_i^p\right) - \sum\nolimits_{i \in Neg} \log\left(\hat{c}_i^0\right). \tag{12}$$

Where, $x_{ij}^{p} = \{1, 0\}$ represents that the $i_{th}$ prediction box matches the $j_{th}$ truth box of category P; P represents the target category; $c_i^p$ represents the confidence that the $i_{th}$ default box is category p.

For location loss in the second part, in order to make the target prediction position more accurate, the loss function Smooth L1 between prediction box l fine-tuned by default box d and truth box g is used, and the formula is as follows:

$$\text{smooth}_{L1}(x) = \begin{cases} 0.5x^2, & |x| < 1 \\ |x| - 0.5, & others \end{cases} \tag{13}$$

$$L_{loc}(x, l, g) = \sum_{i \in Pos}^{N} \sum_{m \in (cx, cy, w, h)} x_{ij}^{k} smooth_{L1}\left(l_i^m - \hat{g}_j^m\right). \tag{14}$$

Where: $\hat{g}_j^{cx} = \left(g_j^{cx} - d_i^{cx}\right)/d_i^w$; $\hat{g}_j^{cy} = \left(g_j^{cy} - d_j^{cy}\right)/d_i^h$; $\hat{g}_j^w = \log\left(\frac{g_j^w}{d_i^w}\right)$; $\hat{g}_j^h = \log\left(\frac{g_j^h}{d_i^h}\right)$; l is the prediction box; g is the truth box; (cx, cy) is the center of the default box; w is the width of the default box; h is the height of the default box; Pos is the positive sample set; d is the corresponding parameter of the $i_{th}$ default box.

### 4.5 Macthing Strategy and Non-maximum Suppression

**Matching Strategy.** During model training, the matching degree between the truth box and the default box is determined by IOU. The formula of IOU is as follows:

$$\text{IOU} = (\text{A} \cap \text{B})/(\text{A} \cup \text{B}) \tag{15}$$

The specific matching steps are as follows:

1. For each truel value in the image, find the default box with the largest IOU to match. The purpose is to ensure that each true value can match with a default box;
2. For the remaining default boxes that do not match, if the IOU of a true value is greater than a certain threshold (which is a critical value, generally set to 0.5), then the default box matches the true value

As for the matching principle, we should note the following points:

1. Positive sample: the default box that matches the true value; Negative sample: The default box that none of the true values match;
2. A true value can match multiple default boxes, but each default box can match only one true value;
3. If the IOU calculated between more than one true value and a default box is greater than the threshold, then the default box will only match the true value of the largest IOU.

**Non-maximum Suppression.** In the final output of the model, there may be multiple target objects, and there will be multiple highly repetitive bounding boxes around each target object. In this case, we need to use non maximum suppression [14] to select the most appropriate bounding box. Non maximum suppression helps to avoid repeatedly detecting the same object. The specific method is: first, all the output bounding boxes are sorted according to the confidence score from large to small, and the bounding box with the highest confidence score is selected from all the bounding boxes. Then, the bounding box and the remaining bounding boxes in the image are used for IOU operation, and the bounding box whose IOU value is greater than a certain threshold is deleted. After deleting all the qualified bounding boxes, the bounding box with the highest confidence is located in the bounding box of the object. Repeat until there is only one bounding box around all target objects. Using the non-maximum suppression method, the most suitable bounding box can be selected for the target object.

# 5 Experiments

## 5.1 Evaluation Criterion and Experimental Data

In the field of object detection in machine learning, the precision-recall (P-R) curve is used to measure the performance of target detection algorithm. In addition, mAP is also commonly used to measure the performance of target detection algorithm. Detection accuracy (AP) refers to the area under the P-R curve, and mAP refers to the average accuracy of different types.

Accuracy P and recall R are defined as follows:

$$P = \frac{Correct\ number\ of\ detected\ targets}{Number\ of\ all\ detected\ targets}. \tag{16}$$

$$\mathrm{R} = \frac{Correct\ number\ of\ detected\ targets}{The\ number\ of\ actual\ targets\ in\ the\ detection\ set}. \tag{17}$$

The algorithm uses tensorflow framework, and the ball detection model is trained and tested on NVIDIA GTX 1060 GPU. The statistical information of the collected data set is shown in Table 1:

**Table 1.** Statistical informatica of data set

| Data set | Basketball | Volleyball | Number of images |
|---|---|---|---|
| Training set | 634 | 612 | 1123 |
| Test set | 55 | 49 | 86 |

## 5.2 Experimental Results and Analysis

Through the experiment, the detection results of basketball and volleyball are shown in Fig. 9.

In the experiment, the recognition rate is above 0.9 for the target with complete target and less background interference. Because the background of the basketball robot competition is simple, the interference is small, the ball target is complete, and the influence of occlusion light is few, it meets the performance requirements of robot detection. The P-R analysis of detection effect is shown in Figs. 10 and 11. We also use the same experimental data to train with Yolo and compare with MobileNet-SSD model in the same test set. The AP and map analysis table of detection effect is shown in Table 2. It can be seen from Table 2 that the mAP value of MobileNet-SSD detection model is 4.1% points higher than that of Yolo model. The detection accuracy of mobilenet SSD for basketball is 3.8% points higher than that of Yolo, and for volleyball is 4.5% points higher than that of Yolo. The experimental results show that the MobileNet-SSD model has the advantages of fast detection speed and high detection accuracy, which realizes the basketball robot's rapid recognition of ball objects and meets the requirements of the competition.

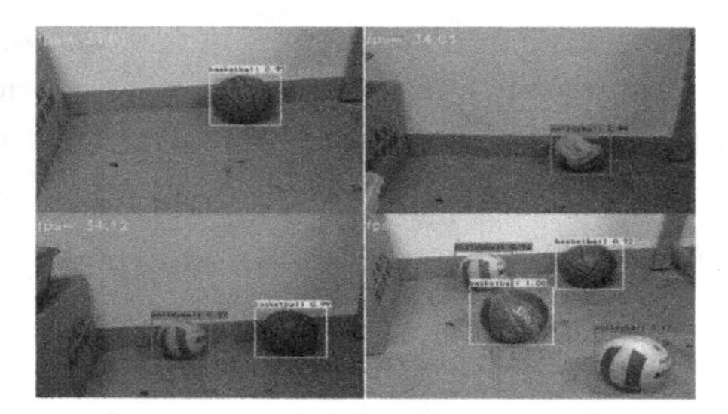

**Fig. 9.** Object detection results

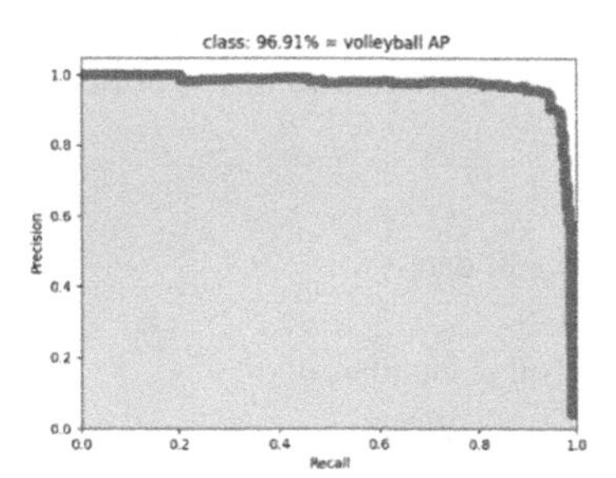


**Fig. 10.** The P-R diagram of volleyballs (MobileNet-SSD)

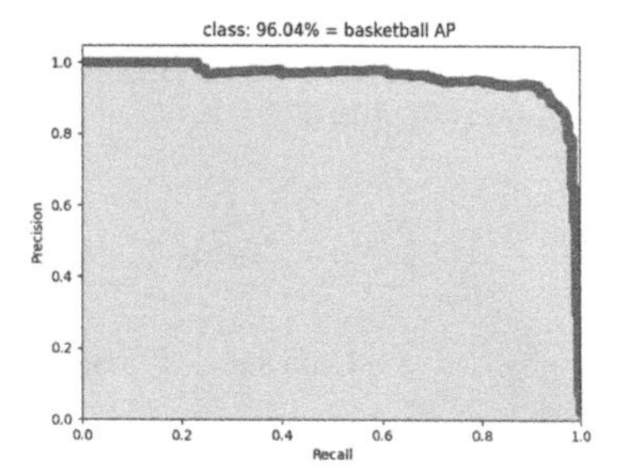


**Fig. 11.** The P-R diagram of basketball (MobileNet-SSD)

**Table 2.** AP and mAP of test set by different models

| Model | AP | | mAP |
|---|---|---|---|
| | Basketball | Volleyball | |
| YOLO | 0.9221 | 0.9245 | 0.9233 |
| MobileNet-SSD | 0.9601 | 0.9691 | 0.9646 |

## 6 Conclusion

In this paper, deep learning is used to solve the problem of ball target detection of basketball robot. We adopt MobileNet-SSD model algorithm. The image features are extracted by using the basic network of the model, and the feature maps of different sizes are added to the network and convolution filtering is performed to obtain the target position coordinates and target categories in the image. Compared with the original robot detection method, it avoids manual extraction of image features and improves the generalization ability of robot target detection. The results show that the mAP value of mobile net SSD model is higher than that of Yolo model, and the detection effect of ball is better. Compared with the original vision scheme based on OpenCV, the sensitivity of the robot to light is reduced. Due to the characteristics of SSD model, the detection speed is fast, but the detection ability of small targets is not strong. Although it does not affect the detection of big balls (basketball and volleyball) by basketball robot, in order to increase the target detection range of basketball robot, such as the detection of table tennis and tennis, on the premise of ensuring the detection speed, improving the detection ability of the target is still the focus of the next step.

## References


1. Xiao, Y., et al.: A review of object detection based on deep learning. Multimedia Tools Appl. **79**(33–34), 23729–23791 (2020). https://doi.org/10.1007/s11042-020-08976-6
2. Girshick, R., Donahue, J., Darrell, T., Malik, J.: Rich feature hierarchies for accurate object detection and semantic segmentation. In: 2014 IEEE Conference on Computer Vision and Pattern Recognition, pp. 580–587. IEEE Press, Columbus (2014)
3. He, K., Gkioxari, G., Dollár, P., Girshick, R.: Mask R-CNN. J. IEEE Trans. Pattern Anal. Mach. Intell. **42**(2), 386–397 (2017)
4. Redmon, J., Divvala, S., Girshick, R., Farhadi.A.: You only look once: unified, real-time object detection. In: 2016 IEEE Conference on Computer Vision and Pattern Recognition, pp. 779–788. IEEE Press, Las Vegas (2016)
5. Redmon, J., Farhadi, A.: YOLO9000: better, faster, stronger. In: 2017 IEEE Conference on Computer Vision and Pattern Recognition, pp. 6517–6525. IEEE Press, Honolulu (2017)
6. Li, C., Wang, R., Li, J., Fei, L.: Face detection based on YOLOv3. In: Jain, V., Patnaik, S., Popenţiu, V.F., Sethi, I. (eds.) Advances in Intelligent Systems and Computing. AISC, vol. 1031, pp. 277–284. Springer, Singapore (2018). https://doi.org/10.1007/978-3-319-45991-2
7. Liu, W., et al.: SSD: single shot multibox detector. In: Leibe, B., Matas, J., Sebe, N., Welling, M. (eds.) ECCV 2016. LNCS, vol. 9905, pp. 21–37. Springer, Cham (2016). https://doi.org/10.1007/978-3-319-46448-0_2

8. Howard, AG., et al.: Mobilenets: efficient convolutional neural networks for mobile vision applications. arXiv preprint arXiv:1704.04861 (2017)
9. Guo, J., Wang, Q., Ren, X.: Target recognition based on kinect combined RGB image with depth image. In: Abawajy, J.H., Choo, K.-K., Islam, R., Xu, Z., Atiquzzaman, M. (eds.) ATCI 2019. AISC, vol. 1017, pp. 726–732. Springer, Cham (2020). https://doi.org/10.1007/978-3-030-25128-4_89
10. Simonyan, K., Andrew Z.: Very deep convolutional networks for large-scale image recognition. arXiv preprint arXiv:1409.1556 (2014)
11. Szegedy, C., et al.: Going deeper with convolutions. In: 2015 IEEE Conference on Computer Vision and Pattern Recognition, pp. 1–9. IEEE Press, Boston (2015)
12. He, K., Ding, L., Wang, L., Cao, F.: Deep residual learning for image recognition. In: 2016 IEEE Conference on Computer Vision and Pattern Recognition, pp. 770–778. IEEE Press, Las Vegas (2016)
13. Lin, T., Dollár, P., Girshick, R., He, K., Hariharan, B., Belongie, S.: Feature pyramid networks for object detection. In: 2017 IEEE Conference on Computer Vision and Pattern Recognition, pp. 770–778. IEEE Press, Honolulu (2017)
14. Gong, M., Wang, D., Zhao, X., Guo, H., Luo, D., Song, M.: A review of non-maximum suppression algorithms for deep learning target detection. In: Seventh Symposium on Novel Photoelectronic Detection Technology and Applications, vol. 11763. International Society for Optics and Photonics (2021)

# Detection of Overlapping and Small-Scale Targets in Indoor Environment Based on Improved Faster-RCNN

Yang Liu(✉) and Wanmi Chen

School of Mechatronic Engineering and Automation,
Shanghai University, Shanghai 200072, China

**Abstract.** This paper proposes an improved Faster-RCNN network to detect small-scale and overlap targets in indoor environment. In the improved Faster-RCNN model, ResNet-50 was used to replace VGG16 as the backbone network to extract multi-layer features of images, and then shallow feature images with rich detailed information were fused with the deep feature images with abstract information. In addition, Soft-NMS is used to replace NMS to improve the performance of overlapping target detection. In self-made dataset, the mAP of the improved model was improved by 1.45% relative to the unimproved model.

**Keywords:** Faster-RCNN · Soft-NMS · Target detection · Multi-features fusion

## 1 Introduction

The accuracy of target recognition and detection greatly affects the subsequent operation of the robot. Since the detection target of a service robot is likely to be the target of small-scale and overlap, the detection accuracy of target of small-scale and overlap is very important for a service robot. Traditional detection methods need to build different mathematical models for detection targets to extract features such as HOG [18] and SURF [19] that are transmitted to traditional machine learning classifiers for detection and recognition. The limitation of traditional methods is that it requires computer vision experts to build mathematical models to extract features from detection targets, and the features extracted from specific mathematical models may only be applicable to certain types of detection targets and are difficult to generalize. In recent years, deep convolutional neural network [14, 15] has made a major breakthrough in image classification. At present, detection methods based on deep learning have strong feature expression ability and semantic expression ability, and its generalization ability is stronger than traditional detection methods. Deep learning-based detection methods are divided into two categories: one stage and two stage. The main difference is whether the network needs to generate candidate borders through the assistance of subnetworks.

The main representative detection algorithms of the one stage include YOLO [8–10] and SSD [12] without generating candidate border through subnetwork. It maps the region of interest on the original image to the feature map generated by the convolutional neural


Q. Han et al. (Eds.): LSMS 2021/ICSEE 2021, CCIS 1469, pp. 23–34, 2021.
https://doi.org/10.1007/978-981-16-7213-2_3

network. Finally, it classifies and regressing the depth features to get the category of the box and the offset.

The main representative detection algorithms of the two stage include SPP-NET [11], Fast-RCNN [3], Fast-RCNN [1], and R-FCN [7]. Such algorithms generate candidate regions that may contain targets through the assistance of subnetworks, and then transmit these candidate regions to the subsequent networks for classification and location regression.

Based on the two-stage Faster-RCNN algorithm, this paper constructs an improved Faster-RCNN model for target detection in indoor environment. We adopt the multi-layer feature fusion strategy which not only enriches the semantic information of the shared feature map, but also makes the shared feature map more contextual information and which is beneficial to detect small-scale targets. In addition, we use ResNet50 to replace Vgg-16 as the backbone network to improve the accuracy of extracting target features and Soft-NMS [16] is used to replace NMS [6] algorithm to improve the overlapping situation of missed items in indoor environment and enhance the generalization ability of item detection. In the self-made dataset, the mAP of the improved model was improved by 1.45% relative to the original model. In particular, the target of small-scale and overlap such as bottles have been improved by 2.14%.

## 2 Faster-RCNN

### 2.1 Faster-RCNN Structure

The network architecture and detection process of Faster RCNN are shown in Fig. 1. Firstly, VGG network generates a Feature Map with abstract information that is transferred to the RPN network to generate a series of proposal regions for detecting targets. And then the proposed regions is mapped on the Feature Map. The regions mapped on the Feature Map will be sent to the follow-up network for classification and regression after the fixed dimensions of ROI Pooling. From the network structure, Faster-RCNN can be thought of as RPN and the Fast RCNN that uses Selective search [4] to generate candidate regions. Compared with Selective search, RPN network greatly reduces the time of generating candidate regions and extracts more accurate proposed regions.

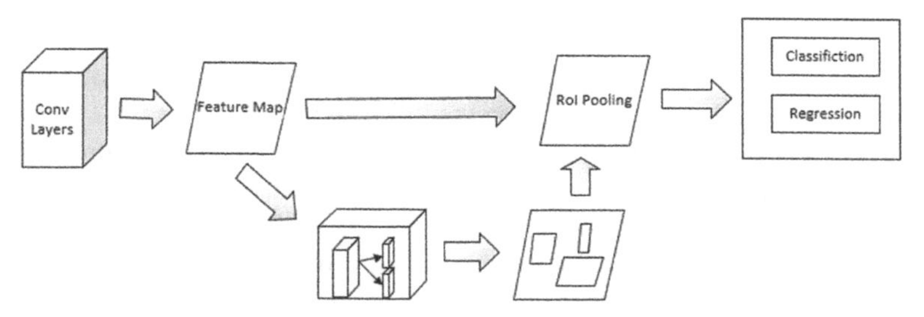


**Fig. 1.** Faster-RCNN structure

## 2.2 RPN

The structure of RPN network is shown in Fig. 2, which is used to assist the main network to generate ROI. In order to better introduce RPN network, we first introduce what is the anchor. Each point on the Feature Map is called an anchor, and each anchor generates different anchor frames in accordance with different scales and aspect ratio in the original image.

Generally, we choose (128, 256, 512) as different scales and (1:1, 1:2, 2:1) as different ratios of width and height. Therefore, a anchor has nine different sizes of anchor boxes on the original picture. The role of RPN is to use these different anchor boxes to predict whether there is a target on the original image. In addition, RPN also corrects the position deviation between the anchor box and the real box, which brings the anchor box close to the real target area.

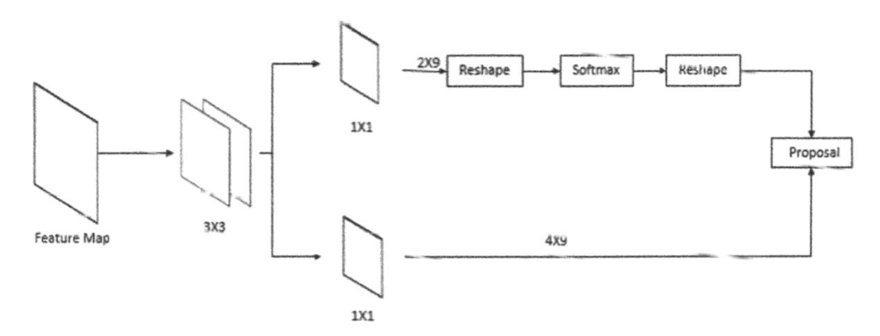


**Fig. 2.** RPN structure

The specific implementation strategy is to use a $3 \times 3$ window to slide on the Feature Map, as shown in Fig. 3. Each sliding window will generate a Feature vector with 512 dimensions. Feature vectors are sent into two different $1 \times 1$ connection layers, one layer is called the classification layer to output the foreground and background confidence, and the other layer is called the regression layer, which outputs the coordinate offset of the reference frame relative to the labeled real frame. Finally, the RPN layer outputs the foreground.

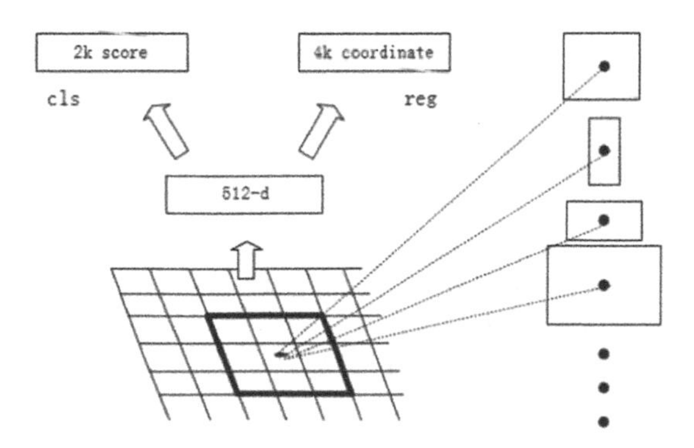


**Fig. 3.** Anchor and anchor boxes

According to the RPN description, the loss function is shown in (1)

$$L = \frac{1}{N_{cls}} \sum_i L_{cls}(p_i, p_i^*) + \lambda \frac{1}{N_{reg}} \sum_i p_i^* L_{reg}(t_i, t_i^*). \tag{1}$$

Where i is the index of the anchor in a batch of data, $L_{cls}$ is a dichotomy (foreground and background) logarithmic loss function, $p_i$ is the predictive classification confidence of anchor[i]. $p_i^* = 1$ when anchor[i] is a positive sample, $p_i^* = 0$ when anchor[i] is a negative sample. $\lambda$ is balancing the equilibrium parameters, $N_{cls}$ is mini-batch size, $N_{reg}$ is the number of anchor Location. $L_{reg}$ is a regression loss function, $t_i$ is the parameterized coordinates of the Anchor prediction border are shown in (2) (3), $t_i^*$ is the parameterized coordinates of the real border of the positive anchor are shown in (4) and (5).

$$t_x = (x - x_a)/\omega_a, t_y = \frac{y - y_a}{h_a}. \tag{2}$$

$$t_\omega = \log(\omega/\omega_a), t_w = \log\left(\frac{h}{h_a}\right). \tag{3}$$

$$t_x^* = (x^* - x_a)/\omega_a, t_y^* = \frac{y^* - y_a}{h_a}. \tag{4}$$

$$t_\omega^* = \log(\omega^*/\omega_a), t_h^* = \log\left(\frac{h^*}{h_a}\right). \tag{5}$$

Where $(x, y)$, $(x_a, y_a)$ and $(x^*, y^*)$ correspond to the central coordinates of the prediction box, anchor point box and real marking box respectively, $\omega$, $\omega_a$, $\omega^*$ correspond to the width of the prediction box, anchor box, and real label box, respectively, h, $h_a$, $h^*$ correspond to the width of the prediction box, anchor box, and real label box, respectively.

In this paper, Anchor is assumed to be a positive sample when the overlap area between anchor box and real region is the largest or is greater than 0.7.When the overlap area o is less than 0.3, it is set as negative sample, and the anchor that does not belongs to positive and negative samples are discarded. The overlap rate of the two regions is calculated by IOU formula, as shown in (6).

$$IoU = \frac{A \cap B}{A \cup B}. \tag{6}$$

## 3 Defects of Network

Although Faster-RCNN, as a general target detection framework, achieved a mAP of 73.2% on VOC2007 and VOC2012 datasets, we found that target detection in the indoor environment still had poor performance in some small-scale target detection and target overlap tasks. We analyze the reasons as follows:

1) In the Faster-RCNN network, the RPN network and the Fast network all use the shared feature layer, so the shared feature layer has some influence on the detection

accuracy of the Faster-RCNN target. With the deepening of the network, information may be lost at each feature layer, and more effective information will be lost when it reaches the deep feature map [17]. The Feature Map extracted by traditional VGG16 has large receptive field and rich abstract semantic information, which has good robustness for large-scale target detection with large resolution. However, as each point of the last layer of Feature Map contains a lot of surrounding information, small-scale target information will be lost. For service robots to detect small-scale targets, it may be difficult to extract the semantic information of the target, so it will lead to missed detection.

2) Faster-RCNN uses NMS algorithm [6] for suppression, and this method has obvious defects. When there are two targets of the same kind in the detection area and the overlap between the two targets is high, the network will reject the target with low confidence, which will lead to the failure of network detection of the target and reduce the mAP, as shown in Fig. 4 [16].

**Fig. 4.** The target in the green box is missed due to the defect of NMS (Color figure online)

## 4 Improvement

### 4.1 Multi-scale Feature Image Fusion

Chen Ze [2] proposed an improved target detection method based on Faster-RCNN to detect small-scale pedestrians, and the experimental results on INRIA and Pascal VOC2012 datasets were 17.58% and 23.78% higher than the original network MAP, respectively. The ResNet [5] improved the problem of network degradation, and using ResNet instead of VGG16 as the extraction feature network was 3.3% higher on average than the traditional Faster RCNN mAP on the VOC2007+VOC2012 dataset. Multi-scale Faster-RCNN [8] to fuse feature images of different feature layers and perform well in detecting multi-scale human faces. Inspired by these method, in order to make the Shared features layer can capture more interested area information, this paper adopts Restnet50 as backbone network and proposes the way of multiple features fusion (deep and shallow features fusion) to enrich the context information, which avoids depth Shared characteristics to the excessive loss of shallow layer of rich details. Fusion is generally divided into two kinds, one is the superposition of feature channels, the other is the sum of feature graphs. As shown in Fig. 5, the shared feature graph is formed by the superposition of the previous several feature graph channels.

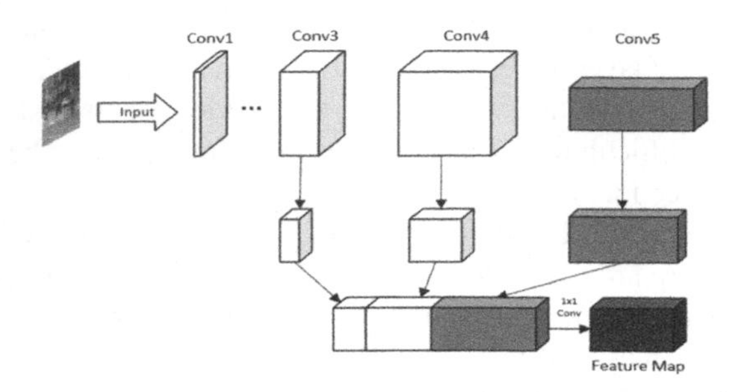


**Fig. 5.** The fusion method is via channel stack

As shown in Fig. 6, the feature diagrams of Conv3,Conv4 and Conv5 were added element by element after adjusting the same scale (that is, the size of the feature graph was consistent with the channel number of the feature graph).

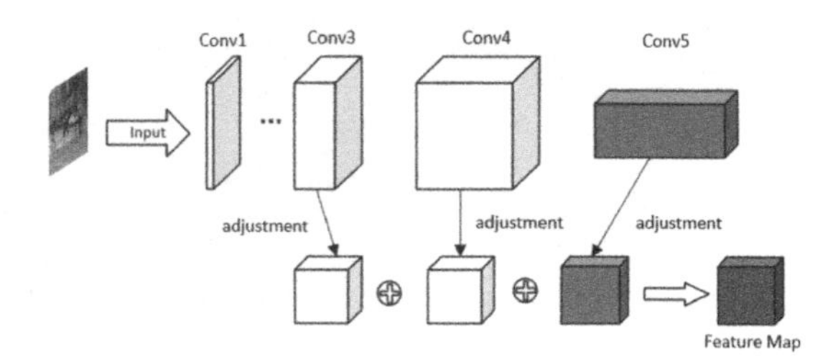


**Fig. 6.** The fusion method is through the addition of feature graphs

The fusion method adopted in this paper is channel fusion. The strategy is to scale the input image to $600 \times 600$ and then get five feature maps through ResNet50 network. And then we adjust the size of the characteristics of Conv3 and Conv4 to be the same as that of Conv5 but the number of channels remains the same. The channel fusion of Conv3, Conv4 and Conv5 after adjustment is carried out. Finally, the features after the superposition of channels are dimensionalized by $1 \times 1$ convolution to 1024 channel number to get the fused feature map. Because the shared feature map adds rich information of the receptive field in the shallow layer, the adjusted feature map has both abstract information and rich detail information in the shallow layer.

## 4.2 Soft-NMS

The defect of NMS algorithm is that when the network deletes the redundant overlapping frame on the same target, it will mistakenly delete the other target next to the target. Aiming at the defect of NMS, Bharat Singh proposed a soft-NMS algorithm [16]. The advantage of this algorithm is that it does not need to retrain the model, and only needs to replace the NMS algorithm directly with the Soft-NMS algorithm. After replacing NMS with Soft-NMS, Bharat Singh used R-FC and Fast-RCNN models to test VOC and COCO [17] and found that the performance was improved. However, Soft-NMS is still a greedy algorithm, which cannot guarantee that the model will leave the target detection box with the optimal global detection. The formula of traditional NMS algorithm is shown in (7):

$$s_i = \begin{cases} s_i, \mathrm{iou}(M, b_i) < N_t \\ 0, \mathrm{iou}(M, b_i) > N_t \end{cases} \tag{7}$$

Where M is the candidate box with the highest confidence in the current stage, and $b_i$ is the remaining candidate box to be processed except the candidate box with the highest confidence in the current stage. When the IOU of the candidate box to be processed is greater than the threshold value, the score of this candidate is set to 0.

The difference between Soft-NMS algorithm and NMS algorithm is that Soft-NMS does not directly remove candidate boxes greater than the threshold value, but gives a attenuation function to attenuate the confidence of candidate boxes greater than the threshold value. Its formula is shown in (8):

$$s_i = s_i F(\mathrm{iou}(M, b_i)). \tag{8}$$

$F(\mathrm{iou}(M, b_i))$ as a attenuation function can be selected as a linear function or a Gaussian function, and its formulas are shown in (9) and (10) respectively:

$$F(\mathrm{iou}(M, b_i)) = e^{\frac{-\mathrm{iou}(M, b_i)^2}{\sigma}}. \tag{9}$$

$$F(\mathrm{iou}(M, b_i)) = \begin{cases} 1, iou(M, b_i) < N_t \\ 1 - iou(M, b_i), iou(M, b_i) > N_t \end{cases} \tag{10}$$

In this paper, the Gaussian function is selected as the attenuation function.

## 5 Datasets and Experiments

### 5.1 Datasets

1) This paper selects images from Pascal VOC2007 that contain the home environment or contain common targets of the home environment. After selection, there are 6418 pictures. Then, the data are randomly scrambled and divided into 6:2:2 ratios of training set, verification set and test set.
2) The self-made test set collection method is to use the service robot platform equipped with cameras (as shown in Fig. 7) in indoor environment to shoot videos for four types of targets (bottle, kettle, fan, chair) in different combinations at different angles for many times. Because the robot is constantly changing its position relative to the target, each target in the data set has a different scale.

**Fig. 7.** Service robot

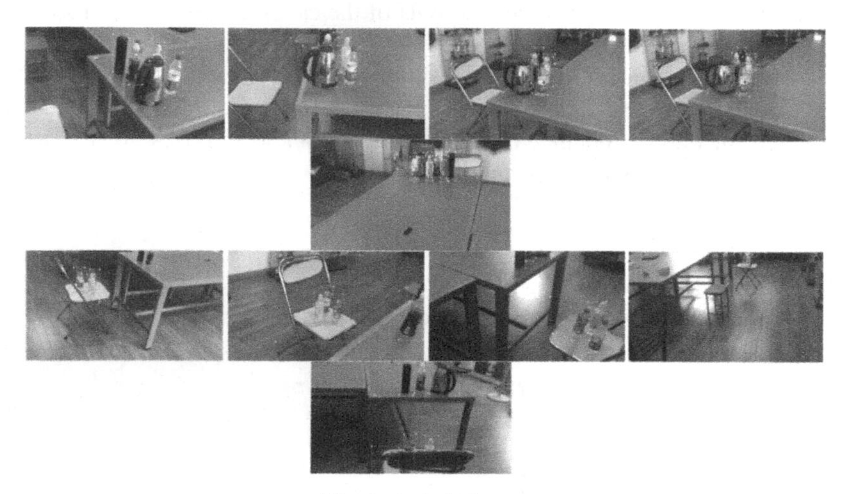

**Fig. 8.** Label pictures

Finally, we intercepted the video to get 600 pictures, and used the LabelImg software to label the targets in the pictures. After the annotation, the data were divided according to the ratio of training set, verifier and test set 6:2:2. Some pictures at different scales after annotation are shown in Fig. 8.

### 5.2 Experiment of Filtered VOC2007

In order to verify the effectiveness of the improved model, the filtered VOC2007 data set is used in this section for verification. The mAP curves of Bike, Person and TV detected by the model before and after improvement are shown in Fig. 9. The mAP values for all targets are shown in Table 1.

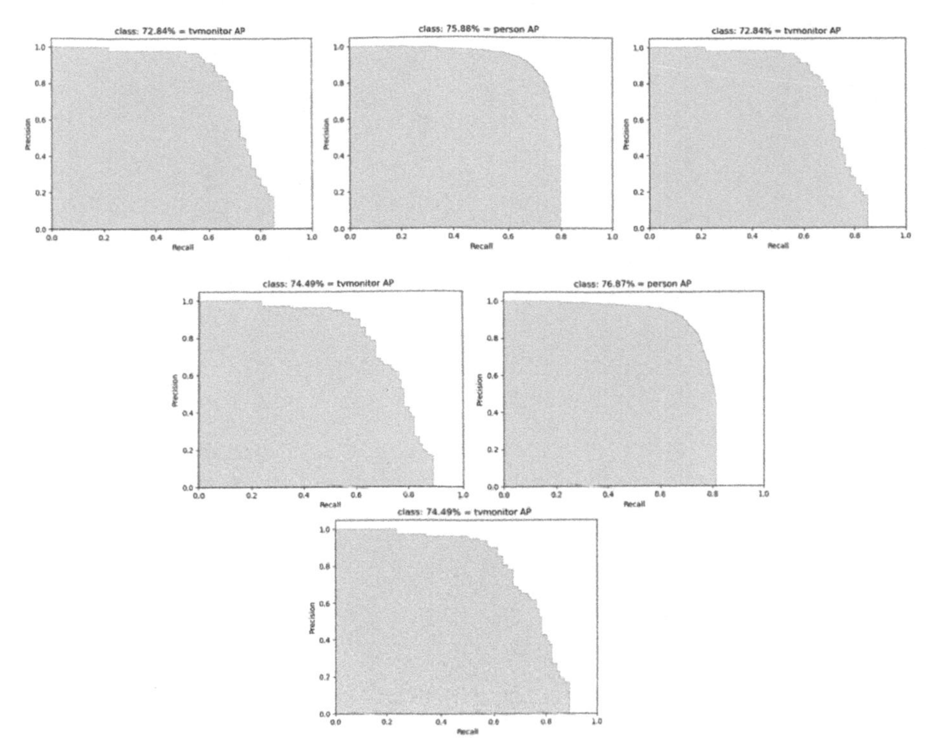


**Fig. 9.** The mAP curves of Bike, Person and TV detected by the model before and after improvement.

**Table 1.** The mAP of all detected by the model before and after improvement

| Method | mAP | Bike | Bottle | Person | TV | Chair | Table |
|---|---|---|---|---|---|---|---|
| R50+Fast | 70.26 | 90.50 | 43.18 | 75.88 | 72.84 | 66.52 | 72.64 |
| SMFR50+Fast | 71.26 | 93.31 | 44.07 | 76.87 | 74.49 | 66.01 | 72.83 |

where the top figure is the detection result of the original model and the bottom figure is the detection result of the improved model. As can be seen from Table 1, the mAP of the original model network is 70.26%.The mAP of model detection using multi-scale feature image fusion and soft-NMS strategy reached 71.26%.Although the overall mAP improvement is not obvious, the mAP of Bike, Person and TV is 1.81% higher than that of the original model.

### 5.3 Self-made Dataset Experiment

The mAP of bottle, fan and kettle of self-made data set is shown in Fig. 10, where the top figure is the detection result of the original model and the bottom figure is the detection result of the improved model. The overall mAP of all targets is shown in Table 2.

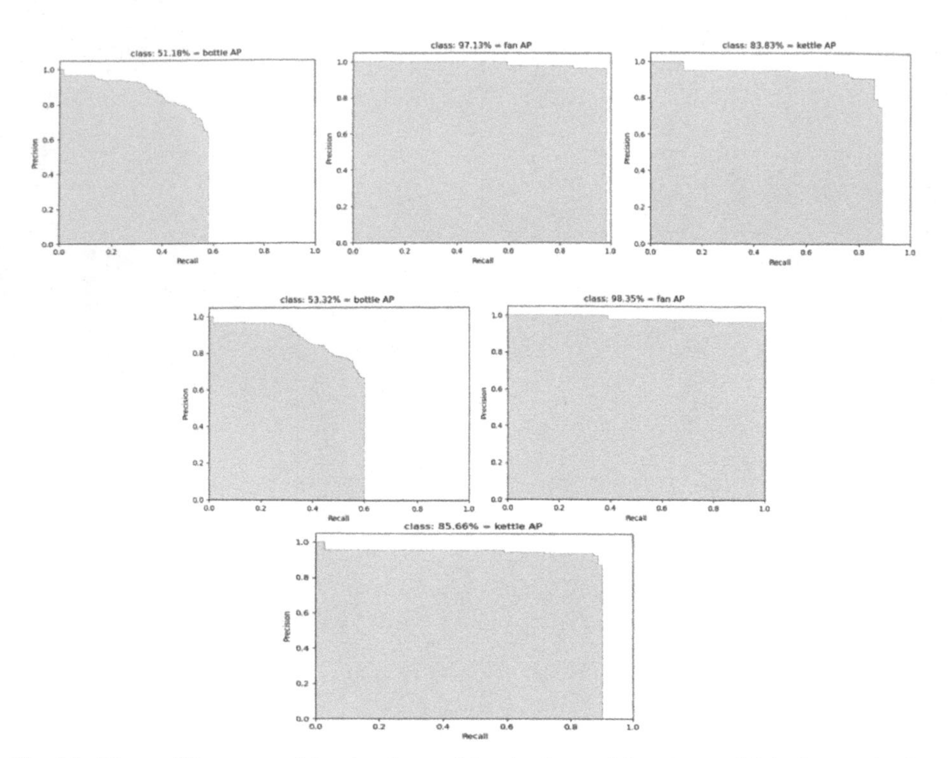


**Fig. 10.** The mAP curves of bottle, fan and kettle detected by the model before and after improvement

**Table 2.** The mAP of all detected by the model before and after improvement

| Method | mAP | Bottle | Fan | Kettle | Chair |
|---|---|---|---|---|---|
| R50+Fast | 82.62 | 51.18 | 97.13 | 83.83 | 98.33 |
| SMFR50+Fast | 84.07 | 53.32 | 98.35 | 85.66 | 98.96 |

As can be seen from Table 2, the improved model has a better detection effect on three types of targets, namely bottle, fan and kettle. In particular, bottle, such target of overlap and small targets, has an increase of 2.14% compared with the original model. Figure 11 shows the test results on this data set before and after the improvement, which visually shows the performance of the improved model in detecting overlapping small targets. In Fig. 11, the top image of the backbone network is the test result of ResNet50 model, and the bottom image is the test result of the improved model.

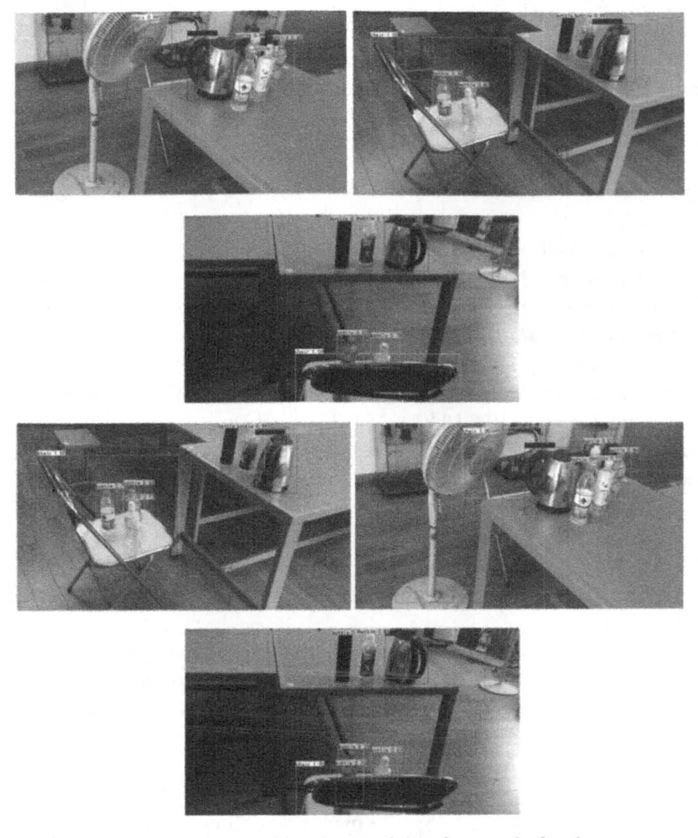

**Fig. 11.** The object detected by the model before and after improvement

As can be seen from Fig. 11, the original model is not as effective as the improved model in detecting bottles with partial occlusion, which indicates that the improved method is robust to a certain extent for the target of small-scale and overlap.

# 6 Conclusion

In order to improve the detection ability of the target of small-scale and overlap in indoor environment, this paper proposes a multi-layer feature fusion strategy aiming at the lack of information in the depth feature map for detection of small-scale targets in indoor environment, which integrates the rich detailed information in the shallow layer with the depth feature map with abstract semantics. Finally, the integrated feature map has both abstract semantic information and abundant shallow information. In addition, Soft-NMS is adopted in this paper to improve the performance of overlapping target detection. Experimental results show that this paper can effectively improve the detection performance of Faster-RCNN network in the target of small-scale and overlap in the indoor environment.

Although the improved network can improve the performance of detecting the small-scale and overlapping targets, there is still the case of missed detection in small-scale target with a high overlapping rate. Therefore, we will continue to study the issueon the basis of this paper to further improve the detection accuracy.

## References


1. Ren, S., He, K., Girshick R., et al.: Faster R-CNN: towards real-time object detection with region proposal networks. J. IEEE Trans. Pattern Anal. Mach. Intell. **39**(6), 1137–1149 (2017)
2. Ze, C., Ye, X., Qian, D., et al.: Small-scale pedestrian detection based on improved faster – RCNN. J. Comput. Eng. **46**(9), 226–232, 241(2020)
3. Girshick, R.: Fast R-CNN. J. Comput. Sci. (2015)
4. Uijlings, J.R., van de Sande, K.E., Gevers, T., Smeulders, A.W.: Selective search for object recognition. J. Int. Comput. Vis. **104**(2), 154–171 (2013)
5. He, K., Zhang, X., Ren, S., et al.: Deep residual learning for image recognition. In: IEEE Conference on Computer Vision and Pattern Recognition, pp. 770–778 (2016)
6. Neubeck, A., Van, Gool, L.: Efficient non-maximum suppression. In: International Conference on Pattern Recognition, pp. 850–855 (2006)
7. Dai, J., Li, Y., He, K., et al.: R-FCN: object detection via region-based fully convolutional networks. In: Proceedings of IEEE Conference on Computer Vision and Pattern Recognition, pp. 379–387 (2016)
8. Redmon, J., Farhadi, A.: YOLOv3: an incremental improvement. In: Proceedings of IEEE Conference on Computer Vision and Pattern Recognition, p. 16 (2018)
9. Redmon, J., Divvala, S., Girshick, R., et al.: You only look once: unified, real-time object detection. In: IEEE Conference on Computer Vision and Pattern Recognition (2016)
10. Redmon, J., Farhadi, A.: YOLO9000: better, faster, stronger. In: IEEE Conference on Computer Vision and Pattern Recognition, pp. 6517–6525 (2017)
11. He, K., Zhang, X., Ren, S., Sun, J.: Spatial pyramid pooling in deep convolutional networks for visual recognition. In: Fleet, D., Pajdla, T., Schiele, B., Tuytelaars, T. (eds.) ECCV 2014. LNCS, vol. 8691, pp. 346–361. Springer, Cham (2014). https://doi.org/10.1007/978-3-319-10578-9_23
12. Liu, W., Anguekiv, D., Erhan, D., et al.: SSD: single shot multibox detector. In: Leibe, B., Matas, J., Sebe, N., Welling, M. (eds.) ECCV 2016. LNCS, vol. 9905, pp. 21–37. Springer, Cham (2016). https://doi.org/10.1007/978-3-319-46448-0_2
13. Simonyan, K., Zisserman, A.: Very deep convolutional networks for large-scale image recognition. arXiv (2014)
14. Krizhevsky, A., Related, S., Technicolor, T.: et al.: ImageNet classification with deep convolutional neural networks. In: Advances in Neural Information Processing Systems, pp. 1097–1105 (2012)
15. Krizhevsky, A., Hinton, G.: Learning multiple layers of features from tiny images (2009)
16. Bodla, N., Singh, B., Chellappa, R., et al.: Soft-NMS – improving object detection with online of code. In: IEEE International Conference on Computer Vision (2017)
17. Zagoruyko, S., Lerer, A., Lin, T., et al.: A multipath network for object detection. J. Comput. Vis. Pattern Recogn. **23**(4), 1604–1607 (2016)
18. Zhu, O., Yeh, M.C., Cheng, K.T., et al.: Fast human detection using a cascade of histograms of oriented gradients. In: Proceedings of IEEE Conference on Computer Vision and Pattern Recognition, pp. 1491–1498 (2016)
19. Li, J., Zhang, Y.: Learning SURF cascade for fast and accurate object detection. In: Proceedings of IEEE Conference on Computer Vision and Pattern Recognition, pp. 3468–3475 (2016)

# Location Analysis and Research of Service Robot Based on Multi-sensor Fusion

Nan Yang(✉) and Wanmi Chen

School of Mechatronic Engineering and Automation, ShangHai University, Shanghai 200072, China

**Abstract.** For mobile service robots, positioning is the premise of all activities. Accurate positioning and path selection can improve service efficiency. Service robots generally operate without a map. It is known that a single sensor will produce relatively large errors in positioning, so a positioning method of multi-sensor fusion is proposed. Based on the integration of the oemometer and lidar, the robot position is estimated and information is fused by dead reckoning and ICP algorithm, and the error is corrected by particle filter.

**Keywords:** Odometer · Multi-sensor information fusion · Laser radar · Particle filter

## 1 Introduction

In service robot positioning technology, can effectively get the current state of the robot's position, is crucial to the next step of the robot positioning, accurate positioning the robot's current position can greatly reduce the system error, through internal sensors, external sensors can obtain the position of the robot, has the different effect of different kinds of sensors, we need to choose the appropriate sensor and filtering mode according to the requirements, and the fusion of multiple sensors can make up for the deficiency of a single sensor. The internal sensors mainly include dead-reckoning method based on odometer, while the external sensors mainly use laser radar to obtain the position and posture information of the robot. Zhao Yibing et al. used Kalman filtering algorithm to integrate odometer and inertial odometry unit, but it was not suitable for long-distance pose estimation [1]. Fan Binbin constructed a global map integrating odometer and lidar based on the ROS platform, and proposed that there was error accumulation over long distances, but this problem was not solved [2]. Zhao Qili used EKF filter to integrate odometer and lidar positioning and applied it to local positioning [3]. Wu Jingyang introduced particle filter Monte Carlo localization algorithm and algorithm experiment in detail [4]. Gerasimosgrigatos compared the advantages and disadvantages of EKF and particle filter in multi-sensor information fusion based on oemometer and sonar positioning [5]. Based on the prediction and update of EKF filter, JB Gao et al. proposed a fusion method based on two state information. Fox et al. used PF for mobile robot localization, and Orton et al. proposed a multi-target tracking and information fusion method based on PF for the disorder measurement from multiple sensors.


Q. Han et al. (Eds.): LSMS 2021/ICSEE 2021, CCIS 1469, pp. 35–43, 2021.
https://doi.org/10.1007/978-981-16-7213-2_4

## 2 Sensor Positioning Principle

### 2.1 Odometer Positioning Based on Dead Reckoning

Odometer as one of basic mobile robot positioning sensors, with high frequency output, simple structure, low cost, in a short period of time has the advantages of high positioning accuracy, if in an ideal world, odometer can accurate estimates of the robot relative to the initial position of the moving distance, as a service robot, the working environment of complex will exist a lot of the system error, such as wheels encounter obstacles or skid will make gyroscope is affected, cause the odometer did a lot of error accumulation with time, the Shanghai university of Liu Zhen and others provide the distinguish method of the system error in [6], through the gyroscope to get around the displacement increment, the actual displacement increment compared with displacement increment theory, if the actual displacement increment is less than the theoretical displacement increment, it indicates that the robot encounters obstacles or skids and other non-systematic errors, otherwise there is no obvious non-systematic errors.

Using dead reckoning [7] to achieve a single positioning of the odometer, dead reckoning is a more widely used means of positioning, in a short period of time to accurate positioning, using the odometer can get the exact orientation, horizontal position, velocity and acceleration, in points, the distance can be calculated at a particular moment as well as the change of direction, so as to analyze the position and posture of the robot (Fig. 1).

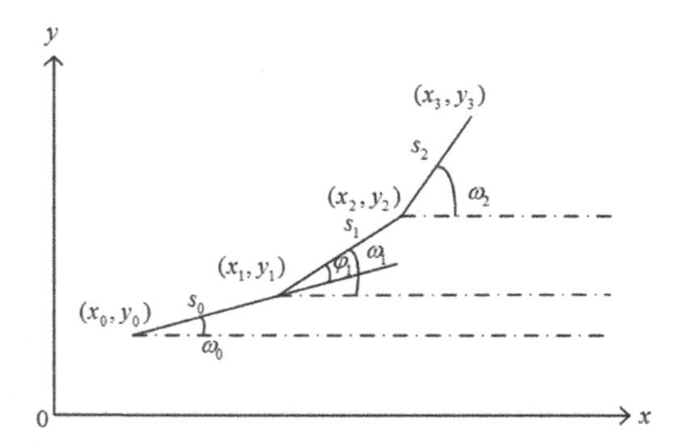


**Fig. 1.** Dead reckoning of circular motion

With $\left(x_t, y_t, w_t\right)$ to represent the robot's pose at time t, the distance traveled at adjacent times is expressed by $S_t$, and the Angle deviation at adjacent times is expressed by $\omega_t$, then the pose at time $t + 1$ can be expressed as:

$$\begin{cases} x_{t+1} = x_t + s_t cos w_t \\ y_{t+1} = y_t + s_t cos w_t \\ \omega_{t+1} = w_t + \varphi_{t+1} \end{cases} \quad (1)$$

In real life, odometers are installed on the left and right axles of the robot to collect the corresponding data and information, and the distance difference between the two wheels can be collected ($|s_l - s_r|$) the Angle change between adjacent distances can be obtained, and the path can be deduced according to the distance difference and Angle change obtained. Modeling and analysis can be carried out according to the actual movement of the service robot. After determining the model, the path calculation can be analyzed, which can be divided into two situations:

## 2.2 The Service Robot Moves in an Arc

Assume that the service robot moves in an arc from origin O to point A. The coordinate of the center position of the robot at time t is (x, y, w), after Δt time to arrive at the point A (x + Δx, y + Δy, w + Δw), The robot refers O to the center of the circle and makes circular motion for the radius; refers $\omega$ to the running Angle (and heading Angle) of the initial position; the left and right wheel spacing is L; the distance traveled by the left and right wheels is used respectively in the robot's motion from O to point A, The distance traveled by the left wheel and the right wheel is used separately $\Delta S_L$, $\Delta S_R$ said, Δx Δy Δw denotes the displacement of the robot from O to point A relative to the abscissa and ordinate and the change of the heading Angle, respectively. Under the ideal condition (Fig. 2):

$$\begin{cases} \Delta S_L = (R - L/2)\Delta W \\ \Delta S_R = (R + L/2)\Delta W \end{cases} \tag{2}$$

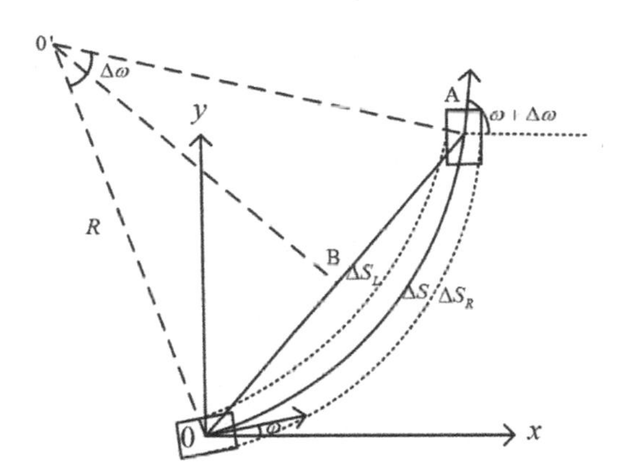


**Fig. 2.** Robot circular trajectory diagram

In the diagram O·O and O·A both are equal to the radius, the triangle ΔO·OA is an isosceles triangle, According to the property of isosceles triangle:

OA = 2OB = 2[sin(Δω/2) · O·O], there are:

$$R = \Delta S/\Delta\omega. \tag{3}$$

$$\Delta x = OA\cos\left(\omega + \frac{\Delta\omega}{2}\right) = 2R\sin\frac{\Delta\omega}{2}\cos\left(\omega + \frac{\Delta\omega}{2}\right) \tag{4}$$

$$= \left[\Delta S / \left(\frac{\Delta\omega}{2}\right)\right] \sin\left(\frac{\Delta\omega}{2}\right) \cos\left(\omega + \frac{\Delta\omega}{2}\right)$$

$$= \Delta S \left[\sin\left(\frac{\Delta\omega}{2}\right) / \left(\frac{\Delta\omega}{2}\right)\right] \cos\left(\omega + \frac{\Delta\omega}{2}\right).$$

$$\Delta y = OA \sin\left(\omega + \frac{\Delta\omega}{2}\right) = 2R \sin\frac{\Delta\omega}{2} \sin\left(\omega + \frac{\Delta\omega}{2}\right). \tag{5}$$

$$= \left[\Delta\omega / \left(\frac{\Delta\omega}{2}\right)\right] \sin\left(\frac{\Delta\omega}{2}\right) \sin\left(\omega + \frac{\Delta\omega}{2}\right)$$

$$= \Delta S \left[\sin\left(\frac{\Delta\omega}{2}\right) / \left(\frac{\Delta\omega}{2}\right)\right] \sin\left(\omega + \frac{\Delta\omega}{2}\right).$$

### 2.3 Robot Moves in a Straight Line

When the sampling interval $\Delta t$ is small enough or the robot is moving in a straight line, in the above for mula $\Delta\omega = 0$, It's given by the limit $\sin\left(\frac{\Delta\omega}{2}\right)/\left(\frac{\Delta\omega}{2}\right) = 1$ there are:

$$\begin{cases} \Delta x = \Delta s \cos\left(\omega + \frac{\omega}{2}\right) \\ \Delta y = \Delta s \sin\left(\omega + \frac{\omega}{2}\right) \end{cases} \tag{6}$$

By combining the equations of circular arc and linear motion, and adding the time constant, we can get:

$$\Delta S = (\Delta S_L + \Delta S_R)/2 \tag{7}$$

$$\begin{cases} x_{t+1} = x_t + \frac{\Delta S_{L,t} + \Delta S_{R,t}}{2} \cos\left(\omega_t + \frac{\Delta S_{R,t} - \Delta S_{L,t}}{2L}\right) \\ y_{t+1} = y_t + \frac{\Delta S_{L,t} + \Delta S_{R,t}}{2} \sin\left(\omega_t + \frac{\Delta S_{R,t} - \Delta S_{L,t}}{2L}\right) \\ \omega_{t+1} = \omega_t + \frac{\Delta S_{R,t} - \Delta S_{L,t}}{L} \end{cases} \tag{8}$$

Then the dead-reckoning formula of the robot is as follows:

$$\begin{bmatrix} x_{t+1} \\ y_{t+1} \\ \omega_{t+1} \end{bmatrix} = \begin{bmatrix} x_{t+1} = x_t + \frac{\Delta S_{L,t} + \Delta S_{R,t}}{2} \cos\left(\omega_t + \frac{\Delta S_{R,t} - \Delta S_{L,t}}{2L}\right) \\ y_{t+1} = y_t + \frac{\Delta S_{L,t} + \Delta S_{R,t}}{2} \sin\left(\omega_t + \frac{\Delta S_{R,t} - \Delta S_{L,t}}{2L}\right) \\ \omega_{t+1} = \omega_t + \frac{\Delta S_{R,t} - \Delta S_{L,t}}{L} \end{bmatrix} \tag{9}$$

It can be seen from the above formula that the positioning of the odometer needs the accumulation of time, under the condition that the speed per unit time remains unchanged and there is no non-mechanical error, the pose in a short time is very accurate. With the accumulation of time, the error will become larger and larger.

Therefore, in order to achieve accurate positioning for a long time, the odometer also needs to rely on other sensors to correct the error.

## 3 LiDAR Positioning is Realized Based on ICP Algorithm [8]

Laser radar transmission laser beam by specific frequency continuously into the surrounding environment, through the imaging laser reflected in sensor, generate two-dimensional graphic environment, so the laser radar, generally installed on the base of high-speed rotation, convenient on the surrounding environment like, usually used with SLAM, laser radar is used to the establishment of a map. The LIDAR scans the same environment at different locations at different times and registers the point cloud data obtained at different locations at different times using ICP algorithm to minimize the mechanical error (Fig. 3).

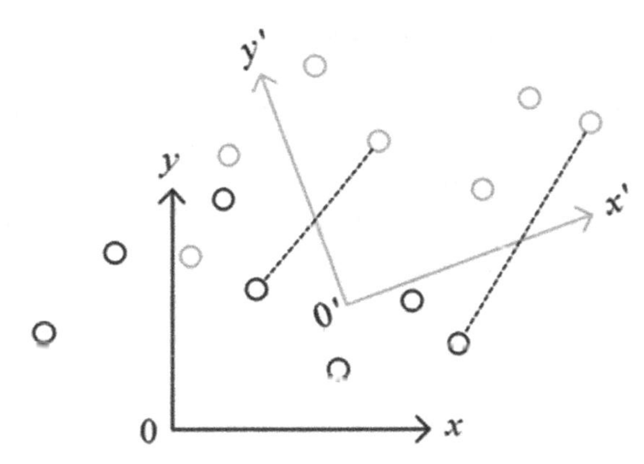


**Fig. 3.** ICP algorithm schematic diagram

The purpose of the iterative nearest point (ICP) is to merge two or more sets of data from different coordinate systems into a single coordinate axis by rotation and translation, which can be accomplished by a set of mappings.

$$H = \begin{bmatrix} a_{11} & a_{12} & a_{13} & t_x \\ a_{21} & a_{22} & a_{23} & t_y \\ a_{31} & a_{32} & a_{33} & t_z \\ v_x & v_y & v_z & s \end{bmatrix}. \tag{10}$$

$H = \begin{bmatrix} a_{11} & a_{12} & a_{13} & t_x \\ a_{21} & a_{22} & a_{23} & t_y \\ a_{31} & a_{32} & a_{33} & t_z \\ v_x & v_y & v_z & s \end{bmatrix}$. is a rotation matrix,You can use $A_{3\times3}$ said, $T\left[t_x\ t_y\ t_z\right]$ T is a translation vector, you can use $T_{3\times1}$ said, $V\left[v_x\ v_y\ v_z\right]$ denote the change vector, S represents the scaling factor. ICP algorithm can only rotate and shift the point cloud array,

Let V be the zero matrix, and S = 1:

$$H = \begin{bmatrix} A_{3\times3} & T_{3\times1} \\ O_{1\times3} & S \end{bmatrix}. \tag{11}$$

And $A_{3\times 3}$ can be expressed as:

$$A_{3\times 3} = \begin{bmatrix} 1 & 0 & 0 \\ 0 & \cos\alpha & \sin\alpha \\ 0 & -\sin\alpha & \cos\alpha \end{bmatrix} \begin{bmatrix} \cos\beta & 0 & -\sin\beta \\ 0 & 1 & 0 \\ \sin\beta & 0 & \cos\beta \end{bmatrix} \begin{bmatrix} \cos\gamma & \sin\gamma & 0 \\ -\sin\gamma & \cos\gamma & 0 \\ 0 & 0 & 1 \end{bmatrix}. \quad (12)$$

Where, $\alpha$, $\beta$ and $\gamma$ represent the rotation Angle of the robot relative to axis x, y and z, z represent the translation distance on axis $t_x$, $t_y$ and $t_z$ respectively, so points in different coordinates can be converted in this way:

$$X' = A_{3\times 3}X + T_{3\times 1}. \quad (12)$$

To find as many corresponding points as possible for conversion, can reduce the error.

Two groups of arrays $P = \{p_1, p_2, \cdots, p_n\}$ and $Q = \{q_1, q_2, \cdots, q_n\}$ in the point cloud data are respectively selected as the source point set and the target point set, and f(A, T) is used to represent the error from the target point set to the source point set, that is, the error between the point sets is converted into the optimal solution of min(f(A, T)).

$$f(A, T) = \sum_{i=1}^{n} \|Ap_i + T - q_i\|^2. \quad (14)$$

The point cloud registration accuracy of ICP is very high, which can be used to correct the errors accumulated by the odometer over time. The lidar makes a picture of the local environment, and the dead-reckoning method is used to estimate the robot's pose according to the data of the odometer, and then the obtained pose is matched with the laser point cloud to fuse the information.The actual pose of the robot measured by the lidar is taken as the element of the target point set, and the estimated pose calculated by the odeometer is taken as the element of the source point set. The standard of minimum error is set according to the actual situation, If $f(A, T) \leq \min$, the point data in the source point set will be retained for error adjustment; if $(A, T) \geq \min$, the points will be directly omitted, and the retained points will form a new point set.

## 4 Odometer and Lidar Positioning Method Based on Particle Filter [9]

In the process of positioning, odometer and laser radar will be affected by the process noise and observation noise, even if the fusion process example of error, but in the laser radar observations are produce process observation error and system error, using particle filter method to nonlinear systems for processing and optimization, to reduce errors as much as possible.

The purpose of particle filtering is to recurse the observed values with noise and estimate the posterior probability density of the samples composed of N particles of the nonlinear system state, and to register the pose calculated by the oedometer with the pose points observed by the lidar by ICP, will points of small errors are calibrated to constitute the new target point set $Q' = \{q'_1, q'_2, \cdots, q'_n\}$, while t times for $X_t = f(X_{t-1}, m_{t-1})$ system of the state, in which $m_{t-1}$ process for noise, t observation vector for $Y_t = f(X_t, n_t)$ moment system, in which $m_{t-1}$ is the observation noise, the posterior probability density of $P(X_t|Y_t)$, the system state sequence and observation sequence of $X_{0:t} = \{X_0, X_1, \cdots, X_t\}$, $Y_{1:t} = \{Y_1, Y_2, \cdots, Y_t\}$, respectively (the state sequence is $Q' = \{q'_1, q'_2, \cdots, q'_n\}$, namely $Q' = X_{0:t}$,Through the state sequence and observation sequence, the posteriori probability density formula is $P(X_t|Y_t) = \frac{P(Y_t|X_{0:t}, Y_{1:t-1})}{P(Y_t|Y_{1:t-1})}$, the sample set is $\{X^i_{0:t}, i = 1, 2, \cdots, N\}$, and the weight set of each particle in the sample is $\{W^i_t, i = 1, 2, \cdots, N\}$, and satisfies $\sum_i^N W^i_t = 1$, then the posteriori probability density at time t is approximately:

$$P(X_{0:t}|Y_{1:t}) \approx \sum_{i=1}^{N} W^i_t \delta\left(X_{0:t} - X^i_{0:t}\right). \tag{15}$$

The integration calculation of the posterior probability density is complex, so it is converted into the summation operation of the weighted samples:

$$\begin{aligned} E(D(X_{0:t})) &= \int D(X_{0:t}) P(X_{0:t}|Y_{1:t}) dX_{0:t} \\ &\approx \sum_{i=1}^{N} W^i_t D\left(X^i_{0:t}\right). \end{aligned} \tag{16}$$

$$\approx \sum_{i=1}^{N} W^i_t D\left(X^i_{0:t}\right).$$

Using particle filter method to extract effective samples in need in the probabilistic, but $P(X_{0:t}|Y_{1:t})$ posteriori probability can be standard, and changeable, it IS difficult to extract effective particle, so the introduction of importance sampling (IS), the purpose IS to avoid the density function of $q(X_{0:t}|Y_{1:t})$ harder to turn to another kind of easy to sample particles, $W^{*i}_t \triangleq W^{*i}_t = \frac{P(X_{0:t}|Y_{1:t})}{q(X_{0:t}|Y_{1:t})}$ of the importance sampling density IS required. $W^{*i}_t$, where $\widetilde{W^i_t} = \frac{W^{*i}_t}{\sum W^{*i}_t}$ is the non-normalized materiality right, and $\widetilde{W^i_t} = \frac{W^{*i}_t}{\sum W^{*i}_t}$ is the normalized materiality right, so $E(D(X_{0:t})) \approx \sum_{i=1}^{N} \widetilde{W^i_t} D\left(X^i_{0:t}\right)$.

After the new observation is obtained, the particle set is updated to obtain the weight of the new particle:

$$W^i_{t-1} \rightarrow W^i_t. \tag{17}$$

$$q(X_{0:t}|Y_{1:t}) = q(X_t|X_{0:t}, Y_{1:t}) q(X_{0:t-1}|Y_{1:t-1}). \tag{18}$$

The weight of the new particle is:

$$W^i_t = W^i_{t-1} \frac{P\left(Y_t|X^i_t\right) P\left(X^i_t|X^i_{t-1}\right)}{q\left(X^i_t|X^i_{t-1}, Y_t\right)}. \tag{19}$$

The process of reducing errors through particle filtering:

(1) Initialize the robot to obtain the observed values and the current calculated pose.

(2) Use ICP formula $f(A, T) = \sum_{i=1}^{n} \|Ap_i + T - q_i\|^2$ to find the optimal solution, and get the new target set, namely the example sample $Q' = \{q'_1, q'_2, \cdots, q'_n\}$.
(3) Convert the integral operation of posterior probability into the weighted summation operation:

$$E(D(X_{0:t})) = \int D(X_{0:t})P(X_{0:t}|Y_{1:t})dX_{0:t} \approx \sum_{i=1}^{N} W_t^i D\left(X_{0:t}^i\right)$$

(4) Update the weight of the particle:

$$W_t^i = W_{t-1}^i \frac{P(Y_t|X_t^i)P(X_t^i|X_{t-1}^i)}{q(X_t^i|X_{t-1}^i, Y_t)}$$

(5) $t = t + 1$
(6) Reduce the variance by changing the probability distribution and repeat steps (2), (3), (4) and (5).

## 5 Conclusion

The positioning of the odometer (source point set) based on dead reckoning has the characteristics of high output frequency and accurate positioning in short time. At the same time, the odometer is also very sensitive to the process error accumulated over time. The positioning over long distances will make the error of the odometer become larger and larger. Laser radar by launching laser beam on the surrounding environment so as to obtain the service robot working environment and the real position of the robot (target), the ICP algorithm from the source point set and target point set registration form particle samples, in global positioning, odometer and laser radar can correct odometer error caused by long distance,Particle filter is used to correct the influence of laser radar observation noise and process noise on the data, and the particles with small weight are screened out to reduce the variance to reduce the error.

## References


1. Zhao, Y., et al.: A novel intelligent vehicle localization method based on multi-sensor information fusion. J. Automot. Eng. **11**(1), 1–10 (2021)
2. Fan, B., Chen, W.: Self-localization of service robot based on inertial navigation and laser. J. Netw. New Media Technol. **6**(2), 46–51 (2017)
3. Zhao, Q.: Research on localization method of mobile robot based on multi-source information fusion. MA thesis, Harbin Institute of Technology (2020)
4. Wu, J.: Development of multi-autonomous mobile robot system based on LiDAR positioning and navigation. MA thesis, Harbin Institute of Technology (2017)
5. Rigatos, G.G.: Extended Kalman and particle filtering for sensor fusion in motion control of mobile robots. J. Math. Comput. Simul. **81**(3) (2010)
6. Liu, Z.: Error analysis and correction of mobile robot log legal position. J. Electron. Meas. Technol. **40**(12), 75–81 (2017)

7. Li, X.: Research on omnidirectional mobile robot localization and magnetic navigation based on dead reckoning. MA thesis, Nanjing University of Science and Technology (2017)
8. Li, Y.X.: Research on LiDar based point cloud data processing algorithm. MA thesis, Changchun University of Science and Technology (2020)
9. Lei, Y.: Particle filtering based service robot localization in dynamic environment. MA thesis, Southwest University of Science and Technology (2018)

# Research on Monocular SLAM of Indoor Mobile Robot Based on Semi-direct Method

Menghao Tian(✉) and Wanmi Chen

School of Mechatronic Engineering and Automation, Shanghai University, Shanghai 200072, China

**Abstract.** In order to improve the real-time performance of the indoor mobile robot equipped with a monocular camera, a monocular SLAM algorithm combining the direct method and the feature-based method is proposed. The algorithm in this paper is mainly divided into three threads: tracking, local mapping and loop closing. In the tracking thread, the initial pose estimation and feature alignment are obtained by minimizing the photometric error of the pixel blocks. If pose estimation through the direct method is not accurate enough, the pose is further optimized by minimizing the reprojection error; In the local mapping thread, the local map is settled and the local BA (Bundle Adjustment) is implemented for the optimization of the local keyframe poses and the spatial position of the local mappoints to improve the local consistency of SLAM; In the loop closing thread, the loop detection and loop optimization for the keyframes are executed to eliminate accumulated error and obtain a globally consistent trajectory. Through comparative experiments on the standard datasets with the mainstream ORB-SLAM algorithm, it is proved that the SLAM system in this paper has better real-time performance and robustness while ensuring the positioning accuracy.

**Keywords:** Indoor mobile robot · Semi-direct method · Monocular visual SLAM

## 1 Introduction

Localization and mapping are of great significance for realizing autonomous navigation of mobile robots. In the absence of prior environmental information, the information is only obtained through the sensors carried by itself. The robot estimates pose and builds the model of the surrounding environment during the movement. The problem is called as Simutaneous Localization and Mapping (SLAM) [1]. Vision sensors are divided into depth cameras, multi-lens cameras and monocular cameras. Because indoor mobile robots are based on embedded systems with weak processing capabilities, considering that the multi-lens cameras and depth cameras are expensive and the algorithm is very complex, this paper selects the monocular SLAM system as the research object.

According to the extraction of visual information in the images and matching methods, visual odometry, as an important part of the SLAM system, can be divided into the feature-based method and the direct method. The feature-based method extracts SIFT


Q. Han et al. (Eds.): LSMS 2021/ICSEE 2021, CCIS 1469, pp. 44–54, 2021.
https://doi.org/10.1007/978-981-16-7213-2_5

[2], SURF [3], FAST [4], ORB [5] from the images, and calculates the reprojection error based on 2D or 3D feature matches to estimate the camera pose and map. MonoSLAM proposed by A.J. Davison in 2007 is the first real-time monocular vision SLAM system [1]. In the same year, Klein et al. proposed PTAM (Parallel Tracking and Mapping) [6] to realize the parallelization of the tracking and mapping thread. ORB-SLAM [7] supports three modes of monocular, stereo and RGB-D to extract and match features. However, the extraction of keypoints and the calculation of descriptors are very time-consuming, and the SLAM system based on the feature-based method cannot be applied to situations with no texture or low texture.

Compared with the feature-based method, the direct method is based on the hypothesis of grayscale invariance solves the camera movement by minimizing the photometric error. Because the direct method eliminates the need to extract image features, the running frame rate is very high. And the system is of great robustness to photometric noise, even can track images in the absence of features. DTAM (Dense Tracking and Mapping) [8] is a monocular SLAM method based on the direct method, which extracts the inverse depth of each pixel to construct a dense depth map. LSD-SLAM (Large Scale Monocular SLAM) [9] marks the successful application of the monocular direct method in SLAM. SVO (Semi-direct Visual Odoemtry) [10], based on the sparse direct method, is very suitable for occasions with limited computing platforms. However, the direct method requires a large overlap of adjacent images, and cannot handle scenes with large changes in visual angle. The constraint based on the hypothesis of grayscale invariance is not strong. It is easy to fall into a local extreme value in the case of ambient luminosity changes.

Aiming at the problems of the feature-based method and the direct method, this paper combines the features and pixel gradient information to compensate for each other's shortcomings, and proposes a monocular SLAM algorithm based on the semi-direct method. This paper uses the direct method to obtain the camera's initial pose estimation and feature matching relationship. Keyframes are selected according to a certain strategy and extracted image features for loop detection and full BA. If the keyframe pose error estimated by the direct method is higher than the set threshold, the pose is further optimized through the feature-based method which interacts with local map.

## 2 Algorithm Framework

As shown in Fig. 1, the system is mainly divided into three parallel threads: tracking thread, local mapping thread, loop closing thread. The main work of each thread is as follows:

1. Tracking thread: Input the image sequences collected by the monocular camera, preprocess the image, including grayscale, distortion correction, and then perform monocular initialization. The tracking is divided into two stages, tracking the adjacent frames and the local map;
2. Local mapping thread: After the tracking thread inserts keyframes, keyframe pose and mappoint coordinates are optimized in the local map. At the same time, some

spatial points and keyframes are culled to maintain a certain map scale. After the keyframe is inserted, new mappoints are created with other keyframes in the local map;

3. Loop closing thread: Loop detection is performed through the bag-of-words model. When the loop is detected, the pose transformation between the loop frame and the current frame is calculated, and the Sim3-constrained pose graph optimization is performed to eliminate the accumulated error in order to a globally consistent trajectory. Finally, the full BA algorithm is executed to optimize all keyframes in the global map.

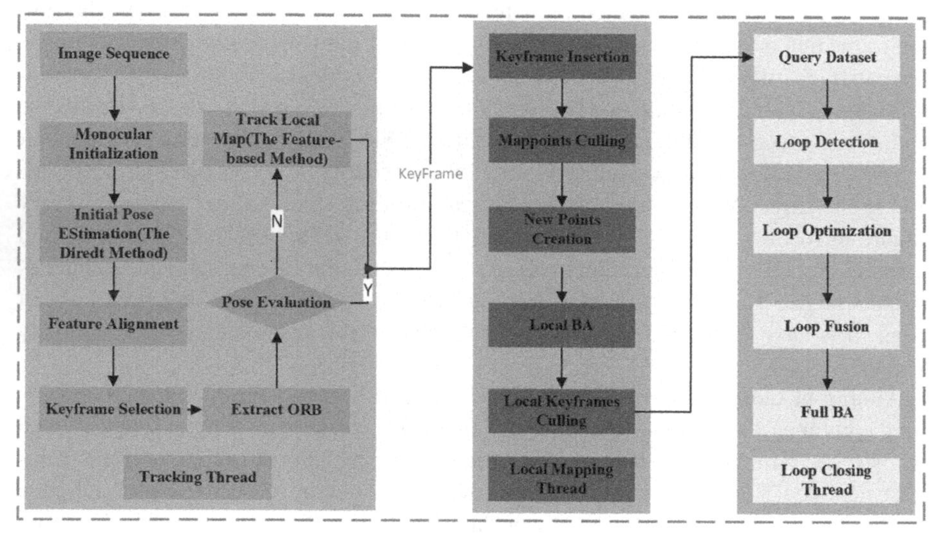


**Fig. 1.** System framework.

## 2.1 Monocular Initialization

The purpose of monocular initialization is to calculate the initial camera pose through the feature-based method and construct the initial three-dimensional feature map, taking the optical center at the initial moment as the origin of the world coordinate system. There are two geometric models to restore the camera pose from 2D matching point pairs: the homography matrix $H$ [11] for planar scenes and the basic matrix $F$ [11] for non-planar scenes. $H$ is solved through DLT (Direct Linear Transform), $F$ is solved by the classic eight-point method [12], and then the two matrices are decomposed by numerical or analytical methods to obtain the camera pose. However, when the feature points are coplanar or the camera is purely rotated, the degree of freedom of the fundamental matrix decreases, which causes so-called degenerate. Therefore, two models are calculated in parallel, and the model with smaller reprojection error is selected as the final motion estimation matrix.

### 2.2 Initial Pose Estimation

After the system collects the images, this paper first uses the direct method to estimate the initial pose by minimizing the photometric error of the same three-dimensional points projected to two adjacent frames. In order to avoid the shortcomings of a single pixel, those are no discrimination and the weak hypothesis of grayscale invariance, this paper utilizes pixel blocks of 4 × 4 for matching, and minimizes the photometric error of all corresponding pixel blocks in two adjacent frames to estimate initial pose. The formula is as follows:

$$T_{cl}^{init} = \arg\min_{T_{cl}} \frac{1}{2} \sum_{i \in \Omega} \|\delta I(\xi, P_{uv})\|^2. \tag{1}$$

where $\Omega$ is the set of pixel blocks, $\delta I(\xi, P_{uv})$ is defined as the grayscale residual of the image blocks,which can be expressed as:

$$\delta I(\xi, P_{uv}) = I_c(\pi^{-1}(T_{cl} P_l^i)) - I_l(\pi^{-1}(T(\xi) P_l^i)). \tag{2}$$

where $P_l^i = T_{lw} \cdot P_w^i$ is the homogeneous representation of the 3D coordinate of the mappoints in the coordinate system of the previous frame, and $\pi^{-1}$ represents the back-projection model from the feature points to the 3D points.

Given an initial relative pose transformation, then continuously update the pose estimation matrix through the incremental update formula. Since the pose transformation matrix is a constrained optimization problem, Lie algebra is used to transform this problem into an unconstrained optimization problem. Using GN (Gauss-Newton) algorithm to solve Eq. (2), after each iteration, update the estimated value $T_{cl}$:

$$T_{cl} = T_{cl} \cdot T^{-1}(\xi). \tag{3}$$

### 2.3 Feature Alignment

In order to obtain the local map keyframes and corresponding local mappoints, the paper use the set of matched mappoints obtained in the initial pose estimation to determine the local keyframes that have covisibility relationship with the current frame. In order to speed up the operation of the algorithm, the mappoints observed in the local keyframes are rasterized according to the pixel coordinates projected to the current frame. Each raster selects up to 5 mappoints with optimal depth to perform pixel matching.

Known the pose estimation of the current frame, project the local mappoints onto the current frame to get the pixel projection position $\pi\left(T_{cw}, P_k^w\right)$. The feature alignment step in this paper uses the iterative method from the high-level pyramid to the low-level pyramid to minimize the grayscale error of the image pixel blocks, and obtains the sub-pixel precision two-dimensional feature coordinates $u_k^{'}$ of the local mappoints projected to the current frame. The calculation formula is:

$$u_k^{'} = \arg\min_{u_k^{'}} \left\| I_c(u_k^{'}) - I_r(A_k \cdot u_k^r) \right\|^2. \tag{4}$$

where $A_k$ is the affine transformation matrix, which is used to adjust the affine transformation of the image blocks caused by the excessive distance between the current frame and the reference keyframe [13]. However, in the process of actually matching the local mappoints and the pixel points of the current frame, the pose transformation between the current frame and the reference keyframe is too large, there will be large differences in the ambient light conditions, and the exposure time of the two image frames is different. These factors lead to different grayscale values of the same local mappoint prejected to the current frame and the reference keyframe. In response to this problem, the gain-exposure model of literature [14] is used to compensate the pixel grayscale of the two frames, but the model is easy to fall into local extreme value. This paper improves it as follow:

$$u_k^{'} = \underset{u_k^{'}}{\arg\min} \left\| I_c(u_k^{'}) - e^{\partial} \cdot I_r(A_k \cdot u_k^r) - \beta \right\|^2. \tag{5}$$

where $e^{\partial}$ is exposure gain compensation, and $\beta$ is exposure offset compensation.

## 2.4 Tracking Local Map

When the keyframe is inserted into the local mapping thread, this paper extracts the ORB features from the keyframes and calculates the BRIEF descriptors [15] for loop detection and tracking the local map. If the direct method cannot obtain a sufficiently accurate pose estimation, then the covisible mappoints of are projected to the current frame, and the local map is tracked by minimizing the reprojection error to obtain a more accurate pose estimation. In order to reduce the interference of moving objects, according to the matching error of 3D-2D relationship, mappoints below the threshold are screened out as effective points to optimaize the camera pose:

$$\hat{T}_{iw} = \underset{T_{iw}}{\arg\min} \frac{1}{2} \sum \rho \left\| \frac{u_c^i - \Pi\pi(T_{iw} P_w^i)}{\sigma_i^2} \right\|^2. \tag{6}$$

where $\hat{T}_{iw}$ is the optimized pose, $T_{iw}$ is the initial pose, $u_c^i$ is the matching pixel coordinates in the current frame image, and $P_w^i$ is the 3D world coordinate of the feature points. $\sigma_i^2$ represents the variance of the feature points, which is related to the scale coefficient of the image pyramid and the number of pyramid layers where the feature points are located [7].

## 2.5 Keyframe Selection

The camera collects data during movement. It is unrealistic to use the estimated pose of each frame as a state variable for back-end optimization. This will lead to the rapid increase of the scale of the graph mode, accordingly pick some representative images as keyframes. The selection of keyframes has a greater impact on the accuracy of the system. When the selected keyframes are dense, there will be too much redundant information, which cannot meet the real-time requirement. When the selected keyframes are sparse, it may increase the difficulty of frames matching, even causes tracking failure. Based on these considerations, the keyframe selection strategy in this paper is as follows:

1. Since the last keyframe is inserted, there are already more than 20 frames of images;
2. The current frame tracks less than 50 mappoints;
3. The covisible information of the current frame and the reference keyframe is less than 80%.

## 2.6 Local Map Management

When a new keyframe is inserted into local map, it is necessary to create and remove mappoints according to a certain strategy, and remove keyframes to update the local map.

1. Create new mappoints: After inserting a new keyframe, in order to ensure that subsequent frames can continue to track, some new mappoints need to be added. For newly inserted keyframes, search for matching points in adjacent keyframes, triangulate the matched point pairs to restore three-dimensional points, and check whether the matching meets the photometric error, scale consistency, and the depth information of the mappoints after triangulation. It can be added into the local map only if the conditions are met;
2. Cull old mappoints: Mappoints obtained by mismatching triangulation, and mappoints that cannot be observed by multiple consecutive keyframes in a local map may bring additional error and increase the burden on the system. Therefore, the mappoints need to be properly culled, and can be retained only meet the following conditions: A stable mappoint can be tracked by at least three keyframes; The ratio of the frame number of the mappoints obtained by actual matching to the frame number of the mappoints that can be observed theoretically is not less than 0.3;
3. Cull key frames: After local map optimization, in order to maintain the scale of the map, it is necessary to detect keyframes with redundant information. If most of the features tracked by a keyframe can be tracked by other keyframes, the keyframe is considered redundant and needs to be eliminated.

## 2.7 Loop Closing Thread

When a new keyframe is inserted, the loop closing thread queries the keyframe database to search for the loop candidate keyframes through the open source library DBoW3 [16]. Each candidate keyframe is detected through time consistency, continuity testing and geometric verification. After detecting the loop information, the visual dictionary is used to match the feature points, and the RANSAC algorithm is used to eliminate the erroneous data association. After obtaining 3D-2D matching relationship, the pose graph optimization algorithm based on Sim3 constraint is executed for loop fusion. In this way, a globally consistent camera movement trajectory without accumulated error can be obtained.

## 3 Experiments

In order to test the performance of the semi-direct SLAM algorithm, comparative experiments with the mainstream ORB-SLAM algorithm are carried out on two standard datasets of TUM and KITTI. The CPU used for processing experimental data is AMD R5 4600U, quad-core 2.10 GHz, memory 16G, and running Ubuntu 20.04 system. The average processing speed of the semi-direct SLAM algorithm is approximately 30 frames per second, which meets the real-time requirements of indoor mobile robots.

### 3.1 Pose Graph Optimization Experiment

In order to verify the effect of the pose graph optimization which is used in the loop closing thread, a pose graph is generated using the simulation program that comes with g2o [17]. The true trajectory of the pose graph is a ball composed of multiple layers from bottom to top. Each layer is a perfect circle, and many circular layers of different sizes form a complete sphere, which contains 2500 pose nodes and 9799 edges.

**Fig. 2.** Pose graph after adding noise.

**Fig. 3.** Pose graph after optimization using g2o's default vertices and edges.

**Fig. 4.** Pose graph after optimization using Lie algebra.

By adding nosie to each edge, a pose graph with accumulated error is obtained (as shown in Fig. 2). Starting from the initial value of these edges and nodes with noise, try to optimize the entire pose graph to obtain data that approximates the true value.

Figure 3 represents the pose graph after graph optimization using g2o's default vertices and edges (g2o uses quaternions and translation vectors to express the posture by default), and Fig. 4 represents the pose graph optimized by Lie algebra. It can be seen that both are almost the same as the pose graph generated by the simulation program before adding noise. However, the former program still has a lower error after the set 30 iterations, while the latter program that uses Lie algebra remains unchanged after 23 iterations. Moreover, the overall error of the latter is 44360 under the SE3 edge measurement, which is slightly smaller than 44811 at 30 iterations of the former. This shows that after using Lie algebra to optimize the pose graph, better results are obtained with fewer iterations.

### 3.2 KITTI Dataset Experiment

The KITTI dataset contains dynamic scenes with moving vehicles such as urban roads and highways. The four sequences 00, 02, 05, and 08 of the KITTI dataset are selected. These sequences contain multiple loops, the path is more complex, the distance between frames is large, the running speed is fast, and the cumulative distance of every loop is long. These can effectively test the loop detection ability of the algorithm. Figure 5 is a comparison diagram of the estimated trajectory with the real trajectory. It can be seen that the trajectory estimated by the semi-direct SLAM algorithm basically coincides with the real trajectory. Table 1 shows the number of loops detected by the algorithm and RMSE (Root Mean Square Error). It can be seen from the table that the semi-direct SLAM algorithm uses the bag-of-words model to construct word vectors and compares them with keyframes in the database, which can effectively detect the loop frames and keep the positioning error within the allowable range. Andbasically there is basically no missing detection.

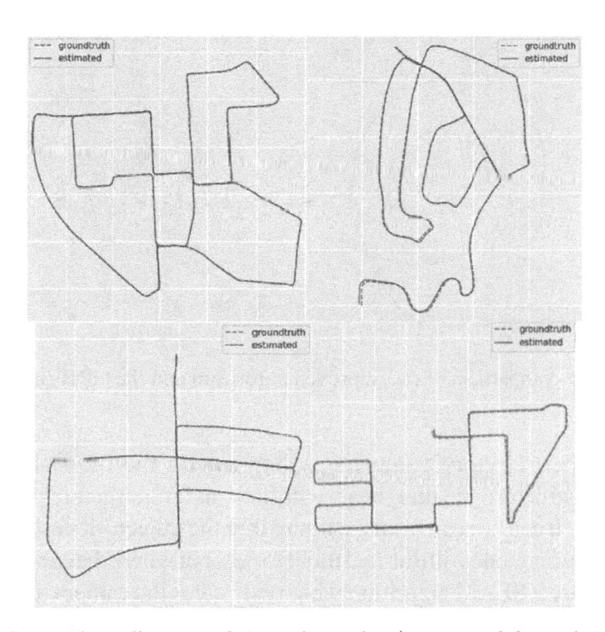


**Fig. 5.** Comparison diagram of the estimated trajectory and the real trajectory.

**Table 1.** The number of detected loops and RSME.

| Sequence | 00 | 02 | 05 | 08 |
|---|---|---|---|---|
| Number | 3 | 2 | 3 | 3 |
| RESE | 0.925 | 1.228 | 0.387 | 0.614 |

### 3.3 TUM Dataset Comparison Experiment

The TUM dataset is obtained by collecting indoor office scenes with a RGB-D camera. It is composed of depth maps and RGB maps. It is a commonly used dataset in various SLAM/VO algorithm experiments. The algorithm performance is further tested on this dataset and compared with the mainstream ORB-SLAM algorithm, as shown in Fig. 6, Tables 2, 3 and 4.

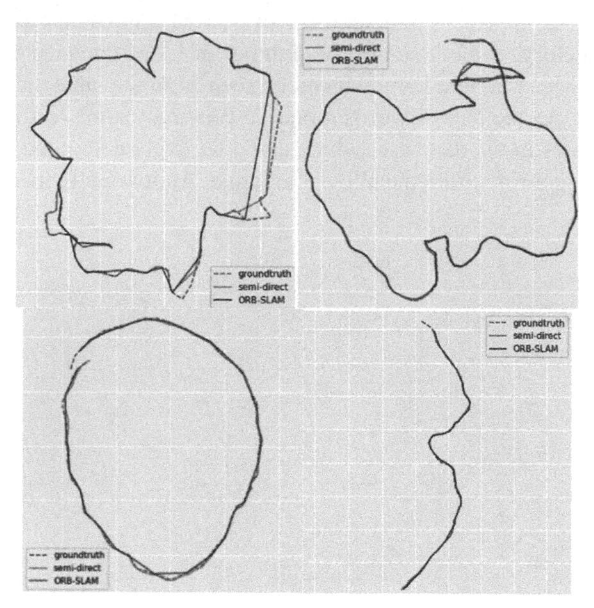


**Fig. 6.** Trajectory comparison of the improved algorithm and the ORB-SLAM algorithm.

Figure 6 are the trajectory comparison diagram on the fr2/desk, fr3/long_office _household, fr3/nostructure_texture_near_withloop and fr3/structure_texture sequences of the TUM dataset. It can be seen from the figure that the trajectories estimated by the two algorithms basically coincide with the real trajectories, but where the camera rotation angle is large, the semi-direct SLAM algorithm which estimates the camera movement through the grayscale gradient information can track image information of adjacent frames faster and has strong robustness, so the tracking effect is better than ORB-SLAM. In the sequence fr2/desk, the camera shakes violently, the tracking effect of the two algorithms is poor.

Tables 2 and 3 are ATE (Absolute Trajectory Error) of the two algorithms respectively. ATE is the difference between the estimated pose and the real pose. The program aligns the real value with the estimated value according to the timestamp of the pose, and calculates the difference between each pair of poses. The four indicators for comparison are: maximum error, average error, median error, and root mean square error. The above three tables shows that the positioning accuracy of the algorithm of the paper is better than the ORB-SLAM algorithm on most sequences, and the running speed is increased by about 20% on average. In fr3/nostructure_texture_near_withloop and

**Table 2.** ATE of the semi-direct SLAM algorithm.

| Sequence | 1 | 2 | 3 | 4 |
|---|---|---|---|---|
| Max | 0.0172 | 0.0418 | 0.0952 | 0.0356 |
| Mean | 0.0080 | 0.0153 | 0.0407 | 0.0166 |
| Median | 0.0079 | 0.0135 | 0.0393 | 0.0168 |
| RESE | 0.0086 | 0.0172 | 0.0471 | 0.0181 |

**Table 3.** ATE of the ORB-SLAM algorithm.

| Sequence | 1 | 2 | 3 | 4 |
|---|---|---|---|---|
| Max | 0.0212 | 0.0581 | 0.0425 | 0.0352 |
| Mean | 0.0095 | 0.0188 | 0.0215 | 0.0190 |
| Median | 0.0091 | 0.0173 | 0.0202 | 0.0179 |
| RESE | 0.0103 | 0.0214 | 0.0230 | 0.0203 |

**Table 4.** Mean tracking time of the two algorithms.

| Sequence | 1 | 2 | 3 | 4 |
|---|---|---|---|---|
| Semi-direct | 0.0211 | 0.0239 | 0.0242 | 0.0255 |
| ORB-SLAM | 0.0295 | 0.0319 | 0.0250 | 0.0304 |

fr3/structure_texture_ sequences, there are obvious textures on the surface of the object, and the feature-based method also has a good tracking effect.

## 4 Conclusion

A monocular vision SLAM algorithm combining the feature-based method and the direct method is proposed. The tracking thread obtains the initial pose estimation and feature correspondence by minimizing the photometric error, and further optimizes the pose by minimizing the reprojection error to track the local map. The local mapping thread maintains a certain scale through reasonable strategies to cull spatial points and keyframes. Finally the bag-of-words model is used to detect loop to obtain a globally consistent trajectory. Experiments on standard datasets show that the semi-direct SLAM algorithm maintains a high positioning accuracy and operating speed. Because the line segment features in indoor scenes are obvious, and have good illumination invariance and rotation invariance. The line segment features can describe the geometric structure information of the environment to build a higher-level environmental map [18, 19]. Therefore, effective use of line segment features to improve the robustness of the system is the focus of the next step.

## References


1. Davison, A.J., Reid, I., Molton, N., et al.: Monoslam: real-time single camera SLAM. J. IEEE Trans. Pattern Anal. Mach. Intell. **29**(6), 1052–1067 (2007)
2. Keltner, D., Sauter, D., Tracy, J., et al.: Emotional expression: advances in basic emotion theory. J. Nonverb. Behav. **43**(2), 133–160 (2019)
3. Bay, H., Tuytelaars, T., Van Gool, L.: SURF: speeded up robust features. In: Leonardis, A., Bischof, H., Pinz, A. (eds.) ECCV 2006. LNCS, vol. 3951, pp. 404–417. Springer, Heidelberg (2006). https://doi.org/10.1007/11744023_32
4. Tsang, V.: Eye-tracking study on facial emotion recognition tasks in individuals with high-functioning autism spectrum disorders. J. Autism. **22**(2), 161–170 (2018)
5. Polikovsky, S., Kameda, Y., Ohta, Y.: Detection and measurement of facial micro-expression characteristics for psychological analysis. Kameda's Publ. **110**, 57–64 (2010)
6. Klein, G., Murray, D.: Parallel tracking and mapping for small AR workspaces. In: 6th IEEE and ACM International Symposium on Mixed and Augmented Reality, ISMAR 2007, pp. 225–234. IEEE (2007)
7. Mur-Artal, R., Montiel, J., Tardós, J.D.: ORB-SLAM: a versatile and accurate monocular SLAM system. J. IEEE Trans. Robot. **31**(5), 1147–1163 (2015)
8. Newcombe, R.A., Lovegrove, S.J., Davison, A.J.: DTAM: dense tracking and mapping in real-time. In: IEEE International Conference on Computer Vision. ICCV 2011, Barcelona, Spain. IEEE (2011)
9. Engel, J., Sturm, J., Cremers, D.: Semi-dense visual odometry for a monocular camera. In: Proceedings of the IEEE International Conference on Computer Vision, pp. 1449–1456 (2013)
10. Forster, C., Pizzoli, M., Scaramuzza, D.: SVO: fast semi-direct monocular visual odometry. In: 2014 IEEE International Conference on (revised edition) Robotics and Automation (ICRA), pp. 15–22. IEEE (2014)
11. Hartley, R., Zisserman, A.: Multiple View Geometry in Computer Vision. Cambridge University Press, Cambridge (2003)
12. Hartley, R.I.: In defense of the eight-point algorithm. J. IEEE Trans. Pattern Anal. Mach. Intell. **19**(6), 580–593 (1997)
13. Klein, G., Murray, D.: Parallel tracking and mapping for small AR workspaces. In: IEEE and ACM International Symposium on Mixed and Augmented Reality, Piscataway, USA, pp. 250–259. IEEE (2008)
14. Szeliski, R.: Image alignment and stitching: a tutorial. Found. Trends Comput. Graph. Vision. **2**(1), 1–104 (2006)
15. Calonder, M., Lepetit, V., Strecha, C., Fua, P.: Brief: binary robust independent elementary features. In: Daniilidis, K., Maragos, P., Paragios, N. (eds.) ECCV 2010. LNCS, vol. 6314, pp. 778–792. Springer, Heidelberg (2010). https://doi.org/10.1007/978-3-642-15561-1_56
16. Gálvez-López, D., Tardos, J.D.: Bags of binary words for fast place recognition in image sequences. J. IEEE Trans. Robot. **28**(5), 1188–1197 (2012)
17. Kümmerle, R., Grisetti, G., Strasdat, H., et al: G2O: a general framework for graph optimization. In: IEEE International Conference on Robotics and Automation, Shanghai, pp. 3607–3613. IEEE (2011)
18. Li, H.F., Hu, Z.H., Chen, X.W.: PLP-SLAM: a visual SLAM method based on point-line-plane feature fusion. J. Robot. **39**(2), 214–220 (2017)
19. Pumarola, A., Vakhitov, A.: PL-SLAM: real-time monocular visual SLAM with points and lines. In: IEEE International Conference on Robotics and Automation, Piscataway, USA, pp. 4503–4508. IEEE (2017)

# Research on Basketball Robot Recognition and Localization Based on MobileNet-SSD and Multi-sensor

Cheng Lin and Wanmi Chen(✉)

School of Mechatronic Engineering and Automation, Shanghai University, Shanghai 200444, China

**Abstract.** The rules of basketball robot game and the structure level of basketball robot system are introduced, and it is clear that the main task requirement of the perception subsystem is to identify and locate the target. The traditional target recognition method is sensitive to natural light, and the generalization ability and robustness are poor. To address the problems of traditional algorithms, this paper implements a target detection model with good robustness and high accuracy based on the application of migration learning on deep learning, using the SSD (Single Shot Multibox Detector) algorithm with a lightweight MobileNet as the feature extraction network. In the target localization method, the localization strategy of "using laser sensors for long distance and short distance" is proposed. Finally, the above research results are integrated and tested in the basketball robot by target recognition experiments, and the fast and accurate target recognition is successfully achieved.

**Keywords:** Basketball robot · Target recognition · Target localization

## 1 Introduction

Traditional ball detection algorithms mainly color features of balls and geometric shape features or a combination of these two features [1] detect the target. Literature [2] uses Houge features for localization detection of basketballs, and literature [3] combines color thresholding and edge detection for basketball detection and localization. The detection algorithm based on color features is affected by the illumination and is prone to misclassification when the background environment is complex. The recognition algorithm based on geometric shape features is more difficult to detect small targets, and the algorithm is not very stable.

The development and maturity of image processing technology and computer vision technology provide new methods for ball detection. In 2016, Liu et al. [4] applied a single deep neural network to image target detection and proposed the SSD (single shot multibox detector) algorithm. In 2018, the literature [5] combined convolutional neural networks (convolutional neural networks) with SSD algorithm for road traffic signal detection. Mobilenet [6] model is a deep learning lightweight classification model proposed by Google for mobile environment with low latency but can maintain certain


Q. Han et al. (Eds.): LSMS 2021/ICSEE 2021, CCIS 1469, pp. 55–66, 2021.
https://doi.org/10.1007/978-981-16-7213-2_6

accuracy. SSD combines the regression idea in YOLO SSD combines the regression idea of YOLO and the anchor mechanism of Faster R-CNN, which not only has the speed of YOLO but also can be as accurate as Faster R-CNN. The SSD-Mobilenet model performs well in terms of recognition accuracy and performance consumption because it combines the advantages of both types of networks. The focus of this paper is to develop a fast target recognition and localization method for basketball robots using the SSD-Mobilenet network model and a multi-sensor system, which can combine real-time and accuracy.

## 2 Robot Sensing Platform Construction

Basketball robots are fully autonomous mobile robots with independent image processing and intelligent decision making capabilities. A complete basketball robot generally consists of a decision subsystem, a perception subsystem, a control subsystem and a communication subsystem. The system structure of the basketball robot is shown in Fig. 1.

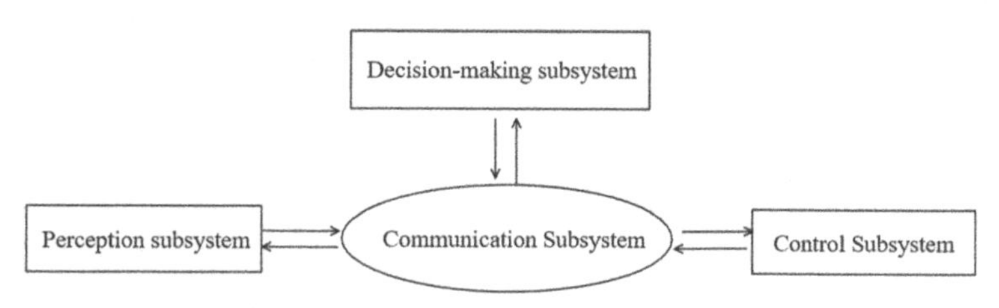


**Fig. 1.** Basketball robot system structure.

### 2.1 The Functions of Each Subsystem

The decision subsystem completes the processing of the data collected by the perception subsystem and then converts them into decision commands to complete the corresponding action process of the robot; the perception subsystem collects the data from each sensor and then transmits them to the decision subsystem through the communication subsystem, and the sensors used are camera sensors for vision and LIDAR for position estimation, odometer and gyroscope, etc.; the control subsystem converts the control subsystem converts the decision commands sent from the decision subsystem into various control signals, such as motion control, ball drop control, and ball pickup control. The motor controllers are commonly used in daily life and industry, and each motor controller contains speed control unit, position control unit and motor drive, etc. The communication subsystem completes the data transfer between the upper and lower computers and between the controller and sensors. Although the basic functions of the system structure of basketball robots are the same, in the specific form of implementation, the robots of different teams have different characteristics and the performance is very different.

The main task of the basketball robot perception subsystem is target identification and localization, and its hardware usually consists of vision sensors. However, in order to improve the speed, accuracy and robustness of target recognition and localization, teams usually add other sensors to assist the perception subsystem to better accomplish the task. In this paper, a combination of monocular camera sensor and laser sensor hardware is used.

## 3 Target Recognition Algorithm

The recognition target of the robot in the basketball robot game is the ball. The traditional target recognition algorithm achieves target recognition by extracting the color features and contour shape features [7] of the ball. The target contour recognition algorithm uses the circular contour feature of the ball, but this type of recognition algorithm is easily disturbed by other background objects in the environment [8], which will not only consume more computation but also reduce the accuracy of recognition. The target color recognition algorithm utilizes the color feature information of basketball and volleyball. The difficulty of this type of algorithm is that the color information collected by the camera will be affected by the ambient lighting and the description of the color features [9].

### 3.1 MobileNet-SSD Recognition Algorithm

With the wide application of CNN in the field of image recognition, deep neural networks have greatly improved the performance of computer vision tasks. For the network structure, the current general trend is to build deeper and more complex network structures in order to build networks with higher accuracy [10]. Although SSD is better than traditional detection algorithms in terms of detection accuracy and detection speed, it still cannot achieve real-time target detection. MobileNet, on the other hand, proposes an efficient lightweight network architecture to build a lightweight convolutional neural network based on depth-separable convolution. By introducing two hyperparameters, width multiplier and resolution multiplier, a small size and low latency model is constructed, which can effectively compromise between latency and accuracy. These hyperparameters allow the selection of a suitably sized model based on constraints [11].

### 3.2 Depth-Separable Convolution

The depth-separable convolution built by MobileNet combines two operations, Depthwise and Pointwise, as shown in Figs. 2 and 3. The combination of the two operations results in a lower number of parameters and lower computational cost than the conventional convolution operation, which allows the MobileNet neural network to combine detection accuracy and detection speed.

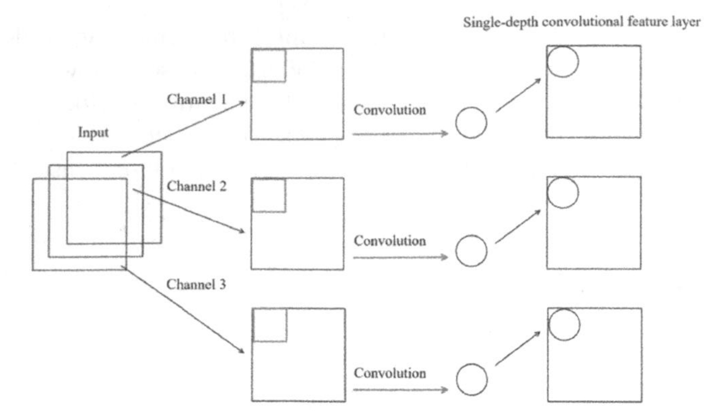


**Fig. 2.** Single-depth convolution process.

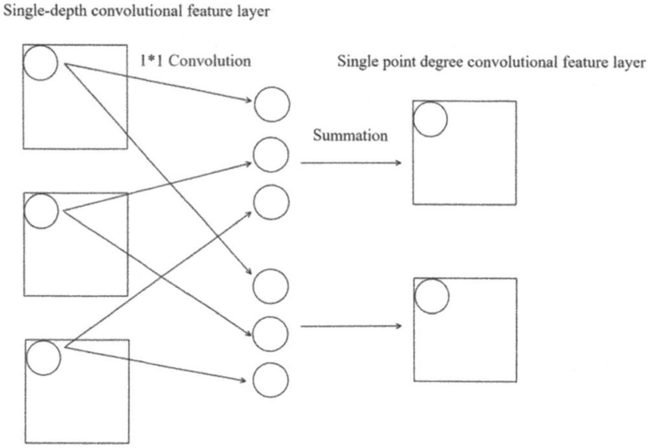

**Fig. 3.** Single point convolution process.

### 3.3 Ball Classification and Position Localization

After getting the input image, MobileNet-SSD needs to localize the specific targets in the image, and then classify the localized targets into basketball class or volleyball class. The loss function of MobileNet-SSD consists of two parts: classification and regression, and the specific process is: defining the input sample $x$, adjusting both the position error loss function and the confidence level according to the specific classification and localization tasks error loss function, and the loss function is shown in Eq. (1),

$$L(x, c, l, g) = \frac{1}{N}\left(L_{conf}(x, c) + \partial L_{loc}(x, l, g)\right). \quad (1)$$

where $N$ is the number of selected predefined boxes; the role of $\partial$ is to adjust the ratio between the position error, and the confidence error, which is set to 1 by default; $c$ is the category confidence prediction value; $l$ is the location of the a priori box of the bounding box; $g$ is the position parameter of the real target; $x$ is an indicator parameter of the specific form $x_{ij}^{p} \in \{1, 0\}$, when $x_{ij}^{p} = 1$, the $i$-th priori box matches $j$-th real target, that is, for the $j$-th real target, the $i$-th priori box is its positive sample, and the class of the real target is $p$; when $x_{ij}^{p} = 0$, for the $j$-th real target, the first prior frame is its negative sample, and the class of the real target is $p$.

where the position loss function $L_{loc}$ uses the SmoothL1 loss function in the form shown in Eqs. (2) to (7),

$$L_{loc}(x, l, g) = \sum_{i \in Pos}^{N} \sum_{m \in \{cx, xy, w, h\}} x_{ij}^{k} smooth_{L1}\left(l_i^m - \hat{g}_j^m\right). \quad (2)$$

$$smooth_{L1}(x) = \begin{cases} 0.5x^2 & if\ |x| < 1 \\ |x| - 0.5 & otherwise \end{cases}. \quad (3)$$

$$\hat{g}_j^{cx} = \frac{g_j^{cx} - d_i^{cx}}{d_i^w}. \quad (4)$$

$$\hat{g}_j^{cy} = \frac{g_j^{cy} - d_i^{cy}}{d_i^h}. \quad (5)$$

$$\hat{g}_j^{w} = \log\left(\frac{g_j^w}{d_i^w}\right). \quad (6)$$

$$\hat{g}_j^{h} = \log\left(\frac{g_j^h}{d_i^h}\right). \quad (7)$$

Where: $l_i^m$, $\hat{g}_j^m$ represent whether the $i$-th prediction box and the $j$-th truth box in the category match for category $k$, the value takes 1 when matching, otherwise takes 0. The coordinates of the center point of the bounding box $(cx, xy)$, the width of the bounding box is $w$ and the height is $h$.

The other part of the loss function $L_{conf}$, is the classification loss, defined as positive and negative case loss:

$$L_{conf}(x, c) = -\sum_{i \in P_w}^{N} x_{ij}^{p} \log \hat{c}_i^0 \quad where\ \hat{c}_i^N = \frac{expc_i^0}{\sum_n expc_i^p}. \quad (8)$$

In the first term, the prediction frame $i$ matches the truth frame $j$ with respect to the category $p$. The higher the probability of $p$ prediction, the smaller the loss. The second term indicates that there is no target in the prediction box, and the higher the probability that the prediction is background, the smaller the loss. Also, the probability $\hat{c}_i^p$ is generated by *softmax*.

### 3.4 Multi-distance Position Localization

Since the positioning of targets at close distances is not accurate using vision sensors, laser sensors can make up for the disadvantage of unsatisfactory positioning results at close distances in order to improve the accuracy of target positioning. In order to give full play to the respective advantages of camera sensors and laser sensors, this paper adopts a "segmented" positioning method, firstly, a sensor-based segmentation: vision sensors are used for long-range (>1 m) target positioning; laser sensors are used for short-range (<1 m) targets. Then the segmentation is based on the positioning process: due to the positioning accuracy and the performance limitation of the basketball robot positioning system, it is almost impossible for the robot to correctly "walk to the sphere" and complete the ball picking action if the movement command is given to the motion control subsystem through the decision subsystem after a one-time positioning. Therefore, a multi-positioning method with "multi-angle and multi-distance levels" is adopted.

# 4 Experimental Results and Analysis

## 4.1 Dataset Construction

There are many datasets for basketball and volleyball categories, and in order to adapt to the official basketball and volleyball types specified by the robotics competition and to select the appropriate distances and angles, the datasets in this paper are derived from real environments and manually labeled. The dataset contains 1021 basketball images, 1211 volleyball images, and 923 mixed basketball and volleyball images, with a total of 8,351 ball targets, as shown in Figs. 4, 5.

The sample dataset is divided into a training set and a test set, where the training set consists of 721 basketball images, 911 volleyball images, and 623 mixed basketball and volleyball images, and the test set consists of 300 basketball images, 300 volleyball images, and 300 mixed basketball and volleyball images.

**Fig. 4.** Single target dataset.

**Fig. 5.** Mixed target dataset.

### 4.2 Experimental Environment

The experimental platform configuration of this paper is shown in Table 1. The model construction, training and testing of the results are all done in the Caffe framework, using the CUDA parallel computing architecture, while the cudnn acceleration library is integrated into the Caffe framework to accelerate the computer computing power.

**Table 1.** Specific environment configuration.

| | | |
|---|---|---|
| Server platform hardware configuration | System | Ubuntu16.04 |
| | GPU | NVIDIA GTX 1650 |
| Software configuration | Programming languages | Python3.7 |
| | Deep learning framework | Caffe |
| | Labeling tools | LabelImg |
| | Accelerated | CUDA10.0 |

### 4.3 Testing Performance Evaluation and Analysis

After 12,000 iterations of the designed model using the above profile, the MobileNet SSD model already has a good effect on the recognition of targets, and the effect is shown in Figs. 6 and 7.

**Fig. 6.** Single target recognition effect.

**Fig. 7.** Mixed target recognition effect.

From Figs. 6 and 7, it can be seen that the constructed MobileNet-SSD fine-tuning model has high accuracy, good robustness to changes in illumination, blur and shadow, and good generalizability to different styles of balls, and it can correctly identify the nearest target ball.

The average detection time obtained from this model is about 0.052 seconds per frame. The maximum frame rate of the field detection video obtained from the camera is 17.8 frames per second, and the detection time needs to be less than 0.056 seconds per frame to complete the real-time detection task. It can be seen that the average detection time of this method is lower than the demanded time, and it is able to detect basketball and volleyball targets in different environments.

(1) Evaluated on a training set

From the loss images of the training process in Figs. 8, 9 and 10, it can be seen that the average level of loss decreases faster before 1000 step in the training process, and then the loss stays around 2.5 with a jump of about 1. At 7000 step, the average level of the loss is nearly stable. It proves that the experiment has selected the appropriate learning parameters to make the whole loss curve drop smoothly.

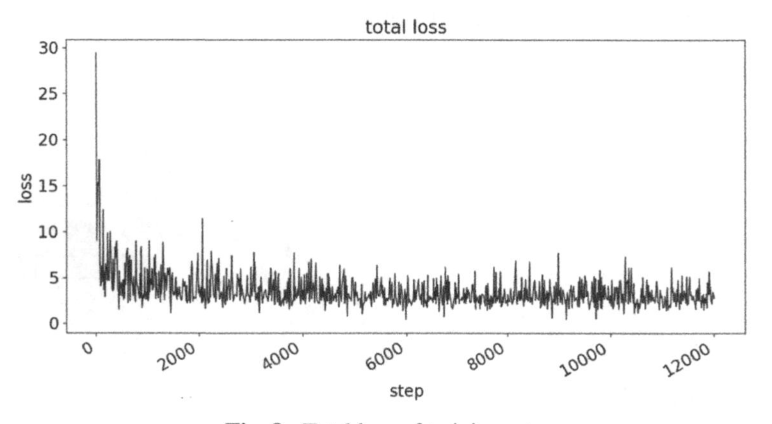


**Fig. 8.** Total loss of training set.

At the same time, although the overall loss curve in the image decreases, the "width" of the curve is larger, which is not conducive to the fitting of the model in the later stage of training, and it is difficult to further improve the accuracy. This indicates that the variance of the samples used in the two adjacent training sessions is too large, and it is necessary to increase the batchsize and sample size to continuously train and improve the model.

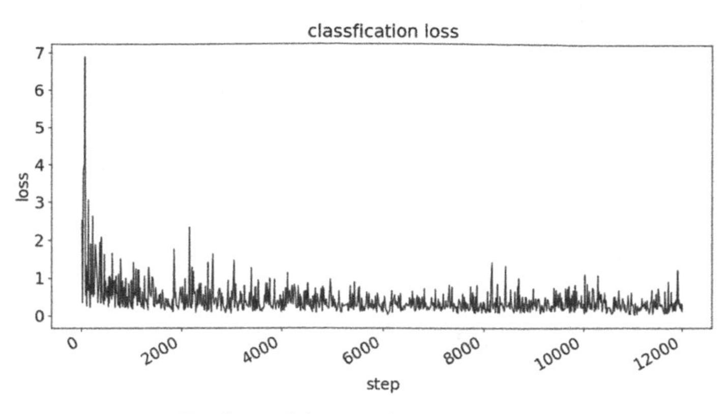


Fig. 9. Training set classification loss.

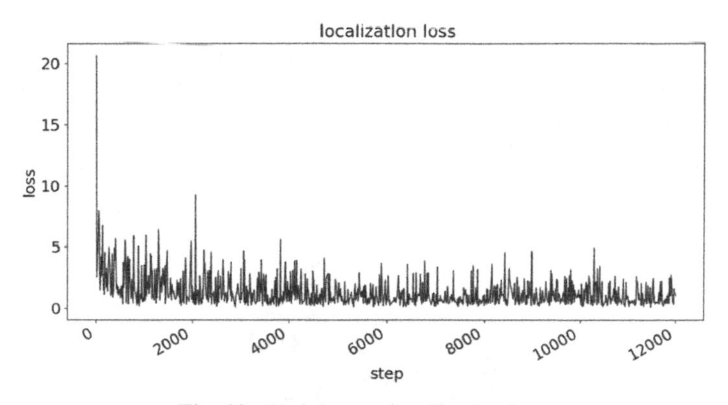


Fig. 10. Training set localization loss.

(2) Evaluated on a test set

The performance of the model training set and test set in Figs. 11 and 12 shows that the model is adequately trained, the loss function converges faster, the recognition accuracy is high, no overfitting is seen, and the training set and test set data meet the expected results.

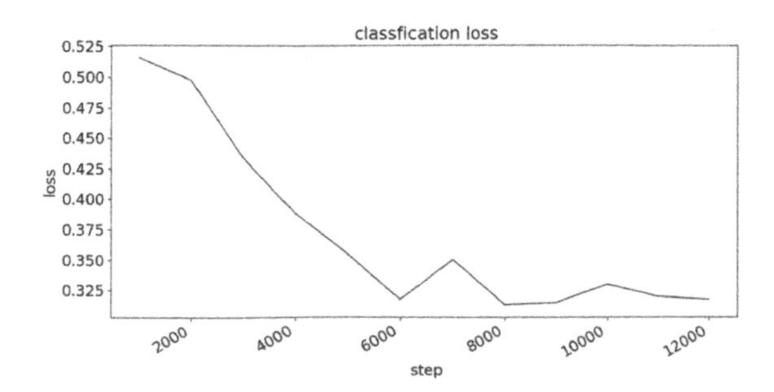


Fig. 11. Test set classification loss.

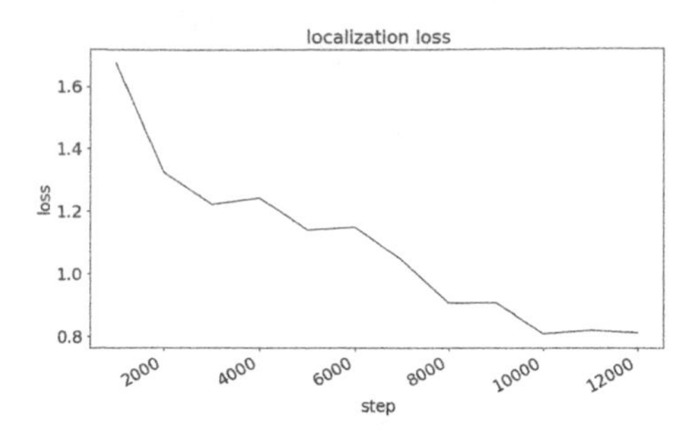


**Fig. 12.** Test set location loss.

(3) Position localization

After the robot identifies the target ball, the control subsystem module is invoked, and when the robot moves to a distance of about 50 cm relative to the target, and the angle is about 270°, the LIDAR vision subsystem module is invoked, and the position of the target ball is at the white point in the point cloud. As shown in Fig. 13.

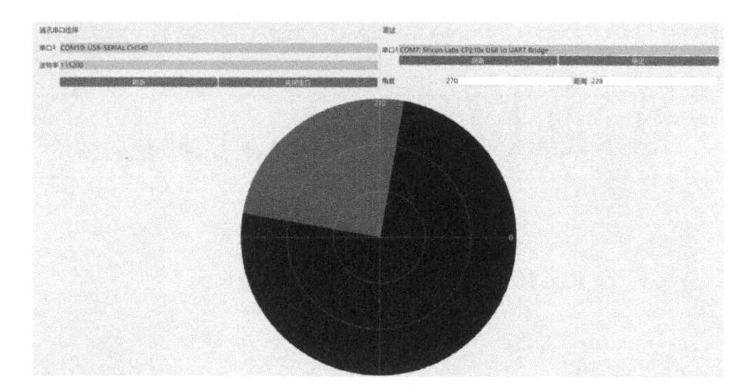

**Fig. 13.** Lidar vision subsystem effect.

Finally, the relative positioning between the robot and the target ball is completed by the LIDAR vision sub-code, as shown in Fig. 14. The target object is accurately picked up, and the identification and localization of the basketball robot target is realized, as shown in Fig. 15.

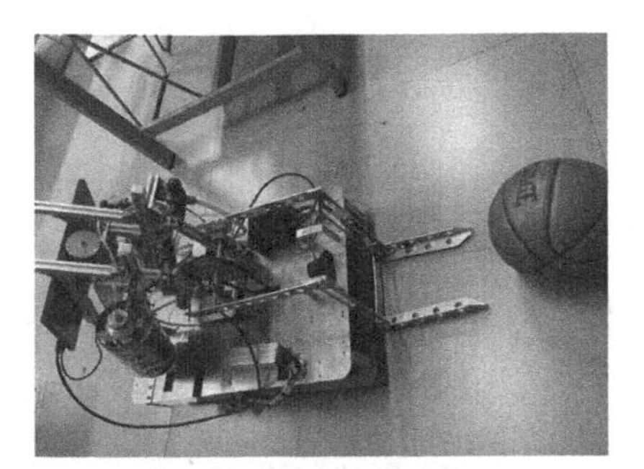

**Fig. 14.** Laser ball finding within 50 cm range.

**Fig. 15.** Pick up the target ball.

## 5 Conclusion

This paper first analyzes the requirements of the ball recognition and localization system based on monocular camera and LIDAR sensor, then introduces the structure and characteristics of MobileNet, constructs a ball detection network model based on MobileNet and SSD according to the task requirements, analyzes the loss function of the model, and realizes a ball detection model based on MobileNet-SSD. The MobileNet-SSD ball detection model is implemented. We also elaborate on each aspect of the MobileNet-SSD model, design and analyze the selection and configuration of the development environment, the preprocessing of the data set, the parameter configuration of the model training and the output results, and demonstrate the performance of the trained model. From the analysis of the structure of the experiments, it can be seen that the MobileNet-SSD model designed in this paper has a good recognition effect on the recognition of balls. In terms of localization by combining multiple sensors and multiple distances, it is able to find the target object accurately.

## References


1. Omachi, M., Omachi, S.: Traffic light detection with color and edge information. In: 2009 2nd IEEE International Conference on Computer Science and Information Technology, pp. 284–287. IEEE Press, Beijing (2009)
2. Li, F.J., Chen, W.M.: Research and implementation of LabVIEW-based vision system for basketball robots. J. Ind. Control Comput. **2014**(8), 10–11 (in Chinese)
3. Zhang, F.Y., Zhou, S., Zhang, T.: Research on basketball localization detection based on improved Hough. J. Autom. Technol. Appl. **38**(8), 118–121 (2019). (in Chinese)
4. Liu, W., et al.: SSD: single shot multiBox detector. In: Leibe, B., Matas, J., Sebe, N., Welling, M. (eds.) ECCV 2016. LNCS, vol. 9905, pp. 21–37. Springer, Cham (2016). https://doi.org/10.1007/978-3-319-46448-0_2
5. Muller, J., Dietmayer, K.: Detecting traffic lights by single shot detection. In: 2018 21st International Conference on Intelligent Transportation Systems, pp. 266–273. IEEE Press, Maui (2018)
6. Howard, A.G., Zhu, M., Chen, B., et al.: MobileNets: efficient convolutional neural networks for mobile vision applications (2017)

7. Lü, W., Wang, J., Yu, W., Bao, X.: Range profile target recognition using sparse representation based on feature space. J. Shanghai Jiaotong Univ. (Sci.) **22**(5), 615–623 (2017). https://doi.org/10.1007/s12204-017-1879-4
8. Feng, Y., Liu, H., Zhao, S.: Moving target recognition and tracking algorithm based on multi-source information perception. Multimed. Tools Appl. **79**(23–24), 16941–16954 (2019). https://doi.org/10.1007/s11042-019-7483-x
9. Zhang, H., Wang, X., Jiang, L., Xu, Y., Jiang, G.: Near color recognition based on residual vector and SVM. Multimed. Tools Appl. **78**(24), 35313–35328 (2019). https://doi.org/10.1007/s11042-019-08164-1
10. Zhang, J., et al.: An improved MobileNet-SSD algorithm for automatic defect detection on vehicle body paint. Multimed. Tools Appl. **79**(31–32), 23367–23385 (2020). https://doi.org/10.1007/s11042-020-09152-6
11. Li, Y., Huang, H., Xie, Q., Yao, L., Chen, Q.: Research on a surface defect detection algorithm based on mobilenet-SSD. J. Appl. Sci. **8**(9), 1678 (2018)

# Artificial Potential Field Method for Area Coverage of Multi Agricultural Robots

Zhonghua Miao, Guo Yang, Chuangxin He, Nan Li, and Teng Sun(✉)

School of Mechatronic Engineering and Automation, Shanghai University, Shanghai 200072, China
sunny315@shu.edu.cn

**Abstract.** The robot area full coverage problem is a particular path planning problem, which requires robots to cover all parts of the area except obstacles. It has been used in the actual production of agriculture, such as weeding, fertilization, spraying. In this paper, an artificial potential field method of multi-robot full coverage path planning algorithm is used to make several agricultural robots spontaneous organization and coordination, to achieve the full coverage of farmland, greenhouse and other agricultural scenes. According to robots' repulsion force to obstacles in the artificial potential field method, the repulsion force between robots is increased. The goal points are generated to lead the robots to move to the direction that is not covered, and the re coverage of the area is minimized as far as possible. Simulations show that the algorithm can achieve more than 95% full coverage of the task area with good performance, adaptability and robustness.

**Keywords:** Full area coverage · Multiple robots · Agricultural robots

## 1 Introduction

Multi-robot cooperative work is an effective means of agricultural production in the future [1]. In larger agricultural environments, a single robot is not competent for crop weeding, fertilizing, spraying, etc. [2], the multi-robot coordination to complete the regional coverage is the most straightforward solution efficiency [3, 4], how to effectively control of multi-robot coordination for the regional coverage tasks, in agriculture has significant application value and practical significance, it has become one of the research hotspots in the field of collaborative control in recent years.

Compared with a single robot, a multi-robot system has many advantages, such as more reliability, more flexibility, stronger robustness and easier expansion [5, 6]. They are more reliable because the failure of any one member of the flock does not bring the whole system down; Multi-robot can reduce errors in ranging, communication, sensing, etc. With the increasing task, the algorithm applied to the multi-robot system is still effective. In view of the many advantages of multi-robots, some scholars have applied multi-robots to the regional coverage task. Therefore, in a certain historical time, multi-robots cooperate to complete the coverage task. Multi-robot area coverage means that multiple robots form a team to effectively cover the entire work area through


Q. Han et al. (Eds.): LSMS 2021/ICSEE 2021, CCIS 1469, pp. 67–76, 2021.
https://doi.org/10.1007/978-981-16-7213-2_7

a cooperative strategy. Thus, the working ability of each robot is effectively improved, and the working time of the overall covering task is reduced. The multi-robot cooperative task can overcome the limitation of the number of robot sensors, make a multi-directional observation of the target area, meet more search needs, and complete the target task more efficiently [3].

### 1.1 Previous Studies

Existing multi-robot region coverage methods are mainly based on task division after environment decomposition [7]. After task division, each robot covers its own region. This method divides the problem into two tasks; the first is the division of the area, the second is the area coverage planning. In [8], the target area is decomposed into a fine grid, and the unified grid unit set represents the entire environment. Each unit has an associated value, the value 0 represents the free space, and the value 1 represents the obstacle space. Reference [9] adopts the off-line algorithm based on trapezoidal decomposition, as shown in the Fig. 1 below, for covering path planning of farmland and agricultural machinery. Based on trapezoidal element decomposition, [10] proposed the double-line scanning method. The double-line scanning method is to cut the target area with two false scan lines, divide the whole area into many blocks by obstacles and double lines, and replace all blocks with nodes respectively to generate the adjacency map. However, when the environment is complex and the edge of the obstacle is irregular, the decomposition of trapezoidal element method is limited.

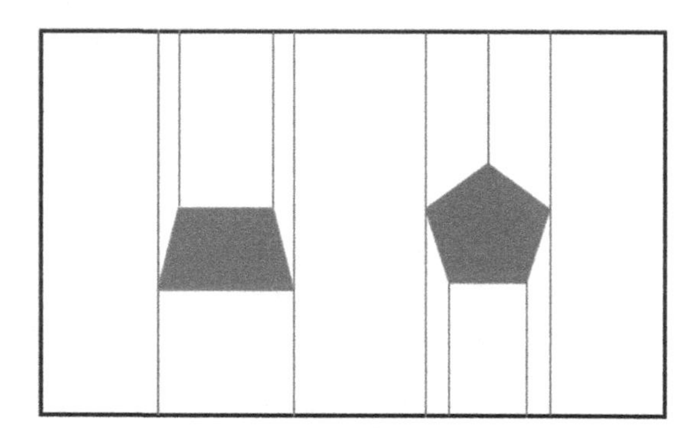

**Fig. 1.** Trapezoidal decomposition diagram.

Since the environmental decomposition in the first step only gets blocks, the next step needs to plan the paths of robots in each partition, and the classic method in robot coverage path planning is to search back and forth [11]. In Reference [12], STC (Spanning Tree Coverage) algorithm was applied to the region coverage for the first time, improving the traditional method and achieving good coverage of the task area. The improved distributed anti-flocking algorithm in reference [13] improves the area coverage of wireless sensor. Besides, emerging methods are increasingly being applied to regional coverage planning. The template-based method considers all motion conditions, customizes the corresponding motion mode for each situation, divides the behavior of the robot into a variety of fixed templates, and selects the appropriate template according to the actual situation. The generated path will consist of multiple templates. The disadvantage is that it takes time to customize the template based on understanding the environment map in advance. Reference [14] combines the template-based method with heuristic path planning and realizes the completeness of coverage by using some predefined templates to realize the robot's real execution behavior.

### 1.2 Contributions and Organization of the Paper

In this paper, a multi-robot full coverage algorithm is designed based on the artificial potential field method. The main work includes: (1) Information map mechanism: The information storage mechanism of the historical trajectory coverage map is designed for each robot so that it can record information with other robots and avoid repeated coverage. (2) Collision avoidance rules: The problem of agricultural robot movement is simplified, and the movement rules are designed according to the repulsive force principle of the artificial potential field method, giving priority to the uncovered area guidance rule and the rule of keeping a certain distance range. (3) In order to prevent agricultural robots from operating in non-work areas, an operating mechanism within the boundary is designed. MATLAB simulation shows that the algorithm can achieve nearly complete coverage of multiple agricultural robots in the work area and has nothing to do with the initial position of each agricultural robot.

The rest of the paper is organized as follows. In the second section, we reviewed the use scenarios of multiple agricultural robots, made a problem description, and introduced information maps. The multi-robot full coverage algorithm based on the artificial potential field method is given and analyzed in the third section. The simulation results of multiple agricultural robots in the obstacle environment are in Sect. 4. The experiment summary and discussion are in Sect. 5.

## 2 Preliminaries

The algorithms presented in this paper consider a group of $N$ agricultural robot moving in a convex region. All agricultural robots are considered isomorphic and isotropic, equipped with the same operating tools, such as weeding tools and spraying tools. The working range of the agricultural robot tool is recorded as $R_w$, it is obvious that the working radius $R_w$ must be greater than the actual size of the agricultural robot. The operating range of the agricultural robot is more prominent than its own radius. The coverage of the agricultural robot is the range that the tool can operate, and does not refer to the range that the actual robot walks through in this paper. In this work, we assume that each agricultural robot is equipped with a wireless communication module with an isotropic range, sensing and communicating with other robots. Each robot can easily get the map information covered by other robots within the given range. Agents are used to name each robot, such as a-agent, b-agent, to distinguish different robots, dynamics of a-agent are given by

$$\begin{cases} \dot{q}_i(t) = p_i(t) \\ \dot{p}_i(t) = u_i(t), \quad i = 1, 2, \ldots, N \end{cases} \tag{1}$$

Where $q_i(t), p_i(t), u_i(t)$ are the position, velocity, and control input of agent $i$ at time $t$. The deviation between the actual position of the robot and the position of the wireless sensor is ignored here, and both are considered to be at the same position.

The working scene of the agricultural robot is discretized into a grid map of $M \times N$ dimensions. The values of M and N depend on the size of the actual working scene and are also limited by precision. Excessive precision will increase computing power and invalidate the algorithm. A map with its own coverage information is created on each robot. The map information exists in the form of a matrix. The value of each unit in the matrix reflects the number of times the agricultural robot traverses it. Obstacles also have unique values in the matrix. Matrix maps allow robots to exchange information about themselves and distinguish between obstacles and workspaces. Figure 2 shows the concept of map discretization. The coordinates, footprints and obstacles of each agricultural robot can be represented in an $M \times N$-dimensional matrix, as shown in the figure. The value 0 in the matrix represents uncovered freedom workspaces, the larger the value represents the free workspaces covered, the larger the value, the more times the robot cover, the unique value represents the obstacles that cannot be explored. The boundary of the map is the boundary of the matrix.

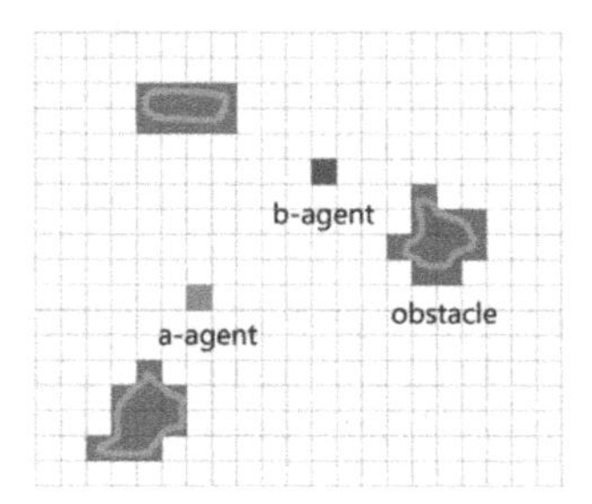


**Fig. 2.** Simple map information.

The coordinates $[x, \ y]$ is used as the agricultural robot's position on the map. As the agricultural robot moves, the map information will be updated continuously, including the robot's current position and the trace of the robot in the past time. At the same time, the information is transmitted in real time to other robots. Through this map update mechanism, each agricultural robot can clearly know the location and coverage information of other robots in the area it has covered. Shared information can reduce the overlap of the overall coverage area. The map information design is used to maximize coverage areas and reduce duplicate coverage areas in the algorithm.

## 3 Multi-robot Covering Algorithm Based on Artificial Potential Field Method

The artificial potential field method is a common method for robot path planning. It assumes that the robot is moving under a virtual force field. The mathematical function of this mechanism is applied to the mobile robot environment model. The robot is attracted by the guide point and the obstacle's repulsive force [15]. The robot is applied with repulsive force from other robots within a certain distance in the paper. The agricultural robot is represented by a square in Fig. 3. The five robots generate trajectories after running for a period. Trajectories are represented by the positions recorded at different times. Obstacles have a fixed range of repulsive force.

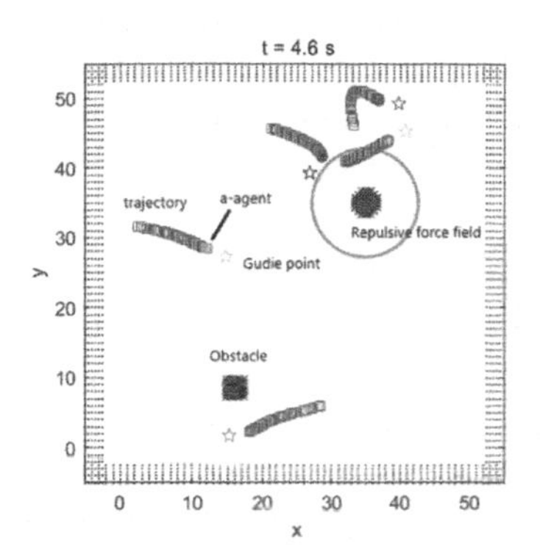


**Fig. 3.** Motion diagram.

In addition, each agricultural robot has its own guide point, which is indicated by a star in the figure. It facilitates the robot to move to uncovered areas. The gravitational force of the robot by the guide point is unique, and the gravitational force guides the robot to the guide point. However, the repulsion force received by the robot is not unique, and the amount of repulsion force received corresponds to the amount within the range of the obstacle repulsion field.

No robot in the picture is repulsed by obstacles. But for the robot marked with yellow, after entering the repulsive force field of the circular obstacle, it is attracted by the attraction from the guide point, the repulsive force from the circular obstacle and the repulsive force from other robots. The a-agent is not in the repulsion field of the obstacle, so it is not affected by the repulsion force of the obstacle. Therefore, the join force of the a-agent will guide the robot to maximize the coverage area and avoid collision with the obstacle in the process of moving to the target point.

### 3.1 Mathematical Model

Artificial potential field method is a path planning method based on virtual force field. The constraint relationship among agricultural robot, guide point and obstacle is expressed by potential field function. The negative gradient of the corresponding potential field function is the expression of the force exerted on the agricultural robot.

The artificial potential field is established by using the potential field function $U$. The initial position coordinate of the robot is $q = (x, y)^T$ in the plane space. The target location coordinate is $q_{goal} = (x_g, y_g)^T$. The potential gravitational field creates a gravitational force on the robot that points to the target point. Also, the potential gravitational field at $q_{obs} = (x_o, y_o)^T$ from the obstacle will produce a repulsive force directed from the obstacle to the robot. The potential field function $U(q)$ at point $q$ can be expressed as

$$U(q) = U_{att}(q) + U_{rep}(q). \tag{2}$$

The gravitational potential field function formula can be defined as

$$U_{att}(q) = \frac{1}{2}\lambda\rho^{2}(q, q_{goal}). \quad (3)$$

Modulus $\lambda$ is a non-negative constant and the gain coefficient of the gravitational potential field, and its value can be changed according to the specific situation to increase or decrease the strength of the gravitational potential field. $\rho(q, q_{\text{goal}})$ represents the distance between the current position of the robot and the guide point to be reached.

The gravitational potential field function formula can be defined as

$$U_{rep}(q) = \begin{cases} \frac{1}{2}\mu\left(\frac{1}{\rho(q, q_{obs})} - \frac{1}{\rho_0}\right)^2, & \rho(q, q_{obs}) < \rho_0 \\ 0, & \rho(q, q_{obs}) \geq \rho_0 \end{cases}. \quad (4)$$

In the formula, $\mu$ is a non-negative constant, representing the gain coefficient of the repulsive potential field, and its effect is similar to that of $\lambda$ in the potential gravitational field. $\rho_0$ is the obstacle distance threshold, and its function is to calculate the influence of the repulsive force within the distance threshold on the robot. $\rho(q, q_{obs})$ represents the distance between the location of the robot and the obstacle within the threshold. Within the robot's communication distance, the mutual repulsion force of two agricultural robots and the above repulsion force is also calculated in the same way, except that the action threshold is different.

## 3.2 Local Minimum Problem

The obstacle avoidance mechanism of the artificial potential field method is to drive the robot to move through the virtual field's force. When the robot is no longer under force, it will stop moving. Suppose the robot at a particular position receives the gravitational force from the guide point and the repulsive force from the obstacle is exactly equal. In that case, it means that the total force received by the robot at this time is 0, and the robot loses the driving force and no longer continues to move.

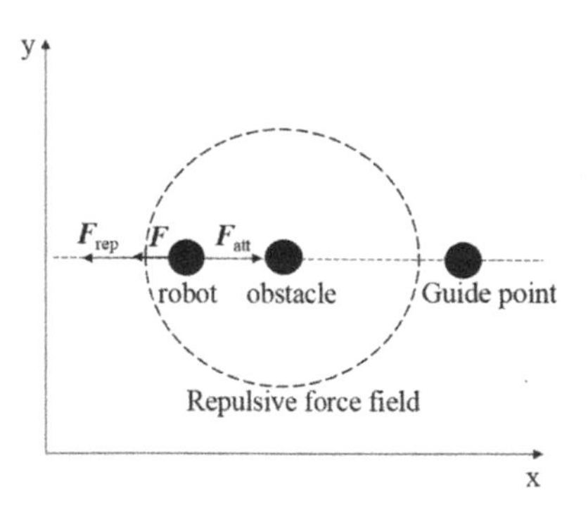


**Fig. 4.** The local minimum problem.

As shown in Fig. 4, this situation leads to a local minimum problem, and the robot does not reach the target point at this time. In order to solve the local minimum problem, the component of gravity on the coordinate axis is adjusted to make the gravity change.

### 3.3 The Guide Point

A function is designed to evaluate the map information. It can increase agricultural robot's coverage and reduce the areas that have not been covered for a long time. Regions with lower evaluations are more likely to increase the probability of the random appearance of guide points.

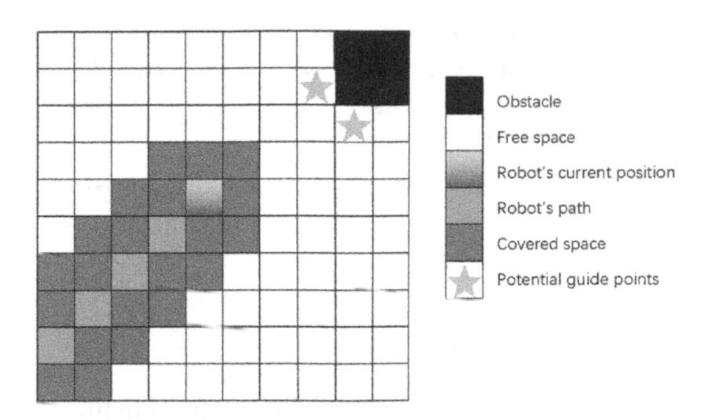


**Fig. 5.** Grid map at guide point.

By querying the unit's value whose value of the map information matrix unit is 0 and its surrounding units, calculate the value of each unit within 5 m around the agricultural robot. The evaluation is the added value, and the target point is generated at the center of the unit with the lowest evaluation. When the agricultural robot cannot find a unit with a value of 0 within a range of 5 m, it means that the surrounding area has been covered and then expand by 5 m until all the units are covered. As shown in Fig. 5, the potential guidance points appear in the current direction of a certain distance from the robot. Due to the existence of obstacles, there are two points that meet the requirements. Finally, the robot will randomly select one of them and move forward.

## 4 Results

We assume that all static obstacles are known before the action starts, so the local information map can be updated to represent the environment's obstacle area. If a unit in the information map coincides with an obstacle in the background, the unit should be unavailable when calculating the guide point's position. The repulsive forces between robots are only activated when they are within the distance threshold. If the map grid is too large, the robot may collide with obstacles before being repelled. A higher resolution information map can embed obstacle information more accurately and ideal performance in obstacle avoidance. However, a high-resolution information map will increase the robot's computational burden, resulting in slow action decisions. The size of the grid needs to be selected according to the actual situation.

In this section, covering based on the multi-robot proposed algorithm's artificial potential field is verified by MATLAB simulation. The covered area is a square area of 50 m * 50 m, and the whole region is discretized into grids of 0.5 m * 0.5 m. The agricultural robot travels at a constant speed of = 1.5 m/s and the working radius $R_w$ = 4 m. The environment with obstacles is tested in MATLAB.

### 4.1 The Evaluation Indexes

The algorithm's evaluation index, the cumulative coverage area rate and task time, are defined as follows:

1) Cumulative coverage area rate: the ratio of the area covered by all agricultural robots at a specific time to the total area.
2) Task completion time: the time it takes to complete all tasks.
3) Obstacle avoidance: Obstacle avoidance performance under obstacles of different sizes, whether robots collide with obstacles or collisions between robots.

The most critical point to evaluate this algorithm is the above three points, and the most crucial point is the third point. Once the agricultural robots have poor obstacle avoidance performance, the loss will be heavy.

### 4.2 Performance Analysis

The obstacle avoidance performance of the algorithm is tested in the case of two obstacles and five robots. In this case, there are more chances of collision. As shown in Fig. 6, the experiment used 5 robots to run under the area. The initial position and orientation of the robot are randomly given. As the robot continues to cover, there is no collision between the robot and the obstacle, and the results show that the algorithm is effective for obstacle avoidance and regional coverage.

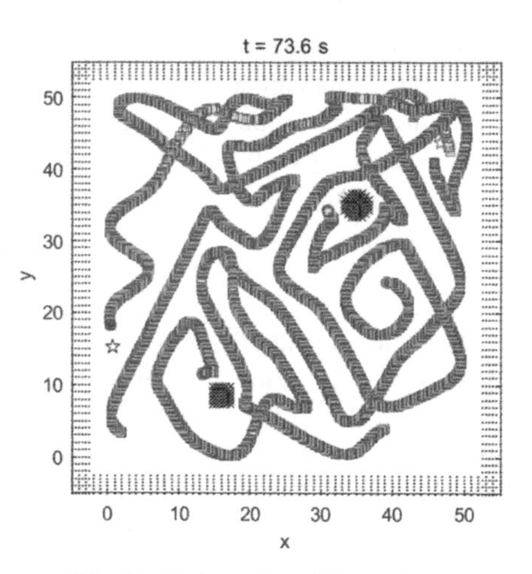


**Fig. 6.** Trajectories of five robots.

The results clearly show that the algorithm can complete the multi-robot full coverage task, and it performs well in the presence of obstacles.

Finally, we analyzed the cumulative coverage area rate when the number of agricultural robots changes. By calculating the average results of 10 simulations, the conclusion shows that the total cumulative coverage area of the robot gradually increases with time, exceeding 95%.

## 5 Conclusion

The multi-robot full coverage algorithm based on artificial potential field achieves adequate coverage of the task area. The algorithm does not need to consider the starting position between the robots. The robots make decisions through local information exchange, which can adapt to different numbers of robots to work together. It has good obstacle avoidance performance, can be used in different environments, and has good adaptability, scalability and robustness. The algorithm can be applied to the full coverage tasks of large-scale agricultural production, and MATLAB simulation verifies the algorithm's performance.

## References


1. Ju, C., Son, H.I.: Modeling and control of heterogeneous agricultural field robots based on Ramadge-Wonham theory. IEEE Robot. Autom. Lett. **5**(1), 48–55 (2020)
2. Zhong, J., Cheng, H., He, L., Ouyang, F.: Decentralized full coverage of unknown areas by multiple robots with limited visibility sensing. IEEE Robot. Autom. Lett. **4**(2), 338–345 (2019)
3. Arezoumand, R., Mashohor, S.: Deploying clustered wireless sensor network by multi-robot system. In: 2014 IEEE International Conference on Control System, Computing and Engineering, pp. 107–111, Penang, Malaysia (2014)
4. Mukherjee, P., Santilli, M., Gasparri, A., Williams, R.K.: Experimental validation of stable coordination for multi-robot systems with limited fields of view using a portable multi-robot testbed - EXTENDED ABSTRACT. In: 2019 International Symposium on Multi-Robot and Multi-Agent Systems (MRS), pp. 4–6, New Brunswick, NJ, USA (2019)
5. Ma, Y., Sun, H., Ye, P., Li, C.: Mobile robot multi-resolution full coverage path planning algorithm. In: 2018 5th International Conference on Systems and Informatics (ICSAI), pp. 120–125, Nanjing, China (2018)
6. Guruprasad, K.R., Dasgupta, P.: Distributed Voronoi partitioning for multi-robot systems with limited range sensors. In: 2012 IEEE/RSJ International Conference on Intelligent Robots and Systems, pp. 3546–3552, Vilamoura-Algarve, Portugal (2012)
7. Miao, X., Lee, J., Kang, B.: Scalable coverage path planning for cleaning robots using rectangular map decomposition on large environments. IEEE Access **6**, 38200–38215 (2018)
8. Lee, H., Jeong, W., Lee, S., Won, J.: A hierarchical path planning of cleaning robot based on grid map. In: 2013 IEEE International Conference on Consumer Electronics (ICCE), pp. 76–77, Las Vegas, NV, USA (2013)
9. Oksanen, T., Visala, A.: Path planning algorithms for agricultural field machines. J. Field Robot. **26**(8), 651–668 (2009)
10. Choset, H.: Coverage of known spaces: the boustrophedon cellular decomposition. J. Auton. Robots **9**(3), 247–253 (2000)
11. Andersen, H.L.: Path planning for search and rescue mission using multicopters. J. Institutt for Teknisk Kybernetikk (2014)
12. Gabriely, Y., Rimon, E.: Spiral-STC: an on-line coverage algorithm of grid environments by a mobile robot. In: IEEE International Conference on Robotics and Automation (ICRA), pp. 954–960. IEEE (2002)
13. Ganganath, N., Yuan, W., Cheng, C.T., Fernando, T., Iu, H.H.C.: Territorial marking for improved area coverage in anti-flocking-controlled mobile sensor networks. In: 2018 IEEE International Symposium on Circuits and Systems (ISCAS), pp. 1–4 (2018)

14. Mao, Y.T., Dou, L.H., Chen, J., et al.: Combined complete coverage path planning for autonomous mobile robot in indoor environment. In: Asian Control Conference 2009, pp. 1468–1473 (2009)
15. Ma, X., Lin, Z., Song, R.: Local path planning for unmanned surface vehicle with improved artificial potential field method. J. Journal of Physics Conference Series. 1634(1), 012125 (2020)

# Collaborative Robot Grasping System Based on Gaze Interaction

Mingyang Li, Yulin Xu(✉), and Aolei Yang

Department of Mechatronic Engineering and Automation,
Shanghai University, Shanghai 200444, China

**Abstract.** The bionic robotic arm has the characteristics of high safety, lightweight and portability, which can assist humans in completing various daily activities in life. Based on the research background of human-machine collaboration and simulating its experimental environment, a vision-assisted grasping framework for collaborative robots serving humans is proposed, and a discrimination scheme based on the combination of the shortest path and gIOU is designed. We can interact with the robotic grasping system by directly gazing at the object of interest. The gaze coordinates are provided by an eye tracker, which is matched with the image of the target detection algorithm to obtain the grasping information, and the signal is transmitted to the grasping system. The grasping system executes the grasping operation through the robotic arm after calculating the target pose, and implements the automatic grasping target by gazing at the target object. The experimental results show that this framework has good positioning accuracy and can effectively complete daily grasping tasks.

**Keywords:** Gaze tracking · Human-machine interaction · Object detection · Robotic grasping

## 1 Introduction

With the rapid development of industrial automation technology, human labor has been gradually replaced by machinery. Industrial robots have begun to develop gradually, and this type of collaborative robot has received intense attention from many researchers. Compared with ordinary industrial robots, a collaborative robot is an automated device that can sense external signals through sensors and perform corresponding actions in real time. This kind of robot is generally used to assist or expand the human movement ability and perception ability, and help us accomplish many complex tasks. In addition, since the clients of collaborative robots are most likely to be the disabled and the elderly, it will be very inconvenient if the collaborative robot needs to be operated with a large number of keys and joysticks like a mouse and keyboard. How to design a new type of human-computer interaction interface to allow target clients to operate more efficiently and conveniently has become a new challenge for researchers.

In order to better improve the interaction between us and collaborative robots, researchers have been exploring various daily interaction signals used by humans in


Q. Han et al. (Eds.): LSMS 2021/ICSEE 2021, CCIS 1469, pp. 77–86, 2021.
https://doi.org/10.1007/978-981-16-7213-2_8

interpersonal communication. Various external sensors collect these signals and analyze commands to convey to the collaborative robot, and present the feedback of the collaborative robot to the client in a simple and clear way. Among these schemes, there are electromyographic signals (EMG) [1–3], brain electrical signals (EEG) [4, 5], gesture signals [6, 7], the speech signal [8], facial expressions [9, 10]. However, as the eye that obtains more than 80% of the information in the human senses, this kind of interaction should be paid more attention.

## 2 Previous Work of Gaze Interaction and Object Detection

Gaze interaction, as a natural way for people to interact with the world, makes it more intuitive, natural and bidirectional, and can help us to be more convenient and flexible, accurately and effectively perceive and communicate with the outside world in daily life. "Eyes are the windows of the soul." Mental activities and eye movements are inseparable. Human gaze information will always naturally reflect what they think and want. This makes gaze interaction possibly the most natural and effortless way of all interactions. The gaze interaction methods can be divided into wearable and non-wearable, contact and non-contact [11]. Among gaze interaction method, since a high precision, non-compulsory, non-contact, non-invasive, etc. pupil - corneal reflection (Pupil Center Corneal Reflection, PCCR) technology become human-computer interaction gaze ideal solution.

Gaze tracking has been used to study human behavior in various disciplines, with a large amount of research existing in the direction of cognitive psychology [12], social psychology [13], assessment and study of student behavior [14], health and medical research [15], linguistics [16], marketing and consumer research [17], and software development client experience [18]. There is less research on using human gaze signals as a control method. Some researchers tried to use gaze signals to control the movement of robots or equipment, and then proposed using gaze to manipulate wheelchairs. In these research work, users are generally provided with a screen, which displays buttons in various directions, such as up, down, left, and right, and by gazing at these buttons to control the wheelchair to drive forward, backward, left or right [19–22]. Researchers have also made different attempts on medical robots [23] and autonomous driving [24].

In order to complete various tasks, collaborative robots need to have the ability to perceive important targets in the environment. Object detection is the basis of many computer vision tasks. Considering the real-time problem, finally adopted the YOLO proposed by Joseph Redmon.

YOLOv1 discards the candidate frame generation part in the two-stage algorithm, but directly implements feature extraction, candidate frame regression and classification in the same branchless deep convolutional network, making the network structure simple. YOLOv2 proposes a new classification backbone Darknet-19, which has 19 convolutional layers and 5 maximum pooling layers.Darknet-19 is faster and more accurate, and can reduce the amount of calculation by 33% without reducing the detection accuracy. Darknet-53, the backbone of YOLOv3, combines the feature pyramid network for multi-scale fusion prediction. Compared to SSD, YOLOv3 is 3 times faster when achieving the same accuracy.

The remainder of this paper consists of two parts. In the first part (Sect. 3) introduces the structure of the whole system, the experimental equipment and the principles of human-computer interaction system. In the second part (Sects. 3, 4, 5 and 6) shows the experimental process and experimental results. We analyze and summarize the experimental results and propose new ideas for improvement.

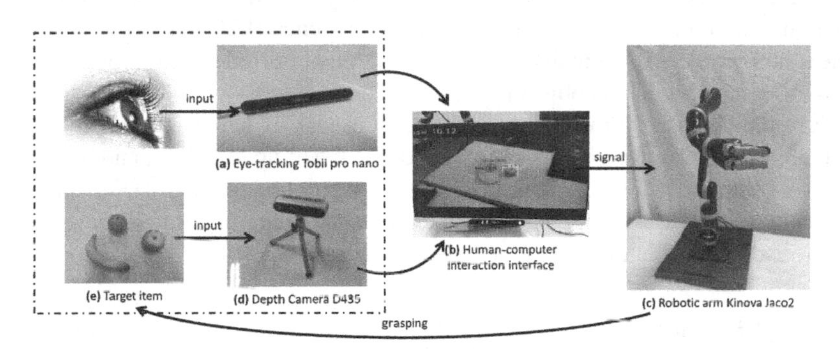


**Fig. 1.** Flow chart of human-computer interaction system experiment

# 3 Materials and Methods

## 3.1 Basic Framework

In order to better collaborate and help humans with mobility impairments achieve an easier and more efficient interaction, a human-machine collaboration system is developed as shown in Fig. 1. The system is mainly divided into two important parts: The recognition detection module is used to communicate with the user, and the capture execution module is used to meet the requirements of the user. The following is a detailed description of the system.

1. The eye recognition component is used to capture gaze data equipped with a Tobii eye tracker.
2. The detection and selection module obtains the video stream of the target area through the Realsense camera. It provides images to the target detection algorithm and gives the user information to grab the object.
3. The human-computer interaction interface part is responsible for analyzing the object that the user wants to grasp by combining the gaze signal with the gaze image information.
4. The robotic arm grasping module executes the grasping of the target through the Kinova robotic arm after obtaining the final grasping intention signal.

In this work, the personal gaze information is obtained by the eye tracker device Tobii pro nano as shown in Fig. 1(a). This is an on-screen eye tracker with a sampling rate of 60 Hz using Tobii Pro's latest technology.

Tobii Pro Nano has a length of 170 mm (6.7") and a weight of 59 g (2.1 oz.). The eye tracker uses video-based pupil-corneal reflection eye tracking technology. It can measure the position of the line of sight, the diameter of the pupil and the position of the eye in the three-dimensional space, and obtain data such as fixation time, pupil size, number of blinks, eye saccades, and eyelid closure.

The recognition unit uses Realsense camera captures the video stream of the target area. It provides input for the target detection algorithm and the robot grabbing pose generation algorithm. As shown in Fig. 1(e), the length, width and height of Realsense are 90 mm × 25 mm × 25 mm, and the resolution reaches 1920 × 1080 in RGB mode, and the frame rate can be maintained at 30 fps. It is the preferred solution for applications such as robot navigation and object recognition.

The robotic arm performing the final grasping task is Kinova Jaco2 in Fig. 1(c). The Jaco2 robotic arm weighs 5.5 kg, the load can reach up to 2.4 kg, and the working radius is 985 mm. The Jaco2 manipulator is a 7-degree-of-freedom (DOF) light-duty manipulator. Its actuator is modular, which enables the manipulator to achieve different functions and be used in different practical scenarios.

## 3.2 Basic Principles

### 3.2.1 Object Identification Module

A human-computer interaction framework is developed based on practical requirements to explain the way to determine the gaze signal to interact with an object, establishing a mathematical model of visual determination as shown in Fig. 2 and select the object to be grasped by considering the gaze and stay time.

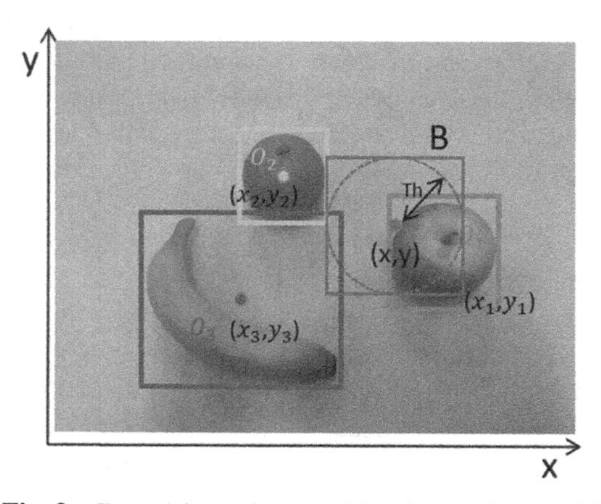


**Fig. 2.** Gaze point and target object interaction model

In this case, the object $O_i$ is considered to be selected if the gaze point stays within the distance threshold *Th* of an object, and the distance between the gaze point and the object $O_i$ is the shortest.

We define the coordinates of the current gaze point by (x, y) and the position of the object by $O_i(x_i, y_i)$, where $1 \leq i \leq n$ and $n$ is the number of identified objects on the image. The selection of object j by the shortest distance is:

$$\mathrm{j} = \arg\min_{\mathrm{i}} \sqrt{(x - x_{\mathrm{i}})^2 + (y - y_{\mathrm{i}})^2}. \tag{1}$$

The experiment is in the actual scene where the objects are placed densely. When the gaze point is at the intersection of the object prediction frame and the critical point of distance determination, the system may make a wrong decision to select the target. So the gIOU method is used to assist in solving this possible problem. This is a way to use the interaction ratio to determine. Since the gaze signal does not provide a window for comparison, we propose an adaptive anchor frame method. This method uses the shape of the prediction frame where the gaze point is located as the anchor frame of the gaze point. It fully and effectively utilizes the prediction frame data, is robust to the scale of the staring target object, and can provide a uniform scalar for comparison of objects of different scales. We indicate the bounding box range of the object by $O^i$, use B to indicate the adaptive anchor box range of the gaze point, determine the intersection and ratio maximum to select object j as

$$j = \arg\max_{i} \frac{O^i \cap B}{O^i \cup B}. \tag{2}$$

Here, the two methods will be combined, and the weight ratio $k_1$, $k_2$ will be set to adapt to different environmental conditions.

$$j = \arg\max_{i} \left( \frac{k_1 \times Th}{\sqrt{(x - x_i)^2 + (y - y_i)^2}} + k_2 \times \frac{O^i \cap B}{O^i \cup B} \right). \tag{3}$$

When developing a graphics-oriented eye tracking application, the most important problem is to map the coordinates of the eye tracker to the appropriate screen coordinates. The eye tracker can calculate the observer's gaze point coordinates on the eye tracker system screen, and we need to match it with the image coordinates displayed on the screen.

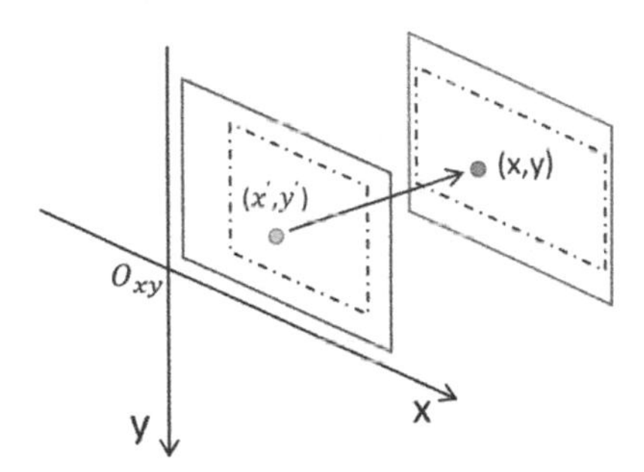


**Fig. 3.** Schematic diagram of gaze point coordinate conversion

In a sampling period, the eye tracker will return a fixation point coordinate pair x and y (for example, for a 60 Hz eye tracker, about 16 ms sample time), this coordinate pair should be mapped into the application window. When using an eye tracker, the data obtained from the eye tracker needs to be mapped to the appropriate area for a given application. As shown in Fig. 3, the two-dimensional eye tracker data (eye tracker system screen coordinates) must be mapped to the two dimensional image display coordinate system. Typically, to map the eye-tracking capture area $\left[(a_x, b_x), (a_y, b_y)\right]$ to the image display area $\left[(c_x, d_x), \left(c_y, d_y\right)\right]$, set the original eye-tracking data coordinates to $I' = \left[x', y', 1\right]$, then:

$$\left[x, y, 1\right] = \left[x', y', 1\right]\begin{bmatrix} k_{1x}k_{2x} & 0 & 0 \\ 0 & k_{1y}k_{2y} & 0 \\ k_{1x}b & k_{1y}b & 1 \end{bmatrix}. \tag{4}$$

Where, $k_{1x}, k_{1y}$ is the scale of eye-movement coordinates scaled to the standard coordinate system.

$$k_{1x} = \frac{1}{b_x - a_x} \tag{5}$$

$k_{2x}, k_{2y}$ is the scale of the standard coordinate system to the image coordinates.

$$k_{2x} = d_x - c_x. \tag{6}$$

$b$ is the offset of the target coordinates relative to the initial coordinates in the standard coordinate system during the scaling process.

$$b = 2ac - ad - bc. \tag{7}$$

Based on (4), we can correctly convert the gaze position from the output of the eye tracker to the image coordinate system.

#### 3.2.2 Object Grasping Module

A robotic arm is a device designed to imitate the structure of a human arm. In the mechanical structure of the robotic arm, the connecting rod imitates the structure of the human upper arm and forearm. The "joint" that connects the connecting rod is a motor, and the motion of the connecting rod can be imitated by the movement of the human arm by the motor. Generally speaking, the position of a rigid body in space can be represented by 6 parameters $(x, y, z, \propto, \beta, \gamma)$. $(x, y, z,)$ represents the position of the object in space, and $(\propto, \beta, \gamma)$ represents the rotation angle of the object in the world coordinate system. In theory, the end of a robotic arm with more than 6 degrees of freedom can make any posture within the movable range.

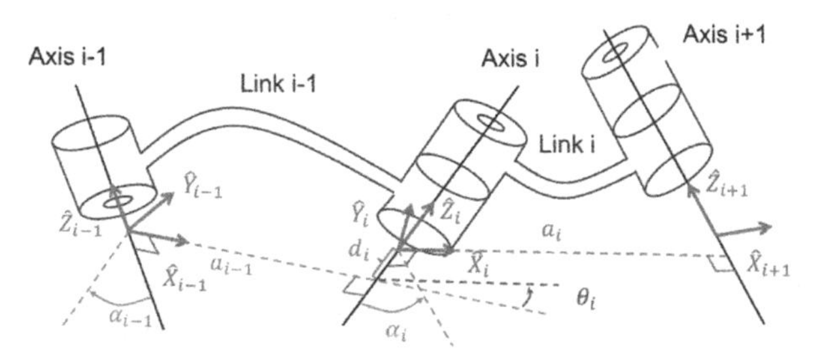


**Fig. 4.** DH parameters model

The forward kinematics and inverse kinematics of the manipulator constitute the kinematics of the manipulator. The DH modeling method of the manipulator (Fig. 4) is currently the most used modeling method. It has a total of four parameters: The length $a_{i-1}$ of the link i–1 represents the vertical distance between the axis of the i–1 joint and the axis of the i joint. The torsion angle $\alpha_{i-1}$ of the link i–1 represents the angle between the joint i–1 and the axis of the joint i. The offset $d_i$ of the link i relative to the link i–1 represents the distance between the axis of the i joint and the common perpendicular of the front and rear joint axes. The joint angle $\theta_i$ represents the rotation angle of the link i relative to the link i–1 around the i axis. The figure shows the angle between two green male perpendiculars.

Through these four parameters, the model can completely and clearly describe the conversion relationship between adjacent coordinate systems, and its conversion matrix T is expressed as:

$$T = \begin{bmatrix} \cos\theta_i & -\sin\theta_i & 0 & a_{i-1} \\ \sin\theta_i\cos\alpha_{i-1} & \cos\theta_i\cos\alpha_{i-1} & -\sin\alpha_{i-1} & -d_i\sin\alpha_{i-1} \\ \sin\theta_i\sin\alpha_{i-1} & \cos\theta_i\sin\alpha_{i-1} & \cos\alpha_{i-1} & d_i\cos\alpha_{i-1} \\ 0 & 0 & 0 & 1 \end{bmatrix}. \tag{8}$$

By using the transformation matrix T for the interpretation of the spatial coordinate system, the problem of positive and inverse kinematics of the robotic arm can be solved.

### 3.2.3 System Structure

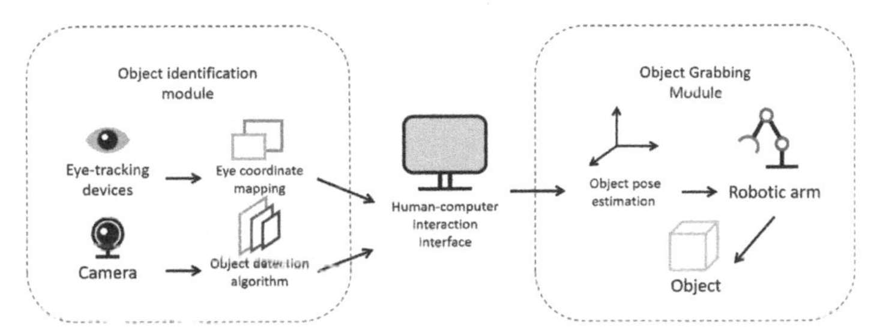


**Fig. 5.** Architecture of a gaze human-robot interaction grasping system for collaborative robots

The basic process of this paper is shown in Fig. 5. It is mainly divided into two parts: the target recognition module and the target capture module: in the target recognition module, the target image is obtained through the Realsense camera and the target detection algorithm is used to distinguish the objects to be captured in the scene. The Object identification module combined with the gaze point information from the eye tracker to confirm that the grasped object is on the human-computer interaction interface. The information of the object to be grasped is transmitted to the target grasping module. After analyzing the pose information of the target object, the grasping parameters are transmitted to the robotic arm to grasp the target, completing the process of the whole human-machine collaboration system.

## 4 Experiments and Results

The whole environment of the collaborative robot grasping system is shown in Fig. 6, which includes the interactive interface and the robotic arm grasping scene. Figure 6(a) is a computer screen running the gaze interactive system. Figure 6(b) shows the scene where the robotic arm reacts to grabbing after receiving the gaze signal. Figure 6(c) shows the scene when the robot is performing a grab after receiving a command. During the experiment, in order to better show the approximate position of the line of sight, we marked the point of sight with a red circle, making the point of sight more intuitive in the experiment.

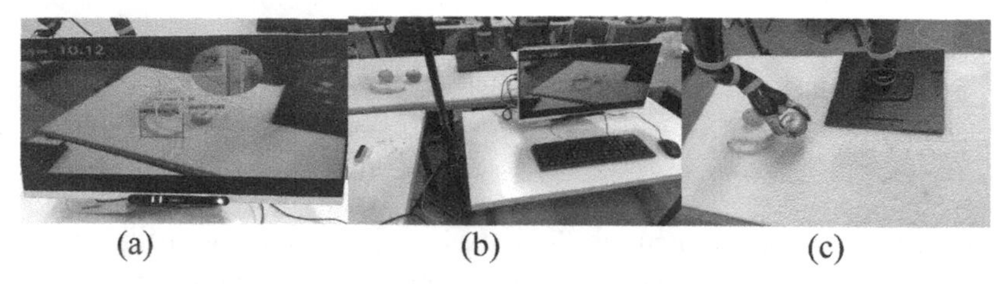

(a) (b) (c)

**Fig. 6.** Experimental scene

In this equipment environment, the collaborative robot grasping system has been used for many simulation experiments, and the environmental conditions have been set (the gaze trigger time unit is 0.001 s), and the interaction distance unit is 50 pixels).

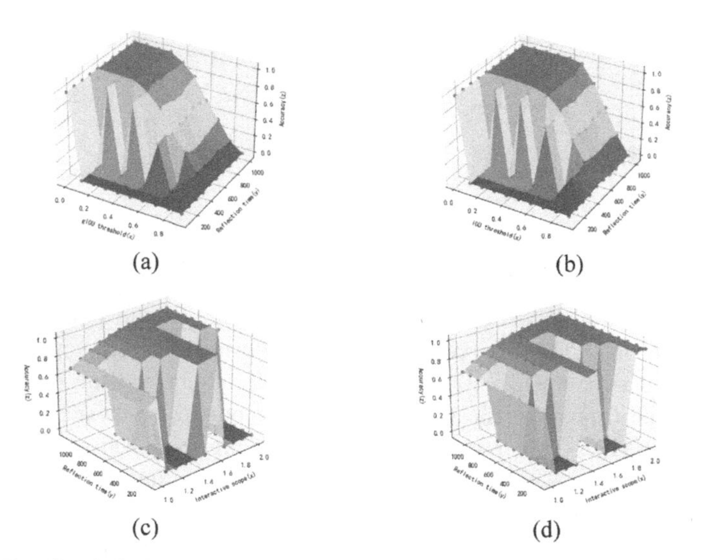

(a) (b)

(c) (d)

**Fig. 7.** Visual display of simulated experimental data with the Gaze HCI system

Figure 7 shows the visualization data diagram of the simulation experiment. Figure 7(a) (b) shows the performance of different scale objects at different gIOU values. Figure 7(c) (d) shows the performance under different distance thresholds. It can be seen from (a) and (b) that the smaller the gIOU value, the higher the accuracy, which means that the smaller the gIOU threshold will make the gripping system more robust. Similarly, in (c) and (d), it can be seen that a larger interaction distance can also increase the robustness of the system. The reflection time means the time for gaze to confirm the target. The shorter the time, the shorter the time from receiving the command to making the grab. The increase in response time can reduce noise interference and enhance the stability of the system. Finally, the gaze trigger time is selected as 0.6 s, the interaction threshold is 80 pixels, and the gIOU threshold is 0.6, which shows good results in the actual test.

## 5 Conclusion

In this paper, a new human-computer interaction framework is developed to match the gaze signal with the robotic arm grasping system. A set of mathematical models are applied in the interaction framework according to the rules of interaction, and its validity is verified through simulated experimental data. We verify the ability to perform grasping tasks using a gaze signal to control the robot's arm in a practical environment. This new interaction framework is expected to better benefit humanity, but there are still many problems that need to be improved, such as the Midas contact problem. In the future work, we will improve the interactive interface and develop reasonable algorithms to solve the Midas contact problem.

**Acknowledgments.** This work is supported by Science and Technology Commission of Shanghai (No. 15411953500).

## References


1. Artemiadis, P.K., Kyriakopoulos, K.J.: An EMG-based robot control scheme robust to time-varying EMG signal features. J. IEEE Trans. Inf. Technol. Biomed. **14**(3), 582–588 (2010)
2. Bisi, S., De Luca, L., Shrestha, B., Yang, Z., Gandhi, V.: Development of an EMG-controlled mobile robot. Robotics **7**(3), 36 (2018)
3. Yang, C., Chang, S., Liang, P., Li, Z., Su, C.: Teleoperated robot writing using EMG signals. In: 2015 IEEE International Conference on Information and Automation, pp. 2264–2269 (2015)
4. Stephygraph, L.R., Arunkumar, N.: Brain-actuated wireless mobile robot control through an adaptive human–machine interface. In: Suresh, L., Panigrahi, B. (eds.) Proceedings of the International Conference on Soft Computing Systems, pp. 537–549. Springer India, New Delhi (2016). https://doi.org/10.1007/978-81-322-2671-0_52
5. Wei, L., Jin, J., Duan, F.: Cognitive-based EEG BCIs and Human Brain-Robot Interactions. J. Comput. Intell. Neurosci. (2017)
6. Chen, S., Ma, H., Yang, C., Fu, M.: Hand gesture based robot control system using leap motion. In: Liu, H., Kubota, N., Zhu, X., Dillmann, R., Zhou, D. (eds.) ICIRA 2015. LNCS (LNAI), vol. 9244, pp. 581–591. Springer, Cham (2015). https://doi.org/10.1007/978-3-319-22879-2_53
7. Liu, J., Luo, Y., Ju, Z.: An interactive astronaut-robot system with gesture control. J. Comput. Intell. Neurosci. **2016**, 1–11 (2016)

8. Zinchenko, K., Wu, C., Song, K.: A study on speech recognition control for a surgical robot. J. IEEE Trans. Ind. Inform. **13**(2), 607–615 (2017)
9. Liu, Z., et al.: A facial expression emotion recognition based human-robot interaction system. J. IEEE/CAA J. Automatica Sinica **4**(4), 668–676 (2017)
10. Yassine, R., Makrem, M., Farhat, F.: A facial expression controlled wheelchair for people with disabilities. J. Comput. Methods Programs Biomed. **165**, 89–105 (2018)
11. Arar, N.M., Thiran, J.P., Theytaz, O.: Eye gaze tracking system and method. J. Econ. Polit. Wkly. **2**(25), 1121–1122 (2016)
12. Gere, A., Kókai, Z., Sipos, L.: Influence of mood on gazing behavior: preliminary evidences from an eye-tracking study. J. Food Qual. Prefer. **61**, 1–5 (2017)
13. Kraines, M.A., Kelberer, L.J., Wells, T.T.: Rejection sensitivity, interpersonal rejection, and attention for emotional facial expressions. J. Behav. Ther. Exp. Psychiatry **59**(1), 31 (2017)
14. Gog, T.V., Kester, L., Nievelstein, F., et al.: Uncovering cognitive processes: different techniques that can contribute to cognitive load research and instruction. J. Comput. Hum. Behav. **25**(2), 325–331 (2009)
15. Melinder, A., Konijnenberg, C., Sarfi, M.: Deviant smooth pursuit in preschool children exposed prenatally to methadone or buprenorphine and tobacco affects integrative visuomotor capabilities. J. Addict. **108**(12), 2175–2182 (2013)
16. Engel, S., Shapiro, L.P., Love, T.: Proform-antecedent linking in individuals with agrammatic aphasia: a test of the Intervener Hypothesis. J. Neurolinguistics **45**, 79–94 (2018)
17. Pham, C., Rundle-Thiele, S., Parkinson, J., et al.: Alcohol warning label awareness and attention: a multi-method study. J. Alcohol Alcohol. **53**(1), 1–7 (2017)
18. Yesilada, Y., Jay, C., Stevens, R., et al.: Validating the use and role of visual elements of web pages in navigation with an eye-tracking study. In: ACM, New York, NY, USA, pp. 11–20 (2008)
19. Araujo, J.M., Zhang, G., Hansen, J., et al.: Exploring eye-gaze wheelchair control. In: ETRA'20: 2020 Symposium on Eye Tracking Research and Applications, New York, NY, USA, Article 16, pp. 1–8 (2020)
20. Eid, M.A., Giakoumidis, N., Saddik, A.E.: A novel eye-gaze-controlled wheelchair system for navigating unknown environments: case study with a person with ALS. J. IEEE Access **4**, 1 (2016)
21. Hyder, R., Chowdhury, S.S., Fattah, S.A.: Real-time non-intrusive eye-gaze tracking based wheelchair control for the physically challenged. In: 2016 IEEE EMBS Conference on Biomedical Engineering and Sciences (IECBES), Kuala Lumpur, Malaysia, pp. 784–787 (2016)
22. Rupanagudi, S.R., et al.: A video processing based eye gaze recognition algorithm for wheelchair control. In: 2019 10th International Conference on Dependable Systems, Services and Technologies (DESSERT), Leeds, UK, pp. 241–247 (2019)
23. Noonan, D.P., Mylonas, G.P., Darzi, A., Yang, G.: Gaze contingent articulated robot control for robot assisted minimally invasive surgery. In: 2008 IEEE/RSJ International Conference on Intelligent Robots and Systems, pp. 1186–1191 (2008)
24. Navarro, J., Osiurak, F., Ovigue, M., et al.: Highly automated driving impact on drivers' gaze behaviors during a car-following task. Int. J. Hum. Comput. Inter. 1–10 (2019)

# Intelligent Data Processing, Analysis and Control

# A Hybrid Recommendation Algorithm Based on Latent Factor Model and Collaborative Filtering

Wenyun Xie(✉) and Xin Sun

School of Mechatronic Engineering and Automation, Shanghai University, Shanghai 200444, China

**Abstract.** The basic recommendation algorithms include item-based CF algorithms and recommendation algorithms based on machine learning. The following article focuses on the system theory and implementation of these two basic algorithms. Based on their core ideas, a new method is proposed, which is a hybrid recommendation algorithm based on the CF algorithm and LFM algorithm. The algorithm uses a latent factor model to generate user preference matrix and item hidden feature matrix, and learns to update the matrix through gradient descent, and then calculates the similarity between items according to the calculation method in collaborative filtering, and finally, after weighting, obtains Top-N recommendation results. The algorithm has been verified, which improves the accuracy of recommendation, improves the shortcomings of collaborative filtering algorithm that only analyzes and displays features, and also alleviates the influence of sparse matrix on recommendation results.

**Keywords:** CF · LFM · Machine learning · Recommendation system · Hybrid recommendation algorithm

## 1 Introduction

In today's society, with the explosive development of Internet technology and communication technology. Faced with massive amounts of information, users spend more and more time on obtaining corresponding information and choices. So, we must address the problem of information inflation, in such situation, search engines, recommendation engines have emerged [1].

The recommendation engine is more inclined to use the system without a clear purpose and unconsciously [2, 3]. The recommendation engine uses the user's historical behavior data and the user's interest number list to recommend suggestions that may be of interest to the user. The long tail theory [4] in the recommender system explains its value differently from search engines. The recommender system provides exposure opportunities for all items in the list, so as to tap the potential profits of the long tail items [5].


Q. Han et al. (Eds.): LSMS 2021/ICSEE 2021, CCIS 1469, pp. 89–97, 2021.
https://doi.org/10.1007/978-981-16-7213-2_9

At present, the mainstream recommendation algorithm is mainly based on CF algorithm. Most of the scoring matrices are sparse matrices, and fails to obtain ideal recommendation results due to the influence of sparse matrices. Liu Yefeng proposed[1] the Slope-one algorithm that populates the scoring matrix with predicted scoring data which extract from users to solve this problem. [6] Vozalis merged the singular value decomposition method and the item-based method into the CF algorithm. [7] Yang proposed an improved CF algorithm that uses a combination of social networks and scoring data to improve accuracy [8].

In order to improve these obstacle, this paper give this proposal that we can use the LFM algorithm to enhance the performance of previous collaborative filtering algorithms. LFM algorithm uses a idea of matrix decomposition to reduce the dimensionality of the score matrix, and uses implicit semantic molecules to explain the user's preferences to make recommendations. Increasingly, A single algorithm usually has limitations and in order to better improve the performance of the algorithm, recommendation systems are now opting for strategies that incorporate multiple recommendation algorithms simultaneously for optimization.

## 2 Analysis of Two Basic Algorithms

### 2.1 Collaborative Filtering

The basic logic of traditional collaborative filtering algorithms is to do calculations on a large number of users' historical behavioral data, use different indicators to stratify users and recommend products to users with the same preferences. This article first uses item-based collaborative filtering. However, this algorithm has a considerable limitation when the number of users is increasing dramatically, the calculation of the user's interest similarity matrix will become more arduous. When a surge in the number of users occurs, the timing complexity of its similarity matrix changes accordingly, and this increasing is exponentially increasing. The traditional item-based collaborative filtering matrix assumes that when two items have a strong similarity in the structure of the algorithm, it can be assumed that users who like one of the items have a high probability of liking the other item as well.. With such an algorithm, the user's historical behavior can also be used to provide a reasonable recommendation explanation for its recommended results.

For Item-CF, In the whole algorithm process, the most characteristic step of the collaborative filtering algorithm lies in its similarity calculation. The most typical similarity calculation method in this algorithm is:

$$w_{ij} = \frac{|N(i) \cap N(j)|}{|N(i)|}. \tag{1}$$

$|N(i)|$ represents the quantity of users who have a superior rating or preference for item i in the user base, and the numerator $|N(i) \cap N(j)|$ is the number of users in the user base who have a higher liking for item i and a higher liking for item j in the same time interval. However, there is a problem with this similarity calculation method. If item j is

1 [1]This work is supported by the National Key Research and Development Program of China (No. 2019YFB1405500).

a popular item, *wij* will be close to 1. In order to punish the popular item, the following is proposed formula:

$$Wij = \frac{|N(i) \cap N(j)|}{\sqrt{|N(i)||N(j)|}}. \tag{2}$$

Considering that items in the popular case have a larger user audience and their similarity will also overlap with most items, this formula penalizes the weight of popular item j in it, reducing the possibility of the above situation [9, 10].

After the similarity calculation for each item, the similarity matrix between items is obtained, but the algorithm also calculates the potential interest of the user for each item, which is calculated as follows:

$$p_{uj} = \sum_{i \in N(u) \cap S(j,k)} w_{ji} r_{ui}. \tag{3}$$

Among them, $N(u)$ represents the collection of items which have high interest to user u; $S(j, k)$ represents the set of k values from the items with high similarity to item j in descending order, taking out the lowest to the highest of them; $w_{ji}$ represents the result of item i and item j under its similarity calculation; $r_{ui}$ represents the user's interest rating for items that have experienced behaviors.

## 2.2 Latent Factor Model

Due to the explosive growth of Internet-based global e-commerce, the number of users and consumer items from around the world in its products is also growing exponentially [11]. For this problem, traditional statistical-based recommendation algorithms are no longer applicable [12]. In 2006, Simon Funk novel approach is to decompose its users' scoring matrix for items into two low-rank matrices, minimize the two matrices based on RMSE coefficients and obtain the training matrix. As a result, the latent factor model was born. LFM successfully solved the problem of the SVD algorithm requiring a large amount of storage memory [13]. The application of the LFM in the recommendation system is that it is expected to connect the potential connection between the item. Simon Funk [14, 15] decompose the high-dimensional matrix of user ratings into two low-rank matrices, as shown below:

$$R = P^T Q. \tag{4}$$

Based on the above, The rating of user u for item i can then be expressed as follows:

$$r_{ui} = \sum_k p_{uk} * q_{ik}. \tag{5}$$

Among them, $p_{uk} = P(u,f)$, $q_{ik} = Q(i,f)$. It can be seen that the method of latent factor model is actually to construct a loss function by minimizing the error RMSE:

$$C(p,q) = \sum_{(u,i) \in train} (r_{ui} - \hat{r_{ui}})^2 = \sum_{(u,i) \in train} (r_{ui} - \sum_{f-1}^{F} p_{uf} q_{if})^2. \tag{6}$$

Considering the over-fitting problem that will occur during the learning process, Based on this situation, we also need to add an extra regularization term to the original loss function to solve the problem, and the regularization formula is as follows, [16] namely:

$$\lambda(||p_u||^2 + ||q_i||^2). \tag{7}$$

Among them $\lambda$ are called regularization parameters.

Finally, the corresponding recurrence formula is:

$$p_{uf} = p_{uf} + \alpha((r_{ui} - \sum_{f-1}^{F} p_{uf} q_{if}) q_{if} - \lambda p_{uf}). \tag{8}$$

$$q_{if} = q_{if} + \alpha((r_{ui} - \sum_{f-1}^{F} p_{uf} q_{if}) p_{uf} - \lambda q_{if}). \tag{9}$$

Among them, $\alpha$ is represented the learning rate or step size among the gradient descent algorithm. Latent factor model no longer directly analyzes the similarity between items or users based on the explicit content of items or users, but uses a large amount of data to study the potential connections between items and users to achieve recommendations. Instead of directly analyzing the similarity between items or users based on their explicit content, the implicit semantic model uses a large amount of data to study the potential connections between items and users to achieve recommendations. This is fundamentally different from the traditional collaborative filtering recommendation algorithm.

## 3 Hybrid Recommendation Algorithm

This paper attempts to combining the respective advantages of the two algorithms, grasping its core ideas, and designing a hybrid recommendation calculation method based on the two, which can alleviate the sparse scoring matrix problem that currently exists in the recommendation system, and it can also improve the defects of sparse scoring matrix and data loss in the process of dimensionality reduction. In the fusion process, both algorithms run simultaneously and independently, and the obtained calculations are weighted and summed. The main variation is the calculation of user preferences for the items.

### 3.1 Comparison of LFM and Item-CF

From a theoretical basis, latent factor modeling is a machine learning based approach, while collaborative filtering is still a traditional method based on statistics. Therefore, the former has certain advantages in solving the problem of the accuracy of the recommendation system. In terms of space complexity, collaborative filtering algorithms need to maintain a memory table to save comment information, So when the quantity of users or items is huge, a lot of storage space is needed, and algorithms based on implicit semantics significant storage space savings based on better solutions for storage. Considering user-friendliness, this algorithm based on collaborative filtering is more convincing. Users can see the recommendation results generated by historical behavior, but the implicit semantic algorithm is an analysis of potential topics, and it is difficult to explain the recommendation in word reason.

### 3.2 Similarity Calculation in Collaborative Filtering

This paper attempts to focus on the similarity calculation method in collaborative filtering and gradient descent iterative calculation in latent factor model. In the Item-CF algorithm, It is a two-step problem, firstly, calculating the similarity between items to generate the item similarity matrix, secondly, predicting the interest of each user for each item, and finally, combining the two to calculate the interest of a user for an item. It is assumed that one of the users u has an evaluation on the item i, and after calculation $r_{ui}$.The result of item i and item j under its similarity calculation is $w_{ij}$, If this issue of user activity is considered, that is, users with more active behaviors should contribute less to the similarity of pairwise items than inactive users. Therefore, a calculation formula called IUF (Inverse User Frequence) is proposed. That is, the inverse parameter of the logarithm of user activity is used to modify the product similarity formula:

$$w_{ij} = \frac{\sum_{u \in N(i) \cap N(j)} \frac{1}{\log 1 + |N(u)|}}{\sqrt{|N(i)||N(j)|}}. \tag{10}$$

After processing the above formula, the interest degree of user i for item j can be more precisely expressed as:

$$p_{uj} = w_{ij} * r_{ui}. \tag{11}$$

### 3.3 Calculation of User's Interest

The LFM model for user interest degree is calculated by using the user's potential interest matrix for the product and the hidden feature matrix of each product as the basis for the interest degree to be learned and calculated. Assuming that the vector corresponding to user u is $P_u$ and the vector corresponding to item j is $Q_j$, so the user's possible interest for item j can be expressed as:

$$p_{uj} = P_u^T * Q_j. \tag{12}$$

It can be seen from the above that the calculation formula of the user interest of the hybrid recommendation algorithm should be:

$$P_{uj} = w_{ij} * r_{ui} + P_u^T * Q_j. \tag{13}$$

According to the formula, the user's interest in each item is specifically calculated, and according to the calculation result, the user's interest is sorted from large to small, then Top-N items are selected to generate a mixed recommendation list.

# 4 Experimental Results and Analysis

## 4.1 Experimental Data Set

The article utilizes the MovieLens data set provided by Grouplens. There are a total of 976 users in the MovieLens data set for a total of 100,000 ratings for 1786 movies, and the ratings range from 1 to 5. To scientifically validate the effectiveness of the algorithm, The data set will be processed into two areas, the training set and the test set, with a ratio of seven to three.

## 4.2 Experimental Evaluation Method Index

In our experiments, we use the following comparison method to verify that the algorithm has certain advantages, four sets of experiments are designed for comparative testing. The four algorithms are: User-CF, Item-CF, LFM, HybridLFM. We need to use some metrics to accurately measure whether these recommendation algorithms have the ability to accurately predict user behavior for a product. This indicator occupies an indispensable position in the evaluation system of the recommendation system. If the user's historical rating records are known, the user's interest model can be learned from it and used to predict the user's rating of an unknown item. The accuracy of prediction is generally calculated by Mean Absolute Error (MAE) or Root Mean Square Error. For a user u and item i in the data set, let $r_{ui}$ is the actual score of user u for item i, but $\hat{r_{ui}}$ is the predicted score given by the recommendation algorithm. The definition of MAE is:

$$MAE = \frac{\sum_{u,i \in T} |r_{ui} - \hat{r_{ui}}|}{|T|}. \tag{14}$$

## 4.3 Experimental Results and Analysis

For this hybrid recommendation algorithm proposed in the article, there are two most important parameters, which are the quantity of hidden factors f in the LFM calculation, also the size of the set of similar items k. The experiment mainly revolves around these two key parameters. The experimental data uses learning rate $\alpha = 0.01$ when calculating the implicit factor, and the regularization coefficient $\lambda = 0.002$, Multiply the learning rate by 0.9 after each iteration and slowly reduce learning rate. The first set of experiments keeps k unchanged for testing (Fig. 1).

Under the condition that k is unchanged (k = 10) in the four algorithms, The magnitude of the number of influence factors f has a considerable impact on the latter two learning model-based algorithms, but has little impact on the collaborative filtering algorithm. Because Item-CF and User-CF algorithms are mainly traditional statistical algorithms, there is no learning process, so the accuracy of the algorithm will not change with the change of f. Within a certain range, the accuracy of the algorithm is significantly improved with the increase of f, but this effect will no longer change as the value of f exceeds a certain amount, and tends to be stable. The hidden factors in each movie are also limited. The second set of experiments controls the hidden factor f unchanged (f = 30), and observes the change in the accuracy of the algorithm brought by the change in the number of neighbor items k (Fig. 2).

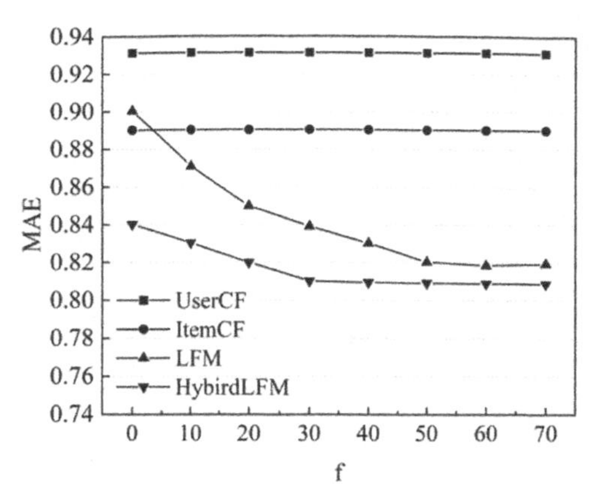


**Fig. 1.** Comparison of MAE with different factor numbers

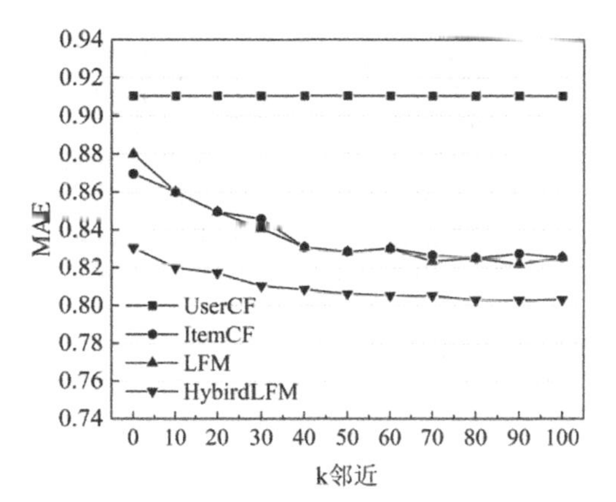


**Fig. 2.** Comparison of MAE with different neighbor numbers

The User-CF algorithm has nothing to do with the number of similar items, so the result data is no change. Item-CF, LFM and the algorithm proposed in this paper, The accuracy of the recommendation increases as the number of similar entries increases, and its variation is relatively consistent. It tends to stabilize when k exceeds 60. This indicates that most users also have a limited range of interests (Table 1).

Chooses two hybrid collaborative filtering algorithms, LICF and LFTRS_CF to compare them to verify that the algorithm proposed in this paper has high accuracy in recommendation results. Under these result with keeping the number of K the same, calculate the accuracy recommended by each algorithm.

Based on the above three experiments in different directions, the HybridLFM algorithm proposed in this paper outperforms the other three algorithms in terms of accuracy of recommendation results Therefore, the recommendation algorithm of item collaborative filtering and implicit semantic fusion improves the accuracy.

**Table 1.** Comparison of MAE with different Algorithms

| K | Algorithms | | |
|---|---|---|---|
| | LICF | LFTRS_CF | HybirdLFM |
| 10 | 0.885 | 0.875 | 0.835 |
| 20 | 0.868 | 0.869 | 0.826 |
| 30 | 0.859 | 0.861 | 0.815 |
| 40 | 0.850 | 0.853 | 0.810 |
| 50 | 0.848 | 0.849 | 0.807 |

## 5 Conclusion

This paper studies the optimization strategy of recommendation algorithm in detail, and proposes a hybrid recommendation algorithm based on Item-CF algorithm and LFM algorithm. Based on the Movielens data set provided by Grouplens as experimental data, we verified the performance of the new algorithm by several methods, such as controlling the number of hidden factors for users as well as products, and the limitation on the number of neighboring items of the item similarity matrix, through the above experiments, we can see that the new hybrid algorithm has high accuracy in terms of recommendation results. A certain improvement has been made to improve the problem that the recommendation accuracy of the basic CF algorithm is not high, because the data sparseness affects the accuracy.

## References

1. Shi, Y., Larson, M., Hanjalic, A.: Collaborative filtering beyond the user-item matrix: a survey of the state of the art and future challenges. J. ACM. Comput. Surv. **47**(1), 1–45 (2014)
2. Wang, J., De Vries, A.P., Reinders, M.J.T.: Unifying user-based and item-based collaborative filtering approaches by similarity fusion. In: Proceedings of the 29th Annual International ACM SIGIR Conference on Research and Development in Retrieval. ACM, Seattle, Washington, USA (2006)
3. Jamali, M., Ester, M.: A transitivity aware matrix factorization model for recommendation in social networks. In: Proceedings of the IJCAI, pp. 2644–2649. AAAI Press (2011)
4. Goel, S., Broder, A., Gabrilovich, E., Pang, B.: Anatomy of the long tail: ordinary people with extraordinary taste. In: ACM WSDM, vol. 10, pp. 201–210 (2012)
5. Luke, A., Johnson, J., Ng, Y.: Recommending long-tail items using extended tripartite graphs. In: IEEE, vol. 1(1), pp. 123–130 (2018)
6. Liu, Y., Chai, T.: An improved slope-one collaborative filtering recommendation algorithm. J. Control Eng. China **24**(2), 257–262 (2017)
7. Vozalis M G,Margaritis K G.:Applying SVD on item-based filtering. In: Proc of International Conference on Intelligent Systems Design and Applications.Washington DC: IEEE Computer Society,pp. 464–469 (2005)
8. Yang, B., Lei, Y., Liu, D., et al.:Social collaborative filtering by trust. J. IEEE Trans. Pattern Anal. Mach. Intell. **39**(8), 1633–1647 (2017)

9. Wang, W., Jing, Y., Liang, H.: An improved collaborative filtering based on item similarity modified and common ratings. In: 2012 International Conference on Cyber Worlds, pp. 231–235. IEEE, Darmstadt (2012)
10. Xin, S., Fan, Y., Zheng, T., Zeng, Y., Li, L., Liu, D.: The improvement and application of collaborative filtering algorithm based on item. J. Front. Artif. Intell. Appl. **309**, 840–845 (2018)
11. Tao, S., Shen, C., Zhu, L., Dai, T., SVDCNN: a convolutional neural network model with orthogonal constraints based on SVD for context aware citation recommendation.J. Comput. Intell. Neurosci. (2020)
12. Qiu, L., Gao, S., Cheng, W., et al.: Aspect-based latent factor model by integrating ratings and review s for recommender system. J. Knowl.-Based Syst. **110**, 233–243 (2016)
13. Zeng, X., Ding, N., Quan, Z.: Latent factor model with heterogeneous similarity regularization for predicting gene-disease associations. In: 2016 IEEE International Conference on Bio Informatics and Bio Medicine, pp.15–18. IEEE Press, Shenzhen (2016)
14. Cheng, Z., Ding, Y., Zhu, L., et al.: Aspect-aware latent factor model: rating prediction with ratings and reviews In: International World Wide Web Conference Committee, pp. 23–27. ACM, Lyon (2018)
15. Luo, X., Zhou, M.C., Li, S., et al.: A nonnegative latent factor model for large-scale sparse matrices in recommender systems via alternating direction method. J. IEEE Trans. Neural Netw. Learn. Syst. **27**(3), 579–592 (2016)
16. Hofmann, T.: Latent semantic models for collaborative filtering. J. ACM Trans. Inf. Syst. **22**(1), 89–115 (2004)

# Multi-plane Slope Stability Analysis Based on Dynamic Evolutionary Optimization

Haiping Ma[1(✉)], Yu Shan[1], Chao Sun[1], Yufei Fan[2], and Zhile Yang[3]

[1] Department of Electrical Engineering, Shaoxing University, Shaoxing, Zhejiang, China
mahp@usx.edu.cn

[2] Rutgers Business School, Rutgers, The State University of New Jersey, Piscataway, NJ, USA
yf193@scarletmail.rutgers.edu

[3] Shenzhen Institute of Advanced Technology, Chinese Academy of Sciences, Shenzhen, Guangzhou, China

**Abstract.** The stability analysis of slope is an important and fundamental problem in geotechnical engineering, which is attracting the attention of many engineers and researchers. This paper proposes a dynamic biogeography-based optimization (BBO), to analyze slope stability problems. First, a real-world three-dimensional (3D) slope is described, and it is modeled as multiple two-dimensional (2D) slope planes, each of which is formulated as a single-objective optimization problem. Then a dynamic biogeography-based optimization algorithm, which is called DBBO, is developed by introducing change detection strategy into BBO. This strategy uses sentry solutions to detect their cost changes in different planes of a 3D slope, to reduce the unnecessary function evaluations. The proposed DBBO is applied to solve a case of classical 2D slope stability analysis and a case of 3D slope stability analysis with multiple planes established in this paper. The numerical results show that the proposed DBBO is a competitive algorithm for analyzing various slope stability problems, compared with other traditional methods.

**Keywords:** Slope stability analysis · Multi-plane · Dynamic · Biogeography-based optimization · Evolutionary computation

## 1 Introduction

Landslide is a complex natural phenomenon due to slope instability, and it usually causes enormous losses for human life and property [1–3]. It is widely understood that slope stability depends on different indicators, such as mechanics parameters, slope height, rainfall, and earthquake. Slope stability analysis becomes one of the most important issues in geotechnical engineering [4, 5]. The limit equilibrium principle commonly used in engineering geology verifies slope stability problem is a NP (non-polynomial) optimization problem, which involves calculation of safety factors for trial sliding surfaces and search for a sliding surface yielding a minimum safety factor. Importantly, safety factor calculation is non-convex, discontinuous with multiple strong local minima [6–8].


Q. Han et al. (Eds.): LSMS 2021/ICSEE 2021, CCIS 1469, pp. 98–108, 2021.
https://doi.org/10.1007/978-981-16-7213-2_10

Several approaches have been proposed for locating a minimum safety factor including classical methods and heuristic optimization techniques. In [9], a hybrid optimization integrating gravitational search algorithm with sequential quadratic programming is used for calculating safety factor, and numerical simulation shows that the proposed algorithm converges faster to accurate solutions for slope stability problems. In [10], swarm intelligence techniques are used for solving some complicated slope stability examples, and results show that the levy flight krill herd algorithm is the most efficient method for this kind of problems. In [11], a scripting system combined the standard landslide analysis program with particle swarm optimization is automatically used to analyze slope stability. The proposed system is tested on a standard soil slope, and results show that it is promising for slope stability analysis.

Previous studies mainly focus on two-dimensional (2D) slope analysis with infinitely wide sliding surfaces. However, all landslides are three-dimensional, and especially a slope is clearly constrained by adjacent rocks or existing structures. Safety factor calculation based on 2D analysis might be overly conservative so the result is unreasonable. To obtain a more accurate solution, it is very significant to perform stability analysis of three-dimensional (3D) slope, which can more truly show the slope collapse. In [12], dynamic time-history stability analysis of a 3D slope is developed, and its reliability is assessed based on the probability density evolution method combined with dimensional-reduction spectrum-random function. Numerical simulations demonstrate that the proposed method can directly get the range and specific location of instability failure sliding surface. In [13], various hybrid intelligent systems, including genetic algorithm, particle swarm optimization combined with artificial neural networks, are introduced to evaluate and predict static and dynamic 3D slope stability, and results show that all methods can approximate a minimum slope safety factor. In [14], a new method combined with numerical manifold is proposed for 3D slope stability analysis, and then genetic algorithm is used to find the critical failure surface. Numerical simulations show that the proposed method is effective and robust.

These methods provide various potential tools to analyze 3D slope stability. Compared with a 2D slope, a 3D slope gives a more global approach to show the effect of boundary conditions and slope width. More importantly, for natural landslides, they are always three-dimensional. So stability analysis of a 3D slope can provide its instability failure range and the corresponding location, which is the important information for landslide prevention and reinforcement. However, for these advanced methods and simulations, the level of expertise is very important. Meanwhile, the difficulties in modeling and validation further prevent their widespread applications. In addition, the gains and tests of complex parameters under 3D conditions pose major challenges for geotechnical engineering.

In light of the above evidence, considering pre-existing 2D slope analysis experience and necessity of development of 3D slope analysis, a multi-plane slope stability analysis method based on evolutionary optimization is proposed in this paper. First, a 3D slope model is established and it is divided into multiple 2D planes. It is in favor of using 2D slope analysis methods to solve a 3D landslide problem. Then, a dynamic biogeography-based optimization (DBBO) algorithm, which uses change detection strategy to detect the solution dynamic change, is proposed to fast search minimum safety factors of each

2D plane and the corresponding critical failure surfaces. Finally, the feasibility and superiority of the proposed method for analyzing slope stability is verified via a 2D slope example and a 3D slope example. The proposed method provides a new idea, which can directly extend various 2D methods for analyzing 3D slope stability, and improves computing efficiency of 3D slope stability analysis.

## 2 Multi-plane Slope Stability Analysis Model

In this section, a 3D slope is cut into multiple 2D planes as shown in Fig. 1. For each plane, it is equivalent to a 2D slope. The element stress of each component of each plane is first calculated using a finite element method. Then the stress component of a point on the sliding surface is further computed, and the surface normal and tangential stresses are determined.

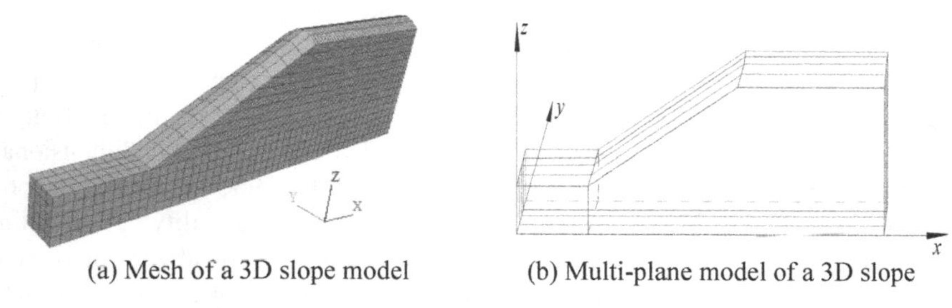


(a) Mesh of a 3D slope model (b) Multi-plane model of a 3D slope

**Fig. 1.** A 3D slope is cut into multiple 2D planes.

Next, for the multi-plane 3D slope, various heuristic methods including evolutionary optimization are used to search a minimum safety factor of a sliding surface, which is corresponding to a critical failure surface of each 2D plane. Here safety factor of a slope is defined as the ratio between the available strength and the strength required for a state of incipient failure along a possible sliding surface. According to the limit equilibrium principle, when a safety factor is less than one, it represents the failure in a slope. In the multi-plane 3D slope, the minimum safety factor of all sliding surfaces of all planes is defined as the minimum safety factor of the whole 3D slope. For the entire time-history ground motion during an earthquake, the process can be repeated to get the minimum safety factor of a 3D slope under dynamic condition.

In the proposed multi-plane 3D slope stability analysis, each plane is taken as a 2D slope. So calculation of safety factor $F_S$ of each plane is based on a 2D slope model, which is represented as follows [12]:

$$F_S = \frac{\sum_{i=1}^{n}(c_i + \sigma_i \tan \varphi_i) l_i}{\sum_{i=1}^{n} \tau_i l_i} \quad (1)$$

where the objective of optimization is to minimize the safety factor $F_S$. $c_i$ and $\varphi_i$ are the adhesion angle and internal friction angle of the $i$th element respectively; $l_i$ is the length of the $i$th element passed by a sliding arc; and $\sigma_i$ and $\tau_i$ are the normal stresses and tangential stresses on the sliding arc surface of the $i$th element.

According to the equilibrium conditions, the normal stresses and tangential stresses of the $i$th element is calculated as.

$$\sigma_i = \frac{\sigma_x + \sigma_y}{2} - \frac{\sigma_x - \sigma_y}{2}\cos 2\alpha - \tau_{xy}\sin 2\alpha \tag{2}$$

$$\tau_i = \frac{\sigma_x - \sigma_y}{2}\sin 2\alpha - \tau_{xy}\cos 2\alpha \tag{3}$$

where $\alpha$ is the solution variable, which is an angle between slice base and horizontal line. $\sigma_x$, $\sigma_y$ and $\sigma_{xy}$ satisfy: $\sigma_x = (\sigma_x^s + \sigma_x^d)$, $\sigma_y = (\sigma_y^s + \sigma_y^d)$, $\tau_{xy} = (\tau_{xy}^s + \tau_{xy}^d)$, $\sigma_x^s$ is the element static horizontal stress, $\sigma_x^d$ is the element dynamic horizontal stress, $\sigma_y^s$ is the element static vertical stress, $\sigma_y^d$ is the element dynamic vertical stress, $\tau_{xy}^s$ is the element static shear stress, and $\tau_{xy}^d$ is the element dynamic shear stress.

## 3 Proposed Approach

To search a minimum safety factor for the multi-plane 3D slope, DBBO is proposed, which is the dynamic version of the original and successful BBO. Some basic operators of BBO including migration and mutation refer to literature [15, 16]. Here we mainly design a new strategy of DBBO to fast search the optimal solutions of the multi-plane 3D slope.

In the proposed multi-plane 3D slope, optimization environment of each plane may change due to geological factors. But for most of planes, the change is very small, not even changed. When there are only slight changes or no changes, there can be a large correlation between the current optimal solutions after a change and the previous ones. In such case, change detection strategy is a very useful method, which makes use of existing optimal solutions in the current plane. In fact, when the change degree is very small, information gained from the other solutions can be exploited and reused to accelerate the convergence speed. In this paper, the change detection strategy is used in the proposed DBBO because most of the potential landslides are relatively stable, and their safety factors do not drastically change for multiple planes in the same 3D slope.

Since the population contains a large number of solutions, it may be time-consuming to evaluate all solutions for each plane. Thus, the change detection strategy uses sentry solutions to detect cost change. That is, the strategy stores good solutions of the current plane into an archive, and then randomly selects a few solutions as sentry solutions to test cost change for other planes. If the cost values of sentry solutions keep unchanged in other planes, the solutions in the archive are considered as the good solutions for other planes, and it does not need to re-evaluate all solutions for other planes. If the cost value change is detected in a certain plane, it re-evaluates all solutions for this plane. In this way, the proposed DBBO can apparently reduce the evaluation numbers and quickly converge to the new prominent regions.

For the selection of sentry solutions, 10% of the solutions in the $k$th plane are randomly chosen from the population to build a sentry set. These sentry solutions are re-evaluated in the $(k + 1)$th plane including the objective values and the constraint values. If the values in the $(k + 1)$th plane are different from the ones in the $k$th plane, a change has taken place. For an optimization problem with a size $N_0$ of sentry set, its change detection process is illustrated as:

$$\underbrace{\begin{array}{llll} \text{sentry solutions} & x_1 = \left(x_1^1, x_1^2, \cdots, x_1^n\right) & x_2 = \left(x_2^1, x_2^2, \cdots, x_2^n\right) \cdots & x_{N_0} = \left(x_{N_0}^1, x_{N_0}^2, \cdots, x_{N_0}^n\right) \\ \text{plane } k & F(x_1, k), G(x_1, k) & F(x_2, k), G(x_2, k) \cdots & F\left(x_{N_0}, k\right), G\left(x_{N_0}, k\right) \\ \text{plane } k+1 & \underbrace{F(x_1, k+1), G(x_1, k+1)}_{test\ a} & \underbrace{F(x_2, k+1), G(x_2, k+1)}_{test\ a} \cdots & \underbrace{F\left(x_{N_0}, k+1\right), G\left(x_{N_0}, k+1\right)}_{test\ a} \end{array}}_{test\ b} \quad (4)$$

In Eq. (4), *Test a* is to examine if the value of objective function $F\left(x_j, k+1\right)$ or the value of constraint functions $G\left(x_j, k+1\right)$ in the $(k+1)$th plane is different from $F\left(x_j, k\right)$ or $G\left(x_j, k\right)$ in the $k$th plane for any sentry solution $x_j$. If this is the case, the objective $F\left(x_j\right)$ or the constraint $G\left(x_j\right)$ has been changed. *Test b* is to consider that a change occurs when at least one objective $F\left(x_j\right)$ or one constraint $G\left(x_j\right)$ has been changed.

The basic DBBO procedures applying to the multi-plane 3D slope stability analysis are summarized as follows:

Step 1. Randomly initialize a population $P$ applying to the first plane, and evaluate objective and constraint values of solutions in $P$.

Step 2. If the criterion of termination is not met, go to step 3; otherwise, it is terminated. Here the maximum function evaluation number is taken as the termination criterion.

Step 3. Use BBO migration operator [15] to create offspring population $O$ for the current plane. Evaluate their values and store the current optimal solutions into the archive set $A$. Then randomly choose some solutions from the offspring population $O$ as a sentry solution set, and re-evaluate their values for the next plane to detect changes.

Step 4. Perform change detection strategy shown in Eq. (4) for the sentry solutions. If cost values of sentry solutions are unchanged, it does not need to re-evaluate all solutions for this plane, and the solutions in the previous plane are considered as the good solutions for this plane. If a change has occurred, it re-evaluates all solutions for this plane.

Step 5. Update optimal solutions in the archive $A$ using the new solutions in this plane if they are the better solutions during the re-evaluation, and then detect the change for the next plane.

## 4 Simulation Results

In this section, to verify the feasibility of the multi-plane 3D slope model and explore the effectiveness of the proposed DBBO, two numerical cases are investigated to search minimum safety factors and their corresponding critical failure surfaces. The results are discussed and compared with previous studies. The first case is a homogeneous 2D slope and the second case is a homogeneous 3D slope.

For the proposed DBBO, some parameters need to be tuned, including the size of population, the rates of migration and mutation. In [17] authors have discussed these parameters in detailed, and give a reasonable parameter set for solving various problems. In this paper, we use population sizes of 20, 50 and 100, immigration and emigration rates based on sine curves, mutation rate of 0.01 per decision variable, and a fixed fitness evaluation numbers of 1000, 5000 and 10000. Each case is optimized over 25 independent runs.

### 4.1 Case 1 – 2D Slope Stability Analysis

The first case is a 2D slope consisting of homogeneous soil, and its mechanical parameters are: unit weight of 17.64 kN/m$^3$, friction angle $\varphi$ of 10° and effective cohesion $c$ of 9.8kPa. The slope geometry parameters are presented in Fig. 2. Dynamic time-history data of seismic load shown in Fig. 3 is added to the slope, and the corresponding displacement nephogram of 2D slope under seismic load is shown in Fig. 4.

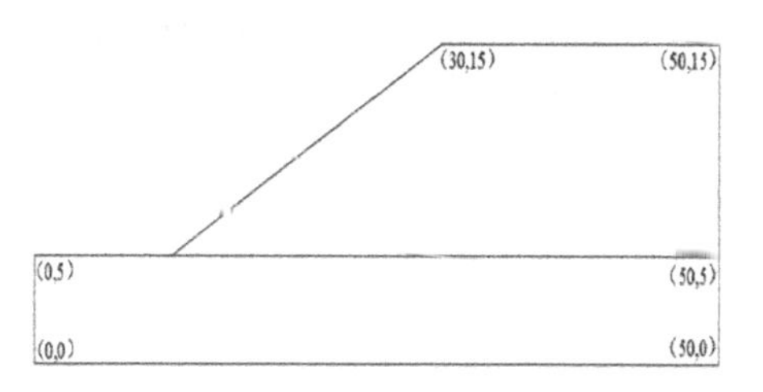


**Fig. 2.** Geometry parameters of a 2D slope.

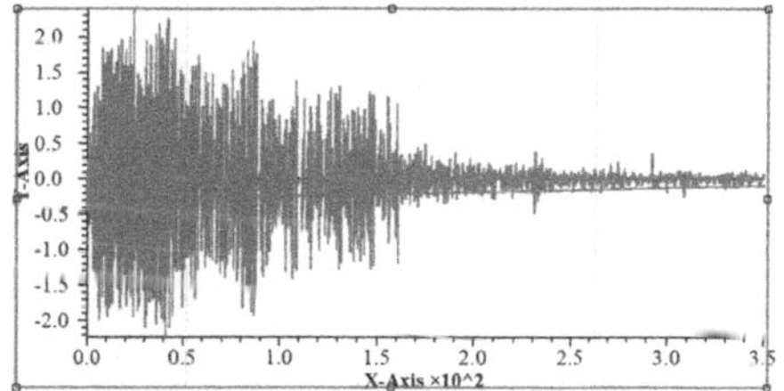


**Fig. 3.** Dynamic time-history data.

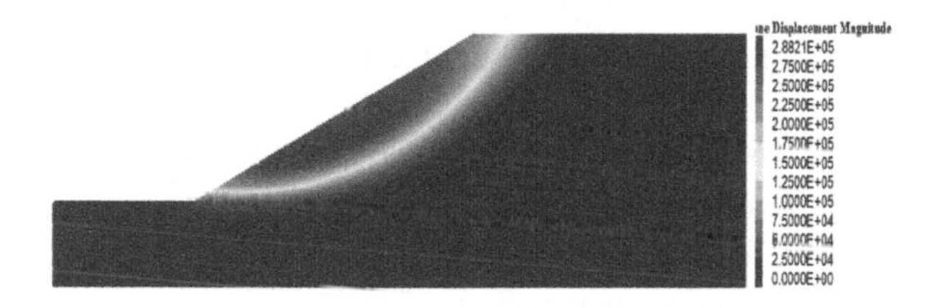


**Fig. 4.** Displacement nephogram of a 2D slope under seismic load.

Table 1 shows the minimum safety factors obtained by the proposed DBBO under the difference tuning parameters. From the table, it has been observed that for the same population size, the mean performance obtained by the proposed DBBO is best when the fitness evaluation numbers are 1000 and 5000, which means that the proposed approach can converge to the optimal solutions after the fitness evaluation numbers are enough. On the other hand, for the same fitness evaluation numbers, the proposed DBBO performs best when the population size is 50, which means that appropriate population size is beneficial to the convergence of the proposed approach.

Furthermore, the performance of the proposed DBBO is compared with other methods in literature. Here we only compare the mean minimum safety factors, which are shown in Table 2. As we see in the table, for case 1 of 2D slope, the proposed DBBO is the second best with a mean minimum safety factor of 1.049. Although opposition-based firefly algorithm is best with a mean minimum safety factor of 0.995, it does not meet the definition requirements of a minimum safety factor more than one in terms of the limit equilibrium principle of slope stability. The empirical results show that the proposed DBBO is a competitive algorithm for the case of 2D slope stability we study.

**Table 1.** Minimum safety factors obtained by DBBO under the difference tuning parameters.

| Pop. size | Eval. number | Best | Worst | Mean | SD |
|---|---|---|---|---|---|
| 20 | 500 | 1.037 | 1.135 | 1.068 | 0.028 |
| | 1000 | 1.032 | 1.137 | 1.062 | 0.016 |
| | 5000 | 1.035 | 1.131 | 1.062 | 0.022 |
| 50 | 500 | 1.021 | 1.127 | 1.051 | 0.007 |
| | 1000 | 1.020 | 1.120 | 1.049 | 0.005 |
| | 5000 | 1.019 | 1.121 | 1.049 | 0.006 |
| 100 | 500 | 1.040 | 1.137 | 1.078 | 0.004 |
| | 1000 | 1.034 | 1.133 | 1.071 | 0.021 |
| | 5000 | 1.034 | 1.131 | 1.071 | 0.016 |

**Table 2.** Performance comparisons of DBBO with other methods.

| Reference | Method | Mean minimum safety factor |
|---|---|---|
| Mohammad et al. [18] | Firefly algorithm (FA) | 1.137 |
| Mohammad et al. [18] | Opposition-based firefly algorithm (OBFA) | 0.995 |
| Li et al. [19] | Real-coded genetic algorithm (RCGA) | 1.321 |
| Raihan et al. [9] | Gravitational search algorithm (GSA) | 1.123 |
| Zhang et al. [20] | Differential evolution (DE) | 1.055 |
| Gandomi et al. [10] | Particle swarm optimization (PSO) | 1.115 |
| Gandomi et al. [10] | Cuckoo search (CS) | 1.064 |
| DBBO (this study) | Dynamic biogeography-based optimization (DBBO) | 1.049 |

Figure 5 shows the minimum safety factors of case 1 obtained by the proposed DBBO under the whole of time-history data of seismic load. It finds that the minimum safety factor is 1.049 when the time is about 40 s. Figure 6 shows the critical failure surface corresponding to the minimum safety factor of 1.049, where the white line represents the critical failure surface obtained by the proposed DBBO, and the interactive curve of the red surface and the blue surface represents the critical failure surface obtained by the traditional strength reduction method.

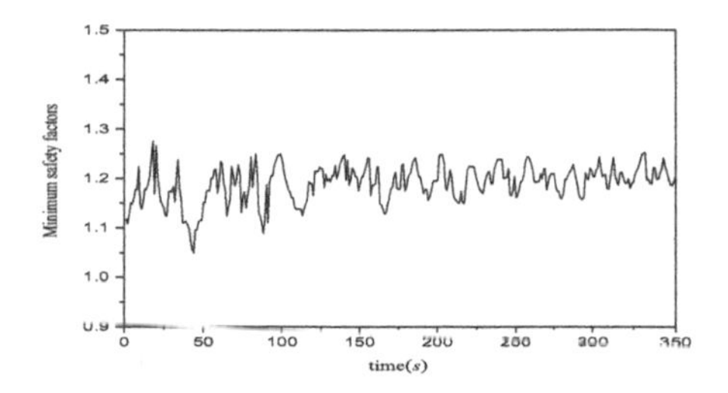


**Fig. 5.** Minimum safety factors.

15
10
5
0 5 10 15 20 25 30 35 40 45 50

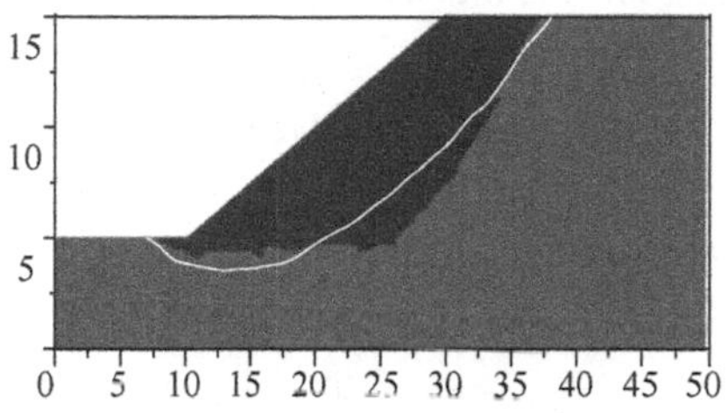

**Fig. 6.** Critical failure surface.

## 4.2 Case 2 – 3D Slope Stability Analysis

The second case is a 3D slope consisting of homogeneous soil. Mechanical parameters of the slope are the same as those of the first case. The width of the slope is set to 4m because it is a 3D slope, and the detailed geometry parameters of the multi-plane 3D model are shown in Fig. 7. To effectively analyze this model, it is divided into five 2D planes based on its width. Namely, in the y-axis direction in Fig. 7, it cuts one plane every meter. In addition, the time-history data of seismic load shown in Fig. 3 is added to this 3D slope, and the corresponding displacement nephogram under seismic load is shown in Fig. 8.

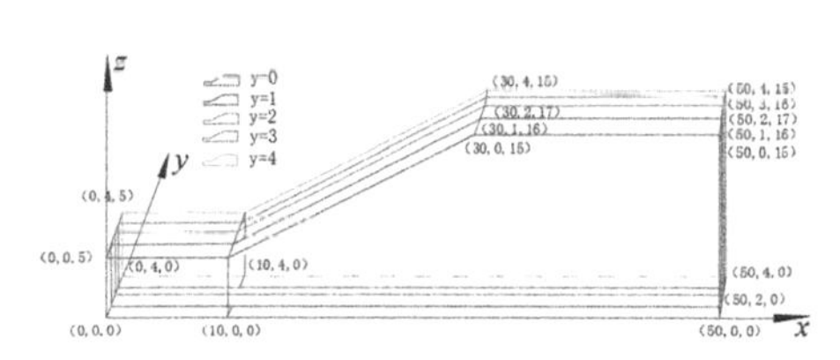


**Fig. 7.** Geometry parameters of a 3D slope.

**Fig. 8.** Displacement nephogram of a 3D slope.

Table 3 shows the minimum safety factors of different planes of the 3D slope obtained by the proposed DBBO. From the table, it has been observed that in the same plane, the proposed DBBO performs best when the population size is 50, which means that the appropriate population size is beneficial to get the optimal solutions for the studied multi-plane 3D slope. On the other hand, for the different planes in the same 3D slope, the obtained minimum safety factors are somewhat different due to the different geometry parameters of each plane, which means that the proposed multiple-plane 3D slope model is feasible, and the proposed DBBO can find the subtle differences of the minimum safety factors inside the 3D slope.

**Table 3.** Minimum safety factors of different planes of the 3D slope obtained by DBBO.

| Plane | Pop. size | Best | Worst | Mean | SD |
|---|---|---|---|---|---|
| Plane 1 (y = 0) | 20 | 1.032 | 1.137 | 1.062 | 0.016 |
| | 50 | 1.020 | 1.120 | 1.049 | 0.005 |
| | 100 | 1.034 | 1.133 | 1.071 | 0.021 |
| Plane 2 (y = 1) | 20 | 1.037 | 1.142 | 1.068 | 0.024 |
| | 50 | 1.025 | 1.105 | 1.054 | 0.020 |
| | 100 | 1.040 | 1.111 | 1.079 | 0.025 |
| Plane 3 (y = 2) | 20 | 1.042 | 1.164 | 1.078 | 0.023 |
| | 50 | 1.041 | 1.156 | 1.070 | 0.019 |
| | 100 | 1.047 | 1.166 | 1.081 | 0.021 |
| Plane 4 (y = 3) | 20 | 1.036 | 1.141 | 1.065 | 0.022 |
| | 50 | 1.025 | 1.107 | 1.051 | 0.019 |
| | 100 | 1.041 | 1.112 | 1.077 | 0.023 |
| Plane 5 (y = 4) | 20 | 1.0321 | 1.138 | 1.060 | 0.018 |
| | 50 | 1.022 | 1.122 | 1.048 | 0.009 |
| | 100 | 1.035 | 1.134 | 1.070 | 0.017 |

Figure 9 shows that the critical failure surface of each plane of the 3D slope corresponding to the minimum safety factors obtained by the proposed DBBO. Table 4 shows that for different planes in the same 3D slope, the depths and locations of the deepest points of critical failure surfaces, and their intersection angles with the boundaries. These results furthermore show the proposed DBBO can discover the differences inside the 3D slope, and it is considered as a promising heuristic method for the multi-plane 3D slope stability analysis we study.

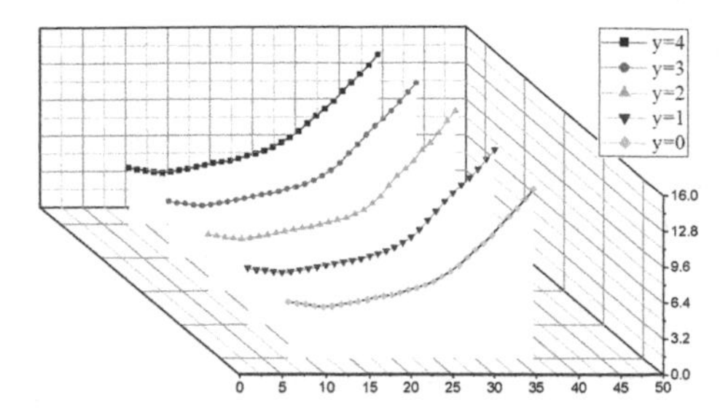


**Fig. 9.** Critical failure surface of each plane of a 3D slope.

**Table 4.** The depth, location, and intersection angle of the deepest point of each plane obtained by DBBO.

| Planes | Minimum safety factor | Depth/m | Location/m | Intersection angle |
|---|---|---|---|---|
| Plane 1 (y = 0) | 1.049 | 5.45 | (26.19, 0, 7.12) | 153.82 |
| Plane 2 (y = 1) | 1.054 | 5.94 | (26.82, 1, 6.90) | 155.37 |
| Plane 3 (y = 2) | 1.070 | 6.35 | (26.91, 2, 6.78) | 157.19 |
| Plane 4 (y = 3) | 1.051 | 5.86 | (26.80, 3, 6.92) | 155.16 |
| Plane 5 (y = 4) | 1.048 | 5.39 | (26.52, 4, 7.10) | 153.01 |

## 5 Conclusions

In this paper a multi-plane 3D slope stability analysis model is built based on complex natural landslide phenomenon, which is posed as a dynamic single-objective optimization problem. Then a DBBO algorithm with cost change detection strategy is proposed, which uses sentry solutions to detect the change of different planes in a same 3D slope. The performance of DBBO is investigated on a case of classical 2D slope stability analysis and a case of multi-plane 3D slope stability analysis, and experimental results show that the proposed DBBO can successfully handle with various slope stability analysis problems studied in this paper.

This paper shows that the proposed DBBO has good results for slope stability analysis problems, but still opens the door for additional development and empirical investigation. First, this paper takes multi-plane 3D slope stability analysis as a dynamic single-objective optimization problem, but future work could formulate it as a constraint dynamic optimization problem, considering various real-world slope environment effects. Second, this paper only uses DBBO to analyze slope stability problems, and future work could employ other types of dynamic evolutionary algorithms to solve the same problems.

**Acknowledgments.** This material was supported in part by the National Natural Science Foundation of China under Grant No. 52077213 and 62003332, and the Zhejiang Provincial Natural Science Foundation of China under Grant No. LY19F030011.

## References


1. Wang, M., Liu, K., Yang, G., Xie, J.: Three-dimensional slope stability analysis using laser scanning and numerical simulation. J. Geomat. Nat. Haz. Risk. **8**(2), 997–1011 (2017)
2. Li, Z., Yang, X.: Stability of 3D slope under steady unsaturated flow condition. J. Eng. Geol. **242**, 150–159 (2018)
3. He, Y., Liu, Y., Zhang, Y., Yuan, R.: Stability assessment of three-dimensional slopes with cracks. J. Eng. Geol. **252**, 136–144 (2019)
4. Singh, J., Banka, H., Verma, A.K.: A BBO-based algorithm for slope stability analysis by locating critical failure surface. J. Neural Comput. Appl. **31**(2), 1–18 (2019)
5. Xu, J., Yang, X.: Seismic stability of 3D soil slope reinforced by geosynthetic with nonlinear failure criterion. J. Soil Dyn. Earth. Eng. **118**, 86–97 (2019)
6. Khajehzadeh, M., Taha, M.R., Shafie, A., Eslami, M.: A modified gravitational search algorithm for slope stability analysis. J. Eng. Appl. Artif. Intel. **25**, 1589–1597 (2012)
7. Kang, F., Li, J., Ma, Z.: An artificial bee colony algorithm for locating the critical slip surface in slope stability analysis. J. Eng. Optimiz. **45**(2), 207–223 (2013)
8. Huang, S., Huang, M., Lyu, Y.: An improved KNN-based slope stability prediction model. J. Adv. Civil Eng. 8894109 (2020)
9. Raihan, T.M., Mohammad, K., Mahdiyeh, E.: A new hybrid algorithm for global optimization and slope stability evaluation. J. Central South Univ. **20**, 3265–3273 (2013)
10. Gandomi, A.H., Kashani, A.R., Mousavi, M., Jalalvandi, M.: Slope stability analyzing using recent swarm intelligence techniques. Int. J. Numer. Anal. Met. **39**(3), 295–309 (2015)
11. Chen, W., Zhao, Y., Liu, L., Wang, X.: A new evaluation method for slope stability based on TOPSIS and MCS. J. Adv. Civ. Eng. 1209470 (2020)
12. Song, L., et al.: Three-dimensional slope dynamic stability reliability assessment based on the probability density evolution method. J. Soil Dyn. Earth. Eng. **120**, 360–368 (2019)
13. Koopialipoor, M., Armaghani, D.J., Hedayat, A., Marto, A.: Applying various hybrid intelligent systems to evaluate and predict slope stability under static and dynamic conditions. J. Soft Comput. **23**, 5913–5929 (2018)
14. Liu, G., Zhuang, X., Cui, Z.: Three-dimensional slope stability analysis using independent cover based numerical manifold and vector method. J. Eng. Geol. **225**, 83–95 (2017)
15. Ma, H.: An analysis of the equilibrium of migration models for biogeography-based optimization. J. Inform. Sci. **180**(18), 3444–3464 (2010)
16. Simon, D.: Biogeography-based optimization. J. IEEE Trans. Evol. Comput. **12**(6), 702–713 (2008)
17. Ma, H., Simon, D.: Analysis of migration models of biogeography-based optimization using markov theory. J. Eng. Appl. Artif. Intel. **24**(6), 1052–1060 (2011)
18. Khajehzadeh, M., Taha, M., Eslami, M.: Opposition-based firefly algorithm for earth slope stability evaluation. J. China Ocean Eng. **28**(5), 713–724 (2014)
19. Li, Y., Chen, Y., Zhan, T., Ling, D., Cleall, P.: An efficient approach for locating the critical slip surface in slope stability analyses using a real-coded genetic algorithm. J. Canadian Geotech. J. **47**(7), 806–820 (2010)
20. Zhang, Z., Chai, J., Zhang, S., Qian, W.: Rearch on slope slide surface search based on differential evolution algorithm with mixed mutation strategy. J. Water Resour. Water Eng. **29** (4), 218--223 (2018)

# Housing Assignment Optimization of Warehouse Breeding

Wenqiang Yang[1], Zhile Yang[2](✉), Zhanlei Peng[1], and Yonggang Chen[1]

[1] School of Mechanical and Electrical Engineering, Henan Institute of Science and Technology, Xinxiang 453003, China

[2] Shenzhen Institute of Advanced Technology, Chinese Academy of Sciences, Shenzhen 518055, China

zl.yang@siat.ac.cn

**Abstract.** This paper aims at optimizing the housing assignment of warehouse breeding. A model of the housing assignment for minimizing the total completion time of feeding in cycle of slaughter is established, and a Gaussian social spider algorithm (GSSO) is proposed. By introducing the subpopulation reconstruction mechanism, Gaussian mating radius and multi-mating operator, GSSO can balance the global exploration and the local exploitation. Simulation results demonstrate that compared with other algorithms, the proposed algorithm achieves higher accuracy and quicker convergence, meanwhile obtains a more satisfactory solution in a shorter period of time.

**Keywords:** Warehouse housing assignment · Social spider optimization algorithm · Subpopulation reconstruction mechanism · Gaussian mating radius · Multi-mating operator

## 1 Introduction

2021 is the first year of China's rural revitalization plan. General Secretary Xi has repeatedly emphasized that a well-off life is not a well-off life, the key is to look at the fellow villagers. It can be seen that the "the work related to agriculture, rural areas and farmers" are the core issue for realizing the China's rural revitalization plan. However, agriculture is the root of the three rural issues, which can be solved effectively by promoting agricultural transformation and developing modern agriculture. In such a context, technologies such as intelligent breeding and intelligent seedling have gradually been applied in agricultural enterprises [1]. In order to maximize profits, breeding companies usually adopt hybrid breeding mode of a multi-variety broiler. Nevertheless, due to the differences in the growth characteristics of different breeds of broiler chickens, the feeding rhythm and the amount of feeding are different. Therefore, the different layout of broiler houses of various breeds will directly affect the timeliness of feeding of various broiler chickens, and then affect their normal growth. Thus it can be seen that carrying out research on the optimization of the housing distribution of multi-breed broiler is only essential but also has good economic and social benefits.


Q. Han et al. (Eds.): LSMS 2021/ICSEE 2021, CCIS 1469, pp. 109–121, 2021.
https://doi.org/10.1007/978-981-16-7213-2_11

There are few research literatures on the house allocation optimization of multi-breed broiler, but relating to it, the research on the optimization of storage space allocation has been more in-depth. De Koster et al. [2] pointed out that the storage location assignment is an effective way to improve picking efficiency, which can save up to 55% of picking time. Lee et al. [3] deemed that the relevant cargo location allocation strategy would lead to certain picking jams, and based on this, proposed an improved related cargo location allocation strategy considering picking delay. Calzavara et al. [4] proposed a storage space allocation model that considers picking posture and picking workload from the perspective of affecting the health of picking workers. Battini et al. [5] based on order data in a certain period of time, counted and quantified the picking probability of goods, and allocated locations close to the storage buffer for goods with high picking frequency. Ozturkoglu et al. [6] proposed an optimization model for location allocation that takes into account both picking distance and storage utilization, and solved it based on a constructive heuristic algorithm. Li et al. [7] established a dynamic cargo location allocation model based on the classification and storage, combining with the degree of correlation between goods, and proposed a greedy genetic algorithm. Roshan et al. [8] introduced energy-saving, green manufacturing and other indicators into the inventory allocation model, and proposed a new type of classified storage strategy. E et al. [9] analyzed the key role played by auto parts storage in production, aiming to improve the efficiency of parts delivery, and shelf stability and safety, and used an improved genetic algorithm to optimize the cargo location. Of course, some scholars have proposed optimization strategies for the allocation of goods in combination with industry characteristics [10, 11].

In summary, firstly, the above-mentioned literatures mainly focus on logistics warehousing, and less consider the field of animal husbandry in terms of application; secondly, storage allocation modeling lays emphasis upon picking efficiency and shelf stability and less considers the influence of goods' own characteristics on the allocation of cargo locations, which will affect the effect of cargo location optimization to a certain extent. In view of these shortcomings, this paper takes the warehouse housing of the multi-breed broiler as the research object, by combining with the feeding characteristics of the broilers throughout the cycle of slaughter, and based on the principle of precision breeding, the research on the optimization of warehouse housing assignment is carried out with the aim of minimizing the delayed growth of the broilers, which is solved by the proposed Gaussian social spider algorithm.

## 2 Problem Description and Modeling

### 2.1 Problem Statement

A breeding enterprise uses automated warehouse to breed multiple varieties of broilers. The automated warehouse system has the following characteristics: $m$ lanes, $n$ layers shelves, each row of shelves has $k$ columns, the distance between adjacent lanes is $D$, and the length and height of each bin are $L$ and $H$ respectively. In order to conveniently research, the start and braking time of the robot are ignored. The horizontal and vertical movements of the robot are independent of each other, which the average speed of are $v_x$ and $v_y$ respectively, and there are $N$ forks. The goods location coordinate is defined as $p_i(x_i, y_i, z_i)$, where $x_i$, $y_i$ and $z_i$ represent the column number, layer number and lane number of the goods location respectively, and the coordinate of warehouse buffer is set to $p_O(0,0,0)$.

**Definition 1:** If the robot passes through the sub-route $r$, then $s_r = 1$; otherwise, $s_r = 0$. If the location $p_i$ belongs to sub-route $r$, then $g_{ir} = 1$; Otherwise, $g_{ir} = 0$.

**Definition 2:** If the robot consecutively accesses $p_i(x_i,y_i,z_i)$, $p_j(x_j,y_j,z_j)$ during the execution of the task, then $e_{ij} = 1$; Otherwise, $e_{ij} = 0$.

Here, the time $t_{ij}$ for successively passing locations $p_i$ and $p_j$ can be expressed as:

$$\mathrm{t}_{ij} = max\left\{\frac{L \times |x_i - x_j|}{v_x}, \frac{H \times |y_i - y_j|}{v_y}\right\}. \tag{1}$$

Taking into account the differences of the growth characteristics of broiler breeds, that is, within the same cycle of slaughter, the rates of weight gain of each breed are different. Therefore, the same feeding delay will have different effects on the growth of each breed. In order to quantify the degree of influence, a delayed growth factor $\varphi_k$ is introduced, where $k$ $(k \in \{1, 2, \cdots, Z\})$ represents the breed number of each broiler. Then the degree of influence $\rho_k$ of feeding delay on the growth of breed $k$ can be expressed as

$$\rho_k = \varphi_k \times T_k. \tag{2}$$

where $T_k$ is the feeding delay time of the k-th breed broiler.

## 2.2 Mathematical Modeling

Assuming that during the entire slaughter cycle, the robot needs to complete $q$ feeding tasks. Through the above analysis, it can be seen that the warehouse housing assignment should minimize the impact on the growth of broilers of various breeds. Therefore, the optimization goal is to minimize the sum of the degree of influence of each breed broiler. The specific mathematical model includes the optimization objective and the constraints are defined as follows,

$$\min f(e) = \sum_{r=1}^{\lceil q/N \rceil} \sum_{i=0}^{q} \sum_{j=0}^{q} t_{ij} \cdot e_{ij} \cdot g_{ir} \cdot g_{jr} \cdot s_r \cdot \varepsilon_{jk} \cdot \varphi_k. \tag{3}$$

$$\text{s.t.} \quad \sum_{i=0}^{q} \sum_{r=1}^{\lceil q/N \rceil} g_{ir} = q. \tag{4}$$

$$\sum_{r=1}^{\lceil q/N \rceil} g_{ir} = 1, \forall i \in \{1, 2, \cdots, q\}. \tag{5}$$

$$\sum_{i=0}^{q} g_{ir} \le N, \forall r \in \{1, 2, \cdots, \lceil q/N \rceil\}. \tag{6}$$

$$\sum_{j=1}^{m} e_{0j} = \lceil q/N \rceil. \tag{7}$$

$$\sum_{i=1}^{m} e_{i0} = \lceil q/N \rceil. \tag{8}$$

where $\varepsilon_{jk}$ represents the feeding times of the broiler with breed number $j$ in the location $p_j$ throughout the slaughter cycle. Equation (3) is the objective to be optimized, Eq. (4) is the constraint on the number of robot tasks, Eq. (5) is that each feeding task appears only once in all paths of the robot, Eq. (6) is the load constraint of the robot, Eq. (7) is the constraint of the starting point of the robot path, Eq. (8) is the constraint of the end point of the robot path.

## 3 Social Spider Optimization Algorithm (SSO)

Social Spider Optimization Algorithm (SSO) is a new type of swarm intelligence optimization algorithm proposed by Cuevas [12] in 2013 by simulating the biological behavior of social spiders, and has been widely adopted in varied fields [13, 14]. The SSO algorithm makes the spider web equal to the solution space of the problem to be solved, and the location of the individual spider corresponds to the potential solution of the optimization problem. The males and females individuals approach the optimal solution through cooperative behavior and mating operator. The basic principles of conventional SSO can be summarized as follows.

### 3.1 Initialization of the Population

The spider population is a partial female population, and the proportion of female individuals in the population is approximately 65–90%. Therefore, the number of female and male spiders can be determined by Eq. (9).

$$\begin{cases} N_f = \lfloor (0.9 - rand \times 0.25) \times NP \rfloor \\ N_m = N - N_f \end{cases}. \tag{9}$$

where $NP$, $N_f$, $N_m$ and are the numbers of the entire population, female spiders and male spiders, $rand$ is a random number between 0 and 1, $\lfloor \cdot \rfloor$ means rounding down. The population of male and female spiders are initialized by Eq. (10).

$$\begin{cases} F_{ij} = h_{jmin} + rand \times \left(h_{jmax} - h_{jmin}\right) \\ M_{ij} = h_{jmin} + rand \times \left(h_{jmax} - h_{jmin}\right) \end{cases}. \tag{10}$$

where $F_{ij}$ represents the $j$-th dimension variable of the i-th female spider, and similarly, $M_{ij}$ represents the male spider, $h_{jmax}$ and $h_{jmin}$ represent the upper and lower bounds of the j-th dimension variable respectively.

### 3.2 Female Cooperative Operator

Female spiders are update by sensing the vibrations on the communal web generated by the superior spiders. Such vibrations are mainly manifested by two behaviors: attraction and repulsion, which are represented by Eq. (11).

$$\mathrm{F}_i^{t+1} = \begin{cases} F_i^t + \alpha Vib_{ci}\left(S_c - F_i^t\right) + \beta Vib_{bi}\left(S_b - F_i^t\right) + \delta(rand - 0.5), r_m \le PF \\ F_i^t - \alpha Vib_{ci}\left(S_c - F_i^t\right) - \beta Vib_{bi}\left(S_b - F_i^t\right) + \delta(rand - 0.5), else \end{cases}. \tag{11}$$

$$Vib_{ij} = \omega_j \times e^{-d_{ij}^2}. \tag{12}$$

$$\omega_j = \frac{J(S_i) - worst_S}{best_S - worst_S}. \tag{13}$$

$$best_s = \min_{k\in\{1,2,\ldots,N\}}\{J(S_k)\}, \quad worst_s = \max_{k\in\{1,2,\ldots,N\}}\{J(S_k)\}. \tag{14}$$

where $r_m, \alpha, \beta$ and $\delta$ are random numbers between [0, 1], $PF$ is the probability factor, $S_c$ is the closest to the individual $i$ possessing the higher weight, $S_b$ is the highest weighted individual in the female population, $Vib_{ku}$ represents the u-th individual's ability to perceive the vibrations generated by the k-th individual, $J(S_i)$ is the objective function value of the individual $i$, and $d_{ij}$ is the Euclidean distance between the individuals $i$ and $j$.

### 3.3 Male Cooperative Operator

From a biological point of view, the male population is composed of dominant and non-dominant male spiders. Herein, dominant individuals are those that have better weight than the median male spider. All the male spiders are sorted in descending order of weights, and the individual located in the middle is considered the median male member $N_f + m$. Instead, other male spiders are non-dominant individuals. Furthermore, the dominant individuals can attract the nearest female spider. However, non-dominant individuals gather around the median male member. Thus, the co-evolution behavior in male population can be expressed as

$$M_i^{t+1} = \begin{cases} M_i^t + \alpha Vib_{fi}(S_f - M_i^t) + \delta(rand - 0.5), \omega_{N_f+i} \geq \omega_{N_f+m} \\ M_i^t + \alpha\left(\frac{\sum_{h=1}^{N_m} M_h^t \omega_{N_f+h}}{\sum_{h=1}^{N_m} \omega_{N_f+h}} - M_i^t\right), \quad else \end{cases}. \tag{15}$$

where $S_f$ is the superior female individual closest to the male dominant individual $i$, and $\frac{\sum_{h=1}^{N_m} M_h^t \omega_{N_f+h}}{\sum_{h=1}^{N_m} \omega_{N_f+h}}$ is the mean weight of the male population.

### 3.4 Mating Operator

Male dominant individuals have strong reproductive ability, for female individuals such male dominated individuals will mate with the female individuals, who are within their mating radius $R$, to reproduce offspring. Suppose that the set including the male dominant individuals and the female individuals within their mating radius is TG. The mating process mainly uses roulette to select individuals dimension by dimension according to Eq. (17) to obtain new individuals. If the new individuals are better than the worst individual in the entire spiders population, then the worst individual is replaced; otherwise, it remains unchanged.

$$R = \sum_{j=1}^{n} \frac{(h_{jmax} - h_{jmin})}{2n}. \tag{16}$$

$$PS_i = \frac{\omega_i}{\sum_{k \in T_G} \omega_k}. \tag{17}$$

where $n$ is the dimension of the problem to be solved.

## 4 Proposed GSSO

The SSO algorithm is mainly applied to continuous optimization problems, and the application research in combination optimization problems such as warehouse housing assignment is still rare. Furthermore, the less information exchange between male and female populations is not conducive to the co-evolution; a fixed mating radius reduces the solving efficiency; a single mating operator limits the optimization space. The above shortcomings will affect the convergence speed and the solution accuracy. To this end, SSO is improved from the following three aspects.

### 4.1 Discrete Encoding Design

To better apply SSO to the discrete optimization problem of warehouse housing allocation, combining the characteristics of the problem, a four-dimensional matrix coding scheme is proposed, in which the four-dimensional matrix corresponds to the column number, layer number, lane number and broiler breed number respectively, that's to say, the four-dimensional matrix intuitively reflects the locations occupied by various breeds of broilers, which not only overcomes the conversion error caused by continuous encoding, but also improves the diversity of the population, thereby improving the optimization performance of the algorithm to a certain extent.

### 4.2 Subpopulation Reconstruction

Although the male and female populations co-evolve within themselves, there are relatively few opportunities for communicating and learning between the populations, which affect the convergence speed of algorithm in some degree. For this purpose, a subpopulation reconstruction mechanism is introduced, that is, the male and female subpopulations are mixed at each iteration, and then the female and male subpopulations are randomly allocated to achieve the purpose of mutual learning between the female and male subpopulations, and thus speed up optimization efficiency.

### 4.3 Gaussian Mating Radius

From the entire evolutionary process of SSO, it can be known that its mating radius has always remained the same. Thus, as the iteration progresses, more and more females and even the worst females will be within the mating radius, which not only increases the complexity of mating, more importantly, it reduces the quality of mating, which will be unfavorable to pass on good genes to the next generation. Therefore, it is necessary to use an adaptive mating radius to flexibly adjust the number of female spider individuals participating in the mating. That is, in the early stage of evolution, a larger mating radius corresponds to a larger solution space, which is conducive to global exploration; conversely, a smaller mating radius in the later stage of evolution is beneficial to local exploitation. The mating radius $r_a$ changes adaptively according to Eq. (18). It can be seen that despite the oscillations, the mating radius $r_a$ generally changes in a decreasing direction, which better balances global exploration and local exploitation.

$$r_a = \left[ \sum\nolimits_{j=1}^{n} \frac{\left(h_{jmax} - h_{jmin}\right)}{2n} \right] \cdot e^{-\left(\frac{t}{MaxGen}\right)} \cdot N(\mu, \sigma). \tag{18}$$

where $N(\mu, \sigma)$ is the random numbers between [0, 1] obeying the Gaussian distribution, $\mu$ and $\sigma$ are 0.5 and 0.16, respectively, $t$ is the number of current generation, and *MaxGen* is the maximum evolutionary generation.

## 4.4 Multi-mating Operator

Mating is an effective way to reproduce offspring. However, the SSO algorithm cannot ensure that the good genes from dominant males can be inherited to the next generation, which will inevitably affects the quality of the offspring. Therefore, in order to increase the chance of the inheritance of good genes, the multi-mating operator is used to guide the mating process to breed good individuals. The principle can be described as: multiple mating operators are involved in mating, and the best offspring individual is selected through comparison. If the best offspring individual is better than the worst individual in the entire population, it is replaced, otherwise it remains unchanged. It can be seen that the multi-mating operator will increase the chance of reproducing good individuals. Assuming that the male and female individuals participating in mating are $S_m = [s_{1m}, s_{2m}, \cdots, s_{nm}]$ and $S_f = \left[s_{1f}, s_{2f}, \cdots, s_{nf}\right]$ respectively, the multi-mating operator can be defined as follows,

(a) Arithmetic mating operator

$$s_{im}^{new} = rand \cdot s_{im} + (1 - rand) \cdot s_{if}. \tag{19}$$

(b) Average mating operator

$$s_{im}^{new} = \left(s_{im} + s_{if}\right)/2. \tag{20}$$

(c) Extreme mating operator

$$s_{im}^{new} = \lfloor rand \rfloor \cdot s_{im} + \left(1 - \lfloor rand \rfloor \cdot s_{if}\right). \tag{21}$$

(d) Boundary mating operator

$$s_{im}^{new} = \lfloor rand \rfloor \cdot h_{imin} + (1 - \lfloor rand \rfloor \cdot h_{imax}). \tag{22}$$

### 4.5 Flow of the Proposed GSSO

**Step1** Initialization of the parameters: population size *NP*, probability factor *PF*, maximum evolution generation *MaxGen*, evolution generation counter *t*.

**Step2** Carry out discrete encoding according to Sect. 3.1, and initialize the positions of $N_f$ female spiders and $N_m$ male spiders. If the generated solution repeats with other solutions, one of the dimensions is chosen to give a random number within the value range of the dimension until it is no longer repeated. If this happens in other steps, follow this method.

**Step3** Global search: $t = t + 1$.

**Step4** Utilize Eq. (13) to calculate the weights of all spider individuals in the population, and record the position $S_{best}$ and the optimal value $J(S_{best})$ of the spider with the highest weight.

**Step5** Perform co-evolution of female populations according to Sect. 2.2.

**Step6** Perform co-evolution of male populations according to Sect. 2.3.

**Step7** Calculate the Gaussian mating radius of the male dominant individual according to Eq. (18).

**Step8** For each male spider dominant individual, it will mate with all female individuals within its Gaussian mating radius according to the multi-mating operator in Sect. 3.4.

**Step9** According to Sect. 3.2, reconstruct the populations of male and female.

**Step10** If $t > MaxGen$, output the optimal solution; otherwise, go to **Step3**.

## 5 Simulations Results

In order to validate the performance of the Gaussian social spider algorithm (GSSO) proposed in this paper, the well-known international standard benchmark library TSPLIB (http://www.iwr.uni-heidelberg.de/groups/comopt/software/TSPLIB95/tsp/) is selected. And compared with the algorithms from references, such as the standard social spider algorithm (SSO) [12], improved teaching algorithm (G-TLBO) [15], improved artificial bee colony algorithm (S-ABC) [16] and improved differential evolution algorithm (ADE) [17]. The experiment are conducted in a PC with Windows 10 system, 3.7 GHz Intel Core, 4 GB RAM, and MATLAB R2014b. The population size *NP* and the maximum evolution generation *MaxGen* of the algorithm GSSO in this paper are 100 and 600 respectively. For the fairness of comparison, SSO, G-TLBO, S-ABC and ADE are adopted the same population size and maximum evolution generation as GSSO, and other parameters are from the references. Each algorithm run 30 times for each trial, and statistics are calculated based on the optimal solution *optimal*, the average value of the optimal solution *mean*, the standard deviation of the optimal solution *std*, and the average running time *atime*. In order to increase the discrimination, the best results are highlighted in boldface.

### 5.1 Sensitivity Analysis of the Probability Factor *PF*

According to the evolutionary principle of the social spider algorithm, female spiders decide whether to approach a better individual according to the probability factor *PF*. Therefore, *PF* becomes an important parameter that affects the performance of the SSO algorithm, and it is necessary to tune it. In order to have universal applicability, we selected representative TSP problems of different scales for testing and analysis. The average value *mean* of the 30 optimal solutions is shown in Table 1. It can be seen from Table 1 that when *PF* is 0.6, all three TSP problems tested can obtain the optimal average value. Besides too large or too small *PF* is not conducive to the optimization of SSO. Too small *PF* results in SSO not being able to guide potential spider individuals to approach better spider individuals, bringing a certain degree of blindness; too large PF will cause more spider individuals to gather to better spider individuals. Based on the experimental results and analysis, the *PF* is set to 0.6 in this paper.

**Table 1.** Sensitivity analysis of the parameter *PF.*

| *PF* | eil51 | eil101 | tsp225 | lin318 |
|---|---|---|---|---|
| 0.1 | 459 | 679 | 4317 | 42806 |
| 0.2 | 472 | 653 | 4285 | 42671 |
| 0.3 | 453 | 660 | 4297 | 42747 |
| 0.4 | 448 | 637 | 4069 | 42488 |
| 0.5 | 440 | 641 | 3950 | 42258 |
| 0.6 | **426** | **629** | **3916** | **42029** |
| 0.7 | 450 | 633 | 3983 | 42114 |
| 0.8 | 437 | 650 | 4149 | 42380 |
| 0.9 | 439 | 661 | 4200 | 42271 |
| 1 | 467 | 672 | 4273 | 42481 |

### 5.2 Benchmark Test and Analysis

To test the performance of GSSO algorithm, compared with SSO, G-TLBO, S-ABC and ADE, the results are shown in Table 2. At the same time, for these four TSP cases, the optimal paths solved by GSSO are also shown in Figs. 1, 2, 3, and 4, respectively.

**Table 2.** Performance evaluation of the algorithms

| Name (Optimum) | | SSO [12] | G-TLBO [15] | S-ABC [16] | ADE [17] | GSSO |
|---|---|---|---|---|---|---|
| eil51 (426) | optimal | 490 | 463 | 429 | 434 | **426** |
| | mean | 517 | 470 | 438 | 447 | **426** |
| | std | 16.82 | 9.20 | 4.73 | 15.05 | **0** |
| | atime/s | 2.85 | 1.68 | 1.80 | 2.11 | **1.37** |
| eil101 (629) | optimal | 659 | 655 | 644 | 635 | **629** |
| | mean | 677.73 | 658.36 | 650.88 | 651.39 | **633.05** |
| | std | 15.86 | 8.47 | 13.21 | 13.59 | **7.37** |
| | atime/s | 4.21 | 2.94 | 3.30 | 3.87 | **2.19** |
| tsp225 (3916) | optimal | 4189 | 4095 | 4036 | 4135 | **3916** |
| | mean | 4259.22 | 4198.30 | 4112.91 | 4179.96 | **3977.31** |
| | std | 105.08 | 113.26 | 46.50 | 68.07 | **24.28** |
| | atime/s | 9.44 | 6.32 | 7.61 | 7.84 | **4.23** |
| lin318 (42029) | optimal | 42284 | 42219 | 42144 | 42189 | **42029** |
| | mean | 42431.92 | 42384.98 | 42268.19 | 42305.54 | **42106.33** |
| | std | 223.47 | 190.88 | 150.42 | 139.66 | **30.61** |
| | atime/s | 17.01 | 14.27 | 16.83 | 16.72 | **5.90** |

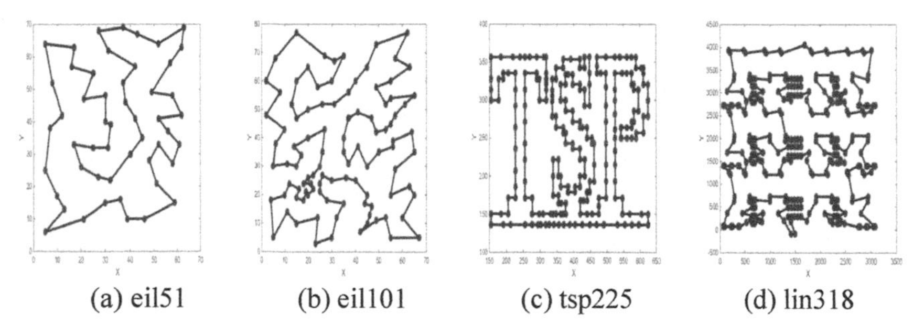
(a) eil51 (b) eil101 (c) tsp225 (d) lin318

**Fig. 1.** Optimal paths obtained by GSSO.

It can be seen from Table 2 that for TSP problems of different scales, GSSO has certain advantages over the other four algorithms in generalization ability. Especially as the scale of the problem increases, the solution quality and solution efficiency become more prominent. It shows that GSSO has strong robustness.

### 5.3 Application to Warehouse Housing Assignment

The parameters of automated warehouse are set as follows: $L = 100$ cm, $H = 50$ cm, $v_x = 1$ m/s, $v_y = 0.5$ m/s, $m = 6$, $n = 10$, $k = 20$, $D = 2.6$ m, $N = 1$. The broiler farm currently breeds 5 types of broiler. The number of locations required for each broiler, the number of feeding times during the breeding cycle, and the delayed growth factors $\varphi$ are shown in Table 3. Each algorithm run 30 times, and the results are shown in Table 4, Figs. 2 and 3. Finally, the warehouse housing assignment results obtained by GSSO is shown in Fig. 4, where X, Y, and Z represent shelf number, column number, and layer number, respectively, and 5 colors from dark to light represent numbers 1 to 5 corresponding to 5 broiler breeds, respectively.

**Table 3.** Parameters related to each breed broiler.

| Breed number | Feeding times | Delayed growth factors | Locations required |
|---|---|---|---|
| 1 | 5 | 2.2 | 470 |
| 2 | 2 | 5.5 | 450 |
| 3 | 2 | 5.7 | 310 |
| 4 | 6 | 1.5 | 700 |
| 5 | 4 | 4.8 | 470 |

**Table 4.** Comparisons of the results among the algorithms.

| Algorithm | optimal | mean | Std | atime/s |
|---|---|---|---|---|
| GSSO | **2.91E + 05** | **2.92E + 05** | **336.51** | **979.50** |
| SSO [12] | 3.26E + 05 | 3.30E + 05 | 2325.4 | 1766.37 |
| G-TLBO [15] | 3.18E + 05 | 3.21E + 05 | 2183.2 | 1306.18 |
| S-ABC [16] | 3.11E + 05 | 3.12E + 05 | 1158.7 | 1315.80 |
| ADE [17] | 3.12E + 05 | 3.14E + 05 | 1779.3 | 1406.55 |

For the optimization problem of warehouse housing assignment, Table 4, Figs. 2 and 3 further verify the superior performance of GSSO. It can be seen from the layout of warehouse housing in Fig. 4 that under the comprehensive influence of the three factors of feeding frequency, delayed growth factor and the required number of locations, the warehouse housing of broiler breed 1 basically presents a descending stair-like distribution in each lane, and the locations are near the aisle. However, the parameter values of the three factors of broiler breed 1 are between the largest and the smallest among all broiler breeds. This also shows that if you simply rely on the maximum or the minimum value to carry out the warehouse housing, the results should be unreasonable, which also

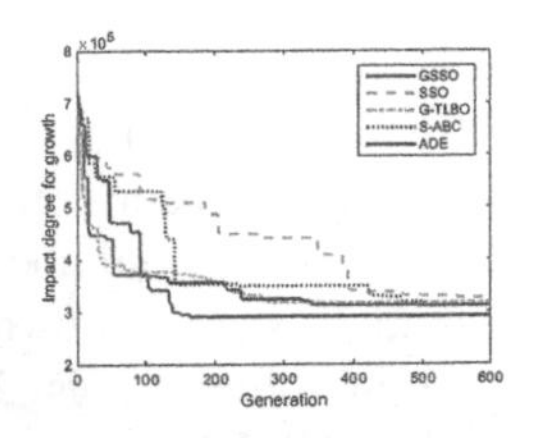


**Fig. 2.** Convergence graph

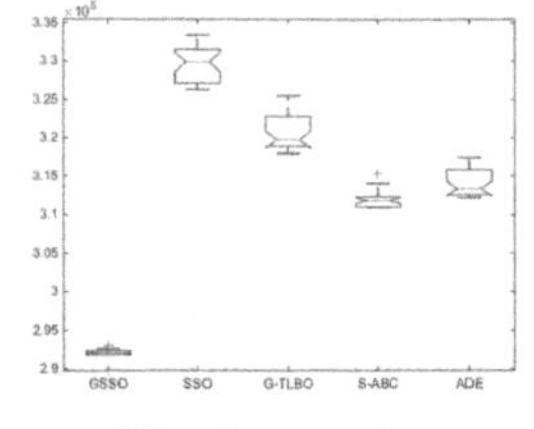


**Fig. 3.** Boxplot

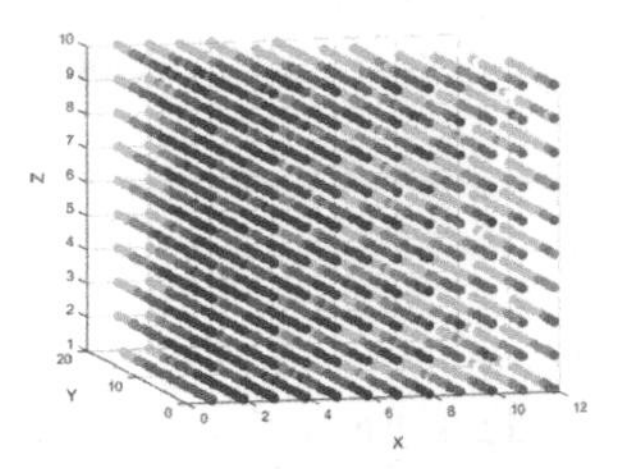


**Fig. 4.** Optimal layout by GSSO

reflects the research value of this paper. From the TSP problem to the warehouse housing assignment, in terms of the quality of the solution, GSSO can find the optimal solution of the TSP for small-scale TSP problems. For larger-scale TSP problems and warehouse housing assignment, although GSSO cannot find the optimal solution, it is very close to it, which is more intuitively confirmed in Fig. 3. In terms of solving efficiency, for all instances, GSSO can converge to the optimal solution or the sub-optimal solution at a faster rate. The above-mentioned excellent performance of GSSO is mainly due to the following three aspects: Firstly, the four-dimensional matrix discrete encoding mechanism not only reduces the conversion error caused by continuous encoding, but also increases the diversity of the population; secondly, the subpopulation is reconstructed through the learning mechanism, which improves the evolutionary ability of the subpopulations. Thirdly, the introduction of the Gaussian mating radius allows the mating radius to be adjusted adaptively and speeds up the convergence. Finally, the multi-mating operator effectively guarantees the opportunity for good genes to reproduce.

## 6 Conclusion

To solve the optimization problem of multi-breed broiler warehouse housing assignment, this paper proposes an improved social spider optimization algorithm (GSSO). The simulation results show that the four-dimensional matrix discrete encoding method, the subpopulation reconstruction mechanism, the Gaussian mating radius and the multi-mating operator makes GSSO achieve a better balance between global exploration and local exploitation. Compared with other algorithms, better results are achieved in terms of efficiency and solution quality. In view of its good optimization performance, in future, it worth considering applying it to related combined optimization problems such as logistics distribution, aviation scheduling, and operating room scheduling.

**Acknowledgments.** This work is supported by the National Key Research and Development Program of China (No. 2018YFB1700500), the National Natural Science Foundation of China (Nos. 52077213, 62003332, 61773156) and the Scientific and Technological Project of Henan Province (No. 202102110281, 202102110282).

## References


1. Zhang, L., Tan, Y., Lyu, H., et al.: Optimization of automatic transplanting path for plug seedlings in greenhouse. J. Trans. Chin. Soc. Agr. Eng. **36**(15), 65–72 (2020)
2. De Koster, R., Rene, T.L.D., Roodbergen, K.J.: Design and control of warehouse order picking: a literature review. J. Eur. Oper. Res. **182**(2), 481–501 (2007)
3. Lee, I.G., Chung, S.H., Yoon, S.W.: Two-stage storage assignment to minimize travel time and congestion for warehouse order picking operations. J. Comput. Ind. Eng. **139**, 1–13 (2020)
4. Calzavara, M., Glock, C.H., Grosse, E.H., et al.: An integrated storage assignment method for manual order picking warehouses considering cost, workload and posture. J. Int. Prod. Res. **57**(8), 2392–2408 (2019)
5. Battini, D., Calzavara, M., Persona, A., et al.: Order picking system design: the storage assignment and travel distance estimation (SA&TDE) joint method. J. Int. Prod. Res. **53**(4), 1077–1093 (2015)
6. Ozturkoglu, O.: A bi-objective mathematical model for product allocation in block stacking warehouses. J. Int. Trans. Oper. Res. **27**(4), 2184–2210 (2020)
7. Li, J., Moghaddam, M., Nof, S.Y.: Dynamic storage assignment with product affinity and ABC classification a case study. J. Int. Adv. Manuf. Technol. **84**(9), 2179–2194 (2016)
8. Roshan, K., Shojaie, A., Javadi, M.: Advanced allocation policy in class-based storage to improve AS/RS efficiency toward green manufacturing. Int. J. Environ. Sci. Technol. **16**(10), 5695–5706 (2018)
9. E, Y., Zu, Q., Cao, M.: Slotting optimization of AS/RS for automotive parts based on genetic algorithm. J. Syst. Simul. **25**(3), 430-435,444 (2013)
10. Mantel, R.J., Schuur, P.C., Heragu, S.S.: Order oriented slotting: a new assignment strategy for warehouses. J. Eur. Ind. Eng. **1**(3), 301–316 (2007)
11. Yuan, R., Graves, S.C., Cezik, T.: Velocity-based storage assignment in semi-automated storage systems. J. Prod. Oper. Manag. **28**(2), 354–373 (2019)
12. Cuevas, E., Cienfuegos, M., Zaldivar, D., et al.: A swarm optimization algorithm inspired in the behavior of the social spider. J. Expert Syst. Appl. **40**(16), 6374–6384 (2013)
13. Elsayed, W.T., Hegazy, Y.G., Bendary, F.M., et al.: Modified social spider algorithm for solving the economic dispatch problem J. . Eng. Sci. Technol. **19**(4), 1672–1681 (2016)
14. Shukla, U.P., Nanda, S.J.: Parallel social spider clustering algorithm for high dimensional datasets. J. Eng. Appl. Artif. Intell. **56**, 75–90 (2016)
15. Güçyetmez, M., Çam, E.: A new hybrid algorithm with genetic teaching learning optimization (G-TLBO) technique for optimizing of power flow in wind thermal power systems. J. Electr. Eng. **98**(2), 145–157 (2016)
16. Sharma, T.K., Pant, M.: Shuffled artificial bee colony algorithm. Soft. Comput. **21**(20), 6085–6104 (2016)
17. Najeh, B.G.: An accelerated differential evolution algorithm with new operators for multi-damage detection in plate-like structures. J. Appl. Math. Model. **80**, 366–383 (2020)

# Automobile Rim Weld Detection Using the Improved YOLO Algorithm

Zhenxin Yuan[1,2], Shuai Zhao[3], Fu Zhao[4], Tao Ma[4], and Zhongtao Li[1,2](✉)

[1] Shandong Provincial Key Laboratory of Network Based Intelligent Computing, University of Jinan, Jinan 250022, China
ise_lizt@ujn.edu.cn

[2] School of Information Science and Engineering, University of Jinan, Jinan 250022, China

[3] Jinan Housing and Urban-Rural Development Bureau, Jinan, China

[4] Shandong Xiaoya Holding Group Co. Ltd., Jinan, China

**Abstract.** At present, the production efficiency of automobile rim in the industrial field is affected by the detection process of automobile rim quality after steel forging. The traditional way is to check welding position manually, which can facilitate the air tightness detection after weld is pressurized. However, this can largely affect production efficiency. By introducing computer vision and image processing, the position of the rim weld can be accurately located, which is more accurate and time-saving. In order to ensure high accuracy and speed of detection, we propose an automobile rim weld detection algorithm YOLOv4-head2-BiFPN on the basis of YOLOv4 algorithm. The experimental results show that, for one thing, it does not affect the detection speed by strengthening feature fusion and removing redundant detection heads. For another, the AP75 of the improved YOLOv4-head2-BiFPN algorithm in the automobile rim weld detection task is 7.7% higher than that of the original YOLOv4 algorithm.

**Keywords:** YOLO · Automobile rim weld · Object detection

## 1 Introduction

With the integration of industrialization and information technology, the development of artificial intelligence, industrial Internet of Things and other technologies have been continuously applied to the industrial field. Technological breakthroughs not only promoting the intelligent transformation and upgrading of industrial production lines, but also upgrading new and old kinetic energy. In the production line of industrial automobile rim, there are several ways of airtightness detection of automobile rim weld: manual methods are often used to calibrate the location of automobile rim weld, and carry out the next step of airtightness detection. The manual method of tightness detection greatly affects production efficiency; another is to use computer vision method to analyze the automobile rim weld through images. Although this method can capture the weld target, it may fail to detect the slight air leakage of the weld. Only by detecting the air flow after pressurization can we accurately grasp the air tightness of the current rim weld.


Q. Han et al. (Eds.): LSMS 2021/ICSEE 2021, CCIS 1469, pp. 122–131, 2021.
https://doi.org/10.1007/978-981-16-7213-2_12

Therefore, the first step is to realize the rapid and accurate location of the rim weld, and deploy the task of detecting and locating the automobile rim weld on the edge computing equipment; and then perform air flow detection to ensure the airtightness of the rim weld. The rapid and precise location of the automobile rim weld is a problem that needs to be solved at the moment.

Therefore, in order to improve the efficiency of automobile rim weld detection and the accuracy of model detection. According to the automobile rim data sets collected in the industrial site, a modified algorithm based on the YOLOv4 model is proposed. For the actual data set, we prune the detection head of the YOLOv4 network structure, that is, the YOLOv4-head2 network structure. On this basis, the feature fusion of spatial feature fusion layer is strengthened to prevent feature loss with the increase number of layers, that is, YOLOv4-head2-BiFPN network structure. After experimental testing, compared with the original YOLOv4 network structure, the upgraded network structure is improved by 0.8% and 4.3% respectively, which improves the performance of the model.

## 2 Related Work

The current object detection algorithms are roughly divided into two categories: two-stage detector represented by faster R-CNN and one-stage detector represented by YOLO. Two-stage generally has the characteristics of high accuracy but slow detection speed, while one stage has the characteristics of fast detection speed and lower accuracy than two-stage. In view of the practical problems, it is necessary to ensure low delay and high precision when the detection of automobile rim weld is applied to the production lines. Therefore, we test the faster R-CNN algorithm and the YOLO algorithm respectively, and make a comparative experiment.

### 2.1 Data Preprocessing

The data set of automobile rim are taken on site. The number of the data set is 5300 and these images are labeled in VOC format.

Target in data sets is located at different positions of the automobile rim. For each frame of image, the target is marked in the form of rectangular box. The rectangular box need to be as close as possible to the target weld. The marked results directly affect the results of model training and model reasoning. The deep learning regression method requires a large number of data sets as the training set. Using various data enhancement strategies on the training set and test set is very important for the training performance of the model. In the YOLO algorithm, the data sets are enhanced by inversion, Cutmix, Mosaic and other transformation methods.

Using Cutmix method, non-information pixels will not appear during the training process, so as to improve the efficiency of training. Cutmix replaces the image area with a patch from another training image and generates more local natural images than Mixup. The cost of training and reasoning remains unchanged.

After flipping, zooming and gamut transformation, four images are combined into one image according to four directions. Mosaic [2] randomly selects four samples from the data set, enriches the background of the detected object, and improves the number of training samples each time by combining images. Four images are processed during batch normalization.

## 2.2 Faster R-CNN Algorithm

Because the task of classification is single classification task, there is only one rim in each image, and there is only one weld on each rim. For this single classification and single object detection task, according to the complexity of the target, we also try to use the less hierarchical backbone feature extraction network than CSPdarknet53 to carry out the experiment.

Faster R-CNN algorithm is the mainstream two-stage detection network. VGG16 and ResNet-50 are the main feature extraction networks of faster R-CNN [16]. The data sets are input into VGG16 and ResNet-50 respectively to compare the detection performance. At the same time, we also try to use ResNet-50 to replace the CSPdarknet53 backbone network of the original YOLOv4 model. Verify the backbone network with a lower number of layers and compare with the faster R-CNN and YOLO model horizontally to observe the effect of data feature extraction, model convergence, and computational reasoning speed.

## 2.3 YOLO Algorithm

Compared with the two-stage model of RCNN series, YOLO model has the advantages of fast reasoning speed and good object detection accuracy. In this paper, we also compare the YOLOv3 and YOLOv4 models. According to the experimental results, YOLOv4 is selected to solve the problem of automobile rim weld detection. The characteristics of network structure of YOLOv4 model are analyzed. Aiming at the practical problems of automobile rim weld detection, the YOLOv4 algorithm is improved. The improved YOLOv4-head2 and YOLOv4-head2-BiFPN network structures are obtained respectively. The detection accuracy is improved compared with YOLOv4.

In order to extract deeper features and reduce feature loss, YOLOv4 adopts multi-scale feature fusion, which makes use of high-level feature map with rich semantic meaning and low precision; low-level feature map with rich semantic meaning and high precision; cross layer connection, after up-sampling, zooming in and splicing with high-level, so that the output feature map not only has high precision, but also has rich semantic meaning. In order to detect targets of different sizes and areas, YOLOv4 [2] adopts multi-scale detection, which detects at three different scales. Each layer is predicted independently, which can improve the accuracy of detection for targets of different sizes.

According to the requirements of the rim weld location task, it is necessary to accurately locate the center point of the automobile rim weld to ensure the accuracy of the next air tightness test results. Therefore, it is necessary to accurately find the position of the weld, including $X_{min}$, $Y_{min}$, $X_{max}$, $Y_{max}$ of the weld target. Compared with the two coordinate values of the target box, the more important is the center position of the weld target, that is, $X_{center}$, $Y_{center}$. Therefore, choosing the appropriate loss function, evaluating the error between the prediction frame and the real frame, and correcting the results of each training directly affect the final performance of the model. IoU [13] is often used to measure the coincidence between the detected target and the real frame, and the intersection and union ratio is used as the evaluation scale in object detection, but IoU cannot accurately reflect the coincidence between the target frame and the real frame. DIoU [14] loss takes into account both the overlapping area of the bounding box (BBox) and the distance between the center points. In our work, DIoU and CIoU [2] as loss function to conduct experiments respectively. In order to minimize the distance between the center points, the two loss calculation methods are compared. The comparison results show that the training convergence of CIoU model is faster than that of DIoU model, and the performance of training results is better.

## 3 Improved YOLO Algorithm

### 3.1 YOLOv4-Head2

According to the characteristics of the location and classification task, in order to get more accurate classification, detection results, and faster detection speed, the YOLO model is modified.

The prior frame provided by the original YOLOv4 algorithm is obtained by clustering the VOC data set, and the VOC data set includes 20 categories. The size of the target in the image of the automobile rim data set is very different from the type of the target in the image of the VOC data set. Therefore, the K-means clustering algorithm is used to regenerate the prior frame [3].

In the detection process, YOLOv4 assigns the large-scale prior frame to the small-scale feature layer to detect large targets, and assigns the small-scale prior frame to the large-scale feature layer to detect small targets. In the industrial production line of automobile rim, the location of image acquisition is fixed, so the proportion of weld target in the image is relatively fixed, which is in a certain range. According to the size of the weld target and the scale of the three detection feature layers of YOLOv4, the redundant detection scale is removed and the detection scale close to the real target scale is retained.

According to the width $w_i$ and height $h_i$ of the prior box after K-means clustering, $S_1, S_2, S_3$ is the detection scale of the three feature layers of YOLOv4 ($S_{wi}$ A is the width of the ith detection scale, $S_{hi}$ is the height of the ith detection scale).

$$S_1 > S_2 > S_3, S_i = S_{wi} \cdot S_{hi}(1) \tag{1}$$

$$\forall w_i \in (S_{w1}, S_{w2}), \forall h_i \in (S_{h1}, S_{h2})(i = 1, 2, \cdot\cdot\cdot, 9)(2) \tag{2}$$

The average area $area_{avg}$ of the prior box is calculated. The formula is as follows.

$$area_{avg} = \frac{\left(\sum_{i=1}^{k} w_i \cdot h_i\right)}{k}(3) \tag{3}$$

$$area_{avg} \in (S_{w1} \cdot S_{h1}, S_{w2} \cdot S_{h2})(4) \tag{4}$$

The core is that during detection, predictions and classifications of different sizes are not well trained, which makes the network incomplete. Therefore, the larger size difference in prediction and classification is removed, the model training and convergence effect are accelerated. Thus, the detector head with detection scale $S_3$ can be removed to optimize the network, prevent the waste of computing resources and accelerate the convergence of the model.

### 3.2 YOLOv4-Head2-BiFPN

Feature fusion between different layers is also very important for the detection task. FPN [4], PANet [8], NAS-FPN [7], BiFPN [9] and other FPN networks have been proposed. Among them, the PANet layer in YOLOv4 adopts the bottom-up and top-down splicing method, and the detection effect is better than that of FPN, which only stitches the bottom features to the shallow layer. Libra-FPN [5] first converges all features to the middle layer, and then fuses with other layers again. Although this method can achieve better fusion effects, the amount of calculation is larger than PANet. NAS-FPN is an FPN based on neural architecture search. It uses reinforcement learning to select the best model structure in the given search space to train the controller. After repeated learning, it finally generates a better structure. Although NAS-FPN automatic search can improve the detection accuracy of the model, the repeated search process will bring some delay problems and increase the model parameters. BiFPN is a balance between speed and accuracy. On the basis of PANet, an edge is added between the input and output nodes of the same layer. In this way, more features can be fused. Compared with Libra-FPN, it has less computation. Therefore, refer to BiFPN to adjust PANet in YOLOv4.

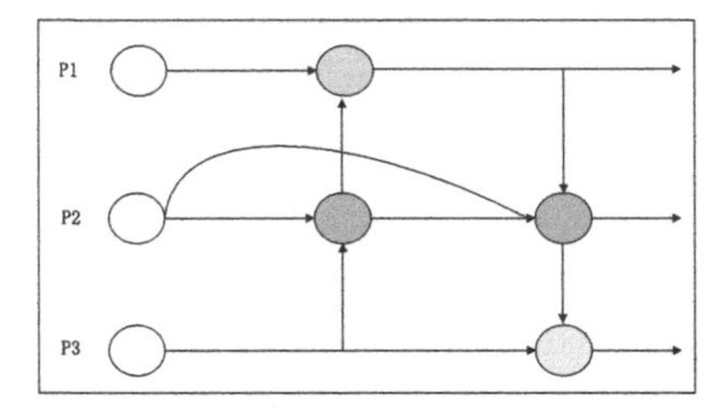


**Fig. 1.** PANet-Alter network structure picture.

Figure 2 shows the improved network structure of YOLOv4-head2-BiFPN based on YOLOv4, On the basis of PANet, it adds feature jump fusion. The strong semantic information of high-level feature maps and detail information of low-level feature maps are fused efficiently, which improved the applicability of the model.

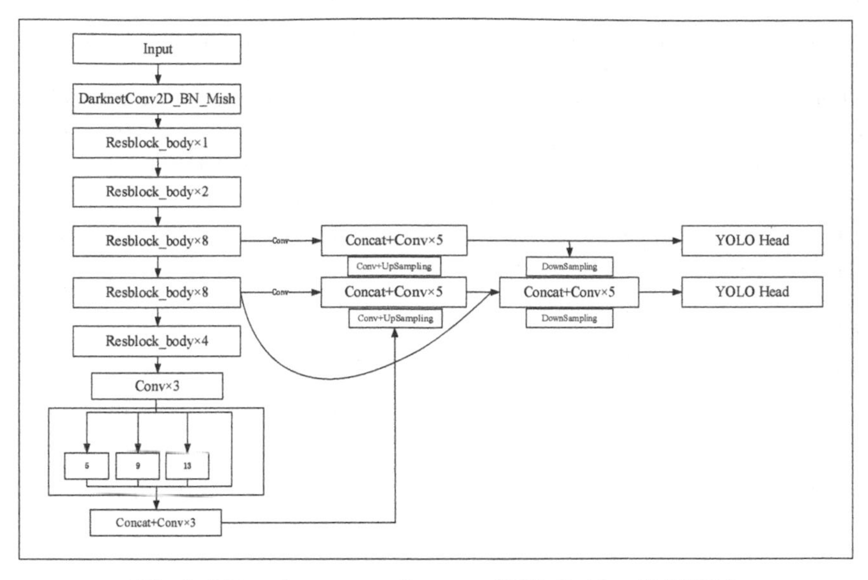


**Fig. 2.** Network structure diagram of YOLOv4-head2-BiFPN.

# 4 Experimental Results and Analysis

## 4.1 Evaluation Index

In order to ensure the consistency of the comparison, the following comparison data are the results of the experiment on 1080Ti platform. The training set is 4240 automobile rim images, and the test set is 1060 automobile rim images. In this paper, IoU is used as the evaluation index of the target positioning task of the automobile rim weld, and the ratio of the area of the two regions to the combined area is defined. In order to ensure the accuracy of the automobile rim weld, IoU $= 0.75$ is taken as the threshold value. If IoU $> 0.75$, it is the true positives, and if IoU $<= 0.75$, it is the false positives. At the same time, the average precision AP is used as the evaluation index of the automobile rim detection task.

Because our classification task is single classification, then AP is mAP. According to different confidence levels, the recall rate is the horizontal coordinate, and the accuracy is the pre-recall curve of the vertical coordinate. AP is the integral of Precision-Recall curve.

## 4.2 YOLOv4-Head2 Effect Experiment

We also test the effect of a single YOLOv4 model in the automobile rim weld detection task and the different image size in the input model. According to the requirements of image acquisition and industrial production site detection, we only test the images with the resolution of $416 \times 416$ and $512 \times 512$, so as to prevent the high resolution of the input image from affecting the efficiency of the detection task after deployment. It can be seen from Table 1 of the experimental results that the AP is improved by 2.1% with 512px × 512px resolution image input.

After modifying the network structure of YOLOv4 and removing the small-scale detection head for detecting large targets, the input data sets of 416 × 416 and 512 × 512 are set respectively for training. The training results are shown in Table 1.

**Table 1.** The AP of the YOLOv4 and YOLOv4-head model in the automobile rim test set When the input image is 416 × 416 and 512 × 512.

| Model | Input size (px × px) | $AP_{75}$ | Model size |
|---|---|---|---|
| YOLOv4 | 416 | 69.5 | 244MB |
| YOLOv4-head2 | 416 | 70.3 | 183MB |
| YOLOv4 | 512 | 71.6 | 244MB |
| YOLOv4-head2 | 512 | 74.7 | 183MB |

According to the experimental results in Table 1, the model size is reduced by 25% after removing the small-scale detection head in YOLOv4 network. When the input image size is 416 × 416, the AP increases by 0.8%, while when the input image size is 512 × 512, the AP increases by 3.1%, and the model parameters are reduced by 64MB. The results show that after removing the small-scale detection head, the calculation of the small-scale detection layer is avoided, and a certain number of parameters are reduced. The performance has also been improved.

### 4.3 YOLOv4-Head2-BiFPN Effect Experiment

In order to compare the detection performance of different algorithms on the automobile rim weld data sets, we also compared Faster R-CNN [11] algorithm with VGG16 and ResNet-50 [12] as the main feature extraction network. YOLOv3 [1] is a common one-stage detection network.

Compared with the faster R-CNN with VGG16 and ResNet-50 as the main feature extraction network, the detection time of YOLOv4 on AP is improved by 25.1% and 22.4% respectively.

The main reason is that the aspect ratio of the prior box of faster R-CNN is fixed in advance. However, the aspect ratio of the automobile rim weld will change according to different positions, which leads faster R-CNN to missing the weld target in the detection. Besides, the confidence of the detection target is higher, but the error between predicted BBox and ground truth BBox is so obvious. The prior box in the YOLOv4 algorithm is obtained by the K-means algorithm clustering, and the size of the prior box is closer to the target size of the data sets, so it is more suitable for detecting the automobile rim weld target. Compared with YOLOv4, the detection speed of faster R-CNN is slower, because the detection problem is divided into two stages. The first step is to generate candidate regions, and the second is to adjust and classify the location of candidate regions. This method has low recognition error rate and slow speed, which cannot meet the effect of real-time detection.

Compared with YOLOv3, YOLOv4 algorithm is superior to YOLOv3 in detection accuracy and detection speed, and AP is increased by 5.9%. The main feature extraction network of YOLOv4 can extract more complex features, increase SPP [4] module, and use different convolution kernel sizes and steps to realize the output of different receptive field features. It can increase the detection accuracy and improve the overall detection effect under the condition of maintaining the same detection time.

In view of the problems studied in our work, we try to use ResNet-50 with lower layers to replace the backbone of YOLOv4 model. After experiments, due to the complexity of the automobile rim weld target, it is easy to cause feature loss when taking ResNet-50 as the backbone network. At the same time, we link the ResNet-50 backbone network layer to feature fusion layer, and input the data set training results which can be seen from Table 2 that the AP of YOLOv4 model with ResNet-50 as the backbone network is 46.1%, which cannot better solve this task.

In order to better extract the features of the rim weld, CSPdarknet53 [15] with deeper feature extraction is used, and for better feature fusion of different levels of feature map. Referring to BiFPN, the PANet in YOLOv4 is modified. The difference between BiFPN and PANet is the cross-layer fusion with low-level feature layer. The complex fusion method is used to enhance the feature fusion extraction. The experiment shows that the effect is better than that of PANet. The experimental results are shown in Table 2 and Table 3.

**Table 2.** AP results on automobile rim test sets under different models.

| Model | $AP_{75}$ | FPS |
|---|---|---|
| Faster R-CNN VGG16 | 44.4 | 15.5 |
| YOLOv4-ResNet-50 | 46.1 | 42 |
| Faster R-CNN ResNet-50 | 47.1 | 11 |
| YOLOv3 | 63.6 | 50 |
| YOLOv4 | 69.5 | 47 |
| YOLOv4-head2 | 70.3 | 49 |
| YOLOv4-head2-BiFPN | 73.8 | 45 |

Compared Table 2 with Table 3, the faster R-CNN detection process is divided into two steps: candidate regions and the adjustment and classification of candidate regions. This method has low recognition error rate, but the speed is very slow, which cannot meet the application requirements of industrial production lines. Compared with faster R-CNN model, it is more suitable for the detection of automobile rim weld. Based on the YOLOv4 model, the method of modifying PANet to strengthen feature fusion and prune the detection head is used in the rim weld detection and location task, which shows better performance. Compared with the original YOLOv4 model, when the input image is 416 × 416, the detection accuracy of YOLOv4-head2 is improved by 0.8% based on the improvement of detection speed. YOLOv4-head2-BiFPN improves the detection accuracy by 4.3% on the basis of little difference between the detection speed

**Table 3.** AP results on automobile rim test sets under YOLOv4, YOLOv4-head2 and YOLO-v4-head2-BiFPN.

| Model | Input Size(px × px) | $AP_{75}$ | FPS |
|---|---|---|---|
| YOLOv4 | 416 × 416 | 69.5 | 47 |
| YOLOv4-head2 | 416 × 416 | 70.3 | 49 |
| YOLOv4-head2-BiFPN | 416 × 416 | **73.8** | **45** |
| YOLOv4 | 512 × 512 | 71.6 | 35 |
| YOLOv4-head2 | 512 × 512 | 74.7 | 40 |
| YOLOv4-head2-BiFPN | 512 × 512 | **79.3** | **35** |

and YOLOv4. When the input image is 512 × 512, the detection accuracy of YOLOv4-head2 is improved by 3.1% based on the improvement of detection speed. YOLOv4-head2-BiFPN improves the detection accuracy by 7.7% on the basis of little difference between the detection speed and YOLOv4. The experimental results show that the image size of the input model has a certain impact on the detection results of the automobile rim weld.

## 5 Conclusion

The end-to-end object detection algorithm can deal with the task of automobile rim weld detection, accurately locate the position of the weld target, and overcome low efficiency and reduce the error rate of manual method. It is feasible to solve the problem of object detection and location of automobile rim weld based on YOLO algorithm. The feature fusion part of YOLO algorithm is modified to strengthen the feature fusion between layers, reduce the consumption of feature information in convolution operation, and reduce the loss of features. At the same time, according to the characteristics of the YOLO algorithm, aiming at the current specific object detection task, according to the size of the target, prune the model detection scale, so as to improve the overall performance of the model, and make it better to solve the problem of automobile rim weld detection in the industrial production lines.

## References

1. Redmon, J., Farhadi, A.: Yolov3: an incremental improvement. arXiv preprint arXiv:1804.02767 (2018)
2. Bochkovskiy, A., Wang, C.Y., Liao H.Y.M.: YOLOv4: optimal speed and accuracy of object detection. arXiv preprint arXiv:2004.10934 (2020)
3. Li, B., Wang, C., Wu, J., et al.: Surface defect detection of aeroengine components based on improved YOLOv4 algorithm. In: Laser & Optoelectronics Progress (2021)
4. He, K., Zhang, X., Ren, S., Sun, J.: Spatial pyramid pooling in deep convolutional networks for visual recognition. J. IEEE Trans. Pattern Anal. Mach. Intell. **37**(9), 1904–1916 (2015)

5. Lin, T., Dollár, P., Girshick, R., He, K., Hariharan, B., Belongie, S.: Feature pyramid networks for object detection. In: 2017 IEEE Conference on Computer Vision and Pattern Recognition (CVPR), pp. 936–944 (2017)
6. Pang, J., Chen, K., Shi, J., Feng, H., Ouyang, W., Lin, D.: Libra R-CNN: towards balanced learning for object detection. In: 2019 IEEE/CVF Conference on Computer Vision and Pattern Recognition (CVPR), pp. 821–830 ( 2019)
7. Yun, S., Han, D., Chun, S., Oh, S.J., Yoo, Y., Choe, J.: CutMix: regularization strategy to train strong classifiers with localizable features. In: 2019 IEEE/CVF International Conference on Computer Vision (ICCV), pp. 6022–6031 (2019)
8. Ghiasi, G., Lin, T., Le, Q.V.: NAS-FPN: learning scalable feature pyramid architecture for object detection. In: 2019 IEEE/CVF Conference on Computer Vision and Pattern Recognition (CVPR), pp. 7029–7038 (2019)
9. Wang, K., Liew, J.H., Zou, Y., Zhou, D., Feng, J.: PANet: few-shot image semantic segmentation with prototype alignment. In: 2019 IEEE/CVF International Conference on Computer Vision (ICCV), pp. 9196–9205 (2019)
10. Tan, M.X., Pang, R.M., Le, Q.: EfficientDet: scalable and efficient object detection. In: The IEEE Conference on Computer Vision and pattern Recognition (2020)
11. Ren, S., He, K., Girshick, R., et al.: Faster R-CNN: towards real-time object detection with region proposal networks. J. IEEE Trans. Pattern Anal. Mach. Intell. **39**(6), 1137–1149 (2017)
12. He, K., Zhang, X.Y., Ren, S.Q., et al.: Deep residual learning for image recognition. arXiv preprint arXiv:1512.03385 (2015)
13. Jiang, B.R., Luo, R.X., Mao, J.Y., et al.: Acquisition of localization confidence for accurate object detection. In: The European Conference on Computer Vision (2014)
14. Zheng, Z., Wang, P., Liu, W., et al.: Distance-IoU loss: faster and better learning for bounding box regression. In: AAAI, pp. 12993–13000 (2020)
15. Wu, J., Ma, J., Zhao, Y.H., et al.: Apple detection in complex scene using the improved YOLOv4 model. J. Agronomy **11**(3), 476 (2021)
16. Hao, R.Y., Lu, B.Y., Cheng, Y., et al.: A steel surface defect inspection approach towards smart industrial monitoring. J. Intell. Manuf. (2020)

# A Novel Algorithm YOLOv4-Mini to Detection Automobile Rim Weld

Linlin Jiang[1,2], Shuai Zhao[3], Fu Zhao[4], Guodong Jian[5], Zhongtao Li[1,2], and Kai Wang[1,2](✉)

[1] Shandong Provincial Key Laboratory of Network Based Intelligent Computing, University of Jinan, Jinan 250022, China
ise_wangk@ujn.edu.cn
[2] School of Information Science and Engineering, University of Jinan, Jinan 250022, China
[3] Jinan Housing and Urban-Rural Development Bureau, Jinan, China
[4] Shandong Xiaoya Holding Group Co., Ltd., Jinan, China
[5] Shandong Xiaoya Precise Machinery Co., Ltd., Jinan, China

**Abstract.** In order to combine the lightweight object detection model with small embedded devices and improve the detection accuracy of automobile rim weld, this paper proposes YOLOv4-mini based on improved YOLOv4-tiny. Firstly, the lightweight network YOLOv4-tiny is adopted as the main architecture. Secondly, the M-SPP structure is added to obtain the main features of the target and utilize the Mosaic method to enhance the dataset. The K-means + + clustering method is utilized to reset the anchor data belonging to the rim weld target. Finally, the parameters of the pre-trained model are fine-tuned to complete the training of the model. Compared with the original algorithm, in the detection task at 50% and 75% threshold of 4240 rim weld datasets, the mAP index can reach 100% and 55.15%, with the detection speed of 173.5FPS, respectively. Therefore, the improved algorithm gains higher accuracy and efficiency for the edge weld detection task.

**Keywords:** Automobile Rim Weld · Object detection · Lightweight network · YOLOv4-tiny · M-SPP

## 1 Introduction

With the development of industrial technology, vacuum tires have become the mainstream of automobile industrial production. In order to ensure good airtightness in the rim, the premise of the production line is to detect the weld of the rim. Early detection usually relies on manual naked eye recognition, which is prone to misjudgment and missed detection due to visual fatigue. Nowadays, in the upsurge of machine vision development, it is of great significance to solve the problem of industrial automation production by leveraging object detection algorithms based on deep learning. Wei et al. used the deep neural network MobileNet combined with the SSD algorithm to detect artificial board defects [1]. Zhu and others constructed a data-enhanced Faster R-CNN algorithm to detect common foreign bodies of the power grid [2]; Guo et al. adopted the radial basis


Q. Han et al. (Eds.): LSMS 2021/ICSEE 2021, CCIS 1469, pp. 132–141, 2021.
https://doi.org/10.1007/978-981-16-7213-2_13

kernel support vector machine to classify steel surface defects. Li et al. utilized image preprocessing to detect the surface defects of PVC pipes [3]. If the object detection model can be transplanted into small embedded equipment and deployed in industrial fields for real-time detection, the production efficiency will be greatly improved and the cost can be effectively reduced. Common small embedded devices include raspberry pie Model 3B+, Jetson Nano, Jetson AGX Xavier, K210, etc. Linux operating system is configured in embedded devices, and the object detection model based on deep learning is embedded into the device to run. Due to the limited computing power of small embedded devices, the detection model with complex network structure cannot be processed, hence the detection accuracy of rim weld is reduced. But its advantages of high reliability and high real-time make it have broad applications and development space in the field of intelligent manufacturing.

As an important branch of computer science, machine vision has a history of more than 20 years. It is not a single application product until now, and has gradually become a necessary part of intelligent manufacturing. Object detection is one of the main tasks of machine vision. By identifying and predicting the position of the target object of the high-resolution image, and associating the position information about the target object, the control system is designed to drive the relevant machine for detection. The object detection network extends from LeNet [4] in 1998 to ResNet [5] in 2015. The depth of the network extends to several layers to thousands of layers. The number of model parameters increases from the improvement of accuracy and depth. In order to achieve the combination of the object detection model and small embedded devices, it is necessary to prune and optimize the network model.

In summary, to improve the robustness of the automobile rim weld detection system, this paper selects the lightweight object detection model combined with small embedded devices. This paper proposes a YOLOv4-mini algorithm, and makes the following improvements: adjust the network structure of YOLOv4-small object detection algorithm based on Darknet framework; add the optimized spatial pyramid pooling structure M-SPP (Mini-Spatial Pyramid Pooling) to promote the ability of the network to extract effective features with minimal computation; K-means + + algorithm is used to cluster the anchor data which is more in line with the weld ratio and reset it. Additionally, the data set of automobile rim weld based on VOC is constructed, and the problem of uneven distribution of weld state is overcome by data enhancement. Meanwhile, the optimized YOLOv4-tiny network is trained, and the network parameters are constantly fine-tuned, and finally the automobile rim weld detection algorithm YOLOv4-mini is obtained. Compared with YOLOv4-tiny, the optimization algorithm YOLOv4-mini in this paper increases the model size by 1.1MB, and the detection accuracy is improved by 12.6% in the $AP_{75}$ index, which enhances the accuracy of rim weld object detection.

The organization is as follows: The Sect. 2 introduces the related work of the object detection algorithm. The Sect. 5 describes the algorithm optimization method in detail. In the Sect. 4, the preparation of experimental data focuses on the comparative analysis of experimental results. Finally, draw conclusions in Sect. 5.

## 2 Related Work

### 2.1 Two-Stage Object Detection Model Based on Candidate Regions

Based on the Two-stage object detection model of RP (Region Proposals), such as R-CNN, Fast R-CNN, Faster R-CNN, etc., the detection problem is divided into two stages: first, the region selection algorithm is adopted to find candidates with rich semantic information Area, and then correct and classify the candidate area by judging the foreground and background.

R-CNN [6] (Regions with CNN features) is a pioneering work based on the candidate region algorithm. For the first time, CNN is introduced into the object detection field, but each candidate region in R-CNN must be sent to the CNN model separately to calculate the feature vector, this operation is very time consuming. Fast R-CNN [7] circumvents the redundant feature extraction operation in R-CNN, and performs a feature extraction on the entire image of the entire region. However, due to utilize the Selective Search algorithm to extract candidate regions, it takes about 2 to 3 s, which cannot fulfill the needs of real-time applications. Faster R-CNN [8] uses RPN (Region Proposal Network) to extract candidate regions directly in the feature map, which is 34 times faster than Fast R-CNN.

### 2.2 Single-Stage Object Detection Model Based on Regression Thinking

Single-stage object detection models based on regression ideas, such as YOLO, SSD, etc., convert the two-stage object detection classification problem into a regression problem. Using a CNN can directly predict the bounding box and class probability of the object, which improves the real-time performance of the detection model.

The YOLO [9–11] (You Only Look Once) object detection model divides the input image into S $\times$ S grids, extracts image features through operations such as convolution, activation, and pooling, detects whether each grid contains an object. Calculate the boundary box area intersection ratio IoU (Intersection over Union), and select the bounding box with the best confidence value through non-maximum suppression (NMS). SSD [12] (Single Shot MultiBox Detector) is based on YOLO and divides the input image into several grids. The anchor mechanism of Faster R-CNN can be used to predict a priori frames with different scales and aspect ratios for each grid. Meanwhile, a priori box explores a small convolution filter combined with a regression method to predict the class score and offset of the bounding box, and the priori box with higher confidence is selected for non-maximum suppression (NMS) to obtain the final result.

In comparison with other object detection algorithms, the YOLO algorithm has become the mainstream algorithm for object detection in the industrial field based on the advantages of high real-time detection. However, due to large weights and a large amount of parameters, it is impossible to complete inference tasks on small embedded devices. Therefore, YOLOv4-tiny is the preferred detection model for developers with relatively tight computing resources. The detection performance of YOLOV4-tiny on GTX 1080Ti for the COCO data set is $AP_{50}$: 40.2%, FPS: 371. In contrast to the previous lightweight network, there is a big improvement in AP and FPS. In order to ensure the balance between the number of parameters and the detection accuracy of the lightweight object detection model, optimizing the network structure has become the research focus of scholars. Shao proposed a lightweight convolutional neural network YOLO-Slim, which modified the main network of YOLOv3 to MobileNet, weakened the network parameters and computation, reduced the size of the model by 90%, and ensured the detection accuracy of 76.42% [13]. Jiang and others added a feature output layer in the YOLOv3 network structure to optimize the loss function. The average accuracy of the optimized ship object detection algorithm is 6.72%, which is higher than that of YOLOv3 [14]. Cao modified the YOLOv4-small backbone network of GhostNet residual structure, the model is only 10% of YOLOv4-small, and the detection speed is increased by 58%, unfortunately, the detection accuracy is not improved [15].

## 3 YOLOv4-Mini Based on Improved YOLOv4-Tiny Network Model

### 3.1 YOLOv4-Tiny Added M-SPP Spatial Pyramid Pooling Structure

The traditional YOLOv4-tiny prunes and compresses the network structure of YOLOv4. The original SPP pyramid pooling structure is deleted, and the feature fusion layer is up-sampled only once, and the output feature layer is reduced to two. Although the reasoning speed of the network is accelerated, the accuracy of object detection is reduced at the same time.

Therefore, this paper improves the ability of extracting effective features by adding the M-SPP (Mini-Spatial Pyramid Pooling) structure, avoids distortion caused by image scaling, and effectively solves the problem of large difference in target size of the image to be detected. The improved network structure is shown in Fig. 1.The structure of spatial pyramid pooling SPP and M-SPP are shown in Fig. 2.

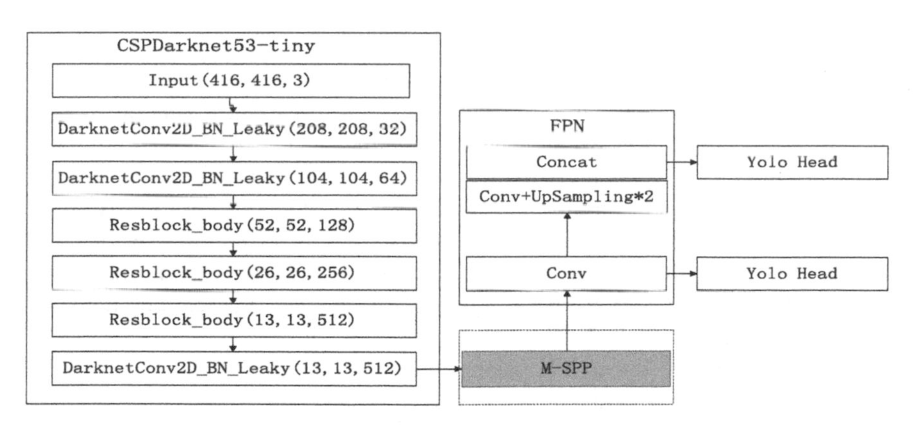


**Fig. 1.** Optimized YOLOv4-mini network structure.

The newly constructed spatial pyramid pooling module M-SPP in this paper is composed of two maximum pooling layers. The design of window size follows the formula:

$$\mathrm{Pool\_size} = \lceil \mathrm{fmap\_ size} \rceil / n_i, \ (n_i \in 1, 3) \tag{1}$$

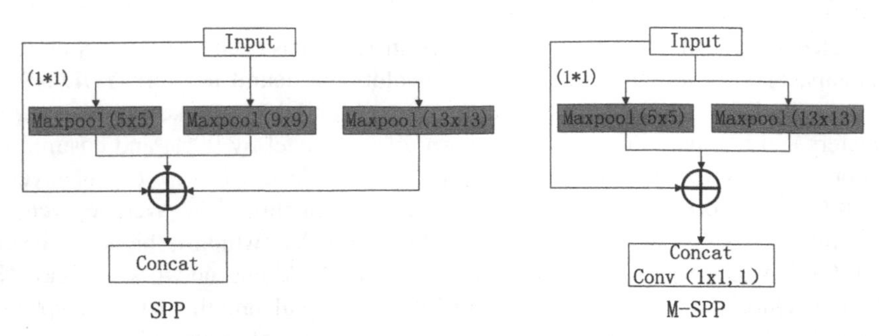


**Fig. 2.** Original SPP structure and modified M-SPP structure.

Therefore, the pooling window is 5 × 5 and 13 × 13, and each dimension extracts the features of the fixed dimension and then fuses them. The first layer of 5 × 5 spatial pyramid divides the feature map into 25 grids, and the width and height of each grid block are (w/5, h/5); The spatial pyramid of the second layer 13 × 13 divides the feature map into 169 grids, and the width and height of each grid block are (w/13, h/13). Furthermore, the third layer space pyramid takes the whole feature map as a block for feature extraction. Finally, all feature layers are merged.

In contrast with the SPP structure, M-SPP reduces the pyramid scale and layer number. For the rim weld target, more appropriate size is designed to extract features and reduce the calculation parameters. The original up-sampling was doubled, and the detection accuracy was improved at the cost of only 1.1 MB.

### 3.2 Resetting Anchor Based on K-means + + Clustering Algorithm

The purpose of anchor clustering is to get a priori box that matches the true box width-to-height ratio. The original YOLOv4-tiny leverages the distance-based K-means clustering algorithm. Six default anchor frames are used in YOLOv4-tiny to adapt to the public data set, and the default anchor frame used is not applicable to the aspect ratio of weld target size.

To avoid the false detection caused by the error between the predicted boundary frame and the real boundary frame, the K-means + + [16] clustering algorithm is adopted in this paper. The basic idea of the K-means + + algorithm to select the initial clustering center is that the distance between the initial clustering centers should be as far as possible. Reset anchor data according to the location of the real box marked by the data set to generate a priori box.

Repeat the K-means + + and K-means algorithm 15 times for the rim weld train set, and the Accuracy results are shown in Fig. 3. Finally, six groups of anchor frame sizes with the highest accuracy of 87.75% were selected: [(29,38), (38,34), (31,46), (48,41), (34,60), (42,54)]. Compared with the default anchor value of YOLOv4-tiny, the detection accuracy of the model is improved by 12%. Accuracy is the average IOU value, which is the coincidence degree between the prior and the true frames.

Comparing the accuracy values of K-means + + and K-means clustering algorithms after 15 calculations, the accuracy value of the K-means + + algorithm is significantly better than that of K-means. In this paper, the model training parameters selected the

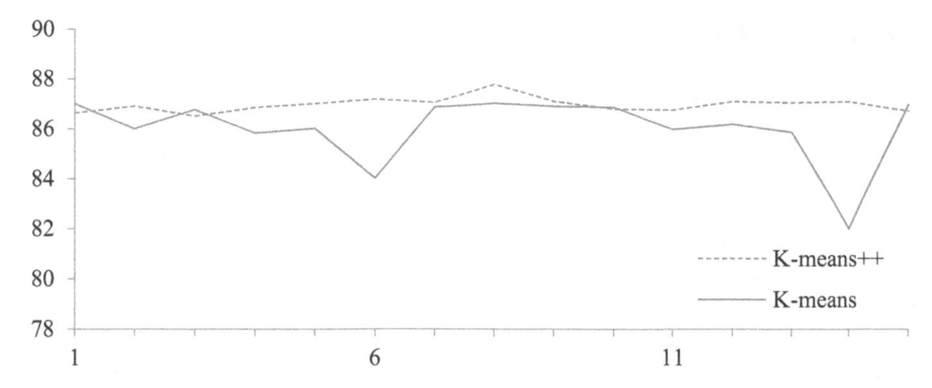


**Fig. 3.** Accuracy comparison between K-means and K-means + + algorithm.

eighth highest accuracy of 87.75%, which means that the obtained anchor frame is closer to the real frame.

### 3.3 Data Enhancement Based on Mosaic

Mosaic [17] data enhancement techniques are capitalizing on to improve the generalization ability of the proposed model. Randomly read 4 training images to flip, zoom, color gamut change, etc., and place them in four directions to combine images and frames, and finally synthesize an image.

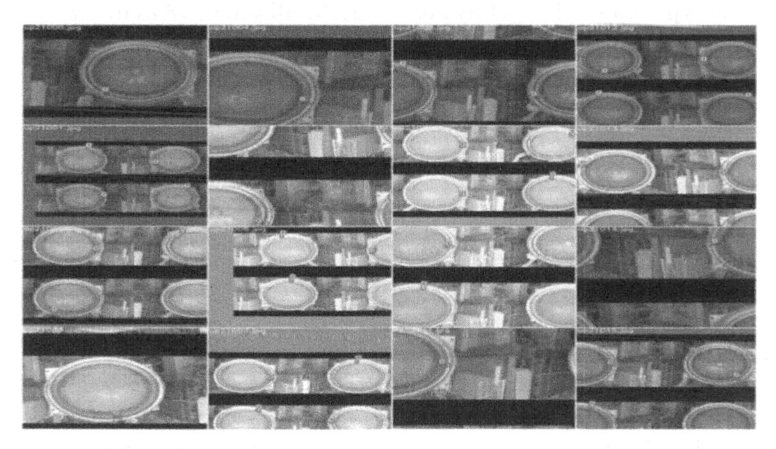

**Fig. 4.** Example of Mosaic-processed car rim weld train data set.

Mosaic can not only enrich the background of detection objects, but also compute 4 images simultaneously, saving the time of batch standardized calculation. Figure 4 is an example of a train data set for car rim welds processed by Mosaic.

# 4 Experiment and Result Analysis

## 4.1 Experimental Data Preparation

In this experiment, an industrial camera was employed for 360-degree rotation shooting, and its resolution was set to 640 × 480, which was used to collect three types of images at the rim production site, a total of 4240 images. Among them, 1240 welds of type A, 1500 welds of type B and C each. The class label of this experiment is weld, 3392 images are randomly selected as the train set, and the remaining 848 images are the test set.

## 4.2 Experimental Configuration and Result Analysis

**Parameter Configuration.** The Jetson AGX Xavier development kit is adopted as the performance verification platform for small embedded devices. The operating system is Ubuntu 18.04, the development language uses C and Python 3.7, and the software environment configuration is CUDA10.2, CUDNN8.0, and OpenCV4.1.1. Algorithms with larger experimental models: YOLOv4 and Faster R-CNN algorithm training and testing platforms use high-performance server, and the GPU model is NVIDIA GeForce GTX 1080 Ti. According to the parameter tuning strategy, the parameters are fine-tuned during the model training process. The initial learning rate is set to 0.00261, momentum is set to 0.9, decay is set to 0.0005, batch size is set to 16, and the object detection model is trained for 30,000 iterations.

**Evaluation Index and Result Analysis.** This paper verifies the algorithm's ability to identify single-type rim weld target by Precision, Recall and mAP as the accuracy evaluation indicators, and the number of FPS as the evaluation indicator of the inference speed. The calculation formulas of Precision, Recall and mAP are shown in (2)-(4).

$$\text{Precision} = \text{TP} / (\text{TP} + \text{FP}) \tag{2}$$

$$\text{Recall} = \text{TP} / (\text{TP} + \text{FN}) \tag{3}$$

$$\text{mAP} = \sum_{C=1}^{C} \text{AP}/\text{C} \tag{4}$$

Where TP represents the correct recognition as a positive sample, FP represents the correct recognition as a negative sample, TN represents a false recognition as a positive sample, and *FN* represents a false recognition as a negative sample. AP is the area under the Precision-Recall curve, which reflects how well the model recognizes a certain category. Where mAP is the average of AP for all classes. The higher the mAP value, the more accurate the detection target and the better the algorithm effect.

In order to verify the effectiveness of the improved algorithm in this paper, 848 rim weld test sets were tested, and the results are shown in Table 1. The Precision and Recall of this algorithm are 71% and 71%, respectively. When the set IoU Threshold is 75%, mAP is 55.15%. The results show that the algorithm in this paper has high detection accuracy and meets the accuracy requirements of automobile rim weld detection.

**Table 1.** Detection results of the improved algorithm.

| Index | IoU% | TP | FP | FN | Precision% | Recall% | AvgIoU% | mAP% ($AP_{75}$) |
|---|---|---|---|---|---|---|---|---|
| Result | 75 | 605 | 243 | 243 | 71 | 71 | 58.55 | 55.15 |

Under the same rim weld train set and test set conditions, this article compares the optimized algorithm YOLOv4-mini with Faster R-CNN, YOLOv4, and YOLOv4-tiny algorithms. The tests are shown in Table 2. In the same configured server, use the common anchor frame for training.

**Table 2.** Detection results of different object detection algorithm.

| Algorithms | Description | Model size(MB) | mAP%($AP_{75}$) | FPS | GPU |
|---|---|---|---|---|---|
| Faster R-CNN | ResNet50 | 113.8 | 47.1 | 11 | Nvidia GTX 1080Ti |
| YOLOv4 | CSPDarknet | 256.2 | 80 | 43.3 | Nvidia GTX 1080Ti |
| YOLOv4-tiny | CSPDarknet-tiny | 23.5 | 42.55 | 168.9 | Nvidia GTX 1080Ti |
| **YOLOv4-mini** | **M-SPP\K-means + + \Mosaic** | **24.6** | **55.15** | **173.5** | **Nvidia GTX 1080Ti** |

The mAP of each object detection algorithm in the table are: 47.1%, 80%, 42.55%, 55.15%; FPS are 11, 43.3, 168.9, and 173.5, respectively. The analysis results show that Faster R-CNN and YOLOv4 networks have high detection accuracy, but the large model size and slow detection speed limit their applications on embedded devices; the detection speed of the YOLOv4-tiny algorithm is fast, by contrast, the accuracy rate is low.

This paper is optimized based on YOLOv4-tiny. Although the model increases the cost of 1.1MB compared with the original YOLOv4-tiny, it boosts the value of mAP. In Table 3, YOLOv4-tiny and the improved algorithm of this paper are designated to compare AP under different thresholds. The algorithm of $AP_{50}$ has increased by 0.12%, the algorithm of $AP_{65}$ has increased by 6.23%, and the one of $AP_{75}$ has increased by 12.6%. The difference in FPS is only 4.6, because the M-SPP structure has only one maximum pooling structure, and the impact on time overhead is negligible, which meets the real-time requirements of industrial production lines.

**Table 3.** AP comparison of YOLO-tiny and YOLO-mini under different thresholds.

| Algorithms | $AP_{50}$% | $AP_{65}$% | $AP_{75}$% | GPU |
|---|---|---|---|---|
| YOLOv4-tiny | 99.88 | 84.88 | 42.55 | Xavier |
| **YOLOv4-mini** | **100.00** | **91.11** | **55.15** | **Xavier** |

In the complex production environment of automobile rims, the improved YOLOv4-mini algorithm in this paper is robust. The detection effect of the rim is shown in Fig. 5. This paper improves the network to take into account the balance of detection speed and accuracy, and can better complete the task of weld detection.

**Fig. 5.** Object detection of YOLO-mini algorithm on rim data set.

## 5 Conclusion

Aiming at the issues of obstruction of computing power and low rim weld detection accuracy of object detection model on small embedded devices, this paper leverages YOLOv4-tiny as the main architecture, and improves the SPP spatial pyramid pooling structure to M-SPP. The ability of the network to extract effective features is enhanced with a very small computational cost. Moreover, the K-means + + clustering algorithm is employed to reset the anchors belonging to the rim weld to avoid the impact on the accuracy of the model due to the difference in the target size of the anchor frame and the weld. The Mosaic data enhancement method is adopted to process the data set to shorten the training time and enhance the generalization ability of the data. In response to the needs of weld detection applications, the detection model optimized in this paper is deployed in small embedded devices. Experiments show that compared with YOLOv4-tiny on the automobile rim weld data set, the improved algorithm in this paper has a 12.6% increase in mAP, which is more accurate than YOLOv4-tiny. Besides, the detection speed reaches 173.5FPS, which meets the real-time requirements of rim weld detection.

## References


1. Wei, Z.F., Xiao, S.H., Jiang, G.Z., et al.: Research on surface defect detection of wood-based panels based on deep learning. J. Forest Ind. **58**(02), 21–26 (2021)
2. Zhu, H.Z., Sun, Z., Lan, Q.Q., et al.: Faster R-CNN based detection of common foreign bodies in power grid lines. J. Electr. Energy Efficiency Manage. Technol. **01**, 58–63 (2021). https://doi.org/10.16628/J.CNKI.2095-8188
3. Li, S.H., Zhou, Y.T., Wang, D., et al.: Surface defect detection of PVC pipe based on machine vision. J. Progress Laser Optoelectron. **56**(13), 100–108 (2019)
4. LeCun, Y., Bottou, L., Bengio, Y.: Gradient-based learning applied to document recognition. J. Proc. IEEE. **86**(11), 2278–2324 (1998)
5. Szegedy, C., Ioffe, S., Vanhoucke, V.: Inception-v4, inception-ResNet and the impact of residual connections on learning (2016)
6. Girshick, R., Donahue, J., Darrell, T., et al.: Rich feature hierarchies for accurate object detection and semantic segmentation. In: Proceedings of the 2014 IEEE Conference on Computer Vision and Pattern Recognition, pp. 580–587 (2013)
7. Girshick, R.: Fast R-CNN. In: Proceedings of the 2015 IEEE International Conference on Computer Vision, pp. 1440–1448 (2015)
8. Ren, S., He, K., Girshick, R., et al.: Faster R-CNN: towards real-time object detection with region proposal networks. J. IEEE Trans. Pattern Anal. Mach. Intell. **39**(6), 1137–1149 (2017)
9. Bochkovskiy, A., Wang, C.Y., Liao H.: YOLOv4: optimal speed and accuracy of object detection. In: Computer Vision and Pattern Recognition. IEEE (2020)
10. Redmon, J., Farhadi, A.: YOLOv3: an incremental improvement. arXiv e-prints (2018)
11. Redmon, J., Divvala, S., Girshick, R., et al.: You only look once: unified, real-time object detection. In: Computer Vision and Pattern Recognition. IEEE (2016)
12. Liu, W., et al.: SSD: single shot multibox detector. In: Leibe, B., Matas, J., Sebe, N., Welling, M. (eds.) ECCV 2016. LNCS, vol. 9905, pp. 21–37. Springer, Cham (2016). https://doi.org/10.1007/978-3-319-46448-0_2
13. Shao, W.P., Wang, X., Cao, Z.R., et al.: Design of lightweight convolutional neural network based on mobilenet and YOLOv3. J. Comput. Appl. **40**(S1), 8–13 (2020).
14. Jiang, W.Z., Li, B.Z., Gu, J.J., et al.: Ship object detection algorithm based on improved YOLO V3. J. Electro-optical Control. 1–5 (2021)
15. Cao, Y.J., Gao, Y.X.: Lightweight beverage recognition based on ghostnet residual structure. J. Netw. Comput. Eng. 1–7 (2021). https://doi.org/10.19678/j.issn.1000-3428.0059966
16. Arthur, D., Vassilvitskii, S.: K-means++: the advantages of carefull seeding. In: Proceedings of the Eighteenth Annual ACM-SIAM Symposiumon Discrete algorithms, Society for Industrial and Applied Mathematics, pp. 1027–1035 (2007)
17. Redmon, J., Farhadi, A.: YOLO9000: better, faster, stronger. In: Computer Vision and Pattern Recognition, pp. 6517–6525. IEEE (2017)

# Detection Method of Automobile Rim Weld Based on Machine Vision

Xin Xiao[1,2], Shuai Zhao[3], Haokun Sun[4], Shuai Li[4], Zhongtao Li[1,2], and Xiangyu Kong[1(✉)]

[1] Shandong Provincial Key Laboratory of Network Based Intelligent Computing, University of Jinan, Jinan 250022, China
ise_kongxy@ujn.edu.cn
[2] School of Information Science and Engineering, University of Jinan, Jinan 250022, China
[3] Jinan Housing and Urban-Rural Development Bureau, Jinan, China
[4] Shandong Xiaoya Precise Machinery Co., Ltd, Jinan, China

**Abstract.** At present, the commonly used automobile tires are vacuum tires, which mainly rely on the rim to seal the gas. The detection of rim air tightness is a key for rim production, and the detection of rim weld is the primary prerequisite for detecting rim air tightness. We introduce the workflow of weld detection and compare three detection methods based on Halcon object detection, YOLOv4 and Faster R-CNN deep learning algorithm, and conduct field verification to prove that the Halcon deep learning method works best. When the IoU is set to 0.75, the detection precision reaches 87.6%, which basically meets the practical demands in the industry.

**Keywords:** Rim Weld · Machine vision · Deep learning · Object detection

## 1 Introduction

The production environment of the rim is complex and harsh, and the welding of the rim will form weld. Due to the complex welding environment and many other factors, there are defects such as sand holes and pores on the weld surface, which greatly affect the quality of the rim. In addition, the rim is a metal product, and the light causes reflection at the rim weld, and factors such as the variety of wheel rims and the inconsistent size of the rim weld make it difficult to identify the rim weld. In recent years, deep learning technology has become a research hotspot and has brought huge economic benefits to all walks of life. It is imperative to apply machine learning technology to automobile rim weld detection.

In order to achieve accurate detection of rim weld, the complex and harsh industrial scene puts forward high requirements on the accuracy and robustness of the recognition algorithm. Therefore, we have tried three different detection methods based on Halcon object detection, YOLOv4 and Faster R-CNN deep learning algorithm. Finally, based on the Halcon deep learning technology, the detection accuracy of rim weld reaches 87.6%, which basically meets the actual production of the automobile industry. Our main structures are as follows.


Q. Han et al. (Eds.): LSMS 2021/ICSEE 2021, CCIS 1469, pp. 142–151, 2021.
https://doi.org/10.1007/978-981-16-7213-2_14

Firstly, we analyzed the research state of automobile rim weld detection and described the workflow of weld detection. What's more, compared algorithms based on Halcon, YOLOv4, Faster R-CNN, and analyzed the performance of each type of algorithm in rim weld detection and location. Finally, we get a conclusion that the method based on Halcon deep learning is more suitable for automobile rim weld detection. However, we cannot modify the encapsulated backbone network, so the performance indicators have reached the upper limit of Halcon.

# 2 Related Work

## 2.1 Research State

At present, manual recognition and non-destructive testing methods are mainly adopted in the automobile rim weld detection. However, due to the complex structure of the rim, the manual recognition efficiency is low, the accuracy rate is low, and the cost is high, which often cannot meet the production needs. Non-destructive testing includes ultrasonic testing, penetrant testing, infrared testing and so on, but the detection speed is slow and the error is large [1]. The automobile rim weld detection method based on machine vision has the characteristics of high accuracy and fast detection speed. Han et al. [2] designed an automobile rim weld detection system based on machine vision, and used an improved Hough transform detection algorithm to perform weld detection on the image; On the basis of Faster R-CNN object detection algorithm, Zhu et al. [3] added SE module and replaced ROI-Pooling with ROI-Align, FPN multi-scale feature fusion network, and finally got the model that can locate and classify wheel surface defects; Zhong et al. [4] used U-Net model and Resnet model as feature extractors to improve the original Faster R-CNN model to detect of automobile weld defects of subway vehicles; On the basis of YOLO object detection algorithm, Wu et al. [5] adjusted appropriate anchor box to suit data set and used multi scale feature fusion to improve the accuracy of defect recognition.

## 2.2 Workflow of Weld Detection

The workflow of weld detection mainly includes six parts: layout of field environment, image acquisition, image processing, weld detection, serial communication, and air tightness detection.

The layout of field environment requires the industrial camera and lens to be fixed on the camera stand, so that the captured image of the rim is centered, and the camera parameters are adjusted to make the captured image clear.

The image acquisition part adopts the hard trigger used in the industry, the acquisition card passively waits for the trigger signal, and each signal corresponds to acquiring an image, and the image acquisition is performed after the signal is received [6].

The image processing part changes the image size, number of channels and other parameters according to the requirements of the deep learning network. We collected 5300 images, including 4240 images for deep learning training and validating, 1060 images for testing.

The rim weld detection part is shown in Fig. 1. The deep learning network is used to locate the weld and get the position parameters of the rectangular box (the coordinates of the upper left corner and the lower right corner of the rectangular box). The coordinates of the center point of the rectangular box are obtained by using the position parameters. Then the angle is obtained according to the center point coordinates of the rim weld, the rim center point coordinates and the preset reference point coordinates. If the weld is not detected, the angle is set to 361°. Among them, the preset reference point is the location of the rim air tightness testing instrument.

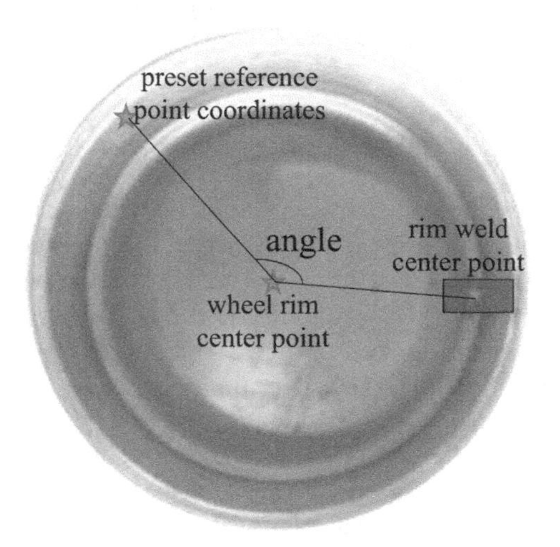


**Fig. 1.** Detection of Rim Weld.

In the serial communication part, the server communicates with PLC through the Modbus protocol, and the angle value is transmitted to PLC. PLC drives the servo motor to control the rotating round table to rotate according to the angle value. When the rotating round table stops rotating, the air-tightness testing instrument performs pressure test on the weld, and the airflow sensor determines whether the rim is leaking. If there is no leakage, the rim enters the next step.

# 3 Experimental Result and Analysis

## 3.1 Evaluation Indicators

We use the average precision mean (mAP), precision, and recall as evaluation indicators. The AP value is the average value of the maximum precision under different recall and it is for a certain category in the data set. The average of the AP values of multiple categories is mAP. The value of mAP is between 0–1, and the larger the value, it means that better performance of the model. We only detect one category, AP is mAP. Among them, the AP value depends on the threshold of IoU. IoU is the ratio of intersection over union between bounding box and ground truth. We take IoU = 0.75 as the threshold, if

IoU > 0.75, it is the true positives, otherwise, it is the false positives. The calculation formula is as follows. Among them, A represents the ground truth labeled with LabelImg software, and B represents the bounding box predicted by the network model.

$$\mathrm{IoU} = (\mathrm{A} \cap \mathrm{B})/(\mathrm{A} \cup \mathrm{B}) \tag{1}$$

The Precision refers to how many of the data predicted to be positive samples are truly positive samples. The calculation formula is as follows. Among them, TP + FP is the number of all positive samples predicted, and TP is the number of positive samples predicted correctly.

$$\mathrm{Precision} = \mathrm{TP}/(\mathrm{TP} + \mathrm{FP}) \tag{2}$$

The Recall refers to the number of positive samples retrieved by the model in the true positive samples. The calculation formula is as follows. Among them, TP + FN is the number of all true positive samples, and TP is the number of positive samples predicted correctly.

$$\mathrm{Recall} = \mathrm{TP}/(\mathrm{TP} + \mathrm{FN}) \tag{3}$$

## 3.2 Experimental Environment

The hardware configuration of the experimental environment is shown in Table 1.

**Table 1.** Experimental environment hardware configuration.

| Processor | Intel Core i7–9700-16G-RAM |
|---|---|
| GPU | RTX2060-6G-GPU |
| Acceleration library | cuda10.2.89 + cudnn7.6.5 |
| Python | Python 3.7 |
| Pytorch | Pytorch 1.5.0 |
| Torchversion | Torchversion 0.6.0 |
| Halcon | Halcon 19.11 Progress |
| OpenCV | 4.0.1 |

## 3.3 Halcon-Based Weld Detection

HALCON is a complete standard machine vision algorithm package developed by MVTec company. It has a widely used machine vision integrated development environment, which saves production costs and shortens the product development cycle [7]. Halcon vision is divided into three categories: the first is the traditional machine vision; the second is machine vision based on classifier; the third is machine vision based on Halcon deep learning.

**Halcon Traditional Vision.** This method must manually extract features and detect the weld by setting a series of rules. Firstly, the collected weld image is preprocessed, the contrast of the image is enhanced, the gray value of the image is linearly transformed, and the preprocessed image is converted into a grayscale image. Then, threshold segmentation of gray image based on gray histogram, and get the general position of the rim weld. Finally, the rim weld is found by the shape selection of feature histogram. This method has many shortcomings. It is only aimed at the situation where the gray value of the target area and the non-target area are very different. With the change of the detected object, all the rules have to be reset, and the adaptability is poor. Therefore, it is not suitable for the detection of rim weld. The original image is shown in Fig. 2 (left), and the weld detection image is shown in Fig. 2 (right).

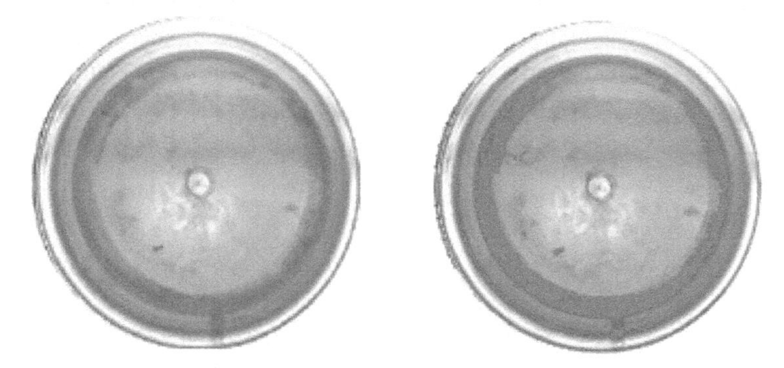

**Fig. 2.** Halcon traditional visual detection.

**Halcon Classifier Vision.** Classification is divided into a separate category based on specific features, such as color, shape, length and other features. Each category through pre-training and learning, and a classifier is obtained. After the object is trained, the classifier compares the object features to identify object. We use color as the basis for classification and use the MLP classifier to classify images. However, this method can only extract the target area based on the relatively shallow edge texture features, so it is

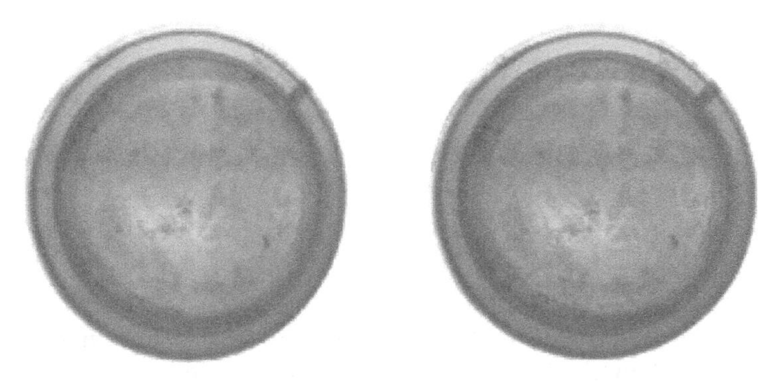

**Fig. 3.** Halcon classifier detection.

not suitable for the detection of rim weld. The original image is shown in Fig. 3 (left), and the weld detection image is shown in Fig. 3 (right).

**Halcon Deep Learning.** Halcon deep learning extracts the features of each level by learning a large number of training sample images, so that the network has the ability to discriminate and reason. It needs to read json format data from a coco file, however, we uses the rectangular box of LabelImg software to get the rim weld, and the image annotation information is saved in the XML file. Therefore, we need to convert the XML format to the json format. The learning rate decay strategy we set is decayed with the number of steps and the average 40 epochs of learning rate decays by 50%. The optimizer chooses the iterative optimization algorithm Stochastic Gradient De-scent (SGD), and the k-means algorithm is used to find the anchor point suitable for the data set. We get the final hyperparameters after repeated parameter adjustment. Parameter settings during training is shown in Table 2.

**Table 2.** Parameter settings during training.

| Parameter variables | Representative meaning | Parameter setting |
|---|---|---|
| batch_size | batch size | 2 |
| Input size | image resolution | 512 × 384 |
| learning_rate | learning rate | 0.0005 |
| momentum | momentum | 0.99 |
| epoch | number of training rounds | 100 |
| weight_prior | regularization parameter | 0.0001 |

The image of the training process is shown in Fig. 4. Figure 4 (above) shows image labeled with LabelImg, and Fig. 4 (bottom) shows the weld detection image.

When evaluating a trained network model, we test four trained network models on the test set, and the experimental results are shown in Table 3.

From the above table, it can be seen that the four network types of Halcon have good detection effects on rim weld. Further analysis, Halcon regards missed detection and false detection as one type, that is, FP and FN are equal, so it gets higher AP and accuracy.

**Fig. 4.** Halcon training process image.

**Table 3.** AP results on test set under Halcon different models.

| Model | $AP_{75}$ |
|---|---|
| pretrained_dl_classifier_alexnet.hdl | 0.83 |
| pretrained_dl_classifier_compact.hdl | 0.809 |
| pretrained_dl_classifier_enhanced.hdl | 0.81 |
| pretrained_dl_classifier_resnet50.hdl | 0.829 |

### 3.4 Weld Detection Based on YOLOv4

YOLOv4 is a one-stage detection algorithm. The classification and location of the target are completed in one detection step. Compared with the previous YOLO series algorithms, it introduces CSP, Mish activation, Mosaic data augmentation, CIoU loss and other tricks to achieve the best trade-off between detection speed and accuracy [8]. YOLOv4 adds a bottom-up feature pyramid to the FPN layer. The FPN layer conveys strong semantic features from top to bottom, while the feature pyramid conveys strong location features from bottom to top. The parameters of different detection layers are fused from different backbone layers to obtain richer feature information. CIoU is used to calculate the loss in training regression, and DIoU is used to apply non-maximum value suppression to the test set, which makes the bounding box regression faster and more accurate, thereby improving the final performance of the model [9].

We train images of 416 × 416 and 512 × 512 respectively, the learning rate is set to 0.001, the learning rate decay strategy is decayed with the number of steps, and the number of training rounds set to 20,000 rounds, batch and subdivision set to 32. The experimental results are shown in Table 4.

**Table 4.** AP results on test set under different size of images.

| Model | Input size | $AP_{75}$ |
|---|---|---|
| YOLOv4 | 416 × 416 | 0.695 |
| YOLOv4 | 512 × 512 | 0.712 |

From the above table, when the image resolution increases, the AP of YOLOv4 increases by 0.017. Therefore, the greater the resolution of the input image, the more accurate recognition of the rim weld.

### 3.5 Weld Detection Based on Faster R-CNN

Faster R-CNN is a two-stage detection algorithm [11]. By adding an RPN network, bounding boxes are generated based on the anchor mechanism, instead of the previous Fast R-CNN [12] and R-CNN [13] selective search algorithm. Finally integrates feature extraction, candidate bounding box selection, bounding box regression and classification in a network, thereby effectively improving the detection accuracy and detection efficiency. The backbone feature extraction networks commonly used in Faster R-CNN are VGG and ResNet networks. We test VGG16 and ResNet-50 network types, and the results are shown in Table 5.

**Table 5.** AP results on test set under different feature extraction networks of Faster R-CNN.

| Model | $AP_{75}$ |
|---|---|
| Faster R-CNN VGG16 | 0.440 |
| Faster R-CNN ResNet-50 | 0.471 |

From the above table, the Faster R-CNN network only detects rim weld on the last layer of feature map, the resolution of the feature map decreases with the deepening of the network depth and continuous convolution and pooling operations, resulting in poor recognition of the rim weld [13]. We can add the FPN network on the basis of VGG16 or ResNet-50. FPN is a fusion of different scale features for detection. It not only adds the location information of the low-level convolution feature, but also integrates the semantic information of the high-level convolution feature. We can also increase the anchor box of different sizes of RPN, so that the anchor box is more suitable for rim weld detection.

### 3.6 Experimental Results

**Table 6.** AP results on test set under different models.

| Model | $AP_{75}$ | Precision | Recall |
|---|---|---|---|
| Faster R-CNN VGG16 | 0.440 | 0.423 | 0.648 |
| Faster R-CNN ResNet-50 | 0.471 | 0.438 | 0.662 |
| YOLOv4 | 0.695 | 0.810 | 0.810 |
| pretrained_dl_classifier_compact.hdl | 0.809 | 0.862 | 0.862 |
| pretrained_dl_classifier_enhanced.hdl | 0.810 | 0.886 | 0.886 |
| pretrained_dl_classifier_resnet50.hdl | 0.829 | 0.881 | 0.881 |
| pretrained_dl_classifier_alexnet.hdl | 0.830 | 0.876 | 0.876 |

The summary of all experimental results are shown in Table 6, it can be seen that the pretrained_dl_classifier_resnet50.hdl network has the best effect on detecting rim weld.

## 4 Conclusion

After the test of different rim weld recognition methods, the experimental results show that the Halcon-based deep learning method can better adapt to the complex conditions of the scene. It gets rid of the traditional manual detection and non-destructive detection method, and the recognition accuracy rate can reach 87.64%, which can meet the needs of the automobile industry. But further analysis, we cannot modify the encapsulated backbone network, so the performance indicators have reached the upper limit of Halcon.

## References

1. Hu, D., Gao, X.D., Zhang, N.F., et al.: Review of status and prospect of weld defect detection. J. Mech. Electr. Eng. **37**(07), 736–742 (2020)
2. Han, X.Y., Duan, J., Dong, S.Q.: Research on automotive weld inspection system based on machine vision. J. Changchun Univ. Sci. Technol. **41**(05), 75–79 (2018)
3. Zhu, C.P., Yang, T.B.: Online detection algorithm of automobile wheel surface defects based on improved faster-RCNN model. J. Surface Technol. **49**(06), 359–365 (2020)
4. Zhong, J.J., He, D.Q., Miao, J., et al.: Weld defect detection of metro vehicle based on improved faster R-CNN. J. Railway Sci. Eng. **17**(04), 996–1003 (2020)
5. Wu, T., Yang, J.C., Liao, R.Y., et al.: Weld detect inspection of battery pack based on deep learning of linear array image. J. Laser Optoelectron. Progress **57**(22), 315–322 (2020)

6. Wang, C., Liu, M.L., Liu, Z.Y., et al.: Influence of trigger mode on acquisition time of low-speed-camera image sequence. J. Trans. Beijing Inst. Technol. **39**(06), 632–637 (2019)
7. Kan, R.F., Yang, L.X., Nan, Y.L.: Research on weld recognition and extraction method based on HALCON image processing. J. Internet of Things Technol. **7**(05), 29–31 (2017)
8. Bochkovskiy, A., Wang, C.Y., Liao H.Y.M.: YOLOv4: Optimal speed and accuracy of object detection. arXiv preprint arXiv: 2004.10934 (2020)
9. Wang, C., et al.: CSPNet: a new backbone that can enhance learning capability of CNN. In: 2020 IEEE/CVF Conference on Computer Vision and Pattern Recognition Workshops (CVPRW), pp. 1571–1580 (2020)
10. Zheng, Z., Wang, P., Liu, W., et al.: Distance-IoU loss: faster and better learning for bounding box regression. In: AAAI, pp. 12993–13000 (2020)
11. Ren, S., He, K., Girshick, R., Sun, J.: Faster R-CNN: towards real-time object detection with region proposal networks. J. IEEE Trans. Pattern Anal. Mach. Intell. **39**(06), 1137–1149 (2017)
12. Girshick, R.: Fast R-CNN. In: 2015 IEEE International Conference on Computer Vision (ICCV), pp. 1440–1448 (2015)
13. Girshick, R., Donahue, J., Darrell, T., Malik, J.: Rich feature hierarchies for accurate object detection and semantic segmentation. In: 2014 IEEE Conference on Computer Vision and Pattern Recognition, pp. 580–587 (2014)
14. Xiang, K., Li, S.S., Luan, M.H., et al.: Aluminum product surface defect detection method based on improved faster RCNN. J. Chin. J. Sci. Instrument **42**(01), 191–198 (2021)

# Event-Triggered Output Feedback Predictive Control of Networked Control Systems Against Denial-of-Service Attacks

Zhiwen Wang[1,2,3](✉), Caihong Liang[2], Xiaoping Wang[3], and Hongtao Sun[4]

[1] Key Laboratory of Gansu Advanced Control for Industrial Processes, Lanzhou University of Technology, Lanzhou, China
[2] College of Electrical and Information Engineering, Lanzhou University of Technology, Lanzhou, China
[3] National Demonstration Centre for Experimental Electrical and Control Engineering Education, Lanzhou University of Technology, Lanzhou, China
[4] College of Engineering, QuFu Normal University, Qufu, China

**Abstract.** This paper studies the event-triggered output feedback predictive control problem of networked control systems (NCSs) subject to Denial-of-Service (DoS) attacks. When the limited energy of attackers is taken into account, the impact of DoS attacks is reasonably assumed to be bounded consecutive packet dropouts. Then, a new-type predictive control sequence is designed by only employing successfully received triggering data at the last time. Furthermore, based on the switching Lyapunov functional approach and linear matrix inequality, the stability criterion and predictive control design are derived in detail. Compared with previous research, when only the latest successfully received output measurements are available, this event-based predictive control strategy can compensate arbitrary bounded packet dropouts under DoS attacks. In addition, a simulation example shows the effectiveness of the proposed control method.

**Keywords:** NCSs · Event-triggered predictive control · Packet dropouts · DoS attacks

## 1 Introduction

With the rapid development of network communication technology, the network control system has been widely concerned by people in the past decades. At present, it has been widely used in unmanned aerial vehicle, spacecraft, smart grid and other fields [1, 2]. Generally speaking, when the sensor or communication network in the network control system is attacked maliciously, the network control system will not be able to control according to the normal system state [3]. In recent years, denial of service attack, as a typical malicious attack means, has been widely used in network systems, and its main way is to block the normal communication of information [4]. Therefore, it is of great significance to study the network control system to meet the information transmission


Q. Han et al. (Eds.): LSMS 2021/ICSEE 2021, CCIS 1469, pp. 152–163, 2021.
https://doi.org/10.1007/978-981-16-7213-2_15

requirements specified in the task and to know how to effectively save communication resources [5]. In addition, considering the communication resource constraints in NCSs, event-triggered communication and control strategies are widely used. Unlike traditional time-triggered strategies, event-triggered strategies can greatly reduce the number of information transmission according to a given event-triggered mechanism, thus improving the utilization of communication resources [6, 7].

The DoS attack has been extensively concerned by many scholars for its immense destructive effects in industrial control systems, such as [8, 9]. In general, the model of DoS attack is established as bounded packet dropouts in discrete systems, the stability result of losses in the transmission of packets occurred in the forward and backward channels is transformed into the robust stability in [10–12]. Furthermore, also in the control of discrete-time constrained nonlinear systems, the random packet dropouts are modeled by a two-state Markov chain and settled unconstrained model predictive control scheme in [13]. Moreover, event-based approach is applied to NCSs with limited communication resources and achieves the purpose of saving network bandwidth in [14–16]. Because of the event-triggered condition which is a function of the instantaneous output of the system excessively raised in [17]. It is necessary to continuously monitor the system output information, which will undoubtedly increase the hardware cost. In addition, the design of an ideal dynamic output feedback controller (DOFC) used before has a certain limitation in solutions which is difficult to implement. The above defects motivate us to introduce the way of a discrete event-triggered mechanism that only needs to place emphasis at the moment of data transmission and does not require extra hardware support in the context of network attacks. Summarized the following two main contributions:

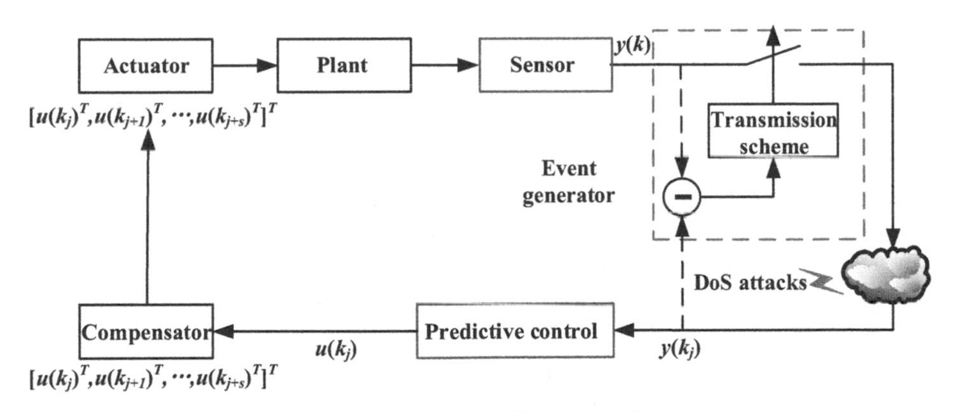


**Fig. 1.** Structure of the NCSs under DoS attack.

1) Aiming at NCSs, a new-type switching system is established, which has good security performance under DoS attacks. Considering that not all the states are available in the practical engineering, it is desirable to make full use of the only output measurements to construct safety controller.

2) Designing the event-triggered predictive control strategy by the latest received triggering data. Different from the traditional model predictive control methods, this way will greatly reduce the dependence on real-time output measurements, thus saving the limited resources of networks while maintaining the desired control performance.

## 2 System Description

The event-based predictive control structure is shown in Fig. 1.Consider the following discrete linear time-invariant system:

$$\begin{cases} x(k+1) = Ax(k) + Bu(k) \\ y(k) = Cx(k) \end{cases}. \tag{1}$$

where $x(k) \in R^n$ is the system state of the controlled plant, $u(k)$ is the control input, and $y(k) \in R^m$ is the measurement output. $A$,$B$, and $C$ are known real constant matrices with appropriate dimensions. It is naturally assumed that the pair $(A, B)$ is controllable and the initial condition for system (1) is given by $x(0) = x_0$.

### 2.1 Event Generator

For the convenience of event-triggered scheme description, we make some assumptions:

*Assumption 1:* In all the data transmissions, each datum has a time stamp.

*Assumption 2:* The sensor is time-driven. It transmits the sampling data to the event trigger with $h > 0$ as the sampling period.

It is supposed that the data packet is collected by the sensor at each sampling instant $kth$, its sampling sequence is described by set $S_1 = \{y_0, y_1, y_2, \cdots, y_k\}$, $k \in R^+$ and $k \to \infty$.

The sampling data packet collected by the sensor will be selectively transmitted to the controller when the pre-designed triggering condition is met. These triggered sampling sequences can be recorded as set $S_2 = \{y_0, y_{k_1}, y_{k_2}, \cdots, y_{k_j}\}$, $k_j \in R^+$ and $\lim\limits_{k\to\infty} k_j = \infty$. It is clear that set $S_2 \subseteq S_1$, which not all datum is sent to the controller, but the packet is judged by the triggering condition.

In order to make full use of communication resources, an event generator is designed to determine whether the data collected by the sensors need to be released to the controller. It consists of a register and a comparator: the former stores the last successful packet sent $y(k_j)$, and the latter determines whether or not the sampled signal $y(k)$ is transmitted to the controller, Clearly, the next triggering release instants of the event generator are denoted by

$$k_{j+1} = k_j + \min_{k\in N, k>k_j} \{k|e^T(k)\Phi e(k) \geq \delta y(k_j)^T \Phi y(k_j)\}. \tag{2}$$

where $\delta > 0$ represents the given triggering parameter. $\Phi$ is the positive definite matrix of appropriate dimension.$e(k)$ is the error between the output measurement $y(k)$ at the current sampling time and $y(k_j)$ as the last triggering update time.

$$e(k) = y(k) - y(k_j). \tag{3}$$

*Remark 1:* The triggering parameter $\delta$ determines the transmission rate of the data packet, that is, the larger the $\delta$ is, the lower the data transmission and the less network bandwidth it consumes.

## 2.2 Predictive Control

*Assumption 3:* The maximum consecutive packet drop- outs caused by DoS attacks, which are denoted by N, is bounded.

Assuming that the trigger state value received by the controller at time instant $k_j$ is $y_{k_j}$, according to the output feedback control law, the corresponding predictive sequences is

$$u(k_{j+s}) = G_s y(k_j). \tag{4}$$

where $k_j$ denotes the instant switching time, $s = 0, 1, 2, \cdots, \sigma_{k_j}$ is a time-varying switching signal, taking the value in a finite subset $s \in Z \triangleq \{0, 1, 2, \cdots, N\}$.

The state $y(k_j)$ is forwarded to the controller, which is packetized together as a set of future control predictions and it sends out the compensator side through the forward channel. Then, the appropriate control value from the event-triggered predictive control sequences $U(k_j) = [u(k_j)^T, u(k_{j+1})^T, u(k_{j+2})^T, \cdots, u(k_{j+N})^T]^T$ is used as the actual control input of the plant is employed to compensate the packet dropouts caused by DoS attacks.

## 2.3 Model Establishment

The output-based control law (4) is applied to the system (1), and the dynamic evolution of the switching system with adjacent trigger time can be described by the following cases:

*Case 1*: DoS-free case

$$\begin{aligned} x(k+1) &= (\mathrm{A} + \mathrm{BG}_0\,\mathrm{C})x(k) - \mathrm{BG}_0\,\mathrm{C}e(k) \\ &= \Psi_0 x(k) - \mathrm{BG}_0\,\mathrm{C}e(k). \end{aligned} \tag{5}$$

where $\Psi_0 = \mathrm{A} + \mathrm{BG}_0\,\mathrm{C}$, $k \in [\mathrm{k}_j,\ \mathrm{k}_{j+1})$ and $\forall j \in N$.

*Case 2*: One-step horizon is jammed by DoS attacks

$$x(k+1) = \Psi_0 x(k). \tag{6}$$

where $k = k_{j+1}$, $e(k) = 0$.

$$\begin{aligned} x(k+1) &= A\Psi_0 x(k) + BG_1 C x(k_j) = (A\Psi_0 + BG_1 C)x(k) - BG_1 Ce(k) \\ &= \Psi_1 x(k) - BG_1 Ce(k). \end{aligned} \tag{7}$$

where $\Psi_1 = A\Psi_0 + BG_1C$, $k \in [k_{j+1}, k_{j+2})$.

*Case 3*: N-steps horizon is jammed by DoS attacks

$$x(k+1) = \Psi_0 x(k). \tag{8}$$

where $k = k_{j+1}$, $e(k) = 0$.

$$x(k+1) = \Psi_1 x(k). \tag{9}$$

where $k = k_{j+2}$, $e(k) = 0$.

$$\begin{aligned} x(k+1) = A\Psi_{N-1}x(k) + BG_N Cx(k_j) &= (A\Psi_{N-1} + BG_N C)x(k) - BG_N Ce(k) \\ &= \Psi_N x(k) - BG_N Ce(k). \end{aligned} \tag{10}$$

where $\Psi_N = A\Psi_{N-1} + BG_N C$, $k \in [k_{j+N}, k_{j+N+1})$.

According to (6)–(10), the following bounded switching control system model subject to DoS attacks is introduced:

$$x(k+1) = \Psi_{\sigma(k_j)} x(k) - BG_{\sigma(k_j)} Ce(k). \tag{11}$$

where $\sigma(k_j) \in \{0, 1, 2, \cdots, N\}$ stands for triggering packet dropout caused by DoS attacks.

## 3 Stability Analysis

**Definition 1**: *If there are positive scalars* c *and* $\lambda > 1$*such that the following inequality*

$$\|x(k)\| \le c\lambda^k \|x(0)\|.$$

*holds, the NCS (1) is said to be exponentially stable, where* $x(0) \in R^n$*is an arbitrary initial value.*

**Lemma 1**:[18] *For an arbitrary matrix* $\Psi \in \mathrm{R}^{n\times n}$*and an arbitrary vector* $x \in R^n$*, the following inequality*

$$\lambda_{\min}\|x\| \le \|\Psi x\| \le \lambda_{\max}\|x\|.$$

*holds, in which* $\lambda_{\max}$*and* $\lambda_{\min}$*are the maximum singular value and the minimum singular value of matrix.*

**Theorem 1**: *For some given scalars* $0 < \lambda_i < 1$, $\mu > 0$ *and* $\forall i \in Z$*, if there are matrices* $P_i$, $\Phi$*, the following inequalities*

$$\begin{bmatrix} \Xi_{11} & * & * \\ \Xi_{21} & \Xi_{22} & * \\ \Xi_{31} & \Xi_{32} & \Xi_{33} \end{bmatrix} < 0. \tag{12}$$

$$P_\alpha < \mu P_\beta, \forall \alpha, \beta \in Z. \tag{13}$$

$$\varepsilon = \mu\overline{\lambda}. \tag{14}$$

*Where*

$$\Xi_{11} = -\lambda_i P_i + \delta C^T \Phi C,\ \Xi_{21} = -\delta\Phi C,\ \Xi_{22} = -\Phi + \delta\overline{\Phi},$$
$$\Xi_{31} = P_i\Psi_i,\ \Xi_{32} = -P_i BG_i C,\ \Xi_{33} = -P_i$$

*hold, then under arbitrary switching laws the system* (11) *is exponentially stable and the decay rate is* $\sqrt[2NM]{\varepsilon}$.

*Proof*: The following Lyapunov function is selected as candidate for the switching system model (11) candidate as

$$V_{\sigma(k_j)}(k) = x^T(k)P_{\sigma(k_j)}x(k). \tag{15}$$

where $P > 0$ is a positive definite matrix.

The inequality (2) is taken into account by taking the difference equations along the trajectory of NCSs (11) considered. It was inferred from that

$$\begin{aligned} \Delta V_{\sigma(k_j)}(k) &= V_{\sigma(k_j)}(k+1) - V_{\sigma(k_j)}(k) \\ \leq x^T(k+1)P_{\sigma(k_j)}x(k+1) &- \lambda_{\sigma(k_i)}x^T(k)P_{\sigma(k_j)}x(k) - e^T(k)\Phi e(k) + \delta y(k_j)^T\Phi y(k_j) \\ &= [\Psi_N x(k) - BG_N Ce(k)]^T P_{\sigma(k_j)}[\Psi_N x(k) - BG_N Ce(k)] \\ -\lambda_{\sigma(k_i)}x^T(k)P_{\sigma(k_j)}x(k) &- e^T(k)\Phi e(k) + \delta\{[y(k) - e(k)]^T\Phi[y(k) - e(k)]\}. \end{aligned} \tag{16}$$

Then, the above inequality (16) can be written as

$$\Delta V_{\sigma(k_j)}(k) = \begin{bmatrix} x^T(k)\ e^T(k) \end{bmatrix}\Xi\begin{bmatrix} x(k) \\ e(k) \end{bmatrix}. \tag{17}$$

Where

$$\Xi = \begin{bmatrix} \Xi_{11} & * \\ \Xi_{21} & \Xi_{22} \end{bmatrix}. \tag{18}$$

$$\Xi_{11} = \Psi^T_{\sigma(k_j)}P_{\sigma(k_j)}\Psi_{\sigma(k_j)} - \lambda_{\sigma(k_j)}P_{\sigma(k_j)} + \delta C^T\Phi C,$$
$$\Xi_{21} = -(BG_{\sigma(k_j)}C)^T P_{\sigma(k_j)}\Psi_{\sigma(k_j)} - \delta\Phi C$$
$$\Xi_{22} = (BG_{\sigma(k_j)}C)^T P_{\sigma(k_j)}BG_{\sigma(k_j)}C - \Phi + \delta\Phi.$$

Decompose (16) and use Shure's complement theorem that

$$\begin{bmatrix} -\lambda_{\sigma(k_j)}P_{\sigma(k_j)} + \delta C^T\Phi C & * & * \\ -\delta\Phi C & -\Phi + \delta\Phi & * \\ P_{\sigma(k_j)}\Psi_{\sigma(k_j)} & -P_{\sigma(k_j)}BG_{\sigma(k_j)}C & -P_{\sigma(k_j)} \end{bmatrix} < 0. \tag{19}$$

Thus, for $k \in [k_j, k_{j+\sigma(k_j)}]$, the following relationship under triggering strategy is guaranteed to be held: $V_{\sigma(k_j)}(k+1) < \lambda_{\sigma(k_j)}V_{\sigma(k_j)}(k)$.

Further, considering its internal switching behavior, we have that

$$V_{\sigma(k_j)}(k) < \lambda_{\sigma(k_j)} V_{\sigma(k_j)}(k_{\sigma(k_j)}) < \cdots < \lambda_{\sigma(k_j)} \lambda_{\sigma(k_{j-1})} \lambda_{\sigma(k_{j-2})} \cdots \lambda_0 V_{\sigma(k_j)}(k_{\sigma(k_j)})$$
$$< \mu \lambda_{\sigma(k_j)} \lambda_{\sigma(k_{j-1})} \cdots \lambda_0 V_{\sigma(k_{j-1})}(k_{\sigma(k_{j-1})}) < \cdots < \mu^{k_j} \lambda_{\sigma(k_j)} \lambda_{\sigma(k_{j-1})} \cdots \lambda_0 \lambda_{\sigma(k_j)} \lambda_{\sigma(k_{j-1})} \cdots \lambda_0 V_{\sigma(k_0)}(k_{\sigma(k_0)}). \tag{20}$$

Define

$$\overline{\lambda} = \lambda_{\sigma(k_j)} \lambda_{\sigma(k_{j-1})} \cdots \lambda_0, \; \varepsilon = \mu \overline{\lambda} < 1. \tag{21}$$

Utilizing (20) and (21) together leads to

$$V_{\sigma(i)}(k) < \varepsilon V_{\sigma(0)}(k_0). \tag{22}$$

By Lemma 1, the following inequality for $\forall \sigma(k_j) \in Z$ will be true that

$$\varsigma = \max\{\sigma(k_j) \in Z | \lambda_{\max}(\Psi_{\sigma(k_j)})\}. \tag{23}$$

$$\eta_1 = \max\{\sigma(k_j) \in Z | \lambda_{\max}(P_{\sigma(k_j)})\}, \eta_2 = \min\{\sigma(k_j) \in Z | \lambda_{\max}(P_{\sigma(k_j)})\}, \tag{24}$$

Meanwhile, define $\chi = \sqrt{\frac{\eta_1}{\eta_2}}$. Therefore

$$\left\| x(k_j) \right\| < \chi \sqrt{\varepsilon}^{k_j} \| x(0) \|. \tag{25}$$

where $k = k_j$ indicates that the real state is adopted to deliver the current controller. Otherwise, the predictive state is employed to compensate the packet dropouts when $k = k_{j+\sigma(k_j)}$. In similar way, we can derive:

$$\left\| x(k_{j+\sigma(k_j)}) \right\| < \varsigma \chi \sqrt{\varepsilon}^{k_j} \| x(0) \|. \tag{26}$$

Meanwhile, $k$, $k_j$ and $k = k_{j+\sigma(k_j)}$ have the following relationships:

$$k < k_j NM, k \le k_{j+\sigma(k_j)} NM \le k_{j+\sigma(k_j)+1} NM. \tag{27}$$

where $M \ge k_{j+1} - k_j$, $M \in N^*$ denotes the upper bound of neighbouring triggering time.

Combining (25) and (27) that

$$\left\| x(k_j) \right\| < \chi \sqrt[2NM]{\varepsilon}^k \| x(0) \|, \left\| x(k_{j+\sigma(k_j)}) \right\| < \varsigma \chi \sqrt[2NM]{\varepsilon}^k \| x(0) \|. \tag{28}$$

It can be issued from (28) that the following inequalities (29) are satisfied for arbitrary instantaneous switching time $k$ after N-steps packet dropouts caused by DoS attacks.

$$\| x(k) \| < \varsigma \chi \sqrt[2NM]{\varepsilon}^k \| x(0) \|. \tag{29}$$

This completes the proof.

## 4 Controller Design

According to Theorem 1, the output predictive control sequences based on event trigger under DoS attack will be derived.

**Theorem 2**: *For given scalars $0 < \lambda_i < 1$, $\mu > 0$ and $\forall i \in Z$, if there exist positive definite matrices $\overline{P}_i$, $\overline{\Phi}$such that the following inequalities*

$$\begin{bmatrix} \tilde{\Xi}_{11} & * & * \\ \tilde{\Xi}_{21} & \tilde{\Xi}_{22} & * \\ \tilde{\Xi}_{31} & \tilde{\Xi}_{32} & \tilde{\Xi}_{33} \end{bmatrix} < 0. \quad (30)$$

*Where*

$$\tilde{\Xi}_{11} = -\lambda_i X_i + \delta \overline{Z},\ \tilde{\Xi}_{21} = -\delta \overline{R},\ \tilde{\Xi}_{22} = -\overline{\Phi} + \delta \overline{\Phi},$$

$$\tilde{\Xi}_{32} = -BG_i CX_i,\ \tilde{\Xi}_{33} = -X_i,\ \overline{A}_i = A^{i+1} + \sum_{l=0}^{i} A^l BG_{i-l} C$$

*holds, then under arbitrary switching laws, the system (11) is exponentially stable and the decay rate is $\sqrt[2NM]{\varepsilon}$.*

*Proof*: Define $X = P^{-1}$, $\Phi = X\Phi X$, $C^T \Psi C = Z$, $\overline{Z} = XZX$, $\Phi C = R$, $\overline{R} - YRX$, then pre- and post-multiplying the inequality (12) by $diag[X, X, X]$ in Theorem 1, we arrive at Theorem 2.

However, because of the coupling non-liner items $BGCX$, the above Theorem 2 still cannot be solved directly. The following theorem is presented to solve such items by LMIs.

**Theorem 3**: *For given scalars $0 < \lambda_i < 1$, $\mu > 0$, $\xi > 0$and $\forall i \in Z$, if there exist positive definite matrices $\overline{P}_i$, $\overline{\Phi}$, full rank matrix M and aleatoric appropriate dimension matrix Qsuch that the following inequalities*

$$\begin{bmatrix} \tilde{\Xi}'_{11} & * & * \\ \tilde{\Xi}'_{21} & \tilde{\Xi}'_{22} & * \\ \tilde{\Xi}'_{31} & \tilde{\Xi}'_{32} & \tilde{\Xi}'_{33} \end{bmatrix} < 0, \begin{cases} \begin{bmatrix} -\xi I & * \\ MC - CX & -I \end{bmatrix} < 0 \\ \xi \to 0 \end{cases}. \quad (31)$$

*Where*

$$\tilde{\Xi}'_{11} = -\lambda_i X_i + \delta \overline{Z},\ \tilde{\Xi}'_{21} = -\delta \overline{R},\ \tilde{\Xi}'_{22} = -\overline{\Phi} + \delta \overline{\Phi}$$

$$\tilde{\Xi}'_{31} = \overline{A}_i X_i + BQ_i C,\ \tilde{\Xi}'_{32} = -BQ_i C,\ \tilde{\Xi}'_{33} = -X_i,\ \overline{A}_i = A^{i+1} + \sum_{l=0}^{i} A^l BQ_{i-l} C.$$

*hold, then under arbitrary switching laws, the addressed NCS* (11) *controlled by $G = QM^{-1}$is exponentially stable and the decay rate is $\sqrt[2QM]{\varepsilon}$.*

*Proof*: It is known from (31) in Theorem 3 that $C = M^{-1}CX$. Then, replacing $GCX$ with $QC$, we can easily get the above result $G = QM^{-1}$, so as to complete this proof.

## 5 Illustrative Simulation

In this part, proposing an example to verify the results under the event-triggered strategy. We consider an inverted pendulum system whose plant model is given by the following formula. The parameters of the inverted pendulum system are listed in Table 1.

$$\begin{cases} u = M\dfrac{d^2y}{dt^2} + m\dfrac{d^2}{dt^2}(y + l\sin\phi) \\ mgl\sin\phi = m\dfrac{d^2}{dt^2}(y + l\sin\phi) \cdot l\cos\phi \end{cases}. \tag{32}$$

The state variables of the system are defined on the basis of the above inequality.

$$x_1 = y, x_2 = \phi, x_3 = \dot{y}, x_4 = \dot{\phi}.$$

Let the sampling period $T = 0.01$ s, then the discrete model of the system is given as

$$\begin{cases} x(k+1) = Ax(k) + Bu(k) \\ y(k) = Cx(k) \end{cases}. \tag{33}$$

Where

$$A = \begin{bmatrix} 0\,1 & 0 & 0 \\ 0\,0 & 0 & 0 \\ 0\,0 & 0 & 1 \\ 0\,0 & 29.43 & 0 \end{bmatrix}, B = \begin{bmatrix} 0\,1\,0\,3 \end{bmatrix}^T, C = \begin{bmatrix} 1 & 0 & 0 & 0 \\ 0 & 0 & 1 & 0 \end{bmatrix}$$

**Table 1.** Parameters of the inverted pendulum system.

| Heading level | Example | Font size and style |
|---|---|---|
| M | Mass of the cart | 1.378 kg |
| l | Length of the pendulum | 0.25 m |
| g | Acceleration of gravity | $9.8\,\text{m/s}^2$ |
| $\phi$ | At an angle to the upright position | – |
| u(t) | Displacement of the cart | – |
| y(t) | Force acted on the cart | – |

By selecting $N = 3$ and applying Theorem 3 with $\delta = 0.85$, $\mu = 1$ and $\lambda_0 = \lambda_1 = \lambda_2 = \lambda_3 = 0.9$, the corresponding control gain $G$ and triggering matrix $\Phi$ can be gained:

$$\Phi = \begin{bmatrix} 1.81 & 1.10 \\ 1.10 & 0.73 \end{bmatrix}, G = \begin{bmatrix} -4.92 & 1.35 \end{bmatrix}$$

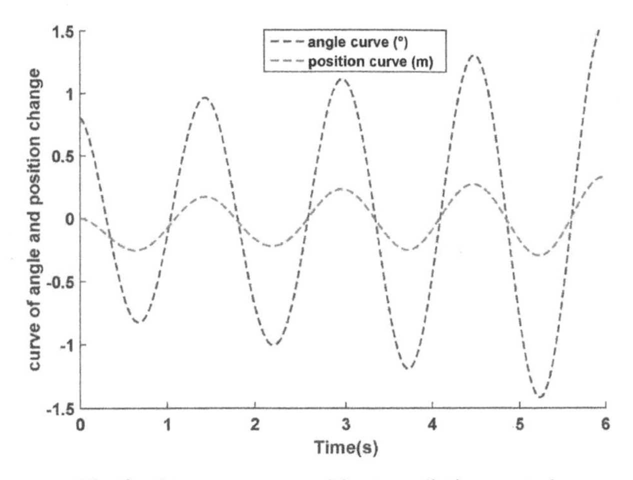


**Fig. 2.** State responses without predictive control.

In the following, $x_0 = [\,8\ 0\ 0\ 0\,]^T$ is the initial condition, the simulation time is. chosen as $t \in [0, 6][0, 6]$. When there is no predictive control strategy (using initial control gain without DoS case), the state responses of the system under the worst attacks are depicted in Fig. 2.

Note that in order to compare the angle curve and position curve more intuitively, the angle value in the following Fig. 2 and Fig. 3 has been reduced by one tenth.

The effect of using predictive control with the event-triggered communication scheme is presented in Fig. 3. From the angle and position curve, it can be seen that this proposed method is effective, which can guarantee that the system with limited packet dropouts is stability. It is found that the system performance is better than the one without predictive control.

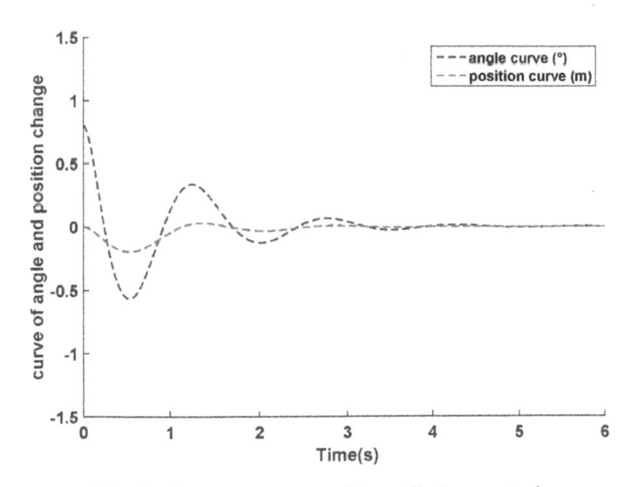


**Fig. 3.** State responses with predictive control.

Furthermore, the statistics in Fig. 4 show that 25 packets are transmitted with an average period of 0.2180 s. These results reveal that the proposed strategy can reduce the transmission data while maintaining the control performance.

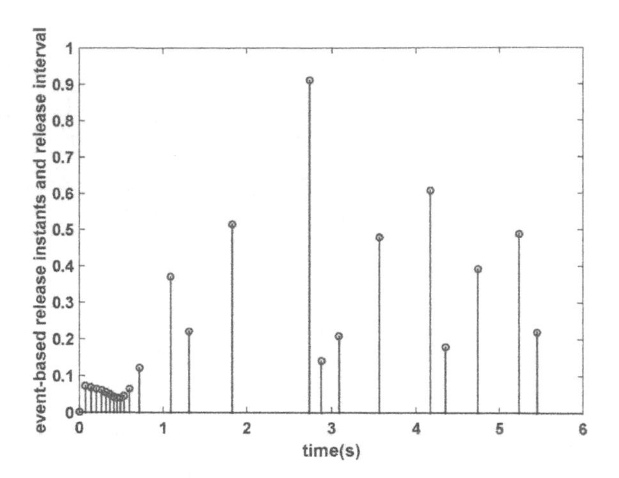


**Fig. 4.** Release instants and intervals with DoS attacks.

## 6 Conclusion

This paper proposed a new-type predictive control strategy for the packet dropouts under the DoS attacks. Firstly, a linear discrete switched system model is established to describe the performance of the NCSs under DoS attacks. Secondly, the stability criterion of the system is derived by LMIs, and then the design method of predictive control sequence is given according to the stability condition of the system. Finally, the simulation results verify the effectiveness of the event-triggered output feedback predictive control method.

## References

1. Gupta, R.A., Chow, M.Y.: Networked control system: overview and research trends. J. IEEE Trans. Ind. Electron. **57**(7), 2527–2535 (2010)
2. Chen, R., Du, D., Fei, M.: A Novel Data Injection Cyber-Attack Against Dynamic State Estimation in Smart Grid. In: Li, K., Xue, Y., Cui, S., Niu, Q., Yang, Z., Luk, P. (eds.) LSMS/ICSEE -2017. CCIS, vol. 763, pp. 607–615. Springer, Singapore (2017). https://doi.org/10.1007/978-981-10-6364-0_61
3. Sundaram, S., Hadjicostis, R.: Distributed function calculation via linear iterative strategies in the presence of malicious agents. J. IEEE Trans, Autom. Control **56**(7), 1495–1508 (2011)
4. Amin, S., Cárdenas, A.A., Sastry, S.S.: Safe and secure networked control systems under denial-of-service attacks. In: Majumdar, R., Tabuada, P. (eds.) HSCC 2009. LNCS, vol. 5469, pp. 31–45. Springer, Heidelberg (2009). https://doi.org/10.1007/978-3-642-00602-9_3
5. Chen, R.: Optimal co-design of control algorithm and bandwidth scheduling for networked control systems. J. Adv. Mater. Res. **314–316**, 2124–2131 (2011)

6. Wei, G., Wang, Z., Alsaadi, F.E., Ding, D.: Event-based security control for discrete-time stochastic systems. J. IET Control Theory Appl. **10**(15), 1808–1815 (2016)
7. Teixeira, A., Perez, D., Sandberg, H., Johansson, K.H.: Attack models and scenarios for networked control systems. In: Proceedings of 1st International Conference on High Confidence Networked Systems, Beijing, China (2012)
8. Lin, H., Antsaklis, P.J.: Stability and persistent disturbance attenuation properties for a class of networked control systems: switched system approach. J. Int. J. Control **78**(18), 1447–1458 (2005)
9. Chen, J., Shi, L., Cheng, P., Zhang, H.: Optimal denial-of-service attack scheduling with energy constraint. J. IEEE Trans. Autom. Control **60**(11), 3023–3028 (2015)
10. Cao, R., Wu, J., Long, C., Li, S.: Stability analysis for networked control systems under denial-of-service attacks. In: Proceedings of the 54th IEEE Conference on Decision and Control, Osaka, Japan (2015)
11. Long, M., Wu, C.H., Hung, J.Y.: Denial of service attacks on network-based control systems: impact and mitigation. J. IEEE Trans. Ind. Inform. **1**(2), 85–96 (2015)
12. Ding, B.: Stabilization of linear systems over networks with bounded packet loss and its use in model predictive control. J. Automatica **47**(11), 2526–2533 (2011)
13. Reble, M., Quevedo, D.E., Allgwer, F.: Control over erasure channels: Stochastic stability and performance of packetized unconstrained model predictive control. J. Int. J. Robust Nonlinear Control **23**(10), 1151–1167 (2013)
14. Fu, W., Yang, S.X., Huang, C., Liu, G.: Predictive triggered control for networked control systems with event-triggered mechanism. Clust. Comput. **22**(4), 10185–10195 (2017). https://doi.org/10.1007/s10586-017-1210-z
15. Zhang, W., Yu, L.: Output feedback stabilization of networked control systems with packet dropouts. J. IEEE Trans. Autom. Control **52**(9), 1705–1710 (2007)
16. Sun, X., Wu, D., Wen, C., Wang, W.: A novel stability analysis for networked predictive control systems. J. IEEE Trans. Circ. Syst. II Express Briefs **61**(6), 453–457 (2014)
17. Lai, S., Chen, B., Lai, T., Yu, L.: Packet-based state feedback control under DoS attacks in cyber-physical systems. J. IEEE Trans. Circ. Syst. II Express Briefs **66**(8), 1421–1425 (2018)
18. Zhang, J., Peng, C.: Event-triggered $H_\infty$ filtering for networked Takagi-Sugeno fuzzy systems with asynchronous constraints. J. IET Signal Process. **9**(5), 403–411 (2015)

# Compression of YOLOv3-spp Model Based on Channel and Layer Pruning

Xianxian Lv[1,2] and Yueli Hu[1,2(✉)]

[1] Shanghai Key Laboratory of Power Station Automation Technology, Shanghai 200444, China
huyueli@shu.edu.cn

[2] School of Mechatronic Engineering and Automation, Shanghai University, Shanghai 200444, China

**Abstract.** Hands is an important medium for human-computer interaction, and it is important for computers to detect human hands in real time. However, the detection algorithm based on deep learning has complex network and requires enormous computation power therefore cannot be deployed to mobile terminals with limited power and memory. To solve this issue, a pruning algorithm combining channel and layer pruning is proposed to compress the network model. The method applies L1 regularization to the scaling factor of BN layer to enhance the network sparsity, and then prunes the channel with smaller scaling factor. Specifically, for a shortcut layer, the previous DBL is evaluated, the $\gamma$ value of each layer is sorted, and the smallest layer is removed. Two groups of experiments were carried out. One compared the effect of the channel pruning and combination of channel and layer pruning, respectively. The other group compared the effect of choosing different thresholds. The results showed that the size of the original model was reduced by 97%, while its inference speed is increased by 1.7 times. Meanwhile, the pruning barely caused loss of precision. Thus, the proposed approach enables hand detection models to be deployed in mobile terminals and make such algorithms more practical.

**Keywords:** YOLOv3-spp · Hand detection · Model pruning

## 1 Introduction

With the rapid development of machine vision and deep learning, object detection technology is gradually applied to all aspects of people's lives, such as security, automatic driving and industrial manufacturing. Among them, human hands become an important way of human-computer interaction, many scenes need computer to realize the detection of human hand [1, 2] rapidly.

Convolutional neural network, as a kind of deep neural network, is superior to traditional algorithms in image classification, object detection [3, 4], pose detection and other computer vision tasks, with higher detection accuracy and faster detection speed. Although the increase of network layers and network structure complexity has greatly improved the accuracy of the algorithm, the drawback is that the number of parameters


Q. Han et al. (Eds.): LSMS 2021/ICSEE 2021, CCIS 1469, pp. 164–173, 2021.
https://doi.org/10.1007/978-981-16-7213-2_16

grows exponentially and the algorithm model is very large. Ordinary CPU has been unable to meet the computational requirements of deep network, so GPU is usually used for training. However, the embedded platform, mobile phone, vehicle, FPGA and other mobile terminals have some limitations of computing power and memory, and the problem of memory bandwidth. Therefore, how to reduce float point operations and training parameters, accelerate forward inference, compress deep learning network model and apply it to resource constrained embedded devices without reducing the detection accuracy has become a new challenge.

The feature expression of general deep learning network is scattered in each layer and each parameter, so deep learning network has certain redundancy in structure and calculation, and the model is usually in a state of over parameter, which is also the premise of deep learning network acceleration and optimization.

Pruning is a method of deep neural network acceleration and compression. It is to cut off some unimportant connections in the neural network. In this way, the complexity of the whole network and the size of the network model will be greatly reduced. For unstructured pruning, the classic three-stage pruning method proposed in NIPS2015 [5], it firstly trains a full precision network, then removes some unimportant nodes, and then trains weights. In ICLR2016, Stanford University [6] proposed a random pruning method called deep compression. However, the random pruning method is very hardware unfriendly. Yoon et al. [7] used the group sparsity method to add sparse regularization to group features to prune some columns of weight matrix, and then used exclusive sparsity enhances the competitiveness of features between different weights to learn more effective filters, and they work together to achieve good pruning results. In an article of ICCV2017, Liu et al. [8] added a scaling factor to each channel, then added sparsity regulation to these scaling factors, and cut off some channels with smaller scaling factor, so as to achieve the slimming effect on the whole network. Sun et al. [9] in ICML2017 analyzed the gradient information in the network training process, and simplified the network back propagation process by removing the gradient with small amplitude, so as to speed up the network training process.

After pruning, it will be more compact in the network model size with little loss precision in detection effect, which provides theoretical support for the deployment of algorithms on mobile terminals.

## 2 YOLOv3-spp Detection Network

YOLO series algorithms [10–12] can be said to be a masterpiece in the history of target detection. As shown in the Table 1 is the performance comparison of various official versions of YOLO on the COCO dataset. It can be seen that the YOLOv3-spp version has improved the mAP by several percentage points compared to the previous versions, both in the speed and accuracy. So this paper mainly chooses YOLOv3-spp network as the research object of model compression.

In 2018, Redmon [11] put forward YOLOv3 algorithm. First it is characterized by the adoption of a new backbone network called Darknet in YOLOv3, and the use of shortcut layer improves the feature extraction ability of the backbone network; Secondly, it uses multiple feature mAPs for prediction, and the accuracy is significantly improved for small targets; As a category loss, the cross entropy loss function works better when the predicted target category is very complex.

**Table 1.** Performance on the COCO dataset.

| Model | Dataset | mAP | FPS |
|---|---|---|---|
| YOLOv2 | COCO | 48.1 | 40 |
| YOLOv3 | COCO | 55.3 | 35 |
| YOLOv3-tiny | COCO | 33.1 | 220 |
| YOLOv3-spp | COCO | 60.6 | 20 |

YOLOv3 body consists of 252 layers. Most convolution structures are composed of convolution layer + BN layer + LeakyRelu, which is called DBL (Darknetconv2d_BN_Leaky). The DBL is the basic component of YOLOv3, as shown in Fig. 1.

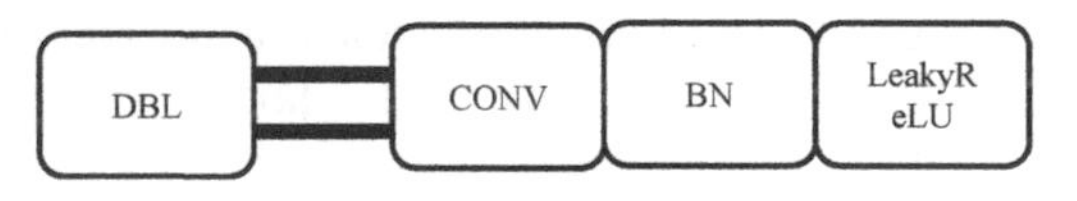


**Fig. 1.** DBL structures.

Compared with the ordinary version of YOLOv3, the spp version adds an spp module between the fifth and sixth convolution layers, which is fully called spatial pyramid pooling structure. This module is composed of four parallel branches, namely the maximum pooling and a concatenation connection with the size of 5 × 5, 9 × 9 and 13 × 13. The module uses the idea of spatial pyramid for reference, and realizes the local feature and global feature through spp module. After the fusion of local feature and global feature, the feature mAP enriches the expression ability of the feature mAP, which is conducive to the situation of large difference in the size of the target in the image to be detected.

## 3 Compression of YOLOv3-spp Model Based on Channel and Layer Pruning

Model pruning can be implemented at different levels, such as weight level [13], core level [14], channel level [15] or layer level. Weight level pruning will have higher flexibility, versatility, and higher compression ratio, but it usually needs to design a special runtime to infer sparse models. On the contrary, layer pruning can run directly under the existing runtime, but it is less flexible. Thus, channel pruning achieves a good balance between flexibility and easy implementation. It can be applied to any typical CNN or fully connected network. After pruning, the model can be trained by using the existing hardware and computing library, and can obtain higher compression ratio and shorter training time.

### 3.1 Overall Plan

We improved the pruning method on Liu et al. [8] basis.The general idea of pruning method can be described as follows: firstly, the scaling factor is introduced for each channel of convolutional layer. And L1 regularization is applied to the scaling factor of BN layer, then these scaling factors are jointly trained to achieve network sparseness. Finally the channels with small scaling factor are pruned and fine-tuned. In particular, the shortcut layer can not prune the channel directly, so it needs to prune the channel separately. Finally, the pruned-yolov3-spp model is obtained by using the channel the layer pruning algorithm. The process of pruning scheme is shown in the Fig. 2.

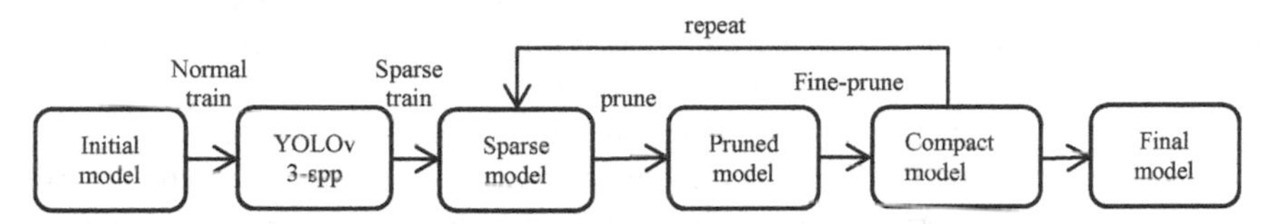


**Fig. 2.** Schematic diagram of pruning process.

**Sparse Training**
The role of sparsity training is to evaluate the importance of each channel. Firstly, a scaling factor λ is introduced for each channel of convolutional network. Then, the weight of the training network and these scaling factors are combined to modify the loss function of the training process:

$$Loss_{prun} = \sum\nolimits_{(x,y)} l(f(x, W), y) + \lambda \sum\nolimits_{\gamma \epsilon \Gamma} g(\gamma). \quad (1)$$

Among them, $Loss_{prun}$ represents the target detection loss of pruned-yoov3-spp based on channel pruning, where (x, y) represents the input image and real label, W represents the weight of the network, $f$ represents the YOLOv3 neural network function. The first sum term is the loss function of CNN's normal training. $\gamma$ is the parameter of all batch normalization layers in the neural network, and g (.) is the penalty term on the scaling factor. We choose g (s) = | s |, namely L1 regularization, which is widely used in sparsification. λ is the balance factor of the two items, which is used to adjust the proportion of object detection loss item and channel scaling factor penalty item in the total loss value. After sparse training, a sparse model can be obtained, in which many scaling factors are close to 0, and the channel pruning can be completed by filtering out the channels with smaller $\gamma$ by threshold.

### 3.2 Using the Scaling Factor of BN Layer

From the learning of YOLOv3 structure in the previous chapter, we can see that YOLOv3 is mainly composed of DBL structure. If the scaling factor is added after the convolution layer, and BN layer is not used, because convolution layer and scaling layer are linear transformations, the overall effect is still linear transformation, so the importance of channel cannot be determined by scaling factor. If the channel factor is added before the BN layer and after the convolution layer, the influence of the channel factor will be eliminated by the nonlinear transformation of the BN layer. If a scaling factor is added behind the BN layer, it is equivalent to two consecutive scaling factors for each channel. The conversion formula from BN layer is as follows:

$$\hat{Z} = \frac{Z_{in} - \mu}{\sqrt{\sigma^2 + \varepsilon}}. \tag{2}$$

$$Z_{out} = \gamma \hat{Z} + \beta. \tag{3}$$

The BN layer has two super parameters $\gamma$ and $\beta$, which $\gamma$ is also called scaling factor, $\beta$ is the bias coefficient, $Z_{in}$, $Z_{out}$ are input and output of BN layer espectively, $\mu$, $\varepsilon$ are the average and standard deviation of the input data.

If $\gamma$ is very small, then the input value to the next layer is very small, which can be ignored. It can also be understood that this channel has little contribution to the network. It also shows that the correlation between scaling factor and convolution channel can be used to evaluate the contribution of convolution channel to the network. Based on the above analysis, the sparse training process directly takes the scaling factor in BN layer as the channel scaling factor for training iteration, which can also save additional network overhead.

### 3.3 Pruning of Shortcut Layer

There are five groups of 23 shortcut connections in YOLOv3, which correspond to the add operation. Pruning can be performed directly between $1 \times 1$ and $3 \times 3$ convolution layers, which is channel pruning. The dimensions of input and output channels of shortcut layer need to be consistent, so pruning cannot be performed directly. If the shortcut is not pruned, the dimension processing problem can be avoided, but the pruning rate is low.

In this paper, we evaluate the DBL before each shortcut layer, rank the $\gamma$ value of each layer, and prune the smallest layer. In order to ensure the integrity of YOLOv3 structure, a shortcut layer and two convolution layers in front of it will be cut at the same time when cutting a shortcut structure. Here, we only consider the shortcut module in the main scissors. There are 23 shortcuts in YOLOv3, with a total of 69 layers of scissors space. In the experiment, we cut out some shortcuts and the accuracy reduce very little.

### 3.4 Fine-Tuning

Whether the pruning effect is good or not depends on the sparsity, and different pruning strategies and threshold settings have different effects with pruning. Sometimes, the accuracy of the model may even rise after pruning, but generally speaking, pruning will damage the accuracy of the model. At this time, it is necessary to fine tune the pruned model to make the accuracy rise. Repeat the above process several times, you can get a more compact model through multi-stage network slimming.

# 4 Experiment and Analysis

In view of the above pruning method, the feasibility is verified by the effect after pruning. The pruned-YOLOv3-spp is obtained by pruning the YOLOv3-spp network with the above pruning method, and is used to detect the human hands.

## 4.1 Selection and Preprocessing of Dataset

In this paper, we select Oxfordhand dataset as the detection dataset. It introduces a comprehensive dataset of hand images collected from various different public image dataset sources. While collecting the data, no restriction was imposed on the pose or visibility of people, nor was any constraint imposed on the environment. In each image, all the hands that can be performed clearly by humans are annotated. There are 4069 pictures in the training set, 738 pictures in the validation set and 821 pictures in the test set.

## 4.2 Experiment Process

The training process of deep neural network has higher requirements on computer hardware, so it needs to call GPU for training, and build the experimental environment on the server, as shown in the Table 2.

**Table 2.** Experimental environment on the server.

| | |
|---|---|
| Operating system | Ubuntu 16.04 |
| GPU | GeForce GTX 1080 Ti *2 |
| CUDA | 10.2 |
| Experimental environment | python3.5 |
| Dependency library | Python numpy opencv Python torch vision Matplotlib pycocotools tqdm tensorboard calculations, etc. |

The experimental steps are as follows:

**Basic Training.** The basic training uses the weight parameters provided on the official website of YOLO as the Initial parameters of network training. The YOLOv3-spp network model is trained on the Oxfordhand dataset, and we set the batch size = 20, epochs = 100.

**Sparsity Training.** First of all, the original model needs to be sparsely trained to obtain a sparse network with partial scaling factor approaching 0, so as to prepare for model pruning. Using batch_Size = 20, epochs = 300, penalty factor scale = 0.001. Large scale generally sparse faster but precision drop faster, small scale generally sparse slower but precision drop slower. This paper uses global constant sparse strategy scale to train YOLOv3-spp network. Other training parameters are the same as normal training.

As shown in the Fig. 3 are the loss curves after basic training and sparse training. The abscissa represents the training epochs, and the ordinate represents the loss.The loss value at the beginning of training decreases greatly, which indicates that the learning rate is appropriate and the gradient decline process is smooth. With the increase of the number of iterations, the loss value gradually flattens and tends to 0, which indicates that the network learning is gradually completed.

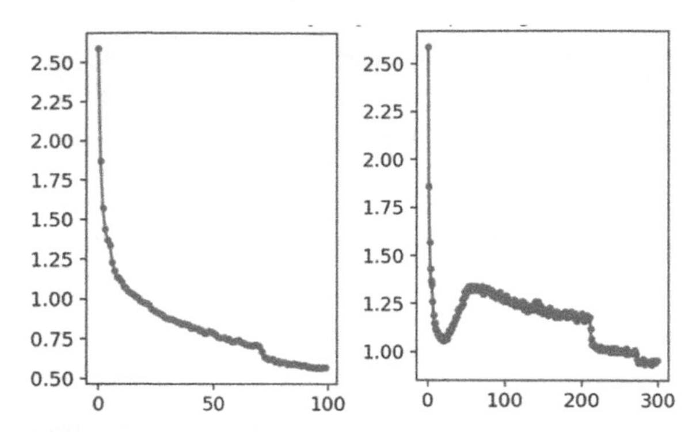


**Fig. 3.** The loss curves after basic training and sparse training and the left one is the loss curve of basic training and the right is the loss curve after sparse training.

**Prune.** In order to verify the performance of pruning algorithm, this paper sets up two groups of comparative experiments.

*The first group is pruning strategy,* which selects channel pruning algorithm and channel and layer pruning algorithm respectively. Add both set the global threshold is 0.85 to verify the pruning effect of different pruning strategies on the network.

*In the second group,* we choose channel and layer pruning algorithm as pruning strategy, but different global thresholds are set for comparison, which are 0, 0.5, 0.7, 0.85, 0.93 and 0.95. In addition, at least 10% of the channels in each convolution layer are reserved, and the whole convolution layer is not removed, so as to verify the pruning effect of different pruning thresholds on the network.

**Fine-Tuning.** Since the model precision drop due to pruning, we make fine-tuning to compensate for the accuracy degradation. After 50 epoch retraining, we get the final pruned-YOLOv3-spp pruning model.

### 4.3 Analysis of Experimental Results

If too many channels are pruned, the model structure will be destroyed, and the detection accuracy will be greatly reduced and the detection function cannot be completed normally; If too few channels are pruned, the network compression effect cannot be achieved. In the experiment, training, pruning and fine-tuning are carried out on Oxford-hand dataset. After different pruning strategies and pruning thresholds, the pruning models with different pruning effects were obtained, and the pruning models were evaluated based on the following indicators: model size, mAP, floating-point operations, inference time and parameters.

**Result of First Comparative Experiment.** Table 3 shows the experimental results of comparing channel and layer pruning algorithm with channel pruning algorithm. The first column is the original model, the second column is channel pruning, and the third column is channel and layer pruning algorithm. The global threshold for pruning is 0.85. Minimum number of channels maintained per layer (layer keep) is 0.01. The cutting number of shortcuts layer is16. The results below show that there are two points missing in the model after clipping the channel (it is found that the fluctuation of mAP up and down 0.02 is a normal, and it can be approximately considered that no loss of accuracy), and the size of model is reduced from 246 M to 33.28 M. After channel and layer pruing, the mAP is reduced to 0.57 and the model size is reduced to 8.05 M, which is 3% of the original, and the model parameters are reduced by 97%.The reasoning speed is reduced from 13.5 ms to 7.9 ms, doubling the speed.

**Table 3.** Results of comparing channel and layer pruning algorithm with channel pruning algorithm.

| Metric | Before | After prune channels | After prune layers (final) |
|---|---|---|---|
| mAP | 0.7905 | 0.7692 | 0.5679 |
| Parameters | 62573334 | 2621963 | 2101335 |
| size | 246.4 | 33.28 | 8.05 |
| Inference time | 0.0135 | 0.0130 | 0.0079 |

In view of the decline of the model accuracy of channel and layer pruned, 50 epochs were fine-tuning, and the accuracy returned to 0.792. The BN layer also is to be normally distributed. The result shows that the loss of accuracy is bare compared with baseline, but the model is greatly compressed, which reduces the resource occupation, improves the running speed, and reduces the computational complexity of the model.

**Result of Second Comparative Experiment.** Table 4 shows the experimental results when the channel and layer pruning algorithm with different global thresholds.

The experimental results show that with the increase of model compression ratio, the parameters and model size decrease significantly, and the image reasoning speed increases significantly. The cost is that the greater the model compression ratio is, the greater the average accuracy of detection and recognition will decline. In the pursuit of speed, when the extreme compression ratio achieves 0.95, the mAP is reduced to 0.005, and the number of parameters and model size are 3.08 M, which is about 1% of the original. Some important convolution kernel structures may be removed, which cannot be recognized, but the accuracy can be improved by fine-tuning for 50 epochs. In actual use of the project, we can flexibly choose the compression ratio of the model according

**Table 4.** Results of the channel and layer pruning algorithm with different global thresholds.

| Global threshold | mAP | Parameters | Size | Inference time |
|---|---|---|---|---|
| 0 | 0.7905 | 6257334 | 246.4 | 0.0135 |
| 0.5 | 0.6772 | 10138185 | 38.75 | 0.0079 |
| 0.7 | 0.5654 | 4667869 | 17.86 | 0.0079 |
| 0.85 | 0.5679 | 2101335 | 8.05 | 0.0079 |
| 0.93 | 0.4749 | 1146748 | 4.40 | 0.0081 |
| 0.95 | 0.0050 | 801331 | 3.08 | 0.0078 |

to the configuration of the specific mobile devices, and seek a balance between the size and accuracy of the model.

In order to intuitively see the detection performance of pruning algorithm, this paper selects some detection result images. In Fig. 4, the results of human hand detection and recognition show that the model has a better effect on human hand detection, can accurately detect human hands, and has a better performance on small target detection and recognition, which is almost the same as the effect of the original model before pruning.

**Fig. 4.** Results of human hand detection.

## 5 Conclusion

In this paper, we propose a channel and layer pruning algorithm for human hand detection in YOLOv3-spp network. By applying L1 regularization to the channel scaling factor enhances the channel sparsity of convolution layer. The convolution layer channels is pruned structurally, and the shortcut layer is pruned separately. The results of two groups of experiments show that the combination of channel and layer pruning has higher compression rate, faster reasoning speed than channel pruning, and the precision drop after fine tuning is acceptable. After pruning, at the cost of a small decline in detection and recognition effect, it will be more compact in the network model size, the memory occupation in the running stage and the amount of calculation, which provides theoretical support for the deployment of hand detection algorithm on mobile terminals.

## References


1. Ge, L., Ren, Z., Li, Y., et al.: 3D hand shape and pose estimation from a single RGB image. J. IEEE (2019)
2. Yang, L., Song, Q., Wang, Z., et al.: Hier R-CNN: instance-level human parts detection and a new benchmark. J. IEEE Trans. Image Process. **30**, 39–54 (2021)
3. Wang, W., He, B., Zhang, L.: High-accuracy real-time fish detection based on self-build dataset and RIRD-YOLOv3. J. Complex. **2021**, 1–8 (2021)
4. Liu, S., Agaian, S.S.: COVID-19 face mask detection in a crowd using multi-model based on YOLOv3 and hand-crafted features. In: Multimodal Image Exploitation and Learning (2021)
5. Han, S., Pool, J., Tran, J., et al.: Learning both weights and connections for efficient neural networks. In: NIPS (2015)
6. Han, S., Mao, H., Dally, W.J.: Deep compression: compressing deep neural networks with pruning, trained quantization and huffman coding. J. Fiber **56**, 3–7 (2015)
7. Yoon, J., Hwan, S.J.: Combined group and exclusive sparsity for deep neural networks. In: Proceedings of the 34th International Conference on Machine Learning, vol. 70, 3958–966 (2017)
8. Liu, Z., Li, J., Shen, Z., Huang, G. Yan, S., Zhang, C.: Learning efficient convolutional networks through network slimming. In: 2017 IEEE International Conference on Computer Vision (ICCV), Venice, Italy, pp. 2755–2763 (2017)
9. Sun, X., Ren, X., Ma, S., Wang, H.: Sparsified back propagation for accelerated deep learning with reduced overfitting. In: ICML (2017)
10. Redmon, J., Divvala, S., Girshick, R., et al.: You only look once: unified, real-time object detection. In: Computer Vision & Pattern Recognition. IEEE (2016)
11. Redmon, J., Farhadi, A.: YOLOv3: An Incremental Improvement. arXiv e-prints (2018)
12. Zhang, P., Zhong, Y., Li, X.: SlimYOLOv3: narrower, faster and better for real-time UAV applications. In: 2019 IEEE/CVF International Conference on Computer Vision Workshop (2019)
13. Li, Y., Zhang, S., Zhou, X., et al.: Build a compact binary neural network through bit-level sensitivity and data pruning. J. Neurocomput. **398**, 45–54 (2020)
14. Kumar, A., Shaikh, A.M., Li, Y., Bilal, H., Yin, B.: Pruning filters with L1-norm and capped L1-norm for CNN compression. Appl. Intell. **51**(2), 1152–1160 (2020). https://doi.org/10.1007/s10489-020-01894-y
15. Wu, D., Lv, S., Jiang, M., et al.: Using channel pruning-based YOLO v4 deep learning algorithm for the real-time and accurate detection of apple flowers in natural environments. J. Comput. Electron. Agric. **178**, 105742 (2020)

# A Chinese Dish Detector with Modified YOLO v3

Mingyu Gao[1], Jie Shi[1], Zhekang Dong[1,2(✉)], Yifeng Han[2(✉)], Donglian Qi[2], Rui Han[2,3], and Yan Dai[3]

[1] School of Electronics and Information, Hangzhou Dianzi University, Hangzhou 310018, China
{mackgao,shij,englishp}@hdu.edu.cn

[2] Department of Electrical Engineering, Zhejiang University, Hangzhou 310027, China
{hanyf,qidl}@zju.edu.cn

[3] Electric Power Research Institute, State Grid Zhejiang Electric Power Co., LTD., Hangzhou 310045, China

**Abstract.** In the field of food recognition, considering the difficulty in feature extraction caused by the diversity of Chinese dish, this paper proposed a Chinese dish recognition method SeDC-YOLO based on Squeeze-and- Excitation (SE) and Deformable Convolution (DC). This method optimized the feature extraction network Darknet53. Specifically, the SE attention mechanism was adopted to model the correlation between the different channels of the feature map as well as strengthen important features. Meanwhile DC was taken to solve the problem that the network is difficult to adapt to geometric deformation caused by regular sampling in standard convolution. Furthermore, the Gradient Harmonizing Mechanism (GHM) was introduced to solve the problem of unbalanced classification samples. Aiming at the local recognition problem in dish recognition, a new Non-Maximum Suppression (NMS) strategy was proposed. The experimental results of 37 types of Chinese dishes showed that the method proposed in this paper has a higher recognition accuracy.

**Keywords:** Chinese dish recognition · Deformable convolution · Squeeze and excitation · Gradient harmonizing mechanism · Non-maximum Suppression

## 1 Introduction

At present, the work of serving food in many fast-food restaurants, canteens of schools and factories is boring, with high labor demand and labor costs. The use of service robots instead of humans [1] to complete the dish serving has become a solution. The first step to complete the dish serving is to recognize and locate dishes, so it is of great significance to research an intelligent identification system for Chinese dishes.

Traditional food recognition algorithms [2–9] mostly adopt SVM, MKL-SVM models, then classify food images based on artificial features. The method of using Convolutional Neural Network to classify food images has achieved good accuracy [10–13].


Q. Han et al. (Eds.): LSMS 2021/ICSEE 2021, CCIS 1469, pp. 174–183, 2021.
https://doi.org/10.1007/978-981-16-7213-2_17

However, our task is to recognize and locate dishes. Then, the object detection algorithm provides us with research ideas. Two-stage algorithm such as Faster R-CNN [14] is more accurate, but with lower speed. While one-stage algorithm such as YOLO v3 [15] is faster, but the accuracy is lower than two-stage algorithm.

The purpose of this research is to design a fast and accurate object detector for the Chinese dish recognition system. The architecture that used for our model is similar to YOLO v3. While, this work has optimized the network and algorithm for the task of Chinese dish recognition, whose contributions are summarized as follow:

- This paper proposes a new and stronger feature extraction network named SeDC-Darknet and modified the scale of the prediction network;
- This paper adapts Gradient Harmonizing Mechanism (GHM) [16] on the classification loss to reduce the impact of sample imbalance;
- This paper raised a new Non-Maximum Suppression (NMS) strategy which is used to solve local repeated recognition for Chinese dishes.

The rest of this paper is organized as follows. Section 2 introduces the proposed algorithm flow. The optimization and adjustment of the network are provided in Sect. 3. In Sect. 4, there are the descriptions of some algorithm improvements. Then this paper explains the experimental environment and analyzes the results in Sect. 5. Finally, Sect. 6 concludes the paper.

## 2 Proposed Method

This paper proposes a detection system which adopts the one-stage object detection algorithm for Chinese dish. The framework of the dish recognition system is shown in Fig. 1. The process can be divided into the following steps:

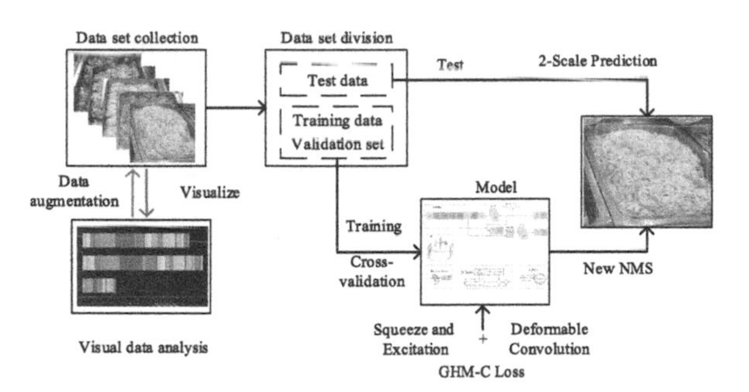


**Fig. 1.** The framework of the dish recognition system

Step 1: the collection of dish images and the production of data sets;
Step 2: data visualization analysis and data augment based on the analysis results;
Step 3: data set division;

Step 4: the optimized model training, the optimization content includes: Squeeze and Excitation [17], Deformable Convolution (DC) [18] and GHM-C Loss;
Step 5: model testing with a new NMS strategy and a 2-scale prediction network.

# 3 Optimization of Network Architecture

Darknet53 contain 5 residual blocks and 2 convolution layers. Specifically, each residual block consists of a 2-fold down-sample convolution layer and a set of repeated residual units [19], the numbers of repeated units for each block are 1, 2, 8, 8, 4. Furthermore, each residual unit contains two convolutional layers. It should be noted that each of the above convolutional layer contains a Convolution, Batch Normalization [20] and Leaky ReLU. And such a convolution layer is named CBL.

## 3.1 New Feature Extraction Network: SeDC-Darknet

Generally, the network pays the same attention to each channel of the feature map. This work introduces an attention mechanism-SE module, which enable the network to pay attention to the relationship between different channel features and learn the importance of different channel features.

The SE module includes two operations. First, the squeeze operation will compress the input feature $X$ with dimension $H \times W \times C$ ($H$ is height, $W$ is width, and $C$ is number of channels). It can be implemented by Global Average Pooling, which is expressed as (1):

$$S_X = \frac{1}{H \times W} \sum_{i=1}^{H} \sum_{j=1}^{W} X(i,j). \tag{1}$$

Then, to make use of the information aggregated in the squeeze operation, excitation operation is followed. This operation opts to employ a gating mechanism by forming a bottleneck with two fully connected (FC) layers around the non-linearity, i.e., a dimensionality-reduction layer with parameters $\mathrm{W}_1$ with reduction ratio r, a ReLU and then a dimensionality increasing layer with parameters $\mathrm{W}_2$:

$$E_S = \sigma(\mathrm{W}_2(\mathrm{ReLU}(\mathrm{W}_1(S_X)))). \tag{2}$$

where $\mathrm{W}_1 \in \mathbb{R}^{C/r \times C}$, $\mathrm{W}_2 \in \mathbb{R}^{C \times C/r}$, $\sigma(.) = \mathrm{sigmoid}(.)$, $S_X$ is the output of the Squeeze.

As shown in Fig. 2, this work incorporates the SE module into each residual unit to form the Se_Resnet. Afterwards, multiple Se_Resnet modules form the Se_ResN block, $N$ represents the number of repetitions of the Se_Resnet module. Finally, the Se-Darknet is formed of 5 Se_ResN blocks, where the values of $N$ are 1, 2, 8, 8, and 4 respectively. At the same time, we denote each Se_ResN block as a stage.

In addition, traditional convolutional neural network structure will be limited by fixed geometry sampling method. In order to improve the ability of our feature extraction network to process unknown deformations, this paper adopts DC to further optimize our Se-Darknet. This work adds a DC after stage5 to adjust the sampling position of the

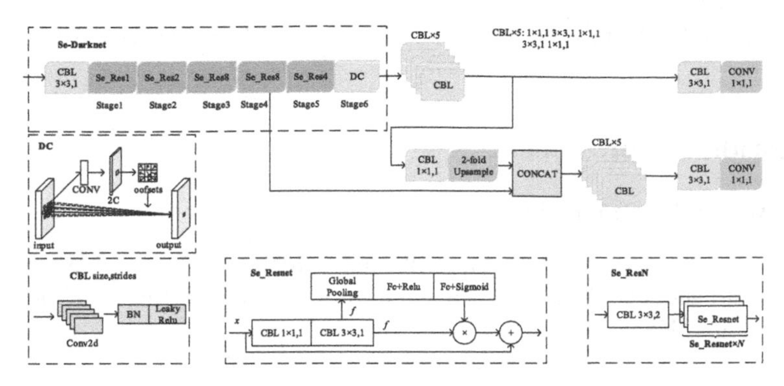

**Fig. 2.** The diagram of the network structure

feature map so that the feature extraction network is able to focus on the area of interest. Then our SeDC-Darknet is formed, and it is shown in Fig. 2.

In the above, DC is based on a parallel network to learn the offset. The convolution kernel in the network generates offsets on the regular sampling points of the input feature map, so as to realize random sampling near the sample center. Moreover, these offsets can be learned and will eventually focus the network on the interested areas and object. The Comparison of standard convolution and deformable convolution sampling is shown in Fig. 3.

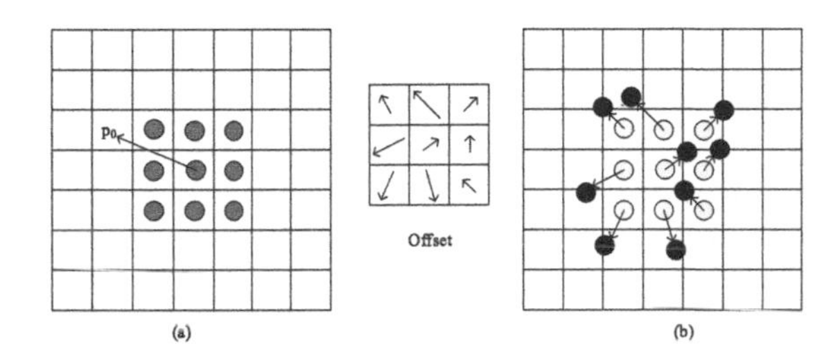

**Fig. 3.** Comparison of standard convolution and deformable convolution sampling

### 3.2 Prediction Network Optimization

This work draws on the method of Feature Pyramid Networks (FPN) [21, 22], and uses the top-down extraction method for the feature map. It performs down-sampling convolution in 5 stages, and each stage is 2-fold down-sampling. Then, the bottom-up method is used for up-sampling convolution during feature fusion. Finally, three feature maps with scales of 52 × 52, 26 × 26, and 13 × 13 are generated in the fusion for prediction.

According to the analysis of the dish data (the analysis process will be discussed in Sect. 5), the sample is mainly composed of medium and large objects. This work deleted the small-scale target prediction network to improve the recognition accuracy of large and medium targets. At the same time, the streamlining of the model improves recognize speed.

# 4 Algorithm Improvement

## 4.1 GHM-Loss for Sample Imbalance

Sample imbalance has always been one of the main problems faced by object detection. This work introduces GHM to reduce the impact of sample imbalance. GHM defines a gradient modulus length $g$, as in (3):

$$g = |p - y| = \begin{cases} 1-p & \text{if } y=1 \\ p & \text{if } y=0 \end{cases} \tag{3}$$

where $p = \text{sigmoid}(x)$, denotes the predicted probability and $y$ is the true label. It can be seen that $g$ is proportional to the difficulty of detection. Simultaneously, $g$ is the gradient module length of the cross-entropy loss, it can be proved as:

$$\begin{aligned} L_{CE} &= \begin{cases} -\log(p) & \text{if } y=1 \\ -\log(1-p) & \text{if } y=0 \end{cases} \\ \frac{\partial p}{\partial x} &= p(1-p) \\ \frac{\partial L_{CE}}{\partial x} &= \frac{\partial L_{CE}}{\partial p}\frac{\partial p}{\partial x} = \begin{cases} p-1 & \text{if } y=1 \\ p & \text{if } y=0 \end{cases} = p - y \\ g &= |p - y| = \left|\frac{\partial L_{CE}}{\partial x}\right|. \end{aligned} \tag{4}$$

In addition, GHM defines the gradient density variable $GD(g)$ to measure the number of samples within a certain gradient range. The excessive attention to outliers is reduced by suppressing samples with high gradient density:

$$\begin{aligned} GD(g) &= \frac{1}{l_\varepsilon(g)} \sum_{k=1}^{N} \delta_\varepsilon(g_k, g) \\ L_{GHM-C} &= \sum_{i=1}^{N} \frac{L_{CE}(p_i, y_i)}{GD(g_i)}. \end{aligned} \tag{5}$$

where, $\delta(g_k, g)$ represents the number of samples whose gradient modulus length is distributed in the range of $(g - \varepsilon/2, g + \varepsilon/2)$ among $N$ samples, $l_\varepsilon(g)$ is the length of $(g - \varepsilon/2, g + \varepsilon/2)$, $\varepsilon$ is an adjustable parameter.

## 4.2 New Non-Maximum Suppression (NMS) Strategy

NMS [24] is used in object detection to extract bounding boxes with high confidence and suppress false detection ones with low confidence. Generally, the model will predict and output many bounding boxes, among which many repeated boxes will locate the same object. NMS can remove these repeated boxes to obtain real prediction results.

The main parameters of NMS include the confidence threshold (score_threshold), the Intersection over Union (IoU) threshold (iou_threshold1), and the maximum number of boxes selected by NMS (max_output_size).

Meanwhile, this paper proposes a new NMS processing strategy, which adds a new threshold parameter (iou_threshold2). The method mainly includes three steps:

First, remove the boxes with confidence lower than score_threshold.

Second, the bounding boxes with the same prediction category are sorted by confidence. This work chooses the one with the highest confidence as the first reserved box, then selects a box in order, and calculates the $IOU$ of the box and each reserved box. It will remove the box when there is an $IOU$ greater than the iou_threshold1, otherwise add the box to the reserved boxes. When the number of reserved boxes is greater than max_output_size or all boxes are processed, this step ends. The $IOU$ calculation method for two boxes (A and B) is as follows:

$$IOU = \frac{area(A \cap B)}{area(A \cup B)}. \tag{6}$$

Finally, repeat the second step, but replace $IOU$ with $nIOU$, eliminate the boxes with $nIOU$ greater than the iou_threshold2, otherwise add the box to the reserved boxes. The calculation method of $nIOU$ is as follows:

$$nIOU = \max(\frac{area(A \cap B)}{area(A)}, \frac{area(A \cap B)}{area(B)}). \tag{7}$$

# 5 Experiments and Analysis

## 5.1 Dish Data Set Description and Analysis

The data images of dishes were taken in the student cafeteria of Hangzhou Dianzi University. Before the dishes were served, these dishes were placed in a large basin, as shown in Fig. 4.

This work has carried out a statistical analysis on the number and size of the ground-truth boxes of each dish. According to the statistical results of the number of samples for each dish, the categories with fewer samples have been expanded.

Then the proportion distribution of all the ground-truth box sizes relative to the original image is shown in Fig. 5. According to the definition of relative size, the area of small object's ground-truth box is less than 1% of the original image. As can be seen from Fig. 5, the proportion is mainly distributed from 4% to 22%. It means the detection targets are mainly medium and large targets in the date set. Therefore, this work removes the predicting network structure for small targets.

Fig. 4. Dish image sample

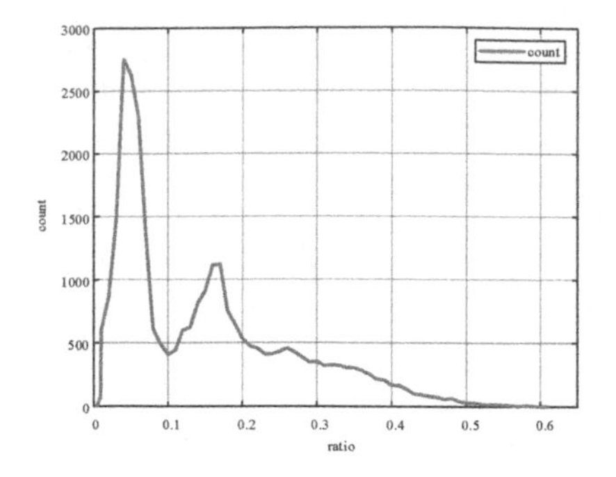


Fig. 5. The distribution of ground-truth box sizes

## 5.2 Environment and Parameter Settings

The experimental environment of the method described in this article is as follows: the CPU is Intel® Core™ i7-6700K @4.00 GHz, the GPU is GTX1080Ti with 11 GB video memory, the operation system is Ubuntu 16.04, the python version is 3.6.8, the deep learning framework is Tensorflow 1.6.0 and Keras 2.1.5.

Afterwards, we set the parameters of the model training as follows:

This work adopts the K-means algorithm to cluster and generate anchors based on the object size distribution of dishes to improve the detection rate of bounding boxes. Finally, the 6 anchors generated are assigned to the predicted feature maps of two sizes: $26 \times 26$: (101,182) (199,186) (284,193), $13 \times 13$: (206,426) (385,394) (497,577).

This work train on a single GPU, and the batch size is set to 10. The Adam optimizer is adopted to update the network weights. And the learning rate, momentum parameter, and weight decay parameter are initialized to 0.001, 0.9, 0.0005 respectively. Then the learning rate adjustment strategy learning rate scaling factor is set to 0.1 and patience is set to 3 epochs. Moreover, the data enhancement parameters are set as follows: angle = 5, saturation = 1.5, exposure = 1.5, hue = 0.1. The above parameter setting refers to the existing paper [15]. In addition, according to the training and test results, we adjust the dimensionality reduction coefficient of Excitation in SeDC-Darknet, the bins number of gradient module length division in GHM-C Loss, the IoU threshold, nIoU threshold, and score_threshold in new NMS. Finally, these five parameters are set to 4, 20, 0.45, 0.55, 0.35 respectively.

## 5.3 Results and Analysis

This work uses the parameters shown in the previous section to train our model. The convergence of the loss curve for 200 iterations of the model is shown in Fig. 6. Since this work has adjusted the structure of the neural network, this work chooses the initialization weights to train our model. At the beginning of the training process, the loss value of the model is very large. But as the training continues and the network weights are adjusted, the loss gradually decreases. After the training is complete for 200 iterations, the loss value tends to stabilize near 14.952. Then, we use the trained model to detect on the test set, and some test results are shown in Fig. 7.

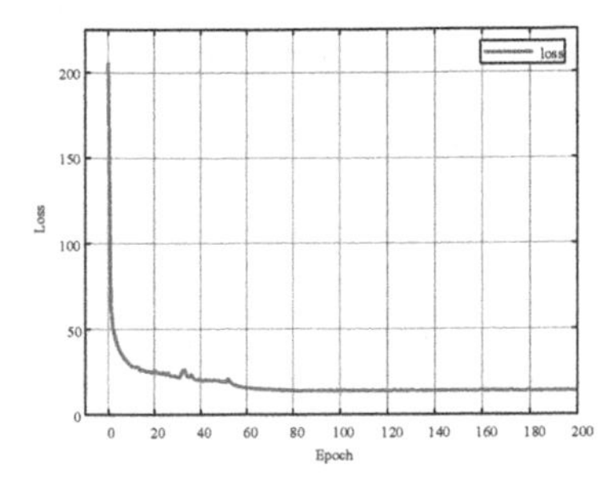


**Fig. 6.** The loss curve

**Fig. 7.** The detection results

At the same time, this work has tested the performance of model using different optimization methods proposed in Sect. 3 and 4 on our data set. The results shown in Table 1 prove the effectiveness of the proposed algorithm.

**Table 1.** Comparison of performance effects

| 2-scale predict | SeDC-Darknet | GHM | New NMS | FPS | mAP (%) | AP50 (%) | AP75 (%) |
|---|---|---|---|---|---|---|---|
| | | | | 16 | 59.0 | 66.5 | 48.0 |
| √ | | | | 18 | 61.6 | 68.3 | 53.9 |
| | √ | | | 15 | 65.5 | 71.5 | 57.6 |
| | | √ | | 12 | 60.0 | 65.3 | 50.1 |
| | | | √ | 16 | 62.0 | 66.9 | 57.4 |
| √ | √ | √ | √ | 14 | 69.9 | 73.5 | 62.1 |

From Table 1, the first row shows the performance of YOLO v3. The other rows are the performance results using proposed methods. It can be seen from the results that the performance impact of each method. Finally, the mAP of the dish recognition model reaches 69.9% by applying all the improved methods on YOLO v3, which is 10% higher than YOLO v3.

In addition, to better illustrate the effectiveness of the proposed method, this paper used the state-of-the-art object detection algorithm for dish recognition and compared with our method. The experimental results are shown in Table 2. From Table 2, the proposed algorithm has higher mAP than traditional dish recognition algorithm and the one-stage algorithm. At the same time, the mAP is only 2.5% lower, but the detection speed is much faster than the two-stage algorithm Faster R-CNN. It can be seen that the recognition algorithm proposed in this paper has a great advantage in the trade-off between speed and accuracy.

**Table 2.** Performance comparison of different recognition algorithm

| M | Backbone | FPS | Recall (%) | mAP (%) | AP50 (%) | AP75 (%) |
|---|---|---|---|---|---|---|
| M1 | - | - | - | 61.3 | - | - |
| M2 | - | - | - | 55.8 | - | - |
| M3 | - | - | - | 62.5 | - | - |
| M4 | Efficient-Net-b0 | - | - | 63.4 | - | - |
| M5 | ResNet-50 | 4 | 71.3 | 72.4 | 79.1 | 64.1 |
| M6 | VGG-16 | 10 | 66.3 | 61.2 | 67.8 | 53.4 |
| M7 | VGG-16 | 9 | 67.1 | 61.6 | 68.3 | 53.9 |
| M8 | ResNet-50 | 7 | 68.2 | 62.0 | 65.3 | 57.1 |
| M9 | DarkNet-53 | 16 | 66.4 | 59.6 | 66.5 | 48.8 |
| Ours | SeDC-Darknet | 14 | 71.0 | 69.9 | 73.5 | 62.1 |

**Note**: M means the method. M1 is MKL-SVM [2], M2 is SVM [6], M3 is MKL [26], M4 is FoRConvD [10], M5 is Faster R-CNN [14], M6 is SSD [25], M7 is RefineDet [26], M8 is RetinaNet [23], and M9 is YOLO v3 [15]. The FPS (Frames Per Second), Recall, mAP (mean Average Precision), and AP are some of the most used metrics for object detection

## 6 Conclusion and Future Work

This paper introduces a method for Chinese dish recognition—SeDC-YOLO. The results show that this method can quickly and efficiently identify various types of dishes. The results of comparison with other state-of-the-art algorithms also show that this method is effective and has advantages in performance. Future work will first improve the quality of the data set. For a good data set can greatly improve the performance of the model. In addition, the introduction of SE and DC increases the amount of model parameters and reduces the model's reasoning FPS. Future work will consider the optimization and pruning of the model to reduce the parameter quantity and model size.

**Acknowledgments.** This research was supported by the National Natural Science Foundation of China (No. 61873077), National Natural Science Foundation of China (No. 62001149), Key R&D Program of Zhejiang Province (No. 2020C01110), and National Science Foundation of Zhejiang Province (No. LQ21F010009).

## References

1. Wirtz, J., Patterson, P.G.: Brave new world: service robots in the frontline. J. Serv. Manag. **29**(5), 907–931 (2018)
2. Joutou, T., Yanai, K.: A food image recognition system with multiple kernel learning. In: 2009 IEEE ICIP, pp. 285–288 (2009)
3. Yang, S., Chen, M.: Food recognition using statistics of pairwise local features. In: 2010 IEEE CVPR, pp. 2249–2256 (2010)

4. Shotton, J., Johnson, M.: Semantic texton forests for image categorization and segmentation. In: 2008 IEEE CVPR, pp. 1–8 (2008)
5. Varma, M., Ray, D.: Learning the discriminative power-invariance trade-off. In: 2007 IEEE ICCV, pp. 1–8 (2007)
6. Matsuda, Y., Hoashi, H.: Recognition of multiple-food images by detecting candidate regions. In: 2012 IEEE ICME, pp. 25–30 (2012)
7. Felzenszwalb, P.F., Girshick, R.B.: Object detection with discriminatively trained part-based models. IEEE TPAMI **32**(9), 1627–1645 (2010)
8. Deng, Y., Manjunath, B.S.: Unsupervised segmentation of color-texture regions in images and video. IEEE TPAMI **23**(8), 800–810 (2001)
9. Sonnenburg, S., Rätsch, G.: Large scale multiple kernel learning. J. Mach. Learn. Res. **7**(57), 1531–1565 (2006)
10. Tomescu, V.I.: FoRConvD: an approach for food recognition on mobile devices using convolutional neural networks and depth maps. In: 2020 IEEE SACI, pp. 000129–000134 (2020)
11. Tan, M., Le, Q.V.: Efficientnet: rethinking model scaling for convolutional neural networks. In: 2019 International Conference on Machine Learning PMLR, pp. 6105–6114 (2019)
12. Chen, X., et al.: Chinesefoodnet: a large-scale image dataset for chinese food recognition. arXiv preprint arXiv:1705.02743 (2017)
13. Fu, Z., Chen, D., Li, H.: ChinFood1000: A Large Benchmark Dataset for Chinese Food Recognition. In: Huang, D.-S., Bevilacqua, V., Premaratne, P., Gupta, P. (eds.) ICIC 2017. LNCS, vol. 10361, pp. 273–281. Springer, Cham (2017). https://doi.org/10.1007/978-3-319-63309-1_25
14. Ren, S., He, K.: Faster R-CNN: towards real-time object detection with region proposal networks. IEEE TMAMI **39**(6), 1137–1149 (2016)
15. Farhadi, A., Redmon, J.: Yolov3: an incremental improvement. In: 2018 IEEE CVPR (2018)
16. Li, B., Liu, Y.: Gradient harmonized single-stage detector. In: Proceedings of the AAAI Conference on Artificial Intelligence, vol. 33(01), pp. 8577–8584 (2019)
17. Hu, J., Shen, L.: Squeeze-and-excitation networks. In: 2018 IEEE CVPR, pp. 7132–7141 (2018)
18. Dai, J., Qi, H.: Deformable convolutional networks. In: 2017 IEEE ICCV, pp. 764–773 (2017)
19. He, K., Zhang, X.: Deep residual learning for image recognition. In: 2016 IEEE CVPR, pp. 770–778 (2016)
20. Ioffe, S., Szegedy, C.: Batch normalization: accelerating deep network training by reducing internal covariate shift. In: International Conference on Machine Learning PMLR, pp. 448–456 (2015)
21. Lin, T.Y., Dollár, P.: Feature pyramid networks for object detection. In: 2017 IEEE CVPR, pp. 2117–2125 (2017)
22. Redmon, J., Farhadi, A.: YOLO9000: better, faster, stronger. In: 2017 IEEE CVPR, pp. 7263–7271 (2017)
23. Lin, T.Y., Goyal, P.: Focal loss for dense object detection. In: 2017 IEEE ICCV, pp. 2980–2988 (2017)
24. Neubeck, A., Van Gool, L.: Efficient non-maximum suppression. In 2006 IEEE ICPR, vol. 3, pp. 850–855 (2006)
25. Liu, W., et al.: SSD: Single Shot MultiBox Detector. In: Leibe, B., Matas, J., Sebe, N., Welling, M. (eds.) ECCV 2016. LNCS, vol. 9905, pp. 21–37. Springer, Cham (2016). https://doi.org/10.1007/978-3-319-46448-0_2
26. Zhang, S., Wen, L.: Single-shot refinement neural network for object detection. In: 2018 IEEE CVPR, pp. 4203–4212 (2018)
27. Hoashi, H., Joutou, T.: Image recognition of 85 food categories by feature fusion. In: 2010 IEEE ISM, pp. 296–301 (2010)

# UJN-Traffic: A Benchmark Dataset for Performance Evaluation of Traffic Element Classification

Yan Li[1,2,3], Shi-Yuan Han[1,2,3,4(✉)], Yuan Shen[1,2,3], Zhen Liu[1,2,3], Ji-Wen Dong[1,2,3], and Tao Xu[1,2,3(✉)]

[1] University of Jinan, Jinan 250022, China
202021200807@mail.ujn.edu.cn,
{ise_hansy,Liuzhen,ise_dongjw,xutao}@ujn.edu.cn
[2] China High-Resolution Earth Observation System Shandong Center of Data and Application, University of Jinan, Jinan 250022, People's Republic of China
[3] Shandong Provincial Key Laboratory of Network Based Intelligent Computing, University of Jinan, Jinan 250022, China
[4] Land Spatial Data and Remote Sensing Technology Institute of Shandong Province, Jinan 250002, China

**Abstract.** The detection of traffic element in remote sensing imagesplays an important role in the construction of traffic infrastructure, and can provide dynamic monitoring and quality evaluation for traffic construction projects. The convolutional neural network (CNN)-based method has achieved good performance in object detection in computer vision, which benefits from the popularity of large-scale datasets. However, the existing dataset can only be used for object detection, and the sample size and diversity need to be improved. These problems have brought challenges to the improvement and application of existing traffic element detection methods. This paper proposes a large-scale highresolution traffic element dataset based on multi-method detection analysis, and validates it based on the CNN-based object detection method. We screened more than 100 satellite images from high-resolution satellite images with a spatial resolution of 0.8 m. Manually labeled 5 categories and 100,000 traffic element samples. In the experiment, three CNN-based detection methods were used as the experimental baseline method. The experimental results proved effectiveness of the UJN-traffic traffic element dataset and provided a useful reference for traffic construction engineering and management.

**Keywords:** Traffic element detection · Roadworks · High-resolution

## 1 Introduction

The purpose of object detection and recognition is to obtain the location and category of a predetermined category target in the image. In recent years, due to the advantages of high-resolution images with high spatial resolution and


Q. Han et al. (Eds.): LSMS 2021/ICSEE 2021, CCIS 1469, pp. 184–193, 2021.
https://doi.org/10.1007/978-981-16-7213-2_18

high temporal resolution, high-resolution remote sensing satellite systems have been applied in many fields of national security, economic construction, and people's livelihood, have produced huge market value. Through the combination of high-resolution remote sensing image data and object detection technology, it is possible to conduct road network status investigation, traffic flow extraction, road construction and construction supervision, road disaster damage assessment, etc., to provide detailed, objective and comprehensive data for road traffic management and maintenance support.

With the increase in the number of satellites and their wide application in land, ocean, transportation and other fields, the object detection of high-resolution satellite data has been developed by leaps and bounds, especially in the field of traffic road construction and supervision. At present, researchers have proposed various datasets based on Google Earth, USGS and SPOT related to transportation, such as the bridge in DOTA dataset [2] and TGRS-HRRSD-Dataset [12]. For the realization of the object detection algorithm, it is not only affected by the algorithm, but also affected by many factors in the remote sensing image, especially for the existing object detection algorithm.The realization of the target detection algorithm is not only affected by the algorithm, but also by many factors in the remote sensing image, especially most of the existing object detection algorithms. In addition, the sample sizes and accuracy of the existing datasets are insufficient, the annotation process of the ground truth is slow, and the annotation accuracy is not satisfactory. These problems have brought challenges to the improvement and application of popular object detection methods such as Faster R-CNN [7] and YOLO [6]. Not only that, the above-mentioned datasets and currently popular datasets only contain the bridge category. The traffic infrastructure construction supervision dataset does not exist.Especially in recent years, governments of various countries have made every effort to promote the construction of transportation infrastructure and reforms in the transportation field. This work is to increase financial support for transportation infrastructure construction, and to provide strong support for national economic and social development, there is an urgent need for a dataset containing multiple categories of transportation elements to be proposed.

In response to the above problems, this paper proposes a large-scale traffic element dataset for remote sensing object detection, and conducts experimental analysis. It is based on China's GF-2 satellite imagery, and Fig. 1 shows an overview of the dataset. We selected more than one hundred remote sensing images for this, and manually labeled all the piers, subgrades, formation samples, and other objects in the images. The total number of labeled samples exceeded 100,000. The dataset has multi-scale and intra-class diversity. And the characteristics of similarity between classes. In the experiment, a variety of classic object detection algorithms are used as the method of traffic element detection, and the traffic element dataset is verified.

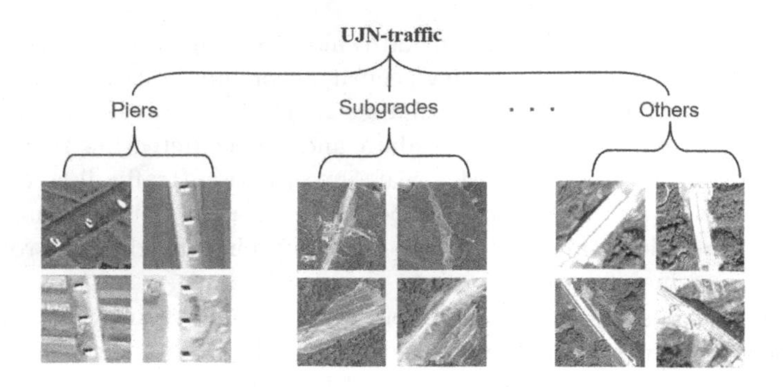


**Fig. 1.** Overview of UJN-traffic.

The main contributions of this article are:

(1) We propose a multi-scale high-resolution traffic element dataset for remote sensing image traffic element object detection and semantic segmentation. Its category and scale exceed existing datasets.
(2) In order to verify the validity of the dataset, the pier is selected as the object to be detected, a multi-means object detection method is proposed, and detailed experimental analysis is carried out.

The second section summarizes existing datasets and detection methods that include traffic element. The third section introduces the traffic element datasets proposed in this paper. The fourth section gives the experimental results and analyzes them. The fifth section gives having reached a conclusion.

## 2 Related Work

Datasets and algorithms for remote sensing object detection using high-resolution satellite images have been proposed and implemented in many fields. Table 1 summarizes the commonly used datasets for bridge detection. As the primary task of remote sensing image processing, remote sensing image object detection has important practical application value in the military and civilian fields, and has received extensive attention and research from international scholars. In particular, the object detection method based on deep learning has made great progress in the field of computer vision. For example, the literature [14] introduced a new two-stage method for CNN training, and improved the effect of the object detection algorithm by fine-tuning the CNN model. Zhang et al. [11] proposed to increase the position constraint and use the spatial relationship of the context to update. It is good to detect road vehicles through high-resolution aerial images.iffalse

As we all know, data sample in deep learning determine the accuracy and diversity of the model to a certain extent. Although existing datasets contain common features in remote sensing images, they are in terms of scale, resolution, image size, and time. There are certain limitations, so that it will have a negative impact on the research and application of many remote sensing object detection and classification. Table 1 compares the classic datasets that has been launched and our dataset in many aspects.

**Table 1.** Traffic datasets

| Dataset | Information | | | | |
|---|---|---|---|---|---|
| | Sources | Image width | Number of traffic elements | Spatial resolution (m) | Year |
| UC merced land use [10] | USGS | 256 | Bridge: 100 | 0.3 | 2010 |
| NWPU VHR-10 [8] | Google earth | 600 | Bridge: 124 | 0.5∼2 | 2014 |
| SIRI-WHU [1] | Google earth | 200 | Bridge: 200 | 2 | 2016 |
| NWPU-RESISC45 [13] | Google earth | 256 | Bridge: 700 | 0.3∼0.2 | 2016 |
| RSOD [5] | Google earth | 1280 | Bridge: 180 | 0.3∼3 | 2017 |
| AID [9] | Google earth | 600 | Bridge: 200 | 0.5∼0.8 | 2017 |
| Pattern-Net [3] | Google earth | 256 | Bridge: 800 | 0.062∼4.693 | 2017 |
| DIOR [3] | Google earth | 800 | Bridge: 2106 | 0.5∼30 | 2019 |
| DOTA | Google earth | 800∼4000 | Bridge: 6000 | 0.1∼5 | 2019 |
| TGRS-HRRSD | Google earth | 250 | Bridge: 4570 | 0.15∼1.2 | 2019 |
| UJN-traffic | GF-2 | 1024 | Bridges: 2000 | 0.8 | 2021 |

## 3 UJN-Traffic

Visual interpretation is a kind of remote sensing image interpretation, which is the inverse process of remote sensing imaging. It refers to the process by which professionals obtain specific target feature information on remote sensing images through direct observation or with the aid of auxiliary interpretation instruments. Because the image has a high resolution. Therefore, we built the UJN-traffic dataset through visual interpretation based on the GF-2 Satellite images. The following will give a detailed introduction to the details of our dataset and the labeling process.

### 3.1 Samples in UJN-Traffic

The UJN-traffic traffic element dataset is composed of five categories based on the construction of road traffic infrastructure, namely, piers, subgrades, formations, bridges, and pavements. The piers are the main supports of the bridge, and the upper building of the bridge is set on the pier. Above; it can be composed of stone, steel, wood or concrete, and built on a solid foundation of the river. Subgrade refers to a belt-shaped structure built according to the location of the route and certain technical requirements as the foundation of the road. It is the foundation of railways and highways. The subgrade is a linear structure built with soil or stone. The formation is below the subgrade surface, and the upper part of the subgrade is greatly affected by train dynamics and hydrological and climate changes. The subgrade soil bears the dynamic stress generated by the train load Under its long-term repeated action, the subgrade is prone to damage or excessive harmful deformation, which affects normal railway transportation. Therefore, the subgrade is the most important key part of the railway subgrade. The dataset contains more than 100,000 images. The sample images have obvious characteristics in remote sensing images. The samples of piers, subgrades, formations, bridges, and road surfaces are shown in Fig. 2.

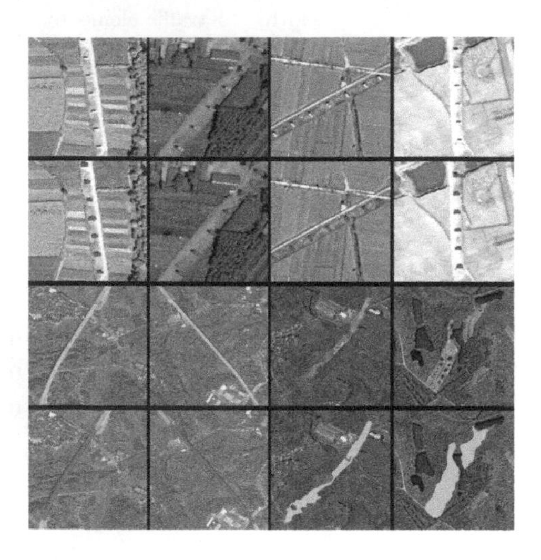

**Fig. 2.** Sample images of UJN-traffic. Piers and labels are on the upper two lines, and subgrade and bridge are on the lower two lines.

Due to the diversity of scene locations, collection times, and weather and lighting conditions, accurate labeling of traffic elements can be a very challenging task. According to our investigation, the UJN-traffic traffic element dataset proposed in this paper is the largest high-resolution remote sensing dataset used for traffic element detection.

# 4 Experimental Result and Analysis

All deep learning experiments are implemented through the Pytorch framework, and experiments are carried out by the Ubuntu 18.04.5 LTS operating system equipped with Intel(R) Xeon(R) Silver 4210 CPU, RTX-2080Ti GPU (12 GB) and 12 GB RAM.

In order to verify the effectiveness of the traffic element dataset, we use the pier sample and three commonly used deep learning object detection models, namely SSD [4], YOLO v5, Faster R-CNN for experimental verification. The experiment uses with the five-fold cross-validation method, the model training images are up to 100 scenes, and the number of samples reaches 90,000.

### 4.1 Experiments of SSD

In our experiment, the main parameters of SSD are as follows, backbone = vgg16_reducedfc, out_channels: (512, 1024, 512, 256, 256, 256, 256), priors_strides: [8, 16, 32, 64, 128, 256, 512], min_sizes: [35.84, 76.8, 153.6, 230.4, 307.2, 384.0, 460.8], max_sizes: [76.8, 153.6, 230.4, 307.2, 384.0, 460.8, 537.65], max_iter: 120000, lr_steps: [80000, 100000], batch_size: 8, lr: 0.001.

The total loss of SSD is:

$$L(x, c, l, g) = \frac{1}{N} L_{conf}(x, c) + \alpha L_{loc}(x, l, g) \tag{1}$$

$L_{conf}$ is the confidence loss, $L_{loc}$ is location loss, $L$ is the number of positive samples of the prior box, $c$ is the predicted value of category confidence, $l$ is the position prediction value of the corresponding bounding box of the a prior box, $g$ is the position parameter of Ground truth.

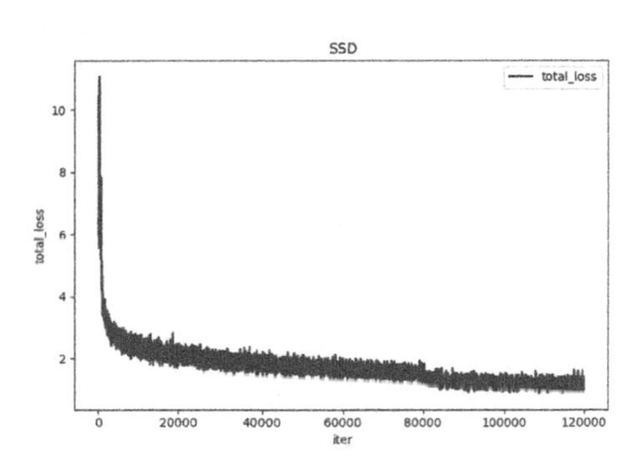


**Fig. 3.** The loss function curve of SSD.

The loss function curve of the SSD training iteration is shown in Fig. 3.

### 4.2 Experiments of Faster R-CNN

In our experiment, the main parameters of SSD are as follows, backbone = vgg16, max_epochs = 20, num_workers = 1, batch_size = 4, lr = 0.001, anchor_ratios = [0.5, 1, 2], lr_decay_step = 5, lr_decay_gamma = 0.1, anchor_scales = [8, 16, 32].

The total loss of Faster R-CNN is:

$$L(p_i, t_i) = \frac{1}{N_{cls}} \sum_i L_{cls}(p_i, t_i^*) + \frac{1}{N_{reg}} \sum_i p_i^* L_{reg}(p_i, t_i^*) \tag{2}$$

$p_i$ is the predicted classification probability of Anchor[$i$], When Anchor[$i$] is a positive sample, $p_i^* = 1$; when Anchor[$i$] is a negative sample, $p_i^* = 0$; $t_i$ is the parameterized coordinates of the Bounding Box predicted by Anchor[$i$], $t_i^*$ is the parameterized coordinates of the Bounding Box of Ground Truth of Anchor[$i$];

$N_{cls}$ is mini-batch size; $N_{reg}$ is the number of Anchor Locations; $L_{reg}$ stands for Smooth L1 function.

The loss function curve of Faster R-CNN training iteration is shown in Fig. 4.

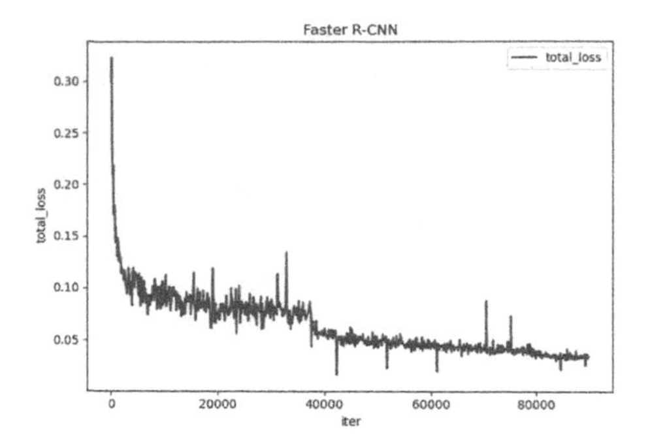


**Fig. 4.** The loss function curve of Faster-RCNN.

### 4.3 Experiments of YOLO V5

In our experiment, the main parameters of SSD are as follows, backbone = yolo5x, max_epochs = 20, depth_multiple: 1.33, batch_size = 4, lr = 0.01.

The total loss of YOLO v5 is:

$$L = L_{obj} + L_{cls} + L_{box} \tag{3}$$

$L_{obj}$ is the object loss used to calculate the mean square error of the confidence score of the object, $L_{cls}$ is the classifier loss used to calculate the cross-entropy loss used to classify the object, $L_{box}$ is the bounding box regression loss based on the mean square error of the predicted border position to judge the penalty for the wrong anchor box detection.

The loss function curve of YOLO v5 training iteration is shown in Fig. 5.

### 4.4 Piers Detection Results

Piers have obvious appearance characteristics, and these goals usually exist on the line of traffic infrastructure construction projects. Therefore, the trained model is used to detect piers in remote sensing images. The test results are shown in Fig. 6.

The ground truth image with bounding box is shown in Fig. 6(a), and the red bounding box represents the pier in the ground real image. Through experiments, the average accuracy of Faster R-CNN reached 90.74%, the average accuracy of SSD was 90.83%, and the average accuracy of YOLO v5 was 87.92%. Figure 6(b) to (d) respectively show the pier detection results of Faster R-CNN, SSD and YOLO v5.

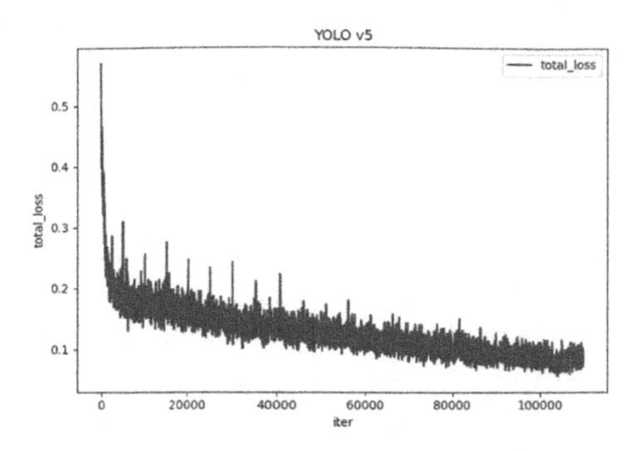


**Fig. 5.** The loss function curve of YOLO v5.

As shown in the figure, the three models can detect traffic elements more accurately. Faster R-CNN for small object detection, it has strong robustness. Although SSD and YOLO v5 are faster than Faster R-CNN in detection speed, the detection accuracy is not ideal. In general, for piers in traffic elements, commonly used object detection methods give satisfactory detection results.

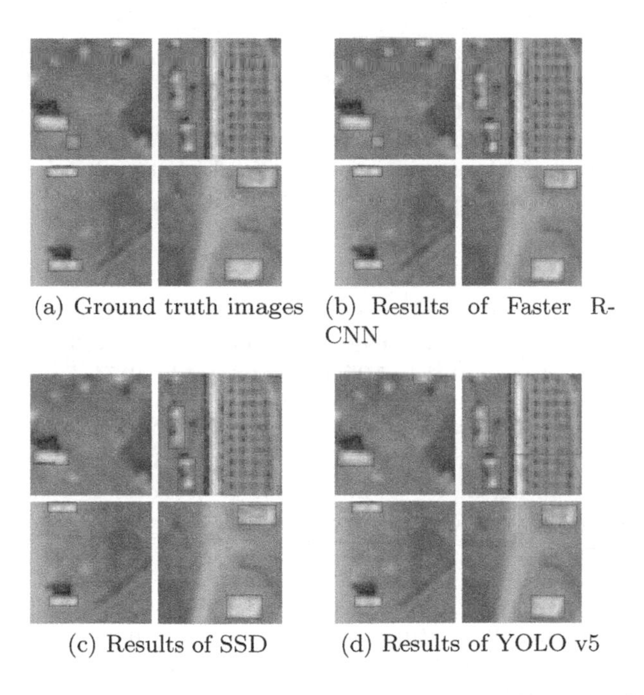

(a) Ground truth images (b) Results of Faster R-CNN

(c) Results of SSD (d) Results of YOLO v5

**Fig. 6.** The piers detection results of Faster R-CNN, SSD and YOLO v5.

## 5 Conclusions

This paper proposes a traffic element dataset named UJN-Traffic for object detection and recognition in high-resolution images. The dataset contains 100,000 traffic element samples, and all traffic elements have been manually annotated. Three common object detection models: Faster R-CNN, SSD, YOLO v5 are used to verify the effectiveness of the pier element in this dataset. The experimental results verify the effectiveness of traffic element piers. The proposed dataset and traffic element object detection method will play an effective role in the dynamic monitoring and quality evaluation of traffic construction projects such as construction progress, land acquisition area, road vegetation, soil erosion, land use and other dynamic monitoring and quality evaluation projects.

In the future, we will continue to verify the effectiveness of other traffic elements in this dataset, so that this traffic dataset can be suitable for the training and testing of semantic segmentation and object classification algorithm models in remote sensing images. The second point is that we plan to release the second edition of the refined dataset to cover more traffic construction scenarios, which will be suitable for various high-resolution remote sensing image data processing tasks.

**Acknowledgment.** This research was supported by the National Natural Science Foundation of China under Grants 61903156 and 61873324, the Natural Science Foundation of Shandong Province for Key Project under Grant ZR2020KF006, and the Taishan Scholar Project of Shandong Province under Grant tsqn201812077, and Natural Science Foundation of University of Jinan (No. XKY1803, XKY1804, XKY1928, XKY2001).

## References

1. Cheng, G., Han, J., Lu, X.: Remote sensing image scene classification: benchmark and state of the art. Proc. IEEE. **105**(10), 1865–1883 (2017)
2. Ding, J., Xue, N., Long, Y., Xia, G.S., Lu, Q.: Learning roi transformer for oriented object detection in aerial images. In: 2019 IEEE/CVF Conference on Computer Vision and Pattern Recognition (CVPR), pp. 2844–2853 (2019)
3. Li, K., Wan, G., Cheng, G., Meng, L., Han, J.: Object detection in optical remote sensing images: a survey and a new benchmark. ISPRS J. Photogram. Remote Sens. **159**, 296–307 (2020)
4. Liu, W., et al.: SSD: single shot multibox detector. In: Leibe, B., Matas, J., Sebe, N., Welling, M. (eds.) ECCV 2016, Part I. LNCS, vol. 9905, pp. 21–37. Springer, Cham (2016). https://doi.org/10.1007/978-3-319-46448-0_2
5. Long, Y., Gong, Y., Xiao, Z., Liu, Q.: Accurate object localization in remote sensing images based on convolutional neural networks. IEEE Trans. Geosci. Remote Sens. **55**(5), 2486–2498 (2017)
6. Redmon, J., Divvala, S., Girshick, R., Farhadi, A.: You only look once: unified, real-time object detection. In: 2016 IEEE Conference on Computer Vision and Pattern Recognition (CVPR), pp. 779–788 (2016)

7. Ren, S., He, K., Girshick, R., Sun, J.: Faster R-CNN: towards real-time object detection with region proposal networks. In: Cortes, C., Lawrence, N., Lee, D., Sugiyama, M., Garnett, R. (eds.) Advances in Neural Information Processing Systems, vol. 28. Curran Associates, Inc. (2015)
8. Su, H., et al.: HQ-ISNET: high-quality instance segmentation for remote sensing imagery. Remote Sens. **12**(6), 989 (2020)
9. Xia, G.S., et al.: AID: a benchmark data set for performance evaluation of aerial scene classification. IEEE Trans. Geosci. Remote Sens. **55**(7), 3965–3981 (2017)
10. Yang, Y., Newsam, S.: Bag-of-visual-words and spatial extensions for land-use classification. In: Proceedings of the 18th SIGSPATIAL International Conference on Advances in Geographic Information Systems, GIS 2010, pp. 270–279. Association for Computing Machinery, New York (2010)
11. Zhang, J., Tao, C., Zou, Z.: An on-road vehicle detection method for high-resolution aerial images based on local and global structure learning. IEEE Geosci. Remote Sens. Lett. **14**(8), 1198–1202 (2017)
12. Zhang, Y., Yuan, Y., Feng, Y., Lu, X.: Hierarchical and robust convolutional neural network for very high-resolution remote sensing object detection. IEEE Trans. Geosci. Remote Sens. **57**(8), 5535–5548 (2019)
13. Zhou, W., Newsam, S., Li, C., Shao, Z.: PatternNet: a benchmark dataset for performance evaluation of remote sensing image retrieval. ISPRS J. Photogramm. Remote Sens. **145**, 197–209 (2018). Deep Learning RS Data
14. ševo, I., Avramović, A.: Convolutional neural network based automatic object detection on aerial images. IEEE Geosci. Remote Sens. Lett. **13**(3), 740–744 (2016)

# Online Adaptive Strategy for Traffic-Light Timing Based on Congestion Pressure

Qi-Wei Sun, Shi-Yuan Han(✉), Ya Fang, and Qiang Zhao

Shandong Provincial Key Laboratory of Network Based Intelligent Computing, University of Jinan, Jinan 250022, China
ise_hansy@ujn.edu.cn, ise_fangy@mail.ujn.edu.cn

**Abstract.** This paper devises a strategy for adjusting automatically the traffic light timing in real time by counting the number of vehicles with detecting devices at both ends of the road. By means of the strategy, based on the data concerning the traffic flow in the previous period, a timing plan firstly for all the cycles in the next period is proposed. At the end of a cycle, the value of congestion pressure in each phase of the following cycle is predicted according to the obtained traffic-flow data. The value is taken as an input variable of the algorithm in order to optimize the timing plan, which will be applied to the next cycle. The proposed strategy, after being verified in the simulation experiment, can be adopted to accelerate the traffic flow in case of unbalanced traffic flow at the intersection.

**Keywords:** Road traffic · Adaptive traffic lights · Traffic congestion · Traffic flow simulation

## 1 Introduction

Traffic congestion has become a common phenomenon, which not only happens in large cities but also in medium and small ones with varying degrees. Solving the problem of traffic congestion has become one of the most important issues in building an urban intelligent transportation system.

In recent years, more attention has been paid to improving the efficiency of traffic flow at intersections and optimizing the settings through the signal-light timing scheme. There are two modes for controlling traffic lights: cycle length being fixed and cycle length being adjusted by traffic conditions. The most widely-used method in allocating signal time is the Webster timing method, which divides a single day into different traffic periods as the signal-timing periods and each period corresponds to a preset signal plan [1]. As to the second mode, sensors are first deployed to detect the traffic flow and other various data, and then the phase difference of signal light is adjusted to optimize the signal timing at the intersection, so as to improve the traffic efficiency or reduce the average delay of vehicles. At present, most of the researches are still focused on


Q. Han et al. (Eds.): LSMS 2021/ICSEE 2021, CCIS 1469, pp. 194–203, 2021.
https://doi.org/10.1007/978-981-16-7213-2_19

the single-point control mode, that is, adjusting the cycle of traffic lights according the traffic-flow data collected by equipment at the intersection, for instance, M. Radivojevi'c [2] adjusts the cycle in real time by analyzing the information of traffic flow through video detectors at the intersection, and [3] adjusts the cycle on the basis of the data collected by the device in real time. Other scholars also adjust the cycle by pre-determining the length of the queue of vehicles in real time [4–7]. But, because the equipment that they are using cannot accurately obtain the number of vehicles in the queue on the road, they can only estimate the length of the queue by analyzing relevant data. So, their research results are still limited in application. To evade the problem of inaccurate data, [8–11] adopt the vehicle-to-vehicle (V2V) or vehicle-to-infrastructure (V2I) method to collect traffic data about each vehicle, and then offer guidance for vehicles on the road or optimize the traffic-light system through cloud computing.

The road net in megalopolises or large cities is mainly designed in the form of radiation. It may facilitate the traffic between the suburbs and the downtown, but also cause the problem that the main roads are filled with too many vehicles during peak hours. To solve this problem, this paper proposes an online adaptive strategy by applying the technology of infrastructure-to-infrastructure(I2I) communication on the road network [12] for urban traffic lights. First, the number of vehicles entering and leaving the road is counted accurately on the road in the previous period by placing detecting devices at both ends of each road. After that, according to the influence from other adjacent intersections, a discrete-time model of dynamic traffic flow is established. Then, an online strategy is proposed to regulate the urban traffic lights by predicting the traffic flow.

The remainder of this paper is organized as follows: The proposed strategy is introduced in Sect. 2. Our algorithm is simulated and analyzed in Sect. 3. Our conclusion is presented in Sect. 4.

## 2 Methodology

### 2.1 Intersection Description

In the system of traffic-signal control, there are often traffic flows from multiple directions at the intersection. To avoid traffic conflicts among these traffic flows, the order of phases of signal lights at the intersection must be determined to ensure that the vehicles will pass in a certain order. At this time, the intersection can be regarded as $N$-phase controlled, where $N$ refers to the total number of phases.

Generally, a typical intersection with four arms is represented below(see Fig. 1), and traffic light phases in the intersection are a straight phase, a left-turning phase, a straight phase for the direction orthogonal to the first one and a left-turning phase for the direction orthogonal to the first one. Figure 2 shows all traffic light phases in the target intersection. The turn to right from all sides are permitted in all phases. The sequence of phase change is $a \to b \to c \to d \to a$.

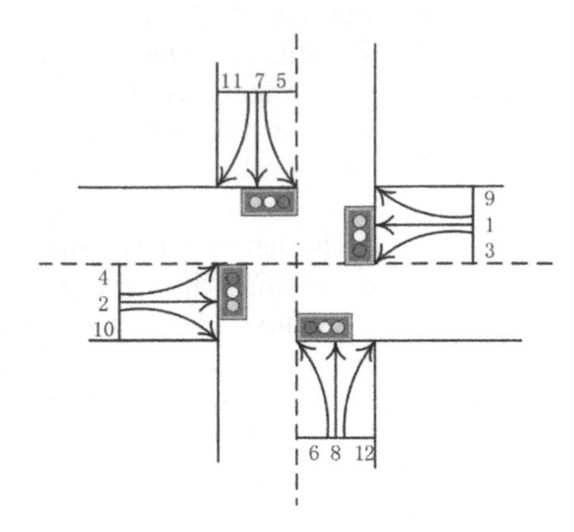


**Fig. 1.** The structure of 4-arm intersection

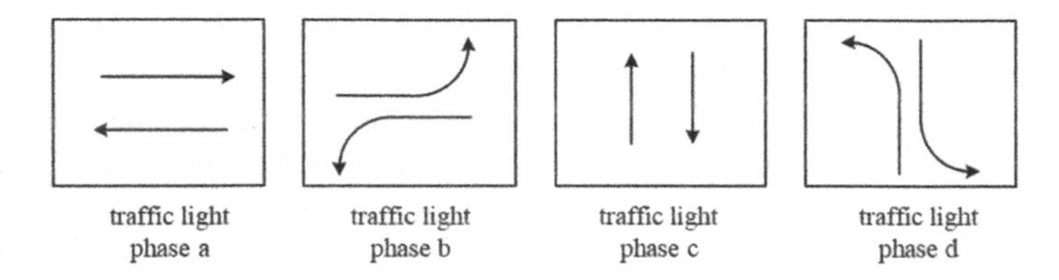


**Fig. 2.** All traffic-light phases at the target intersection

## 2.2 Framework

In this paper, the proposed method is under the following, as shown in Fig. 3, which mainly consists of two parts: the strategy to get the timing plan and the strategy of online adaptive planning.

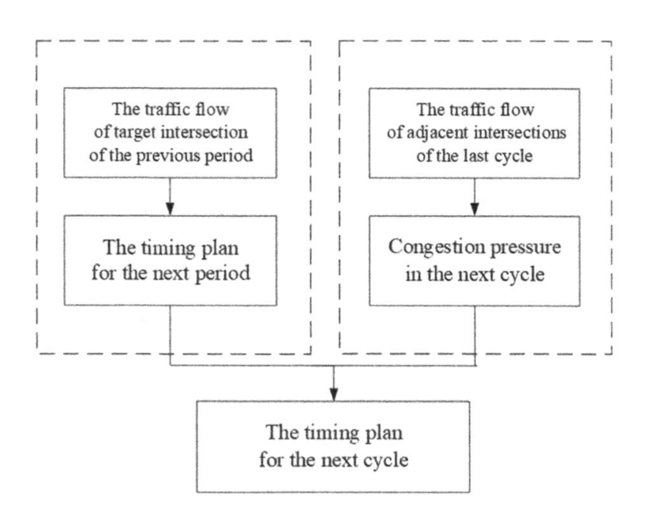


**Fig. 3.** Framework of timing plan

The traffic flow collected of the previous period provides basic data for the timing plan for the next period; as the traffic flow changes with time, the signal cycle may become too long in the next period so as to cause a waste of resources; or become too short to stop the traffic flow. At this time, the signal cycle needs to be adjusted before the approach of the next period. Thus, the traffic flow of each phase can be used to regulate the length of green light and to design a balanced regulation plan; and a reasonable period for each phase can be obtained on the basis of the overall situation.

## 2.3 Online Adaptive Strategy for Traffic-Light Timing

Based on the traffic timing plan in this period, and according to the influence of adjacent intersections on the target intersection, a discrete-time model of dynamic traffic flow is established and priority ranks of regulation of each phase are devised, and then signal plan is adjusted according to the priority ranks so as to obtain a new signal plan. The plan for the next cycle will be adjusted according to the plan in this period, not the last one, as is illustrated in this part.

According to the impact of adjacent intersections on the target intersection, the road network can be mathematically represented; taking an intersection of typical four-phase control as example, we obtain the relationship between of traffic network and flow, which is described in Fig. 4.

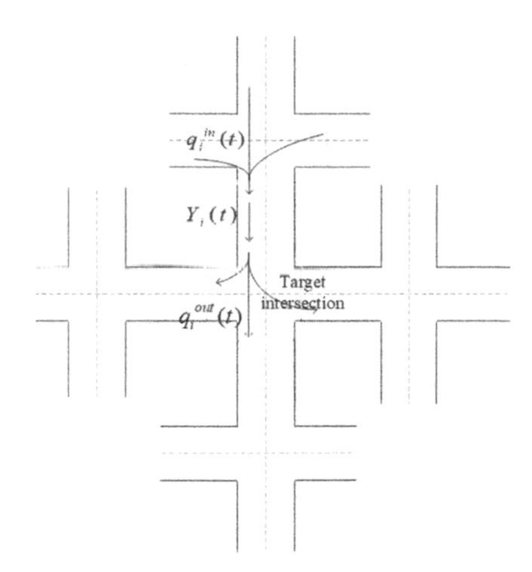


**Fig. 4.** Road network and traffic flow

In Fig. 4, $i$ represents the number of traffic-light phases that the traffic flow enters from other adjacent intersections. $Y_i(t)$ represents the number of vehicles in the $i$ phase of the target intersection at the beginning of the $t$ time interval, ${q_i}^{in}(t)$, ${q_i}^{out}(t)$ respectively represents the number of inbound and outbound vehicles at the intersection, and $N$ represents the total number of phases at the intersection.

Assuming that the maximum flow $Q_i$ of each phase and the turning probability $\tau_i$ in each direction are measurable constant variables, $L_i$ is the $i$ phase lost time, then the inflow ${q_i}^{in}(t)$ and outflow ${q_i}^{out}(t)$ in the $i$ phase can be obtained by the following two formulas:

$$ {q_i}^{in}(t) = \sum_{j \in {I_i}^{in}} \tau_{j,\, i} \cdot Y_j(t) \tag{1} $$

$$ {q_i}^{out}(t) = Q_i \cdot \frac{\sum_{i=1}^{N} T_i}{\sum_{i=1}^{N} (T_i + L_i)} \tag{2} $$

In the formulas aforesaid, ${I_i}^{in}$ represents the traffic flow set including all the vehicles entering the target road at the intersection from other adjacent intersections, and $\tau_{j,\, i}$ represents the steering probability of the traffic flow entering from the $j$ intersection during the $i$ phase. Then, the dynamic traffic flow of each phase at the target intersection can be obtained; $Y_i(t+1)$ represents the number of vehicles in the $i$ phase of the target intersection in $t+1$ time interval can be represented as the following discrete-time model:

$$ Y(t+1) = Y(t) + q^{in}(t) - q^{out}(t) \tag{3} $$

$$ q^{in}(t) = \begin{bmatrix} {q_1}^{in}(t) \\ \vdots \\ {q_i}^{in}(t) \\ \vdots \\ {q_N}^{in}(t) \end{bmatrix}; q^{out}(t) = \begin{bmatrix} {q_1}^{out}(t) \\ \vdots \\ {q_i}^{out}(t) \\ \vdots \\ {q_N}^{out}(t) \end{bmatrix}; $$

$$ Y(t) = \begin{bmatrix} y_1(t) \\ \vdots \\ y_i(t) \\ \vdots \\ y_N(t) \end{bmatrix}; Y(t+1) = \begin{bmatrix} y_1(t+1) \\ \vdots \\ y_i(t+1) \\ \vdots \\ y_N(t+1) \end{bmatrix}. $$

According to the number of vehicles entering the target intersection and that of vehicles leaving the intersections which upstream the target intersection at a fixed moment, the number of vehicles passing through the target intersection at the previous moment is predicted to optimize the signal plan and increase the traffic efficiency.

As to the problem of congestion disparity between the green-light phases at the intersection of the main road and the branch, a better resolution of traffic-light cycle is proposed on the basis of the different congestion pressure of phases. This resolution is to regulate the length of green light according to the number of vehicles at the upper and lower sections of the road and also according to the traffic-flow data forecasted in real time. On the basis of timing plan obtained before, the length of each green-light phase in every traffic-light cycle is regulated by applying the discrete-time model of dynamic traffic flow. The congestion-pressure index of each green-light phase at the intersection is mathematically represented as the following formula:

$$w_i(t+1) = \frac{1}{M}\sum_{j=1}^{M}(\frac{y_j(t+1)}{y_{j_{max}}} + \frac{1}{F}\sum_{k=1}^{F}\beta_j(1-\frac{y_k(t+1)}{y_{k_{max}}})) \tag{4}$$

In the above formula, $w_i(t+1)$ stands for the degree of congestion pressure to the $i$ phase of green light in the next timing cycle; $y_j$ (t + 1) represents the number of vehicles entering the intersection from the upper section of the road $j$ during the $i$ phase of green light in the next timing cycle; and $y_k$ (t + 1) presents the number of vehicles leaving the intersection for the lower section of the road $k$ during the $i$ phase of green light in the next timing cycle. $M$ represents the number of upper section of the road, and $F$ presents the number of the lower section of the road. $y_{j_{max}}$ and $y_{k_{max}}$ represent the maximum number of vehicles on the upper and lower sections of the road respectively. $\beta_i$ represents the probability that the traffic on the upper section of the road will enter the lower section of the road $k$, and $\sum \beta_i = 1$ .

From the congestion pressure obtained from the above formula, the priority rank for adjusting the length of each green light at the intersection is used to optimize the length of each green light, and the difference between congestion-pressure index of each green-light phase can be determined according to the S obtained from the following formula.

$$S\left(t+1\right) = \frac{\sum_{i=1}^{N}\left(w_i\left(t+1\right)-\bar{w}\left(t+1\right)\right)^2}{N} \tag{5}$$

$$\bar{w}\left(t+1\right) = \frac{\sum_{i=1}^{N} w_i\left(t+1\right)}{N} \tag{6}$$

$\bar{w}\left(t+1\right)$ is the value to represent the difference in road congestion in each cycle. In this paper, we set two fixed values $S_{\min}$ and $S_{\max}$, so $S\left(t+1\right)$ can be divided into three intervals. If $S\left(t+1\right) \leq S_{\min}$, it means that the adjustment priority ranks of each green-light phase are roughly the same, and so is the load condition; under such circumstances, only the length of the green-light phase highest in the priority rank is slightly regulated. If $S\left(t+1\right) \geq S_{\min}$ and $S\left(t+1\right) \leq S_{\max}$, they illustrate the wide gap between the priority ranks for adjusting the length corresponding to each green-light phase, and also illustrate the disparity of traffic conditions of each road. So, the length of each phase highest or lowest in the priority rank is adjusted. If $S\left(t+1\right) \leq S_{\max}$, it shows a wider gap between the priority ranks for adjusting the length corresponding to each green-light phase. Hence the length of each green-light phase must be adjusted.

## 3 Simulation and Analysis

Based on the SUMO traffic simulation software, a road network model for multiple intersections is established to simulate the traffic flow so as to facilitate testing the online optimization model [13,14]. In the experiment conducted in this part, based on Stefan Krauß's microscopic modeling of traffic flow, the variables of the main-road delay, the vehicle-stopping times and the volume of traffic flow in a certain moment are analyzed to test the feasibility of the model established in this paper. Also in this experiment, the collected traffic data are used to obtain the traffic flow in the next cycle. Then according the traffic flow, the adjustment of the time length of each green-light phase in the next cycle. The overall structure of the simulation environment is illustrated in Fig. 5.

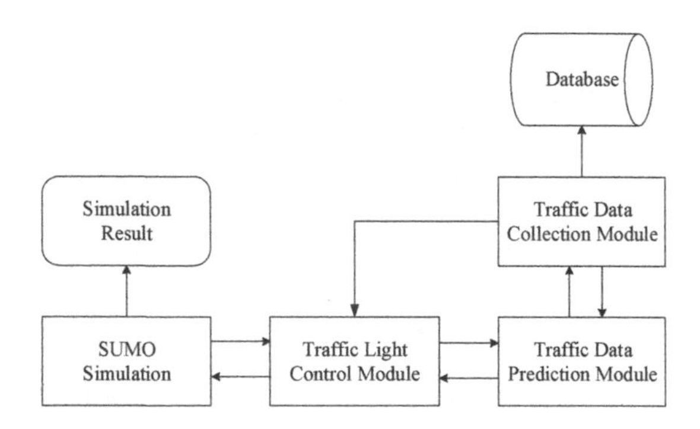


**Fig. 5.** The overall structure of the simulation environment

In this part, a model of a target intersection with other four adjacent intersections is established, on the basis of SUMO traffic software, to form a road net. In this road net, the main road is the middle one from south to north, and all others are regarded as branches (Fig. 6).

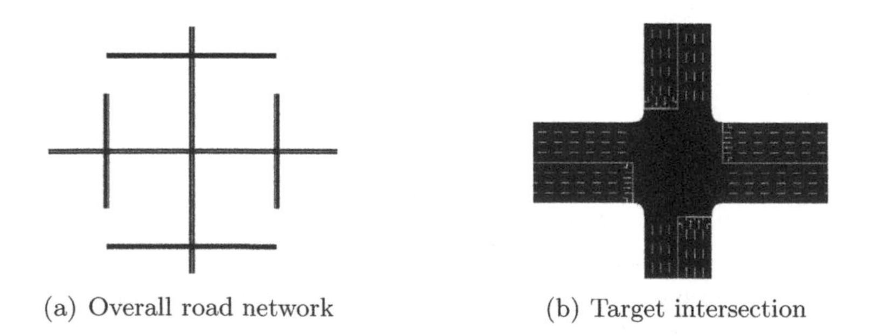

(a) Overall road network (b) Target intersection

**Fig. 6.** Road network used for simulation

Each road has four entrance lanes in one direction. Let's take one of the roads running from south to north as an example, the lanes in one direction are divided from left to right into one left-turning lane, two straight-going lanes and one right-turning lane. The lane is 3.5 m wide. Each vehicle follows a fixed route, which is engendered at random. See Table 1 for vehicle-speed interval and vehicle proportion.

**Table 1.** Vehicle proportion and speed interval

| Vehicle type | Proportion | Speed interval (km/h) |
|---|---|---|
| Car | 96% | (40, 60) |
| Bus | 4% | (30, 45) |

The indicators used to evaluate the performance of the model come from the output files in the SUMO simulation. In these files are recorded for every step the data of the traffic, the traffic lights and other factors involved in the simulation process. The evaluation indicators in this paper mainly include the total time length of vehicle delay, the queue length on all roads in the real time, and the average queue length.

In this simulation experiment conducted under different congestion conditions, a comparison is made between the proposed timing strategy and the Webster timing strategy according to the parameters set in the previous part.

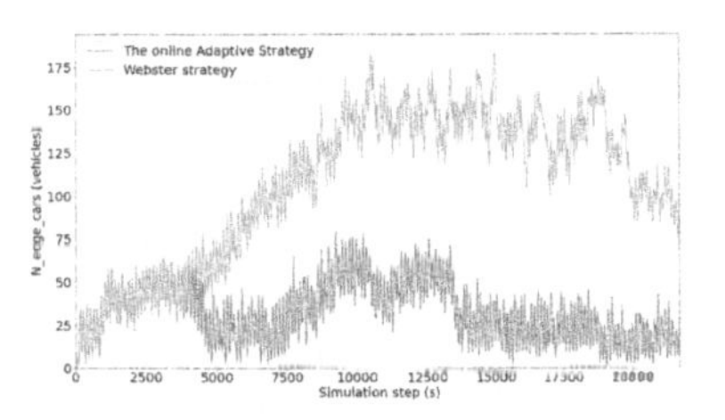


(a) the vehicles on north edge

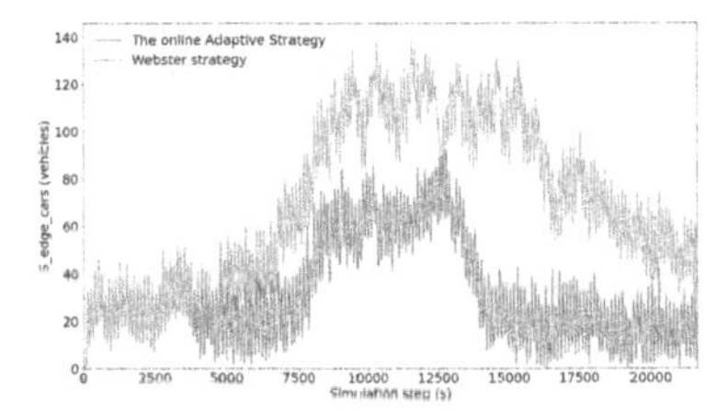


(b) the vehicles on south edge

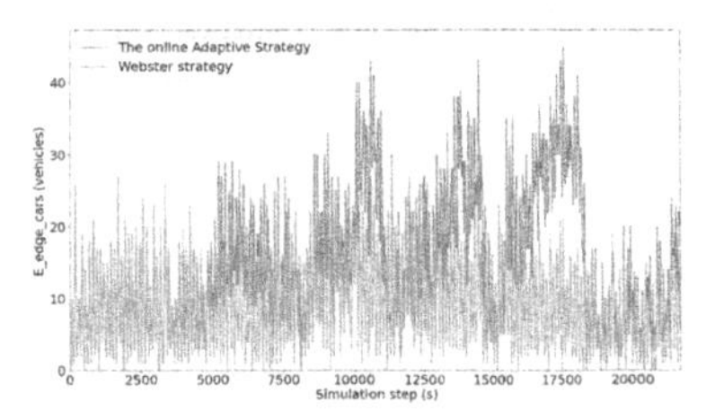


(c) the vehicles on east edge

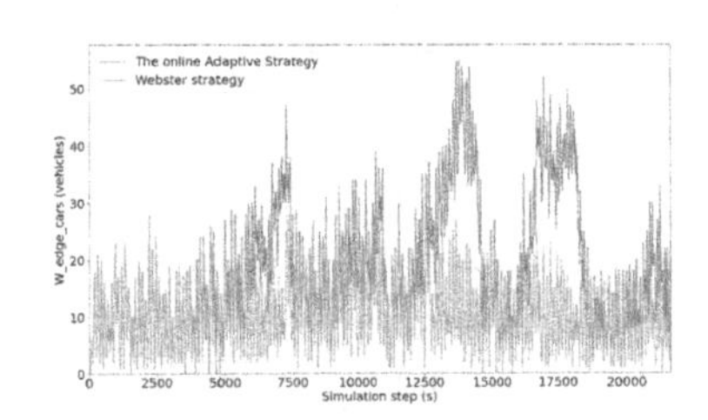


(d) the vehicles on west edge

**Fig. 7.** Performance comparison for cumulative delay

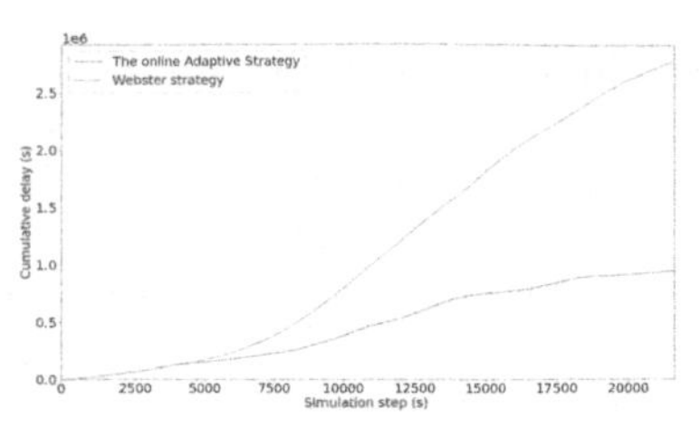


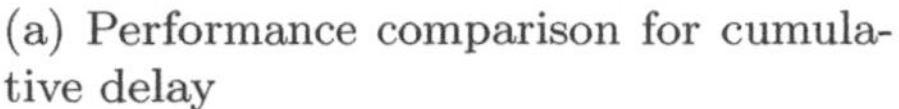
(a) Performance comparison for cumulative delay

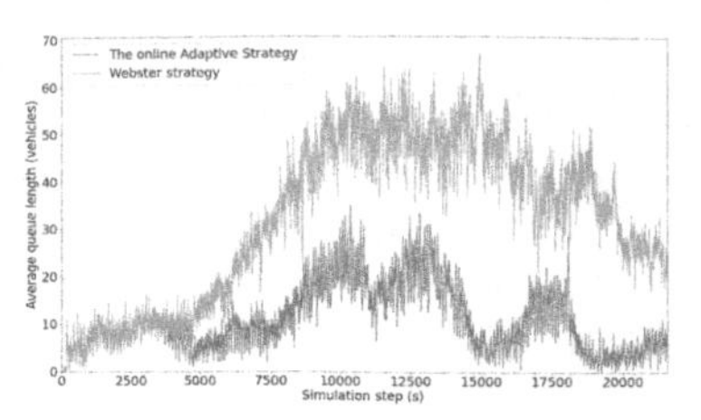


(b) Performance comparison for average queue length

**Fig. 8.** Performance comparison in simulation

In this part, the main road and the branch are divided, so the scene can be simulated in an unbalanced traffic-flow situation. After the tests by comparing the indicators, the proposed timing strategy is obviously more effective than the Webster timing strategy in that the average queue length decreases by 66.11% compared with that of the Webster timing strategy and the cumulative delay decreases by 58.81%. In Fig. 7, it can be seen that, in case of unbalanced traffic flow, on the main road in the north-south direction, the timing strategy proposed in the previous part is obviously more effective than the Webster timing strategy, while on the branch road in the east-west direction, the proposed timing strategy is as effective as the Webster timing strategy, or maybe a little worse. As the traffic flow on the main road is the focus of this research, the traffic flow on the branches is a little less smooth. But the overall performance of the proposed timing strategy is more effective, as is illustrated in Fig. 8; because the total length of traffic delay is shorter and the average queue length is also much shorter than the corresponding indicators in Webster timing strategy.

## 4 Conclusion

In this paper, an online adaptive regulation strategy for signal plan was proposed. The traffic flow is represented mathematically as a congestion-pressure index, as so to obtain a priority rank for each green-light cycle. Then, according to the priority rank, the timing scheme is optimized. Thus, the traffic flow on the main roads will be reduced first of all, to ensure the connecting link between all the districts of the city.

The proposed strategy, after being verified in the simulation experiment, can accelerate the traffic flow on the main road at the intersection even in rush hours and therefore cut down the traffic accidents to some extent. At the same time, the proposed strategy can be adopted to improve the overall traffic efficiency so as to obtain good economic benefits. Therefore, this strategy is effective in solving the problem of traffic congestion at the intersection of the main road.

**Acknowledgment.** This research was supported by the National Natural Science Foundation of China under Grants 61903156 and 61873324, the Natural Science Foundation of Shandong Province for Key Project under Grant ZR2020KF006, and the Taishan Scholar Project of Shandong Province under Grant tsqn201812077.

## References

1. Webster, F.V.: Traffic signal settings. Road Research Technique Paper 39. Road Research Laboratory, London (1958)
2. Liang, X., Guler, S.I., Gayah, V.V.: An equitable traffic signal control scheme at isolated signalized intersections using connected vehicle technology. Transp. Res. Part C Emerg. Technol. **110**, 81–97 (2020)
3. Stevanovic, A., Stevanovic, J., So, J., Ostojic, M.: Multi-criteria optimization of traffic signals: mobility, safety, and environment. Transp. Res. Part C Emerg. Technol. **55**, 46–68 (2015)
4. Tiaprasert, K., Zhang, Y., Wang, X.B., Zeng, X.: Queue length estimation using connected vehicle technology for adaptive signal control. IEEE Trans. Intell. Transp. Syst. **16**, 2129–2140 (2015)
5. Liu, G., Li, P.F., Qiu, T.Z., Han, X.: Development of a dynamic control model for oversaturated arterial corridor. Procedia-Soc. Behav. Sci. **96**, 2884–2894 (2013)
6. Comert, G.: Simple analytical models for estimating the queue lengths from probe vehicles at traffic signals. Transp. Res. Part B: Methodol. **55**, 59–74 (2013)
7. Hao, P., Sun, Z.B., Ban, X.G., Guo, D., Ji, Q.: Vehicle index estimation for signalized intersections using sample travel times. Transp. Res. Part C Emerg. Technol. **36**, 513–529 (2013)
8. Briesemeister, L., Schafers, D., Hommel, G.: Disseminating messages among highly mobile hosts based on inter-vehicle communication. In: IEEE Intelligent Vehicle Symposium, pp. 522–527. IEEE Press, Detroit (2000)
9. Zhang, Y., Zhou, Y.: Distributed coordination control of traffic network flow using adaptive genetic algorithm based on cloud computing. J. Netw. Comput. Appl. **119**, 110–120 (2018)
10. Cao, Z., Lu, L., Chen, C., Chen, X.: Modeling and simulating urban traffic flow mixed with regular and connected vehicles. IEEE Access. **9**, 10392–10399 (2021)
11. Pandit, K., Ghosal, D., Zhang, H.M., Chuah, C.: Adaptive traffic signal control with vehicular ad hoc networks. IEEE Trans. Veh. Technol. **62**(4), 1459–1471 (2013)
12. Ott, J., Kutscher, D.: Drive-thru internet: IEEE 802.11b for "automobile" users. In: Twenty-third Annual Joint Conference of the IEEE Computer and Communications Societies, pp. 373. IEEE Press, Hong Kong, China (2004)
13. Lopez, P. A., Behrisch, M.: Microscopic traffic simulation using SUMO. In: 21st International Conference on Intelligent Transportation Systems, pp. 2575–2582. IEEE Press, Maui (2018)
14. Sommer, C., Yao, Z., German, R., Dressler, F.: Simulating the influence of IVC on road traffic using bidirectionally coupled simulators. In: IEEE INFOCOM Workshops 2008, pp. 1–6. IEEE Press, Phoenix (2008)

# UJN-Land: A Large-Scale High-Resolution Parcel of Land of Multi-temporal Dataset with CNN Based Semantic Segmentation Analysis

Yuan Shen[1,2,3], Yulin Wang[4], Shiyong Yang[5](✉), Yan Li[1,2,3], Shiyuan Han[1,2,3], Zhen Liu[1,2,3], and Tao Xu[1,2,3](✉)

[1] School of Information Science and Engineering, University of Jinan, Jinan 250022, China
ise_xut@ujn.edu.cn

[2] Highresolution Earth Observation System Shandong Center of Data and Application, University of Jinan, Jinan 250022, China

[3] Shandong Provincial Key Laboratory of Network Based Intelligent Computing, University of Jinan, Jinan 250022, People's Republic of China

[4] Shandong Institutes of Industrial Technonlogy, Jinan 250102, China

[5] Land Spatial Data and Remote Sensing Technology Institute of Shandong Province, Jinan 250002, China

**Abstract.** Semantic segmentation of land in remote sensing images plays an important role in urban management and rural planning, and can provide intelligent analysis for urban development. Convolutional neural network (CNN) based methods have achieved good performance in object detection in computer vision, which benefits from the popularity of large-scale datasets. However, most of segmentation of land dataset were constructed based on various satellite images and samples that lack the number and diversity of existing datasets. These problems have brought challenges to the improvement and application of CNN based land parcel segmentation methods. This paper presented a large-scale high-resolution land dataset based on CNN analysis. More than 200,000 images of 256*256 pixels were obtained from Gaofen satellite images with a spatial resolution of 0.8 m, in which four category samples were manually annotated. In the experiment, three CNN based semantic segmentation methods were used as experimental baseline methods. The experimental results showed that the availability of the UJN-Land dataset is beneficial to the planning and management of cultivated land.

**Keywords:** Semantic segmentation · UJN-Land dataset · Multi-temporal

## 1 Introduction

In recent years, the application of high-resolution remote sensing satellite images has made great progress, and great progress has been made in the protection and planning of cultivated land. High-resolution remote sensing satellite images have the advantages of high spatial resolution and high temporal and spatial resolution, and are convenient


Q. Han et al. (Eds.): LSMS 2021/ICSEE 2021, CCIS 1469, pp. 204–213, 2021.
https://doi.org/10.1007/978-981-16-7213-2_20

for applications in land use, regional planning, food security, environmental monitoring. In the above-mentioned research fields, semantic segmentation and classification of multi-temporal land is a key step in the application of remote sensing images, and it plays an important role in land and resources surveys, land functional area planning, and land category change detection. With the increase in the number of satellites and the improvement of the spatial resolution of satellite images, the semantic segmentation of high-resolution satellite image dataset has attracted the attention of researchers, especially in the field of crops.

Many methods [1–7] of land and crop datasets and land functional block segmentation are proposed by research. At present, researchers have proposed datasets containing land based on Google Earth and the United States Geological Survey, such as the DeepGlobe Land Cover Classification Challenge dataset, GID dataset, CCF's BDCI 2020 competition dataset and NWPU Dataset. Large-scale annotated datasets are very important for convolutional neural network based semantic segmentation methods, as is the segmentation of land parcels in remote sensing images. However, the existing land segmentation datasets are usually constructed based on various satellites with different spatial resolutions. In addition, the number and diversity of existing datasets are insufficient, the annotation process of ground truth is slow, and the annotation accuracy is not high. These problems have brought challenges to the improvement and application of the CNN based land segmentation method. In response to the above problems, this paper proposes a large-scale high resolution dataset with experimental analysis of semantic segmentation of multi-temporal remote sensing land. The proposed dataset is named UJN-Land dataset, which is based on the high-resolution satellite image of China Gaofen-2. Figure 1 shows an overview of the keyway of the dataset. Acquired 200,000 remote 256*256 sensing images, and manually annotated some samples in the images. UJN-Land dataset has the characteristics of large scale, in-cabin diversity and similarity between classes. In the experiment, the algorithm based on semantic segmentation of multi-temporal land was used as the baseline method of land segmentation to verify UJN-Land dataset.

The main contributions of this paper are: (1) A large-scale high-resolution multi-temporal land segmentation dataset for remote sensing images is proposed. (2) Based

**Fig. 1.** Overview of UJN-lands.

on the convolutional neural network method, detailed experimental analysis was carried out to prove the validity of the proposed dataset.

The rest of the paper is organized as follows: the existing land datasets and semantic segmentation methods are summarized in Sect. 2. the proposed UJN-Land dataset is introduced in Sect. 3. Section 4 presents the experimental results with analysis and conclusions are given in Sect. 5.

## 2 Related Work

Datasets and algorithms [2, 3] for remote sensing semantic segmentation with high resolution satellite images have been proposed and implemented in many fields. The commonly used datasets for semantic segmentation are summarized in Table 1.

**Table 1.** Parcel of land datasets.

| Dataset | Sources | Image size | Number of samples | Number of sample type | Spatial Resolut-ion/m | Year |
|---|---|---|---|---|---|---|
| EvLab-SS | World-View-2/GeoEye/Quick Bird GF2/Aerial image | 4500*4500 | 60 | 11 | 0.1–2 | 2013 |
| RSSCN7 | – | 400*400 | 2800 | 7 | - | 2015 |
| SIRI-WHU | Google Earth | 200*200 | 2400 | 12 | 2 | 2016 |
| NWPU Dataset | – | 256*256 | 31500 | 45 | 0.2 | 2017 |
| DeepGlobe Land Cover Classification Challenge | DigitalGlobe | 2448*2448 | 803 | 7 | 0.5 | 2018 |
| DLRSD | USGS National Map | 256*256 | 2100 | 21 | – | 2018 |
| GID | GF-2 | 7200*6800 | 150 | 15 | 0.8–10 | 2018 |
| DeepGlobe Road Detection Challenge | DigitalGlobe | 1024*1024 | 6226 | 1 | – | 2018 |
| SEN12MS | Sentinel-1/Sentinel-2/MODIS | – | 180748 | – | 20/10/500 | 2019 |
| BDCI2020 | – | 256*256 | 145981 | 7 | – | 2020 |
| FUSAR-Map | GaoFen-3 | 1024*1024 | 610 | 4 | 3 | 2021 |
| UJN-Land | GaoFen-2 | 256*256 | 200000 | 4 | 0.8 | 2021 |

Researchers have also proposed many semantic segmentation methods for remote sensing images. The semantic segmentation methods based on deep learning has achieved great success in computer vision and CNN based semantic segmentation method for remote sensing images has been proposed. For example, Deeplab V3+ [4] adapt the Xception model for the segmentation task and apply depthwise separable convolution to both ASPP module and decoder module. There can arbitrarily control the resolution of extracted encoder features by convolution to compromise precision and runtime. A Multi-Path Refinement Networks for High-Resolution Semantic Segmentation called RefineNet was proposed by Lin et al. [6]. Their method explicitly leverages all available information from the down-sampling process, allowing for high-resolution predictions using remote residual connections.

For semantic segmentation and classification of remote sensing images, researchers have built many datasets and made great progress in semantic segmentation and classification algorithms. However, challenges such as insufficient number and diversity of samples and the cumbersome labeling process, which will lead to a negative impact on research and the semantic segmentation of remote sensing images based on convolutional neural networks. Solving the above problems, this paper firstly constructs a land dataset named UJN-Land based on GaoFen satellite. Compared with the other existing datasets, UJN-Land dataset has the advantages of high-resolution, sufficient number of samples, and diverse samples, which is convenient for the research of semantic segmentation in remote sensing images. Secondly, CNN based methods on detailed analysis is proposed in experiments. Figure 2 show the framework of our work, in which an efficient method of constructing the dataset is also illustrated.

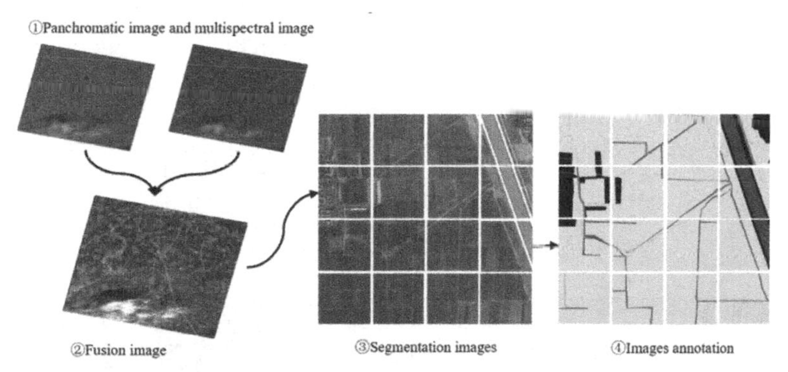


**Fig. 2.** The framework of our work. First, the high resolution panchromatic and multispectral images are selected from GaoFen satellite images. Obviously, fusion image is formed by fusion of two images, and fusion image is divided into 256*256 images and labeled manually.

## 3 UJN-Land

UJN-Land is a multi-temporal land parcel dataset constructed by manual annotation based on Gaofen-2 satellites which is equipped with two high-resolution 1-m panchromatic and 4-m multispectral cameras, with sub-meter spatial resolution, hyperspectral, high spatial resolution, all-weather, all-weather, real-time/quasi-real-time Earth observation capability, high positioning accuracy and rapid attitude maneuverability. The following is a detailed description and efficient labeling process of UJN-Land Dataset.

### 3.1 Efficient Labeling

In our work, UJN-Land Dataset currently contains 200,000 images of 256*256 pixels of the main winter wheat producing areas in Shandong Province. The train dataset categories are now divided into wheat, road, soil and water.

Firstly, the 1 m panchromatic and 4 m multispectral images of Gaofen-2 are respectively orthorectified, and then the two corresponding images are fused into a satellite image with a resolution of 0.8 m. Next, we will handle fusion image with color correction and color balance, which is divided into images of 256*256 pixel. After the divided image format is converted to jpg format, it is pixel-level annotated to create annotated items and sample type attributes and store the semantic information of each pixel. UJN-Land dataset labels and colors are shown in Table 2. After the image is annotated, a json format file will be generated, and the json format file will be uniformly converted into a png format file, which named corresponding to the jpg format file and saved. Part of the image of the annotated dataset is shown in Fig. 3.

**Table 2.** Labels and colors of UJN-land datasets.

| Label | Color | RGB |
|---|---|---|
| Road | Red | [255,0,0] |
| Wheat | Green | [0,255,0] |
| Soil | Gold | [255,215,0] |
| Water | Blue | [0,0,255] |

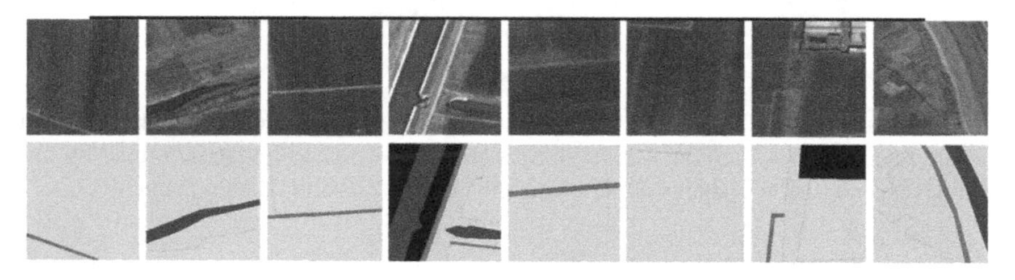

**Fig. 3.** Part of the image of the annotated dataset.

### 3.2 Sample in UJN-Land Dataset

UJN-Land dataset contains more than 200,000 images, which selects GaoFen-2 satellite images under different conditions under different weather, seasons and imaging conditions for sample labeling. The sample image has a spatial resolution of 0.8m, covering suburban, rural, mountainous, and grain growing areas. The sample images of UJN-Land datasets are displayed in in Fig. 4, which are selected images of winter wheat in different seasons and different months in Shandong Province for fusion images labeling. In terms of multi-temporal land segmentation, UJN-Land datasets have the characteristics of large-scale, intra-class diversity and similarity between classes. UJN-Land remote sensing images in different months are showed in Fig. 4. Although the existing remote sensing datasets contain more scenes, the number of samples cannot meet the needs of model training for semantic segmentation or classification of multi-temporal land based on CNN. To the best of our knowledge, the UJN-land dataset proposed in this paper is the largest high-resolution remote sensing dataset currently used for semantic segmentation of multi-temporal land parcels.

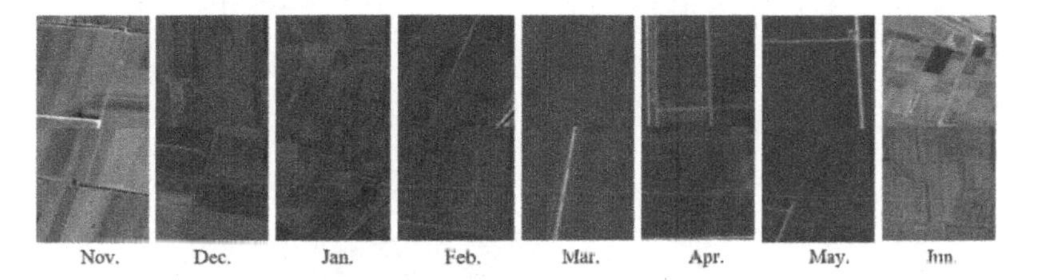


**Fig. 4.** UJN-Land images in different months.

## 4 Experimental Result and Analysis

In order to verify the usability of UJN-Land dataset, three commonly used deep convolutional neural networks for semantic segmentation. Segnet [10], Unet [11] and PsPnet [12] models are used for verification experiments. Ten-fold cross-validation is used in the experiment. The number of images for model training is 3000. The hardware environment uses 2080Ti GPU server, TensorFlow and Pytorch deep learning framework. The following is an introduction to our experiment.

### 4.1 Experimental of Segnet

In our experiments, the main parameters of Segnet are set as follows. We use the backbone network mobile [5] model to reduce the amount of calculation. It uses the 11 layers of the Encoder's 13-layer network structure to extract multiple features of f4 for processing. Mobile modle uses 2*2 window and stride 2 (non-overlapping window) to perform maximum pooling, and output the sub-sample twice. Maximum pooling is used to achieve translation invariance on small spatial displacements in the input image. Sub-sampling generates a large input image context (spatial window) for each pixel in the feature map and where alpha = 1.0, depth_multiplier = 1. Mobile modle uses decoder to upsampling Upsampling2D multiple times. Finally, a specific number of layers is obtained, and the result is transformed into reshape.

The loss function for Segnet is:

$$L(Y,\ P(Y|X)) = -\log P(Y|X)\,. \tag{1}$$

Where L(Y, P(Y|X)) means that the sample X makes the probability P(Y|X) reach the maximum value in the case of classification Y.

The accuracy and loss function graphs of the Segnet training set based on the Mobile backbone network, as well as the predicted accuracy and loss function are shown in the Fig. 5.

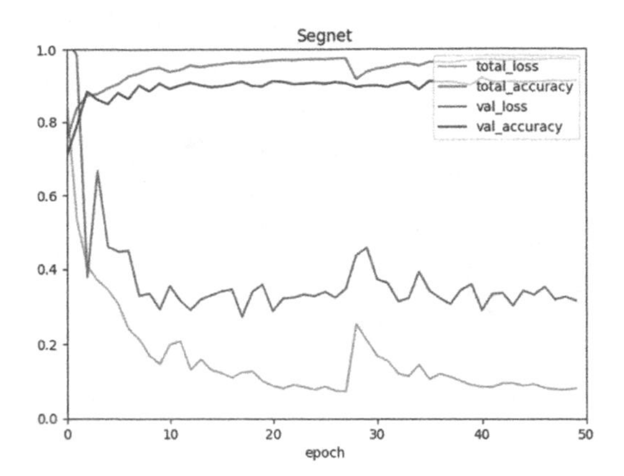


**Fig. 5.** The accuracy and loss function graphs of the Segnet training set based on the Mobile backbone network, as well as the predicted accuracy and loss function.

## 4.2 Experimental of Unet

The Unet network structure consists of three parts. The first part is the backbone feature extraction part. We can use the backbone part to obtain one feature layer after another. The backbone feature extraction part of Unet is similar to VGG, which is a stack of convolution and maximum pooling. By using the main feature extraction part, we can obtain five preliminary effective feature layers. In the second step, we will use these five effective feature layers to perform feature fusion. The second part is to strengthen the feature extraction part. We can use the five preliminary effective feature layers obtained from the main part to perform up-sampling and perform feature fusion to obtain a final effective feature layer that combines all features. The third part is the prediction part. We will use the last effective feature layer finally obtained to classify each feature point, which is equivalent to classifying each pixel.

The energy function is computed by a pixel-wise soft-max over the final feature map combined with the cross entropy loss function.

The soft-max is defined as:

$$p_k(x) = \frac{\exp(a_k(\mathbf{x}))}{\sum_{k'=1}^{K} \exp\left(a_{k'}(x)\right)} \tag{2}$$

Where $a_k$(x) denotes the activation in feature channel k at the pixel position x ∈ Ω with Ω ∈ $Z^2.K$ is the number of classes and $p_k(x)$ is the approximated maximum-function.

I.e. $p_k(x) \approx 1$ for the k that has the maximum activation $a_k(x)$ and $pk(x) \approx 0$ for all other k.

The accuracy and loss function graphs of the Unet training set based on the backbone network, as well as the predicted accuracy and loss function are shown in the Fig. 6.

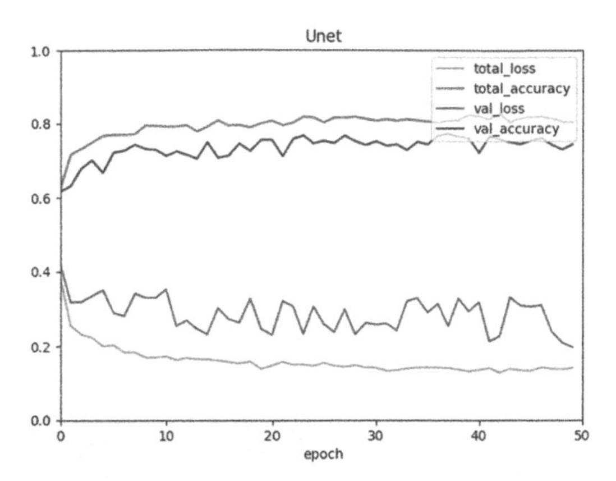


**Fig. 6.** The accuracy and loss function graphs of the Unet training set based on the Mobile backbone network, as well as the predicted accuracy and loss function.

## 4.3 Experimental of PsPnet

They divide the acquired feature layer into grids of different sizes through our pyramid pool module and the proposed pyramid scene analysis network (PsPNet), and each grid is internally averaged and pooled through context based on different regions. To take advantage of the function of global context information. Our global a priori representation can effectively produce high-quality results on scene parsing tasks, and PsPNet provides an excellent framework for pixel-level prediction.

In our experiments, the main parameters of PsPnet are set as follows: base learning rate = 0.01, power = 0.9, momentum = 0.9, weight decay = 0.0001, batchsize = 16.

The accuracy and loss function graphs of the PsPnet training set based on the backbone network, as well as the predicted accuracy and loss function are shown in the Fig. 7.

## 4.4 Parcel Semantic Segmentation Results

The wheat information in the land has obvious appearance characteristics, these wheat fields usually exist in rural areas or near towns. These models can accurately identify these three types of label information in UJN-LandDataset, which can be accurately segmented, and Segnet has achieved better detection results. Partial semantic segmentation prediction result are shown in the Fig. 8.

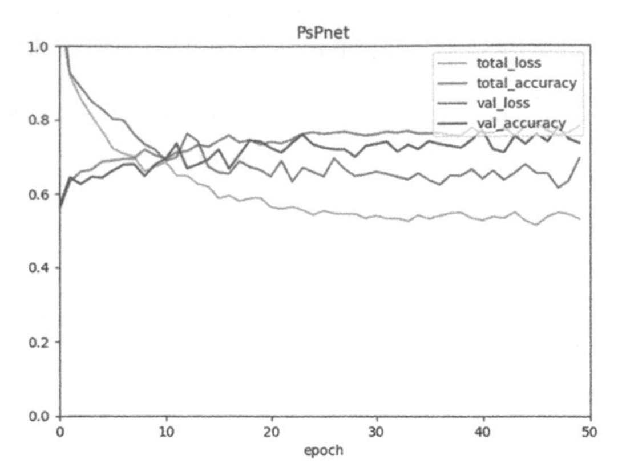


**Fig. 7.** The accuracy and loss function graphs of the PsPnet training set based on the Mobile backbone network, as well as the predicted accuracy and loss function.

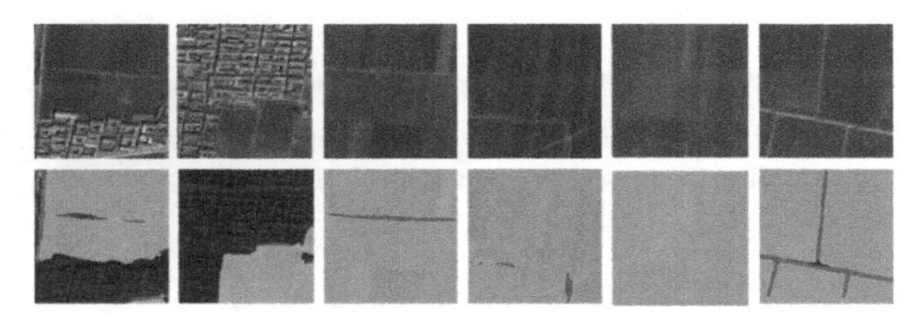

**Fig. 8.** Partial semantic segmentation prediction result.

# 5 Conclusion and Discussion

This paper proposes a large-scale multi-temporal land parcel segmentation dataset named UJN-Land. Segnet, Unet and PsPnet were utilized for validation of UJN-Land Dataset. The proposed dataset contains more than 1 million land samples, and some of the data set samples have been manually annotated. The proposed data set contains more than 200,000 sample images of land, and some of the data set samples have been manually annotated. The land samples in the dataset are rich and diverse, with the characteristics of large-scale, multi-scale, intra-class diversity and inter-class similarity in land. The experimental results verify the validity of UJN-Land. The proposed dataset and semantic segmentation method from UJN-Land dataset will play an important role in the application of urban and rural planning, such as land management and planning. In the future work, we will further expand the category and quantity of UJN-Land, and propose effective segmentation methods for the semantic segmentation of multi-temporal satellite images and multi-category. The detection of the semantic information segmentation of the same plot in different months will be the focus of the next step.

**Acknowledgments.** This research was supported by the National Natural Science Foundation of China under Grants 61903156 and 61873324, the Natural Science Foundation of Shandong Province for Key Project under Grant ZR2020KF006, and the Taishan Scholar Project of Shandong

Province under Grant tsqn201812077, and Natural Science Foundation of University of Jinan (XKY1928, XKY2001).

## References


1. Islam, M.A., Rochan, M., Bruce, N., Yang, W.: Gated feedback refinement network for dense image labeling. In: Computer Vision Pattern Recognition. IEEE Press (2017)
2. Jegou, S., Drozdzal, M., Vazquez, D., Romero, A., Bengio, Y.: The one hundred layers tiramisu: Fully convolutional densenets for semantic segmentation. In: 2017 IEEE Conference on Computer Vision and Pattern Recognition Workshops (CVPRW). IEEE Press (2016)
3. Long, J., Shelhamer, E., Darrell, T.: Fully convolutional networks for semantic segmentation. J. IEEE Trans. Pattern Anal. Mach. Intell. **39**(4), 640–651 (2015)
4. Chen, L.C., Zhu, Y., Papandreou, G., Schroff, F., Adam, H.: Encoder-decoder with atrous separable convolution for semantic image segmentation. In: Ferrari, V., Hebert, M., Sminchisescu, C., Weiss, Y. (eds.) ECCV 2018. LNCS, vol. 11211, pp. 833–851 Springer, Cham (2018). https://doi.org/10.1007/978-3-030-01234-2_49
5. Howard, A.G., et al: Mobilenets: Efficient convolutional neural networks for mobile vision applications (2017)
6. Lin, G., Milan, A., Shen, C., Reid, I.: Refinenet: multi-path refinement networks for high-resolution semantic segmentation. In: 2017 IEEE Conference on Computer Vision and Pattern Recognition (CVPR) (2017)
7. Yi, Z., Chang, T., Li, S., Liu, R., Hao, A.: Scene-aware deep networks for semantic segmentation of images. IEEE Access **7**, 69184–69193 (2019)
8. Paszke, A., Chaurasia, A., Kim, S., Culurciello, E.: Enet: A deep neural network architecture for real-time semantic segmentation (2016)
9. Wang, M., Ma, Y., Li, F., Guo, Z.: A hybrid semantic segmentation based on levelset evolution driven by fully convolutional networks. IEEE Access **9**, 42556–42567 (2021)
10. Badrinarayanan, V., Kendall, A., Cipolla, R.: Segnet: a deep convolutional encoder-decoder architecture for image segmentation. J. IEEE Trans. Pattern Anal. Mach. Intell. **39**(12), 2481–2495 (2017)
11. Ronneberger, O., Fischer, P., Brox, T.: U-net: Convolutional networks for biomedical image segmentation. In: Navab N., Hornegger J., Wells W., Frangi A. (eds.) Medical Image Computing and Computer-Assisted Intervention – MICCAI 2015. MICCAI 2015. LNCS, vol. 9351, pp. 234–241. Springer, Cham (2015). https://doi.org/10.1007/978-3-319-24574-4_28
12. Zhao, H., Shi, J., Qi, X., Wang, X., Jia, J.: Pyramid scene parsing network. J. IEEE Comput. Soc. (2016)

# An End-to-End Feature-Complementing Method for Building Classification with Residual Network

Zhongyi Zhang[1,2], Kun Liu[1,2(✉)], Rui Ding[3], Tao Xu[1,2], Jinguang Yao[1,2], and Tao Sun[1,2]

[1] School of Information Science and Engineering, University of Jinan, Jinan 250022, People's Republic of China

[2] China High-Resolution Earth Observation System Shandong Center of Data break and Application, University of Jinan, Jinan 250022, People's Republic of China
ise_liuk@ujn.edu.cn

[3] School of Comprehensive, Shandong Vocational University of Foreign Affairs, Weihai 264504, People's Republic of China

**Abstract.** With the development of deep learning, deep learning algorithms are gradually applied to remote sensing image interpretation. However, most of the current algorithms are solely using panchromatic images or multispectral images for processing, which has certain limitations. This paper proposed an end-to-end feature-complementing artificial building classification method and did experiments on remote sensing images captured by the domestic remote sensing satellite Gaofen-2. The results show that this method can improve classification accuracy. Through the analysis of the results, the guidance for improving the performance of the classifier is summarized, which provides a reference for the selection and optimization of the feature extraction network in the remote sensing image interpretation algorithm based on deep learning.

**Keywords:** Remote sensing · Building classification · Feature complementary · Residual network

## 1 Introduction

Remote sensing technology is a very important technology for earth observation. By launching unmanned aerial vehicles or artificial satellites with various electromagnetic wave sensors, capturing electromagnetic waves of different frequencies reflected on the earth's surface and generating remote sensing images to observe the earth.

In recent years, as the deep learning technology rising, deep learning methods are widely used in remote sensing. Deep learning algorithms have achieved significant success on many image analysis tasks [12], and it is applied in many aspects of social production and activity, such as traffic [17], building extraction [16], land use monitoring [6], forestry [1], ocean monitoring [9], etc. Deep learning algorithms can be used in most tasks of remote sensing image processing,


Q. Han et al. (Eds.): LSMS 2021/ICSEE 2021, CCIS 1469, pp. 214–223, 2021.
https://doi.org/10.1007/978-981-16-7213-2_21

such as classification, image fusion, image segmentation, change monitoring and image registration, etc.

However, compared with the three-channel RGB images processed in computer vision, remote sensing images have their particularities such as spatial and spectral resolution, making it difficult to directly apply algorithms in computer vision to remote sensing images. There are also differences in attributes of different types of remote sensing images. For panchromatic images and multispectral images acquired by the same remote sensing satellite, the spatial resolution of panchromatic images is higher than that of multispectral images. However, panchromatic images are single-channel images that only contain a single band of ground object information. But the multispectral image contains feature information of multiple bands. Based on the characteristics of panchromatic images and multispectral images, in order to make better use of the information of these two types of images, image fusion methods are indispensable.

This paper proposed an end-to-end feature-complementing artificial building classification method. By combining Gram-Schmidt pan sharpening fusion algorithm [8] and Resnet-101 as the backbone network, it reached a higher accuracy of artificial buildings classification on domestic remote sensing satellite images. The experiments were done on the 5M-Building data set [11], an artificially annotated data set based on the remote sensing images taken by domestic remote sensing satellite Gaofen-2. The results show that the fusion of panchromatic and multispectral images can improve the recognition accuracy, and it also verifies that Resnet-101 is a better feature extractor of remote sensing images compared with other commonly used convolutional neural networks in remote sensing image classification tasks. By doing experiments the effectiveness of this method was verified.

The rest of the paper is organized as follows: Sect. 2 introduces the related work of remote sensing image classification methods, and summarizes the classic remote sensing image fusion methods and image classification methods. Section 3 introduces the model structure and algorithm steps of the method proposed in this paper in detail. Section 4 introduces the experiments of longitudinal comparison of different types of data and the lateral comparison of different neural network models and the evaluation of experimental results. Finally, Sect. 5 concludes the full paper.

## 2 Related Work

In recent years, with the explosive growth of data and computing power, image classification methods based on deep learning have emerged one after another, and excellent results have been achieved on many data sets. The classification method based on RGB image provides a reference for remote sensing image classification. For example, the classic image classification model AlexNet [7], Inception V3 [15], etc. Many researchers have improved the classification method based on RGB images combined with the characteristics of remote sensing images, and achieved good results. Such as remote sensing image scene classification method based on AlexNet and pyramid pooling [4], unknown ship classification method based on Inception V3 network [10], etc. In order to make better use of panchromatic and multispectral images, it is necessary to perform fusion operations on remote sensing images.

### 2.1 Image Fusion Methods

Remote sensing image fusion methods can be divided into the following categories: 1) fusion methods based on component replacement, 2) fusion methods based on multi-resolution analysis, 3) hybrid fusion methods, 4) model-based fusion methods. The method based on component replacement uses mathematical transformation to map the multi-spectral image to another space to separate the spatial structure and spectral information into different components. This type of method includes methods based on principal component analysis [3,13], Gram-Schmidt method [8] and Brovey transformation method, etc. Methods based on multi-resolution analysis mainly include fusion methods based on wavelet transform, least squares support vector machines, etc. The hybrid method is a fusion method that combines the advantages of component replacement and multi-resolution analysis, such as the fusion method based on independent component analysis [2] and wavelet decomposition. Model-based fusion methods mainly include methods based on Markov random fields, methods based on statistics, methods based on coefficient matrix decomposition, etc.

### 2.2 Image Classification Methods

The image classification methods can be divided into two types: traditional methods and methods based on deep learning. With the continuous development of deep learning, image classification algorithms based on deep learning emerge in an endless stream, and their classification accuracy far exceeds traditional image classification algorithms. In 2012, AlexNet [7] won the championship in the ILSVRC-2012 competition, proved that the deep learning method has huge potential for processing image classification tasks. In 2014, LeCun and others proposed the Inception (GoogLeNet) network [14], which once again increased the optimal classification accuracy rate on the ImageNet dataset to 69.8%. After continuous improvement, Inception's newer version of Inception V3 Achieved a Top1 accuracy rate of 78.8%. Resnet [5] is one of the most breakthrough methods in the field of computer vision in recent years. It has solved the deep neural network to a certain extent by introducing residual connections. The problem of network gradient dispersion makes the output of the network contain more input information and shallow information, which further improves the classification accuracy.

## 3 Proposed Method

In this paper, we propose an end-to-end feature-complementing artificial building classification method. As shown in the architecture diagram of the method of Fig. 1, this method accepts panchromatic and multispectral image as input, and uses the Gram-Schmidt pan sharpening method [8] to complement the information of the panchromatic image and the multispectral image and send them directly into Resnet network for classification.

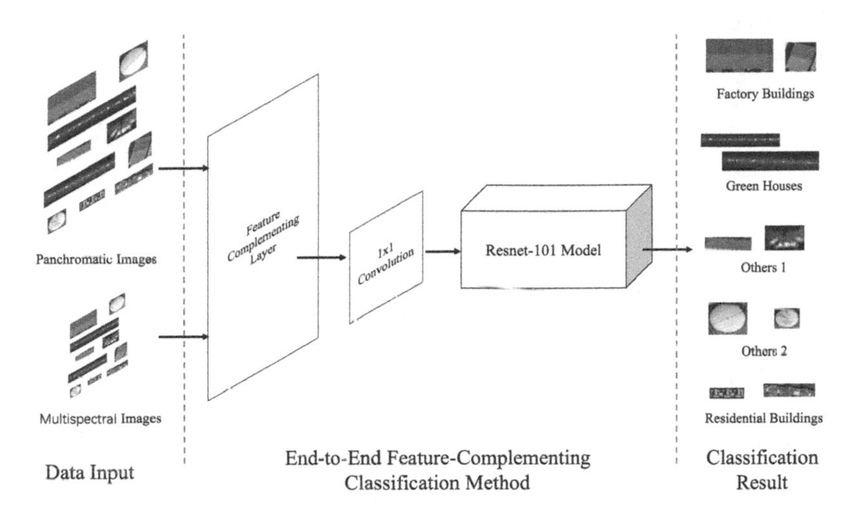


**Fig. 1.** Architecture of proposed method

In order to make better use of the information about the panchromatic image and the multispectral image and improve the classification effect, the algorithm used the fusion method to complement the image features. The algorithm accepts the registered panchromatic images and multispectral images, and uses Gram-Schmidt pan sharpening method for fusion.

The detailed steps of the Gram-Schmidt fusion method are as follows: The first step is to simulate a panchromatic band from the low-resolution spectral band. In order to simulate this band, each band of the multispectral image is multiplied by an appropriate coefficient, and then added together to obtain the simulated panchromatic image $PAN_{sim}$ with low resolution.

The second step is to take the simulated panchromatic band as the first variable of GS transform to carry out Gram Schmidt transform for the simulated panchromatic band and multispectral band which can be expressed by:

$$I_{GS1}(i,j) = PAN_{sim}(i,j) \tag{1}$$

where $i$, $j$ are the row and column number of the image. $I_{GS1}$ is the first band of Gram Schmidt transform. The $T$-th GS variable is constructed from the previous $T-1$ GS variables by:

$$GS_T(i,j) = (MSS_T(i,j) - \mu_T) - \sum_{l=1}^{T-1} \phi(MSS_T, GS_l) \times GS_l(i,j) \tag{2}$$

where $GS_T$ is the $T$-th variable generated after GS transform, $MSS_T$ is the $T$-th band affected by the original multi-spectrum. $\mu_T$ is the mean of the grayscale value of the $T$-th original multispectral band image. By adjusting the statistical value of the high-resolution panchromatic image, that can match the first variable after the GS transform $GS1$, and the modified high-resolution panchromatic images can be produced. This modification helps to maintain the spectral characteristics of the original multispectral band image, the modification is as follows:

$$\mu_T = \frac{\sum_{j=1}^{W}\sum_{i=1}^{H} MSS_T(i,j)}{W \times H} \tag{3}$$

$$\phi(MSS_T, GS_l) = \left[\frac{\sigma(MSS_T, GS_l)}{\sigma(GS_l, GS_l)^2}\right] \tag{4}$$

$$\sigma_t = \sqrt{\frac{\sum_{j=1}^{W}\sum_{i=1}^{H}(MSS_T(i,j) - \mu_T)}{W \times H}} \tag{5}$$

where $W$, $H$ are the width and height of the images respectively, $\phi$ is the covariance, $\sigma_T$ is the standard deviation of the $T$-th band. Finally, the modified high-resolution panchromatic image is used to replace the first band after GS transform, generating a new set of transform bands, forms the fused high-resolution multi-spectral image which applies in the inversion of GS transform. The inversion method of GS transform can be expressed by:

$$MSS_T(i,j) = (GS_T(i,j) + \mu_T) + \sum_{l=1}^{T-1} \phi(MSS_T, GS_l) \times GS_l(i,j) \tag{6}$$

After the above methods, a fused image can be obtained. Further, the size of fused image is adjusted to make it correspond to the network input size. The proposed method added a $1 \times 1$ convolution layer in front of Resnet. This operation can reduce the dimension of the input data and reduce the computational complexity. It also gives the ability to accept images of different channel numbers as input to enhance the robustness of the algorithm.

The proposed method uses residual neural network as backbone network for image classification. In 2015, He et al. proposed residual network [5], which added residual connections on the traditional CNN, make it easier to optimize, and can gain accuracy from considerably increased depth. Therefore, the deep structure of the network can get shallow information, and the network can converge easily, avoid the network degradation and achieve higher correct rates. Resnet consists of several residual blocks, which can be represented by the following formula:

$$x_{l+1} = h(x_l) + \mathcal{F}(x_l, W_l) \tag{7}$$

A residual block has two parts: residual part and identity mapping part. $h(x_l)$ represents identity mapping part, $\mathcal{F}(x_l, W_l)$represents residual part, consists of two or three convolution operations typically. When $x_l$ and $x_{l+1}$ have different number of feature maps, $1 \times 1$ convolution is needed to increase or reduce the dimension of $x_l$, makes $h(x_l) = W_l'x$, where $W_l'$ is a $1 \times 1$ convolution operation. When $x_l$ and $x_{l+1}$ have the same number of feature maps, $h(x_l) = x_l$.

Experiments have proven that Resnet networks with residual connection have higher classification accuracy compared with other networks, which will be described in detail in Sect. 4.

# 4 Experiments

The experiment was conducted on 5M-Building data set [11]. Classical neural network models such as AlexNet, VGG-16, Inception V3 and four variants of Resent: Resnet-18, Resnet-34, Resnet-50, Resnet-101 are used for comparison.

Images of the 5M-building data set is derived from domestic remote sensing satellite Gaofen-2, in which the spatial resolution of panchromatic image was 0.8 m, and the spatial resolution of multispectral image was 3.2 m. This data set contains a large number of artificially labeled building samples, including residential buildings, factory buildings, greenhouses and two types of buildings labeled as other 1 and other 2, a total of five classes of buildings. The panchromatic and multispectral images of the dataset are aligned by geographic location information registration. In this paper, some annotated regions of 5M-building dataset are randomly extracted and formed a sub-dataset. The category of the sub-dataset is the same as that of the original dataset, and the same number of samples are randomly selected from each category to balance the dataset. Each category contains 1595 samples, with a total of 7975 building samples.

All of models use stochastic gradient descent (SGD) as the optimizer and use cross entropy loss function. Different learning rates are set for each model to make their performance as good as possible. $1 \times 1$ convolution is used for dimension reduction before all models. The weight initialization is the same as that of the model convolution. Each model modified the number of neurons in the last classification layer to match the number of categories. The weight initialization is the same as that of the full connection layer of the model. In this experiment, the pre-training model on ImageNet dataset is used to speed up the training process. We used PyTorch deep learning framework and 2 NVIDIA RTX 2080 Ti cards to complete this experiment.

For convenience, three schemes were defined. Scheme A classifies the multispectral images without feature complementary, the scheme B classifies the panchromatic images and without feature complementary as well, and the scheme C uses the end-to-end feature-complementing method to classify the two images of the panchromatic and multispectral images. Table 1 shows the accuracy of classification under three schemes using different backbone networks.

Figure 2 shows the RoC curves of Resnet-101 under scheme A, B and C. Figure 3 shows the confusion matrices of Resnet-101 under scheme A, B and C. Figure 4 shows the accuracy and macro-average RoC curves under scheme C.

Table 1 indicates that the Gram-Schmidt pan sharping fusion method combined with Resnet-101 model achieved the highest classification accuracy. By comparing different models, it can be seen that the residual connection introduced by Resnet transfers the shallow features to the deep layer of the network,

**Table 1.** The accuracy under three schemes using different backbone networks.

| Model | Scheme A | Scheme B | Scheme C |
|---|---|---|---|
| AlexNet | 85.87% | 87.12% | 91.87% |
| VGG-16 | 81.63% | 83.75% | 88.13% |
| Inception V3 | 84.62% | 86.62% | 90.87% |
| Resnet-18 | 84.75% | 86.75% | 90.75% |
| Resnet-34 | 85.25% | 86.62% | 90.63% |
| Resnet-50 | 86.12% | 87.88% | 92.37% |
| Resnet-101 | 86.75% | 88.25% | **93.50%** |

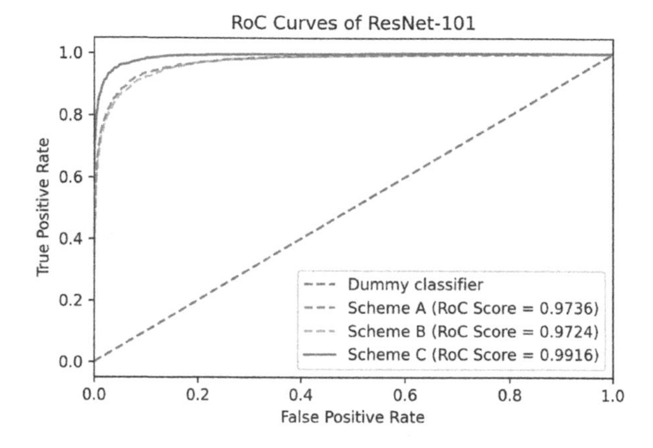


**Fig. 2.** RoC curves of Resnet-101 under scheme A, B and C.

the final extracted features contain more shallow features compared to other models. By using different sizes of convolution kernels in the Inception V3 network, the features of each layer contain information from a variety of receptive fields. Comparing Resnet-18, Resnet-34, Resnet-50 and Resnet-101, it can be found that as the network depth increases, the classification accuracy tends to rise. Hence, it can be concluded that for high-resolution remote sensing images, the extraction of image features should not only consider deep semantic features, but also need to combine the shallow detail features, which are both important for feature extraction. However, the effect on improving the feature extraction network is restricted by combining the feature information under different receptive fields.

From the perspective of the image type, it can be seen from Table 1 that the accuracy of the end-to-end feature-complementing classification method proposed in this paper is 4%–5% higher than that of the solely use of panchromatic

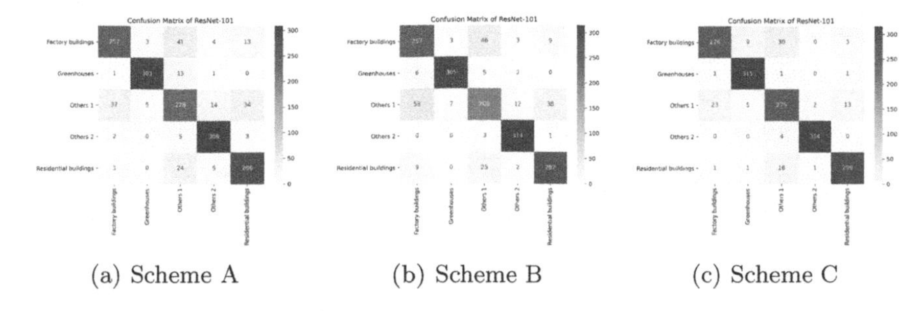


(a) Scheme A (b) Scheme B (c) Scheme C

**Fig. 3.** Confusion matrices of Resnet-101 under scheme A, B and C.

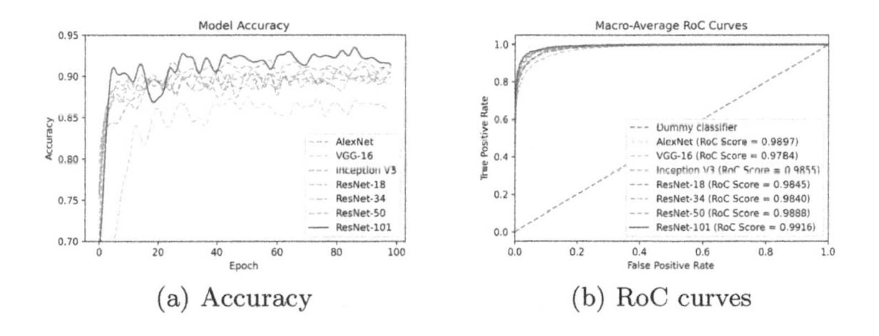


(a) Accuracy (b) RoC curves

**Fig. 4.** Accuracy and macro-average RoC curves under scheme C.

images and 5%–7% higher than that of multispectral images, respectively, which indicates the classification accuracy can be significantly improved by complementing the features of panchromatic and multispectral images.

From the perspective of the sample category, it can be seen from the Fig. 5(g) that the accuracy of samples of class Greenhouses and Other 2 is high, which is due to the shape, size and color of the Greenhouses samples on the image are relatively similar. The samples of class Other 2 are mainly circular buildings, which have obvious characteristics. The Factory buildings and the Other 1 category samples have a relatively low recognition rate due to their complex shapes and diverse colors. Figure 5 shows that the difference in the recognition ability of these two types of samples is the main reason for the difference in accuracy. This reflects that an appropriate increase in the number of network layers within a certain range has a large difference. When the appearance of samples are diverse, it has a positive effect. But for the category with obvious characteristics of the sample appearance, it does not benefit much.

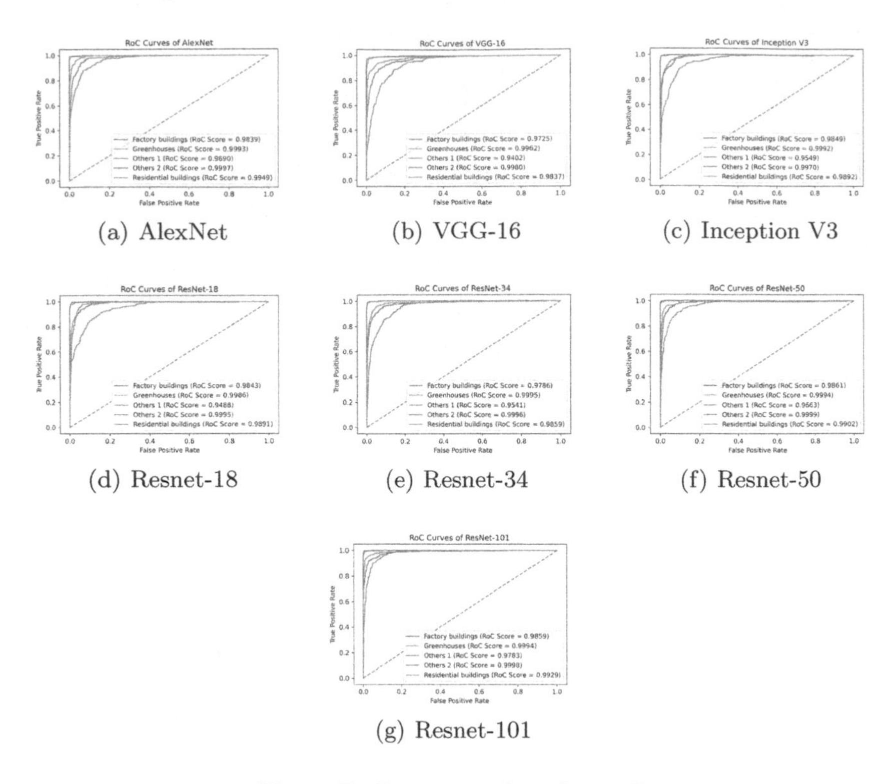


(a) AlexNet (b) VGG-16 (c) Inception V3

(d) Resnet-18 (e) Resnet-34 (f) Resnet-50

(g) Resnet-101

**Fig. 5.** RoC curves under scheme C.

# 5 Conclusion

This paper proposed an end-to-end feature-complementing artificial building classification method. The method combines the features of panchromatic images and multispectral images, adds $1 \times 1$ convolution to reduce the dimension of the fused image, and uses the Resnet-101 model for image classification. Finally, experiments prove that compared to using panchromatic images or multispectral images separately, the method proposed in this paper has a better performance on the 5M-Building data set based on the domestic Gaofen-2 remote sensing satellite. Through experiments, conclusion could be drawn that the methods that help to improve the classification effect of remote sensing images are as follows: 1) Combine panchromatic and multispectral image features; 2) Combine shallow and deep features; 3) For classes which the distribution of features within the class are relatively scattered, it is worth considering to deepen the network depth appropriately. In fact, the classification accuracy reflects the performance of the model as a feature extractor. This conclusion also has value for the selection of the backbone network of the deep learning model in the interpretation algorithm based on domestic remote sensing satellite images.

## References

1. Bai, B., Tan, Y., Guo, D., Xu, B.: Dynamic monitoring of forest land in fuling district based on multi-source time series remote sensing images. ISPRS Int. J. Geo-Inf. **8**(1), 36 (2019)
2. Chen, F., Qin, F., Peng, G., Chen, S.: Fusion of remote sensing images using improved ICA mergers based on wavelet decomposition. Procedia Eng. **29**, 2938–2943 (2012)
3. Green, A.A., Berman, M., Switzer, P., Craig, M.D.: A transformation for ordering multispectral data in terms of image quality with implications for noise removal. IEEE Trans. Geosci. Remote Sens. **26**(1), 65–74 (1988)
4. Han, X., Zhong, Y., Cao, L., Zhang, L.: Pre-trained AlexNet architecture with pyramid pooling and supervision for high spatial resolution remote sensing image scene classification. Remote Sens. **9**(8), 848 (2017)
5. He, K., Zhang, X., Ren, S., Sun, J.: Deep residual learning for image recognition. In: Proceedings of the IEEE Computer Society Conference on Computer Vision and Pattern Recognition, December 2016, pp. 770–778 (2016)
6. Joshi, N., et al.: A review of the application of optical and radar remote sensing data fusion to land use mapping and monitoring. Remote Sens. **8**(1), 70 (2016)
7. Krizhevsky, A., Sutskever, I., Hinton, G.E.: ImageNet classification with deep convolutional neural networks. Commun. ACM. **60**(6), 84–90 (2017)
8. Laben, C.A., Brower, B.V.: United States Patent 19. (19) (2000)
9. Lee, Z., Marra, J., Perry, M.J., Kahru, M.: Estimating oceanic primary productivity from ocean color remote sensing: a strategic assessment. J. Marine Syst. **149**, 50–59 (2015)
10. Liu, K., Yu, S., Liu, S.: An improved InceptionV3 network for obscured ship classification in remote sensing images. IEEE J. Sel. Top. Appl. Earth Obs. Remote Sens. **13**, 4738–4747 (2020)
11. Lu, Z., Xu, T., Liu, K., Liu, Z., Zhou, F., Liu, Q.: 5M-building: a large-scale high-resolution building dataset with CNN based detection analysis. In: Proceedings - International Conference on Tools with Artificial Intelligence, ICTAI, November 2019, pp. 1385–1389 (2019)
12. Ma, L., Liu, Y., Zhang, X., Ye, Y., Yin, G., Johnson, B.A.: Deep learning in remote sensing applications: A meta-analysis and review. ISPRS J. Photogram. Remote Sens. **152**, 166–177 (2019)
13. Shahdoosti, H.R., Ghassemian, H.: Combining the spectral PCA and spatial PCA fusion methods by an optimal filter. Inf. Fusion **27**, 150–160 (2016)
14. Szegedy, C., et al.: Going deeper with convolutions. J. Chem. Technol. Biotechnol. **91**(8), 2322–2330 (2014)
15. Szegedy, C., Vanhoucke, V., Ioffe, S., Shlens, J., Wojna, Z.: Rethinking the inception architecture for computer vision. In: Proceedings of the IEEE Computer Society Conference on Computer Vision and Pattern Recognition, vol. 2016, pp. 2818–2826. IEEE Computer Society, December 2016
16. Xu, Y., Wu, L., Xie, Z., Chen, Z.: Building extraction in very high resolution remote sensing imagery using deep learning and guided filters. Remote Sens. **10**(1), 144 (2018)
17. Zhang, Z., Liu, Q., Wang, Y.: Road extraction by deep residual U-net. IEEE Geosci. Remote Sens. Lett. **15**(5), 749–753 (2018)

# Optimal-Performance-Supervised Vibration Control for Nonlinear Discrete-Time System with Delayed Input Under Sinusoidal Disturbance

Shi-Yuan Han[1], Qi-Wei Sun[1], Xiao-Fang Zhong[2(✉)], and Jin Zhou[1]

[1] Shandong Provincial Key Laboratory of Network Based Intelligent Computing, University of Jinan, Jinan 250022, China
{ise_hansy,ise_zhouj}@ujn.edu.cn

[2] School of Data and Computer Science, Shandong Women's University, Jinan 250300, China

**Abstract.** In this paper, an optimal-performance-supervised vibration control strategy is developed for a class of discrete-time nonlinear systems subject to delayed input and sinusoidal disturbance, which makes up the optimal-trajectory-based vibration controller (OTVC) and the optimal-performance-supervised iterative algorithm (OPSIA). More specifically, by employing the reference optimal trajectories obtained from a closed-loop augmented system under the typical optimal state feedback controller, the original vibration control problem is reconstructed as an optimal-trajectory-guided tracking control (OTGTC) problem. After that, OTVC is derived from a sequence of nonhomogeneous linear two-point boundary value (TPBV) problem, which consists of the feedback term with system state, the feedforward terms with system states of sinusoidal disturbance and reference closed-loop system, and the compensation term with an infinity vector sequence for nonlinear dynamic and delayed input. Meanwhile, by defining the terminal condition involving with the desired minimum performance value, OPSIA is proposed to realize the computability of OTVC. Finally, the effectiveness of the proposed strategy is verified by employing a simple nonlinear discrete-time vehicle active suspension under different scenarios.

**Keywords:** Nonlinear control · Vibration control · Delayed input · Sinusoidal disturbance · Iterative algorithm

## 1 Introduction

Strictly speaking, the nonlinearities, external disturbances, and time delays are unavoidable in the real-world practical systems, such as ocean platforms [1,2], vehicle active suspensions [3,4], power systems [5,6], and so on. Unfortunately, while attempting to meet the classic optimal control objectives, these effective solutions for optimal control problem with linear systems are invalid for nonlinear


Q. Han et al. (Eds.): LSMS 2021/ICSEE 2021, CCIS 1469, pp. 224–233, 2021.
https://doi.org/10.1007/978-981-16-7213-2_22

systems under external disturbances and time delays, such as neither guaranteeing the system stability nor minimizing the related performance index [7,8]. As a result, many control strategies have been proposed to improve the control performance of nonlinear systems, such as observer-based control [8,9], adaptive control [1,4], $H\infty$ control [10,11], approximation optimal control [12,13], and so on.

Typically, the analytical theories and numerical methods are viewed as the basic mathematical tools for the analysis and controller design for nonlinear systems. In retrospect, various effective analytical methods were proposed, such as variational calculus [14,15] and Hamilton-Jacobi-Bellman (HJB) equation [16,17] for continuous-time systems, and dynamic programming involving with Bellman equation [18] for discrete-time systems. However, while discussing the nonlinear systems under the constrains of external disturbances and time delays, it is still difficult to directly gain the exact optimal solution [19,20].

Motivated by the aforementioned analyses, this paper proposes an iterative vibration control strategy for discrete-time nonlinear system with sinusoidal disturbances and delayed input, in which the iterative process is supervised by the reference optimal trajectory and desired minimum performance value.

## 2 Problem Description

Consider a nonlinear discrete-time system with delayed input under persistent sinusoidal disturbances

$$\begin{aligned} x(k+1) &= Ax(k) + Bu(k-h) + Dv(k) + f(x(k)), \\ x(k) &= x_0, \quad k = -h, -h+1, \ldots, 0, \\ y(k) &= Cx(k), \end{aligned} \tag{1}$$

where $x(k) \in \mathbb{R}^n$ is system state; $v(k) \in \mathbb{R}^r$ denotes the sinusoidal disturbance; $u(k) \in \mathbb{R}^m$ is control force with input delay $h > 0$; $y(k) \in \mathbb{R}^l$ denotes system output; $x_0$ is the initial state vector. $f(x(k)) : C^1(\mathbb{R}^n) \to U \subset \mathbb{R}^n$ is nonlinear term involved system state $x(k)$; $U$ is bounded open set and $0 \subset U$. $A, B, C$ and $D$ are the real constant matrices with appropriate dimensions.

The dynamic characteristics of the sinusoidal disturbance $v(k)$ can be expressed as

$$v(k) = \begin{bmatrix} v_1(k) \\ \vdots \\ v_r(k) \end{bmatrix} = \begin{bmatrix} \alpha_1 \sin(\omega_1 kT + \varphi_1) \\ \vdots \\ \alpha_r \sin(\omega_r kT + \varphi_r) \end{bmatrix}, \tag{2}$$

where $T$ denotes the sampling period; $\alpha_i$ and $\varphi_i$ are the unknown amplitude and phase, respectively; the known frequency $\omega_i$ satisfies the following condition

$$-\pi < \omega_1 \leq \omega_2 \leq \ldots \leq \omega_r \leq \pi. \tag{3}$$

The classic aim of vibration control problem is to offset or eliminate hazards of vibration caused by sinusoidal disturbance $v(k)$ in (2) for system (1) with small energy consumption $u(k)$. Beyond that, the additional objective of this paper is to compensate the negative influences from delayed input and nonlinear dynamic for improving the control performance of system (1).

# 3 Transformation of Vibration Control Problem

## 3.1 Reconstructions of Delay System and Sinusoidal Disturbances

First, an input delay-free form of delay system (1) will be derived from a designed transformed vector. By defining the following transformed vector as

$$z(k) = x(k) + \Gamma(h,k), \tag{4}$$

where $\Gamma(h,k) = \sum_{i=k-h}^{k-1} A^{k-i-1-h} Bu(i)$, the equivalent input-delay-free form of (1) can be obtained, which is given by

$$\begin{aligned} z(k+1) &= Az(k) + B_z u(k) + Dv(k) + f(z(k) - \Gamma(h,k)), \\ y(k) &= Cz(k) - C\Gamma(h,k), \end{aligned} \tag{5}$$

where $B_z = A^{-h}B$, $z(0) = x_0$.

For sinusoidal disturbances, by introducing the following discrete vector $v_\omega(k)$ with matrices of $\psi_1$ and $\psi_2$ as

$$\begin{aligned} v_\omega(k) &= \left[ v_1(k - \tfrac{\pi}{2\omega_1}), \cdots, v_r(k - \tfrac{\pi}{2\omega_r}) \right]^T, \\ \psi_1 &= diag\{\cos\omega_1, \cos\omega_2, \ldots, \cos\omega_r\}^T, \\ \psi_2 &= diag\{\sin\omega_1, \sin\omega_2, \ldots, \sin\omega_r\}^T, \end{aligned} \tag{6}$$

and defining a disturbance state vector as $w(k) = [v(k) \;\; v_\omega(k)]^T \in \mathbb{R}^{2r}$, based on the periodic properties, the sinusoidal disturbance $v(k)$ in (2) can be reformulated as the output of the following exosystem

$$w(k+1) = Gw(k), v(k) = Ew(k), \tag{7}$$

where

$$G = \begin{bmatrix} \psi_1 & -\psi_2 \\ \psi_2 & \psi_1 \end{bmatrix}, \; E = \begin{bmatrix} I_r \\ 0 \end{bmatrix}. \tag{8}$$

## 3.2 Formulation of OTGTC Problem

First, while ignoring the nonlinear term $f(x(k))$ and input delay $h$ in (1) and combining with sinusoidal disturbance expressed as (7), an augmented system with system state $x_d(k) = [x(k) \;\; w(k)]^T$ is constructed as

$$\begin{cases} x_d(k+1) = A_d x_d(k) + B_d u_d(k), \\ \bar{y}(k) = C_d x_d(k), \end{cases} \tag{9}$$

where

$$A_d = \begin{bmatrix} A & DE \\ 0 & G \end{bmatrix}, B_d = \begin{bmatrix} B \\ 0 \end{bmatrix}, C_d = \begin{bmatrix} C & 0 \end{bmatrix}. \tag{10}$$

Based on the classic optimal control theory, an optimal state feedback controller is given in the following proposition to make the following quadratic performance index reach minimum value, which is described as

$$J_d = \lim_{N\to\infty} \frac{1}{2N} \sum_{k=1}^{N} \left[\bar{y}^T(k)Q\bar{y}(k) + {u_d}^T(k)Ru_d(k)\right], \tag{11}$$

where $Q \in \mathbb{R}^{l\times l}$ is a positive semi-definite matrix and $R \in \mathbb{R}^{m\times m}$ is a positive definite matrix.

**Proposition 1.** Consider the augmented system (9) with respect to the quadratic performance index (11), there exists the unique optimal state feedback controller

$$u^*(k) = -Kx_d(k) = -(R + B_d^T PB_d)^{-1}B_d^T PA_d x_d(k), \tag{12}$$

where $P$ is the uniquc solution of the Riccati equation

$$P = \begin{bmatrix} P_{11} & P_{12} \\ P_{12}^T & P_{22} \end{bmatrix} = C_d^T QC_d^T + A_d^T P(I + B_d R^{-1}B_d^T P)^{-1}A_d. \tag{13}$$

Meanwhile, (9) can be rewritten as

$$x_d(k+1) = Fx_d(k), \bar{y}(k) = C_d x_d(k), \tag{14}$$

where $F = A_d - B_d K$, and the minimum value of (14) is obtained as

$$J_{\min}{}^* - \tfrac{1}{2}(x^T(0)P_{11}x^T(0) + w^T(0)P_{22}w^T(0) + 2x^T(0)P_{12}w(0)). \tag{15}$$

By introducing the tracking error as $e(k) = \bar{y}(k) - y(k)$, the following performance index is employed to design the tracking controller, which is given by

$$J = \lim_{N\to\infty} \frac{1}{2N} \sum_{k=1}^{N} \left[e^T(k)Qe(k) + u^T(k)Ru(k)\right]. \tag{16}$$

After that, by employing (1), (7) and (9), (11) can be with the following form

$$\begin{aligned} J = \lim_{N\to\infty} \tfrac{1}{2N} \sum_{k=1}^{N} \{&z^T(k)Q_{zz}z(k) + x_d(k)C_d^T QC_d x_d(k) + 2x_d(k)Q_{x_d\Gamma}u(k) - \\ &2w^T(k)Q_{wu}u(k) - 2x_d(k)C_{zx_d}^T z(k) + u^T(k)\tilde{R}u(k) - \\ &2z^T(k)Q_{z\Gamma}u(k) - 2f^T(z(k) - \Gamma(h,k))Q_{zz}\Gamma(h,k)\}, \end{aligned} \tag{17}$$

where

$\tilde{R} = R + R_{uu}$, $Q_{zz} = C^T QC$, $Q_{zx_d} = C^T QC_d$, $Q_{x_d\Gamma} = \sum_{i=1}^{h} (F^i)^T C_{zx_d}^T A^{i-1}B_z$,

$R_{uu} = \sum_{i=1}^{h} (A^{h-i}B_z)^T Q_{zz}A^{h-i}B_z$, $Q_{z\Gamma}\Gamma = \sum_{i=1}^{h} (A^i)^T Q_{zz}A^{i-1}B_z$,

$Q_{wu} = \sum_{i=1}^{h} \sum_{j=0}^{h-i} (A^j DEG^{i-1})^T Q_{zz}A^{j+i-1}B_z$.

Thus, the original vibration control problem is reconstructed as OTGTC problem for designing the controller to make the system output $y(k)$ of (1) track the reference optimal trajectory $\bar{y}(k)$ in (14), and minimize performance index (16) under the constrains of nonlinear system (5), sinusoidal disturbance (7), and reference closed-loop system (14). According to the maximum principle theory, the tracking control law can be expressed as

$$u(k) = -\tilde{R}^{-1}\{Q_{x_d\Gamma}^T x_d(k) - Q_{wu}^T w(k) - Q_{z\Gamma}^T z(k) + B_z^T \lambda(k+1)\}, \tag{18}$$

where $\lambda(k)$ satisfies the following nonlinear TPBV problem

$$\begin{cases} z(k+1) = A_z z(k) + Q_w w(k) - B_z \tilde{R}^{-1} Q_{x_d\Gamma}^T x_d(k) \\ -B_z \tilde{R}^{-1} B_z^T \lambda(k+1) + f\left(z(k) - \Gamma\left(h,k\right)\right), \\ \lambda(k) = Q_{zz} z(k) - Q_{zx_d} x_d(k) - Q_{z\Gamma} u\left(k\right) - f_z^T\left(z(k) - \Gamma\left(h,k\right)\right) Q_{zz} \Gamma\left(h,k\right) \\ \qquad + \left(A^T + f_z^T\left(z(k) - \Gamma\left(h,k\right)\right)\right) \lambda(k+1), \\ z(0) = x_0, \quad \lambda(\infty) = 0, \end{cases} \tag{19}$$

where $A_z = A + B_z \tilde{R}^{-1} Q_{z\Gamma}^T$, $Q_w = DE + B_z \tilde{R}^{-1} Q_{wu}^T$; $f_z\left(z(k) - \Gamma\left(h,k\right)\right)$ denotes the Jacobian matrix of $f\left(z(k) - \Gamma\left(h,k\right)\right)$ along with $z(k)$.

# 4 Optimal-Performance-Supervised Vibration Control Strategy

## 4.1 Design of OTVC

First, the following lemmas are given to design the OTVC.

**Lemma 1.** *([21]) Consider the nonlinear discrete-time system under persistent disturbances*

$$\begin{cases} \xi(k+1) = L\xi(k) + \delta(\xi(k), \upsilon(k), k), \\ \xi(0) = \xi_0, \end{cases} \tag{20}$$

*where $\xi \in \mathbb{R}^n$ is the system state with initial state value $\xi_0$; $\upsilon \in \mathbb{R}^m$ denotes the disturbances; $\delta$: $C^l(\mathbb{R}^n) \to U \subset \mathbb{R}^n$ is nonlinear term involved with system state $\xi(k)$, disturbance $\upsilon(k)$ and $k$, and satisfies the Lipschitz condition on $\mathbb{R}^{n+m+1}$.*

*The following vector sequence $\{\xi^{(i)}(k)\}$ uniformly converges to the solution of (20), which is expressed as*

$$\begin{aligned} &\xi^{(0)}(k) = \Phi(k,0)\xi_0, \quad k = 0, 1, \cdots, \\ &\xi^{(i)}(k) = \Phi(k,0)\xi_0 + \sum_{i=0}^{k-1} \left\{\Phi\left(k, i+1\right) \delta\left(\xi^{(i-1)}(i), \upsilon(i)\right)\right\}, \\ &\xi^{(i)}(0) = \xi_0, \quad i = 0, 1, 2, \cdots, \end{aligned} \tag{21}$$

*where $\Phi(k,m)$ is the state transition matrix of (25) with $m = 0, 1, \cdots, k$ .*

**Lemma 2.** *([22]) Let $A_1 \in \mathbb{R}^{n\times n}, B_1 \in \mathbb{R}^{m\times m}, C_1 \in \mathbb{R}^{n\times m}$ and $X \in \mathbb{R}^{n\times m}$. The following matrix equation $A_1 X B_1 - X = C_1$ has a unique solution $X$ if and only if*

$$\lambda_i(A_1) \times \lambda_j(B_1) \neq 1, i = 1, 2, \cdots, n, \quad j = 1, 2, \cdots, m, \tag{22}$$

*where $\lambda(\bullet)$ denotes eigenvalues of matrix of $\bullet$.*

In order to show the designed OTVC with a concise description, let

$$
\begin{aligned}
&S=\tilde{R}+B_z^T P_z B_z, \quad K_w=B_z^T P_w G-Q_{wu}^T+B_z^T P_z D E, \\
&K_z=B_z^T P_z A-Q_{z\Gamma}^T, \quad K_{x_d}=Q_{x_d\Gamma}^T+B_z^T P_{x_d} F, \\
&T=I+B_z \tilde{R}^{-1} B_z^T P_z, \quad Q_{x_d}=Q_{x_d\Gamma}^T+B_z^T P_{x_d} F.
\end{aligned} \tag{23}
$$

Thus, OTVC is described in the following theorem.

**Theorem 1.** Consider the nonlinear discrete-time system (1) with delayed input under sinusoidal disturbances (2), OTVC exists with the following form, which is described as

$$
\begin{aligned}
u^*(k)=-S^{-1}\{&K_1 z(k)+B_z^T P_z f\left(z(k)-\Gamma(h, k)\right) \\
&+K_{x_d} x_d(k)+\lim_{j \rightarrow \infty} B_z^T g^{(j)}(k+1)+K_w w(k)\},
\end{aligned} \tag{24}
$$

where $P_z$ is the unique positive-definite solution of the following Riccati matrix equation

$$
P_z=Q_{zz}-Q_{z\Gamma} \tilde{R}^{-1} Q_{z\Gamma}^T+A_z^T P_z T^{-1} A_z, \tag{25}
$$

$P_{x_d}$ and $P_w$ are the unique solutions of the following Stein matrices equations

$$
\begin{aligned}
P_{x_d} &= Q_{z\Gamma} \tilde{R}^{-1} Q_{x_d\Gamma}^T \\
&\quad -Q_{zx_d}-A_z^T P_z T^{-1} B_z \tilde{R}^{-1} Q_{x_d}+A_z^T P_{x_d} F, \\
P_w &= A_z^T P_z T^{-1} Q_w+ \\
&\quad A_z^T\left(I-P_z T^{-1} B_z \tilde{R}^{-1} B_z^T\right) P_w G-Q_{z\Gamma} \tilde{R}^{-1} Q_{wu}^T.
\end{aligned} \tag{26}
$$

$g^{(j)}(k)$ is the solution of the following adjoint vector sequences

$$
\begin{aligned}
&g^{(0)}(k)=0, \quad f^{(0)}(k)=0, \quad \Gamma^{(0)}(h, k)=0, \\
&g^{(j)}(k)=f_z^{T(j-1)}(k)\{P_w G w(k)+P_{x_d} F x_d(k) \\
&\quad -Q_{zz} \Gamma^{(j-1)}(h, k)+P_z T^{-1} \times\{A_z z^{(j)}(k) \\
&\quad +\left(Q_w-B_z \tilde{R}^{-1} B_z^T P_w G\right) w(k)-B_z \tilde{R}^{-1} Q_{x_d} x_d(k)\} \\
&\quad +\left(A_z^T+f_z^{T(j-1)}(k)\right) \times\{g^{(j)}(k+1) \\
&\quad \left.+P_z T^{-1}\left(f^{(j-1)}(k)-B_z \tilde{R}^{-1} B_z^T g^{(j)}(k+1)\right)\right\}, \\
&g^{(j)}(\infty)=0, \quad j=1,2,3, \cdots, \quad k=0,1,2, \cdots,
\end{aligned} \tag{27}
$$

in which $f^{(j)}(k)=f\left(z^{(j)}(k)-\Gamma^{(j)}(h, k)\right)$, $f_z^{T(j)}(k)=f_z^T\left(z^{(j)}(k)-\Gamma^{(j)}(h, k)\right)$, and $z^{(j)}(k)$ is obtained from the following vector sequences

$$
\begin{aligned}
&z^{(j)}(k+1)=T^{-1}\{A_z z^{(j)}(k)+Q_w w(k) \\
&\quad +f^{(j-1)}(k)-B_z \tilde{R}^{-1}(B_z^T g^{(j)}(k+1) \\
&\quad \left.+B_z \tilde{R}^{-1}\left(Q_{x_d} x_d(k)+B_z^T P_w G w(k)\right)\right\}, \\
&z^{(j)}(0)=x_0, \quad j=0,1, \cdots, \quad k=1,2, \cdots.
\end{aligned} \tag{28}
$$

### 4.2 Design of OPSIA

In this subsection, an iterative algorithm is designed to realize the computability of proposed OTVC (24), which is supervised by the desired minimum performance value $J^*_{min}$ in (15).

First, by replacing $\lim\limits_{j\to\infty} B_z g^{(j)}(k+1)$ by a finite-step term of $B_z^T g^{(M)}(k+1)$ in (24), the computable form of OTVC (24) can be written as

$$\begin{aligned} u^*(k) = -S^{-1}\{&K_1 z(k) + K_{x_d} x_d(k) + K_w w(k) \\ &+ B_z^T P_z f\left(z(k) - \Gamma\left(h, k\right)\right) + B_z^T g^{(M)}(k+1)\}, \end{aligned} \tag{29}$$

where the finite positive integer $M$ denotes the iteration time. Meanwhile, by bringing the calculated $u^{(j)}(k)$ from (24) at the $j$th iteration procedures into (1), the performance index value $J^{(j)}$ can be calculated from

$$J^{(j)} = \lim_{N\to\infty} \tfrac{1}{2N} \sum_{k=1}^{N} \left(x^{(j)T}(k) C^T Q C x^{(j)}(k) + u^{(j)T}(k) R u^{(j)}(k)\right). \tag{30}$$

Thus the terminal condition of iterative process is determined by a discrepancy rate with respect to the desired minimum performance value $J^*_{min}$ in (18), which is described as

$$\left(\left|J^{(M)} - J^*_{\min}\right| / J^*_{\min}\right) < \varepsilon, \tag{31}$$

where $\varepsilon$ is a positive threshold constant. If inequality (31) holds, the iterative process could be broken.

## 5 Simulation Results

### 5.1 Vehicle Active Suspension Under Road Disturbances and Delayed Input

By employing an active suspension from [13] and using the reconstructed method for sinusoidal disturbances given in Sect. 3, road disturbance $v(k)$ can be calculated under vehicle velocity $v_0 = 20\ m/s$, road length $l = 400\ m$, and road roughness ($B$ Grade, good) with $G_d(\Omega_0) = 64 \times 10^{-6}\ m^3$. After that, the effectiveness of proposed vibration control strategy is verified by comparing with relation results, including open-loop suspension system, desired reference optimal trajectories in (14) with minimum performance value of (15), approximation optimal vibration controller (AOVC) in [13], and $H_\infty$ control scheme in [23].

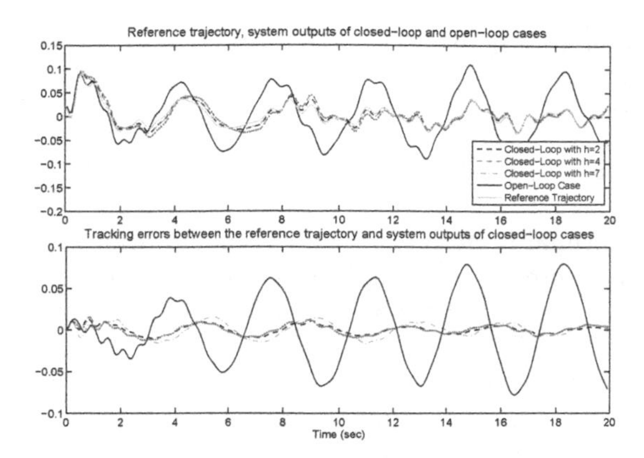


**Fig. 1.** Curves of suspension deflection and tracking error under different scenarios.

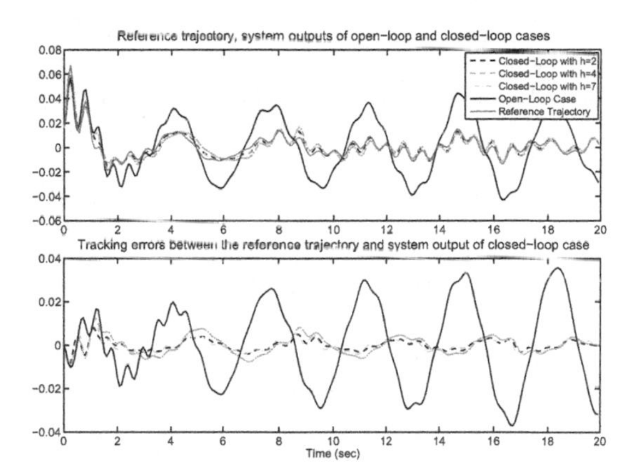


**Fig. 2.** Curves of tire deflection and tracking error under different scenarios.

Considering augment system (9) and performance index (11), the performance values of the desired minimum case and open-loop case are $J^*_{min} = 1.49 \times 10^5$ and $J^{ol} = 9.05 \times 10^5$, respectively. The threshold constant of discrepancy rate sets as $\varepsilon = 15\%$ in Algorithm 1. In order to show the effectiveness of proposed vibration control strategy with different input delays more intuitive, the responses and tracking error of suspension deflection and tire deflection are displayed in Figs. 1 and 2, including the open-loop system, system output of closed-loop system (1) under (29), and reference optimal trajectories $\bar{y}$ in (14). Obviously, system outputs under proposed vibration control strategy could track the reference optimal trajectories with low energy consumption and small tracking error. Thus, under the supervision and guidance from the reference optimal trajectories and desired minimum performance value, the proposed strategy behaves better for offsetting the road disturbances with sinusoidal features, and compensating the nonlinear dynamic and delayed input for nonlinear discrete-time suspension system.

## 6 Conclusion

In this paper, an optimal-performance-supervised vibration control strategy has been developed for discrete-time nonlinear systems with delayed input and sinusoidal disturbances. Compared to existing methods, one salient feature of the proposed OPSIA with OTVC lies in that the designed control strategy is supervised by the reference optimal trajectory and the desired minimum performance value. Furthermore, the gap in performance between OPSIA with OTVC and reference optimal performance is small by using less iteration time.

**Acknowledgments.** This research was supported by the National Natural Science Foundation of China under Grants 61903156 and 61873324, the Natural Science Foundation of Shandong Province for Key Project under Grant ZR2020KF006, and the Foundation of Shandong Women's University under Grants 2019GSPSJ07.

## References

1. Kim, G., Hong, K.: Adaptive sliding-mode control of an offshore container crane with unknown disturbances. IEEE ASME Trans. Mechatron. **24**(6), 2850–2861 (2019)
2. Zhang, B., Cai, Z., Gao, S., Tang, G.: Delayed proportional-integral control for offshore steel jacket platforms. J. Franklin Inst. **356**(12), 6373–6387 (2019)
3. Han, S., Zhou, J., Zhang, Y., Chen, Y., Tang, G., Wang, L.: Active fault tolerant control for discrete vehicle active suspension via reduced-order observer. IEEE Trans. Syst. Man Cybern. Syst. 1–11 (2020)
4. Li, H., Zhang, Z., Yan, H., Xie, X.: Adaptive event-triggered fuzzy control for uncertain active suspension systems. IEEE Trans. Cybern. **49**(12), 4388–4397 (2019)
5. Zhao, X., Wang, X., Zhang, S., Zong, G.: Adaptive neural backstepping control design for a class of nonsmooth nonlinear systems. IEEE Trans. Syst. Man Cybern. Syst. **49**(9), 1820–1831 (2019)
6. Vafamand, N., Khooban, M., Dragicevic, T., Blaabjerg, F.: Networked fuzzy predictive control of power buffers for dynamic stabilization of DC microgrids. IEEE Trans. Industr. Electron. **66**(2), 1356–1362 (2019)
7. Sariyildiz, E., Oboe, R., Ohnishi, K.: Disturbance observer-based robust control and its applications: 35th anniversary overview. IEEE Trans. Industr. Electron. **67**(3), 2042–2053 (2020)
8. Lin, L., Liu, Z., Kao, Y., Xu, R.: Observer-based adaptive sliding mode control of uncertain switched systems. IET Control Theory Appl. **14**(3), 519–525 (2020)
9. Ren, X., Yang, G., Li, X.: Sampled observer-based adaptive output feedback fault-tolerant control for a class of strict-feedback nonlinear systems. J. Franklin Inst. **356**(12), 6041–6070 (2019)
10. Yuan, Y., Wang, Z., Guo, L.: Event-triggered strategy design for discrete time nonlinear quadratic games with disturbance compensations: the noncooperative case. IEEE Trans. Syst. Man Cybern. Syst. **48**(11), 1885–1896 (2018)

11. Jiang, X., Xia, G., Feng, Z., Li, T.: Non-fragile $H\infty$ consensus tracking of nonlinear multi-agent systems with switching topologies and transmission delay via sampled-data control. Inf. Sci. **509**, 210–226 (2020)
12. Abidi, B., Elloumi, S.: Decentralized observer-based optimal control using two-point boundary value successive approximation approach application for a multi-machine power system. Trans. Inst. Meas. Control. **42**(9), 1641–1653 (2020)
13. Han, S., Zhang, C., Tang, G.: Approximation optimal vibration for networked nonlinear vehicle active suspension with actuator delay. Asian J. Control **19**(3), 983–995 (2017)
14. Berkani, S., Manseur, F., Maidi, A.: Optimal control based on the variational iteration method. Comput. Math. Appl. **64**(4), 604–610 (2012)
15. Bourdin, L., Trelat, E.: Pontryagin maximum principle for finite dimensional nonlinear optimal control problem on time scales. SIAM J. Control. Optim. **51**(5), 3781–3813 (2013)
16. Chilan, C., Conway, B.: Optimal nonlinear control using Hamilton-Jacobi-Bellman viscosity solutions on unstructured grids. J. Guid. Control. Dyn. **43**(2), 30–38 (2020)
17. Song, R., Zhu, L.: Optimal fixed-point tracking control for discrete-time nonlinear systems via ADO. IEEE-CAA J. Automatica Sinica **6**(3), 657–666 (2019)
18. Wang, Z., Wei, Q., Liu, D.: Event-triggered adaptive dynamic programming for discrete-time multi-player games. Inf. Sci. **506**, 457–470 (2020)
19. Mirhosseini-Alizamini, S., Effati, S.: An iterative method for suboptimal control of a class of nonlinear time-delayed systems. Int. J. Control **92**(12), 2869–2885 (2019)
20. Yang, X., Wei, Q.: An off-policy iteration algorithm for robust stabilization of constrained-input uncertain nonlinear systems. Int. J. Robust Nonlinear Control **28**, 5747–5765 (2018)
21. Tang, G., Wang, H.: Successive approximation approach of optimal control for nonlinear discrete-time systems. Int. J. Syst. Sci. **3**, 153–161 (2005)
22. Sujit, K.: The matrix equation $axb + cxd = e$. SIAM J. Appl. Math. **32**, 823–825 (1997)
23. Du, H., Zhang, N.: $H\infty$ control of active vehicle suspensions with actuator time delay. J. Sound Vib. **307**, 236–252 (2007)

# A Novel PID Control for MIMO Systems in the DPLS Framework with Gaussian Process Model

Wenli Zhang[1], Xuxia Zhang[2], Mifeng Ren[1(✉)], Junghui Chen[3], and Lan Cheng[1]

[1] College of Electrical and Power Engineering, Taiyuan University of Technology, Taiyuan, China
[2] Taiyuan Branch, China Merchants Bank Company Limited, Taiyuan, China
[3] Department of Chemical Engineering, Chung-Yuan Christian University, Chung-Li, Taoyuan City 320, Taiwan ROC

**Abstract.** For multi-input and multi-output (MIMO) systems, its nonlinearity and strong coupling make the control problem more complicated. This paper introduces a new PID control strategy based on Gaussian process (GP) model to solve the problem of dynamic nonlinearity and strong coupling in complex industrial fields. This processing method is a combination of partial least squares (PLS) as an external framework and GP for internal processing. Firstly, the dynamic PLS (DPLS) is used to extract the input and output variables to eliminate the correlation, and the obtained feature matrix is used as the input of the GP to establish the model. Therefore, the MIMO control problem in the original space is transformed into multiple single-input and single-output (SISO) control problems in the latent space. Finally, a numerical example are used to illustrate the efficiency of the proposed control method.

**Keywords:** MIMO system · Partial least squares · Gaussian process · PID control

## 1 Introduction

The PID controller has been extensively used in the industry due to its simplicity, easy on-line tuning and robustness. In the early days, Yusof R et al. proposed a multivariable self-tuning PID controller based on estimation strategy [1]. A tuning method for self-dispersing relay feedback test of fully cross-coupled multivariable PID controller was proposed in [2]. For the first time, the diagonal of multi-variable PID controller independent of non-diagonal line was designed. During the controller design process, the most difficult point is the mutual coupling within the multivariable system. In order to obtain satisfactory control effects, the decoupling problem of the multivariable system must be studied. Many scholars have proposed various effective control theories and control methods, such as internal model control (IMC) [2], generalized predictive control [3], etc. Although the generalized predictive control can improve the control performance, it is required to solve the Diophantine equation with the huge computational burden.


Q. Han et al. (Eds.): LSMS 2021/ICSEE 2021, CCIS 1469, pp. 234–244, 2021.
https://doi.org/10.1007/978-981-16-7213-2_23

For the strongly coupled multivariable system, the PID parameter tunning is difficult. To solve this problem, PLS regression framework is used to decouple the multivariable PID control. The basic idea of transferring the traditional control design to the PLS framework was firstly proposed by Kaspar and Ray [4],which has the advantage of automatic decoupling, ease of loop pairing. And many researchers have further developed this framework. Chen et al. [5,6] designed a multi-loop adaptive PID controller based on PLS decoupling structure. Considering the dynamic modeling of PLS, a feedforward control scheme was proposed for multivariate linear and nonlinear systems using a DPLS framework [7]. A multi-loop IMC scheme combined with a feedforward strategy is proposed in [8]based on a linear DPLS framework. Then multi-loop nonlinear IMC design based on ARX neural network model for nonlinear DPLS framework is proposed in [9]. Zhao et al.extended the multi-loop IMC solution to adaptive control to address serious plant model changes [10]. However, when the process changes or the environment changes, because the control performance depends on the quality of the model, the accuracy of the model is very crucial, and the information about the uncertainty of the model is very valuable. GP in machine learning can solve this problem well.

The prediction variance provided by the GP model can be used to measure the uncertainty of the model. If the prediction variance of the model is high, it means that the uncertainty of the model is high; otherwise, the uncertainty of the model is low and the prediction model is more accurate. GP model is widely used in different nonlinear dynamic systems. The curve-fitting GP model method was firstly introduced by O'Hagan [11] and later compared with the widely used model of Rasmussen [12], which led to the rapid expansion of GP model research. Likar and Kocijan [13] applied predictive control of the GP model in gas-liquid separation equipment. Considering the advantages of PLS and GP, a dynamic GP model was built for the MIMO system in the framework of PLS [14]. Motivated by this idea, a GP-based PID controller is designed in this paper for the MIMO system in the DPLS framework. By doing that, the MIMO control problem can be simplified to the multi-loop SISO control problem.

The rest of this paper is arranged as follows. Section 2 introduces the GP-DPLS modelling for MIMO systems. The algorithm for implementing PID controller design in the DPLS framework based on the GP model is given in Sect. 3. Section 4 illustrates the validity of the proposed control method. Section 5 gives the conclusion of this paper.

## 2 GP-DPLS Framework

Considering a MIMO system with $m$-inputs and $l$-outputs:$\boldsymbol{u}_k = [u_{1k}, u_{2k}, \cdots, u_{mk}]^T$, $\boldsymbol{y}_k = [y_{1k}, y_{2k}, \cdots, y_{lk}]^T$. Given the set-point as $\boldsymbol{y}_k^{set} = [y_{1k}^{set}, y_{2k}^{set}, \cdots, y_{lk}^{set}]^T$, the goal of this paper is to design a PID controller to make the system output $\boldsymbol{y}_k$ approach to $\boldsymbol{y}_k^{set}$as closely as possible. In order to solve the problems of model uncertainty and strong coupling of the MIMO system, the GP-DPLS framework is used.

The GP-DPLS method extracts the feature information of the dynamic process through DPLS, eliminates the collinearity of the data, and reduces the dimension of the input variable. It inherits the advantages of DPLS and GP, namely the robustness of DPLS and nonlinear processing capabilities of GP. Compared with other inner nonlinear model in the DPLS framework, the parameters of GP model are significantly less, and their optimization is much easier.

### 2.1 Outer DPLS Framework

DPLS is used as the outer framework to extract pairs of latent variables (score vectors) from the collinearity problem. At the same time, the feature information is obtained, and the data set is normalized and the mean value is concentrated. Use two datasets $\boldsymbol{U}$ (an input data matrix with size $n \times m$) and $\boldsymbol{Y}$ (an output data matrix with size $n \times l$), where $n, m, l$ each represent the number of observations, predictor variables and predicted variables.

$$\boldsymbol{U} = \begin{bmatrix} u_{1k} & u_{2k} & \cdots & u_{mk} \\ u_{1,k-1} & u_{2,k-1} & \cdots & u_{m,k-1} \\ \vdots & \vdots & \ddots & \vdots \\ u_{1,k-n+1} & u_{2,k-n+1} & \cdots & u_{m,k-n+1} \end{bmatrix} \boldsymbol{Y} = \begin{bmatrix} y_{1k} & y_{2k} & \cdots & y_{lk} \\ y_{1,k-1} & y_{2,k-1} & \cdots & y_{l,k-1} \\ \vdots & \vdots & \ddots & \vdots \\ y_{1,k-n+1} & y_{2,k-n+1} & \cdots & y_{l,k-n+1} \end{bmatrix}$$

The outer relations for $\boldsymbol{U}$, $\boldsymbol{Y}$ can be formulated as

$$\begin{gathered} \boldsymbol{U} = \boldsymbol{t}_1 \boldsymbol{p}_1^T + \boldsymbol{t}_2 \boldsymbol{p}_2^T + \cdots + \boldsymbol{t}_a \boldsymbol{p}_a^T + \boldsymbol{E}_{a+1} = \boldsymbol{T}\boldsymbol{P}^T + \boldsymbol{E}_{a+1} \\ \boldsymbol{t}_i = \left[t_{i,k}, t_{i,k-1}, \cdots, t_{i,k-n+1}\right]^T, \boldsymbol{p}_i = \left[p_{i,1}, p_{i,2}, \cdots, p_{i,m}\right]^T \end{gathered} \tag{1}$$

$$\begin{gathered} \boldsymbol{Y} = \boldsymbol{v}_1 \boldsymbol{q}_1^T + \boldsymbol{v}_2 \boldsymbol{q}_2^T + \cdots + \boldsymbol{v}_a \boldsymbol{q}_a^T + \boldsymbol{F}_{a+1} = \boldsymbol{V}\boldsymbol{Q}^T + \boldsymbol{F}_{a+1} = \hat{\boldsymbol{Y}} + \boldsymbol{F}_{a+1} \\ \boldsymbol{v}_i = \left[v_{i,k}, v_{i,k-1}, \cdots, v_{i,k-n+1}\right]^T, \boldsymbol{q}_i = \left[q_{i,1}, q_{i,2}, \cdots, q_{i,l}\right]^T \end{gathered} \tag{2}$$

where $a$ is the number of principal components which can be determined by statistical techniques, such as the cross-validation or heuristic techniques,e.g.Both $\boldsymbol{T}$ and $\boldsymbol{V}$ represent the score matrices,$\boldsymbol{P}$ and $\boldsymbol{Q}$ are the loading matrices of $\boldsymbol{T}$ and $\boldsymbol{V}$. In addition,$\boldsymbol{t}_i$ and $\boldsymbol{v}_i$ are the $i^{th}$ $(i \leq a)$ orthogonal column vectors of the score matrices $\boldsymbol{T}$ and $\boldsymbol{V}$, $\boldsymbol{p}_i$ and $\boldsymbol{q}_i$ are the $i^{th}$ vectors of the loading matrices $\boldsymbol{P}$ and $\boldsymbol{Q}$. $\boldsymbol{E}_{a+1}$ and $\boldsymbol{F}_{a+1}$ are the residual matrices of $\boldsymbol{U}$ and $\boldsymbol{Y}$, respectly.

Then, we can decouple the $m$-input and $l$-output multivariable system into $a$ independent loops. And the multivariable control problem can be converted to be the univariate ones.

Through the nonlinear iterative partial least-squares (NIPALS) algorithm, then can get the following formula

$$\mathbf{p}_i^T = \mathbf{t}_i^T \mathbf{X} / \mathbf{t}_i^T \mathbf{t}_i \tag{3}$$

$$\mathbf{t}_i = \mathbf{X}\mathbf{w}_i \tag{4}$$

where $\mathbf{w}_i$ is the weight vector.

## 2.2 Inner GP Model

In the previous section, the original spatial data matrix is mapped to the latent space through the DPLS outer framework to obtain a one-to-one corresponding input score vector $\mathbf{t}_i$ and output score vector $\mathbf{v}_i$, forming multiple single loops in the latent variable space. Then for each pair of score vectors, GP model is used to establish the dynamic relationship between the input and output variables in the latent space.

GP model is a set of random variables with simultaneous distribution:

$$P\left(\boldsymbol{v}_i \mid \boldsymbol{C}_i, \boldsymbol{Z}_i\right) = \frac{1}{L_i} \exp\left(-\frac{1}{2}\left(\boldsymbol{v}_i - \boldsymbol{\mu}_i\right)^T \boldsymbol{C}_i^{-1}\left(\boldsymbol{v}_i - \boldsymbol{\mu}_i\right)\right) \tag{5}$$

where $\boldsymbol{Z}_i = \{\boldsymbol{z}_{i,1}, \cdots, \boldsymbol{z}_{i,k}\}$ refers to the input vectors and $\boldsymbol{v}_i = \{\boldsymbol{v}_{i,k}, \cdots, \boldsymbol{v}_{i,k-n+1}\}$ with mean vector $\boldsymbol{\mu}_i$ is the output of GP model. In the dynamic modeling process, $\boldsymbol{z}_{i,k}$ can be seen as containing past manipulated latent variables and output latent variables.

$$\boldsymbol{z}_{i,k} = \left[t_{i,k-1}, \cdots, t_{i,k-b+1}, v_{i,k-1}, \cdots, v_{i,k-d+1}\right]^T \tag{6}$$

where $b$ and $d$ are the past $b$ manipulated latent variables and the past $d$ output latent variables. $L_i$ is a proper normalizing constant. $\boldsymbol{C}_i$ is the covariance matrix of the data defined by the parameterized covariance function $\boldsymbol{C}\left(\boldsymbol{z}_{i,r_1}, \boldsymbol{z}_{i,r_2}\right)\left(r_1, r_2 = 1, 2, \ldots, k\right)$.

Since the joint density $P\left(v_{i,k+1}, \boldsymbol{v}_i\right)$is also Gaussian, the inference on $v_{i,k+1}$can be readily obtained for the given $\boldsymbol{D}_i = \left(\boldsymbol{Z}_i, \boldsymbol{v}_i\right)$. According to Bayes' theorem, the expression of posterior distribution is obtained:

$$P\left(v_{i,k+1} \mid \boldsymbol{v}_i\right) = \frac{1}{L_i} \exp\left[-\frac{\left(v_{i,k+1} - \mu_{i,k+1}\right)^2}{2\sigma_{i,k+1}^2}\right] \tag{7}$$

where $\mu_{i,k+1} = g^T\left(z_{i,k+1}\right) C_i^{-1} v_i$ is the mean prediction at the time k+1, and $\sigma_{i,k+1}^2 = \boldsymbol{C}\left(\boldsymbol{z}_{i,k+1}, \boldsymbol{z}_{i,k+1}\right) - g^T\left(\boldsymbol{z}_{i,k+1}\right) \boldsymbol{C}^{-1} g\left(\boldsymbol{z}_{i,k+1}\right)$ is the variance of this prediction,$g^T\left(\boldsymbol{z}_{i,k+1}\right) = \left[\boldsymbol{C}\left(\boldsymbol{z}_{i,k+1}, \boldsymbol{z}_{i,1}\right) \cdots \boldsymbol{C}\left(\boldsymbol{z}_{i,k+1}, \boldsymbol{z}_{i,k}\right)\right]$. The vector $g^T\left(\boldsymbol{z}_{i,k+1}\right) \boldsymbol{C}^{-1}$ can be regarded as a smoothing term, which weights the training output to predict the new input vector $\boldsymbol{z}_{i,k+1}$. The variance of this prediction provides a confidence level.

Covariance function is very crucial and the general choice is

$$\boldsymbol{C}\left(\boldsymbol{z}_{i,r_1}, \boldsymbol{z}_{i,r_2}\right) = a_{i0} + a_{i1} \sum_{s=1}^{d} \boldsymbol{z}_{i,r_1;s} \boldsymbol{z}_{i,r_2;s} + v_{i0} \exp\left(-\sum_{s=1}^{d} w_{i,s}\left(\boldsymbol{z}_{i,r_1;s} - \boldsymbol{z}_{i,r_2;s}\right)\right) + q_i \delta_{i;r_1 r_2} \tag{8}$$

where $a_{i0}$, $a_{i1}$, $w_{i,s}$, $q_i$ and $v_{i0}$ are the hyper-parameters.$v_{i0}$ controls the global scale of local correlation, $a_{i1}$ can take different distance measurements in every input dimension,$s$ and $q_i$ are the estimated value of the noise variance. $\delta_i$ is a Kronecker delta parameter, if $r_1 = r_2$, then $\delta_i = 1$, otherwise $\delta_i = 0$.

Under the GP model, the prior obeys the Gaussian distribution. The hyper-parameters can be estimated by maximizing the log-likelihood

$$L(\theta_i) = -\frac{1}{2}\log(|\boldsymbol{C}_i|) - \frac{1}{2}\boldsymbol{v}_i^T\boldsymbol{C}_i^{-1}\boldsymbol{v}_i - \frac{N}{2}\log(2\pi) \tag{9}$$

where $\theta_i = [a_{i0}, a_{i1}, w_{is}, q_i, v_{i0}]$.

Maximizing the log-likelihood function can be achieved by deriving the hyper-parameters from the log-likelihood function:

$$\frac{\partial L}{\partial \theta_i} = -\frac{1}{2}tr\left(C_i^{-1}\frac{\partial C_i}{\partial \theta_i}\right) + \frac{1}{2}\boldsymbol{v}_i^T\boldsymbol{C}_i^{-1}\frac{\partial \boldsymbol{C}_i}{\partial \theta_i}\boldsymbol{C}_i^{-1}\boldsymbol{v}_i \tag{10}$$

Then, the covariance function of $v_{i,k+1}$, $\sigma^2_{i,k+1}$ can be obtained.

# 3 PID Control Strategy

Based on the GP-DPLS model, a multivariate PID control strategy is proposed. The framework of PID control based on GP-DPLS is shown in Fig. 1.

Using the PLS compensation structure, the MIMO control design problem can be transformed into multiple univariate control loop designs in the latent subspace. Therefore, only the PID control strategy of the SISO for each loop will be in detail. The process is post-multiplied by the matrix $\mathbf{W}_u\mathbf{P}$ as shown in Fig. 1 Where $\mathbf{P}^+$ and $\mathbf{Q}^+$ are the generalized inverse of $\mathbf{P}$ and $\mathbf{Q}$, respectively; $\mathbf{W}_u$ and $\mathbf{W}_Y$ are the input and output scaling matrixes, respectively.

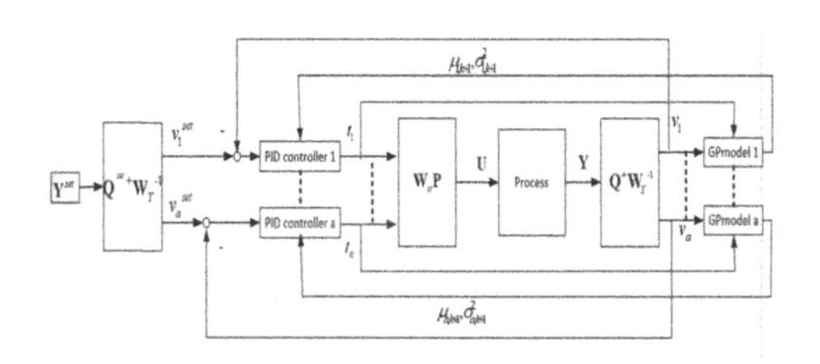


**Fig. 1.** PID control strategy in the DPLS framework by using GP model

We can independently design the PID controller for each SISO loop of the considered multivariate system in the latent space. For the $i^{th}(i = 1, 2, \cdots, a)$ loop, the PID controller algorithm is

$$t_i = k_{iP}\left[\tilde{e}_i + \frac{1}{k_{iI}}\int \tilde{e}_i\ \mathrm{d}t + k_{iD}\frac{\mathrm{d}\tilde{e}_i}{\mathrm{d}t}\right] \tag{11}$$

where $k_{iP}$, $k_{iI}$ and $k_{iD}$ are each refer to the proportional gain, integral time constant and the derivative time constant.$\tilde{e}_i = v_i^{\text{set}} - v_i$ is the tracking error in the latent space. So the discrete PID can be described as

$$\begin{aligned}\Delta t_{i,k} &= t_{i,k} - t_{i,k-1} \\ &= \underbrace{k_{iP}\left[1 + (\Delta k / k_{iI}) + (k_{iD}/\Delta k)\right]}_{k_{ik}^0} \tilde{e}_{i,k} + \underbrace{k_{iP}\left[-1 - 2\left(k_{iD}/\Delta k\right)\right]}_{k_{ik}^1} \tilde{e}_{i,k-1} \\ &+ \underbrace{k_{iP}\left(k_{iD}/\Delta k\right)}_{k_{ik}^2} \tilde{e}_{i,k-2} = \tilde{e}_{i,k}{}^T \mathbf{k}_{ik}\end{aligned} \tag{12}$$

where $\mathbf{k}_{ik} = \left[k_{ik}^0 \; k_{ik}^1 \; k_{ik}^2\right]^T$, $\tilde{e}_{i,k}^T = \left[\tilde{e}_{i,k} \; \tilde{e}_{i,k-1} \; \tilde{e}_{i,k-2}\right]^T$.

The PID controller design can straightly use the identified GP model. And minimize the objective function to get the best control action. Using the PLS framework, the relationship between the original objective function and the latent space objective function is

$$\begin{aligned}J_k &= E\left\{\left\|\boldsymbol{y}_{k+1}^{set} - \boldsymbol{y}_{k+1}\right\|_2^2\right\} + \lambda \left\|\Delta \boldsymbol{u}_k\right\|_2^2 \\ &= \sum_{i=1}^{a} \left\{E\left[\left\|\boldsymbol{v}_{i,k+1}^{set} - v_{i,k+1}\right\|_2^2 \left\|\boldsymbol{q}_i^T\right\|_2^2\right] + \lambda \left\|\Delta t_{i,k}\right\|_2^2 \left\|\boldsymbol{p}_i^T\right\|_2^2\right\}\end{aligned} \tag{13}$$

The objective function in the original space is slightly less than the sum of the objective functions in the latent variable space. It means that the control performance will decrease in the latent variable space.

For a loops in the latent space, we can design the PID controller for each loop one by one independently. For the $i^{th}$ loop, using the fact that $\mathrm{Var}\{v_{i,k+1}\} = E\left\{v_{i,k+1}^2\right\} - E^2\{v_{i,k+1}\}$ and the predictive variance obtained from GP, the objective function can be written as

$$J_{ik} = \left(v_{i,k+1}^{set} - \mu_{i,k+1}\right)^2 + \sigma_{i,k+1}^2 + \lambda \Delta t_{i,k}^2 \tag{14}$$

Minimize the objective function $J_{ik}$ to adjust the PID parameters. The purpose is to design control actions to minimize the error between the system output value and the expected value. The variance of GP is considered during the optimization process. The predicted variance represents model uncertainty information, resulting in a more robust control system.

A gradient based optimization algorithm is used to minimize the objective function $J_{ik}$ in order to adjust the parameters.

$$t_{ik}^* = \frac{\partial J_{ik}}{\partial k_{ik}} = \frac{\partial J}{\partial t_{i,k}} \frac{\partial t_{i,k}}{\partial k_{ik}} = 0 \tag{15}$$

$$\frac{\partial J_{ik}}{\partial t_{i,k}} = 2\left(\mu_{i,k+1} - v_{i,k+1}^{set}\right) \frac{\partial \mu_{i,k+1}}{\partial t_{i,k}} + \frac{\partial \sigma_{i,k+1}^2}{\partial t_{i,k}} + 2\lambda \Delta t_{i,k} \tag{16}$$

$$\frac{\partial t_{i,k}}{\partial k_{ik}} = \tilde{e}_{i,k} \tag{17}$$

The partial derivative terms of Eq. (16) is expressed by the following formula:

$$\frac{\partial \mu_{i,k+1}}{\partial t_{i,k}} = \frac{\partial \mathbf{g}^T(\mathbf{z}_{i,k+1})}{\partial t_{i,k}} \mathbf{C}_i^{-1} \mathbf{v}_i \tag{18}$$

$$\frac{\partial \sigma_{i,k+1}^2}{\partial t_{i,k}} = \frac{\partial \mathbf{C}(\mathbf{z}_{i,k+1}, \mathbf{z}_{i,k+1})}{\partial t_{i,k}} - 2\mathbf{g}^T(\mathbf{z}_{i,k+1}) \mathbf{C}^{-1} \frac{\partial \mathbf{g}(\mathbf{z}_{i,k+1})}{\partial t_{i,k}} \tag{19}$$

Moreover, the derivative of the covariance function (Eq. (8)) is

$$\begin{aligned}\frac{\partial C(z_{i,k+1}, z_{i,r_1})}{\partial t_{i,k}} =& a_{i1} z_{i,r_i,1} + v_{i0} \exp\left(-\sum_{s=1}^{d} w_{i,s}(z_{i,r_1;s} - z_{i,k+1;s})\right) \\ & \times 2w_{i,1}(z_{i,r_1,1} - t_{i,k})\end{aligned} \tag{20}$$

If $z_{i,r_1} = z_{i,k+1}$, we have $\frac{\partial C(z_{i,k+1}, z_{i,k+1})}{\partial t_{i,k}} = a_{i1} t_{i,k}$. Thus,

$$\frac{\partial \mathbf{g}}{\partial t_{i,k}} = \left[\frac{\partial \mathbf{C}(\mathbf{z}_{i,k+1}, \mathbf{z}_{i,1})}{\partial t_{i,k}} \quad \frac{\partial \mathbf{C}(\mathbf{z}_{i,k+1}, \mathbf{z}_{i,2})}{\partial t_{i,k}} \quad \cdots \quad \frac{\partial \mathbf{C}(\mathbf{z}_{i,k+1}, \mathbf{z}_{i,k})}{\partial t_{i,k}}\right]^T \tag{21}$$

After obtaining the control algorithm in the latent space, the PID controller in the original space can be formulated according to (1). The system is decoupled into more separate control loops in the potential space of DPLS. It avoids the difficulty of loop matching between the controlled variable and the manipulated variable in the traditional decoupling control design. In addition, there is no need for centralized control design for MIMO systems, and the PID controller can be designed separately for each control loop in the potential space. Thus, the design of the controller is no longer a tedious and time-consuming task.

## 4 Case Study

Consider the following non-linear systems with 3-input and 3-output:

$$\begin{aligned}y_1(t+1) =& -0.2y_1(t) + 0.3u_1(t) + 1.5\sin(u_2(t)) + u_3(t) - 0.5u_1(t-1) \\ & - 0.8u_2(t-1) + u_3(t-1) + \frac{u_2(t-1)}{1+u_2(t-1)^2}\end{aligned}$$

$$\begin{aligned}y_2(t+1) =& 0.6y_2(t) + 0.2\sin(u_1(t)) + u_2(t) + 0.5\sin(u_3(t)) + 0.3u_1(t-1) \\ & + 1.5u_2(t-1) + u_3(t-1) + \frac{u_3(t-1)}{1+u_3(t-1)^2}\end{aligned}$$

$$\begin{aligned}y_3(t+1) =& -0.4y_3(t) + 0.2u_1(t) + \sin(u_2(t)) + 0.5\sin(u_3(t)) + u_1(t-1) \\ & + 0.5u_2(t-1) + u_3(t-1) + \frac{u_1(t-1)}{1+u_1(t-1)^2}\end{aligned}$$

Taking the above multivariable nonlinear discrete time system as an example, three reference input signals are given: $u_1(t) = \text{sign}(1.5\sin(\pi t/40))$, $u_2(t) =$

sign(sin($\pi t/30$)) and $u_3(t) = 0$. Compared to other control methods, our latent space PID control strategy can be regarded as a relatively simple control scheme, and it can also achieve good control performance.

Before implementing the control scheme, we need to obtain the GP-DPLS model. Before the predictive model is obtained, the input signal is added to the system to generate the input data and output data for GP-DPLS modeling. The obtained input and output data are normalized to zero and the variance is one. After the sampled data is decoupled by the DPLS method, the system dimension decreases, and the number of principal elements is 2. The two PID controllers are designed in the latent space independently, and then they are reconstructed back into the original space.

### 4.1 GP Model with Sufficient Data

The training results of the first pair of latent variables and the second pair of latent variables are shown in Fig. 2. It shows the comparison results between the predicted output of the GP-DPLS model and the actual system model. The variance of $y_1$ is between 0.03 and 0.036 and the variance of $y_2$ is between 0.005 and 0.02. It shows the prediction power of the GP-DPLS model, and the predictive output is more consistent with the actual process output.

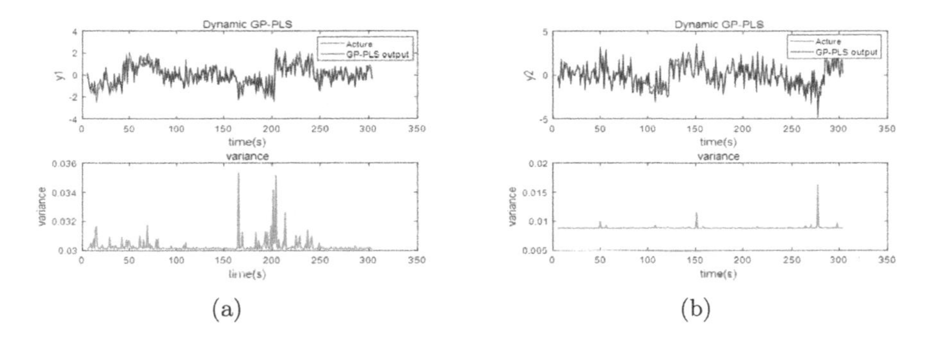


**Fig. 2.** The predictive output and variance of the GP model under system accuracy.

Based on the obtained GP-DPLS model, a single-loop PID controller in the latent space with the DPLS framework for each loop is designed. The control output results are shown in Fig. 3. When the model is accurate, the control results obtained by using the GP model are similar when the variance $\sigma_{i,k+1}^2$ is considered and not considered in (14).

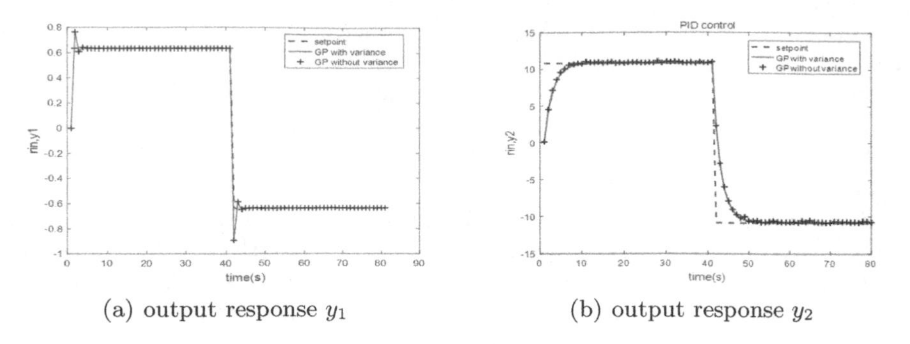


(a) output response $y_1$

(b) output response $y_2$

**Fig. 3.** Output responses under GP-DPLS-PID control

## 4.2 GP Model with Insufficient Data

However, there may be lack of data during actual operation. The figure below shows the corresponding prediction variances obtained from the internal GP model. Compared with Fig. 2, the prediction variance showed in Fig. 4 is much larger, indicating that the model is more uncertain.

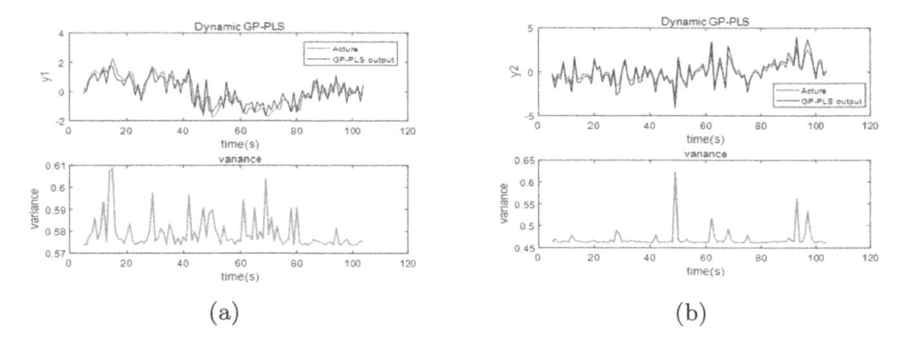


(a)

(b)

**Fig. 4.** The prediction output and variance of the GP model under system inaccuracy.

There may be insufficient data in the actual operation process, the resulting model is not accurate, and the prediction effect is not very good. Output responses under GP-DPLS-PID control algorithm are shown in Fig. 5. On the one hand, without considering the variance term, the unreliable predicted value is used without any extra checks in the control process. Due to the large forecast error in the manipulated variable, it will cause a large variation when tracking the set point. On the other hand, when considering the variance term, the system's response is optimized during this process, preventing the system from reaching a more variable area and becoming more robust. Correspondingly, Figs. 6 and 7 show the variation of the control inputs, respectively. The fluctuations in Figs. 6(a) and 7(a) are less than the fluctuations in Figs. 6(b) and 7(b). All the experimental results show the effectiveness of the variance term in the objective function (14), and demonstrates the excellent control performance of the proposed control method.

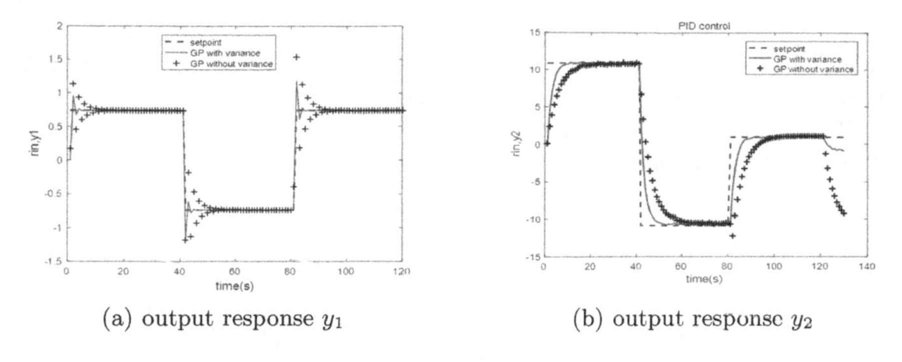


(a) output response $y_1$ (b) output response $y_2$

**Fig. 5.** Output responses under GP-DPLS-PID control

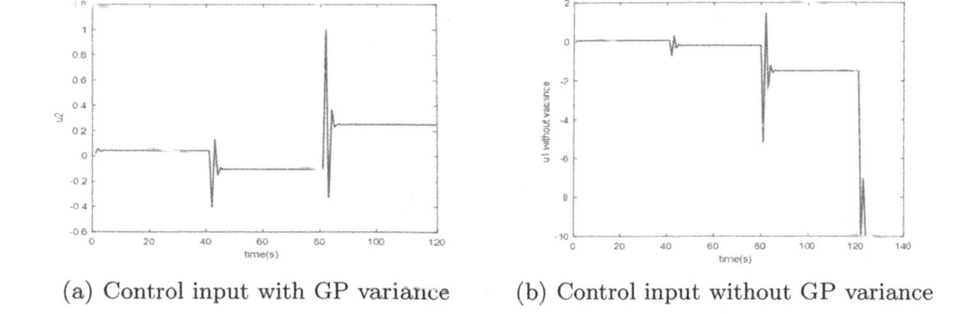


(a) Control input with GP variance (b) Control input without GP variance

**Fig. 6.** Comparison of control input for $y_1$.

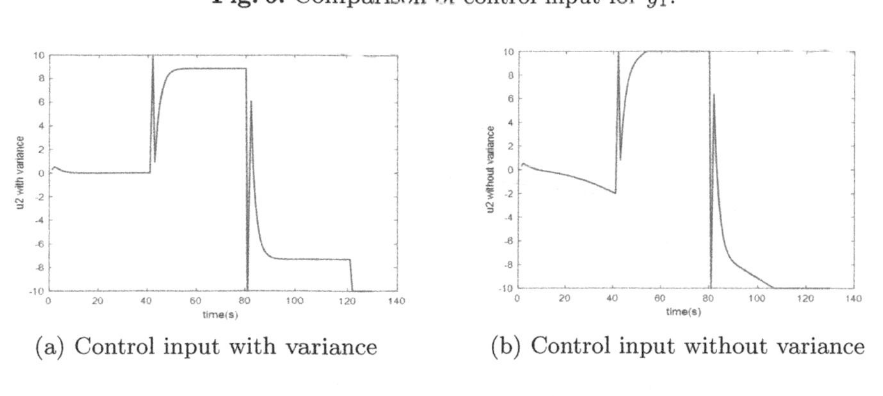


(a) Control input with variance (b) Control input without variance

**Fig. 7.** Comparison of control input for $y_2$.

## 5 Conclusion

This research uses the DPLS framework to design the PID controller, and the internal model is the GP model. It combines the advantages of GP model and DPLS framework to effectively deal with the problems of nonlinearity and strong coupling of the considered multivariable systems. Compared with the PID control algorithm directly designed in the original space, the presented latent space PID controller can also achieve the expected effect. And the proposed GP-DPLS based PID control method is much simpler during the design process, which could be used in real complex industrial systems).

## References

1. Yusof, R., Omatu, S., Khalid, M.: Self-tuning PID control: a multivariable derivation and application. IFAC Proc. Vol. **26**(2), 1055–1061 (1993)
2. Wang, Q.G., Zou, B., Lee, T.H., Bi, Q.: Auto-tuning of multivariable PID controllers from decentralized relay feedback. Automatica **33**(3), 319–330 (1997)
3. Xue, M.S., Fan, D., Wei, H.H.: Decoupling design of generalized predictive control for multivariable systems. Control Eng. China **18**(1), 39–42 (2011)
4. Kaspar, M.H., Ray, W.H.: Partial least squares modelling as successive singular value decompositions. Comput. Chem. Eng. **17**(10), 985–989 (1993)
5. Chen, J., Cheng, Y.C., Yea, Y.: Multiloop PID controller design using partial least squares decoupling structure. Korean J. Chem. Eng. **22**(2), 173–183 (2005)
6. Chen, J., Cheng, Y.C.: Applying partial least squares based decomposition structure to multiloop adaptive proportional-integral-derivative controllers in nonlinear processes. Ind. Eng. Chem. Res. **43**(18), 5888–5898 (2004)
7. Lakshminarayanan, S., Shah, S.L., Nandakumar, K.: Modeling and control of multivariable processes: dynamic pls approach. AIChE J. **43**, 2307–2322 (1997)
8. Liang, B.: Multi-loop internal model controller design based on a dynamic PLS framework. Chin. J. Chem. Eng. **18**(2), 277–285 (2010)
9. Hu, B., Zhao, Z., Liang, J.: Multi-loop nonlinear internal model controller design under nonlinear dynamic PLS framework using ARX-neural network model. J. Process Control **22**(1), 207–217 (2012)
10. Zhao, Z., Bin, H.U., Liang, J.: Multi-loop adaptive internal model control based on a dynamic partial least squares model. J. Zhejiang Univ.-Sci. A (Appl. Phys. Eng.) **12**(03), 190–200 (2011)
11. O'Hagan, A.: Curve fitting and optimal design for prediction. J. Roy. Stat. Soc.: Ser. B (Methodol.) **40**(1), 1–24 (1978)
12. Rasmussen, C.E.: Evaluation of gaussian processes and other methods for non-linear regression. Ph.D. thesis, University of Toronto (1999)
13. Likar, B., Kocijan, J.: Predictive control of a gas-liquid separation plant based on a gaussian process model. Comput. Chem. Eng. **31**(3), 142–152 (2007)
14. Liu, H., Yang, C., Carlsson, B., Qin, S.J., Yoo, C.: Dynamic nonlinear pls modeling using gaussian process regression. Ind. Eng. Chem. Res. **58**(36), 16676–16686 (2019)

# An Enhanced Discrete Human Learning Optimization for Permutation Flow Shop Scheduling Problem

Ling Wang[1], Mengzhen Wang[1], Jun Yao[1], and Muhammad Ilyas Menhas[2](✉)

[1] Shanghai Key Laboratory of Power Station Automation Technology, School of Mechatronics Engineering and Automation, Shanghai University, Shanghai 200444, China
{wangling,Wangmengzhen,grandone0529}@shu.edu.cn

[2] Department of Electrical Engineering, Mirpur University of Science and Technology (MUST), Mirpur 10250, AJ&K, Pakistan

**Abstract.** Human learning optimization (HLO) is a promising meta-heuristic algorithm inspired by human learning mechanisms. As a binary-coded algorithm, HLO can be used for discrete problems without modifying the encoding scheme. However, as the scale of the problem grows, the "combinatorial explosion" phenomenon inevitably occurs, which increases the computational complexity and reduces the optimization efficiency of HLO. Therefore, this paper extends HLO and proposes an enhanced discrete HLO (EDHLO) algorithm for the Permutation Flow Shop Scheduling Problem (PFSSP), in which the original three learning operators are reformed and new improvement strategies for PFSSP are introduced. First, the Nawaz-Ensco-Ham (NEH) algorithm is utilized to initialize partial population to improve the quality of initial solutions. Second, two crossover operators are introduced into the individual learning of EDHLO to improve the diversity of the population and enhance the quality of the solution. Third, a local search approach is applied to manage escape from the local minimum. A total of 21 benchmark problems have been taken into account to evaluate the performance of EDHLO. The experimental outcomes and the comparison with other metaheuristics verify the effectiveness and superiority of the EDHLO algorithm.

**Keywords:** Human learning optimization · Permutation flow shop scheduling problem · Discrete human learning optimization

## 1 Introduction

In recent years, to break through the bottleneck limitation of traditional optimization algorithms, various intelligent optimization algorithms based on animal foraging or other learning behaviors, such as Artificial Bee Colony, Shuffled Frog Leaping Algorithm and Firefly Algorithm, have emerged. Compared with other animals in nature, human has evolved a higher level of intelligence and learning ability and is able to solve complex problems that other animals are unable to solve. Inspired by a simple yet general human learning mechanism, Wang et al. proposed Human Learning Optimization (HLO) [1],


Q. Han et al. (Eds.): LSMS 2021/ICSEE 2021, CCIS 1469, pp. 245–257, 2021.
https://doi.org/10.1007/978-981-16-7213-2_24

in which the random learning operator, individual learning operator and social learning operator were designed through simulating the learning mechanism of humans to search out the optimal solution of problems.

To further improve the optimization performance of HLO, Wang et al. proposed an adaptive simplified human learning optimization algorithm (ASHLO), which set the linearly increasing and linearly decreasing adaptive strategy for the parameters pr and pi, respectively, to maintain the balance between algorithmic exploration and exploitation [2]. Based on the phenomenon that human IQ follows normal distribution on the whole and shows an upward trend, a diverse human learning optimization algorithm (DHLO) was proposed, where the social learning capability of all individuals was subject to normal distribution and tuned dynamically [3]. The simulation results indicate that DHLO possesses better global optimization performance. Yang et al. adopted an adaptive strategy based on cosine and sine functions to propose a sine-cosine adaptive human learning optimization algorithm (SCHLO) [4]. Later, An improved adaptive human learning optimization algorithm (IAHLO) [5] which used a two-stage adaptive strategy to dynamically adjust the execution probability of random learning operators is proposed, so that the algorithm focused on exploration to diversify the population at the beginning of iterations, and focused on local search at the later phase of iterations to enhance the mining ability of the algorithm. Nowadays, many various practical problems, such as extraction of text abstracts [6], financial market forecasts [4], image processing [7], mixed variable engineering optimization problems [8], and intelligent control [9–11], have been successfully solved by HLO.

As a binary algorithm, HLO is capable of solving discrete optimization problems directly. Fan W et al. adopted the ASHLO to take place local search in standard VNS to solve scheduling problems with multiple constraints [12]. Li et al. took advantage of HLO to solve the actual production scheduling problem of a dairy plant [13]. Ding et al. combined an improved PSO and some scheduling strategies to solve the flexible job shop scheduling problem under the algorithm architecture of HLO [14]. A. Shoja et al. combined ASHLO with GA and PSO, respectively, to solve the two-stage supply chain network design problem aiming at minimizing cost [15].

However, with the scale of the problem grows, the size of the feasible solution set grows exponentially, and the phenomenon of "combinatorial explosion" inevitably occurs, which results in a significant decrease in the optimization efficiency of binary HLO for discrete problems. Therefore, this paper extends HLO and proposes an enhanced discrete HLO (EDHLO) algorithm for the PFSSP, which is a classical discrete problem and plays an essential part in ensuring the steady progress of production and improving resource utilization. To solve PFSSP more efficiently, the learning operators of EDHLO are improved and the efficient heuristic algorithm NEH [16] is combined with random initialization to improve the quality of the initial solutions. Besides, a local search strategy is utilized to help EDHLO algorithm get rid of the local optima.

The structure of this paper is as follows: Sect. 2 is a description of the PFSSP and the related definitions. Section 3 describes the proposed algorithm EDHLO. The experimental results of EDHLO and other metaheuristics are given in Sect. 4. Section 5 summarizes this paper.

## 2 Permutation Flow Shop Scheduling Problem

PFSSP which is a simplified model of assembly-line production in many manufacturing enterprises, one of the most well-known classes of scheduling problems, has been investigated extensively and intensively due to its importance in the aspect of academy and engineering applications [17].

The current approaches for solving PFSSP mainly can be divided into 3 classifications, i.e., precision methods, heuristic rule-based methods and meta-heuristic algorithms [18]. PFSSP, as an NP-hard combinatorial optimization problem, grows exponentially in computational complexity as the problem scale increases. Although the precision methods can find the exact solution to the problem, they are only suitable for solving small-scale PFSSP considering the computational resources. Constructive heuristics are methods for solving specific problems by inductive reasoning and experimental analysis according to experience, which can obtain nearly optimal solutions. However, some of them are difficult to obtain satisfactory solutions. Compared with the other two categories, meta-heuristic algorithms are more advantageous in terms of stability, convergence speed and computational efficiency. Thus, currently, meta-heuristic algorithms have become the most effective and efficient method to solve scheduling problems, and the main hotspots of research are also concentrated on the improvement of the meta-heuristics and the development of new algorithms.

The description of PFSSP is as following: $n$ jobs require $m$ processes on different machines sequentially, and the processing sequence of $n$ jobs on all machines is the same, with the goal to find the best permutation to minimize the makespan. It should be noted that each machine at the same time only one job is allowed to be machined, accordingly, each job can be handled on only one machine at a time. The processing time $H(p_j,k)$ is the time for job $p_j$ to be processed on the $k$-th machine. $C(p_j,k)$ represents the processing completion time of job $p_j$ on the $k$-th machine. PFSSP can be mathematically formulated as follows Eq. (1) [19]:

$$\begin{cases} C(p_1, 1) = H(1, 1) \\ C(p_j, 1) = H(j, 1) + C(p_{j-1}, 1)\, j = 2, 3, \ldots, n \\ C(p_1, k) = H(1, k) + C(p_1, k-1)\, k = 2, 3, \ldots, m \\ C(p_j, k) = H(j, k) + \max(C(p_{j-1}, k), C(p_j, k-1))\, j = 2, 3, \ldots, n;\ k = 2, 3, \ldots, m \\ C_{\max} = C(p_n, m) \end{cases} \tag{1}$$

The optimal permutation $p^*$ can be found by the optimization in Eq. (2):

$$p^* = \arg\min C(p_n, m),\ \forall p \in \Omega. \tag{2}$$

## 3 Enhanced Discrete Human Learning Optimization for PFSSP

### 3.1 Initialization

The solution of PFSSP is a single linked list, which focuses on the position of each element in the whole permutation. Therefore, different from the standard HLO, EDHLO uses the integer encoding framework, where the individual solution in EDHLO is represented by an integer array as Eq. (3),

$$x_i = [x_{i1}\, x_{i2} \cdots x_{ij} \cdots x_{iM}],\ x_{ij} \in \{1, 2, \cdots, M\}, 1 \le i \le N, 1 \le j \le M \tag{3}$$

where $x_i$ is the $i$-th individual, $x_{ij}$ is the $j$-th element of the $i$-th individual, $N$ is the size of population and $M$ denotes the length of solutions, i.e. the number of jobs. Under the assumption that the prior knowledge of problems does not exist at first, each variable of an individual in EDHLO is initialized with an integer between 1 and $M$ stochastically to present the number of jobs.

Aiming at solving PFSSP efficiently, the NEH algorithm, an efficient heuristic algorithm, is used to generate 10% of the individuals, so as to enhance the quality of the initial population and maintain the diversity of the population.

## 3.2 Learning Operators

The standard HLO generates new candidates to search out the optimal solution by the random learning operator, the individual learning operator and the social learning operator, but these three learning operators are not applicable to PFSSP. Thus, EDHLO redesigns and introduces the learning operators to solve PFSSP more efficiently.

### 3.2.1 Random Learning Operator for PFSSP-Like Problems

In reality, humans usually have to solve a problem by using the random learning strategy at the beginning due to a lack of prior knowledge. Moreover, since a person is easy to be affected by various factors like interference and forgetting, the human cannot fully replicate the previous experiences. And consequently, learning processes are always accompanied by randomness [20]. To imitate the random learning behavior for the PFSSP, the random learning operator is designed in EDHLO as Eq. (4),

$$x_{ij} = R(M) \tag{4}$$

where *R(M)* represents the random learning operator which randomly chooses a job not scheduled yet.

### 3.2.2 Individual Learning Operators for PFSSP-Like Problems

Individual learning refers to the ability of an individual to construct his own knowledge by reflecting on the extrinsic stimuli and sources [21]. Drawing on previous experience is conducive to efficient learning of humans. Set up an Individual Knowledge Database (IKD) to store individual optimum solutions to simulate the individual learning mechanism described above, as shown in Eqs. (5–6),

$$IKD = \begin{bmatrix} ikd_1 \\ ikd_2 \\ \vdots \\ ikd_i \\ \vdots \\ ikd_N \end{bmatrix}, \ 1 \le i \le N \tag{5}$$

$$ikd_i = \begin{bmatrix} ikd_{i1} \\ ikd_{i2} \\ \vdots \\ ikd_{ip} \\ \vdots \\ ikd_{iK} \end{bmatrix} = \begin{bmatrix} ik_{i11} & ik_{i12} & \cdots & ik_{i1j} & \cdots & ik_{i1M} \\ ik_{i21} & ik_{i22} & \cdots & ik_{i2j} & \cdots & ik_{i2M} \\ \vdots & \vdots & & \vdots & & \vdots \\ ik_{ip1} & ik_{ip2} & \cdots & ik_{ipj} & \cdots & ik_{ipM} \\ \vdots & \vdots & & \vdots & & \vdots \\ ik_{iK1} & ik_{iK2} & \cdots & ik_{iKj} & \cdots & ik_{iKM} \end{bmatrix}, 1 \le p \le K. \tag{6}$$

where $ikd_i$ represents the IKD of individual $i$, $K$ is the size of IKD, $ikd_{ip}$ is the p-th best solution of individual $i$, and each $ik_{ipj}$ is an integer between 1 and M.

EDHLO uses its previous knowledge in the IKD to generate a high-quality candidate by conducting the individual imitation learning operator (IILO) as Eq. (7).

$$x_{ij} = ik_{ipj} \tag{7}$$

Besides, to diversify the population and further enhance the quality of the solution, the individual swap learning operator (ISLO) and the individual reversal-insertion learning operator (IRLO), inspired and borrowed from two cross operators, i.e. the swap exploration operator and reversal-insertion operator, are introduced in the individual learning process. The swap exploration operator and reversal-insertion operator are successful ways of solving PFSSP in the previous studies, and their effectiveness has been verified in [22] and [23] respectively.

As shown in Fig. 1, the individual swap learning operator refers to exchanging the positions of two random jobs. The number of executions of the individual swap learning operator is thirty percent of the number of jobs. The individual reversal-insertion learning operation is described in Fig. 2 where it exchanges the position of two adjacent jobs, removes them from the original permutation and then inserts them in any other ($n$-2) positions. These two operators only need individual knowledge and therefore they are designed as individual learning operators. The newly generated permutation will replace the original one if and only if it has a better fitness.

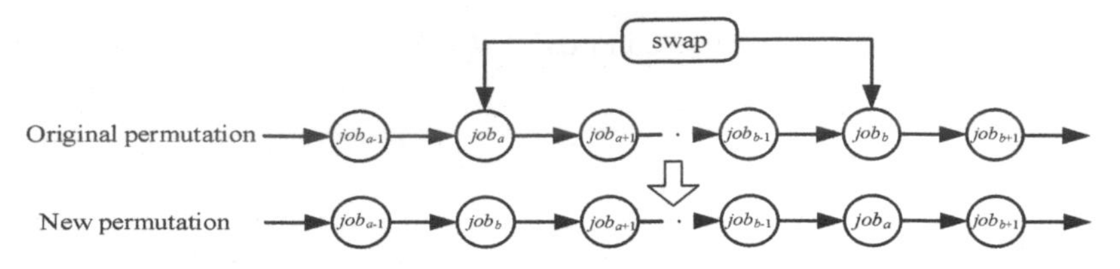


**Fig. 1.** The individual swap learning operator

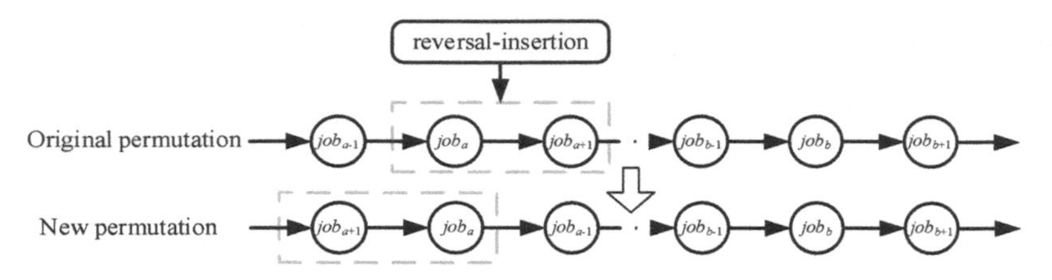


**Fig. 2.** The individual reversal-insertion learning operator

### 3.2.3 Social Learning Operator for PFSSP-Like Problems

The process of tackling complicated problems through individual learning alone can be quite slow and inefficient. Therefore, as social animals, human beings naturally learn from others in the collective to gain experience and expand their knowledge [24]. By directly or indirectly transferring experiences, humans can improve their competence and learning efficiency. To utilize the social experience productively, Social Knowledge Database (SKD) which stores the optimal solution found by the whole population in the search process is established in EDHLO as Eq. (8),

$$SKD = \begin{bmatrix} skd_1 \\ skd_2 \\ \vdots \\ skd_q \\ \vdots \\ skd_H \end{bmatrix} = \begin{bmatrix} sk_{11} & sk_{12} & \cdots & sk_{1j} & \cdots & sk_{1M} \\ sk_{21} & sk_{22} & \cdots & sk_{2j} & \cdots & sk_{2M} \\ \vdots & \vdots & & \vdots & & \vdots \\ sk_{q1} & sk_{q2} & \cdots & sk_{qj} & \cdots & sk_{qM} \\ \vdots & \vdots & & \vdots & & \vdots \\ sk_{H1} & sk_{H2} & \cdots & sk_{Hj} & \cdots & sk_{HM} \end{bmatrix}, 1 \leq q \leq H \tag{8}$$

where $skd_q$ denotes the $q$-th optima solution in the SKD and $H$ is the size of the SKD. When an individual performs social learning, it operates as Eq. (9).

$$x_{ij} = sk_{qj} \tag{9}$$

### 3.3 Updating of IKD and SKD

Although the individual learning operators and the social learning operator can learn the order of jobs scheduled for processing from the better solutions efficiently, they may produce duplicate job numbers and yield infeasible solutions. Therefore, after the new population is generated, EDHLO will find the duplicate job numbers from the random position of the individual sequence and replace them randomly with the job numbers that have not been assigned. Then the new candidates are substituted into the fitness function to calculate the fitness values. Only when the number of candidates stored in IKD is fewer than the set value $K$ or the new candidate solution is superior to the worst in IKD, the new candidate solution will be saved in the IKD directly, otherwise, it will be discarded. For the SKD, the same updating strategy is applied. Moreover, to obtain a superior SKD and help the EDHLO escape from the local minimum, a local search approach is applied in EDHLO.

The local search method operates as follows: move each job in the optimal permutation with dimension M to the remaining (M-1) positions sequentially except the current position, and keep the rest of the permutation unchanged. After every movement, calculate the fitness of the new permutation. If the fitness is better, update the SKD. To avoid redundant computations, the local search is activated every 50 iterations.

### 3.4 Implementation of EDHLO

In summary, EDHLO performs the random learning operator, individual learning operators, and social learning operator with specified probabilities to search for the optimal solution. The implementation of EDHLO can be described as below:

***Step 1:*** set the parameters of EDHLO, such as the population size, maximum number of iterations, *pr*, *pi;*
***Step 2:*** initialize 10% of individuals by NEH and randomly yield the rest initial population;
***Step 3:*** evaluate the fitness of the whole population and generate the initial IKDs and SKD;
***Step 4:*** yield new candidates by executing the learning operators of EDHLO as Eq. (4), Eq. (7) and Eq. (9) with the probabilities of *pr*, (*pi-pr*) and (1-*pi*), respectively;
***Step 5:*** fix the infeasible solutions and execute the ISLO and the IRLO for 20% of individuals;
***Step 6:*** update the IKDs and SKD according to the calculated fitness of new individuals.
***Step 7:*** perform the local search on the SKD every 50 iterations;
***Step 8:*** Determine whether the termination condition is satisfied, if so, output the optimal solution found, otherwise go to step 4.

## 4 Experimental Results and Discussion

To estimate the performance, the proposed EDHLO, together with six meta-heuristic algorithms, i.e. the particle swarm optimization based memetic algorithm (PSOMA) [25], hybrid genetic algorithm (HGA) [26], discrete bat algorithm (DBA) [27], hybrid differential evolution algorithm with local search (L-HDE) [22], hybrid backtracking search algorithm (HBSA) [19] and chaotic whale algorithm (CWA) [23], are adopted to solve 21 benchmark problems proposed by Reeves. Referring to the literature [1], the IKD and SKD sizes of EDHLO are 1, and the remaining parameters of EDHLO are specified as given below: population size $popsize = 60$, maximum iterations $Gmax = 2000$, $pr = 0.1$, $pi = 0.45$. Run each instance 30 times independently to compare. The computations were carried out using a PC with AMD Ryzen 5 2600 Six-Core Processor CPU and 32GB RAM on Windows 10, 64-bit operating system.

The best relative error (BRE), the average relative error (ARE) and the worst relative error (WRE) of the optimal solution are selected as the evaluation metrics, which are defined as Eqs. (10–12),

$$BRE = \frac{S_{Best} - BKS}{BKS} \times 100\%. \tag{10}$$

$$ARE = \frac{S_{Average} - BKS}{BKS} \times 100\%. \tag{11}$$

$$WRE = \frac{S_{Worst} - BKS}{BKS} \times 100\%. \tag{12}$$

where $S_{Best}$, $S_{Average}$ and $S_{Worst}$ represent the best solution, the average solution and the worst solution found by the algorithm respectively, and BKS represents the optimal solution value known so far of the benchmark problem.

Table 1 lists the experimental data of EDHLO. The corresponding data of compared algorithms are obtained from the original literature and the experimental results not given in the original literature are indicated by "–".

Figures 3 and 4 show the average values of the three evaluation metrics on the Reeves benchmark problems. Since the original literature of CWA contains results for only 19 Reeves benchmarks, for the sake of fairness, CWA is compared with EDHLO separately. It can be seen in Figs. 3 and 4 that EDHLO achieves the best results. Compared with PSOMA, HGA, DBA, L-HDE and HBSA, EDHLO ranks first in terms of the mean of BRE, ARE and WRE, with the lowest values of 0.258, 0.564 and 0.628, respectively. According to the values of the mean of BRE, ARE and WRE, EDHLO is also superior to CWA. The above experimental outcomes prove the effectiveness and stability of EDHLO.

**Table 1.** Comparison of EDHLO and other algorithms on Reeves benchmarks

| Problem | n*m | BK | | PSOM | HG | DB | L- | HBS | CW | EDHL |
|---|---|---|---|---|---|---|---|---|---|---|
| REC0 | 20*5 | 124 | BR | **0** | **0** | **0** | **0** | **0** | **0** | **0** |
| | | | AR | 0.144 | 0.14 | 0.08 | **0** | 0.14 | **0** | **0** |
| | | | WR | 0.16 | – | 0.16 | **0** | 0.16 | **0** | **0** |
| REC0 | 20*5 | 110 | BR | **0** | **0** | **0** | **0** | **0** | **0** | **0** |
| | | | AR | 0.189 | 0.09 | 0.08 | **0** | 0.08 | **0** | 0.018 |
| | | | WR | 0.721 | – | 0.18 | **0** | 0.18 | **0** | 0.180 |
| REC0 | 20*5 | 124 | BR | 0.242 | **0** | 0.24 | 0.242 | 0.24 | **0** | **0** |
| | | | AR | 0.249 | 0.29 | 0.24 | 0.242 | 0.24 | **0** | **0** |
| | | | WR | 0.402 | – | 0.24 | 0.242 | 0.24 | **0** | **0** |
| REC0 | 20*1 | 156 | BR | **0** | **0** | **0** | **0** | **0** | **0** | **0** |
| | | | AR | 0.986 | 0.69 | 0.57 | **0** | 0.46 | 0.16 | **0** |
| | | | WR | 1.149 | – | 1.14 | **0** | 1.15 | 0.83 | **0** |
| REC0 | 20*1 | 153 | BR | **0** | **0** | **0** | **0** | **0** | **0** | **0** |
| | | | AR | 0.621 | 0.64 | 0.63 | 0.026 | 0.07 | 0.04 | **0** |
| | | | WR | 1.691 | – | 2.40 | 0.260 | 0.65 | 0.32 | **0** |
| REC1 | 20*1 | 143 | BR | **0** | **0** | **0** | **0** | **0** | **0** | **0** |
| | | | AR | 0.129 | 1.1 | 1.16 | **0** | **0** | **0** | **0** |
| | | | WR | 0.978 | – | 2.65 | **0** | **0** | **0** | **0** |
| REC1 | 20*1 | 193 | BR | 0.259 | 0.36 | 0.41 | **0** | 0.1 | **0** | **0** |
| | | | AR | 0.893 | 1.68 | 1.46 | 0.275 | 0.53 | 0.45 | **0.140** |
| | | | WR | 1.502 | – | 3.78 | 0.777 | 1.14 | 0.82 | **0.259** |
| REC1 | 20*1 | 195 | BR | 0.051 | 0.56 | 0.15 | **0** | 0.05 | **0** | **0** |
| | | | AR | 0.628 | 1.12 | 1.22 | 0.523 | 0.64 | 0.57 | **0.179** |
| | | | WR | 1.076 | – | 2.10 | 1.180 | 1.18 | 1.02 | **0.462** |
| REC1 | 20*1 | 190 | BR | **0** | 0.95 | 0.36 | **0** | **0** | **0** | **0** |
| | | | AR | 1.33 | 2.32 | 1.27 | 0.363 | 1 | 0.67 | **0.074** |
| | | | WR | 2.155 | – | 2.15 | 0.946 | 2.16 | 1.41 | **0.368** |

*(continued)*

**Table 1.** (*continued*)

| Problem | n*m | BK | | PSOM | HG | DB | L- | HBS | CW | EDHL |
|---|---|---|---|---|---|---|---|---|---|---|
| REC1 | 30*1 | 209 | BR | 0.43 | 0.62 | 0.57 | 0.287 | 0.29 | 0.28 | **0.143** |
| | | | AR | 1.313 | 1.32 | 0.92 | 0.702 | 0.81 | 0.53 | **0.397** |
| | | | WR | 2.102 | – | 2.02 | 1.242 | 1.29 | 1.05 | **0.860** |
| REC2 | 30*1 | 201 | BR | 1.437 | 1.44 | 1.43 | 0.645 | 0.69 | 0.64 | **0.406** |
| | | | AR | 1.596 | 1.57 | 1.67 | 1.279 | 1.5 | 1.47 | **1.234** |
| | | | WR | 1.636 | – | 2.23 | 1.438 | 1.83 | 1.63 | **1.438** |
| REC2 | 30*1 | 201 | BR | 0.596 | 0.40 | 0.79 | 0.348 | 0.45 | **0.34** | **0.348** |
| | | | AR | 1.31 | 0.87 | 1.17 | 0.428 | 1.28 | 0.85 | **0.417** |
| | | | WR | 2.038 | – | 2.38 | 0.497 | 3.08 | 1.93 | **1.044** |
| REC2 | 30*1 | 251 | BR | 0.835 | 1.27 | 1.63 | 0.557 | 0.4 | – | **0.279** |
| | | | AR | 2.085 | 2.54 | 2.92 | 1.082 | 1.29 | – | **0.712** |
| | | | WR | 3.233 | – | 3.94 | 1.632 | 2.43 | – | **1.353** |
| REC2 | 30*1 | 237 | BR | 1.348 | 1.10 | 1.01 | 0.253 | 0.25 | **0.01** | 0.253 |
| | | | AR | 1.605 | 1.83 | 1.41 | 0.851 | 1.27 | 1.00 | **0.839** |
| | | | WR | 2.402 | – | 2.29 | 1.222 | 2.57 | 1.72 | **1.054** |
| REC2 | 30*1 | 228 | BR | 1.442 | 1.4 | 1.04 | 0.831 | 0.57 | 0.49 | **0.087** |
| | | | AR | 1.888 | 2.7 | 2.58 | 1.049 | 1.42 | 1.24 | **0.822** |
| | | | WR | 2.492 | – | 3.93 | 1.443 | 2.97 | 2.31 | **1.443** |
| REC3 | 50*1 | 304 | BR | 1.51 | 0.43 | 2.29 | 0.427 | 0.43 | 0.42 | **0.263** |
| | | | AR | 2.254 | 1.34 | 3.39 | **0.644** | 1.91 | 0.99 | 0.785 |
| | | | WR | 2.692 | – | 4.53 | **0.920** | 2.66 | 1.47 | 1.478 |
| REC3 | 50*1 | 311 | BR | **0** | **0** | 0.61 | **0** | **0** | **0** | **0** |
| | | | AR | 0.645 | 0.78 | 0.72 | 0.244 | 0.59 | **0** | 0.328 |
| | | | WR | 0.834 | – | 1.73 | 0.835 | 1.28 | **0** | 0.739 |
| REC3 | 50*1 | 327 | BR | **0** | **0** | **0** | **0** | **0** | **0** | **0** |
| | | | AR | **0** | **0** | 0.03 | **0** | **0** | **0** | **0** |
| | | | WR | **0** | – | 0.09 | **0** | **0** | **0** | **0** |
| REC3 | 75*2 | 495 | BR | 2.101 | 3.75 | 3.37 | 2.565 | 1.92 | – | **1.827** |
| | | | AR | 3.537 | 4.9 | 4.87 | 3.001 | 2.93 | – | **2.467** |
| | | | WR | 4.039 | – | 5.97 | 3.555 | 4.2 | – | **3.272** |
| REC3 | 75*2 | 508 | BR | 1.553 | 2.2 | 2.28 | 1.730 | 0.9 | **0.08** | 0.806 |
| | | | AR | 2.426 | 2.79 | 3.85 | 1.832 | 1.88 | 1.57 | **1.246** |
| | | | WR | 2.83 | – | 5.34 | 2.005 | 3.38 | 1.98 | **1.553** |
| REC4 | 75*2 | 496 | BR | 2.641 | 3.64 | 3.81 | 2.661 | 1.69 | 1.45 | **1.008** |
| | | | AR | 3.684 | 4.92 | 5.09 | 3.350 | 2.72 | 2.34 | **2.185** |
| | | | WR | 4.052 | – | 6.53 | 3.770 | 3.55 | 3.04 | **2.742** |

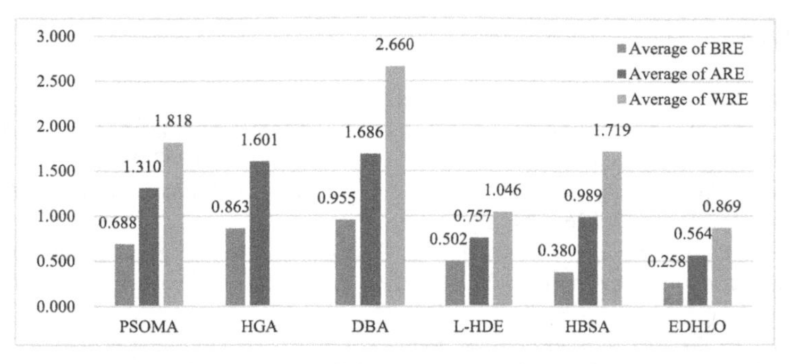


**Fig. 3.** The comparison of PSOMA, HGA, DBA, L-HDE, HBSA and EDHLO on Reeves benchmarks

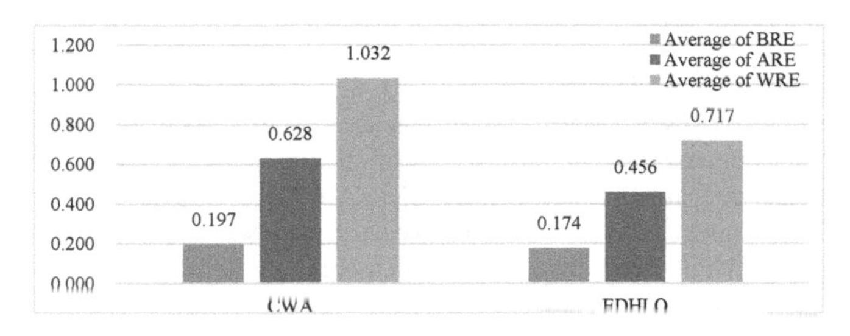


**Fig. 4.** The comparison of CWA and EDHLO on Reeves benchmarks

## 5 Conclusions

To further study HLO and extend it to solve discrete problems more efficiently, this paper proposed an enhanced discrete human learning optimization (EDHLO) algorithm for solving PFSSP. EDHLO is based on the standard learning mechanisms of HLO, but it redesigns and introduces the learning operators to solve PFSSP efficiently. In the proposed EDHLO, the efficient heuristic algorithm NEH is integrated with random initialization to improve the original population quality and maintain the diversity of the original population, two cross operators are introduced into the individual learning of EDHLO to upgrade the local exploitation capacity, and a local search method is used to facilitate EDHLO to get rid of the local optimum. The experimental results on 21 benchmark problems and the comparisons with other meta-heuristics verify the effectiveness and stability of the EDHLO.

**Acknowledgments.** This work is supported by National Key Research and Development Program of China (No. 2019YFB1405500), National Natural Science Foundation of China (Grant No. 92067105 & 61833011), Key Project of Science and Technology Commission of Shanghai Municipality under Grant No. 19510750300 & 19500712300, and 111 Project under Grant No. D18003.

## References

1. Wang, L., Ni, H.Q., Yang, R.X., et al.: A simple human learning optimization algorithm. In: Fei, M., Peng, C., Su, Z., Song, Y., Han, Q. (eds.) LSMS/ICSEE 2014, Part II. CCIS, vol. 462, pp. 56–65. Springer, Heidelberg (2014). https://doi.org/10.1007/978-3-662-45261-5_7
2. Wang, L., Ni, H.Q., Yang, R.X., et al.: An adaptive simplified human learning optimization algorithm. Inf. Sci. **320**, 126–139 (2015)
3. Wang, L., An, L., Pi, J., Fei, M., Pardalos, P.M.: A diverse human learning optimization algorithm. J. Global Optim. **67**(1–2), 283–323 (2016). https://doi.org/10.1007/s10898-016-0444-2
4. Yang, R., Xu, M., He, J., Ranshous, S., Samatova, N.F.: An intelligent weighted fuzzy time series model based on a sine-cosine adaptive human learning optimization algorithm and its application to financial markets forecasting. In: Cong, G., Peng, W.-C., Zhang, W.E., Li, C., Sun, A. (eds.) ADMA 2017. LNCS (LNAI), vol. 10604, pp. 595–607. Springer, Cham (2017). https://doi.org/10.1007/978-3-319-69179-4_42
5. Wang, L., Pei, J., Wen, Y.L., et al.: An improved adaptive human learning algorithm for engineering optimization. Appl. Soft Comput. **71**, 894–904 (2018)
6. Alguliyev, R., Aliguliyev, R., Isazade, N.: A sentence selection model and HLO algorithm for extractive text summarization. In: 2016 IEEE 10th International Conference on Application of Information and Communication Technologies (AICT), pp. 206--209 (2016)
7. Bhandari, A.K., Kumar, I.V.: A context sensitive energy thresholding based 3D Otsu function for image segmentation using human learning optimization. Appl. Soft Comput. **82**, 105570 (2019)
8. Wang, L., Pei, J., Menhas, M.I., et al.: A Hybrid-coded human learning optimization for mixed-variable optimization problems. J. Knowl-Based Syst. **127**, 114–125 (2017)
9. Han, Z., Qi, H., Wang, L., Menhas, M.I., Fei, M.: Water level control of nuclear power plant steam generator based on intelligent virtual reference feedback tuning. In: Li, K., Zhang, J., Chen, M., Yang, Z., Niu, Q. (eds.) ICSEE/IMIOT -2018. CCIS, vol. 925, pp. 14–23. Springer, Singapore (2018). https://doi.org/10.1007/978-981-13-2381-2_2
10. Menhas, M.I., Wang, L., Ayesha, N.U., et al.: Continuous human learning optimizer based PID controller design of an automatic voltage regulator system. In: 2018 Australian & New Zealand Control Conference (ANZCC), pp. 148–153 (2018)
11. Wen, Y., Wang, L., Peng, W., Menhas, M.I., Qian, L.: Application of intelligent virtual reference feedback tuning to temperature control in a heat exchanger. In: Li, K., Fei, M., Du, D., Yang, Z., Yang, D. (eds.) ICSEE/IMIOT -2018. CCIS, vol. 924, pp. 311–320. Springer, Singapore (2018). https://doi.org/10.1007/978-981-13-2384-3_29
12. Fan, W., Pei, J., Liu, X., et al.: Serial-batching group scheduling with release times and the combined effects of deterioration and truncated job-dependent learning. J. Glob. Optim. **71**(1), 147–163 (2017)
13. Li, X.Y., Yao, J., Wang, L., et al.: Application of human learning optimization algorithm for production scheduling optimization. In: Fei, M., Ma, S., Li, X., Sun, X., Jia, L., Su, Z. (eds.) Advanced Computational Methods in Life System Modeling and Simulation. ICSEE 2017, LSMS 2017. Communications in Computer and Information Science, vol. 761, pp. 242--252. Springer, Singapore (2017). https://doi.org/10.1007/978-981-10-6370-1_24
14. Ding, H.J., Gu, X.S.: Hybrid of human learning optimization algorithm and particle swarm optimization algorithm with scheduling strategies for the flexible job-shop scheduling problem. Neurocomputing **414**, 313–332 (2020)
15. Shoja, A., Molla-Alizadeh-Zavardehi, S., Niroomand, S.: Hybrid adaptive simplified human learning optimization algorithms for supply chain network design problem with possibility of direct shipment. Appl. Soft Comput. **96**, 106594 (2020)

16. Nawaz, M., et al.: A heuristic algorithm for the m-machine, n-job flow-shop sequencing problem. Omega **11**, 91–95 (1983)
17. Reeves, C.R.: A genetic algorithm for flowshop sequencing. Comput. Oper. Res. **22**(1), 5–13 (1995)
18. Zwied, A.N.H., Ismal, M.M., Mohamed, S.S.: Permutation flow shop scheduling problem with makespan criterion: literature review. J. Theor. Appl. Inf. Technol. **99**, 4 (2021)
19. Lin, Q., Gao, L., Li, X., et al.: A hybrid backtracking search algorithm for permutation flow-shop scheduling problem. Comput. Ind. Eng. **85**, 437–446 (2015)
20. Ali, M.Z., Awad, N.H., Suganthan, P.N., et al.: A modified cultural algorithm with a balanced performance for the differential evolution frameworks. J. Knowl.-Based Syst. **111**, 73–86 (2016)
21. Forcheri, P., Molfino, M.T., Quarati, A.: ICT driven individual learning: new opportunities and perspectives. J. Educ. Technol. Soc. **3**, 51–61 (2000)
22. Liu, Y., Yin, M., Gu, W.: An effective differential evolution algorithm for permutation flow shop scheduling problem. Appl. Math. Comput. **248**, 143–159 (2014)
23. Li, J., Guo, L.H., Li, Y., et al.: Enhancing whale optimization algorithm with chaotic theory for permutation flow shop scheduling problem. Int. J Comput. Intell. Syst. **14**(1), 651–675 (2021)
24. Roediger, H.L.: Reflections on intersections between cognitive and social psychology: a personal exploration. J. Eur. J. Soc. Psychol. **40**(2), 189–205 (2010)
25. Liu, B., Wang, L., Jin, Y.H.: An effective PSO-based memetic algorithm for flow shop scheduling. IEEE Trans. Syst. Man Cybern. B Cybern. 37(1), 18–27 (2007)
26. Tseng, L.-Y., Lin, Y.-T.: A hybrid genetic algorithm for no-wait flowshop scheduling problem. Int. J. Prod. Econ. **128**(1), 144–152 (2010)
27. Luo, Q., Zhou, Y., Xie, J., et al.: Discrete bat algorithm for optimal problem of permutation flow shop scheduling. Sci. World J. **2014**, 630–280 (2014)

# Key Technology for Service Sharing Platform Based on Big Data

Jialan Ding, Wenju Zhou(✉), Xian Wu, Yayu Chen, and Minrui Fei

School of Mechatronic Engineering and Automation,
Shanghai University, Shanghai 200444, China
{zhouwenju,xianwu,chenyayu}@shu.edu.cn, mrfei@staff.shu.edu.cn

**Abstract.** To solve the problems of low conversion rate of the sharing service platform, this paper proposes an open science and technology sharing service platform which integrates resource retrieval, technology consulting, technology finance, qualification certification, entrepreneurial incubation etc. based on traditional technology sharing service model. The distributed shared collaborative technology service resource pool and independent service transaction and review system are constructed based on big data technology. By promoting the rapid connection and matching the demand and supply under precise services, this platform provides more standardized and efficiently technological services for small and medium-sized enterprises.

**Keywords:** Big data · Service sharing · Achievement transformation

## 1 Introduction

Science and technology service capability is one of the important ways to promote the optimization and upgrading of industrial structure, cultivate strategic emerging industries, and improve the competitiveness of high-tech industries. In the era of big data, the demand for regional innovation is increasingly complicated [1], and the modern technology service system is gradually developing towards informatization [2], which is making technology service innovation encounter more opportunities and challenges. The existing Scientific Research Facilities and Instrument Sharing Service Platform only integrates instrument and equipment resource services in the province such as instrument inquiry, testing appointment, technical consultation, and member unit publicity [3]. The platform exists problems such as single service items, too few resources in all regions, and lack of online transaction and review functions, which are not conducive to promoting the transfer and transformation of scientific research results.

In order to improve the deficiencies of the existing technology service sharing platform, this paper designs the overall functional architecture of the platform based on the analysis of the characteristics of the existing science and technology service sharing platform and investigating the urgent needs of the science and technology service industry. On the one hand, this platform enhances the comprehensive scientific and technological service capabilities by means of gathering domestic and foreign resources by building


Q. Han et al. (Eds.): LSMS 2021/ICSEE 2021, CCIS 1469, pp. 258–267, 2021.
https://doi.org/10.1007/978-981-16-7213-2_25

a comprehensive technology service platform and multi-source heterogeneous resource pool based on big data technology [4]. On the other hand, this platform is equipped with an independent service transaction and review system to promote industry-university-research cooperation between enterprises and universities and scientific research units, and promote rapid docking under the matching of supply and demand, and ultimately increase the actual benefit conversion rate of scientific and technological resources.

## 2 Key Technology

### 2.1 Microservice-Based Application System Architecture

Microservice is a new architectural model. Its core idea is to decouple complex systems into micro-applications to complete business closed loops in various scenarios [5]. In order to makes up for the shortcomings of traditional platforms, microservices communicate with each other through a lightweight communication mechanism and can be independently deployed on single or multiple servers, which have the characteristics of openness, autonomy and growth [6].

According to different service characteristics, the platform adds modules such as log monitoring, load balancing, authorization authentication, and service fusing in the configuration on the basis of using Spring Cloud tools for distributed configuration management of microservices, in particular, it uses Docker. The platform is centered on user needs to design multiple independent subsystem businesses, and aggregated to form a complete scientific and technological service system by creating microservices separately for each subsystem. Each resource library can not only maintain its own integrity and security, but also participate in the implementation of sharing, thereby improving resource utilization.

### 2.2 Component-Based Module Development

The component-based development idea is to split the page into multiple components. The CSS, JS, templates, logic and other resources that each component depends on are developed and maintained together. Components can be reused within the system and be nested between components, which effectively increases the reusability and flexibility of components, improves system development efficiency, and reduces project development costs.

The web front-end part of the platform is developed based on the Vue.js -framework, which is a component-based MVVM architecture model. The front-end interface is divided into multiple page modules through business functions. Developers first define hook functions by constructing Vue instances. Each Vue page is a Vue instance. Then they mount common objects into the Vue prototype, and use component libraries or self-encapsulated business components for functional module design. Finally, ESLint can be used to check code specifications to ensure code consistency.

### 2.3 Multi-source Heterogeneous Resource Pool Based on Big Data

The service sharing platform gathers all kinds of science and technology resource information needed for the development of small and medium-sized enterprises on the basis of the traditional platform [7]. It has the characteristics of extensive data sources, messy data standards, various fields involved, and unstructured. According to the unified data specification, data resources are cleaned and summarized through ETL tools under big data technology, and are stored by relying on Hadoop distributed file system and Spark computing engine. In order to form a hierarchical and clearly structured data classification, massive resources which are used to sort out the data by extracting data tags are extracted for statistical analysis and correlation analysis. Unstructured data is processed by text content mining and analysis, audio and video content structure transformation, and is transformed into structured data by algorithms. Through the construction of a multi-source heterogeneous data resource pool, it provides good data resource support for the service sharing platform.

## 3 Design of System Architecture

The platform adopts a component-based development model with separation of front-end and background to build an information system, which is based on B/S architecture and J2EE development standard specifications. The front section adopts techniques such as Vue.js framework+Element-UI+jQuery+AJAX+CSS and JS encapsulation. The background is mainly Spring Boot framework+MyBatis and MySQL+MINIO database [8]. A screenshot of the homepage of the platform is shown in Fig. 1.

**Fig. 1.** Screenshot of the platform

In order to achieve high cohesion and low coupling between functional modules, the platform adopts a multi-layer structure design method and microservice architecture. According to the demand, the platform architecture can be divided into several main functional levels such as terminal access, user layer, application layer, application support layer, data layer, and base layer, as shown in Fig. 2.

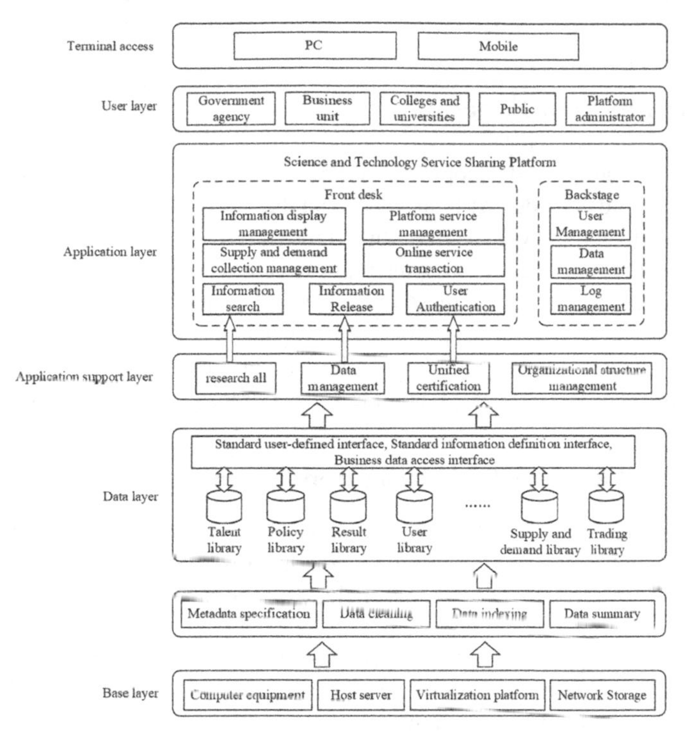


**Fig. 2.** Platform system architecture diagram

The base layer mainly includes basic hardware devices such as computer equipment, host servers, and storage devices. The layer builds private cloud services such as virtualization platforms and network storage, so that the platform can receive highly scalable and highly available basic support services. The data layer cleans, indexes, and summarizes large amounts of metadata, and uses Redis as cache middleware, and then provides data storage services for upper-level applications through standard user-defined interfaces, standard information-defined interfaces, and business data access interfaces. In order to provide users with a big data sharing platform and achieve the principle of separation of front-end and background, the heterogeneous data is logically and physically integrated through XML technology and each microservice exposes an API interface to the outside. In order to reduce the functional coupling of the platform, the application support layer adopts organizational structure management to build public components such as full-text search, data management, and unified authentication. The application layer includes the front-end portal system and the back-end resource pool management system. In order to provide more comprehensive services to the various user roles of the platform, the front-end and background use API interfaces for data transmission, and Ajax technology for achieving data requests between the client and the server. A business process corresponds to a microservice which is independently deployed and has its own separate storage database, thereby improving the fault tolerance rate of the platform [9]. The user layer includes government agencies, enterprises and units, colleges and universities, the public and platform administrators, which is suitable for the multi-role characteristics of platform users.

# 4 Design of System Function

The entire system adopts a development model with front-end and background separation, the background provides API interfaces and the front-end is developed based on the Vue.js framework of the MVVM model. In order to improve the efficiency of science and technology resource sharing services, the platform also adopts a micro-service architecture to make each subsystem business relatively independent. Integrating the needs of small and medium-sized enterprises for technological services, as shown in Fig. 3, the platform is divided into three major parts: the background resource pool management system, the front-end portal system, and the service transaction and review system.

## 4.1 Background Resource Pool Management System

The background resource pool management system is used to quickly operate and visually manage all databases and files on the platform. It is based on the Element-UI component library to build a framework and includes three modules: system management, user management, and data management.

The system management module is used to store system logs and perform message management. The system logs record all user operation data behaviors, and realize log subcontracting records, which is convenient for quick positioning and analysis when platform problems occur; message management function is used by administrators to release, modify and delete platform announcement information.

The user management module provides functions such as user registration, audit authorization, information maintenance, and authority management. Platform users can obtain the permission to log in to the platform by registering. After the platform administrator completes the review and authorization, the service provider users can maintain the information of their own services in the personal center of the platform, and the users of institutional units can maintain the information of their own units, scientific research talents, and scientific research results in the personal center of the platform. At the same time, the platform administrator can open or close multi-role user accounts through the authority management function based on the results of the credit review.

The data management module provides functions such as module management, supply and demand management, information maintenance, and transaction management. In the module management, the data of all independent modules of the system is stored in the form of forms, and the administrator can add, delete, modify, query, import and other functions of the data content. The supply and demand management includes two parts: supply management and demand management. The supply management stores the service entry application and the demand management stores the service demand application submitted by the user at the front end.

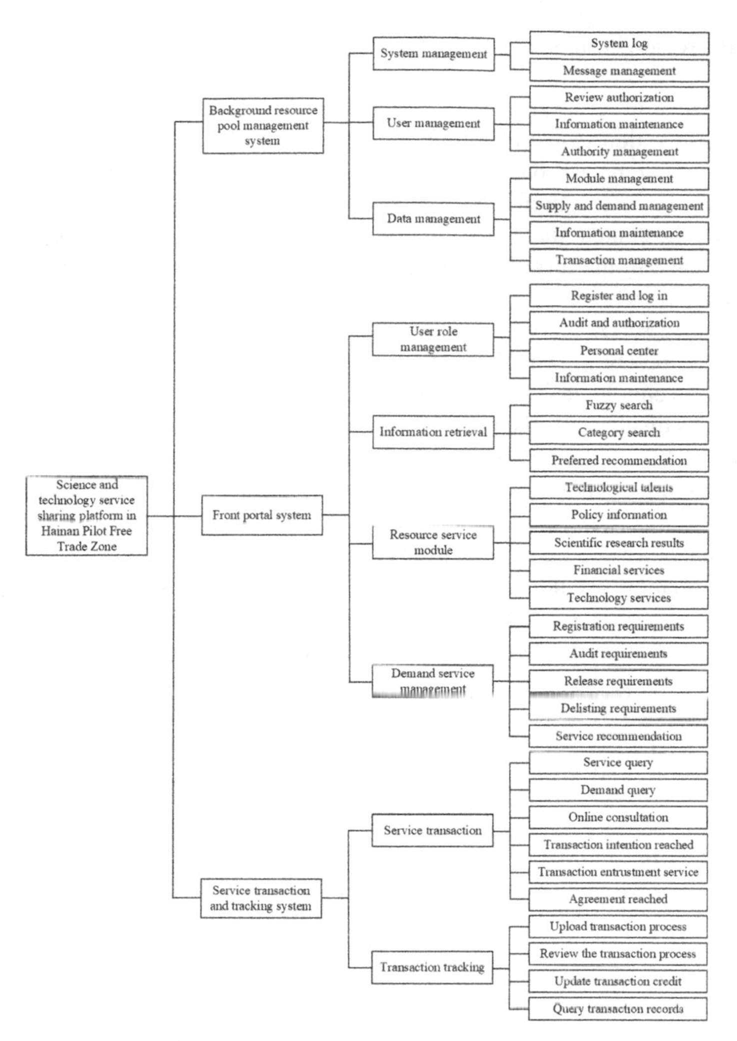


**Fig. 3.** System function structure diagram

## 4.2 Front-End Portal System

The front-end portal system is used to clearly and visually display and apply all the functions and services of the platform, including user role management, information retrieval, resource service module, demand service management and other sub-function modules.

The user role management module is presented in the form of a personal center. After the user submits the registration application information, the platform administrator reviews and authorizes it in the background, and the user can get more platform service functions for login authorization and enter the personal center module. The user can manage and maintain the personal center according to their respective role positioning, and view the review results of the application content by the platform administrator in real time.

The information retrieval module provides three methods: site-wide fuzzy search, keyword classification search and preference recommendation. The preference recommendation is implemented based on a recommendation algorithm which is analyzing and predicting the user's preferences according to the user's historical behavior. And then filter out the content that meets their preferences in the resource pool, and present them on the page in the form of keyword hot words, which is convenient for users to click and search.

The resource service module includes sub-modules such as scientific and technological talents, policy information, scientific research results, financial services, and scientific and technological services. The corresponding scientific and technological resource data is displayed in the form of a form in the sub-module, which is dividing the keyword catalog according to different business types for users to perform conditional search and making preference recommendations according to the user's historical clicks and filters.

The demand service management module performs process management on user demand information, including demand registration, demand review, demand release, demand removal, service recommendation and other functions. Users can fill in the demand information in the demand release section when they want to make a demand to the platform after the search failed. The platform administrator conducts demand review, the reviewed demand is imported into the demand database of the platform, and be published to the front desk display page. When the demand is solved, the demand can be removed from the shelf and deleted in the demand library.

### 4.3 Service Transaction and Review System

The service transaction and review system is an independent system that provides transaction functions for all users on the site, and supports the sustainable operation of the platform. It is the last step after the user confirms the purchase service intention and mainly includes a service transaction module and a follow-up review module.

The business process of the service transaction module is divided into service inquiry, demand inquiry, online consultation, intention accomplishment and transaction accomplishment. Users can inquire about the scientific and technological services and needs they are interested in in the search engine on the site, and initially determine the services or needs they want to purchase. To further understand the scientific and technological information of service providers, the online consultation function provides a real-time chat tool which is convenient for users in need to consult with experts such as scientific and technological talents, teams, and project teams. When the demand side and the service side reach a transaction intention, the transaction is reached after confirmation by the platform. The business process of the follow-up review module is divided into transaction process upload, transaction process review, transaction credit update and transaction record query. After the demand side and the service side reach a transaction, the supporting materials need to be uploaded to the platform and be reviewed and credited by the platform administrator.

## 5 Design of Database

E-R diagrams provide a way to represent entity types, attributes and connections. The entity model in the system mainly includes service side, demand side, platform administrator, service, demand, settle in, release, scientific research results, order, transaction and transaction review. The E-R diagram shown in Fig. 4 shows the relationship between entities.

There are three kinds of connections between entities, namely one-to-one, one-to-many, and many-to-many. The service side can provide multiple services and the demand side can put forward multiple requirements. When the supply and demand match, the transaction is concluded through the platform review, and the transaction status is continuously tracked and reviewed to realize the transformation of scientific and technological achievements between small and medium-sized enterprises and expert institutes and other teams.

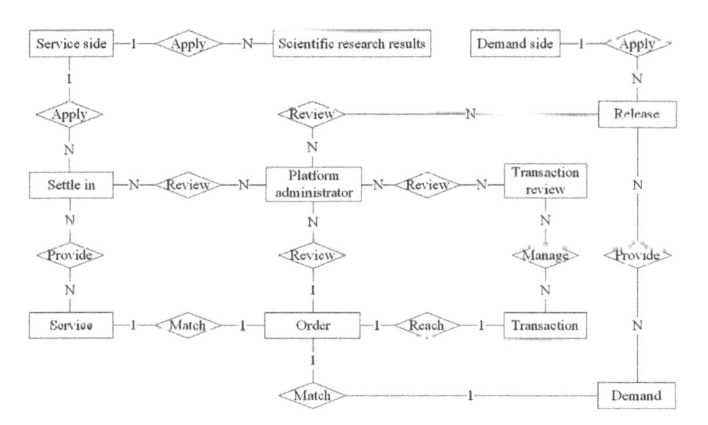


**Fig. 4.** Database E-R diagram

## 6 Design of Business Process

According to the system function structure diagram proposed above, platform users are divided into service side and demand side to respond to different types of user needs, and a business flowchart of the technology service sharing platform is given, as shown in Fig. 5.

Users can inquire about the corresponding services or requirements through the registration and login platform. If no matching information is retrieved, the service side can fill in their own scientific and technological service results to apply for the platform and the demand side can fill in their own scientific and technological service requirements to apply for publication on the platform. The both sides conduct consultation and exchanges through online chat tools and reach a deal intention if the supply and demand match, and then the transaction is concluded after the platform is reviewed and confirmed. In the process of business processing, both parties strictly implement

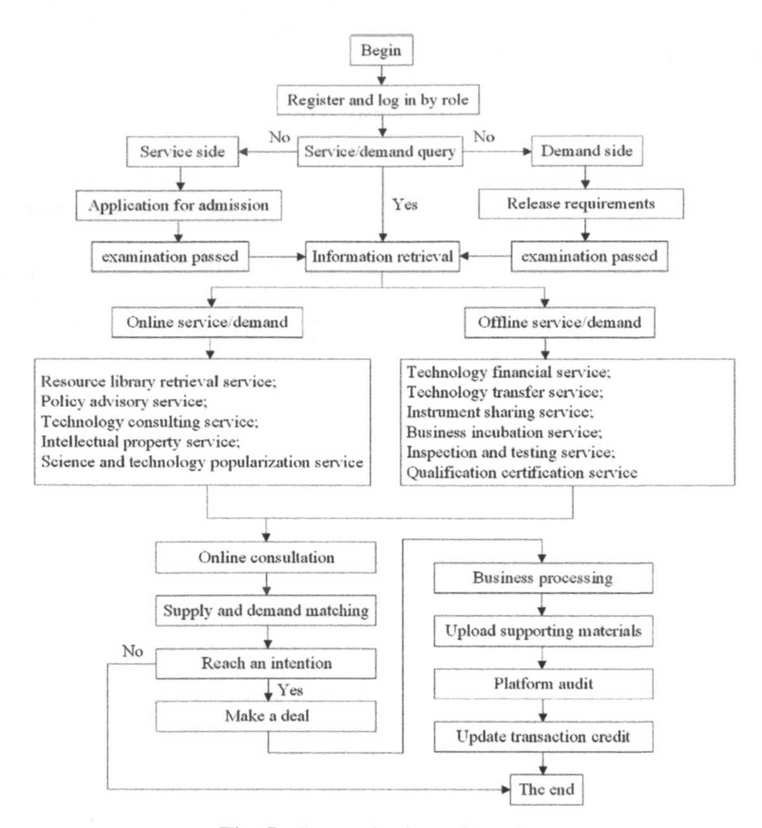


**Fig. 5.** System business flow chart

the service transaction and review process and upload supporting materials at any time. Once the platform administrator discovers untrustworthy and non-compliance with the transaction agreement during the review process, the both sides will immediately terminate the transaction, be updated the transaction credit and pulled into the platform's untrustworthy blacklist.

## 7 Conclusion

This paper aims to design an open service sharing platform to improve the regional comprehensive science and technology service capabilities and promote the transfer and transformation of scientific research results. Through the use of big data technology and component development ideas, this paper constructs a multi-source heterogeneous data resource pool and splits business logic to complete the design of functional modules, so that SMEs and scientific research institutes can share scientific and technological service resources and needs. Finally, the platform is equipped with an independent service transaction and review system to realize reliable and safe transaction services and sustainable operation of the platform between the parties to the transaction, and ultimately improve the actual benefit conversion rate of scientific and technological resources.

**Acknowledgements.** This research is financially supported by the National Key R&D Program of China (No. 2019YFB1405500).

## References

1. Zhao, Q.: Resource sharing strategy of regional scientific and technological innovation based on big data analysis platform. In: 2018 International Conference on Virtual Reality and Intelligent Systems (ICVRIS), Hunan, China, 2018, pp. 134--138 (2018)
2. Chen, M., Chen, Y., Liu, H.F., Xu, H.: Influence of information technology capability on service innovation in manufacturing firms. Ind. Manag. Data Syst. **121**(2) (2020)
3. Hainan Scientific Research Facilities and Instrument Sharing Service Platform. https://www.hidyw.com/
4. Su, H., Lin, X., Xie, Q., Chen, W., Tang, Y.: Research on the construction of tourism information sharing service platform and the collection of tourist satisfaction. In: 2018 3rd International Conference on Smart City and Systems Engineering (ICSCSE), Xiamen, China, pp. 640--643 (2018)
5. Miao, K.H., Li, J., Hong, W.X., Chen, M.T., Huang, C.X.: A microservice-based big data analysis platform for online educational applications. Sci. Program. (2020)
6. Razzaq, A.: A systematic review on software architectures for IoT systems and future direction to the adoption of microservices architecture. SN Comput. Sci. **1**(6), 1–30 (2020)
7. Zhang, Y., Wang, Y.G., Ding, H.W., Li, Y.Z., Bai, Y.P.: Deep well construction of big data platform based on multi-source heterogeneous data fusion. Int. J. Internet Manufact. Serv. **6**(4), 371–388 (2019)
8. Lu, D., Qiu, Y.P., Qian, C., Wang, X., Tan, W.: Design of campus resource sharing platform based on SSM framework. IOP Conf. Ser.: Mater. Sci. Eng. 490(6), 062043 (2019)
9. Isak, S., Endrit, M., Blend, B., Tonit, B.: Design of modern distributed systems based on microservices architecture. Int. J. Adv. Comput. Sci. Appl. (IJACSA) **12**(2), 153–159 (2021)

# A Novel YOLO Based Safety Helmet Detection in Intelligent Construction Platform

Meng Yang[1], Zhile Yang[1](✉), Yuanjun Guo[1], Shilong Su[2], and Zesen Fan[2]

[1] Shenzhen Institute of Advanced Technology, Chinese Academy of Sciences, Shenzhen 518055, Guangdong, China
zl.yang@siat.ac.cn
[2] China Construction Science and Technology Group Cooperation, Shenzhen 518118, China

**Abstract.** Safety is the foremost important issue on the construction site. Wearing a safety helmet is a compulsory issue for every individual in the construction area, which greatly reduces the injuries and deaths. However, though workers are aware of the dangers associated with not wearing safety helmets, many of them may forget to wear helmet at work, which leads to significant potential security issues. To solve this problem, we have developed an automatic computer-vision approach based on Convolutional Neural Network (YOLO) to detect wearing condition. We create a safety helmet image dataset of people working on construction sites. The corresponding images are collected and labeled and are used to train and test our model. The YOLO based model is adopted and parameters are well tuned. The precision of the pro-posed model is 78.3% and the accuracy rate is 20 ms. The results demonstrate that the proposed model is an effective method and comparatively fast for the recognition and localization in real-time helmet detection.

**Keywords:** Object detection · Construction safety · Helmet detection · Computer vision

## 1 Introduction

The construction industry is considered to be one of the most dangerous industries. There is a high accident rate within the field [1]. According to the Health and Safety Executive, construction workers suffer 34% of fatal injuries. The unique, dynamic, and complex working environment of construction site increase workers' exposure to dangers [3]. Wearing safety helmet can greatly reduce the possibility of injury and death when workers are in danger. Checking whether workers wear safety helmets or not is a major issue in the construction industry.

At present, all construction sites require workers to wear safety helmets before entering the construction area. Most construction sites use manual inspection to confirm whether workers wear safety helmets. The situation in construction site is very complex, and manual inspection cannot easily cover all areas. Moreover, although humans have high accuracy in most detection tasks, it is impossible for humans to focus on these tasks all the time.


Q. Han et al. (Eds.): LSMS 2021/ICSEE 2021, CCIS 1469, pp. 268–275, 2021.
https://doi.org/10.1007/978-981-16-7213-2_26

In recent years, object detection, as the basic task of computer vision, has been developed rapidly. In fact, Helmet detection is a special case of Object detection, their purpose is to identify the type of object and the location information contained in it [4]. In the past, intelligent system like DPM uses the sliding window method [5] for object detection: the sliding window slides evenly on the image, in every position the system will classify the object in the sliding window. In 2012, the work of Krizhevsky et al. [6] laid the foundation for the development of CNN-based object detection. In this work, CNN has developed rapidly because it makes good use of the close relationship between pixels in neighboring areas of pictures [7] and become the main method of object detection. In particular, YOLO [8] and R-CNN [9] are two representative CNN methods in object detection. YOLO is a one-step method the image only needs to pass the model once to output location and classification information. R-CNN is a two-step method that first outputs the position of the prediction box, and then classifies the object on the basis of the prediction box.

## 2 Research Method

### 2.1 You Only Look Once (YOLO)

In object detection, You Only Look Once (YOLO) is a famous and excellent model. The Architecture of YOLO has 24 convolutional layers and 2 fully connected layers. The architecture is shown in Fig. 1.

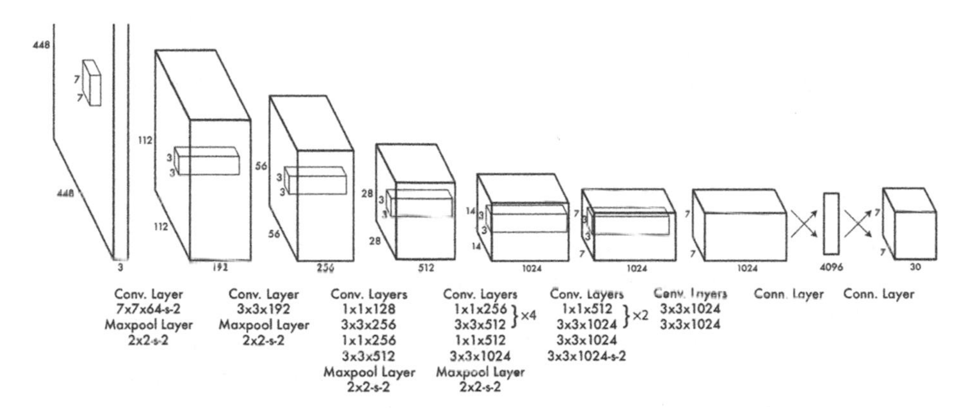


**Fig. 1.** The architecture of YOLO [8].

YOLO take an input image and resize it to 448 × 448 and divides the image into a 7 × 7 grid. Each grid will predict 2 bounding boxes. Therefore, YOLO will predict 98 bounding boxes. Each bounding box consists of 5 predictions, including confidence score, x, y, w, h. The confidence score is the IOU of predicted bounding box and actual bounding box. The IOU is an elimination in object detection, can be defined as:

$$IOU = Area\ of\ Overlap/Area\ of\ Union$$

If the IOU > 0.5 It is assumed that the object has been detected by this predicted bounding box. If the grid detects an object, the grid is responsible for classifying the object. Then YOLO will use non-max-suppressing method to get the final boxes from the 98 boxes. NMS is widely used in object detection and is aimed to eliminate redundant candidate bounding boxes and find the best detection boxes. The whole process is shown in Fig. 2.

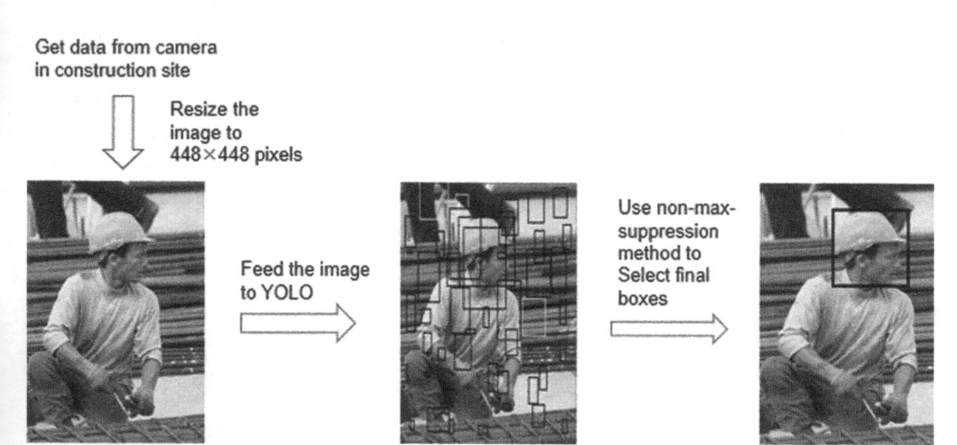
Get data from camera in construction site
Resize the image to 448×448 pixels
Feed the image to YOLO
Use non-max-suppression method to Select final boxes

**Fig. 2.** The procedure of YOLO in dealing with an image.

Compare with other models, the detection speed of YOLO is faster because the image only needs to go through the YOLO model once to get the information of location and classification. Instead of using two step method for classification and localization of object [10].

## 2.2 Dataset Establishment

In object detection tasks, there are many well-known data sets such as PASCAL VOC [11], COCO [12] and ImageNet [13]. Fully consider the needs of our actual applications, for helmet detection task, we only need to classify two categories (helmet, person). Up to date, there is no existing public safety helmet dataset. In light of this, we build the data set by ourselves. The Images are collected from websites and taken by our construction site cameras. In order to get more images of PERSON, we also add some images of the person in PASCAL VOC data set. Finally, we get a data set includes 1600 pictures in the form of JPG file and labels in the form of XML file. The details of datasets are described in Table 1.

**Table 1.** Detailed rules of PASCAL VOC, COCO, ImageNet and our dataset.

| Label setting | Classes | Label form | Number of images | Number of Bounding boxes |
|---|---|---|---|---|
| Our dataset | 2 | XML | 1,633 | 3,789 |
| PASCAL VOC | 20 | XML | 12,000 | 27,000 |
| COCO | 91 | JOSN | 328,000 | 2,500,000 |
| Imagenet | 21841 | XML | 1,034,908 | |

## 2.3 Experiments

All the tasks described in this article are performed on the system having following specifications: OS: Ubuntu (64 bit). CPU: Intel i9-10920X @ 3.5 GHz. CPU RAM: 128 GB. GPU: NVIDIA GeForce RTX 3090. GPU Ram size: 24 GB.

We divide the datasets into three parts: 1200 images as the training set and 200 images as the validation set and 200 images as the testing set. We feed our images and labels to the model and then detect the test image with trained model. The workflow we implemented for this paper is explained in Fig. 3.

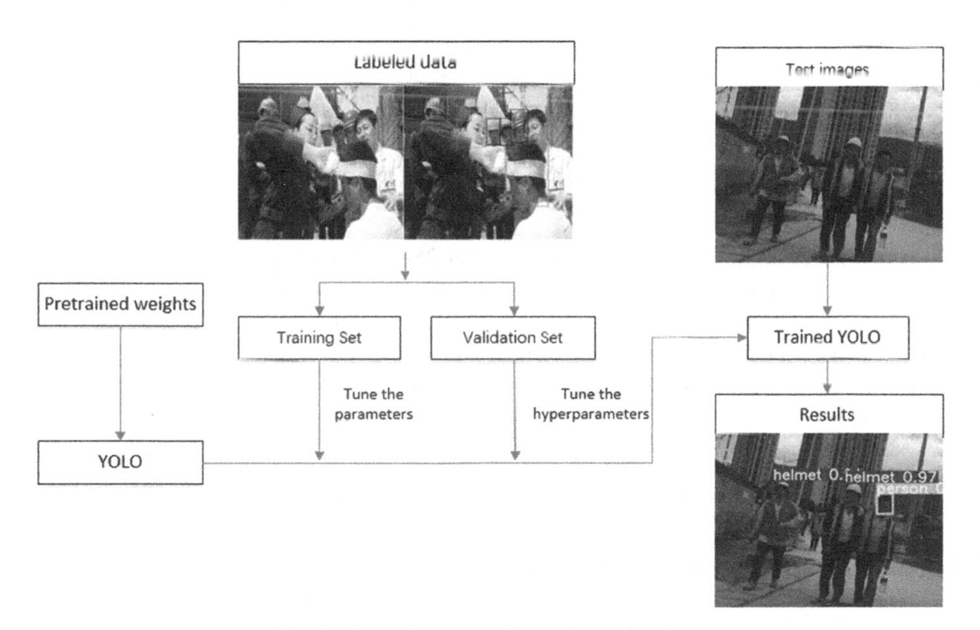


**Fig. 3.** The whole workflow of our algorithm.

In training process, we set the batch size of 32, a learning rate of 0.0001, a momentum of 0.9 and a decay of 0.0005. To avoid overfit, we use dropout = 0.5, other hyperparameters are the same in YOLO paper. We train the model 200 iterations and each iteration costs 58.32s. The loss, precision, mean average precision (mAP) during training is shown in Fig. 4.

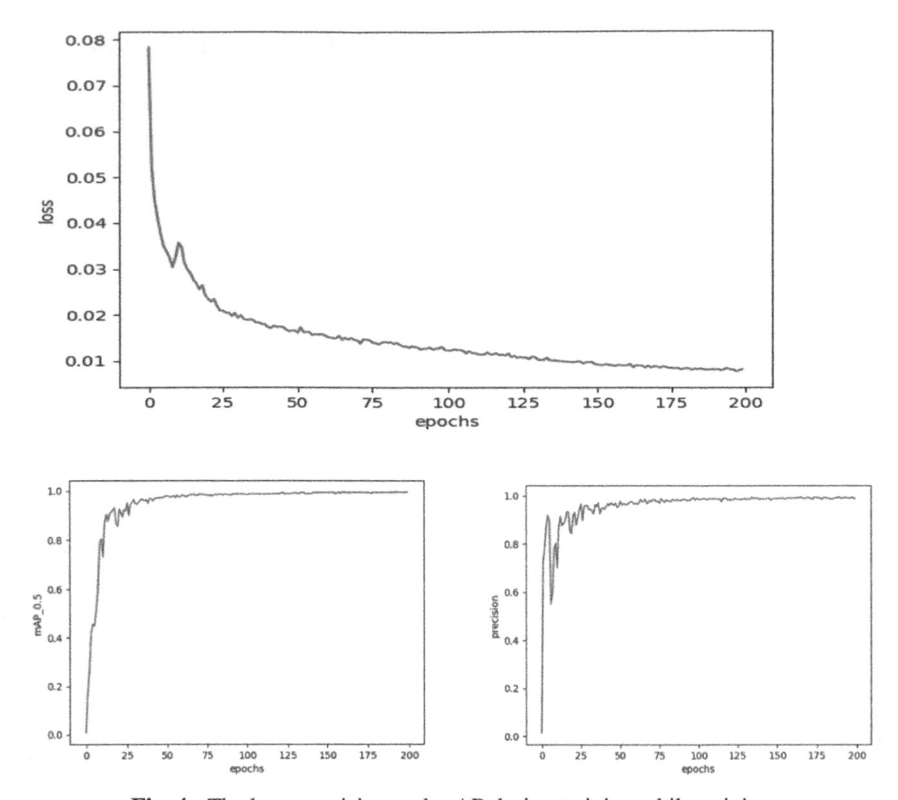


**Fig. 4.** The loss, precision and mAP during training while training.

At the beginning, the loss drops quickly, after 25 iterations, the loss come to 0.02087 and decrease slowly in future iterations. After 200 iterations, the loss is Infinitely close to 0 while the precision is 0.9897, the Recall is 0.9914 and the mAP is 0.9955. That means the model has already reached a steady state after training 200 epochs. For more accuracy and to reduce the average loss, we can train our model with much number of iterations.

## 3 Results

In Object detection, the final predicted bounding boxes are divided into four types: true positive (TP), false positive (FN), true negative (TN) and false negative (FN) [14]. TP means that the system accurately recognizes whether a worker is wearing a helmet. When the system recognizes a worker who is not wearing a helmet as a person wearing a helmet, a FP will occur. More information is shown in Table 2. The commonly used evaluation indicators are precision ($P = \frac{TP}{TP+FP}$) and recall ($R = \frac{TP}{TP+FN}$).

**Table 2.** The confusion matrix of classification task

| Truth | Prediction | |
|---|---|---|
| | Positive | Negative |
| Positive | TP | FN |
| Negative | FP | TN |

For each image, it only takes about 20ms to get the result. We finally get the precision on our testing data set is 78.3%. Some results of helmet detection are shown in Fig. 5. The categories and the confidence score are shown up the bounding box. Different class is shown in different color. We demonstrate that YOLO is an effective method and comparatively fast for recognition and localization in helmet detection.

**Fig. 5.** Result of YOLO method for processing our dataset

## 4 Conclusions

This study focuses on the application of computer vision technology to detect whether workers wear helmet on construction sites. In order to real-time detection, we use YOLO as our basic model. According to our result, we find that YOLO is very fast and the accuracy is acceptable. Using experiments, our research has proved that computer vision can complete the task of identifying helmets on construction sites in real time with the precision of 78.3%.

Though sufficient results are obtained, lots of challenges remain to be solved. First, the situation on the construction site is relatively complicated, and the camera may not be able to capture the complete picture of the worker, which adds great difficulty to helmet detection. Second, our data set has only 1600 images, which is very small for training a model. Future work will focus on updating our dataset and update the model. It is believed that computer vision has great potential in automatically identifying people's unsafe behaviors such as FFH (Fall From Height), safety equipment inspection in construction sites.

## References

1. Li, H., Li, X., Luo, X., Siebert, J.: Investigation of the causality patterns of non-helmet use behavior of construction workers. Autom. Constr. **80**, 95–103 (2017)
2. Health and Safety in Construction Sector in Great Britain (2019). https://www.hse.gov.uk/statistics
3. Seo, J.O., Han, S.U., Lee, S.H., Kim, H.: Computer vision techniques for construction safety and health monitoring. Adv. Eng. Inform. **29**(2), 239–251 (2015)
4. Fang, W., Love, P.E.D., Luo, H., Ding, L.: Computer vision for behaviour-based safety in construction: a review and future directions. Adv. Eng. Inform. **43**, 100980 (2020)
5. Felzenszwalb, P.F., Girshick, R.B., McAllester, D., Ramanan, D.: Object detection with discriminatively trained part-based models. IEEE Trans. Pattern Anal. Mach. Intell. **32**(9), 1627–1645 (2009)
6. Krizhevsky, A., Ilya, S., Geoffrey, E.H.: Imagenet classification with deep convolutional neural networks. Adv. Neural Inf. Process. Syst. **25**, 1097–1105 (2012)
7. Hadji, I., Richard, P.W.: What Do We Understand about Convolutional Neworks? (2018)
8. Redmon, J., Divvala, S., Girshick, R., Farhadi, A.: You only look once: unified, real-time object detection. In: Proceedings of the IEEE Conference on Computer Vision and Pattern Recognition (2016)
9. Girshick, R., Donahue, J., Darrell, T., Malik, J.: Rich feature hierarchies for accurate object detection and semantic segmentation. In: Proceedings of the IEEE Conference on Computer Vision and Pattern Recognition (2014)
10. Shinde, S., Kothari, A., Gupta, V.: YOLO based human action recognition and localization. Procedia Comput. Sci. **133**, 831–838 (2018)
11. Everingham, M., Eslami, S.A., Van Gool, L., Williams, C.K., Winn, J., Zisserman, A.: the pascal visual object classes challenge: a retrospective. Int. J. Comput. Vis. **111**(1), 98–136 (2015)
12. Lin, T.-Y., et al.: Microsoft coco: common objects in context. In: Fleet, D., Pajdla, T., Schiele, B., Tuytelaars, T. (eds.) Computer Vision – ECCV 2014, pp. 740–755. Springer, Cham (2014). https://doi.org/10.1007/978-3-319-10602-1_48

13. Deng, J., Dong, W., Socher, R., Li, L.J., Li, K., Fei-Fei, L.: Imagenet: a large-scale hierarchical image database. In: Proceedings of the IEEE Conference on Computer Vision and Pattern Recognition (2009)
14. Liu, Z., Cao, Y., Wang, Y., Wang, W.: Computer vision-based concrete crack detection using u-net fully convolutional networks. Autom. Constr. **104**, 129–139 (2019)

# Advanced Neural Network Theory and Algorithms

# Hierarchical Finite-Horizon Optimal Control for Stackelberg Games via Integral Reinforcement Learning

Jiayu Chen[1], Xiaohong Cui[1,2](✉), Yao Lu[1], Wenjie Chen[1], and Binbin Peng[1]

[1] College of Mechanical and Electrical Energineering, China Jiliang University, Hangzhou 310018, Zhejiang, People's Republic of China

[2] Key Laboratory of Intelligent Manufacturing Quality Big Data Tracing and Analysis of Zhejiang Province, China Jiliang University, Hangzhou 310018, People's Republic of China

**Abstract.** This paper presents the hierarchical optimal multi-control algorithm for the partially unknown finite-horizon linear systems. The key to the multi-control problem is to find the solution to the coupled leader-follower Hamilton-Jacobi (HJ) equations. The coupled leader-follower HJ equations are derived and we define the Stackelberg equilibrium of the games. To solve the HJ equations for the follower, an integral reinforcement learning (IRL) approach is given on the basis of the PI algorithm, which uses partial knowledge of the dynamic systems. The Finite-Neural-Network (FNN) structure is used to approximate the value functions in the proposed algorithm. Simulation results are shown to verify the effectiveness of the scheme.

**Keywords:** Finite-horizon optimal control · Stackelberg games · Integral reinforcement learning

## 1 Introduction

Optimal control problems of multiplayer behaviors and decision-making have been widely concerned [1,2]. Compared with the single control problem, the different players not only optimize their performances but also consider other players' profits. In Nash games [3], the players take actions synchronously, and the coupled leader-follower Hamilton-Jacobi (HJ) equations are solved. For the nonzero-sum games, players have taken different strategies to achieve the Nash equilibrium [3]. Stackelberg game is deemed as hierarchical optimal control. All players achieve the Stackelberg equilibrium, and the Stackelberg game is a constrained two-level optimal problem different from the Nash games [4,5].

Hierarchical control and optimization approach is applied in energy microgrids [6], and electric vehicles [7]. The hierarchical decision-making process is Stackelberg games, [4] and [5] respectively deduced the opened-loop and closed-loop Stackelberg policies. Vamvoudakis, Lewis, and Johnson [8] presented an


Q. Han et al. (Eds.): LSMS 2021/ICSEE 2021, CCIS 1469, pp. 279–288, 2021.
https://doi.org/10.1007/978-981-16-7213-2_27

online adaptive optimal control algorithm based on policy iteration for Stackelberg games. The necessary and sufficient condition of existence of the Stackelberg-Nash equilibrium is obtained in [9]. The continue-time two-play input hierarchical optimal control policy with quadratic cost functions for the nonlinear systems was studied in [10]. Most of the above-mentioned approximate equilibrium strategy methods require completely known knowledge of system dynamics. The optimal control problem of unknown dynamic systems by IRL with off-policy methods [11,12]. For the finite-horizon optimal control problem, optimal control policy is difficult to obtain owing to the time-varying HJB equation [13]. The HJB was solved by NN least-squares approximation in [14]. The finite-horizon problem for nonlinear discrete-time systems was considered in [15,16].

The main contribution of this paper is that find the solution to the coupled leader-follower Hamilton-Jacobi (HJ) equations for multi-control finite-horizon problem via integral reinforcement learning (IRL).

The formulation of the finite-horizon multi-player control problem is provided in Sect. 2. Section 3 firstly deduces the coupled leader-follower HJ equations anNd then presents the PI algorithm. Section 4 proposes the IRL method based on the PI algorithm, followed by the description of the FNN value function approximation. Section 5 presents the simulation example that demonstrates the effectiveness of the algorithm.

## 2 Problem Formulation

Consider the following continuous-time dynamic systems described as

$$\dot{x} = Ax + B_1 u_1 + B_2 u_2, \tag{1}$$

where $x(t) \in R^n$ denotes the state of the dynamic systems. $u_i \in R^{p_i}, i = 1, 2$ is the control signals and $u_1$ is the strategy controlled by the follower and $u_2$ is controlled by the leader. $Ax + B_1 u_1 + B_2 u_2$ is Lipschitz continuous on a compact set $\Omega \subset R^n$ which contains the origin. $Ax$ is the internal dynamics and it is assumed that matrix $A$ is unknown. The input matrices $B_1$ and $B_2$ can be obtained.

For the finite-horizon multi-control problem, the key issue is to find the minimum of the following performance functions. They are illustrated below.

$$\begin{aligned} J_1(x(t_0), u_1, u_2) &= \psi_1(x(t_f), t_f) + \int_{t_0}^{t_f} \frac{1}{2}(x^T(t)Q_1 x(t) + 2\theta u_1^T(t) u_2(t) \\ &+ \|u_1(t)\|^2)dt = \psi_1(x(t_f), t_f) + \int_{t_0}^{t_f} r_1(x, u_1, u_2)dt, \end{aligned} \tag{2}$$

$$\begin{aligned} J_2(x(t_0), u_1, u_2) &= \psi_2(x(t_f), t_f) + \int_{t_0}^{t_f} \frac{1}{2}(x^T(t)Q_2 x(t) + 2\theta u_2^T(t) u_1(t) \\ &+ \|u_2(t)\|^2)dt = \psi_2(x(t_f), t_f) + \int_{t_0}^{t_f} r_2(x, u_1, u_2)dt. \end{aligned} \tag{3}$$

The terminal state $x(t_f)$ is penalized by $\psi_i(x(t_f), t_f)$, $t_f$ is a fixed terminal time, where $\theta > 0$ denotes the coupled coefficient of the different players. While the state $x(t)$ and the control strategies $u = [u_1, u_2]$ are penalized by $r_i(x, u_1, u_2)$from $t_0$ to $t_f$, $Q_i, i = 1, 2$ are positive definite functions. If the players do not take actions simultaneously and they make the strategies in different levels, the multi-player control problem is deemed as a hierarchical optimal problem. Therefore, the definition of Stackelberg equilibrium should be given.

**Definition 1 (Stackelberg equilibrium** [8,10]**).** The pair$(u_1^*, u_2^*) \in U_F \times V_L$ is regarded as providing a unique Stackelberg equilibrium for the game if the following conditions are satisfied:

1. For each fixed control $\overline{u}_2 \in V_L$, there is always a unique control $\overline{u}_1 \in U_F$ that minimizes $J_1(u_1, \overline{u}_2)$. Then it means that there exists a unique mapping $T : V_L \to U_F$ such that $J_1(Tu_2, u_2) \leq J_1(u_1, u_2)$ for all $u_1$ and for each $u_2$.
2. The leader seeks optimal solution $u_2^*$ from space $V_L$ to minimize $J_2(Tu_2, u_2)$ such that, for all $u_2$, it has $J_2(Tu_2^*, u_2^*) \leq J_2(Tu_2, u_2)$.
3. $u_2^*$ is the optimal Stackelberg leadership-control policy, and $u_1^* = Tu_2^*$ is the optimal follow-control policy.

## 3 Policy Iteration (PI) for Finite-Horizon Stackelberg Games

In this part, we introduce an improved policy iteration (PI) algorithm for the finite-horizon Stackelberg games, which relaxes the requirements of the complete system knowledge. The corresponding value functions under the admissible control policies are defined as follows.

$$\begin{aligned} V_1(x(t), \mu_1, \mu_2) &= \psi_1(x(t_f), t_f) + \int_t^{t_f} \frac{1}{2}\left(x^T Q_1 x + 2\theta \mu_1^T \mu_2 + \|\mu_1\|^2\right) d\tau \\ &= \psi_1(x(t_f), t_f) + \int_t^{t_f} r_1(x, \mu_1, \mu_2) d\tau, \end{aligned} \tag{4}$$

$$\begin{aligned} V_2(x(t), \mu_1, \mu_2) &= \psi_2(x(t_f), t_f) + \int_t^{t_f} \frac{1}{2}\left(x^T Q_2 x + 2\theta \mu_2^T \mu_1 + \|\mu_2\|^2\right) d\tau \\ &= \psi_2(x(t_f), t_f) + \int_t^{t_f} r_2(x, \mu_1, \mu_2) d\tau. \end{aligned} \tag{5}$$

The Hamilton function for the follower is defined below

$$H_1(x, \nabla V_1, u_1, u_2) = r_1(x, u_1, u_2) + \nabla V_1^T (Ax + B_1 u_1 + B_2 u_2) \tag{6}$$

where $\nabla V_1 = \frac{\partial V_1}{\partial x}$ denotes the gradient vector. For the fixed leader policy $u_2$, the follower's response can be obtained by the stationary condition.

$$\frac{\partial H_1}{\partial u_1} = 0 \Rightarrow \mu_1^*(u_2) = \arg\min_{u_1} H_1(x, u_1, u_2, \nabla V_1) = -(B_1^T \nabla V_1 + \theta u_2). \tag{7}$$

The leader should make its strategy to optimize the following performance index function, which is constrained by the follower's rational selection as the leader's anticipation. Then, the Hamilton function associated with the leader is defined as

$$H_2(x, \nabla V_2, \mu_1^*, \mu_2) = r_2(x, \mu_1^*, \mu_2) + \nabla V_2^T (Ax + B_1\mu_1^* + B_2\mu_2) \tag{8}$$

Applying the stationary condition for the leader, the leader's optimal control policy is derived below.

$$\begin{aligned} \frac{\partial H_2}{\partial u_2} = 0 \Rightarrow \mu_2^* &= \arg\min_{u_2} H_2(x, \mu_1^*, u_2, \nabla V_2) \\ &= -\frac{1}{(1-2\theta^2)}\left((B_2 - \theta B_1)^T \nabla V_2 - \theta B_1^T \nabla V_1\right) \end{aligned} \tag{9}$$

The following time-varying leader-follower Lyapunov-like equations are

$$-\frac{\partial V_1}{\partial t} = r_1(x, \mu_1, \mu_2) + (\nabla V_1)^T (Ax + B_1\mu_1 + B_2\mu_2) \tag{10}$$

$$-\frac{\partial V_2}{\partial t} = r_2(x, \mu_1^*, \mu_2) + (\nabla V_2)^T (Ax + B_1\mu_1^* + B_2\mu_2). \tag{11}$$

Therefore, the coupled leader-follower Hamilton-Jacobi(HJ) equations are obtained by substituting the optimal responses to the two leaders (7) and (9) into the time-varying Lyapunov-like equations.

$$\begin{aligned} &(\nabla V_1^*)^T Ax + \frac{1}{2}x^T Q_1 x + \frac{1}{1-2\theta^2}\theta(\nabla V_1^*)^T B_1 B_1^T \nabla V_1^* \\ &-\frac{1}{2}\frac{1}{(1-2\theta^2)^2}\|(1-\theta^2)(\nabla V_1^*)^T B_1 - \theta(\nabla V_2^*)^T (B_2 - \theta B_1)\|^2 \\ &-\frac{1}{1-2\theta^2}(\nabla V_1^*)^T B_2 (B_2 - \theta B_1)^T \nabla V_2^* = -\frac{\partial V_1^*}{\partial t} \end{aligned} \tag{12}$$

$$\begin{aligned} &(\nabla V_2^*)^T Ax + \frac{1}{2}x^T Q_2 x - (\nabla V_2^*)^T B_1 B_1^T (\nabla V_1^*) \\ &-\frac{1}{2}\frac{1}{1-2\theta^2}\|(\nabla V_2^*)^T (B_2 - \theta B_1) - \theta(\nabla V_1^*)^T B_1\|^2 = -\frac{\partial V_2^*}{\partial t} \end{aligned} \tag{13}$$

Then the Stackelberg equilibrium of the games are defined by the continuous-time systems and the corresponding performance index functions with $u_1$ as the follower and $u_2$ as the leader are obtained below.

$$\mu_2^* = -\frac{1}{(1-2\theta^2)}\left((B_2 - \theta B_1)^T \nabla V_2^* - \theta B_1^T \nabla V_1^*\right) \tag{14}$$

$$\mu_1^*(u_2) = -(B_1^T \nabla V_1 + \theta \mu_2^*). \tag{15}$$

We focus on solving the coupled leader-follower HJ equations to obtain the time-varying value functions. However, the coupled leader-follower HJ equations is a time-varying and nonlinear partial differential equations.

**Remark 1.** It is noted that the leader makes its decision prior to the follower, and it is different from the simultaneous games such as a standard nonzero-sum games. Considering the finite-horizon leader-follower Stackelberg games, we provide the following improved policy iteration algorithm. The terminal condition is added into the standard policy iteration algorithm, so it updates the value functions and control policies iteratively subject to the terminal constraints.

**Algorithm 1 (Improved PI Algorithm).** Given the initial admissible control $\mu_1^0(x(t),t)$, $\mu_2^0(x(t),t)$, then the iterations begin as the two steps.

1. (Policy evaluation) Calculate the value functions $V_i^j$ by

$$\begin{aligned} -\frac{\partial V_1}{\partial t} &= r_1(x,\mu_1^j,\mu_2^j) + (\nabla V_1^j)^T(Ax + B_1\mu_1^j + B_2\mu_2^j) \\ V_1^j(x(t_f),t_f) &= \psi_1(x(t_f),t_f) \end{aligned} \tag{16}$$

$$\begin{aligned} -\frac{\partial V_2}{\partial t} &= r_2(x,u_1^j,u_2^j) + (\nabla V_2^j)^T(Ax + B_1\mu_1^j + B_2\mu_2^j) \\ V_2^j(x(t_f),t_f) &= \psi_2(x(t_f),t_f) \end{aligned} \tag{17}$$

2. (Policy improvement) The updated control input is obtained by

$$\mu_2^{j+1}(x,t) = -\frac{1}{(1-2\theta^2)}((B_2 - \theta B_1)^T\nabla V_2^j - \theta B_1^T\nabla V_1^j) \tag{18}$$

$$\mu_1^{j+1}(x,t) = -(B_1^T\nabla V_1 + \theta\mu_2^{j+1}) \tag{19}$$

Theorem 1 shows that the iterative values converge to the optimal values.

**Theorem 1.** Let $\mu_1^0(x(t))$, $\mu_2^0(x(t))$ be the initial admissible control policies, and the iteration steps are implemented according to (16) and (17). Then the sequence $V_i^j$ is convergent, i.e., $\lim V_i^j = V_i^*$ as $j \to \infty$.

*Proof.* The results can be obtained by the similar derivation as the procedure in [17].

**Remark 2.** The iterative method that provides a framework to obtain the approximate stackelberg equilibrium is represented by (14) and (15). In Algorithm 1, the internal dynamic matrix $A$ should be known in advance. In the next section, we mainly deal with the Stackelberg games with finite-horizon based on reinforcement learning scheme.

## 4 Integral Reinforcement Learning (IRL) and the Implement of NN

To relax the requirement of the integral dynamics, the following Algorithm 2 is given to deal with the Stackelberg games. Moreover, the finite-horizon terminal conditions are also considered.

### 4.1 Finite-Horizon Reinforcement Learning Algorithm

**Algorithm 2 (IRL Algorithm for finite-horizon Stackelberg games).** Let's begin with initial admissible controls $\mu_i^{(0)}, i = 1, 2$ and then apply the iteration steps below.

1. Use the following integral equations subject to the terminal constraints to update the value functions $V_i^j(x(t), t)$

$$V_1^j(x(t)) = \int_t^{t+T} r_1(x, \mu_1^j, \mu_2^j)d\tau + V_1^j(x(t+T)), V_1^j(x(t_f), t_f) = \psi_1(x(t_f), t_f) \tag{20}$$

$$V_2^j(x(t)) = \int_t^{t+T} r_2(x, \mu_1^j, \mu_2^j)d\tau + V_2^j(x(t+T)), V_2^j(x(t_f), t_f) = \psi_2(x(t_f), t_f) \tag{21}$$

2. Apply the received value functions to improve the control responses by the updated equations below.

$$\mu_2^{j+1}(x, t) = -\frac{1}{(1 - 2\theta^2)}((B_2 - \theta B_1)^T \nabla V_2^j - \theta B_1^T \nabla V_1^j) \tag{22}$$

$$\mu_1^{j+1}(x, t) = -(B_1^T \nabla V_1 + \theta \mu_2^{j+1}) \tag{23}$$

The following Theorem 2 will show equivalence between the two different iteration algorithms.

**Theorem 2.** The solution solved by Algorithm 1 is equivalent to the solution obtained by the iterative IRL scheme. For avoiding repetition we omit the derivation and the similar proof can be referenced by literature [17].

Compared with the PI algorithm, the IRL algorithm is more intelligent, and only needs to know the hierarchical finite-horizon optimal control problem of partial dynamics information.

### 4.2 Description of Finite-Neural Network (FNN)

In this section, we give the framework based on NN to approximate the time-varying value functions. Then the online IRL algorithm is implemented iteratively. The optimal value functions are time-based, so the structure of NN is different from the standard differential games with infinite-horizon. To facilitate the analysis, we apply the BP NN [11] and [17] with weights being linear.

$$V_i(x, t) = \sum_{k=1}^{\infty} w_{ik}^* \phi_{ik} = W_{iL}^{*T} \varphi_i(x(t), t_f - t) + \varepsilon_i(x, t_f - t), i = 1, \ldots, N, \tag{24}$$

where $W_{iL}^* = [w_{i1}^* \quad w_{i2}^* \quad \ldots \quad w_{iL}^*]^T, i = 1, 2$ denote the ideal weight vectors of the NNs. $L$ is the number of neurons in the hidden layer, $\varphi_i(x, t_f - t) = [\phi_{i1}, \ldots, \phi_{iL}]^T \in R^L, i = 1, \ldots, N$ are time-varying basis function vectors,

$\varepsilon_i(x, t_f - t)$ are the reconstruction errors. Because the weight vectors $W_{iL}^*, i = 1, 2$ can not be obtained, the outputs of the NNs are given as the follows.

$$\hat{V}_i(x, t) = W_i^T \varphi_i(x(t), t_f - t) \tag{25}$$

where $W_i$ denotes the estimate value of the unknown weight vector $W_{iL}^*$.

Then, the obtained control responses are written by

$$\mu_2^{j+1}(x, t) = -\frac{1}{(1 - 2\theta^2)}\left((B_2 - \theta B_1)^T \nabla \hat{V}_2^j - \theta B_1^T \nabla \hat{V}_1^j\right) \tag{26}$$

$$\mu_1^{j+1}(x, t) = -(B_1^T \nabla \hat{V}_1 + \theta \mu_2^{j+1}) \tag{27}$$

Substituting the approximate value functions and the control responses (25)–(27) into the integral equations (20) and (21), we obtain the following residual errors

$$\begin{aligned} \zeta_i(x(t), t) &= \int_t^{t+T} r_i(x, \mu_1^{j+1}, \mu_2^{j+1}) d\tau + W_i^{j+1T}(\varphi_i(x(t+T), t_f - (t+T)) \\ &- \varphi_i(x(t), t_f - t)) = \tilde{W}_i^T(\varphi_i(x(t), t_f - t) - \varphi_i(x(t+T), t_f - (t+T))) \\ &+ \varepsilon_i(x(t), t_f - t) - \varepsilon_i(x(t+T), t_f - (t+T)), i = 1, 2. \end{aligned} \tag{28}$$

where $\tilde{W}_i = W_{iL}^* - W_i^{(j+1)}, i = 1, 2$ are the weight estimation errors.

Note that $V_i(x(t_f), t_f) = \psi_i(x(t_f), t_f), i = 1, 2$. During the implementation of IRL on the basis of NNs, the terminal constraints should be satisfied. To deal with the constraints, another terminal error is defined below.

$$\varsigma_i = W_i^{(j+1)T} \varphi_i(x(t_f), 0) - \psi_i(x(t_f), t_f), i = 1, 2. \tag{29}$$

Let $\delta_i$, $\overline{\psi}_i$ and $\overline{V}_i$ be the augmented errors and the augmented terms. Where $\delta_i = [\zeta_i, \quad \varsigma_i], i = 1, \ldots, N$.

$$\overline{\psi}_i = [\varphi_i(x(t+T), t_f - (t+T)) - \varphi_i(x(t), t_f - t), \varphi_i(x(t_f), 0)], \tag{30}$$

$$\overline{V}_i = [\int_t^{t+T} r_i(x, \mu_1^{(j+1)}, \mu_2^{(j+1)}) d\tau, -\psi_i(x(t_f), t_f)], i = 1, \ldots, N. \tag{31}$$

Therefore, we have $\delta_i = W_i^{(j+1)T} \overline{\psi}_i + \overline{V}_i, i = 1, 2$. The weight vectors are updated to minimize both two errors. The least squares method is applied to obtain the ideal values iteratively.

Note

$$N_i = \int_{\Omega} \delta_i \delta_i^T dx, i = 1, \ldots, N. \tag{32}$$

For the Lebesgue integral, we adopt the inner product to notate, then we have $< f, g >= \int_{\overline{\Omega}} f g^T dx$. Utilizing the least-squares method, we compute the NN weight vectors to force the errors $\delta_i$ to be zero in an average sense. In other words, let $\nabla N_i = 0$, thus the following equation is obtained.

$$\langle \delta_i, \frac{d\delta_i}{dW_i^{(j+1)}} \rangle_{\overline{\Omega}} = 0, i = 1, 2. \tag{33}$$

By further derivation, we have the updating formula for NN weight vectors as follows.

$$W_i^{j+1} = -\Gamma_i^{-1}\langle \overline{\psi}_i, \overline{V}_i(W_i^j)\rangle_{\overline{\Omega}}, i = 1, \ldots, N. \tag{34}$$

where $\Gamma_i = \langle \overline{\psi}_i, \overline{\psi}_i \rangle_{\overline{\Omega}}$.

**Remark 3.** In this paper, a single critic NN is applied for each player, and the time-varying is also considered. The burden of NN learning is lessened when compared with the dual NNs. The selection of NN basis functions is usually by experiences. There is not a standard method to obtain the NN basis function, the trials are required until the best performance is received.

### 4.3 Convergence of NN-Based Learning Algorithm

We now discuss the convergence of the NN-based approximation theorem algorithm. Firstly, we supply Theorem 3 to show the approximation relation between the approximated value functions and the iterative value function.

**Theorem 3.** If the basis functions $\{\phi_{ik}\}_1^L, i = 1, 2$ are linearly independent on the compact $\overline{\Omega}$, $V^j = [V_1^j, V_2^j]$ is solved by the IRL algorithm, and the activation functions $\{\phi_{ik}\}_1^L, i = 1, 2$ are selected to ensure the solution $V_i$ and $\nabla V_i$ can be uniformly approximated when $L \to \infty$. Then the following results can be obtained.

$$\sup_{(x,t)\in\overline{\Omega}} |\hat{V}_i^j - V_i^j| \to 0, i = 1, 2; \tag{35}$$

$$\sup_{(x,t)\in\overline{\Omega}} |\hat{\mu}_2^{j+1}(x,t) - \mu_2^{j+1}(x,t)| \to 0, \ \sup_{(x,t)\in\overline{\Omega}} |\hat{\mu}_1^{j+1}(\mu_2) - \mu_1^{j+1}(\mu_2)| \to 0 \tag{36}$$

*Proof.* For the space problem, we will give the proof procedure in our future paper. The key is to show the convergence of the control response, which is different from the standard differential games. The following Theorem 4 provides that the NN-based solution solved by the least-squares method converges to the optimal Stackelberg equilibrium.

**Theorem 4.** Assumed that the conditions of Theorem 3 hold, then for $\forall \varepsilon > 0$, $\exists j_0, L_0$, when $j \geq j_0, L \geq L_0$, we have the following results.

$$\sup_{(x,t)\in\overline{\Omega}} |\hat{V}_i^j - V_i^*| < \varepsilon, i = 1, 2; \ \sup_{(x,t)\in\overline{\Omega}} |\hat{\mu}_2^j - \mu_2^*| < \varepsilon; \ \sup_{(x,t)\in\overline{\Omega}} |\hat{\mu}_1^j - \mu_1^*| < \varepsilon \tag{37}$$

*Proof.* The results are obtained by Theorem 2 and Theorem 3 directly.

## 5 Simulation

In this section, a simulation example is given to illustrate the theory. Consider the continuous-time dynamics below.

$$\dot{x} = Ax + B_1 u_1 + B_2 u_2 \tag{38}$$

where $A = 0$, $B_1 = B_2 = I$, and the matrices $Q_1 = diag\{9, 4\}$, $Q_2 = diag\{18, 8\}$. Here, $u_1$ is deemed as the follower and $u_2$ is the leader who makes the policy firstly. The terminal time $t_f = 3$, and the terminal penalized functions $\psi_1(x(t_f), t_f) = x(t_f)^T x(t_f)$, $\psi_2(x(t_f), t_f) = 2x(t_f)^T x(t_f)$.

Select the $2 \times 3 \times 1$ time-varying NN activation function vector. It has been seen that the weights for both the follower and the leader converge to the constants along with the time steps in Fig. 1. States $x1$ and $x2$ are converge to zero, and the approximate optimal control $u11, u12, u22, u23$ for the leader and follower is figured in Fig. 1. Moreover, the figure also describes the state curves under the approximate optimal control inputs, which the knowledge of the internal dynamics avoid.

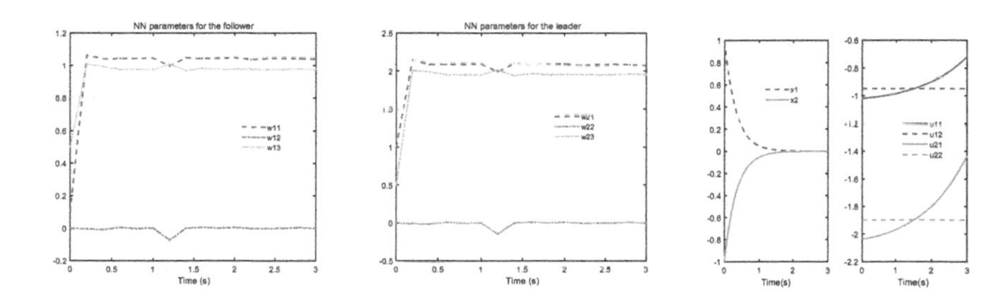


**Fig. 1.** NN weight, state, and control curves of the follower and leader

## 6 Conclusions

This paper presents a hierarchical finite-horizon multi-control algorithm for Stackelberg games, which adapts online to learn the optimal control without completely using knowledge of the dynamic systems via IRL method. Besides, FNN-based online iterative learning scheme and the proof of convergence of the algorithm are given. Finally, the simulation example is provided to demonstrate the validity of the proposed algorithm.

**Acknowledgment.** The authors would like to thank the National Natural Science Foundation of China under Grant (61903351).

## References

1. Jiang, H., Zhang, H., Luo, Y., Han, J.: Neural-network-based robust control schemes for nonlinear multiplayer systems with uncertainties via adaptive dynamic programming. IEEE Trans. Syst. Man Cybern. Syst. **49**(3), 579–588 (2019)
2. Song, R., Du, K.H.: Mix-zero-sum differential games for linear systems with unknown dynamics based on off-policy IRL. Neurocomputing **398**, 280–290 (2020)
3. Jiao, Q., Modares, H., Xu, S.: Multi-agent zero-sum differential graphical games for disturbance rejection in distributed control. Automatica **69**, 24–34 (2016)

4. Abou-Kandil, H., Betrand, P.: Analytical solution for an open-loop Stackelberg game. IEEE Tran. Automat. Control **30**(12), 1222–1224 (1985)
5. Xu, J., Zhang, H., Chai, T.: Necessary and sufficient condition for two-player Stackelberg strategy. IEEE Trans. Autom. Control **60**(5), 1356–1361 (2015)
6. Castro, M.V., Moreira, C., Carvalho, L.M.: Hierarchical optimisation strategy for energy scheduling and volt/var control in autonomous clusters of microgrids. IET Renew. Power Gener. **14**(1), 27–38 (2020)
7. Qiu, S., et al.: Hierarchical energy management control strategies for connected hybrid electric vehicles considering efficiencies feedback. Simul. Model. Pract. Theory **90**, 1–15 (2019)
8. Vamvoudakis, K., Lewis, F., Johnson, M.: Online learning algorithm for Stackelberg games in problems with hierarchy. In: 51st IEEE Annual Conference on Decision and Control (CDC), Maui, Hawaii, pp. 10–13 (2012)
9. Kebriaei, H., Iannelli, L.: Discrete-time robust hierarchical linear-quadratic dynamic games. IEEE Trans. Autom. Control **63**(3), 902–909 (2018)
10. Mu, C., Wang, K., Zhang, Q.: Hierarchical optimal control for input-affine nonlinear systems through the formulation of Stackelberg game. Inf. Sci. **517**, 1–17 (2020)
11. Zhang, H., Cui, X., Luo, Y., Jiang, H.: Finite-Horizon $H\infty$ tracking control for unknown nonlinear systems with saturating actuators. IEEE Trans. Neural Netw. Learn. Syst. **29**(4), 1200–1212 (2018)
12. Ren, H., Zhang, H.G., Wen, Y.L., Liu, C.: Integral reinforcement learning off-policy method for solving nonlinear multi-player nonzero-sum games with saturated actuator. Neurocomputing **335**, 96–104 (2019)
13. Heydari, A., Balakrishnan, S.: Finite-horizon control-constrained nonlinear optimal control using single network adaptive critics. IEEE Trans. Neural Netw. Learn. Syst. **24**(1), 145–157 (2013)
14. Zhao, Q., Xu, H., Dierks, T., Jagannathan, S.: Finite-horizon neural network-based optimal control design for affine nonlinear continuous-time systems. In: The 2013 International Joint Conference on Neural Networks (IJCNN), Dallas, TX, USA, pp. 1–6 (2013)
15. Qiming, Z., Hao, X., Sarangapani, J.: Neural network-based finite-horizon optimal control of uncertain affine nonlinear discrete-time systems. IEEE Trans. Neural Netw. Learn. Syst. **26**(3), 486–499 (2015)
16. Hao, X., Sarangapani, J.: Neural network-based finite horizon stochastic optimal control design for nonlinear networked control systems. IEEE Trans. Neural Netw. Learn. Syst. **26**(3), 472–485 (2015)
17. Cui, X., Zhang, H., Luo, Y., Zu, P.: Online finite-horizon optimal learning algorithm for nonzero-sum games with partially unknown dynamics and constrained inputs. Neurocomputing **185**, 37–44 (2016)

# Activity Recognition Through Micro-Doppler Image Based on Improved Faster R-CNN

Zhetao Sun, Kang Liu, Aihong Tan(✉), Tianhong Yan, and Fan Yang

Research Center for Sino-German Autonomous Driving, College of Mechanical and Electrical Engineering, China Jiliang University, Hangzhou 310018, China
Tanah@cjlu.edu.cn

**Abstract.** A method for activity recognition through Micro-Doppler image based on improved Faster R-CNN is presented. The data captured by millimeter wave radar has excellent environmental resistance and robustness even in bad weather. Firstly, the data captured by the radar is converted into Micro-Doppler image to realize the invisible and non-contact perception of object distance and motion. Secondly, the Faster-RCNN is used to recognize the Micro-Doppler image and automatically extract the features of the images. So as to improve the detection accuracy. Feature extraction network, Region Proposal Networks (RPN) and ROI of the model in Faster-RCNN have been improved. The experimental results show that the mAP of the improved model is increased by 11% and 9.1% on the test set and the measured data respectively under nearly same computing time.

**Keywords:** Micro-Doppler · Faster-RCNN · Activity recognition · Region Proposal Network · Feature extraction

## 1 Introduction

The radar was invented at the beginning of last century, which was originally used for detection and identification for military purposes. With the development of modern radar technology, radar, as a sensor, has been applied more and more in the civilian field, not just in the military field, and the target-recognition technology of radar has attracted widespread attention from scientists all over the world [1–3].

To recognize the movement state of the target by detecting some typical features in the movement process of the target. Compared with imaging the overall movement of the target, it only needs to collect some typical features in the movement process and completes the recognition by extracting these typical features. The process has more efficient features, among which the research on micro-motions features has received extensive attention from the academic and industrial circles in recent years [4–7].

Micro-Doppler images are the main manifestation of micro-motion features, and feature extraction on the Micro-Doppler images is the key to complete activity recognition. The classic feature extraction methods based on empirical and statistical features have certain limitations in applicability and flexibility, and both methods rely on target


Q. Han et al. (Eds.): LSMS 2021/ICSEE 2021, CCIS 1469, pp. 289–298, 2021.
https://doi.org/10.1007/978-981-16-7213-2_28

prior information heavily and cannot be adaptive to change extraction method according to the respective characteristics of each feature [8–12]. The use of deep learning technology can improve the precision of sample test recognition greatly, and the feature extraction method based on deep learning does not depend on the input data form and human-prescribed prior knowledge, which obtains weights suitable for network feature extraction through a large amount of data training. Li et al. [13] proposed to use NN classifier to analyze Micro-Doppler features to realize dynamic hand gesture recognition. However, the NN classifier needs to recalculate the distance to all training samples for each test sample, which results in the calculation becoming large, taking a long time, and also increasing the memory requirements. Moreover, when the number of various training samples is distributed unevenly, the classification precision may decrease. Erol et al. [14] proposed an activity recognition method that combines convolutional neural network (CNN) and Micro-Doppler images. This method inputs the Micro-Doppler images to be tested into the trained recognition model to complete the classification, which solves the problem that the NN classifier has a large amount of calculation and the precision is affected by the distribution of training samples. However, this method has strict requirements on input data and training data, which can't allow other activity signals in the data. If the radar captures an activity signal and then captures some other activity signals by mistake, it will cause serious impact on recognition and reduce the recognition performance of the model.

This paper proposes an activity recognition method through Micro-Doppler image based on improved Faster R-CNN [15]. First, the FMCW radar signal is subjected to short-time Fourier transform (STFT) to obtain Micro-Doppler images. Then, the Faster-RCNN deep learning network is used to identify the precise location of the action waveform in the Micro-Doppler images, thereby improving the performance of the network against clutter interference. The category output by Faster-RCNN classifies the action, and the position of the recognition box on the horizontal axis reflects the time period when the action occurs. In addition, in order to improve recognition realization, we performed data enhancement and improved the feature extraction network, RPN and ROI of Fasters-RCNN. The experimental results show that the Micro-Doppler image activity recognition method based on the improved Faster R-CNN has a good performance.

## 2 System Overview

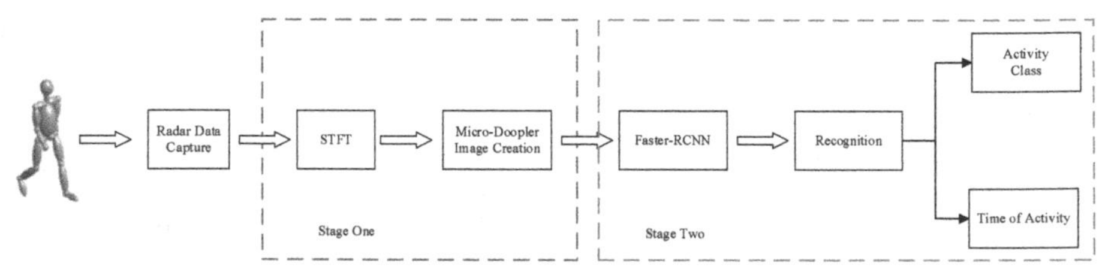


**Fig. 1.** System overview of activity recognition.

The activity recognition through Micro-Doppler image based on improved Faster R-CNN system is shown in Fig. 1. The system consists of two stages: the creation of Micro-Doppler images and the recognition of human activities based on Micro-Doppler images.

In the first stage, the Micro-Doppler image is obtained by performing Range FFT, ROI, MIT filter, and STFT on the data captured by the radar sensor.

In the second stage, the Micro-Doppler image generated in the first stage is sent to the Faster-RCNN network. To output its category and the recognition frame reflecting the specific position of the action waveform in the Micro-Doppler image through feature extraction, proposal extraction, bounding box regression and classification. Then the time when the action occurred can be calculated by the horizontal axis coordinates of the recognition box.

## 3 Micro-Doppler Image Creation and Data Enhancement

In this paper, the most commonly used short-time Fourier transform method in time-frequency analysis is used to convert radar sampling data into Micro-Doppler images. The specific operation steps are shown in Fig. 2.

Step 1: Performing Range FFT on the original radar data to obtain the Range-time image.

Step 2: Enlarging the features in the Range -time image to get the ROI image.

Step 3: To get the static clutter suppression image by performing the MTI filter on the ROI image.

Step 4: A short-time Fourier transform (STFT) with a 0.2 s Hanning window and 95% overlap is used to process each row of the data matrix, and the result data of each row after the STFT processing is sequentially accumulated.

Finally, the Micro-Doppler image is obtained.

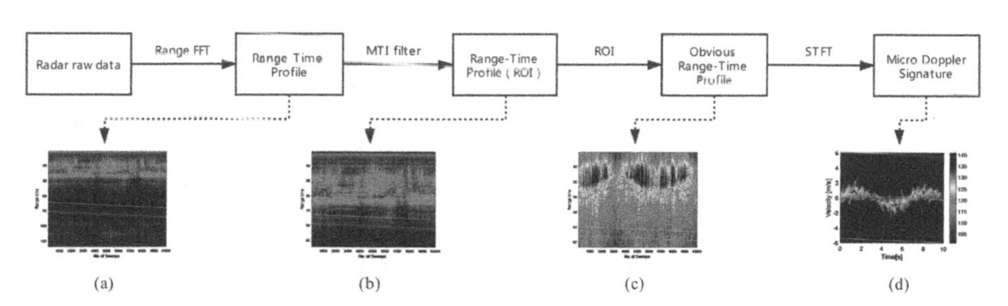


**Fig. 2.** Micro-Doppler image generation flow chart and its corresponding visualization results, of which the upper part is the flow chart, (a) Range-Time Image, (b) ROI Image, (c) Static Clutter Suppression Image, (d) Micro-Doppler Image.

Figure 3 is an example of a radar Micro-Doppler images converted from radar sampling data. It can be seen that the Micro-Doppler images has the ability to describe the micro-motion feature of the target finely. The horizontal axis of the figure is the time axis and the vertical axis is the velocity component of the scattering points of the human body, where a positive value corresponds to the velocity component of the scattering point far away from the radar and a negative value corresponds to the velocity component of the scattering point close to the radar. The gradation value corresponding to the color in the figure represents the strength of the echo signal reflected by the scattering points of the human body. Figure 4 shows the Micro-Doppler images of the six types of human activities researched in this paper.

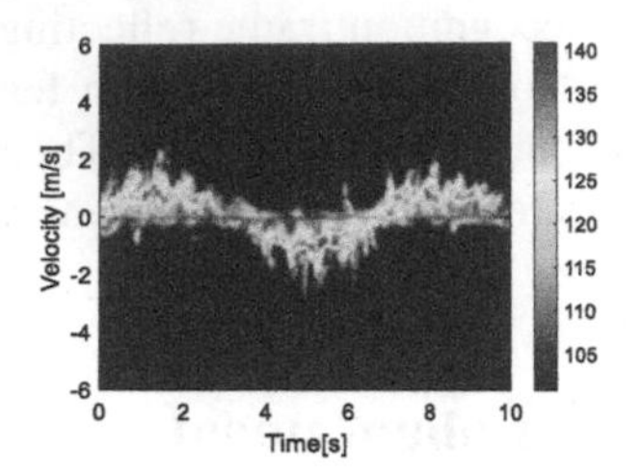


**Fig. 3.** Example of a radar Micro-Doppler images

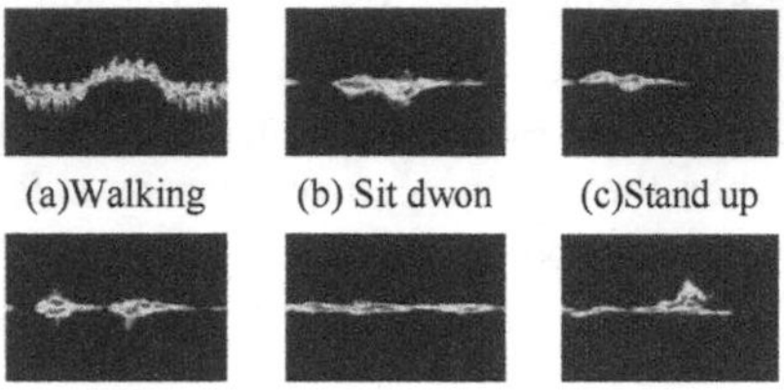

(a)Walking (b) Sit dwon (c)Stand up

(d)Pick up (e)Drink (f)Fall

**Fig. 4.** Micro-Doppler images of six types of human activities

In practice, especially in health care, some quick actions are often more dangerous. In order to enhance the recognition ability and generalization performance of the model, data enhancement is carried out in this paper. On the premise of keeping the overall profile of the vertical axis waveform unchanged, the Micro-Doppler image waveform is compressed along the time axis to simulate the Micro-Doppler image during fast motion. On this basis, the original Micro-Doppler waveform is compressed twice and three times respectively. In addition, for Faster-RCNN, it is to perform feature learning on the micro-Doppler waveform, which has nothing to do with the position of the waveform in the image, so this paper compresses the waveform on the left side of the image uniformly.

## 4 Improvements of Faster R-CNN

### 4.1 Improvement of Feature Extraction Network

For deep learning, the deeper the network, the stronger the nonlinear expression ability, and the stronger the learning ability of the network. At present, the Faster R-CNN algorithm uses VGG16 [16] as a feature extraction network commonly, but the network depth of the VGG16 network is only 16 layers, and due to the phenomenon of gradient disappearance and gradient explosion, it simply superimposing the hidden layer on the basis of the VGG16 network does not improve the fitting ability of the model. Moreover, this kind of shallow network can only extract low-level features such as image edges, colors, and textures, which cannot extract abstract features with conceptual significance highly.

ResNet (Residual Neural Network) [17] uses Residual Units when constructing deep neural networks, as shown in Fig. 5. Among them, relu is the activation function, $x$ is the input of the neural network, $H(x)$ is the feature expected to be learned when the network inputs $x$, and $F(x)$ is the residual. In the residual learning unit, $x$ is copied into two copies, where one is input to the Weight Layer to learn the new residual feature

*F(x)* and the other is output to the final result through the shortcut connection directly. Then the original expected output *H(x)* can be expressed as *F(x)* + *x*. When *F(x)* = *0*, then *H(x)* = *x*. At this time, the depth residual unit only does the identity mapping, while the network performance will not decrease. When *F(x)* ≠ *0*, the deep residual unit learns new features based on the input features. Therefore, the goal of residual learning has been changed. It is no longer to learn a complete output *H(x)*, but *H(x)-x*, which is the residual *F(x)*. This feedforward short connection form neither introduces new parameters nor increases computational complexity, thus simplifying the difficulty of learning objectives and avoiding the problem of gradually reaching saturation with the continuous increase of network depth. In addition, the optimization residual *F(x)* is easier than the optimized expected output *H(x)*, and it also avoids gradient disappearance and gradient explosion. So, this paper uses ResNet101 as the feature extraction network instead of VGG16.

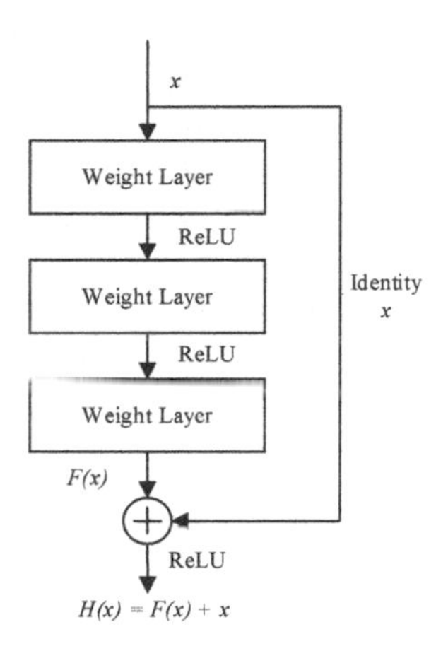


**Fig. 5.** Schematic diagram of Residual unit

### 4.2 Improvement of RPN

In the Faster R-CNN target recognition algorithm, after RPN receives the input of Feature maps, it uses feature points as reference points to generate three scales ($8^2$, $16^2$, $32^2$) and three ratios (1:1, 1:2, 2:1), a total of 9 anchors, as shown in Fig. 6.

The preset anchor size of Faster R-CNN is set for most of the targets in the Pascal VOC and MS COCO data sets, and most of the action waveforms in the Micro-Doppler images that need to be identified in this paper are elongated. The original RPN setting Anchor size cannot recall the action waveform efficiently in the Micro-Doppler image. Figure 7 is a statistical histogram of the ground truth (GT) of the data set cited in this paper on the feature map. The GT scales of all radar waveforms are less than 900. It can be seen that the anchor with a scale of $32^2$ (1024) in the original RPN is not suitable for target recognition in the data set studied in this paper.

Figure 8 is a scatter plot of the GT size and anchor size distribution of the data set cited in this paper. It can be seen from the figure that the size of the GT does not fit well

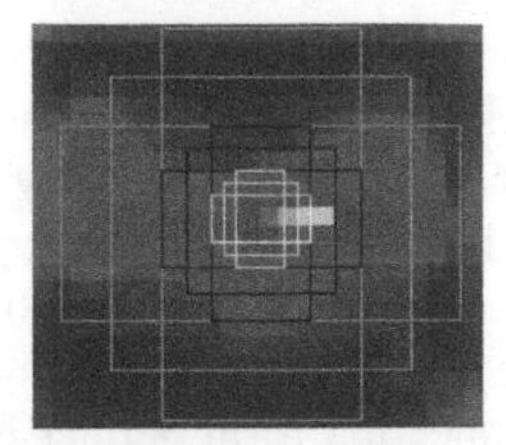

Fig. 6. Schematic diagram of Anchors

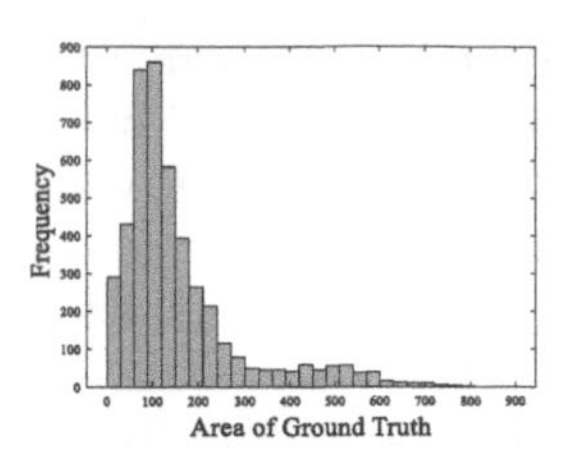


Fig. 7. GT area statistics histogram

with the size of the anchors. Figure 9 and Fig. 10 reflect that the preset Anchor size fully in Faster R-CNN is not reasonable for target recognition in this paper, and the pertinence is not strong.

This paper proposes targeted improvements based on the characteristics of the action waveforms in the data set. First, the scale of the GT is in ascending order. Then, the GT is divided into three categories, large, medium, and small, according to their scale. Through counting the average areas of these three types separately, we get $S_{min}= 69 \approx 8 \times 8$, $S_{med}= 123 \approx 11 \times 11$, $S_{max}= 308 \approx 17 \times 17$. Therefore, in order to improve the recognition performance of the model, this paper sets the scale of the anchors to ($8^2$, $11^2$, $17^2$).

This paper uses statistics on the ratio of the GT, and divides the data into three categories, large, medium, and small according to the proportion, and calculates the average of these three categories. The results are as follows: $R_{min}\approx 0.22$, $R_{med}\approx 0.39$, $R_{max}\approx 0.83$. Therefore, this paper sets the ratio of the anchor to (0.22, 0.39, 0.83).

Figure 9 is a scatter point distribution diagram of the improved anchor size and GT size. It can be seen that the fit between the two is strong, and the pertinence of the model is improved significantly.

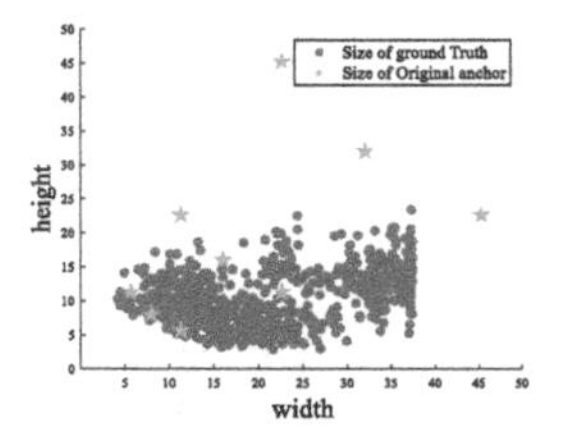


Fig. 8. Scatter plot of GT size and anchor size distribution

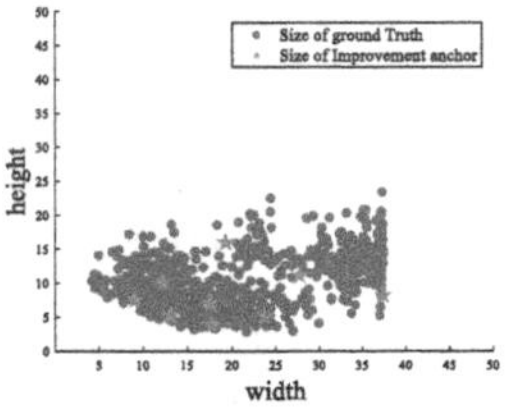

Fig. 9. Scatter plot of GT size and anchor size distribution

### 4.3 Improvement of ROI

In the process of obtaining ROI, Faster R-CNN rounded the calculation results of the two coordinate values, so the positioning results obtained cannot be aligned with the corresponding pixels in the original image. In order to solve this problem, ROI Align is used to calculate and use floating point numbers to record the coordinate results when the size of the feature map is fixed, so as to avoid errors in the calculation results and achieve precise positioning of small targets. Figure 10 is a schematic diagram of the ROI align in this paper. The black cell is a feature maps with a size of 4 × 6, and the red solid line cell is the ROI. First, dividing the ROI into 2 × 2 cells evenly. When the number of sampling points is 4, each cell needs to be divided into 4 small squares, and the center of each square is the sampling point position. The coordinates of these sampling points are usually floating-point types.

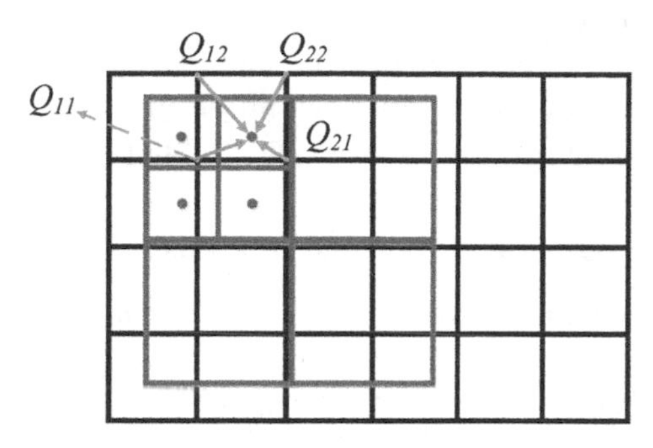


**Fig. 10.** ROI align

As shown by the four arrows in Fig. 10, the pixel values $Q_{11}= (x_1, y_1)$, $Q_{12}= (x_1, y_2)$, $Q_{21}= (x_2, y_1)$ and $Q_{22}= (x_2, y_2)$, are used for bilinear interpolation to calculate the pixels of the sampling point. and the bilinear interpolation formula is

$$F(R_1) \approx \frac{x_2 - x}{x_2 - x_1} F(Q_{11}) + \frac{x - x_1}{x_2 - x_1} F(Q_{21}). \tag{1}$$

$$F(R_2) \approx \frac{x_2 - x}{x_2 - x_1} F(Q_{12}) + \frac{x - x_1}{x_2 - x_1} F(Q_{22}). \tag{2}$$

$$F(P) \approx \frac{x_2 - x}{x_2 - x_1} F(R_1) + \frac{x - x_1}{x_2 - x_1} F(R_2). \tag{3}$$

According to formula (3), the pixel value of the sampling point can be obtained, and finally each sampling point is subjected to max pooling, and a uniform size feature map can be obtained.

## 5 Experiment and Analysis

### 5.1 Experiment Environment

The computer system used in this paper is Ubuntu 18.04; Nvidia TITAN RTX is used for GPU acceleration; the video memory is 24G; the programming language is python; the deep learning framework is pytorch1.0.

Considering the specific application scenarios of activity recognition, the data used in this paper comes from the radar sampling data set of the University of Glasgow [18]. This data set collected 1420 radar sampling data from nearly a hundred volunteers. Based on this, this paper uses data enhancement to expand the data set to 4320 samples. Then we draw 60% as the training set, 20% as the validation set, and the remaining 20% as the test set randomly.

Training parameter setting: the number of training epoch is 100 steps, the total number of iterations is 42,000, and the initial learning rate is 0.0010. The batch-size for training is set to 16.

This paper uses average precision (AP), mean average precision (mAP) and the time to recognize a single image as the evaluation metrics of the model. Among them, AP characterizes the recognition precision of a certain category and is used to evaluate the recognition performance of the model for a certain category; mAP is the average of AP values of all categories and is used to evaluate the overall performance of the model.

### 5.2 Experimental Design Comparison and Analysis

In order to verify the performance improvement of the improved algorithm proposed in this paper compared with the original Faster R-CNN, this paper sets up the following four experimental groups under the same training set:

M1: Traditional Faster R-CNN algorithm does not make any improvements;
M2: Use ResNet101 as the feature extraction network on the basis of M1;
M3: Improve the RPN on the basis of M2;
M4: Improve the ROI layer on the basis of M3.

The experimental comparison results are shown in Table 2. It can be seen from Table 1 that under the same training and test data sets, after using ResNet101 as the feature extraction network, mAP is increased by 3.4%. On this basis, after using the improved RPN method proposed in this paper, the mAP value has increased by 10.9%, and the model performance has been improved significantly. Since the action waveform recognized in this paper belongs to a larger target in the recognition, a slight deviation has little effect on the recognition, so after further switching to ROI align as the ROI layer, the mAP improvement is not significant. It can also be seen from the table that the improvements in this paper have little effect on the processing time of a single image, and there is no significant increase in computing power.

**Table1.** Comparison of experimental results Evaluation indicators.

| Evaluation index | M1 | M2 | M3 | M4 |
|---|---|---|---|---|
| mAP(%) | 84.5 | 87.9 | 95.4 | 95.5 |
| Time(s) | 0.072 | 0.073 | 0.073 | 0.075 |

### 5.3 Performance Comparison of Measured Data

In order to further study the feasibility and effectiveness of deep learning-based Micro-Doppler image recognition in practical applications, we use FMCW radar to collect 100 brand-new measured data as test samples under the configuration of C-band (5.8 GHz), bandwidth of 400 MHz, sweep interval of 1ms, and output power of about + 18 dBm.

To input the measured samples into the trained original Faster R-CNN and the improved algorithm in this paper to count the categories of the AP and the overall mAP to evaluate the model's recognition performance for each category and its overall performance. The recognition results are shown in Table 2. The test results show that compared with the original algorithm, the improved algorithm has a slight decrease in AP values of all types except sitting up, and the AP values of other actions have been improved. The mAP value is also 9.1% higher than that of the traditional recognition algorithm. The improved algorithm maintains a high recognition performance on the measured data, which proves that it has good feasibility and effectiveness in practical applications.

**Table 2.** Comparison of measured data test results.

| Model | Walking | Sit down | Stand up | Pickup |
|---|---|---|---|---|
| Faster R-CNN | 88.5 | 100 | 100 | 100 |
| Our method | 100 | 100 | 95.2 | 100 |
| Model | Drink | Fall | mAP | |
| Faster R-CNN | 37.2 | 94.4 | 86.7 | |
| Our method | 80.26 | 99.3 | 95.8 | |

## 6 Conclusion

This paper proposes a method for activity recognition through Micro-Doppler image based on improved Faster R-CNN. Performing Range FFT, ROI, MIT filter, and STFT on the radar data to get the Micro-Doppler image, and using the Faster R-CNN to recognize the image. Then expanding the data set to increase the generalization ability of the model through data enhancement. In addition, this paper recognizes Micro-Doppler images more efficiently through improving the feature extraction network, RPN, and ROI layers of Faster R-CNN. Experiments show that the improved network improves the mAP of the test set by 11% and the mAP of the measured data by 9.1%, which reflect excellent recognition capabilities. The noise interference experiment proves that the method in this paper still has good recognition ability and good robustness under the influence of noise.

**Acknowledgements.** This work was supported by the Provincial Natural Science Foundation of Zhejiang (Y21F010057).

## References


1. Yu, X., Chen, X., Huang, Y., Guan, J.: Fast detection method for low-observable maneuvering target via robust sparse fractional Fourier transform. IEEE Geosci. Remote. Sens. Lett. **17**(6), 978–982 (2019)
2. Mou, X., Chen, X., Guan, J., Chen, B., Dong, Y.: Marine target detection based on improved faster R-CNN for navigation radar PPI images. In: 2019 International Conference on Control, Automation and Information Sciences (ICCAIS), Chengdu, pp. 1–5. IEEE (2019)
3. Chen, X., Yu, X., Huang, Y., Guan, J.: Adaptive clutter suppression and detection algorithm for radar maneuvering target with high-order motions via sparse fractional ambiguity function. Appl. Earth Obs. Remote Sens. **13**, 1515–1526 (2020)
4. Zhao, R., Ma, X., Liu, X., Li, F.: Continuous human motion recognition using Micro-Doppler signatures in the scenario with micro motion interference. IEEE Sens. J. **21**(4), 5022–5034 (2021)
5. Lang, Y., Wang, Q., Yang, Y., Hou, C., Huang, D., Xiang, W.: Unsupervised domain adaptation for Micro-Doppler human motion classification via feature fusion. IEEE Geosci. Remote. Sens. Lett. **16**(3), 392–396 (2018)
6. Zeng, Z., Amin, M.G., Shan, T.: Automatic arm motion recognition based on radar Micro-Doppler signature envelopes. IEEE Sens. J. **20**(22), 13523–13532 (2020)
7. Yang, Y., Hou, C., Lang, Y., Guan, D., Huang, D., Xu, J.: Open-set human activity recognition based on Micro-Doppler signatures. Pattern Recognit. **85**, 60–69 (2019)
8. Saho, K., Fujimoto, M., Masugi, M., Chou, L.-S.: Gait classification of young adults, elderly non-fallers, and elderly fallers using Micro-Doppler radar signals: simulation study. IEEE Sens. **17**(8), 2320–2321 (2017)
9. Rabbani, M., Feresidis, A.: Wireless health monitoring with 60 GHz-band beam scanning Micro-Doppler radar. In: 2020 IEEE MTT-S International Microwave Biomedical Conference (IMBioC), Toulouse, pp. 1–3 (2020)
10. Erol, B., Amin, M.G., Gurbuz, S.Z.: Automatic data-driven frequency-warped cepstral feature design for Micro-Doppler classification. IEEE Trans. Aerosp. Electron. Syst. **54**(4), 1724–1738 (2018)
11. Gu, F.-F., Fu, M.-H., Liang, B.-S., Li, K.-M., Zhang, Q.: Translational motion compensation and Micro-Doppler feature extraction of space spinning targets. IEEE Geosci. Remote. Sens. Lett. **15**(10), 1550–1554 (2018)
12. Du, L., Li, L., Wang, B., Xiao, J.: Micro-Doppler feature extraction based on time-frequency spectrogram for ground moving targets classification with low-resolution radar. IEEE Sens. J. **16**(10), 3756–3763 (2016)
13. Li, G., Zhang, R., Ritchie, M., Griffiths, H.: Sparsity-driven Micro-Doppler feature extraction for dynamic hand gesture recognition. IEEE Trans. Aerosp. Electron. Syst. **54**(2), 655–665 (2017)
14. Erol, B., Gurbuz, S.Z., Amin, M.G.: Motion classification using kinematically sifted acgan-synthesized radar Micro-Doppler signatures. IEEE Trans. Aerosp. Electron. Syst. **56**(4), 3197–3213 (2020)
15. Ren, S., He, K., Girshick, R., Sun, J.: Faster R-CNN: towards real-time object detection with region proposal networks. IEEE Trans. Pattern Anal. Mach. Intell. **39**(6), 1137–1149 (2017)
16. Simonyan, K., Zisserman, A.: Very deep convolutional networks for large-scale image recognition. In: 3rd International Conference on Learning Representations (2014)
17. He, K., Zhang, X., Ren, S., Sun, J.: Deep residual learning for image recognition. In: 2016 IEEE Conference on Computer Vision and Pattern Recognition (CVPR), Las Vegas, pp. 770–778 (2016)
18. Fioranelli, F., Shah, S.A., Li, H., Shrestha, A., Yang, S., Le Kernec, J.: Radar sensing for healthcare. Electron. Lett. **55**(19), 1022–1024 (2019)

# Prediction of Landslide Risk Based on Modified Generalized Regression Neural Network Algorithm

Di Zhang, Qing Li(✉), Renwang Ge, Fenglin Li, and Wencai Tian

National Key Laboratory of Disaster Detection Technology and Instruments, China Jiliang University, Hangzhou 310018, China

**Abstract.** As a natural disaster, landslide causes immeasurable losses. The particle swarm optimization algorithm (PSO) and genetic algorithm (GA) are used to modify the smoothing factor of the generalized regression neural network (GRNN), which improves the prediction efficiency of GRNN. By building a landslide monitoring platform, the rainfall, shallow soil moisture content, deep soil moisture content, soil glide stress, and surface displacement are used as five landslide factors for landslide risk analysis, and the modified landslide models are applied to the processing of landslide data. These two models are used to predict the landslide risk, and compared with the models of BP, Elman neural network and RBF neural network for landslide prediction. The results illustrate that the modified GRNN landslide models have better prediction effects of landslide risk than BP neural network model, Elman neural network model, and Radial basis function neural network model, which provides a reference for engineering practice.

**Keywords:** Prediction of landslide risk · Generalized regression neural network (GRNN) · Particle swarm optimization (PSO) · Genetic algorithm (GA) · Landslide monitoring platform

## 1 Introduction

Geological disasters are prone to happen on earth, and the disastrous losses caused by landslides are immeasurable. Therefore, international scholars have made endless researches on landslide risk prediction. Researching out a reliable landslide risk prediction method owns great application value and practical significance.

Scholars have made certain developments in the study of the landslide warning. Fengshan Wang et al. proposed a case based reasoning method for landslide risk early warning in the emergency maintenance process. In the risk early warning case of landslide, a method of using the relation matrix to retrieve the local and global similarity is proposed. It can be realized that the early warning and disposal of the landslide risk in the emergency repair process of the underground engineering damaged by the earthquake [1]. Huaxi Gao et al. studied the historical data of rainfall and regional landslides in Shenzhen, and analyzed the relationship between landslides and rainfall [2]. Crissa D.


Q. Han et al. (Eds.): LSMS 2021/ICSEE 2021, CCIS 1469, pp. 299–308, 2021.
https://doi.org/10.1007/978-981-16-7213-2_29

Femandez et al. proposed a landslide warning system that uses rain gauges and smart accelerometers and inclinometers to intercept possible landslides.

This system is based on a multi-sensor warning system and can effectively anticipate possible landslides [3]. Aiming at handling the difficulties of landslide warning and communication in the Internet of things in mountainous areas, Amrita Joshi et al. proposed a method based on edge computing to process and analyze landslide data such as rainfall, pore pressure, humidity, and displacement [4]. Using geographic information system technology (GIS), Mujie Li et al. proposed a quantitative model of landslide probability composed of rainfall intensity-duration threshold model, combined with historical landslide disaster data [5]. Ji Weiwei et al. proposed the landslide geological hazard prediction model using BP neural network, which can reduce the dimensionality of different influencing factors of different landslide bodies [6]. Cui Yangyang et al. used an ensemble learning algorithm to study the risk assessment of geological hazards, and compared the prediction accuracy of the three models of Bagging, Boosting and random forest, which were used to determine the geological hazard risk assessment model in the study area [7].

Most of the above studies involving landslide warning have taken into account the data of a single landslide warning signal. In the actual process, there is a definite coupling relationship between different landslide warning signals. The landslide warning model based on multiple landslide influence factors is more important which is close to the real situation. Because the final cause of the landslide is precisely due to the combined action of multiple landslide factors in the actual landslide process. In this paper, a generalized neural network model is utilized to process multiple landslide signals, and finally landslide risk prediction models are generated. Generalized regression neural network has the advantages of fast convergence speed and good nonlinear approximation performance. The modified GRNN risk models are proposed.

## 2 Landslide Monitor Platform

In consideration of monitoring the landslide process, the landslide monitoring platform is built, by which the corresponding landslide factors are measured. Mainly landslide factors involve rainfall, shallow soil moisture content, deep soil moisture content, soil sliding stress and surface displacement, as well as the inclination angle of the landslide body.

Landslide monitor platform is built by three important parts, as shown in Fig. 1. The disaster simulation test system is divided into a soil-carrying box, water pump, simulated rainfall sprinkler head, and a hydraulic lifting system. The size of the box body is 4.4 m $\times$ 4 m $\times$ 1.5 m, and it can load about 60 tons of soil at most. The box body can be lifted and lowered between the degree of 0 and 60, as shown in Fig. 2. The simulated rainfall system is composed of a water pump, a regulating valve and a rainfall sprinkler, and the rainfall change is realized by the size of the control valve.

In specific experiments, landslide data are collected by monitoring sensors to collect five kinds of experimental data, surface displacement, soil glide stress, rainfall, deep water content and shallow water content, and build a host computer platform to receive the experimental data. Among them, the surface displacement sensor adopts ZLP-S-1000MM-V pullwire sensor, the soil sliding stress data collection adopts MASSPG

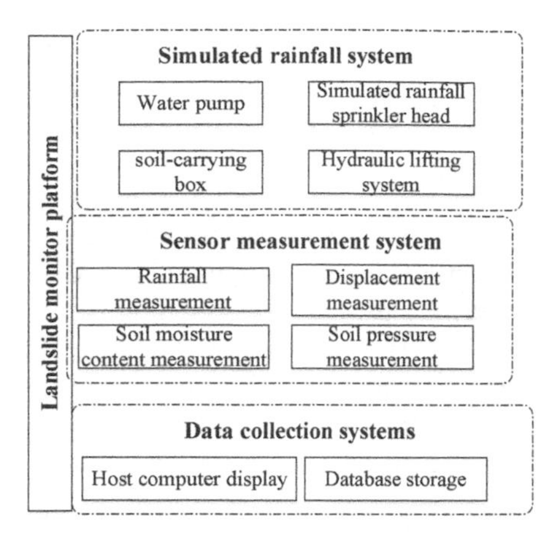


**Fig. 1.** The construction of landslide monitor platform.

**Fig. 2.** The front and back of the landslide disaster test site.

vibrating wire earth pressure gauge, and the measurement of deep and shallow water content uses MS-10 water content sensor. The measurement of the amount adopts the tipping bucket type rainfall FDY-02 sensor. The above sensors are all transmitted to the upper computer program through the RS485 protocol, and it is best to receive the computer terminal display through data collection system.

## 3 Modified Generalized Regression Neural Network Algorithm

In this section, the traditional GRNN and modified GRNN algorithms will be introduced.

### 3.1 Generalized Regression Neural Network

Generalized regression neural network (GRNN) is a kind of radial basis function neural network. GRNN achieves strong nonlinear mapping ability and learning speed [8]. It has stronger advantages than traditional back propagation neural network. GRNN network finally converges to the sample size. Optimal regression with more agglomeration and less sample data has a good prediction effect and can also handle unstable data [9]. Although GRNN does not seem to have radial basis accuracy, it actually has great advantages in classification and fitting, especially when the accuracy of the data is relatively poor.

GRNN is a modification of the Radial Basis Function Neural Network (RBF) [10]. The weight link between the hidden layer and the output layer of the RBF is disappeared. The least squares superposition of Gaussian weights [11]. And GRNN network is divided into four layers, as an input layer, a pattern layer, a summation layer and an output layer.

Input data $X = [x_1, x_2, \ldots, x_n]^T$, enter the test sample, and the number of nodes is equal to the feature dimension of the sample. Output data $Y = [y_1, y_2, \ldots, y_n]^T$, the number of key neurons in the output layer is equal to the dimensionality of the output vector in the training sample. The pattern layer neuron function is obtained by Eq. (1).

$$p_i = \exp[-\frac{(\mathbf{x} - \mathbf{x}_i)^T(\mathbf{x} - \mathbf{x}_i)}{2\sigma^2}], i = 1, 2, \ldots, n \tag{1}$$

Calculate the value of the result of each sample in the test sample and the training sample is obtained by Eq. (2).

$$S_D = \sum_{i=1}^{n} p_i \tag{2}$$

Another type of calculation formula is to perform a weighted summation of the neurons in all the model layers. The weight in the i-th node in the model layer and the j-th molecular sum node in the summation layer is the i The j-th element in the output sample Y, the transfer function is obtained by Eq. (3).

$$S_{N_j} = \sum_{i=1}^{n} y_{ij} p_j, j = 1, 2, \ldots, k \tag{3}$$

The output of neuron j is refers to the j-th element of the outcome, that is, the output layer function is obtained by Eq. (4).

PSO is essentially an efficient algorithm, compared with other optimization algorithms, group search is faster. And many scholars have studied it and expanded the use of the algorithm.

$$y_j = \frac{S_{N_j}}{S_D}, j = 1, 2, \ldots, k \tag{4}$$

### 3.2 Particle Swarm Optimization Algorithm

PSO is essentially an efficient algorithm [12], compared with other optimization algorithms, group search is faster. And many scholars have studied it and expanded the use of the algorithm.

The algorithm steps are as follows:

Step 1. Initialize the initial state of each particle;

Step 2. Calculate fitness value;

Step 3. According to the function, the individual optimal value 'pbest' is obtained;

Step 4. The optimal individual value 'pbest' is compared with the global optimal value 'gbest';

Step 5. Update the state quantity of each particle, this is the update of speed and position;

Step 6. Boundary condition processing;

Step 7. Judge whether the algorithm meets the condition, return to step 2 if the condition is not met; output the optimal value directly if the condition is met.

The speed and position update in step 5 satisfy the following Eqs. 5 and 6.

$$v_{ij}(t+1) = w \cdot v_{ij}(t) + c_1 r_1(t)[p_{ij}(t) - x_{ij}(t)] + c_2 r_2(t)[p_{ij}(t) - x_{ij}(t)] \tag{5}$$

$$x_{ij}(t+1) = x_{ij}(t) + v_{ij}(t+1) \tag{6}$$

In the PSO, the inertia weight w, acceleration coefficients c1 and c2, and boundary condition processing strategies affect the performance and efficiency of the algorithm. The size of the w indicates how much the particle inherits the current velocity. The greater the inertial weight, the stronger the global optimization ability, but the weaker the local optimization ability. The acceleration constant adjusts the maximum step length of particles flying in the direction of 'pbest' and 'gbest' respectively. And $c1 = 0$ is easy to fall into the local optimum. It means that the probability of finding the optimal solution is small, when $c2 = 0$. Generally, $c1 = c2$ is set so that individual experience and group experience have the same importance, and the optimization result is relatively optimized.

### 3.3 Genetic Algorithm

Genetic algorithm (GA) is an adaptive global optimization search algorithm formed by using the genetic and evolution process of living individuals in the natural environment. It performs a series of operations such as selection, crossover, and mutation on the population to produce a new generation of the population, and gradually approaches or reach optimal solution [13].

The mutation probability is essentially random mutation, which may cause the loss of excellent genes or fall into pure random search.

The algorithm steps are as follows:

Step 1. Initialize the population $P_0$ randomly;

Step 2. Calculate the function of all individuals in the population;

Step 3. The selected is sorted, and good individuals are selected to be inherited to the next generation;

Step 4. For the selected individual, exchange part of the chromosomes between them in the light of the crossover probability to generate a new individual;

Step 5. Mutation: For the selected individual, change the gene according to the mutation probability. After the group is selected, crossed, and mutated, the next generation group is obtained, then the fitness value is calculated, and sorted according to the fitness value;

Step 6. Judgment: If the number of evolution is met or the accuracy meets the requirements, the optimal solution will be output, otherwise skip back to step 2.

### 3.4 Modified Generalized Regression Neural Network Algorithm

In the GRNN training process, the smoothing factor is often difficult to determine. Aiming at the difficulty that the smoothing factor is hard to value in the generalized regression neural network, the test sample of the PSO algorithm is constructed based on the GRNN model, and the test error is used as the fitness function. The smoothing factor of the GRNN is optimized by the PSO algorithm, and the error of the model output value is reached. The smallest smoothing factor belongs to the optimal smoothing factor, which constitutes the modified PSO-GRNN algorithm, as shown in Fig. 3.

Like the above-mentioned modified algorithm, in order to locate the optimal size of the smoothing factor, we constructed a GRNN smoothing factor optimization method based on GA and used it in the landslide model. Refer to Sect. 3.3 for the specific GA algorithm steps. The algorithm block diagram of GA-GRNN is shown in Fig. 3. The first is to optimize the smoothing factor of GRNN to achieve the error of the model output value. The obtained optimal smoothing factor is substituted into the original GRNN algorithm to form a modified GRNN algorithm.

The modified algorithm has the characteristics. Firstly, the selection of the smoothing factor is as optimal as possible. Secondly, the calculation speed of the algorithm is not greatly affected. This is more scientific than the traditional GRNN based on experience to select the smoothing factor.

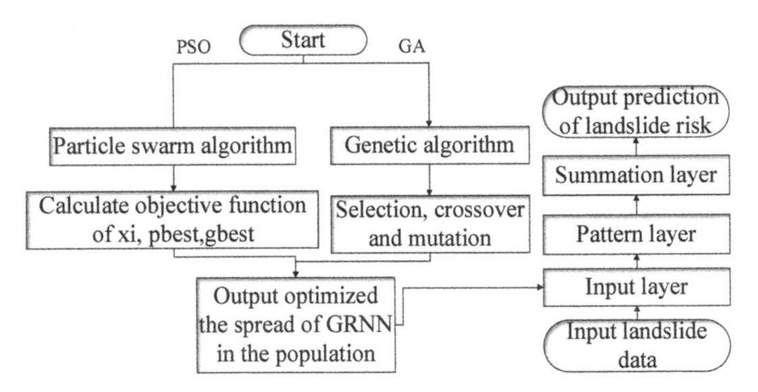


**Fig. 3.** The structure of PSO-GRNN and GA-GRNN.

## 4 Experimental Data and Predictive Performance Analysis

### 4.1 Experimental Data

The data collected in the landslide disaster experiment are sorted out and analyzed, and the prediction analysis is carried out by improving the PSO-GRNN algorithm. The data collected in the experiment have 139 sets of data, 10 typical sets of which as shown in Table 1, where the stability margin refers to the probability of landslide occurrence. The stability margin close to 0 indicates that the area remains relatively stable, and stability margin close to 1 indicates the possibility of a landslide risk is more likely greater. The sample data are randomly divided into two groups in the neural network training, namely the training set and the test set.

**Table 1.** Sample data collected in landslide experiment.

| Surface displacement /mm | Shallow soil moisture content /% | Deep soil moisture content /% | Rainfall /mm | Soil glide stress /Kpa | Landslide stability margin |
|---|---|---|---|---|---|
| 0 | 17.05 | 15.50 | 1.98 | 11.13 | 0.0165 |
| 0 | 34.20 | 25.40 | 11.14 | 11.26 | 0.0707 |
| 4.09 | 34.35 | 25.70 | 11.66 | 11.26 | 0.0774 |
| 20.12 | 35.85 | 26.75 | 14.95 | 11.64 | 0.1072 |
| 113.89 | 38.95 | 27.15 | 43.50 | 14.94 | 0.2800 |
| 131.63 | 37.90 | 26.95 | 46.52 | 15.45 | 0.3064 |
| 297.01 | 39.65 | 27.00 | 49.70 | 15.83 | 0.5194 |
| 570.15 | 39.25 | 28.60 | 51.92 | 16.47 | 0.8662 |
| 593.00 | 38.40 | 28.95 | 52.20 | 16.72 | 0.8954 |
| 641.76 | 39.15 | 28.55 | 52.72 | 17.10 | 0.9576 |

### 4.2 Entropy Method and Stability Score of Data

The scale of the landslide probability is analyzed based on the data obtained by the landslide monitor platform. Aiming at ensuring the objective and fairness of the evaluation results, this study mainly uses the entropy method to evaluate the landslide probability [14]. The entropy approach is an objective weighting method, which is based primarily on the scale of the information which is provided by each indicator to affect the weight. When the degree of dispersion of the information is greater, its role in the assessment is greater.

According to the landslide evaluation method made by previous researchers, the comprehensive index obtained by the entropy method is used as the score of landslide stability in this article.

The entropy method is obtained by Eqs. 7–9. Firstly, standardize the obtained landslide data, and then determine the entropy value according to the Eq. 7, then determine the weight according to the Eq. 8, and calculate the comprehensive performance score of the landslide data, which is expressed by Eq. 9.

$$e_j = -k\sum_{i=1}^{n} p_{ij} \log p_{ij} \tag{7}$$

$$w_i = \frac{g_i}{\sum_{j=1}^{m} g_i} \tag{8}$$

$$score_i = \sum_{j=1}^{m} w_j x_{ij}, i = 1, 2, \cdots, n \tag{9}$$

Through experiments, the collected sensor data including surface displacement, soil glide stress, rainfall, deep water content and shallow water content are used as input, and the stability margin of the landslide is used as output. Note that in the above data, when the stability margin is smaller than 0.2, it is absolutely safe. And when the stability margin is greater than 0.7, the risk of the landslide is higher. When the stability margin of the landslide is between 0.2 and 0.7, the risk of the landslide is between two extreme situations. This is the key interval that we focus on forecasting.

Through the sample data in Table 1, 100 sets of sample data are randomly selected to train the GRNN. The input in the sample is first eliminated from the overlapping information between the input, and the input variables are normalized.

### 4.3 Performance Comparison and Error Analysis

Aiming at verifying the feasibility of the modified models, the modified algorithms are compared with traditional GRNN for landslide risk prediction model. The comparison performance uses the root mean square error (RMSE). The RMSE is the square root of the sum of squared distances from the true value of each data. The deviation between the observed value and the true value is measured by RMSE.

The lower error and root mean square error means (RMSE) that the deviation of the algorithm is smaller and the stability is better. Another thing to emphasize is that compared with the neural network model of previous researchers, we used BP, Elman and RBF landslide risk model as the control group, which is compared with the modified GRNN landslide prediction models.

Through these algorithms, the output stability margins are obtained. Since the stability factor is between (0, 1), there is no need to renormalize the output result. The output result is the same as the test sample. The error rate of the stability coefficient is shown in Table 2.

**Table 2.** 100 sets of RMSE and error comparison experimental results for different algorithms in landslide risk prediction model.

| BP | Elman | RBF | GRNN | PSO-GRNN | GA-GRNN |
|---|---|---|---|---|---|
| Mean of error | | | | | |
| 0.0126 | 0.0562 | 0.0499 | 0.0169 | **0.0115** | **0.0117** |
| Mean of RMSE | | | | | |
| 0.0305 | 0.0938 | 0.5514 | 0.0201 | **0.0142** | **0.0130** |

It can be seen from Table 2 that the RMSE between the landslide stability margin predicted by the neural network and the real landslide stability margin is the smallest, showing that the error is the smallest in the training model for landslide risk prediction. The GRNN neural network after the neural network model training is useful for predicting the probability of occurrence of landslide risk (Fig. 4).

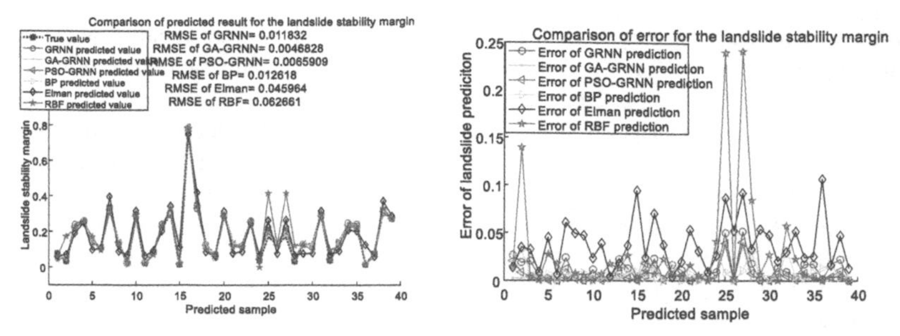


**Fig. 4.** Comparison of performance for different algorithms.

Comparing the traditional GRNN and the modified GRNN models, from a statistical point of view, the RMSE of GA-GRNN is smaller, followed by PSO-GRNN, and the traditional GRNN model is inferior in dealing with landslide risk prediction.

## 5 Conclusion

The modified GRNN landslide models are proposed, which use the PSO and GA to obtain the GRNN optimized smoothing factor. Through the fast search of the minimum value of the PSO and irritation selective search of GA, the most suitable GRNN smoothing factor is found, and the two modified GRNN algorithms are used for the prediction and evaluation of landslide disasters. Compared with the landslide risk prediction results of the BP, the Elman and RBF, the modified GRNN models have a high predictive effect in the actual landslide risk prediction. What needs to be pointed out is that GA-GRNN landslide model is slightly better than PSO-GRNN algorithm.

The future of landslide risk prediction is the study of deep learning models based on big data, which will have great significance for high-precision landslide risk prediction.

**Acknowledgments.** The authors would like to acknowledge the financial support given to this work, with the support of National Key Research and Development Program Project 2017YFC0804604, Zhejiang Key Research and Development Program Project 2018C03040.

## References

1. Wang, F., He, L., Liu, M., Zhang, H.: A landslide risk early-warning method in urgent repair operations for earthquake-damaged underground engineering. In: Proceedings of 2010 IEEE International Conference on Intelligent Computing and Intelligent Systems, ICIS 2010, vol. 2, pp. 576–579 (2010)
2. Gao, H.: Analyses on coupling correlation between landslides and rainfall and application in the early-warning and forecast. In: 2011 Electric Technology and Civil Engineering, ICETCE 2011 – Proceedings, pp. 643–6436 (2011)
3. Fernandez, C.D., Mendoza, K.J.A., Tiongson, A.J.S., Mendoza, M.B.: Development of microcontroller-based landslide early warning system. In: IEEE Region 10 Conference (TENCON) Annual International Conference Proceedings/TENCON, pp. 3000–3005 (2017)

4. Joshi, A., Grover, J., Kanungo, D.P., Panigrahi, R.K.: Edge assisted reliable landslide early warning system. In: 2019 IEEE 16th India Council International Conference Indicon 2019, Symposium Proceedings, pp. 2019–2022 (2019)
5. Li, M., Zhu, M., He, Y., He, Z., Wang, N.: Warning of rainfall-induced landslide in Bazhou District. School of Resources and Environment. University of Electronic Science and Technology of China. Department of Natural Resources of Sichuan Province, Chengdu, Sichuan, PRC 610072 Sichuan Research. In: IGARSS 2020 - 2020 IEEE International Geoscience and Remote Sensing Symposium 2020, Waikoloa, HI, USA, 2020, pp. 6879–6882 (2020)
6. Weiwei, J., Qing, L., Chuanqi, L., Maojie, W., Zizhang, Y.: Study on stability margin of rock slope based on comprehensive measurement. J. Bull. Sci. Technol. **35**, 24–28 (2019)
7. Yangyang, C., Niandong, D., Xiaofan, C., Yi, D., Congcong, X.: Geological disaster risk assessment based on ensemble learning algorithm. J. Water Power. **46**, 37–41 (2020)
8. Yidirim, T., Cigizoglu, H.K.: Comparison of generalized regression neural network and MLP performances on hydrologic data forecasting. In: Proceedings of the 9th International Conference on Neural Information Processing, 2002. ICONIP 2002, Computer Intelligence E-Age, vol. 5, pp. 2488–2491 (2002)
9. Celikoglu, H.B.: Application of radial basis function and generalized regression neural networks in non-linear utility function specification for travel mode choice modelling. J. Math. Comput. Model. **44**, 640–658 (2006)
10. Kumar, G., Malik, H.: Generalized regression neural network based wind speed prediction model for Western Region of India. J. Procedia Comput. Sci. **93**, 26–32 (2016)
11. Wang, N., Chen, C., Yang, C.: A robot learning framework based on adaptive admittance control and generalizable motion modeling with neural network controller. J. Neurocomput. **390**, 260–267 (2020)
12. Salameh, T., Kumar, P.P., Sayed, E.T., Abdelkareem, M.A., Rezk, H., Olabi, A.G.: Fuzzy Modeling and Particle Swarm Optimization of Al2O3/SiO2 nanofluid. J. Int. J. Thermofluids. **10**, 100084 (2021)
13. Velliangiri, S., Karthikeyan, P., Arul Xavier, V.M., Baswaraj, D.: Hybrid electro search with genetic algorithm for task scheduling in cloud computing. J. Ain Shams Eng. J. **12**, 631–639 (2021)
14. Zhang, N., Li, Q., Li, C., He, Y.: Landslide early warning model based on the coupling of limit learning machine and entropy method. J. Phys. Conf. Ser. **1325**, 012076 (2019)

# Rough K-means Algorithm Based on the Boundary Object Difference Metric

Pengda Zhong, Tengfei Zhang(✉), Xiang Zhang, Xinyuan Hu, and Wenfeng Zhang

College of Automation and College of Artificial Intelligence, Nanjing University of Posts and Telecommunications, Nanjing 210023, Jiangsu, China

**Abstract.** Rough k-means algorithm can effectively deal with the problem of the fuzzy boundaries. But traditional rough k-means algorithm set unified weight for boundary object, ignoring the differences between individual objects. Membership degree method of rough fuzzy k-means algorithm is used to measure the membership degree of boundary object to the clusters that they may belong to, ignoring the distribution of neighbor points of the boundary object. So, according to the distribution of neighbor points of the boundary object, we put forward a new rough k-means algorithm to measure the weight of boundary objects. The proposed algorithm considers the distance from boundary objects to their neighbor points and the number of neighbor points of boundary objects together to dynamically calculate the weights of boundary object to clusters that may belong to. Simulation and experiment, through examples verify the effectiveness of the proposed method.

**Keywords:** Rough k-means · Rough-fuzzy k-means · Boundary objects · Neighbor point

## 1 Introduction

Clustering analysis [1] is a kind of unsupervised learning methods, and widely applied to various fields. Clustering can be considered as a process of dividing an object set into several clusters according to similarity standard so that objects of high similarity in the same clusters while objects of high dissimilarity are separated into different clusters [2]. Here are some main clustering approaches: clustering based on density; clustering based on partition; clustering based on hierarchy; clustering based on model; clustering based on grid [3].

K-means [4–6] is a very classic partition-based clustering algorithm. Traditional k-means algorithms belong to hard clustering, namely each sample object is strictly assigned to a certain cluster.

Rough set theory [7] is a mathematical tool of dealing with uncertainty knowledge. Lingras et al. [8] applied rough set theory to k-means clustering, and put forward the rough k-means (RKM) algorithm. In RKM algorithm, each cluster can be divided into the lower approximation and the border area. Lower approximation and boundary region constitute upper approximation. Objects in lower approximation certainly belong to the


Q. Han et al. (Eds.): LSMS 2021/ICSEE 2021, CCIS 1469, pp. 309–318, 2021.
https://doi.org/10.1007/978-981-16-7213-2_30

cluster while objects in border area possibly belong to the cluster. Since RKM algorithm can effectively deal with uncertainty objects, it has received widespread attention and become the focus of research scholars [9, 10]. Peters et al. [11] put forward a new formula of mean computation which is more in line with the concept of upper and lower approximation set. In addition, they also used relative distance instead of absolute distance. Literature [12] proposed a rough k-means clustering algorithm based on unbalanced space distance metric which comprehensive considered the effect of both density and distance of each data object on cluster centers computation.

Only set up unified weights for lower approximation and boundary objects, ignoring the individual differences. π Rough K-means algorithm [13] calculated weight of the boundary object by the numbers of clusters that the boundary object may belong to according to the indifference principle, but ignored the importance of distance. Membership degree function is used in fuzzy k-means (FKM) algorithm [14, 15] to calculate membership degree of objects to a cluster. But the algorithm calculated membership degree of one object to only one cluster relative to all clusters. Actually membership of objects to a cluster only associated with the clusters that the object may belong to and has nothing to do with the rest of the clusters. Rough Fuzzy k-means algorithm (RFKM) [16–19] introduced the concept of membership degree on the basis of rough k-means algorithm and calculated membership degree of boundary object to clusters that it may belong to. But only according to the distance from the boundary object to cluster centers to calculate membership degree, ignoring the distribution of neighbor points of boundary object. Literature [20, 21] calculated the membership of boundary object to the cluster according to the distance from the boundary object to its nearest object in lower approximation of the cluster, but ignored the distribution of other neighboring points in the lower approximation.

For the above problems, this paper proposes a new rough k-means algorithm based on boundary object difference measure. The new algorithm calculates the weight of boundary object in the center computation based on the distribution of neighbor points of boundary object. Through the simulation experiments on UCI data sets, verify the validity of the algorithm in this paper.

## 2 Basic Knowledge of Rough K-means

### 2.1 Lingras and West Rough K-means

Lingras et al. improved the means formula in which the weight coefficient shows the importance of lower approximation and boundary area on the mean calculation. The formula of mean calculation is:

$$m_i = \begin{cases} \omega_l \sum\limits_{x_n \in \underline{C_i}} \frac{x_n}{|\underline{C_i}|} + \omega_b \sum\limits_{x_n \in \hat{C}_i} \frac{x_n}{|\hat{C}_i|} & \underline{C}_i \neq \varphi \wedge \hat{C}_i \neq \varphi \\ \sum\limits_{x_n \in \underline{C_i}} \frac{x_n}{|\underline{C_i}|} & \underline{C}_i \neq \varphi \wedge \hat{C}_i = \varphi \\ \sum\limits_{x_n \in \hat{C}_i} \frac{x_n}{|\hat{C}_i|} & \underline{C}_i = \varphi \wedge \hat{C}_i \neq \varphi \end{cases} \tag{1}$$

In our paper, the upper approximation of cluster $C_i$ is represented by $\overline{C}_i$, the lower approximation of cluster $C_i$ is represented by $\underline{C}_i$ and the boundary area of cluster $C_i$ is represented by $\hat{C}_i$. The weight of lower approximation is $w_l$ and the weight of boundary area is $w_b$.

## 2.2 π Rough K-means

Traditional rough clustering algorithm calculates the sub-mean of lower approximation and boundary area and then weighted sum of the two sub-means, ignoring the difference of objects in lower approximation and boundary area. It is based on directly weighted object to compute means function in π RCM:

$$v_i = \frac{\sum_{x_n \in \overline{C}_i} \frac{x_n}{|B_{x_n}|}}{\sum_{x_n \in \overline{C}_i} \frac{1}{|B_{x_n}|}}. \tag{2}$$

Here $B_{x_n}$ represents set of clusters that the boundary object $x_n$ may belong to and $|B_{x_n}|$ is the number of clusters in the set.

It does not contain any parameters in Eq. (2), but it is necessary to add weight of lower approximation and boundary area in some special cases. So the mean function with weights in π RCM as follows:

$$v_i = \frac{w_l \sum_{x_n \in \underline{C}_i} x_n + w_b \sum_{x_n \in \hat{C}_i} \frac{x_n}{|B_{x_n}|}}{w_l |\underline{C}_i| + w_b \sum_{x_n \in \hat{C}_i} \frac{1}{|B_{x_n}|}}. \tag{3}$$

## 2.3 Rough Fuzzy K-means

Rough Fuzzy k-means (RFKM) algorithm introduced the concept of membership degree in Rough k-means algorithm [15]. Membership function can reflect membership degree of an object to a certain cluster relative to the other clusters. The mean formula as follows:

$$v_i = \begin{cases} w_l \frac{\sum_{x_n \in \underline{C}_i} u^m{}_{in} x_n}{\sum_{x_n \in \underline{BU}_i} u^m{}_{in}} + w_b \frac{\sum_{x_n \in \hat{C}_i} u^m{}_{in} x_n}{\sum_{x_n \in \hat{C}_i} u^m{}_{in}} & if \underline{C}_i \neq \phi \wedge \hat{C}_i \neq \phi \\ \frac{\sum_{x_n \in \hat{C}_i} u^m{}_{in} x_n}{\sum_{x_n \in \hat{C}_i} u^m{}_{in}} & if \underline{C}_i = \phi \wedge \hat{C}_{i_i} \neq \phi \\ \frac{\sum_{x_n \in \underline{C}_i} u^m{}_{in} x_n}{\sum_{x_n \in \underline{C}_i} u^m{}_{in}} & otherwise \end{cases}. \tag{4}$$

Where the calculation formula of membership degree is:

$$u_{in} = \frac{1}{\sum_{j=1}^{c} \left(\frac{d_{in}}{d_{jn}}\right)^{\frac{2}{m-1}}}. \tag{5}$$

# 3 Rough K-means Algorithm Based on the Boundary Object Difference Metric

## 3.1 Analysis of Boundary Object Neighbor Point Distribution

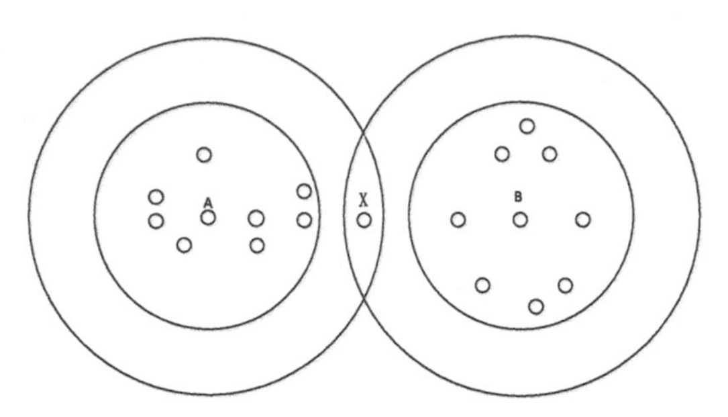


**Fig. 1.** Boundary object neighbor point distribution.

As shown in Fig. 1, the distance from boundary object X to the center of cluster A and cluster B is almost the same, but the distribution of the neighbor points of lower approximation in cluster A and cluster B is different. In lower approximation of cluster A, there are more neighbor points of X and these neighbor points are close to X; But lower approximation of cluster B, there are less neighbor points of X and these neighbor points are far from X when compared with cluster A. So, boundary object X more likely belongs to cluster A relative to the cluster B.

In conclusion, for a cluster if there are more neighbor points of a boundary object in its lower approximation and these neighbor points are more close to the boundary object than other clusters, the boundary object is more likely belong to the cluster compared with other clusters.

## 3.2 Weight Measurement of Boundary Objects

For a cluster $C_i$, one of its boundary objects is $x_n$, threshold of neighbor degree is $\eta$. In lower approximation of cluster $C_i$, the set of neighbor points of boundary objects $x_n$ are as follows:

$$F_p = \{p : |x_p - x_n| \leq \eta | x_p \in \underline{C_i}\}. \quad (6)$$

For lower approximation of cluster $C_i$, the number of neighbor points of boundary object $x_n$ is s. The 'neighbor degree' of boundary object $x_n$ to cluster $C_i$ is:

$$\begin{aligned} L_{in} &= \sum\nolimits_{p \in F_p} \frac{1}{d_{pn}} \\ &= \sum_{p=1}^{s} \frac{1}{\|x_p - x_n\|} \\ &= \frac{1}{d_{1n}} + \frac{1}{d_{2n}} + \ldots + \frac{1}{d_{sk}}. \end{aligned} \quad (7)$$

As shown in Eq. (7), the larger the number s of objects in lower approximation that are close to boundary objects $x_n$, the larger 'neighbor degree', $L_{in}$; and the closer the distance $d(p, n)$ between boundary object and neighbor points in lower approximation, the larger 'neighbor degree'. The larger neighbor degree, boundary objects are more similar with lower approximation and more likely belong to cluster $C_i$. Conversely, the smaller the neighbor degree, the less similar between boundary object and the lower approximation set, and less likely that boundary objects belong to the cluster $C_i$.

The set of clusters that may contain boundary object $x_n$ as follows:

$$F_n = \{f : x_n \in \overline{C_f}\}. \tag{8}$$

The weight of boundary object $x_n$ in mean calculation of the cluster $C_i$ is:

$$w_{in} = \frac{1}{\sum_{j \in F_n} \left(\frac{L_{jn}}{L_{in}}\right)^2}. \tag{9}$$

As shown in Eq. (9), the importance of boundary object to a cluster in mean calculation not only has something to do with neighbor degree to the cluster, but also relates to neighbor degree of clusters that the boundary object may belong to.

### 3.3 BODM_RKM Algorithm

According to the weight calculation method of boundary objects proposed by Eq. (8), the new mean calculation formula is:

$$V_i = \begin{cases} \frac{\sum_{x_n \in \underline{C_i}} x_n + \sum_{x_n \in \hat{C}_i} w_{in} x_n}{|\underline{C_i}| + \sum_{x_n \in \hat{C}_i} w_{in}}, & if\, \underline{C_i} \neq \phi \wedge \hat{C}_i \neq \phi \\ \frac{\sum_{x_n \in \hat{C}_i} w_{in} x_n}{\sum_{x_n \in \hat{C}_i} w_{in}}, & if\, \underline{C_i} = \phi \wedge \hat{C}_i \neq \phi \\ \frac{\sum_{x_n \in \underline{C_i}} x_n}{|\underline{C_i}|}, & otherwise. \end{cases} \tag{10}$$

As shown in Eq. (9), the objects in lower approximation determinately belongs to the cluster while objects in boundary area may belong to the cluster. The weight of boundary object is $w_{in}$.

With N data, a new algorithm of processing steps as follows:

Input: data set $\{x_n\}$, the number of clusters k, relative distance threshold $\zeta$, neighbor degree threshold $\eta$.

Output: k rough class cluster.

## 4 Experiments

### 4.1 Experiment Environment and Data Sets

We carry out experiments on UCI data sets to verify the validity of the algorithm in this paper by compared with other algorithms. Experimental environment: operating system is Windows 7, the CPU frequency of 2.20 GHz, 4 cores, memory is 2 GB, simulation software MATLAB is 7.10.0 (Table 1).

**Table 1.** Data set information.

| Data set | Number of samples | Number of features | Number of clusters |
|---|---|---|---|
| Iris | 150 | 4 | 3 |
| Wine | 178 | 13 | 3 |
| Seeds | 210 | 7 | 3 |
| Ionosphere | 351 | 34 | 2 |

## 4.2 Evaluation Index

In order to assess the quality of the clustering results, effective clustering evaluation index is very important. There are many evaluation index, such as Xie-Beni [22], Silhoutte [23], Davies - Bouldin (DB) and Dunn [24] and other indicators. Among them, the DB and Dunn index are widely used. Davies-Bouldin index is to minimize the distance within clusters $S(v_i)$ and maximize the distance between the clusters $d(v_i, v_k)$. The more similar within the cluster and the more separate between clusters, the lower the DB index i.e. the better clustering effect. The higher the Dunn index, the better the clustering effect.

**Davies-Bouldin Index:**

$$DB = \frac{1}{c}\sum_{i=1}^{c}\max_{k\neq i}\left\{\frac{S(v_i) + S(v_k)}{d(v_i, v_k)}\right\}. \tag{11}$$

Where the distance within the cluster is:

$$\begin{aligned} S(v_i) &= \frac{\underline{w}\sum_{x_n\in\underline{C_i}}\|x_n - v_i\|^2 + \hat{w}\sum_{x_n\in\hat{C}_i}\frac{\|x_n - v_i\|^2}{|B_{x_n}|}}{\underline{w}\left|\underline{C_i}\right| + \hat{w}\sum_{x_n\in\hat{C}_i}\frac{1}{|B_{x_n}|}} \\ &= \frac{\sum_{x_n\in\overline{C_i}}\frac{\|x_n - v_i\|^2}{|B_{x_n}|}}{\sum_{x_n\in\overline{C_i}}\frac{1}{|B_{x_n}|}}. \end{aligned} \tag{12}$$

**Dunn Index:**

$$Dunn = \min_i\left\{\min_{k\neq i}\left\{\frac{d(v_i, v_k)}{\max_l S(v_l)}\right\}\right\}. \tag{13}$$

Literature [11] used four criteria to evaluate the clustering quality.

CQ1: Ok, Number of objects that be assigned to correct lower approximations, the more the better.

CQ2: πOk, Number of objects that be assigned to correct lower approximations and correct boundaries. Objects be assigned to correct boundaries account for $\frac{1}{|B_{xn}|}$, the higher the better.

CQ3: Bd, Number of boundary areas objects, the lower the better.

CQ4: ¬Ok, Number of objects that be assigned to incorrect lower approximations, the lower the better.

### 4.3 DB and DUNN Index Comparison

The proposed algorithm is compared and analyzed with πRKM and RFKM on Iris, Wine and Seeds data sets as mentioned in Sect. 4.1.

These data sets are normalized to interval [0,1] and same data set choose the same initial centers. πRKM algorithm run in standard mode, namely $w_l = w_b = 0.5$. My RKM algorithm does not need to set the weight parameters of the upper and lower approximation set. In order to comprehensively compare these algorithms, threshold parameter is $\zeta = 1.2, 1.4, 1.6, 1.8$. Comparison results of different algorithms are shown in Table 2.

**Table 2.** DB and Dunn index comparison.

| Algorithm | | πRKM | | | | BOBD-RKM | | | |
|---|---|---|---|---|---|---|---|---|---|
| ζ | | 1.2 | 1.4 | 1.6 | 1.8 | 1.2 | 1.4 | 1.6 | 1.8 |
| Iris | DB | 0.1885 | 0.2071 | 0.2144 | 0.2953 | 0.1861 | 0.1904 | 0.2105 | 0.2369 |
| | Dunn | 8.5895 | 7.1981 | 6.9604 | 5.2913 | 8.8688 | 8.6857 | 7.0841 | 6.9168 |
| Wine | DB | 0.7828 | 0.8666 | 1.1990 | – | 0.7710 | 0.8419 | 0.9703 | 1.2140 |
| | Dunn | 2.2378 | 2.0856 | 1.4950 | – | 2.2932 | 2.1917 | 1.8885 | 1.4715 |
| Seeds | DB | 0.3120 | 0.3418 | 0.3701 | 0.4130 | 0.3118 | 0.3417 | 0.3643 | 0.3958 |
| | Dunn | 5.5287 | 4.8748 | 4.5147 | 4.1114 | 5.5375 | 5.1187 | 4.6448 | 4.3504 |

Results in Table 2 show that on the IRIS data set, DB index values of put forwarded algorithm are smaller than π RKM algorithm for all of the threshold value of ζ and Dunn index values are all greater than that of π RKM. It indicates that the proposed algorithm is better than π RKM algorithm on IRIS data set. And algorithm put forwarded in this paper clustering effect is better still when $\zeta = 1.8$. On Seeds dataset, the DB and Dunn indexes of my RKM algorithm are also better than ΠRKM.

In general, the DB and Dunn index of myRKM algorithm is better than that of πRKM algorithm. It indicates that myRKM makes more similar within clusters and further separated between clusters. As a result, the cluster effect of myRKM is better than that of πRKM.

### 4.4 Clustering Results Comparison

We use the iris, wine and Ionosphere data sets and the data have been normalized to interval [0,1]. We choose the same initial centers for all algorithms on same data set.

Threshold parameter $\zeta = 1.2, 1.4, 1.6, 1.8$. For RFKM, $w_l = 0.7$, $w_b = 0.3$. For myRKM and RFKM, exponent parameter m = 2. Table 3 shows clustering results of different algorithms.

**Table 3.** Clustering results comparison.

| Algorithm | | πRKM | | | | RFKM | | | | BOBD_RKM | | | |
|---|---|---|---|---|---|---|---|---|---|---|---|---|---|
| Data set | ζ | Ok | πOk | ¬Ok | Bd | Ok | πOk | ¬Ok | Bd | Ok | πOk | ¬Ok | Bd |
| Iris | 1.2 | 131 | 133.50 | 14 | 5 | – | – | – | – | 130 | 132.5 | 15 | 5 |
| | 1.4 | 124 | 134.50 | 5 | 21 | – | – | – | – | 126 | 131.5 | 13 | 11 |
| | 1.6 | 119 | 132 | 5 | 26 | 126 | 135 | 6 | 18 | 118 | 132 | 4 | 28 |
| | 1.8 | 113 | 130.83 | 1 | 36 | 126 | 135 | 6 | 18 | 114 | 130.5 | 3 | 33 |
| Wine | 1.2 | 150 | 162.33 | 3 | 25 | – | – | – | – | 152 | 163.33 | 3 | 23 |
| | 1.4 | 135 | 154.17 | 1 | 42 | – | – | – | – | 136 | 155 | 1 | 41 |
| | 1.6 | 105 | 137.17 | 1 | 72 | 144 | 159.17 | 3 | 31 | 114 | 142.17 | 1 | 63 |
| | 1.8 | – | – | – | – | 137 | 154.5 | 3 | 38 | 87 | 126.17 | 0 | 91 |
| Ionosphere | 1.2 | 202 | 231 | 91 | 58 | – | – | – | – | 202 | 230.5 | 92 | 57 |
| | 1.4 | 156 | 227.50 | 52 | 143 | – | – | – | – | 169 | 225.5 | 69 | 113 |
| | 1.6 | – | – | – | – | 180 | 223 | 85 | 86 | 151 | 226.5 | 49 | 151 |
| | 1.8 | – | – | – | – | 171 | 224.5 | 73 | 107 | 142 | 227 | 39 | 170 |

As we can see from Table 3, for different data sets and different relative threshold, the results of different algorithms are different. For Iris data set, clustering effect of π RKM algorithm is slightly better than myRKM and RFKM. RFKM has no meaningful results when threshold is small. For Wine data set, myRKM has the best clustering effect while π RKM generates meaningless result with the larger threshold. On Ionosphere data set, clustering results of myRKM and π RKM algorithms are similar when $\zeta = 1.2$; when $\zeta = 1.4$, myRKM is better than that of πRKM on Ok and Bd criterions while not as good as πRKM on πOk and ¬Ok criterions; With the increase of threshold, more and more objects are assigned to boundary areas, π RKM algorithm gets meaningless results and myRKM still has better clustering effect.

## 5 Conclusion

According to the distribution of boundary object's neighbor points in clusters that boundary object may belong to, this paper proposes a new boundary object weight calculation method based on neighbor degree. Through the simulation on UCI data sets, it shows that the proposed algorithm have advantage in making the more similar within the cluster and the more separate between clusters compared with other algorithms.

**Acknowledgments.** This work was supported by the National Natural Science Foundation of China under Grant 62073173 and 61833011, the Natural Science Foundation of Jiangsu Province, China under Grant BK20191376, the University Natural Science Foundation of Jiangsu Province under Grant TJ219022.

## References

1. Huang, D., Lai, J.H., Wang, C.D.: Combining multiple clusterings via crowd agreement estimation and multi-granularity link analysis. J. Neurocomput. **170**, 240–250 (2015)
2. Xu, R., Wunsch, D.: Survey of clustering algorithms. J. IEEE Trans. Neural Netw. **16**, 645–678 (2005)
3. Zhou, S., Xu, Z., Tang, X.: A method for determining the optimal number of clusters based on affinity propagation clustering. J. Control Decis. **26**, 1147–1152, 1157 (2011)
4. Hartigan, J.A., Wong, M.A.: A k-means clustering algorithm. J. Appl. Stat. **28**, 100–108 (1979)
5. Amorim, R.C.: A survey on feature weighting based k-means algorithms. J. Classif. **33**, 210–242 (2016)
6. Khandare, A., Alvi, A.S.: Survey of improved k-means clustering algorithms: improvements, shortcomings and scope for further enhancement and scalability. In: Satapathy, S.C., Mandal, J.K., Udgata, S.K., Bhateja, V. (eds.) Information Systems Design and Intelligent Applications. AISC, vol. 434, pp. 495–503. Springer, New Delhi (2016). https://doi.org/10.1007/978-81-322-2752-6_48
7. Pawlak, Z.: Rough sets. Int. J. Comput. Inf. Sci. **11**, 341–356 (1982)
8. Lingras, P., West, C.: Interval set clustering of web users with rough k-means. J. Intell. Inf. Syst. **23**, 5–16 (2004)
9. Dou, C., Zhang, Z., Yue, D., Song, M.: Improved droop control based on virtual impedance and virtual power source in low-voltage microgrid. J. LET Gener. Transm. Distrib. **11**(4), 1046–1054 (2017)
10. Dou, C., Yue, D., Li, X., Xue, Y.: MAS-based management and control strategies for integrated hybrid energy system. J. IEEE Trans. Ind. Inform. **12**(4), 1332–1349 (2016)
11. Peters, G.: Some refinements of rough k-means clustering. J. Pattern Recogn. **39**, 1481–1491 (2006)
12. Zhang, T., Chen, L., Ma, F.: A modified rough c-means clustering algorithm based on hybrid imbalanced measure of distance and density. J. Int. J. Approx. Reasoning **55**, 1805–1818 (2014)
13. Peters, G.: Rough clustering utilizing the principle of indifference. J. Inf. Sci. **277**, 358–374 (2014)
14. Jiang, Z., Li, T., Min, W., et al.: Fuzzy c-means clustering based on weights and gene expression programming. J. Pattern Recogn. Lett. **90**, 1–7 (2017)
15. Chen, H.P., Shen, X.J., Lv, Y.D., et al.: A novel automatic fuzzy clustering algorithm based on soft partition and membership information. J. Neurocomputing **236**, 104–112 (2017)
16. Maji, P., Pal, S.K.: RFCM. A hybrid clustering algorithm using rough and fuzzy sets. J. Fundamenta Informaticae. **80**, 475–496 (2007)
17. Mitra, S., Banka, H., Pedrycz, W.: Rough–fuzzy collaborative clustering. J. IEEE Trans. Syst. Man Cybern. Part B (Cybern.) **36**, 795–805 (2006)
18. Paul, S., Maji, P.: A new rough-fuzzy clustering algorithm and its applications. In: Babu, B.V., et al. (eds.) Proceedings of the Second International Conference on Soft Computing for Problem Solving (SocProS 2012), December 28-30, 2012. AISC, vol. 236, pp. 1245–1251. Springer, New Delhi (2014). https://doi.org/10.1007/978-81-322-1602-5_130

19. Shi, J., Lei, Y., Zhou, Y., et al.: Enhanced rough–fuzzy c-means algorithm with strict rough sets properties. J. Appl. Soft Comput. **46**, 827–850 (2016)
20. Li, F., Liu, Q.: An extension to rough c-means clustering algorithm based on boundary area elements discrimination. In: Peters, J.F., Skowron, A., Ramanna, S., Suraj, Z., Wang, X. (eds.) Transactions on Rough Sets XVI. LNCS, vol. 7736, pp. 17–33. Springer, Heidelberg (2013). https://doi.org/10.1007/978-3-642-36505-8_2
21. Weng, S., Dong, Y., Dou, C., Shi, J., Huang, C.: Distributed event-triggered cooperative control for frequency and voltage stability and power sharing in isolated inverter-based microgrid. J. IEEE Trans. Cybern. **49**(4), 1427–1439 (2018)

# Reducing the Width of a Feed-Forward Neural Network

Xian Guo[1], Jianing Yang[1(✉)], Gaohong Wang[2], Keyu Chen[1], and Libao Yang[1]

[1] China Industrial Control Systems Cyber Emergency Response Team, Beijing 100040, China
[2] Shanghai Institute of Process Automation and Instrumentation, Shanghai 200233, China

**Abstract.** We study feed-forward neural networks with linear rectification in this paper. Inspired by the idea of "neuron coverage" and "equivalence decomposition", we show that the FFNN's have a natural equivalent relation on them, and can be applied to decide whether the width of the network can be reduced.

## 1 Introduction

We study feed-forward neural networks with linear rectification in this paper. Inspired by the idea of "neuron coverage" [4] and "equivalence decomposition" [3], we show that there is a natural equivalence relation on the FFNN's, and the relation can be applied to decide whether the width of the network can be reduced. The main theoretical foundation of our work is that the space of network parameters forms a dual space of the data points, which is explained in Sect. 3. In Sect. 2, we review some works related to the study of FFNN's. In Sect. 3, we explain the main idea of the paper with examples. We also define the equivalence relation and show how the idea can be applied to reduce width of neural networks. The result of experiments of the proposed method is presented in Sect. 4.

## 2 Related Works

In this section, we introduce related works on neuron coverage and an equivalence relation of data points. These works focus on the understanding of neural networks. In [4], the authors introduced the idea of neuron coverage. They also produced adversarial examples by computing a gradient descent on data points. This inspires the idea that the network paramters themselves form a dual space of the data points.

In [3], a similar idea is applied and an equivalence class is defined on the points of a data cloud for a given FFNN with ReLU activation. Two points $x, y \in X$ are equivalent if and only if the points $x$ and $y$ activate the same set of neurons of the given FFNN. Note that if $x$ and $y$ are equivalent in $X$, then we can see that the points in the same equivalence class are classified by the


Q. Han et al. (Eds.): LSMS 2021/ICSEE 2021, CCIS 1469, pp. 319–326, 2021.
https://doi.org/10.1007/978-981-16-7213-2_31

same linear separater. It follows that an FFNN actually provides us with an equivalence relation on the data points. Moreover, the FFNN is linear on each of the equivalence classes. However, as the structure of the neural network gets more and more complicated, it is in general not easy to see what the equivalence relation and the linear classifiers are like.

In [5], this parameter space is systematically studied, and statistical learning theory is built based on the idea. From this point of view, we wish to setup a theory to choose a better network structure.

## 3 FFNN's with ReLU Activation

The related works from Sect. 2 have shown efforts to understand how neural networks work in general. Among all these works, the study has mainly focused on the inter-relationship between the data cloud and the neural network. We make this idea clear in our paper, i.e., the FFNN's form a dual space of the data points.

### 3.1 The FFNN's Form a Dual Space of the Data Points

In statistical learning, we usually start with the distribution of the data points, and look for the right classifier or estimator of the data. The training is usually processed as an optimizer to minimize certain loss function of the parameters. Conversely, in the work of Generating Adversarial Networks and DeepXplore [4], the training process involves the computation of derivatives of data points. On one hand, this process of generating examples enriches the data cloud, which prevents the network from being over-fitting as well as provides a method to test the robustness of the network. On the other hand, the data clouds that present in practice may not always be well-labeled, hence we hope to learn more about the 'manifold' structure of the data cloud.

Combining both views of the networks, we see that the loss function is a function of both the network parameters and the data points, and the spaces are dual to each other. By minimizing the loss function over fixed data samples, we can train classifiers of the data cloud in the usual way. Conversely, we can study the behaviour of data samples for fixed classifiers on the data cloud, as we shall see in the following examples.

**Example 1 (Multiple classifier voter).** *The most common application of multiple classifiers of a data cloud is to use them as voters for classification. The multiple classifiers can be of the same or different structures. Combining the results of multiple classifiers, we can ensure a higher confidence of the classification result.*

**Example 2 (Semi-supervised learning).** *In a recent paper [2], the authors introduced the Mixmatch algorithm for semi-supervised learning by using the classifiers of different approaches to generate labels of newcomers in the data. In daily examples, it is very common situation that we have to investigate small* *datasets or deal with ambiguous labels. The method enriches the data samples and labels so that a better result can be expected. The idea is to look into the joint distribution of the samples over different classes and classifiers. The distribution is then sharpened for unlabelled data samples to setup labels for training.*

**Example 3 (The class "others").** *We can take the idea of semi-supervised learning from [2] further to deal with the "others" class. In practice, we may encounter data samples with ambiguous labels. For example, in a classification problem for quality assurance or quality control, some NG examples can be very close to OK examples, which makes it difficult to tell the difference. Usually in such problems, there are more OK examples than NG examples. Moreover, for NG examples, it is often the case that the product has a quality problem for various reasons, which is again hard to differentiate, and further classification of NG examples into subclasses would make the dataset more unbalanced. Applying the idea of the MixMatch approach, we expect the ambiguities to present themselves as anomolies in the joint distribution of samples. In such cases, it is helpful to introduce an extra class called "others" to deal with the anomolies, and a newly discovered extra class may correspond to some product flaw that is previously unclassified.*

### 3.2 Equivalence of FFNN's

Recall that in [3], an equivalence relation is defined on the data points, given an FFNN as its classifier. The authors assume that the FFNN's have ReLU activators. The ReLU function was first introduced into neural networks for its faster performance of convergence than the sigmoid function. In FFNN's, the ReLU function activates a neuron according to the sign of the output, and the equivalence relation is defined for data points to activate the same neurons.

We investigate the idea further in this section and see how it can be applied to give an equivalence relation on the FFNN's. Let's look into the example of a two dimensional plane embedded in the three dimensional space. For simplicity, we assume that the plane passes through the origin, hence the embedding can be represented by a $2 \times 3$ matrix. We view this matrix as a linear map $L : X \to Y$, where we think of $X$ as the space of data points and $Y$ as the space of the first layer of neurons. Since the space $Y$ has eight quadrants, one might at first glance guess that the data cloud $X$ has eight equivalent classes according to the definition in [3]. But a close look will show that we can have at most six equivalence classes, depending on the embedding map $L$. It follows that different ways of the embedding will decide different equivalence relations on the data points.

**Example 4.** *Now let's take the example from above and make it more explicit. Let*

$$L = \begin{bmatrix} 1 & 0 & 2 \\ 0 & 1 & 3 \end{bmatrix}$$

*represent an embedding of the plane in $\mathbb{R}^3$. Then the image of L passes through six octants in $\mathbb{R}^3$, namely $(+++), (+--), (+-+)$ and their opposites, as shown*

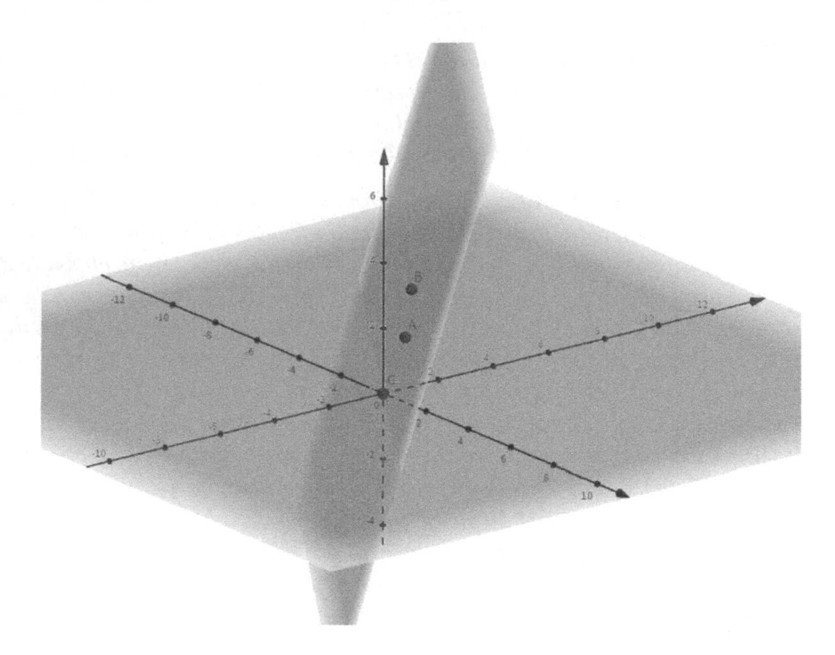

**Fig. 1.** Embedding of a plane in $R^3$

*in Fig. 1. For $x = (x_0, x_1) \in X$, and $y = L(x) = (y_0, y_1, y_2) \in Y$, if $x_0, x_1 > 0$, then $y_0, y_1, y_2 > 0$ and $L(x)$ is in the octant $(+ + +)$. If $x_0 > 0$ and $x_1 < 0$, then $y_0 > 0$, $y_1 < 0$, and the sign of $y_2$ can be either positive or negative, hence the octants $(+ - -)$ and $(+ - +)$ intersect the image of $L$. Since the map $L$ is linear, it also passes through the opposite of the three octants. The reader should observe that the image of $L$ does not intersect the octants $(+ + -)$ and $(- - +)$.*

*Assume that the bias is 0, we can apply the definition from [3], and see that the points in the domain $\mathbb{R}^2$ of $L$ can be divided into six classes, each in correspondence to the named octants above.*

If we fix the first two columns of $L$ and allow the third column to vary, we can see that whether the image of $L$ passes through a different octant depends exactly on the signs of the entries of the third column. Correspondingly, we can have four types of classifications on $\mathbb{R}^2$. Note that the statement we get for the embedding $L$ is generally correct. Indeed, as long as the matrix formed by the first two columns are invertible, we can always perform a row reduction to make the linear embedding into the form of $L$.

**Example 5.** *Now let us expand the dimension of the target space and consider*

$$L = \begin{bmatrix} 1 & 0 & 1 & 2 \\ 0 & 1 & 3 & 7 \end{bmatrix}.$$

*We have already learned from the above example that the signs of entries in the last two columns effect the embedding. But there is more. It is easy to see that*

*$L$ passes through the orthoplex $(+-+-)$ and, by computing the inverse of the matrix formed by the last two columns of $L$, we can show that this is the only possibility of the form $( \_ \_ +-)$ for $L$. More specifically, if a point $x$ is in the image of $L$ with $x_3 > 0$ and $x_4 < 0$, it follow that $x_1 > 0$ and $x_2 < 0$.*

*Now Let*

$$L' = \begin{bmatrix} 1\,0\,1\,2 \\ 0\,1\,3\,5 \end{bmatrix}.$$

*It is not hard to see that $L'$ passes through the orthoplex $(-++-)$, although the entries in $L$ and $L'$ are all positive. The reader may have already found out that it is the determinant of the last two columns that causes the difference. And in this example, this determinant and the signs of the relevant entries together determine the form of the embedding $L$.*

Now we are ready to state our main theorem, which generalizes the examples above. The proof is completely parallel to the cases in low dimensions and can be done by induction.

**Theorem 1.** *The images of two linear maps $L_i : X \to Y$ for $i = 1, 2$ pass through the same orthoplexes if and only if their "sub-determinants" are all of the same or opposite signs.* □

Following the theorem, we define two FFNN's to be equivalent when their sub-determinants are all of the same or opposite signs.

**Definition 1.** *Let $L$ and $L'$ be matrices of shape $m \times n$ and $l = \min(m, n)$. We say that the matrices $L$ and $L'$ are* **equivalent** *if and only if for each pair of $l \times l$ submatrices of $L$ and $L'$, the signs of the determinants of the corresponding submatrices are all the same or all opposite to each other. Two FFNN's of the same structure are* **equivalent** *if and only if for each linear layer of network, the corresponding matrices are equivalent.*

With this definition, we can say that the matrices

$$L = \begin{bmatrix} 1\,0\,1\,2 \\ 0\,1\,3\,7 \end{bmatrix} \text{ and } L' = \begin{bmatrix} 1\,0\,1\,2 \\ 0\,1\,3\,5 \end{bmatrix}$$

are not equivalent since

$$det \begin{pmatrix} 1\,2 \\ 3\,7 \end{pmatrix} = 1, \text{ but } det \begin{pmatrix} 1\,2 \\ 3\,5 \end{pmatrix} = -1.$$

**Remark 1.** *As a remark, we also note that if we switch two rows of the embedding matrix, then the determinants of the matrices change signs. In other words, the orientation of the embedding has been changed, and the equivalence relation we defined in Definition 1 is up to orientation of the embedding.*

### 3.3 Reducing Width of FFNN's

We have defined an equivalence relation on FFNN's in the previous section. Note that if two FFNN's are equivalent, then the FFNN's will define similar equivalence classes on the data points and they should determine similar classification results.

As an application, in the training process of an FFNN, we should require not only the minimization of the loss function, but also the convergence of the equivalent class of the FFNN. We present some experiments in Sect. 4. There are cases in which the training process converges to the minimum value of the loss function but the equivalent class of the FFNN still varies. This suggests that there is extra space for the variables of the FFNN to move around without changing the performance of the network and we can reduce the width of the network in such cases.

The problem with tracking the signs of the determinants is that the complexity of the computation of determinants, as well as the number of determinants to be compared grow dramatically as the size of the matrices increases. In order to make the comparison between to linear layers easier, we compute *the sum of the signs of determinants* as an indicator, rather than compare all the signs. The change of the sum indicator implies that some of the determinants has switched sign. On the other hand, it is rare that two signs of the determinants should change simultaneously to cancel the sign switch of the other. Hence the sum indicator is sufficient to track the change of signs when the gradient descent process moves the layer embedding in a continuous manner, in which case the sum changes by a number of 2. In particular, the stochastic gradient descent optimizer changes the coordinates continuously. Other optimizers, such as the Adam optimizer, may change the coordinates rapidly in order to avoid local minima, but are still continuous most of the time, and the sum indicator works in these cases as well.

We present an experiment of how the process of reducing network width work in Sect. 4.

## 4 Experiments

We carry out the experiment on the Iris dataset, and show how the idea in Sect. 3 can be applied. The Iris dataset consists of 150 data points, and we split them into training samples and validation samples by a factor of 0.8.

We build the model using the Tensorflow [1] framework, starting with an FFNN with two hidden layers of dimension 16 and 4, respectively, the number of features and classes being 4 and 3. We train the model for 5000 epochs with the Adam optimizer, choosing batch size to be 64 and learning rate to be 0.01. The result of the training classifies most training/validation samples correctly, with an exception of less than or equal to two errors. The two misclassified samples are $x = (6.3, 2.8, 5.1, 1.5)$ from Iris virginica and $x = (6.0, 2.7, 5.1, 1.6)$ from Iris versicolor.

While the training loss converges to 0 during the training process, the sum indicator of the first hidden layer varies at the last 100 steps, as is shown in Fig. 2.

```
[30 28 28 32 34 34 26 30 30 30 30 30 26 34 32 28 28 32 30 28 28 30 30 32
 32 32 32 34 30 32 32 30 30 30 30 28 30 30 30 32 30 28 30 28 30 32 32 34
 32 32 28 28 28 28 26 26 28 28 28 28 28 28 28 28 28 26 30 30 30 32 30 32
 30 32 34 30 32 28 30 34 32 32 30 30 34 34 36 36 36 36 34 34 30 32 30 30
 30 30 26 28]
```

**Fig. 2.** Values of the sum indicator for a wide FFNN

We can see from the figure that the sum indicator varies quickly during the training process. One should also observe that the sum indicator changes by a number of 2 most of the time, as we mentioned in Sect. 3.3

Given the variance in Fig. 2, we reduce the width of the first hidden layer to 8. The training result reaches the same precision and misclassifies at most the two samples mensioned above. The sum indicator becomes quite steady in this case, as shown in Fig. 3

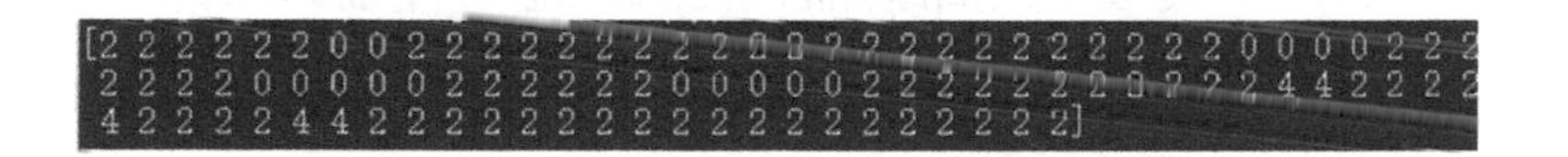
```
[2 2 2 2 2 2 0 0 2 2 2 2 2 2 2 2 2 0 0 2 2 2 2 2 2 2 2 2 2 2 2 0 0 0 0 2 2 2
 2 2 2 2 0 0 0 0 0 2 2 2 2 2 2 2 0 0 0 0 0 2 2 2 2 2 2 2 0 2 2 2 4 4 2 2 2 2
 4 2 2 2 2 4 4 2 2 2 2 2 2 2 2 2 2 2 2 2 2 2 2 2 2 2]
```

**Fig. 3.** Values of the sum indicator for a medium-sized FFNN

We further reduce the width of the first hidden layer to 5. Note that we did not try the number 4, which would force the embedding to be an isomorphism from the first hidden layer to the second. In the case of width 5, we have again managed to reach the same precision, and the sum indicator becomes constant as the training loss converges. The result is show in Fig. 4.

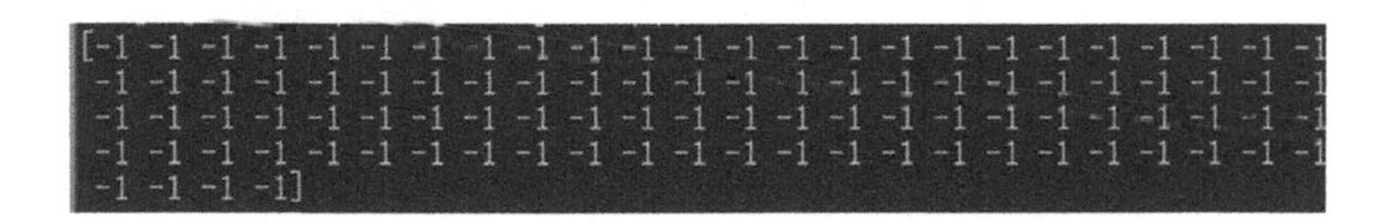
```
[-1 -1 -1 -1 -1 -1 -1 -1 -1 -1 -1 -1 -1 -1 -1 -1 -1 -1 -1 -1 -1 -1 -1 -1
 -1 -1 -1 -1 -1 -1 -1 -1 -1 -1 -1 -1 -1 -1 -1 -1 -1 -1 -1 -1 -1 -1 -1 -1
 -1 -1 -1 -1 -1 -1 -1 -1 -1 -1 -1 -1 -1 -1 -1 -1 -1 -1 -1 -1 -1 -1 -1 -1
 -1 -1 -1 -1 -1 -1 -1 -1 -1 -1 -1 -1 -1 -1 -1 -1 -1 -1 -1 -1 -1 -1 -1 -1
 -1 -1 -1 -1]
```

**Fig. 4.** Values of the sum indicator for a narrow FFNN

In conclusion, we applied the sum indicator as a measure of the complexity of the model and have succeeded in narrowing the model to a much smaller size while keeping the model's precision. We have carried out the process for the first hidden layer. The process can be applied to the other hidden layer in a similar manner.

## 5 Conclusion

We presented an idea of reducing the width of FFNN's in this paper. The key observation is that the FFNN's form a dual space of the data space (Sect. 3.1). Inspired by the equivalence relation on data space defined in [3], we then defined an equivalence relation on the FFNN's. We used the sum indicator from Sect. 3.3 to track the equivalence relation of the FFNN's during a training process. And in Sect. 4, we carried out experiments to show that the idea of the sum indicator can be applied to reduce the width of an FFNN.

## References

1. Abadi, M.: Tensorflow: a system for large-scale machine learning. In: 12th USENIX Symposium on Operating Systems Design and Implementation, pp. 265–283 (2016)
2. Berthelot, D., Carlini, N., Goodfellow, I., Oliver, A., Papernot, N., Raffel, C.: MixMatch: a holistic approach to semi-supervised learning. In: Advances in Neural Information Processing Systems, vol. 32 (2019)
3. Lei, N., Kehua, S., Cui, L., Yau, S.-T., Gu, X.D.: A geometric view of optimal transportation and generative model. Comput. Aided Geom. Des. **68**, 1–21 (2019)
4. Pei, K., Cao, Y.: DeepXplore: automated whitebox testing of deep learning systems. In: Proceedings of the 26th Symposium on Operating Systems Principles, pp. 1–18 (2017)
5. Watanabe, S.: Algebraic Geometry and Statistic Learning Theory. Cambridge University Press, Cambridge (2009)

# Decision Tree as Neural Network

Xian Guo[1], Jianing Yang[1(✉)], Gaohong Wang[2], Keyu Chen[1], and Libao Yang[1]

[1] China Industrial Control Systems Cyber Emergency Response Team, Beijing 100040, China
[2] Shanghai Institute of Process Automation and Instrumentation, Shanghai 200233, China

**Abstract.** We show that given a dataset and a decision-tree classifier for the dataset, we can construct a neural network from the decision-tree such that the neural network classifies the data points exactly as the decision-tree does. This provides a point of view to partially explain neural networks.

## 1 Introduction

We show that given a dataset and a decision-tree classifier for the dataset, we can construct a neural network from the decision-tree such that the neural network classifies the data points exactly as the decision-tree does. This provides a point of view to partially explain neural networks. For example, pruning trees has the effect of reducing the width of neural networks.

The decision-tree classifier is a traditional classifier, which uses a determinant algorithm to classify data points into leaves of a binary tree. Each non-leaf node is divided into two subclasses such that the entropy of the division is minimized. On the other hand, deep neural networks have gained much popularity in recent years, but they have always been suffering from the problem of lacking interpretability.

In this paper, we present a method to reproduce the consequence of a decision tree with a neural network. Given a decision tree, we implement the corresponding neural network with one hidden layer. While the hidden layer is a fully connected layer implementing the linear conditions of the decision tree, the prediction layer is a three-axis layer, with the extra axis mimicking the reward gains on different leaves of tree.

Sections 2 and 3 are devoted to explaining the main idea of the paper. In Sect. 2, we introduce the idea with an example. The algorithm and more details of implementation are given in Sect. 3. The theorems and propositions in these sections show the effort to characterize the matrices in the derived neural networks. In Sect. 4, we provide some comments on how the work of this paper is related and can be applied to the understanding of neural networks. For example, in an earlier paper [3], the author tried to give some criteria for reducing the width of neural networks. We can see that, in the work of transforming from decision trees to neural networks, reducing the width of the network corresponds to pruning the decision tree. We conclude the paper with future expectations to more understanding of neural networks.


Q. Han et al. (Eds.): LSMS 2021/ICSEE 2021, CCIS 1469, pp. 327–336, 2021.
https://doi.org/10.1007/978-981-16-7213-2_32

# 2 Decision Tree as Neurel Network

## 2.1 Condition Matrix, Weight Matrix, and Tree Matrix

We start the introduction of our result by an example. For simplicity, we work with the Iris dataset. The Iris dataset consists of three iris species with 50 samples each. The samples are described by a tuple of four numbers indicating its properties such as the sepal-length.

We first construct a decision tree to do the prediction, as shown in Fig. 1.

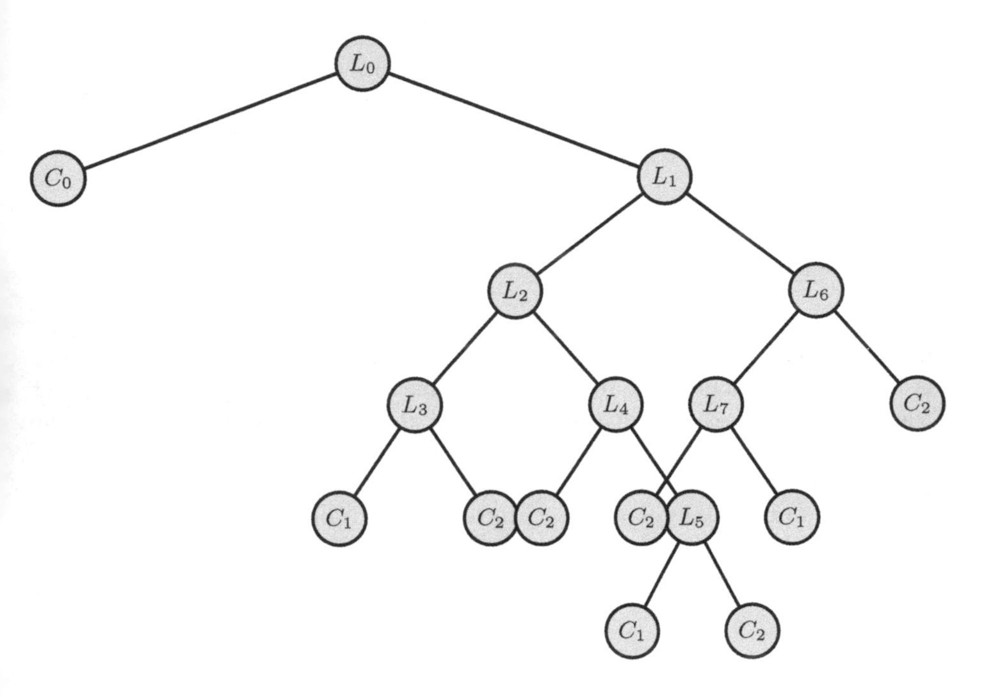


**Fig. 1.** Decision tree for the classification of iris dataset

The decision tree contains 8 non-leaf nodes and 9 leaves, and has depth 6. The leaves are named as "$C_i$", indicating that samples in the leaf "$C_i$" belong to the $i$-th class. The non-leaf nodes are named from "$L_0$" to "$L_7$", each implementing a linear condition for separation. The linear conditions form the following coefficient matrix

$$\mathtt{dt} = \begin{bmatrix} 0 & 0 & 0 & 1 & -0.8 \\ 0 & 0 & 0 & 1 & -1.75 \\ 0 & 0 & 1 & 0 & -4.95 \\ 0 & 0 & 0 & 1 & -1.65 \\ 0 & 0 & 0 & 1 & -1.55 \\ 1 & 0 & 0 & 0 & -6.95 \\ 0 & 0 & 1 & 0 & -4.85 \\ 0 & 1 & 0 & 0 & -3.1 \end{bmatrix},$$

with the $i$-th row corresponding to the condition "$L_i$", respectively. We call this matrix the **condition matrix**. For example, if $L_0(x) \leq 0$ for some sample point $x = (x_0, x_1, x_2, x_3)$, i.e., $x_3 - 0.8 \leq 0$, then $x$ belongs to the left child $C_0$ of $L_0$. Otherwise, we further test $x$ on the condition $L_1$.

In general, a binary decision tree is a strict binary tree, i.e., each node of the tree has either 2 or 0 sub-nodes. It follows that a decision tree consists of $2n+1$ nodes, where $n$ is the number of linear conditions for separation, corresponding to the non-leaf nodes of the tree. The other $n+1$ nodes are leaves of the tree and encode the result of classification.

Next, we need an implementation that transforms the output of the linear conditions into credits of the sample points being in different classes. In order to do this, we collect the following information for each node $L_i$:

1. the priority of the node,
2. the sample classes for $x$ with $L_i(x) > 0$, and
3. the sample classes for $x$ with $L_i(x) \leq 0$.

We collect the information into weight matrices. Firstly, we have a priority matrix:

$$\texttt{priority} = \begin{bmatrix} 16 & 8 & 4 & 2 & 2 & 1 & 4 & 2 \end{bmatrix}.$$

The nodes $L_i$ have different priorities, from top to bottom. The higher nodes in the tree have higher priorities, and the priority matrix implements it with a bit at the position of the priority, e.g., $L_0$ has the highest priority and $L_5$ has the lowest.

To implement the sample classes for $x$ with $L_i(x) > 0$, we use the matrix:

$$\texttt{\_weight\_r} = \begin{bmatrix} 0 & 0 & 0 & 0 & 0 & 0 & 0 & 0 \\ 1 & 1 & 1 & 0 & 1 & 0 & 0 & 1 \\ 1 & 1 & 1 & 1 & 1 & 1 & 1 & 0 \end{bmatrix}$$

Each column of the matrix represents whether or not the right child of the corresponding node contains sample points from various classes.

Similarly, for $L_i(x) \leq 0$, we have

$$\texttt{\_weight\_l} = \begin{bmatrix} 1 & 0 & 0 & 0 & 0 & 0 & 0 & 0 \\ 0 & 1 & 1 & 1 & 0 & 1 & 1 & 0 \\ 0 & 1 & 1 & 0 & 1 & 0 & 1 & 1 \end{bmatrix}$$

Finally, updating the weight matrices with priority, we get

$$\texttt{weight\_r} = \texttt{\_weight\_r} * \texttt{priority},$$

$$\texttt{weight\_l} = \texttt{\_weight\_l} * \texttt{priority},$$

where the asterial stands for scaler multiplication along each column of the weights.

Having created the weight matrices to encode the classification information, we now need to find out how to sum up the credits for classification.

We build the following **tree matrix** for tracing down the decision tree.

**Definition 1.** *Let $T$ be a decision tree with $n+1$ leaves. The corresponding tree matrix is an $(n+1) \times n$ matrix, with $1$, $-1$, and $0$ as its entries. For a fixed leaf and a non-leaf node of the tree, either the leaf is on the left branch, on the right branch, or cannot be reached from the node. We mark the corresponding entry of the matrix by $-1$, $1$, or $0$, respectively. The matrix is unique up to the order of the rows and columns.*

As an example, the following matrix correspond to the decision tree shown in Fig. 1:

$$\texttt{tree} = \begin{bmatrix} -1 & 0 & 0 & 0 & 0 & 0 & 0 & 0 \\ 1 & -1 & -1 & -1 & 0 & 0 & 0 & 0 \\ 1 & -1 & -1 & 1 & 0 & 0 & 0 & 0 \\ 1 & -1 & 1 & 0 & -1 & 0 & 0 & 0 \\ 1 & -1 & 1 & 0 & 1 & -1 & 0 & 0 \\ 1 & -1 & 1 & 0 & 1 & 1 & 0 & 0 \\ 1 & 1 & 0 & 0 & 0 & 0 & -1 & -1 \\ 1 & 1 & 0 & 0 & 0 & 0 & -1 & 1 \\ 1 & 1 & 0 & 0 & 0 & 0 & 1 & 0 \end{bmatrix}.$$

There are different views of the tree matrix. We can see that the rows and columns of the matrix correspond to the leaves and non-leaf nodes of the tree, respectively. Indeed, each row of the matrix describes the trace to get from the root to the corresponding leaf, with $-1$ indicating left and $1$ indicating right at each node. The entry being $0$ indicates that the leaf cannot be reached from the node. Similarly, each column of the matrix marks whether a leaf is on its left or right, or cannot be reached.

We setup the following proposition and theorems that describe the process to recover the decision tree from the matrix.

**Proposition 1.** *The matrix corresponding to a decision tree has the following properties:*

1. *Since the decision tree is a strict binary tree, for each column of the matrix, there is at least one $-1$ and one $1$ in the column.*
2. *For each row of the matrix, there is a non-zero entry that is the unique $1$ or $-1$ of its column. The column corresponds to the parent of the leaf represented by the row.*

**Theorem 1.** *With a choice of the ordering of rows and columns, the tree matrix can be decomposed into a block matrix*

$$\begin{bmatrix} -1 & L & 0 \\ 1 & 0 & R \end{bmatrix}.$$

*The first column block is the unique column of the tree matrix consisting of non-zero entries, corresponding to the root of the tree. The blocks L and R correspond to the left and right subtree of the root, respectively. We can build the tree from the matrix by finding the root of L and R recursively.*

**Theorem 2.** *There is at least one column of the matrix with all zeros, except for one* $-1$ *and one* $1$*. Such a column* $C$ *corresponds to a node whose children are both leaves. The two rows with the non-zero entries of* $C$ *are indentical except for* $C$*. Removing the column* $C$ *and collapsing the indentical rows produces a new matrix. This process corresponds to collapsing the leaves of* $C$ *in the decision tree, and the result gives a new decision tree.*

## 2.2 Decision Tree as Neural Network

In the previous section, we have created the condition matrix `dt`, weight matrices `weight_l` and `weight_r`, and tree matrix `tree` to encode all the information we have from the decision tree. Now we show how to piece all the information together in a neural network.

Recall that the two weight matrices `weight_l` and `weight_r` are of size $m \times n$ and store classification and priority information. The tree matrix is of size $(n+1) \times n$ and implements the structure of the tree. In our example, we have $m = 8$ and $n = 3$.

The key to combine the information together is to build the matrix `dt2` as presented in the code below. In this paper, we present the code in Python style and we assume that the reader is familiar with the Python Numpy package [4].

```
1  weight_l = np.expand_dims(weight_l, 0)
2  weight_r = np.expand_dims(weight_r, 0)
3
4  dt2 = np.expand_dims(tree, 1)
5  weight_l = dt2 * weight_l
6  weight_r = dt2 * weight_r
7  dt2 = np.where(dt2 < 0, weight_l, weight_r)
```

The result is a matrix `dt2` of shape $(n+1) \times m \times n$, which has three axes. The first axis has dimension $n+1$, corresponding the leaves of the decision tree.

For example, the first slice of `dt2` is an $8 \times 3$ matrix:

$$\begin{bmatrix} -16 & 0 & 0 & 0 & 0 & 0 & 0 & 0 \\ 0 & 0 & 0 & 0 & 0 & 0 & 0 & 0 \\ 0 & 0 & 0 & 0 & 0 & 0 & 0 & 0 \end{bmatrix}.$$

It has only one non-trivial column, since the corresponding leaf has only one ancestor, namely $L_0$. Since the leaf is a left child of $L_0$, we choose the column from `weight_l`, with the minus sign indicated by the tree matrix. Similarly, the reader can check that the second slice of `dt2` is

$$\begin{bmatrix} 0 & 0 & 0 & 0 & 0 & 0 & 0 & 0 \\ 16 & -8 & -4 & -2 & 0 & 0 & 0 & 0 \\ 16 & -8 & -4 & 0 & 0 & 0 & 0 & 0 \end{bmatrix},$$

being a right descendant of $L_0$ and a left descendant of $L_1$, $L_2$, and $L_3$.

The input data multiplied with the condition matrix gives the result whether the data should be classified to the left or right child of the corresponding node, indicated by the sign of the result. Taking the signs of the result and multiplying them with each slice of dt2 evaluates the data on the corresponding leaf, i.e., if a sample point were to track along the tree and be classified to the leaf, the credit would be computed. The credits then form a prediction matrix of shape $9 \times 3$. The first axis represents credits on each leaf and the second axis represents credits for each class. The argument with maximum value predicts the class of a data sample and which leaf the sample belongs to.

In conclusion, the prediction process can be implemented by the following code:

```
1  h0 = np.matmul(data, dt[:, :4].T) + dt[:, 4].T
2  h0 = np.maximum(np.sign(h0), 0) - 1/2
3  pred = np.matmul(dt2, h0.T)
4  nn_pred = np.argmax(np.max(pred, 0), 0)
```

We can see that the highest credit is achieved along the path to the correct leaf for classification. Indeed, along the wrong directions, the data point will either receive negative or zero credits. If we get zero credits, the data point will no longer receive any more credits from the sub-tree, or, if we get some negative credits, the negative credits on each node have higher priorities than the credits from its sub-tree and we will end up with a negative credit. It follows that the correct path gets the highest credit and we have proved that the neural network reproduces the consequences of the decision tree.

## 3 Algorithm

In the previous section, we have actually proved that we can construct a neural network to reproduce the consequences of the decision tree. We summarize the algorithm in the following theorem. We will also provide the code to construct the tree matrix.

**Theorem 3.** *Given a decision tree classifier, we can construct a neural network that reproduces the consequences of the decision tree. Assume that the data has $l$ features and $m$ classes, and that the decision tree has $n$ non-leaf nodes and $n+1$ leaves. The neural network has two layers, with*

1. *the condition matrix of size $l \times n$ that implements the linear conditions in the decision tree. The columns of the condition matrix correspond to the linear conditions at each non-leaf node of the tree.*
2. *a matrix of shape $(n+1) \times m \times n$ that contains the priority and classification information of the tree. The axes correspond to leaves, classes, and non-leaf nodes of the tree. Fixing a leaf $L$ and a non-leaf node $N$, the $m$-vector indicates the priority and the classes of a sub-node $S$ of $N$. The sub-node $S$ is chosen to be the left or right child of $N$ depending on whether $L$ is to the left or right*

*of $N$, and the $m$-vector has a sign correspondingly. If $L$ is not a descendant of $N$, then the $m$-vector will be zero.*

3. *the hidden layer is computed as the sign of the conditions at each node, and the prediction is given as the highest credit along the axes of leaves and classes.*

In Sect. 2, we have shown how to create the matrices of the neural-network from the weight, priority, and tree matrices. The scikit-learn package [] stores the tree structure as two arrays of left and right children. In the rest of this section, we will show how the priority and tree matrices can be computed, as presented in the following code.

```
1  l = np.vstack([range(n_node), self.tree.children_left])
2  r = np.vstack([range(n_node), self.tree.children_right])
3
4  for _ in range(n_node):
5      if not (_ in self.tree.children_left or
6              _ in self.tree.children_right):
7          root = _
8          break
9
10 a = coo_matrix(([1]*self.n, l[:, ind_non_leaf]),
11                shape = (n_node, n_node))
12 b = coo_matrix(([1]*self.n, r[:, ind_non_leaf]),
13                shape = (n_node, n_node))
14
15 z_bar = (b - a).todense()
16 z = (b + a)
17
18 _priority = np.linalg.inv(np.eye(n_node) - z/2)
19 priority = np.array(_priority[root, ind_non_leaf])
20
21 tree_matrix = np.linalg.inv(np.eye(n_node) - z)
22 tree_matrix = np.matmul(z_bar, tree_matrix)
23 tree_matrix = tree_matrix[ind_non_leaf][:, ind_leaf]
```

In the code above, `n` is number of all nodes of the tree. `ind_non_leaf` and `ind_leaf` are indices of the non-leaf nodes and the leaves, respectively. In our case `n` is 17 and `n_node` is 8.

**Lemma 1.** *The matrix* `z` *constructed in the above code is a sparse matrix indicating whether a pair of nodes are parent and child. It follows that the matrix* `z^2` *indicates whether a pair of nodes are grandparent and grandchild. The same statement holds for* `z^n` *with $n \in \mathbb{N}$, and* `z^n` *is zero for $n$ large.*

Recall that the tree matrix `tree` encodes information whether a pair of nodes are ancestor and descendant. Hence we can partially recover the tree matrix from `z`, up to sign.

**Proposition 2.** *Up to sign, the tree matrix is the sum of powers of* `z`

$$\mathtt{z} + \mathtt{z}^2 + \mathtt{z}^3 + \cdots .$$

*Since* `z` *is nilpotent, we can compute that*

$$\mathtt{z}^0 + \mathtt{z}^1 + \mathtt{z}^2 + \cdots . = \mathtt{(I + z)}^{-1},$$

*where* `I` *is the identity matrix.*

Similarly, for the matrix `z_bar`, we have:

**Proposition 3.** *The matrix* $\mathtt{(I + z\_bar)}^{-1}$ *differs from the matrix* $\mathtt{(I + z)}^{-1}$ *by signs. The signs in* $\mathtt{(I + z\_bar)}^{-1}$ *indicate whether an odd or even times of switching directions is needed to get from one node to another.*

The key point to the proof of these results is that the non-zeros elements of the sparse matrices $\mathtt{z}^n$ do not overlap with each other. With this in mind, we can actually recover the tree matrix and the priority matrix.

**Theorem 4.** *The tree matrix can be computed as slices of the matrix product*

$$\mathtt{z\_bar} * \mathtt{(I + z)}^{-1},$$

*and the priority matrix can be computed as slices of the matrix*

$$\mathtt{(I + z/2)}^{-1},$$

*as shown in the code above.*

## 4 Comments and Conclusion

In this paper, we have presented a way to reproduce the consequence of a decision tree with a neural network. The idea is to provide some understanding of the neural network structure through the comparison between neural networks and decision trees.

Given a decision tree, we implement the corresponding neural network with one hidden layer. While the hidden layer is a fully connected layer implementing the linear conditions of the decision tree, the prediction layer is a three-axis layer, with the extra axis mimicking the reward gains on different leaves of tree.

Before we conclude the paper, we provide several comments on the relation and application of our work to the study of neural networks.

### 4.1 Comments

**The Extra Axis and Multiple Voters.** In our neural network, we needed an extra axis in the prediction layer to mimic the reward gains on different leaves of decision tree. In a neural network, we do not know which leaf is the right stop for the input sample, hence we had to compute the results on all the leaves together. This gave rise to an extra axis in the prediction layer, making it different from the usual fully connected layers. However, the extra axis can be very useful. We can think of the extra axis as multiple voters, e.g., the mnist classifier as a convolutional neural network committee [2]. In models of localisation such as "yolo", the extra axis provides local information of the image for detection jobs.

**Width of the Neural Network.** In [3], the authors tried to give some criteria for reducing the width of neural networks. In the work of transforming from decision trees to neural networks, we know that the width of the neural network corresponds to the number of nodes of decision tree. It follows that reducing the width of the network corresponds to pruning the decision tree. It will be interesting see how the pruning of the tree can be implemented as part of the structure of the neural network.

**Training the Neural Network.** Given a decision tree and the corresponding neural network, we can replace the sign function with some differentiable activation function and train the neural network.

In order to make the classification result visible, we carried out the experiment on the 0-th and 2-nd dimension of the Iris dataset. The picture shows the classification results of the original decision tree, the decision neural network, and the decision neural network after training. The reader can see from the figure that the decision tree and the constructed neural network have the same classification boundaries. The trained neural network, however, with the sigmoid function replacing linear rectification no longer has a piecewise linear classification boundary (Fig. 2).

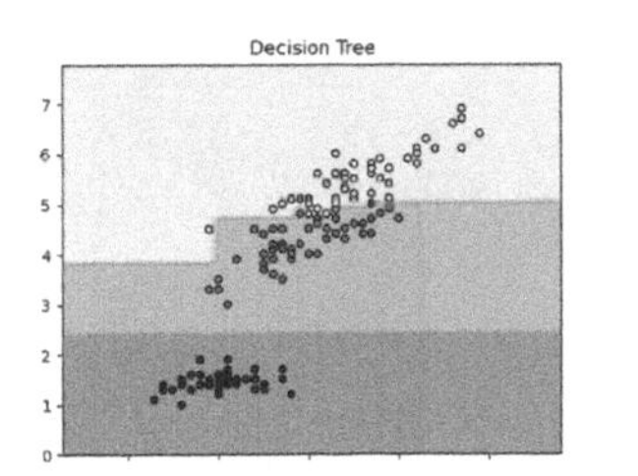

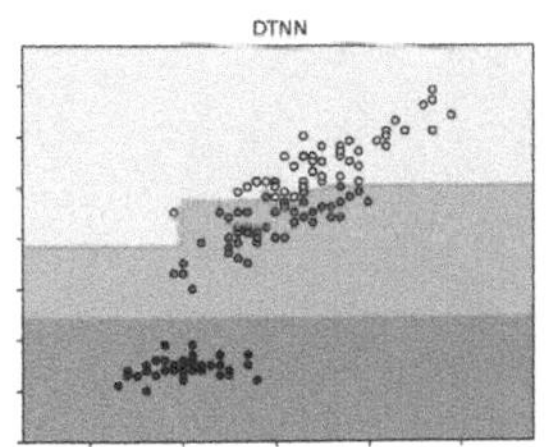

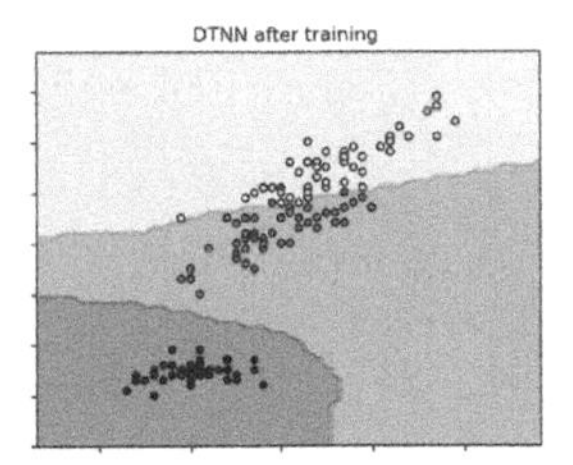


**Fig. 2.** Decision boundary of decision tree, decision tree neural network, and decision tree neural network after training

Note that the parameters of the decision tree neural network is sparse by construction. However, the training process always tend to adjust the coefficients of the network. In particular, the parameters of the neural network fail to be sparse after the training process. It will be great help in the understanding of networks to know which role the loss function, such as the entropy loss, plays in the correspondence between neural networks and decision trees. In the cases when the decision tree or the neural network is overfitting, the regularisation term in the loss function for training can also be interesting.

### 4.2 Conclusion

In conclusion, we have presented a way to reproduce the consequence of a decision tree with a neural network. But this is just the beginning of the effort to better understanding the structure of neural networks. We want more understanding of the loss functions in the network and which roles they play in the correspondence between neural networks and decision trees. Some effort has been shown in Sect. 3 trying to characterize the tree matrix. In future work, we wish to see more criteria of the properties of the matrices, thus providing interpretability to the neural networks.

## References

1. Buitinck, L., Louppe, G., Blondel, M., Pedregosa, F., Mueller, A., Grisel, O., et al.: API design for machine learning software: experiences from the scikit-learn project. In: ECML PKDD Workshop: Languages for Data Mining and Machine Learning, pp. 108–122 (2013)
2. Ciresan, D.C., Meier, U., Gambardella, L.M., Schmidhuber, J.: Convolutional neural network committees for handwritten character classification. In: Proceedings of the 2011 International Conference on Document Analysis and Recognition, pp. 1135–1139 (2011)
3. Guo, X., Yang, J., Wang, G., Chen, K., Yang, L.: Reducing the width of a feed-forward neural network. In: 2021 International Conference on Intelligent Computing for Sustainable Energy and Environment (2021, to appear)
4. Harris, C.R., et al.: Array programming with NumPy. Nature **585**, 357–362 (2000)

# Analysis and Prediction of Business and Financial Data Based on LSTM Neural Network

Yong Ji, Mingze Song, Jin Wen, Xiaoting Zhao(✉), and Yufeng Wang

State Grid Liaoning Integrated Energy Service Co., Ltd., ShenYang, China

**Abstract.** In the context of today's informatization and big data, the traditional financial management model of power grid projects has fallen behind. The financial department is unable to calculate and control key indicators in the process of engineering projects, but can only calculate the results and solidify assets. In response to such problems, in order to advance the role of financial to important nodes in the process of projects, this paper proposes a method for analyzing and predicting business financial data based on LSTM neural network, which uses multi-source heterogeneous data at the current stage of the project to affect future costs Forecast with income, and the financial department calculates indicators based on the forecast results. When the indicators are poor, the project decision of the current stage can be adjusted to realize the closed-loop monitoring of the business process. The results show that the method is accurate and effective.

**Keywords:** LSTM neural network · Business and financial data · Prediction

## 1 Introduction

With economic development, the scale of development of power companies is expanding, the national grid system framework has been put into construction on a large scale, and large amounts of funds have been invested in power grid projects. In the era of Informa ionization and big data, the drawbacks of the original engineering financial management model are gradually exposed [1]. At this stage, the financial management of power grid projects is within a regional framework. It is necessary to combine regional development trends to predict revenue and overall costs, strengthen comprehensive budget control, and accurately control project costs and progress data, and adjust deviations in a timely and rapid manner. Process-based management and control mode, rather than the traditional statistical accounting of results and asset consolidation. The financial management of the industry-financial integration project is a process management that is accurate to each step [2]. The significance of project financial management is not to count the final cost of the project, but to integrate into the process control of the project. During the implementation of the business, starting from the financial process indicators, the track of the project budget is not deviated, and there is "correction" of the business. The shortcomings of the project, the sequence process of the project is turned into a closed-loop control system, so that while controlling the cost of the project, the project progress


Q. Han et al. (Eds.): LSMS 2021/ICSEE 2021, CCIS 1469, pp. 337–348, 2021.
https://doi.org/10.1007/978-981-16-7213-2_33

can be reversely managed to achieve the expected effect of the project, and the project management and financial management can be perfectly integrated [3, 4].

Literature [5] mainly elaborates that power grid companies carry out dynamic forecasts of project funding needs, and accurately predict the amount and time of project payment according to the scale of project investment and construction progress, and form the results of long-term, medium- and short-term project funding demand forecasts, and improve financing. arrangement. Literature [6] analyzed that in the context of industry-finance integration, power grid companies actively constructing a new budget management model is the key path to optimize the effect of budget management. Based on this, first analyze the characteristics of the integrated budget management of industry and finance, and discuss the current situation and predicaments of the budget management of power grid enterprises. Literature [7] uses the system dynamics analysis method to predict the financial data of the grid company. First, the overall structure of the financial forecasting system is analyzed, and then combined with the business characteristics of the grid company, each subsystem is analyzed separately, and the grid company's financial data forecast model is established. Literature [8] predicts the purchase and sale of electricity indicators. The purchase and sale of electricity indicators is an important indicator that affects the economic benefits of power grid enterprises. Scientific forecasts of electricity purchase and sale, income, cash flow, etc. are important to ensure the accuracy of the financial budget of power grid enterprises. premise.2 Analysis and prediction of business and financial data based on LSTM.

## 1.1 Model Design

1) Network selection

Long short-term memory network (LSTM) is a new type of network developed on the basis of RNN, which can store and access historical sequence information better than ordinary RNN, thereby effectively solving the phenomenon of gradient disappearance and explosion [10, 11]. Under the condition that the time feedback mechanism is basically unchanged, the memory unit and the gate mechanism are introduced to realize the storage and control of the information flow [12]. This mechanism guarantees the long-term screening and retention of information in its special neuron structure, thereby effectively avoiding the problem of information loss caused by lateral depth. In addition, the gating mechanism of this memory unit also relieves the attenuation of the gradient when the gradient drops, and solves the problem of gradient disappearance during training to a certain extent. In summary, this article chooses LSTM network to implement the prediction model.

2) Network structure

LSTM, the long and short-term memory unit, is an improved recurrent neural network [13]. The hidden layer has a cyclic chain structure. In a standard cyclic neural network, as shown in the figure, the hidden layer is very simple. For example, there is only one tanh function in Fig. 1.

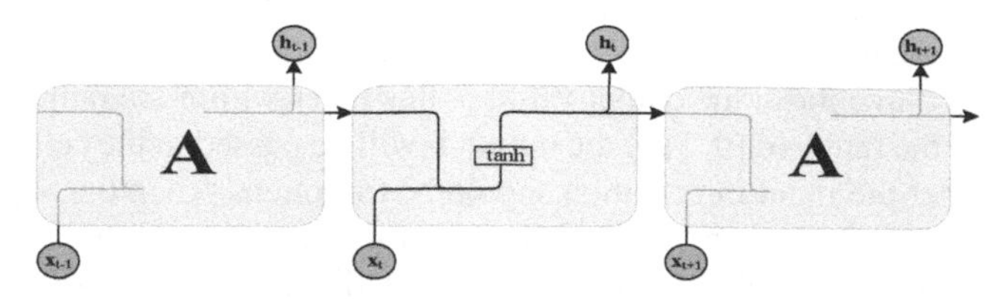

**Fig. 1.** Recurrent module structure of recurrent neural network

Similar to RNN, LSTM also has a cyclic module structure. The difference is that its internal structure is more complex, as shown in Fig. 2. As can be seen from the figure, different from the RNN's internal structure, which has only one tanh layer, LSTM has four interactive layers.

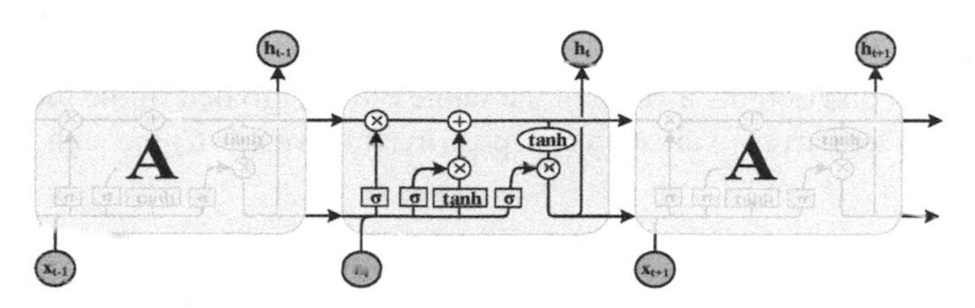

**Fig. 2.** LSTM's loop module structure

The key point of LSTM is that cell states are added to the hidden layer, and each neuron in the hidden layer contains four interactive layers to protect and control the cell state. They are the forget gate layer, the input gate layer, the update gate layer and the output gate layer.

(1) Forgotten door layer

This layer determines what information we discard from the cell state. As shown in Fig. 3, the figure describes the specific structure of the door layer. It can be seen in Fig. 3 that the input data of the forget gate layer is $x_t$ and $h_{t-1}$, that is, the input data at the

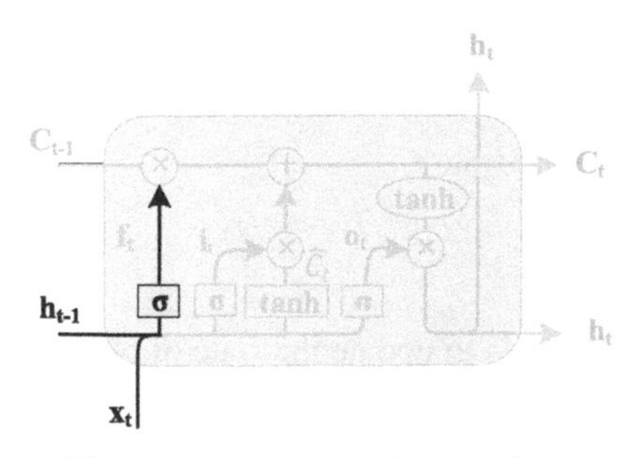

**Fig. 3.** LSTM forgets the door layer

current time and the output data of the hidden layer at the previous time. Then a sigmoid function $\sigma$ is used to control the output value of this layer within the range of [0, 1].

The value in the range of [0, 1] of the output $f_t$ will be passed to the cell state $C_{t-1}$ to determine whether the input information should be forgotten. When the output value is 1, it means that the information will be fully retained; when it is 0, on the contrary, the information will be completely forgotten. The calculation formula of this layer is shown in 1, where $W_f$ is the parameter that the LSTM network model needs to learn, and $b_f$ is the bias term of the forget gate.

$$f_t = \sigma\left(W_f * \left[h_{t-1}, x_t\right] + b_f\right). \tag{1}$$

(2) Enter the door layer

The structure of this layer is shown in Fig. 4, which consists of two layers. The first part is similar to the forget gate layer. It also processes the input $h_{t-1}$ and $x_t$ through the sigmoid function and outputs a value in the range of [0, 1] to determine the input value at a time. The other part is to create a new post-location vector $\tilde{C}_t$ through *tanh*.

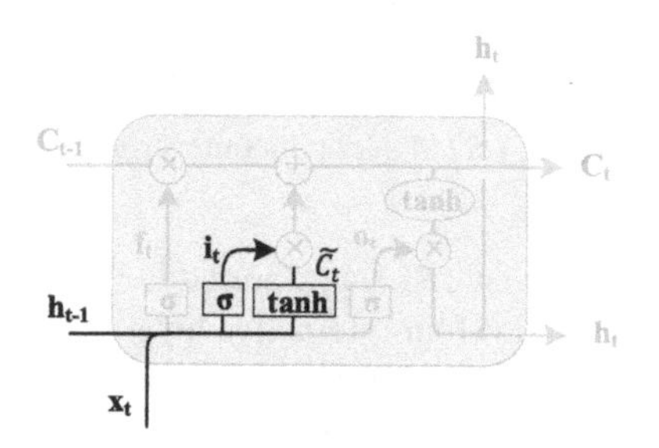


**Fig. 4.** LSTM input gate layer

The formulas 2 and 3 of this process are shown, where $b_i$ is the bias term of the input gate layer, and $b_c$ is the bias term of the post-address vector.

$$i_t = \sigma\left(W_t * \left[h_{t-1}, x_t\right] + b_i\right). \tag{2}$$

$$\tilde{C}_t = tanh\left(W_c * \left[h_{t-1}, x_t\right] + b_c\right). \tag{3}$$

$\tilde{C}_t$ will then be added to the state $C_t$.

(3) Update the door layer

The main function of this layer is to update the state of old cells. About to update $C_{t-1}$ to $C_t$, as shown in Fig. 5.

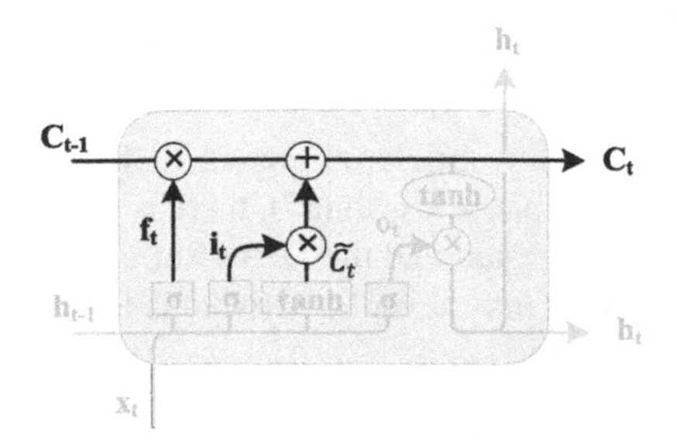


**Fig. 5.** LSTM update gate layer

The calculation formula for this process is shown in 4:

$$C_t = f_t * C_{t-1} + i_t * \tilde{C}_t. \tag{4}$$

(4) Output gate layer

As shown in the structure of the LSTM output gate layer in Fig. 6, this layer is also composed of two parts. Part of it is the same as the forget gate layer and the input gate layer, and the information that needs to be output is determined according to the input of this time and the output value of the previous loop body. Then multiply the output $o_t$ of the first part with the cell state $C_t$ filtered by the *tanh* layer to obtain the output value $h_t$.

The calculation formula of this process is shown in 5 and 6, $b_o$ is the offset term of the output $o_t$ of the first part.

$$o_t = \sigma\left(W_o * \left[h_{t-1}, x_t\right] + b_o\right). \tag{5}$$

$$h_t = o_t * tanh(C_t). \tag{6}$$

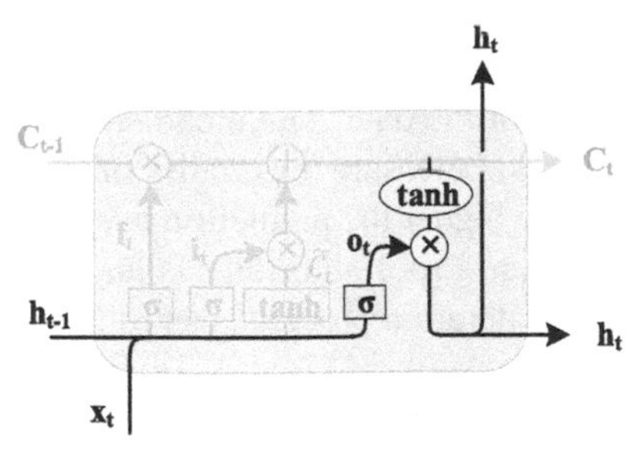


**Fig. 6.** LSTM output gate layer

3) Select activation function

Combining the background of this article, choosing RELU as the activation function, its advantage is that when the input is less than 0, there is no problem of gradient disappearance, the learning speed is faster, and the Dead ReLu Problem can be solved at the same time, that is, some neurons may never be activated., Resulting in the corresponding parameters can never be updated.

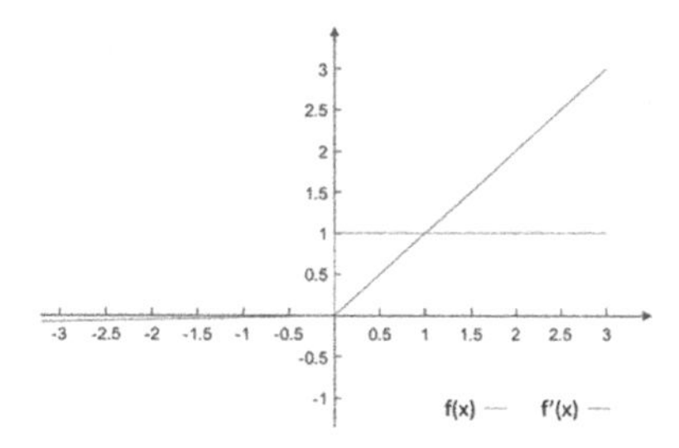


**Fig. 7.** ReLu activation function

4) Loss function and optimization method

The problem in this thesis is a predictive regression problem. The output, which is different from the classification problem, is a discrete variable. The predictive model is a continuous variable, which is one or more specific values, such as housing price prediction, sales volume prediction, etc. Combining the background of the problem and its characteristics, the loss function chosen in this thesis is the mean square error loss function (MSE). The calculation method is shown in 7:

$$MSE(y, y') = \frac{\sum_{i=1}^{n}\left(y_i - y_i'\right)^2}{n}. \tag{7}$$

Where $y_i$ is the output of the neural network, $y_i'$ is the label value of the training data, and $i$ represents the dimension of the data. The mean square error between the output value and the label value is used to measure the quality of the output value. The smaller the mean square error value, the higher the prediction accuracy.

RMSprop (Root Mean Square Prop) algorithm is an improved stochastic gradient descent algorithm. The specific iteration process is as follows:

- First calculate the size of the gradient $g$, where $L(f(x_i;\theta), y_i)$ is the loss function, $\nabla_\theta$ represents the derivative of the weight $\theta$, and $m$ is the number of samples for small batch training. The formula is shown in 8.

$$g = \frac{1}{m}\nabla_\theta \sum_i L(f(x_i;\theta), y_i). \tag{8}$$

- The last gradient cumulant $r$ is attenuated by the attenuation coefficient $\rho$, and the new cumulant is calculated with the gradient $g$ to update $r$. The formula is shown in 9.

$$r = \rho r + (1 - \rho) g \odot g. \tag{9}$$

- The calculated weight is updated to $\Delta\theta$, where $\eta$ is the learning set, and $\delta$ is a constant usually taken as $10^{-6}$, which is used to stabilize the value when divided by a decimal. The formula is shown in 10.

$$\Delta\theta = -\frac{\eta}{\sqrt{\delta + r}} \odot g. \tag{10}$$

- Use $\Delta\theta$ to update the value of weight $\theta$. The specific formula is shown in 11.

$$\theta = \theta + \Delta\theta \tag{11}$$

5) Dropout technology

Dropout technology, as a technology to prevent over-fitting, can greatly reduce the error rate of over-fitting models, improve its generalization ability, and improve the overall effect of the model. Dropout is different from the general regularization method. It does not start from the cost function of the model, but achieves the effect of preventing overfitting by changing the method of the model itself. The working principle of Dropout is shown in Fig. 8. It randomly disconnects the information transmission between certain neurons in a certain proportion during the model training process, so that the weight values corresponding to some neurons will not be updated in this training. While the Dropout method acts as a certain regularization function, it also achieves the effect of simplifying network training and improving a certain training speed.

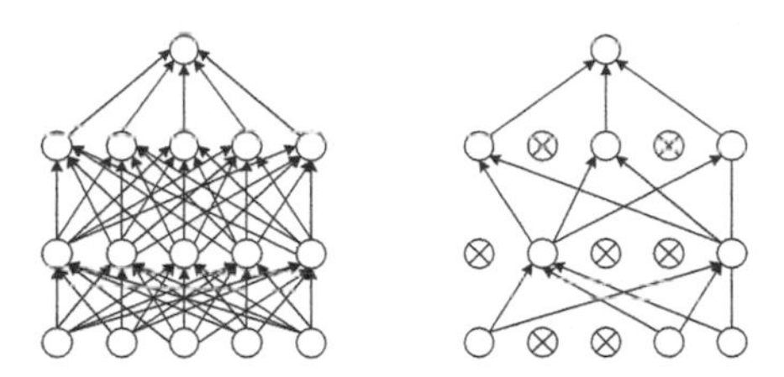

**Fig. 8.** Working principal diagram of Dropout

6) Xavier initialization method

The weight initialization of the neural network can improve the training process of the model. If the initial value of the weight is too small, it will be difficult for the data information to be transmitted in the network, and if the weight is too large, the gradient cannot be effectively backpropagated. The Xavier initialization weight algorithm initializes the weights randomly to a relatively good distribution range, which is beneficial to speed up the convergence of the deep neural network and optimize the network weights in a more correct direction.

The initial weight range and distribution are mainly to solve the problem of the activation function derivation during gradient descent. The activation function mainly includes Sigmoid and *tanh*, etc. The function and derivative form are shown in 12 and 13:

$$y = sigmoid(x) = \frac{1}{1+e^{-x}} \Rightarrow sigmoid'(x) = y(1-y). \tag{12}$$

It can be seen that the gradient of these two functions near the origin of the independent variable is larger, and the network weight update speed is faster at this time, and it will show better characteristics. When the independent variable is in the saturation region at both ends of the function, the corresponding gradient value will become too small, so as to reduce the efficiency and accuracy of optimization. Use the Xavier initialization method to limit the randomization interval to a visible range. This range depends on the number of nodes in the layer network and the type of activation function, as shown in the initialization formulas 13 and 14 for *tanh* and Sigmoid:

$$tanh : \theta \sim Uniform\left(-\frac{\sqrt{6}}{\sqrt{fan_{in}+fan_{out}}}, \frac{\sqrt{6}}{\sqrt{fan_{in}+fan_{out}}}\right). \tag{13}$$

$$sigmoid : \theta \sim Uniform\left(-4\cdot\frac{\sqrt{6}}{\sqrt{fan_{in}+fan_{out}}}, 4\cdot\frac{\sqrt{6}}{\sqrt{fan_{in}+fan_{out}}}\right). \tag{14}$$

Among them, $fan_{in}$ and $fan_{out}$ correspond to the number of input nodes and output nodes in a certain layer of the network, and *Uniform* represents a uniformly distributed initialization method.

## 1.2 Model Implementation and Training

Model implementation and training the training process of the LSTM interval prediction model based on the construction of the loss function method is shown in the figure. The main steps are as follows, and the specific process is shown in Fig. 9:

- First, perform data preprocessing on the data set, which mainly includes input data, project time series progress, completed budget, loan interest, material cost, construction investment, insurance premium, labor cost, project scale, project location, construction period, and commissioning period, Depreciation period, electricity structure, load development trend, etc., to obtain a training set for the model;
- Randomly initialize the model weights through the Xavier initialization method;
- Input the training set for forward calculation of the model, and calculate the value of the loss function $f$;
- In the minimum batch training update mode, the loss function $f$ is back-propagated through the RMSprop gradient descent algorithm, and the corresponding gradient value is calculated;

- The weight is updated by the gradient value obtained by the loss function $f$;
- If the last iteration training of the sample is completed, the model training is stopped, and the model training is now complete.

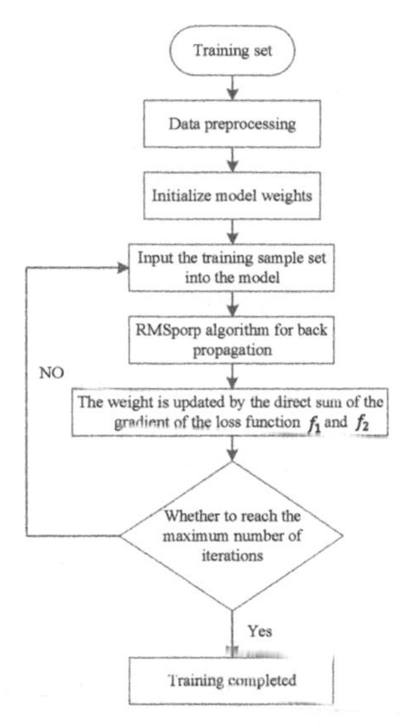


**Fig. 9.** LSTM interval prediction flowchart

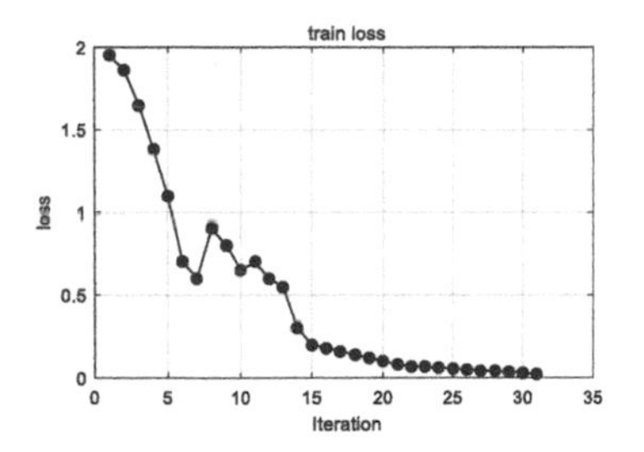


**Fig. 10.** LSTM training result graph

First prepare the multi-source heterogeneous data of the engineering project, then build the model, and finally train the model and test. The output of the neural network is the engineering cost and electricity sales revenue. According to the model audit above, the model training and testing are carried out. The picture shows the passed data After preprocessing and training through the LSTM neural network model, the test results on the test data set are shown in Figs. 2, 3, 4, 5, 6, 7, 8, 9 and 10. It can be seen that as the number of iterations increases, the model converges relatively quickly. When the iteration reaches the sixth round, the loss value has already fallen below 1. Although the loss value fluctuates slightly at the beginning of training, as the number of iterations increases during the training process, the final convergence is achieved, which is less than 0.05.

## 2 Case Analysis

L Province Electric Power Co., Ltd. was established in 1999 and is a wholly-owned subsidiary of the State Grid Corporation of China. Its core business is the construction and operation of power grids, and its power supply business area covers the whole province. The company's headquarters has 21 departments, 34 subordinate units, and 65,500 full-caliber employees.

The data information between the upper and lower levels, between internal units, between business and financial in the L company system is relatively independent, and it cannot realize the interconnection and sharing and timely transmission of information. Various information including financial information is divided vertically and horizontally., Which restricts the deep-level processing and comprehensive utilization of information. Financial personnel spend a lot of energy on daily accounting, and the depth and breadth of data analysis and mining are limited, failing to establish a multi-dimensional and all-round financial data analysis system, and financial information has limited support for decision-making.

In this thesis, the current financial evaluation indicators are commonly used in power grid projects to demonstrate the analysis and forecasting methods of industry financial data based on LSTM proposed in this thesis.

At present, power grid engineering projects mainly adopt the following financial indicators in the NPV-based financial evaluation method: NPV (Net Present Value), IRR (Internal Rate of Return), PP (Payback Period), ROI (Return On Investment).

This article selects the completed engineering projects of L Province Electric Power Co., Ltd. as the samples for this analysis. The following two projects are listed for specific analysis, and all the data can be derived from the grid system.

Project A is a 110 kV power transmission and transformation project, and Project B is a 220 kV power transmission and transformation project. Using this article, the cost and income of projects A and B are predicted based on LSTM. The results are shown in Table 1.

**Table 1.** Project A, B projected cost and income.

| Item Number | Cost (ten thousand yuan) | Average income (ten thousand yuan) |
|---|---|---|
| A | 4232.80 | 5459.46 |
| B | 9763.25 | 16512.33 |

The results of project A and B forecast indicators are fed back to the financial department, and the financial department provides guidance on the project's engineering conditions, and uses financial indicators to comprehensively evaluate the effect of the

**Table 2.** Guide the financial indicators of the first two projects.

| Item number | NPV | IRR | PP | ROI |
|---|---|---|---|---|
| A | 1533.21 | 9.12% | 13.81 year | 7.32% |
| B | 3112.07 | 8.23% | 11.55 year | 11.22% |

**Table 3.** The financial indicators of the two projects after the guidance.

| Item number | NPV | IRR | PP | ROI |
|---|---|---|---|---|
| A | 1652.08 | 9.79% | 12.11 year | 8.56% |
| B | 3319.46 | 8.66% | 10.13 year | 11.79% |

guidance. Table 2 provides guidance on the financial indicators of the first two projects, and Table 3 provides guidance Financial indicators of the latter two projects.

From the comparison of the financial indicators of the two projects before and after the guidance in the table, we can see that using the data analysis method of this article, the real-time data in the project is used to predict the cost and revenue through LSTM during the implementation of the project. The financial department uses the forecast Indicators After the targeted knowing of the construction projects in progress, the financial indicators of the project have been improved, and the financial guidance business has been realized, which can make the project create greater benefits. The big data analysis results of business financial are used to analyze the results. To tap the potential of the project's income and make the financial work forward. The financial department is not only constrained on the settlement after the event, but also participates in the whole project by predicting the indicators in the whole project, and knows the project from the perspective of the financial profession. The progress of the project is conducive to the prevention and control of potential risks in engineering projects, and has further improved the company's benefits.

## 3 Conclusion

This thesis takes the multi-source heterogeneous data in power grid engineering projects as the research object, and proposes a method of analyzing and predicting business financial data based on LSTM neural network. After many comparative experiments, it is proved that the business financial data based on LSTM neural network is proposed in this thesis. The analysis and forecasting method can accurately predict the cost and profit of the engineering project. At the same time, the financial department will conduct a financial professional analysis on the forecast data. This article mainly evaluates the project situation through the forecast results on four key financial indicators. It enables financial to advance to the process of engineering projects, through the analysis of predictive indicators, accurately guides the engineering work beforehand, verifies the applicability and rationality of the method proposed in this thesis, and enables the financial The department further changed the sequence of the project into a closed-loop control system, so that while controlling the cost of the project, it reversely manages the progress of the project to achieve the expected effect of the project, so that the project management and financial management are perfectly integrated.

## References


1. Zhang, Y.Q.: Research on the operation mode of enterprise industry-financial integration under cloud computing—taking the State Grid as an example. J. Friends Acc. **24**, 58–60 (2018)
2. Zhang, C.P., Wang, X.C.: Research on the budget management mode of the integration of business and financial of power grid enterprises. J. East China Electric Power. **42**(12), 2728–2731 (2014)
3. Su, L.F.: Research on the budget management model of the integration of business and financial. J. Friends Acc. **12**, 66–68 (2012)
4. Xie, A., Lu, M.X., Liu, Y.Q.: Research on the construction and application of new enterprise management accounting information system from the perspective of industry and financial integration. J. Contemp. Acc. **03**, 12–13 (2019)
5. Chen, H.: Research on dynamic forecast of power grid project fund demand. J. Oper. Manag. **05**, 112–116 (2020)
6. Wei, N.: Analysis of the budget management model of the integration of industry and financial in power grid enterprises. J. Contemp. Acc. **20**, 62–63 (2019)
7. Wan, M.Y., Wang, C., Chen, J.Y., Zhang, J.C., Zhao, W.B.: Research on the financial forecast model of power grid companies based on system dynamics. J. Modern Econ. Inf. **19**, 153–154 (2016)
8. Zhou, Z.: Research on forecasting modeling of electricity purchase and sale of power grid companies based on financial data. J. Fin. Econ. (05), 238--239+242 (2016)
9. Lu, J., Zhang, Q., Yang, Z., Tu, M., Lu, J., Peng, H.: Short-term load forecasting method based on CNN-LSTM hybrid neural network model. J. Autom. Electr. Power Syst. **43**(08),131–137 (2019)
10. Dai, J.J., Song, H., Sheng, G.H., Jiang, X.C., Wang, J.Y., Chen, Y.F.: Research on the prediction method of power transformer operation state using LSTM network. J. High Voltage Technol. **44**(04), 1099–1106 (2018)
11. Peng, W., Wang, J.R., Yin, S.Q.: Attention-LSTM-based short-term load forecasting model in power market. J. Power Syst. Technol. **43**(05), 1745–1751 (2019)
12. Wang, Z.P., Zhao, B., Ji, W.J., Gao, X., Li, X.B.: Short-term load forecasting method based on GRU-NN model. J. Autom. Electr. Power Syst. **43**(05), 53–58 (2019)
13. Yuan, Y.F.: The application of LST M model in free cash flow forecasting in enterprise value evaluation. J. China Collective Econ. **28**, 85–86 (2018)

# Advanced Computational Methods and Applications

# A Compound Wind Speed Model Based on Signal Decomposition and LSSVM Optimized by QPSO

Sizhou Sun[1,2](✉), Jingqi Fu[2], and Lisheng Wei[1]

[1] School of Electrical Engineering, Anhui Polytechnic University, Wuhu 241000, China
[2] School of Mechatronics Engineering and Automation, Shanghai University, Shanghai 200072, China
jqfu@staff.shu.edu.cn

**Abstract.** A new compound model based on wavelet packet decomposition (WPD) and quantum particle swarm optimization algorithm (QPSO) tuning least squares support vector machine (LSSVM), namely WPD-QPSO-LSSVM, is developed in this study for forecasting short-term wind speed. In the developed model, WPD is firstly applied to preprocess the raw volatile wind speed data test samples to obtain relatively stable different components. Then, LSSVMs are utilized to predict short-term wind speed by these stable subseries components after the input variables are reconstructed by partial autocorrelation function (PACF), and the final short term wind speed forecasting results can be obtained by aggregation of each prediction of different components. In the end, the actual historical wind speed data are applied to evaluate the forecasting performance of the proposed WPD-QPSO-LSSVM model. Compared with the recent developed methods, the proposed compound WPD-QPSO-LSSVM approach can effectively improve the forecasting accuracy.

**Keywords:** Wavelet packet decomposition · Quantum particle swarm optimization algorithm · Least squares support vector machine · Wind speed forecasting

## 1 Introduction

With the development of industrial technology and computing technology, wind power has experienced rapid development over the past few decades. The global new installed wind turbine capacity reaches 60.4 GW and the overall installed wind power capacity approaches to approximately 651 GW in 2019, which can effectively reduce greenhouse gas emissions [1]. However, this rapidly growing wind power has yielded great negative effects on the operation dispatching, economic analysis and power system stability for complex nonlinear characteristics of wind speed. Wind power forecasting modeling is one of the simplest methods to overcome these problems [2].

During last few decades, numerous reliable models have been made for improving wind speed or wind power forecasting accuracy, which mainly includes physical methods, time series forecasting approaches, artificial intelligent methods and compound


Q. Han et al. (Eds.): LSMS 2021/ICSEE 2021, CCIS 1469, pp. 351–360, 2021.
https://doi.org/10.1007/978-981-16-7213-2_34

models [1, 3, 4]. Physical methods make wind power prediction using massive physical data of wind farm and numerical weather prediction model. High computational cost and low accuracy hinder physical methods in the practical real-time application. Statistical methods are skilled in capturing linear information within wind speed data.

To overcome the limitations of physical methods and statistical approaches, artificial intelligent based models and Extreme Learning Machine (ELM) [11], Echo state network (ESN) [5, 8], artificial neural networks (ANN) [2], wavelet neural network (WNN) [10], back-propagation neural network (BPNN) [9] and LSSVM [13], have been developed for their potential capacity in extracting nonlinear feature in the testing samples [5, 6]. For single individual model always suffer from over-fitting or under-fitting, high sensitivity to inappropriate initial parameters, and instability, compound models based on signal preprocess, parameter optimization and artificial intelligent approach have been developed to enhance the wind speed forecasting performance. Reference [5] proposed a novel hybrid wind speed prediction model, in which ESN is employed to make forecasting of each decomposed subseries after three important parameters of ESN are tuned by differential evolution (DE). In Ref. [9], fast ensemble empirical mode decomposition (FEEMD) was developed for preprocessing the empirical wind speed samples to obtain more reliable subseries, and improved BPNN was utilized to make forecasting of each decomposed subseries. Reference [2] developed a novel compound wind speed forecasting model combining ANN optimized by crisscross optimization algorithm with wavelet packet decomposition (WPD). Liu et al. [17] developed a compound model combining support vector machine (SVM) model tuned by genetic algorithm (GA) with WT to make short-term wind speed prediction. SVM is one of widely used machine learning methods which realizes its function based on structure risk minimization theory and it can do well in processing small sample and non-linearity signal with high dimension. However, SVM method is often time-consuming for its complex computation. To overcome this shortage, LSSVM is developed by solving linear equations substituting for a quadratic programming problem, thus, improving the computing performance.

Inspired by above literatures, a compound short-term wind speed forecasting model WPD-QPSO-LSSVM is proposed in this study. In the developed model, LSSVM method is adopted as the core forecasting engine to construct wind speed forecasting model. To improve the regression performance of LSSVM, WPD is utilized to decompose the empirical wind speed time series into relatively stable components, and quantum particle swarm optimization algorithm (QPSO) is employed to optimize the parameters in LSSVM for avoiding falling into local optimum.

The remains of this study are arranged as follows. Wind speed forecasting modeling is described in Sect. 2. Case study is presented in Sect. 3. Finally, the conclusions are obtained in Sect. 4.

# 2 Wind Speed Forecasting Modeling

## 2.1 Wind Speed Preprocess Method

WPD, well-known signal processing method, is developed by improvement of wavelet translate (WT). Compared with WT method, WPD breaks down not only the approximation coefficient components but also the detail coefficient components, therefore, WPD approach can obtain more stable subseries for forecasting engine, the process of signal decomposition at three level is displayed as follows (Fig. 1).

(a) Signal process by WT at three level

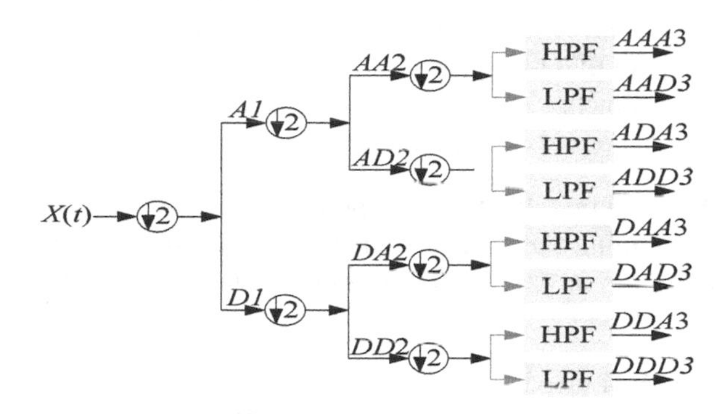


(b) Signal process by WPD at three level

**Fig. 1.** Signal decomposition by WT and WPD at three level

## 2.2 LSSVM Optimized by QPSO

LSSVM model is skilled in solving small sample and nonlinear signal with high dimension, however, the regression performance of LSSVM is affected by kernel function and kernel parameters. As radial basis function (RBF), express as Eq. (1), with less parameters has good local exploitation capacity in signal process at fast speed, it is adopted as kernel function for LSSVM.

$$f_{RBF}(x_i, x_j) = \exp(-\frac{||x_i - x_j||}{2\delta^2}). \tag{1}$$

Penalty factor $\gamma$ and kernel parameter $\delta$ are two important parameters in LSSVM model that influence the wind speed forecasting results, thus, they are tuned by artificial intelligent method QPSO to obtain optimal parameter combination for LSSVM. Mean absolute percent error (MAPE), expressed as Eq. (2), is used as the fitness function to evaluate the forecasting results.

$$MAPE = \frac{1}{N}\sum_{i=1}^{N}\frac{|s(i) - \hat{s}(i)|}{s(i)} \times 100\%. \tag{2}$$

Where $s(i)$ and $\hat{s}(i)$ mean are actual wind speed data and forecasting wind speed value, respectively.

### 2.3 The Working Mechanism of the Proposed WPD-QPSO-LSSVM Model

a. The Proposed WPD-QPSO-LSSVM Model

The historical wind speed data from a wind farm located in East of Anhui province are used to evaluate the proposed compound WPD-QPSO-LSSVM model, shown in Fig. 2, where $P_m$, $P_g$ and $S$ denote personal optimum, global optimum and mean optimal location center of population, respectively. The wind speed forecasting modeling process can be divided into following stage.

*Stage I:* Wind speed preprocessing. Apply WPD method to decompose the training wind speed data into a few components with different frequency. Prior to wind speed forecasting obtained by QPSO-LSSVM method, PACF is utilized to calculate time lag of each subseries for construction of the input candidate matrix. To lower learning difficulty of LSSVM, each decomposed component is translated linearly into [0, 1].

*Stage II:* LSSVM tuned by QPSO. Inappropriate parameters may cause LSSVM into over-fitting or under-fitting state when applied into wind speed prediction. This study utilizes QPSO algorithm to optimize the model parameters for improving generation ability.

*Stage III:* Train LSSVM tuned by QPSO with every decomposed component, the 1st–576th wind speed time series are employed as the training samples, the subsequent 576st–672th wind speed time series are utilized as the testing samples to evaluate the forecasting results.

*Stage IV:* Apply the well-trained QPSO-LSSVM model to carry out wind speed forecasting and utilize the statistical index RMSE, MAPE and MAE to analyze the forecasting results.

b. Evaluation Index for Wind Speed Forecasting

To evaluate the forecasting performance of the proposed WPD-QPSO-LSSVM model, three statistical indices, expressed as Eq. (3) and (4), are used to analyze the forecasting results.

$$RMSE = \sqrt{\frac{1}{N}\sum_{i=1}^{N}|s(i) - \hat{s}(i)|^2}. \quad (3)$$

$$MAE = \frac{1}{N}\sum_{i=1}^{N}|s(i) - \hat{s}(i)|. \quad (4)$$

Where RMSE and MAE denote root mean square error, mean absolute error and, respectively. $s(i)$ and $\hat{s}(i)$ mean are actual wind speed data and forecasting wind speed value, respectively.

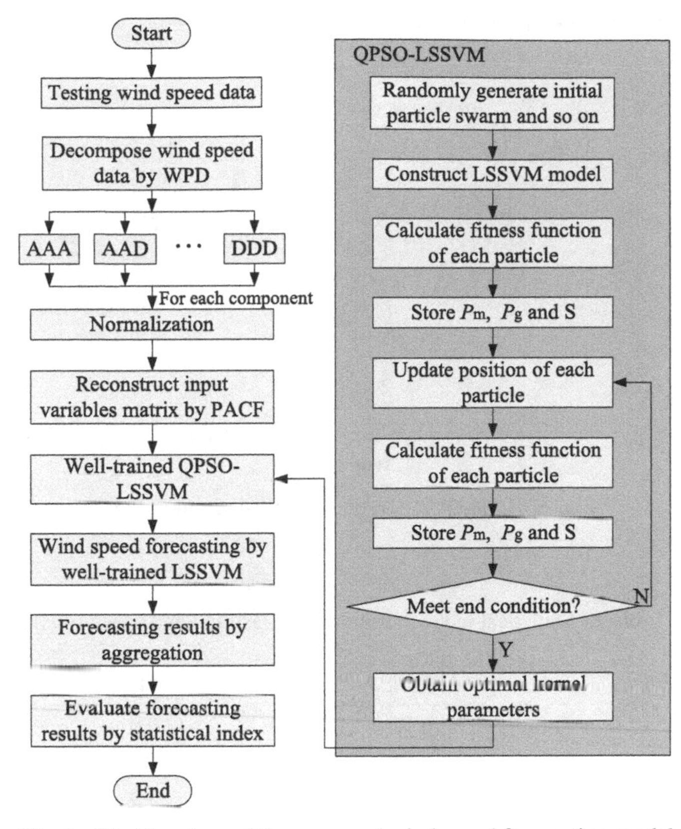


**Fig. 2.** Working flow of the proposed wind speed forecasting model

## 3 Case Study

### 3.1 Testing Wind Speed Time Series

In this study, 15-min interval wind speed time series collected from a wind farm of East Anhui are utilized to test the proposed WPD-QPSO-LSSVM model. The wind speed time series are shown as Fig. 3, and its statistical properties are listed in Table 1. From the figure and table, wind speed time series fluctuate highly.

### 3.2 Wind Speed Preprocessing

In the study, WPD approach is applied to decompose the training samples to eliminate the high volatility and obtain relatively stable subseries. The Daubechies of order4 (db4) is adopted as the mother wavelet function for WPD. The decomposed results obtained by WT and WPD are shown in Fig. 4 and Fig. 5, respectively.

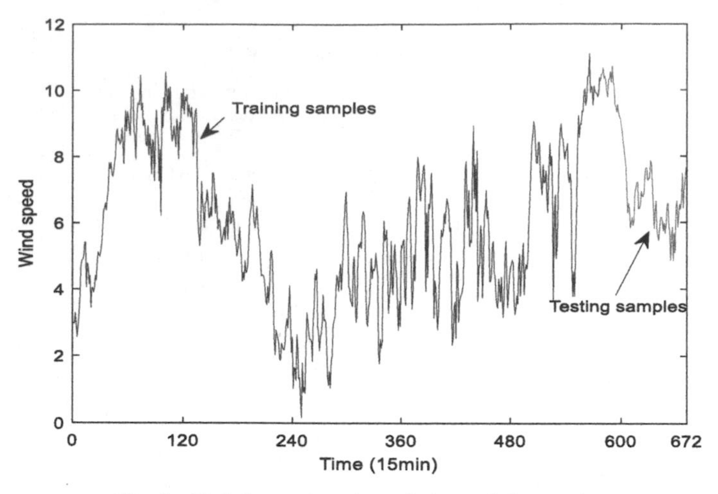


**Fig. 3.** Training and testing wind speed time series

**Table 1.** Statistical index of training and testing wind speed (m/s)

| Data set | Max. | Min. | Median | Mean | St. dev. |
|---|---|---|---|---|---|
| Total samples | 11.09 | 0.17 | 5.93 | 6.01 | 2.33 |
| Training samples | 11.09 | 0.17 | 5.47 | 5.76 | 2.34 |
| Testing samples | 10.72 | 4.85 | 6.92 | 7.44 | 1.66 |

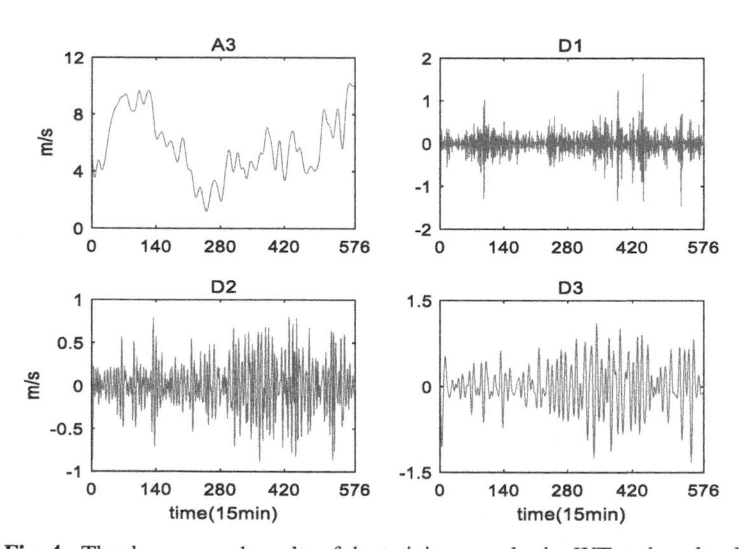


**Fig. 4.** The decomposed results of the training samples by WT at three level

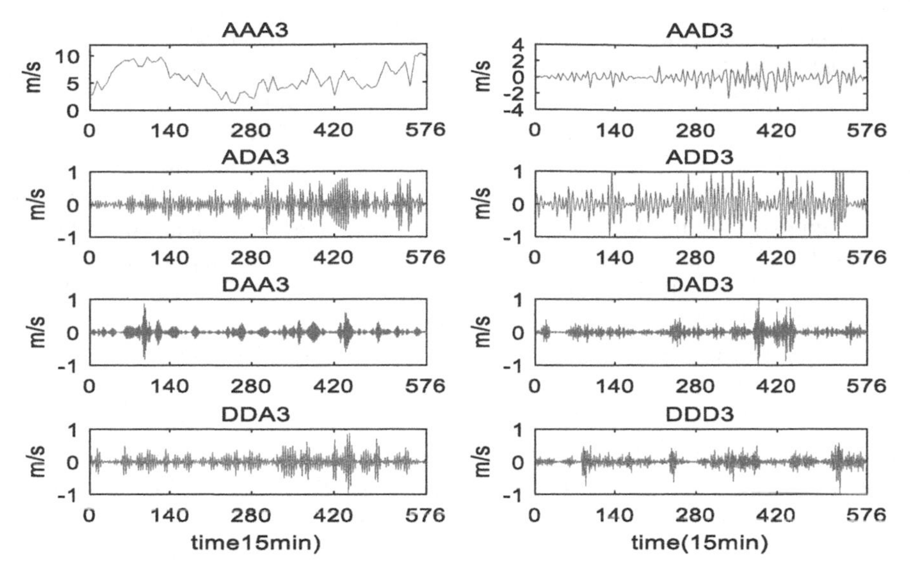


**Fig. 5.** The decomposed results of the training samples by WPD at three level

After decomposition, each subseries is linearly normalized into interval [0, 1] to reduce the forecasting difficulties. Prior to wind speed forecasting by LSSVM, the input variables matrixes are reconstructed according to the lag values of each decomposed variables determined by PACF, if lag value is 8, the reconstruction of input matrix for LSSVM by PACF is shown in Fig. 6.

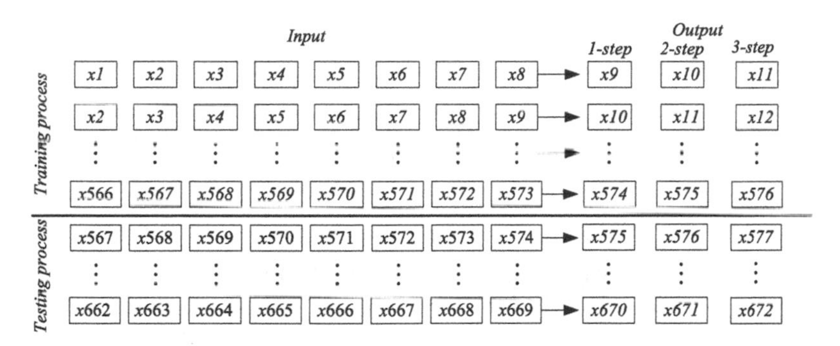


**Fig. 6.** Reconstruction of input matrix for LSSVM by PACF

### 3.3 Numerical Results

The first $1^{st}$–$576^{th}$ wind speed time series are used to train the QPSO-LSSVM model and the subsequent $577^{th}$–$672^{th}$ wind speed time series are applied to test the QPSO-LSSVM model. The testing process are divided into two parts: Experiment I and Experiment II. In

Experiment I, compare the developed WPD-QPSO-LSSVM model with single forecasting models including Persistence, ARMA, BPNN, WNN, and LSSVM. In Experiment II, compare the developed WPD-QPSO-LSSVM model with other hybrid forecasting models including QPSO-LSSVM, WT-GA-SVM [17], and EMD-ANN [18].

Statistical indices of the proposed and single forecasting models are shown in Table 2. Persistence is always used as benchmark model to evaluate the new developed forecasting approach. In the same way, Persistence is applied to compare with the new developed WPD-QPSO-LSSVM model. From Table 2, the one-step MAPE values of Persistence, ARMA, BPNN, WNN, LSSVM and the proposal are 13.13%, 11.03%, 9.03%, 8.46%, 7.88% and 4.75%, respectively. Compared with Persistence model, the forecasting performance of the proposal improves obviously. The proposal also outperforms single model ARMA, BPNN, WNN and LSSVM in not only one-step forecasting, but also two-step and three-step forecasting. The reasons of this comparisons are: single forecasting models utilize the original wind speed time series to make prediction, which shows highly nonlinear characteristics and improve forecasting difficulties. In addition, QPSO algorithm is applied to tune LSSVM which avoid over-fitting or under-fitting.

**Table 2.** Statistical indices of the proposed and single forecasting models

| Horizon | Index | Persistence | ARMA | BPNN | WNN | LSSVM | Proposal |
|---|---|---|---|---|---|---|---|
| 1-step | MAE (m/s) | 0.99 | 0.84 | 0.67 | 0.65 | 0.61 | 0.35 |
| | MAPE (%) | 13.13 | 11.03 | 9.03 | 8.46 | 7.88 | 4.75 |
| | RMSE (m/s) | 0.99 | 0.92 | 0.82 | 0.80 | 0.78 | 0.59 |
| 2-step | MAE (m/s) | 1.13 | 0.93 | 0.79 | 0.7 | 0.68 | 0.41 |
| | MAPE (%) | 15.09 | 12.82 | 10.46 | 9.69 | 9.11 | 5.56 |
| | RMSE (m/s) | 1.06 | 0.96 | 0.89 | 0.84 | 0.82 | 0.64 |
| 3-step | MAE (m/s) | 1.35 | 1.05 | 0.92 | 0.89 | 0.85 | 0.5 |
| | MAPE (%) | 18.23 | 14.24 | 12.31 | 11.82 | 11.11 | 6.79 |
| | RMSE (m/s) | 1.16 | 1.03 | 0.96 | 0.94 | 0.92 | 0.71 |

Statistical indices of the proposed and other hybrid forecasting models are shown in Table 3. WT-GA-SVM [17] and EMD-ANN [18] are recently developed hybrid wind speed forecasting models, which are employed to further evaluate the developed WPD-QPSO-LSSVM method. The parameters in WT-GA-SVM and EMD-ANN models are set according to Refs. [17] and [18], respectively. Compared with WT-GA-SVM and EMD-ANN models, the fitness function MAPE values of the proposal are cut by 0.63% and 0.88% in on-step forecasting, respectively. Compared with QPSO-LSSVM, the MAPE values are cut by 2.17% in one-step forecasting. From statistical indices in the tables, it can be obviously seen that the new developed WPD-QPSO-LSSVM model obtains smaller forecasting errors than WT-GA-SVM, EMD-ANN, the reasons are that WPD can yield more stable decomposed subseries than WT and EMD, and the regression performance of LSSVM outperforms SVM and ANN. Thus, the comparisons substantiate the developed WPD-QPSO-LSSVM model can carry out wind speed prediction effectively.

**Table 3.** Statistical indices of the proposed and other hybrid forecasting models

| Horizon | Index | QPSO-LSSVM | WT-GA-SVM | EMD-ANN | Proposal |
|---|---|---|---|---|---|
| 1-step | MAE | 0.52 | 0.41 | 0.42 | 0.35 |
| | MAPE | 6.92 | 5.38 | 5.63 | 4.75 |
| | RMSE | 0.72 | 0.64 | 0.65 | 0.59 |
| 2-step | MAE | 0.61 | 0.47 | 0.51 | 0.41 |
| | MAPE | 8.15 | 6.39 | 6.71 | 5.56 |
| | RMSE | 0.78 | 0.69 | 0.71 | 0.64 |
| 3-step | MAE | 0.71 | 0.55 | 0.59 | 0.5 |
| | MAPE | 9.58 | 7.41 | 7.86 | 6.79 |
| | RMSE | 0.85 | 0.74 | 0.77 | 0.71 |

## 4 Conclusion

A new hybrid WPD-QPSO-LSSVM model is developed for multi-step ahead wind speed forecasting in this study. Some conclusions can be drawn from above analysis and comparisons: (1) the single forecasting models BPNN, WNN and LSSVM obtain higher forecasting accuracy than Persistence and ARMA for the intelligent methods can better deal with the nonlinear time series than the statistical approaches. (2) The compound models WT-GA-SVM, EMD-ANN and the proposal model can obtain higher prediction accuracy than the single forecasting methods Persistence, ARMA, BPNN, WNN and LSSVM because original wind speed time series exhibit high nonlinearity and volatility that require preprocessing. (3) The proposed WPD-QPSO-LSSVM mothed outperforms the recently developed EMD-ANN and WT-GA-SVM models in that WPD method can obtain more stable and smaller subseries than WT and EMD approaches, and the regression performance of LSSVM also outperforms SVM and ANN. Thus, the new developed WPD-QPSO-LSSVM model is an effective wind speed approach.

**Acknowledgments.** This work was Supported by Natural Science Research Projects of Colleges and Universities in Anhui Province (KJ2020A0348 and KJ2020ZD39); the Key Laboratory of Electric Drive and Control of Anhui Higher Education Institutes, Anhui Polytechnic University (DQKJ202003); the Open Research Fund of Anhui Key Laboratory of Detection Technology and Energy Saving Devices, Anhui Polytechnic University under grant DTESD2020A02 and DTESD2020A04.

## References


1. Tahmasebifar, R., Moghaddam, M.P., Sheikh-El-Eslami, M.K., et al.: A new hybrid model for point and probabilistic forecasting of wind power. J. Energy. **211**, 119016 (2020)
2. Meng, A., Ge, J., Yin, H., et al.: Wind speed forecasting based on wavelet packet decomposition and artificial neural networks trained by crisscross optimization algorithm. J. Energy Convers. Manage. **114**, 75–88 (2016)
3. Wang, J., Li, Q., Zen, B.: Multi-layer cooperative combined forecasting system for short-term wind speed forecasting. J. Sustain. Energy Technol. Assess. **43**, 22–33 (2021)
4. Luo, L., Li, H., Wang, J., et al.: Design of a combined wind speed forecasting system based on decomposition-ensemble and multi-objective optimization approach. J. Appl. Math. Model. **89**, 49–72 (2021)
5. Hu, H., Wang, L., Tao, R.: Wind speed forecasting based on variational mode decomposition and improved echo state network. J. Renew. Energy **164**, 729–751 (2021)
6. Hu, J., Heng, J., Wen, J., et al.: Deterministic and probabilistic wind speed forecasting with denoising-reconstruction strategy and quantile regression-based algorithm. J. Renew. Energy **162**, 1208–1226 (2020)
7. Moreno, S.R., Mariani, V.C., Coelho, L.S.: Hybrid multi-stage decomposition with parametric model applied to wind speed forecasting in Brazilian Northeast. J. Renew. Energy **164**, 1508–1526 (2021)
8. Tang, G., Wu, Y., Li, C., et al.: A novel wind speed interval prediction based on error prediction method. IEEE Trans. Ind. Inf. **16**(11), 6806–6815 (2020)
9. Sun, W., Wang, Y.: Short-term wind speed forecasting based on fast ensemble empirical mode decomposition, phase space reconstruction, sample entropy and improved back-propagation neural network. J. Energy Convers. Manage. **157**, 1–12 (2018)
10. Xiao, L., Qian, F., Shao, W.: Multi-step wind speed forecasting based on a hybrid forecasting architecture and an improved bat algorithm. J. Energy Convers. Manage. **143**, 410–430 (2017)
11. Zhang, C., Zhou, J., Li, C.: A compound structure of ELM based on feature selection and parameter optimization using hybrid backtracking search algorithm for wind speed forecasting. J. Energy Convers. Manage. **143**, 360–376 (2017)
12. Wang, D., Luo, H., Grunder, O., et al.: Multi-step ahead wind speed forecasting using an improved wavelet neural network combining variational mode decomposition and phase space reconstruction. J. Renew. Energy **113**, 1345–1358 (2017)
13. Wang, Y., Wang, J., Wei, X.: A hybrid wind speed forecasting model based on phase space reconstruction theory and Markov model: a case study of wind farms in northwest China. J. Energy **91**, 556–572 (2015)
14. Naik, J., Dash, S., Dash, P.K., et al.: Short-term wind power forecasting using hybrid variational mode decomposition and multi-kernel regularized pseudo inverse neural network. J. Renew. Energy **118**, 180–212 (2018)
15. Chen, X., Zhao, J., Jia, X., et al.: Multi-step wind speed forecast based on sample clustering and an optimized hybrid system. J. Renew. Energy **165**, 595–611 (2021)
16. Tian, Z., Li, H., Li, F.: A combination forecasting model of wind speed based on decomposition. Energy Rep. **7**, 1217–1233 (2021)
17. Liu, D., Niu, D., Wang, H., et al.: Short-term wind speed forecasting using wavelet transform and support vector machines optimized by genetic algorithm. J. Renew. Energy **62**, 592–597 (2014)
18. Guo, Z., Zhao, W., Lu, H.: Multi-step forecasting for wind speed using a modified EMD-based artificial neural network model. J. Renew. Energy **37**, 241–249 (2012)
19. Wang, S., Zhang, N., Lei, W., et al.: Wind speed forecasting based on the hybrid ensemble empirical mode decomposition and GA-BP neural network method. J. Renew. Energy **94**, 629–636 (2016)

# Robot Path Planning Based on Improved RRT Algorithm

Xiuyun Zeng(✉), Huacai Lu, Hefeng Lv, and Peiyang Li

Anhui Key Laboratory of Electric Drive and Control, Anhui Polytechnic University, Wuhu, Anhui, China

**Abstract.** Path planning is an important research content of mobile robot. Because of the successful application of Rapidly-Exploring Random Tree (RRT) algorithm in robot path planning, it has been greatly studied and developed since it was proposed. In this paper, the robot path planning simulation experiment is carried out for two two-dimensional maze environment, whose sizes are 860 * 770 and 1250 * 1000 respectively. One of the mazes is circular, and the other is composed of many square obstacles. In the case of unknown maze environment, any two points in the maze are selected as the starting point and end point to plan the path between them. In this paper, RRT algorithm is used, and then the algorithm is improved. Dijkstra algorithm is used to find the optimal path from the beginning to the end. In this paper, the random search tree algorithm is used to start from the starting point and search the random number until the target point is reached. Then Dijkstra algorithm is used to search the shortest path from the target point to the end point. Record the nodes and paths of the fast search tree, and then find the optimal path through Dijkstra algorithm. Through the implementation of fast search tree algorithm by Visual Studio (VS), the optimal path is obtained. The optimal path is shown by Matlab.

**Keywords:** Robot path planning · RRT · Dijkstra · Maze environment

## 1 Introduction

With the continuous development of artificial intelligence, robots are widely used in various fields, such as military use, intelligent transportation, etc. Intelligent becomes a new navigation mark in the future robot field, among which path planning has become an important research direction of robot technology [1]. Path planning refers to how to find a proper path from the starting point to the end point in the obstacle environment space, so that the robot can safely and without collision to bypass all obstacles in the course of movement, and then plan an optimal feasible path according to an optimized target.

Robot path planning algorithm is one of the core contents of mobile robot research. It started in 1970s [3]. So far, there have been a lot of research results, which have made great contributions to the wide application of robots [4]. However, with the development of industrial and social needs, the traditional robot path planning algorithm is slow to


Q. Han et al. (Eds.): LSMS 2021/ICSEE 2021, CCIS 1469, pp. 361–369, 2021.
https://doi.org/10.1007/978-981-16-7213-2_35

meet the planning conditions in complex environment. At present, the algorithms of robot path planning include A* algorithm, Probabilistic Roadmaps algorithm (PRM), Particle Swarm Optimization algorithm (PSO), Neural Network algorithm and Rapid-exploration Random Tree algorithm (RRT) [2]. Among them, the A* algorithm has strong search ability and can converge to the global optimal path, but the computation is complex and depends on the integrity of grid resolution, which is not suitable for high-dimensional environmental path planning. Although the PRM algorithm can solve the problem of high-dimensional spatial path planning, the search rate is low in complex environment. PSO algorithm [8] is simple and easy to realize and has fast convergence speed, but it is easy to fall into local optimal solution, and it is not suitable for path planning in local and high-dimensional environment. The neural network algorithm has excellent learning ability, and it can be combined with other algorithms with strong robustness, but its generalization ability is poor. RRT algorithm [7] has the advantages of less parameters, simple structure, strong search ability, easy to combine with other algorithms, which can effectively solve high dimensional space and complex problems, and is more suitable for path planning of mobile robot and multi degree of freedom industrial robot, so it is widely used and improved [5].

In this paper, RRT algorithm and Dijkstra algorithm are used to find the optimal path from the beginning to the end. The fast search algorithm of random tree starts from the beginning, and searches the random number until it reaches the target point [6]. Then, Dijkstra algorithm is used to search for the shortest path from target point to end point. Record the nodes and paths of fast search tree, find the optimal path by Dijkstra algorithm, and realize the fast search random tree algorithm through Visual Studio (VS) to get the optimal path. Finally, the optimal path is displayed by Matlab.

## 2 RRT Algorithm

RRT algorithm is a path planning algorithm to search high-dimensional space by randomly constructing space filling tree. The tree is gradually formed by random sampling in the search space in an incremental way. In essence, it tends to grow to the direction of the unexplored area. It can quickly search the configuration space by finding the nearest node in the tree in the direction of the random point and expanding with a specified step size. RRT algorithm, as a technique of generating trajectories for nonholonomic systems with state constraints, can be directly applied to the study of nonholonomic systems. RRT algorithm is a random algorithm, and its expansion mode is different from the general expansion methods.

The general steps of RRT algorithm is to define a starting point as $q$start in the environment, and then randomly scatter a point in the environment to get the point $q$rand. If $q$rand is not in the obstacle area, then connect $q$start and $q$rand, we get a line L, If I is not in the obstacle, it moves a certain distance along L from $q$start to $q$rand gets a new point $q$new. Then $q$start, $q$new and the line segments between them form the simplest tree. On this basis, continue to repeat, scatter points in the environment to get the point $q$rand of the obstacle free area, and then find a point $q$near nearest to the $q$rand on the existing tree, Connect two points. If the line has no obstacles, move a certain distance along the line from $q$near to $q$rand get a new point $q$new, which is added to the existing tree, so as to realize the expansion of the tree. Repeat the above process until the target point (or nearby point) is added to the tree, then we can find a path from the starting point to the target point on the tree.

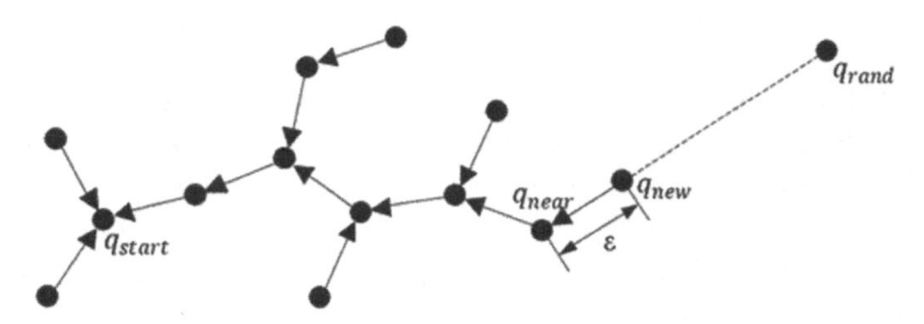


**Fig. 1.** Schematic diagram of RRT algorithm extension process.

As shown in Fig. 1, it is the basic RRT algorithm expansion process schematic diagram. Firstly, in the initialization process, the initial state node $q_{start}$ is used to initialize the random tree. At this time, there is only one node in the tree. Then, a random point $q$rand is randomly sampled in the barrier free space. After traversing in the random tree through the specified distance evaluation function, the node nearest to $q_{rand}$ is selected, which is called the node $q_{near}$, a straight line connects $q_{rand}$ and $q_{near}$, and then in the direction from $q_{near}$ to $q_{rand}$, take $q_{near}$ as the starting point to intercept the step length of ε, at this time, the other end of ε generates a new node $q_{new}$, which can be called $q_{new}$ as the child node of $q_{near}$, and judge the distance between $q_{near}$ and $q_{near}$. If the path between $q_{new}$ collides with the obstacle, the newly generated node $q_{new}$ will be discarded and the random tree will not make any changes; if not, the new node $q_{new}$ will be added to the random tree. When the shape space and the specified time permit, the above process is iterated until the distance between the newly generated node and the target point is within the set maximum tolerance distance. Finally, the random tree is returned and the path from the initial node $q_{start}$ to the target node $q_{goal}$ in the tree is connected, and the path planning ends.

## 3 Improved RRT Algorithm

Due to the randomness of the traditional RRT extended leaf nodes, a large number of leaf nodes are blind, and the expansion speed may be slow. Therefore, this paper proposes an extension point selection strategy. The core idea is to simulate the gravity effect of the target point in the artificial potential field method, and improve the random sampling function in RRT algorithm [9], so that the algorithm can guide the growth direction of the extended tree, so as to explore the target as soon as possible, and make the extended tree grow more tendentious without a lot of invalid expansion, so as to shorten the time of path planning and save resources. The schematic diagram of expansion point selection is shown in Fig. 2.

The new leaf node is jointly determined by the random node $q_{rand}$ and the target point $q_{goal}$ to guide the tree to expand to the target point. Define a constant $p_{threshold}$, when the random value temp is less than $p_{threshold}$, use the above method to guide the tree expansion, otherwise it is still expanding towards the random direction. In this way, the randomness of the extended tree is preserved, and it can be extended to the target faster. Adaptive step size strategy solves this problem, it can use the information collected in the process of exploration to adapt to the expansion of the tree, so that the tree can cover less blocked space area faster. Its basic idea is: using two variables E1 and E2 (initially 1) to represent the expansion length factor of random direction and guidance direction respectively. When a given node expands successfully in the above two directions, the E value in this direction is increased, and the new node is expanded by multiplying the new e value by the step size. If the expansion fails in this direction, that is, it encounters obstacles, the E value is initialized. The combination of extension point selection strategy and adaptive step size strategy can quickly explore the unknown space and guide the extension tree to the final target point.

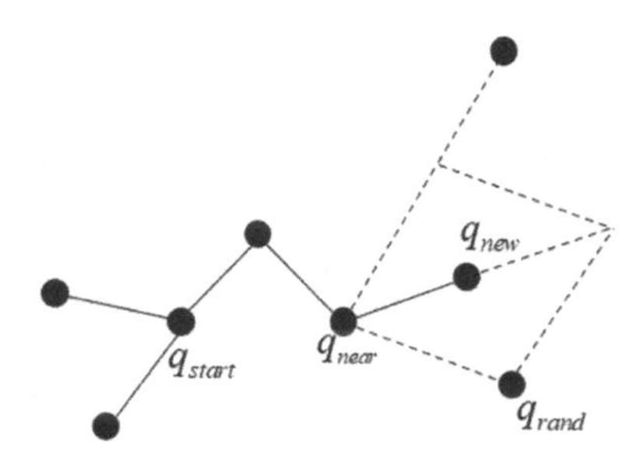


**Fig. 2.** Extension point selection strategy.

Combination Dijkstra algorithm optimization, Dijkstra algorithm is first proposed by Dijkstra, which is the shortest path algorithm from one node to other nodes. The main feature of the algorithm is that it expands gradually from the starting point to the outside, until the shortest distance is found at the end. The basic idea of Dijkstra algorithm is to find the shortest path from the source point to other nodes under the joint action of two sets. The first set s is the set of solved shortest path vertices, and the set S only contains the source point S at the initial time; the second set U is the set of undetermined shortest path points, in which the distance between the nodes connected with the source point is the corresponding weight, and the distance between the nodes not directly connected is set to $\infty$. Select the shortest distance point in U to add to S, and remove the point in U. Taking the point with the minimum distance from the source point as the source point, the distance value from the source point is updated gradually. Repeat the above two steps until set s covers all nodes and the algorithm ends.

First of all, build the maze environment, define the size of the maze, the location of the obstacles in the maze, the coordinates of the starting point and the ending point, and define the path cost. Build a fast search tree algorithm. RRT algorithm is a search method based on probability sampling. Figure 3 (a) shows the program flow chart of RRT algorithm, Fig. 3 (b) is the program flow chart of the key extension process of random tree.

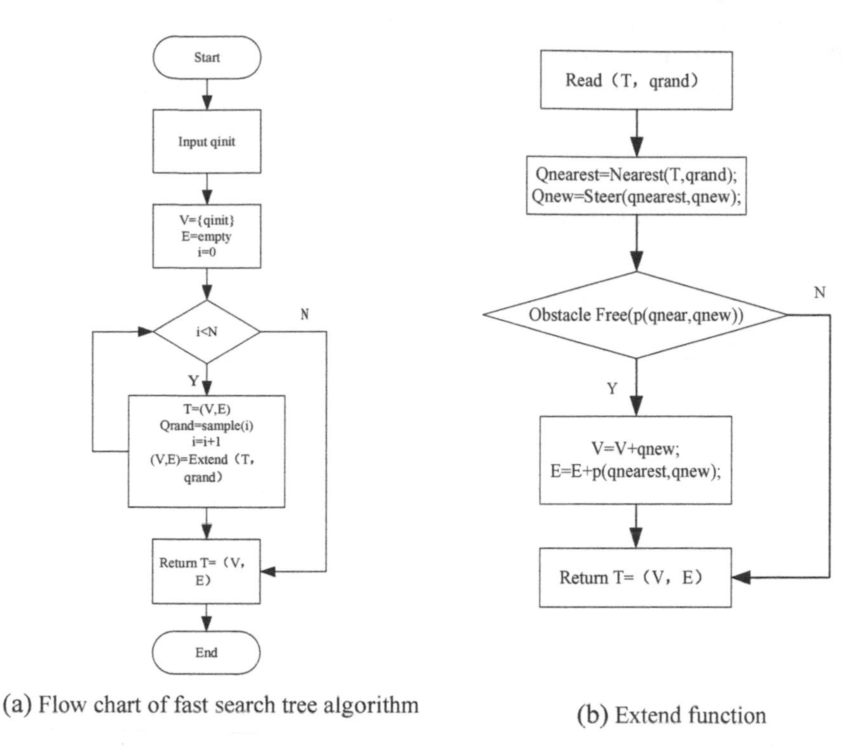


(a) Flow chart of fast search tree algorithm (b) Extend function

**Fig. 3.** Improve RRT flow chart.

## 4 Experiment and Simulation

In order to verify the effectiveness of RRT algorithm, experiments are carried out on the visual studio simulation platform, and the path and cost obtained after optimization are drawn in Matlab to verify the effectiveness of the algorithm. Simulation experiments are carried out in different starting point maps before and after the improvement. The starting point position of the extended tree is represented by * and the target point position coordinates are represented by circles. The green areas in the environmental map represent obstacles.

When the starting point is set to (810, 0) and the focus is set to (380, 380), as shown in the figure below, it is an extension of the basic RRT algorithm in two-dimensional space. In order to understand the RRT algorithm more clearly, the black point in the center of the map is set as the starting point, and the target point and obstacles are not set. The size of the map is 860 * 770, and the step size is 50 cm. The blue line represents the extended structure of the random tree, the simulation diagram is shown in Fig. 4 (a), (b), (c) and (d). The four pictures are the effect pictures of RRT algorithm expansion of 200, 1000, 2000 steps. As can be seen from the figure, RRT With the continuous sampling, the number of nodes in the tree increases, and the expansion of the main body of the tree also conforms to the law of probability and extends to the four vertices. At the same time, the other nodes also expand to the four vertices as the main body. When the number of sampling points reaches a certain number, the number of nodes increases. The machine tree will cover the whole free space, which indirectly proves the probabilistic completeness of RRT algorithm.

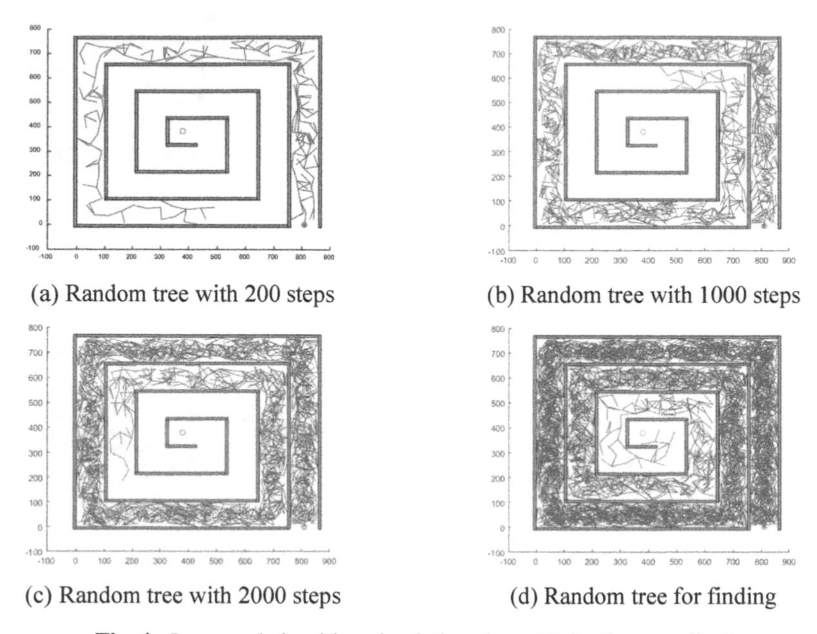


(a) Random tree with 200 steps (b) Random tree with 1000 steps

(c) Random tree with 2000 steps (d) Random tree for finding

**Fig. 4.** Improved algorithm simulation chart. (Color figure online)

The shortest path obtained by Dijkstra is as follows (Fig. 5).

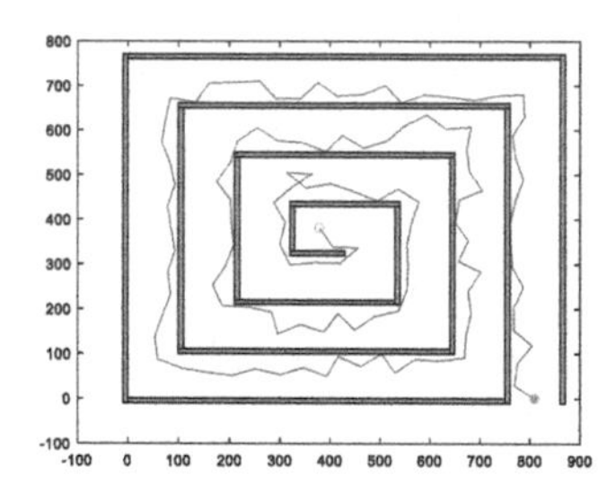


**Fig. 5.** The shortest path obtained by Dijkstra.

When the starting point and the end point are interchanged, the operation result chart is shown in Fig. 6.

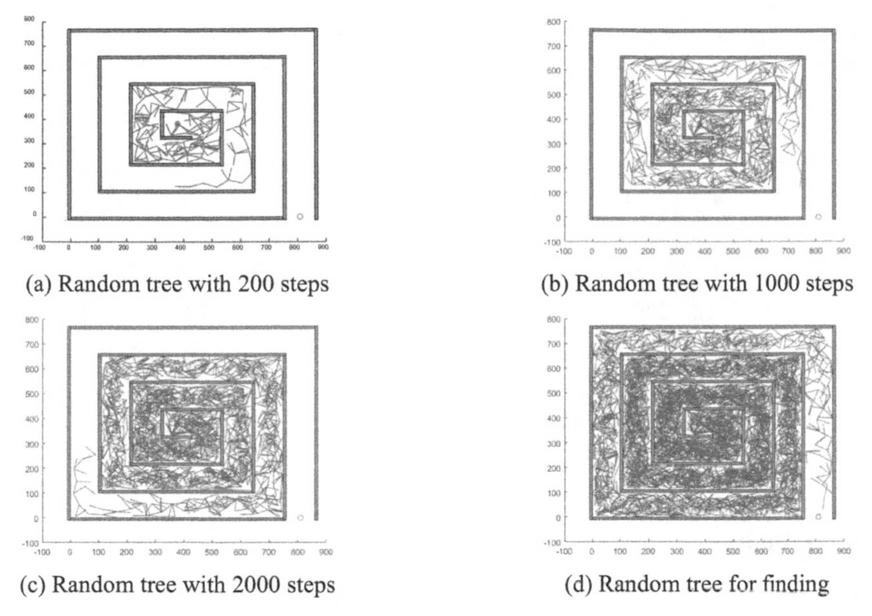


(a) Random tree with 200 steps (b) Random tree with 1000 steps

(c) Random tree with 2000 steps (d) Random tree for finding

**Fig. 6.** Improved algorithm simulation chart.

The shortest path obtained by Dijkstra is as follows (Fig. 7).

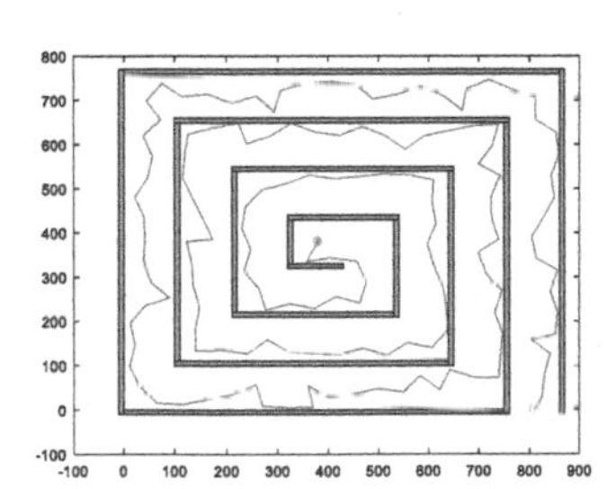


**Fig. 7.** The shortest path obtained by Dijkstra.

When the starting point is set to (1100, 60) and the key point is set to (150, 900), as shown in the figure below, it is an extension of the basic RRT algorithm in two-dimensional space. In order to understand the RRT algorithm more clearly, the black point in the center of the map is set as the starting point, and no target point and obstacles are set. The map size is 1250 * 1000, and the step size is 20 cm. The blue line represents the extended structure of the random tree. Figure 8 and Fig. 9 shows the RRT algorithm extends the rendering of 200, 1000, 2000 steps. It can be seen from the figure that RRT algorithm is random and uniform sampling in the unexpanded free space in the geometric space. With the continuous sampling, the number of nodes in the tree increases, and the expansion of the main body of the tree also conforms to the law of probability and extends to the four vertices. At the same time, other nodes also expand to the surrounding with the branches generated in the direction of the four vertices When the number reaches a certain value, the random tree will cover the whole free space, which indirectly proves the probabilistic completeness of RRT algorithm.

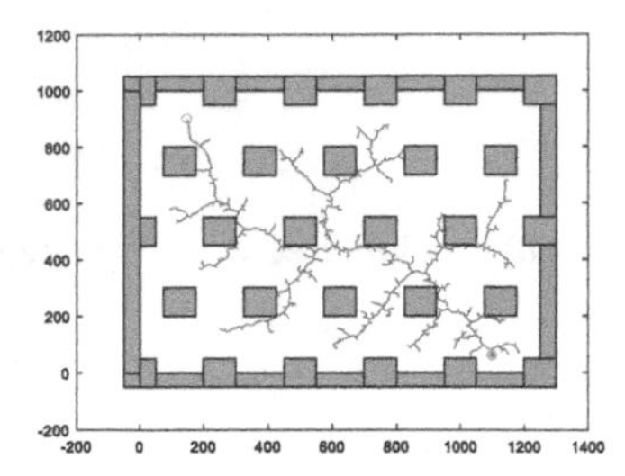


**Fig. 8.** Tree diagram of random number search.

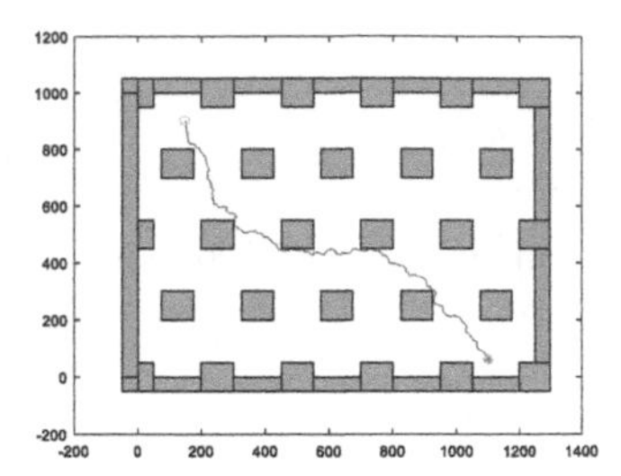


**Fig. 9.** The shortest path obtained by Dijkstra.

## 5 Conclusion

RRT algorithm has an important position in the path planning field of mobile robot because of its own advantages. In this paper, the two-dimensional maze environment is simulated by RRT algorithm, and then Dijkstra algorithm is used to find the optimal path from the beginning to the end. Although RRT algorithm has many advantages, its own defects also limit its application in practice. In the face of particularly complex environment, the number of nodes is very large, the calculation amount will be multiplied, and the planning speed will be slower and slower. Further improvement is needed in the future research.

**Acknowledgments.** We thank for support from Wanjiang Collaborative Innovation Center for High-end Manufacturing Equipment Open Fund (No. GCKJ2018013), and Research Projects of Anhui University of Technology (No. Xjky2020022).

## References


1. Luo, D., Li, D., Yin, L.J.: Review of typical intelligent algorithms in robot path planning. Comput. Knowl. Technol. **16**(26), 180–181 (2020)
2. Feng L.C.: Research on the algorithm of parametric RRT unmanned vehicle motion planning based on guidance domain. University of Science and Technology of China, Hefei (2017)
3. Zhang, G.L., Hu, X.M., Chai, J.F.: Overview of path planning algorithm and application. Mod. Mach. **18**(5), 85–90 (2011)
4. Noreen, I., Khan, A., Habib, Z.: Optimal path planning using RRT* based approaches: a survey and future directions. Int. J. Adv. Comput. Sci. Appl. **7**(11), 11–16 (2016)
5. Shi, Z.: Research on PSO optimization algorithm in mobile robot path planning. J. Guiyang Univ. **16**(01), 1–6 (2021)
6. Jiang, H.Y.: Research on improved path planning algorithm based on RRT. Guangxi University (2020)

7. Chen, Q.L., Jiang, H.Y., Zheng, Y.J.: Survey of rapidly-exploring random tree for robot path planning. Comput. Eng. Appl. **55**(16), 10–17 (2019)
8. Liu, Y.Q., Zhao, H.C., Liu, X., Xu, Y.: An improved RRT algorithm for path obstacle avoidance planning of industrial robots. Inf. Control **50**(02), 235–246 (2021)
9. Liang, Z.Y., Cheng, F.X., Wei, W.: Improved RRT* path planning algorithm. J. Changchun Univ. Technol. **41**(06), 602–607 (2020)

# Robot Dynamic Collision Detection Method Based on Obstacle Point Cloud Envelope Model

Aolei Yang[1], Quan Liu[1], Wasif Naeem[2](✉), and Minrui Fei[1]

[1] School of Mechatronic Engineering and Automation, Shanghai University, Shanghai 200444, China

[2] School of Electronics, Electrical Engineering and Computer Science, Queen's University Belfast, University Road, Belfast BT7 1NN, U.K.
w.naeem@qub.ac.uk

**Abstract.** This paper presents an unsupervised dynamic collision detection approach based on obstacle point cloud envelope model. Based on the generated robot envelope model using the D-H parameters, an invalid point cloud filtering method is designed to filter the point cloud data out of the reachable space of the robot. The oriented bounding box is then adopted to generate an approximate obstacle model. The relevant simplex is further calculated by fusing the robot envelope model, and the Gilbert-Johnson-Keerthi algorithm is employed to take the collision detection between model cells. A collision detection experimental platform was finally constructed to verify the feasibility and effectiveness of the proposed approach.

**Keywords:** Clustering segmentation · Collision detection · Obstacle model approximation · Robot envelope

## 1 Introduction

Robots are gradually entering people's daily lives from the traditional industrial field, such as brain-computer interface [1], assisted grasping [2], autonomous obstacle avoidance [3] and so on. The dynamic collision detection between the robot and the working space is the premise of the normal operation of the robot in the unstructured environment. Traditional collision detection is mainly based on the fusion of multiple sensors, such as torque sensor [4], current or voltage sensor [5], which analyzes the value change after collision. However, in unstructured complex dynamic environment, these traditional methods have a high possibility of a secondary collision because they cannot detect the surrounding environment in real time, and it is therefore difficult to perform tasks efficiently.

At present, researchers have done some work in this area. Lee et al. [6] designed a cascade classifier to detect conical obstacles based on the Haar feature obtained by Adaboost algorithm training. Abiyev's team [7] also employs the supervised support vector machine to classify the collected images and obtain obstacle areas of the world map for collision detection. Heo, YJ et al. [8] proposed a collision detection framework


Q. Han et al. (Eds.): LSMS 2021/ICSEE 2021, CCIS 1469, pp. 370–378, 2021.
https://doi.org/10.1007/978-981-16-7213-2_36

based on a deep learning approach to learn robot collision signals and recognize any occurrence of a collision. Wu, F et al. [9] developed a compact obstacle detection and avoidance system based on magnetic field for small spherical robots.

Collision detection based on lidar or depth camera is another hot research direction. Wang, X et al. [12] used a depth camera to acquire 3D point cloud data, and presented a self-recognition method to make the self-collision detection of a dual-arm robot. Germi et al. [13] used the depth map data collected by RGB-D camera to calculate the pose of obstacles, and estimated the dynamical parameters of moving obstacles. Based on 3D point cloud, Han, D et al. [14] proposed a fast convex hull approach to approximate spatial obstacles. Chen, W et al. [15] adopted lidar to collect data, extracting edge features of obstacles and clustering their point cloud, recognizing the obstacle through neural networks, and finally employing the point cloud post-processing algorithm to localize the obstacle.

This paper proposes an unsupervised robot collision detection approach based on obstacle point cloud envelope model. Based on the generated robot envelope model, an invalid point cloud filtering method is designed to filter the point cloud data. The oriented bounding box (OBB) is then adopted to generate an approximate obstacle model. The relevant simplex is further calculated by fusing the robot envelope model, and the Gilbert-Johnson-Keerthi (GJK) algorithm is employed to detect the collision between the robot and the obstacle models. An experimental platform also is constructed to verify the proposed method.

This paper is organized as follows. First, Sect. 2 explains how to construct the robot envelop model and filter the point cloud data out of the reachable space of the robot. In Sect. 3, we discuss some details of the approximate generation of obstacle model and collision detection. In Sect. 4 we construct a robot collision detection verification platform to verify the feasibility and effectiveness of the proposed approach. Finally, Sect. 5 provides the concluding remarks.

## 2 Problem Description and Method Architecture

### 2.1 Coordinate System Definition and Collision Detection Scene

In this paper, the Franka robot with 7 degrees of freedom (DOF) is used to describe the method, and a typical collision detection scene is shown in Fig. 1.

In the figure, {B} represents the basic coordinate system (CS) of the robot, which is defined as the global world coordinate system {W} in the collision detection scene. {C} is the depth camera coordinate system, its *Z*-axis is aligned with the shooting axis of the depth camera, and the other axis meet the right-hand screw rule. The relative coordinate transformation matrix ${}_C^W T$ of {W} and {C} can be represented by Eq. (1) achieved by hand-eye calibration.

$$ {}_C^W T = \{\ R(r, p, y), T(x, y, z)\} \tag{1} $$

where $R(r, p, y)$ is the rotation vector and $T(x, y, z)$ is the translation vector. Camera placement should consider the reachable space or actual workspace of the entire robot, which should be exposed to the depth camera field of view.

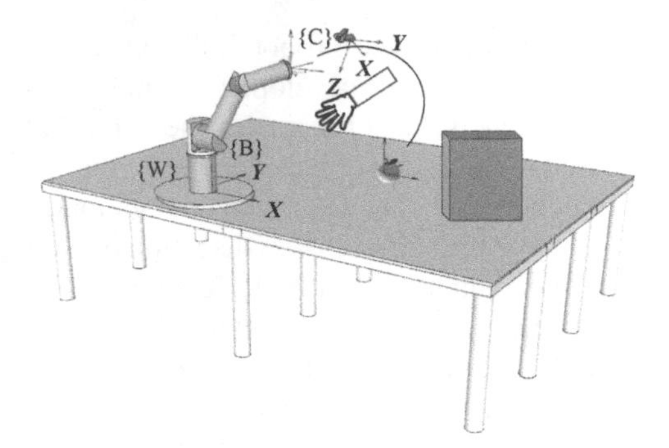


**Fig. 1.** Typical scene of robot collision detection.

### 2.2 Overall Architecture of the Proposed Approach

The proposed architecture is shown in Fig. 2, which consists of the following three parts: the construction of robot real-time envelope model $\boldsymbol{M}_{\boldsymbol{Rob}}$, the implementation of approximate generation of obstacle model (AGOM) $\boldsymbol{M}_{\boldsymbol{Obs}}$, and the collision detection.

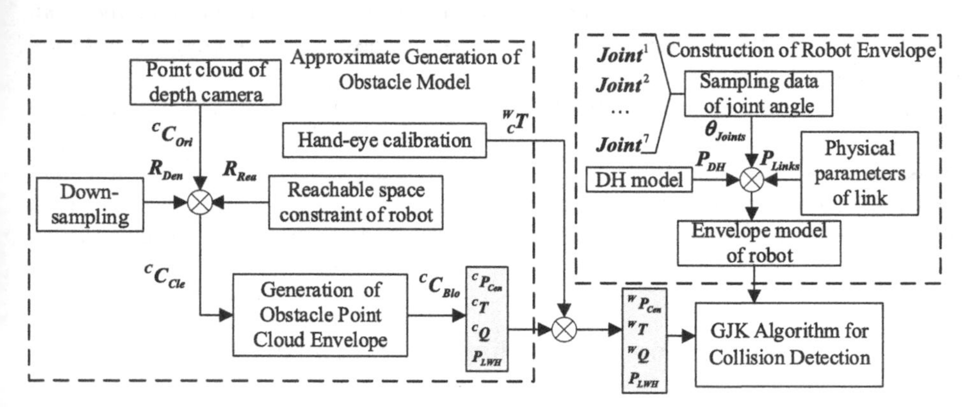


**Fig. 2.** Architecture of the proposed approach.

## 3 Robot Envelope Model and Invalid Point Cloud Filtering

### 3.1 Construction of the Robot Envelope Model

In this paper, the relative transformation relationship between the coordinate systems $\{L_n\}$ and $\{L_m\}$ of any two adjacent links calculated by D-H parameters. Collecting real-time joint angle sensor data $\boldsymbol{\theta}_{\boldsymbol{Joints}}$ of the robot in task execution, for the joint angle $\theta_{L_n}$ of any link $L_n$ at a certain moment, it has a $3 * 3$ rotation transformation matrix as follows:

$$ {}_{L_n}^{W}Q = \begin{bmatrix} cos\theta_{L_n} & sin\theta_{L_n} & 0 \\ -sin\theta_{L_n} & cos\theta_{L_n} & 0 \\ 0 & 0 & 1 \end{bmatrix} \tag{2} $$

Since the base of the robot coincides with the world coordinate system, the transformation matrix between {D} and {W} of any link $L_d$ can be expressed by Eq. (3). It should be noted that ${}^{W}_{L_n}Q$ needs to be transformed into a 4 * 4 homogeneous matrix.

$$ {}^{W}_{L_d}T = {}^{W}_{L_1}T^{W}_{L_1}Q \cdots {}^{L_n}_{L_m}T^{W}_{L_m}Q \cdots {}^{L_{d-1}}_{L_d}T^{W}_{L_d}Q \tag{3} $$

Considering the actual physical parameters of the Franka robot and the environment of the GJK algorithm is a convex hull object, this paper adopts a cube model to construct the robot envelope model. The final robot model is shown in Fig. 3.

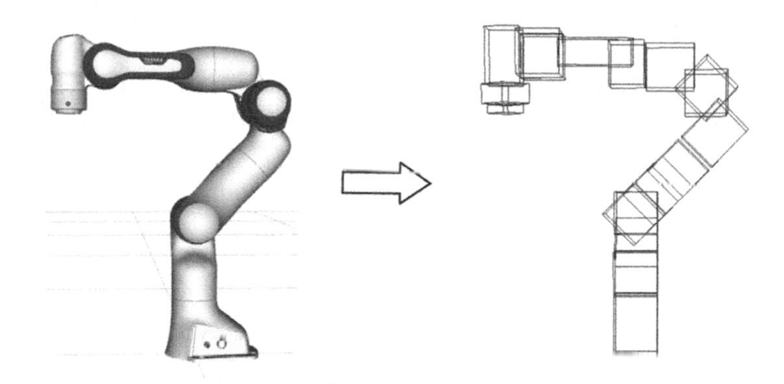

**Fig. 3.** Robot envelope model.

### 3.2 Invalid Point Cloud Filtering

Combined with the robot envelope model constructed in 3.1, this paper designs a Monte Carlo based invalid point cloud filtering method to filter the point cloud data out of the reachable space of ssssssssthe robot. In the upper and lower limits of the joint angle, random multiple sets of joint angle, the point cloud of the robot end position can be received by recording the space coordinate $(x, y, z)$ of the robot end position.

According to the distribution of 3D point cloud, the fitting equation of space ellipsoid is proposed as follows where the variables $[a, b, c, d]$ are the unknown parameters to be solved.

$$ \frac{x^2}{a^2} + \frac{y^2}{b^2} + \frac{(z+d)^2}{c^2} = 1 \tag{4} $$

Following is the variable solution. The covariance matrix can be employed to analyze the spatial correlation of multi-dimensional variables, and the covariance matrix $C$ can be constructed by Eq. (5):

$$ C = \begin{Bmatrix} \mathrm{cov}(X,X) & \mathrm{cov}(X,Y) & \mathrm{cov}(X,Z) \\ \mathrm{cov}(Y,X) & \mathrm{cov}(Y,Y) & \mathrm{cov}(Y,Z) \\ \mathrm{cov}(Z,X) & \mathrm{cov}(Z,Y) & \mathrm{cov}(Z,Z) \end{Bmatrix} \tag{5} $$

The above Eq. (5) has the following relationship:

$$\mathrm{cov}(X,Y) = \frac{\sum_{i=1}^{n}(\overline{X}_i - \overline{X})(\overline{Y}_i - \overline{Y})}{n-1}$$
$$\mathrm{cov}(X,Y) = \mathrm{cov}(Y,X) \tag{6}$$

The eigenvalues and eigenvectors of the covariance matrix $C$ of the point cloud are employed as the reference and the eigenvalues are sorted according to their magnitude. Three representative directions $[v_1, v_2, v_3]$ are selected where, $\left[\vec{v_1}, \vec{v_2}, \vec{v_3}\right]$ are calculated by normalizing to determine the equatorial and polar directions of the ellipsoid.

Suppose that the calculated maximum and minimum points in the direction of three orthogonal vector are $\left[P_{v1}^{max}, P_{v2}^{max}, P_{v3}^{max}\right]$ and $\left[P_{v1}^{min}, P_{v2}^{min}, P_{v3}^{min}\right]$ respectively, then $a = 0.5\left(P_{v1}^{max} - P_{v1}^{min}\right)$, $b = 0.5\left(P_{v2}^{max} - P_{v2}^{min}\right)$, $c = 0.5\left(P_{v3}^{max} - P_{v3}^{min}\right)$, $d = -0.5\left(P_{v3}^{max} + P_{v3}^{min}\right)$ exists for the parameters required to fit the ellipsoid.

# 4 AGOM Algorithm and Collision Detection

## 4.1 AGOM Algorithm

The original point cloud data ${}^{C}\boldsymbol{C}_{Ori}$ provided by the depth camera is down-sampled, and the dense point cloud is transformed into sparse point cloud while preserving the effective spatial feature. The point cloud after down-sampling is applied to the constraint of the robot reachable space in 3.2, and the point cloud ${}^{C}\boldsymbol{C}_{Cle}$ in the robot reachable space can be screened out.

Assuming that any point ${}^{C}P = [{}^{C}P_X, {}^{C}P_Y, {}^{C}P_Z]$ in ${}^{C}\boldsymbol{C}_{Ori}$ is transformed to {W} coordinate system to receive ${}^{W}P = [{}^{W}P_X, {}^{W}P_Y, {}^{W}P_Z]$, only the point cloud satisfying the following Eq. (7) is retained as the obstacle point cloud ${}^{C}\boldsymbol{C}_{Cle}$ in the robot workspace.

$$\frac{{}^{W}P_X^2}{a^2} + \frac{{}^{W}P_Y^2}{b^2} + \frac{({}^{W}P_Z^2 + d)^2}{c^2} - 1 \le 0 \tag{7}$$

The AGOM algorithm adopts supervoxel to cluster the point cloud of obstacle, evaluate the concavity of the boundary point cloud of each supervoxel cluster after clustering, and mark it. Then the constraint plane cutting algorithm is employed to segment the obstacles, and the OBB is applied to envelope the cluster after the point cloud of obstacle is divided, which effectively solves the above problem.

${}^{C}\boldsymbol{C}_{Cle}$ is clustered into point cloud regions with similar attributes by supervoxel clustering. The color information of the point cloud is not considered in this paper. After receiving the supervoxel cluster ${}^{C}\boldsymbol{C}_{Blo}$, the constrained planar cuts (CPC) based on the concavity-convexity is adopted to segment ${}^{C}\boldsymbol{C}_{Blo}$. Firstly, the CPC algorithm is employed to mark the concavity of the supervoxel clusters boundary point cloud, then the RANSAC algorithm with direction weighting and local constraints is applied for the local concave point to fit the cutting plane and combined with the score of concavity evaluation function to determine whether a local plane cutting is required. The OBB is finally employed to envelope the point cloud cluster after the obstacle segmentation.

For the parameter matrix ${}^{C}\boldsymbol{R}_{\boldsymbol{OBB}}$ of OBB in any {C} coordinate system, it can be transformed into {W} coordinate system achieved by Eq. (1), and the result is as follows.

$$ {}^{W}R_{OBB} = \{{}^{W}P_{Cen}, {}^{W}T, {}^{W}Q, P_{LWH}\} \tag{8} $$

where ${}^{W}P_{Cen}$ is the position of the centroid point of ${}^{W}R_{OBB}$, $P_{LWH}$ is the length, width and height parameters of ${}^{W}R_{OBB}$, ${}^{W}Q$ is the rotation matrix of ${}^{W}R_{OBB}$ in {W} coordinate system, and ${}^{W}T$ is the translation matrix of ${}^{W}R_{OBB}$ in {W} coordinate system. The directions of the coordinate axis of the OBB coordinate system are determined by the three orthogonalized eigenvectors.

### 4.2 Collision Detection Between Obstacle and Robot

The real-time robot envelope model $\boldsymbol{M}_{\boldsymbol{Rob}}$ and obstacle model $\boldsymbol{M}_{\boldsymbol{Obs}}$ can be achieved from 3.1 and 4.1, both envelope cells are convex bodies, which meet the conditions of GJK algorithm. How to find the appropriate support point in the convex body is very important to the results of GJK collision detection algorithm and to find the fastest escape direction. The collision detection process of any obstacle envelope cell M and robot envelope cell N can be described with Fig. 4.

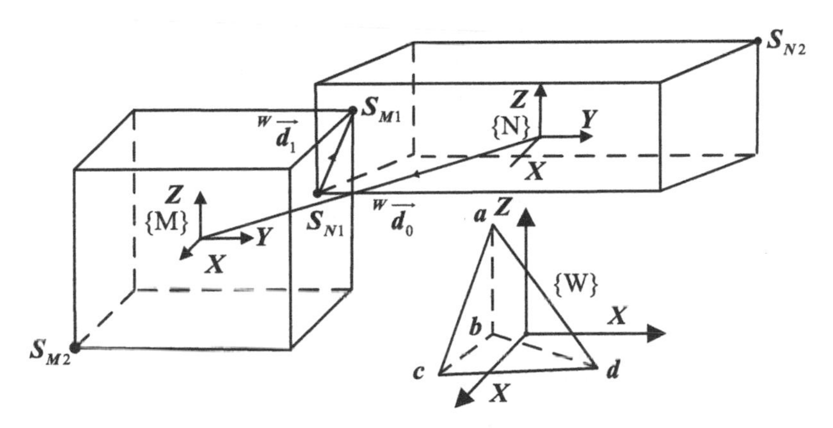


**Fig. 4.** Simplex solution for collision detection between cells.

The collision of obstacle envelope cell M and robot envelope N can be determined by the existence of a simplex containing the origin of {W} coordinate system. If it exists, the collision occurs, and the fastest escape direction is the normal vector direction to the plane $\boldsymbol{P}_{\boldsymbol{Clo}}$ nearest to the origin away from the origin.

## 5 Collision Detection Experiment and Analysis

The robot adopted in the experiment was the Franka Panda robot with 7 DOF, the obstacle point cloud of environment was provided by RealSense D435 depth camera, and dynamic obstacles were simulated by human arms. The approach proposed in this paper was realized by C++ programming, robot control and obstacle release were implemented based on the Moveit! plugin in the ROS platform. The state of the robot envelope and obstacle models calculated by the algorithm in real time was visualized based on the PCL tool. The experiment was carried out under the Ubuntu 16.04 system. The computing platform CPU adopted is Intel core i7-7700HQ, no graphics card is required, and the ROS version is Kinetic.

### 5.1 Collision Detection Effect Verification

This paper transforms the obstacle point cloud and the obstacle envelope model calculated by AGOM algorithm to {W} coordinate system by hand-eye calibration. Figure 5 demonstrates the collision detection process between the robot envelope model and the obstacle model, the whole process intercepted four states to illustrate the effectiveness of the collision detection algorithm, including: Initial state, End collision, Enamel collision, and Recovery state. It can be seen in Fig. 5 that the collision between the obstacle and the robot models can be effectively detected and the collision position can be successfully perceived.

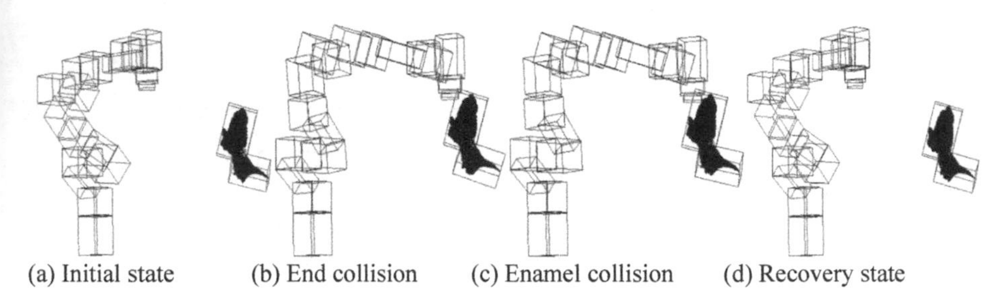

(a) Initial state (b) End collision (c) Enamel collision (d) Recovery state

**Fig. 5.** Collision detection process between robot and obstacle models.

### 5.2 Reliability Verification of AGOM

In order to verify the reliability of the AGOM algorithm, rely on the ROS platform to adopt the calculated obstacle model as the reference obstacle for Moveit! motion planning, and update it in Rviz timely. It is verified on the Franka robot, in which the planned trajectory and joint angle at the end is shown in Fig. 6 when robot executes the task, and the physical scene is shown in Fig. 7.

The experimental results show that the proposed AGOM algorithm can successfully detect the obstacles in the depth camera field of view and achieve the obstacle avoidance, which verifies the effectiveness of the proposed AGOM algorithm for the obstacle envelope.

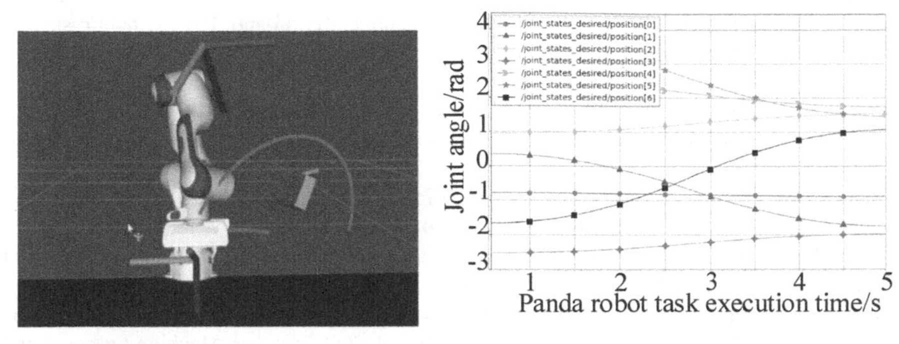


**Fig. 6.** Robot end trajectory and real-time joint angle during task execution.

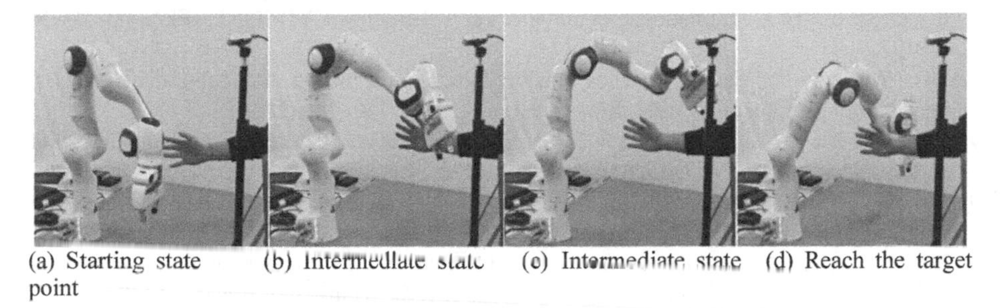

(a) Starting state point (b) Intermediate state (c) Intermediate state (d) Reach the target

**Fig. 7.** Experiments of Franka robot avoiding obstacles.

# 6 Conclusion

The robot dynamic collision detection method based on obstacle point cloud envelope model is proposed to overcome the limitation of traditional robot collision detection. The robot envelope model is constructed, and the random sampling based invalid point cloud filtering method is designed to filter the point cloud data. At the same time, the AGOM algorithm is proposed to provide a more accurate description approach of the obstacle spatial pose considering the concavity-convexity and connectivity of obstacle point cloud. The experimental results verify the feasibility and effectiveness of the proposed algorithm.

**Acknowledgments.** This research was supported by Natural Science Foundation of Shanghai (18ZR1415100).

# References

1. Li, Z., Li, J., Zhao, S., Yuan, Y., Kang, Y., Chen, C.L.P.: Adaptive neural control of a kinematically redundant exoskeleton robot using brain–machine interfaces. J. IEEE Trans. Neural Netw. Learn. Syst. **30**(12), 3558–3571 (2019)

2. Abi-Farraj, F., Pacchierotti, C., Arenz, O., Neumann, G., Giordano, P.R.: A haptic shared-control architecture for guided multi-target robotic grasping. J. IEEE Trans. Haptics, **13**(2), 270–285 (2020)
3. Yang, A., Naeem, W., Fei, M., et al.: A cooperative formation-based collision avoidance approach for a group of autonomous vehicles. J. Int. J. Adapt. Control Signal Proc. **31**, 489–506 (2016)
4. Lou, Y., Wei, J., Song, S.: Design and optimization of a joint torque sensor for robot collision detection. J. IEEE Sensors J. **19**, 6618–6627 (2019)
5. Indri, M., Trapani, S., Lazzero, I.: Development of a virtual collision sensor for industrial robots. J. Sensors Basel, **17**(5), 1148 (2017)
6. Lee, C.J., Tseng, T.H., Huang, B.J., et al.: Obstacle detection and avoidance via cascade classifier for wheeled mobile robot. In: International Conference on Machine Learning and Cybernetics. IEEE, pp. 403–407 (2015)
7. Abiyev, R.H., Arslan, M., Gunsel, I., et al.: Robot pathfinding using vision based obstacle detection. In: IEEE International Conference on Cybernetics CYBCONF. IEEE, pp. 29–34 (2017)
8. Heo, Y.J., Kim, D., Lee, W., et al.: Collision detection for industrial collaborative robots: a deep learning approach. J. IEEE Robot. Autom. Lett. **4**, 740–746 (2019)
9. Fang, W., Akash, V., Song, S.G., et al.: A compact magnetic field-based obstacle detection and avoidance system for miniature spherical robots. J. Sensors **17**, 1231 (2017)
10. Ball, D., Upcroft, B., Wyeth, G., et al.: Vision-based obstacle detection and navigation for an agricultural robot. J. Field Robot. **33**, 1107–1130 (2016)
11. Luo, R.C., Lai, C.C.: Multisensor fusion-based concurrent environment mapping and moving object detection for intelligent service robotics. J. IEEE Trans. Ind. Electr. **61**, 4043–4051 (2014)
12. Wang, X., Yang, C., Ju, Z., et al.: Robot manipulator self-identification for surrounding obstacle detection. J. Multimedia Tools Appl. **76**, 6495–6520 (2016)
13. Germi, S.B., Zamanian, A., Arzati, M.A., et al.: Estimation of moving obstacle dynamics with mobile RGB-D camera. In: RSI International Conference on Robotics and Mechatronics (ICRoM), pp. 156–161 (2017)
14. Han, D., Nie, H., Chen, J., et al.: Dynamic obstacle avoidance for manipulators using distance calculation and discrete detection. J. Robot. Comput. Integ. Manuf. **49**, 98–104 (2018)
15. Chen, W., Liu, Q., Hu, H., et al.: Novel laser-based obstacle detection for autonomous robots on unstructured terrain. J. Sensors **20**, 5048 (2020)

# RSS/AOA Indoor Target Location Method Based on Weighted Least Squares Algorithm

Tiankuan Zhang, Jingqi Fu[✉], and Yiting Wang

Shanghai University, Shanghai, China
jqfu@shu.edu.cn

**Abstract.** Due to there are large deviations and uncertainties in determining the coordinates of the target position in the indoor environment solely relying on the received signal strength (RSS), and the emergence of a new type of antenna array provides the possibility to measure the angle of arrival (AOA). Therefore, a target location method that combines RSS and AOA measurements in an indoor environment is proposed. Aimed at the problem of large positioning deviation caused by the mixed positioning of AOA and RSS under the least square solution method due to no consideration of noise, a weighted least square method based on approximate measurement noise is proposed. Then a set of indoor positioning verification system is designed. The influence of the introduction of the new weighting mechanism on the target position coordinate estimation is analyzed, and the experimental results show the superior performance of the new algorithm.

**Keywords:** Indoor target positioning · Received signal strength (RSS) · Angle of arrival (AOA) · Weighted least squares (WLS)

## 1 Introduction

With the growth of the society's demand for indoor positioning, such as: catering services, environmental monitoring, medical monitoring, construction and many other fields, indoor positioning has extremely broad application prospects [1–5]. With the development of wireless communication technology, sensor technology, and signal processing technology, electronic devices equipped with various wireless communication methods have also shown explosive growth. Therefore, indoor positioning technology has increasingly become a focus of attention in the field of target positioning.

The traditional indoor positioning mostly relies on the method of measuring distance or angle to locate the target. Methods of measuring distance usually include received signal strength (RSS), time of arrival (TOA), and time difference of arrival (TDOA) [6]. The methods of measuring angle usually include angle of arrival (AOA) and angle of departure (AOD) [7]. In [8], an error self-correction


Q. Han et al. (Eds.): LSMS 2021/ICSEE 2021, CCIS 1469, pp. 379–387, 2021.
https://doi.org/10.1007/978-981-16-7213-2_37

positioning method based on RSS is proposed. The error self-correction factor of the correction node replaces the ranging error factor of the unknown node to compensate for the ranging error, which improves the positioning accuracy to a certain extent, but when the multipath attenuation and background noise are serious, a sharp rise in measurement error cannot be avoided. In [9], an ultra-wideband AOA estimation method based on a single base station antenna array is proposed. Under the condition that a single base station is equipped with a directional antenna array, it avoids the problem of inaccurate clock synchronization caused by the arrival time difference of multi-base station measurement signal pulses. However, this method puts forward strict requirements on the antenna array design of a single base station, and the hardware implementation is relatively difficult. In [10], a target positioning method of RSS and AOA mixed measurement is proposed, and the least square model of measured value and target position is derived through coordinate transformation. Then the distance between the nodes and the signal transmission power are introduced into the weight parameters respectively, deriving the weighted least squares solution of the target position.

In order to minimize reduce the error caused by RSS and AOA in the measurement process, this paper further studies the hybrid measurement method based on RSS and AOA. In this paper, the position of the target obtained by the least square algorithm is corrected by introducing the noise weighting factor. The noise parameters are estimated without knowing the real noise, and the estimated position is compensated. Finally, a set of actual indoor positioning system is designed to verify the effectiveness of the algorithm.

## 2 System Model

Consider that there are N sensors (anchors) distributed indoors, and their coordinates of position are respectively $a_i = [a_{ix}, \mathrm{a}_{iy}]^T$ for $i = 1, 2, ..., N$. The coordinates of the node to be positioned are $\mathrm{x} = [x_x, x_y]^T$. For the measurement of RSS, the logarithmic path distance loss model [11] is usually used :

$$L_{\mathrm{i}} = L_0 - 10\gamma \log_{10}\frac{\mathrm{d}_i}{d_0} + n_i,\ \text{for}\ i = 1, 2, ..., N \tag{1}$$

Where $L_0$ (dB) is the reference power at distance $d_0$ (m), $\gamma$ is the path loss exponent (PLE), $\mathrm{d}_i = \|\mathrm{x} - a_i\|$ indicates the distance between the node to be located and the i-th anchor, $L_i$ (dB) is the received power at distance $d_i$ (m), and $n_i$ represents the measurement noise of RSS between the node to be located and the i-th anchor. Suppose $n_i$ obeys a Gaussian distribution with mean zero and variance $\delta_{\mathrm{n}_i}^2$, denoted as $\mathrm{n}_i \in N(0, \delta_{n_i}^2)$.

In a complex indoor environment, the measurement of RSS usually has intense noise, such as signal interference in the same frequency band, indoor obstacle interference, and human measurement interference, etc. In order to reduce the positioning error, while measuring the RSS, it is also necessary to measure the direction angle of the node to be located relative to the anchor node, i.e., AOA.

The essence of obtaining the angle of arrival is the processing of the phase difference of the bluetooth signal. At the beginning of 2019, the Bluetooth Alliance released the latest specification of Bluetooth 5.1, in which the new direction finding function introduces angle of arrival (AOA) and angle of departure (AOD) into the bluetooth [12]. Suppose there is a fixed-frequency bluetooth beacon signal spreading in an open area (ignoring obstacles and other 2.4G signal interference in the air). If two receivers are on the same radius of the same transmitting end at a certain moment, the phase difference received by the receiver should be zero, but if the two receivers are at positions with different radii, the phase difference received is not zero [13].

In the light of this principle, it is assumed that the unknown node sends a special data packet through a single antenna, and the anchor node has multiple antennas arranged in an array as shown in Fig. 1(a). Due to the different distances between the antennas and the unknown node, the signals received by different antennas will have phase differences. When switching between antennas, the amplitude and phase are obtained from the received signal, and the relative orientation of the target is calculated by the phase difference.

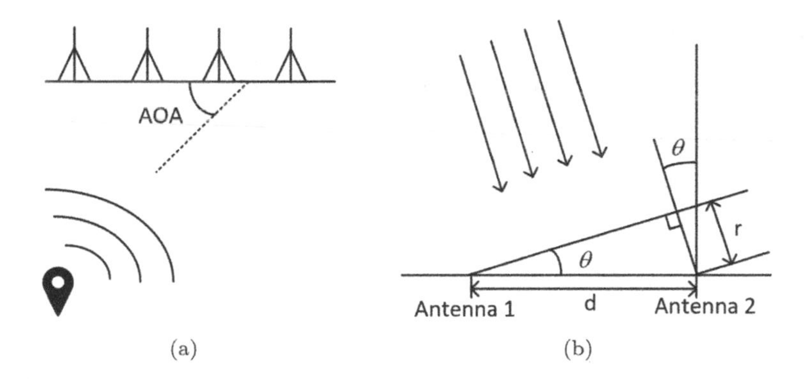


**Fig. 1.** Illustration of antenna array configuration and the relationship of arrival angle and phase difference

Assuming that the bluetooth signal transmitted by the unknown node comes in the form of a plane wave, as shown in Fig. 1(b). $\theta$ is the angle of arrival. According to the geometric relationship, $\theta = \sin^{-1}\frac{r}{d}$, $r = \frac{\lambda\phi}{2\pi}$. Where $r$ is the distance difference between the signal transmitted by the unknown node to reach antenna 1 and antenna 2, $d$ is the actual distance between antenna 1 and antenna 2, $\lambda$ is the wavelength of the signal, $\phi$ is the phase difference between the signal reaching antenna 1 and antenna 2, the angle of arrival is calculated as follows:

$$\theta = \sin^{-1}\frac{\lambda\phi}{2\pi \mathrm{d}} \tag{2}$$

Now, if the angle between the target and the anchor node is $\theta_{\mathrm{i}}$, for $i = 1, 2, ..., N$, according to the position relationship between the nodes, the angel is modeled as:

$$\theta_{\rm i} = \arctan\frac{x_y - a_{iy}}{x_x - a_{ix}} + m_i,\ \text{for}\ i = 1, 2, ..., N \tag{3}$$

Where $m_i$ is the measurement noise of angel between the target and the i-th anchor , suppose $m_i$ obeys a Gaussian distribution with mean zero and variance $\delta^2_{\mathrm{m}_i}$, denoted as $\mathrm{m}_i \in N(0, \delta^2_{m_i})$.

The position estimation of the node to be located is usually obtained by minimizing the following expression [14]:

$$\hat{x} = \arg\min_x \sum_{i=1}^{N} (\frac{1}{\delta^2_{\mathrm{n}_i}}\varepsilon^2_{L_i} + \frac{1}{\delta^2_{\mathrm{m}_i}}\varepsilon^2_{\theta_i}) \tag{4}$$

Where $\varepsilon_{\mathrm{L_i}}{=}L_i - L_0{+}10\gamma\log_{10}\frac{\mathrm{d}_i}{d_0}$ and $\varepsilon_{\theta_{\rm i}}{=}\theta_{\rm i} - \tan^{-1}\frac{x_y - a_{iy}}{x_x - a_{ix}}$ represent the measurement error of RSS and AOA respectively.

# 3 Least Squares Estimation and Weighted Least Squares Estimation Method

## 3.1 Least Squares Estimation

The mean square error of the node position of the maximum likelihood (ML) estimation is the smallest. However, since the ML in (4) is non-convex, it is difficult to find the global optimal solution. Therefore, by transforming the coordinates of formulas (1) and (3), the least squares (LS) estimator can be used to approximate (4). In the case of ignoring noise, formula (1) can be transformed into:

$$\lambda_{\rm i}\mathrm{f}^T_i(x - a_i) - \eta d_0 \approx 0,\ \text{for}\ i = 1, 2, ..., N \tag{5}$$

Where $\lambda_{\rm i} = 10^{\frac{L_{\rm i}}{10\gamma}}$, $\mathrm{f}_i = \left[\frac{1}{2\cos\theta_i}\ \frac{1}{2\sin\theta_i}\right]^T$, $\eta{=}10^{\frac{L_0}{10\gamma}}$. Similarly, without considering noise, formula (3) can be transformed into:

$$h^T_i(x - a_i) \approx 0,\ \text{for}\ i = 1, 2, ..., N \tag{6}$$

Where $\mathrm{h}_i = \left[-\sin\theta_i\ \cos\theta_i\right]^T$. The following estimator can be obtained through the least squares criterion:

$$\hat{\mathrm{x}}_{LS} = \arg\min_x \sum_{i=1}^{N} \left(\lambda_{\rm i}\mathrm{f}^T_i(x - a_i) - \eta d_0\right)^2 + \sum_{i=1}^{N}\left(h^T_i(x - a_i)\right)^2 \tag{7}$$

It can be written in matrix form:

$$\underset{\mathrm{x}}{\text{minimize}}\ \|Ax - b\|^2 \tag{8}$$

Where $A=\begin{bmatrix} \vdots \\ \lambda_{\mathrm{i}} \mathrm{f}_i^T \\ \vdots \\ h_i^T \\ \vdots \end{bmatrix}$, $\mathrm{b}=\begin{bmatrix} \vdots \\ \lambda_{\mathrm{i}} \mathrm{f}_i^T a_i + \eta d_0 \\ \vdots \\ h_i^T a_i \\ \vdots \end{bmatrix}$. Therefore, the closed-form solution of the target position by LS is:

$$\hat{\mathrm{x}}_{LS} = (A^T A)^{-1}(A^T b) \tag{9}$$

### 3.2 Weighted Least Squares Estimation

In order to give more importance to the different links of each measurement, a weighting mechanism needs to be introduced. In [10], a distance-based weighting mechanism is introduced. The author believes that the short-range measurement of RSS and AOA is more accurate than the long-range measurement. Therefore, when the distance between the anchor node and the target is close, the measured value is given a large weight to improve its credibility in the positioning process. Conversely, when the anchor node is far away from the target, the measured value is given a small weight, which reduces its credibility in the positioning process. At present, some articles have studied the weighting mechanism based on path-loss exponent (PLE). In [15], the position of the node and the PLE of each link are jointly estimated for positioning. In [16], PLE is assumed for each communication link. However, these assumptions are simplifications of the actual environment PLE and are not accurate.

In this section, we study a least square method based on noise weighting. First, the approximate position of the target is calculated by LS in Sect. 3.1. Then this position is approximated to the real position of the target, and the noise of the measured values of the target is estimated. The estimated noise is introduced into the weight, and the following estimator can be obtained by weighted least squares(WLS):

$$\hat{\mathrm{x}}_{WLS} = \arg\min_{x} \sum_{i=1}^{N} \mathrm{w}_{n_{\mathrm{i}}} (\lambda_{\mathrm{i}} f_i^T (x - a_i) - \eta d_0)^2 + \sum_{i=1}^{N} \mathrm{w}_{m_i} (h_i^T (x - a_i))^2 \tag{10}$$

Where $\mathrm{w}_{n_{\mathrm{i}}} = 1 - \frac{\hat{n}_i}{\sum_{i=1}^{N} \hat{n}_i}$, $\mathrm{w}_{\mathrm{m}_{\mathrm{i}}} = 1 - \frac{\hat{m}_i}{\sum_{i=1}^{N} \hat{m}_i}$, the noise terms $\hat{n}_i$ and $\hat{m}_i$ are solved by the following formula:

$$\hat{\mathrm{n}}_i = L_{\mathrm{i}} - L_0 + 10\gamma \log_{10} \frac{\|\hat{x}_{LS} - a_i\|}{d_0} \tag{11}$$

$$\hat{\mathrm{m}}_i = \theta_{\mathrm{i}} - \tan^{-1} \frac{\hat{x}_{LS_y} - a_{iy}}{\hat{x}_{LS_x} - a_{ix}} \tag{12}$$

Formula (10) can be abbreviated as:

$$\underset{\mathrm{x}}{\mathrm{minimize}} \, \|W(Ax - b)\|^2 \tag{13}$$

Where $W = I_2 \otimes \text{diag}(w)$, with $\otimes$ denoting the Kronecker product, the closed-form solution of the target position by WLS is:

$$\hat{\mathbf{x}}_{WLS} = (A^T W^T W A)^{-1}(A^T W^T b) \tag{14}$$

# 4 Experiment and Analysis

So as to verify the RSS and AOA joint positioning method based on the weighted least squares algorithm, a set of wireless sensor network indoor positioning verification system based on Bluetooth 5.1 is constructed. The system mainly includes the mobile node to be located, anchor nodes, the gateway and related processing softwares. The gateway is used to receive and transmit data. The mobile node to be located provides the gateway with information such as RSS and AOA in real time. The anchor node provides its own physical coordinates and is equipped with an antenna array to measure phase information. The gateway, the node to be located and the anchor node adopt the RF chip CC2642R produced by TI. In addition, the anchor node carries the BOOSTXL-AOA antenna array board produced by TI for use with CC26X2R.

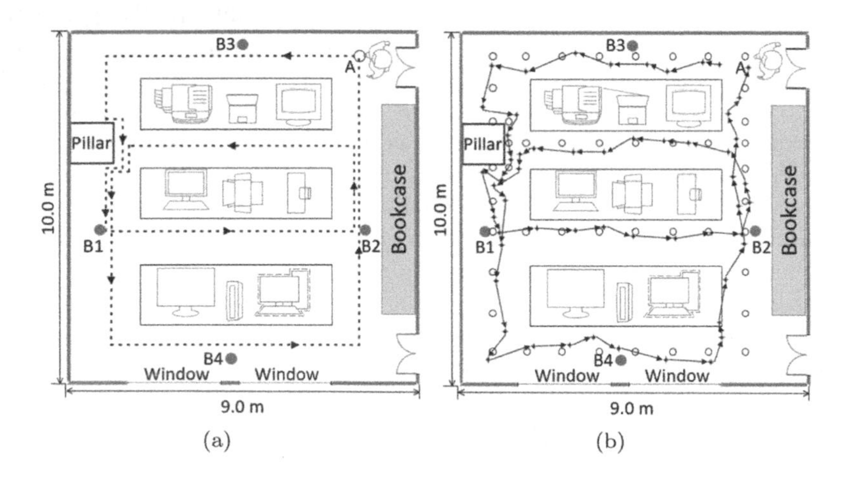


**Fig. 2.** Illustration of location area and location tracking results

The area to be positioned is shown in Fig. 2(a), which is a complex indoor environment. Four anchor nodes are installed on the walls and corners of the experimental area and are marked as B1, B2, B3, B4 in Fig. 2(a). The tester takes the target node and starts at point A according to the trajectory in Fig. 2(a), walks clockwise through the room back to point A, and marks 56 test points along the trajectory. The least square method combining RSS and AOA (LS), the least square method based on distance weighted RSS and AOA fusion (WLS-D), and the least square method based on noise weighted RSS and AOA fusion (WLS-N) are used for positioning. Among them, WLS-D is a distance-weighted fusion method mentioned in[10], and the calculation of its weight is $w_i = 1 - \frac{\hat{d}_i}{\sum_{i=1}^{N} \hat{d}_i}$, with $\hat{d}_i = d_0 10^{\frac{L_0 - L_i}{10\gamma}}$. Figure 2(b) shows the target location tracking results under the WLS-N solution. Figure 3 shows the positioning error of LS, WLS-D and WLS-N, and the positioning error analysis is shown in Table 1.

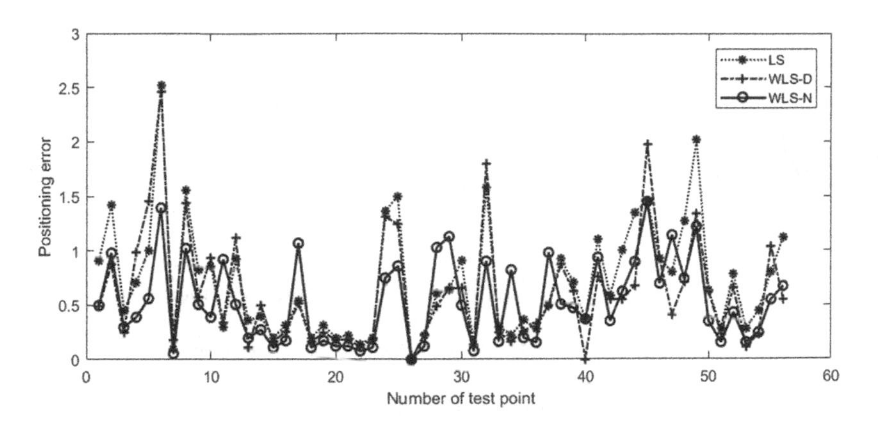


**Fig. 3.** Illustration of the positioning error of LS,WLS-D and WLS-N

**Table 1.** The positioning effect of three indoor positioning algorithms

| Algorithm | Average error (m) | Maximum error (m) | Variance |
|---|---|---|---|
| LS | 0.72 | 2.51 | 0.42 |
| WLS-D | 0.62 | 2.46 | 0.39 |
| WLS-N | 0.53 | 1.45 | 0.33 |

It can be seen from Table 1 and Fig. 3 that compared with WLS-D, the average error of WLS-N has dropped by 0.09 m, and the maximum error has dropped by 1.01 m. Compared with traditional LS, the positioning accuracy of WLS-N is improved by 26%, and the variance of the average error is reduced to 0.33. All these show that the fusion method based on noise weighting can effectively improve the positioning accuracy.

On the basis of the appeal experiment, the testers used LS, WLS-D, and WLS-N to repeat the experiment 30 times along the trajectory of Fig. 2. After recording the positioning error of each point, the cumulative distribution function(CDF) diagram of the positioning error of multiple repeated position experiments is obtained, as shown in Fig. 4. At the same time, count the proportion of each algorithm's positioning error within 1.2 m, as shown in Table 2.

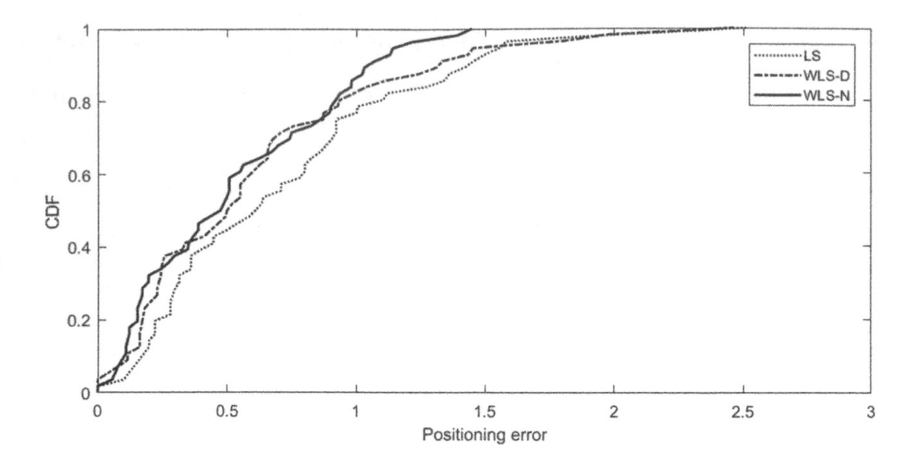


**Fig. 4.** The CDF graph of three positioning methods

**Table 2.** Probability of realizing positioning error within 1.2 m

| Algorithm | Probability/% |
|---|---|
| LS | 82.14 |
| WLS-D | 85.71 |
| WLS-N | 92.86 |

It can be from Fig. 4 and Table 2 that the WLS-N proposed in this paper has a greater probability of achieving a positioning error within 1.2 m than the traditional LS and WLS-D. The probabilities of LS, WLS-D, and WLS-N are 82.14%, 85.71%, and 92.86%, respectively. Experimental results show that the WLS-N proposed in this paper can achieve fast and accurate indoor positioning.

## 5 Conclusion

This paper proposes a joint positioning algorithm of RSS and AOA based on noise weighting. Through coordinate transformation, the least squares solution of the target position is derived, and the real target position is replaced by this to estimate the noise of each measurement value. According to the size of the noise, different weights are assigned to each error term of the target least squares solution. A set of indoor positioning verification system is designed. In the positioning experiments at multiple locations, the average positioning error of LS and WLS-D both reach more than 0.6 m. The average positioning error of WLS-N is reduced to 0.53 m. In addition, the WLS-N has a high probability of 92.86% to achieve precise positioning with a positioning error within 1.2 m. Through experimental verification, the WLS-N positioning algorithm proposed in this paper has a good indoor positioning effect.

## References


1. Ullah Khan, I., et al.: An improved hybrid indoor positioning system based on surface tessellation artificial neural network. Meas. Control **53**, 9–10 (2020)
2. Alhomayani, F., Mahoor, M.H.: Deep learning methods for fingerprint-based indoor positioning: a review. J. Locat. Based Serv. **14**(3), 129–200 (2020)
3. Cheng, K.C., et al.: Using indoor positioning and mobile sensing for spatial exposure and environmental characterizations: pilot demonstration of PM2.5 mapping. Environ. Sci. Technol. Lett. **6**(3), 153–158 (2019)
4. Prince Gregary, B., Little Thomas, D.C.: Two-phase framework for indoor positioning systems using visible light. Sensors (Basel, Switzerland) **18**(6), 1917 (2018)
5. Qiu, C., Mutka, M.W.: Walk and learn: enabling accurate indoor positioning by profiling outdoor movement on smartphones. Pervasive Mob. Comput. **48**, 84–100 (2018)
6. Coluccia, A., Fascista, A.: Hybrid TOA/RSS range-based localization with self-calibration in asynchronous wireless networks. J. Sensor Actuator Netw. **8**(2), 31 (2019)
7. Angeline Beulah, V., Venkateswaran, N.: Sparse linear array in the estimation of AoA and AoD with high resolution and low complexity. Trans. Emerging Telecommun. Technol. **31**(4), e3840 (2020)
8. Zheng, L.: Error self-correction positioning algorithm based on received signal strength indication. J. Sensor Technol. **27**(7), 970–975 (2014)
9. Hao, Z., et al.: Ultra-wideband positioning AOA estimation method based on single base station antenna array. J. Electron. Inf. Technol. **35**(8), 2024–2028 (2013)
10. Tomic, S., et al.: A closed-form solution for RSS/AoA target localization by spherical coordinates conversion. IEEE Wirel. Commun. Lett. **5**(6), 680–683 (2016)
11. Chen, J., Chen, L.: Overview of research on wireless sensor network positioning. Comput. Knowl. Technol. **17**(1), 41–42+51 (2021)
12. Woolley, M.: Bluetooth direction finding function technology overview. Single Chip Microcomput. Embed. Syst. Appl. **19**(7), 90 (2019)
13. Chen, W.: Discussion and test of bluetooth positioning technology. Foreign Electron. Meas. Technol. **39**(10), 143–146 (2020)
14. Yan, C., Ma, J.: Accurate positioning method for 3D sensor network spatial RSS and AOA hybrid measurement. J. Sensor Technol. **30**(3), 450–455 (2017)
15. Salman, N., Kemp, A.H., Ghogho, M.: Low complexity joint estimation of location and path-loss exponent. IEEE Wirel. Commun. Lett. **1**(4), 364–367 (2012)
16. Khan, M.W., et al.: Localisation of sensor nodes with hybrid measurements in wireless sensor networks. Sensors **16**(7), 1143 (2016)

# Research on Vacuum Detection of Thermal Cup Based on Infrared Camera

Haiwei Huang(✉), Hongwei Xu, and Jian Sun

College of Mechanical and Electrical Engineering, China Jiliang University, Hangzhou 310018, China

**Abstract.** Vacuum insulation cup mainly relies on the vacuum of its internal vacuum layer to improve the insulation performance, so you can detect the vacuum of the vacuum layer of the vacuum insulation cup to detect the insulation performance. In this paper, the temperature change of the outer wall of the vacuum insulation cup under different vacuum degrees is taken as the research object, and the heat transfer process of the vacuum insulation cup is simulated by using COMSOL Multiphysics 5.4 (hereinafter referred to as COMSOL), and the experimental verification is carried out based on the infrared camera [1]. The paper simulates and experimentally verifies the temperature trend of the outer wall of the insulation cup under different vacuum degrees, the relationship between the insulation performance of the thermal cup and its vacuum degree was obtained, it promotes the development of vacuum degree detection technology of thermos cup [2].

**Keywords:** Online inspection · Heat transfer · Infrared images · COMSOL

## 1 Introduction

With the development and improvement of living standards, insulation cup more and more people's lives as a necessity [3]. It is made based on the principle of vacuum insulation, by maximizing the vacuum in the vacuum layer to achieve the isolation of heat transfer, thus achieving the purpose of insulation. But the insulation cup insulation performance testing has been a problem, its performance depends mainly on the vacuum layer within the vacuum degree high or low [4]. In the current vacuum testing process, there is generally low efficiency, energy consumption, high cost and other problems, resulting in uneven insulation effect of the insulation cup on the market.

The insulation cups used in this research are made of stainless steel inside and outside, the stainless insulation cup has a pagoda head at the bottom, and is connected to the vacuum pump by a vacuum gas tube. The vacuum degree of the inner vacuum layer was changed by vacuum pump, and its air tightness was checked. Compared with the traditional glass, it is refined with advanced vacuum technology and has good sealing performance [5]. The vacuum level of the vacuum layer of the insulated cup is expressed by its air pressure, which varies from $10^5$ Pa to $10^6$ Pa, the detection time of infrared camera is 1 min. The method uses the principle of infrared thermography by slowly


Q. Han et al. (Eds.): LSMS 2021/ICSEE 2021, CCIS 1469, pp. 388–395, 2021.
https://doi.org/10.1007/978-981-16-7213-2_38

injecting 100 °C hot water into a thermos cup and letting it stand for 1min, to detect, process and calculate the infrared images of the insulation cup under different vacuum degrees, and get the trend relationship between the temperature change of the outer wall of the insulation cup and the change of the size of its vacuum layer air pressure, so as to detect the vacuum degree of the insulation cup [6, 7].

## 2 Detection Principle of Infrared Camera

The principle of infrared camera detection is based on the fact that all objects in nature above absolute zero will radiate infrared radiation that contains information about the characteristics of the object, the infrared radiation in the scene captured by the infrared camera is mapped into grayscale values and converted into infrared images, so that the temperature level in the scene being photographed can be determined based on the magnitude of the infrared radiation value. To be precise, the greater the radiation intensity of a part of the scene, the higher the gray value of this part reflected in the image, and the brighter and hotter it is. Infrared image detection methods are divided into active and passive according to the way they cause temperature differences. Active infrared image detection means that the target under test is injected by heating, so that the target under test loses its thermal equilibrium, and does not need to make its internal temperature reach a uniform and stable state, but in its internal temperature is not uniform, with thermal conductivity process of infrared detection. Passive infrared image detection is the infrared detection in the process of heat exchange between the target under test and the environment, using the condition that the temperature of the target under test is different from the temperature of the surrounding environment [8]. Currently, active heating methods are often used.

The main principle of infrared camera-based vacuum detection for thermal cups:

There are three types of heat exchange in the insulated cup: conduction, convection, and radiation. The 100 °C hot water slowly injected into the thermos cup first has heat convection with the inner wall of the thermos cup, and the stainless inner wall absorbs the heat of the hot water, and heat conduction occurs inside the thermos cup [9]. However, due to the small thickness of the inner wall, heat conduction consumption of heat is basically negligible. The air pressure inside the vacuum layer is low, and there is basically no air flow, so the heat transfer from the inner wall to the outer wall is mainly heat radiation. Radiation heat exchange can be calculated by Eq. (1)

$$\phi = \varepsilon A\sigma\left(T_1^4 - T_2^4\right). \tag{1}$$

Where $T_1$, $T_2$ are the temperatures of the inner and outer walls of the vacuum layer respectively, $\varepsilon$ is the stainless steel 304 emissivity, $A$ is the surface area of the cup wall, $\sigma$ is the Stefan-Boltzmann constant. In the same environment, injected hot water temperature and the same time conditions, due to the pressure in the vacuum layer is different, then the insulation cup outside the performance of the temperature is different. According to this characteristic phenomenon, the infrared camera is used to detect the change of thermal radiation around the outer wall of the insulation cup and map this change on top of the temperature value, and finally the temperature value is used to judge whether the vacuum level inside the vacuum layer is in accordance with the requirements [10].

# 3 COMSOL Simulation of Vacuum Insulation Cup

## 3.1 Structural Model of Stainless Insulation Cup

Due to the need to change the vacuum of the experimental stainless steel insulation cup, while taking into account that the vacuum layer can't be destroyed, but also to facilitate its vacuum operation, so will be in the bottom of the stainless steel insulation cup shell welding a pagoda head, keep the liner intact, pagoda head end welded to the insulation cup shell, and the insulation cup vacuum layer connected to the other end of the pagoda head with a vacuum gas tube and vacuum gas pump connected between the two and a valve, the control the process of vacuuming. The air tightness test was carried out before the experiment, the structure is shown in Fig. 1.

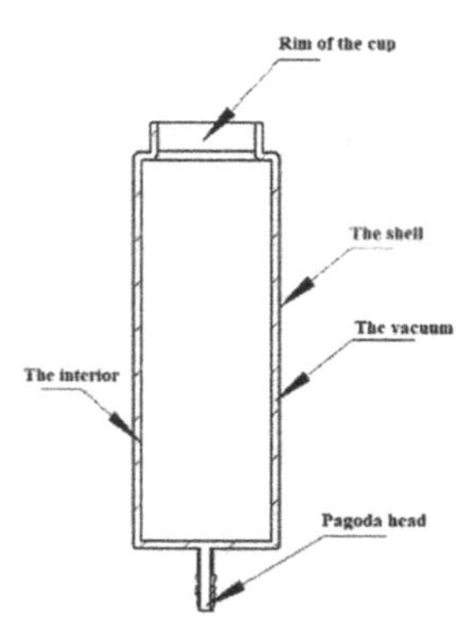


**Fig. 1.** Stainless steel insulation cup structure diagram.

## 3.2 COMSOL Geometric Model Construction

The first step in conducting COMSOL simulation is to establish an accurate and suitable geometric model for the heat transfer process of the thermal cup, before proceeding with

**Table 1.** Actual study of insulation cup parameters.

| Measurement points | Numerical value (mm) |
|---|---|
| Height of cup wall | 215 |
| Effective cup wall height | 200 |
| Inside diameter of cup body | 31 |
| Outer diameter of cup body | 37 |
| Thickness of vacuum layer | 4 |
| Cup radius | 26 |
| Cup bottom thickness | 10 |

the model building, it is necessary to measure various data of the insulation cup and the measured data of the actual thermal cup used in the study are shown in Table 1.

When choosing the modeling dimension, the two-dimensional axisymmetric structure is selected, which is consistent with the thermos cup being an axisymmetric structure and saves the modeling time. According to the measured data in Table 1, the thermal cup model structure is constructed by COMSOL with a physical field control mesh, and the mesh density is hyperfine, and the mesh density satisfies the mesh irrelevance. The finite element geometry and network model of the thermos cup are shown in Figs. 2 and 3.

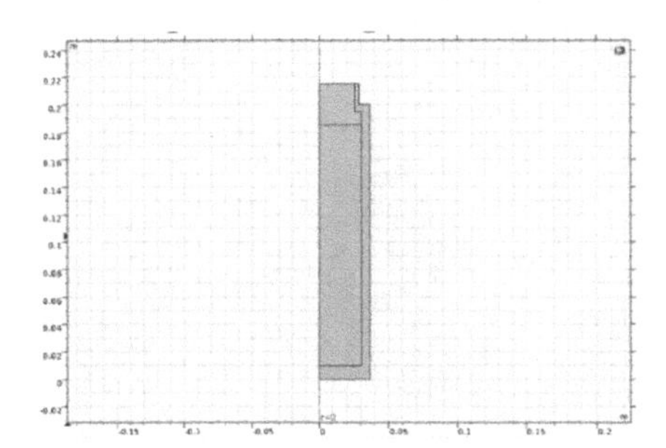

**Fig. 2.** Geometric model of insulation cup.

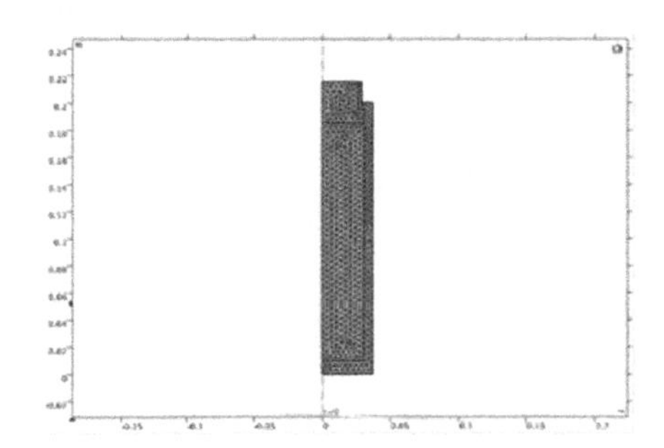

**Fig. 3.** Network model of insulation cup.

Since the injected hot water is essentially stationary with respect to the thermos, the effect of fluid flow on heat transfer can be ignored. COMSOL does not have a specific description of the concept of "vacuum layer", and the impact of air pressure changes in the vacuum layer of the insulation cup is described by the characteristics of this new material, the most intuitive is the size of the thermal conductivity to describe the impact of this change in air pressure.

### 3.3 Model Simulation Calculation Results and Analysis

The boundary condition settings in the finite element calculation should be the same as the actual experimental test conditions, the temperature of the outer wall of the thermos cup is the same as the ambient temperature, which is 20 °C. The material in the cup is defined as water, and the initial temperature is set at 100 °C, because the temperature change of vacuum cup during heat transfer is studied, the transient solver is used to solve the problem, and the heat transfer time is set to 60 s, and the time step is set to 2 s. In addition, in order to facilitate the experimental verification, it is necessary to set up a group of one-dimensional drawing group. The drawing area of this group is the outer wall of the thermos cup, as shown by the blue line in Fig. 4.

During the simulation process, the air pressure in the vacuum layer is taken to be $10^5$–$10^6$ pa in several values for simulation experiments. Due to the limitation of space, the following three-dimensional diagrams of the heat transfer process on the surface of the insulation cup when the internal pressure of the vacuum layer is $8*10^5$ pa and $2*10^5$ pa are selected for comparison (Fig. 5 (a) and (b)).

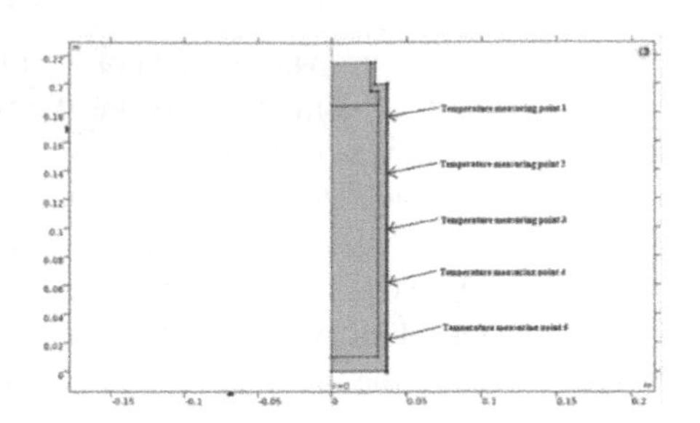

**Fig. 4.** Data acquisition area for 1D mapping group.

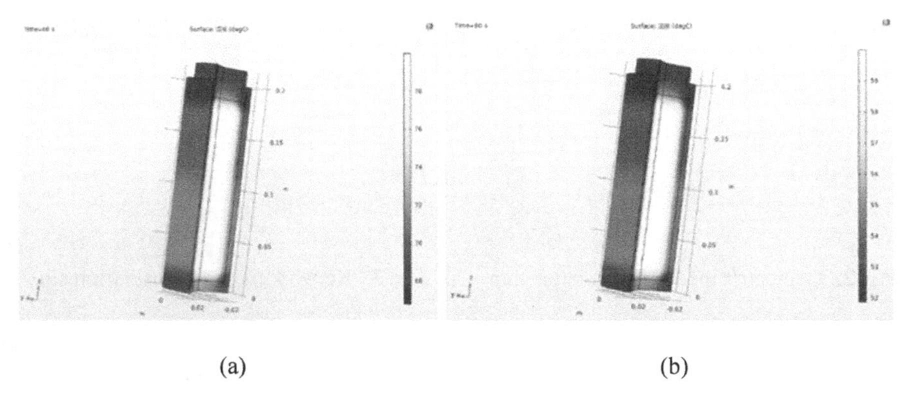

(a) (b)

**Fig. 5.** 3D simulation results of thermal cups under two vacuum levels.

Figure 6 (a) and (b) are the temperature change of the outer surface of the wall of the vacuum insulation cup after 1 min of heat transfer when the air pressure inside the vacuum layer of the insulation cup is $8*10^5$ pa and $2*10^5$ pa, respectively, where the curve change inside Fig. 6 is based on the time change [11].

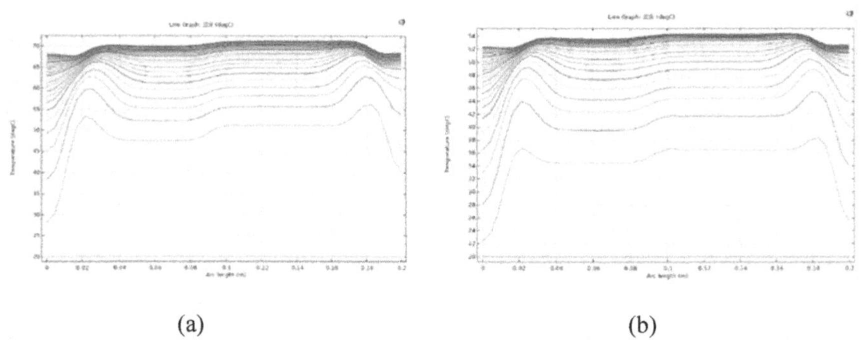

(a) (b)

**Fig. 6.** Simulation of the temperature change of the wall of a thermal cup under two vacuum levels.

From the above CMOSOL simulation of the cup wall temperature change graph with time can be seen, in the same conditions with the insulation layer air pressure increases, the insulation cup heat dissipation process the greater the temperature change, the most obvious temperature change in the heat transfer process between the mouth and the bottom of the cup of the thermal insulation cup, at the temperature measuring point 1 and the temperature measuring point 5, a peak value of the temperature value is reached and the change rate of the two temperature measuring points is the most obvious under different air pressure. It can be concluded that the infrared image of the outer wall of the thermos will change differently under different vacuum levels, so the simulation results show that the infrared camera-based thermos vacuum detection is feasible.

## 4 The Experimental Test

According to the detection principle, the experimental platform shown in Fig. 7 was built to carry out the experimental verification of infrared camera-based vacuum detection of thermal insulation cups.

**Fig. 7.** Vacuum insulation cup vacuum degree experimental test platform.

In the process of the experiment, the insulation cups were tested under different vacuum degrees, i.e., different air pressures in the vacuum layer, and the experimental insulation cups were pumped using a vacuum pump to determine the current air pressure in the vacuum layer by the vacuum gauge readings on the platform. After controlling the air pressure in the vacuum layer, close the pumping valve, and then slowly inject 100 °C hot water into the insulation cup to be tested, and let it stand for 1 min, infrared camera connected to the processing software, acquisition of infrared thermal images [12], and after processing the effective cup wall height is divided into five temperature measurement points, display each temperature measurement point of the cup wall temperature and record, the number of experiments conducted five times. So in the actual vacuum testing, the vacuum inside the insulation layer can be determined by testing the surface temperature change of the insulation cup.

As shown in Figs. 8 and 9, the temperature change trend of each temperature measurement points of the insulation cup with the air pressure of $8*10^5$ pa and $2*10^5$ pa in the vacuum layer is injected with hot water and left for 1 min.

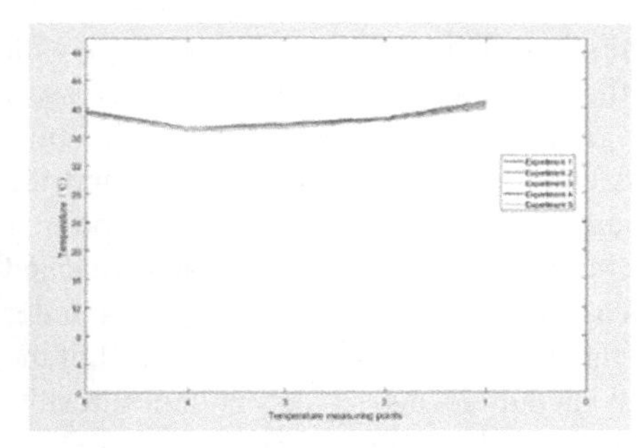

Fig. 8. Temperature change under $8*10^5$ pa.

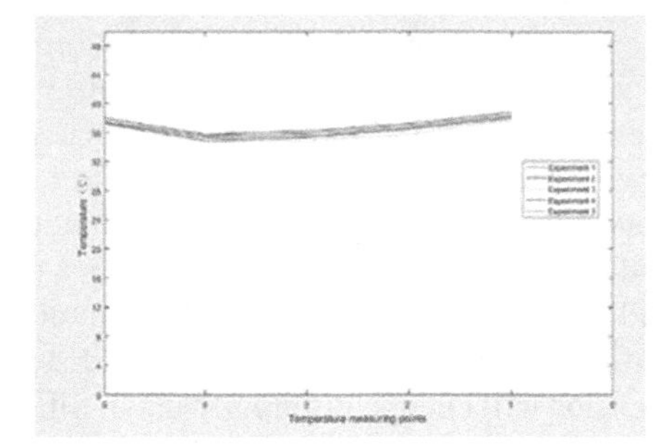

Fig. 9. Temperature change under $2*10^5$ pa.

It can be seen that, with the reduction of air pressure in the vacuum layer, the surface temperature of the insulation cup wall is gradually decreasing trend, the heat transferred to the surface of the insulation cup becomes smaller, the insulation cup surface temperature becomes relatively low. The mouth and bottom of the cup, i.e., temperature measurement point 1 and temperature measurement point 5, reached the peak temperature, and the temperature change was more obvious, while the body of the cup, i.e., temperature measurement points 2, 3 and 4, had a more moderate temperature change. Therefore, in the actual insulation cup vacuum detection, the use of infrared camera to shoot the insulation cup bottom or cup mouth detection effect is most obvious.

## 5 Conclusion

This paper conducts an exploratory study on infrared camera-based vacuum detection method for thermal insulation cups, the heat transfer process of the vacuum insulation cup is simulated by COMSOL software, and the temperature is obtained by using the infrared camera to shoot and process the calculation, and the feasibility of the thermal infrared technology detection method is verified through experiments and data analysis, and certain results are achieved [13]. The higher the vacuum layer of the insulation cup, the higher the vacuum, the lower the surface temperature, that is, the insulation effect is better. In the simulation process, it was found that when the vacuum degree changed, the temperature of the mouth and the wall of the vacuum cup changed greatly in the heat transfer process, and it was verified by experiments. By analyzing the experimental data, it was found that the temperature of the bottom and the mouth of the vacuum cup changed most obviously in the heat transfer process. Therefore, in the actual insulation cup vacuum testing, the cup mouth and cup bottom are more suitable as the insulation cup testing parts.

## References

1. Yuan, H., Liu, D., Wang, X.H., Liu, P., Rong, M.Z., Ding, H.B.: Preliminary study on vacuum degree online detection of vacuum circuit breaker based on laser induced breakdown spectroscopy. In: 2016 International Conference on Condition Monitoring and Diagnosis (CMD), pp. 135–138 (2016)

2. Liu, J.B.: Research on structure design of thermocouple junction for spacecraft thermal vacuum tests. J. Phys. Conf. Ser. **1754**, 012067 (2021)
3. Wang, S.J., Feng, Y.J., Huang, Y.Q., Wu, J.C., Yu, L.F.: Pressure measurement of vacuum insulation panel with infrared spectroscopy. J. Chin. J. Vacuum Sci. Technol. **11**, 1074–1079 (2013)
4. Li, D.: Research on online inspection method of vacuum degree of squash automatic packaging. J. China Condiment. **8**, 110–113 (2020)
5. Zhou, F., Liu, Q.Y.: COMSOL-based numerical simulation of heat and mass transfer for fructose dehydration. J. Acta Energ. Solar. Sinica **6**, 1677–1683 (2019)
6. Yu, S.X.: Research on detection of micro cracks on metal surface based on infrared thermal wave technology. J. Mach. Des. Manuf. Eng. **50**, 101–106 (2021)
7. Zhao, J., Wu, M.Y., Song, G., Li, T.R.: Measurement and application of non-contact infrared temperature measurement device in vacuum drying oven. J. Sci. Technol. Innov. Herald **170**, 55–56 (2020)
8. Bao, X.D., Yu, X.L., Mao, H.X.: Research on fluid field and infrared radiation of vacuum plume based on theoretical analytical method. J. Infrared Laser Eng. **7**, 49 (2020)
9. Chen, Y.H., Lin, W.Y., Teng, C.C., Su, G.D.: Detection of vacuum level inside sealed field emission displays by infra spectroscopy. In: 2005 International Vacuum Nanoelectronics Conference, UK (2005)
10. Jeon, G., Kim, W.Y., Lee, H. C.: Thin-film vacuum packaging based on porous anodic alumina (PAA) for infrared (IR) detection. In: SENSORS, 2012. IEEE (2012)
11. Ito, T., Tokuda, T., Kimata, M., Tokashiki, N.: Vacuum packaging technology for mass production of uncooled IRFPAs. In: Proceedings of SPIE, vol. 7298, pp. 72982A-1–10 (2009)
12. Zhang, L.Y., Du, Y.X., Ding, L.: Research on threshold segmentation algorithm and its application on infrared small target detection algorithm. In: 2014 12th International Conference on Signal Processing (ICSP), pp. 67–682 (2014)
13. Zhang, Z.P., Song, X.Y., Li, Y.F.: Application of infrared thermography in the study of vacuum pre-cooling characteristics of different fruits and vegetables. J. Sci. Technol. Bullet. **237**, 15–17 (2009)

# Application of Blockchain Technology in the Assessment System

Yinyi Zheng and Xiai Chen(✉)

China Jiliang University, 258 Xueyuan Street, Xiasha Higher Education Park, Hangzhou, China
xachen@cjlu.edu.cn

**Abstract.** This paper is aimed at the data security needs of the assessment system, and uses the characteristics of blockchain technology such as distributed storage, tamper-proof modification, and consensus mechanism, and proposes a certificate assessment system based on blockchain technology. The data of the system is stored in a distributed structure, that is, multiple databases are distributed in various regions of the country, and data synchronization is carried out between each node. After the user obtains the certificate, the system will hash the certificate data and store it in the database in the area where it is located, and finally synchronize it to other nodes through broadcasting to complete the data storage. Compared with the assessment system designed by the existing centralized scheme, the assessment system based on the blockchain distributed storage scheme has the advantages of traceability, non-tampering, and decentralization, which improves the security of certificate data records.

**Keywords:** Assessment system · Blockchain · Distributed · Security

## 1 The Background and Significance of the Research

Blockchain originated from Bitcoin, which is a technical solution proposed by Satoshi Nakamoto in 2008 [1]. The blockchain is essentially a decentralized distributed database based on encryption algorithms. It has distinctive features such as decentralization, traceability, unforgeability, full recording, openness and transparency.

Blockchain technology provides a solution to the problem of trust [2–4]. Nowadays, financial institutions have invested a lot of cost in the study of the traditional solution of trust issues, and the technical solution of blockchain provides new ideas for solving trust issues at low cost in the future. The assessment system based on blockchain technology improves the security of system data from the technical solution and reduces the cost of maintaining data security in traditional solutions [8].

In 2019, the central government raised blockchain technology to the level of national strategy. Therefore, blockchain technology has great development prospects in many aspects such as national economic development, information technology innovation, and people's livelihood. Our country's current economy is turning to high-quality development, and blockchain technology can provide certain technical support for this transformation.


Q. Han et al. (Eds.): LSMS 2021/ICSEE 2021, CCIS 1469, pp. 396–403, 2021.
https://doi.org/10.1007/978-981-16-7213-2_39

## 2 The Core Technology of Blockchain

### 2.1 Hash and Digital Signature

The relevant principles of cryptography mainly used in the blockchain: 1. Hash; 2. Digital signature.

Hash. The hash function can transform an input of any length into a fixed-length output, and this output is the hash value of the previous input. However, because the input space is much larger than the hash value space, it may have the same hash output under completely different inputs, that is, if input $A \neq B$, $H(A) = H(B)$, this phenomenon is called "hash collision". Although the hash function will have "collision", "collision" is difficult to create artificially, that is, given A, find H(A). It is difficult to find an $A1 \neq A$ such that $H(A1) = H(A)$. In the case that the input space is large enough and the output distribution of the hash function is sufficiently uniform, given an input B, H(B) is calculated, and it is almost impossible to calculate B according to H(B) inversely.

Digital signature. The digital signature uses the public key to encrypt (decrypt) the digital digest information, and uses the private key to decrypt (encrypt) the operation. The public key is publicly distributed, while the private key is only known by each owner. If the information is signed with a private key, then only the public key corresponding to the private key can be used for signature verification to prove that the sender of the information is the owner of the private key. This encryption algorithm, which uses different keys for encryption and decryption, is called an asymmetric encryption algorithm. In summary, the digital signature mechanism uses a pair of corresponding public and private keys to verify the signature to ensure the authenticity and integrity of the data and prevent data from being tampered with.

### 2.2 Data Structure

Blockchain, as its name suggests, is composed of blocks connected by hash value. As shown in Fig. 1 below, this is the structure of two adjacent blocks in the blockchain.

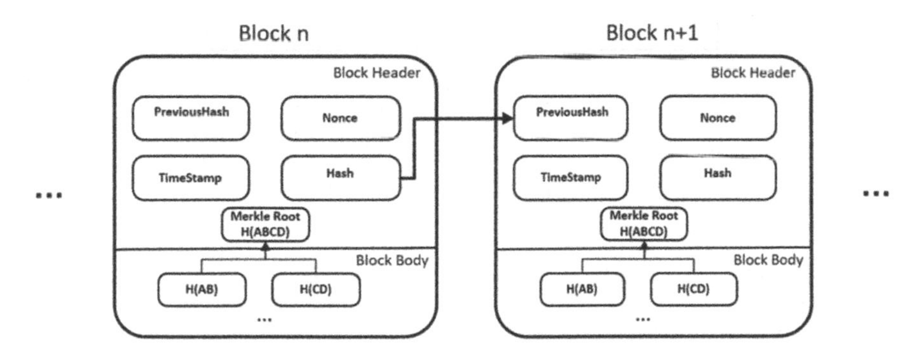


**Fig. 1.** Blockchain data structure diagram

A single block is the data storage unit of the entire block chain, and a block is divided into a block header and a block body.

The block header mainly contains the basic information of the current block such as the previous hash, random number, timestamp, current hash, and the root node of the Merkel tree. The previous hash refers to the hash value of the previous block. The random number is counted and recorded in the process of implementing the proof-of-work (POW) mechanism. The timestamp is used to record the creation time of the block. The current hash, which is obtained by hashing all the information of the current block plus the previous block hash value. The root node of the Merkel tree summarizes and records all transaction data in the block body.

The information recorded in the block body is the task data carried by the block, which specifically contains a list of all transaction data. In the blockchain, as long as the information of one of the blocks is tampered with, it will affect all blocks starting from this block.

## 2.3 Merkle Tree

Merkel tree was proposed by Ralph Merkle in 1979. It is essentially a binary tree data structure, which is composed of data blocks, leaf nodes, intermediate nodes, and root nodes. As shown in Fig. 2 below.

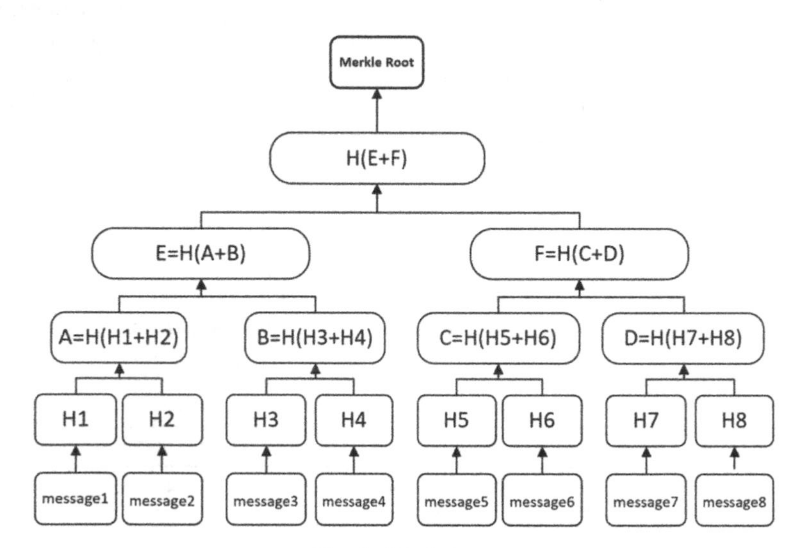


**Fig. 2.** Merkel tree

If want to build a Merkel tree, we must first start from the lowest level of the original data. Hash the original data (message1–message8) at the bottom layer separately to obtain the hash values (H1–H8). Starting from H1–H8, the hash values of adjacent leaf nodes are formed into a new string for hash operation. For example, H1 + H2 is hashed to get A, and A + B is hashed to get E, In the same way, F is obtained, and then E + F is calculated to finally get the root node of the Merkel tree.

### 2.4 Consensus Mechanism

The blockchain consensus mechanism is to formulate a plan to reach a certain consensus between all bookkeeping nodes [9, 11, 12]. This is not only a means of identification, but also a means of preventing tampering. The following mainly describes several common consensus mechanisms [5–7, 10].

### 2.5 Proof of Work

The proof-of-work (POW) mechanism is a consensus mechanism applied in the Bitcoin system. The Bitcoin system uses a proof-of-work mechanism to allow everyone to compete for the right to bookkeeping. The user who is the first to calculate the corresponding block according to the established rules can get the reward of the system-Bitcoin. This is "mining", and the people who participate in the mining are called "miners". And this speed depends on the computing power of the computer used for mining.

### 2.6 Proof of Stake

In general, the working principle of the proof-of-stake (POS) is: the verifier pledges the resources it holds as a guarantee of the ability to create a block. The more resources pledged, the greater the probability that the system will assign the right to create a block to the verifier.

Compared with POW, POS does not require a lot of unnecessary calculations, does not consume excessive resources, and accelerates the verification speed. But the problem is that this proof mechanism will lead to "the rich get richer".

### 2.7 Delegated Proof of Stake

DPOS improves the verification speed by reducing the number of verifiers on the basis of POS. The validator is selected by the resource holder through voting, and the weight of each vote is determined by the voter's holding of resources. Voters can vote for verifiers at any time. DPOS is more fair than POS.

## 3 Classification of Blockchain

### 3.1 Public Blockchain

The public blockchain, as the name implies, is public and open, allowing anyone to participate in and view all the data that has occurred on the chain, without any permission distribution, so it is the chain with the highest degree of openness and decentralization in the blockchain. Data operations, updates, and storage on the public blockchain do not rely on third-party (centralized) servers, it depends on every node on the system network. For example, the Bitcoin system is a public blockchain network system. Any node on the network has the right to participate in the packaging of blocks. Based on the POW consensus mechanism, it is determined who will package the latest block. However, due to this characteristic of the public blockchain, there are too many participating nodes, and its transaction response speed is very low, resulting in low efficiency. Furthermore, the openness and openness of the public blockchain makes data without privacy.

### 3.2 Private Blockchain

Private blockchain. It is a concept relative to the public blockchain. The private blockchain is not open to the outside world, and is only used in the internal application of the organization, such as the bill management and accounting audit of various large enterprises. Every member who participates in the operation of private blockchain data has submitted an identity certificate in advance and has been authenticated, so the possibility of malicious attacks is relatively small. Since the number of nodes on the private blockchain is small, and each node has a high degree of trust, the transaction does not need to be confirmed by all network nodes, so the transaction speed between nodes is very fast. Transactions on private blockchains only need to be confirmed by a few recognized high computing power nodes, and their transaction costs are much lower than public blockchains and consortium blockchains.

### 3.3 Consortium Blockchain

Consortium blockchain is a model for achieving alliances between institutions or between organizations. Consortium blockchains have different permission operation requirements for different scenarios. Therefore, like private blockchains, consortium blockchains also require certain identity certification and permission authentication. The number of nodes in the consortium blockchain is often determined, and authorization is required to join or exit the consortium network. Various institutions (organizations) form an consortium block chain through a certain common interest appeal to jointly maintain the data of the blockchain and ensure data security.

## 4 Application in the Assessment System

### 4.1 Basic Structure of the System

The assessment system based on the blockchain solution is suitable for the assessment records of certificates. When a person obtains a certificate after an assessment in a certain area, the data needs to be recorded in the database, in order to be able to conveniently query in various regions of the country, and to ensure that the data cannot be tampered with, so the system adopts distributed storage Structure, there are multiple ledgers used for data storage of certificate information in various regions of the country. As shown in Fig. 3 below, when the ledgers (also called nodes) of each region are updating data, data transmission is carried out to the remaining nodes through broadcasting to ensure that the data of each node is consistent.

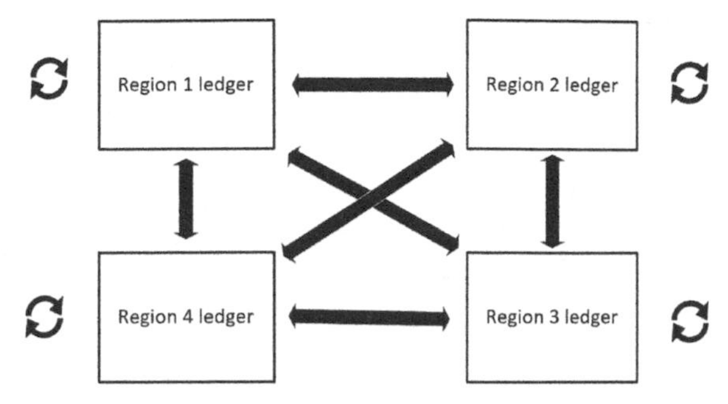


**Fig. 3.** Structure of the assessment system

## 4.2 Basic Principles of the System

Each node in the assessment system has the right to update data, and all other nodes in the system are kept synchronized with data through broadcasting. As shown in Fig. 4 below, Region X ledger first uses the SHA-256 algorithm to obtain the digital digest information of the certificate information data that needs to be updated. After that, the digital digest information is encrypted with the private key of Region X ledger to obtain the encrypted data, and the encrypted data is broadcasted with the digital digest information. After Region Y ledger receives the data transmitted by Region X ledger, it uses the public key of Region X ledger (published by Region X ledger) to decrypt the encrypted data. The decrypted digital digest information is compared with the received digital digest information. If the comparison is successful, it can be proved that the certificate information has not been illegally modified during the transfer process. The legality of the transmitted data is verified through digital signatures, and only legal and valid data are packaged into blocks to complete the update of the certificate information data.

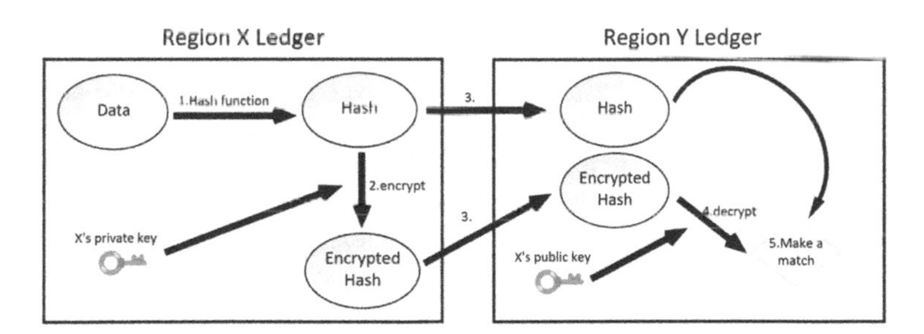


**Fig. 4.** Digital signature process

Each node stores the certificate information data in the form of blockchain, as shown in Fig. 5 above. As mentioned in 2.2, If you want to illegally tamper with the nth block on the blockchain in the node, you must modify all blocks from the nth block to the latest block, so as to ensure that the tampering action of the block is on this blockchain is

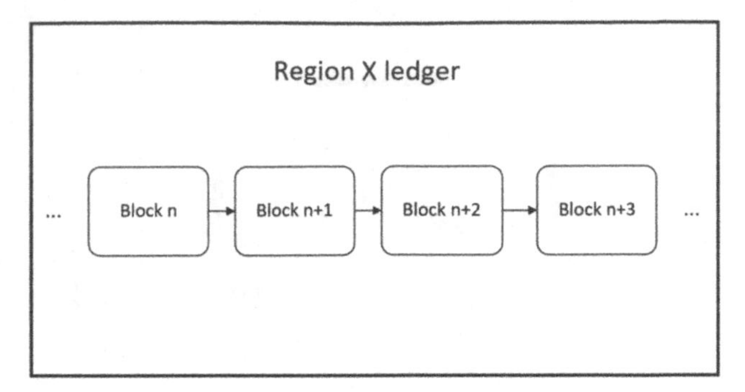


**Fig. 5.** Storage structure of a single node

legal. But in the entire system, the data modification of this node cannot be recognized by the system.

The assessment system uses a distributed storage structure. Distributed storage ensures the consistency of all node data. If the data of a certain node is tampered with due to the above reasons, the data of this node is inconsistent with nodes in other regions. Therefore, it can be concluded that the node data has been illegally tampered with by comparing the data of each node. Set the number of nodes in the system to an odd number ($2n + 1$). Under the premise that the rate of node tampering in the entire system does not exceed 50%, the system can accurately find the location of the tampered node, the location of the illegal data on the internal blockchain of the node, and the time of the tampering by comparing the data between the nodes, ensuring that the data in the system cannot be tampered with.

## 5 Summary

In summary, blockchain technology guarantees data security from three levels of the system: 1. Data structure level. At the data structure level, the blocks on the blockchain of each node are connected by hash values. The irreversible algorithm of hash ensures that the block connection has a certain degree of security. 2. Data transmission level. The system uses digital signature technology in the blockchain to ensure the security of data transmission. First, the digital digest obtained by the hash operation is encrypted with a private key. After transmission to another node, the node uses the corresponding public key to decrypt, and finally confirms whether the data has been tampered with during transmission, and only package and record data that has been confirmed to be valid (not tampered with). 3. System consensus level. All nodes of the entire assessment system adopt a distributed storage method, under the consensus that the data of each node is consistent, the tampering of the data of a few nodes will not be recognized by the system.

## References

1. Kaushal, P.K., Bagga, A., Sobti, R.: Evolution of bitcoin and security risk in bitcoin wallets. In: 2017 International Conference on Computer, Communications and Electronics, pp. 172–177. IEEE Press, India (2017)
2. Frauenthaler, P., Sigwart, M., Spanring, C., Sober, M., Schulte, S.: ETH relay: a cost-efficient relay for ethereum-based blockchains. In: 2020 IEEE International Conference on Blockchain (Blockchain), pp. 204–213. IEEE Press, Greece (2020)
3. Davenport, A., Shetty, S.: Air gapped wallet schemes and private key leakage in permissioned blockchain platforms. In: 2019 IEEE International Conference on Blockchain, Atlanta, GA, USA, pp. 541–545. IEEE Press (2019)
4. Lunardi, R.C., Alharby, M., Nunes, H. C., Zorzo, A.F., Dong, C., Moorsel, A.v.: Context-based consensus for appendable-block blockchains. In: 2020 IEEE International Conference on Blockchain, pp. 401–408. IEEE Press, Greece (2020)
5. D., Kim, R., Ullah, B., Kim.: RSP Consensus Algorithm for Blockchain. In: 2019 20th Asia-Pacific Network Operations and Management Symposium, pp.1--4. IEEE Press, Matsue, Japan (2019)
6. Pahlajani, S., Kshirsagar, A., Pachghare, V.: Survey on private blockchain consensus algorithms. In: 2019 1st International Conference on Innovations in Information and Communication Technology, pp. 1–6. IEEE Press, India (2019)
7. Watanabe, H., Fujimura, S., Nakadaira, A., Miyazaki, Y., Akutsu, A., Kishigami, J.J.: Blockchain contract: a complete consensus using blockchain. In: 2015 IEEE 4th Global Conference on Consumer Electronics, pp. 577--578. IEEE Press, Japan (2015)
8. Natoli, C., Gramoli, V.: The blockchain anomaly. In: 2016 IEEE 15th International Symposium on Network Computing and Applications, Cambridge, MA, USA, pp. 310–317. IEEE Press (2016)
9. Watanabe, H., Fujimura, S., Nakadaira, A., Miyazaki, Y., Akutsu, A., Kishigami, J.: Blockchain contract: securing a blockchain applied to smart contracts. In: 2016 IEEE International Conference on Consumer Electronics, Las Vegas, NV, USA, pp. 467–468. IEEE Press (2016)
10. Sankar, L.S., Sindhu, M., Sethumadhavan, M.: Survey of consensus protocols on blockchain applications. In: 2017 4th International Conference on Advanced Computing and Communication Systems, pp. 1–5. IEEE Press, India (2017)
11. Sukhwani, H., Martínez, J.M., Chang, X., Trivedi, K.S., Rindos, A.: Performance modeling of PBFT consensus process for permissioned blockchain network (Hyperledger Fabric). In: 2017 IEEE 36th Symposium on Reliable Distributed Systems, pp. 253–255. IEEE Press, China (2017)
12. Zheng, Z., Xie, S., Dai, H., Chen, X., Wang, H.: An overview of blockchain technology: architecture, consensus, and future trends. In: 2017 IEEE International Congress on Big Data (BigData Congress), Honolulu, HI, USA, pp. 557–564. IEEE Press (2017)

# A Multi-Populations Human Learning Optimization Algorithm

Jiaojie Du, Ling Wang(✉), and Minrui Fei

Shanghai Key Laboratory of Power Station Automation Technology, School of Mechatronics Engineering and Automation, Shanghai University, Shanghai 200072, China
wangling@shu.edu.cn

**Abstract.** Due to the greediness of the individual learning operator and the social learning operator, the standard HLO is likely to fall into local optimum. Inspired by the fact that the multi-populations mechanism can increase the diversity, this paper proposes a multi-population human optimization algorithm (MPHLO), which divides the population into the elite subpopulation and ordinary subpopulation according to the fitness value. The elite population uses a global topology to fasten the convergence of the algorithm, while the ordinary population uses a ring topology to maintain the diversity of the algorithm. In addition, the two subpopulations both learn from the global optimal individual to guarantee the global search ability of the algorithm. The presented MPHLO is applied to solve CEC14 benchmark function and its results are compared with other state-of-art metaheuristic algorithms to evaluate its performance. The experimental results demonstrate that MPHLO has advantages and significantly outperforms the compared algorithms.

**Keywords:** Human learning optimization · Multi-populations · Global topology · Ring topology

## 1 Introduction

Human learning optimization (HLO) [1] is an emerging metaheuristic algorithm which is motivated by the learning process of human beings to improve the equality of solutions and achieve the purpose of searching for optimization. HLO consists of three learning operators, namely random learning operator (RLO), individual learning operator (ILO) and social learning operator (SLO), that imitate the random learning strategy, the individual learning strategy and the social learning strategy in the learning activities of humans, respectively. Considering the ease of implementation and excellent global search ability, HLO is a very promising optimization algorithm.

To improve the performance of HLO, different kinds of variants of HLO have been proposed. An adaptive simplified human learning optimization (ASHLO) [2] algorithm, which adopts the linear adaptive strategies for *pr* and *pi*, is proposed to balance the exploration and exploitation. Inspired by the fact of the IQ of human follows a normal distribution curve [3], a diverse human learning optimization algorithm (DHLO) [4] is proposed to dynamically adjust the value of *pi* for enhancing the global search ability of


Q. Han et al. (Eds.): LSMS 2021/ICSEE 2021, CCIS 1469, pp. 404–421, 2021.
https://doi.org/10.1007/978-981-16-7213-2_40

the algorithm. Later, a new adaptive mechanism based on the sine-cosine functions [5] is designed to strengthen the search efficiency and reduce the workload of the parameter settings. Recently, a hybrid-coded human learning optimization [6] is proposed to solve mixed-variable optimization problems, in which the binary and discrete variables are optimized by the standard HLO while the real variables are searched out by a continuous HLO. In [7], ASHLO is hybridized with Genetic Algorithms as well as Particle Swarm Optimization (PSO) to solve the supply chain network design problem with possibility of direct shipment, and the obtained results reveal the effectiveness of the hybrid strategy. Besides, the hybrid of HLO and PSO [8] is used to solve groups of the flexible job-shop scheduling problems (FJSP), and the results prove that HLO-PSO can tackle most of single-objective FJSP more efficiently. Nowadays, the HLO algorithms have been successfully applied to knapsack problems [1, 2, 4], multi-dimensional knapsack problems [4, 9] scheduling optimization [10], extractive text summarization [11], financial markets forecasting [5], optimal power flow calculation [12], mixed-variable engineering optimization problems [6], and etc.

The original HLO has advantages of simple structure, few control parameters and easy understanding. Because HLO performs individual learning and social learning by copying the individual optima and social optima, it has an excellent search efficiency and high convergence speed. However, on the other hand, HLO may suffer from the premature convergence problem and fall into local optimum due to these characteristics. Multi-swarm technique has attracted more attention during the recent decade [13] as it can maintain the diversity of the population effectively [14]. Therefore, a multi-population human learning optimization algorithm (MPHLO) is proposed in this paper. Specifically, the population is divided into two sub-populations according to the finesses of individuals, which are named as the elite subpopulation and ordinary subpopulation, respectively, and the information transfers between the subpopulations is achieved by learning the best optima of the population along with the evolutionary process. Besides, two different neighborhood topologies are used in the subpopulations to improve the population diversity and sear efficiency simultaneously.

This paper is arranged as follows. Section 2 presents the proposed multi-populations human learning optimization algorithm in details. Then, MPHLO is applied to tackle CEC14 benchmark functions to evaluate its performance, and results and discussion are provided in Sect. 3. Section 4 concludes this paper.

## 2 Multi-Populations Human Learning Optimization

The multi-population approaches are useful for maintaining the population diversity, because various sub-populations can be situated in different regions of search space [15]. Inspired by this fact, the population of MPHLO is partitioned into two subpopulations according to the fitness values, i.e. the elite sub-population and ordinary sub-population. The neighborhood topology significantly affects the information interaction between individuals, and therefore controls the exploration and exploitation of the algorithms [13]. To enable MPHLO to maintain diversity and have the fast convergence, different topologies are used for the two subpopulations to generate the new solutions. To maintain the fast convergence of HLO, the global topology is used in the elite subpopulation, and the ring topology is used in the ordinary subpopulation to increase the diversity of the algorithm. In addition, each subpopulation learns from the global optimal individual to prevent the two subpopulations from falling into local optimum due to lack of information exchange. The learning mechanism MPHLO is depicted as Fig. 1 where the red dot represents the individual with the global optimal value of the whole population, the blue dots represent the individual with the optimal fitness value in each subpopulation, and the black dots represent the other individuals of the subpopulations. The blue lines in the two subpopulations represent the connections of individuals. Note that the individuals in the elite subpopulation are fully connected and those in the ordinary population are connected in a ring topology. The two dotted lines with two arrows indicate that the social optimal of the whole population is obtained by comparing the socially optimal values of two subpopulations. Two solid lines with black arrows indicate that the individuals of the two subpopulations learn from the global optima with a certain probability.

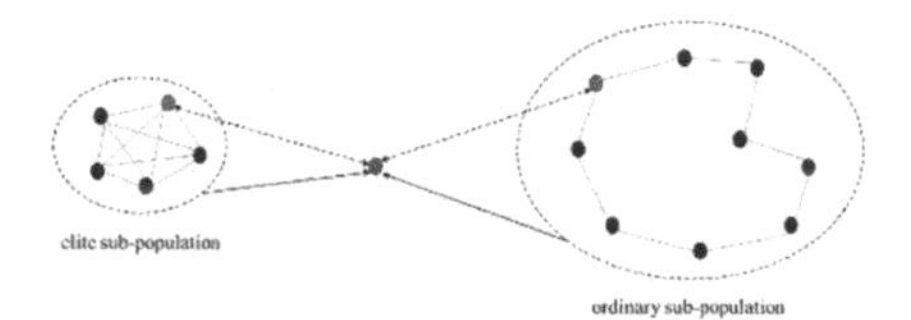


**Fig. 1.** The learning mechanism of MPHLO

MPHLO adopts the binary-coding framework in which each bit corresponds to a basic component of knowledge of problems. Therefore, an individual is initialized as "0" or "1" randomly assuming that there is no prior-knowledge of problems, which is represented by a binary string as Eq. (1)

$$x_i = \left[ x_{i1}\ x_{i2}\ \cdots\ x_{ij}\ \cdots\ x_{iM} \right], 1 \leq i \leq N, 1 \leq j \leq M \tag{1}$$

where $x_{ij}$ is the $j$-th bit of $i$-th individual, $N$ is the number of the population, and $M$ is the length of solution. After initializing all the individuals, the initial population of MPHLO is generated as Eq. (2)

$$X = \begin{bmatrix} x_1 \\ x_2 \\ \vdots \\ x_i \\ \vdots \\ x_N \end{bmatrix} = \begin{bmatrix} x_{11} & x_{12} & \cdots & x_{1j} & \cdots & x_{1M} \\ x_{21} & x_{22} & \cdots & x_{2j} & \cdots & x_{iM} \\ \vdots & \vdots & & \vdots & & \vdots \\ x_{i1} & x_{i2} & \cdots & x_{ij} & \cdots & x_{iM} \\ \vdots & \vdots & & \vdots & & \vdots \\ x_{N1} & x_{N2} & \cdots & x_{Nj} & \cdots & x_{NM} \end{bmatrix}, x_{ij} \in \{0, 1\}, 1 \leq i \leq N, 1 \leq j \leq M \tag{2}$$

## 2.1 Topologies of MPHLO

As mentioned above, two different topologies, the global topology and the ring topology, are used in the elite subpopulation and ordinary subpopulation, respectively.

#### 2.1.1 Global Topology of Elite Sub-population

The individuals with the fitness value ranked in the top $\lceil \lambda N \rceil$ are selected to constitute the elite subpopulation and the remaining $(N - \lceil \lambda N \rceil)$ individual constitute the ordinary subpopulation, where $\lambda$ is the proportion coefficient. To guarantee the search efficiency, the global topology is adopted in the elite sub-population. In global topology, all the individuals are connected by each other and therefore the optimal information can spread very fast, which can fasten the convergence.

#### 2.1.2 Ring Topology of Ordinary Sub-population

It is known that population topology can increase the diversity of population. Ring topology in which the individuals could interact with its neighbors is one of the most commonly used topologies. Inspired by this, the ordinary subpopulation employs a ring topology to maintain the diversity and explore solution space. In the ring topology, it is a point-to-point closed structure and each individual is only connected with its left and right neighbors. Specifically, each individual is directly connected to $k$ individuals on the left and right adjacent sides according to the number during initialization where $k$ is set to 1 in this paper based on trial-and-error methods for better diversity.

### 2.2 Learning Operators of MPHLO

In MPHLO, the four learning operators, i.e. the random learning operator, individual learning operator and two social learning operators, are used to generate new candidates, which are described as follows.

#### 2.2.1 Random Learning Operator

Random learning is very important and is always accompanying with the learning process. In the early stage of learning process, people always learn by their random acts due to the lack of understanding of a new problem. In the following study, humans still learn randomly because of interference, forgetting, only knowing partial knowledge of problems and other factors [2, 16]. In addition, in the process of learning, people constantly explore new ways to better solve problems. Thus, MPHLO uses the random operator to simulate random learning as Eq. (3)

$$x_{ij} = \begin{cases} 0, & 0 \leq r_1 \leq 0.5 \\ 1, & else \end{cases} \quad (3)$$

where $r_1$ is a stochastic number generated in the interval [0,1].

#### 2.2.2 Individual Learning Operator

Individual learning refers to the individual acquiring new skills and knowledge in the course of behavior and through the results of behavior [17]. People remember useful experience and learn from previous experience, when encountering similar problems, people can avoid mistakes and learn more efficiently. To mimic individual learning of human, MPHLO store personal best experience in an individual knowledge database (IKD), which can be described as Eq. (4)

$$ikd_i = \begin{bmatrix} ikd_{i1} \\ ikd_{i12} \\ \vdots \\ ikd_{ip} \\ \vdots \\ ikd_{iL} \end{bmatrix} = \begin{bmatrix} ik_{i11} & ik_{i12} & \cdots & ik_{i1j} & \cdots & ik_{i1M} \\ ik_{i21} & ik_{i22} & \cdots & ik_{i2j} & \cdots & ik_{i2M} \\ \vdots & \vdots & & \vdots & & \vdots \\ ik_{ip1} & ik_{ip2} & \cdots & ik_{ipj} & \cdots & ik_{ipM} \\ \vdots & \vdots & & \vdots & & \vdots \\ ik_{iL1} & ik_{iL2} & \cdots & ik_{iLj} & \cdots & ik_{iLM} \end{bmatrix}, 1 \leq p \leq L \tag{4}$$

where $ikd_i$ denotes the IKD of person *i*, *L* is the number of solutions saved in the IKD, and $ikd_{ip}$ stands for the *p-th* best solution of person *i*.

The individual learning of MPHLO generates new solutions according to the knowledge in the IKD as Eq. (5)

$$x_{ij} = ikd_{ipj} \tag{5}$$

### 2.2.3 Social Learning Operators

In a social environment, people can further develop their abilities and achieve more efficient and effective through learn from each other. In MPHLO, the social learning operators include the social learning operator in the elite subpopulation, the social network learning operator in the ordinary subpopulation and the global social learning operator.

a) Social learning operator in the elite subpopulation (SLOES)
   To possess the efficient search ability, like HLO, the best solution of elite subpopulation is also stored in the social knowledge database of elite subpopulation (SKDE) which can be described as Eq. (6)

$$SKDE = \begin{bmatrix} skde_1 \\ skde_2 \\ \vdots \\ skde_s \\ \vdots \\ skde_Q \end{bmatrix} = \begin{bmatrix} ske_{11} & ske_{12} & \cdots & ske_{1j} & \cdots & ske_{1M} \\ ske_{21} & ske_{22} & \cdots & ske_{2j} & \cdots & ske_{2M} \\ \vdots & \vdots & & \vdots & & \vdots \\ ske_{s1} & ske_{s1} & \cdots & ske_{sj} & \cdots & ske_{sM} \\ \vdots & \vdots & & \vdots & & \vdots \\ ske_{Q1} & ske_{Q2} & \cdots & ske_{Qj} & \cdots & ske_{QM} \end{bmatrix}, 1 \leq s \leq Q \tag{6}$$

   where $skde_s$ denotes the *s-th* solution in the SKDE and Q is the size of the SKDE. Based on the knowledge in the SKDE, MPHLO can perform the social learning operator in the elite subpopulation to generate new solutions as Eq. (7).

$$x_{ij} = ske_{sj} \tag{7}$$

b) Social network learning operator in the ordinary subpopulation (SLOOS)
In the ordinary subpopulation, each individual stores the best solution of its neighbors in the social knowledge database of neighbors (SKN) for social learning, which can be indicated as Eq. (8)

$$SKN = \begin{bmatrix} skn_1 \\ skn_2 \\ \vdots \\ skn_l \\ \vdots \\ skn_G \end{bmatrix} = \begin{bmatrix} sk_{i11} & sk_{i12} & \cdots & sk_{i1j} & \cdots & sk_{i1M} \\ sk_{21} & sk_{22} & \cdots & sk_{2j} & \cdots & sk_{2M} \\ \vdots & \vdots & & \vdots & & \vdots \\ sk_{l1} & sk_{l2} & \cdots & sk_{lj} & \cdots & sk_{lM} \\ \vdots & \vdots & & \vdots & & \vdots \\ sk_{i1} & sk_{G2} & \cdots & sk_{Gj} & \cdots & sk_{GM} \end{bmatrix}, 1 \le l \le G \tag{8}$$

where $skn_l$ denotes the *l-th* solution in the SKN and G is the size of the SKN.

When the ordinary sub-population executes the SLOOS, it chooses the best solution in the SKN and then copies the corresponding value as Eq. (9).

$$x_{ij} = skn_{lj} \tag{9}$$

c) Global social learning operator
To enhance the information interaction between subpopulations and improve the search result, both of subpopulations learn the social optimum of the whole population with certain probability, which is stored in social knowledge database (SKD) as Eq. (10).

$$SKD = [skd_1] = \begin{bmatrix} sk_{11} & sk_{12} & \cdots & sk_{1j} & \cdots & sk_{1M} \end{bmatrix} \tag{10}$$

MPHLO performs the global social learning operator to generate a new solution as Eq. (11)

$$x_{ij} = sk_{1j} \tag{11}$$

### 2.3 Implementation of MPHLO

In summary, the individual in the elite subpopulation yields a new solution by performing the random learning operator, individual learning operator, social learning operator in the elite subpopulation and global social learning operator as Eq. (12)

$$x_{ij} = \begin{cases} Rand(0, 1) & 0 \le r_2 \le pr \\ ik_{ipj} & pr \le r_2 \le pi_1 \\ ske_{sj} & pi_1 \le r_2 \le pm_1 \\ sk_{1j} & else \end{cases} \tag{12}$$

where $r_2$ is a random number generated between 0 and 1 using a uniform distribution. $pr$, $pi_1$ and $pm_1$ are three control parameters used to determine the probability of running the different operators. Specifically, $pr$ is the probability of random learning. In addition to individual learning, social learning of elite subpopulation and the global social learning are represented by the values of $(pi_1 - pr)$, $(pm_1 - pi_1)$ and $(1 - pm_1)$ respectively.

The process of the ordinary subpopulation generating new solutions can be described as Eq. (13)

$$x_{ij} = \begin{cases} Rand(0, 1) & 0 \le r_3 \le pr \\ ik_{ipj} & pr \le r_3 \le pi_2 \\ skn_{lj} & pi_2 \le r_3 \le pm_2 \\ sk_{1j} & else \end{cases} \tag{13}$$

where $r_3$ is a stochastic number between 0 and 1. $pr$, $pi_2$ and $pm_2$ are three control three control parameters deciding the probability of running the operators. Specifically, $pr$ is the probability of random learning. In addition to individual learning, social learning of ordinary subpopulation and the global social learning are represented by the values of $(pi_2 - pr)$, $(pm_2 - pi_2)$ and $(1 - pm_2)$, respectively.

The implementation of MPHLO is presented as Algorithm 1.

**Algorithm 1:** The framework of MPHLO algorithm

**Algorithm 1: MPHLO**

1: Set the parameter values, such as $N, \lambda, pi_1, pm_1, pi_2, pm_2, k, G\max$;

2: Initialize the population randomly;

3: Calculate the fitness of each individual and initialize the IKD and SKD;

4: Divide the population into two subpopulations according to the fitness;

5: Initialize the ring topology of ordinary subpopulation;

6: Initialize the SKDE of the elite subpopulation and the SKN of the ordinary subpopulation;

7: while $cn < G\max$ do

8: while $i < \lceil \lambda N \rceil$ do

9: Update the population according to Eq. (12);

10: End while

11: while $i < (N - \lceil \lambda N \rceil)$ do

12: Update the population according to Eq.(13);

13: End while

14: Calculate the fitness of each new candidate;

15: Update the IKD, SKDE, SKN and SKD according to the updating rules

16: $cn = cn + 1$

17: End while

18: Output the global best solution

## 3 Experimental Results and Discussions

The proposed algorithm MPHLO is compared with a simple human learning optimization (SHLO) [1], moreover, other recent binary optimization algorithms, i.e. artificial algae algorithm (AAA) [18], adaptive harmony search (ABHS) [19], improved binary differential evolution (IBDE) [20], modified binary bat algorithm (MBBA) [21] and time-varying mirrored S-shaped transfer function for binary particle swarm optimization (TVMS-BPSO) [22], were also used in the comparison. The CEC14 benchmark functions [23] were selected to evaluate the performance of these algorithms. A set of parameters of MPHLO were obtained by trail-and -error. For a fair comparison, SHLO, AAA, IBDE, MBBA, ABHS and TVMS-BPSO tales the recommended parameters to tackle these problems. Table 1 illustrates the parameters used in all the algorithms. In addition, all the algorithms ran 100 times on all the function independently. And the population size was set to 50 and the maximum iterations was 3000 in the 10-dimensional functions, for all algorithms. For the 30-dimensional functions, the population size and the maximum iterations were set to 100 and 6000, respectively. Each decision variable was coded by 30 bits.

**Table 1.** Parameters setting of MPHLO, SHLO, AAA, ABHS, IBDE, MBBA and TVMS-BPSO

| Algorithm | Parameters |
|---|---|
| MPHLO | $pr = 5/M, pi_1 = 0.88, pm_1 = 0.92, pi_2 = 0.87, pm_2 = 0.90$ |
| SHLO | $pr = 5/M, pi = 0.85 + 2/M$ |
| AAA | $DSP = 0.66$ |
| ABHS | $NGC = 20, PAR = 0.2, C = 15$ |
| IBDE | $\delta = 0.05, \alpha = 1.0, p_{sm} = 0.008, b = 0.5$ |
| MBBA | $F_{max} = 2.0, F_{min} = 0, \omega_0 = 0.5, a = 0.4, \alpha = 0.9, \gamma = 0.9, r_0 = 0.5$ |
| TVMS-BPSO | $c1 = c2 = 2.0, \mu_{max} = 1.0, \mu_{min} = 0.1, \omega = 1.0, v_{max} = 10, v_{min} = -10$ |

The numerical results, the t-test and the Wilcoxon signed-rand test (W-test) results on 10-dimensinal and 30-dimensinal functions are given in Table 2 and Table 4, respectively. The values equal to "1" or "−1" denotes the results obtained by MPHLO is significantly better or worse than the compared algorithms at 95% confidence, while the value equal to "0" represents that the achieved results by MPHLO and the compared algorithm are not statistically different. For clearing analyzing and comparing the performance, the summary results of the t-test and W-test on the 10-dimensinal and 30-dimensinal functions are listed in Table 3 and Table 5, respectively.

**Table 2.** Results of all the algorithms on the 10-dimensional benchmark functions

| Function | Metric | MPHLO | SHLO | AAA | ABHS | IBDE | MBBA | TVMS-BPSO |
|---|---|---|---|---|---|---|---|---|
| F1 | Best | 1.24E + 03 | 7.83E + 03 | 2.83E + 06 | 1.76E + 04 | 9.21E + 04 | 9.53E + 05 | 5.41E + 03 |
| | Mean | **4.60E + 04** | 1.06E + 05 | 5.30E + 07 | 7.34E + 06 | 5.68E + 06 | 3.41E + 07 | 1.44E + 05 |
| | Std | **4.17E + 04** | 7.72E + 04 | 2.12E + 07 | 8.78E + 06 | 5.90E + 06 | 2.48E + 07 | 1.22E + 05 |
| | t-test | / | 1 | 1 | 1 | 1 | 1 | 1 |
| | W-test | / | 1 | 1 | 1 | 1 | 1 | 1 |
| F2 | Best | **8.34E + 01** | 3.12E + 03 | 2.47E + 09 | 3.97E + 05 | 6.58E + 06 | 4.69E + 08 | 1.19E + 04 |
| | Mean | **8.13E + 05** | 1.33E + 07 | 8.77E + 09 | 1.66E + 08 | 3.01E + 08 | 2.42E + 09 | 2.16E + 07 |
| | Std | **3.45E + 06** | 1.86E + 07 | 2.31E + 09 | 1.94E + 08 | 3.05E + 08 | 1.37E + 09 | 2.03E + 07 |
| | t-test | / | 1 | 1 | 1 | 1 | 1 | 1 |
| | W-test | / | 1 | 1 | 1 | 1 | 1 | 1 |
| F3 | Best | **2.02E + 00** | 1.55E + 01 | 1.05E + 04 | 1.96E + 02 | 7.43E + 01 | 2.18E + 03 | 1.40E + 01 |
| | Mean | **2.06E + 02** | 7.69E + 02 | 2.17E + 04 | 6.06E + 03 | 2.90E + 03 | 3.15E + 04 | 1.31E + 03 |
| | Std | **2.54E + 02** | 4.81E + 02 | 5.44E + 03 | 5.60E + 03 | 2.27E + 03 | 5.44E + 04 | 1.16E + 03 |
| | t-test | / | 1 | 1 | 1 | 1 | 1 | 1 |
| | W-test | / | 1 | 1 | 1 | 1 | 1 | 1 |
| F4 | Best | **3.05E-02** | 2.30E-01 | 4.82E + 02 | 1.57E + 00 | 1.07E + 00 | 5.17E + 01 | 7.04E-01 |
| | Mean | **5.89E + 00** | 1.71E + 01 | 1.68E + 03 | 4.62E + 01 | 3.49E + 01 | 2.14E + 02 | 2.47E + 01 |
| | Std | 8.27E + 00 | 1.16E + 01 | 5.55E + 02 | 2.07E + 01 | 1.78E + 01 | 1.92E + 02 | 1.45E + 01 |
| | t-test | / | 1 | 1 | 1 | 1 | 1 | 1 |
| | W-test | / | 1 | 1 | 1 | 1 | 1 | 1 |
| F5 | Best | 5.55E + 00 | 1.43E + 01 | 2.01E + 01 | 7.26E + 00 | 2.00E + 01 | 1.95E + 01 | **3.95E + 00** |
| | Mean | **1.96E + 01** | 1.99E + 01 | 2.04E + 01 | 1.99E + 01 | 2.00E + 01 | 2.04E + 01 | 1.97E + 01 |
| | Std | 2.16E + 00 | 6.75E-01 | 7.64E-02 | 1.27E + 00 | **1.09E-02** | 1.93E-01 | 2.09E + 00 |
| | t-test | / | 1 | 1 | 0 | 0 | 1 | 0 |
| | W-test | / | 1 | 1 | 1 | 1 | 1 | 0 |

*(continued)*

**Table 2.** (*continued*)

| Function | Metric | MPHLO | SHLO | AAA | ABHS | IBDE | MBBA | TVMS-BPSO |
|---|---|---|---|---|---|---|---|---|
| F6 | Best | 1.03E-01 | 5.26E-01 | 8.02E + 00 | 7.36E-01 | 3.10E + 00 | 5.01E + 00 | 3.20E-01 |
| | Mean | **1.20E + 00** | 1.93E + 00 | 1.01E + 01 | 4.32E + 00 | 4.94E + 00 | 8.78E + 00 | 2.01E + 00 |
| | Std | 6.90E-01 | 7.88E-01 | **5.48E-01** | 1.27E + 00 | 9.61E-01 | 1.40E + 00 | 9.80E-01 |
| | t-test | / | 1 | 1 | 1 | 1 | 1 | 1 |
| | W-test | / | 1 | 1 | 1 | 1 | 1 | 1 |
| F7 | Best | **2.54E-02** | 6.15E-02 | 3.35E + 01 | 2.80E-01 | 2.12E-01 | 5.58E + 00 | 3.47E-02 |
| | Mean | **1.77E-01** | 4.76E-01 | 7.74E + 01 | 2.44E + 00 | 2.92E + 00 | 3.22E + 01 | 6.26E-01 |
| | Std | **1.00E-01** | 4.14E-01 | 1.51E + 01 | 1.57E + 00 | 2.03E + 00 | 1.99E + 01 | 6.86E-01 |
| | t-test | / | 1 | 1 | 1 | 1 | 1 | 1 |
| | W-test | / | 1 | 1 | 1 | 1 | 1 | 1 |
| F8 | Best | **2.21E-04** | 4.01E-03 | 6.24E + 01 | 3.21E-01 | 5.02E + 00 | 1.25E + 01 | 2.15E-02 |
| | Mean | **7.98E-01** | 1.69E + 00 | 1.03E + 02 | 7.49E + 00 | 1.57E + 01 | 4.28E + 01 | 3.63E + 00 |
| | Std | 8.05E-01 | 1.54E + 00 | 1.35E + 01 | 3.48E + 00 | 3.90E + 00 | 1.33E + 01 | 2.04E + 00 |
| | t-test | / | 1 | 1 | 1 | 1 | 1 | 1 |
| | W-test | / | 1 | 1 | 1 | 1 | 1 | 1 |
| F9 | Best | **1.17E + 00** | 1.32E + 00 | 6.38E + 01 | 6.98E + 00 | 1.11E + 01 | 3.37E + 01 | 4.06E + 00 |
| | Mean | **4.45E + 00** | 7.58E + 00 | 1.05E + 02 | 1.82E + 01 | 2.38E + 01 | 5.83E + 01 | 1.03E + 01 |
| | Std | **1.47E + 00** | 2.68E + 00 | 1.00E + 01 | 6.62E + 00 | 6.81E + 00 | 1.44E + 01 | 3.72E + 00 |
| | t-test | / | 1 | 1 | 1 | 1 | 1 | 1 |
| | W-test | / | 1 | 1 | 1 | 1 | 1 | 1 |
| F10 | Best | **7.13E-02** | 1.60E-01 | 9.07E + 02 | 2.61E-01 | 7.63E + 00 | 2.13E + 02 | 2.56E-01 |
| | Mean | **4.86E-01** | 2.36E + 00 | 1.27E + 03 | 3.77E + 01 | 1.67E + 02 | 9.11E + 02 | 6.45E + 00 |
| | Std | **9.61E-01** | 3.08E + 00 | 1.25E + 02 | 6.28E + 01 | 9.86E + 01 | 2.97E + 02 | 1.29E + 01 |
| | t-test | / | 1 | 1 | 1 | 1 | 1 | 1 |
| | W-test | / | 1 | 1 | 1 | 1 | 1 | 1 |
| F11 | Best | **6.46E + 00** | 6.66E + 00 | 1.20E + 03 | 6.66E + 01 | 9.44E + 01 | 7.50E + 02 | 8.12E + 00 |

(*continued*)

**Table 2.** (*continued*)

| Function | Metric | MPHLO | SHLO | AAA | ABHS | IBDE | MBBA | TVMS-BPSO |
|---|---|---|---|---|---|---|---|---|
| | Mean | **8.55E + 01** | 1.38E + 02 | 1.62E + 03 | 5.41E + 02 | 6.82E + 02 | 1.51E + 03 | 2.51E + 02 |
| | Std | **6.83E + 01** | 9.29E + 01 | 1.55E + 02 | 2.32E + 02 | 1.90E + 02 | 2.88E + 02 | 1.63E + 02 |
| | t-test | / | 1 | 1 | 1 | 1 | 1 | 1 |
| | W-test | / | 1 | 1 | 1 | 1 | 1 | 1 |
| F12 | Best | 3.06E-02 | 4.02E-02 | 7.13E-01 | 3.00E-02 | 6.90E-02 | 4.80E-01 | **2.60E-02** |
| | Mean | **1.18E-01** | 1.47E-01 | 1.29E + 00 | 1.94E-01 | 2.30E-01 | 1.23E + 00 | 1.33E-01 |
| | Std | **4.57E-02** | 5.62E-02 | 1.93E-01 | 1.14E-01 | 8.73E-02 | 4.26E-01 | 7.22E-02 |
| | t-test | / | 1 | 1 | 1 | 1 | 1 | 0 |
| | W-test | / | 1 | 1 | 1 | 1 | 1 | 0 |
| F13 | Best | **2.53E-02** | 2.54E-02 | 2.60E + 00 | 9.06E-02 | 3.31E-02 | 2.06E-01 | 3.51E-02 |
| | Mean | **6.03E-02** | 1.30E-01 | 3.91E + 00 | 3.43E-01 | 2.27E-01 | 1.13E + 00 | 1.24E-01 |
| | Std | **1.96E-02** | 4.91E-02 | 5.36E-01 | 1.39E-01 | 8.98E-02 | 6.75E-01 | 5.53E-02 |
| | t-test | / | 1 | 1 | 1 | 1 | 1 | 1 |
| | W-test | / | 1 | 1 | 1 | 1 | 1 | 1 |
| F14 | Best | **1.57E-02** | 4.16E-02 | 6.05E + 00 | 1.10E-01 | 1.41E-01 | 2.37E-01 | 3.57E-02 |
| | Mean | **4.36E-02** | 1.73E-01 | 1.76E + 01 | 5.14E-01 | 3.53E-01 | 6.18E + 00 | 2.32E-01 |
| | Std | **1.89E-02** | 1.05E-01 | 5.15E + 00 | 2.62E-01 | 2.46E-01 | 4.35E + 00 | 1.87E-01 |
| | t-test | / | 1 | 1 | 1 | 1 | 1 | 1 |
| | W-test | / | 1 | 1 | 1 | 1 | 1 | 1 |
| F15 | Best | **2.29E-01** | 3.99E-01 | 6.98E + 02 | 8.78E-01 | 1.44E + 00 | 1.15E + 01 | 4.24E-01 |
| | Mean | **8.92E-01** | 1.36E + 00 | 7.53E + 03 | 1.43E + 01 | 3.83E + 00 | 1.80E + 03 | 1.34E + 00 |
| | Std | **3.83E-01** | 5.95E-01 | 5.77E + 03 | 5.74E + 01 | 1.56E + 00 | 2.94E + 03 | 5.30E-01 |
| | t-test | / | 1 | 1 | 1 | 1 | 1 | 1 |
| | W-test | / | 1 | 1 | 1 | 1 | 1 | 1 |

### 3.1 Low-Dimensional Benchmark Functions

From Table 2 and Table 3, it is evident that MPHLO is superior to the other algorithms on 10-dimensional functions. Specifically, MPHLO obtains the optimal Mean values on all the functions. The t-test in Table 3 clearly show that MPHLO is outperform than SHLO, AAA, ABHS, IBDE, MBBA and TVMS-BPSO on 15, 15, 14, 14, 15, 15 out of 30 functions, respectively, at the same time it is inferior to them on none. In addition, the W-test results represent that MPHLO significantly surpasses SHLO, AAA, ABHS, IBDE, MBBA and TVMS-BPSO on all the benchmark functions.

**Table 3.** Summary result of the t-test and W-test on the 10-dimensional functions

| | SHLO | AAA | ABHS | IBDE | MBBA | TVMS-BPSO |
|---|---|---|---|---|---|---|
| t-test | | | | | | |
| 1 | 15 | 15 | 14 | 14 | 15 | 15 |
| 0 | 0 | 0 | 1 | 1 | 0 | 0 |
| −1 | 0 | 0 | 0 | 0 | 0 | 0 |
| W-test | | | | | | |
| 1 | 15 | 15 | 15 | 15 | 15 | 15 |
| 0 | 0 | 0 | 0 | 0 | 0 | 0 |
| −1 | 0 | 0 | 0 | 0 | 0 | 0 |

### 3.2 High-Dimensional Benchmark Functions

Table 4 indicates that MPHLO seek out the best Mean solutions on 13 out of 15 functions, while it is inferior to SHLO, AAA, ABHS, IBDE, MBBA and TVMS-BPSO on 0, 0, 0, 1, 0 and 2 functions. And t-test results of Table 5 show that MPHLO is superior than SHLO, AAA, ABHS, IBDE, MBBA and TVMS-BPSO on 14, 15, 15, 14, 15, and 14 functions, respectively, while it is inferior to them on 0, 0, 0, 1, 0 and 1 functions. Besides, the results of the W-test display that MPHLO is better than AAA, ABHS and MBBA on all the functions. MPHLO is obviously outperform than SHLO, IBDE and TVMS-BPSO on 14, 14 and 14 functions while is worse than them on 0, 1 and 1 functions, respectively. According to the results of Table 4 and Table 5, MPHLO can obtain superior or similar results than SHLO, AAA, ABHS, IBDE, MBBA and TVMS-BPSO on the 30-dimensional benchmark functions. The results indicates that the proposed algorithm can achieve better solutions compared to other binary algorithms.

**Table 4.** Results of all the algorithms on the 30-dimensional benchmark functions

| Function | Metric | MPHLO | SHLO | AAA | ABHS | IBDE | MBBA | TVMS-BPSO |
|---|---|---|---|---|---|---|---|---|
| F1 | Best | 7.58E + 05 | 1.77E + 06 | 6.51E + 08 | 1.48E + 07 | 3.04E + 07 | 1.27E + 08 | 3.01E + 06 |
| | Mean | 6.20E + 06 | 1.23E + 07 | 1.10E + 09 | 1.00E + 08 | 9.08E + 07 | 4.44E + 08 | 1.64E + 07 |
| | Std | 3.45E + 06 | 5.99E + 06 | 1.82E + 08 | 7.53E + 07 | 4.58E + 07 | 2.24E + 08 | 1.01E + 07 |
| | t-test | / | 1 | 1 | 1 | 1 | 1 | 1 |
| | W-test | / | 1 | 1 | 1 | 1 | 1 | 1 |
| F2 | Best | 4.57E + 06 | 2.03E + 07 | 7.75E + 10 | 3.17E + 08 | 5.81E + 08 | 1.21E + 10 | 1.12E + 08 |
| | Mean | 9.63E + 07 | 4.10E + 08 | 1.27E + 11 | 2.10E + 09 | 2.83E + 09 | 2.74E + 10 | 5.77E + 08 |
| | Std | 8.34E + 07 | 2.94E + 08 | 1.41E + 10 | 1.41E + 09 | 1.40E + 09 | 8.33E + 09 | 3.73E + 08 |
| | t-test | / | 1 | 1 | 1 | 1 | 1 | 1 |
| | W-test | / | 1 | 1 | 1 | 1 | 1 | 1 |
| F3 | Best | 2.62E + 02 | 4.21E + 02 | 1.28E + 05 | 8.02E + 02 | 1.74E + 03 | 4.35E + 04 | 4.14E + 02 |
| | Mean | 1.30E + 03 | 3.89E + 03 | 1.77E + 05 | 1.98E + 04 | 7.81E + 03 | 9.01E + 04 | 4.76E + 03 |
| | Std | 7.42E + 02 | 2.34E + 03 | 1.95E + 04 | 1.58E + 04 | 3.22E + 03 | 2.62E + 04 | 3.00E + 03 |
| | t-test | / | 1 | 1 | 1 | 1 | 1 | 1 |
| | W-test | / | 1 | 1 | 1 | 1 | 1 | 1 |
| F4 | Best | 5.39E + 01 | 9.39E + 01 | 1.55E + 04 | 1.64E + 02 | 1.69E + 02 | 8.84E + 02 | 1.14E + 02 |
| | Mean | 1.40E + 02 | 1.77E + 02 | 3.55E + 04 | 3.66E + 02 | 2.94E + 02 | 3.04E + 03 | 1.74E + 02 |
| | Std | 3.31E + 01 | 3.89E + 01 | 7.15E + 03 | 1.73E + 02 | 7.12E + 01 | 1.50E + 03 | 3.23E + 01 |
| | t-test | / | 1 | 1 | 1 | 1 | 1 | 1 |
| | W-test | / | 1 | 1 | 1 | 1 | 1 | 1 |
| F5 | Best | 2.00E + 01 | 2.00E + 01 | 2.09E + 01 | 2.00E + 01 | 2.00E + 01 | 2.05E + 01 | 2.00E + 01 |
| | Mean | 2.01E + 01 | 2.01E + 01 | 2.10E + 01 | 2.01E + 01 | 2.00E + 01 | 2.08E + 01 | 2.01E + 01 |
| | Std | 6.91E-02 | 5.91E-02 | 4.72E-02 | 5.17E-02 | 2.02E-02 | 1.12E-01 | 4.97E-02 |
| | t-test | / | 0 | 1 | 0 | −1 | 1 | 0 |
| | W-test | / | 0 | 1 | 0 | −1 | 1 | 0 |

*(continued)*

**Table 4.** (*continued*)

| Function | Metric | MPHLO | SHLO | AAA | ABHS | IBDE | MBBA | TVMS-BPSO |
|---|---|---|---|---|---|---|---|---|
| F6 | Best | 4.33E + 00 | 7.66E + 00 | 3.93E + 01 | 1.35E + 01 | 1.54E + 01 | 2.63E + 01 | 6.42E + 00 |
| | Mean | 9.36E + 00 | 1.27E + 01 | 4.26E + 01 | 2.02E + 01 | 2.04E + 01 | 3.34E + 01 | 1.18E + 01 |
| | Std | 1.77E + 00 | 2.07E + 00 | 9.95E-01 | 2.89E + 00 | 1.86E + 00 | 2.80E + 00 | 2.41E + 00 |
| | t-test | / | 1 | 1 | 1 | 1 | 1 | 1 |
| | W-test | / | 1 | 1 | 1 | 1 | 1 | 1 |
| F7 | Best | 5.45E-01 | 1.09E + 00 | 5.85E + 02 | 5.54E + 00 | 5.61E + 00 | 1.24E + 02 | 1.37E + 00 |
| | Mean | 1.87E + 00 | 4.90E + 00 | 8.56E + 02 | 1.92E + 01 | 2.14E + 01 | 2.30E + 02 | 5.60E + 00 |
| | Std | 8.66E-01 | 2.72E + 00 | 7.35E + 01 | 1.02E + 01 | 9.48E + 00 | 6.09E + 01 | 3.25E + 00 |
| | t-test | / | 1 | 1 | 1 | 1 | 1 | 1 |
| | W-test | / | 1 | 1 | 1 | 1 | 1 | 1 |
| F8 | Best | 4.50E + 00 | 4.55E + 00 | 4.13E + 02 | 1.62E + 01 | 3.94E + 01 | 1.20E + 02 | 8.87E + 00 |
| | Mean | 9.51E + 00 | 1.54E + 01 | 4.93E + 02 | 3.36E + 01 | 5.79E + 01 | 1.96E + 02 | 2.01E + 01 |
| | Std | 3.64E + 00 | 4.13E + 00 | 2.99E + 01 | 7.69E + 00 | 7.58E + 00 | 2.79E + 01 | 5.48E + 00 |
| | t-test | / | 1 | 1 | 1 | 1 | 1 | 1 |
| | W-test | / | 1 | 1 | 1 | 1 | 1 | 1 |
| F9 | Best | 2.33E + 01 | 3.77E + 01 | 4.08E + 02 | 6.37E + 01 | 7.75E + 01 | 2.00E + 02 | 3.47E + 01 |
| | Mean | 4.27E + 01 | 5.91E + 01 | 5.93E + 02 | 1.27E + 02 | 1.44E + 02 | 2.83E + 02 | 6.37E + 01 |
| | Std | 9.29E + 00 | 1.16E + 01 | 4.41E + 01 | 3.16E + 01 | 1.92E + 01 | 4.05E + 01 | 1.29E + 01 |
| | t-test | / | 1 | 1 | 1 | 1 | 1 | 1 |
| | W-test | / | 1 | 1 | 1 | 1 | 1 | 1 |
| F10 | Best | 7.05E-01 | 3.49E + 00 | 6.46E + 03 | 1.83E + 01 | 4.41E + 02 | 3.10E + 03 | 3.64E + 00 |
| | Mean | 1.84E + 01 | 9.00E + 01 | 7.48E + 03 | 4.10E + 02 | 1.08E + 03 | 4.79E + 03 | 1.20E + 02 |
| | Std | 2.69E + 01 | 8.89E + 01 | 2.61E + 02 | 2.07E + 02 | 2.37E + 02 | 6.19E + 02 | 1.02E + 02 |
| | t-test | / | 1 | 1 | 1 | 1 | 1 | 1 |
| | W-test | / | 1 | 1 | 1 | 1 | 1 | 1 |
| F11 | Best | 7.37E + 02 | 6.22E + 02 | 6.58E + 03 | 1.52E + 03 | 2.26E + 03 | 4.45E + 03 | 1.13E + 03 |

(*continued*)

**Table 4.** (*continued*)

| Function | Metric | MPHLO | SHLO | AAA | ABHS | IBDE | MBBA | TVMS-BPSO |
|---|---|---|---|---|---|---|---|---|
| | Mean | 1.46E + 03 | 2.07E + 03 | 7.46E + 03 | 3.18E + 03 | 2.96E + 03 | 6.27E + 03 | 2.31E + 03 |
| | Std | 3.44E + 02 | 5.25E + 02 | 3.04E + 02 | 5.64E + 02 | 3.37E + 02 | 5.52E + 02 | 4.55E + 02 |
| | t-test | / | 1 | 1 | 1 | 1 | 1 | 1 |
| | W-test | / | 1 | 1 | 1 | 1 | 1 | 1 |
| F12 | Best | 8.41E-02 | 6.76E-02 | 1.68E + 00 | 1.03E-01 | 9.37E-02 | 7.73E-01 | 5.91E-02 |
| | Mean | 2.30E-01 | 2.28E-01 | 2.68E + 00 | 2.53E-01 | 2.39E-01 | 1.65E + 00 | 1.89E-01 |
| | Std | 9.94E-02 | 7.86E-02 | 2.73E-01 | 7.19E-02 | 5.98E-02 | 4.53E-01 | 7.63E-02 |
| | t-test | / | 0 | 1 | 0 | 0 | 1 | −1 |
| | W-test | / | 0 | 1 | 0 | 0 | 1 | −1 |
| F13 | Best | 1.21E-01 | 2.13E-01 | 7.97E + 00 | 3.06E-01 | 2.44E-01 | 2.62E + 00 | 1.69E-01 |
| | Mean | 2.06E-01 | 3.75E-01 | 1.05E + 01 | 5.71E-01 | 4.27E-01 | 3.86E + 00 | 3.27E-01 |
| | Std | 4.17E-02 | 8.02E-02 | 1.17E + 00 | 1.35E-01 | 7.48E-02 | 6.22E-01 | 7.17E-02 |
| | t-test | / | 1 | 1 | 1 | 1 | 1 | 1 |
| | W-test | / | 1 | 1 | 1 | 1 | 1 | 1 |
| F14 | Best | 4.86E-02 | 1.51E-01 | 2.21E + 02 | 1.81E-01 | 1.52E-01 | 2.70E + 01 | 1.61E-01 |
| | Mean | 2.30E-01 | 4.26E-01 | 3.26E + 02 | 1.78E + 00 | 2.82E + 00 | 7.51E + 01 | 3.49E-01 |
| | Std | 1.89E-01 | 2.58E-01 | 2.87E + 01 | 2.79E + 00 | 3.94E + 00 | 2.38E + 01 | 1.98E-01 |
| | t-test | / | 1 | 1 | 1 | 1 | 1 | 1 |
| | W-test | / | 1 | 1 | 1 | 1 | 1 | 1 |
| F15 | Best | 5.31E + 00 | 4.71E + 00 | 1.03E + 06 | 2.29E + 01 | 1.80E + 01 | 5.99E + 03 | 5.88E + 00 |
| | Mean | 1.04E + 01 | 1.64E + 01 | 4.17E + 06 | 1.25E + 03 | 1.77E + 02 | 2.50E + 05 | 1.45E + 01 |
| | Std | 3.99E + 00 | 7.45E + 00 | 1.49E + 06 | 3.41E + 03 | 2.49E + 02 | 3.78E + 05 | 7.96E + 00 |
| | t-test | / | 1 | 1 | 1 | 1 | 1 | 1 |
| | W-test | / | 1 | 1 | 1 | 1 | 1 | 1 |

**Table 5.** Summary results of the t-test and W-test on the 30-dimensional functions

| | SHLO | AAA | ABHS | IBDE | MBBA | TVMS-BPSO |
|---|---|---|---|---|---|---|
| t-test | | | | | | |
| 1 | 14 | 15 | 15 | 14 | 15 | 14 |
| 0 | 1 | 0 | 0 | 0 | 0 | 0 |
| −1 | 0 | 0 | 0 | 1 | 0 | 1 |
| W-test | | | | | | |
| 1 | 14 | 15 | 15 | 14 | 15 | 14 |
| 0 | 1 | 0 | 0 | 0 | 0 | 0 |
| −1 | 0 | 0 | 0 | 1 | 0 | 1 |

# 4 Conclusion

As HLO generates new solutions mainly through copying the individual and social best solutions, it is likely to lose the diversity quickly, and therefore it may easily fall into local optimum. Inspired by the fact that the multiple-populations mechanism can increase the diversity, a new multi-population human optimal learning algorithm is proposed in this paper. According to the fitness value, the population is divided into two subpopulations, which are named as the elite subpopulation and ordinary subpopulation, respectively. The elite population uses a global topology to fasten the convergence of the algorithm, while the ordinary population uses a ring topology to maintain the diversity of the algorithm. Besides, the two subpopulations have a certain probability to learn the global optimal individual of the whole population to guarantee the global search ability of the algorithm. The CEC14 benchmark functions were adopted to test MPHLO for evaluating its performance. The results were compared with those of the other six state-of-art optimization algorithms, i.e. SHLO AAA, ABHS, IBDE, MBBA and TVMS-BPSO, which demonstrate that MPHLO is significantly better than the compared algorithms.

**Acknowledgments.** This work is supported by the National Key Research and Development Program of China No. 2019YFB1405500, Key Project of Science and Technology Commission of Shanghai Municipality under Grant No. 16010500300, and 111 Project under Grant No. D18003.

# References

1. Wang, L., Ni, H., Yang, R., Fei, M., Ye, W.: A simple human learning optimization algorithm. In: Fei, M., Peng, C., Su, Z., Song, Y., Han, Q. (eds.) LSMS/ICSEE 2014. CCIS, vol. 462, pp. 56–65. Springer, Heidelberg (2014). https://doi.org/10.1007/978-3-662-45261-5_7
2. Wang, L., Ni, H., Yang, R., Pardalos, P.M., Du, X., Fei, M.: An adaptive simplified human learning optimization algorithm. Inf. Sci. **320**, 126–139 (2015)
3. Holden, W.: The bell curve: intelligence and class structure in american life. Transform. Anthropol. **6**(5), 87–89 (2010)

4. Wang, L., An, L., Pi, J., Fei, M., Pardalos, P.M.: A diverse human learning optimization algorithm. J. Global Optim. **67**(1–2), 283–323 (2016). https://doi.org/10.1007/s10898-016-0444-2
5. Yang, R., Xu, M., He, J., Ranshous, S., Samatova, N.F.: An intelligent weighted fuzzy time series model based on a sine-cosine adaptive human learning optimization algorithm and its application to financial markets forecasting. In: Cong, G., Peng, W.-C., Zhang, W.E., Li, C., Sun, A. (eds.) ADMA 2017. LNCS (LNAI), vol. 10604, pp. 595–607. Springer, Cham (2017). https://doi.org/10.1007/978-3-319-69179-4_42
6. Wang, L., Pei, J., Menhas, M.I., Pi, J., Fei, M., Pardalos, P.M.: A hybrid-coded human learning optimization for mixed-variable optimization problems. Knowl.-Based Syst. **127**, 114–125 (2017)
7. Shoja, A., Molla-Alizadeh-Zavardehi, S., Niroomand, S.: Hybrid adaptive simplified human learning optimization algorithms for supply chain network design problem with possibility of direct shipment. Appl. Soft Comput. **96**, 106594 (2020)
8. Ding, H., Gu, X.: Hybrid of human learning optimization algorithm and particle swarm optimization algorithm with scheduling strategies for the flexible job-shop scheduling problem. Neurocomputing **414**, 313–332 (2020)
9. Wang, L., Yang, R., Ni, H., Ye, W., Fei, M., Pardalos, P.M.: A human learning optimization algorithm and its application to multi-dimensional knapsack problems. Appl. Soft Comput. **34**, 736–743 (2015)
10. Li, X., Yao, J., Wang, L., Menhas, M.I.: Application of human learning optimization algorithm for production scheduling optimization. In: Fei, M., Ma, S., Li, X., Sun, X., Jia, L., Su, Z. (eds.) LSMS/ICSEE -2017. CCIS, vol. 761, pp. 242–252. Springer, Singapore (2017). https://doi.org/10.1007/978-981-10-6370-1_24
11. Alguliyev, R., Aliguliyev, R. Isazade, N.: A sentence selection model and HLO algorithm for extractive text summarization. In: 2016 IEEE 10th International Conference on Application of Information and Communication Technologies (AICT) (2017)
12. Cao, J., Yan, Z., Xu, X., He, G., Huang, S.: Optimal power flow calculation in AC/DC hybrid power system based on adaptive simplified human learning optimization algorithm. J. Mod. Power Syst. Clean Energy **4**(4), 690–701 (2016)
13. Wang, S., Liu, G., Gao, M., Cao, S., Wang, J.: Heterogeneous comprehensive learning and dynamic multi-swarm particle swarm optimizer with two mutation operators. Inf. Sci. **540**, 175–201 (2020)
14. Xia, X., Tang, Y., Li, J., Hua, C., Guan, X.: Dynamic multi-swarm particle swarm optimizer with cooperative learning strategy. Appl. Soft Comput. **29**, 169–183 (2015)
15. Venkata, R., Saroj, R.A.: A self-adaptive multi-population based Jaya algorithm for engineering optimization. Swarm Evol. Comput. **37**, 1–26 (2017)
16. Nangle, D.W., Erdley, C.A., Adrian, M., Fales, J.: A conceptual basis in social learning theory. In: Nangle, D., Hansen, D., Erdley, C., Norton, P. (eds.) Practitioner's Guide to Empirically Based Measures of Social Skills, pp. 37–48. Springer, New York (2009). https://doi.org/10.1007/978-1-4419-0609-0_3
17. Forcheri, P., Molfino, M., Quarati, A.: ICT driven individual learning: new opportunities and perspectives. Educ. Technol. Soc. **3**, 51–61 (2000)
18. Korkmaz, S., Kiran, M.S.: An artificial algae algorithm with stigmergic behavior for binary optimization. Appl. Soft Comput. **64**, 627–640 (2018)
19. Guo, Z., Yang, H., Wang, S., Zhou, C., Liu, X.: Adaptive harmony search with best-based search strategy. Soft. Comput. **22**(4), 1335–1349 (2016). https://doi.org/10.1007/s00500-016-2424-3
20. Qian, S., Ye, Y., Liu, Y., Xu, G.: An improved binary differential evolution algorithm for optimizing PWM control laws of power inverters. Optim. Eng. **19**(2), 271–296 (2017). https://doi.org/10.1007/s11081-017-9354-5

21. Meraihi, Y., Acheli, D., Ramdane-Cherif, A.: QoS multicast routing for wireless mesh network based on a modified binary bat algorithm. Neural Comput. Appl. **31**(7), 3057–3073 (2017). https://doi.org/10.1007/s00521-017-3252-9
22. Beheshti, Z.: A time-varying mirrored S-shaped transfer function for binary particle swarm optimization. Inf. Sci. **512**, 1503–1542 (2020)
23. Liang, J.J., Qu, B.Y., Suganthan, P.N.: Problem definitions and evaluation criteria for the CEC 2014 special session and competition on single objective real-parameter numerical optimization (2013)

# Fuzzy, Neural, and Fuzzy-Neuro Hybrids

# Output-Based Ultimate Boundedness Control of T-S Fuzzy Systems Under Deception Attacks and Event-Triggered Mechanism

Fuqiang Li[1], Lisai Gao[2], Xiaoqing Gu[1](✉), and Baozhou Zheng[1]

[1] College of Sciences, Henan Agricultural University, Zhengzhou 450002, China
[2] School of Mechatronics Engineering and Automation, Shanghai University, Shanghai 200444, China

**Abstract.** This paper studies ultimate boundedness output feedback control of T-S fuzzy systems under event-triggered mechanism and stochastic deception attacks. Firstly, using both state-relative information and state-irrelative information, a general event-triggered mechanism (ETM) is proposed, which can save system resources such as network bandwidth, and exclude Zeno behavior absolutely. Secondly, a closed-loop system model is established, which integrates parameters of the ETM, random deception attacks, network-induced delays and fuzzy controller in a unified framework. Thirdly, exponentially ultimately bounded stability conditions in mean square are obtained, and design method for a fuzzy dynamic output feedback controller is presented. Finally, example confirms effectiveness of the proposed methods.

**Keywords:** T-S fuzzy system · Event-triggered mechanism · Deception attacks · Ultimately bounded stability

## 1 Introduction

Although networked control systems (NCSs) benefit from communication networks, they also face serious threat from cyber attacks [1]. For instance, deception attacks destroy data authenticity and integrity by injecting false data, while DoS attacks prevent data transmission by jamming communication network [2]. Deception attacks often destroy important infrastructures in an unpredictable and stealthy way, which motivates us to study deception attacks here.

Considering effects of deception attacks, secure control of NCSs has been drawing more and more attentions. For instance, by modeling stealthy deceptive attack and analysing its perniciousness, the work [3] studies distributed secure optimal frequency regulation of power grid. Considering deception attacks and Round-Robin protocol scheduling in controller-to-actuator channels, the work [4] designs a sliding mode controller for networked switched systems, and presents sufficient conditions for input-to-state stability in probability. Some works introduce the ETM into secure control. For instance, the work [5] studies path tracking secure control of networked autonomous vehicles under a learning-based


Q. Han et al. (Eds.): LSMS 2021/ICSEE 2021, CCIS 1469, pp. 425–434, 2021.
https://doi.org/10.1007/978-981-16-7213-2_41

ETM and deception attacks. Considering deception attacks obeying independent Bernoulli processes, the work [6] studies decentralized event-triggered control for uncertain systems. Considering random deception attacks, the work [7] designs a $H_\infty$ load frequency controller for networked power systems under a memory-based ETM. Using a switching method, the work [8] studies dynamic event-triggered $\mathcal{L}_\infty$ security control for NCSs under Bernoulli stochastic deception attacks. Although these works present many important results, they often assume systems are linear and system states are measurable. However, in practice, most systems are nonlinear, and it is often difficult to measure system states directly [9], which motivates us to study output-based security control of nonlinear systems.

To address the problems mentioned above, this paper focuses on output-based ultimate boundedness control of T-S fuzzy systems under the ETM and deception attacks. Main contributions are listed as follows. Firstly, using state-relative information and a constant, a discrete ETM is proposed, which can save system resources and avoid Zeno behavior in theory. Secondly, a closed-loop system is established, which characterizes effects of the ETM, network-induced delays and deception attacks in a unified framework. Thirdly, exponentially ultimately bounded stability criteria in mean square are obtained, and a fuzzy dynamic output feedback (FDOF) controller is further designed.

## 2 Problem Formulation

### 2.1 System Description

Consider a T-S fuzzy system with $r$ plant rules as:

*Plant rule* $i$: IF $\theta_1(t)$ is $M_{i1}$, $\theta_2(t)$ is $M_{i2}$ and ... and $\theta_g(t)$ is $M_{ig}$, THEN

$$\begin{cases} \dot{x}(t) = A_i x(t) + B_i u(t) \\ y(t) = C_i x(t), \ i = 1, \dots, r \end{cases} \tag{1}$$

where $x(t) \in \mathbb{R}^n$, $u(t) \in \mathbb{R}^{n_u}$ and $y(t) \in \mathbb{R}^{n_y}$ denote plant state, control input and measurement output, respectively. $r$ indicates number of IF-THEN rules, $\theta(t) = [\theta_1(t), \theta_2(t), \dots, \theta_g(t)]$ is the premise variable vector, $M_{ij} (j = 1, \dots, g, i = 1, \dots, r)$ is the fuzzy set, and $A_i, B_i$ and $C_i$ are gain matrices.

By using product fuzzy inference, center-average defuzzifier, and a singleton fuzzifier [10], dynamics of the T-S fuzzy system (1) can be expressed as

$$\begin{cases} \dot{x}(t) = \sum_{i=1}^{r} \mu_i(\theta(t))[A_i x(t) + B_i u(t)] \\ y(t) = \sum_{i=1}^{r} \mu_i(\theta(t)) C_i x(t) \end{cases} \tag{2}$$

where $\mu_i(\theta(t)) = \frac{\varphi_i(\theta(t))}{\sum_{i=1}^{r} \varphi_i(\theta(t))}$ is normalized membership function with $\varphi_i(\theta(t)) = \Pi_{j=1}^{g} M_{ij}(\theta_j(t))$, and $\mu_i(\theta(t))$ satisfies $\mu_i(\theta(t)) \geq 0$ and $\sum_{i=1}^{r} \mu_i(\theta(t)) = 1$.

### 2.2 Event-Triggered Mechanism

Using periodically-sampled measure output, a discrete ETM is proposed as [11]

$$\begin{aligned} t_{k+1}h =& t_k h + \min_{j\in\mathbb{N}}\{jh|[y(t_k h) - y(t_k h + jh)]^T \Omega \\ & [y(t_k h) - y(t_k h + jh)] \geq \delta y(t_k h)^T \Omega y(t_k h) + \delta_0\} \end{aligned} \tag{3}$$

where scalars $\delta \in [0, 1), \delta_0 > 0$, matrix $\Omega > 0$, $t_k h$ denotes the latest triggering instant, $t_{k+1}h$ indicates the next triggering instant, $t_k h + jh$ and refers to current sampling instant.

### 2.3 Closed-Loop System Modeling

Without considering attacks, based on characteristics of the ZOH, controller input can be described as

$$\tilde{y}(t) = y(t_k h),\ t \in [t_k h + \tau_k, t_{k+1}h + \tau_{k+1}) \tag{4}$$

where $\tau_k, \tau_{k+1} \in [\underline{\tau}, \bar{\tau}]$ denote network-induced delays.

Defining $\epsilon_k = t_{k+1} - t_k - 1$, holding intervals of the ZOH can be divided as $[t_k h + \tau_k, t_{k+1}h + \tau_{k+1}) = \bigcup_{\ell_k=0}^{\epsilon_k} \phi_{\ell_k}^{t_k}$, where

$$\phi_{\ell_k}^{t_k} = \begin{cases} [t_k h + \tau_k + \ell_k h, t_k h + \tau_k + (\ell_k + 1)h),\ \ell_k = 0, 1, \ldots, \epsilon_k - 1 \\ [t_k h + \tau_k + \ell_k h, t_{k+1}h + \tau_{k+1}),\ \ell_k = \epsilon_k \end{cases} \tag{5}$$

Define functions $e(t)$ and $\eta(t)$ as

$$e(t) = y(t_k h) - y(t_k h + \ell_k h),\quad \eta(t) = t - (t_k h + \ell_k h), t \in \phi_{\ell_k}^{t_k} \tag{6}$$

where $\underline{\tau} \leq \eta(t) \leq h + \bar{\tau}$.

Considering effects of deception attacks, it follows from (4) that

$$\hat{y}(t) = \alpha(t)\mathscr{D}(\tilde{y}(t)) + \tilde{y}(t),\ t \in \phi_{\ell_k}^{t_k} \tag{7}$$

where $\mathscr{D}(\tilde{y}(t))$ is attack signal, and $\alpha(t) \in \{0, 1\}$ is a Bernoulli distribution with mathematical expectation $\mathbb{E}\{\alpha(t)\} = \bar{\alpha}$. To avoid being detected by security defense system in practice, attack energy is often limited as [12]

$$\mathscr{D}^T(\tilde{y}(t))\mathscr{D}(\tilde{y}(t)) \leq \tilde{y}^T(t)G^T G\tilde{y}(t) \tag{8}$$

where $G$ stands for upper bound of the nonlinearity.

Design an output-based fuzzy controller with $r$ rules:

*Controller rule j*: IF $\theta_1(t)$ is $M_{j1}$, $\theta_2(t)$ is $M_{j2}$ and ... and $\theta_g(t)$ is $M_{jg}$, THEN

$$\begin{cases} \dot{x}_c(t) = A_{c_j}x_c(t) + B_{c_j}x_c(t - \eta(t)) + C_{c_j}\hat{y}(t) \\ u(t) = D_{c_j}x_c(t),\quad t \in \phi_{\ell_k}^{t_k} \end{cases} \tag{9}$$

where $x_c(t) \in \mathbb{R}^n$ is controller state, $A_{c_j}$, $B_{c_j}$, $C_{c_j}$ and $D_{c_j}$ are gain matrices. Then, the fuzzy controller can be obtained as

$$\begin{cases} \dot{x}_c(t) = \sum_{j=1}^{r} \mu_j(\theta(t))[A_{c_j} x_c(t) + B_{c_j} x_c(t-\eta(t)) + C_{c_j} \hat{y}(t)] \\ u(t) = \sum_{j=1}^{r} \mu_j(\theta(t)) D_{c_j} x_c(t), \quad t \in \phi_{\ell_k}^{t_k} \end{cases} \tag{10}$$

Using the plant (2) and the FDOF controller (10), the closed-loop system can be obtained as

$$\dot{\mathscr{X}}(t) = \sum_{i=1}^{r}\sum_{j=1}^{r} \mu_i\mu_j[\bar{A}\mathscr{X}(t) + \bar{A}_d\mathscr{X}(t-\eta(t)) + \bar{B}_e e(t) + \alpha(t)\bar{B}_a\mathscr{D}(\tilde{y}(t))] \tag{11}$$

where $\mathscr{X}(t) = \begin{bmatrix} x(t) \\ x_c(t) \end{bmatrix}$, $\bar{A} = \begin{bmatrix} A_i & B_i D_{c_j} \\ 0 & A_{c_j} \end{bmatrix}$, $\bar{A}_d = \begin{bmatrix} 0 & 0 \\ C_{c_j} C_i & B_{c_j} \end{bmatrix}$, $\bar{B}_e = \begin{bmatrix} 0 \\ C_{c_j} \end{bmatrix}$ and $\bar{B}_a = \begin{bmatrix} 0 \\ C_{c_j} \end{bmatrix}$. To simplify representation, $\mu_i$ and $\mu_j$ are used to represent $\mu_i(\theta(t))$ and $\mu_j(\theta(t))$, respectively.

## 3 Stability Analysis

**Theorem 1.** Given a sampling period $h$, parameters of network-induced delay $\bar{\tau}, \underline{\tau}$, attack parameter $G$ and triggering parameter $\delta \in (0,1), \delta_0 > 0$, if there exist positive definite matrices $\Omega > 0, P > 0, R > 0, S > 0, Q_i > 0 (i = 1,2,3)$, and matrices $U_2$, $U_3$ satisfying

$$\begin{bmatrix} Q_2 & * \\ U_2 & Q_2 \end{bmatrix} > 0, \quad \begin{bmatrix} Q_3 & * \\ U_3 & Q_3 \end{bmatrix} > 0 \tag{12}$$

$$\begin{bmatrix} \Pi_{11}^l & * & * & * & * \\ \Pi_{21} & \Pi_{22} & * & * & * \\ \Pi_{31} & 0 & \Pi_{33} & * & * \\ \Pi_{41} & 0 & 0 & \Pi_{44} & * \\ \Pi_{51} & 0 & 0 & 0 & \Pi_{55} \end{bmatrix} < 0, \ l = 2,3 \tag{13}$$

where

$\Pi_{11}^l = \Xi^l - \Lambda_1^T(\eta_{10}^2 Q_1 + \eta_{21}^2 Q_2 + \eta_{32}^2 Q_3)\Lambda_1 - \bar{\alpha}(1-\bar{\alpha})\Lambda_2^T(\eta_{10}^2 Q_1 + \eta_{21}^2 Q_2$
$+\eta_{32}^2 Q_3)\Lambda_2 - e_6^T \Omega e_6 - e_7^T e_7$, $\Pi_{21} = col\{\eta_{10}\Lambda_1, \eta_{21}\Lambda_1, \eta_{32}\Lambda_1\}$,
$\Pi_{31} = col\{\eta_{10}\Lambda_2, \eta_{21}\Lambda_2, \eta_{32}\Lambda_2\}, \Pi_{33} = diag\{-\beta Q_1^{-1}, -\beta Q_2^{-1}, -\beta Q_3^{-1}\}$,
$\Pi_{22} = diag\{-Q_1^{-1}, -Q_2^{-1}, -Q_3^{-1}\}, \Pi_{41} = \Lambda_3, \ \Pi_{51} = \Lambda_4$,
$\Pi_{44} = -\delta^{-1}\Omega^{-1}, \Pi_{55} = -I, \beta = (\bar{\alpha}(1-\bar{\alpha}))^{-1}$,
$\Lambda_1 = \bar{A}e_1 + \bar{A}_d e_3 + \bar{B}_e e_6 + \bar{\alpha}\bar{B}_a e_7, \Lambda_2 = \bar{B}_a e_7, \Lambda_3 = C_i E_1 e_3 + e_6, \Lambda_4 = G\Lambda_3$,
$\Xi^l = \sigma e_1^T P e_1 + He\{e_1^T P \Lambda_1\} + e_1^T R e_1 - e^{-\sigma\eta_1} e_2^T R e_2 + col\{e_2, e_4\}^T S col\{e_2, e_4\}$
$-e^{-\sigma\rho}[e_4^T \ \ e_5^T] S [e_4^T \ \ e_5^T]^T + \Lambda_1^T(\eta_{10}^2 Q_1 + \eta_{21}^2 Q_2 + \eta_{32}^2 Q_3)\Lambda_1 + \bar{\alpha}(1-\bar{\alpha})\Lambda_2^T(\eta_{10}^2 Q_1$
$+\eta_{21}^2 Q_2 + \eta_{32}^2 Q_3)\Lambda_2 - e^{-\sigma\eta_1} e_{12}^T Q_1 e_{12} - e^{-\sigma\eta_2}(i-2) e_{24}^T Q_2 e_{24} - e^{-\sigma\eta_3}(i-2) e_{43}^T Q_3 e_{43}$
$-e^{-\sigma\eta_3}(i-2) e_{35}^T Q_3 e_{35} - e^{-\sigma\eta_3}(i-2) He\{e_{35}^T U_3 e_{43}\} - e^{-\sigma\eta_3}(3-i) e_{45}^T Q_3 e_{45} - e^{-\sigma\eta_2}(3$
$-i) e_{23}^T Q_2 e_{23} - e^{-\sigma\eta_2}(3-i) e_{34}^T Q_2 e_{34} - e^{-\sigma\eta_2}(3-i) He\{e_{34}^T U_2 e_{23}\}$,
$e_1 = [I\ 0\ 0\ 0\ 0\ 0\ 0], e_2 = [0\ I\ 0\ 0\ 0\ 0\ 0], e_3 = [0\ 0\ I\ 0\ 0\ 0\ 0], e_4 = [0\ 0\ 0\ I\ 0\ 0\ 0]$,
$e_5 = [0\ 0\ 0\ 0\ I\ 0\ 0], e_6 = [0\ 0\ 0\ 0\ 0\ I\ 0], e_7 = [0\ 0\ 0\ 0\ 0\ 0\ I], e_{12} = e_1 - e_2, e_{23} =$
$e_2 - e_3, e_{34} = e_3 - e_4, e_{24} = e_2 - e_4, e_{43} = e_4 - e_3, e_{35} = e_3 - e_5, e_{45} = e_4 - e_5$,
$E_1 = [I\ 0], E_2 = [0\ I]$, $He\{\}$ refers to sum of a matrix and the matrix transpose,
$diag\{\}$ denotes diagonal matrix and $col\{\}$ indicate column matrix.

Then, the T-S fuzzy system (11) under the ETM and deception attacks is exponentially ultimately bounded in mean square.

*Proof.* Construct a Lyapunov functional as

$$\begin{aligned} V(t) =& \mathscr{X}^T(t)P\mathscr{X}(t) + \int_{t-\eta_1}^{t} e^{\sigma(s-t)}\mathscr{X}^T(s)R\mathscr{X}(s)ds \\ &+ \int_{t-\rho}^{t} e^{\sigma(s-t)}\varsigma^T(s)S\varsigma(s)ds \\ &+ \eta_{10}\int_{-\eta_1}^{-\eta_0}\int_{t+\theta}^{t} e^{\sigma(s-t)}\dot{\mathscr{X}}^T(s)Q_1\dot{\mathscr{X}}(s)dsd\theta \\ &+ \eta_{21}\int_{-\eta_2}^{-\eta_1}\int_{t+\theta}^{t} e^{\sigma(s-t)}\dot{\mathscr{X}}^T(s)Q_2\dot{\mathscr{X}}(s)dsd\theta \\ &+ \eta_{32}\int_{-\eta_3}^{-\eta_2}\int_{t+\theta}^{t} e^{\sigma(s-t)}\dot{\mathscr{X}}^T(s)Q_3\dot{\mathscr{X}}(s)dsd\theta \end{aligned} \tag{14}$$

where $P > 0, R > 0, S > 0, Q_1 > 0, Q_2 > 0, Q_3 > 0, \eta_0 = 0, \eta_1 = \underline{\tau}, \eta_3 = h+\bar{\tau}, \eta_2 = (\eta_3+\eta_1)/2, \eta_{10} = \eta_1 - \eta_0, \eta_{21} = \eta_2 - \eta_1, \eta_{32} = \eta_3 - \eta_2, \rho = (\eta_3 - \eta_1)/2$ and $\varsigma(t) = \text{col}\{\mathscr{X}(t-\eta_1), \mathscr{X}(t-\eta_2)\}$.

Taking the derivative of $V(t)$ [13], and then taking the expectation, we have

$$\mathbb{E}\{\dot{V}(t)\} \le -\sigma\mathbb{E}\{V(t)\} + \sum_{i=1}^{r}\sum_{j=1}^{r}\mu_i\mu_j[\xi^T(t)\Xi^l\xi(t)] \tag{15}$$

where $\xi(t) = col\{\mathscr{X}(t), \mathscr{X}(t-\eta_1), \mathscr{X}(t-\eta(t)), \mathscr{X}(t-\eta_2), \mathscr{X}(t-\eta_3), e(t), \mathscr{D}(\tilde{y}(t))\}$.

Using (3) and (8), it follows from (15) that

$$\mathbb{E}\{\dot{V}(t)\} \le -\sigma\mathbb{E}\{V(t)\} + \sum_{i=1}^{r}\sum_{j=1}^{r}\mu_i\mu_j[\xi^T(t)\bar{\Xi}^l\xi(t)] + \delta_0 \tag{16}$$

where $\bar{\Xi}^l = \Xi^l - e_6^T\Omega e_6 - e_7^T e_7 + \delta(C_iE_1e_3 + e_6)^T\Omega(C_iE_1e_3 + e_6) + (C_iE_1e_3 + e_6)^T G^T G(C_iE_1e_3 + e_6)$.

Applying Schur complement lemma to (13), we obtain $\bar{\Xi}^l < 0$. Substituting this inequality into (16) yields

$$\mathbb{E}\{\dot{V}(t)\} \le -\sigma\mathbb{E}\{V(t)\} + \delta_0 \Rightarrow \mathbb{E}\{V(t)\} \le e^{-\sigma t}\mathbb{E}\{V(0)\} + \frac{\delta_0}{\sigma}(1 - e^{-\sigma t}) \tag{17}$$

Due to $\mathscr{X}^T(t)P\mathscr{X}(t) \le V(t)$, we can derive from (17) that

$$\mathbb{E}\{\|\mathscr{X}(t)\|^2\} \le \frac{1}{\lambda_{min}\{P\}}[\mathbb{E}\{V(0)\} - \frac{\delta_0}{\sigma}]e^{-\sigma t} + \frac{\delta_0}{\sigma\lambda_{min}\{P\}} \tag{18}$$

where $\lambda_{min}\{P\}$ indicates the minimum eigenvalue, and the bounded set that the state $\mathscr{X}(t)$ converges to as $t \to \infty$ is obtained as $\mathscr{B} = \sqrt{\frac{\delta_0}{\sigma\lambda_{min}\{P\}}}$.

Therefore, one derives that the T-S fuzzy system (11) under the ETM and deception attacks is exponentially ultimately bounded in mean square [14].

*Remark 1.* In Theorem 1, reciprocally convex method and delay partitioning technique are employed to reduce conservativeness [15]. For instance, using the mass-spring-damping system in [16] with $A_1 = [0\,1; -10.88\ -2]$, $A_2 = [0\,1; -8\ -2]$, $B_1 = B_2 = [0\ 1]^T$, $C_1 = C_2 = [1\ 0]$, for given delay lower bound $\underline{\tau} = 0.01\,\mathrm{s}$, the method in [17] achieves a maximum allowable upper bound $\bar{\tau} = 0.46\,\mathrm{s}$, while Theorem 1 can obtain a larger $\bar{\tau} = 0.68\,\mathrm{s}$.

# 4 Design of the FDOF Controller

**Theorem 2.** Given a sampling period $h$, parameters of network-induced delay $\bar{\tau}, \underline{\tau}$, attack parameter $G$ and triggering parameter $\delta \in (0, 1), \delta_0 > 0$, if there exist positive definite matrices $\Omega > 0, \bar{R} > 0, \bar{S} > 0, \bar{Q}_i > 0 (i = 1, 2, 3)$, $H = \begin{bmatrix} X & * \\ I & Y \end{bmatrix} > 0$, and matrices $\bar{U}_2$, $\bar{U}_3$ such that

$$\begin{bmatrix} \bar{Q}_2 & * \\ \bar{U}_2 & \bar{Q}_2 \end{bmatrix} > 0, \quad \begin{bmatrix} \bar{Q}_3 & * \\ \bar{U}_3 & \bar{Q}_3 \end{bmatrix} > 0 \tag{19}$$

$$\begin{bmatrix} \bar{\Pi}_{11}^{l} & * & * & * & * \\ \bar{\Pi}_{21} & \bar{\Pi}_{22} & * & * & * \\ \bar{\Pi}_{31} & 0 & \bar{\Pi}_{33} & * & * \\ \bar{\Pi}_{41} & 0 & 0 & \bar{\Pi}_{44} & * \\ \bar{\Pi}_{51} & 0 & 0 & 0 & \bar{\Pi}_{55} \end{bmatrix} < 0, \ l = 2, 3 \tag{20}$$

where

$$\begin{aligned}
&\bar{\Pi}_{11}^{i} = \sigma e_1^T H e_1 + He\{e_1^T \bar{\Lambda}_1\} + e_1^T \bar{R} e_1 - e^{-\sigma\eta_1} e_2^T \bar{R} e_2 + col\{e_2, e_4\}^T \bar{S} col\{e_2, e_4\} \\
&- e^{-\sigma\rho}[e_4^T \quad e_5^T] \bar{S} [e_4^T \quad e_5^T]^T - e^{-\sigma\eta_1} e_{12}^T \bar{Q}_1 e_{12} - e^{-\sigma\eta_2}(i-2) e_{24}^T \bar{Q}_2 e_{24} - e^{-\sigma\eta_3}(i \\
&- 2) e_{43}^T \bar{Q}_3 e_{43} - e^{-\sigma\eta_3}(i-2) e_{35}^T \bar{Q}_3 e_{35} - e^{-\sigma\eta_3}(i-2) He\{e_{35}^T \bar{U}_3 e_{43}\} - e^{-\sigma\eta_3}(3 \\
&- i) e_{45}^T \bar{Q}_3 e_{45} - e^{-\sigma\eta_2}(3-i) e_{23}^T \bar{Q}_2 e_{23} - e^{-\sigma\eta_2}(3-i) e_{34}^T \bar{Q}_2 e_{34} - e^{-\sigma\eta_2}(3-i) He\{e_{34}^T \bar{U}_2 e_{23}\} \\
&- e_6^T \Omega e_6 - e_7^T e_7, \\
&\bar{\Pi}_{21} = col\{\eta_{10}\bar{\Lambda}_1, \eta_{21}\bar{\Lambda}_1, \eta_{32}\bar{\Lambda}_1\}, \bar{\Pi}_{31} = col\{\eta_{10}\bar{\Lambda}_2, \eta_{21}\bar{\Lambda}_2, \eta_{32}\bar{\Lambda}_2\}, \\
&\bar{\Pi}_{41} = \bar{\Lambda}_3, \bar{\Pi}_{51} = \bar{\Lambda}_4, \bar{\Pi}_{22} = \mathrm{diag}\{\bar{Q}_1 - 2H, \bar{Q}_2 - 2H, \bar{Q}_3 - 2H\}, \\
&\bar{\Pi}_{33} = \mathrm{diag}\{\bar{Q}_1 - 2\sqrt{\beta}H, \bar{Q}_2 - 2\sqrt{\beta}H, \bar{Q}_3 - 2\sqrt{\beta}H\}, \bar{\Pi}_{44} = \Omega - 2\bar{\delta}, \bar{\Pi}_{55} = -I, \\
&\bar{\Lambda}_1 = \tilde{A} e_1 + \tilde{A}_d e_3 + \tilde{B}_e e_6 + \bar{\alpha}\tilde{B}_a e_7, \bar{\Lambda}_2 = col\{0, \omega_2\} e_7, \bar{\Lambda}_3 = [C_i X \ C_i] e_3 + e_6, \\
&\bar{\Lambda}_4 = G\bar{\Lambda}_3, \tilde{A} = \begin{bmatrix} A_i X + B_i \omega_1 & A_i \\ \omega_4 & Y A_i \end{bmatrix}, \tilde{A}_d = \begin{bmatrix} 0 & 0 \\ \omega_3 & \omega_2 C_i \end{bmatrix}, \tilde{B}_e = \begin{bmatrix} 0 \\ \omega_2 \end{bmatrix}, \tilde{B}_a = \begin{bmatrix} 0 \\ \omega_2 \end{bmatrix}, \\
&\omega_1 = D_{c_j} N^{-1}(I - YX), \omega_3 = \omega_2 C_i X + N B_{c_j} N^{-1}(I - YX), \\
&\omega_2 = N C_{c_j}, \omega_4 = Y A_i X + Y B_i \omega_1 + N A_{c_j} N^{-1}(I - YX), \\
&\bar{R} = \Phi_1^T R \Phi_1, \bar{S} = \Psi_1^T S \Psi_1, \bar{Q}_1 = \Phi_1^T Q_1 \Phi_1, \bar{Q}_2 = \Phi_1^T Q_2 \Phi_1, \bar{Q}_3 = \Phi_1^T Q_3 \Phi_1, \bar{U}_2 \\
&= \Phi_1^T U_2 \Phi_1, \bar{U}_3 = \Phi_1^T U_3 \Phi_1 \text{ and } \bar{\delta} = \delta^{-\frac{1}{2}},
\end{aligned}$$

then, the T-S fuzzy system (11) under the ETM and deception attacks is exponentially ultimately bounded in mean square, and gain matrices of an equivalent FDOF controller are obtained as

$$\begin{cases} \bar{A}_{c_j} = (\omega_4 - Y A_i X - Y B_i \omega_1)(I - YX)^{-1}, & \bar{C}_{c_j} = \omega_2 \\ \bar{B}_{c_j} = (\omega_3 - \omega_2 C_i X)(I - YX)^{-1}, & \bar{D}_{c_j} = \omega_1 (I - YX)^{-1} \end{cases} \tag{21}$$

*Proof.* Decompose matrix $P$ and define matrices $\Phi_1, \Phi_2 = P\Phi_1$ as

$$P = \begin{bmatrix} Y & * \\ N^T & N^T Y^{-1} N - N^T X N \end{bmatrix}, \ \Phi_1 = \begin{bmatrix} X & I \\ N^{-1}(I - YX) & 0 \end{bmatrix}, \ \Phi_2 = \begin{bmatrix} I & Y \\ 0 & N^T \end{bmatrix} \tag{22}$$

where $X > 0, Y > 0$, and $X, Y, N \in \mathbb{R}^{n \times n}$.

Defining $\Psi_1 = \text{diag}\{\Phi_1, \Phi_1\}$, $\Psi_2 = \text{diag}\{\Psi_1, \Psi_1, \Phi_1, I, I, \Psi_3, I, I\}$ and $\Psi_3 = \text{diag}\{\Phi_2, \Phi_2, \Phi_2, \Phi_2, \Phi_2, \Phi_2\}$, transform the conditions in Theorem 1 as

$$\begin{bmatrix} \bar{Q}_2 & * \\ \bar{U}_2 & \bar{Q}_2 \end{bmatrix} = \Psi_1^T \begin{bmatrix} Q_2 & * \\ U_2 & Q_2 \end{bmatrix} \Psi_1 > 0, \ \begin{bmatrix} \bar{Q}_3 & * \\ \bar{U}_3 & \bar{Q}_3 \end{bmatrix} = \Psi_1^T \begin{bmatrix} Q_3 & * \\ U_3 & Q_3 \end{bmatrix} \Psi_1 > 0 \tag{23}$$

$$\begin{bmatrix} \bar{\Pi}_{11}^l & * & * & * & * \\ \bar{\Pi}_{21} & \hat{\Pi}_{22} & * & * & * \\ \bar{\Pi}_{31} & 0 & \hat{\Pi}_{33} & * & * \\ \bar{\Pi}_{41} & 0 & 0 & \Pi_{44} & * \\ \bar{\Pi}_{51} & 0 & 0 & 0 & \bar{\Pi}_{55} \end{bmatrix} = \Psi_2^T \begin{bmatrix} \Pi_{11}^l & * & * & * & * \\ \Pi_{21} & \Pi_{22} & * & * & * \\ \Pi_{31} & 0 & \Pi_{33} & * & * \\ \Pi_{41} & 0 & 0 & \Pi_{44} & * \\ \Pi_{51} & 0 & 0 & 0 & \Pi_{55} \end{bmatrix} \Psi_2 < 0, \ l = 2, 3 \tag{24}$$

where $\hat{\Pi}_{22} = \text{diag}\{-H\bar{Q}_1^{-1}H, -H\bar{Q}_2^{-1}H, -H\bar{Q}_3^{-1}H\}$ and $\hat{\Pi}_{33} = \text{diag}\{-\beta H\bar{Q}_1^{-1}H, -\beta H\bar{Q}_2^{-1}H, -\beta H\bar{Q}_2^{-1}H\}$. If $\mathscr{M} \geq 0$, one has $(\mathscr{N} - \mathscr{M})\mathscr{M}^{-1}(\mathscr{N} - \mathscr{M}) \geq 0$, which implies $-\mathscr{N}\mathscr{M}^{-1}\mathscr{N} \leq \mathscr{M} - 2\mathscr{N}$. Applying the inequality to $\hat{\Pi}_{22}, \hat{\Pi}_{33}$ and $\Pi_{44}$ in (24), we obtain $\bar{\Pi}_{22}, \bar{\Pi}_{33}$ and $\bar{\Pi}_{44}$ in (20). To deal with the unknown $N$, $x_c(t) = N^{-1}\bar{x}_c(t)$ is used to introduce an equivalent FDOF controller with gain matrices in (21).

## 5 Example

Consider the following T-S fuzzy system in [18] with gain matrices

$$A_1 = \begin{bmatrix} 0.12 & -0.13 \\ 0.06 & -0.20 \end{bmatrix}, \quad B_1 = \begin{bmatrix} -0.21 \\ -0.02 \end{bmatrix}, C_1 = \begin{bmatrix} 0.14 \\ -0.01 \end{bmatrix}^T,$$

$$A_2 = \begin{bmatrix} -0.32 & -0.21 \\ -0.16 & 0.20 \end{bmatrix}, B_2 = \begin{bmatrix} -0.22 \\ 0.18 \end{bmatrix}, C_2 = \begin{bmatrix} 0.20 \\ -0.30 \end{bmatrix}^T,$$

$$\mu_1 = 0.5 - 0.5sin(x_2(t)), \quad \mu_2 = 0.5 + 0.5sin(x_2(t))$$

Other parameters are given as: $h = 100\,\text{ms}$, $\underline{\tau} = 10\,ms, \bar{\tau} = 30\,\text{ms}$, $\sigma = 0.1$, $\delta = 0.003$, $\delta_0 = 1 \times 10^{-9}$, $\bar{\alpha} = 0.13$, $G = 0.4$ and $\mathcal{H}(\tilde{y}(t)) = tanh(0.4\tilde{y}(t))$. Based on Theorem 2, parameters of the ETM and FDOF controller are obtained, which are omitted due to page limitation.

As shown in Fig. 1, although the system is affected by both of deception attacks and the ETM, the designed controller can still stabilise the system. As shown in Fig. 2, the average triggering interval of the designed ETM is 0.43 s, which is larger than sample period 0.1 s, and thus system resources can be saved. The minimum interval between consecutive event-triggering instants is sample period 0.1 s, and thus Zeno behavior can be excluded absolutely. Although 12 out of 94 triggering packets are polluted by deception attacks, the system can still be stabilized, which illustrates validity of the proposed method.

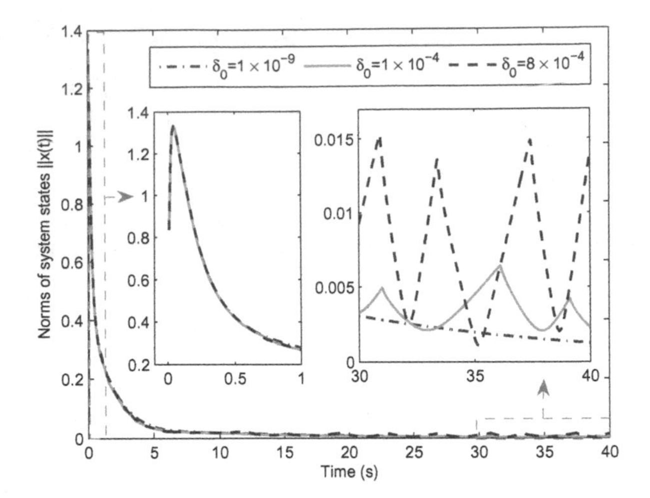


**Fig. 1.** Norms of system states of the T-S fuzzy system.

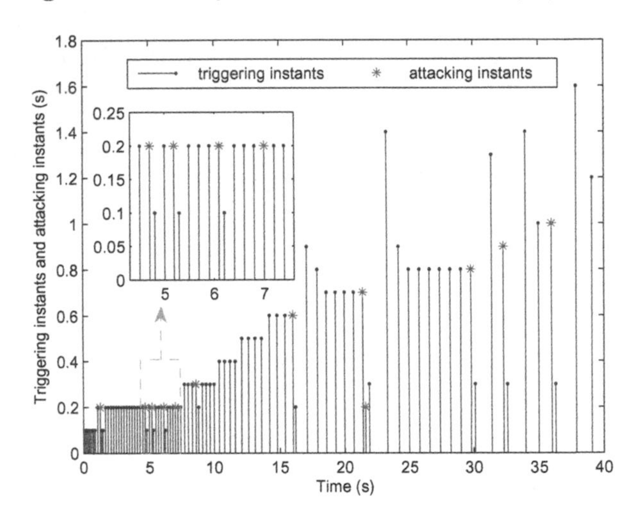


**Fig. 2.** Triggering instants and attacking instants.

## 6 Conclusion

This paper studies output-based ultimate boundedness control of T-S fuzzy systems under the ETM and Bernoulli-distribution deception attacks. First, using sampled output and a constant, an ETM is introduced, which can reduce unnecessary usage of system resources and exclude Zeno behavior absolutely. Then, a closed-loop fuzzy system model is built, and exponentially ultimately bounded stability criteria in mean square are obtained. Finally, a fuzzy output-based controller is designed, and example illustrates validity of the proposed methods.

**Funding.** This work was supported by National Natural Science Foundation of China [61703146]; Scientific Project in Henan Province [202102110126]; Backbone Teacher in Henan Province [2020GGJS048]; Scientific Project in Henan Agricultural University [KJCX2016A09, KJCX2015A17]; Key Project of Henan Education Department [19A413009].

## References

1. Peng, C., Sun, H., Yang, M., Wang, Y.L.: A survey on security communication and control for smart grids under malicious cyber attacks. IEEE Trans. Syst. Man Cybern. Syst. **49**(8), 1554–1569 (2019)
2. Hu, S., Yuan, P., Yue, D., Dou, C., Cheng, Z., Zhang, Y.: Attack-resilient event-triggered controller design of DC microgrids under DoS attacks. IEEE Trans. Circ. Syst. I Regul. Pap. **67**(2), 699–710 (2020)
3. Weng, S., Yue, D., Dou, C.: Secure distributed optimal frequency regulation of power grid with time-varying voltages under cyber attack. Int. J. Robust Nonlinear Control **30**(3), 894–909 (2020)
4. Zhao, H., Niu, Y., Jia, T.: Security control of cyber-physical switched systems under Round Robin protocol: input-to-state stability in probability. Inf. Sci. **508**, 121–134 (2020)
5. Gu, Z., Yin, T., Ding, Z.: Path tracking control of autonomous vehicles subject to deception attacks via a learning-based event-triggered mechanism. IEEE Trans. Neural Netw. Learn. Syst. (2021)
6. Sun, Y., Yu, J., Yu, X., Gao, H.: Decentralized adaptive event-triggered control for a class of uncertain systems with deception attacks and its application to electronic circuits. IEEE Trans. Circuits Syst. I Regul. Pap. **67**(12), 5405–5416 (2020)
7. Tian, E., Peng, C.: Memory-based event-triggering $H_\infty$ load frequency control for power systems under deception attacks. IEEE Trans. Cybern. **50**(11), 4610–4618 (2020)
8. Wu, Z., Xiong, J., Xie, M.: Dynamic event-triggered $\mathcal{L}_\infty$ control for networked control systems under deception attacks: a switching method. Inf. Sci. **561**, 168–180 (2021)
9. Hu, S., Yue, D., Han, Q.L., Xie, X., Chen, X., Dou, C.: Observer-based event-triggered control for networked linear systems subject to denial-of-service attacks. IEEE Trans. Cybern. **50**(5), 1952–1964 (2020)
10. Liu, J., Yang, M., Zha, L., Xie, X., Tian, E.: Multi-sensors-based security control for T-S fuzzy systems over resource-constrained networks. J. Franklin Inst. **357**(7), 4286–4315 (2020)
11. Yue, D., Tian, E., Han, Q.L.: A delay system method for designing event-triggered controllers of networked control systems. IEEE Trans. Autom. Control **58**(2), 475–481 (2013)
12. Liu, J., Yang, M., Xie, X., Peng, C., Yan, H.: Finite-time $H_\infty$ filtering for state-dependent uncertain systems with event-triggered mechanism and multiple attacks. IEEE Trans. Circuits Syst. I Regul. Pap. **67**(3), 1021–1034 (2020)
13. Gao, L., Li, F., Fu, J.: Event-triggered output feedback resilient control for NCSs under deception attacks. Int. J. Control Autom. Syst. **18**(9), 2220–2228 (2020)
14. Zou, L., Wang, Z., Han, Q.L., Zhou, D.: Ultimate boundedness control for networked systems with try-once-discard protocol and uniform quantization effects. IEEE Trans. Autom. Control **62**(12), 6582–6588 (2017)

15. Zhang, X.M., Han, Q.L., Seuret, A., Gouaisbaut, F., He, Y.: Overview of recent advances in stability of linear systems with time-varying delays. IET Control Theory Appl. **13**(1), 1–16 (2019)
16. Zhao, T., Dian, S.: Fuzzy dynamic output feedback $H_\infty$ control for continuous-time T-S fuzzy systems under imperfect premise matching. ISA Trans. **70**, 248–259 (2017)
17. Zhang, X.M., Han, Q.L.: Event-triggered dynamic output feedback control for networked control systems. IET Control Theory Appl. **8**(4), 226–234 (2014)
18. Li, Y., Zhang, Q., Luo, X.: Robust $L_1$ dynamic output feedback control for a class of networked control systems based on T-S fuzzy model. Neurocomputing **197**, 86–94 (2016)

# Vibration Attenuation for Jacket Stuctures via Fuzzy Sliding Mode Control

Zhihui Cai(✉) and Hua-Nv Feng

College of Science, China Jiliang University, Hangzhou 310018, China

**Abstract.** In this paper, fuzzy sliding mode control scheme is designed for offshore jacket stuctures with parameter perturbations and external waves. Firstly, the Takagi-Sugeno fuzzy model is proposed via considering the parameter perturbation of the system. Secondly, based on the state signals of the jacket stuctures with current and delayed states, a novel fuzzy sliding mode control method that is established from state signals of the system with current and delay to reduce oscillations of jacket stuctures. Thirdly, based on the obtained model, a sufficient condition for stability of offshore platfrom is derived via Lyapunov-Krasovskii functional approach. Finally, several simulation results show that the mixed delayed robust fuzzy sliding mode $H_\infty$ control strategy can effectively stabilize offshore structures and make better performance of the controlled system.

**Keywords:** Jacket platform · T-S model · Fuzzy sliding mode control · Active control · Time-delay

## 1 Introduction

Offshore jacked platforms are the basic infrastructures for explorating oil and gas resources in different depths [1]. These platforms, which suffer excessive oscillations, are inescapably affected by wave forces, nonlinear self-excited wave forces, earthquakes, and other perturbations [2–4]. In order to reduce the impact of structural vibration on system properties and improve the reliability of the offshore structure, passive control [5,6], semi-active control [7] and active control [8] are investigated and implemented extensively. Due to the flexibility and potential to satisfy the performance requirements, active control strategies have attracted intensive attention over the past decades [9]. To mention a few, sampled-data control [10] and sliding mode control [11] strategies for offshore structures are presented.

Note that the offshore platform is large in size and complex in structure. It is a hard work to design an accurate dynamic model. Actually, the simplified model of the offshore platform still presents nonlinear characteristics. Takagi-Sugeno(T-S) fuzzy model has been widely applied to deal with the modeling problems for complex nonlinear systems in recent years [12,13]. T-S fuzzy model has also been


Q. Han et al. (Eds.): LSMS 2021/ICSEE 2021, CCIS 1469, pp. 435–445, 2021.
https://doi.org/10.1007/978-981-16-7213-2_42

used as an effective model to approximate nonlinear system, and the stabilization of T-S fuzzy models has been widely studied [14,15]. For example, in [16], an adaptive sliding mode control model was established using fuzzy control method. To cope with the nonlinear control problem of active suspension systems, incorporating the adaptive sliding mode control and the T-S fuzzy-based control is suggested by [17]. As ones of the important structural parameters of the offshore structure, damping ratios and natural frequencies are usually approximated as constants in the analysis. Nevertheless, the two important parameters are actually time-varying. For this reason, it is great significance to built fuzzy model directly based on the above two system variables, and then study the active control for such offshore structures.

Sliding mode control has been regarded as a effective tool and applied to active control for offshore structures recently. For example, a novel robust integral sliding mode control strategy has been descripted to ensure the stability of offshore platform under parameter perturbations and nonlinear wave forces in [18]. From [19], a novel general dynamic model for the jacket structure is established via applying a chosen time delay to the controller. Although many research results on sliding mode control theory of the offshore structure have been reported in recent years, most of the aforementioned researches are still based on the constant parameters of offshore structure [20]. However, in practical, the parameter of offshore structure such as natural frequencies are not constant due to the offshore structures are used to exploration and storage the marine oil and gas resources. Thus, it is great significance to extend the previous research results to this case that the parameter of offshore structure are not constant. Although there are a great number of research results on fuzzy control and sliding mode control that can stimulate the active control for the vibration systems. As far as the author knows, the fuzzy sliding mode controllers for jacket structure have not yet been well investigated, there are still challenges facing platform dynamics and improving platform uncertainty simultaneously, which motivates this paper.

Driven by the above discussions, our aim in this research is to develop a new fuzzy sliding mode control scheme of the offshore jacket structures. Firstly, considering jacket structure with external wave force, a T-S fuzzy model is created via using T-S fuzzy method to approximate the time-varying parameters such as damping ratios and natural frequencies. Secondly, applying a chosen time delay to the control loop, a mixed delayed robust fuzzy sliding mode $H_\infty$ controller is proposed. Thirdly, by use of the chosen Lyapunov-Krasovskii candidate, a sufficient stability condition of the system is obtained. Finally, some results of numerical simulations show the usefulness of the designed method.

## 2 Problem Formulation

A classic offshore jacket structure with an tuned mass damper (TMD) [1] which is shown in Fig. 1 is considered. Considering the uncertainty of the platform, the model of the system can be formulated as:

$$\begin{cases} \ddot{z}_1 = -2(\xi_1^1 + \Delta\xi_1^1)(\omega_1^1 + \Delta\omega_1^1)\dot{z}_1 - (\omega_1^1 + \Delta\omega_1^1)^2 z_1 \\ \quad -\phi_1 K_T(\phi_1 z_1 + \phi_2 z_2) + \phi_1 K_T z_T + \phi_1(\zeta_1 - u) \\ \quad +\phi_1 C_T \dot{z}_T - \phi_1 C_T(\phi_1 \dot{z}_1 + \phi_2 \dot{z}_2) + f_1 + f_2 \\ \ddot{z}_2 = -2(\xi_2^1 + \Delta\xi_2^1)(\omega_2^1 + \Delta\omega_2^1)\dot{z}_2 - (\omega_2^1 + \Delta\omega_2^1)^2 z_2 \\ \quad -\phi_2 K_T(\phi_1 z_1 + \phi_2 z_2) + \phi_2 K_T z_T + \phi_2(\zeta_2 - u) \\ \quad +\phi_2 C_T \dot{z}_T - \phi_2 C_T(\phi_1 \dot{z}_1 + \phi_2 \dot{z}_2) + f_3 + f_4 \\ \ddot{z}_T = -2(\xi_T^1 + \Delta\xi_T^1)(\omega_T^1 + \Delta\omega_T^1)(\dot{z}_T - \phi_1 \dot{z}_1 - \phi_2 \dot{z}_2) \\ \quad -(\omega_T^1 + \Delta\omega_T^1)^2(z_T - \phi_1 z_1 - \phi_2 z_2) + u/m_T \end{cases} \tag{1}$$

where $z_1$, $\xi_1^1$, $\omega_1^1$ and $\phi_1$ are the generalized coordinate, damping ratio, natural frequency and shapes vector of the first vibration modes, respectively; $z_2$, $\xi_2^1$, $\omega_2^1$ and $\phi_2$ are the generalized coordinate, damping ratio, natural frequency and shapes vector of the second vibration modes, respectively; $z_T$, $\omega_T^1$, $\xi_T^1$, $C_T$, $K_T$ and $m_T$ are the horizontal displacement, natural frequency, damping ratio, damping, stiffness and mass of the tuned mass damper, respectively; $\Delta\xi_1^1$, $\Delta\xi_2^1$, $\Delta\xi_T^1$, $\Delta\omega_1^1$, $\Delta\omega_2^1$ and $\Delta\omega_T^1$ are the perturbations of parameters $\xi_1^1$, $\xi_2^1$, $\xi_T^1$, $\omega_1^1$, $\omega_2^1$ and $\omega_T^1$ , respectively; $f_i$ $(i = 1, \cdots, 4)$ and $u$ are the self-excited wave forces and control force of the system, respectively.

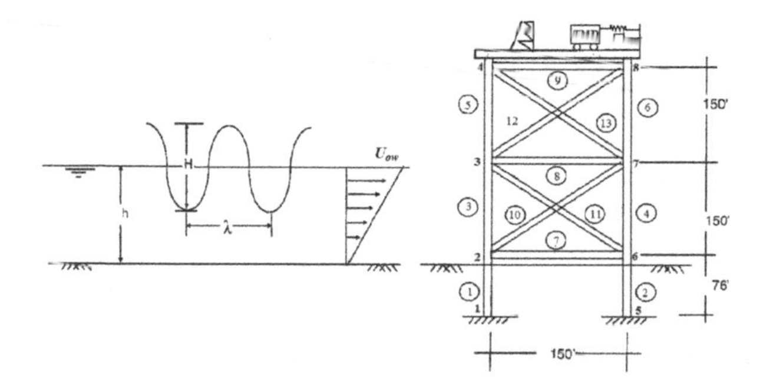


**Fig. 1.** The sketch of jacket platform with a TMD [1]

Let

$$\begin{cases} \xi_1 = \xi_1^1 + \Delta\xi_1^1, \quad \xi_2 = \xi_2^1 + \Delta\xi_2^1, \quad \xi_T = \xi_T^1 + \Delta\xi_T^1, \\ \omega_1 = \omega_1^1 + \Delta\omega_1^1, \omega_2 = \omega_2^1 + \Delta\omega_2^1, \omega_T = \omega_T^1 + \Delta\omega_T^1. \end{cases} \tag{2}$$

Then, the Eq. (1) can be expressed as:

$$\begin{cases} \ddot{z}_1 = -2\xi_1\omega_1\dot{z}_1 - \omega_1^2 z_1 - \phi_1 K_T(\phi_1 z_1 + \phi_2 z_2) + f_1 + f_2 \\ \quad +\phi_1 K_T z_T + \phi_1(\zeta_1 - u) + \phi_1 C_T \dot{z}_T - \phi_1 C_T(\phi_1 \dot{z}_1 + \phi_2 \dot{z}_2) \\ \ddot{z}_2 = -2\xi_2\omega_2\dot{z}_2 - \omega_2^2 z_2 - \phi_2 K_T(\phi_1 z_1 + \phi_2 z_2) + f_3 + f_4 \\ \quad +\phi_2 K_T z_T + \phi_2(\zeta_2 - u) + \phi_2 C_T \dot{z}_T - \phi_2 C_T(\phi_1 \dot{z}_1 + \phi_2 \dot{z}_2) \\ \ddot{z}_T = -2\xi_T\omega_T(\dot{z}_T - \phi_1\dot{z}_1 - \phi_2\dot{z}_2) - \omega_T^2(z_T - \phi_1 z_1 - \phi_2 z_2) + u/m_T. \end{cases} \tag{3}$$

Let $x = [x_1, \ x_2, \ \cdots, \ x_6]^{\mathrm{T}}$, $f = [f_1 + f_2 \ , f_3 + f_4]^{\mathrm{T}}$ and $\zeta = [\zeta_1 \ \ \zeta_2]^{\mathrm{T}}$.

Where $x_1 = z_1$, $x_2 = \dot{z}_1$, $x_3 = z_2$, $x_4 = \dot{z}_2$, $x_5 = z_T$, $x_6 = \dot{z}_T$.

Then, the Eq. (3) can be approximated to a state space model:

$$\dot{x}(t) = Ax(t) + Bu(t) + Df(x,t) + D_0\zeta(t), \tag{4}$$

where

$$\begin{cases} A = \begin{bmatrix} 0 & 1 & 0 & 0 & 0 & 0 \\ \alpha_{21} & \alpha_{22} & \alpha_{23} & \alpha_{24} & \alpha_{25} & \alpha_{26} \\ 0 & 0 & 0 & 1 & 0 & 0 \\ \alpha_{41} & \alpha_{42} & \alpha_{43} & \alpha_{44} & \alpha_{45} & \alpha_{46} \\ 0 & 0 & 0 & 0 & 0 & 1 \\ \alpha_{61} & \alpha_{62} & \alpha_{63} & \alpha_{64} & \alpha_{65} & \alpha_{66} \end{bmatrix} \\ B = \begin{bmatrix} 0 & -\phi_1 & 0 & -\phi_2 & 0 & 1/m_T \end{bmatrix}^{\mathrm{T}} \\ D = \begin{bmatrix} 0\,1\,0\,0\,0\,0 \\ 0\,0\,0\,1\,0\,0 \end{bmatrix}^{\mathrm{T}} \quad D_0 = \begin{bmatrix} 0 & \phi_1 & 0 & 0 & 0 & 0 \\ 0 & 0 & 0 & \phi_2 & 0 & 0 \end{bmatrix}^{\mathrm{T}}, \end{cases} \tag{5}$$

with

$$\begin{aligned}
&\alpha_{21} = -\omega_1^2 - K_T\phi_1^2, && \alpha_{22} = -2\xi_1\omega_1 - C_T\phi_1^2, && \alpha_{23} = -K_T\phi_1\phi_2, && \alpha_{24} = -C_T\phi_1\phi_2, \\
&\alpha_{25} = \phi_1 K_T, && \alpha_{26} = \phi_1 C_T, && \alpha_{41} = -K_T\phi_1\phi_2, && \alpha_{42} = -C_T\phi_1\phi_2, \\
&\alpha_{43} = -\omega_2^2 - K_T\phi_2^2, && \alpha_{44} = -2\xi_2\omega_2 - C_T\phi_2^2, && \alpha_{45} = \phi_2 K_T, && \alpha_{46} = \phi_2 C_T, \\
&\alpha_{61} = \omega_T^2\phi_1, && \alpha_{62} = 2\xi_T\omega_T\phi_1, && \alpha_{63} = \omega_T^2\phi_2, && \alpha_{64} = 2\xi_T\omega_T\phi_2, \\
&\alpha_{65} = -\omega_T^2, && \alpha_{66} = -2\xi_T\omega_T.
\end{aligned}$$

Here uniformly bounded function $\|f(x,t)\| \leqslant \mu \|x(t)\|$, where $\mu > 0$ is a given value.

Note that, $\xi_i$ and $\omega_i$ of system have time-varying characteristics. In order to obtain a more generalized offshore sturcture model, the fuzzy dynamic model of offshore sturcture will be established on the above time-varying variables. Based on the analysis above, we give the fuzzy premise variable $\chi(t) = \{\xi_1(t), \xi_2(t), \xi_T(t), \omega_1(t), \omega_2(t), \omega_T(t)\}$, $r$ represents the number of fuzzy rules, $\xi_v^i(t)$ and $\omega_v^i(t)$ $(i = 1, 2, \cdots, r;\ v = 1, 2, T)$ represent fuzzy sets. Then the fuzzy rules are defined as follows:

IF $\xi_1(t)$ is $\xi_1^i(t)$ and $\xi_2(t)$ is $\xi_1^2(t)$ and $\cdots$ and $\omega_T(t)$ is $\omega_T^i(t)$, THEN

$$\dot{x}(t) = A_i x(t) + Bu(t) + Df(x,t) + D_0\zeta(t). \tag{6}$$

Then, applying the center-average defuzzifier approach, the overall fuzzy dynamical model can be descripted in the following.

$$\dot{x}(t) = \sum_{i=1}^{r} \psi_i(\chi(t))[A_i x(t) + Bu(t) + Df(x,t) + D_0\zeta(t)], \tag{7}$$

where $\sum_{i=1}^{r} \psi_i(\chi(t)) = 1$ and $\psi_i(\chi(t)) \in [0,1]$.

The output equation of the system is given by the following expression:

$$y(t) = C_1 x(t) + D_1\zeta(t), \tag{8}$$

where

$$C_1 = \begin{bmatrix} 1\,0\,0\,0\,0\,0 \\ 0\,0\,1\,0\,0\,0 \end{bmatrix}, \quad D_1 = \begin{bmatrix} 0.1 & 0 \\ 0 & 0.1 \end{bmatrix}.$$

In what follows, a mixed delayed robust fuzzy sliding mode $H_\infty$ control scheme will be developed and the following results will be obtained.

1) The closed-loop system is robustly stable in the developed sliding surface, the output system shows that $\|y(t)\| \leq \gamma \|\zeta(t)\|$, for $\zeta(t) \in L_2[0, \infty)$ under the zero initial condition, where $\mu > 0$ is a given value.
2) The state trajectory of system (7) can be driven by the designed delayed mixed fuzzy sliding mode $H_\infty$ controller into the sliding surface in finite time and maintain the state thereafter.

## 3 Fuzzy Sliding Mode Control

A fuzzy sliding mode surface with mixed delayed states is designed firstly. Then, a mixed delayed robust fuzzy sliding mode $H_\infty$ control strategy will be provided for the system.

### 3.1 Design of Mixed Delayed Fuzzy Sliding Surface

The mixed delayed fuzzy sliding surface function is given by

$$s(t) = \sum_{i=1}^{r}\sum_{j=1}^{r} \psi_{ij}(\chi(t))[Gx(t) - \int_0^t G(A_i + BK_j)x(\theta)d\theta - \int_0^{t-\tau} GBK_{\tau j}x(\theta)d\theta - Gx_0]. \tag{9}$$

Where $G$, $K_j$ and $K_{\tau j}$ are $1 \times 6$ real matrix to be determined such that $GB$ is nonsingular matrix, $\psi_{ij}(\chi(t)) := \psi_i(\chi(t))\psi_j(\chi(t))$ and $\tau$ is a given positive scalar.

When the system state trajectory enters the sliding mode, we have $s(t) = 0$ and $\dot{s}(t) = 0$. Let $\dot{s}(t) = 0$, the equivalent control law $u_{eq}(t)$ from (9) is given by

$$u_{eq}(t) = \sum_{j=1}^{r} \psi_j(\chi(t))[K_j x(t) + K_{\tau j}x(t-\tau) - (GB)^{-1}G(Df(x,t) + D_0\zeta(t))]. \tag{10}$$

Subscribing (10) into (7) yields the sliding motion as

$$\dot{x}(t) = \sum_{i=1}^{r}\sum_{j=1}^{r} \psi_{ij}(\chi(t))[(A_i + BK_j)x(t) + BK_{\tau j}x(t-\tau) + \bar{D}f(x,t) + \bar{D}_0\zeta(t)]. \tag{11}$$

where $\bar{D} = \bar{G}D$, $\bar{D}_0 = \bar{G}D_0$ and $\bar{G} = I - B(GB)^{-1}G$.

A sufficient condition with the gain matrix $K_j$ and $K_{\tau j}(j = 1, 2, \cdots, r)$ is shown in the following proposition.

**Proposition 1.** *The sliding motion (11) with $\zeta(t)=0$ is robustly stable and the level $\gamma$ of $H_\infty$ attenuation is ensured, if $\tau>0$, $\mu>0$ and $\gamma>0$ are given scalars and there exist $6\times 6$ matrices $P>0$, $Q>0$, $R>0$, $P_2>0$, $P_3>0$ and $1\times 6$ matrices $K_j$, $K_{\tau j}$ such that*

$$\begin{bmatrix} \Phi_{11} & \Phi_{12} & \Phi_{13} & P_2\bar{D} & P_2\bar{D}_0 & C_1^{\mathrm{T}} & \mu I \\ * & -Q-R & \Phi_{23} & 0 & 0 & 0 & 0 \\ * & * & \Phi_{33} & P_3\bar{D} & P_3\bar{D}_0 & 0 & 0 \\ * & * & * & -I & 0 & 0 & 0 \\ * & * & * & * & -\gamma^2 I & D_1^{\mathrm{T}} & 0 \\ * & * & * & * & * & -I & 0 \\ * & * & * & * & * & * & -I \end{bmatrix} < 0, \tag{12}$$

*where*

$$\begin{cases} \Phi_{11}=A_i^{\mathrm{T}}P_2+P_2A_i+P_2BK_j+K_j^{\mathrm{T}}B^{\mathrm{T}}P_2-R+Q, \ \Phi_{12}=R+P_2BK_{\tau j}, \\ \Phi_{13}=P-P_2+A_i^{\mathrm{T}}P_3+K_j^{\mathrm{T}}B^{\mathrm{T}}P_3, \Phi_{23}=K_{\tau j}^{\mathrm{T}}B^{\mathrm{T}}P_3, \quad \Phi_{33}=\tau^2R-2P_3. \end{cases}$$

*Proof.* Choose the Lyapunov-Krasovskii candidate as follows

$$V(t,x_t)=x^{\mathrm{T}}(t)Px(t)+\int_{t-\tau}^{t}x^{\mathrm{T}}(s)Qx(s)ds+\tau\int_{-\tau}^{0}ds\int_{t+s}^{t}\dot{x}^{\mathrm{T}}(\theta)R\dot{x}(\theta)d\theta, \tag{13}$$

where $P>0$, $Q>0$, $R>0$.

Differentiating $V(t,x_t)$ and applying the descriptor method, we obtain

$$\begin{aligned} \dot{V}(t,x_t)=&\dot{x}^{\mathrm{T}}(t)Px(t)+x^{\mathrm{T}}(t)P\dot{x}(t)+x^{\mathrm{T}}(t)Qx(t)-\tau\int_{t-\tau}^{t}\dot{x}^{\mathrm{T}}(s)R\dot{x}(s)ds \\ &+\tau^2\dot{x}^{\mathrm{T}}(t)R\dot{x}(t)-x^{\mathrm{T}}(t-\tau)Qx(t-\tau)+\sum_{i=1}^{r}\sum_{j=1}^{r}\psi_{ij}(\chi(t))[2(x^{\mathrm{T}}(t)P_2 \\ &+\dot{x}^{\mathrm{T}}(t)P_3)][-\dot{x}(t)+(A_i+BK_j)x(t)+BK_{\tau j}x(t-\tau)+\bar{D}f(x,t)+\bar{D}_0\zeta(t)], \end{aligned} \tag{14}$$

where $P_2>0$ and $P_3>0$. By applying further Jensen's inequality, we have

$$-\tau\int_{t-\tau}^{t}\dot{x}^{\mathrm{T}}(s)R\dot{x}(s)ds\le-\int_{t-\tau}^{t}\dot{x}^{\mathrm{T}}(s)dsR\int_{t-\tau}^{t}\dot{x}(s)ds. \tag{15}$$

To obtain the stability conditions of the considered system, let $\zeta(t)=0$ in (11) and denote

$$\eta_1^{\mathrm{T}}(t)=[x^{\mathrm{T}}(t)\ x^{\mathrm{T}}(t-\tau)\ \dot{x}^{\mathrm{T}}(t)\ f^{\mathrm{T}}(x,t)]. \tag{16}$$

Then from (14) and (15), we have

$$\dot{V}(t,x_t)\le\sum_{i=1}^{r}\sum_{j=1}^{r}\psi_{ij}(\chi(t))\eta_1^{\mathrm{T}}(t)\Xi_1(t)\eta_1(t), \tag{17}$$

where

$$\begin{cases} \Lambda_{11} = A_i^{\mathrm{T}} P_2 + P_2 A_i + P_2 B K_j + K_j^{\mathrm{T}} B^{\mathrm{T}} P_2 - R + Q + \mu^2 I, \\ \Lambda_{12} = R + P_2^{\mathrm{T}} B K_{\tau j}, \ \ \Lambda_{13} = P - P_2 + A_i^{\mathrm{T}} P_3 + K_j^{\mathrm{T}} B^{\mathrm{T}} P_3, \\ \tilde{\Xi}_1(t) = \begin{bmatrix} \Lambda_{11} & \Lambda_{12} & \Lambda_{13} & P_2 \bar{D} \\ * & -R-Q & K_{\tau j}^{\mathrm{T}} B^{\mathrm{T}} P_3 & 0 \\ * & * & \tau^2 R - 2P_3 & P_3 \bar{D} \\ * & * & * & -I \end{bmatrix}. \end{cases}$$

It is clear that the inequality $\Xi_1(t) < 0$ can be guaranteed by (12), which implies that the sliding motion (11) is robustly stable with $\zeta(t) = 0$.

Next, for nonzero $\zeta(t) \in L_2[0, \infty)$, it will be proven that the $H_\infty$ performance is guaranteed.

Note that

$$\dot{V}(t, x_t) + y^{\mathrm{T}}(t) y(t) - \gamma^2 \zeta^{\mathrm{T}}(t) \zeta(t) \le \sum_{i=1}^{r} \sum_{j=1}^{r} \psi_{ij}(\chi(t)) \eta_2^{\mathrm{T}}(t) \Xi(t) \eta_2(t) \qquad (18)$$

where

$$\begin{cases} \eta_2^{\mathrm{T}}(t) = [x^{\mathrm{T}}(t) \ x^{\mathrm{T}}(t-\tau) \ \dot{x}^{\mathrm{T}}(t) \ f^{\mathrm{T}}(x,t) \ \zeta^{\mathrm{T}}(t)] \\ \Xi_2(t) = \left[P_2^{\mathrm{T}} \bar{D}_0 \ 0 \ P_3^{\mathrm{T}} \bar{D}_0 \ 0\right]^{\mathrm{T}}, \Xi_3(t) = \left[C_1^{\mathrm{T}} \ 0 \ 0 \ 0\right]^{\mathrm{T}} \\ \Xi(t) = \begin{bmatrix} \Xi_1(t) & \Xi_2(t) & \Xi_3(t) \\ * & -\gamma^2 I & D_1^{\mathrm{T}} \\ * & * & -I \end{bmatrix} \end{cases}$$

By the Schur Complement, $\Xi(t) < 0$ is equivalent to the inequality (12). Notice that the initial condition $V(0) = 0$. Integrating inequality (18) from 0 to $\infty$ shows $\int_0^\infty y^T(t) y(t) dt \le \gamma^2 \int_0^\infty f^T(t) f(t) dt$.

Due to nonlinear term $P_2 B K_j$, $K_j^{\mathrm{T}} B^{\mathrm{T}} P_3$, $P_2 B K_{\tau j}$, $K_{\tau j}^{\mathrm{T}} B^{\mathrm{T}} P_3$, the inequality (12) is nonlinear. Denote $\hat{P} = \mathrm{diag}\{P_2^{-1}, P_2^{-1}, P_2^{-1}, I, I, I, I\}$. To obtain $K_j$ and $K_{\tau j}$ in (9), pre- and post-multiply (12) by $\hat{P}$ and $\hat{P}^T$, respectively. Setting $P_3 = \sigma P_2$, $\bar{P}_2 = P_2^{-1}$, $\bar{K}_j = K_j P_2^{-1}$, $\bar{K}_{\tau j} = K_{\tau j} P_2^{-1}$, $\bar{P} = P_2^{-1} P P_2^{-1}$, $\bar{Q} = P_2^{-1} Q P_2^{-1}$ and $\bar{R} = P_2^{-1} R P_2^{-1}$. The equivalent version of Proposition 1 can be obtained.

**Proposition 2.** *The sliding motion* (11) *with* $\zeta(t) = 0$ *is robustly asymptotically stable and the level* $\gamma$ *of* $H_\infty$ *attenuation is ensured, if* $\tau > 0$, $\sigma > 0$, $\mu > 0$ *and* $\gamma > 0$ *are given values and there exist* $6 \times 6$ *matrices* $\bar{P} > 0$, $\bar{Q} > 0$, $\bar{R} > 0$, $\bar{P}_2 > 0$ *and* $1 \times 6$ *matrices* $\bar{K}_j$, $\bar{K}_{\tau j}$ *such that*

$$\begin{bmatrix} \Phi_1 & \Phi_2 & \Phi_3 & \bar{D} & \bar{D}_0 & \bar{P}_2 C_1^{\mathrm{T}} & \mu \bar{P}_2 \\ * & -\bar{Q} - \bar{R} & \Phi_4 & 0 & 0 & 0 & 0 \\ * & * & \Phi_5 & \sigma \bar{D} & \sigma \bar{D}_0 & 0 & 0 \\ * & * & * & -I & 0 & 0 & 0 \\ * & * & * & * & -\gamma^2 I & D_1^{\mathrm{T}} & 0 \\ * & * & * & * & * & -I & 0 \\ * & * & * & * & * & * & -I \end{bmatrix} < 0, \qquad (19)$$

*where*

$$\begin{cases} \Phi_1 = A_i\bar{P}_2 + \bar{P}_2A_i^{\mathrm{T}} + B\bar{K}_j + \bar{K}_j^{\mathrm{T}}B^{\mathrm{T}} - \bar{R} + \bar{Q}, \ \ \Phi_2 = \bar{R} + B\bar{K}_{\tau j}, \\ \Phi_3 = \bar{P} - \bar{P}_2 + \sigma\bar{P}_2A_i^{\mathrm{T}} + \sigma\bar{K}_j^{\mathrm{T}}B^{\mathrm{T}}, \ \ \Phi_4 = \sigma\bar{K}_{\tau j}^{\mathrm{T}}B^{\mathrm{T}}, \ \ \Phi_5 = \tau^2\bar{R} - 2\sigma\bar{P}_2. \end{cases}$$

*Moreover,* $K_j = \bar{K}_j\bar{P}_2^{-1}$ *and* $K_{\tau j} = \bar{K}_{\tau j}\bar{P}_2^{-1}$, *respectively.*

### 3.2 Mixed Delayed Fuzzy Sliding Mode Control Law

Now, a mixed delayed robust fuzzy sliding mode $H_\infty$ control (MDRFSMHC) law can be designed to ensure the reachability of the specified sliding surface.

The MDRFSMHC law is given as

$$u(t) = \sum_{j=1}^{r} \psi_j(\chi(t))[K_jx(t) + K_{\tau j}x(t-\tau) - \rho(x(t))\ sgn(s(t))], \tag{20}$$

where $\phi > 0$, $\rho(x(t)) = (GB)^{-1}\left[\mu \|GD\|\,\|x(t)\| + \|GD_0\|\,\|\zeta(t)\| + \phi\right]$, $sgn(\cdot)$ is the sign function.

Now, we give a proposition as follow.

**Proposition 3.** *The state trajectory of the system* (7) *under the hybrid delay robust fuzzy sliding mode control law* (20) *can be driven to the sliding surface* $s(t) = 0$ *in a finite time and maintain this state.*

*Proof.* Choose the Lyapunov function as

$$V_2(s(t)) = \frac{1}{2}s^2(t). \tag{21}$$

From Eqs. (7), (9) and (20), it follows that

$$\dot{s}(t) = \sum_{i=1}^{r}\sum_{i=1}^{r} \psi_{ij}(\chi(t))G[Df(x,t) + D_0\zeta(t) - B\rho(x(t))sgn(s(t))]. \tag{22}$$

Hence, combining (20), (21) with (22) yields $\dot{V}_2(s(t)) \le -\phi|s(t)|$.

## 4 Simulation Results

First, the parameters of TMD and offshore platforms are taken from [19], where the depth of water, height of wave and the length of wave are 76.2 m, 12.19 m, 182.88 m, respectively. The shape vectors $\phi_1 = -0.003445$, $\phi_2 = 0.00344628$. The mass, stiffness and damping coefficient of TMD are supposed as $m_T = 469.4836$ kg, $K_T = 1551.5$ N/m and $C_T = 256\ N_s/m$.

For simplicity, we will choose $\omega_1(t)$, $\omega_2(t)$ and $\omega_T(t)$for fuzzification in this case. And $\xi_1 = \xi_2 = 0.005$, $\xi_T = 0.15$. The natural frequencies of system are supposed as $\omega_1(t) = 1.8180 + 0.1sin(t)$, $\omega_2(t) = 10.8683 + 0.1sin(t)$ and $\omega_T(t) =$

$1.8180+0.1sin(t)$. The fuzzy sets of $\omega_1(t)$, $\omega_2(t)$ and $\omega_T(t)$ are defined as $\{B, S\}$. They are $\omega_1^i = \{1.7180, 1.9180\}$ rad/s, $\omega_2^i = \{10.7683, 10.9683\}$ rad/s and $\omega_T^i = \{1.7180, 1.9180\}$ rad/s, separately. According to the above parameter values, the system matrix of the platform can be calculated by (5).

Then, we will show several advantages of the fuzzy sliding mode control law. When the jacket stucture subject to external wave is under no controller, the peak values of displacement response curves of the three floors are 1.4180 m, 1.5371 m and 1.6181 m, respectively. Then, we focus on designing the mixed delayed robust fuzzy sliding mode $H_\infty$ controller (MDRFSMHC). For given $\mu = 0.8$, $\sigma = 0.1$, $\tau = 0.08$ s and $\gamma = 9.8$. According to the proposition 2, the gain matrix $K_j$ and $K_{\tau j}$ can be obtained as

$$K_1 = [-56190\ \ 27970\ \ 11953\ \ -2390\ \ -34560\ \ -71110]\,,\ K_2 = [-55960\ \ 27850\ \ 11916\ \ -2360\ \ -32770\ \ -70670]$$
$$K_3 = [-55380\ \ 27560\ \ 11854\ \ -2270\ \ -32410\ \ -69940]\,,\ K_4 = [-55590\ \ 27670\ \ 11888\ \ -2300\ \ -34180\ \ -70350]$$
$$K_5 = [-55530\ \ 27220\ \ 11758\ \ -2200\ \ -32030\ \ -69180]\,,\ K_6 = [-55870\ \ 27430\ \ 11824\ \ -2250\ \ -33880\ \ -69790]$$
$$K_7 = [-56180\ \ 27540\ \ 11844\ \ -2300\ \ -32410\ \ -70000]\,,\ K_8 = [-55590\ \ 27670\ \ 11888\ \ -2300\ \ -34180\ \ -70350]$$
$$K_{1\tau} = [428.74\ \ -235.81\ \ -898.82\ \ 52.30\ \ 291.01\ \ 548.54]\,,\ K_{2\tau} = [408.19\ \ -225.45\ \ -861.73\ \ 50.29\ \ 285.08\ \ 523.19]$$
$$K_{3\tau} = [421.23\ \ -232.38\ \ -807.37\ \ 51.52\ \ 294.40\ \ 539.14]\,,\ K_{4\tau} = [445.89\ \ -245.35\ \ -940.76\ \ 53.75\ \ 303.20\ \ 570.65]$$
$$K_{5\tau} = [531.70\ \ -284.42\ \ -1114.35\ \ 58.30\ \ 356.95\ \ 669.23]\,,\ K_{6\tau} = [447.98\ \ -247.03\ \ -950.18\ \ 54.21\ \ 301.76\ \ 573.48]$$
$$K_{7\tau} = [470.83\ \ -253.24\ \ -984.01\ \ 55.29\ \ 315.62\ \ 594.24]\,,\ K_{8\tau} = [445.89\ \ -245.35\ \ -940.76\ \ 53.75\ \ 303.20\ \ 570.65]$$

Under the MDRFSMHC, oscillation amplitudes of the first, second, third floors and required control force are about 0.2307 m, 0.2484 m, 0.2611 m, $3.5564 \times 10^5$ N, respectively. After calculation, the MDRFSMHC reduces the average amplitude of the three-layer displacement of the offshore platform to about 16%. Then, the curves of the first, second, third floors and control force required of the uncertain system under MDRFSMHC are depicted in Fig. 2. Clearly, we can observe that the vibration amplitudes of the three floors are effectively mitigated under this scheme.

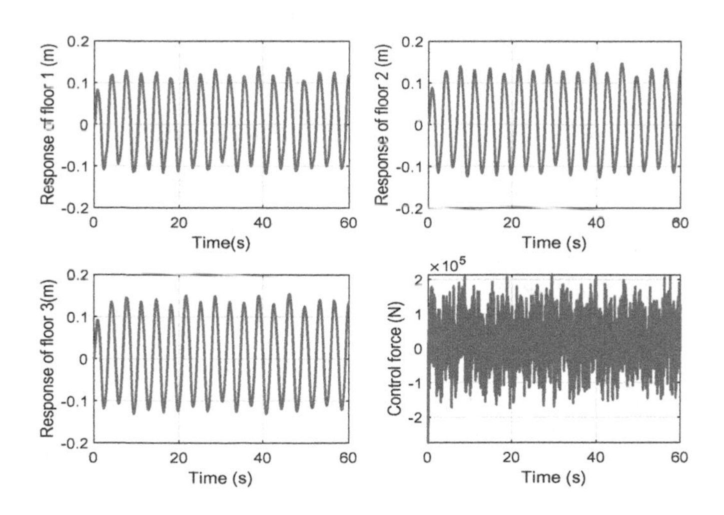


**Fig. 2.** The response of control performance of system under MDRFSMHC.

## 5 Conclusions

This paper is concerned with the fuzzy sliding mode control law for offshore structure with external wave has been solved. Considering the impact of time-varying parameters damping ratios and natural frequencies, the T-S fuzzy model is descripted. A new sufficient condition of system stability has been derived. Finally, several simulation results show the effectiveness of the designed controller.

## References


1. Zribi, M., Almutairi, N., Abdel-Rohman, M., et al.: Nonlinear and robust control schemes for offshore steel jacket platforms. Nonlinear Dyn. **35**, 61–80 (2004)
2. Hirdaris, S.E., Bai, W., Dessi, D., et al.: Loads for use in the design of ships and offshore structures. Ocean Eng. **78**, 131–174 (2014)
3. Kim, D.H.: Neuro-control of fixed offshore structures under earthquake. Eng. Struct. **31**, 517–522 (2009)
4. Abdel Raheem, S.E., Arab, S.E.: Study on nonlinear response of steel fixed offshore platform under environmental loads. Arab. J. Sci. Eng. **39**, 6017–6030 (2014)
5. Ou, J., Long, X., Li, Q.S., et al.: Vibration control of steel jacket offshore platform structures with damping isolation systems. Eng. Struct. **29**, 1525–1538 (2007)
6. Moharrami, M., Tootkaboni, M.: Reducing response of offshore platforms to wave loads using hydrodynamic buoyant mass dampers. Eng. Struct. **81**, 162–174 (2014)
7. Li, H.N., Huo, L.S.: Semi-active TLCD control of fixed offshore platforms using artificial neural networks. China Ocean Eng. **17**, 277–282 (2003)
8. Zhang, B.-L., Huang, Z.W., Han, Q.-L.: Delayed non-fragile $H_\infty$ control for offshore steel jacket platforms. J. Vib. Control **21**, 959–974 (2015)
9. Zhang, B.-L., Han, Q.-L., Zhang, X.-M.: Recent advances in vibration control of offshore platforms. Nonlinear Dyn. **89**(2), 755–771 (2017). https://doi.org/10.1007/s11071-017-3503-4
10. Zhang, W.B., Han, Q.-L., Tang, Y., Liu, Y.R.: Sampled-data control for a class of linear time-varying systems. Automatica **103**, 126–134 (2019)
11. Nourisola, H., Ahmadi, B., Tavakoli, S.: Delayed adaptive output feedback sliding mode control for offshore platforms subject to nonlinear wave-induced force. Ocean Eng. **104**, 1–9 (2015)
12. Liu, Y.J., Tong, S.: Adaptive neural network tracking control of uncertain nonlinear discrete-time systems with nonaffine dead-zone input. IEEE Trans. Cybern. **45**, 497–505 (2017)
13. Li, J., Zhou, S., Xu, S.: Fuzzy control system design via fuzzy Lyapunov functions. IEEE Trans. Syst. Man Cybern. Part B Cybern. **38**, 1657–1661 (2008). A Publication of the IEEE Systems Man and Cybernetics Society
14. Kamal, E., Aitouche, A., Ghorbani, R., Bayart, M.: Robust fuzzy fault-tolerant control of wind energy conversion systems subject to sensor faults. IEEE Trans. Sustain. Energy **3**, 231–241 (2012)
15. Liu, Y.J., Tong, S.: Adaptive fuzzy control for a class of nonlinear discrete-time systems with backlash. IEEE Trans. Fuzzy Syst. **22**, 1359–1365 (2014)
16. Li, H., Wang, J., Shi, P.: Output-Feedback based sliding mode control for fuzzy systems with actuator saturation. IEEE Trans. Fuzzy Syst. **24**, 1282–1293 (2015)

17. Li, H.Y., Yu, J.Y., Liu, H.H.: Adaptive sliding-mode control for nonlinear active suspension vehicle systems using T-S fuzzy approach. IEEE Trans. Industr. Electron. **60**, 3328–3338 (2012)
18. Zhang, B.-L., Han, Q.-L., Zhang, X.-M., et al.: Integral sliding mode control for offshore steel jacket platforms. J. Sound Vib. **331**, 3271–3285 (2012)
19. Zhang, B.-L., Han, Q.-L., Zhang, X.-M., et al.: Sliding mode control with mixed current and delayed states for offshore steel jacket platforms. IEEE Trans. Control Syst. Technol. **22**, 1769–1783 (2014)
20. Zhang, B.-L., Han, Q.-L., Zhang, X.-M., Tang, G.-Y.: Active Control of Offshore Steel Jacket Platforms. Springer, Singapore (2019). https://doi.org/10.1007/978-981-13-2986-9

# Real-Time Multi-target Activity Recognition Based on FMCW Radar

He Pan, Kang Liu, Lei Shen, and Aihong Tan(✉)

Research Center for Sino-German Autonomous Driving, College of Mechanical and Electrical Engineering, China Jiliang University, Hangzhou 310018, China
Tanah@cjlu.edu.cn

**Abstract.** Since multi-target activity recognition solutions based on wearable devices require that sensors be worn at all times, as well as vision-based multi-target activity recognition solutions are susceptible to weather or lighting conditions and are prone to revealing personal privacy, so this paper proposes a real-time multi-target activity recognition solution based on FMCW (Frequency Modulated Continuous Wave) radar to realize non-contact real-time multi-target activity recognition. Firstly, the point cloud data collected by FMCW radar are pre-processed. Secondly, each target information is obtained by multi-target tracking algorithm, and the activity key points and activity feature are extracted. Thirdly, according to the activity characteristic quantity, the data is classified by the classifier to realize the indoor real-time multi-target activity recognition. The test results show that the real-time activity recognition accuracy of this solution for falling, sitting and walking can reach 100%, 87.2% and 100%, respectively.

**Keywords:** Activity recognition · Multi-target tracking · Millimeter wave radar · Feature extraction · Real time

## 1 Introduction

In recent years, with the popularization and development of artificial intelligence, human activity recognition has gradually become a hot topic that fits real life, and it is also a hot topic in current academic and industrial research [1].

At present, the typical human activity recognition solutions can be divided into wearable and non-contact types according to the types of sensors. The common types of wearable sensors include wristbands, ankle monitors and so on, but it must be worn with yourself and the cost is high, so its application prospects are poor. The most studied non-contact human activity recognition is vision-based activity recognition, the core of which is to input the single frame image or video data of human activity collected by camera, and realize multi-object activity recognition through machine learning or deep learning [2]. The camera is a traditional non-contact sensor, which has high requirements for storage space, information processing capabilities and working environment. In addition, the public's awareness of the protection of personal privacy is becoming stronger, and the unrestricted use of cameras in privacy-sensitive areas will also cause controversy [3–5].


Q. Han et al. (Eds.): LSMS 2021/ICSEE 2021, CCIS 1469, pp. 446–455, 2021.
https://doi.org/10.1007/978-981-16-7213-2_43

In recent years, with the improvement of precision manufacturing processes, radar measurement accuracy is getting higher and higher, and the cost is gradually decreasing. Low-power radar electromagnetic waves will not affect the target. The radar is minimally affected by the external environment. It has very stable working performance, and there is no hidden danger of infringement of privacy at all, which meets the requirements of human activity recognition scenes. At present, many radar-based activity recognition solutions are single target activity recognition, and can not judge in real time. Literature [6] uses the micro-Doppler map generated from radar data to extract features, and uses support vector machines for classification, but it cannot be judged in real time. Literature [7] recognizes human activity micro-Doppler images generated from radar data through deep learning. Although its classification accuracy is high, it requires a large amount of data and a computer with excellent performance. Using micro-Doppler information only recognizes high specificity activities, which is not suitable for daily life [8–10].

This paper proposes a real-time indoor multi-target tracking and activity recognition solution based on FMCW radar to realize real-time multi-target activity detection. At the same time, it has been tested, and its accuracy can meet the designs of people's fall in the intelligent health care system. First, the point cloud data output by the FMCW radar in real time is detected by a multi-target tracking algorithm frame by frame to generate point cloud data and target information for each frame. Then extract the activity characteristics of each frame data. Finally, the classifier is used for activity recognition. The rest of this article is organized as follows. The second section summarizes the indoor real-time multi-target activity recognition system based on FMCW radar. The Sect. 3 describes the data processing and multi-target tracking algorithm. The Sect. 4 describes the real-time activity recognition algorithm. The Sect. 5 tests the accuracy of the real-time multi-target activity recognition solution through actual scene experiments. Section 6 summarizes the article.

## 2 Overview of Real-time Multi-target Activity Recognition System Based on FMCW Radar

The personnel activity detection system designed in this paper is shown in Fig. 1, which consists of two parts: data acquisition and data processing. The data collection environment is a room, the floor of which is relatively flat, so when walking target height will not change significantly due to external factors. The FMCW radar module is installed at a height of 1.8 m from the ground through a tripod and placed close to the wall. The data collected by the radar are transmitted to the computer through the USB port. The acquisition rate is $5\text{frame} \cdot s^{-1}$. The data processing part includes three parts: data preprocessing, multi-target tracking algorithm, and activity recognition algorithm. The FMCW millimeter wave radar module uses TI's IWR6843ISK. The module's working frequency band is 60–64 GHz. The IWR6843ISK includes an on-board etched long-distance antenna, four receivers and three transmitters, and a radar acquisition system with 12 virtual antenna elements is constructed. The maximum output power is 10 dBm, and the measured azimuth angle is 120°, the measured elevation angle is 60° [11]. The radar antenna distribution is shown in Fig. 2.

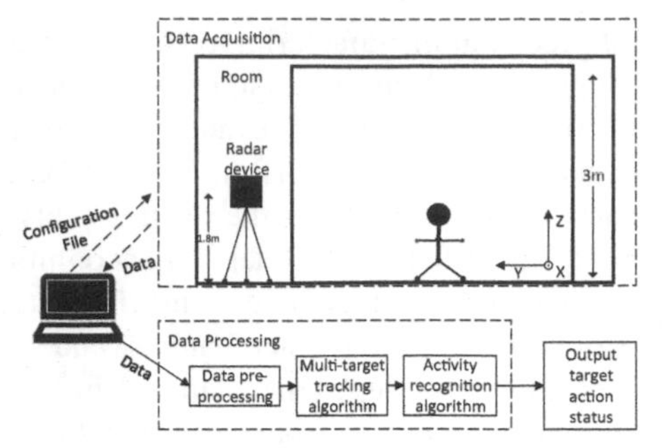


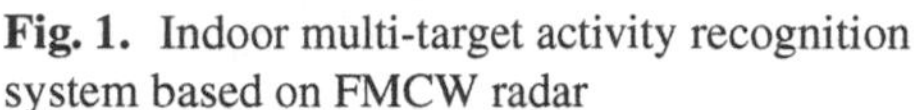

**Fig. 1.** Indoor multi-target activity recognition system based on FMCW radar

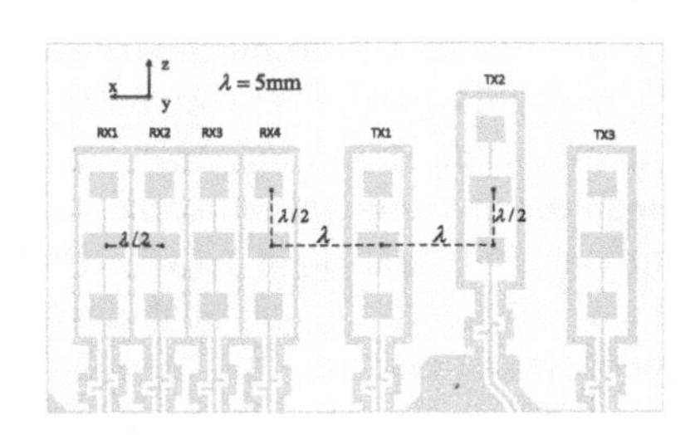


**Fig. 2.** Radar antenna layout

The propagation time between the detected target and the radar is obtained by the frequency difference of the reflected and transmitted wave of FMCW radar, and then the distance $R$ between the target and the radar is estimated. As shown in Fig. 3, the phase of the signal sent by TX1 on the 4 receiving antennas is $\left[\ 0\ \omega\ \ 2\omega\ 3\omega\ \right]$, where $\omega = \lambda/2 * \sin\theta$, Since the distance between TX1 and TX3 is $2\lambda$, the phase of the signal sent by TX3 on the four receiving antennas is $\left[\ 4\omega\ 5\omega\ \ 6\omega\ 7\omega\ \right]$. Since the horizontal distance between TX1 and TX2 is $\lambda$ and the vertical distance is $\lambda/2$, the phase of the signal from TX2 on the 4 receiving antennas are $-\sqrt{5}\omega+\lambda/2*\sin\varphi$, $1-\sqrt{5}\omega+\lambda/2*\sin\varphi$, $2-\sqrt{5}\omega+\lambda/2*\sin\varphi$ and $3-\sqrt{5}\omega+\lambda/2*\sin\varphi$. Therefore, by using MIMO virtual antenna array, the antenna with 3 transmit and 4 receive is constructed into 1 transmit and 12 receive [9].

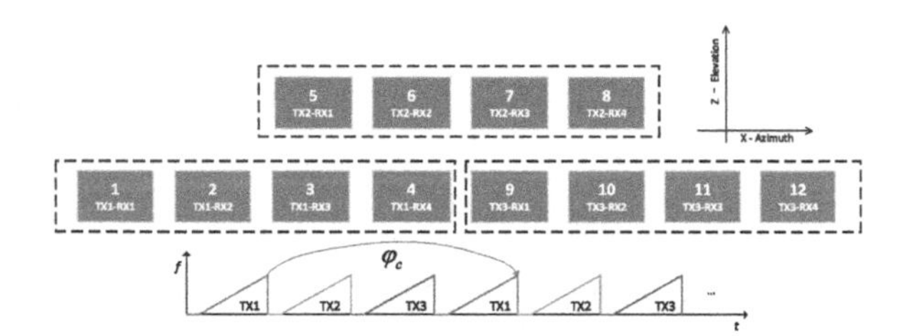


**Fig. 3.** MIMO virtual antenna array

# 3 Data Preprocessing, Multi-target Tracking Algorithm and Parameter Configuration

## 3.1 Data Preprocessing

According to the distance, azimuth and elevation angle $(R, \theta, \varphi)$ of the real-time output of the radar, the rectangular coordinates $(x, y, z)$ where the target is located can be calculated by formula (1).

$$x = R\cos\phi\sin\theta,\ y = R\cos\phi\cos\theta,\ z = R\sin\phi. \tag{1}$$

In addition, the phase difference $\phi_c$ generated by the two signals transmitted by the same antenna in a short time interval $T_i$ is detected, and the moving speed $v$ of the detection target is obtained from formula (2).

$$v = \frac{4\pi T_i}{\lambda \cdot \phi_c}. \tag{2}$$

In order to identify the reflected signals of targets, the background elimination method is used to eliminate the reflected signals of static objects in the environment. This paper uses linear interpolation to process the data. After data preprocessing, the point cloud information generated by the reflection of the human body in the detection environment can be obtained [13]. The point cloud is a set of rich measurement vectors which contains position, speed and signal-to-noise ratio (SNR).

### 3.2 Multi-target Tracking Algorithm

The multi-target tracking algorithm takes point cloud data as input, performs target detection and positioning, and outputs target information. We use a constant acceleration model for multi-target tracking. The state vector $s(n)$ of the constant acceleration model is defined as the following formula (3). The state of the Kalman filter in the algorithm at time n is formula (4), the state vector of the target in the previous frame is used to predict the state vector of the target in the next frame, $w(n)$ is the process noise vector of the constant acceleration model.

$$s(n) \triangleq \left[ x(n)\ y(n)\ z(n)\ \ \dot{x}(n)\ \dot{y}(n)\ z(n)\ \ (n)\ \ddot{y}(n)\ \ddot{z}\ (n) \right]^T. \tag{3}$$

$$s(n) = Fs(n-1) + w(n). \tag{4}$$

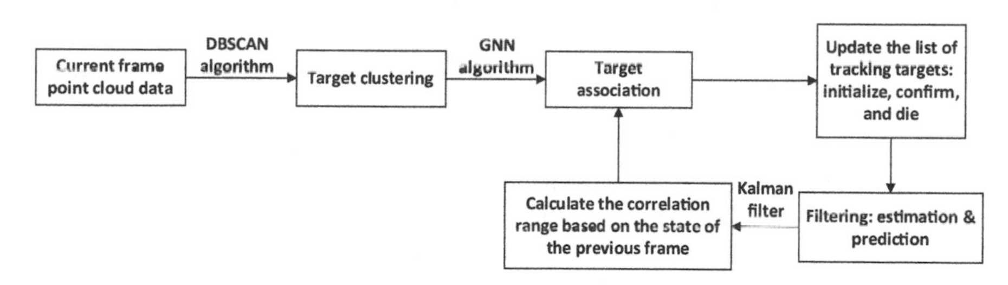


**Fig. 4.** Multi-target tracking flowchart

The process of the multi-target tracking algorithm adopted in this paper is mainly divided into three links: clustering, target association and filtering. Clustering uses DBSCAN algorithm. Target association adopts GNN strategy for association. The filtering link adopts the optimal Kalman filter in the sense of least squares, as shown in Fig. 4. The actual effect is shown in Fig. 5.

An example of tracking two targets is shown in Fig. 6, where at time n − 1, the centroids of the two tracking targets are marked as $G_1(n-1)$ and $G_2(n-1)$, respectively.

Firstly, predict the centroid of the tracking target. The predicted centroid of the target is $G_{1,apr}(n)$ and $G_{2,apr}(n)$. Secondly, point cloud detection is carried out around each centroid. At time n, a set of point clouds $[u_1, u_2, \cdots\cdots u_9]$ is obtained from the data, and associate it with the corresponding trajectory. It can be seen from Fig. 6 that the relevant point cloud at time n of target 1 has $[u_1, u_4, u_5, u_8, u_9]$, and the average value is $\overline{u_1}(n)$. We call the difference between $G_{1,apr}(n)$ and $\overline{u_1}(n)$ innovation, which can evaluate the difference between our predicted results and the observed results. Finally, we move the target centroid at time n to $G_1(n)$ by the value of innovation [14].

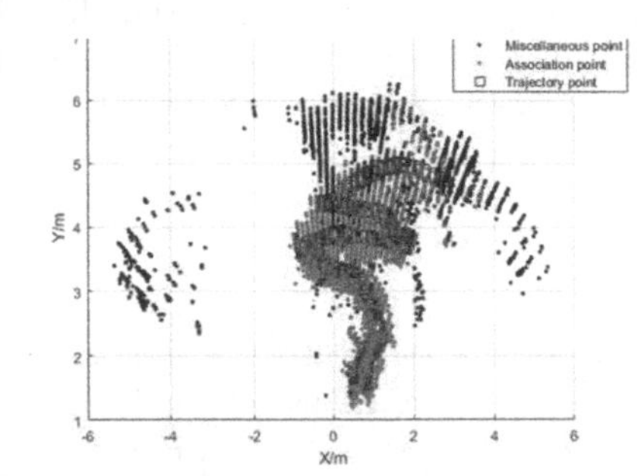


Fig. 5. Single target tracking effect

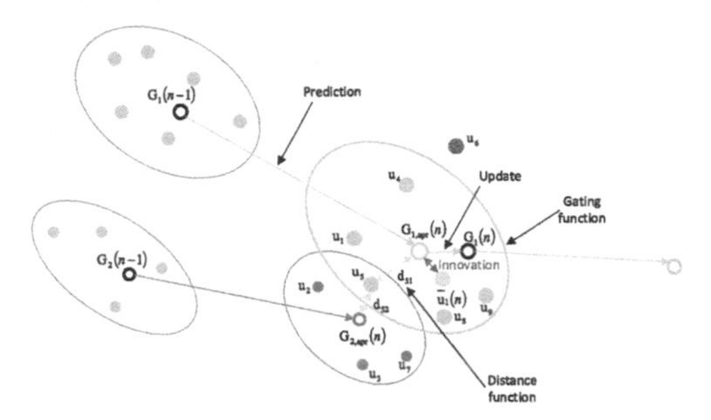


Fig. 6. Example of multi-target tracking algorithm

### 3.3 Parameter Configuration

Firstly, we configure the size parameters of the detection area. After the test, it is found that the number of point clouds at a distance of 6 m from the radar decreases sharply. When the radar is level with the height of the person, the detection speed is more accurate. Therefore, we set the detection area size to 6 m*6 m and the radar installation height to 1.8 m. The detection direction of the radar is parallel to the wall. Secondly, for each tracking trajectory, we need to cluster around each tracking point. The range of this clustering is considered from the physical size of the target, the expected target activity trajectory, the expected tracking error and so on. After the test, when the range of clustering is a circle of 1 m, the tracking sensitivity is higher. Then, for points within the clustering range, if the number of points, the value of the point's SNR, and the value of the point's velocity all exceed the threshold, we start to track a target. After the test, it is found that when the number of points in the clustering range exceeds 15, the SNR exceeds 20, and the speed value exceeds 0.002 m/s, the tracking effect is relatively good. Finally, set the maximum amount of change in the acceleration of the tracking target in each direction to 10 m/s to make the effect of multi-target tracking more accurate.

## 4 Key Point Extraction, Feature Extraction and Activity Recognition

In order to recognize the activity state of the human body, it is necessary to further extract the key points of its activity, and then perform feature extraction. After extracting the features, we configure the parameters of each activity recognition in the classifier. The more accurate the parameter configuration, the higher the success rate of activity recognition.

## 4.1 Key Point Extraction

The activity recognition accuracy rate of the solution with the maximum, average, and minimum value of the activity key point data is compared in Fig. 7. It can be seen from the Fig. 7 that the solution based on the average value of the activity key point data is better than the other two solutions in the overall situation. Therefore, this article chooses the mean value as the key point of the activity.

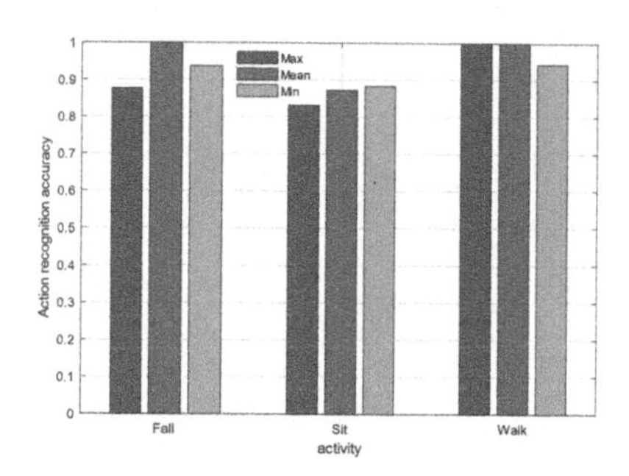


**Fig. 7.** The effect of key point data on the accuracy of activity recognition

After that, moving average filtering is used to process the unsmooth activity key points, which is helpful for us to analyze the data. It can be seen from Fig. 8 that falling corresponds to a rapid decrease in height on the z axis, and at the same time there is a significant change in the speed. When the speed is negative, it means that the direction of movement is far away from the radar detection module.

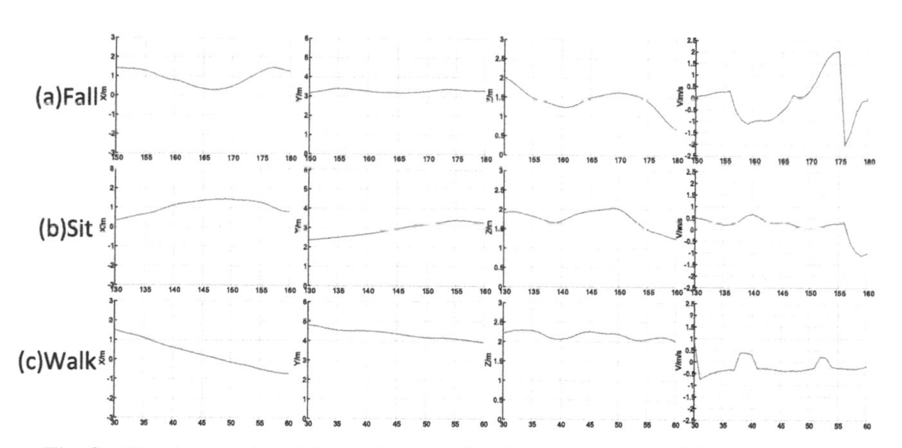


**Fig. 8.** The change of position and speed of action key points in different behaviors

## 4.2 Feature Extraction

This paper proposes the use of four features: the average value of data Z ($Z_{mean}$), the average value of data X ($X_{mean}$), the average value of data Y ($Y_{mean}$), and the speed

($V_{obj}$). When the target goes from standing to sitting, gravity center of the target will decrease slightly and the size of the target will change from a tall and thin state to roughly the same three-dimensional state. When the target is falling, gravity center of the target will be greatly reduced and the size of the target will become wide and flat. We compare the recognition accuracy rates corresponding to different combinations of these four features in Fig. 9. It can be seen from the figure that when using all 4 features the recognition accuracy is the highest.

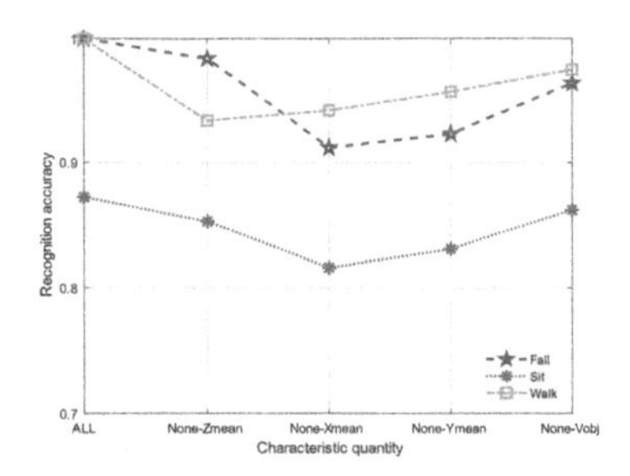


**Fig. 9.** The effect of key point data on the accuracy of activity recognition

## 4.3 Activity Recognition

Based on the above test results, this paper extracts all the four features mentioned above to form a feature vector, and uses a naive Bayes classifier for classification to realize the recognition of multi-target activities [15]. The classifier process is shown in Fig. 10.

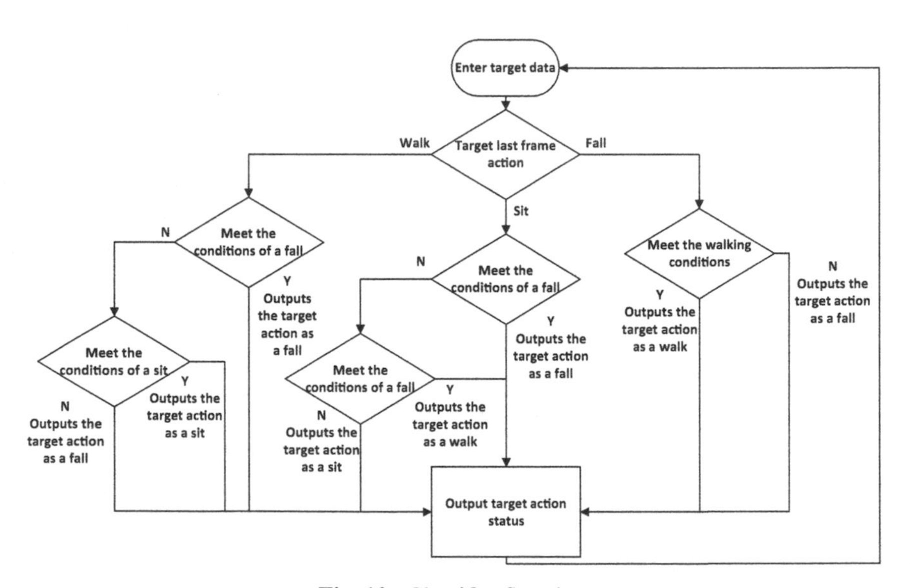


**Fig. 10.** Classifier flowchart

## 5 Real-Time Testing

The test environment of this article and part of the action test is shown in Fig. 11. After the tester enters the room, he performs some activities required for the test within 6 m from the radar, such as walking-falling, walking-sitting, sitting-falling and other activities. A total of 10 testers (9 males and 1 female) were invited to participate in the test experiment. The testers' heights ranged from 160 to 185 cm, tall, short, fat and thin, and weighed between 45 to 90 kg. First, each person performs 5 tests individually, a total of 50 tests are performed in a single test, and each test is tested for 3 types of actions, and then two tests are conducted in groups of two, and a total of 10 tests are conducted in a two-person test. At last, a group of 4 people conducted 2 tests, and a total of 4 people conducted 10 tests.

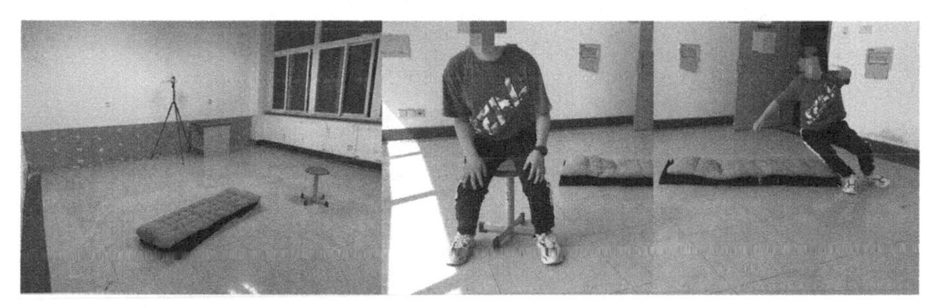

**Fig. 11.** Test environment and part of the action test

This article uses the recognition accuracy of the activity as the evaluation criterion. After test, the accuracy of the indoor multi-target activity recognition system based on FMCW radar proposed in this paper is shown in Table 1. The data statistics in the table include single, double and multiple people. From the table, it can be seen that the recognition accuracy of the three activities of falling, sitting, and walking are 100%, 87.2%, and 100%, respectively. The system can identify fall and non-fall events very accurately. The misjudgment of the activity is mainly caused by the misjudgment of the two activities of walking and sitting.

**Table 1.** Activity recognition accuracy table

| Activity | Fall(recognition) | Sit(recognition) | Walk(recognition) |
|---|---|---|---|
| Fall | 1 | 0 | 0 |
| Sit | 0 | 0.872 | 0.128 |
| Walk | 0 | 0 | 1 |

The activity recognition solution adopted in this paper has the accuracy rate of activity recognition for falling, sitting and walking under the scenes of different target numbers, as shown in Tables 2 and 3. It can be seen from the table that the wrong judgment rate of activities in single-player scenes is low, and the wrong judgment rate of activities in multiplayer scenes is higher. In a scene with more than two persons, due to mutual occlusion between the test targets, the point cloud of the occluded target is reduced, resulting in the inaccurate data of the occluded target detected by the FMCW radar, which in turn leads to misjudgment of the target's activity, which affects the accuracy of activity recognition rate.

**Table 2.** Single-person scene activity recognition accuracy table

| Activity | Fall(recognition) | Sit(recognition) | Walk(recognition) |
|---|---|---|---|
| Fall | 1 | 0 | 0 |
| Sit | 0 | 0.953 | 0.047 |
| Walk | 0 | 0 | 1 |

**Table 3.** Multiplayer scene activity recognition accuracy table

| Activity | Fall(recognition) | Sit(recognition) | Walk(recognition) |
|---|---|---|---|
| Fall | 1 | 0 | 0 |
| Sit | 0.059 | 0.807 | 0.134 |
| Walk | 0 | 0 | 1 |

## 6 Conclusion

This paper proposes a scheme of indoor multi-target activity recognition based on FMCW radar. Through multi-target tracking of the data collected by FMCW radar, the key points of the target action are extracted, the appropriate action features are selected, and the detected data is classified, thereby Realize multi-target activity recognition suitable for indoor scenes. Through the test results, it is found that the multi-target tracking algorithm used in this paper can effectively detect the information of the test target from the point cloud data collected by the FMCW radar. For the action data of each target, we take the average value of the action key point data as the action judgment basis, which can effectively reduce the data processing capacity and has a better accuracy rate of activity recognition. At the same time, we select the four feature quantities of the height mean $Z_{mean}$, X mean $X_{mean}$, Y mean $Y_{mean}$, and speed mean $V_{obj}$ in the action data as the feature quantities for activity recognition. The test has a good activity recognition accuracy. The experimental results show that the activity recognition scheme used in this paper has achieved 100%, 87.2%, and 100% accuracy for the activity recognition of falling, sitting and walking, respectively. However, in the scene of more than two persons, the action is blocked by the targets. The recognition accuracy rate has declined.

**Acknowledgements.** This work was supported by the Provincial Natural Science Foundation of Zhejiang (Y21F010057).

## References

1. Park, S., Aggarwal, J.: Recognition of two-person interactions using a hierarchical Bayesian network. In: First ACM SIGMM International Workshop on Video Surveillance, pp. 65–76 (2003)
2. Ji, S., Xu, W., Yang, M., Yu, K.: 3D convolutional neural networks for human action recognition. IEEE Trans. Pattern Anal. Mach. Intell. **35**, 221–231 (2012)
3. Li, T., Fan, L., Zhao, M., Liu, Y., Katabi, D.: Making the invisible visible: Action recognition through walls and occlusions. In: Proceedings of the IEEE/CVF International Conference on Computer Vision, pp. 872–881 (2019)
4. Banerjee, D., et al.: Application of spiking neural networks for action recognition from radar data. In: 2020 International Joint Conference on Neural Networks (IJCNN), pp. 1–10. IEEE (2020)
5. Park, J., Cho, S.H.: IR-UWB radar sensor for human gesture recognition by using machine learning. In: 2016 IEEE 18th International Conference on High Performance Computing and Communications; IEEE 14th International Conference on Smart City; IEEE 2nd International Conference on Data Science and Systems (HPCC/SmartCity/DSS), pp. 1246–1249. IEEE (2016)
6. Vandersmissen, B., et al.: Indoor person identification using a low-power FMCW radar. IEEE Trans. Geosci. Remote Sens. **56**, 3941–3952 (2018)
7. Chuma, E.L., Roger, L.L.B., de Oliveira, G.G., Iano, Y., Pajuelo, D.: Internet of Things (IoT) privacy–protected, fall-detection system for the elderly using the radar sensors and deep learning. In: 2020 IEEE International Smart Cities Conference (ISC2), pp. 1–4. IEEE (2020)
8. Li, H., et al.: Multisensor data fusion for human activities classification and fall detection. In: 2017 IEEE SENSORS, pp. 1–3. IEEE (2017)
9. Luo, D., Du, S., Ikenaga, T.: Multi-task and multi-level detection neural network based real-time 3D pose estimation. In: 2019 Asia-Pacific Signal and Information Processing Association Annual Summit and Conference (APSIPA ASC), pp. 1427–1434. IEEE (2019)
10. Ugolotti, R., Sassi, F., Mordonini, M., Cagnoni, S.: Multi-sensor system for detection and classification of human activities. J. Ambient Intell. Humanized Comput. **4**, 27–41 (2013)
11. Will, C., Vaishnav, P., Chakraborty, A., Santra, A.: Human target detection, tracking, and classification using 24-GHz FMCW radar. J. IEEE Sens. J. **19**, 7283–7299 (2019)
12. Ren, A., Wang, Y., Yang, X., Zhou, M.: A dynamic continuous hand gesture detection and recognition method with FMCW radar. In: 2020 IEEE/CIC International Conference on Communications in China (ICCC), pp. 1208–1213. IEEE (2020)
13. Chen, M.-C., Wu, H.-Y.: Design of a fall detection system. In: 2011 International Conference on Electric Information and Control Engineering, pp. 3148–3151. IEEE (2011)
14. Livshitz, M.: Tracking radar targets with multiple reflection points. J. TI Internal Document (2017)
15. Baird, Z., Rajan, S., Bolic, M.: Classification of human posture from radar returns using ultra-wideband radar. In: 2018 40th Annual International Conference of the IEEE Engineering in Medicine and Biology Society (EMBC), pp. 3268–3271. IEEE (2018)

# Hand Gesture Recognition in Complex Background Based on YOLOv4

Li Wang, Shidao Zhao, Fangjie Zhou(✉), Liang Hua, Haidong Zhang, Minqiang Chen, and Lin Zhou

Department of Electrical Engineering, Nantong University, Nantong 226019, China

**Abstract.** Traditional methods often use a monochrome background such as green as the background so that the recognition process is not affected by the background, but this operation makes recognition limited to the use of the background plate, greatly reducing flexibility and losing the original gesture recognition to improve the level of human-computer interaction. In this paper, we propose the use of background subtraction to first eliminate the background so that recognition is not affected by the background, solving the problem of complex backgrounds affecting gesture recognition; then use YOLOv4 for gesture recognition to recognize Chinese hand gestures. After using background removal, the gesture recognition accuracy of YOLOv4 can be as high as 80%.

**Keywords:** Hand gesture recognition · Background subtraction · YOLOv4

## 1 Introduction

Machine vision-based gesture recognition is important for intelligent human-computer interaction. However, vision-based gesture recognition can be easily influenced by the background environment. The background often contains other information that interferes with computer vision, reducing the accuracy of gesture image recognition. In the past, people have used skin color distribution models [1, 2] to accomplish hand image or other parts of person's image extraction, but this can lead to objects with similar colors to hand skin color, such as human faces, entering the image recognition, increasing the difficulty of image recognition or the complexity of the image recognition procedure [3]. In response, we propose to use background subtraction to pre-process the image.

We want to make our work easier to accept by people, different other papers which use hand gestures that are regulated by themselves. Although by regulating hand gestures, the difficulty of recognition will drop and the accuracy will grow, it is not easy to use since this kind of hand gesture regulation is not widely accepted by people. And that means people need to learn the regulation before they use, which does not meet the requirement of intelligent human interaction. So, we need a dataset that is widely accepted by people. However, many papers [4, 5] chose to use public datasets that are made by big company or some universities. But these kinds of datasets do not include the one we want. So, we made our own dataset to satisfy the need from Chinese hand gesture.


Q. Han et al. (Eds.): LSMS 2021/ICSEE 2021, CCIS 1469, pp. 456–461, 2021.
https://doi.org/10.1007/978-981-16-7213-2_44

Since the image to be recognized is a black background filled gesture, the dataset is also acquired using background subtraction to acquire a black background filled gesture image. In addition, the color images are converted to black and white in order to reduce the amounts of operations required for image recognition and to increase the speed of model training and gesture recognition.

Due to the camera and the ambient light source, there is inevitably a lot of noise in the images after capture. We also use Gaussian filtering to smooth out the images. This reduces the interference of noise, etc. with image recognition.

In order to achieve better target detection, we use YOLOv4 as the network for gesture recognition.

## 2 Background Subtraction

Background Subtraction is a relatively basic tool in visual surveillance [6]. This method is simple to implement, easy to compute and therefore responsive. It is a more effective method for extracting areas of change.

There are many handcraft models used to handle the effects from background [7, 8]. They did excellent performance in their fields. However, handcraft methods can not handle the conditions not prepared, which are common in real practice.

Background Subtraction starts by selecting an image as the background, after which each pixel of the subsequent image is subtracted from the background image to eliminate the background.

After eliminating the background the gesture comes within the visual range of the camera to obtain an image containing only the gesture.

The formula follows as:

$$d = |I_l(x, y, i) - B_l(x, y)|. \tag{1}$$

$$ID_L(x, y, i) = \begin{cases} d, if\ d \geq T \\ 0, if\ d < T \end{cases}. \tag{2}$$

where $ID_L$ is the background frame difference map, $B_L$ is the background luminance component, $i$ denotes the number of frames ($i = 1, \ldots N$), $N$ is the total number of frames in the sequence, and $T$ is the threshold value [9].

The advantage of this method is the simplicity of the algorithm design. However, it also poses some problems. When the background light source emission changes, the background under this method will be different from the original condition, leading to a simple background subtraction to determine that the background is different from the original, thus feeding the background information into the new image. Thus, the gesture recognition does not achieve the desired effect.

## 3 Gaussian Filtering

Gaussian filtering is a common computer vision image processing function [10] Gaussian pre-filtering can significantly reduce the noise level of an image, thus eliminating the systematic error caused by noise in the DVC displacement measurement results [11].

Therefore, we use Gaussian filtering to filter the images and reduce the interference caused by noise on recognition.

## 4 Dataset

**Fig. 1.** Hand gestures used in our dataset.

The Fig. 1 shows the examples of the hand gestures used in our dataset. Each hand gesture represents a number, means 0, 1, 2, 3, 4, 5, 6, 7, 8, 9. Fist can mean 10 but we think 0 is more widely used in computer, so we use fist to represent 0, not 10.

Our dataset is made by ourselves. It contains over 10,000 images in this dataset. These hand gestures are all Chinese hand gestures. They are widely accepted by Chinese. We chose these hand gestures after we see other's work [12, 13] use hand gestures which are widely accepted in India. So, we want to make the hand gesture recognition we made more useful and easier for many people to accept.

Our images in the dataset are all pretreated by the methods mentioned above, in order to reduce net's work and have a better accuracy.

## 5 YOLOv4 Network

YOLOv4 is a network model proposed by Alexey Bochkovskiy et al. It is faster and more accurate than YOLOv3.

YOLOv4 was theoretically justified and tested for comparison using CSPDarkNet53 [14] as the backbone network, using SPP, PAN [15] as the neck and YOLOv3 as the head.

Alexey Bochkovskiy et al. compared YOLOv4 with networks such as LRF and SSD on Maxwell, Pascal, and Volta GPUs, and showed that YOLOv4 outperformed LRF and SSD.

In addition, YOLOv4 has lower GPU requirements and can be trained and used on regular GPUs with 8-16GB-VRAM, this makes YOLOv4 portable for use on a wider range of platforms and has high practical and commercial value.

However, since the dataset images we used were all 384 * 320 images, we decided that using the 608 * 608 input was a waste of resources and not conducive to faster training and recognition, so we used an input image size of 416 * 416. This will make training and recognition faster and more efficient.

Before the images are sent into the net, we use the methods mentioned above to pretreat the images in order to let the net analysis the images easier.

The loss and accuracy curves of the YOLOv4 network during training are shown in Figs. 2, 3.

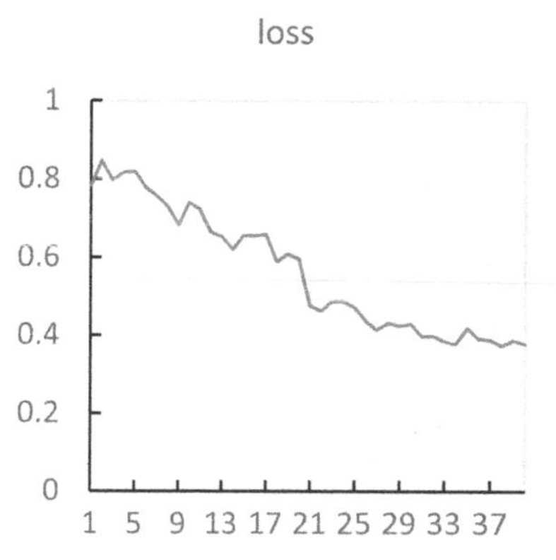


**Fig. 2.** The curve of loss will drop with the epochs' increase.

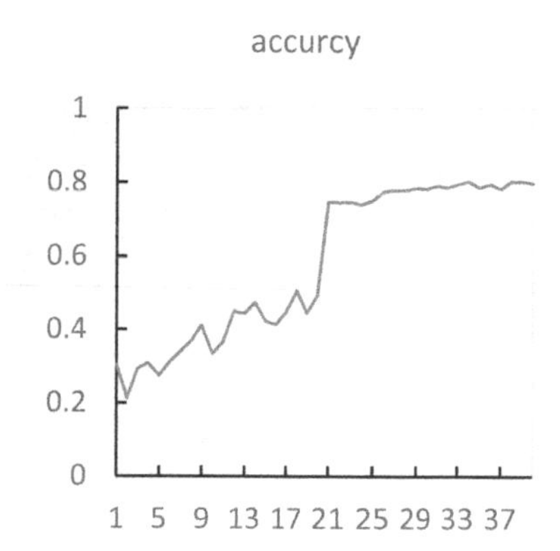


**Fig. 3.** The curve of accuracy will increase and then converge to 0.8.

# 6 Experimental Results

The actual experimental results show that the effect of the background on gesture recognition can be effectively eliminated in complex backgrounds. The final accuracy is 0.8 (Fig. 4).

Figures 5 show the black noise that results on the image of the hand when objects of the same color as the hand are present in the background. This is due to the fact that the background color matches the color of the part of the hand, causing a simple background differential to incorrectly subtract the hand from the background color.

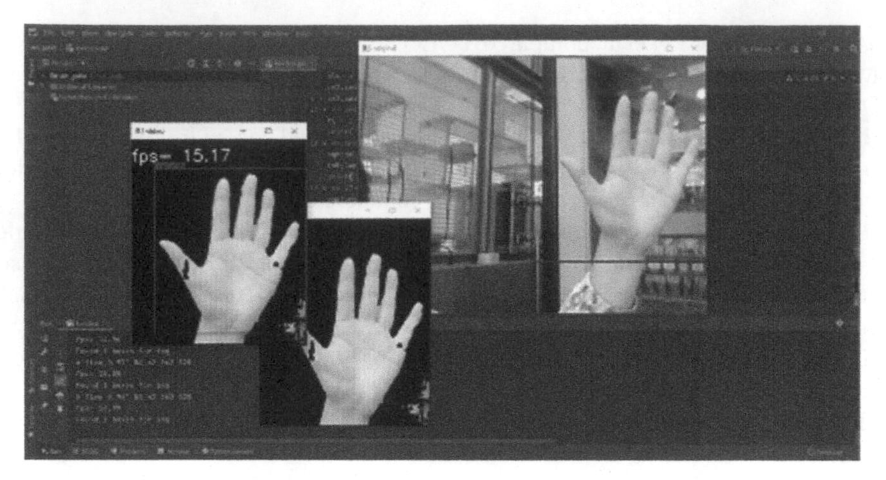


**Fig. 4.** Detection results against a complex background.

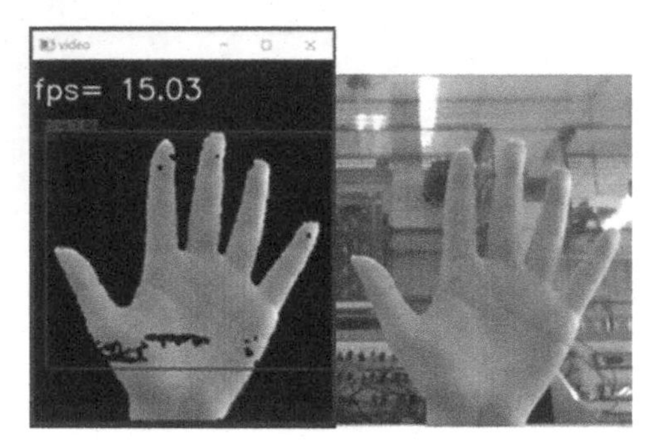


**Fig. 5.** The noise affects the result.

## 7 Conclusion

The aim of the solution in this thesis is gesture recognition in complex contexts. For complex backgrounds, we used background subtraction to eliminate the background directly to a black background using subtraction. Afterwards we used YOLOv4 as a neural network model for gesture recognition. The YOLOv4 network shows good recognition performance without interference from the background.

In future work, we will use more advanced and effective background elimination methods to compensate for the current simple background differencing methods affected by light, taking into account the overall performance. We will also consider the introduction of logical judgements into the recognition process.

## References

1. Zhou, K., et al.: Research on recognition and application of hand gesture based on skin color and SVM. J. Comput. Meth. Sci. Eng. **20**(1), 269–278 (2019)

2. Biplab, K.C., et al.: Combining image and global pixel distribution model for skin colour segmentation. Pattern Recogn. Lett. **88**, 33–40 (2017)
3. Rajkumar, C., Mahendran, S.K.: Enhanced key frame identification and background subtraction methods for automatic license plate recognition of motorcyclists without helmets. Mater. Today: Proc. ISSN 2214-7853 (2021)
4. Na, I.S., Kim, S.H., Lee, C.W., Duong, N.H.: Hand component decomposition for the hand gesture recognition based on FingerPaint dataset. In: Eleventh International Conference on Ubiquitous and Future Networks (ICUFN) (2019)
5. Himamunanto, A.R., Rustad, S., Soeleman, M.A., Sidhik, G.F.: Silhouette analysis of hand gesture dataset using histogram profile feature extraction. In: 2020 International Seminar on Application for Technology of Information and Communication (2020)
6. Iszaidy, I., et al: An analysis of background subtraction on embedded platform based on synthetic dataset. In: Journal of Physics: Conference Series, vol. 1755, no. 1, p. 012042 (2021)
7. Barnich, O., Droogenbroeck, M.V.: A powerful random technique to estimate the background in video sequences. In: IEEE International Conference on Acoustics, pp. 1–10. IEEE Press (2004)
8. Yoshinaga, S., Shimada, A., Nagahara, H., Taniguchi, R.-I.: Background model based on statistical local difference pattern. In: Park, J.-I., Kim, J. (eds.) ACCV 2012. LNCS, vol. 7728, pp. 327–332. Springer, Heidelberg (2013). https://doi.org/10.1007/978-3-642-37410-4_30
9. Alamgir, H.M., Dung, N.V., Nam, H.E.: The trade-off between accuracy and the complexity of real-time background subtraction. IET Image Process. **15**(2), 350–368 (2020)
10. Zhang, Y.B., Bie, X.B.: Research on target detection and extraction algorithm in static background. Laser J. **42**(02), 94–97 (2021)
11. Li, G., Qiu, S., Lin, L., Zeng, R.: New moving target detection method based on background differencing and coterminous frames differencing. Chin. J. Sci. Instr. **27**(8), 961–964 (2006)
12. Kharate, G.K., Ghotkar, A.S.: Vision based multi-feature hand gesture recognition for Indian sign language manual signs. Int. J. Smart Sens. Intell. Syst. **9**(1), 124–147 (2016)
13. Chu, X.Z., Liu, J., Shimamoto, S.: A sensor-based hand gesture recognition system for Japanese sign language. In: 2021 IEEE 3rd Global Conference on Life Sciences and Technologies (LifeTech) (2021)
14. Wang, C. Y., Liao, H., Wu, Y. H., Chen, P.Y., Yeh, I. H.: CSPNet: a new backbone that can enhance learning capability of CNN. In: 2020 IEEE/CVF Conference on Computer Vision and Pattern Recognition Workshops (CVPRW), vol. 37, no. 9, pp. 1904–1916. IEEE Press (2020)
15. Liu, S., Qi, L., Qin, H., Shi, J., Jia, J.: Path aggregation network for instance segmentation. In: Proceedings of the IEEE Conference on Computer Vision and Pattern Recognition (CVPR), pp. 8759–8768. IEEE Press (2018)

# Design and Implementation of Security Labeling System Based on Neighborhood Rough Set

Ping Yu[1], Shiyong Yang[3](✉), Yuan Shen[2], Yan Li[2], Zhen Liu[2], and Tao Xu[2](✉)

[1] College of Information Science and Technology, Zibo Normal College, Zibo 255130, China
[2] School of Information Science and Engineering, University of Jinan, Jinan 250022, China
ise_xut@ujn.edu.cn
[3] Land Spatial Data and Remote Sensing Technology Institute of Shandong Province, Jinan 250002, China

**Abstract.** At present, "smart security" as a produsction of modern science and technology, it provides an important technical guarantee for personal safety, family property safety and even the safety of the whole society. In this paper, a neighborhood rough set model system based on annotated result set is proposed to solve the problem that the manual operation and maintenance cost of "smart security" system is huge, which can not meet the requirements of security system construction. Experiments show that the system platform based on neighborhood rough set can greatly reduce the cost of manual operation and improve the accuracy of security video annotation.

**Keywords:** Intelligent security · Rough set · Neighborhood rough set

## 1 Introduction

In recent years, a kind of "smart city" construction initiated by the security industry is in full swing. Among them, "smart security" as a production of modern science and technology, it provides an important technical guarantee for personal safety, family property safety and even the safety of the whole society "Smart security" [1] is a high-tech product of the combination of computer technology, electronic science and technology, Internet of things technology and sensor technology. The monitoring of security system [2] can effectively crack down on man-made crimes and prevent natural disasters in time. At present, in all walks of life, "smart security" has played a very important technical support, security systems are also all over the city. A large number of video real-time uninterrupted generation, not only makes the security system manual operation and maintenance cost is huge, the quality of video and other factors or the security system manual operation and maintenance can not meet the construction requirements of the security system.

With the rapid development of the new generation of information technology represented by artificial intelligence, it provides a feasible scheme for the automatic operation and maintenance of "smart security" system. Deep learning algorithms such as neural network calculation [3], fuzzy reasoning [4], genetic algorithm [5] can be used in the


Q. Han et al. (Eds.): LSMS 2021/ICSEE 2021, CCIS 1469, pp. 462–471, 2021.
https://doi.org/10.1007/978-981-16-7213-2_45

automatic operation and maintenance of "smart security" system. Among them, deep learning method can extract data features from unstructured data, more clearly express the meaning of data, and make it easier for people to understand the deep meaning brought by rich data. For the training samples of deep learning, the amount of training set is extremely important, but how to structure, level and efficient the data is also essential. Therefore, it is of great practical significance and value to design a reliable, sustainable and iterative annotation platform, which is particularly important for how to clean and extract the experimental data from the massive actual security data in line with the video quality algorithm research.

## 2 Research on Neighborhood Based Rough Set

### 2.1 Traditional Rough Set Theory

Classical rough set theory [6–8] was put forward by Professor Z. Pawlak in 1982, which is a kind of expression and application of incomplete or imprecise knowledge. Pawlak rough set is a very useful mathematical tool for knowledge representation, especially in the field of describing data uncertainty.

In Pawlak rough set model, if the category is equivalent, then the object x will be classified into the corresponding category X. Then [x] is defined as a subset of X, that is, $P(x|[x]) = |x \cap [x]|/|[x]|=1$. Therefore, Pawlak rough set model can be regarded as a qualitative classification model. However, this classification model has some limitations, which can be divided into the following two categories:

1) The attribution probability p (x|[x]) must be equal to 1, so it is very sensitive to noise data.
2) Equivalence relation class [x] is defined based on indiscernibility, which can not be directly used as digital data processing.

For the first limitation, decision-making rough sets (DTRs) [9] introduces a generalized decision framework to solve it. The general idea is to avoid the influence of noise data on rough set model by decision tree. In fact, it is very difficult for Pawlak rough set model to obtain effective results from noise data without classification error tolerance. In order to overcome this problem, many researchers have studied different probabilistic rough set (PRS) models. Compared with Pawlak rough set model, all PRS models achieve probability tolerance, $P(x|[x]) \geq \alpha$. It is used to classify object x into category x, and the threshold value is not used $\alpha$ Between 0 and 1.

### 2.2 Neighborhood Rough Set Model

Pawlak's classical rough set theory and probabilistic rough set based on minimum Bayes are usually based on absolute equivalence. For some discrete numerical data, important information may be lost in the process of processing, and different discretization strategies will affect the final processing results.

Neighborhood rough set can divide the universe into several equivalent segments. For dealing with discrete data, we can approach the fact by upper approximation and lower approximation. At the same time, we can reduce the influence and distortion of noise data on the spatial structure by compressing the mapping relationship.

**Definition 1:** given a decision table s, for any object Xi ∈ u and a ⊆ at, for Xi in a.
The domain set on can be defined as:

$$\delta A(x_i) = \{x_j | x_j \in U, A(x_i, x_j) \leq \delta\}. \tag{1}$$

among Δ Minkowski distance is selected as its metric function.

$$M(x_i, x_j) = \left(\sum_{K=1}^{M} |I_{aK}(X_i) - I_{aK}(X_j)|^M\right)^{\frac{1}{M}}. \tag{2}$$

xi and xj are two objects in the n-dimensional space At = {A1, A2,..., an}, It has the following characteristics.

1) If M = 1, it is Manhattan distance;
2) If M = 2, it is Euclidean distance;
3) If M approaches positive infinity, it is expressed as Chebyshev distance.

Given the metric space <U, Δ>, the neighborhood system consists of neighborhood particles{δ(xi)|xi ∈ U}. In other words, <U, NA> can be called neighborhood approximation space defined by attribute set. C is the covering relation of domain relation N of U. The basic domain subset caused by covering object C is the neighborhood rough set approximation set of constructing domain description system. Then for any subset X ⊆ U, the definitions of lower approximation and upper approximation of X with respect to C can be expressed as follows:

$$\underline{N}_c(X) = \{x_i \subseteq X, x_i \in U\}. \tag{3}$$

$$\overline{N}_c(X) = \{x_i \cap X \neq \varnothing, x_i \in U\}. \tag{4}$$

Through the definition of the lower approximation and the upper approximation of X with respect to C, the rough approximation of X with respect to C, the positive range, the boundary region and the negative range of X can be expressed as follows:

$$POS_c(X) = \underline{N}_c(X). \tag{5}$$

$$\overline{N}_c(X) = \{x_i \cap X \neq \varnothing, x_i \in U\}. \tag{6}$$

$$NEG_c(X) = U - POS_c(X) \cup BND_c(X) = U - \overline{N}_c(X). \tag{7}$$

**Theorem 1:** Given a set of space metrics <U, N> and two nonnegative neighborhood parameters δ1 and δ2, and defining δ1 < δ2, we can get the following results:

$$\forall x_i \in U : \delta_1(x_i) \subseteq \delta_2(x_i). \tag{8}$$

$$\forall x_i \in U : \delta_1(x_i) \subseteq \delta_2(x_i). \tag{9}$$

In the general neighborhood rough set model, the object x can usually be described by the corresponding equivalent description class, that is: [x]. So that we can define.

$$p(x|\delta(x)) = \frac{|\delta(x) \cap X|}{\delta(x)}). \tag{10}$$

On the basis of this metric (<U,N>), for subspace B ⊆ n, the upper and lower approximations of probability about X can be defined as the following formula:

$$\underline{N}^{(\alpha,\beta)} = \{x_i|P(X|\delta B(X_I \leq \alpha), x_i \in U\}. \tag{11}$$

$$\overline{N}^{(\alpha,\beta)} = \{x_i|P(X|\delta B(X_I \leq \alpha), x_i \in U\}. \tag{12}$$

If we define a decision table in which the attribute set is divided by subclasses, then we define the cutting attribute set as d = {D1, D2,..., DN}. Then in the neighborhood rough set, the positive, negative and related boundary regions of any a(8).

Ttribute sub unit can be expressed as:

$$POS_B^{(\alpha,\beta)}(X) = \underline{N}_B^{(\alpha,\beta)}(x). \tag{13}$$

$$BND_B^{(\alpha,\beta)}(X) = \overline{N}_B^{(\alpha,\beta)} - \underline{N}_B^{(\alpha,\beta)}(x). \tag{14}$$

$$NEG_B^{(\alpha,\beta)}(X) = U - (POS_B^{(\alpha,\beta)}(x) \cup BND_B^{(\alpha,\beta)}(X) = U - \overline{N}_B^{(\alpha,\beta)}. \tag{15}$$

From the semantic level, the objects in the positive domain are classified into "certain" categories with high probability. Generally, the larger positive region often has smaller boundary domain, which means less fuzzy and uncertain objects. In the platform, this feature is often used to test the size of the "ambiguous" areas of these annotations, The quality of classification can be defined by the formula:

$$\gamma_B^{(\alpha,\beta)}(D) = \frac{\left|POS_B^{(\alpha,\beta)}(d)\right|}{|U|}. \tag{16}$$

Because it is a NP hard problem to calculate all attribute reduction, in many studies, the general idea can be summarized into two main directions: heuristic and search strategy. The general idea of heuristics is to reduce or extend the reserved attributes in the positive field, so as to achieve the goal of preserving or deleting the attributes in this way. The other is non directional search strategy. Directional search strategy can further classify deletion method, addition method and addition deletion method. Non directional search strategy is usually applied to evolutionary algorithm, swarm algorithm and other group based meta heuristic algorithms.

### 2.3 Neighborhood Rough Set Model Based on Annotation Result Set

In the annotation system, we adopt the multi-level arbitration annotation model to classify the video data manually. Each video usually needs three annotation workers to classify it manually at the same time. According to the classification results, arbitrators intervene in the arbitration according to the following rules.

1) If the evaluation results of the three evaluators are consistent, the video does not need to be arbitrated by an arbitrator.
2) If the result of one of the three evaluators is inconsistent with that of other evaluators, the video needs to be arbitrated by an evaluator with rich evaluation experience.

    2.a) If the arbitrators' arbitration results are not consistent with the majority of the evaluation results, the evaluation experts will intervene in the arbitration.
    2.b) If the arbitrator's arbitration result is consistent with the majority evaluation, it will be classified as the mode evaluation result.

3) If the evaluation results of the three evaluators are inconsistent, the evaluation experts directly intervene, and the evaluation results are regarded as the classification results of the video. According to this classification model, the evaluation results are reliable in the quality of classification, and each video is assigned to its own classification.

However, the classification result of video is often not a single distortion type, two or more distortion types are more likely to appear in a video. After labeling about 22 W channels of video, we found that about half of the videos were intervened by experienced arbitrators, and about 1/7 channels were intervened by evaluation experts. The first mock exam shows that the objective factors affect the labeling of this model. After costly tagging cost, some co existing distorted videos lose their valuable background value in some sense. If these videos are organized efficiently and efficiently, they can be organized into related video sets for a more meaningful and valuable work. If we simply retrieve the videos marked with different results, some objective factors, such as manual misoperation or inexperience of reviewers, will affect the reliability and maintainability of the data to a certain extent. Therefore, the platform expects to use the basic theory of neighborhood rough set, to construct a neighborhood model, to show the degree of association between classes and the degree of membership between instances and classes mathematically, and to deeply mine and analyze the relationship between the macro video sets, so that the expensive annotation cost can get more value.

## 3 System Design Based on Neighborhood Rough Set

After each batch of video is labeled, the corresponding neighborhood rough set is constructed according to the neighborhood rough set theory. The main work is divided into two steps:

1) Calculate the correlation degree between distortion classification words.
2) The membership between each image and distortion classification is calculated.

### 3.1 Calculate the Degree of Association Between Classifiers

The classification of distorted image is mainly based on discrete model. The label in the label information table may be composed of multiple markers, and the final label result after arbitration by the arbitrator is the decision attribute of the label. Therefore, it is necessary to process and decompose the data when constructing the rough set model.

Firstly, the tagging information table is divided according to the tagging words.

1) If the video vi has been annotated by a certain annotation word lj by two ordinary taggers, then the record vi is divided into the constituent elements of the equivalence relation class of the annotation word lj;
2) If the video vi arbitrator annotates a label word lj, then the record vi is divided into the component elements of the equivalent relation class of the label word lj;
3) If there are more than three tagging words in the video, it will be included in the exception class without adding any tagging word equivalent class.

Then, according to the types of video distortion, the universe of Discourse (tag information table) can be cut into different sets of equivalence classes. There are intersection relations between equivalence classes and equivalence classes. These intersection relations are the sets marked by more than two kinds of tag words at the same time, that is, the video sets with certain "divergence". The specific results are shown in Table 1.

**Table 1.** Neighborhood rough set category association degree table.

| Dataset label name | Normal video | Fuzzy anomaly | Abnormal color | Abnormal occlusion | Abnormal interference | Decoding exception | Abnormal brightness | Signal loss | PTZ jitter |
|---|---|---|---|---|---|---|---|---|---|
| Normal video | 1.00 | 0.15 | 0.37 | 0.65 | 0.00 | 0.32 | 0.07 | 0.15 | 0.03 |
| Fuzzy anomaly | 0.15 | 1.00 | 0.09 | 0.63 | 0.05 | 0.00 | 0.03 | 0.22 | 0.00 |
| Abnormal color | 0.37 | 0.09 | 1.00 | 0.00 | 0.78 | 0.00 | 0.78 | 0.00 | 0.00 |
| Abnormal occlusion | 0.65 | 0.63 | 0.00 | 1.00 | 0.15 | 0.00 | 0.13 | 0.00 | 0.00 |
| Abnormal interference | 0.00 | 0.05 | 0.01 | 0.15 | 1.00 | 0.15 | 0.34 | 0.07 | 0.00 |
| Decoding exception | 0.32 | 0.00 | 0.00 | 0.00 | 0.15 | 1.00 | 0.00 | 0.80 | 0.11 |
| Abnormal brightness | 0.07 | 0.03 | 0.78 | 0.13 | 0.34 | 0.00 | 1.00 | 0.01 | 0.02 |
| Signal loss | 0.15 | 0.22 | 0.00 | 0.09 | 0.07 | 0.80 | 0.01 | 1.00 | 0.00 |
| PTZ jitter | 0.03 | 0.00 | 0.00 | 0.00 | 0.00 | 0.00 | 0.02 | 0.00 | 1.00 |

According to the actual situation, the platform selects $\delta = 0.25$, that is, when $y > 0.25$, the support relationship between li and lj is considered, and the support relationship between li and lj is calculated.

### 3.2 Calculate the Membership of Image and Distortion Type

According to the support relation of the association degree between the segmentation words, the upper approximation set and the lower approximation set of each tag word about the tag information table can be obtained. According to the situation, the membership calculation formula is transformed as follows:

$$\mathrm{SIM}(\mathrm{li}, \mathrm{gi}, \mathrm{n}, \mathrm{m}) = \beta\underline{\mathrm{SIM}}(\mathrm{li}, \mathrm{gi}) + (1 + \beta)\overline{\mathrm{SIM}}(\mathrm{li}, \mathrm{gi}) \times \frac{n}{m} \tag{17}$$

Where n is the number of times the image gi is hit and marked as li, and m is the total number of times the image gi is marked. Then the images are classified according to Table 2.

**Table 2.** Result processing rules of neighborhood rough set.

| Condition | Conclusion | SIM value range |
|---|---|---|
| SIM (li, gi) > 0.5 | The image gi belongs to the marker li | (0.50, 1] |
| SIM (li, gi) < 0.5<br>SIM (li, gi) > 0.25 | The image g is not sure whether it belongs to the marker li | (0.25, 0.5] |
| SIM (li, gi) < 0.25 | The image gi does not belongs to the marker li | (0.00, 0.25] |

In the system, the platform constructs a neighborhood rough set suitable for video distortion annotation based on the annotation data of 22 W channel video, and classifies the video according to the distortion types according to the above rules. The number of videos of each category and relevant SIM information are shown in the Table 3.

**Table 3.** Result of neighborhood rough set processing.

| Distortion type | Number | Average value of SIM | Maximum value of SIM | Minimum value of SIM |
|---|---|---|---|---|
| Normal video | 17572 | 0.74 | 1.00 | 0.51 |
| Fuzzy anomaly | 38811 | 0.84 | 1.00 | 0.76 |
| Abnormal occlusion | 3343 | 0.87 | 1.00 | 0.73 |
| Abnormal interference | 1872 | 0.93 | 1.00 | 0.90 |
| Abnormal color | 2382 | 0.94 | 1.00 | 0.78 |

*(continued)*

**Table 3.** (*continued*)

| Distortion type | Number | Average value of SIM | Maximum value of SIM | Minimum value of SIM |
|---|---|---|---|---|
| Decoding exception | 4902 | 0.65 | 0.93 | 0.50 |
| Abnormal brightness | 1846 | 0.73 | 1.00 | 0.52 |
| PTZ jitter | 310 | 0.95 | 1.00 | 0.90 |
| Signal loss | 636 | 0.78 | 1.00 | 0.81 |

In the later spot check, according to the previous spot check information table records, the platform tracks the samples used in the previous spot check, searches and verifies the label classification results according to the rough set processing category, and calculates the new label accuracy rate. The specific results are shown in Table 4.

**Table 4.** Quality table of neighborhood rough set.

| Distortion type | Number | Recall | Accuracy | F1-score |
|---|---|---|---|---|
| Normal video | 1201 | 73.82% | 67.32% | 69.34% |
| Fuzzy anomaly | 1397 | 65.32% | 72.14% | 70.25% |
| Abnormal occlusion | 181 | 61.32% | 75.66% | 62.87% |
| Abnormal interference | 51 | 93.78% | 85.25% | 89.92% |
| Abnormal color | 92 | 78.25% | 64.23% | 62.38% |
| Abnormal brightness | 152 | 63.87% | 75.23% | 69.16% |

## 4 Practical Application Results of the System

In the security video monitoring of a certain area, the platform contains about 33 W actual security video. Among them, about 280000 videos are classified by platform annotation, and these videos are successfully divided into nine abnormal types and normal videos. In addition to normal videos, fuzzy abnormal accounts for the largest proportion, reaching 78294. Through the actual inspection, the manual marking records are shown in Table 5.

**Table 5.** Marking record.

| No. | Marked quantity | Time/day | Total number of taggers | Spot check marking accuracy | Number of arbitrations |
|---|---|---|---|---|---|
| 1 | 87632 | 41 | 9 | 64.32% | 42031 |
| 2 | 46321 | 18 | 13 | 80.13% | 19342 |
| 3 | 58216 | 21 | 13 | 76.57% | 39323 |
| 4 | 32019 | 14 | 13 | 76.78% | 21397 |
| 5 | 28413 | 14 | 12 | 70.32% | 11323 |
| 6 | 42310 | 21 | 12 | 87.19% | 19235 |
| 7 | 14291 | 7 | 21 | 77.61% | 7394 |

In the process of each batch labeling task, the system monitors the quality according to the labeling situation. The system will generate the corresponding statistical report, which will contain the following data: video sampling, quality inspection confusion matrix, the number of videos through arbitration icon, video annotation time statistics. The report of a certain period is shown in the Table 6.

**Table 6.** Statistical report.

| | Normal video | Abnormal occlusion | Fuzzy anomaly | Night video | Abnormal interference | Abnormal brightenss | Changing scenes | Abnormal color | Signal loss | DVR freeze | Unknown fault |
|---|---|---|---|---|---|---|---|---|---|---|---|
| Normal video | 808 | 17 | 274 | 13 | 1 | 25 | 2 | 25 | 0 | 38 | 15 |
| Abnormal occlusion | 31 | 380 | 26 | 0 | 0 | 5 | 0 | 2 | 0 | 0 | 6 |
| Fuzzy anomaly | 134 | 26 | 1134 | 11 | 6 | 28 | 3 | 35 | 2 | 20 | 11 |
| Night video | 0 | 0 | 1 | 265 | 0 | 3 | 0 | 12 | 0 | 0 | 1 |
| Abnormal interference | 2 | 3 | 10 | 1 | 74 | 2 | 0 | 3 | 0 | 1 | 3 |
| Abnormal brightness | 14 | 2 | 51 | 7 | 2 | 185 | 0 | 20 | 0 | 3 | 1 |
| Changing scenes | 0 | 0 | 1 | 0 | 0 | 2 | 11 | 0 | 0 | 0 | 1 |
| Abnormal color | 20 | 5 | 100 | 3 | 9 | 31 | 0 | 197 | 0 | 3 | 7 |
| Signal loss | 1 | 1 | 6 | 2 | 0 | 1 | 0 | 0 | 24 | 1 | 3 |
| DVR freeze | 0 | 0 | 0 | 0 | 0 | 0 | 0 | 0 | 0 | 0 | 0 |
| Unknown fault | 8 | 2 | 12 | 0 | 4 | 3 | 0 | 1 | 5 | 2 | 110 |

657 cases passed the arbitration at one time, 689 cases were awarded by arbitration, 110 cases were awarded by second level arbitration, and 1456 cases were awarded. In general, the accuracy rate of annotation results based on neighborhood rough set in distortion category increases slightly. The accuracy rate of normal video increased by 5.44%,

and the recall rate increased by 3.5%; The accuracy rate of fuzzy anomaly increased by 9.06%, and the recall rate decreased by 12.00%; The accuracy of occlusion increased; The accuracy rate of occlusion increased by 8.15%, and the recall rate almost unchanged; In the interference category, the accuracy rate decreased by 6.08%, and the recall rate increased by 8.45%; The accuracy rate of abnormal brightness increased by 11.86%, and the recall rate increased by 3.87%. For some subjective evaluation of color distortion categories, the accuracy rate increases obviously, such as fuzzy abnormal, brightness abnormal, but the effect of the recall rate is not obvious, and even in some categories has a downward trend. The application of neighborhood rough set in annotation is feasible and can improve the quality of annotation obviously.

## 5 Summary

This paper proposes a neighborhood rough set model system based on annotated result set, and designs and implements a neighborhood rough set system platform. Experiments show that the system platform based on neighborhood rough set can greatly reduce the cost of manual operation and improve the accuracy of security video annotation.

**Acknowledgments.** This research was supported by the Natural Science Foundation of University of Jinan (XKY1928, XKY2001).

## References

1. Murray, A.T., Kim, K., Davis, J.W., Machiraju, R., Parent, R.: Coverage optimization to support security monitoring. Comput. Environ. Urban Syst. **31**(2), 133–147 (2007)
2. Balamurugan, N.M., Babu, T.K.S.R., Adimoolam, M., John, A.: A novel effificient algorithm for duplicate video comparison in surveillance video storage systems. J. Ambient Intell. Hum. Comput.1–12 (2021)
3. Bosch, A., Zisserman, A., Munoz, X.: Scene classifification via plsa. In: Leonardis Computer Vision – ECCV 2006, pp. 517–530. Springer, Heidelberg (2006). https://doi.org/10.1007/11744085_40
4. Cao, L., Fei-Fei, L.: Spatially coherent latent topic model for concurrent segmentation and classifification of objects and scenes. In: 2007 IEEE 11th International Conference on Computer Vision, pp. 1–8. IEEE Press (2007)
5. Fei-Fei, L., Perona, P.: A Bayesian hierarchical model for learning natural scene categories. In: 2005 IEEE Computer Society Conference on Computer Vision and Pattern Recognition (CVPR), vol. 2, pp. 524–531 (2005)
6. Dai, J.H., Li, Y.X.: Heuristic genetic algorithm for minimal reduction decision system based on rough set theory. In: Proceedings. International Conference on Machine Learning and Cybernetics, vol. 2, pp. 833–836 (2002)
7. Yuan, Z., Chen, H., Xie, P., Zhang, P., Liu, J., Li, T.: Attribute reduction methods in fuzzy rough set theory: an overview, comparative experiments, and new directions. Appl. Soft Comput. **107**, 107353 (2021)
8. Rehman, N., Ali, A., Hila, K.: Note on tolerance-based intuitionistic fuzzy rough set approach for attribute reduction. Expert Syst. Appl. **175**, 114869 (2021)
9. Yin, J., Opoku, L., Miao, Y.H., Zuo, P.P., Yang, Y., Lu, J.F.: An improved site characterization method based on interval type-2 fuzzy C-means clustering of CPTu data. Arab. J. Geosci. **14**(14), 1–11 (2021)

# Continuous Human Learning Optimization with Enhanced Exploitation

Ling Wang[1], Bowen Huang[1], Xian Wu[1], and Ruixin Yang[2](✉)

[1] Shanghai Key Laboratory of Power Station Automation Technology, School of Mechatronics Engineering and Automation, Shanghai University, Shanghai 200444, China
{wangling,huangbowen,xianwu}@shu.edu.cn

[2] Department of Computer Science, North Carolina State University, Raleigh, NC, USA
ryang9@ncsu.edu

**Abstract.** Human Learning Optimization (HLO) is an emerging meta-heuristic with promising potential. Although HLO can be directly applied to real-coded problems as a binary algorithm, the search efficiency may be significantly spoiled due to "the curse of dimensionality". To extend HLO, Continuous HLO (CHLO) is developed to solve real-values problems. However, the research on CHLO is still in its initial stages, and further efforts are needed to exploit the effectiveness of the CHLO. Therefore, this paper proposes a novel continuous human learning optimization with enhanced exploitation (CHLOEE), in which the social learning operator is redesigned to perform global search more efficiently so that the individual learning operator is relieved to focus on performing local search for enhancing the exploitation ability. Finally, the CHLOEE is evaluated on the benchmark problem and compared with CHLO as well as recent state-of-the-art meta-heuristics. The experimental results show that the proposed CHLOEE has better optimization performance.

**Keywords:** Continuous human learning optimization · Human learning optimization · Social learning operator · Individual learning operator

## 1 Introduction

Human Learning Optimization (HLO) [1] is an emergent meta-heuristic algorithm inspired by the learning mechanisms of humans, in which three learning operators are developed by mimicking the general learning behaviors of humans to search out the optimal solution, including the random learning operator (RLO), the individual learning operator (ILO) and the social learning operator (SLO).

To improve the performance of HLO, various enhanced variants of HLO have been subsequently proposed. Wang et al. [2] propose an adaptive simplified human learning algorithm (ASHLO), which adopts linear adaptive strategies for *pr* and *pi*, two control parameters determining the probabilities of running RLO, ILO and SLO. With this strategy, ASHLO can achieve a better trade-off between exploration and exploitation. Later, Yang et al. [3] present a sine-cosine adaptive human learning optimization (SCHLO)


Q. Han et al. (Eds.): LSMS 2021/ICSEE 2021, CCIS 1469, pp. 472–487, 2021.
https://doi.org/10.1007/978-981-16-7213-2_46

algorithm in which *pr* and *pi* utilize the adaptive strategies based on sine and cosine functions to avoid premature convergence. According to the fact that the IQ scores of humans follow gaussian distribution and increase with the development of society and technology, a diverse human learning optimization (DHLO) algorithm is designed in which the values of *pi* are randomly initiated by a gaussian distribution and periodically updated to enhance the global search ability [4]. Recently, an improved adaptive human learning optimization (IAHLO) algorithm [5] with a two-stage adaptive strategy is designed in which the control parameter of the random learning, i.e. *pr*, linearly increases at first and then decreases after a certain number of iterations. With this new adaptive strategy, IAHLO can efficiently maintain the diversity at the early of iterations and perform an accurate local search in the late stages of iterations, and therefore the performance is significantly improved.

Attracted by its promising search ability, HLO has been applied to various problems. For example, Wang et al. apply the HLO algorithm to knapsack problems [5] and engineering design problems [9]. Li et al. [6] designs a discrete HLO to solve the production scheduling problem. Cao et al. adopt HLO to solve the power flow calculation problem of the AC / DC hybrid power system and then further use the Multi-Objective HLO to solve the multi-objective power flow calculation optimization problem [8]. Ashish et al. [10] combine the HLO and 3D Ostu functions for image segmentation. Nowadays, the HLO and its variants have been successfully used in intelligent control [11], job-shop scheduling [12], supply chain network [13], extractive text summarization [14] and financial markets forecasting [15].

As a binary-coding algorithm, HLO can be directly used to solve continuous problems. However, the curse of dimensionality may rise when HLO is used to solve the high-dimensional continuous problems. Thus, a continuous HLO (CHLO) is firstly presented in [16] to efficiently optimize the real variables of problems. However, the research on CHLO is still in its very initial stage, and further efforts are needed to exploit the effectiveness to attain better performance. Therefore, this paper proposes a new continuous human learning optimization with enhanced exploitation (CHLOEE), in which the social learning operator is redesigned to perform global search more efficiently, and therefore the individual learning operator can be relieved to focus on performing local search for enhancing the exploitation ability. In this way, the exploitation capability and the final performance of CHLOEE are improved.

The remaining content of this paper is organized as follows. The CHLOEE algorithm is introduced in Sect. 2. A comprehensive parameter study is performed in Sect. 3 to analyze why the CHLOEE can achieve better exploitation capability. Section 4 presents the experimental results of the CHLOEE and compared algorithms on the benchmark problem. Finally, the conclusion is drawn in Sect. 5.

## 2 Continuous Human Learning Optimization with Enhanced Exploitation

Unlike the standard HLO, an individual in CHLOEE is directly represented as a set of continuous variables instead of a binary string as Eq. (1).

$$x_i = \left[x_{i1} x_{i2} \cdots x_{ij} \cdots x_{iD}\right], 1 \leq i \leq N, 1 \leq j \leq D \tag{1}$$

where $x_i$ is the $i$-th individual, $N$ is the number of individuals and $D$ denotes the dimension of solutions.

Under the assumption that the prior knowledge of problems does not exist initially, each variable of individuals is randomly initialized between the lower bound and the upper bound of problems as Eq. (2).

$$x_{ij} = x_{min,j} + r_1 \times \left(x_{max,j} - x_{min,j}\right), 1 \leq i \leq N, 1 \leq j \leq D \tag{2}$$

where $x_{min,j}$ and $x_{max,j}$ are the lower bound and upper bound of variable $j$ and $r_1$ is a stochastic number between 0 and 1.

Consistent with CHLO, CHLOEE iteratively generates new candidates to search out the optimal solution by three learning operators, i.e. random learning operator, individual learning operator and social learning operator. The individual learning operator and social learning operator are performed according to the best personal knowledge and social experiences. Hence, CHLOEE establishes the individual knowledge database (IKD) as Eqs. (3&4) and social knowledge database (SKD) as Eq. (5) to reserve the personal best solutions and the best knowledge of the population respectively,

$$IKD = \begin{bmatrix} ikd_1 \\ ikd_2 \\ \vdots \\ ikd_i \\ \vdots \\ ikd_N \end{bmatrix}, 1 \leq i \leq N \tag{3}$$

$$ikd_i = \begin{bmatrix} ikd_{i1} \\ ikd_{i2} \\ \vdots \\ ikd_{ip} \\ \vdots \\ ikd_{iQ} \end{bmatrix} = \begin{bmatrix} ik_{i11} & ik_{i12} & \cdots & ik_{i1j} & \cdots & ik_{i1D} \\ ik_{i21} & ik_{i22} & \cdots & ik_{i2j} & \cdots & ik_{i2D} \\ ik_{ip1} & ik_{ip2} & \cdots & ik_{ipj} & \cdots & ik_{ipD} \\ \vdots & \vdots & \cdots & \vdots & \cdots & \vdots \\ ik_{iQ1} & ik_{iQ2} & \cdots & ik_{iQj} & \cdots & ik_{iQD} \end{bmatrix}, 1 \leq p \leq Q, 1 \leq j \leq D \tag{4}$$

$$SKD = \begin{bmatrix} skd_1 \\ skd_2 \\ \vdots \\ skd_q \\ \vdots \\ skd_M \end{bmatrix} = \begin{bmatrix} sk_{11} & sk_{12} & \cdots & sk_{1j} & \cdots & sk_{1D} \\ sk_{21} & sk_{22} & \cdots & sk_{2j} & \cdots & sk_{2D} \\ \vdots & \vdots & \vdots & \vdots & \vdots & \vdots \\ sk_{q1} & sk_{q2} & \cdots & sk_{qj} & \cdots & sk_{qD} \\ \vdots & \vdots & \vdots & \vdots & \vdots & \vdots \\ sk_{M1} & sk_{M2} & \cdots & sk_{Mj} & \cdots & sk_{MD} \end{bmatrix}, 1 \leq q \leq M, 1 \leq j \leq D \tag{5}$$

where $ikd_i$ is the individual knowledge database of individual $i$, $ikd_{ip}$ is the $p$-th best solutions of individual $i$, $skd_q$ is the $q$-th solution in $SKD$ and $M$ denotes the size of the $SKD$.

### 2.1 Learning Operators

Human learning is a highly complicated process, of which the study is part of neuropsychology, learning theory, etc. However, humans can use the random learning strategy, individual learning strategy and social learning strategy to solve problems effeciently. Hence, CHLOEE adoptes the random learning operator, individual learning operator and social learning operator to generate new candidates.

#### 2.1.1 Random Learning Operator

Random learning always occurs in human learning. At the beginning, people may adopt random strategies for comprehending the problem because of inadequate prior knowledge. With the progress of learning, the random learning strategy is remained to maintain the peculiar creativity [4]. The CHLOEE simulates those behaviors of humans through the random learning operator, as shown in Eq. (6).

$$x_{ij} = x_{min,j} + r_2 \times (x_{max,j} - x_{min,j}) \tag{6}$$

where $r_2$ is a random number between 0 and 1.

#### 2.1.2 Individual Learning Operator

Individual learning refers to building knowledge through individual reflection about external stimuli and sources [2]. With the individual learning strategy, humans can learn from their own previous experience and knowledge, and therefore they can avoid mistakes and improve the learning efficiency when they encounter similar problems. To mimic this learning ability, CHLOEE runs the individual learning operator as Eq. (7) to generate new solutions.

$$x_{ij} = ik_{ipj} \pm F_i \times r_3 \times \left(sk_{qj} - ik_{ipj}\right) \tag{7}$$

where $F_i$ is the individual learning factor, $r_3$ is a random number between 0 and 1, $ik_{ipj}$ is the *j-th* variable of the *p-th* solution in the IKD of individual $i$ and $sk_{qj}$ is the *j-th* corresponding variable of the *p-th* solution in the social knowledge database.

#### 2.1.3 Social Learning Operator

Social learning is an effective way for a person to absorb helpful knowledge from the collective experience of the population, and consequently they can grow their capabilities and realize more excellent proficiency with advanced knowledge. To imitate the social learning of humans, the original CHLO designs the social learning operator as Eq. (8).

$$x_{ij} = sk_{qj} + S_l \times r_4 \times \left(sk_{qj} - ik_{ipj}\right) \tag{8}$$

where $r_4$ stands for a random number between 0 and 1 and $S_l$ is a social learning factor.

To better enhance the performance of CHLO, CHLOEE makes an improvement on the social learning operator, which is re-defined as Eq. (9).

$$x_{ij} = sk_{qj} \pm F_s \times r_5 \times \left(sk_{qj} - ik_{ipj}\right) \tag{9}$$

where $r_5$ stands for a random number between 0 and 1 and $F_s$ is a social learning factor.

With the new social learning operator, the CHLOEE can perform global search more efficiently, and thus the individual learning operator can be relieved to focus on performing a local search for enhancing the exploitation ability. More details will elaborate in the parameter study and analysis part.

In summary, CHLOEE generates new solutions by performing the random learning operator, the individual learning operator and the social learning operator as follows,

$$x_{ij} = \begin{cases} x_{min,j} + r_2 \times \left(x_{max,j} - x_{min,j}\right), & 0 \le r \le pr \\ ik_{ipj} \pm F_i \times r_3 \times \left(sk_{qj} - ik_{ipj}\right), & pr < r \le pi \\ sk_{qj} \pm F_s \times r_5 \times \left(sk_{qj} - ik_{ipj}\right), & pi < r \le 1 \end{cases} \tag{10}$$

where $r$ is a stochastic number between 0 and 1; $pr$, ($pi$-$pr$) and (1-$pi$) decide the probabilities to execute random learning operator, individual learning operator and social learning operator, respectively.

### 2.2 Updating of the IKD and SKD

After all the individuals of CHLOEE generate new candidate solutions, the fitness of each solution is calculated according to the pre-defined fitness function. The new candidate will be stored in the IKD when the number of solutions stored in IKD is less than the pre-defined size of the IKD. Otherwise, the worst solution in the IKD will be substituted by the new candidate solution only if the fitness value of the new candidate solution is better. The SKD updates in the same way. Note that the sizes of the IKD and SKD should be set to 1 for a better balance between the performance and computational complexity for non-Pareto multi-objective optimization according to previous works on HLO [1].

Besides, if an individual does not find a better solution in the successive 100 generations, its IKD will be re-initialized to help the individual start a new search and avoid being trapped in the local optima.

### 2.3 Implementation of CHLOEE

The implementation of CHLOEE can be summarized as follows.

Step 1: set the control parameters of CHLOEE, such as the population size, the maximal generation number, $pr$, $pi$, $F_i$ and $F_s$;

Step 2: initialize the population randomly;

Step 3: evaluate the fitness of the initial population and set the initial IKD and SKD;

Step 4: yield new candidate solutions through performing the learning operators as Eq. (10);

Step 5: evaluate the new candidates and update the IKD and SKD;

Step 6: if conditions meet the termination conditions, output optimal solutions; otherwise, go to Step 4.

## 3 Parameter Study and Analysis

To gain a deep understanding of the proposed CHLOEE, a comprehensive parameter study is carried out in this section, in which *pr*, *pi*, $F_i$ and $F_s$ are considered. The first two parameters determine the execution probabilities of three learning operators while the latter two are the individual learning and social learning range factors. The Taguchi method [17] is adopted to investigate the influence of those four parameters on the performance of CHLOEE based on the 10-dimensional and 30-dimensional F2 and F6, which belong to the CEC05 benchmark functions. All the combinations of parameter values are given in Table 1. The population size and maximal generation number on the 10-dimensional functions are set to 50 and 3000, and they are increased to 100 and 6000 on the 30-dimensional functions. The mean of the best values (Mean) over 100 runs is used to evaluate the performance of different combinations of parameter values. The results are given in Table 2, where the best numbers are in bold. To conveniently choose a set of fair parameter values and analyze the characteristics of CHLOEE, the trials of which the results all ranking in the top 20% on 4 problems, i.e., 10–D F2, 10–D F6, 30–D F2 and 30–D F6, are selected and listed in Table 3.

**Table 1.** Factor levels

| Parameters | Factor levels | | | | | | | | |
|---|---|---|---|---|---|---|---|---|---|
| | 1 | 2 | 3 | 4 | 5 | 6 | 7 | 8 | 9 |
| *pr* | 0.001 | 0.005 | 0.01 | 0.015 | 0.017 | 0.02 | 0.023 | 0.027 | 0.035 |
| *pi* | 0.1 | 0.3 | 0.45 | 0.5 | 0.55 | 0.6 | 0.7 | 0.8 | 0.9 |
| $F_i$ | 0.1 | 0.3 | 0.5 | 0.8 | 1.0 | 1.2 | 1.5 | 2 | 2.5 |
| $F_s$ | 0.1 | 0.3 | 0.5 | 0.8 | 1.0 | 1.2 | 1.5 | 2 | 2.5 |

**Table 2.** The results of the parameter study

| Trial | Parameter Factors | | | | 10–D | | 30–D | |
|---|---|---|---|---|---|---|---|---|
| | *pr* | *pi* | $F_i$ | $F_s$ | F2 | F6 | F2 | F6 |
| 1 | 0.035 | 0.9 | 0.1 | 2 | 3.2982E–03 | 1.3509E + 00 | 5.5568E + 03 | 8.1273E + 01 |
| 2 | 0.027 | 0.5 | 1 | 0.8 | 8.5265E–14 | 6.3206E + 00 | 5.5800E–01 | 8.8663E + 00 |
| 3 | 0.01 | 0.1 | 0.5 | 0.5 | 2.4308E + 02 | 2.1343E + 03 | 8.5299E + 03 | 4.3236E + 02 |
| 4 | 0.023 | 0.55 | 1.2 | 2 | 1.9059E + 00 | 1.5592E + 01 | 2.1081E + 02 | 1.2132E + 02 |
| 5 | 0.015 | 0.55 | 0.5 | 1 | 4.7748E–14 | **1.0522E + 00** | 1.4217E–07 | **5.1033E + 00** |
| 6 | 0.027 | 0.9 | 0.5 | 1.5 | 1.0637E–09 | 4.5429E + 00 | 1.4133E + 02 | 5.6296E + 01 |
| 7 | 0.017 | 0.45 | 1.2 | 0.5 | 2.4881E + 02 | 1.1845E + 03 | 8.9882E + 02 | 2.1596E + 02 |
| 8 | 0.017 | 0.55 | 0.1 | 1.2 | 4.3201E–14 | 1.1280E + 00 | 1.1113E + 00 | 1.7160E + 01 |

*(continued)*

**Table 2.** *(continued)*

| Trial | Parameter Factors | | | | 10−D | | 30−D | |
|---|---|---|---|---|---|---|---|---|
| | *pr* | *pi* | $F_i$ | $F_s$ | F2 | F6 | F2 | F6 |
| 9 | 0.001 | 0.6 | 2 | 0.8 | 5.3433E−14 | 1.5186E + 02 | 3.3254E + 01 | 1.2458E + 03 |
| 10 | 0.02 | 0.55 | 0.3 | 0.8 | 8.5265E−14 | 4.8083E + 00 | 1.1165E + 00 | 1.8936E + 01 |
| 11 | 0.02 | 0.5 | 0.5 | 0.3 | 4.2728E + 02 | 1.7867E + 03 | 8.9001E + 03 | 1.5320E + 04 |
| 12 | 0.035 | 0.6 | 0.8 | 0.5 | 1.0565E + 02 | 2.9037E + 01 | 8.8085E + 00 | 7.5907E + 01 |
| 13 | 0.02 | 0.7 | 2.5 | 1.5 | 1.4886E + 02 | 1.9010E + 02 | 4.5507E + 03 | 1.1583E + 07 |
| 14 | 0.005 | 0.8 | 0.8 | 2 | 4.2064E−14 | 4.8912E + 00 | 9.9698E−01 | 5.3195E + 01 |
| 15 | 0.023 | 0.7 | 0.1 | 0.3 | 5.7103E + 02 | 5.0219E + 02 | 3.4913E + 03 | 8.9158E + 03 |
| 16 | 0.005 | 0.5 | 2 | 1.5 | 2.6019E + 01 | 3.0662E + 01 | 9.6560E + 02 | 6.5404E + 05 |
| 17 | 0.015 | 0.7 | 1.5 | 2 | 5.5800E−01 | 2.3220E + 01 | 3.1214E + 02 | 1.3608E + 02 |
| 18 | 0.001 | 0.1 | 0.1 | 0.1 | 3.4027E + 02 | 1.5887E + 03 | 2.8364E + 04 | 3.8943E + 05 |
| 19 | 0.001 | 0.7 | 0.8 | 1 | 4.9454E−14 | 3.6597E + 00 | **9.8545E−11** | 1.1005E + 01 |
| 20 | 0.02 | 0.8 | 2 | 0.5 | −1.4049E−11 | 1.6216E + 01 | 9.7448E + 01 | 1.4746E + 02 |
| 21 | 0.01 | 0.45 | 0.1 | 1.5 | 1.0173E−03 | 1.0049E + 01 | 8.8839E + 01 | 4.0403E + 01 |
| 22 | 0.02 | 0.45 | 0.8 | 0.1 | 4.2848E + 02 | 1.5301E + 03 | 1.5643E + 04 | 6.8798E + 03 |
| 23 | 0.035 | 0.5 | 1.2 | 1 | 6.7075E−14 | 2.6315E + 00 | 5.5812E−01 | 1.3165E + 01 |
| 24 | 0.005 | 0.3 | 0.1 | 0.8 | 1.2150E−02 | 1.4205E + 02 | 5.5811E + 01 | 1.2045E + 02 |
| 25 | 0.017 | 0.1 | 1 | 1 | 7.1049E−04 | 2.4393E + 02 | 3.3869E + 02 | 7.8437E + 01 |
| 26 | 0.027 | 0.6 | 1.2 | 0.3 | 2.9874E + 02 | 1.1779E + 03 | 1.4553E + 03 | 1.3102E + 04 |
| 27 | 0.001 | 0.3 | 0.5 | 1.2 | 4.3201E−14 | 2.1185E + 01 | 3.5750E + 01 | 4.7133E + 01 |
| 28 | 0.005 | 0.55 | 1.5 | 0.5 | 2.1553E + 01 | 7.3975E + 02 | 8.4869E + 00 | 8.6182E + 05 |
| 29 | 0.035 | 0.55 | 1 | 1.5 | 4.4338E−14 | 3.7172E + 00 | 1.1464E + 01 | 5.4171E + 01 |
| 30 | 0.01 | 0.55 | 2 | 0.1 | 6.8935E + 00 | 9.2222E + 01 | 1.1435E + 01 | 1.6571E + 07 |
| 31 | 0.02 | 0.9 | 1.5 | 1 | 4.4906E−14 | 1.1400E + 01 | 4.8352E + 00 | 8.3767E + 01 |
| 32 | 0.02 | 0.6 | 0.1 | 2.5 | 3.3676E + 01 | 3.6618E + 01 | 7.2738E + 03 | 1.1895E + 07 |
| 33 | 0.023 | 0.6 | 1 | 0.1 | 3.9998E + 02 | 1.2004E + 03 | 1.3387E + 04 | 1.6318E + 06 |
| 34 | 0.02 | 0.3 | 1 | 2 | 4.0733E + 01 | 2.2100E + 01 | 1.5289E + 03 | 1.3102E + 06 |
| 35 | 0.017 | 0.6 | 0.5 | 2 | 3.5811E−06 | 1.4192E + 01 | 6.3226E + 02 | 2.5858E + 04 |
| 36 | 0.01 | 0.7 | 1.2 | 0.8 | 7.3328E−14 | 1.1533E + 01 | 6.0805E−09 | 3.1716E + 01 |
| 37 | 0.023 | 0.45 | 2 | 1 | 5.5800E−01 | 2.7141E + 01 | 7.0795E + 01 | 6.9447E + 01 |
| 38 | 0.001 | 0.9 | 1 | 0.5 | 2.5514E + 01 | 9.6001E + 02 | 4.2772E−03 | 3.0161E + 02 |
| 39 | 0.023 | 0.9 | 0.3 | 2.5 | 1.2908E−06 | 2.8068E + 00 | 3.5458E + 03 | 6.5232E + 01 |
| 40 | 0.027 | 0.3 | 1.5 | 0.1 | 5.7794E + 02 | 2.4498E + 03 | 1.6123E + 04 | 1.4987E + 09 |
| 41 | 0.017 | 0.9 | 2.5 | 0.8 | 1.2521E + 02 | 3.9053E + 02 | 5.7435E + 03 | 1.5413E + 07 |
| 42 | 0.01 | 0.6 | 1.5 | 1.2 | 4.7748E−14 | 1.0378E + 01 | 1.4824E + 01 | 7.2156E + 01 |

*(continued)*

**Table 2.** (*continued*)

| Trial | Parameter Factors | | | | 10–D | | 30–D | |
|---|---|---|---|---|---|---|---|---|
| | *pr* | *pi* | $F_i$ | $F_s$ | F2 | F6 | F2 | F6 |
| 43 | 0.027 | 0.55 | 0.8 | 2.5 | 1.1161E + 00 | 1.7000E + 01 | 2.6285E + 03 | 2.9748E + 05 |
| 44 | 0.005 | 0.1 | 0.3 | 0.3 | 1.1399E + 02 | 1.5236E + 03 | 1.7808E + 04 | 2.3526E + 03 |
| 45 | 0.023 | 0.3 | 2.5 | 0.5 | 6.3577E–01 | 1.3985E + 02 | 4.9172E + 01 | 1.2939E + 04 |
| 46 | 0.027 | 0.45 | 2.5 | 1.2 | 4.5412E + 01 | 3.3311E + 01 | 8.5934E + 02 | 5.3049E + 05 |
| 47 | 0.02 | 0.1 | 1.2 | 1.2 | 7.7875E–14 | 1.6113E + 01 | 1.9230E + 02 | 9.1414E + 01 |
| 48 | 0.015 | 0.1 | 0.8 | 0.8 | 1.9888E + 00 | 1.3330E + 02 | 2.6675E + 02 | 1.3868E + 02 |
| 49 | 0.023 | 0.1 | 1.5 | 1.5 | 3.7498E + 01 | 3.8509E + 01 | 9.9279E + 02 | 7.7435E + 04 |
| 50 | 0.035 | 0.8 | 0.3 | 1.2 | 8.8676E–14 | 2.8730E + 00 | 2.7536E + 01 | 6.2786E + 01 |
| 51 | 0.015 | 0.6 | 0.3 | 1.5 | 2.5751E–10 | 4.8977E + 00 | 1.0253E + 01 | 4.2195E + 01 |
| 52 | 0.035 | 0.1 | 2.5 | 2.5 | 4.2838E + 02 | 5.8549E + 06 | 9.4390E + 03 | 1.7296E + 08 |
| 53 | 0.001 | 0.5 | 1.5 | 2.5 | 6.2891E + 01 | 4.0307E + 02 | 3.7711E + 03 | 3.3476E + 06 |
| 54 | 0.01 | 0.9 | 0.8 | 0.3 | 2.2057E + 02 | 1.2862E + 03 | 1.6868E + 01 | 7.4624E + 03 |
| 55 | 0.01 | 0.3 | 0.3 | 1 | 7.1623E–14 | 1.4151E + 02 | 3.0202E + 02 | 3.9228E + 01 |
| 56 | 0.005 | 0.7 | 1 | 1.2 | 5.1159E–14 | 5.0973E + 00 | 9.8054E–08 | 1.4587E + 01 |
| 57 | 0.017 | 0.7 | 2 | 2.5 | 1.3105E + 02 | 1.6632E + 02 | 4.8031E + 03 | 1.4981E + 07 |
| 58 | 0.015 | 0.3 | 1.2 | 2.5 | 1.4202E + 02 | 1.7965E + 02 | 4.8597E + 03 | 2.8610E + 07 |
| 59 | 0.035 | 0.7 | 0.5 | 0.1 | 3.3308E + 02 | 3.9551E + 04 | 1.1088E + 04 | 2.9133E + 07 |
| 60 | 0.005 | 0.9 | 1.2 | 0.1 | 4.1053E + 02 | 4.0953E + 03 | 1.5300E + 01 | 1.4716E + 06 |
| 61 | 0.01 | 0.5 | 2.5 | 2 | 2.6628E + 02 | 1.0955E + 03 | 6.4545E + 03 | 5.8844E + 07 |
| 62 | 0.001 | 0.55 | 2.5 | 0.3 | 7.3328E–14 | 1.5570E + 05 | 1.2694E + 01 | 1.4450E + 05 |
| 63 | 0.005 | 0.6 | 2.5 | 1 | 2.2121E + 01 | 7.3567E + 01 | 1.2415E + 03 | 1.0422E + 05 |
| 64 | 0.01 | 0.8 | 1 | 2.5 | **4.0359E–14** | 6.8930E + 00 | 3.3575E + 00 | 7.4433E + 01 |
| 65 | 0.017 | 0.3 | 0.8 | 1.5 | 4.0927E–14 | 1.4603E + 01 | 1.3746E + 02 | 5.7748E + 01 |
| 66 | 0.015 | 0.8 | 2.5 | 0.1 | 2.2817E–10 | 5.3768E + 01 | 1.6503E + 02 | 1.8194E + 02 |
| 67 | 0.035 | 0.45 | 1.5 | 0.8 | 8.1855E–14 | 1.1614E + 02 | 5.2513E + 00 | 3.0542E + 01 |
| 68 | 0.001 | 0.45 | 0.3 | 2 | 2.2479E + 01 | 3.9165E + 01 | 1.6835E + 03 | 1.8678E + 06 |
| 69 | 0.017 | 0.5 | 0.3 | 0.1 | 2.0900E + 02 | 7.4486E + 02 | 1.4536E + 04 | 1.5237E + 04 |
| 70 | 0.005 | 0.45 | 0.5 | 2.5 | 4.5119E + 01 | 4.6787E + 01 | 6.0573E + 03 | 1.4875E + 07 |
| 71 | 0.015 | 0.45 | 1 | 0.3 | 1.2855E + 02 | 1.6840E + 03 | 1.2696E + 04 | 2.0115E + 05 |
| 72 | 0.027 | 0.1 | 2 | 2 | 2.3256E + 02 | 4.6168E + 01 | 3.6456E + 03 | 2.8280E + 07 |
| 73 | 0.027 | 0.8 | 0.1 | 1 | 1.0334E–12 | 2.0125E + 00 | 3.6156E + 00 | 6.1364E + 01 |
| 74 | 0.035 | 0.3 | 2 | 0.3 | 1.3033E + 02 | 1.2684E + 03 | 6.3221E + 02 | 1.9111E + 02 |
| 75 | 0.001 | 0.8 | 1.2 | 1.5 | 5.3433E–14 | 9.3673E + 00 | 9.0297E–05 | 4.6238E + 01 |
| 76 | 0.023 | 0.8 | 0.5 | 0.8 | 7.0656E–13 | 5.1318E + 00 | 2.3803E–02 | 4.3803E + 01 |
| 77 | 0.015 | 0.5 | 0.1 | 0.5 | 3.8767E + 01 | 4.0183E + 01 | 7.2529E + 02 | 1.9895E + 02 |

(*continued*)

**Table 2.** *(continued)*

| Trial | Parameter Factors | | | | 10–D | | 30–D | |
|---|---|---|---|---|---|---|---|---|
| | *pr* | *pi* | $F_i$ | $F_s$ | F2 | F6 | F2 | F6 |
| 78 | 0.017 | 0.8 | 1.5 | 0.3 | 7.2191E–14 | 1.3907E + 02 | 1.6580E–04 | 5.0803E + 01 |
| 79 | 0.023 | 0.5 | 0.8 | 1.2 | 5.1159E–14 | 1.1032E + 00 | 3.3989E–06 | 1.1257E + 01 |
| 80 | 0.015 | 0.9 | 2 | 1.2 | 1.0731E + 01 | 4.8533E + 01 | 1.7146E + 03 | 1.4285E + 05 |
| 81 | 0.027 | 0.7 | 0.3 | 0.5 | 1.8397E + 01 | 1.2939E + 02 | 2.1540E + 01 | 1.8776E + 02 |

**Table 3.** The trials all ranking in the top 20% on the four problems

| Trial | Parameter Factors | | | | 10–D F2 | | 10–D F6 | | 30–D F2 | | 30–D F6 | |
|---|---|---|---|---|---|---|---|---|---|---|---|---|
| | *pr* | *pi* | *Fi* | *Fs* | Mean | Rank | Mean | Rank | Mean | Rank | Mean | Rank |
| 5 | 0.015 | 0.55 | 0.5 | 1 | 4.77E–14 | 7 | **1.05E + 00** | 1 | 1.42E–07 | 4 | **5.10E + 00** | 1 |
| 8 | 0.017 | 0.55 | 0.1 | 1.2 | 4.32E–14 | 4 | 1.13E + 00 | 3 | 1.11E + 00 | 13 | 1.72E + 01 | 7 |
| 19 | 0.001 | 0.7 | 0.8 | 1 | 4.95E–14 | 8 | 3.66E + 00 | 9 | **9.85E–11** | 1 | 1.10E + 01 | 3 |
| 23 | 0.035 | 0.5 | 1.2 | 1 | 6.71E–14 | 11 | 2.63E + 00 | 6 | 5.58E–01 | 11 | 1.32E + 01 | 5 |
| 56 | 0.005 | 0.7 | 1 | 1.2 | 5.12E–14 | 9 | 5.10E + 00 | 15 | 9.81E–08 | 3 | 1.46E + 01 | 6 |
| 79 | 0.023 | 0.5 | 0.8 | 1.2 | 5.12E–14 | 9 | 1.10E + 00 | 2 | 3.40E–06 | 5 | 1.13E + 01 | 4 |

The results of Table 2 and Table 3 show that the social learning operator has the most significant impact on the performance of the CHLOEE as the high-ranking trials are highly dependent on the certain values of *pi* and $F_s$. The values of *pi* are between 0.5 and 0.7 can achieve better performance, far smaller than the default values of *pi* (0.8) in the original CHLO, which means CHLOEE tends to execute more social learning to guarantee the global search ability as the individual learning operator in CHLOEE only focus on the local search to enhance the exploitation ability, which is be further analyzed. Meanwhile, $F_s$' values concentrate between 1 and 1.2 for high performance trials. The inappropriate values of $F_s$ will weaken the algorithm performance since too big $F_s$ will miss the possible optimal solutions during the search process, whereas the diversity of the CHLOEE cannot be maintained effectively when too small $F_s$ is set. In contrast, as for *pr* and $F_i$, the values can be more flexible.

Table 3 indicates that the comprehensive performance of trial 5 is the best as its total ranking is the highest. Therefore, the parameter values of trial 5 are used as the default values, in which *pr*, *pi*, $F_i$ and $F_s$ are set to 0.015, 0.55, 0.5, 1, respectively.

The search behaviors of CHLO and CHLOEE are drawn in Fig. 1 and Fig. 2, respectively, in which it assumes that CHLO and CHLOEE have the same the individual optimal solution ($ik_{ipj}$) and the social optimal solution ($sk_{qj}$). The search range of the random learning operator of CHLO is the same as CHLOEE, which is depicted by the blue line. The search ranges of the individual learning operator and social learning operator are marked by the black arrow line and the orange arrow line, respectively. Note that the ranges of all the learning operators are determined by the default parameters of the original CHLO ($I_L = 1$, $S_L = 2$) and CHLOEE ($F_i = 0.5$, $F_s = 1$).

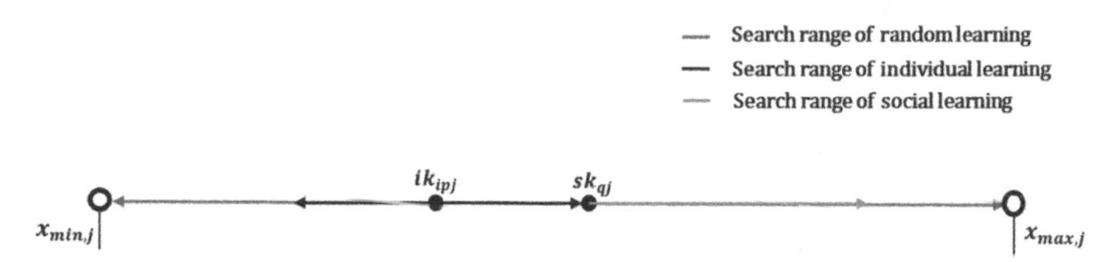


**Fig.1.** The diagram of CHLO learning operator operation

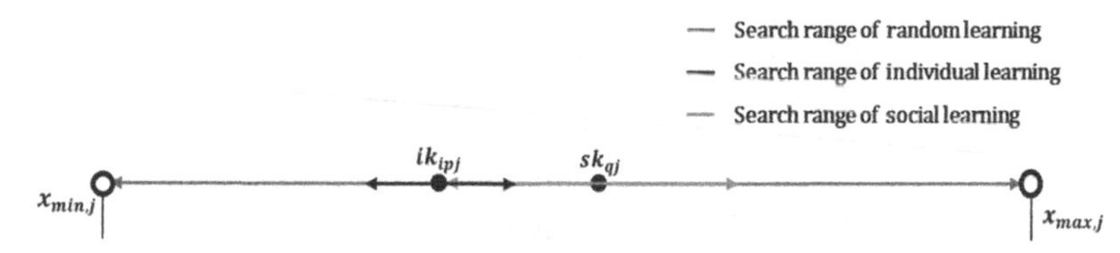


**Fig.2.** The diagram of CHLOEE learning operator operation

As the best solution is likely between $ik_{ipj}$ and $sk_{qj}$, it can be seen from Fig. 1 and Fig. 2 that CHLO and CHLOEE with the recommended parameter values both cover this area. However, the search strategies of these two algorithms for this sensitive area are totally different. CHLO only uses its individual learning operator to search this range and therefore its individual learning factor, $I_L$, has to be set to 1. On the contrary, the proposed CHLOEE mainly relies on the social learning operator to search this area, and therefore the individual learning factor of CHLOEE can be set as a much smaller value to perform local search. Considering that $ik_{ipj}$ will approach $sk_{qj}$ gradually and the individual learning operator of CHLOEE also covers part of the sensitive area, CHLOEE can search this important area much more efficiently than CHLO. Thus, the exploitation ability of CHLOEE is significantly enhanced.

## 4 Experimental Results and Discussions

To verify the performance, the proposed CHLOEE is adopted to solve the CEC20 benchmark functions, and the results are compared with those of the CHLO [16] and recent metaheuristic algorithms, including the Grasshopper Optimization Algorithm (GOA)

[18], Hybrid Harmony Search and Simulated Annealing algorithm (HS-SA) [19], Adaptive Learning-based Particle Swarm Optimization (ALPSO) [20] and Seagull Optimization Algorithm (SOA) [21]. For a fair comparison, the recommended parameter values are adopted for all the algorithms, which are listed in Table 4. All the algorithms run 100 times independently on each function.

**Table 4.** Parameter settings of all the algorithms

| Algorithm | Parameters |
|---|---|
| CHLOEE | $pr = 0.015, pi = 0.5, I_l = 0.5, S_l = 1, rl = 100$ |
| CHLO (2017) | $pr = 0.1, pi = 0.85, I_L = 1, S_L = 2, rl = 100$ |
| GOA (2017) | $cmin = 0.00004, cmax = 1, c = cmax - l \times (cmax - cmin)/L$ |
| HS-SA (2018) | $HMS = 5, HMCR = 0.9, PAR = 0.3, BW = 0.01, \alpha = 0.99, L = 3D, \beta = 0.95$ |
| ALPSO (2018) | $w = [0.4,0.9], c1 = c2 = 2$ |
| SOA (2019) | $N\_agents = 100,\ f\_c = 2,\ A = f\_c - (x \times (f\_c/Max\_iteration))$ |

The simulation results of CHLOEE and other competitors on the 10-dimensional and 30-dimensional functions are provided in Table 5 and Table 6. The W-test results in Table 5 indicate that the CHLOEE significantly surpasses CHLO, GOA, ALPSO and SOA on all the 10-dimensional functions while it is better than HS-SA on 8 out of 10 functions. The t-test results are consistent with the W-test results.

Table 6 shows that the proposed CHLOEE obtains the best mean numerical results on all the 30-dimensional functions. The results of the t-test in Table 6 show that the CHLOEE is superior to CHLO, GOA, HS-SA, ALPSO and SOA on 10, 9, 9, 10 and 10 out of 10 functions, and the W-test results reveal that CHLOEE outperforms these compared algorithms on 9, 9, 9, 10 and 10 out of 10 functions.

Therefore, based on the results of the CEC20 benchmark problems, it is fair to claim that the CHLOEE has a better search performance.

**Table 5.** Results of all the algorithms on the 10-dimensional CEC20 benchmarks

| Function | Metric | CHLOEE | CHLO | GOA | HS-SA | ALPSO | SOA |
|---|---|---|---|---|---|---|---|
| F1 | Best | 3.48E−01 | 1.04E + 01 | 1.18E + 07 | **1.07E−01** | 2.49E + 00 | 7.38E + 04 |
| | Mean | **2.21E + 03** | 7.27E + 03 | 3.11E + 07 | 2.32E + 03 | 1.37E + 08 | 1.91E + 08 |
| | Std | 3.13E + 03 | 4.09E + 03 | 7.74E + 06 | **2.85E + 03** | 4.13E + 08 | 2.49E + 08 |
| | t-test | / | 1 | 1 | 0 | 1 | 1 |
| | W-test | / | 1 | 1 | 0 | 1 | 1 |

*(continued)*

**Table 5.** (*continued*)

| Function | Metric | CHLOEE | CHLO | GOA | HS-SA | ALPSO | SOA |
|---|---|---|---|---|---|---|---|
| F2 | Best | **3.60E + 00** | 2.68E + 01 | 1.25E + 02 | 6.89E + 00 | 1.02E + 01 | 9.12E + 01 |
| | Mean | 2.06E + 02 | 4.73E + 02 | 5.62E + 02 | **1.75E + 02** | 4.08E + 02 | 6.66E + 02 |
| | Std | 1.31E + 02 | 2.30E + 02 | 1.81E + 02 | **1.26E + 02** | 2.17E + 02 | 2.20E + 02 |
| | t−test | / | 1 | 1 | 0 | 1 | 1 |
| | W−test | / | 1 | 1 | 0 | 1 | 1 |
| F3 | Best | **3.71E + 00** | 6.60E + 00 | 2.23E + 01 | 5.03E + 00 | 1.58E + 01 | 1.77E + 01 |
| | Mean | **1.55E + 01** | 1.96E + 01 | 3.80E + 01 | 2.53E + 01 | 3.45E + 01 | 4.81E + 01 |
| | Std | **3.79E + 00** | 5.85E + 00 | 4.10E + 00 | 6.47E + 00 | 9.47E + 00 | 1.26E + 01 |
| | t−test | / | 1 | 1 | 1 | 1 | 1 |
| | W−test | / | 1 | 1 | 1 | 1 | 1 |
| F4 | Best | **1.68E−01** | 2.61E−01 | 1.67E + 00 | 4.06E−01 | 7.44E−01 | 4.03E−01 |
| | Mean | **9.84E−01** | 2.00E + 00 | 2.74E + 00 | 2.38E + 00 | 1.58E + 01 | 2.36E + 00 |
| | Std | 4.49E−01 | 1.22E + 00 | **4.25E−01** | 1.63E + 00 | 6.88E + 01 | 1.38E + 00 |
| | t−test | / | 1 | 1 | 1 | 1 | 1 |
| | W−test | / | 1 | 1 | 1 | 1 | 1 |
| F5 | Best | **4.25E + 01** | 5.62E + 01 | 4.05E + 02 | 6.71E + 03 | 1.66E + 02 | 1.08E + 03 |
| | Mean | **2.43E + 03** | 1.66E + 04 | 2.59E + 03 | 3.44E + 05 | 5.54E + 03 | 8.74E + 03 |
| | Std | 1.40E + 03 | 1.98E + 04 | **1.11E + 03** | 3.68E + 05 | 5.47E + 03 | 4.99E + 03 |
| | t−test | / | 1 | 1 | 1 | 1 | 1 |
| | W−test | / | 1 | 1 | 1 | 1 | 1 |
| F6 | Best | **2.98E−01** | 6.80E−01 | 5.80E + 00 | 4.56E−01 | 1.36E + 00 | 1.77E + 00 |
| | Mean | **2.76E + 01** | 9.61E + 01 | 1.44E + 02 | 9.05E + 01 | 1.02E + 02 | 1.34E + 02 |
| | Std | 4.98E + 01 | 7.60E + 01 | 7.91E + 01 | 8.74E + 01 | 9.03E + 01 | **4.06E + 01** |
| | t-test | / | 1 | 1 | 1 | 1 | 1 |
| | W-test | / | 1 | 1 | 1 | 1 | 1 |
| F7 | Best | **1.25E−01** | 6.43E−01 | 3.53E + 02 | 1.19E + 02 | 4.45E + 01 | 4.06E + 02 |
| | Mean | **3.75E + 01** | 2.87E + 03 | 1.23E + 03 | 6.14E + 04 | 1.14E + 03 | 6.59E + 03 |
| | Std | **4.54E + 01** | 7.39E + 03 | 6.12E + 02 | 1.03E + 05 | 4.12E + 03 | 5.64E + 03 |
| | t-test | / | 1 | 1 | 1 | 1 | 1 |
| | W-test | / | 1 | 1 | 1 | 1 | 1 |
| F8 | Best | 2.44E + 01 | 2.30E + 01 | 2.23E + 01 | 2.71E + 01 | 1.00E + 02 | **1.56E + 01** |
| | Mean | **1.05E + 02** | 1.50E + 02 | 1.13E + 02 | 2.79E + 02 | 1.12E + 02 | 4.32E + 02 |
| | Std | 3.53E + 01 | 1.05E + 02 | **1.29E + 01** | 3.23E + 02 | 1.68E + 01 | 5.05E + 02 |
| | t-test | / | 1 | 1 | 1 | 1 | 1 |

(*continued*)

**Table 5.** (*continued*)

| Function | Metric | CHLOEE | CHLO | GOA | HS-SA | ALPSO | SOA |
|---|---|---|---|---|---|---|---|
| | W-test | / | 1 | 1 | 1 | 1 | 1 |
| F9 | Best | **5.84E−06** | 1.00E + 02 | 1.13E + 02 | 1.00E + 02 | 1.00E + 02 | 3.00E + 02 |
| | Mean | **2.97E + 02** | 3.59E + 02 | 3.32E + 02 | 3.37E + 02 | 3.53E + 02 | 3.51E + 02 |
| | Std | 1.03E + 02 | 4.65E + 01 | 8.70E + 01 | 6.96E + 01 | 4.54E + 01 | **1.12E + 01** |
| | t-test | / | 1 | 1 | 1 | 1 | 1 |
| | W-test | / | 1 | 1 | 1 | 1 | 1 |
| F10 | Best | **3.98E + 02** | 3.98E + 02 | 3.99E + 02 | 3.98E + 02 | 3.98E + 02 | 3.98E + 02 |
| | Mean | **4.20E + 02** | 4.36E + 02 | 4.23E + 02 | 4.31E + 02 | 4.36E + 02 | 4.32E + 02 |
| | Std | 2.37E + 01 | 2.88E + 01 | 2.20E + 01 | 2.31E + 01 | 2.83E + 01 | **1.86E + 01** |
| | t-test | / | 1 | 1 | 1 | 1 | 1 |
| | W-test | / | 1 | 1 | 1 | 1 | 1 |

**Table 6.** Results of all the algorithms on the 30-dimensional CEC20 benchmarks

| Function | Metric | CHLOEE | CHLO | GOA | HS-SA | ALPSO | SOA |
|---|---|---|---|---|---|---|---|
| F1 | Best | **2.53E−01** | 1.14E + 01 | 2.84E + 08 | 2.11E + 06 | 4.34E + 06 | 1.73E + 09 |
| | Mean | **6.57E + 03** | 1.17E + 04 | 3.92E + 08 | 8.38E + 06 | 2.99E + 09 | 6.74E + 09 |
| | Std | **6.94E + 03** | 7.66E + 03 | 4.63E + 07 | 3.24E + 06 | 2.98E + 09 | 2.57E + 09 |
| | t-test | / | 1 | 1 | 1 | 1 | 1 |
| | W-test | / | 1 | 1 | 1 | 1 | 1 |
| F2 | Best | **9.04E + 02** | 1.57E + 03 | 3.70E + 03 | 1.32E + 03 | 2.74E + 03 | 2.27E + 03 |
| | Mean | **2.00E + 03** | 2.78E + 03 | 4.76E + 03 | 2.23E + 03 | 5.23E + 03 | 4.41E + 03 |
| | Std | **4.05E + 02** | 4.84E + 02 | 5.04E + 02 | 4.29E + 02 | 9.58E + 02 | 8.26E + 02 |
| | t-test | / | 1 | 1 | 1 | 1 | 1 |
| | W-test | / | 1 | 1 | 1 | 1 | 1 |
| F3 | Best | **6.09E + 01** | 8.13E + 01 | 1.82E + 02 | 1.02E + 02 | 2.16E + 02 | 2.03E + 02 |
| | Mean | **9.56E + 01** | 1.28E + 02 | 2.11E + 02 | 1.49E + 02 | 3.18E + 02 | 3.21E + 02 |
| | Std | 1.95E + 01 | 2.30E + 01 | **9.04E + 00** | 2.13E + 01 | 4.87E + 01 | 5.00E + 01 |
| | t-test | / | 1 | 1 | 1 | 1 | 1 |
| | W-test | / | 1 | 1 | 1 | 1 | 1 |
| F4 | Best | **2.71E + 00** | 4.22E + 00 | 1.55E + 01 | 8.91E + 00 | 2.00E + 01 | 1.95E + 01 |
| | Mean | **7.04E + 00** | 1.10E + 01 | 1.86E + 01 | 1.57E + 01 | 1.38E + 03 | 1.25E + 03 |
| | Std | 3.24E + 00 | 5.54E + 00 | **9.93E−01** | 4.29E + 00 | 3.46E + 03 | 2.23E + 03 |

(*continued*)

**Table 6.** (*continued*)

| Function | Metric | CHLOEE | CHLO | GOA | HS-SA | ALPSO | SOA |
|---|---|---|---|---|---|---|---|
| | t-test | / | 1 | 1 | 1 | 1 | 1 |
| | W-test | / | 1 | 1 | 1 | 1 | 1 |
| F5 | Best | **1.86E + 04** | 2.50E + 04 | 8.20E + 04 | 5.05E + 05 | 3.52E + 04 | 1.39E + 05 |
| | Mean | **1.22E + 05** | 7.97E + 05 | 2.90E + 05 | 3.62E + 06 | 1.86E + 06 | 6.26E + 05 |
| | Std | **8.10E + 04** | 6.61E + 05 | 1.29E + 05 | 2.08E + 06 | 1.58E + 06 | 4.51E + 05 |
| | t-test | / | 1 | 1 | 1 | 1 | 1 |
| | W-test | / | 1 | 1 | 1 | 1 | 1 |
| F6 | Best | **1.81E + 01** | 6.24E + 01 | 2.74E + 02 | 3.56E + 01 | 3.01E + 02 | 1.11E + 02 |
| | Mean | **1.24E + 02** | 3.51E + 02 | 4.65E + 02 | 2.46E + 02 | 7.61E + 02 | 6.59E + 02 |
| | Std | **8.70E + 01** | 1.49E + 02 | 1.09E + 02 | 1.16E + 02 | 2.57E + 02 | 2.54E + 02 |
| | t-test | / | 1 | 1 | 1 | 1 | 1 |
| | W-test | / | 1 | 1 | 1 | 1 | 1 |
| F7 | Best | **2.95E + 03** | 1.79E + 04 | 3.51E + 04 | 2.59E + 04 | 1.05E + 04 | 4.53E + 04 |
| | Mean | **3.91E + 04** | 3.06E + 05 | 9.72E + 04 | 1.11E + 06 | 5.08E + 05 | 4.93E + 05 |
| | Std | **2.66E + 04** | 3.64E + 05 | 4.47E + 04 | 8.88E + 05 | 3.44E + 05 | 3.12E + 05 |
| | t-test | / | 1 | 1 | 1 | 1 | 1 |
| | W-test | / | 1 | 1 | 1 | 1 | 1 |
| F8 | Best | **1.00E + 02** | 1.00E + 02 | 1.79E + 02 | 1.16E + 02 | 2.23E + 02 | 7.89E + 02 |
| | Mean | **1.53E + 03** | 2.97E + 03 | 1.64E + 03 | 2.43E + 03 | 3.97E + 03 | 4.49E + 03 |
| | Std | 1.27E + 03 | 1.31E + 03 | **8.09E + 02** | 1.13E + 03 | 2.07E + 03 | 9.28E + 02 |
| | t-test | / | 1 | 0 | 1 | 1 | 1 |
| | W-test | / | 1 | 0 | 1 | 1 | 1 |
| F9 | Best | 4.60E + 02 | 5.03E + 02 | 5.27E + 02 | **4.60E + 02** | 5.33E + 02 | 4.68E + 02 |
| | Mean | **5.09E + 02** | 6.06E + 02 | 5.52E + 02 | 5.13E + 02 | 6.33E + 02 | 5.38E + 02 |
| | Std | 2.94E + 01 | 3.83E + 01 | **1.01E + 01** | 2.68E + 01 | 5.64E + 01 | 2.89E + 01 |
| | t-test | / | 1 | 1 | 0 | 1 | 1 |
| | W-test | / | 1 | 1 | 0 | 1 | 1 |
| F10 | Best | **3.82E + 02** | 3.83E + 02 | 4.00E + 02 | 3.87E + 02 | 3.89E + 02 | 4.58E + 02 |
| | Mean | **3.88E + 02** | 3.89E + 02 | 4.04E + 02 | 4.11E + 02 | 5.06E + 02 | 5.62E + 02 |
| | Std | 7.78E + 00 | 9.62E + 00 | **3.25E + 00** | 2.18E + 01 | 1.21E + 02 | 5.78E + 01 |
| | t-test | / | 1 | 1 | 1 | 1 | 1 |
| | W-test | / | 0 | 1 | 1 | 1 | 1 |

## 5 Conclusion

To improve the performance of the CHLO, a new CHLOEE algorithm is proposed in this work. By re-designing the social learning operator, the individual learning operator of CHLOEE is relieved to focus on performing local search to enhance the exploitation ability. The parameter study and performance analysis on CHLOEE demonstrate that the duty of searching for the sensitive area, where the best solution may exist, is transferred from the individual learning operator in CHLO to the social learning operator in CHLOEE, and the individual learning operator of CHLOEE act as a local search operator with a small learning factor. Moreover, unlike that the social learning operator and individual learning operator in CHLO search different areas, the social learning operator in CHLOEE works together with the individual learning operator to search the sensitive area, which further strengthens the exploitation ability of CHLOEE. The comparison results with CHLO and recent state-of-the-art meta-heuristics show that the proposed CHLOEE can search for the optimal solution more efficiently and has significantly better performance.

**Acknowledgments.** This work is supported by National Key Research and Development Program of China (No. 2019YFB1405500), National Natural Science Foundation of China (Grant No. 92067105 & 61833011), Key Project of Science and Technology Commission of Shanghai Municipality under Grant No. 19510750300 & 19500712300, and 111 Project under Grant No. D18003.

## References

1. Wang, L., Ni, H., Yang, R., Fei, M., Ye, W.: A simple human learning optimization algorithm. In: Fei, M., Peng, C., Su, Z., Song, Y., Han, Q. (eds.) LSMS/ICSEE 2014. CCIS, vol. 462, pp. 56–65. Springer, Heidelberg (2014). https://doi.org/10.1007/978-3-662-45261-5_7
2. Wang, L., Ni, H., Yang, R., et al.: An adaptive simplified human learning optimization algorithm. J. Inf. Sci. **320**, 126–139 (2015)
3. Yang, R., Xu, M., He, J., Ranshous, S., Samatova, N.F.: An intelligent weighted fuzzy time series model based on a sine-cosine adaptive human learning optimization algorithm and its application to financial markets forecasting. In: Cong, G., Peng, W.C., Zhang, W.E., Li, C., Sun, A. (eds.) ADMA 2017. LNCS (LNAI), vol. 10604, pp. 595–607. Springer, Cham (2017). https://doi.org/10.1007/978-3-319-69179-4_42
4. Wang, L., An, L., Pi, J., Fei, M., Pardalos, P.M.: A diverse human learning optimization algorithm. J. Global Optim. **67**(1–2), 283–323 (2016). https://doi.org/10.1007/s10898-016-0444-2
5. Wang, L., Pei, J., Wen, Y., et al.: An improved adaptive human learning algorithm for engineering optimization. J. Appl. Soft Comput. **71**, 894–904 (2018)
6. Li, X., Yao, J., Wang, L., Menhas, M.I.: Application of human learning optimization algorithm for production scheduling optimization. In: Fei, M., Ma, S., Li, X., Sun, X., Jia, L., Su, Z. (eds.) LSMS/ICSEE -2017. CCIS, vol. 761, pp. 242–252. Springer, Singapore (2017). https://doi.org/10.1007/978-981-10-6370-1_24
7. Cao, J., Yan, Z., He, G.: Application of multi-objective human learning optimization method to solve AC/DC multi-objective optimal power flow problem. J. International J. Emerg. Electr. Power Syst. **17**(3), 327--337 (2016)

8. Cao, J., Yan, Z., Xu, X., et al.: Optimal power flow calculation in AC/DC hybrid power system based on adaptive simplified human learning optimization algorithm. J. Modern Power Syst. Clean Energy. **4**(4), 690--701 (2016)
9. Wang, L., Yang, R., Ni, H., et al.: A human learning optimization algorithm and its application to multi-dimensional knapsack problems. J. Appl. Soft Comput. **34**, 736–743 (2015)
10. Bhandari, A.K., Kumar, I.V.: A context sensitive energy thresholding based 3D Otsu function for image segmentation using human learning optimization. J. Appl. Soft Comput. **82**, 105–570 (2019)
11. Wen, Y., Wang, L., Peng, W., Menhas, M.I., Qian, L.: Application of Intelligent Virtual Reference Feedback Tuning to Temperature Control in a Heat Exchanger. In: Li, K., Fei, M., Du, D., Yang, Z., Yang, D. (eds.) ICSEE/IMIOT -2018. CCIS, vol. 924, pp. 311–320. Springer, Singapore (2018). https://doi.org/10.1007/978-981-13-2384-3_29
12. Ding, H., Gu, X.: Hybrid of human learning optimization algorithm and particle swarm optimization algorithm with scheduling strategies for the flexible job-shop scheduling problem. J. Neurocomput. **414**, 313–332 (2020)
13. Shoja, A., Molla-Alizadeh-Zavardehi, S., Niroomand, S.: Hybrid adaptive simplified human learning optimization algorithms for supply chain network design problem with possibility of direct shipment. J. Appl. Soft Comput. **96**, 106–594 (2020)
14. Alguliyev, R., Aliguliyev, R., Isazade, N.: A sentence selection model and HLO algorithm for extractive text summarization. In: 2016 IEEE 10th International Conference on Application of Information and Communication Technologies (AICT). pp. 1--4 (2016)
15. Yang, R., He, J., Xu, M., Ni, H., Jones, P., Samatova, N.: An Intelligent and Hybrid Weighted Fuzzy Time Series Model Based on Empirical Mode Decomposition for Financial Markets Forecasting. In: Perner, P. (ed.) ICDM 2018. LNCS (LNAI), vol. 10933, pp. 104–118. Springer, Cham (2018). https://doi.org/10.1007/978-3-319-95786-9_8
16. Wang, L., Pei, J., Menhas, M.I., et al.: A hybrid-coded human learning optimization for mixed-variable optimization problems. J. Knowl. Based Syst. **127**, 114–125 (2017)
17. Ghani, J.A., Choudhury, I.A., Hassan, H.H.: Application of Taguchi method in the optimization of end milling parameters. J. Mater. Proc. Technol. **145**(1), 84–92 (2004)
18. Saremi, S., Mirjalili, S., Lewis, A.: Grasshopper optimisation algorithm: theory and application. J. Adv. Eng. Softw. **105**, 30–47 (2017)
19. Assad, A., Deep, K.: A hybrid harmony search and simulated annealing algorithm for continuous optimization. J. Inf. Sci. **450**, 246–266 (2018)
20. Wang, F., Zhang, H., Li, K., et al.: A hybrid particle swarm optimization algorithm using adaptive learning strategy. J. Inf. Sci. **436**, 162–177 (2018)
21. Dhiman, G., Kumar, V.: Seagull optimization algorithm: theory and its applications for large-scale industrial engineering problems. J. Knowl. Based Syst. **165**, 169–196 (2019)

# Observer-Based Fuzzy Sliding Mode Control for Networked Control Systems Under Cyber Attacks

Jin Zhang, Dajun Du(✉), Changda Zhang, Baoyue Xu, and Minrei Fei

Shanghai Key Laboratory of Power Station Automation Technology,
School of Mechanical Engineering and Automation, Shanghai University,
Shanghai 200072, People's Republic of China
zhangjin_@shu.edu.cn, ddj@i.shu.edu.cn

**Abstract.** Cyber attacks and disturbance in networked control systems (NCSs) will inevitably decline the control system performance even cause the system crash. This paper investigates observer-based fuzzy sliding mode control (FSMC) for NCSs under cyber attacks and disturbance. Firstly, to improve the control performance, an observer is employed to decline the effect of cyber attacks and disturbance, and the estimation errors are proved to converge asymptotically to zero. Secondly, an observer-based FSMC controller is proposed to maintain the system stability and decline chattering, and the stability of the closed-loop system is proved. Finally, simulation results demonstrate the effectiveness of the proposed method in comparison with other approaches.

**Keywords:** Networked control systems · Cyber attacks · Observer-based FSMC · Stability

## 1 Introduction

With the development of network communication technology, the network is applied to connect the controllers, actuators and sensors deployed in different locations, forming networked control systems (NCSs). NCSs are employed to solve the problems of wiring difficulties, high costs, and difficult maintenance in traditional control systems. However, there are also some factors in NCSs that affect the control performance such as network uncertainties issues, disturbance [1], cyber attacks [2] and so on. Therefore, how to investigate control approaches for NCSs under these cases is an open issue.

Since disturbance or cyber attacks affect the control performance, the control methods with strong robustness need to be usually chosen. Due to strong robustness and fast response characteristics, sliding mode control (SMC) method is employed to suppress disturbance [3]. For example, a robust SMC method is proposed for robot control systems to decline unknown disturbances [4]. However, there is a chattering problem in the real SMC, which affects the system


Q. Han et al. (Eds.): LSMS 2021/ICSEE 2021, CCIS 1469, pp. 488–497, 2021.
https://doi.org/10.1007/978-981-16-7213-2_47

performance. If it is not solved properly, it may cause the instability of the whole system. To improve the performance of conventional SMC, the improved SMC are studied by reducing chattering effects [5]. For example, a fuzzy sliding mode control (FSMC) method is proposed by suppression of disturbance for quadrotor positioning and tracking control [6]. Moreover, there exist some observer-based methods to reduce the effect of disturbance [7].

Except for disturbance, NCSs are vulnerable to cyber attacks. For example, Iran's nuclear facilities were attacked in 2010, Venezuelan's power grid was attacked in 2019 and so on. There exist different cyber attacks such as denial of service (DoS) attacks [8] and false data injection attacks (FDIAs) [9]. Since great harm of cyber attacks, the security of NCSs has been paid more and more attention. For example, a secure controller is proposed for cyber-physical systems under DoS attacks [10]. To provide locally reliable state estimations and detect FDIAs, resilient attack detections are constructed [11]. More research can be found in some review literatures [12–14]. Therefore, it is of great practical significance to design a secure controller which can maintain system stability under cyber attacks.

The above mentioned studies do not consider cyber attacks and disturbance together. However, cyber attacks and disturbance in NCSs will decline the control performance even cause system unstable. There exist two difficulties: 1) The accuracy of the signals is affected by cyber attacks and disturbance. How to obtain accuracy estimation signals by designing an observer is the first challenge. 2) The estimation errors will affect the performance of the controller. How to design a controller with strong robustness is another challenge.

To address these challenges, this paper investigates observer-based FSMC for NCSs under cyber attacks and disturbance. The main contributions include: 1) To improve the control performance, an observer is employed to decline the effect of cyber attacks and disturbance, and the estimation errors are proved to converge asymptotically to zero. 2) To maintain the system stability and decline chattering, an observer-based FSMC controller is designed, and the stability of the closed-loop system is proved.

The rest of this paper is organized as follows. NCSs under cyber attacks and disturbance are described in Sect. 2. An observer is designed to compensate the impact of cyber attacks, and an observer-based FSMC are presented in Sect. 3. Simulation results and discussion are provided to show the effectiveness of the proposed method in Sect. 4 and conclusions are given in Sect. 5.

## 2 Problem Formulation

The structure of NCSs under cyber attacks and disturbance is shown in Fig.1. The system output $y(t)$ is firstly sampled by sensors and sent to the controller. Then the control signal $u(t)$ is calculated and transmitted to the actuator via the network, which may be attacked by the attack signal $u_a(t)$. Finally, the attacked control signal (i.e., $u(t) + u_a(t)$) is employed to the plant by the actuator.

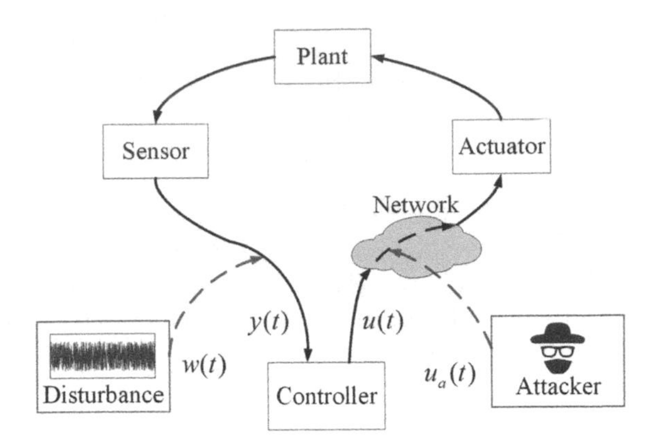


**Fig. 1.** Structure of NCSs under cyber attacks and disturbance.

Considering cyber attacks and disturbance against NCSs, the system model is expressed as

$$\dot{x}(t) = Ax(t) + Bu(t) + Eu_a(t), \tag{1}$$
$$y(t) = Cx(t) + Dw(t), \tag{2}$$

where $x(t) \in \mathbb{R}^n$ is the system state, $y(t) \in \mathbb{R}^p$ is the system output, $u(t) \in \mathbb{R}^m$ is the control signal, $w(t) \in \mathbb{R}^q$ is the bounded disturbance signal, $u_a(t) \in \mathbb{R}^l$ is the bounded attack signal, $A$, $B$ and $C$ are known matrices of appropriate dimensions, as well as $D$ and $E$ are known matrices with full column rank.

For the matrix $D$, singular value decomposition is performed, i.e., there exists a nonsingular matrix $U$ such that $\tilde{D} = UD = [0; D_2]$, where $D_2 \in \mathbb{R}^{q \times q}$ is invertible. Therefore, the nonsingular transformation $\tilde{y}(t)$ of $y(t)$ can be given by

$$\tilde{y}(t) = Uy(t) = \tilde{C}x(t) + \tilde{D}w(t), \tag{3}$$

where $\tilde{C} = UC = [C_1; C_2]$, $C_1 \in \mathbb{R}^{(p-q)\times n}$, $C_2 \in \mathbb{R}^{q\times n}$. Define $\tilde{y}(t) = [y_1(t); y_2(t)]$ and it follows that

$$y_1(t) = C_1x(t), \tag{4}$$
$$y_2(t) = C_2x(t) + D_2w(t). \tag{5}$$

By decomposing $y(t)$ into $y_1(t)$ and $y_2(t)$, the effect of cyber attacks and disturbance on the system can be separated, which brings great convenience for the design of observers. To improve the control performance under cyber attacks and disturbance, a secure controller is necessary to be designed.

## 3 Observer-Based FSMC for NCSs

After the above model is established, the observer-based FSMC controller is designed as shown in Fig. 2, where the observer is designed to obtain the estimate $\hat{x}(t)$, $\hat{u}_a(t)$ and $\hat{w}(t)$ of $x(t)$, $u_a(t)$ and $w(t)$ respectively. These estimated results can be used to decline the effect of cyber attacks and disturbance.

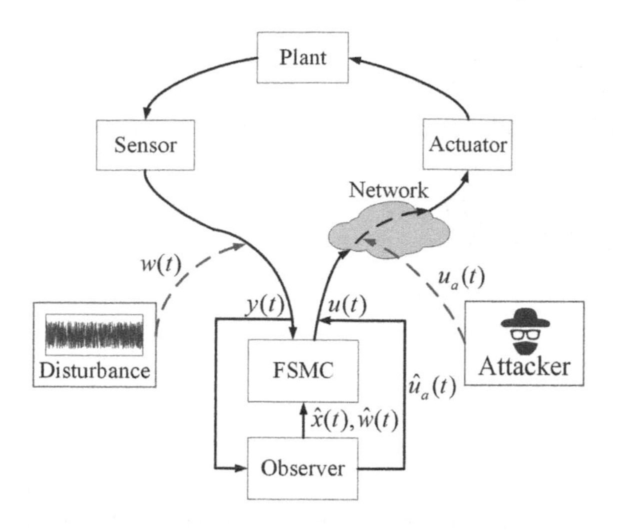


**Fig. 2.** Structure of NCSs with Observer-Based FSMC.

### 3.1 Design and Convergence Analysis of Observer

To obtain the estimated results, the observer is designed by the following three parts:

1) To obtain estimation of cyber attacks, $u_a(t)$ needs be decoupled from the state estimation error equation. From (4) and (5), if $rank(C_1E) = rank(E)$, there exists a matrix $M \in \mathbb{R}^{l\times(p-q)}$ satisfies

$$MC_1E = I. \tag{6}$$

From (6), the particular solution of $M$ can be obtained by $M = (C_1E)^+ := \left[(C_1E)^T(C_1E)\right]^{-1}(C_1E)^T$.

Multiplying both sides of (1) by $MC_1$, it follows that

$$MC_1\dot{x}(t) = MC_1Ax(t) + MC_1Bu(t) + MC_1Eu_a(t). \tag{7}$$

Substituting (4) and (6) into (7), $u_a(t)$ can be re-written as

$$u_a(t) = M\dot{y}_1(t) - MC_1Ax(t) - MC_1Bu(t). \tag{8}$$

To reconstruct $u_a(t)$ and according to (8), $\hat{u}_a(t)$ is designed as

$$\hat{u}_a(t) = M(\dot{y}_1(t) - C_1A\hat{x}(t) - C_1Bu(t)). \tag{9}$$

2) To obtain $\hat{u}_a(t)$ by (9), $\hat{x}(t)$ can be estimated by

$$\dot{\hat{x}}(t) = A\hat{x}(t) + Bu(t) + E\hat{u}_a(t) + L(y(t) - \hat{y}(t) - D\hat{w}(t)),$$

where $\hat{y}(t) = C\hat{x}(t)$, and $L$ is the observer gain. Define $z(t) := \hat{x}(t) - EMy_1(t)$ and $T = I - EMC_1$ ($TE = 0$), $\hat{x}(t)$ can be further expressed as

$$\begin{cases} \dot{z}(t) = (TA - LC)z + TBu(t) - LD\hat{w}(t) + ((TA - LC)EM)y_1(t) + Ly(t) \\ \hat{x}(t) = z(t) + EMy_1(t). \end{cases} \tag{10}$$

3) Based on the nonsingular transformation of $y(t)$ in (3), the observer of $w(t)$ can be designed as

$$\hat{w}(t) = Ge_y(t), \tag{11}$$

where $e_y(t) = y(t) - \hat{y}(t)$, $G = \left[G_1, D_2^{-1}\right]$, and $G_1 \in \mathbb{R}^{q\times(p-1)}$ is a weight matrix.

After the design of the observer, the convergence of the observer is analyzed by Theorem 1. For the convenience of the proof of Theorem 1, the estimation errors are defined as $e(t) = x(t) - \hat{x}(t)$, $\tilde{u}_a(t) = u_a(t) - \hat{u}_a(t)$, and $\tilde{w}(t) = w(t) - \hat{w}(t)$.

**Theorem 1.** *Consider the system (1) with the observer (9), (10) and (11), if there exist the matrices $Z \in R^{n\times p}$, $P = P^T > 0 \in R^{n\times n}$ satisfying*

$$PTA - ZC^* + (PTA - ZC^*)^T < 0,$$

*and let $L = ZP^{-1}$ in (11), where $C^* = C - DGC$, then $e(t)$, $\tilde{u}_a(t)$ and $\tilde{w}(t)$ will converge asymptotically to zero.*

*Proof.* The error of state estimation is expressed as $e(t) = x(t) - z(t) - EMy_1(t) = Tx(t) - z(t)$.

From (10), $\dot{e}(t)$ can be given by $\dot{e}(t) = T\dot{x}(t) - \dot{z}(t) = (TA - LC)e(t) + LD\tilde{w}(t)$. From (4), (5) and (11), $\tilde{w}(t)$ can be given by $\tilde{w}(t) = GCe(t)$, and $\dot{e}(t)$ becomes $\dot{e}(t) = (TA - LC^*)e(t)$.

Choose a Lyapunov function as $V(t) = e^T(t)Pe(t)$. Taking its derivative with respect to $t$ and let $PL = Z$, yields

$$\dot{V}(t) = e^T(t)(PTA - ZC^* + (PTA - ZC^*)^T)e(t) < 0.$$

Therefore, $e(t)$ will converge asymptotically to zero. It completes the proof.

### 3.2 Design and Stability Analysis of Sliding Mode Controller

The above designed can obtain $\hat{x}(t)$, $\hat{u}_a(t)$ and $\hat{e}(t)$. Therefore, $u_a(t)$ and $w(t)$ can be offset by $\hat{u}_a(t)$ and $\hat{w}(t)$, respectively. Using the estimated results to offset cyber attacks and disturbance, the system (1) becomes

$$\dot{x}(t) = Ax(t) + Bu(t) + E(u_a(t) - \hat{u}_a(t)), \tag{12}$$

$$y(t) = Cx(t) + D(w(t) - \hat{w}(t)). \tag{13}$$

Considering the effect of $u_a(t) - \hat{u}_a(t)$ and $w(t) - \hat{w}(t)$, (13) can be re-written as

$$\dot{x}(t) = Ax(t) + B(u(t) + f(x,t)), \tag{14}$$

where $f(x,t)$ is the error from the estimation, and $\|f(x,t)\| \leq \delta_f$.

To decline the effect of estimation error, the sliding mode controller for (14) is designed as

$$u(t) = u_{eq}(t) + u_n(t). \tag{15}$$

After the controller is designed, the sliding mode variable is defined as

$$s(t) = B\bar{P}x(t), \tag{16}$$

where $\bar{P} > 0 \in \mathbb{R}^{n \times n}$, $s(t) = 0$ is achieved by designing the appropriate $\bar{P}$.

According to equivalent control principle [15], when the system state reaches the sliding mode surface, let $f(x,t) = 0$. From (14) and $\dot{s}(t) = 0$, it follows that $\dot{s}(t) = B\bar{P}\dot{x}(t) = B\bar{P}(Ax(t) + Bu(t)) = 0$. For $u_{eq}(t)$ in (15), it is designed as $u_{eq}(t) = -(B^T\bar{P}B)^{-1}B\bar{P}Ax(t)$. To ensure $s\dot{s} < 0$, $u_n(t)$ in (15) is designed as $u_n(t) = -(B^T\bar{P}B)^{-1}\left[\left|B^T\bar{P}B\right|\delta_f + \varepsilon_0\right] sgn(s)$.

Choose a Lyapunov function as $V(t) = \frac{1}{2}s^2$. From (14), (15) and (16) yields

$$\dot{V}(t) = s\dot{s} = -\left[\left|B^T\bar{P}B\right|\delta_f + \varepsilon_0\right]|s| + B^T\bar{P}Bf(x,t) < -\varepsilon_0\,|s|.$$

Therefore, the accessibility of the sliding surface $s\dot{s} < 0$ is satisfied.

Using linear matrix inequality to design an appropriate $\bar{P}$ in (16), and according to [16], the controller is written as

$$u(t) = -Kx(t) + v(t), \tag{17}$$

where $v(t) = Kx(t) + u_{eq} + u_n$.

Substituting (17) into (14), the closed-loop system can be expressed as

$$\dot{x}(t) = (A - BK)x(t) + B(v(t) + f(x,t)), \tag{18}$$

where $K$ is designed to make $A - BK$ be Hurwitz, which can ensure the stability of the closed-loop system. The design of $\bar{P}$ and $K$ is given in Theorem 2.

**Theorem 2.** *If there exist matrix $Y \in R^{m \times n}$ and symmetric positive definite matrix $X \in R^{n \times n}$, the following linear matrix inequality*

$$AX - BY + XA^T - Y^TB^T < 0,$$

*holds. Let $\bar{P} = X^{-1}$ and $K = Y\bar{P}$, then the designed controller can satisfy the stability of the closed-loop system (18).*

*Proof.* Choose Lyapunov function as $V(t) = x^T(t)\bar{P}x(t)$, it follows that

$$\dot{V}(t) = 2x^T(t)\bar{P}(A - BK)x(t) + 2x^T(t)\bar{P}B(v(t) + f(x,t)). \tag{19}$$

From (15), there exists $t \geq t_0$ and $s(t) = B\bar{P}x(t) = 0$ holds, which results in $s^T(t) = x^T(t)\bar{P}B = 0$. Furthermore, (19) becomes

$$\dot{V}(t) = x^T(t)(\bar{P}(A - BK) + (A - BK)^T\bar{P})x(t). \tag{20}$$

Let $\bar{P} = X^{-1}$ and $K = Y\bar{P}$, (20) becomes

$$\dot{V}(t) = x^T(t)(X^{-1}(A - BYX^{-1}) + (A - BYX^{-1})^T X^{-1})x(t). \tag{21}$$

If both sides of (21) are multiplied by $X$, yields

$$\dot{V}(t) < x^T(t)(AX - BY + XA^T - Y^T B)x(t) < 0.$$

Therefore, the stability of (18) can be ensured by the above designed controller. It completes the proof.

### 3.3 Design of Fuzzy Rules

Since it is difficult to eliminate completely the chattering in the process of sliding mode control, fuzzy rules are used to adjust $s(t)$ for mitigating chattering. The structure of the FSMC is shown in Fig. 3, where $s(t)$ represents the input variable, $\mu(t)$ represents the output variable. The fuzzy domain of $s(t)$ is $d_f$, and the fuzzy subsets of the language values of input and output variables are $s = \{NB\ NM\ NS\ Z\ PS\ PM\ PB\}$, and $\mu = \{NB\ NM\ NS\ Z\ PS\ PM\ PB\}$. Fuzzy rules are designed as

If $s$ is NS then $\mu$ is NS,
If $s$ is NM then $\mu$ is NM,
If $s$ is NB then $\mu$ is NB,
If $s$ is Z then $\mu$ is Z,
If $s$ is PS then $\mu$ is PS,
If $s$ is PM then $\mu$ is PM,
If $s$ is PB then $\mu$ is PB.

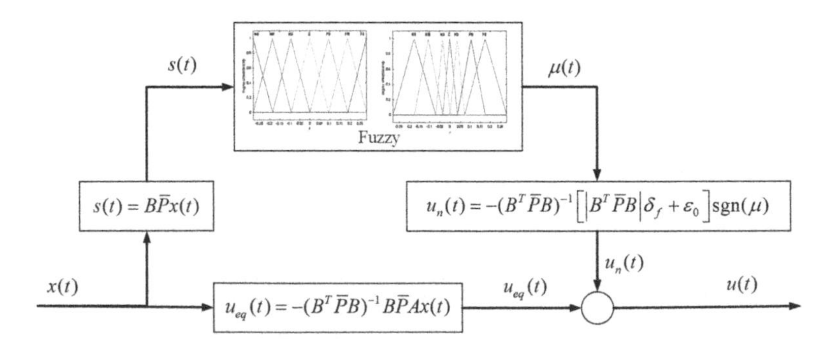


**Fig. 3.** Structure of the FSMC.

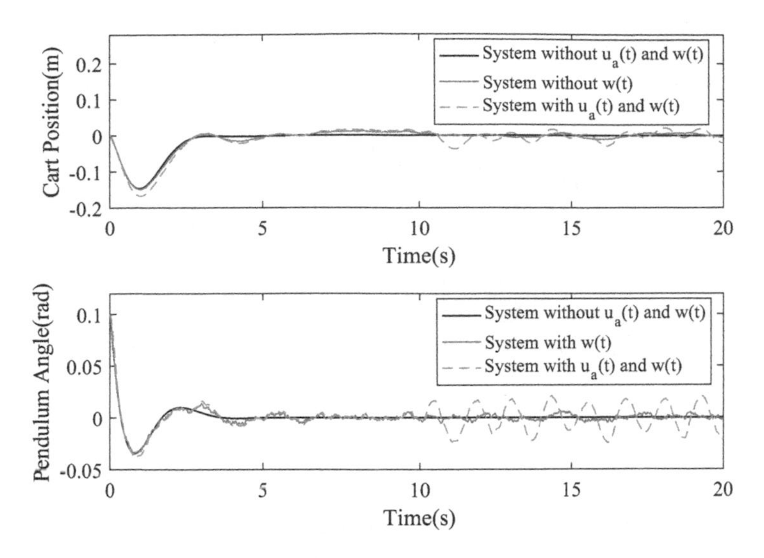


**Fig. 4.** Comparison the outputs between the system without $u_a(t)$ and $w(t)$, the system with $w(t)$ and the system with $u_a(t)$ and $w(t)$.

Using fuzzy rules, the change trend of $s(t)$ can be reduced, and $u_n(t)$ becomes $\tilde{u}_n(t) = -(B^T\bar{P}B)^{-1}\left[\left|B^T\bar{P}B\right|\delta_f + \varepsilon_0\right] sgn(\mu)$, and the convergence of $s(t)$ is not changed according to the fuzzy rule: If $s$ is $Z$, then $\mu$ is $Z$. Therefore, the FSMC still meets the requirements of sliding mode surface accessibility and closed-loop system stability.

Using the above designed controller to control the NCSs under cyber attacks and disturbance can improve the control performance.

# 4 Numerical Simulation and Discussion

The networked inverted pendulum model [17] is used to verify the effectiveness of the proposed control method. The system parameters are given by $A = \begin{bmatrix} 0 & 0 & 1\,0 \\ 0 & 0 & 0\,1 \\ 0 & 0 & 0\,0 \\ 0 & 29.431 & 0\,0 \end{bmatrix}$, $B = \begin{bmatrix} 0 \\ 0 \\ 1 \\ 3 \end{bmatrix}$, $C = \begin{bmatrix} 1\,0\,0\,0 \\ 0\,1\,0\,0 \\ 0\,0\,1\,0 \\ 0\,0\,0\,1 \end{bmatrix}$, $D = B$, $E = \begin{bmatrix} 1\,1\,1\,1 \end{bmatrix}^T$, $w(t) \in [-0.02, 0.02]$, $d_f \in [-0.28, 0.28]$, $u_a(t) = 0.5sin(5t)$ and starts from $t = 10s$.

To analyze the effect of cyber attacks and disturbance on the control performance. Figure 4 shows the comparison of control performance between networked inverted pendulum system without cyber attacks and disturbance, networked inverted pendulum system with $w(t)$, and networked inverted pendulum system with $u_a(t)$ and $w(t)$. It can be seen from Fig. 4 that NCSs under $w(t)$ and $u_a(t)$ has poor control performance.

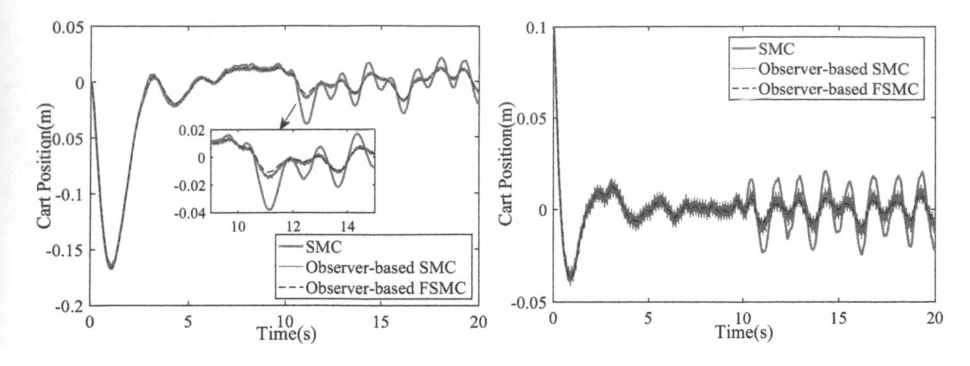


**Fig. 5.** Comparison the outputs among the system using SMC, observer-based SMC and observer-based FSMC methods.

To verify the proposed control approach can effectively improve the system control performance and decline chattering, several control methods including SMC, observer-based SMC and observer-based FSMC are operated for networked inverted pendulum system under cyber attacks and disturbance, respectively. Figure 5 shows cart position and pendulum angle of different controller approaches respectively. It can be found from Fig. 5 that: 1) Three kinds of control methods can realize the stability of networked inverted pendulum system, but the results of SMC fluctuates greatly after the system is attacked. However, the performances of observer-based SMC and observer-based FSMC are better than SMC method. 2) Comparing the control results of observer-based SMC and observer-based FSMC, it can be seen that there is obvious chattering phenomenon of cart position and pendulum angle by the observer-based SMC, and the chattering can be effectively reduced by observer-based FSMC. According to the above analysis, the proposed observer-based FSMC can achieve reasonably good effectiveness of declining chattering.

## 5 Conclusion

This paper proposes an observer-based FSMC method for NCSs under cyber attacks and disturbance. A continuous linear time-invariant model is firstly used to describe NCSs under cyber attacks and disturbance. An observer is then designed to estimate the cyber attacks and disturbance, and its convergence is proved. Furthermore, an observer-based FSMC method is employed to decline the effect of cyber attacks, disturbance, and chattering. Finally, numerical simulation is given to show the effectiveness of the proposed method. The future work is to consider unbounded attack and even some attacks with concealment.

**Acknowledgment.** This work is supported by the National Natural Science Foundation of China (92067106) and Project of Science and Technology Commission of Shanghai Municipality, China (20JC1414000, 19500712300, 19510750300).

## References

1. Wu, X.Q., Zhao, Y.J., Xu, K.X.: Nonlinear disturbance observer based sliding mode control for a benchmark system with uncertain disturbances. ISA Trans. **110**, 63–70 (2021)
2. Mo, Y.L., Sinopoli, B.: On the performance degradation of cyber-physical systems under stealthy integrity attacks. IEEE Trans. Autom. Control **61**(9), 2618–2624 (2016)
3. Wang, Q., Yu, H., Wang, M., Qi, X.: An improved sliding mode control using disturbance torque observer for permanent magnet synchronous motor. IEEE Access **7**, 36691–36701 (2019)
4. Wang, H., Pan, Y., Li, S., Yu, H.: Robust sliding mode control for robots driven by compliant actuators. IEEE Trans. Control Syst. Technol. **27**(3), 1259–1266 (2019)
5. Zaihidee, F.M., Mekhilef, S., Mubin, M.: Application of fractional order sliding mode control for speed control of permanent magnet synchronous motor. IEEE Access **7**, 101765–101774 (2019)
6. Huaman Loayza, A.S., Pérez Zuñiga, C.G.: Design of a fuzzy sliding mode controller for the autonomous path-following of a quadrotor. IEEE Latin America Trans. **17**(6), 962–971 (2019)
7. Chen, H., Fan, Y., Chen, S., Wang, L.: Observer-based event-triggered optimal control for linear systems. In: Chinese Control And Decision Conference (CCDC), pp. 1234–1238 (2019)
8. De, P.C., Tesi, P.: Input-to-state stabilizing control under denial-of-service. IEEE Trans. Autom. Control **60**(11), 2930–3944 (2015)
9. Yao, L., Peng, N., Reiter, M.K.: False data injection attacks against state estimation in electric power grids. ACM Trans. Inf. Syst. Secur. **14**(1), 1–33 (2011)
10. Wang, Z., Li, L., Sun, H., Zhu, C., Xu, X.: Dynamic output feedback control of cyber-physical systems under DoS attacks. IEEE Access **7**, 181032–181040 (2019)
11. Guan, Y., Ge, X.: Distributed attack detection and secure estimation of networked cyber-physical systems against false data injection attacks and jamming attacks. IEEE Trans. Sig. Inf. Process. Netw. **4**(1), 48–59 (2018)
12. Zhang, H., Liu, B., Wu, H.: Smart grid cyber-physical attack and defense: a review. IEEE Access **9**, 29641–29659 (2021)
13. Zhou, C., Hu, B., Shi, Y., Tian, Y.C., Li, X., Zhao, Y.: A unified architectural approach for cyberattack-resilient industrial control systems. Proc. IEEE **109**(4), 517–541 (2021)
14. Liang, G., Zhao, J., Luo, F., Weller, S.R., Dong, Z.Y.: A review of false data injection attacks against modern power systems. IEEE Trans. Smart Grid **8**(4), 1630–1638 (2017)
15. Bonnans, J.F., Tiba, D.: Equivalent control problems and applications. In: Lecture Notes in Control and Information Sciences, vol. 97, no. 1, pp. 154–161 (1987). https://doi.org/10.1007/BFb0038749
16. Gouaisbaut, F., Dambrine, M., Richard, J.P.: Robust control of delay systems: a sliding mode control, design via LMI. Syst. Control Lett. **46**(4), 219–230 (2002)
17. Du, D.J., et al.: Real-time H control of networked inverted pendulum visual servo systems. IEEE Trans. Cybern. **50**(12), 5113–5126 (2020)

# Intelligent Modelling, Monitoring, and Control

# Path Planning of Manipulator Based on Improved Informed-RRT* Algorithm

Qicheng Li, Nan Li, Zhonghua Miao, Teng Sun(✉), and Chuangxin He

School of Mechanical and Electrical Engineering and Automation, Shanghai University, Shanghai 200444, China

**Abstract.** To solve the problems of low efficiency and slow convergence of traditional RRT algorithm and RRT* algorithms, an improved informed-RRT* algorithm is proposed in this paper. The algorithm keeps the probability completeness and path optimality of RRT algorithm, improves the speed of iterative convergence and the quality of the generated path. After the final path is obtained, the problem of sharp and burr in the generated trajectory is solved by trajectory smoothing strategy. Finally, the comparison experiment shows that the performance of the proposed algorithm in three-dimensional space is better than RRT* algorithm, and the algorithm is applied to real manipulator, which verifies the feasibility of the algorithm.

**Keywords:** Informed-RRT* · Trajectory smooth · Path planning · Manipulator1

## 1 Introduction

The Manipulator is one of the earliest industrial robots, the most common of which is 6-DOF articulated manipulator. At present, 6-DOF manipulator has been widely used in various fields of industry, such as assembly, painting, welding, etc., but the second generation of traditional manipulators is still used in industry. This kind of manipulators still uses the traditional teaching method in path planning, and integrates some high-precision sensors, which can work in a slightly more complex environment. However, with the gradual application of manipulators in various fields, such as agricultural picking, medical treatment, aviation and space station, the working environment is becoming more and more complex and the functions are becoming more and more diversified. The demand for the third generation of intelligent manipulators that can automatically analyze and avoid obstacles is increasing. Correspondingly, compared with the more mature two-dimensional path planning problem, the path planning problem in high-dimensional space also needs to be studied.

Path planning problem refers to a collision free optimal path or suboptimal path from the known starting position to the target position according to a certain optimal index (such as time and path length) under certain motion constraints. At present, global path planning methods can be divided into environmental modeling method, path planning method based on graph searching and path planning method based on sampling. However, because the manipulator usually has more than six degrees of freedom, and


Q. Han et al. (Eds.): LSMS 2021/ICSEE 2021, CCIS 1469, pp. 501–510, 2021.
https://doi.org/10.1007/978-981-16-7213-2_48

the manipulator itself is in a high-dimensional space in configuration, the complexity of the graph based algorithm for the state discretization of the manipulator configuration space is too high. Therefore, in the path planning of the manipulator, the sampling based planning algorithm is usually used to avoid the discretization of the state space. In the sampling based algorithm, RRT (rapid exploring random trees) algorithm is usually used in the path planning of manipulator.

The RRT algorithm was first proposed by Lavalle [1]. The algorithm fills the whole space randomly through a random tree whose root is the starting point of the path, and finally finds a path from the starting point to the end point from the random tree. Because the RRT algorithm is based on the sampling principle, it is complete in probability first, that is, when the number of iterations of the algorithm is enough, the probability of finding a feasible solution will tend to 1 [2], so the RRT algorithm is suitable for solving the path planning problem of robots under complex and dynamic conditions [3]. But at the same time, because of the sampling characteristics of RRT algorithms, there are too many invalid searches, the generated path quality is not high, and the path solutions are quite different each time. This will lead to a large number of unnecessary movements of the manipulator, which will shorten the joint life of the manipulator. In order to solve the above problems, Karaman and Frazzoli proposed RRT* algorithm [4], which made the search path reach the asymptotic optimization by sacrificing time cost. In 2014, the Informed-RRT* [5] algorithm proposed by Gammell not only keeps the completeness of the solution, but also reduces the search area and improves the search speed. Informed-RRT* algorithm has also become an important research direction of RRT algorithms.

In addition, if the path generated by RRT algorithm is not smoothed, when the robot arm moves to the sharp point near the node, the sudden increase and sudden drop of speed and acceleration is also a challenge to the joint drive of the robot arm. In order to solve this problem, the generated path is usually smoothed, so this paper proposes to optimize the path generated by Informed-RRT* algorithm by B-spline interpolation method [6].

# 2 Analysis of RRT* and Informed-RRT*

## 2.1 Principle Analysis of RRT* Algorithm

The basic principle of RRT algorithm is as follows:

1) Generate a random search tree from the initial point, with the root vertex of the tree as the starting point.
2) Each iteration randomly generates a sampling point $X_{rand}$ in the unexplored area with a certain probability to guide the random tree growth.
3) Select a vertex closest to the sampling point in the grown random tree $X_{nearest}$.
4) Use the extended function Steer to advance from $X_{nearest}$ to $X_{rand}$ in a certain step to generate new nodes. If there is a collision, stop this growth If there is no collision, the newly generated nodes are added to the random tree.
5) If there is no collision, the newly generated vertexes are added to the random tree.

However, RRT algorithm has two problems: firstly, the search is too redundant because of the blindness of random sampling; Secondly, because of the rapidity of search, the search results are not asymptotically optimal. In order to solve the above problems, RRT* algorithm is produced. RRT* algorithm is based on RRT algorithm, adding the process of reselecting parent vertexes [10], taking the temporary node as the center, selecting a circle with radius R, selecting the node with the lowest path cost in the circle as the parent node, and reconnecting the nodes on the tree after each iteration, thus ensuring the asymptotic optimality of the solution. The pseudo code of RRT* algorithm is shown in Table 1:

**Table 1.** RRT* algorithm process.

| RRT* Algorithm |
|---|
| 1. T. init($X_{source}$) |
| 2. For i = 1 to n do |
| 3. $X_{rand}$ = Sample (X) |
| 4. $X_{nearest}$ = findnearestnode (T , $X_{rand}$) |
| 5. $X_{new}$ = Steer ($X_{nearest}$ , $X_{rand}$) |
| 6. if no collision （$X_{nearest}$ , $X_{rand}$） then |
| 7. V = add_vertex ($X_{new}$) |
| 8. E = add_egde ($X_{nearest}$ , $X_{new}$) |
| 9. $X_{near}$ = nearnodes (T , $X_{new}$ , r) |
| 10. for all ($X_{near}$ ,$X_{near}$) |
| 11. rewire ($X_{near}$ ,$X_{near}$ ) |
| 12. for all ($X_{near}$ ,$X_{new}$) |
| 13. rewire ($X_{near}$ ,$X_{new}$) |
| 14. if near_goal($X_{new}$) then |
| 15. Success ( ) |

The values to be input are starting point $X_{source}$, end point $X_{goal}$, iteration times n, search radius r and spanning tree T. The first eight steps of the algorithm are exactly the same as the RRT algorithm, except that from the ninth step to the thirteenth step, a process of reselecting the parent vertex is added. The specific method is as follows: all the nodes on the random tree in the circle with radius R around the new node are entered into $X_{near}$ (a sphere if it is a three-dimensional space), and then the path cost of all $X_{new} \rightarrow X_{near} \rightarrow X_{source}$ is iteratively calculated. Select $X_{near}$ of the shortest path to connect with $X_{new}$, and then we will check whether the cost of reaching other points in $X_{near}$ through $X_{new}$ is smaller than the original one. if so, we will remove the original path and add a new one. Figure 1(a), (b) and (c) are the Rewire process of two-dimensional plane.

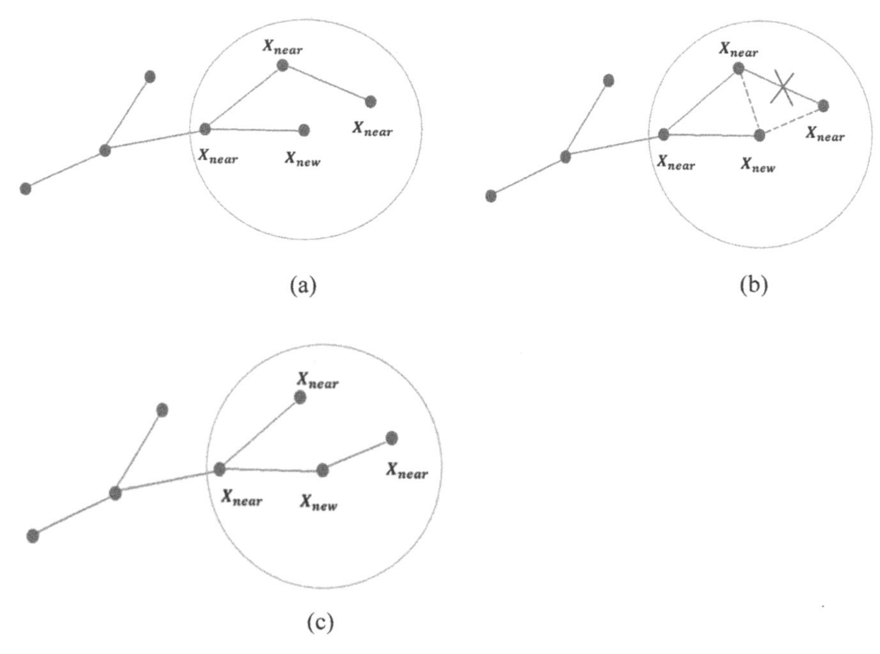


**Fig. 1.** The two-dimensional plane reselects the parent vertex process.

## 2.2 Principle Analysis of Informed-RRT * Algorithm

In order to solve the problem that RRT* algorithm wastes computing resources and has low efficiency in high-dimensional space. Gammell proposed an RRT* algorithm based on prior knowledge, called Informed-RRT* algorithm. On the basis of RRT* algorithm, the algorithm improves the sampling function Sample (). After finding the first feasible path, before each iteration, the sampling space is reduced to an ellipse (in three-dimensional space, it is a super ellipsoid), the starting point and the end point are the two focuses of the ellipse, the long axis A of the ellipse is the first solved path length $l_{best}$, the Euclidean distance $l_{min}$ from the starting point to the end point, and the short axis of the ellipse is:

$$b = \sqrt{l_{best}^2 - l_{min}^2} \tag{1}$$

The method to make the sampling point fall within the ellipse is to randomly sample a point $X_{ball}$ in a unit circle in the world coordinate system, and then use the (2):

$$X_{rand} = CLX_{ball} + X_{center} \tag{2}$$

Generate a random point $X_{rand}$ in the ellipsoid, where c is a rotation matrix, which is used to transform the world coordinate system into the elliptical coordinate system, and $L$ is a diagonal matrix, $L = diag\{r_1, r_2, r_3, \cdots, r_n\}_n$, in this paper, n is taken as 3, and $X_{center}$ is the midpoint from the starting point to the target point.

As shown in Fig. 2, we can know that the sum of the distances from the point on the boundary of the ellipse to the starting point and the end point is $l_{best}$, while any point other than the ellipse cannot generate a shorter path length. This improved method greatly improves the efficiency of progressive optimization, and the higher the dimension of search space, the more computing resources this method can save [8]. At the same time, every time we find a new path, we will update the length of $l_{best}$ again and reduce the sample space again until the end of iteration.

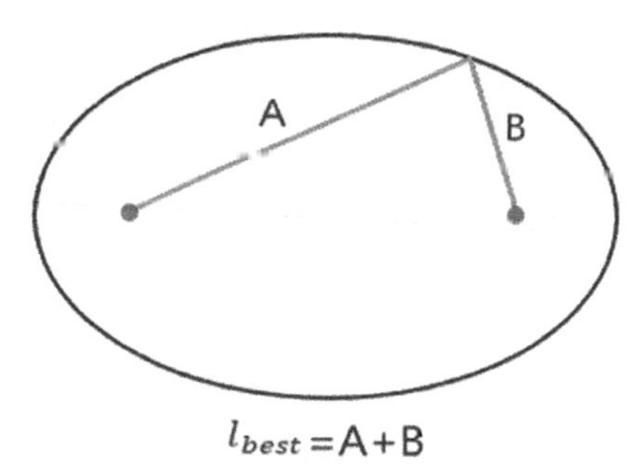


**Fig. 2.** The distance from one point to two focuses on the ellipse is the long axis, so the distance from one point outside the ellipse to two focuses is longer than $l_{best}$.

## 3 Trajectory Smoothing of Manipulator

Although the Informed-RRT* algorithm has been able to obtain a better path, when it is applied to a real robot arm, because the generated path is a time-stamped path, it is necessary to optimize the path to avoid the jitter of speed or acceleration during motion. At present, the commonly used trajectory planning method for robots is polynomial interpolation method, including cubic polynomial interpolation and quintic polynomial interpolation. The principle of interpolation method is to provide the constraint conditions of speed and acceleration, and then calculate the interpolation polynomial through the coordinate simultaneous equation of path points. However, if the complex curve is interpolated by polynomial, there will be a problem of too much calculation, and the change of single point of polynomial interpolation method will not change the interpolation result [7].

Therefore, in this paper, B-spline interpolation method is used for smoothing. The principle of B-spline interpolation method is to form polygons with type points to generate curves. The advantage of spline interpolation method is that even if one value point is changed, the whole trajectory will change greatly, which can improve the optimization accuracy.

The general form of b-spline function is as follows:

$$P(u) = \sum\nolimits_{i=0}^{n} Q_i N_{i,k}(u) \tag{3}$$

In which $Q_i = (i = 1, 2, 3, \cdots n)$ is a type-valued point, and $N_{i,k}(u) = (i = 1, 2, 3, \cdots n)$ is the canonical basis function of k-th order b-spline. meanwhile, we can find the basis function $N_{i,k}(u)$ from the de Boer recursion formula [8]:

$$\begin{cases} N_{i,0}(u) = \begin{cases} 1, (u_i \leq u \leq u_{i+1}) \\ 0 \end{cases} \\ N_{i,k}(u) = \frac{u-u_i}{u_{i+k}-u_i} N_{i,k-1}(u) + \frac{u_{i+k+1}-u}{u_{i+k+1}-u_{i+1}} N_{i+1,k-1}(u) \end{cases} \tag{4}$$

Another advantage of B-spline is that we can specify the fixed points through which the curve passes. Because the path planning of mechanical arm must pass through the starting point and the target point, we can fix the starting point and the target point in the type value point for the path planning of mechanical arm [9].

# 4 Experiments

## 4.1 Analysis of Algorithm Simulation Experiment

This experiment is run on ubuntu16.04 system by installing the corresponding configuration files and using cpp files to generate nodes. In simulation, the manipulator's end effector is considered as a point, and the attitude of the manipulator and the solution of forward and inverse kinematics are not studied, we only study the point-to-point spatial path planning problem.

Firstly, the program defines three-dimensional space in a cube of 100*100*100. At the same time, a cylinder with a radius of 10 and a height of 100 is set at the center of the cube as an obstacle. The starting point is set to (2,2,2), and the end point is set to (95,95,95), We use Rviz to visualize the simulation results. The simulation results are used to evaluate the performance of Informed-RRT* algorithm and RRT* algorithm. The two algorithms are identical in the basic code part, and the program running results are shown in Fig. 3 and Fig. 4:

We set the iterative search times of nodes to 10,000 times and the iterative search step size to 5. When the algorithm finds a feasible path, the algorithm will not stop, but will continue to find better paths. When ten better paths are found, the algorithm stops, and the shortest path output is shown in blue. From the result diagram, although the generated path is still the suboptimal path, as long as the number of iterations is enough, both algorithms can find the optimal path in theory. Also we can see that although the paths generated by the two algorithms are not much different, the random trees generated by RRT* algorithm are evenly distributed in the whole space, resulting in a lot of invalid

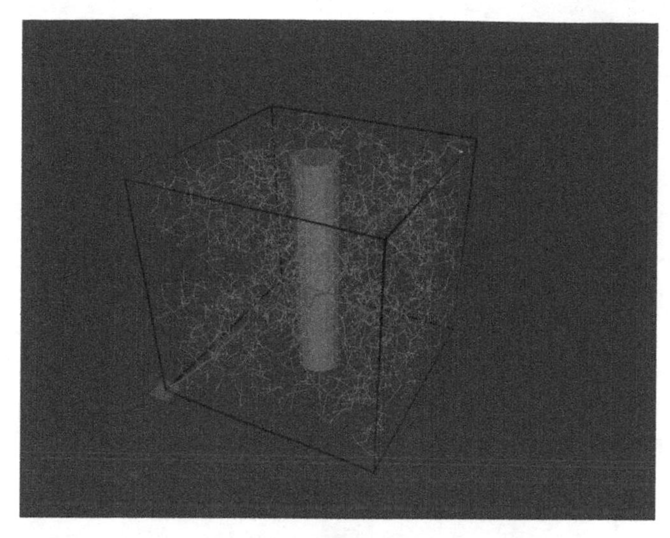

**Fig. 3.** RRT* algorithm result diagram.

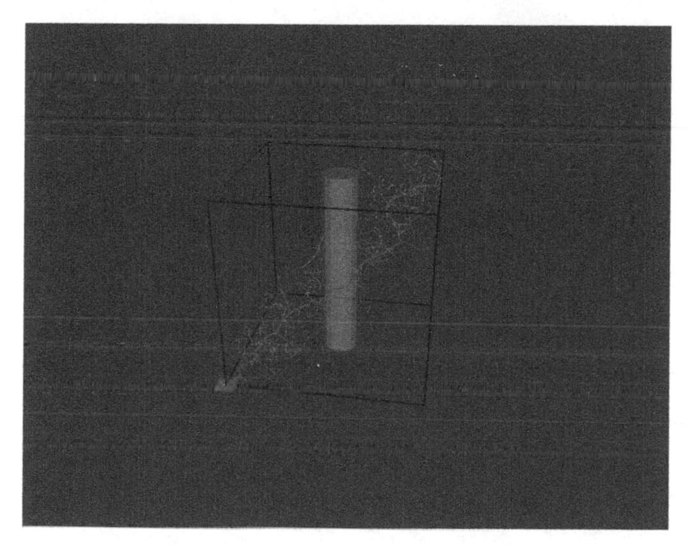

**Fig. 4.** Informed-RRT* algorithm result diagram.

searches, while the random trees generated by Informed-RRT* algorithm are basically surrounded by ellipsoids. When the algorithm ends, record the number of nodes on the random tree, the shortest path length generated and the time needed to optimize the ten paths, as shown in the Fig. 5, Fig. 6 and Fig. 7:

It can be seen from the comparison graph that although the average shortest path generated in the end is not much different, the time required to update the shortest path and the number of vertexes generated by the Informed-RRT* algorithm are much smaller than that of the RRT* algorithm, and Informed-RRT* is stable in terms of time

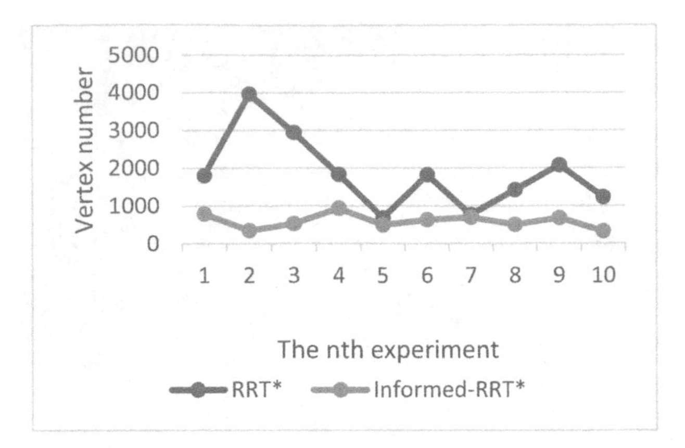


**Fig. 5.** The comparison graph of two algorithms' random tree size.

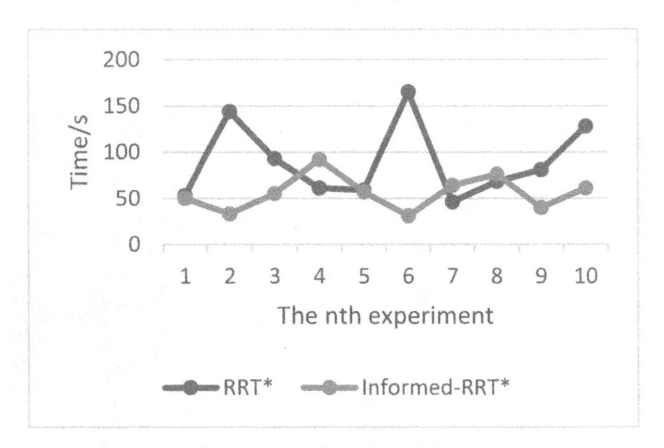


**Fig. 6.** The comparison graph of time-consuming of updating ten paths between two algorithms.

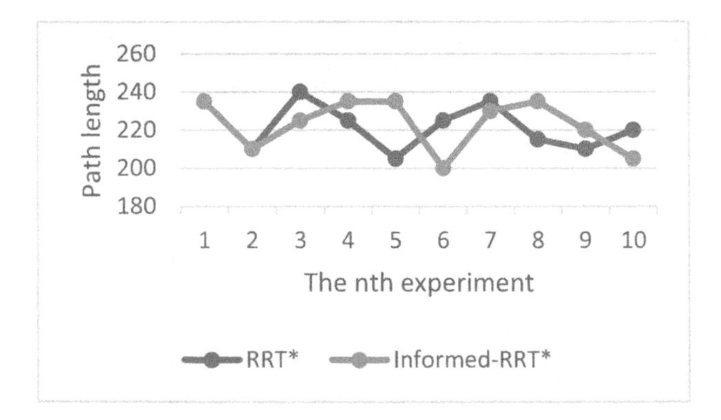


**Fig. 7.** The comparison graph of shortest path length between two algorithms.

consumption and number of nodes, while RRT* is random.Therefore, the experimental analysis shows that the performance of the Informed-RRT* algorithm is obviously better than that of the RRT* algorithm in three-dimensional state.

### 4.2 Real Manipulator Experiment

The manipulator used in this experiment is QDD-Lite series desktop robot arm. Different from traditional large industrial manipulators, this robot arm has the advantages of small size, low energy consumption, flexible joints and so on, and each joint is equipped with a separate driver, which is suitable for verifying the path planning algorithm of manipulator.

The robot arm is developed using Moveit based on ROS, Moveit integrates the functions of modeling manipulators, motion path planning, collision detection and solving forward and inverse kinematics of manipulators. The platform is connected to an open source motion planning algorithm library OMPL through the motion planning interface. Many current mainstream motion planning algorithms (such as RRT* algorithm) are integrated in the algorithm library for developers to use, but at the same time, Moveit also provides a user-defined planning algorithm interface for developers to use their own path planning algorithms.

After importing the improved algorithm into the motion planning database, we plan for motion obstacle avoidance, the manipulator smoothly bypasses obstacle without any collision, and the end moves in a relatively smooth curve. The angular displacement curve of each joint during movement is as shown in the Fig. 8.

It can be seen that the slope of the joint curve is not obviously sharp, indicating that there is no sudden increase or decrease in speed or acceleration, which meets our design requirement.

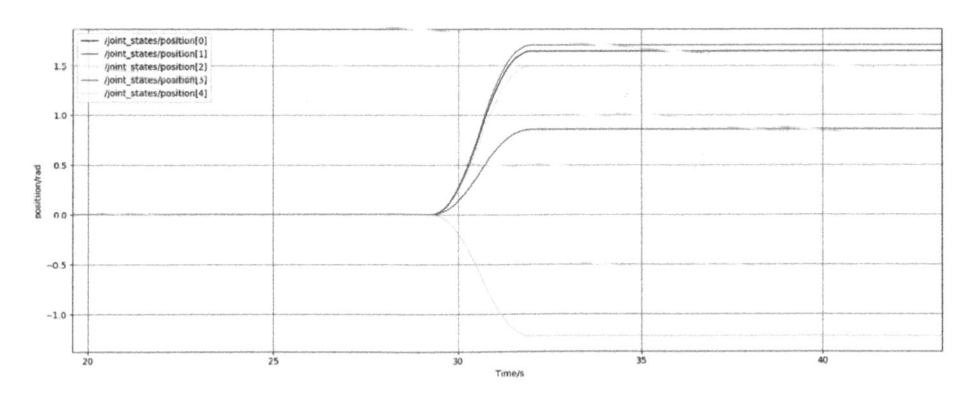


**Fig. 8.** The position of each joints.

## 5 Conclusion

In this paper, an improved Informed-RRT* algorithm is proposed to solve the problems of slow planning speed and low efficiency when the traditional RRT* algorithm is applied to the manipulator in three-dimensional space. This algorithm improves the search speed of RRT algorithm by reducing the search area, and optimizes the trajectory with B-spline interpolation function.

The algorithm is verified on the simulation platform, and the results show that the algorithm proposed in this paper is obviously superior to RRT* algorithm in search time and search efficiency when the planned path lengths are similar. Moreover, through the experimental verification, it is proved that the algorithm proposed in this paper can be effectively applied to the real manipulator, and the feasibility of the algorithm is verified.

## References


1. LaValle, S.M.: Rapidly-exploring random trees: a new tool for path planning. 98--11 (1988)
2. Kavraki, L.E., Kolountzakis, M.N., Latombe, J.C.: Analysis of probabilistic roadmaps for path planning. IEEE Trans. Robot. Autom. **14**(1), 166–171 (1998)
3. Kuffner, J., Lavalle, S.: RRT-connect: an efficient approach to single-query path planning. In: Proceedings Millennium Conference IEEE Interantional Conference Robotics Automation Symposia, vol. 2, pp. 995--1001, San Francisco (2000)
4. Karaman, S., Frazzoli, E.: Sampling-based algorithms for optimal motion planning. Int. J. Robot. Res. **30**(7), 846–894 (2011)
5. Gammell, J.D., Srinivasa, S.S., Barfoot, T.D.: Informed RRT*: Optimal sampling-based path planning focused via direct sampling of an admissible ellipsoidal heuristic. In Proceedings of IEEE/RSJ International Conference Intelligent Robots System, pp. 2997--3004 (2014)
6. Shan, E., Dai, B., Song, J., Sun, Z.: A dynamic RRT path planning algorithm based on B-Spline. In: 2009 Second International Symposium on Computational Intelligence and Design, pp. 25--29, Changsha, China (2009)
7. Koyuncu, E., Inalhan, G.: A probabilistic B-Spline motion planning algorithm for unmanned helicopters flying in dense 3d environment. In: IEEE/RSJ International Conference on Intelligent Robots and Systems, pp. 815--821 (2008)
8. Hutton, D.M.: Numerical analysis and optimization. Oxford University Press (2007)
9. Kim, M., Song, J.: Informed RRT* towards optimality by reducing size of hyperellipsoid. In: 2015 IEEE International Conference on Advanced Intelligent Mechatronics (AIM). pp. 244–248, Busan, Korea (2015)
10. Chen, L., Shan, Y., Tian, W., Li, B., Cao, D.: A fast and efficient double-tree RRT-like sampling-based planner applying on mobile robotic systems. IEEE/ASME Trans. Mech. **23**(6), 2568–2578 (2018)

# Dynamics of Cyber-Physical Systems for Intelligent Transportations

Zhuoran Chen and Dingding Han(✉)

School of Information Science and Technology, Fudan University, Shanghai 200433, China
{zrchen20,ddhan}@fudan.edu.cn

**Abstract.** Cyber-Physical System is a complex system that integrates perception, computing, communication and control to construct the mapping, interaction and coordination of human, machine, object, environment and information in cyber and physical space. In transportation cyber-physical system, the topology of road network is extremely complex, vehicles are widely distributed, and information changes rapidly. With the development of embedded systems and communication technology, interactions between vehicles and other devices are becoming faster and more reliable, making dynamic route guidance possible. This paper proposes a dynamic route guidance method named Q-learning Rerouting (QRR), which updates Q-value table by periodically acquiring real-time average travel time as penalty of actions and guides vehicles according to maximized value function strategy. Two common scenarios in intelligent transportation are built: congestion caused by vehicle failure and speed limit on waterlogged roads. Compared with static route guidance, QRR makes full use of road capacity and allows vehicles move faster.

**Keywords:** Cyber-physical system · Dynamic route guidance · Q-learning

## 1 Introduction

Cyber-physical System (CPS) has aroused extensive attention since it was proposed by National Science Foundation in 2006 [1]. CPS is a complex system that integrates the technology of sensing, computing, communication and control, constructing the mutual mapping, timely interaction and cooperation between human, machine, object, environment and information in cyber and physical space [2]. In CPS, multi-agents cooperate with each other to maintain orderly operation. For example, literature [3] constructs a multi-agent grid defense system that joins cyber criminals and defenders. Literature [4] proposes a multi-agent framework to prevent and detect CPS hardware failure. CPS has a wide range of applications such as smart grid, energy distribution, medical care, environmental monitoring, robot collaboration, industrial automation and so on [5]. Literature [6] introduces typical cases of CPS in manufacturing, electronics, automobile, aviation and other industries.

In the field of intelligent transportation, the transportation CPS (T-CPS) realizes the overall coordination of the transportation cyber system and transportation physical system through communication, interaction and adaptation [7]. Vehicles obtain information,


Q. Han et al. (Eds.): LSMS 2021/ICSEE 2021, CCIS 1469, pp. 511–520, 2021.
https://doi.org/10.1007/978-981-16-7213-2_49

estimate traffic conditions, and plan in advance by communicating with other vehicles and road side units (RSU) [8], so as to reduce congestion, and improve traffic conditions. Literature [9] proposed an on-board CPS to dynamically assess collision risk according to vehicle movement position, driver behavior and road geometry. Literature [10] put forward a CPS method for large-scale transportation based on communication between vehicles and infrastructure as well as cloud and on-board computing.

With the booming expansion of the fifth Generation of mobile communication technology (5G) and Vehicle-to-Everything (V2X), a number of test demonstration area has been established, such as Wuxi C-V2X project, which contains abnormal vehicle remind, dangerous road conditions remind, road situation remind and other V2X scenarios. Thanks to reliable and effective real-time traffic information transmission, dynamic route guidance will also become one of the key services of T-CPS [11].

Given a starting and ending pair, the easiest way to find the shortest or fastest path between two points is to use shortest path algorithm such as Dijkstra [12] or A* [13], which is adopted by traditional route guidance. However, due to the dynamic change of traffic conditions, the shortest route will not always lead to be shortest travel time. While combined with real-time traffic information, a path with shorter travel time can be found [14]. Considering that drivers' perception may be subjective and fuzzy, analytic hierarchy process based on fuzzy reasoning is used to process real-time traffic information in Literature [15]. Literature [16] actively looks for signs of congestion in road network and selects one of the five proposed lane change strategies to guide vehicles. Literature [17] proposed a dynamic route guidance method based on the maximum flow theory to balance the traffic load of the road network. The author of literature [18] believes that reinforcement learning can learn and improve performance through interaction with the environment itself, and the strategy of continuous optimization can provide solutions to traffic congestion.

In view of the above analysis, this paper proposes a Q-learning based dynamic route guidance method (QRR), which uses average travel time (ATT) of the road segment as punishment for actions, updates Q-value table and guides vehicles according to the maximization value strategy. Our approach reduces ATT and can identify abnormal scenarios based on real-time information.

## 2 Method

Q-learning is an approach in Reinforce Learning (RL) that combines Monte Carlo approach with dynamic programming [19]. The task scenario of RL is as follows: the description of machine-aware environment constitutes state space $X$ and the actions can be taken constitute action space $A$. When an action $a \in A$ acts on the current state $x \in X$, $x$ will be transferred to another state according to the potential transfer function $P$ and receive the feedback from the potential reward function $R$.

Q-learning uses the state-action value function $Q^{\pi}(x, a)$ to represent the cumulative reward for using the strategy $\pi$ after performing the action $a$ starting with state $x$. Monte Carlo method averages multiple attempts as an approximation to $Q^{\pi}(x, a)$. Suppose the estimate based on t times sampling is.

$$Q_t^{\pi}(x, a) = \frac{1}{t}\sum_{i=1}^{t} r_t. \tag{1}$$

When sampling $r_{t+1}$ at $t+1$, through incremental update, then

$$Q_{t+1}^{\pi}(x,a) = Q_t^{\pi}(x,a) + \frac{1}{t+1}\left(r_{t+1} - Q_t^{\pi}(x,a)\right). \tag{2}$$

RL can usually be described by Markov Decision Process (MDP). In MDP, the next time state of the system is determined only by the current time state, so the expression can be written as a recursive form using the idea of dynamic programming, considering the accumulate reward discount $\gamma$, and replacing $1/(1+\mathrm{t})$ with a smaller positive value $\alpha$, and then $Q^{\pi}(x,a)$ finally is updated as follows.

$$Q_{t+1}^{\pi}(x,a) = Q_t^{\pi}(x,a) + \alpha\left(R_{x\rightarrow x'}^{a} + \gamma Q_t^{\pi}(x',a') - Q_t^{\pi}(x,a)\right). \tag{3}$$

Where $x'$ is the state transferred after state $x$ performs action $a$, and $a'$ is the action selected on $x'$ based on strategy $\pi$.

Q-learning is an off-policy approach, that is, different execution strategies and evaluation strategies are adopted. The execution strategy interacts with the environment during training to generate data; the evaluation strategy learning data generated by execution strategy.

The execution strategy uses $\varepsilon$-greedy strategy.

$$\pi^{\varepsilon}(x) = \begin{cases} \frac{1}{|\mathrm{A}|}, & \epsilon; \\ argmax_{a'}Q(x,a'), & 1-\epsilon. \end{cases} \tag{4}$$

In strategy $\pi^{\varepsilon}(x)$, the probability of choosing non-optimal action is $\epsilon/|\mathrm{A}|$ and the probability of choosing current optimal action is $1-\epsilon+\epsilon/|\mathrm{A}|$, ensuring the training process exploratory, every action could be discovered and not easy to fall into local optimum.

The evaluation strategy adopts maximized value function to give full play to its utilization.

$$\pi(x) = argmax_{a''}Q(x,a''). \tag{5}$$

## 3 Model

Based on Q-learning, we propose a dynamic route guidance method Q-learning re-routing (QRR) based on real-time traffic information. We take the road where the vehicle is as the state, the choice of the next road as the action, the average travel time of action as the penalty, and reward the action of reaching the destination.

We maintain a Q-value table for each vehicle destination, update the action reward $R$ with real-time traffic information, update the Q-value table with the greed strategy, and finally select the optimal route according to the vehicle's current position with the maximized value function.

The flow of QRR is shown in Fig. 1. It can be seen that QRR realizes closed loop of perception, analysis, decision and execution in CPS.

1) Perception: we obtain vehicle's destination, vehicle's current position and ATT through the road, which is defined as follows.

$$ATT = \frac{length}{\overline{v}}. \quad (6)$$

2) Analysis: we take $ATT_j$ as the penalty of the action from any road $i$ to road $j$ and take a reward large enough for the action of reaching the destination,

$$R^a_{i \to j} = -ATT_j. \quad (7)$$

$$R^a_{i \to terminal} = reward\ big\ enough. \quad (8)$$

3) Decision: we maintain a Q-value table for each destination and update each Q-value table with the updated reward functions according to Eq. (3).
4) Execution: Based on the updated Q-value table, we replace the original path with the new route obtained according to Eq. (5).

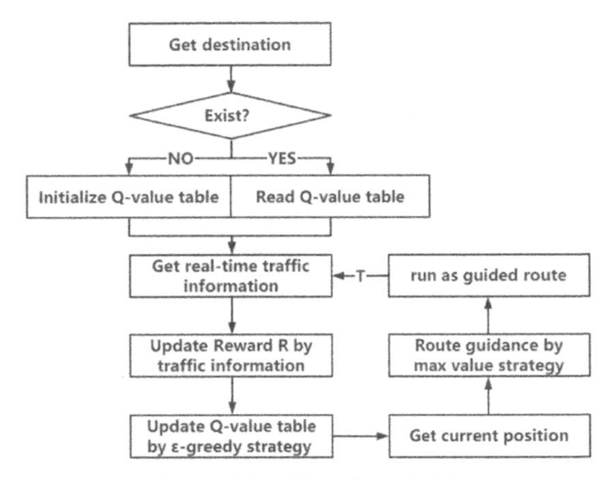


**Fig. 1.** Algorithm flow of QRR

# 4 Experiment and Results

## 4.1 Simulation Settings

Simulation of Urban Mobility (SUMO) [20] is a spatially-continuous and time-discrete microscopic traffic simulator which uses Cartesian Coordinates and updates object status at equal time intervals. SUMO also provides an API named Traffic Control Interface (TraCI) to control the simulation, we can retrieve and modify the attribute values of simulation objects such as roads, vehicles and so on.

#### 4.1.1 Guidance Settings

We compare three modes of QRR with the static guidance.

1) Static guidance: Use Dijkstra Algorithm to generate a path with the shortest time according to the information of the static road network;
2) QRR-oneShot: Once a vehicle enters the road network, conduct QRR once for it with the information of the current road network;
3) QRR-together: Conduct QRR for all vehicles in the road network at the same time with time interval T;
4) QRR-step: Conduct QRR for each vehicle from departure with time interval $\tau$.

#### 4.1.2 Discrete Event Settings

We consider the following two common scenarios of intelligent transportation:

1) Speed limit on waterlogged roads.
   We simulate the speed limit by setting the maximum permissible speed of the road. As is shown in Fig. 2, after entering the flooded road, the vehicle moves at a speed not exceeding the specified speed in the red circle.

**Fig. 2.** Smart Transportation Scene1: Speed limit on waterlogged roads

2) Congestion caused by vehicle failure.

We simulate the congestion by stopping a vehicle for a period of time. As is shown in Fig. 3, a vehicle travelling along the yellow line stops for a period of time after entering the rectangular area with a red opening.

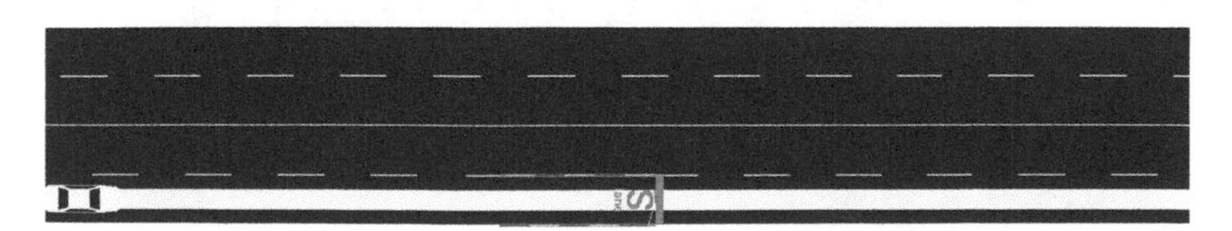

**Fig. 3.** Smart Transportation Scene2: Congestion caused by vehicle failure

### 4.1.3 Road Network

The regular road network is generated by 6 * 3 rectangles, as is shown in Fig. 4, 200 m on the long side and 100 m on the short side, single lane and no signal lights. Waterlogged segment is set in C1D and two congestions are set in B0C0 and C2D2.

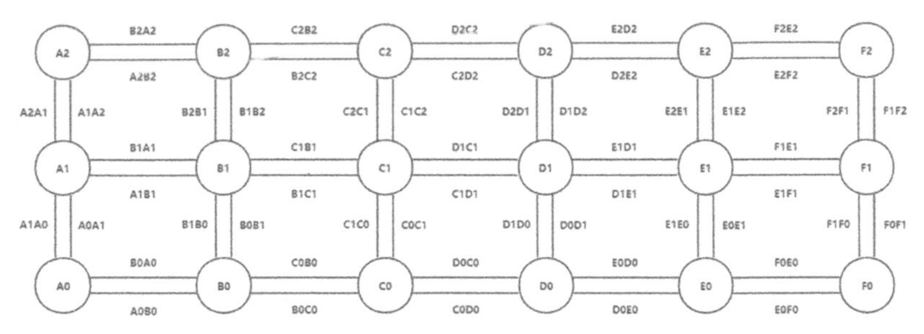


**Fig. 4.** Road network

### 4.1.4 Traffic Demand

We send 100 cars from A1B1 to E1F1 by the same time interval. The shortest path under static guidance is <A1B1 B1C1 C1D1 D1E1 E1F1>.

## 4.2 Results

### 4.2.1 QRR Functional Verification

For functional verification each vehicle is guided in QRR-oneShot mode. Vehicle trajectory is shown in Fig. 5. Each vehicle is able to choose the most reasonable path <B1C1 C1C0 C0D0> to avoid waterlogged and congested roads.

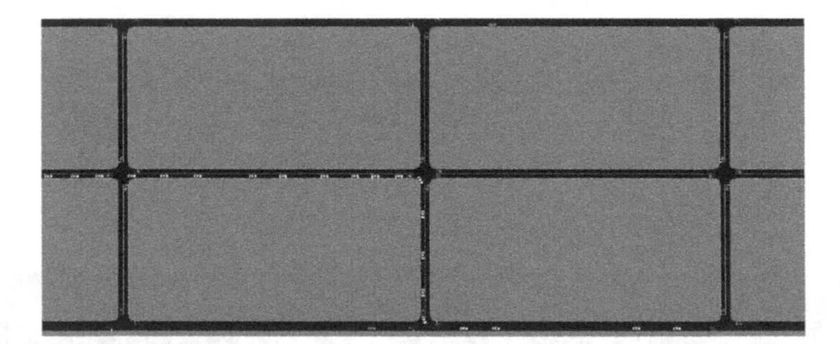

**Fig. 5.** Vehicles under QRR are able to avoid waterlogged or congested roads

In the simulation, we also find the emergence of game: $ATT_{B1C1}$ increases when vehicles going through < B1C1 C1C0 > slow down at intersection C1, making the reward of choosing B1C1 smaller than that of B1B2. Therefore, subsequent vehicles adopt game behavior and choose another way < B1B2 B2C2 C2C1 > as shown in Fig. 6(a). The lateral traffic flow from B1C1 is cut off due to the priority of the intersection as shown in Fig. 6(b). Increasing $ATT_{B1C1}$ makes more vehicles behave like 6(a) as shown in Fig. 6(c) and the game forms a positive feedback. However, when the condition of B1C1 improves, vehicles enter the lateral traffic flow again as shown in Fig. 6(d).

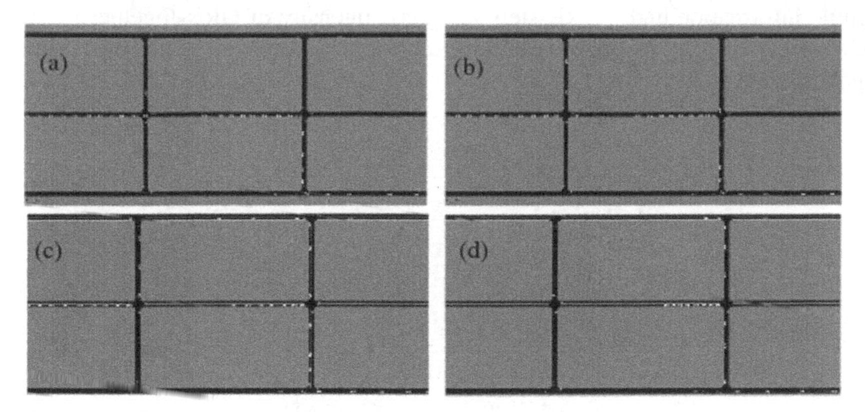


**Fig. 6.** Game under QRR-oneShot

Vehicles that adopt game behavior have a longer route length and the trend of game's positive feedback is as shown in Fig. 7.

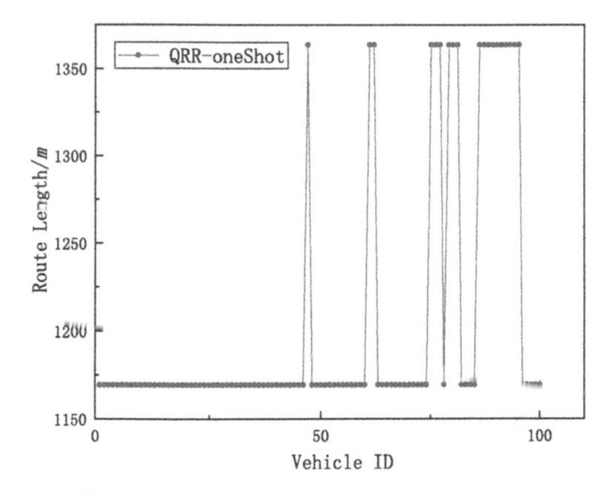


**Fig. 7.** Route Length under QRR-oneShot

### 4.2.2 QRR Modes Comparison

On the basis of road network in 4.3.1, we remove two congestions to compare four modes of guidance with a freer road network.

Firstly, we consider the effect of time interval $\tau$ in QRR-together and QRR-step. With a smaller $\tau$, guidance is more frequent and vehicle stays shorter on its original route. We take $\tau = 1, 2, 3..., 30$ for comparison according to the size of road network.

Results are shown in Fig. 8. Average route length represents the intensity of the game, which decreases with the increase of $\tau$. Due to the fact that vehicles in the same state are more possible to take the same action under QRR-together while more sensitive to traffic information under QRR-step, the game intensity of QRR-together is smaller than that of QRR-step. Average travel time and waiting time of these two modes change irregularly with $\tau$.

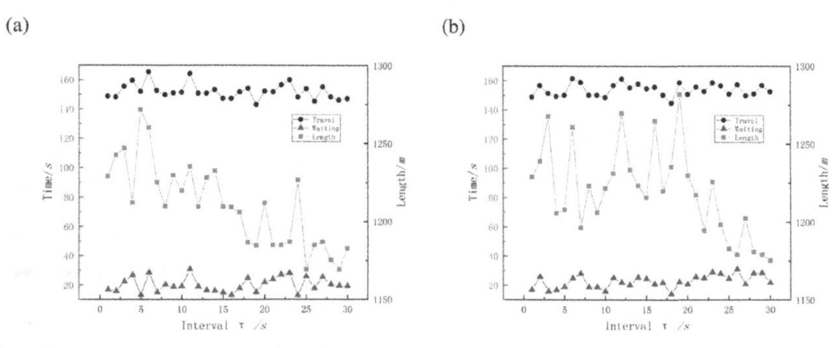


**Fig. 8.** Average travel time, waiting time and route length of different time intervals: (a) QRR-together; (b) QRR-step

Secondly, we compare the number of vehicles in the network with simulation steps. As shown in Fig. 9, QRR can significantly reduce the number of vehicles in the road network and it takes a shorter time for all vehicles to pass the road network.

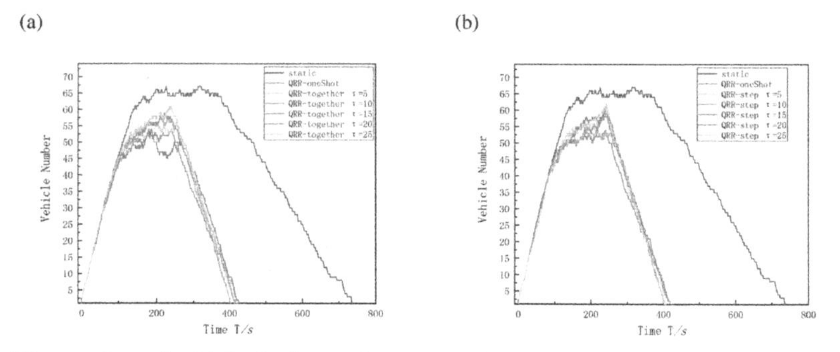


**Fig. 9.** The number of vehicles in the road network with simulation steps: (a) QRR-together; (b) QRR-step

## 5 Conclusion

In this paper, a dynamic route guidance method QRR is proposed. For two common scenarios in intelligent transportation: road congestion and waterlogged roads speed limits, QRR is able to guide vehicles to circumvent adaptively under the framework of CPS, suggesting that real-time information such as road average travel time is able to identify the special situations and help decision making. However, the resulting game behavior also partly reflects individual indicators in the long road cannot fully represent the road conditions and the reward function needs to be optimized.

Compared with static route guidance, QRR can significantly improve road capacity and make full use of road network resources. However, the road network used in the experiment is relatively simple. More realistic configuration of network should be considered, and other machine learning methods can be considered to conduct dynamic guidance for vehicles.

**Acknowledgments.** Supported by the National Key R&D Program of China under Grant No.2018YFB2101302, the National Natural Science Foundation of China under Grant No. 11875133 and Grant No.11075057.

## References

1. Lun, Y.Z., D'Innocenzo, A., Smarra, F., Malavolta, I., Di Benedetto, M.D.: State of the art of cyber-physical systems security: an automatic control perspective. J. Syst. Softw. **149**, 174–216 (2019)
2. Cyber-Physical Systems Whitepaper. http://www.cesi.cn/201703/2251.html
3. Rege, A., Ferrese, F., Biswas, S., Bai, L.: Adversary dynamics and smart grid security: a multiagent system approach. In: 7th International Symposium on Resilient Control Systems, pp. 1–7. IEEE Press, Colorado (2014)
4. Sanislav, T., Zeadally, S., Mois, G., Fouchal, H.: Multi-agent architecture for reliable Cyber-Physical Systems (CPS). In: 2017 IEEE Symposium on Computers and Communications, pp. 170–175. IEEE Press, Heraklion (2017)
5. Han, D., Ma, Y.: An Empirical Study on Topological and Dynamic Behavior of Complex Networks (in Chinese). Peking University Press, Beijing (2012)
6. China Electronics Standardization Institute: Typical application cases of Cyber-Physical Systems (CPS) (in Chinese). Publishing House of Electronics Industry, Beijing (2017)
7. Li, Y., et al.: Nonlane-discipline-based car-following model for electric vehicles in transportation-cyber-physical systems. J. IEEE Trans. Intell. Transp. Syst. **19**, 38–47 (2017)
8. Kim, K., Kumar, P.R.: Cyber–physical systems: a perspective at the centennial. J. Proc. IEEE **100**, 1287–1308 (2012)
9. Wu, C., Peng, L., Huang, Z., Zhong, M., Chu, D.: A method of vehicle motion prediction and collision risk assessment with a simulated vehicular cyber physical system. J. Transp. Res. Pt. C-Emerg. Technol. **47**, 179–191 (2014)
10. Besselink, B., et al.: Cyber–physical control of road freight transport. J. Proc. IEEE. **104**, 1128–1141 (2016)
11. Lin, J., Yu, W., Zhang, N., Yang, X., Ge, L.: Data integrity attacks against dynamic route guidance in transportation-based cyber-physical systems. J. Model. Anal. Defense. IEEE Trans. Veh. Technol. **67**, 8738–8753 (2018)
12. Dijkstra, E.W.: A note on two problems in connexion with graphs. J. Numer. Math. **1**, 269–271 (1959)
13. Hart, P.E., Nilsson, N.J., Raphael, B.: A formal basis for the heuristic determination of minimum cost paths. J. IEEE Trans. Syst. Sci. Cybern. **4**, 100–107 (1968)

14. Ding, J., Wang, C., Meng, F., Wu, T.: Real-time vehicle route guidance using vehicle-to-vehicle communication. J. IET Commun. **4**, 870–883 (2010)
15. Li, C., Anavatti, S.G., Ray, T.: Analytical hierarchy process using fuzzy inference technique for real-time route guidance system. J. IEEE Trans. Intell. Transp. Syst. **15**, 84–93 (2013)
16. Pan, J., Popa, I.S., Zeitouni, K., Borcea, C.: Proactive vehicular traffic rerouting for lower travel time. J. IEEE Trans. Veh. Technol. **62**, 3551–3568 (2013)
17. Ye, P., Chen, C., Zhu, F.: Dynamic route guidance using maximum flow theory and its MapReduce implementation. In: 14th International IEEE Conference on Intelligent Transportation Systems, pp. 180–185. IEEE Press, Washington D C (2011)
18. Koh, S.S., Zhou, B., Yang, P., Yang, Z., Fang, H., Feng, J.: Reinforcement Learning for Vehicle Route Optimization in SUMO. In: 4th International Conference on Data Science and Systems, pp. 1468–1473. IEEE Press, Exeter (2018)
19. Zhou, Z.: Machine Learing (in Chinese). Tsinghua University Press, Beijing (2016)
20. Lopez, P.A., et al.: Microscopic Traffic Simulation using SUMO. In: 21st International Conference on Intelligent Transportation Systems, pp. 2575–2582. IEEE Press, Maui (2018)

# Modeling and Active Control of Offshore Structures Subject to Input Delays and Earthquakes

Yue-Ting Sun[1], Yan-Dong Zhao[1(✉)], Xian-Li Chen[2], and Chao Li[3]

[1] College of Automation and Electronic Engineering, Qingdao University of Science and Technology, Qingdao 266061, China
ydzhao@qust.edu.cn
[2] College of Information Science and Technology, Qingdao University of Science and Technology, Qingdao 266061, China
chenxl@qust.edu.cn
[3] State Key Laboratory of Coastal and Offshore Engineering, Faculty of Infrastructure Engineering, Dalian University of Technology, Dalian 116024, China
chaoli@dlut.edu.cn

**Abstract.** This paper addresses the dynamic modeling and active controller design of offshore structures subject to control delays and earthquakes. First, by considering the time-varying control delays and earthquakes, a delayed dynamic model of the marine structure subject to earthquake is established. Then, the asymptotic stability are ensured by some sufficient conditions and the $H_\infty$ performance index of closed-loop system are derived, and the design scheme of an active damping $H_\infty$ controller for the offshore structure is presented. The simulation results show that the vibration of marine structures caused by earthquakes can be effectively reduced through the designed control scheme.

**Keywords:** Active control · Offshore structure · Time delay · $H_\infty$ control · Earthquakes

## 1 Introduction

Offshore structures are essential platforms for marine resources exploration such as oil and gas extraction from deep sea [1,2]. It is known that the offshore platforms are commonly suffered long-term action of wind, wave, current and tide load, ice, and earthquakes, which cause the platforms vibration and threaten the lives of staff. The problem of vibration reduction has become an attractive research topic. To reduce the vibration of various marine structures, there are normally four types of methods, i.e., passive control [3], active control [4,5], semi-active control [6], and hybrid control. Under the influence of external wave loads [7,8] and self-excited hydrodynamics [9,10], most of control plans are presented to improve the performance of offshore platforms.

Supported by the National Natural Science Foundation of China under Grant 61773356.


Q. Han et al. (Eds.): LSMS 2021/ICSEE 2021, CCIS 1469, pp. 521–530, 2021.
https://doi.org/10.1007/978-981-16-7213-2_50

As a common load of marine structures, earthquake can also cause vibration of offshore platforms [11–14]. To attenuate the earthquake-induced vibration of the offshore structure, several research results have been reported. Under four types of seismic ground action, a damping isolation system is developed in [11]. In [12], a decentralized sliding mode method is presented to reduce the impact of earthquakes on offshore platforms. By considering the nonlinear dynamic model of fluid solid interaction, a neural-control method is studied for the marine structure subject to earthquakes [14]. Generally speaking, time-delay exits in almost all real control systems [15–17]. However, time-delays are not considered yet in designing the active control schemes for the offshore structures affected by earthquakes, which prompted us to conduct this research.

In this paper, by considering the time-varying input delays and earthquakes simultaneously, and discusses the dynamic modeling of offshore structures and active control problem. First, a dynamic model of offshore structure subject to time-varying input delays and earthquakes is established. Then, a robust $H_\infty$ controller is designed to suppress the vibration of the structure, and the asymptotic stability and $H_\infty$ performance index of the resulting closed-loop system are obtained. Third, according to the simulation results, it is verified that the proposed active control plan is effective.

## 2 Problem Statement

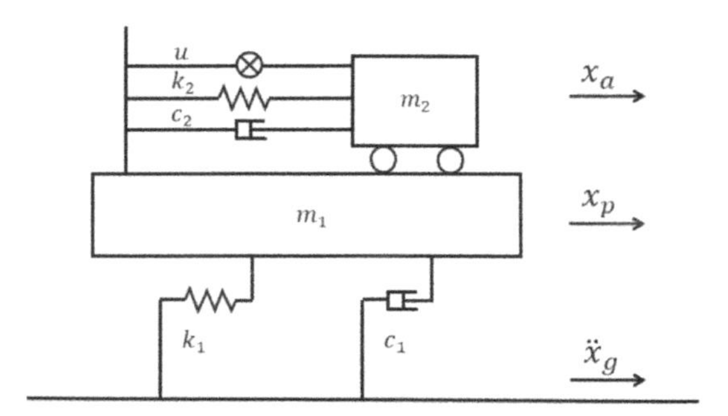


**Fig. 1.** An AMD-based offshore structure under the earthquake.

The marine platform with an active mass damper (AMD) mechanism considered in this paper is shown in Fig. 1. By considering the input delays of the system under seismic, the offshore structure model is established, which can be expressed as

$$\begin{cases} m_1\ddot{x}_p + c_1\dot{x}_p + k_1x_p - k_2(x_a - x_p) \\ -c_2(\dot{x}_a - \dot{x}_p) + m_1\ddot{x}_g + u(t-d) = 0 \\ m_2\ddot{x}_a + c_2(\dot{x}_a - \dot{x}_p) + m_2\ddot{x}_g + k_2(x_a - x_p) - u(t-d) = 0. \end{cases} \quad (1)$$

where the displacement, mass, damping and stiffness of the offshore structure and the AMD are represented by $x_p$, $m_1$, $c_1$, $k_1$, $x_a$, $m_2$, $c_2$ and $k_2$, respectively;

the earthquake acceleration is represented by $\ddot{x}_g$, the control input is denoted by $u$, and $d$ is time-varying input delay satisfying $0 \leq \tau_1 \leq d(t) \leq \tau_2$.

Let $x = \begin{bmatrix} x_1 \; x_2 \; x_3 \; x_4 \end{bmatrix}^T$, $\zeta = \ddot{x}_g$, where $x_1 = x_p$, $x_2 = x_a$, $x_3 = \dot{x}_p$, $x_4 = \dot{x}_a$. Note that $c_i = 2m_i\xi_i\omega_i$ and $k_i = m_i\omega_i^2$, where $i = 1, 2$, $\omega_i$ and $\xi_i$ are natural frequencies and damping ratios, respectively. Then, from (1), one has

$$\dot{x}(t) = Ax(t) + Bu(t - d(t)) + D_e\zeta(t) \tag{2}$$

where

$$\begin{cases} A = \begin{bmatrix} 0 & 0 & 1 & 0 \\ 0 & 0 & 0 & 1 \\ -\frac{m_1\omega_1^2+m_2\omega_2^2}{m_1} & \frac{m_2\omega_2^2}{m_1} & \frac{-2(m_1\omega_1\xi_1+m_2\omega_2\xi_2)}{m_1} & \frac{2m_2\omega_2\xi_2}{m_1} \\ \omega_2^2 & -\omega_2^2 & 2\omega_2\xi_2 & -2\omega_2\xi_2 \end{bmatrix} \\ B = \begin{bmatrix} 0 \; 0 \; -\frac{1}{m_1} \; \frac{1}{m_2} \end{bmatrix}^T, \; D_e = \begin{bmatrix} 0 \; 0 \; -1 \; -1 \end{bmatrix}^T. \end{cases} \tag{3}$$

The controlled output equation is given as

$$y(t) = C_1x(t) + W_1\zeta(t) \tag{4}$$

where $C_1$ and $W_1$ are given matrices.

To design a control law $u(t)$ as

$$u(t) = Kx(t) \tag{5}$$

where the gain matrix $K$ needs to be determined.

Designing the $H_\infty$ control law (5) is the purpose of this article such that (i) the resulting closed-loop system (2) with $\zeta = 0$ is asymptotically stable; and (ii) the closed-loop system with $H_\infty$ performance

$$\|y(t)\| \leq \gamma\|\zeta(t)\| \tag{6}$$

can be ensured for the earthquakes $\zeta(t) \in L_2[0, \infty)$ and a defined parameter $\gamma > 0$.

## 3 Design of the Active Damping $H_\infty$ Controller

In this section, the sufficient conditions for the asymptotic stability of the closed-loop system are discussed, and the active $H_\infty$ controller is presented.

From (2) and (5), the equation of the closed-loop system can be derived as

$$\dot{x}(t) = Ax(t) + BKx(t - d(t)) + D_e\zeta(t). \tag{7}$$

$x(\iota) = \varsigma(\iota)$ is the supplement of the initial condition of state $x(t)$, where $\iota \in [-\tau_2, 0]$, and $\varsigma(0) = x_0$.

Let $e_1 = [I\ 0\ 0\ 0\ 0]$, $e_2 = [0\ I\ 0\ 0\ 0]$, $e_3 = [0\ 0\ I\ 0\ 0]$, $e_4 = [0\ 0\ 0\ I\ 0]$, $e_5 = [0\ 0\ 0\ 0\ I]$, and

$$\begin{cases} \nu(t) = \mathrm{col}\left\{x(t), x(t-d(t)), x(t-\tau_1), x(t-\tau_2), \frac{1}{\tau_1}\int_{t-\tau_1}^{t} x(\iota)d\iota\right\} \\ \mu(t) = \mathrm{col}\left\{\nu(t), \zeta(t)\right\} \end{cases} \tag{8}$$

Then rewrite the closed-loop system (7) as

$$\dot{x}(t) = \Gamma\nu(t) + D_e\zeta(t) \tag{9}$$

where $\Gamma = Ae_1 + BKe_2$.

**Proposition 1.** *$\tau_1$ and $\tau_2$ are given the scalars with $\tau_2 > \tau_1 \geq 0$, and $\gamma > 0$, the closed-loop system (9) is asymptotically stable with the prescribed $H_\infty$ disturbance attenuation level $\gamma$ while $\zeta(t) = 0$, if there exist $4 \times 4$ real matrices $P_1 > 0, P_2 > 0$, $H_0 > 0, H_1 > 0, H_2 > 0$, $Q_1 > 0, Q_2 > 0, Q_3 > 0$, and a $1 \times 4$ gain matrix $K$ such that*

$$\begin{bmatrix} \Lambda & e_1^T P_1 D_e & e_1^T C_1^T & \Lambda_4^T & \Lambda_5^T & \sqrt{\tau_1}\Gamma^T H_1 & \delta\Gamma^T H_2 \\ * & -\gamma^2 I & W_1^T & 0 & 0 & \sqrt{\tau_1}D_e^T H_1 & \delta D_e^T H_2 \\ * & * & -I & 0 & 0 & 0 & 0 \\ * & * & * & -H_1 & 0 & 0 & 0 \\ * & * & * & * & -3H_1 & 0 & 0 \\ * & * & * & * & * & -H_1 & 0 \\ * & * & * & * & * & * & -H_2 \end{bmatrix} < 0 \tag{10}$$

$$\begin{bmatrix} H_2 & H_0 \\ * & H_2 \end{bmatrix} \geq 0 \tag{11}$$

*where $\Lambda = \Lambda_1 + \Lambda_2 + \Lambda_3 + \Lambda_6$, and*

$$\begin{aligned}
\Lambda_1 &= (Ae_1 + BKe_2)^T P_1 e_1 + e_1^T P_1 (Ae_1 + BKe_2) + \tau_1 e_5^T P_2 (e_1 - e_3) + \tau_1 (e_1 - e_3)^T P_2 e_5 \\
\Lambda_2 &= e_1^T Q_1 e_1 - e_3^T Q_1 e_3 + e_3^T Q_2 e_3 - e_4^T Q_2 e_4 + e_3^T Q_3 e_3 \\
\Lambda_3 &= -(e_2 - e_4)^T H_2 (e_2 - e_4) - (e_3 - e_2)^T H_2 (e_3 - e_2) \\
&\quad - (e_2 - e_4)^T H_0 (e_3 - e_2) - (e_3 - e_2)^T H_0^T (e_2 - e_4) \\
\Lambda_4 &= \sqrt{\tau_1}(M_1 e_1 + M_2 e_3 + M_3 e_5), \quad \Lambda_5 = \sqrt{\tau_1}(\aleph_1 e_1 + \aleph_2 e_3 + \aleph_3 e_5) \\
\Lambda_6 &= e_1^T (M_1 + M_1^T + \aleph_1 + \aleph_1^T) e_1 + e_3^T (-M_2 - M_2^T + \aleph_2 + \aleph_2^T) e_3 \\
&\quad + e_3^T (-M_1^T + M_2 + \aleph_1^T + \aleph_2)^T e_1 + e_1^T (-M_1^T + M_2 + \aleph_1^T + \aleph_2) e_3 \\
&\quad + e_5^T (M_3 - 2\aleph_1^T + \aleph_3)^T e_1 + e_1^T (M_3 + \aleph_3 - 2\aleph_1^T) e_5 + e_5^T (-2\aleph_3 - 2\aleph_3^T) \\
&\quad e_5 + e_5^T (-M_3 - 2\aleph_2^T + \aleph_3)^T e_3 + e_3^T (-M_3 - 2\aleph_2^T + \aleph_3) e_5.
\end{aligned}$$

*Proof.* The Lyapunov functional is chosen as:

$$V(t, x_t) = V_1(t, x_t) + V_2(t, x_t) + V_3(t, x_t) \tag{12}$$

where

$$V_1(t,x_t) = x^T(t)P_1x(t) + \int_{t-\tau_1}^{t} x^T(\iota)d\iota P_2 \int_{t-\tau_1}^{t} x(\iota)d\iota$$
$$V_2(t,x_t) = \int_{t-\tau_1}^{t} x^T(\iota)Q_1x(\iota)d\iota + \int_{t-\tau_2}^{t-\tau_1} x^T(\iota)Q_2x(\iota)d\iota + \int_{t-d(t)}^{t-\tau_1} x^T(\iota)Q_3x(\iota)d\iota$$
$$V_3(t,x_t) = \int_{-\tau_1}^{0}\int_{t+\varrho}^{t} \dot{x}^T(\iota)H_1\dot{x}(\iota)d\iota d\varrho + \delta\int_{-\tau_2}^{-\tau_1}\int_{t+\varrho}^{t} \dot{x}^T(\iota)H_2\dot{x}(\iota)d\iota d\varrho$$

with $\delta = \tau_2 - \tau_1$, $x_t = x(t+\varrho), \varrho \in [-\tau_2, 0]$, $P_1 > 0, P_2 > 0$, $H_1 > 0, H_2 > 0$, $Q_1 > 0, Q_2 > 0$ and $Q_3 > 0$.

Set $\zeta(t) \equiv 0$ in (9). Then one obtains

$$\dot{x}(t) = \Gamma\nu(t). \tag{13}$$

Denote

$$\eta_1(t) = -\int_{t-\tau_1}^{t} \dot{x}^T(\iota)H_1\dot{x}(\iota)d\iota, \quad \eta_2(t) = -\delta\int_{t-\tau_2}^{t-\tau_1} \dot{x}^T(\iota)H_2\dot{x}(\iota)d\iota. \tag{14}$$

Then, taking the derivative of the Lyapunov-Krasovskii functional (12) for $t$ along the trajectory of the system (13) gives

$$\dot{V}(t,x_t) = \dot{V}_1(t,x_t) + \dot{V}_2(t,x_t) + \dot{V}_3(t,x_t) \tag{15}$$

where

$$\dot{V}_1(t,x_t) = \nu^T(t)\Lambda_1\nu(t), \quad \dot{V}_2(t,x_t) = \nu^T(t)\Lambda_2\nu(t)$$
$$\dot{V}_3(t,x_t) = \nu^T(t)(\tau_1\Gamma^TH_1\Gamma + \delta^2\Gamma^TH_2\Gamma)\nu(t) + \eta_1(t) + \eta_2(t). \tag{16}$$

By applying Lemma 2 in [18], one yields

$$\eta_1(t) \le \nu^T(t)[V_1^T(\Omega_1 + \Psi)V_1]\nu(t) \tag{17}$$

where $V_1 = [e_1^T\ e_2^T\ e_3^T]^T$, $\Omega_1 = \tau_1(MH_1^{-1}M^T + \frac{1}{3}\aleph H_1^{-1}\aleph^T)$, $M = [M_1, M_2, M_3]^T, \aleph = [\aleph_1, \aleph_2, \aleph_3]^T$, and $\Psi = (\Psi_{ij})$ with

$$\begin{cases} \Psi_{11} = M_1^T + M_1 + \aleph_1 + \aleph_1^T, \Psi_{12} = -M_1^T + M_2 + \aleph_1^T + \aleph_2 \\ \Psi_{13} = M_3 + M_3 - 2\aleph_1^T, \Psi_{22} = -M_2 - M_2^T + \aleph_2 + \aleph_2^T \\ \Psi_{23} = -M_3 - 2\aleph_2^T + \aleph_3, \Psi_{33} = -2\aleph_3 - 2\aleph_3^T. \end{cases}$$

Notice $V_1^T\Omega_1V_1 = \Lambda_4^TH_1^{-1}\Lambda_4 + \Lambda_5^T\frac{H_1^{-1}}{3}\Lambda_5$, $V_1^T\Psi V_1 = \Lambda_6$. Then (17) can be rewritten as

$$\eta_1(t) \le \nu^T(t)(\Lambda_4^TH_1^{-1}\Lambda_4 + \Lambda_5^T\frac{H_1^{-1}}{3}\Lambda_5 + \Lambda_6)\nu(t)\ . \tag{18}$$

By applying Lemma 1 in [19], one yields $\eta_2(t) \leq \nu^T(t)\Lambda_3\nu(t)$. Then we have

$$\dot{V}(t, x_t) \leq \nu^T(t)\Sigma\nu(t) \tag{19}$$

where $\Sigma = \Lambda_1 + \Lambda_2 + \Lambda_3 + \Lambda_6 + \Lambda_4^T H_1^{-1}\Lambda_4 + \Lambda_5^T \frac{H_1^{-1}}{3}\Lambda_5 \tau_1 \Gamma^T H_1 \Gamma + \delta^2 \Gamma^T H_2 \Gamma$.

By the Schur complement $\Sigma < 0$ is equivalent to (10). Note that if $\Sigma < 0$, system (13) is asymptotically stable.

To prove that the inequality $y^T(t)y(t) < \gamma^2\zeta^T(t)\zeta(t)$ is guaranteed, taking the derivative of the Lyapunov-Krasovskii functional (15) for $t$ along the trajectory of the system (9) yields

$$\dot{V}(t, x_t) + y^T(t)y(t) - \gamma^2\zeta^T(t)\zeta(t) \leq \beta^T(t)\Xi\beta(t) \tag{20}$$

where

$$\Xi = \begin{bmatrix} \Sigma + e_1^T C_1^T C_1 e_1 & e_1^T P_1 D_e + e_1^T C_1^T W_1 + \tau_1 \Gamma^T H_1 D_e + \delta^2 \Gamma^T H_2 D_e \\ * & W_1^T W_1 - \gamma^2 I + \tau_1 D_e^T H_1 D_e + \delta^2 D_e^T H_2 D_e \end{bmatrix}. \tag{21}$$

Applying Schur complement yields

$$\begin{aligned} \Xi < 0 \Leftrightarrow & \begin{bmatrix} \Sigma & e_1^T P_1 D_e + \tau_1 \Gamma^T H_1 D_e + \delta^2 \Gamma^T H_2 D_e & e_1^T C_1^T \\ * & -\gamma^2 I + \tau_1 D_e^T H_1 D_e + \delta^2 D_e^T H_2 D_e & W_1^T \\ * & * & -I \end{bmatrix} < 0 \\ & \Leftrightarrow \Phi + \Pi < 0 \end{aligned} \tag{22}$$

where

$$\Phi = \begin{bmatrix} \Lambda + \Lambda_4^T H_1^{-1}\Lambda_4 + \Lambda_5^T \frac{H_1^{-1}}{3}\Lambda_5 & e_1^T P_1 D_e & e_1^T C_1^T \\ * & -\gamma^2 I & D_1^T \\ * & * & -I \end{bmatrix}$$

$$\Pi = \begin{bmatrix} \tau_1 \Gamma^T H_1 \Gamma + \delta^2 \Gamma^T H_2 \Gamma & \tau_1 \Gamma^T H_1 D_e + \delta^2 \Gamma^T H_2 D_e & 0 \\ * & \tau_1 D_e^T H_1 D_e + \delta^2 D_e^T H_2 D_e & 0 \\ * & * & 0 \end{bmatrix}$$

Note that the inequality (22) is equivalent to (10). Thus from (20), we have $\|y(t)\| < \gamma \|\zeta(t)\|$, $\zeta(t) \in L_2[0, \infty)$. The proof is complete.

To solve the gain matrix $K$, let

$\bar{P}_1 = P_1^{-1}, \bar{P}_2 = P_2^{-1}, \bar{K} = KP_1^{-1}, \bar{Q}_i = P_1^{-1}Q_iP_1^{-1}, \bar{M}_i = P_1^{-1}M_iP_1^{-1}, \Im = \mathrm{diag}\{P_1,\ P_1\}$,
$\bar{\aleph}_i = P_1^{-1}\aleph_iP_1^{-1},\ i = 1, 2, 3, \bar{H}_0 = P_1^{-1}H_0P_1^{-1}, \bar{H}_j = P_1^{-1}H_jP_1^{-1},\ j = 1, 2$,
$\hbar = \mathrm{diag}\{P_1,\ P_1,\ P_1,\ P_1,\ P_1\}, \pounds = \mathrm{diag}\{\hbar^{-1},\ I,\ I,\ P_1^{-1},\ P_1^{-1},\ H_1^{-1},\ H_2^{-1}\}$.

$\pounds$ and its transpose are multiplied on both sides of (10), $\Im$ and its transpose are multiplied on both sides of (11), the following proposition is proposed.

**Proposition 2.** *$\tau_1$ and $\tau_2$ are given scalars with $\tau_2 > \tau_1 \geq 0$, and $\gamma > 0$, the closed-loop system* (9) *is asymptotically stable with the prescribed $H_\infty$ disturbance attenuation level $\gamma$, if there exist $4 \times 4$ real matrices $\bar{P}_1 > 0$, $\bar{P}_2 > 0$, $\bar{H}_0 > 0$, $\bar{H}_1 > 0$, $\bar{H}_2 > 0$, $\bar{Q}_1 > 0$, $\bar{Q}_2 > 0$, $\bar{Q}_3 > 0$, and a $1 \times 4$ gain matrix $\bar{K}$ such that*

$$\begin{bmatrix} \bar{\Lambda} & e_1^T D_e & e_1^T \bar{P}_1 C_1^T & \bar{\Lambda}_4 & \bar{\Lambda}_5 & \bar{\Lambda}_7 & \bar{\Lambda}_8 \\ * & -\gamma^2 I & W_1^T & 0 & 0 & \sqrt{\tau_1} D_e^T & \delta D_e^T \\ * & * & -I & 0 & 0 & 0 & 0 \\ * & * & * & -\bar{H}_1 & 0 & 0 & 0 \\ * & * & * & * & -3\bar{H}_1 & 0 & 0 \\ * & * & * & * & * & -H_1^{-1} & 0 \\ * & * & * & * & * & * & -H_2^{-1} \end{bmatrix} < 0 \tag{23}$$

$$\begin{bmatrix} \bar{H}_2 & \bar{H}_0 \\ * & \bar{H}_2 \end{bmatrix} \geq 0 \tag{24}$$

*where $\bar{\Lambda} = \bar{\Lambda}_1 + \bar{\Lambda}_2 + \bar{\Lambda}_3 + \bar{\Lambda}_6$ and*

$$\begin{aligned}
\bar{\Lambda}_1 &= e_1^T \bar{P}_1 A^T e_1 + e_2^T \bar{K}^T B^T e_1 + e_1^T A \bar{P}_1 e_1 + e_1^T B \bar{K} e_2 \\
&\quad + \tau_1 e_5^T \bar{P}_2 (e_1 - e_3) + \tau_1 (e_1 - e_3)^T \bar{P}_2 e_5 \\
\bar{\Lambda}_2 &= e_1^T \bar{Q}_1 e_1 - e_3^T \bar{Q}_1 e_3 + e_3^T \bar{Q}_2 e_3 - e_4^T \bar{Q}_2 e_4 + e_3^T \bar{Q}_3 e_3 \\
\bar{\Lambda}_3 &= -(e_2 - e_4)^T \bar{H}_2 (e_2 - e_4) - (e_3 - e_2)^T \bar{H}_2 (e_3 - e_2) \\
&\quad - (e_2 - e_4)^T \bar{H}_0 (e_3 - e_2) - (e_3 - e_2)^T \bar{H}_0^T (e_2 - e_4) \\
\bar{\Lambda}_4 &= \sqrt{\tau_1} (\bar{M}_1 e_1 + \bar{M}_2 e_3 + \bar{M}_3 e_5), \ \bar{\Lambda}_5 = \sqrt{\tau_1} (\bar{\aleph}_1 e_1 + \bar{\aleph}_2 e_3 + \bar{\aleph}_3 e_5) \\
\bar{\Lambda}_6 &= e_1^T (\bar{\aleph}_1 + \bar{\aleph}_1^T + \bar{M}_1 + \bar{M}_1^T) e_1 + e_3^T (\bar{\aleph}_2 + \bar{\aleph}_2^T - \bar{M}_2 - \bar{M}_2^T) e_3 \\
&\quad + e_3^T (\bar{\aleph}_1^T + \bar{\aleph}_2 - \bar{M}_1^T + \bar{M}_2)^T e_1 + e_1^T (\bar{\aleph}_1^T + \bar{\aleph}_2 - \bar{M}_1^T + \bar{M}_2) e_3 \\
&\quad + e_5^T (\bar{M}_3 - 2\bar{\aleph}_1^T + \bar{\aleph}_3)^T e_1 + e_1^T (\bar{M}_3 + \bar{\aleph}_3 - 2\bar{\aleph}_1^T) e_5 + e_5^T (-2\bar{\aleph}_3 - 2\bar{\aleph}_3^T) \\
&\quad e_5 + e_5^T (-\bar{M}_3 - 2\bar{\aleph}_2^T + \bar{\aleph}_3)^T e_3 + e_3^T (-\bar{M}_3 - 2\bar{\aleph}_2^T + \bar{\aleph}_3) e_5 \\
\bar{\Lambda}_7 &= \sqrt{\tau_1} (e_1^T \bar{P}_1 A^T + e_2^T \bar{K}^T B^T), \ \bar{\Lambda}_8 = \delta (e_1^T \bar{P}_1 A^T + e_2^T \bar{K}^T B^T).
\end{aligned}$$

In addition, $K = \bar{K} \bar{P}_1^{-1}$ determines the controller gain.

## 4 Simulation

The relevant parameters of the marine structure are taken from [20]. The masses, damping ratios, and natural frequencies of the structure and the AMD are set as $m_1 = 7825307$ kg,$\xi_1 = 2\%$, $\omega_1 = 2.0466$ rad/s, $m_2 = 78253$ kg,$\xi_2 = 20\%$, $\omega_2 = 2.0074$ rad/s. Thus, the matrices in (2) can be obtained

$$A = \begin{bmatrix} 0 & 0 & 1 & 0 \\ 0 & 0 & 0 & 1 \\ -4.2290 & 0.0403 & -0.0899 & 0.0080 \\ 4.0297 & -4.0297 & 0.8030 & -0.8030 \end{bmatrix}$$

$$B = 10^{-4} \times \begin{bmatrix} 0 \ 0 \ -0.0013 \ 0.1278 \end{bmatrix}^T, \ D_e = \begin{bmatrix} 0 \ 0 \ -1 \ -1 \end{bmatrix}^T.$$

The matrices $C_1$ and $W_1$ in (4) are given as $C_1 = \begin{bmatrix} 1\,0\,0\,0 \\ 0\,0\,1\,0 \end{bmatrix}, W_1 = \begin{bmatrix} 0.5 \\ 0.5 \end{bmatrix}$. The earthquake data are taken form [21], the curve of acceleration of earthquake is shown in Fig. 2.

First, the performance of the uncontrolled offshore structure is given. The peak of the displacement, velocity, accelerate vibration amplitude are calculated with 0.3948 m, 0.8000 m/s, 1.6703 m/s$^2$, respectively. Their root mean square (RMS) values are 0.1462 m, 0.3001 m/s, 0.6374 m/s$^2$, respectively.

Then, we design an active state feedback $H_\infty$ controller (SFHC). For this, let $\gamma = 1.5$, step size $h = 0.01$ s. Set $\tau_1 = 0.01$ s and $\tau_2 = 0.06$ s. According to Proposition 2, the gain matrix $K$ of an SFHC as

$$K_{SFHC} = 10^7 \times \begin{bmatrix} 2.0204\ 0.0019\ 0.2992\ -0.0112 \end{bmatrix}.$$

When the marine platform applies the designed SFHC, the control force required and the peak and RMS values of displacement, velocity, and acceleration of the structure are listed in Table 1, and the curves of control force, responses of the structure system are depicted in Fig. 3. It is found from Table 1 and Fig. 3 that the designed SFHC can significantly reduce the vibration amplitude of offshore structures caused by earthquakes.

Now, we analyze the maximum tolerable upper bounds $\tau_{max}$ of the input delays of the marine structure. For given values of lower bounds of time-delays $\tau_1$, we compute the maximum tolerable upper bounds $\tau_{max}$. For the different values of $\tau_1$, i.e., 0, 0.01, 0.02, 0.03, and 0.04 s, the values of $\tau_{max}$ are 0.182, 0.173, 0.163, 0.152, and 0.139. This indicates that if the input delays are taken vales on the intervals as [0, 0.182], [0.01, 0.173], [0.02, 0.163], [0.03, 0.152], and [0.04, 0.139], the performance of the system can be guaranteed based on the proposed $H_\infty$ controller.

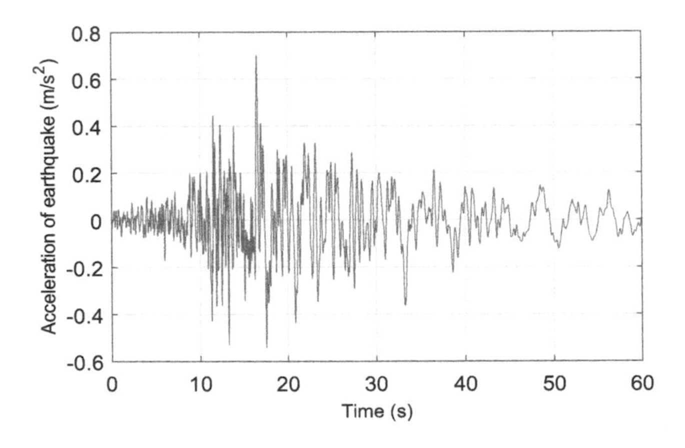


**Fig. 2.** Acceleration of earthquake.

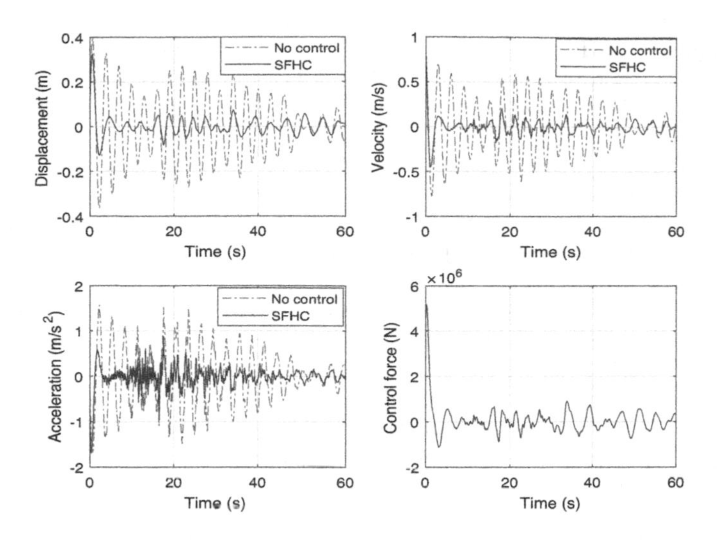


**Fig. 3.** Curves of responses and control force of the offshore structure.

**Table 1.** RMS and Peak values of system and control forces under different controllers.

| | Controllers | $x_1$ (m) | $\dot{x}_1$ (m/s) | $\ddot{x}_1$ (m/s$^2$) | $u$ ($10^6$ N) |
|---|---|---|---|---|---|
| RMS | No control | 0.1462 | 0.3001 | 0.6374 | – |
| | SFHC | 0.0483 | 0.0879 | 0.2375 | 0.6500 |
| Peak | No control | 0.3948 | 0.8000 | 1.6703 | – |
| | SFHC | 0.3254 | 0.8000 | 1.6852 | 5.1774 |

## 5 Conclusions

The vibration control issue of the marine structures affected by earthquakes has been studied in this paper. By considering the time-varying delays on the actuator, the offshore platform with delayed dynamic model has been developed. A state feedback $H_\infty$ controller has been proposed and its existence conditions has been derived based on constructing a delay-dependent Lyapunov function. The effectiveness of this control scheme has been proved by the simulation results.

## References

1. Si, Y.-L., Karimi, H.-R., Gao, H.-J.: Modelling and optimization of a passive structural control design for a spar-type floating wind turbine. Eng. Struct. **69**, 168–182 (2017)
2. Zhang, B.-L., Han, Q.-L., Zhang, X.-M., Tang, G.-Y.: Active Control of Offshore Steel Jacket Platforms. Springer, Singapore (2019). https://doi.org/10.1007/978-981-13-2986-9

3. Wang, S.-Q., Li, H.-J., Ji, C.-Y., Jiao, G.-Y.: Energy analysis for TMD-structure systems subjected to impact loading. China Ocean Eng. **16**(3), 301–310 (2002)
4. Som, A., Das, D.: Sliding mode control of fixed jacket offshore platform structures. J. Basic Appl. Eng. Res. **2**(10), 883–886 (2015)
5. Zhang, B.-L., Han, Q.-L., Zhang, X.-M.: Recent advances in vibration control of offshore platforms. Nonlin. Dyn. **89**, 755–771 (2017)
6. Wu, B., Shi, P.-F., Wang, Q.-Y., Guan, X.-C., Qu, J.-P.: Performance of an offshore platform with MR dampers subjected to ice and earthquake. Struct. Control Health Monit. **18**, 682–697 (2011)
7. Kazemy, A., Lam, J., Li, X.-W.: Finite-frequency $H_\infty$ control for offshore platforms subject to parametric model uncertainty and practical hard constraints. ISA Trans. **83**, 53–65 (2018)
8. Moharrami, M., Tootkaboni, M.: Reducing response of offshore platforms to wave loads using hydrodynamic buoyant mass dampers. Eng. Struct. **81**, 162–174 (2014)
9. Zhang, B.-L., Wei, H.-M., Cai, Z.-H., Li, Q., Tang, G.-Y.: Resilience analysis and design of event-triggered offshore steel jacket structures. Neurocomputing **400**, 429–439 (2020)
10. Ma, H., Hu, W., Tang, G.-Y.: Networked predictive vibration control for offshore platforms with random time delays, packet dropouts and disordering. J. Sound Vib. **441**, 187–203 (2019)
11. Ou, J.-P., Long, X., Li, Q.-S., Xiao, Y.-Q.: Vibration control of steel jacket offshore platform structures with damping isolation systems. Eng. Struct. **29**(7), 1525–1538 (2007)
12. Som, A., Das, D.: Seismic vibration control of offshore jacket platforms using decentralized sliding mode algorithm. Ocean Eng. **152**, 377–390 (2018)
13. Bargi, K., Hosseini, S.R., Tadayon, M.H., Sharifian, H.: Seismic response of a typical fixed jacket-type offshore platform (SPD1) under sea waves. Open J. Mar. Sci. **1**(2), 36–42 (2011)
14. Kim, D.H.: Neuro-control of fixed offshore structures under earthquake. Eng. Struct. **31**(2), 517–522 (2009)
15. He, Y., Wu, M., She, J.-H., Liu, G.-P.: Delay-dependent robust stability criteria for uncertain neutral systems with mixed delays. Syst. Control Lett. **51**(1), 57–65 (2004)
16. Park, P.: A delay-dependent stability criterion for systems with uncertain time-invariant delays. IEEE T. Automat. Contr **44**(4), 876–877 (1999)
17. Wu, L., Shi, P., Wang, C., Gao, H.: Delay-dependent robust $H_\infty$ and $L_2 - L_\infty$ filtering for LPV systems with both discrete and distributed delays. IEE P-Contr. Theor. Ap. **153**(4), 483–492 (2006)
18. Zhang, X.-M., Wu, M., She, J.-H., He, Y.: Delay-dependent stabilization of linear systems with time-varying state and input delays. Automatica **41**, 1405–1412 (2005)
19. Peng, C., Fei, M.-R.: An improved result on the stability of uncertain T-S fuzzy systems with interval time-varying delay. Fuzzy Sets Syst. **212**, 97–109 (2013)
20. Li, H.-J., Hu, S.-L.J., Jakubiak, C.: $H_2$ active vibration control for offshore platform subjected to wave loading. J. Sound Vib. **263**(4), 709–724 (2003)
21. Li, C., Hao, H., Li, H.-N., Bi, K.-M., Chen, B.-K.: Modeling and simulation of spatially correlated ground motions at multiple onshore and offshore sites. J. Earthq. Eng. **21**(3–4), 359–383 (2017)

# Design of Distributed Monitoring System for Nitrogen Generators Based on Edge Gateway

Kai Liu, Yongyan Zhong(✉), Juan Chen, Zhen Zhu, Jintian Ge, and Yuanzhu Xiang

School of Electrical Engineering, Nantong University, Nantong, Jiangsu, China
zhong.yy@ntu.edu.cn

**Abstract.** Aiming at the monitoring system for the distributed nitrogen generator devices, a remote distributed monitoring system for mine nitrogen generator devices based on the edge gateway is designed. Three parts are involved in the monitoring system, they are local devices, cloud server and monitoring side. Local devices are typical industrial control system with PLC controller and network communicator to realize remote monitoring. More than common remote monitoring system, an edge gateway is added near the local controller, helping handle data, store data and compress data, communicating with cloud server for remote monitoring. The cloud server communicates with the nitrogen generator devices with MQTT protocol. The monitoring side is programmed using JQuery and Echarts to build Web interface. Two tests were conducted to confirm the time delay and control reliability of the remote monitoring system, it showed that time delay was less than 400ms with no packet loss and success control rate is more than 99.9%.

**Keywords:** Mine nitrogen generator device · MQTT · Distributed monitoring · Edge gateway

## 1 Introduction

It has rich reserves in coal and big coal producing in China, and the producing of coal is related to the safety of miners, but the development of national economy. Due to the existence of a large amount of leftover coal in the goaf of coal mines, spontaneous combustion of leftover coal often occurs, which will seriously affects the safety production of coal mines [1, 2]. At present, method of nitrogen injection is adopted to prevent the fire in most coal mines, which relies on chemical inertia of nitrogen [3, 4]. Usually, nitrogen is injected in the form of liquid into the goaf, which can be used for prevention and fire suppression during the gasification process, absorbs a large amount of heat as well. Nitrogen is directly separated from air with pressure swing adsorption principle using nitrogen generators and then cooled and pressed into liquid nitrogen. Therefore, nitrogen generator device is an important mine equipment, and it is important to supervise the nitrogen generator device. Usually, remote monitoring system for nitrogen generators is with the client-server mode, in which clients monitoring single nitrogen generator device

This work was supported by the National Natural Science Foundation of China (62073154).


Q. Han et al. (Eds.): LSMS 2021/ICSEE 2021, CCIS 1469, pp. 531–540, 2021.
https://doi.org/10.1007/978-981-16-7213-2_51

are arranged distributed, communicate with the monitoring center, also called the main station. Nowadays, it has realized that more and more monitoring systems are based on 4G communication technology [5], which is more flexible due to the wireless and reliable connection. The communication protocol includes RS-485, field control bus, TCP and so on, and twisted cable, optical fiber, and Wi-Fi are involved in the communication medium. With remote monitoring system, status and data of nitrogen generators are transferred to the monitoring center in real time, which is helpful for maintenance, avoiding fault exists and guaranteeing the production efficiency of mining [6, 7]. However, this client-server monitoring mode is mainly centralized monitoring, it leads to disastrous consequences, once the monitoring center or any of the client node occurs fault [8].

Furthermore, for the manufacturing enterprises of nitrogen generator devices, it is suit to establish the distributed monitoring system for nitrogen generator devices, in order to achieve better remote maintenance, and then enhances the competitiveness of products, finally promotes the benign development of enterprises.

Because of the distribution of nitrogen generator devices, the idea of distributed monitoring system should be introduced. Also, with the Internet of Things and cloud-computing involved, distributed monitoring system become more easily realization. That cloud server is acted as the monitoring center, the client nodes connected the cloud server through the IoT. Monitoring system application is realized through browser, app and wechat, this mode is called B/S (browser/server) mode.

A distributed monitoring system of mine nitrogen generator devices based on edge gateway is introduced in this paper. Herein, the edge gateway is used as gateway connects the cloud server and client controller, which can handle more complex task including data storage, data reduction and data transmission.

## 2 Architecture of System

Three parts are included in the distributed monitoring system for mining nitrogen generators, which are device, cloud and monitoring side, as shown in Fig. 1.

In Fig. 1, there are two towers, tower A and tower B, in the device side, which are used for nitrogen producing separately, one for nitrogen generation with pressure desorption and one for regeneration at a time, then alternately work, produce nitrogen continuously [9, 10]. Programmable logic controller, also named PLC, is selected as controller in the system for its reliable function. The combination of Raspberry Pi 4B and touch screen is selected as the edge gateway, in which Raspberry Pi handles the data storage and data reduction, while the touch screen is responsible for the human-machine-interface and data transferring for its wireless communication function. On the one hand, the human-machine interaction of field equipment can be realized, on the other hand, the monitoring data and alarm information can be sent to the cloud server, also receives the instruction command from the cloud server [11].

The cloud server is developed with B/S architecture, which is divided into access layer, service layer and application layer. The access layer authenticates the client Nodes with the protocol of MQTT (Message Queuing Telemetry Transport). The service layer provides micro-service and micro-applications, which includes client-Node-Access, device-monitoring and alarm-service etc. The application layer consists of

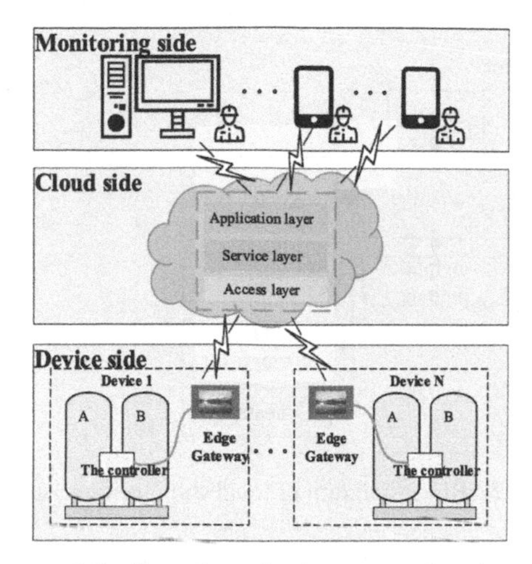


**Fig. 1.** Diagram of distributed monitoring system for nitrogen generators

several application pages, which contain real-time monitoring, historical data display, remote operation, device information and system-configuration.

Monitoring side is also based on publish/subscribe protocol mode MQTT, monitoring tools such as Browsers, APPs and applets subscribe specific topics, are serving as MQTT client; MQTT server, known as the broker, publishes data to different MQTT client according to specific topics, is established in cloud server. Monitoring interface display the information and the status of the mine nitrogen generator devices, remotely control the nitrogen generator devices as well.

# 3 Design of Local Device

Local device of the monitoring system is to monitor and control the nitrogen generator device, to communicate the cloud server in the local site. This paper mainly introduces the design of hardware of local system and the design of edge gateway.

## 3.1 Design of Local Monitoring System

The hardware of local monitoring system includes controllers, sensors and actuators. Two kinds of controller are involved in the system, one is programmable logic controller (PLC) which is for the nitrogen generator control, and the other is Raspberry Pi 4B which is for the edge gateway. Sensors include temperature sensor, pressure sensor, nitrogen concentration sensor and switches. Actuators include valves, air compressor and heater, which is driven by output module connected to the controller. The block diagram of the local monitoring system is shown in Fig. 2.

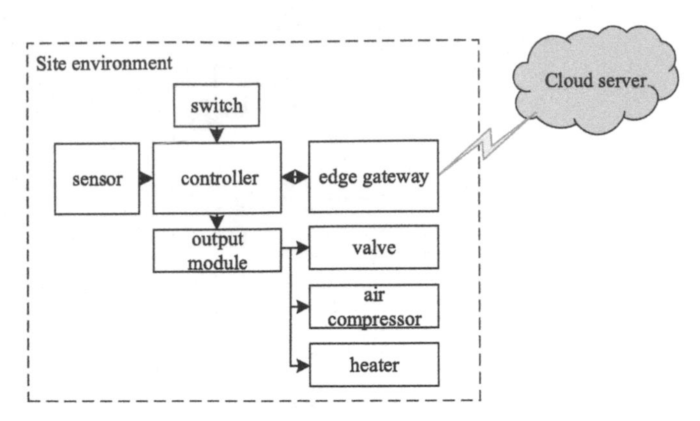


**Fig. 2.** Block diagram of local monitoring system

Siemens S7-1200 series PLC is selected as the controller, it supports communication protocol of TCP, with which PLC connects edge gateway and touch Panel. Also, the PLC supports analog input module for data acquisition of the sensors. Temperature sensor is selected as PT-100 thermal resistance, pressure sensor type is selected as GPD60, and FMA 900A type of nitrogen concentration sensor is settled in the system. Raspberry Pi 4B is chosen as the edge gateway, data acquisition, data reduction is handled by Raspberry Pi, and the data is transfer to the touch panel, which is acted as the HMI and the communicator to the cloud server.

### 3.2 Design of the Edge Gateway

The nitrogen generator device's real-time data is sent to cloud server, to show the intact and lossless information, less sampling interval, more precise information, besides, more dimensional data, more intact information. Meanwhile, large amount of data will occupy enough bandwidth and deduce time-delay as well. Data compression is the most common method to solve these problem. Furthermore, JSON (JavaScript Object Notation) is commonly used in transmission of Internet of Things, combination method of JSON and data-Compressing based IoT (abbr. JCIoT) is introduced in the paper.

The JCIoT compression method compresses JSON real-time data in two ways, which are: time stamp-based compression and data dictionary-based compression. The compression based on time stamp is realized by the data filter design, which filters the static data and the dynamic data within the reasonable fluctuation range set by the data filter, and then reduces the transmission frequency. The compression method based on data dictionary mainly relies on Huffman-coding principle, which uses short code to map the string data with high probability, and long code to map the string data with low probability, thus it will realize the content compression.

Traditionally, data encoding is based on the static dictionary, which is deduced by the experts' experience. When facing complex environment, there is usually a lack of mapping relationship between relevant strings, it is impossible to achieve efficient compression ratio, so it is necessary to establish dynamic dictionary with self-updating. Bayesian analysis algorithm is introduced in the paper to count the longer string data in the JSON string and thus predict the future reuse probability, according to the occurrence frequency by

$$P(B|A) = \frac{P(B)P(A|B)}{P(A)}. \quad (1)$$

Where A is long string events, B is high frequency string event and P(B|A) represents the probability that string is a high frequency string in the case of long string. The high probability string is encoded by the short string and added into the data dictionary.

Firstly, the long JSON string commonly used in nitrogen generator system is used as the training sample set, and the initial classifier is obtained by training through Bayesian network. Secondly, the classifier is optimized based on the string data parsed in the JSON data stream to generate a new data dictionary. Finally, the new data dictionary is synchronized with the cloud data dictionary to achieve server-client (also known as north-south) synchronization.

# 4 Design of Remote Monitoring

## 4.1 Communication Protocol Design

Generally, communication protocols in the Internet of Things include REST/HTTP, COAP, MQTT, DDS, AMQP, XMPP, and JMS. Among these, MQTT is a kind of lightweight communication protocol designed specifically for low bandwidth, high latency or unreliable networks [12]. It is based on the publish/subscribe mode, the protocol has the advantages of low power consumption, high efficiency and open-source [13, 14], which is fit for the application scenario of mining nitrogen generator device.

The MQTT broker of the monitoring system is set in the cloud server, then the edge gateway in the device side, browser, APP and applet in the monitoring side, are set as MQTT clients. The clients publish data with different MQTT topic, also these clients subscribe specific MQTT topics, therefore, it is important to design the communication topic, which can specific the communication contents.

MQTT supports hierarchical topic, which is suit for protocol parsing. In this monitoring system, the hierarchical format of MQTT theme is designed with "project Name/unit ID/device ID/ device Type/function Code". In which "project Name" is the abbreviation of the enterprise name, "unit ID" presents the working unit ID of the mining nitrogen generator device, and "device ID" presents the ID of the mining nitrogen generator device, "device Type" presents the type of the nitrogen generator device.

Different message of mining nitrogen generator device is distinguished from the arrangement of message topic "project Name/ unit ID/ device ID". Also, specific function Code message should be subscribe according to the topic of "function Code", herein "function Code" is referred as the Code of function. Table 1 lists the main MQTT topic content function codes in the paper.

The function codes RTV, STV, FTC and ACK in the table represents the message topic of publish messages on the edge gateway, which realize the data and instruction communication. Meanwhile, The edge gateway receives device control instructions by subscribing message topic of function Code CMD to realize remote control.

**Table 1.** Function code of MQTT topic

| No | Function code | Definition |
|---|---|---|
| 1 | RTV | Real-time data |
| 2 | STV | Static configuration |
| 3 | FTC | Equipment fault code |
| 4 | ACK | Control instruction reply |
| 5 | CMD | Device control instruction |

### 4.2 Design of Cloud Server

Design of cloud server includes three parts, the design of access layer, service layer and application layer. That Access layer is built with MQTT Broker as the core to realize the data access of the nitrogen processor at the side of the device. Service layer includes four parts, that are core database, instruction module, rule engine and message distribution, which realizes the basic functions of cloud server. Application layer mainly includes web application, which provides interface of visual monitoring. The cloud server architecture is shown in Fig. 3.

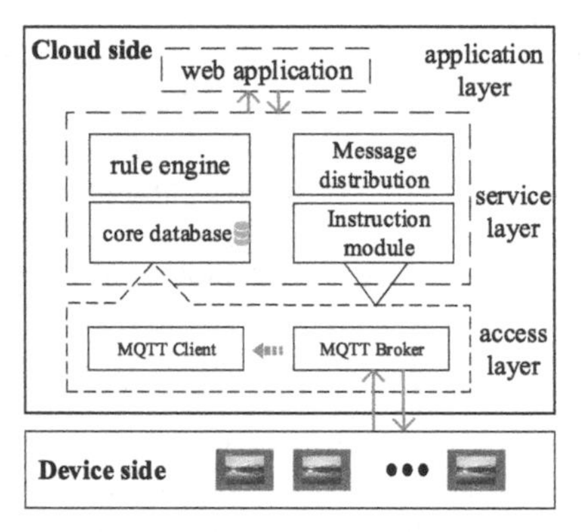


**Fig. 3.** Architecture of cloud server

In the monitoring system, EMQ is selected as MQTT broker in Access layer, which is associated with the core database through MQTT client, who is responsible for authenticating the client devices to ensure system security. Furthermore, the MQTT Client is programmed to parse the data from the device side, to store the data in the core database.

Service layer is developed using Python language to realize data storage, data transferring and remote control functions. User identification data is stored in core database, the web users accesse by the rule engine, also, their authority level determinates the level of core database, instruction module and message distribution. Besides, real-time data, historical data and instruction are also stored in core database, which will be provided to the application layer.

Application layer is designed for web application based on B/S architecture, and the interface is developed with jQuery and Echarts. By calling the interface provided by the service layer, it realizes the function of administrator login and the remote monitoring of the mining nitrogen generator devices.

## 4.3 Design of Monitoring Interface

Application layer presents visualization of data and control of nitrogen generator devices, pages include user login page, main page, real-time monitoring page, historical data display page, and device details page etc.

In the login page, user is authenticated with user name and password, then is assigned with permission level. Overall information of the nitrogen generator devices is shown in the main page, includes geographical distribution information, online status, device type, and overview of alarm. The basic information of the nitrogen generator devices is shown in the device details page, which includes type of device and device online time, real-time curves (purity, flow rate and pressure of nitrogen) are also shown in the page, as shown in Fig. 4.

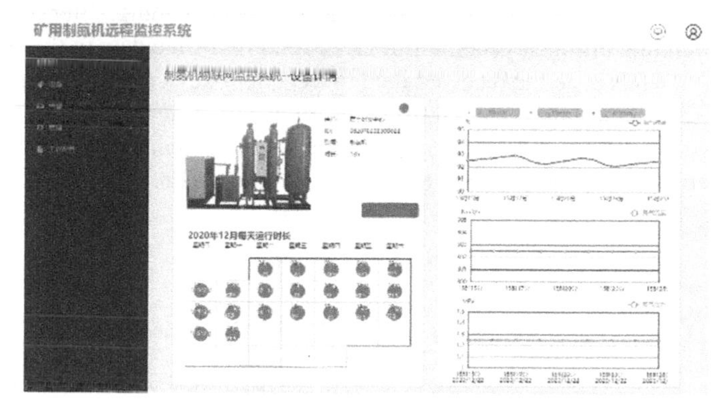


**Fig. 4.** The device details page

Real-time monitoring of the nitrogen generator devices is shown in real-time monitoring page, it shows the status of device. Also, valves are remotely switched on the page, the instructions are sent to the nitrogen generator device to realize the remote control, as shown in Fig. 5.

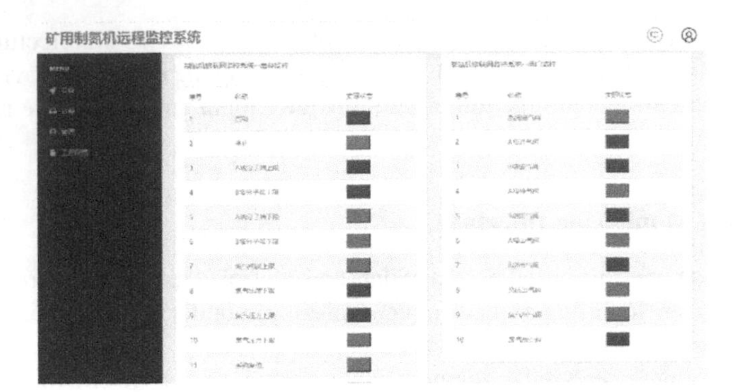


**Fig. 5.** Real-time monitoring page

# 5 Performance Evaluation

## 5.1 Time-Delay Test

To test time-delay of the monitoring system, a large number of mining nitrogen generator devices are simulated using software. The cloud server is running, after accessing the server, the simulating nitrogen generator devices send real-time data to the server per 2 s, and the server sends back the received data to the client devices. Then time-delay can be calculated from timestamp interval, to be more accurate, test will last at least 30 s, and the average time-delay is recorded, the number of the simulating nitrogen generator devices range from 100 to 10,000, the time-delay is seen in Table 2.

**Table 2.** Time-delay test

| Test group | Concurrency | Time delay (ms) |
|---|---|---|
| No. 1 | 100 | 22 |
| No. 2 | 500 | 30 |
| No. 3 | 1000 | 49 |
| No. 4 | 5000 | 125 |
| No. 5 | 10000 | 400 |

The test results show that with the increase of the concurrent number of analog devices, the communication delay of data is relatively increased. When the concurrent amount reaches 10000, time-delay is still below 400 ms, which means it basically meets the requirement of real time.

## 5.2 Reliability Test

In order to test the reliability of the remote monitoring system, especially the reliability of remote control, reliability test was conducted in the paper. Three nitrogen generator devices were taken as the samples, they are labeled as No. 1, No. 2 and No. 3. Each one was connected to the cloud server of monitoring system, Testers login the system to control the nitrogen generator device. Three groups of test were laid on the three nitrogen generator devices, switched the valves repeatedly for 100, 500 and 1000 times respectively. And the test results were recorded in Table 3.

**Table 3.** Reliablity test of control

| Test group | Test 1 (100) | Test 2 (500) | Test 3 (1000) |
|---|---|---|---|
| No. 1 | 100 | 500 | 1000 |
| No. 2 | 100 | 500 | 999 |
| No. 3 | 100 | 500 | 1000 |

Reliability test results show that when the test group is small, the control success rate is 100%, and when the test group sample increases to 1000, the success rate still reach 99.9%. so, it comes to the conclusion that the cloud server has a good control reliability.

## 6 Conclusion

In this paper, a distributed monitoring system based on edge gateway is designed for the mining nitrogen generator devices. The system can realize the real-time detection of parameters such as heating temperature, nitrogen pressure and nitrogen concentration of the mining nitrogen generator device, remote control can be done on the website or APP. More than common remote monitoring system, an edge gateway is added near the local controller, helping handle data, store data and compress data, communicating with cloud server for remote monitoring. With the monitoring system, the distributed miming nitrogen devices could be monitored on the browsers, and be checked anytime and anywhere, which provides a guarantee for the safety production of the coal mine.

## References

1. Hao Changsheng, X., Ren, Y.X., et al.: Study on prevention and fire extinguishing technology of open nitrogen injection in goaf of fully mechanized caving face. J. Coal Technol. **37**(11), 150–152 (2018)
2. Cong, Y., Mingyun, T., Ruiqing, Z., et al.: Optimization of dynamic nitrogen injection parameters in fully-mechanized mining face of Liuzhuang Mine. J. Min. Saf. Environ. Prot. **45**(04), 32–36 (2018)
3. Wenjin, Z.: Design and implementation effect analysis of three-zone prevention and control scheme of goaf spontaneous combustion. J. Coal Chem. Ind. **41**(06), 126–128 (2018)
4. Xu Shiqing, W., Huabang, T.S., et al.: Optimization simulation of fire prevention and extinguishing method of nitrogen injection in deep high gas fully mechanized mining face. J. Coal Technol. **37**(10), 234–237 (2018)
5. Guanglu, Y.: Research on coal mine safety monitoring system architecture under the environment of internet of things. J. China Equip. Eng. **23**, 111–112 (2020)

6. Zhaoyan, X., Wang Kailiang, W., Zhuofan. : Design of mine ventilator remote monitoring system based on internet of things. J. Ind. Min. Autom. **43**(01), 81–84 (2017)
7. Xiaojia, X.: Research on centralized monitoring and monitoring system of +973 horizontal belt conveyor in dongqu coal mine. J. Coal Chem. Ind. **41**(06), 81–84 (2018)
8. Wei, J.: Multi-level drainage monitoring system of coal mine based on distributed control. J. Electron. World. **19**, 152–168 (2016)
9. Huang, L., Fu, Q., Hou, P., et al.: High purification of nitrogen from PSA. J. Low Temp. Spec. Gas **30**(02), 19–22 (2012)
10. Hua, Y.U., Tianzhao, F.E.N.G.: Comparison and selection of nitrogen production technology. J. Chem. Fertilizer Des. **50**(01), 13–15 (2012)
11. Jing-feng, Y.A.N.G.: Research and application of 4G+WiFi wireless communication substation based on multi-system convergence. J. Shaanxi Coal. **39**(01), 127–130 (2020)
12. Hongli, W.A.N., Yuchen, L.I.: Design and architecture analysis of internet of things server based on MQTT protocol. J. Softw. Eng. **23**(06), 39–41 (2020)
13. Guo, L., Hu, W., Zhang, Z.: Application of MQTT protocol in internet of things. J. Comput. Knowl. Technol. **15**(28), 31–32 (2019)
14. Jiang, S., Fang, Y.: A data acquisition control system based on MQTT protocol. J. Inf. Commun. **08**, 80–82 (2019)

# Event-Based $H_\infty$ Filtering for Descriptor Systems

Lei Qing and Jinxing Lin(✉)

College of Automation and Artificial Intelligence, Nanjing University of Posts and Telecommunications, Nanjing 210023, China

**Abstract.** This article is concerned with a kind of descriptor system $H_\infty$ filtering problem based on event-triggered mechanism which is proposed to abate the limited network bandwidth in network channel. By using the method of input delay, the closed-loop system is transformed into a class of time-delay system, which is convenient to use Lyapunov function theory for stability analysis. On this basis, A sufficient criterion for a system with $H_\infty$ performance to be regular, stable and impulse free is given, and the co-design condition of the filter is be taken into account. At length a simulation is showed to demonstrate the reliability of the method.

**Keywords:** Event-triggered communication · $H_\infty$ filtering · Descriptor system

## 1 Introduction

Filtering problem is very important in the field of control. For dynamic systems with noise, the purpose of filtering is to use noisy output measurements to estimate the state of the system or their linear combination. At first, Kalman filtering was proposed to deal with the problem of filtering. However, Kalman filtering is sensitive to uncertainty of the external noise signal. If statistical characteristics of the noise are unknown, so it is difficult for us to use Kalman filter to handle. In view of this, scholars have proposed a $H_\infty$ filtering method, whose objective is to project a filter to ensure the obtained filter error system is stable, and the $L_2$ induced norm from input disturbance to filter error output meets the specified $H_\infty$ performance index. Different from the Kalman filter, the $H_\infty$ filter can ignore the statistical characteristics of the external noise.

Descriptor systems, also known as differential algebraic systems and implicit systems, are kind of more general dynamic system which are widely used in practical systems. Descriptor system includes not only the differential equations describing dynamic characteristics of the system, but also algebraic equations describing the static constraints of the system, which shows that it is a more natural and expressive mathematical model. In the past several years, a large number of research has been made on $H_\infty$ filtering for descriptor systems, see for example [5–8].

With the development of digital control system, the intermittent sampling data can be used for remote estimation. Data is sampled periodically and then transmitted periodically. In the actual networked system, the communication resources and the power


Q. Han et al. (Eds.): LSMS 2021/ICSEE 2021, CCIS 1469, pp. 541–551, 2021.
https://doi.org/10.1007/978-981-16-7213-2_52

of sensors are limited. To cut down the communication density and the use of network bandwidth, Event-triggered mechanism has gradually appeared in the vision of scholars and has been widely developed and applied. Now, scholars have proposed some event-triggered methods and among them the discrete-time triggered communication scheme [1] is widely favored. In the past several years, the research of event-based $H_\infty$ filtering has received extensive attention, and fruitful achievement have been emerged in [2–4]. Continuous systems are considered in reference [2], by introducing a new event-triggered mechanism and using the method of output time delay, the closed-loop disturbance system model with specific time-delay is obtained. At the same time, Cooperative design of communication delay and event-triggered method in a unified network system. The discrete-time system with packet loss is taken into account in [3]. By the condition of event-triggered mechanism, the packet loss is described by Bernoulli process. Under the condition of ensuring the exponential stability of the system, a new sufficient condition is given. Reference [4] deals with the switching field of normal systems.

As is known to all, the research on $H_\infty$ filtering with event-triggered mechanism is still in the shallow stage, and the existing research also has more or less deficiencies. In reference [9], the overfilting problem of a class of descriptor systems based on event-triggered is studied by using the output delay method. However, the sawtooth characteristics of the output delay are not considered in the subsequent analysis, and the stability proof of the algebraic system is not considered in the process of proving the stability which will bring some conservatism. Therefore, this paper will expand the research based on the above two matters.

This article we present the event-triggered $H_\infty$ filtering for descriptor systems. We first use the event-triggered method in [1] and the output delay method to model the system as an error perturbed system with transmission delay. Second, a Lyapunov-Krasovskii function with lower conservatism is constructed to get the $H_\infty$ performance. Then, by using the method of differential-algebraic decomposition, we decompose the descriptor system into differential system and algebraic system, and analyze the stability of algebraic system by using the iterative method in [10], finally, we give the design of the filter and simulation.

**Notations:** $R^l$ represents $l$-dimensional Euclidean space; $R^{l \times p}$ represents the set of all $l \times p$ real matrix; $Q > 0$ indicate that $Q$ is positive define; $I$ represents the identity matrix and $I_r$ means $r$ order identity matrix; $L_2[0, \infty)$ represents the space of square-integrable vector functions over $[0, \infty)$; $*$ represents the symmetric term; $\lambda_{\max} R$ and $\lambda_{\min} R$ represent the maximal and minimal eigenvalues of matrix $R$.

## 2 Problem Formulation

The descriptor system plant is represented by:

$$\begin{cases} E\dot{x}(t) = Ax(t) + Bw(t) \\ y(t) = Cx(t) + Dv(t) \\ z(t) = Lx(t) \end{cases} . \tag{1}$$

where $x(t) \in R^n$ is the state; $y(t) \in R^m$ is the measurements vector; $w(t) \in R^l$ and $v(t) \in R^q$ are external disturbance which are part of $L_2[0, \infty)$; $z(t) \in R^p$ indicate output signal which can be estimated; $E \in R^{n \times n}$ represent a singular matrix with $rank(E) = r < n$. $A,\ B,\ C,\ D,\ L$ are constant matrices.

Here, we employ a class of a filter as follows.

$$\begin{cases} \dot{x}_f(t) = A_f x_f(t) + B_f \tilde{y}(t) \\ z_f(t) = C_f x_f(t) + D_f \tilde{y}(t) \end{cases}. \tag{2}$$

where $x_f(t) \in R^n$ and $\tilde{y}(t) \in R^m$ represent the filter state and the input of the filter; $z_f(t) \in R^p$ is used to estimate $z(t)$; $A_f,\ B_f,\ C_f$ and $D_f$ are appropriately dimensioned filter matrices which will been designed.

To cut down the communication density and the use of network bandwidth, we put forward an event-triggered mechanism instead of the traditional time-triggered mechanism. Similar to [1], assuming that the sampling period is a constant. The sampling data is processed by the event generator before the transmission, and the data is transmitted in the form of a single packet in the communication channel without packet loss. This paper uses the following event- triggered function [1].

$$[y((m+o)h) - y(mh)]^T \Omega [y((m+o)h) - y(mh)] \le \sigma y^T((m+o)h) \Omega y((m+o)h). \tag{3}$$

where $h$ is the sampling cycle, $\Omega$ represents a matrix with $\Omega > 0$, $o = 1, 2, \ldots,$ and $\sigma \in [0, 1)$.

On the basis of function (3), by assuming that the released time are $t_0 h,\ t_1 h,\ t_2 h...$, where $t_0 = 0$ represents the initial time. $s_i h = t_{i+1} h - t_i h$ denoting the $ith$ release period. Considering the existence of transmission delay in the system, we define the time-varying transmission delay in the network channel as $\tau_k$ and $\tau_k \in [0, \widehat{\tau})$. $\widehat{\tau}$ is a real constant and $\widehat{\tau} > 0$. Then, the measurements $y(t_0 h),\ y(t_1 h),\ y(t_2 h)...$ will arrive at the the filter $\tilde{y}(t)$ at the instant $t_0 h + \tau_0,\ t_1 h + \tau_1,\ t_2 h + \tau_2...$, respectively. Therefore, it is easily conclude that $\tilde{y}(t) = y(t_k h)$, $t \in [t_k h + \tau_{t_k}, t_{k+1} h + \tau_{t_{k+1}})$. We can divide the interval as $[t_k h + \tau_{t_k}, t_{k+1} h + \tau_{t_{k+1}}) = \bigcup_{i=0}^{d_M} I_i$, $\widehat{\tau} = \max\{\tau_{t_k}\}$.

By using the treatment in [1], we finally get that for interval $[t_k h + \tau_{t_k},\ t_{k+1} h + \tau_{t_{k+1}})$

$$e_k^T(t) \Omega e_k(t) \le \sigma y^T(t - \tau(t)) \Omega y(t - \tau(t)). \tag{4}$$

So the filter (2) can be translated into

$$\begin{cases} \dot{x}_f(t) = A_f x_f(t) + B_f y(t - \tau(t)) + B_f e_k(t) \\ z_f(t) = C_f x_f(t) + D_f y(t - \tau(t)) + D_f e_k(t) \end{cases}. \tag{5}$$

**Define** $\overline{z}(t) = z(t) - z_f(t)$, $\zeta^T(t) = [x^T(t)\ x^T_f(t)]$, $\omega^T(t) = [w^T(t)\ v^T(t - \tau(t))]$. According to (1) and (5), we have the fully filtering error system

$$\begin{cases} \overline{E}\dot{\zeta}(t) = \overline{A}\zeta(t) + \overline{B}H\zeta(t - \tau(t)) + \overline{B}_e e_k(t) + \overline{B}_w \omega(t) \\ \overline{z}(t) = \overline{C}_1 \zeta(t) + \overline{F}H\zeta(t - \tau(t)) - \overline{D}_f e_k(t) + \overline{D}\omega(t) \end{cases}. \tag{6}$$

where

$$\overline{E} = \begin{bmatrix} E & 0 \\ 0 & I_n \end{bmatrix}, \overline{A} = \begin{bmatrix} A & 0 \\ 0 & A_f \end{bmatrix}, \overline{B} = \begin{bmatrix} 0 \\ B_f C \end{bmatrix}, H = [I_n \; 0], \overline{B}_e = \begin{bmatrix} 0 \\ B_f \end{bmatrix},$$

$$\overline{B}_w = \begin{bmatrix} B & 0 \\ 0 & B_f D \end{bmatrix}, \overline{C}_1 = [L \; -C_f], \overline{F} = -\overline{D}_f C, \overline{D} = [0 \; -\overline{D}_f D].$$

After building the above model, the $H_\infty$ filtering problem of descriptor systems under event-triggered mechanism can be transformed into designing a filter such as

1) The closed-loop system (10) with $\omega(t) = 0$ is regular, stable and impulse free.
2) Under the initial condition of zero, the filtering error $\overline{z}(t)$ satisfies $\|\overline{z}(t)\|_2 \leq \gamma \|\omega(t)\|_2$, for any nonzero $\omega(t) \in L_2[0, \infty)$, where $\gamma$ is a $H_\infty$ performance index.

# 3 Main Result

In this part, we first put forward the $H_\infty$ performance analysis of the system (6), and give the proof in the meantime. Then, the design of filter gains is proposed.

**Theorem 1.** For given scalars $\lambda > 0$, $\gamma > 0, 0 \leq \sigma < 1$ and $\alpha > 0$, system (6) is regular, impulse free and exponentially stable with a $H_\infty$ disturbance attenuation level $\gamma$ if there exist matrices $Q = Q^T > 0$, $R = R^T > 0$, $\Omega = \Omega^T > 0$ and a non-singular matrix $P$ such as

$$P^T \overline{E} = \overline{E}^T P \geq 0. \tag{7}$$

$$\coprod_0 = \Omega_1 + \sum\nolimits_{11}^T \sum\nolimits_{11} + \lambda^2 \sum\nolimits_{12}^T Q \sum\nolimits_{12}^T < 0. \tag{8}$$

$$\coprod_h = \Omega_1 + \sum\nolimits_{11}^T \sum\nolimits_{11} + \sum\nolimits_{22}^T R \sum\nolimits_{22} < 0. \tag{9}$$

where

$$\Omega_0 = \begin{bmatrix} \Omega_{11} & \Omega_{12} & P^T \overline{B}_e & P^T B_w \\ * & \Omega_{22} & 0 & \sigma C^T \Omega [0 \; D] \\ * & * & -\Omega & 0 \\ * & * & * & -\gamma^2 I + \sigma [0 \; D]^T \Omega [0 \; D] \end{bmatrix}$$

$$\Omega_1 = \begin{bmatrix} \Omega_{11} & \Omega_{12} & P^T \overline{B}_e & P^T B_w \\ * & \Omega_{22}' & 0 & \sigma C^T \Omega [0 \; D] \\ * & * & -\Omega & 0 \\ * & * & * & -\gamma^2 I + \sigma [0 \; D]^T \Omega [0 \; D] \end{bmatrix}$$

$$\Omega_{11} = P^T \overline{A} + \overline{A}^T P + 2\alpha \overline{E}^T P - e^{-2\alpha\lambda} \overline{E}^T Q \overline{E}, \; \Omega_{12} = P^T \overline{B} H + e^{-2\alpha\lambda} \overline{E}^T Q \overline{E}$$

$$\Omega_{22} = -e^{-2\alpha\lambda}\overline{E}^T Q\overline{E} - (2\alpha\lambda - 1)e^{-2\alpha\lambda}R + \sigma C^T \Omega' C$$

$$\Omega_{22}' = -e^{-2\alpha\lambda}\overline{E}^T QE - e^{-2\alpha\lambda}R + \sigma C^T \Omega' C, \sum{}_{12} = [\overline{A}\ \overline{B}H\ \overline{B}_e\ \overline{B}_w]$$

$$\sum{}_{22} = [I\ 0\ 0\ 0], \sum{}_{11} = [\overline{C}_1\ \overline{F}H\ -D_f\ \overline{D}],\ \Omega' = \begin{bmatrix} \Omega & 0 \\ 0 & 0 \end{bmatrix}.$$

**Proof.** The regular and impluse free are first proved. The proof of this part is similar to [11] and the specific steps can refer to the proof method in [11].

Through the proof in [11], we know that system (6) satisfies the condition of regular and impulse-free, so we can find two matrices $U$ and $W$ satisfying $U\overline{A}W = \begin{bmatrix} \overline{A}_1 & 0 \\ 0 & I \end{bmatrix}$, Moreover, pre-and post-multiply $\Omega_2 < 0$ by $[-\overline{B}_4^T\ I]$ and its transpose, we get

$$\overline{B}_4^T R_4 \overline{B}_4 - e^{-2\alpha\lambda} R_4 < 0$$

which implies $\rho(e^{\alpha\lambda}\overline{B}_4) < 1$, By using lemma 2 in literature [10], there exist

$$\left\| e^{i\alpha\lambda} B_4^i \right\| \leq \beta\gamma_2^i,\ i = 1, 2\cdots,\ \beta > 1,\ \gamma_2 \in (0, 1). \tag{10}$$

The above deduction will be helpful for the later proof of algebraic subsystem.

Second, we prove stability of the differential subsystem. We propose a discontinuous Lyapunov-Krasovskii function:

$$V(t) = V_1(t) + V_2(t) + V_3(t). \tag{11}$$

where

$$V_1(t) = \zeta^T(t)\overline{E}^T P\dot{\zeta}(t),\ V_2(t) = \lambda(\lambda - \tau(t)) \int_{t-\tau(t)}^{t} e^{2\alpha(s-t)}\dot{\zeta}^T(s)\overline{E}^T Q\overline{E}\dot{\zeta}(s)ds$$

$$V_3(t) = (\lambda - \tau(t))e^{-2\alpha\lambda}\zeta^T(t-\tau(t))R\zeta(t-\tau(t)) + \frac{1}{\lambda}\int_{-\tau(t)}^{0}\int_{t+\theta}^{t} e^{2\alpha(s-t)}\zeta^T(s)R\zeta(s)dsd\theta$$

Here by (3), through simple calculation (4) is equivalent to

$$\sigma\{\ [\mathrm{C}\ 0]\zeta(t-\tau(t)) + [0\ \mathrm{D}]\omega(t)\}^{\mathrm{T}}\Omega\{\ \cdots\} \geq e_k^T(t)\Omega e_k(t). \tag{12}$$

Combining (11) with (12) and using Jensen inequality, it is easily to calculate that

$$\begin{aligned}
\dot{V}(t) + 2\alpha V(t) \leq\ & 2\zeta^T(t)\overline{E}^T P\dot{\zeta}(t) + 2\alpha\zeta^T(t)\overline{E}^T P\zeta(t) + \lambda(\lambda - \tau(t))\dot{\zeta}^T(t)\overline{E}^T Q\overline{E}\dot{\zeta}(t) \\
& + \frac{\tau(t)}{\lambda}\zeta^T(t)R\zeta(t) - e^{-2\alpha\lambda}(\zeta(t) - \zeta(t-\tau(t)))^T\overline{E}^T Q\overline{E}(\zeta(t) - \zeta(t-\tau(t))) \\
& + [2\alpha(\lambda - \tau(t)) - 1]e^{-2\alpha\lambda}\zeta^T(t-\tau(t))R\zeta(t-\tau(t)) + e^T(t)e(t) - e^T(t)e(t) \\
& + \sigma\{\ [\mathrm{C}\ 0]\xi(t-\tau(t))\}^{\mathrm{T}}\Omega\{\ [\mathrm{C}\ 0]\xi(t-\tau(t))\}\ - e_k^T(t)\Omega e_k(t)
\end{aligned}$$

**Define** $\Delta(t) = [\zeta^T(t)\ \zeta^T(t-\tau(t))\ e_k^T(t)]^T$, according to the method in [8], we conclude that $\dot{V}(t) \le \Delta^T(t) \coprod'_{\tau(t)} \Delta(t)$, $\forall t \in [t_k h + \tau_{t_k}, t_{k+1}h + \tau_{t_{k+1}})$.

where

$$\coprod'_{\tau(t)} = \begin{bmatrix} \Omega_{11} & \Omega_{12} & P^T\overline{B}_e \\ * & \Omega''_{22} & 0 \\ * & * & -\Omega \end{bmatrix} + \begin{bmatrix} \overline{C}_1^T \\ H^T\overline{F}^T \\ -D_f^T \end{bmatrix} [\overline{C}_1\ \overline{F}H\ -D_f]$$

$$+ \lambda(\lambda - \tau(t)) \begin{bmatrix} \overline{A}^T \\ H^T\overline{B}^T \\ \overline{B}_e^T \end{bmatrix} Q[\overline{A}\ \overline{B}H\ \overline{B}_e] + \frac{\tau(t)}{\lambda}\xi^T R\xi(t)$$

$$\Omega''_{22} = -e^{-2\alpha\lambda}\overline{E}^T Q\overline{E} - [2\alpha(\lambda - \tau(t) - 1]e^{-2\alpha\lambda}R + \sigma C^T\Omega' C$$

Here we find when $\tau(t) \to 0$ and $\tau(t) \to h$, $\coprod'_{\tau(t)} = \frac{h-\tau(t)}{h}\coprod'_0 + \frac{\tau(t)}{h}\coprod'_h$, when $\coprod'_0 < 0$ and $\coprod'_h < 0$ hold simultaneously, we can prove that $\dot{V}(t) < 0$. $\coprod'_0 < 0$ and $\coprod'_h < 0$ can be derived from (8) and (9), so it is easily to obtain that $\dot{V}(t) + 2\alpha V(t) \le 0$, which leads to $V(t) \le e^{-2\alpha t}V(\varphi(t))$. Now we define

$$\theta(t) = \begin{bmatrix} \theta_1(t) \\ \theta_2(t) \end{bmatrix} = J^{-1}\zeta(t). \tag{13}$$

It is clear that we can get the following estimation condition:

$$\lambda_1\|\theta_1(t)\|_2 \le V(t) \le e^{-2\alpha t}V(\varphi(t)) \le \lambda_2 e^{-2\alpha t}\|\varphi\|_c^2. \tag{14}$$

This can lead to

$$\|\theta_1(t)\|_2 \le \sqrt{\frac{\lambda_2}{\lambda_1}}\|\varphi\|_c e^{-\alpha t}. \tag{15}$$

According to Definition 1 in [12], the proof is completed.

Third, we testify the stability of algebraic subsystem. Combining with (6) and (13), then using the similar treatment in [11], by defining $\overline{B}_e = [B_1\ B_2]^T$.

So we can transfrom system (6) into

$$\begin{cases} \dot{\theta}_1(t) = \overline{A}_1\theta_1(t) + \overline{B}_1\theta_1(t-\tau(t)) + \overline{B}_2\theta_2(t-\tau(t)) + B_1 e_k(t) \\ 0 = \theta_2(t) + \overline{B}_3\theta_1(t-\tau(t)) + \overline{B}_4\theta_2(t-\tau(t)) + B_2 e_k(t) \end{cases}. \tag{16}$$

By using an iterative method similar in [10], we have

$$\theta_2(t) = (-\overline{B}_4)^{k(t)}\theta_2(t_{k(t)}) - \sum_{i=0}^{k(t)-1}(-\overline{B}_4)^i\overline{B}_3\theta_1(t_{i+1}) - \sum_{i=0}^{k(t)-1}(-\overline{B}_4)^i B_2 e_{k(t)-i}(t_i)$$

Here $k(t)$ and $i$ are positive integers, Then, according to Eq. (12) we have

$$\left\|e_{k(t)-i}(t_i)\right\| \le \psi\|\theta_1(t_{i+1})\|,\ \forall i = 1, 2, \ldots. \tag{17}$$

where $\psi = \sqrt{\frac{\lambda_{\max}(\sigma C^T \Omega C)}{\lambda_{\min}\Omega}}$, Using the similar method in [10] and combining with (10), (15) and (17), it can be easily calculated.

$$\begin{aligned}\|\theta_2(t)\| &\le \left\|\overline{B}_4^{k(t)}\right\| \|\varphi\|_c + \left\|\overline{B}_3\right\| \sum_{i=0}^{k(t)-1} \left\|\overline{B}_4^i\right\| \|\theta_1(t_{i+1})\| + \psi \|B_2\| \sum_{i=0}^{k(t)-1} \left\|\overline{B}_4^i\right\| \|\theta_1(t_{i+1})\| \\ &\le [X + (\left\|\overline{B}_3\right\| + \psi \|B_2\|)\sqrt{\frac{\lambda_2}{\lambda_1}} e^{\alpha\lambda} M] \|\varphi\|_c e^{-\alpha t}\end{aligned}$$

where $X = \sqrt{\frac{\lambda_{\max} R_4}{\lambda_{\min} R_4}}$, $M = \frac{\beta}{1-\gamma_2}$. According to Definition 1 in [12], the proof is completed.

Last, we prove the $H_\infty$ performance. For nonzero $\omega(t) \in L_2[0, \infty)$, according to the part 2 and defining $\Delta_2(t) = [\,\zeta^T(t)\ \zeta^T(t-\tau(t))\ e_k^T(t)\ \omega^T(t)\,]^T$. We have

$$\begin{aligned}&\dot{V}(t) + 2\alpha V(t) \le \frac{h-\tau(t)}{h} \Delta_2^T \coprod_0 \Delta_2 + \frac{\tau(t)}{h} \Delta_2^T \coprod_h \Delta_2 - \overline{z}^T(t)\overline{z}(t) + \gamma^2 \omega^T(t)\omega(t) \\ &\le -\overline{z}^T(t)\overline{z}(t) + \gamma^2 \omega^T(t)\omega(t).\end{aligned} \tag{18}$$

Integrating both sides of (18) from $(0, +\infty)$, we get

$$V(+\infty) - V(0) \le \int_0^{+\infty} [-\overline{z}^T(t)\overline{z}(t) + \gamma^2 \omega^T(t)\omega(t)] d(t). \tag{19}$$

For $\omega(t) \in L_2[0, \infty)$, Under zero initial condition $V(0) = 0$ and $V(+\infty) \ge 0$, it is easily to obtain $\int_0^{+\infty} \overline{z}^T(t)\overline{z}(t)dt \le \int_0^{+\infty} \omega^T(t)\omega(t)dt$. Which implies that $\|\overline{z}(t)\|_2 \le \gamma \|\omega(t)\|_2$, so we completed the proof.

**Remark 1.** It can be seen from the definition of (5) that the time-delay $\tau(t)$ is a piecewise linear function satisfying sawtooth characteristics, and its derivative is equal to 1. However, these characteristics are not considered in [9], which will bring some conservatism. Motivated by the idea in [11], we propose an exponential Lyapunov-Krasovskii functional which uses the sawtooth characteristic of $\tau(t)$.

**Remark 2.** The stability proof of singular system includes not only the proof of differential subsystem, but also the proof of algebraic subsystem. However, literatures [9] do not consider the proof of algebraic subsystem, which is not rigorous. In Theorem 1, by using an iterative method, we give the proof of the stability of the algebraic subsystem.

**Remark 3.** Due to the existence of singular matrix in descriptor system, there will exit equality constraints in filter design, which will bring some numerical problems to solve LMIs. To solve these problems, we refer to the ideas in [11], by finding a matrix $P_1 > 0$ and $S_1$ which is a matrix with full column rank and satisfies $E^T S_1 = 0$. So, $P$ in (7) can be substituted for $P_1 \overline{E} + S\Theta$.

**Theorem 2.** For given parameters $\lambda > 0,\ \gamma > 0, 0 \le \sigma < 1$ and $\alpha > 0$. There exists a filter in the form of (2) such that the closed-loop system (6) under the event-triggered scheme (3) is solvable, if there exist matrices $P_{11} > 0, U > 0, Q_{11} > 0, \Omega > 0, R_{11} > 0$ and $M_j (j = 1,2,\ 3,\ 4)$, $\Theta_1$ with appropriate dimensions, $S_1$ is represented in *remark 3* such that

$$U - P_{11} < 0. \tag{20}$$

$$\begin{bmatrix} X_{11} & X_{12} & X_{13} & M_2 & X_{15} & M_2D & X_{17} & L^T \\ * & X_{22} & X_{23} & M_2 & U^TB & M_2D & 0 & -M_3^T \\ * & * & X_{33} & 0 & 0 & X_{36} & 0 & -C^TM_4^T \\ * & * & * & -\Omega & 0 & 0 & 0 & -M_4^T \\ * & * & * & * & -\gamma^2 I_l & 0 & X_{57} & 0 \\ * & * & * & * & * & X_{55} & 0 & -D^TM_4^T \\ * & * & * & * & * & * & -\lambda^2 Q_{11} & 0 \\ * & * & * & * & * & * & * & -I \end{bmatrix} < 0. \tag{21}$$

$$\begin{bmatrix} Y_{11} & X_{12} & X_{13} & M_2 & X_{15} & M_2D & L^T \\ * & X_{22} & X_{23} & M_2 & U^TB & M_2D & -M_3^T \\ * & * & Y_{33} & 0 & 0 & X_{36} & -C^TM_4^T \\ * & * & * & -\Omega & 0 & 0 & -M_4^T \\ * & * & * & * & -\gamma^2 I_l & 0 & 0 \\ * & * & * & * & * & X_{66} & -D^TM_4^T \\ * & * & * & * & * & * & -I \end{bmatrix} < 0. \tag{22}$$

where

$$X_{11} = (P_{11}E + S_1\Theta_1)^T A + A^T(P_{11}E + S_1\Theta_1) + 2\alpha E^T P_{11}E \ - e^{-2\alpha\lambda} E^T Q_{11} E$$

$$X_{12} = M_1 + A^T U + 2\alpha E^T U,\ X_{13} = M_2 C + e^{-2\alpha\lambda} E^T Q_{11} E$$

$$X_{15} = (P_{11}E + S_1\Theta_1)^T B,\ X_{17} = \lambda^2 A^T Q_{11},\ X_{22} = M_1 + M_1^T + 2\alpha U,\ X_{23} = M_2 C$$

$$X_{33} = -e^{-2\alpha\lambda}[E^T Q_{11} E - R_{11} + 2\alpha\lambda R_{11}] + \sigma C^T \Omega C,\ X_{36} = \sigma C^T \Omega D$$

$$X_{57} = \lambda^2 B^T Q_{11},\ X_{66} = -\gamma^2 I_q + \sigma D^T \Omega D,\ Y_{33} = -e^{-2\alpha\lambda}[E^T Q_{11} E + R_{11}] + \sigma C^T \Omega C$$

$$Y_{11} = (P_{11}E + S_1\Theta_1)^T A + A^T(P_{11}E + S_1\Theta_1) + 2\alpha E^T P_{11} - e^{-2\alpha\lambda} E^T Q_{11} E + R_{11}$$

We can design the filter parameters as follows:

$$A_f = M_1 U^{-1},\ B_f = M_2,\ C_f = M_3 U^{-1},\ D_f = M_4$$

**Proof.** Define $Q = \begin{bmatrix} Q_{11} & 0 \\ 0 & Q_{22} \end{bmatrix}$, $P_1 = \begin{bmatrix} P_{11} & P_{12} \\ * & P_{22} \end{bmatrix}$, $R = \begin{bmatrix} R_{11} & 0 \\ 0 & R_{22} \end{bmatrix}$, $\Theta = [\,\Theta_1\ \Theta_2\,]$.

Substituting $P$ in (8) into $P_1\overline{E} + S\Theta$, It can be seen from simple calculation that $P_1 = P_1^T > 0$, by using schur complement we can get (20). Then by defining

$$Y_1 = diag\{I\ P_{12}^{-T}P_{22}\ I\ I\ I\ I\ I\ I\},\ Y_2 = diag\{I\ P_{12}^{-T}P_{22}\ I\ I\ I\ I\ I\}$$

and using the method in [13], we can easily get (21) and (22).

## 4 An Illustrative Example

For the sake of ensure the availability of this method. we give the descriptor system in (1) with

$$E = \begin{bmatrix} 1\,0\,0 \\ 0\,1\,0 \\ 0\,0\,1 \end{bmatrix},\ A = \begin{bmatrix} -3 & 1 & 0 \\ 0.3 & -2.5 & 2 \\ -0.1 & 0.3 & -3.8 \end{bmatrix},\ B = \begin{bmatrix} 1 \\ 0 \\ 1 \end{bmatrix},\ C = \begin{bmatrix} 0.8\ 0.3\ 0 \end{bmatrix},$$
$$D = 0.2,\ L = \begin{bmatrix} 0.1\ -1\ 0.2 \end{bmatrix}$$

letting $\alpha = 0.1$, $\lambda = 0.2, \gamma = 0.3$ and $S = \begin{bmatrix} 0\ 0\ 1 \end{bmatrix}^T$, by using Theorem 2, we get

$$A_f = \begin{bmatrix} -17.5680 & -3.7125 & -7.2390 \\ 4.6520 & -6.0957 & 9.0852 \\ -10.7101 & 0.6332 & -17.6369 \end{bmatrix},\ B_f = \begin{bmatrix} -0.0999 \\ -0.0432 \\ -0.0033 \end{bmatrix},$$
$$C_f = \begin{bmatrix} 2.2015\ 1.5847\ -0.2213 \end{bmatrix},\ D_f = 0.0036.$$

Then the event-triggered parameters can be calculated as $\Omega = 0.9680$, Assume the external disturbance is $w(t) = \begin{cases} 0.7\sin(1.5\pi t), & if\ t \in [1.5, 8] \\ 0, & otherwise \end{cases}$.

Under a initial condition $x(t) = \begin{bmatrix} 0.3\ 0.1\ 0 \end{bmatrix}^T$, the response of $z(t)$ and $z_f(t)$ are indicated in Figs. 1 and 2 shows the data release instants and release intervals. We can clearly find that the system of substitution is stable.

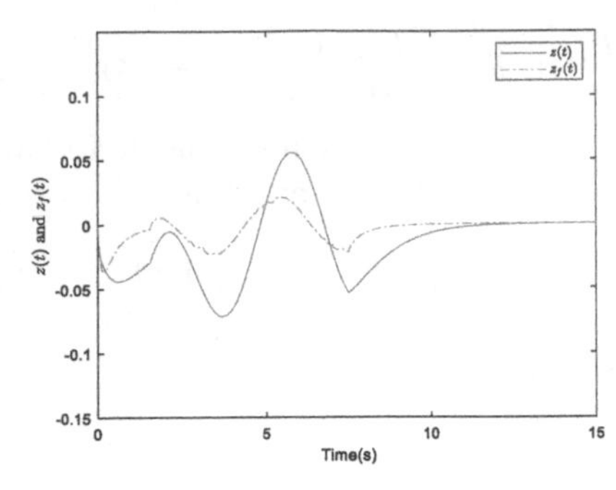

**Fig. 1.** The output $z(t)$ and $z_f(t)$.

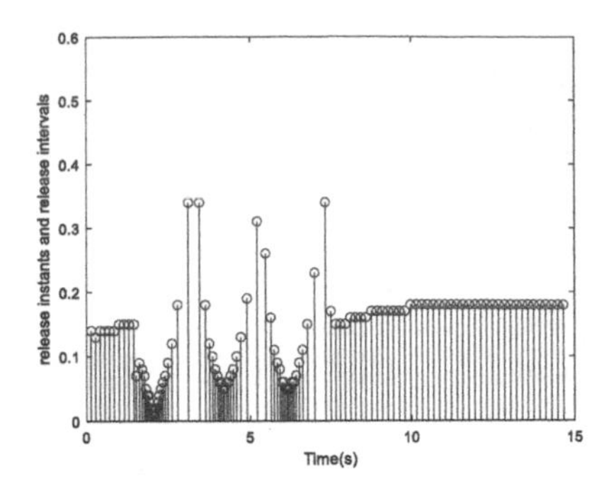


**Fig. 2.** The release instants and intervals.

# 5 Conclusion

In this article, the event-triggered $H_\infty$ filtering for descriptor system has been taken into account. By considering the sawtooth structure of time-delay and costructing the Lyapunov-krasovskii functional, a suffcient condition in terms of LMI has been obtained to make sure that the singular closed-loop system is admissible and have a $H_\infty$ performance. In the mean time, we give a collaborative design idea of the filter gain parameters. Finally, we show a simulation to indicate the availability of the method which we propose in this article.

# References

1. Yue, D., Tian, E.G., Han, Q.L.: A delay system method for designing event-triggered controls of networked control systems. J. IEEE Trans. Automat. Control **58**(2), 475–481 (2013)
2. Hu, S.L., Yue, D.: Event-based filtering for networked system with communication delay. J. Signal Process. **92**(9), 2029–2039 (2012)
3. Zhou, L., Pan, R., Xiao, X.Q., Sun, T.T.: Event-triggered filtering for distrete-time systems over unreliable networks with package dropouts. J. Neurocomput. **218**(19), 346–353 (2016)
4. Xiao, X.Q., Park, J.H., Zhou, L.: Event-triggered filtering of distrete-time switched linear systems. J. ISA Trans. **77**, 112–121 (2018)
5. Wu, Z., Su, H., Chu, J.: $H\infty$ filter design for singular time-delay systems. Proc. Inst. Mech. Eng. Part I: J. Syst. Control Eng. **223**(7), 1001–1015 (2009)
6. Wang, G.L., Xu, S.Y.: Robust $H\infty$ filtering for singular time-delay systems with uncertain Markovian switching probabilities. Int. J. Robust Nonl. Control **25**(3), 376–393 (2013)
7. Wu, Z.G., Su, H.Y., Xue, A.K.: Delay-dependent filtering for singular markovian jump time-delay system. J. Signal Process. **90**, 1240–1248 (2010)
8. Yue, D., Han, Q.L.: Robust $H\infty$ filter design of uncertain descriptor systems with discrete and distributed delays. J. IEEE Trans. Automat. Control **52**(11), 3200–3212 (2004)
9. Song, W.X., Sun, J., Chen, J.: Event-based $H\infty$ filtering for networked singular systems IEEE. In: Proceedings of the 35th Chinese Control Conference, pp. 7503–7507. IEEE, Chengdu (2016)
10. Haidar, A., Boukas, E.K., Xu, S., Lam, J.: Exponential stability and static output feedback stabilization of singular time-delay systems with saturating actuators. J. IET Control Theory Appl. **3**(9), 1293–1305 (2009)

11. Xiao, X.Q., Sun, T.T., Lu, G.P., Zhou, L.: Event-triggered stabilization for singular systems based on sampled-data. In: Proceedings of the 33rd Chinese Control Conference, pp. 5690–5694. IEEE, Nanjing (2014)
12. Wu, Z.G., Su, H.Y., Chu, J.: $H\infty$ filtering for singular systems with time-varying delay. Int. J. Robust Nonl. Control **20**(11), 1269–1284 (2009)
13. Zhang, X.M., Han, Q.L.: Robust $H\infty$ filtering for a class of uncertain linear systems with time-varying delay. J. Automat. **44**, 157–166 (2008)

# Modeling Thermal Sensation Prediction Using Random Forest Classifier

Linyi Jin[1], Tingzhang Liu[1], and Jing Ma[2](✉)

[1] School of Mechatronic Engineering and Automation, Shanghai University Shanghai, Shanghai, China
jly23@formix.com, liutzhcom@oa.shu.edu.cn

[2] Department of Computer Science and Software Engineering, Auckland University of Technology, Auckland, New Zealand
jing.ma@aut.ac.nz

**Abstract.** Indoor thermal comfort has been getting more attention while in growing demand. To achieve the optimal coordination of energy consumption and resident comfort, thermal sensation predition models are needed to guide HVAC (heating, ventilation and air conditioning) settings. This article is to develop a new thermal sensation vote (TSV) prediction model using random forest classifier, and experiments are carried out in data preprocessing, feature engineering, hyperparameter selection, training ratio and other aspects to improve the performance of the prediction model. After optimization, the final accuracy of prediction model reaches 67.1% for 7-point TSV, 72.7% for 5-point TSV and 98.6% for 3-point TSV. Compared with similar experiments conducted on ASHRAE PR-884, the model developed in this article has promising results, which proves the effectiveness of the optimization methods.

**Keywords:** Thermal sensation prediction · Random forest classifier · Thermal sensation vote

## 1 Introduction

People spend up to 80% of their time indoors [1]. The comfort of the indoor environment directly affects their working efficiency [2]. Heating, ventilation and air conditioning (HVAC) systems can directly change the indoor thermal environment and meet people's demands for thermal comfort. However, deviations in people's perception of the comfortable temperature range can lead to an uncomfortable environment and waste of energy, that is, the indoor temperature is set too high or too low. Therefore, a thermal sensation prediction model is required to advise the related parameters, so as to make the indoor environment comfortable.

With the development of machine learning, researchers have widely applied adaptive methods to thermal sensation prediction [3]. The self-adaptive characteristics of machine learning methods and the complexity of comfort prediction allow machine learning methods to achieve good results in thermal sensation


Q. Han et al. (Eds.): LSMS 2021/ICSEE 2021, CCIS 1469, pp. 552–561, 2021.
https://doi.org/10.1007/978-981-16-7213-2_53

prediction. Researches show that compared to the PMV model, almost every ML algorithm was able to achieve better prediction performance [4].

As an integrated learning method with superior performance, random forest is often used for thermal comfort modeling. Siliang Lu etc. developed a thermal comfort model with RP-884 based on k-nearest neighbor (KNN), random forest (RF), and support vector machine (SVM) [5]. In the absence of feature selection and hyperparameter adjustment, the prediction accuracy of KNN, SVM and RF was 49.3%, 48.7% and 48.7%. Tanaya Chaudhuri et al. [4] used random forest to model gender-specific physiological parameters obtained from wearable sensors. The results showed significant sex differences in a number of subjective and physiological responses. The accuracy of male and female thermal state prediction was 92.86% and 94.29%. Previous work has hinted at the importance of model feature selection, parameter selection and data set partitioning in the process of modeling. Therefore, in order to find the best predictive model, this article further studies the influence of feature selection methods in the modeling process and the influence of various modeling methods on model performance.

## 2 A Brief Theory of Random Forest Classifier

### 2.1 Ensemble Learning

The idea of ensemble learning is to combine many algorithms that are suitable for different scopes and have different functions for complex tasks. Random forest is a special bagging algorithm. Bagging is based on bootstrap resampling technique, which divides the data of the original training set into M different data sets and obtains M basic learners based on each data set. Then, each data set is fitted to model $\varphi_j$ and the predicted value $\varphi_j(x)$ is obtained. Finally, the predicted value is averaged as the final output of the model, so a more stable classifier $f(x)$ can be obtained.

$$f(x) = \frac{1}{m}\sum_{j=1}^{m}\varphi_j(x) \tag{1}$$

### 2.2 Random Forest Construction Steps

Random forest uses decision trees as the base classifier model for bagging. The steps to build a random forest model are as follows:

(1) Firstly, the Bootstrap resampling method is used to randomly extract M new sample sets from the original data set, and M classification decision trees are constructed with the new samples. The number of trees M refers to the same variable as the hyperparameter of model n_estimators mentioned later.
(2) Secondly, assuming there are n features, $n_{try}$. Features are randomly selected at each node of each tree, where $n_{try} < n$.

(3) The information entropy of each feature is calculated, and the feature with the strongest classification ability is selected for node splitting by probability.
(4) Finally, the generated multiple trees are formed into a random forest to classify the new data, the result adopts majority voting. The equation to get the final classification result is as follows:

$$H(X) = argmax \sum_{i=I}^{M} I(h_i(X) = Y) \tag{2}$$

$h_i$ represents the i-th decision tree, Y represents the target output, $I()$ represents the indicator function.

## 3 Dataset and Modeling Methods

To fulfill the aims, Six aspects related to modeling thermal sensation via Machine Learning (ML) were examined.

### 3.1 Data Source

For the training of the random forest classifier, the RP-884 Adaptive Model Project dataset was used [6]. Among the 52 tables, there are about 20,000 thermal voting observation results under different conditions from 46 buildings around the world. Locations include places in Europe, Asia, Australia or North America. A classification modeling method is applied in this paper, so the sample whose thermal comfort voting value is discrete value is chosen for the study. After selection, 19 tables were selected from 52 tables, with a total of 8001 thermal comfort data. Table 1 presents a description of the variables in the dataset.

### 3.2 Predictor

The current thermal environment evaluation methods are roughly divided into four categories: (1) Direct indicators: environmental indicators that can be directly measured by instruments, such as air temperature. (2) Experimental indicators: through the characteristics of heat exchange between humans and the environment, multiple variables are summarized into a single index, which can only be obtained through experiments, such as effective temperature and operating temperature. (3) Indicators derived from theory: considers the level of human activity and clothing Circumstances, indicators synthesized by integrating multiple indicators based on the heat balance equation, such as PMV, PPD, etc. (4) Questionnaire indicators: indicators obtained by directly asking the subject's thermal sensation status, such as thermal sensation voting value (TSV). "Feeling" can be used as a concept of "satisfaction with thermal environment", so thermal sensation is often used as a method to evaluate thermal comfort.

For a long time, researchers have used questionnaires to obtain people's subjective feelings about the thermal environment in the research of human thermal comfort. Therefore, using TSV is the most intuitive way to evaluate the current thermal environment.

### 3.3 Data Cleaning

Before modeling with machine learning algorithms, there may be partial missing, irrelevant, duplicated and abnormal data in the original data set. Although this part of the data may not affect the overall trend of large-volume data, it can seriously affect the execution efficiency during the modeling process, and may even cause deviations in the modeling results of some machine learning algorithms that are sensitive to abnormal data [7].

According to theoretical analysis and subsequent feature selection, the 12 variables PRXY_TSA, MET, CLO, UPHOLST, TAAV, TRAV, VELAV, RH, day06_ta, day06_rh, day15_ta and day15_rh are considered important for TSV prediction modeling. The specific meaning and scope of these variables can be found in Table 1. If any of these variables is missing, this sample is deleted. The total number of datasets reduced from 8001 to 7222.

**Table 1.** Variable descriptions and correlation with TSV

| Abbreviation | Feature name | Range | Correlation coefficient with TSV |
|---|---|---|---|
| TSV | Thermal sensation votes | 7-point scale | 1 |
| TAAV | Average of three heights' air temperature | [6.2, 42.7] | 0.6 |
| TRAV | Average of three heights' mean radiant temperature | [3.69, 46.7] | 0.6 |
| CLO | Ensemble clothing insulation | [0.174, 2.287] | 0.41 |
| day06_ta | Outdoor 6 a.m. air temp on day of survey | [−27.2, 29.5] | 0.34 |
| day06_rh | Outdoor 6 a.m. rel humid on day of survey | [43, 100] | 0.33 |
| day15_ta | Outdoor 3 p.m. air temp on day of survey | [−22.6, 40.5] | 0.26 |
| RH | Relative humidity | [2, 100] | 0.16 |
| MET | Average metabolic rate of subject | [0.64, 4.5] | 0.12 |
| VELAV | Average of three heights' air speed | [−0.01, 5.83] | 0.089 |
| PRXY_TSA | Thermal acceptability | [0.1, 2.0] | 0.073 |
| day15_rh | Outdoor 3 p.m. rel humid on day of survey | [30.6, 100] | 0.04 |
| UPHOLST | Insulation of the subject's chair | [0.0, 0.33] | 0.023 |

### 3.4 Normalization and Dataset Partitioning

It is common practice to divide the available data into subsets [8]. 7222 samples were randomly divided into three parts: training set (80%), validation set (10%), test set (10%). Training set is used to train the model, validation set used to adjust the hyperparameters of the model and to make a preliminary assessment of the model's ability. Test set is used to verify the generalization ability of the model. How exactly these subsets were used is described in the Sect. 4.2.

Features in the data set have different dimensions and large differences between each other, which may affect the accuracy of the model. In the data set, the value range of relative humidity RH is [2, 100], and the value range of average metabolic rate of subject MET is [0.64, 1.5]. In the process of calculating distance and correlation, the impact of RH can be much greater than that of MET. Eventually, RH is given a larger weight due to a larger value range, and the weight of MET is weakened. Therefore, normalizing the original features is required to eliminate the influence of the dimension and scope between the features.

All features $x_i$ of the dataset were normalized using maximum and minimum method Eq. 3.

$$x_i = \frac{x_i - x_{min}}{x_{max} - x_{min}} \tag{3}$$

# 4 Experiments

### 4.1 Feature Selection

Feature selection is an important part of feature engineering, its purpose is to find the optimal feature subset. Feature selection can eliminate irrelevant or redundant features, so as to reduce the number of features, improve model accuracy, and reduce running time. This section first studies the relationship between features and TSV based on correlation analysis, and then looks at the impact of the number of input features on the model, finally determines the 10 input variables of the predict model.

Correlation analysis was used to study the relationships between variables and TSV. The covariance between two variables is the theoretical basis of correlation analysis, the covariance equation is as follows:

$$cov(X_i, X_j) = \frac{\sum_{k=1,k=1}^{n} (X_{ik} - \overline{X_i})(X_{jk} - \overline{X_j})}{n-1} \tag{4}$$

$X_i$ and $X_j$ represent the i-th and j-th variables;
$\overline{X_i}$ and $\overline{X_j}$ stand for the average value of $X_i$ and $X_j$;
$n$ represents the number of samples.

The greater the covariance value between two variables, the more relevant they are. The correlation between the 12 variables and TSV was examined.

Table 1 lists the correlations between each variable and TSV in order from highest to lowest. Those variables that are strongly related to TSV should be emphatically considered as model inputs.

It worth noting that the correlation coefficient between TAAV and TRAV is 0.99, which shows that TAAV and TRAV have a strong correlation. Therefore, in order to avoid input redundancy, only TRAV is selected as the input variable.

An increase in the number of input variables can improve the performance of the thermal prediction model, but it will not increase indefinitely. Too many variables may cause the model to overfit. In order to find the optimal number of input features, the random forest classifier was trained with different numbers of features as input step by step according to the relevance ranking (only TRAV is used between TAAV and TRAV), and the accuracy of the obtained model is shown in Table 2.

**Table 2.** Model performance with different input features.

| Number of features | Training accuracy | Validation accuracy | Number of features | Training accuracy | Validation accuracy |
|---|---|---|---|---|---|
| 11 | 0.999 | 0.7136 | 6 | 0.997 | 0.567 |
| 10 | 0.999 | 0.7136 | 5 | 0.951 | 0.501 |
| 9 | 0.999 | 0.695 | 4 | 0.950 | 0.493 |
| 8 | 0.999 | 0.603 | 3 | 0.947 | 0.459 |
| 7 | 0.999 | 0.547 | 2 | 0.900 | 0.429 |

Table 2 shows that using the top 11 variables of the correlation has the same performance as the model using the top 10 variables. As the input feature quantity increase, the accuracy of the validation set gradually increased from 42% to 71.3%. Therefore, the top 11 related variables other than TAAV are selected as input variables, which are: TRAV, CLO, day06_ta, day06_rh, day15_ta, RH, MET, VELAV, PRXY_TSA and day15_rh.

### 4.2 Parameters Adjustment

In the random forest algorithm, there are three main hyper-parameters: n_estimators, max_features, max_depth. n_estimators indicates the number of trees in the forest, it is a typical influencing factor in which model performance is inversely proportional to model efficiency. Even so, it is more appropriate to try to increase the number of trees to improve the stability of the model. max_features indicates the number of features to consider when looking for the best split. max_depth indicates the maximum depth of the tree.

In this research, the parameters max_depth and n_estimators have been optimally selected by the "Grid Search" method [8]. The range of n_estimators considers [1, 201], and the tuning step is set to 10. Correspondingly, the phenomenon of model over-fitting suggests that a simpler model should be used, so parameter max_depth is adjusted within the range of [1, 41], with a step size of 2. Each time a pair (max_depth, n_estimators) is automatically selected and then applied to the training model. If there are multiple (max_depth, n_estimators) pairs for best performance, the smaller pairs is preferred to improve model efficiency.

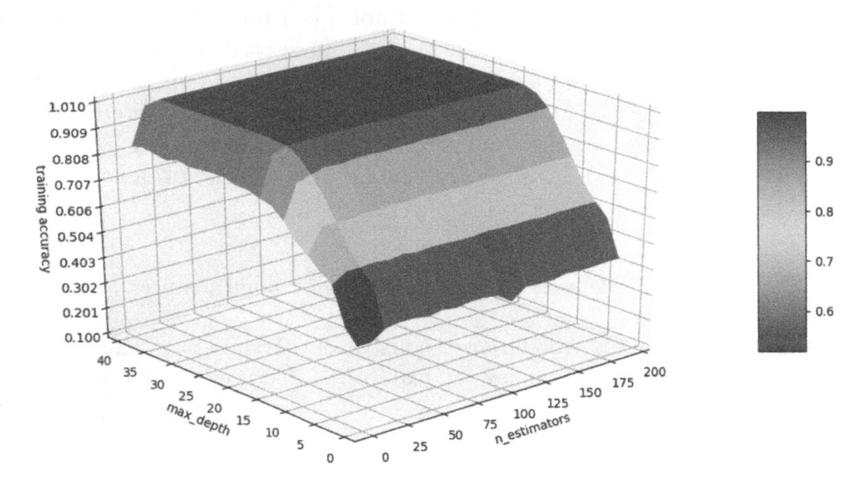


**Fig. 1.** Training accuracy.

It can be seen from Fig. 1 and Fig. 2 that as max_depth increases, the accuracy of both the training set and the validation set has improved. When max_depth > 15, the accuracy of the validation set will no longer increase significantly as the depth changes; In terms of n_estimators, the trend is alike. When n_estimators < 21, the accuracy of the model continues to increase with n_estimators. When the number of trees > 21, the accuracy of the verification set remains basically unchanged. Therefore, n_estimators = 25 and max_depth = 15 are chosen as the optimal parameters of the RF algorithm, accuracy of the model on the validation set reached 71.3%. Compared with the default parameters (n_estimators=100, max_depth = 5) for the training set accuracy rate of 60% and the validation set accuracy rate of 59%, appropriate parameters further improve the generalization performance of the model.

In general, the proportion of data set partition has little influence on the model effect. When the training set accounts for 60%–90% of the total data, the model has the best performance, and the accuracy rate of test set reaches 70%.

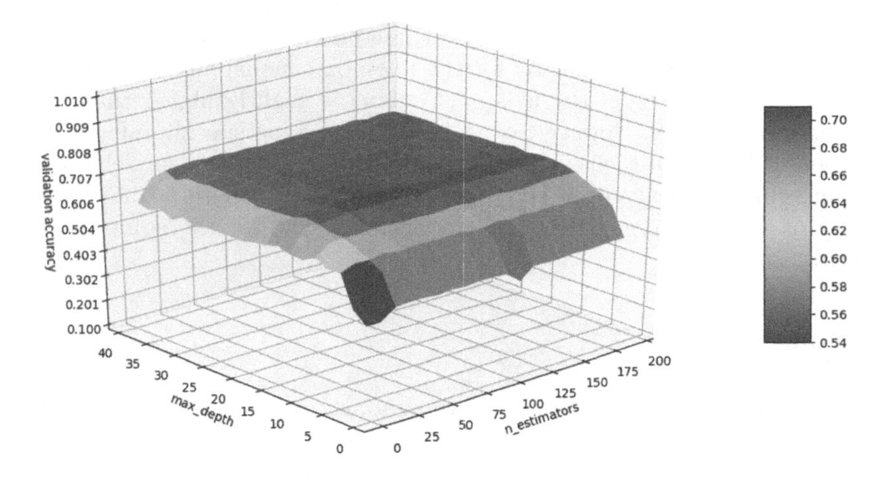


**Fig. 2.** Validation accuracy.

### 4.3 TSV Forms

The scale of TSV is other important consideration. ASHRAE 7-point thermal sensation scale is an ordinal scale that comprises the following categories: 3 (hot); 2 (warm); 1 (slightly warm); 0 (neutral); −1 (slightly cool); −2 (cool); −3 (cold). Different applications require different model accuracy. Many scholars have proposed in the field research on thermal comfort that the ASHRAE thermal sensation vote can be appropriately scaled according to the actual situation. Some scholars scale the original −3, −2, −1, 0, +1, +2, +3 interval to −2, −1, 0, +1, +2, using five-point TSV to make thermal comfort predictions [9]. Others like [1,10–12] divide the original interval into cold discomfort [−3, −1), comfortable/acceptable [−1, 1], hot discomfort (1, 3] in three intervals to better adapt to the on-site situation. In real practice, one should determine whether to pursue higher TSV prediction accuracy or to capture more detailed information on occupants' thermal response. Both pursuits have practical meaning [7] (Table 3).

**Table 3.** Model performance for different TSV forms

| TSV forms | Training | Validation | Testing |
|---|---|---|---|
| 3-point | 1.0 | 0.968 | 0.986 |
| 5-point | 0.999 | 0.757 | 0.727 |
| 7-point | 0.999 | 0.7136 | 0.671 |

Not surprisingly, the 3-point TSV is easier to predict, and the three-point prediction model can almost predict thermal sensation exactly. In situations where slight discomfort is acceptable, the three-point TSV can be used to ensure the accuracy of the prediction. As for the five-point model and the seven-point model, the performance of the five-point model is slightly higher than that of the seven-point model. However, considering that some specific information is lost when scaling from the seven-point model to the five-point model, based on the comprehensive evaluation, the seven-point model is more recommended in the case of lower requirements for model accuracy and higher requirements for comfort information.

## 5 Conclusion

This article shows the process of predictive modeling of thermal sensation votes using random forest classification. Comparisons of machine learning methods were conducted to determine the best modeling method. The data analyzed in the article was derived from the well-known and commonly used data set ASHRAE RP-884. 8001 samples with thermal sensation votes of standard 7-point scale are selected from the project. After data cleaning, the samples with missing important variables are removed and finally obtained 7222 samples. The main results of the analysis can be summarized as follows.

(1) The correlation matrix was used for feature selection, and the variables that highly correlated with TSV were tested in order from high to low. The model achieved the best effect when the top 11 variables were selected as the input. It is worth noting that since TAAV is highly correlated with TRAV, only TRAV is used between them to avoid redundancy.
(2) Adjusting the hyperparameters of the model can greatly affect the performance of the model. The grid search method is used to optimize max_depth and n_estimators, the model performance reaches the best when n_estimators = 25 and max_depth = 15. (99% for training set and 71.3% for validation). Compared with the default parameter n_estimators = 100 and max_depth = 5, the accuracy of training set and test set increased by 39% and 14% respectively.
(3) The performance of models with different training set ratios was tested, and the results shows that the most commonly used 8:1:1 data set partition is the most appropriate for modeling thermal sensation predition.
(4) The output target is thermal sensation votes, the form of TSV should be selected according to the actual situation. For 3-point TSV, the accuracy of the model in the test set reached 98.6%, which is suitable for the situation with low requirements on comfort level; for 5-point TSV, the test accuracy was 72.7%; for the 7-point TSV, the final test accuracy rate is 67.1%. Compared with the three-point TSV, the 7-point TSV model has rather poor performance, but the prediction results contain more information.

## References

1. Klepeis, N.E., Nelson, W.C., et al.: The national human activity pattern survey (NHAPS): a resource for assessing exposure to environmental pollutants. J. Expo. Anal. Environ. Epidemiol. **11**(3), 231–252 (2001)
2. Zhong, C., Liu, T. Zhao, J.: Modeling the thermal prediction using the fuzzy rule classifier. In: Chinese Automation Congress (CAC 2019), IEEE, Hangzhou, China, 22–24 November, pp. 3184–3188 (2019)
3. de Dear, R.J., Brager, G.S.: Developing an adaptive model of thermal comfort and preference. AHRAE Trans. (1998)
4. Fanger, P.O.: Thermal Comfort. Danish Technical Press, Copenhagen (1970)
5. Lu, S., Wang, W., Lin, C., Hameen, E.: Data-driven simulation of a thermal comfort-based temperature set-point control with ASHRAE RP884. Build. Environ. **156**, 137–146 (2019)
6. De Dear, R.: Macquarie university's ASHRAE RP-884 adaptive model project data downloader (2012)
7. Liao, Q.: Data Mining and Data Modeling, vol. 36. National Defense Industry Press, Beijing (2010)
8. Maier, H.R., Dandy, G.C.: Neural networks for the prediction and forecasting of water resources variables: a review of modelling issues and applications. Environ. Modell. Softw. **15**, 101–24 (2000)
9. Han, J.: Thermal comfort model of natural ventilation environment and its application in the Yangtze River Basin. Hunan University (2008)
10. Chaudhuri, T., Soh, Y., Li, H., Xie, L.: A feedforward neural network based indoor-climate control framework for thermal comfort and energy saving in buildings. Appl. Energy. **248**, 44–53 (2019)
11. Wang, Z., Yu, H., Luo, M., et al.: Predicting older people's thermal sensation in building environment through a machine learning approach: modelling, interpretation, and application. Build. Environ. **161** (2019)
12. Megri, A., Naqa, I.: Prediction of the thermal comfort indices using improved support vector machine classifiers and nonlinear kernel functions. Indoor Built Environ. **25**, 6–16 (2014)

# Design and Implementation of Airborne Fire Control Radar Simulation System

Xianfa Ji[1], Congcong Zhang[2], Zhanyuan Chang[2,3], and Chuanjiang Li[2,3](✉)

[1] Aviation Maintenance NCO Academy, AFEU, Xinyang 464000, Henan, China
[2] The College of Information, Mechanical and Electrical Engineering, Shanghai Normal University, Xuhui 200234, Shanghai, China
licj@shnu.edu.cn
[3] Shanghai Engineering Research Center of Intelligent Education and Bigdata, Shanghai Normal University, Xuhui 200234, Shanghai, China

**Abstract.** In this article, the design of airborne fire control radar simulation system is introduced. The simulation hardware platform of the airborne fire control radar is established by simulation computer. In order to realize the full function simulation software of the airborne fire control radar, we combined the theory of radar equation and radar jamming equation to construct the scene of airborne fire control radar, the radar target model and the typical EMI environment. The whole process of searching, intercepting and tracking target of airborne fire control radar is simulated by digital method. Finally, the simulation system also has the functions of theoretical learning and evaluation and the application shows that the simulation system achieves a satisfactory learning and training effect.

**Keywords:** Airborne fire control radar · Radar simulation · Modeling · Assessment

## 1 Introduction

Airborne fire control radar training simulator is an important application field of radar simulation technology, which is widely used in military field. It is a necessary device for fire control system operators to carry out basic training and advanced training. Its advantages are safety, economy, controllability, repeatability, risk-free, and freedom from climatic conditions and space constraints. It also can highlight the unique advantages of training such as high efficiency and high benefit [1–3]. The fire control radar simulation training system can not only carry out the routine operation training of fire control radar in searching, intercepting and tracking targets, but also train the operators to take appropriate measures to counter the jamming signals in the complex electromagnetic environment [4–6]. In this paper, we mainly introduce the design technical ideas and implementation of an airborne fire control radar simulation training system and complete the full function digital simulation of an airborne fire control radar. The system simulates the whole process of fire control radar detection target by constructing


Q. Han et al. (Eds.): LSMS 2021/ICSEE 2021, CCIS 1469, pp. 562–571, 2021.
https://doi.org/10.1007/978-981-16-7213-2_54

the scene model of airborne fire control radar, and establishes a typical radar countermeasure application environment with fire control radar as the core to make the operation more convenient, so that operators can be in a more realistic fire control radar training environment, and the training effect is improved. On the other hand, the system adds theoretical learning and assessment modules to further improve the use efficiency of the simulation training system, so that the trainees can complete the whole process of theoretical learning, theoretical assessment, actual operation and actual assessment, and improve the training effect [7].

## 2 System Structure

Airborne fire control radar simulation system is mainly composed of simulation system and test evaluation system. In the simulation part, the simulation function of airborne fire control radar is completed by establishing aircraft flight motion model and fire control radar function model; In the test evaluation system, the evaluation contents of theory and actual operation are obtained flexibly by establishing an open interface test system, and the operation and theory of fire control radar are evaluated. Finally, the evaluation results are given.

### 2.1 System Hardware Structure

The fire control radar simulation system belongs to the desktop simulation system. The system hardware is simple and easy to carry. It is mainly composed of high-performance simulation computer, radar operation function simulation, Head-Up Display (HUD) and Multi-Function Display (MFD), as shown in Fig. 1.

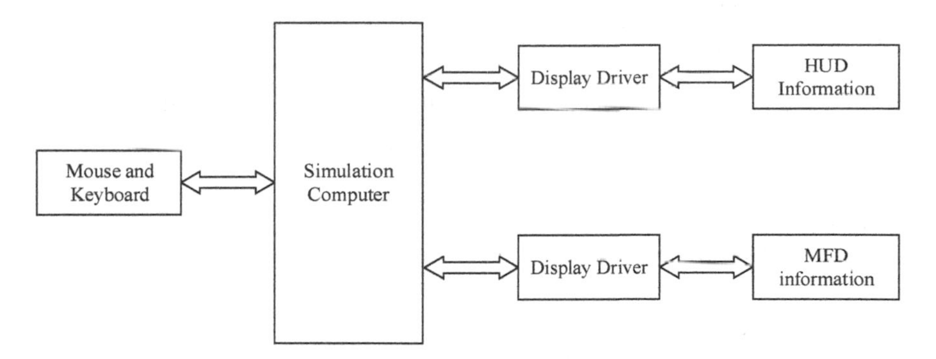


**Fig. 1.** Block diagram of system.

Simulation computer is the core of the whole simulation training system which controls the operation of the entire simulation training system. It is responsible for setting the initial conditions of simulation training and radar counter electromagnetic environment, etc. It can also conduct human-computer dialogue and has the ability to guide the trainees' operation and evaluate the training situation. The simulation training system includes the following three parts. Firstly, through the display the information of the fire control radar and relevant display information of HUD are displayed. Secondly, through the simulated radar operation function keys, the fire control radar state transition signal is obtained, and the state transition information is sent to the simulation computer. Thirdly, through calling the corresponding training model of the fire control radar for calculation, the calculation result is calculated by the display program to control the display of MFD, so as to realize the conversion of radar working state and the synchronization of display.

HUD is a touch screen LCD, which is mainly used to display the synchronous changes of HUD information with the change of radar working state.

## 2.2 System Software Structure

The software of airborne fire control radar simulation system adopts the object-oriented programming method. Through the modular and hierarchical design of the objects, enhance the operability, reliability, versatility, maintainability and upgrade ability of the system. This system mainly consists of five parts: operation control module, airborne fire control radar system module, display control module, radar counter electromagnetic environment model and record evaluation module.

The overall structure of the software is shown in Fig. 2.

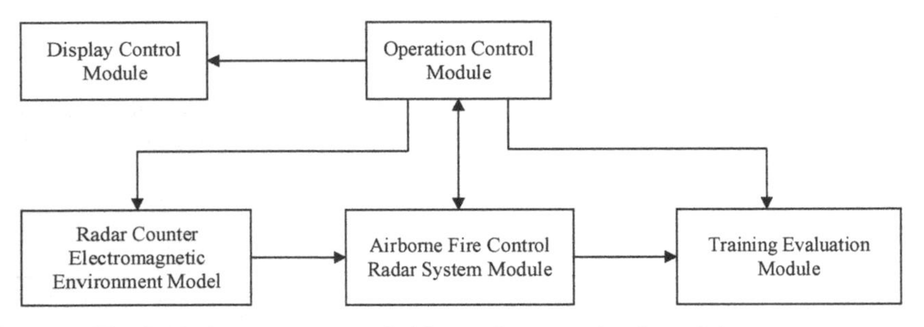


**Fig. 2.** Software structure of airborne fire control radar training system.

### Operation Control Module

The operation control module is the core part of airborne fire control radar simulation system. This part controls the operation process of the entire training software. It realizes the comprehensive scheduling of the working mode model in the radar simulation system and simulates the data of each software sub-module of the training system. Through this module, the interaction and sharing function are realized.

### Airborne Fire Control Radar System Module

The airborne fire control radar system module establishes the models of various radar working modes. This part mainly simulates the working modes of radar in function. The idea of modeling includes building radar target model, simulating the display information and state transition mode of various working modes of airborne fire control radar searching and tracking target.

**Display Control Module**
The display control module includes a radar terminal display module, a radar control module, and a scanning display module. The module control part can simulate the functions of the airborne fire control radar operating parts, and the display part can display the various working screens of the radar.

**Radar Counter Electromagnetic Environment Module**
In the part of radar countermeasures model, the working state of the airborne fire control radar and the working picture of the radar after taking the corresponding anti-jamming measures are established respectively under the condition of active jamming and passive jamming. The simulation training software can simulate and display the radar display screen when the fire control radar is affected by noise interference, range deception interference, speed deception interference, angle deception interference, narrow-band aiming noise interference, and man-made negative interference. According to the theory of radar countermeasures, the radar interference model is established. And at the same time the characteristics of the corresponding interference and treatment measures are given.

**Training Evaluation Module**
The training evaluation module mainly includes the recording and performance evaluation of the training process of the trainees. It mainly consists of theoretical study, theoretical assessment, simulation training and operational assessment. The training software can realize the simulation training of the fire control radar. Both the theoretical learning and the assessment part adopt an open interface structure, flexibly import the theoretical learning question library, randomly set the theory and actual operation assessment questions, and realize the system assessment and assessment function. The theoretical learning module part can be set with a variety of question types, which can update the input question library and objective test question library in time. And the assessment module part can record and score the test situation of personnel.

## 3 Modeling of Simulation System

### 3.1 Flight Situation Environment Model

The flight situation model is mainly composed of aircraft parameter settings, top view of flight situation, side view of flight situation and scenario of combat scene.

The aircraft parameter setting module mainly includes setting the speed and altitude of the carrier aircraft, setting the heading, speed and altitude of the target aircraft, and selecting the number and type of the corresponding target aircraft. The aircraft number is represented by an ID number. The heading includes east and west. The flight speed can be adjusted and the speed unit is $v$ m/s. The flight altitude can be set as low altitude, medium altitude and high altitude, so as to distinguish the radar up and down looking detection target. The side view of the flight situation is used to view the flight situation from the side in order to observe the changes of the aircraft's flight speed, heading and altitude in real time. The flight situation top view is used to look down on the flight situation in order to observe the changes of the aircraft's flight speed, heading and position in real time. When the altitude of the aircraft changes, the position of the corresponding aircraft in the side view changes. At the same time, different radar usage scenarios can be set according to simulation needs.

### 3.2 Mathematical Model of Aircraft Movement

In the radar simulation system, the carrier plane and the target plane move in a straight line at a constant speed, and the specific movement scene is carried out according to the parameters set by the flight situation. In the process of movement, the speed, heading and altitude can be changed at any time. The simulation program is designed according to the idea of circular operation, that is, the process of aircraft motion is a process of circular flight. When the aircraft flies out of the boundary, it will fly again from the initial flight position to ensure that the simulation process continues.

In the simulation system, the mathematical model for calculating the moving distance of the aircraft in the view is as follows:

$$\begin{aligned} \frac{v * t}{\Delta x} &= \frac{X}{w} \\ \Delta x &= \frac{v * t * X}{w} \end{aligned} \tag{1}$$

Where $\Delta x$ is the displacement of aircraft in view, $v$ and $t$ are the aircraft speed and flight time respectively, $X$ and $w$ are the range and width of the view.

### 3.3 Radar Target and Jamming Model

The main theoretical basis for realizing the functional simulation of radar electronic warfare system is the radar range equation [8, 9]. According to the equation, the target echo reflected from the target with an oblique distance of $R$, the power of the echo signal received by the radar is:

$$P_{\mathrm{R}} = \frac{P_t G^2 \lambda^2 \sigma D}{(4\pi)^3 R^4 L} \tag{2}$$

Where $P_t$ is transmit signal power (W). $G$ is the one-way antenna power gain in the target direction. $\lambda$ is the working wavelength of the radar (m). $\sigma$ is the target radar cross-sectional area RCS (m2). $D$ is the improvement factor of radar anti-jamming. $L$ is the comprehensive loss factor of radar and atmospheric transmission loss factor.

The target echo signal received by radar is interweaved with jamming. The forms of interference are receiver noise, clutter and electronic interference. Signal noise ratio (SNR) is a measure of the ability to detect targets in the presence of interference. When the SNR reaches a certain threshold, the radar can find the target. After the SNR is given, the detection probability can be calculated by making appropriate assumptions about the fluctuation statistical characteristics of the target and interference. If there is no target, a "discovery" (exceeding the threshold) may be obtained due to the fluctuation characteristics of the interference signal, which is called false alarm [10–12]. Therefore, in the function simulation of airborne fire control radar, according to the above theoretical requirements, and the specific radar equipment parameters, we simulate the situation when the radar finds the target. Considering the existence of interference, there is a certain probability of false targets when the target appears, that is, the false alarm of radar detection. The function simulation can simulate the time and position of radar finding target and whether the target really exists.

The jamming received by radar generally includes cover jamming and deceptive jamming. Covert jamming mainly aims at the working state of radar searching target, which usually affects the SNR of radar, covers the target information, and reduces the detection ability of radar. Deceptive jamming mainly aims at the tracking working state of radar, whose main function is to destroy the tracking system of radar, make the radar get the wrong target data, so as to protect the real target [13, 14]. In radar function simulation system, considering the influence of jamming on radar simulation, we introduce the influence coefficient of jamming on radar. Firstly, the relative jamming and anti-jamming performance indexes are analyzed mathematically, and then the quantitative performance evaluation is carried out. Finally, the comprehensive influence of jamming on radar detection and tracking performance is determined.

### 3.4 Radar Target Detection and Validation Model

To confirm the target by airborne fire control radar, it is necessary to judge whether the target is in the scanning range [15]. After setting the target parameters, the movement is carried out according to the above flight model. At $t$ time, the position of the aircraft is $(x_1, y_1)$ and the target position is $(x_2, y_2)$. The radar range is $R$ and the beam scanning range is $\theta$. When the target position satisfies inequality (3), the current target should appear in the simulation screen, that is, the radar has detected the target. In the actual simulation process, we should also consider whether the radar is interfered, the strength of the interference, the RCS of the target and other factors to make a comprehensive judgment to get the target conclusion. When the targets meet the conditions of appearance, targets will appear in the training screen and the radar beam simulation scan.

$$\begin{cases} (x_2 - x_1)^2 + (y_2 - y_1)^2 < R^2 \\ y_2 - y_1 < \tan\frac{\theta}{2}(x_2 - x_1) \\ y_2 - y_1 > -\tan\frac{\theta}{2}(x_2 - x_1) \end{cases} \quad (3)$$

The schematic diagram of the calculation method when the target appears in the beam scanning range is shown in Fig. 3.

### 3.5 Mathematical Model of Tracking Target Motion

When tracking a target, the radar needs to continuously calculate the real-time position information of the target. From the above analysis, when the target meets inequality (3), the target appears in the radar display screen. After intercepting the target, it is assumed

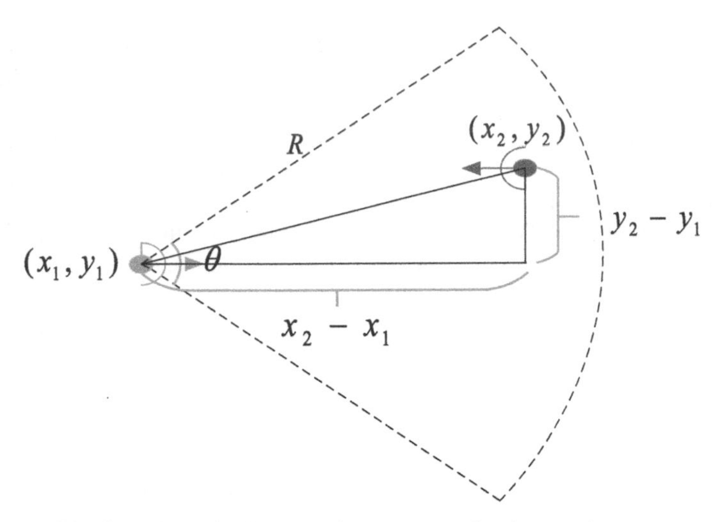


**Fig. 3.** Schematic diagram of target appearing in scanning range.

that the current position of the carrier is $(x_1, y_1)$, the current position of the target is $(x_2, y_2)$, the range of the current radar is $R$ and the azimuth angle of the target is $\alpha$. Then the real-time position of the target $(x, y)$ can be calculated.

According to the scale of the parameter setting interface, the actual horizontal distance between the carrier and the target is calculated as formula (4).

$$\frac{X}{w} = \frac{d}{x_2 - x_1} \tag{4}$$

Where $d$ is actual horizontal distance between two aircraft, $X$ and $w$ are the range and width of the view respectively.

The scale of parameter interface and beam scanning is shown as formula (5).

$$\frac{R}{h} = \frac{d}{y} \tag{5}$$

Where $h$ is the height of the view. y is one of the coordinate values of the target real-time position $(x, y)$.

Assuming that the position of the carrier plane is the coordinate (0,0), $y$ can be obtained from Eq. (4) and Eq. (5).

The actual angle between the carrier and the target is known to be formula (6).

$$\alpha = \arctan \frac{x_2 - x_1}{y_2 - y_1} \tag{6}$$

And then we can get the other coordinate value of the target real-time position, that is, $x = y \tan \alpha$.

## 4 Operation Control of Simulation System

The specific system operation control flow of simulation training system is as follows. After trainees enter the radar simulation training system, they can select the training contents which include theoretical learning module and simulation training module. Trainees can complete the learning of radar and other related theory in theoretical learning module. The specific learning content can be imported by the volume library generating program and the import interface is open. There are three parts in simulation training module, Air-to-Air, Air-to-Surface and Assessment System. Air-to-Air and Air-to-Surface are two training modes. The assessment system is divided into two categories: theoretical assessment and simulation operation assessment. The interface is open and the way of importing test questions is flexible and convenient.

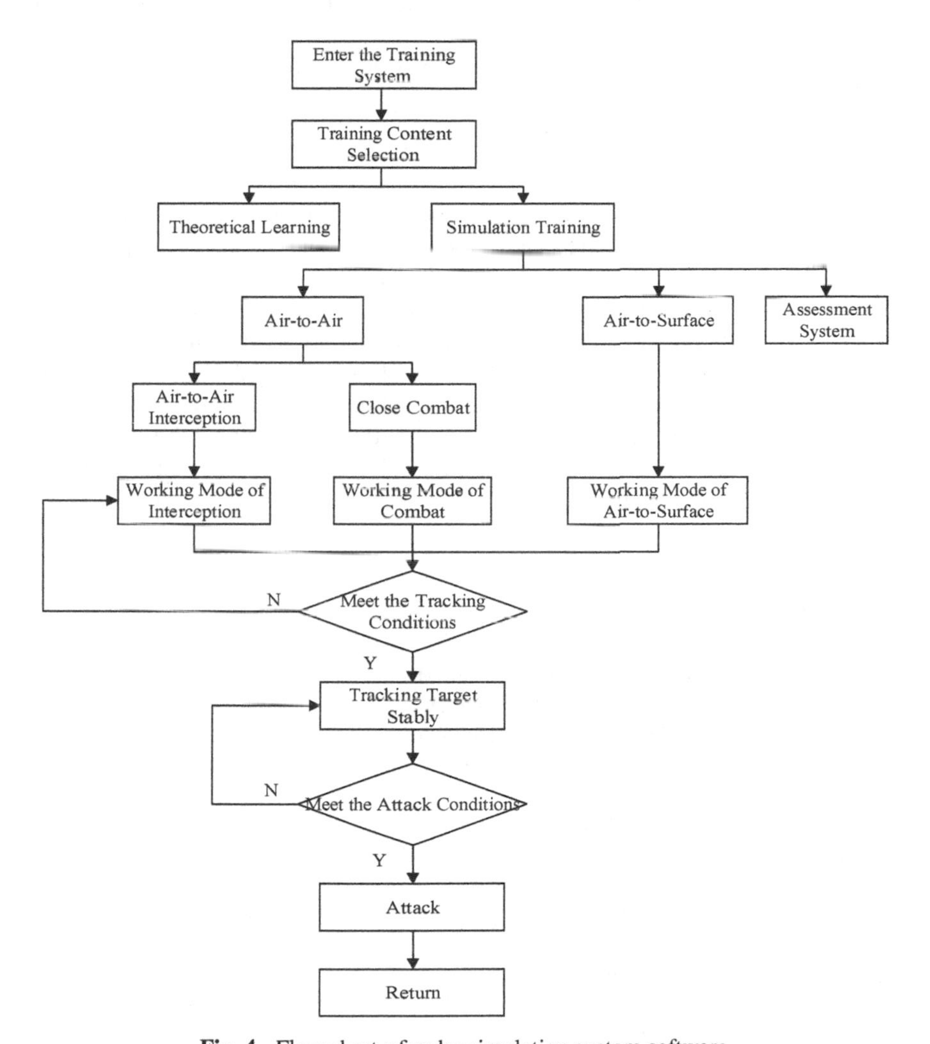


**Fig. 4.** Flow chart of radar simulation system software.

In the process of simulation training, trainees should set the aircraft flight motion parameters and flight situation environment parameters firstly, and then click the confirm button to enter the radar simulation training interface. The position of the target and the aircraft is set in the parameter setting interface and the beam scanning is displayed and changed synchronously on the display and the beam scanning simulation area. When the speed and heading of the aircraft or the target have changed, it is updated synchronously in the simulation training interface to realize the overall synchronous real-time update.

The simulation training part can complete all kinds of functional training of fire control radar, including the entry and exit of the Working modes which are air-to-air interception, close combat, air-to-surface mode, dynamically displaying relevant information on the simulated MFD and so on. In this part, trainees also can simulate the whole process of airborne fire control radar which includes searching for targets, intercepting and tracking targets, entering weapon attack envelope, forming attack conditions and completing weapon launch Attack.

The operation control flow chart of the simulation system is shown in Fig. 4.

## 5 Conclusion

The innovation of the airborne fire control radar simulation training system lies in the whole process of the airborne fire control radar's searching target, intercepting target and weapon launching guidance. In addition, the basic process of fire control calculation is simulated. Compared with the related meteorological radar simulation system and navigation radar simulation system, the system has comprehensive function, outstanding characteristics of fire control calculation and strong practicability. The simulation training system has been used in many training institutions, and more than 100 people have been operated and used. The results show that the system runs stably and reliably, the software interface is standard and the trainees also have achieved better operational training effect. The operational loss of the actual radar is effectively reduced and the economic benefit is obvious.

## References


1. Adamy, D.: Introduction to Electronic Warfare Modeling and Simulation (2006)
2. Hu, Y.H., Luo, J., Jian, F.S., et al.: RadarDL-A radar simulation description language. In: 2012 International Conference on System Simulation (ICUSS 2012) (2012)
3. Barton, D.K.: Radar system analysis and modeling. J. IEEE Aerospace Electron. Syst. Mag. **20**(4), 23–25 (2005)
4. Song, K., Liao, J.B., Yang, Q.: Research of modeling and simulation of aviation ignition equipment. J. Adv. Mater. Res. **690**, 2906–2911 (2013)
5. Hu, Z.Z., Xia, X.Y., Liu, Z.T., et al.: Radar countermeasure model of simulation training system oriented to command. J. Command Control Simul. **04**, 34–38 (2020)
6. Cao, L.Y., Dong, Y., Guo, W.N.: Development trend analysis of airborne fire-control radars. J. Aeronaut. Sci. Technol. **32**(06), 1–8 (2021)
7. Yang, H.D., Liu, S.X.: The reform and practice of the radar principle courses experiment teaching. In: 2015 International Conference on Social Science, Education Management and Sports Education (2015)

8. Skolnik, M.: Radar Handbook. Third Edition (2008)
9. Stove, A.G.: Radar Basics and Applications. John Wiley & Sons, Hoboken (2010)
10. Jiang, Q.: Introduction to Network Radar Countermeasure Systems. Springer, Berlin Heidelberg (2016). https://doi.org/10.1007/978-3-662-48471-5_1
11. Ji, X.F.: The technique characteristic and interference research of airborne phased array fire control rada. J. Dev. Innov. Mach. Electr. Prod. **028**(003), 13--15 (2015)
12. Hong, B., Wang, W.Q., Liu, C.C.: Mutual interference alignment for co-existing radar and communication systems. J. Digit. Sign. Process. **112**(4), 103–107 (2021)
13. Bonacci, D., Vincent, F., Gigleux, B.: Robust DoA estimation in case of multipath environment for a sense and avoid airborne radar. J. Let Radar Sonar Navig. **11**(5), 797–801 (2017)
14. Luong, D., Rajan, S., Balaji, B.: Estimating correlation coefficients for quantum radar and noise radar: a simulation study. In: 2020 IEEE Radar Conference, pp. 1--5. IEEE Press (2020)
15. Wang, Y., Cheng, S., Zhou, Y., et al.: Air-to-air operation modes recognition of airborne fire control radar based on DS evidence theory. J. Modern Radar. **39**(05), 79–84 (2017)

# Event-Triggered Distributed Maximum Correntropy Filtering in Non-Gaussian Environments

Haifang Song, Derui Ding(✉), Xueli Wang, and Meiling Xie

Department of Control Science and Engineering, University of Shanghai for Science and Technology, Shanghai 200093, China
deruiding2010@usst.edu.cn

**Abstract.** This paper investigates the distributed filtering issue for general stochastic nonlinear systems with non-Gaussian noises. A novel adaptive event-triggered scheme is designed to reduce the communication consumption. A new event-triggered distributed maximum correntropy filter (E-DMCC) is proposed by means of the maximum correntropy criterion, and the corresponding explicit expressions of desired filter gains and the upper bound of filtering error covariance are given in light of fixed-point iterative update rules and some typical matrix inequalities. Finally, a ballistic object tracking system is utilized to show the merit of the constructed distributed filtering in non-Gaussian noise environments.

**Keywords:** Distributed filtering · Maximum correntropy criterion · Fixed-point iterative update rules · Non-Gaussian noises

## 1 Introduction

Distributed filtering is widely utilized in industrial research fields such as fault detection, target tracking and intelligent transportation system monitoring [1]. In the Gaussian noise environments, Kalman filter (KF) and its extended algorithm are often used to handle the distributed filtering problems, and it is not difficult to obtain the effective solutions [2]. However, it is well known that the noise suffered by the system in actual industrial applications is often in the form of non-Gaussian noises (such as impulse noises, shot noises and cauchy noises).

The maximum correntropy criterion (MCC), as a new method which can capture and utilize the high-order statistical characteristics of data signals to quantify the similarity between different variables, is widely employed in the estimation of non-Gaussian systems [1–3]. What needs to be further clarified is that due to technical limitations, the closed form of the analytical solution for the MCC algorithm is difficult to achieve in terms of acquisition. Therefore, people usually in light of iterative update algorithms (such as gradient-based schemes, fixed-point iterative update methods, etc.) to solve the analytic solution

This work was supported in part by the National Natural Science Foundation of China under Grants 61973219 and 61933007.


Q. Han et al. (Eds.): LSMS 2021/ICSEE 2021, CCIS 1469, pp. 572–582, 2021.
https://doi.org/10.1007/978-981-16-7213-2_55

acquisition problem in the entropy filtering issues [4–6]. It is worth mentioning that the existing research on MCC algorithms is mostly limited to centralized filtering, while the further expansion and application of corresponding research in the field of distributed state estimation is still an open and challenging issue.

In addition, since most filters use a time-triggered scheme to realize communication scheduling, which means that it will consume a lot of communication resources. In recent years, the development and application of many event-triggered scheme (ETS) have brought hope to solve a problem [7,8]. The existing event-triggered scheme usually adopt a constant threshold, which means that these schemes cannot improve the system performance with the help of dynamically adjusting the communication frequency. Therefore, the adaptive ETS has attracted more and more attention in recent years [9,10]. However, since the adaptive rules introduced in the developed adaptive event-triggered scheme are usually decrease monotonously, which will inevitably lead to the event-triggered threshold becoming smaller and smaller, and the information transmission becoming more and more frequent. Therefore, how to design an adaptive scheme to eliminate the monotonicity of adaptive rules in existing schemes has become more and more important.

In summary, our aim in this paper is to develop a event-triggered distributed maximum correntropy filtering to handle the estimation problem of stochastic nonlinear systems with non-Gaussian noises. The main contributions of this article are highlighted as follows: 1) the weighted MCC instead of the traditional MSE has been employed to construct the cost function for improving the filtering performance; 2) a novel distributed filter is proposed by considering the communication topology and applying a adaptive ETS to reduce communication consumption; 3) the explicit expression of filter gains and the filter error covariance upper bound are achieved via the solution of a set of Riccati-type equations, which is solved by the fixed-point iterative update rules.

## 2 Problem Formulation and Preliminaries

In this article, the physical relationship of $n$ interconnected nodes composing the network can be described via a undirected graph $\mathcal{G} = (\mathcal{V}, \mathcal{E})$, where a finite vertex set $\mathcal{V} = \{1, 2, \ldots, n\}$ and an edge set $\mathcal{E} \subseteq \mathcal{V} \times \mathcal{V}$ denote the set of nodes and the set of communication links between nodes, respectively. If edge $(i, j) \in \mathcal{E}$, then the $j$th node is called the neighbor of the $i$th node, i.e. the $i$th node can obtain the information of $j$-th node. The nonnegative matrix $\mathcal{H} = [a_{ij}] \in \mathbb{R}^{n*n}$ stand for the adjacency matrix, which is set as $a_{ij} > 0$ if edge $(i, j) \in \mathcal{E}$, and $a_{ij} = 0$ otherwise. Furthermore, select $\mathcal{N}_i = \{j : (i, j) \in \mathcal{E}\}$ to describe the set of neighbors of the $i$-th node, and set the corresponding number of neighbors is $d_i$.

Consider the following class of discrete uncertain time-varying nonlinear systems:

$$\begin{aligned} x_{k+1} &= (A_k + \Delta A_k)x_k + f(x_k) + B_k\omega_k, \\ y_{i,k} &= C_{i,k}x_k + v_{i,k}, \quad i \in \mathcal{V}, \end{aligned} \tag{1}$$

where $x_{i,k} \in \mathbb{R}^{n_x}$ indicates the system state vector and $y_{i,k} \in \mathbb{R}^{n_y}$ denotes the *ideal* measurement output. $\omega_k \in \mathbb{R}^{n_w}$ and $v_{i,k} \in \mathbb{R}^{n_v}$ stands for the process and the measurement noise, which are assumed to be mutually independent arbitrary noise sequences with zero-mean and variance $Q_k$ and $R_{i,k}$, respectively. $A_k$, $D_k$, $B_k$ and $C_{i,k}$ are known matrices. The uncertain matrix $\Delta A_k$ satisfying

$$\Delta A_k = \mathcal{T}_k U_k \mathcal{S}_k, \quad U_k U_k^T \leq I_{n_x}, \tag{2}$$

where $U_k$ stands for the time-varying uncertainty, $\mathcal{T}_k$ and $\mathcal{S}_k$ are known time-varying matrices. The function $f(x_k)$ is a known nonlinear function and meets

$$\|f(u) - f(v)\| \leq \|F(u - v)\|, \quad \forall u, v \in \mathbb{R}^{n_x}, \tag{3}$$

where $F$ represent a known matrix.

Fully considering the topology structure, a distributed filter in *ideal communication* is constructed as follows:

$$\begin{cases} \widehat{x}_{i,k}^- = A_{k-1}\widehat{x}_{i,k-1}^+ + f(\widehat{x}_{i,k-1}^+) + \varepsilon \sum_{j\in\mathcal{N}_i} a_{ij}(\widehat{x}_{i,k-1}^+ - \widehat{x}_{j,k-1}^+), \\ \widehat{x}_{i,k}^+ = \widehat{x}_{i,k}^- + K_{i,k}(y_{i,k} - C_{i,k}\widehat{x}_{i,k}^-), \end{cases} \tag{4}$$

where $\widehat{x}_{i,k}^-$ and $\widehat{x}_{i,k}^+$ respectively denote the one-step prediction and estimate of the $i$-th filter at time $k$. $\varepsilon \in (0, 1/\rho)$ and the matrix $K_{i,k}$ are the consensus gain and the filter gain with $\rho = \max\{d_i\}$. Assuming that the initial estimated state meets $\widehat{x}_{i,0}^+ = \widehat{x}_{i,0}^- = \left[\mu_0^T\ 0\right]^T$ $(i \in \mathcal{V})$.

In this paper, an adaptive ETS is introduced to reduce the communication burden. Specifically, the estimator $i$-th broadcasts the local estimate $\widehat{x}_{i,k}^+$ to its neighbors only when it meets the following event-triggered condition:

$$\aleph(g_{i,k}, \delta_{i,k}) = g_{i,k}^T g_{i,k} - \delta_{i,k} > 0, \tag{5}$$

where $g_{i,k} = \widehat{x}_{i,k}^+ - \widehat{x}_{i,k_c^i}^+$, $k_c^i$ stands for the latest transmission time instant of $i$-th estimator and the sequence of event-triggered instants $0 \leq k_0^i < k_1^i < \ldots < k_c^i < \ldots$ can be described as

$$k_{c+1}^i = \min\{k | k > k_c^i, \quad \aleph(g_{i,k}, \delta_{i,k}) > 0\}. \tag{6}$$

The adaptive rule on the time-varying trigger threshold $\delta_{i,k}$ is designed as follows

$$\delta_{i,k+1} = \max\left\{\delta_m, \quad \min\left\{\delta_M,\ \delta_{i,k}\big(1 - (2a_i/\pi)\text{arccot}(g_{i,k}^T g_{i,k} - b_i)\big)\right\}\right\}, \tag{7}$$

where "arccot(·)" denotes the arc cotangent function, $a_i \in (-1, 0)$ and $b_i > 0$ are given constants to adjust the output of arccot(·), $\delta_m$ with $\delta_M$ are two predetermined scalars and meet $\delta_m \leq \delta_M$. In this paper, we assume $\delta_{i,0} = \delta_m$.

**Lemma 1.** *For the adaptive event-triggered scheme (5)–(7) with the given initial condition $\delta_{i,0} = \delta_m > 0$, the scalar $\delta_{i,k}$ satisfies $\delta_m \leq \delta_{i,k} \leq \delta_M$ for all $k \geq 0$, and the condition $g_{i,k}^T g_{i,k} \leq \delta_M$ always holds too.*

With the help of the introduction of the adaptive ETS, we can reconstruct the *actual* estimator structure as follows:

$$\begin{cases} \widehat{x}_{i,k}^{-} = A_{k-1}\widehat{x}_{i,k-1}^{+} + f(\widehat{x}_{i,k-1}^{+}) + \varepsilon \sum_{j\in\mathcal{N}_i} a_{ij}\left(\widehat{x}_{i,(k-1)_c^i}^{+} - \widehat{x}_{j,(k-1)_c^j}^{+}\right), \\ \widehat{x}_{i,k}^{+} = \widehat{x}_{i,k}^{-} + K_{i,k}^{*}(y_{i,k} - C_{i,k}\widehat{x}_{i,k}^{-}), \end{cases} \tag{8}$$

where $K_{i,k}^{*}$ is the filter gain to be designed.

Furthermore, correntropy is often employed to describe the similarity of random variables $X$ and $Y$: $V(X,Y) = \mathbb{E}\{\kappa(X,Y)\} = \int \kappa(x,y)dF_{X,Y}(x,y)$, where $\mathbb{E}$, $\kappa(\cdot,\cdot)$ and $F_{X,Y}(x,y)$ are the expectation operator, the kernel function and the joint distribution function of random variables $X$ and $Y$, respectively. The Gaussian function is utilized as the kernel function in this paper: $G_\sigma(x,y) = \exp(-\|x-y\|^2/2\sigma^2)$, where $\sigma$ is the kernel size. In addition, since $F_{X,Y}(x,y)$ is usually unknown and the available samples are limited in practice, the sample mean estimator rather than the expectation operator is usually utilized to construct the cost function $\hat{V}(X,Y) = \frac{1}{N}\sum_{i=1}^{N} G_\sigma(x_i,y_i)$. With the help of the above analysis, the following cost function based on MCC criterion is employed to evaluate the filtering performance of the constructed filter (8):

$$\mathcal{J}(\widehat{x}_{i,k}^{+}) = G_\sigma\left(\|\widehat{x}_{i,k}^{+} - \widehat{x}_{i,k}^{-}\|_{(\Xi_{ii,k}^{-})^{-1}}\right) + G_\sigma\left(\|y_{i,k} - C_{i,k}\widehat{x}_{i,k}^{+}\|_{R_{i,k}^{-1}}\right), \tag{9}$$

where $R_{i,k}$ and $\Xi_{ii,k}^{-}$ are the measurement noise variance and the upper bound of one-step prediction errors covariance for filter $i$ at time step $k$ to be defined in the following section, respectively.

In summary, the main aim of this article is to propose a feasible distributed filtering to obtain the desired filter gain $K_{i,k}^{*}$ by maximizing the cost function $\mathcal{J}(\widehat{x}_{i,k}^{+})$.

## 3 Main Results

For simplicity, define

$$e_{i,k}^{-} = x_{i,k} - \widehat{x}_{i,k}^{-}, e_{i,k}^{+} = x_{i,k} - \widehat{x}_{i,k}^{+}, P_{ii,k}^{-} = \mathbb{E}\{e_{i,k}^{-}(e_{i,k}^{-})^T\}, P_{ii,k}^{+} = \mathbb{E}\{e_{i,k}^{+}(e_{i,k}^{+})^T\}.$$

where $e_{i,k}^{-}$ and $e_{i,k}^{+}$ are the one-step prediction errors and the filter errors, respectively, $P_{ii,k}^{-}$ and $P_{ii,k}^{+}$ are their corresponding variance matrices. In accordance with the event-triggered protocol (5)–(7) and the filter structure (8), we can rewritten $\widehat{x}_{i,k}^{-}$ as follows

$$\begin{aligned} \widehat{x}_{i,k}^{-} = {} & A_{k-1}\widehat{x}_{i,k-1}^{+} + f(\widehat{x}_{i,k-1}^{+}) + \varepsilon \sum_{j\in\mathcal{N}_i} a_{ij}(\widehat{x}_{i,k-1}^{+} - \widehat{x}_{j,k-1}^{+}) \\ & - \varepsilon \sum_{j\in\mathcal{N}_i} a_{ij}(g_{i,k-1} - g_{j,k-1}). \end{aligned}$$

Then one can achieve the one-step prediction errors $e_{i,k}^-$:

$$\begin{aligned} e_{i,k}^- &= A_{k-1}e_{i,k-1}^+ + \Delta A_{k-1}x_{k-1} + f(x_{k-1}) - f(\widehat{x}_{i,k-1}^+) + B_{k-1}\omega_{k-1} \\ &\quad + \varepsilon \sum\nolimits_{j\in\mathcal{N}_i} a_{ij}(e_{i,k-1}^+ - e_{j,k-1}^+) + \varepsilon \sum\nolimits_{j\in\mathcal{N}_i} a_{ij}(g_{i,k-1} - g_{j,k-1}). \end{aligned}$$

Since $\mathbb{E}\{\omega_{k-1}\} = 0$, then the covariance matrices $P_{ii,k}^-$ can be achieved by

$$\begin{aligned} P_{ii,k}^- &= A_{k-1}P_{ii,k-1}^+A_{k-1}^T + \Delta A_{k-1}\mathbb{E}\{x_{k-1}x_{k-1}^T\}\Delta A_{k-1}^T + B_{k-1}Q_{k-1}B_{k-1}^T \\ &\quad + \mathbb{E}\{\big(f(x_{k-1}) - f(\widehat{x}_{i,k-1}^+)\big)\big(f(x_{k-1}) - f(\widehat{x}_{i,k-1}^+)\big)^T\} \\ &\quad + \varepsilon^2 \sum\nolimits_{j\in\mathcal{N}_i} a_{ij}^2\mathbb{E}\{(e_{i,k-1}^+ - e_{j,k-1}^+)(e_{i,k-1}^+ - e_{j,k-1}^+)^T\} \\ &\quad + \varepsilon^2 \sum\nolimits_{j\in\mathcal{N}_i} a_{ij}^2\mathbb{E}\{(g_{i,k-1} - g_{j,k-1})(g_{i,k-1} - g_{j,k-1})^T\} \\ &\quad + \mathbb{E}\{\mathscr{A}_{i,k-1} + \mathscr{A}_{i,k-1}^T\} + \mathbb{E}\{\mathscr{B}_{i,k-1} + \mathscr{B}_{i,k-1}^T\} + \mathbb{E}\{\mathscr{C}_{i,k-1} + \mathscr{C}_{i,k-1}^T\} \\ &\quad + \mathbb{E}\{\mathscr{D}_{i,k-1} + \mathscr{D}_{i,k-1}^T\} + \mathbb{E}\{\mathscr{E}_{i,k-1} + \mathscr{E}_{i,k-1}^T\} + \mathbb{E}\{\mathscr{F}_{i,k-1} + \mathscr{F}_{i,k-1}^T\} \\ &\quad + \mathbb{E}\{\mathscr{G}_{i,k-1} + \mathscr{G}_{i,k-1}^T\} + \mathbb{E}\{\mathscr{H}_{i,k-1} + \mathscr{H}_{i,k-1}^T\} + \mathbb{E}\{\mathscr{I}_{i,k-1} + \mathscr{I}_{i,k-1}^T\} \\ &\quad + \mathbb{E}\{\mathscr{J}_{i,k-1} + \mathscr{J}_{i,k-1}^T\}, \end{aligned} \tag{10}$$

where

$$\begin{aligned} \mathscr{A}_{i,k-1} &= A_{k-1}e_{i,k-1}^+(\Delta A_{k-1}x_{k-1})^T, \\ \mathscr{B}_{i,k-1} &= A_{k-1}e_{i,k-1}^+\big(f(x_{k-1}) - f(\widehat{x}_{i,k-1}^+)\big)^T, \\ \mathscr{C}_{i,k-1} &= \varepsilon A_{k-1}e_{i,k-1}^+ \sum\nolimits_{j\in\mathcal{N}_i} a_{ij}(e_{i,k-1}^+ - e_{j,k-1}^+)^T, \\ \mathscr{D}_{i,k-1} &= \varepsilon A_{k-1}e_{i,k-1}^+ \sum\nolimits_{j\in\mathcal{N}_i} a_{ij}(g_{i,k-1} - g_{j,k-1})^T, \\ \mathscr{E}_{i,k-1} &= \Delta A_{k-1}x_{k-1}\big(f(x_{k-1}) - f(\widehat{x}_{i,k-1}^+)\big)^T, \\ \mathscr{F}_{i,k-1} &= \varepsilon \Delta A_{k-1}x_{k-1} \sum\nolimits_{j\in\mathcal{N}_i} a_{ij}(e_{i,k-1}^+ - e_{j,k-1}^+)^T, \\ \mathscr{G}_{i,k-1} &= \varepsilon \Delta A_{k-1}x_{k-1} \sum\nolimits_{j\in\mathcal{N}_i} a_{ij}(g_{i,k-1} - g_{j,k-1})^T, \\ \mathscr{H}_{i,k-1} &= \varepsilon\big(f(x_{k-1}) - f(\widehat{x}_{i,k-1}^+)\big) \sum\nolimits_{j\in\mathcal{N}_i} a_{ij}(e_{i,k-1}^+ - e_{j,k-1}^+)^T, \\ \mathscr{I}_{i,k-1} &= \varepsilon\big(f(x_{k-1}) - f(\widehat{x}_{i,k-1}^+)\big) \sum\nolimits_{j\in\mathcal{N}_i} a_{ij}(g_{i,k-1} - g_{j,k-1})^T, \\ \mathscr{J}_{i,k-1} &= \varepsilon^2 \sum\nolimits_{j\in\mathcal{N}_i} a_{ij}^2(e_{i,k-1}^+ - e_{j,k-1}^+)(g_{i,k-1} - g_{j,k-1})^T. \end{aligned}$$

Furthermore, with the help of (1) and (8), one can obtain the filter errors $e_{i,k}^+$:

$$e_{i,k}^+ = [I_{n_x} - K_{i,k}^*C_{i,k}]e_{i,k}^- - K_{i,k}^*v_{i,k}.$$

Hence, the variance matrices $P_{ii,k}^+$ can be obtained as

$$P_{ii,k}^+ = [I_{n_x} - K_{i,k}^*C_{i,k}]P_{ii,k}^-[I_{n_x} - K_{i,k}^*C_{i,k}]^T + K_{i,k}^*R_{i,k}K_{i,k}^{*T}. \tag{11}$$

Noting that due to the existence of parameter uncertainty and nonlinearity, some unknown terms exist in (10) and (11). Thus, it is difficult to calculate the accurate value of $P^-_{ii,k}$ and $P^+_{ii,k}$. Fortunately, we can achieve their corresponding upper bound values as follows.

**Lemma 2.** *For the given gain sequences $\{K^*_{i,k}\}$ $(i \in \mathcal{N})$, and positive scalars $\tau_{i,s}$, $s = 1, \ldots 13$, if the following two coupled Riccati-type equations:*

$$\begin{aligned}
\Xi^-_{ii,k} &= l_{i,1}A_{k-1}\Xi^+_{ii,k-1}A^T_{k-1} + l_{i,2}\mathrm{tr}\{\mathcal{S}_{k-1}\big((1+\tau_{i,11})\Xi^+_{ii,k-1} + (1+\tau^{-1}_{i,11}) \\
&\quad \times \mathbb{E}\{\widehat{x}^+_{i,k-1}\widehat{x}^{+T}_{i,k-1}\}\big)\mathcal{S}^T_{k-1}\}\mathcal{T}_{k-1}\mathcal{T}^T_{k-1} + l_{i,3}FF^T\Xi^+_{ii,k-1} + B_{k-1}Q_{k-1}B^T_{k-1} \\
&\quad + l_{i,4}\varepsilon^2\sum\nolimits_{j\in\mathcal{N}_i} a^2_{ij}\big((1+\tau_{i,12})\Xi^+_{ii,k-1} + (1+\tau^{-1}_{i,12})\Xi^+_{jj,k-1}\big) \\
&\quad + l_{i,5}\varepsilon^2\sum\nolimits_{j\in\mathcal{N}_i} a^2_{ij}(2+\tau_{i,13}+\tau^{-1}_{i,13})\delta_M,
\end{aligned} \tag{12}$$

$$\Xi^+_{ii,k} = [I_{n_x} - K^*_{i,k}C_{i,k}]\Xi^-_{ii,k}[I_{n_x} - K^*_{i,k}C_{i,k}]^T + K^*_{i,k}R_{i,k}K^{*T}_{i,k}, \tag{13}$$

*are solvable with positive definite solutions $\Xi^-_{ii,k}$ and $\Xi^+_{ii,k}$ subject to*

$$P^-_{ii,0} \le \Xi^-_{ii,0}, \quad P^+_{ii,0} \le \Xi^+_{ii,0}.$$

*where*

$$\begin{aligned}
&l_{i,1} = 1+\tau_{i,1}+\tau_{i,2}+\tau_{i,3}+\tau_{i,4}, \quad l_{i,2} = 1+\tau^{-1}_{i,1}+\tau_{i,5}+\tau_{i,6}+\tau_{i,7}, \\
&l_{i,3} = 1+\tau^{-1}_{i,2}+\tau^{-1}_{i,5}+\tau_{i,8}+\tau_{i,9}, \quad l_{i,4} = 1+\tau^{-1}_{i,3}+\tau^{-1}_{i,6}+\tau^{-1}_{i,8}+\tau_{i,10}, \\
&l_{i,5} = 1+\tau^{-1}_{i,4}+\tau^{-1}_{i,7}+\tau^{-1}_{i,9}+\tau^{-1}_{i,10}.
\end{aligned} \tag{14}$$

*Then the covariance matrices $P^-_{ii,k}$ and $P^+_{ii,k}$ in (10) and (11) satisfy*

$$P^-_{ii,k} \le \Xi^-_{ii,k}, \quad P^+_{ii,k} \le \Xi^+_{ii,k}.$$

With the aid of the above analysis, we can achieve the following theorem.

**Theorem 1.** *For the system (1) with non-Gaussian noises, the MCC cost function $\mathcal{J}(\widehat{x}^+_{i,k})$ is maximized when the constructed E-DMCC (8) utilizes the gain*

$$K^*_{i,k} = \big((\Xi^-_{ii,k})^{-1} + Z_{i,k}C^T_{i,k}R^{-1}_{i,k}C_{i,k}\big)^{-1}Z_{i,k}C^T_{i,k}R^{-1}_{i,k}, \tag{15}$$

*with*

$$Z_{i,k} = \frac{G_\sigma(\|y_{i,k} - C_{i,k}\widehat{x}^+_{i,k}\|_{R^{-1}_{i,k}})}{G_\sigma(\|\widehat{x}^+_{i,k} - \widehat{x}^-_{i,k}\|_{(\Xi^-_{ii,k})^{-1}})}, \tag{16}$$

*where $R_{i,k}$ is the measurement noise variance, and the upper bound of one-step prediction errors covariance $\Xi^-_{ii,k}$ for filter $i$ at time step $k$ is set as (12).*

It is worth mentioning that $Z_{i,k}$ depends on $\widehat{x}^+_{i,k}$ by according to (16), and $K^*_{i,k}$ depends on $Z_{i,k}$ with the help of (15). Hence, we can conclude that the required filter gain $K^*_{i,k}$ depends on $\widehat{x}^+_{i,k}$. Additionally, in light of $G_\sigma(x,y) = \exp(-\|x-y\|^2/2\sigma^2)$, it is not difficult to know that the right side of $\widehat{x}^+_{i,k} = \widehat{x}^-_{i,k} + K^*_{i,k}(y_{i,k} - C_{i,k}\widehat{x}^-_{i,k})$ can be regarded as a nonlinear function of $\widehat{x}^+_{i,k}$, i.e., the optimal solution can be rewritten as $\widehat{x}^+_{i,k} = g(\widehat{x}^+_{i,k})$, which is difficult to solve directly. Furthermore, we use the fixed-point iterative algorithm to overcome this shortcoming.

## 4 Illustrative Example

In this section, a recursive robust state estimation problem for a ballistic object tracking system is exploited to illustrate the effectiveness of the designed E-DMCC.

The considered ballistic object tracking systems is described by the following discrete-time nonlinear stochastic equations:

$$x_{k+1} = (A_k + \Delta A_k)x_k + f(x_k) + B_k\omega_k$$

with

$$A_k = \begin{bmatrix} 1 & T & 0 & 0 \\ 0 & 1 & 1 & 0 \\ 0 & 0 & 1 & T \\ 0 & 0 & 0 & 1 \end{bmatrix}, \quad F = \begin{bmatrix} T^2/2 & 0 \\ T & 0 \\ 0 & T^2/2 \\ 0 & T \end{bmatrix}, \quad \mathcal{T}_k = \begin{bmatrix} 0.1 \\ 0.3 \\ 0.2 \\ 0.1 \end{bmatrix}, \quad \mathcal{S}_k^T = \begin{bmatrix} 0.1 \\ 0 \\ 0.2 \\ 0.1 \end{bmatrix},$$

$$U_k = \sin(3k), \qquad B_k = \text{eye}(4), \qquad \xi(x_{2,k}) = \eta_1 \exp(-\eta_2 x_{2,k}),$$

$$h(x_k) = -\frac{\mathrm{g}\xi(x_{2,k})}{2\gamma}\sqrt{\dot{x}_{1,k}^2 + \dot{x}_{2,k}^2}\begin{bmatrix} \dot{x}_{1,k} \\ \dot{x}_{2,k} \end{bmatrix}, \quad H = \begin{bmatrix} 0 \\ -\mathrm{g} \end{bmatrix}, \quad f(x_k) = F(h(x_k) + H).$$

where $x_k = [x_{1,k} \quad \dot{x}_{1,k} \quad x_{2,k} \quad \dot{x}_{2,k}]$ is the state vector, where $x_{1,k}$ and $x_{2,k}$ stand for the target abscissa and the target ordinate; $T = 0.2\,\text{s}$, $\mathrm{g} = 9.81\text{m/s}$ and $\gamma = 4 \times 10^4 \text{kg/ms}^2$ are the sampling period, the gravity acceleration, the ballistic coefficient; $\xi(\cdot)$ is the air density which usually expressed as an exponentially decaying function of the object height, and we set $\eta_1 = 1.227$, $\eta_1 = 1.093 \times 10^{-4}$.

The sensor network composed of 5 sink nodes, and its communication topology is shown in Fig. 1. The weighted adjacency matrix and the measurement matrices are set as follows ($i = 1, 2 \ldots 5$).

$$\mathcal{H} = \begin{bmatrix} 0 & 0.1 & 0.1 & 0 & 0 \\ 0.1 & 0 & 0.1 & 0 & 0 \\ 0.1 & 0.1 & 0 & 0.1 & 0 \\ 0 & 0 & 0.1 & 0 & 0.1 \\ 0 & 0 & 0 & 0.1 & 0 \end{bmatrix}, \quad C_{i,k} = \begin{bmatrix} 1+0.1i & 0 & 0 & 0 \\ 0 & 0 & 1+0.1i & 0 \end{bmatrix}.$$

Furthermore, the non-Gaussian noises adopted in this paper are set as follows:

$$\omega_k = \begin{bmatrix} Ga(\alpha_{x_1}, \beta_{x_1}) \\ Ga(\alpha_{\dot{x}_1}, \beta_{\dot{x}_1}) \\ Ga(\alpha_{x_2}, \beta_{x_2}) \\ Ga(\alpha_{\dot{x}_2}, \beta_{\dot{x}_2}) \end{bmatrix}, \quad v_{i,k} = \begin{bmatrix} Ga(\alpha_{y_{i,1}}, \beta_{y_{i,1}}) \\ Ga(\alpha_{y_{i,2}}, \beta_{y_{i,2}}) \end{bmatrix}.$$

where $Ga(\alpha, \beta)$ represents the Gamma distribution, $\alpha$ and $\beta$ are the shape parameter and the scale parameter. In particular, we set

$$\begin{aligned}
&\alpha_{x_1} = 0.10, \quad \alpha_{\dot{x}_1} = 0.11, \quad \alpha_{x_2} = 0.12, \quad \alpha_{\dot{x}_2} = 0.13, \\
&\alpha_{y_{1,1}} = 0.11, \; \alpha_{y_{1,2}} = 0.12, \; \alpha_{y_{2,1}} = 0.13, \; \alpha_{y_{2,2}} = 0.14, \alpha_{y_{3,1}} = 0.15, \; \alpha_{y_{3,2}} = 0.16, \\
&\alpha_{y_{4,1}} = 0.17, \; \alpha_{y_{4,2}} = 0.18, \; \alpha_{y_{5,1}} = 0.19, \; \alpha_{y_{5,2}} = 0.2, \\
&\beta_{x_1} = 1.0, \quad \beta_{\dot{x}_1} = 1.1, \quad \beta_{x_2} = 1.2, \quad \beta_{\dot{x}_2} = 1.3, \quad \beta_{y_{1,1}} = 1.1, \quad \beta_{y_{1,2}} = 1.2, \\
&\beta_{y_{2,1}} = 1.3, \quad \beta_{y_{2,2}} = 1.4, \quad \beta_{y_{3,1}} = 1.5, \quad \beta_{y_{3,2}} = 1.6, \quad \beta_{y_{4,1}} = 1.7, \quad \beta_{y_{4,2}} = 1.8, \\
&\beta_{y_{5,1}} = 1.9, \quad \beta_{y_{5,2}} = 2.0.
\end{aligned}$$

Additionally, the trigger threshold parameter are set as $\delta_m = 10$, $\delta_M = 15$, $a_i = -0.2$, $b_i = 1$. Here, we only plot the event trigger moment and the trigger function of filters 1–3 in Figs. 2–3. The kernel bandwidth, the estimated accuracy parameter, the maximum number of single iterations, are set as $\sigma = 10$, $\varsigma^* = 10^{-6}$, $t^* = 10^4$, respectively. The consensus gain is selected as $\varepsilon = 0.01$, and other parameters are set as $\tau_{i,s} = 1$ $(s = 1, 2, \ldots, 13)$. Furthermore, for all $i$, we set the initial conditions are

$$x_0 = [1, \; 1, \; 1, \; 1]^T, \qquad \hat{x}^+_{i,0} = \hat{x}^-_{i,0} = [0, \; 0, \; 0, \; 0]^T,$$

$$\Xi^+_{ii,0} = \Xi^-_{ii,0} = P^+_{ii,0} = P^-_{ii,0} = \text{diag}\{1, 1, 1, 1\}, \quad K^*_{i,0} = \begin{bmatrix} 1 \; 0 \; 1 \; 0 \\ 0 \; 1 \; 1 \; 1 \end{bmatrix}^T.$$

We execute the algorithm and draw the simulation results in Figs. 4–5. It is worth noting that we only analyze the test results of filters 1–3 to facilitate graphical deployment. Specifically, Fig. 2 and Fig. 3 stand for the real value $x_{1,k}$ (the black solid line) and its estimation $\hat{x}^+_{1,ik}$ (the red dashed line), the real value $\dot{x}_{1,k}$ (the black solid line) and its estimation $\hat{\dot{x}}^+_{1,ik}$ (the red dashed line) on filter $i$, respectively. In these subgraphs, one can see that the deviation between the black solid line and the red dashed line is relatively small, which means that the algorithm proposed in this paper can estimate the system state very well. As such, we can conclude that the developed filtering algorithm can track the target well in no-Gaussian noise environments.

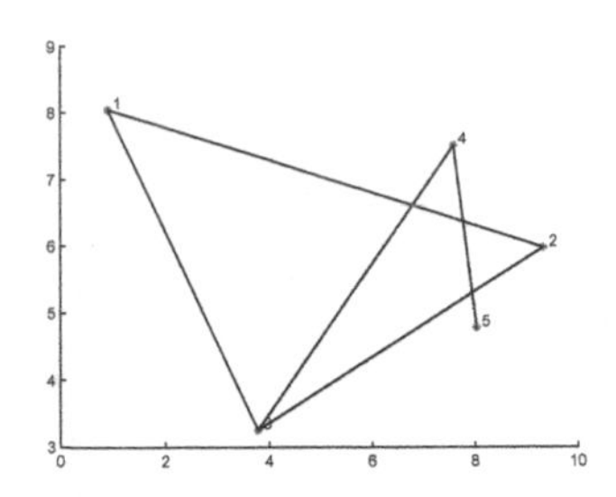

**Fig. 1.** The communication topology.

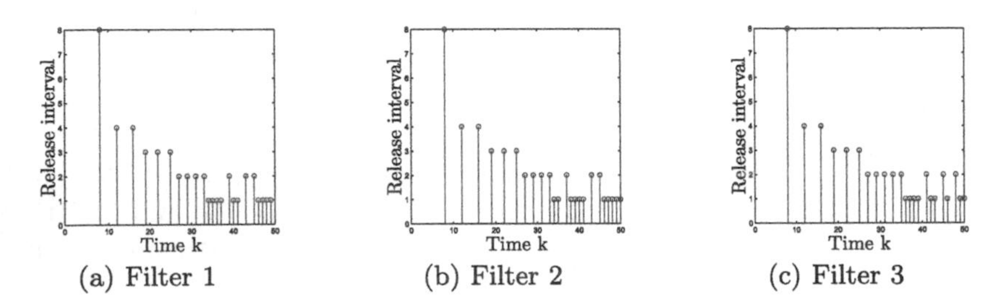


(a) Filter 1 (b) Filter 2 (c) Filter 3

**Fig. 2.** The released instants and intervals of ETS.

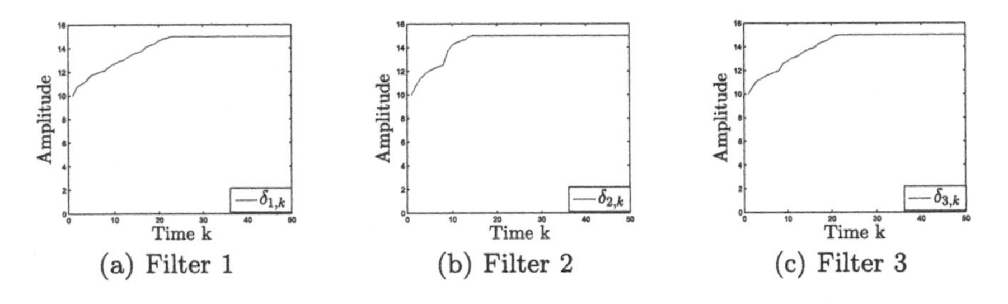


(a) Filter 1 (b) Filter 2 (c) Filter 3

**Fig. 3.** The trigger threshold function $\delta_{i,k}$.

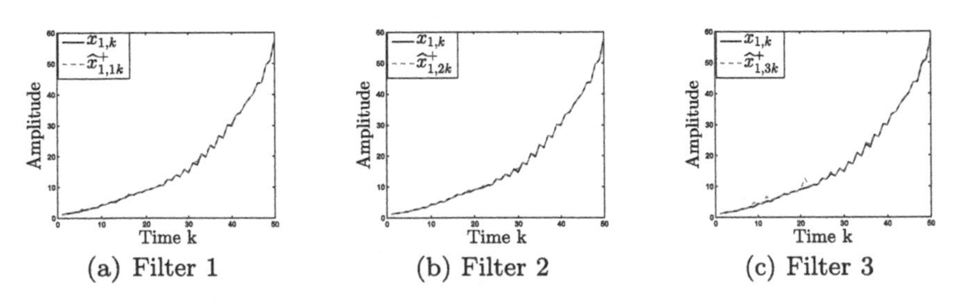


(a) Filter 1 (b) Filter 2 (c) Filter 3

**Fig. 4.** The actual system state $x_{1,k}$ and its estimation $\widehat{x}^{+}_{1,ik}$.

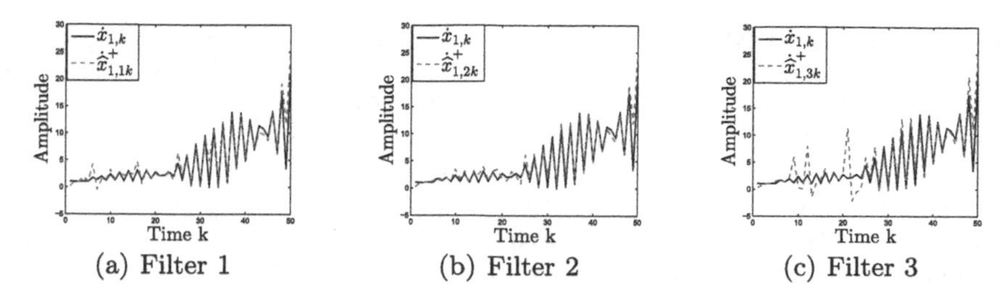

(a) Filter 1 (b) Filter 2 (c) Filter 3

**Fig. 5.** The actual system state $\dot{x}_{1,k}$ and its estimation $\hat{\dot{x}}^{+}_{1,ik}$.

## 5 Conclusions

This paper has designed a event-triggered distributed maximum correntropy for stochastic nonlinear systems with non-Gaussian noises. A weighted MCC instead of MSE has been used to design the cost function for improving the filtering performance. In addition, a adaptive ETS has been employed to reduce the communication consumption. The corresponding explicit expressions of desired filter gains and the upper bound of filtering error covariance for the constructed filtering (E-DMCC) are given by means of fixed-point iterative update rules and some typical matrix inequalities. At last, a ballistic object tracking system is employed to show the merit of the designed distributed filtering in non-Gaussian noise environments.

## References


1. Ding, L., Wang, L., Yin, G., Zheng, W., Han, Q.-L.: Distributed energy management for smart grids with an event-triggered communication scheme. IEEE Trans. Contr. Syst. T. **27**(5), 1950–1961 (2019)
2. Ding, D., Han, Q.-L., Wang, Z., Ge, X.: A survey on model-based distributed control and filtering for industrial cyber-physical systems. IEEE Trans. Ind. Informat. **15**(5), 2483–2499 (2019)
3. Chen, B., Liu, X., Zhao, H., Príncipe, J.C.: Maximum correntropy Kalman filter. Automatica **76**, 70–77 (2017)
4. Chen, B., Wang, J., Zhao, H., Zheng, N., Príncipe, J.C.: Convergence of a fixed-point algorithm under maximum correntropy criterion. IEEE Signal Proc. Let. **22**(10), 1723–1727 (2015)
5. Song, H., Ding, D., Dong, H., Wei, G., Han, Q.-L.: Distributed entropy filtering subject to DoS attacks in non-Gauss environments. Int. J. Robust. Nonlin. **30**(3), 1240–1257 (2020)
6. Song, H., Ding, D., Dong, H., Han, Q.-L.: Distributed maximum correntropy filtering for stochastic nonlinear systems under deception attacks. IEEE Transactions on Cybernetics to be Published. https://doi.org/10.1109/TCYB.2020.3016093
7. Ding, L., Han, Q.-L., Ge, X., Zhang, X.-M.: An overview of recent advances in event-triggered consensus of multiagent systems. IEEE Trans. Cybern. **48**(4), 1110–1123 (2018)

8. Tian, E., Peng, C.: Memory-based event-Triggering $H_\infty$ load frequency control for power systems under deception attacks. IEEE Transactions on Cybernetics to be Published. https://doi.org/10.1109/TCYB.2020.2972384
9. Ge, X., Han, Q.-L.: Distributed formation control of networked multi-agent systems using a dynamic event-triggered communication mechanism. IEEE Trans. Ind. Electron. **64**(10), 8118–8127 (2017)
10. Girard, A.: Dynamic triggering mechanisms for event-triggered control. IEEE Trans. Autom. Control **60**(7), 1992–1997 (2015)

# Guaranteed Cost Control of Networked Inverted Pendulum Visual Servo System with Computational Errors and Multiple Time−Varying Delays

Qianjiang Lu[1], Dajun Du[1](✉), Changda Zhang[1], Minrei Fei[1], and Aleksandar Rakić[2]

[1] Shanghai Key Laboratory of Power Station Automation Technology, School of Mechanical Engineering and Automation, Shanghai University, Shanghai 200072, People's Republic of China
ddj@i.shu.edu.cn

[2] School of Electrical Engineering, University of Belgrade, Bulevar Kralja Aleksandra 73, 11000 Belgrade, Serbia

**Abstract.** Computational errors and multiple time−varying delays will inevitably affect control performance of networked inverted pendulum visual servo system (NIPVSS). This paper investigates guaranteed cost control of NIPVSS. Firstly, computational errors is converted into parameter uncertainties and NIPVSS is then modeled as a closed-loop system with parameter uncertainties and multiple time−varying delays. Secondly, a guarantee cost controller is designed to maintain the asymptotic stability of NIPVSS and meanwhile minimize the performance index. Finally, experimental results demonstrate the feasibility of the proposed method while the effectiveness of parameter uncertainties and multiple time−varying delays on the system stability and performance index are analyzed.

**Keywords:** Networked inverted pendulum visual servo system · Guaranteed cost control · Parameter uncertainties · Multiple time-varying delays

## 1 Introduction

Due to its nonlinear, multi-variable and strongly coupled characteristics [1], inverted pendulum system (IPS) ha always been typical experimental platform to verify new control theory and algorithms for further applying to practical control fields [2]. With the rapid development of visual sensing and network technology, they are gradually being used in IPS, making IPSs further become into NIPVSS.

Compared with traditional IPS, due to limited communication bandwidth and unbalanced network load [3,4], NIPVSS inevitably has network − induced time delay, which will affect control performance. For example, network − induced delay


Q. Han et al. (Eds.): LSMS 2021/ICSEE 2021, CCIS 1469, pp. 583–592, 2021.
https://doi.org/10.1007/978-981-16-7213-2_56

is converted into parameter uncertainty and a memoryless state feedback controller is designed to achieve system stability [5,6]. For visual servo control systems, sliding mode variable structure control is employed to solve image processing delay [7]. Furthermore, network-induced delay and image processing delay are considered and a switching controller is designed to make the system stability [8,9]. It is worth finding that a multiple time–varying delays system with disturbances is considered and $H_\infty$ controller is designed to suppress the disturbance while solve the effect of multiple time-varying delay on NIPVSS [10].

Computational errors are often neglected in visual servo control systems [11]. Computational errors are regarded as an external disturbance [10]. For errors caused by hinge moment of inertia and friction coefficient, the system is modeled as parameter uncertainties [12]. Angle measurement error, which caused by low resolution limit of the sensor, is also described as parameter uncertainy [13].

The above mentioned work has achieved some results of the stability, which ignore error characteristics and studies system performance index. In particular, there exist two problems: 1) Computational errors and multiple time–varying delays will affect system stability, how to combine into system model is the first challenge problem; 2) Considering computational errors and multiple time–varying delays, how to design a controller with tolerance ability and evaluate system performance is another challenge.

To address these challenges, this paper investigates guaranteed cost control under parameter uncertainties and multiple time–varying delays. The main contributions include:1) Considering the effects of computational errors and multiple time–varying delays, a closed-loop model is established with parameter uncertainties and multiple time–varying delays. 2) To improve the performance of the NIPVSS, a guaranteed cost controller is designed, which can tolerate a certain computational errors and the upper bound on the performance index of the closed-loop system is given.

The rest of this paper is organized as follows. In Sect. 2, parameter uncertainties and multiple time–varying delays model of NVSIPS is established. Sufficient conditions for guaranteed cost controller is presented in Sect. 3. Section 4 demonstrates experimental results and discusses the influence of uncertainty on NIPVSS stability and performance index. Conclusion and future work are given in Sect. 5.

## 2 Problem Formulation

The structure of NIPVSS is shown in Fig. 1. The movement image of inverted pendulum is captured by industrial camera at time instant $t_k$ and the state variable $x(t)$ is computed in image processing unit. After computing $x(t)$, the camera is triggered to take the next image, while $x(t)$ is transmitted to the remote controller through the network. The controller uses the received information to compute the corresponding control signal, and it is sent to the actuator. Finally, inverted pendulum can run stably.

In Fig. 1, it can be found that network-induced delay $\tau_k^{sc}$ (from sensor to controller), $\tau_k^{ca}$ (from controller to actuator), image processing delay $d_k$ and

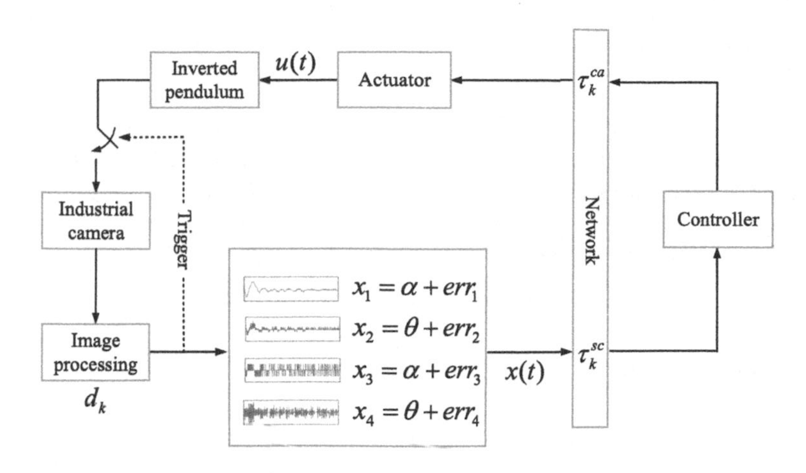


**Fig. 1.** The structure of NIPVSSs.

state errors $err(t)$ affect the performance of NIPVSS. Although $H_\infty$ disturbance attenuation level $\gamma$ is employed to achieve the stability of NIPVSS [10], how to further restrain performance index of NIPVSS and explore an optimal control is not considered. Therefore, the idea of guaranteed cost control is adopted to design a guaranteed cost controller.

Consider the following model of inverted pendulum

$$\begin{aligned} \dot{x}(t) &= Ax(t) + Bu(t) \\ y(t) &= Cx(t) + Du(t) \end{aligned} \tag{1}$$

where $x(t) \in \mathbb{R}^4$ is the system state; $x_1 = \alpha, x_2 = \theta, x_3 = \dot{\alpha}$ and $x_4 = \dot{\theta}$ are cart position, pendulum angle, cart and angular velocity respectively; $A$, $B$, $C$ and $D$ are constant matrices.

Considering control signal $u(t) = Kx(t - d(t) - \tau(t))$, the closed-loop model of NIPVSS with multiple time−varying delays can be given by

$$\begin{aligned} &\dot{x}(t) = Ax(t) + BKx(t - d(t) - \tau(t)) \\ &t \in \left[t_k + d_k + \tau_k^{sc} + \tau_k^{ca}, t_{k+1} + d_{k+1} + \tau_{k+1}^{sc} + \tau_{k+1}^{ca}\right). \end{aligned} \tag{2}$$

Due to unbalanced changes in environmental background and illumination, image processing inevitably contains noise, which further leads to the computational errors of state variables. These computational errors are converted into parametric uncertainties. Define the conversion equation as follows

$$err(t) = \Delta A x(t) \tag{3}$$

where $\Delta A = diag\left\{ \Delta_1 \; \Delta_2 \; \Delta_3 \; \Delta_4 \right\}$ represents parameter uncertainty matrix; $\Delta_1$, $\Delta_2$, $\Delta_3$ and $\Delta_4$ represent parameter uncertainties of above four states.

According to (3), parameter uncertainty matrix can usually be expressed as

$$\varDelta A = DF(t)E \tag{4}$$

Substituting (4) into (2), the closed-loop system model of NIPVSS with parameter uncertainties and multiple time−varying delays can be given by

$$\begin{aligned} &\dot{x}(t) = (A + DF(t)E)x(t) + BKx(t - d(t) - \tau(t)) \\ &t \in \left[t_k + d_k + \tau_k^{sc} + \tau_k^{ca}, t_{k+1} + d_{k+1} + \tau_{k+1}^{sc} + \tau_{k+1}^{ca}\right). \end{aligned} \tag{5}$$

**Control Objective:** Define performance index of the system as

$$J = \int_0^{\infty} [x^T(t)Qx(t) + u^T(t)Ru(t)]d(t) \tag{6}$$

where $Q$ and $R$ are the given symmetric weighting matrices. For the system (5), a memoryless state feedback controller gain $K$ needs to be designed, so that the system is asymptotically stable and $J$ is minimized.

# 3 Guaranteed Cost Control of NIPVSSs

The above has presented the closed-loop model of NIPVSS. Theorems of guaranteed cost control of NIPVSSs are presented below.

**Theorem 1.** *For the given constants* $0 < \tau$, $0 < d_1 < d_2$, $0 < \mu_1$, $0 < \mu_2$, $\varepsilon_i (i = 1, 2, 3, 4)$, *matrices* $Q$ *and* $R$, *if there exist* $0 < \lambda$ *and real symmetric matrices* $X$, $\tilde{Q}_i$ $(i = 1, 2, \cdots, 7)$, $\tilde{Z}_i$ $(i = 1, 2, 3, 4)$ *satisfying*

$$\begin{bmatrix} \varPsi_{11} & Z_1 & 0 & 0 & Z_3 & 0 & BY & 0 & \varPsi_{19} & \varPsi_{110} & \varPsi_{111} & \varPsi_{112} & XE^T & X & Y^T \\ * & \varPsi_{22} & Z_2 & 0 & 0 & 0 & Z_4 & 0 & 0 & 0 & 0 & 0 & 0 & 0 & 0 \\ * & * & \varPsi_{33} & Z_2 & 0 & 0 & 0 & 0 & 0 & 0 & 0 & 0 & 0 & 0 & 0 \\ * & * & * & \varPsi_{44} & 0 & 0 & 0 & 0 & 0 & 0 & 0 & 0 & 0 & 0 & 0 \\ * & * & * & * & \varPsi_{55} & Z_3 & 0 & 0 & 0 & 0 & 0 & 0 & 0 & 0 & 0 \\ * & * & * & * & * & \varPsi_{66} & 0 & 0 & 0 & 0 & 0 & 0 & 0 & 0 & 0 \\ * & * & * & * & * & * & \varPsi_{77} & Z_4 & \varPsi_{79} & \varPsi_{710} & \varPsi_{711} & \varPsi_{712} & 0 & 0 & 0 \\ * & * & * & * & * & * & * & \varPsi_{88} & 0 & 0 & 0 & 0 & 0 & 0 & 0 \\ * & * & * & * & * & * & * & * & \varPsi_{99} & \varPsi_{910} & \varPsi_{911} & \varPsi_{912} & 0 & 0 & 0 \\ * & * & * & * & * & * & * & * & * & \varPsi_{1010} & \varPsi_{1011} & \varPsi_{1012} & 0 & 0 & 0 \\ * & * & * & * & * & * & * & * & * & * & \varPsi_{1111} & \varPsi_{1112} & 0 & 0 & 0 \\ * & * & * & * & * & * & * & * & * & * & * & \varPsi_{1212} & 0 & 0 & 0 \\ * & * & * & * & * & * & * & * & * & * & * & * & -\lambda I & 0 & 0 \\ * & * & * & * & * & * & * & * & * & * & * & * & * & Q^{-1} & 0 \\ * & * & * & * & * & * & * & * & * & * & * & * & * & * & R^{-1} \end{bmatrix} < 0 \tag{7}$$

*where* $d_{12} = d_2 - d_1, d_3 = d_2 + \tau, d_{13} = d_3 - d_1$,

$$\Psi_{11} = AX + XA^T + (\tilde{Q}_1 + \ldots + \tilde{Q}_7) - \tilde{Z}_1 - \tilde{Z}_3 + \lambda DD^T,$$
$$\Psi_{22} = -\tilde{Q}_1 - \tilde{Z}_1 - \tilde{Z}_2 - \tilde{Z}_4, \Psi_{33} = -(1-\mu_1)\tilde{Q}_2 - 2\tilde{Z}_2, \Psi_{44} = -\tilde{Q}_3 - \tilde{Z}_2,$$
$$\Psi_{55} = -(1-\mu_2)\tilde{Q}_4 - 2\tilde{Z}_3, \Psi_{66} = -\tilde{Q}_5 - \tilde{Z}_3, \Psi_{77} = -2\tilde{Z}_4, \Psi_{88} = -\tilde{Q}_7 - \tilde{Z}_4,$$
$$\Psi_{99} = -2\varepsilon_1 X + \varepsilon_1^2\tilde{Z}_1 + \lambda d_1^2 DD^T, \Psi_{1010} = -2\varepsilon_2 X + \varepsilon_2^2\tilde{Z}_2 + \lambda d_{12}^2 DD^T,$$
$$\Psi_{1111} = -2\varepsilon_3 X + \varepsilon_3^2\tilde{Z}_3 + \lambda\tau^2 DD^T, \Psi_{1212} = -2\varepsilon_4 X + \varepsilon_4^2\tilde{Z}_4 + \lambda d_{13}^2 DD^T,$$
$$\Psi_{19} = d_1 XA^T + \lambda d_1 DD^T, \Psi_{110} = d_{12} XA^T + \lambda d_{12} DD^T,$$
$$\Psi_{111} = \tau XA^T + \lambda\tau DD^T, \Psi_{112} = d_{13} XA^T + \lambda d_{13} DD^T, \Psi_{79} = d_1 Y^T B^T,$$
$$\Psi_{710} = d_{12} Y^T B^T, \Psi_{711} = \tau Y^T B^T, \Psi_{712} = d_{13} Y^T B^T, \Psi_{910} = \lambda d_1 d_{12} DD^T,$$
$$\Psi_{911} = \lambda d_1 \tau DD^T, \Psi_{912} = \lambda d_1 d_{13} DD^T, \Psi_{1011} = \lambda d_{12}\tau DD,$$
$$\Psi_{1012} = \lambda d_{12} d_{13} DD^T, \Psi_{1112} = \lambda\tau d_{13} DD^T,$$

*then the closed-loop system (5) is asymptotically stable for parameter uncertainties satisfying* $\|F(t)\| \leqslant 1$ *and multiple time−varying delays. The gain of controller is* $K = YX^{-1}$ *and there exists an performance upper bound* $J^*$.

*Proof* Define Lyapunov−Krasovskii function as

$$V(x(t)) = V_1 + V_2 + V_3 + V_4 + V_5 + V_6 + V_7, \tag{8}$$

where

$$V_1 = x^T(t)Px(t), V_2 = \int_{t-d_1}^{t} x^T(s)Q_1x(s)ds + \int_{t-d(t)}^{t} x^T(s)Q_2x(s)ds$$
$$+ \int_{t-d_2}^{t} x^T(s)Q_3x(s)ds, V_3 = \int_{-d_1}^{0}\int_{t+\theta}^{t} d_1\dot{x}^T(s)Z_1\dot{x}(s)dsd\theta + \int_{-d_2}^{-d_1}\int_{t+\theta}^{t} d_{12}$$
$$\dot{x}^T(s)Z_2\dot{x}(s)dsd\theta, V_4 = \int_{t-\tau(t)}^{t} x^T(s)Q_4x(s)ds + \int_{t-\tau}^{t} x^T(s)Q_5x(s)ds,$$
$$V_5 = \int_{-\tau}^{0}\int_{t+\theta}^{t} \tau\dot{x}^T(s)Z_3\dot{x}(s)dsd\theta, V_6 = \int_{t-d(t)-\tau(t)}^{t} x^T(s)Q_6x(s)ds$$
$$+ \int_{t-d_3}^{t} x^T(s)Q_7x(s)ds, V_7 = \int_{-d_3}^{-d_1}\int_{t+\theta}^{t} d_{13}\dot{x}^T(s)Z_4\dot{x}(s)dsd\theta.$$

Then, taking the derivative of $V(x(t))$ with respect to $t$ and integrating $\dot{V}(x(t))$ in the interval $0 \to \infty$ lead to

$$\int_0^\infty \dot{V}(x(t)) < -\int_0^\infty x^T(t)Qx(t) + u^T(t)Ru(t) \Rightarrow J < V(x(0)) \tag{9}$$

Furthermore, using *Schur* lemma, premultiplying and postmultiplying with $diag\{P^{-1}\ P^{-1}\ P^{-1}\ P^{-1}\ P^{-1}\ P^{-1}\ P^{-1}\ P^{-1}\ Z_1^{-1}\ Z_2^{-1}\ Z_3^{-1}\ Z_3^{-1}\ \mathrm{I}\}$ and setting $\tilde{Q}_i = P^{-1}Q_iP^{-1}\ i = 1, 2, \cdots, 7$, $\tilde{Z}_j = P^{-1}Z_jP^{-1}\,(i = 1, 2, \cdots, 4)$, $X = P^{-1}$, $Y = KX$ and $-Z_j^{-1} < -2\varepsilon_j P^{-1} + \varepsilon_j^2 P^{-1}Z_jP^{-1}\,(j = 1, 2, \cdots, 4)$, we can get Theorem 1. Hence, the closed-loop system (5) are asymptotically stable and there is an upper bound on the performance index. The proof is thus completed.

Theorem 1 provides a sufficient condition for the existence of guaranteed cost control law. Furthermore, the specific upper bound $J^*$ of the performance index is given by Theorem 2.

**Theorem 2.** *For the given constants* $0 < \tau$, $0 < d_1 < d_2$, $0 < \mu_1$, $0 < \mu_2$, $\varepsilon_i (i = 1, 2, 3, 4)$, $\delta_j (j = 1, 2, \cdots, 7)$, $N_i (i = 1, 2, \cdots, 7)$, $Q$ *and* $R$, *if there exist* $0 < \lambda$, $0 < \alpha$, *real symmetric positive definite matrices* $X$, $\tilde{Q}_i$ $(i = 1, 2, \cdots, 7)$, $\tilde{Z}_i$ $(i = 1, 2, \cdots, 7)$, $M_i$ $(i = 1, 2, \cdots, 7)$ *satisfying*

$$\begin{aligned} &\min J^* = \alpha + Trace(M_1 + M_2 + \cdots + M_7) \\ &s.t. \ (i) \qquad (6) \\ &\quad (ii) \begin{bmatrix} -\alpha & x^T(0) \\ * & -X \end{bmatrix} < 0 \\ &\quad (iii) \begin{bmatrix} -M_i & N_i^T \\ N_i & \frac{\tilde{Q}_1 - 2\varepsilon_j X}{\varepsilon_j^2} \end{bmatrix} < 0 \\ &\quad j = 1, 2, \cdots 7 \end{aligned} \tag{10}$$

*then the closed-loop system (5) is asymptotically stable for any parameter uncertainties satisfying* $\|F(t)\| \leqslant 1$ *and multiple time−varying delays. The gain of the guaranteed cost controller is* $K = YX^{-1}$ *and the upper bound is* $J^* = \alpha + Trace(M_1 + M_2 + M_3 + M_4 + M_5 + M_6 + M_7)$.

*Proof.* According to (10), the upper bound of the system performance index is $V(x(0))$. According to $V(x(t))$, $V(x(0))$ is can be expressed as

$$\begin{aligned} V(x(0)) < x^T(0)\,Px(0) + \int_{-d_1}^{0} x^T(s)\,Q_1 x(s)\,ds + \int_{-d_2}^{0} x^T(s)\,Q_2 x(s)\,ds \\ + \int_{d_2}^{0} x^T(s)\,Q_3 x(s)\,ds + \int_{\tau}^{0} x^T(s)\,Q_4 x(s)\,ds + \int_{-\tau}^{0} x^T(s)\,Q_5 x(s)\,ds \\ + \int_{0-d_2-\tau}^{0} x^T(s)\,Q_6 x(s)ds + \int_{-d_3}^{0} x^T(s)\,Q_7 x(s)ds \end{aligned} \tag{11}$$

The upper bound on $x^T(0)Px(0)$ is constrained by the following inequality

$$\begin{bmatrix} -\alpha & x^T(0) \\ * & -X \end{bmatrix} < 0$$

Taking $\int_{-d_1}^{0} x^T(s)\,Q_1 x(s)\,ds$ for example and the process of finding the upper bound of other integral term is the same as it. The upper bound on $\int_{-d_1}^{0} x^T(s)\,Q_1 x(s)\,ds$ is constrained by the following inequality

$$\begin{bmatrix} -M_1 & N_1^T \\ N_1 & \frac{\tilde{Q}_1 - 2\varepsilon_1 X}{\varepsilon_1^2} \end{bmatrix} < 0$$

where $\int_{-d_1}^{0} x^T(s)\,x(s)\,ds = N_1 N_1^T$.

According to the above analysis, the upper bound of performance index $J^*$ can be obtained by minimizing $\alpha + Trace(M_1 + M_2 + M_3 + M_4 + M_5 + M_6 + M_7)$ to ensure the upper bound of the performance index given by (6).

## 4 Experiments Analysis and Discussion

The above two theorems have presented controller design methods and performance index upper bound. To verify the above proposed methods, according to statistical analysis of experimental results, the values of the relevant parameters are given as follow

$$d_1 = \min_{k \in N} (d_k) = 0.017s, d_2 = \max_{k \in N} (d_k + d_{k+1}) = 0.038s,$$

$$\delta_j = 1.1, j = 1, 2, \cdots, 7, \mu_1(t) = 0.7, \mu_2(t) = 0.3, x(0) = \begin{bmatrix} 0 \; 0.1 \; 0 \; 0 \end{bmatrix}^T,$$

$$\tau = \max_{k \in N} \left( \tau_{k+1}^{sc} + \tau_{k+1}^{ca} \right) = 0.01s, \sigma_1 = 0.01, \sigma_2 = 0.75, \sigma_1 = 1.2, \sigma_1 = 0.022,$$

$$A = \begin{bmatrix} 0 & 0 & 1 & 0 \\ 0 & 0 & 0 & 1 \\ 0 & 0 & 0 & 0 \\ 0 & 29.4 & 0 & 0 \end{bmatrix}, B = \begin{bmatrix} 0 \\ 0 \\ 1 \\ 3 \end{bmatrix}, D \begin{bmatrix} 1 & 0 & 0 & 0 \\ 0 & 1 & 0 & 0 \\ 0 & 0 & 1 & 0 \\ 0 & 0 & 0 & 1 \end{bmatrix}, F(t) = \begin{bmatrix} r_1(t) & 0 & 0 & 0 \\ 0 & r_2(t) & 0 & 0 \\ 0 & 0 & 0 & 0 \\ 0 & 0 & 0 & 0 \end{bmatrix},$$

$$E = \begin{bmatrix} 0.4 & 0 & 0 & 0 \\ 0 & 0.82 & 0 & 0 \\ 0 & 0 & 0 & 0 \\ 0 & 0 & 0 & 0 \end{bmatrix}, -1 \leqslant r_1(t) \leqslant 1, -1 \leqslant r_2(t) \leqslant 1$$

and the $N_i, i = 1, 2, \cdots, 7$ can be solved according to $\varphi(t), t \in \begin{bmatrix} \varphi \; 0 \end{bmatrix}$.

From the previous analysis, the parameter uncertainties range of the cart position and angle pendulum are $-0.4 \leqslant \Delta_1 \leqslant 0.4$ and $-0.82 \leqslant \Delta_2 \leqslant 0.82$. Under these parameter uncertainties, we choose weighting matrices $Q = I_4$, $R = 1$ and solve (7) and (11), the controllers $K_1$ and $K_2$ can be obtained as follows:

$$K_1 = \begin{bmatrix} 4.60 \; -38.41 \; 5.22 \; -7.03 \end{bmatrix}, K_2 = \begin{bmatrix} 4.35 \; -38.49 \; 5.06 \; -6.95 \end{bmatrix}.$$

The states of cart position and pendulum angle are shown in Figs. 2 and 3 under $K_1$ and $K_2$, respectively. As can be seen from Figs. 2 and 3 that the controllers have better control over the system. Furthermore, according to Theorems 1 and 2, it is found that the upper bound of system guaranteed performance index in Theorem 1 is 16.57, while the upper bound of optimal guaranteed performance index in Theorem 2 is 15.61, which confirms the existence of optimal feasible solution in Theorem 2.

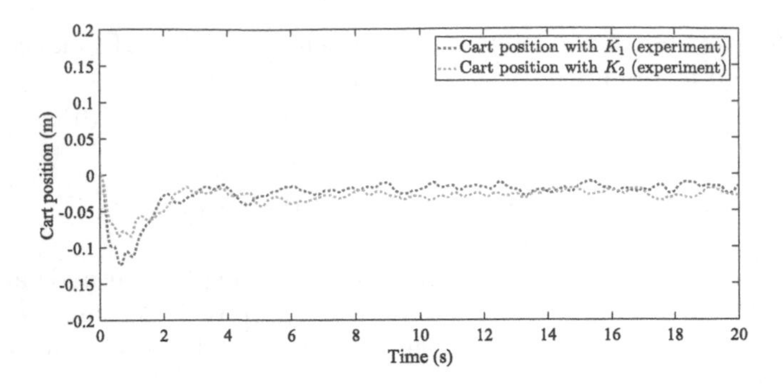


**Fig. 2.** Cart position under $K_1$ and $K_2$.

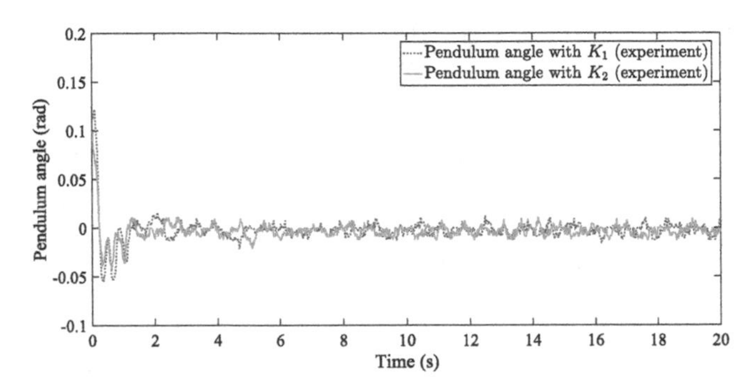


**Fig. 3.** Pendulum angle under $K_1$ and $K_2$.

**Table 1.** Given maximum absolute range of $\Delta_1$, the maximum absolute range of $\Delta_2$.

| Given $\lvert\Delta_1\rvert_{max}$ | 0.66 | 0.63 | 0.55 | 0.33 | 0.11 | 0.0 |
|---|---|---|---|---|---|---|
| $\lvert\Delta_2\rvert_{max}$ | 0.0 | 0.5 | 1 | 2 | 3 | 3.4 |

### 4.1 The Influence of Parameter Uncertainties on NIPVSS Stability and Performance Index

Given absolute range of $\Delta_1$, we can infer the absolute range of $\Delta_2$, which can ensures the closed-loop system (5) to be stable and minimize performance index. The corresponding relationship between $\Delta_1$ and $\Delta_2$ is presented in Table 1. From Table 1, two conclusions can be obtained: 1) The experimental range $\Delta_1 \in \left[-0.4\ 0.4\right]$ and $\Delta_2 \in \left[-0.82\ 0.82\right]$ is a subset of the theoretical range $\Delta_1 \in \left[-0.55\ 0.55\right]$ and $\Delta_2 \in \left[-1\ 1\right]$, which verifies the effectiveness of the proposed method; 2) when the range of parameter uncertainties is changed, as long as it is within the range of Table 1, there is still a guarantee cost controller to make the system stable and there still exists an upper bound of performance index.

**Table 2.** Given $|\Delta_1|$ and $|\Delta_2|$, the corresponding performance upper bound $J^*$.

| Given $\|\Delta_1\|$ | 0 | 0.2 | 0.3 | 0.4 | 0.55 |
|---|---|---|---|---|---|
| Given $\|\Delta_2\|$ | 0 | 0.2 | 0.4 | 0.82 | 1 |
| $J^*$ | 15.07 | 15.11 | 15.18 | 15.61 | 18.76 |

**Table 3.** Given network–induced delay $\tau(t)$ or image processing delay $d(t)$, the corresponding performance upper bound $J^*$.

| Given $\tau(t)$ | 0.01 | 0.012 | 0.014 | 0.016 | 0.018 |
|---|---|---|---|---|---|
| $J^*$ | 15.61 | 15.69 | 15.94 | 16.39 | 17.87 |
| Given $\tau(t)$ | 0.01 | 0.012 | 0.014 | 0.016 | 0.018 |
| $J^*$ | 15.61 | 15.69 | 15.94 | 16.39 | 17.87 |

The above analyzes parameter uncertainties range on the stability of NIPVSS, following by analysis of the influence of parameter uncertainty on performance index. We select $\Delta_1 \in [-0.55\ 0.55]$ and $\Delta_2 \in [-0.1\ 0.1]$ and the performance indices under the corresponding subsets are computed as shown in Table 2. It can be seen from Table 2 that to overcome the increase of parameter uncertainties range, the performance index of NIPVSS also increases, which means NIPVSS need greater performance cost to overcome greater parameter uncertainties.

### 4.2 The Influence of Multiple Time–varying Delays on NIPVSS Stability and Performance Index

The previous subsection has analyzed the influence of parameter uncertainties on NIPVSS including stability and performance index. This subsection analyzes the influence of multiple time–varying delays on NIPVSS. For network–induced delay or image processing time, we increase delays and compute corresponding performance index upper bound as shown in Table 3. It is obvious that the performance index upper bound increases as network–induced delay or image processing time $d(t)$ increases, which means the increase of time delays reduces system performance. It is worth noting that the controllers in network–induced delay range in Table 3 enable the system to run stably.

## 5 Conclusion

This paper has proposed a design method of guaranteed cost controller for NIPVSS with model uncertainty and multiple time-varying delay. For computational errors and time–varying delays, NIPVSS is modeled combined with parameter uncertainties and multiple time–varying delays. A guaranteed cost controller is then designed to tolerate computational errors and time-varying delay and performance index has an upper bound. Experimental results confirms that the effectiveness of the proposed controller on NIPVSS platform. In the future, it is a challenge to reduce parameter uncertainties range and ensure the security of NIPVSS.

**Acknowledgments.** This work is supported by Project of Science and Technology Commission of Shanghai Municipality, China (19510750300, 20JC1414000, 195007 12300), the National Natural Science Foundation of China (92067106), and 111 Project (D18003).

## References

1. Li, Z., Zhang, Y.: Robust adaptive motion/force control for wheeled inverted pendulums. Automatica **46**(8), 1346–1353 (2010)
2. Žilić, T., Pavković, D., Zorc, D.: Modeling and control of a pneumatically actuated inverted pendulum. ISA Trans. **48**(3), 327–335 (2009)
3. Du, D., Li, X., Li, W., Chen, R., Fei, M., Wu, L.: ADMM-based distributed state estimation of smart grid under data deception and denial of service attacks. IEEE Trans. Syst. Man Cybern. Syst. **49**(8), 1–14 (2019)
4. Du, D., Chen, R., Li, X., Wu, L., Zhou, P., Fei, M.: Malicious data deception attacks against power systems: a new case and its detection method. Trans. Inst. Measurement Control **41**(6), 1590–1599 (2019)
5. Xie, G., Wang, L.: Stabilization of NCSs with time−varying transmission period. In: 2005 IEEE International Conference on Systems, vol. 4, pp. 3759–3763 (2005)
6. Xie, G., Wang, L.: Stabilization of networked control systems with time−varying network−induced delay. In: 2004 43rd IEEE Conference on Decision and Control, vol. 4, pp. 3551–3556 (2004)
7. Song, Y., Du, D., Sun, Q., Zhou, H., Fei, M.: Sliding mode variable structure control for inverted pendulum visual servo systems. IFAC-PapersOnLine **52**(11), 262–267 (2019)
8. Chen, C., Wu, H., Kühnlenz, K., Hirche, S.: Switching control for a networked vision−based control system. Automatisierungstechnik **59**(2), 124–133 (2011)
9. Wu, H., Lou, L., Chen, C., Hirche, S., Kühnlenz, K.: A framework of networked visual servo control system with distributed computation. In: International Conference on Control Automation Robotics and Vision, pp. 1466–1471 (2010)
10. Du, D., Zhang, C., Song, Y., Zhou, H., Li, W.: Real-time $H_\infty$ control of networked inverted pendulum visual servo systems. IEEE Trans. Cybern. **50**(12), 1–14 (2019)
11. Wu, H., Lou, L., Chen, C., Hirche, S., Kühnlenz, K.: Cloud−based networked visual servo control. IEEE Trans. Ind. Electron. **60**(2), 554–566 (2013)
12. Dong, J., Liu, D., Liu, Z.: Design of the triple inverted-pendulum $H_\infty$ robust controller with uncertain factors. Industrial Instrumentation and Automation (2017)
13. Takahashi, N., Sato, O., Kono, M.: Robust control method for the inverted pendulum system with structured uncertainty caused by measurement error. Artif. Life Robot. **14**(4), 574–577 (2009)

# Intelligent Manufacturing, Autonomous Systems, Intelligent Robotic Systems

# Double Generating Functions Approach to Quadrupedal Trot Gait Locomotion

Chuliang Xie, Dijian Chen(✉), Tao Xiang, Shenglong Xie, and Tao Zeng

School of Mechanical and Electrical Engineering, China Jiliang University, Hangzhou 310018, China
djchen@cjlu.edu.cn

**Abstract.** This paper introduces a double generating functions based method for the trot gait locomotion of the quadruped robot. A virtual leg technique is employed to simplify the quadrupedal model to a bipedal model, the double generating functions based algorithm is given to generate the energy optimal trot gaits, and finally a comparison between the introduced method and CPG based method illustrates the substantial advantage of the method. The double generating functions method belongs to the model based locomotion method, but due to its particular algorithm structure it is efficient in online computation. furthermore, it can give more detailed planning such as optimal energy consumption and gait pattern control than that of the popular bionic based methods. Though the simulation is for the linear dynamics, the introduced method also works well for the nonlinear system by involving the PD tracking control technique.

**Keywords:** Double generating functions · Trot gait locomotion · Virtual leg · CPG

## 1 Introduction

Among legged robot, the quadruped robot possesses a moderate number of feet, good motion performance and high carrying capacity, hence has a broad application prospect [1]. The quadrupedal gait locomotion has been one of the focus research in the robot field for years. At present, model based and bionic based methods are the two common control strategies applied to quadruped robot gait locomotion [2]. Model based method is a traditional quadrupedal locomotion method, its control idea is modeling-planning-control [3]. M. Raibert proposes the application of a virtual leg model to control the gait of a quadruped robot [4], and his team develops the first jumping robot in the world with a spring-loaded inverted pendulum model as a robot dynamics model [5]. D. Gong builds a relative complete three-dimensional model of a quadruped robot, and the inverse kinematics of the model is used to plan the leg trajectory [6]. A. W. Winkler uses a simplified centroidal dynamics model to represent the leg movement of a quadruped robot,

This work is supported by the National Natural Science Foundation of China (Grant No. 62003321), and Natural Science Foundation of Zhejiang Province, China (Grant No. LQ19F030006 and LQ20E050017).


Q. Han et al. (Eds.): LSMS 2021/ICSEE 2021, CCIS 1469, pp. 595–606, 2021.
https://doi.org/10.1007/978-981-16-7213-2_57

which is affected by the position and force of the foot [7]. X. Chen employs the spiral theory to analyze the mobility of quadruped robot [8]. In the process of establishment of the kinematic model, the robot body is divided into translation and rotation of the body center. The bionic based method via CPG (central pattern generator) is that researchers conduct engineering simulation, simplification and improvement on the CPG, neural vibrator, biological reflection and so on, gradually form a more scientific motion control method and theory, which can be used to realize the control of robot rhythmic motion [9]. Among the existing CPG models, the Matsuoka differential oscillator model proposed by Matsuoka has simple mathematical expression and clear biological significance, and is widely used in rhythmic motion control of robots. Then H. Kimura introduces feedback signal based on this model, and simulates the influence of external interference on the stability of the system [10]. Y. Fukuoka verifies the effectiveness and accuracy of Kimura's improved CPG model through the gait control experiment of tekken quadruped robot [11]. H. Zheng modifies and refines Mastuoka's oscillator model, and verifies the effectiveness of robot rhythmic motion control method based on biological CPG control mechanism through simulation [12]. A nonlinear oscillator is introduced as the mathematical model of CPG, which can generate the rhythmic signal output when CPG lacks sensor feedback [13]. Based on the nonlinear oscillator model, Hopf model and Van der Pol model are widely used in practical engineering [14]. Compared with Kimura CPG model, the parameters of Hopf oscillator correspond to amplitude, phase and frequency one by one. It has the characteristic of limit cycle and is suitable for the gait planning of the legged robot.

In the field of robot, the energy consumption in the process of robot walking is an important problem to be studied. This problem can be formalized as a standard optimal control problem, which can be solved by Riccati transformation method [15], shooting method, parameter optimization and other methods [16, 17]. However, when dealing with a large number of different boundary conditions, these methods ignore the importance of online computing, which makes the efficiency of online gait generation very low. In recent years, the problem of optimal trajectory generation considering energy consumption has been widely studied in the field of robotics. Recently, C. Park proposes a single generating function (GF) method based on optimal trajectory generation, which can generate their optimal trajectory under different boundary conditions [18]. Z. Hao applies the method of double GFs to the gait planning of bipedal walking robot when walking on flat ground. By using this method, the finite time linear quadratic (LQ) optimal control problem is solved and the optimal gait and input torque are generated under the condition of considering the optimal energy [19]. In this method, the numerical integration required in the traditional GF method is implemented offline, and only simple algebraic operation is carried out online, which reduces the effort of online calculation of the system. In this paper, the GF based method is extended to the trot gait locomotion of the quadruped robot. The virtual leg technology is used to build the dynamic model of the diagonal gait of quadruped robot. By simplifying the model, the complex and tedious derivation can be reduced. The problem of energy optimal robot locomotion problem is transformed to the optimal control problem by using the energy related performance index. At the same time, a pair of suitable GFs are introduced into the regular transformation of Hamiltonian system to solve the optimal control problem, and the optimal trajectory generated by

GF is derived. Finally, the real time performance and energy consumption index of the algorithm are studied by comparing the two different types of robot locomotion methods.

## 2 Dynamic Model

### 2.1 Virtual Leg Technique

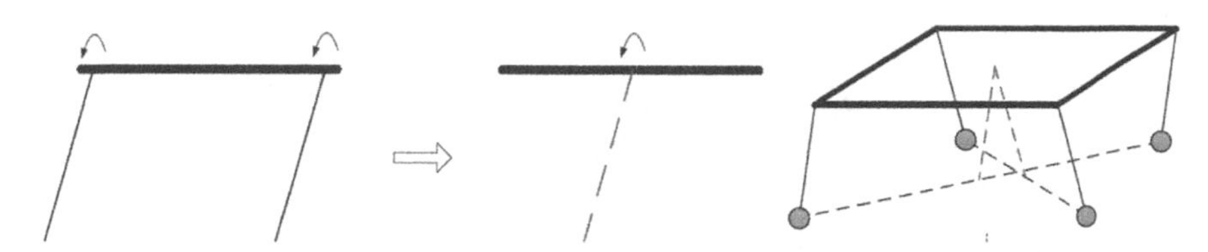

**Fig. 1.** Diagram of Virtual leg technique and virtual leg (the red dotted line) of quadruped robot in trot gait (Color figure online)

As shown in Fig. 1(a), virtual leg technique is a structural simplification of the gait of the quadruped robot with the legs moving in pairs [20]. Compared with the original model, virtual legs have exactly the same function. With the help of the technique, the quadrupedal trot gait can be treated as the bipedal gait. In detail, the robot diagonal legs in each motion cycle have the same phase, which meets the required condition of the technique that the diagonal legs in trot gait can be replaced by the virtual legs at the center of mass of the robot body. Thus, the quadruped robot can be regarded as a biped robot, and its model can be treated as a bipedal model, which is shown in Fig. 1(b).

### 2.2 Dynamic Model of Robot

In the process of modeling, the size and shape of the robot body are ignored, but its mass is preserved, and the robot is simplified as a particle with mass. Each leg of the robot is equivalent to a link connecting the common body, each leg has two rotational degrees of freedom of hip joint and knee joint. In addition, the counterclockwise direction is the positive direction selected in this paper. As shown in Fig. 2, the physical parameters and variables in the modeling process are: $m_{II}$ is the mass of the body of the robot, $m_1$ the mass of the leg, $m_2$ the mass of the thigh, $a_1$ the distance from the center of gravity of the leg to the foot end, $b_1$ the distance from center of gravity of the leg to the knee joint, $b_2$ the distance from the center of gravity of the thigh to the hip joint, $L_1$ the length of the leg, $L_2$ the length of the thigh, $g$ the acceleration of gravity, $\theta_1$ and $\theta_4$ are the angles of equivalent knee joints, $\theta_2$ and $\theta_3$ are the angles of equivalent hip joints, $u_1$ and $u_4$ are the input torques of knee joints, $u_2$ and $u_3$ are the input torques of hip joints. Moreover, it is assumed that the transition of the supporting leg occurs instantaneously when the swinging leg touches the ground and previous supporting leg leaves the ground, the collision of the swinging leg with the ground is assumed to be inelastic and without sliding, and there is no sliding between the support leg and the ground.

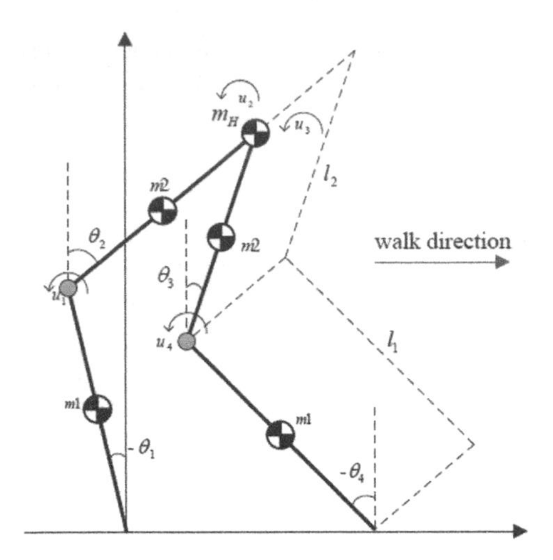


**Fig. 2.** Model of biped walking robot

The virtual leg technique is employed here to establish a simplified dynamic model of the quadruped robot with trot gait. Because the shape of the body is ignored, the motion of the robot is mainly described by the action of the legs. Such a system can be seen as a four-bar linkage system. We use the Euler–Lagrange method to build the dynamic model of the quadruped robot.

$$M(\theta)\ddot{\theta} + N(\theta, \dot{\theta})\dot{\theta} + G(\theta) = u. \tag{1}$$

where $\theta := (\theta_1, \theta_2, \theta_3, \theta_4)^{\mathrm{T}}$ is the generalized coordinate vector, the vector of the joint torques is denoted as $u: = (u_1,u_2,u_3,u_4)^{\mathrm{T}}$, $M \in R^{4\times4}$ is the generalized mass matrix, $N \in R^{4\times4}$ the Coriolis forces matrix, and $G \in R^4$ the gravity term and stiffness matrix.

Although the biped robot model reduces the complexity and redundancy of the model, it is still a multivariate and highly nonlinear equation. In order to start the following algorithm research, we linearize it firstly. We define the state variable $x := (\theta_1, \theta_2, \theta_3, \theta_4, \dot{\theta}_1, \dot{\theta}_2, \dot{\theta}_3, \dot{\theta}_4)^{\mathrm{T}}$, where the variables in the right hand side are denoted by $x_1, x_2, x_3, x_4, x_5, x_6, x_7$, and $x_8$ respectively. By combining with

$$\dot{x}_1 = x_5,\ \dot{x}_2 = x_6,\ \dot{x}_3 = x_7,\ \dot{x}_4 = x_8,$$

we obtain the dynamic equation

$$\dot{x} = f(x, u) \tag{2}$$

where the nonlinear function $f: = R^8 \times R^4 \rightarrow R^8$.

In order to get the zero position of the robot, we use Taylor expansion of the nonlinear function $f(x,u)$. When $(x,u) = (0,0)$, the nonlinear function $\dot{x} = f(x, u) = (0, 0)$ representing that the state variable $x(t)$ is constant for all t. At this time, the robot belongs to the static state, the angular velocity and the input torque of each joint are all zero, so *(x,u)* = *(0,0)* is called as the balance point of the robot. Finally, the linearized version of the dynamic Eq. (2) is

$$\dot{x} = Ax + Bu \tag{3}$$

where the parameters

$$A = \left.\frac{\partial f(x, u)}{\partial x}\right|_{(x,u)=(0,0)},$$

$$B = \left.\frac{\partial f(x, u)}{\partial u}\right|_{(x,u)=(0,0)}$$

# 3 LQ Optimal Control Problem and Double GFS Based Method

## 3.1 Energy Optimal Control Problem

For the robot locomotion, the problem of energy consumption has always been a research hotspot. In fact, this problem is to generate the input torque and the walking pattern (joint angle) of the robot during walking. It can be expressed by minimizing the cost function of the angle and the torque during the trajectory generation. Here, we set a cost function with the consideration of the energy

$$J(u; x(t_0)) = \frac{1}{2}\int_{t_0}^{t_f}\left(x^{\mathrm{T}}Qx + u^{\mathrm{T}}Ru\right)dt, \tag{4}$$

where the constant $Q \in \mathbb{R}^{8\times8}$ is positive definite, $R \in \mathbb{R}^{4\times4}$ is positive definite, and $t_0$ and $t_f$ are the initial and terminal times respectively.

Based on the above, then the energy-optimal trot gait locomotion problem of the quadruped robot (biped model equivalently) can be formulated as

*Problem 1.*

$$\min_{u} J(u; x(t_0)),$$

$$s.t.\, \dot{x} = Ax + Bu,$$

$$x(t_0) = x_0,\ x\left(t_f\right) = x_f.$$

on the fixed time interval $t \in \left[t_0, t_f\right]$. The vectors $x_0$ and $x_f \in \mathbb{R}^8$ are the given values of the fixed initial and terminal configuration states, respectively.

The optimal trajectories of the joint angles and input torques are obtained by minimizing the cost function (4). The state equation in Problem 1 is the linearized version of the robot dynamics derived in the previous section. In the fixed time interval $[t_0, t_f]$, the time values $x_0$ and $x_f$ representing the initial and terminal values of the body posture of a robot in a gait cycle respectively, are the boundary conditions for solving the energy-optimal trot gait locomotion problem of the quadruped robot (Problem 1).

### 3.2 GF Method and Optimal Trajectory Generation

Problem 1 is solved via double GFs based method. A pair of GFs is introduced and their expressions are exhibited. Finally, the optimal state and input trajectories are given by such a method. The detailed algorithms are as follows.

---

**Algorithm 1** Off-line part, calculate GF coefficients.

---

**Input:** constants, $A$, $Q$, and $G$; initial and terminal time, $t_0$ and $t_f$; time step, $\Delta t$.
**Output:** GF coefficients, $X_f(t), Y_f(t), X_b(t)$, and $Y_b(t), \forall t \in [t_0, t_f]$.
1: $X_f(t_0) \leftarrow 0; Y_{f(t0)} \leftarrow -I; X_{b(t_f)} \leftarrow 0; Y_{b(t_f)} \leftarrow -I$ ;
2: **for** $t = t_0$ **to** $t_f$ (step $\Delta t$) **do**
3: $\quad \mathrm{Xf}(t+\Delta t) \leftarrow \mathrm{Xf}(t) + (\mathrm{AXf}(t)\text{-}\mathrm{Xf}(t)^\top \mathrm{A}^\top\text{-}\mathrm{Xf}(t)^\top \mathrm{QXf}(t) + \mathrm{G}) \cdot \Delta t$ ;
4: $\quad Yf(t+\Delta t) \leftarrow Yf(t) - (Yf(t)QXf(t) - Yf(t)A^\top) \cdot \Delta t$ ;
5: **end for**
6: **for** $t = t_f$ **to** $t_0$ (step $\Delta t$) **do**
7: $\quad X_b(t) \leftarrow X_b(t-\Delta t) + (AX_b(t) - Xb(t)^\top A^\top - Xb(t)^\top QX_b(t) + G) \cdot \Delta t$ ;
8: $\quad Y_b(t) \leftarrow Y_b(t-\Delta t) - (Y_b(t)QX_b(t) - Y_b(t)A^\top) \cdot \Delta t$ ;
9: **end for**

---

---

**Algorithm 2** On-line part, generate optimal trajectories.

---

**Input:** constants, $B$ and $R$; initial and terminal time, $t_0$ *and* $t_f$ ; initial and terminal states, $x_0$ and $x_f$; time step, $\Delta t$; GF coefficients, $X_f(t), Y_f(t), X_b(t)$, and $Y_b(t), \forall t \in [t_0, t_f]$.
**Output:** optimal state and input, $x^*(t)$ and $u^*(t)$.
1: **if** there is a computational demand of BCs($x_0$, $x_f$) then ;
2: $\quad x(t_0) \leftarrow x_0; x(t_f) \leftarrow x_f$
3: $\quad$ **for** $t = t_0$ **to** $t_f$ (step $\Delta t$) **do**
4: $\quad\quad x^*(t) \leftarrow Xb(t)(Xf(t) - Xb(t))^{-1} Yf(t)^\top x(t0) - Xf(t)(Xf(t) - Xb(t))^{-1} Yb(t)^\top x(tf)$;
5: $\quad\quad u^*(t) \leftarrow R^{-1}B^\top (Xf(t) - Xb(t))^{-1} Yf(t)^\top x(t0) + R^{-1}B^\top (Xf(t) - Xb(t))^{-1} Yb(t)^\top x(tf)$;
6: $\quad$ **end for**
7: **else**
8: $\quad$ **goto** 1;
9: **end for**

---

Algorithms 1 and 2 represent the online part and offline part of the GF algorithm respectively. In the case of complex model, it takes a lot of efforts to solve ordinary differential matrix equations, while this part is solved offline by double GFs method according to Algorithm 1, which greatly improves the efficiency of the computation. In Algorithm 2, the boundary conditions $x_0$ and $x_f$, the initial time $t_0$, and the terminal time $t_f$ can be easily changed to generate different optimal trajectories according to the task requirements. The reason is that in Algorithm 2, all of them are algebraic operations, when the robot needs to change the boundary condition parameters when switches the gait, the effort of online calculation brought by this change is almost zero.

# 4 Trot Gait Locomotion via Double Gfs and CPG Based Methods

This section compares the simulation results of the double *GFs* based algorithm with the CPG based bionic control algorithm. According to the comparison, this section exhibits the particular advantage of the double *GFs* based method.

The parameters of the quadruped robot in this section are chosen as $l_1 = l_2 = 0.16$ [m], $a_1 = a_2 = 0.08$ [m], $b_1 = b_2 = 0.08$ [m], $m_1 = m_2 = 1/6$ [kg], $m_H = 4/3$ [kg], $g = 9.8$ [m/s$^2$]. Based on some tentative simulations, we select the parameters $Q = I_{8\times 8}$ and $R = \text{diag}\ (1, 1, 2, 3)$.

## 4.1 Locomotion via Double Gfs Based Method

After the parameters of the robot are selected, the matrices A and B in the linearized dynamic equation can be determined. Now we only need to determine the boundary conditions $x_0$, $x_f$, and the gait period T to get the optimal state and the input trajectories.

In our previous equivalent bipedal model of the quadruped robot, we assumed that the time period for the robot to walk one step started at the moment when the swinging leg currently raises, and ends at the next transition moment when the swinging leg falls and the supporting leg alternates. As shown in Fig. 3, the dotted line in the figure is the transition position of the support leg during the robot walking one step.

In fact, due to the limitations of the physical prototype in the design of the mechanism, the rotation angles of the hip and knee joints of the quadruped robot can only be constrained within a specific range. In this paper, the robot hip joint angle range $\theta_h \in [105°, 165°]$ and the knee joint angle range $\theta_k \in [-60°, -120°]$. The configuration of the initial joint of a leg of the quadruped robot is exhibited in Fig. 3.

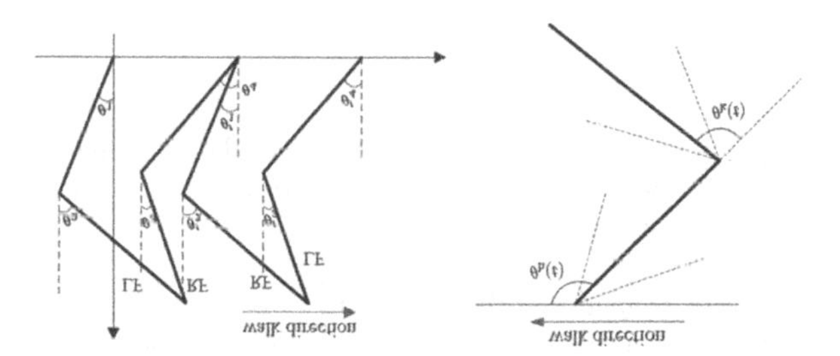

**Fig. 3.** Diagram of gait switching and initial configuration.

For convenience, we set the initial configuration of the robot as

$$x_0 = \left(-30°, 60°, 45°, -48.78°, 0, 0, 0, 0\right)^{\mathrm{T}}$$

According to a gait cycle (T = 1 [s]) of the CPG, the end state can be obtained

$$x_f = \left(-48.78°, 45°, 60°, -30°, 0, 0, 0, 0\right)^{\mathrm{T}}$$

Assume that the initial and end configuration of a robot walking step are stationary, that is, the angular velocity of each joint is zero. In order to maintain the principle of control variables, we set the gait cycle of the robot to take a step to be the same as the CPG method to plan a step. That is $T \in \left[t_0, t_f\right] = [0, 1]$.

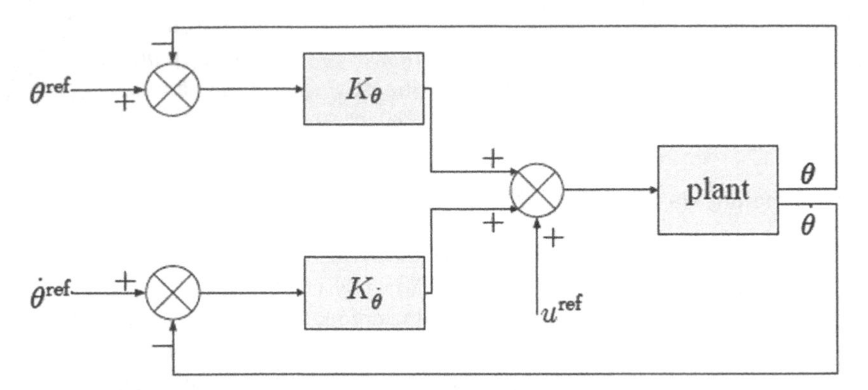


**Fig. 4.** PD tracking control system block diagram.

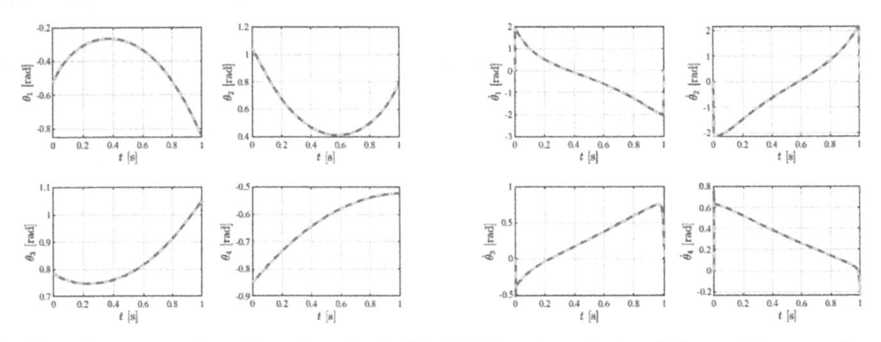


**Fig. 5.** Reference angle trajectories via double GFs based method and the tracking ones by the PD tracking control and reference angular velocity trajectories via double GFs based method and the tracking ones by the PD tracking control.

At the same time, we considered the modeling error caused by the linearization of the nonlinear robot model. A PD controller is built to track the trajectory. The system block diagram is shown in Fig. 4.

Figures 5 show the reference optimal angle and angular velocity trajectories obtained by the double GFs based method and also the tracking ones of the nonlinear system by the PD tracking control technique. As can be seen from the two figures, the tracking trajectories of the original nonlinear dynamic system are close to the reference ones generated by the double GFs based algorithm. The errors are small and can be almost ignored. It implies that the introduced double GFs based algorithm does also work for the nonlinear dynamics via PD tracking control, which widens the application of the method.

Figure 6 gives the corresponding input torque trajectories in 1 [s] via double GFs. It can be seen from the figure that the changes of the joint torques are relatively large at

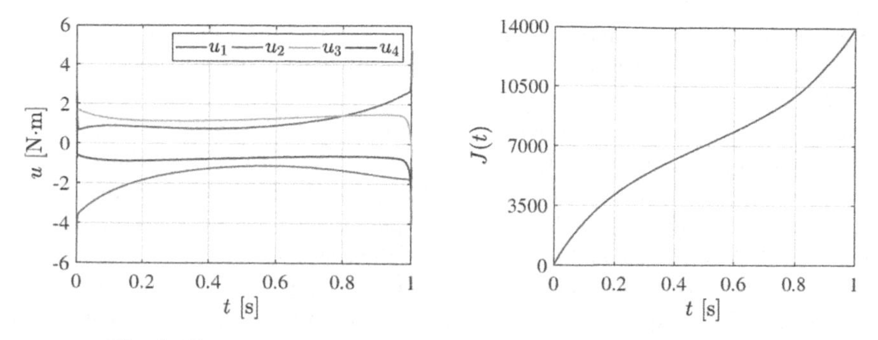


**Fig. 6.** The generated joint torque u and the cost function value $J$ ($t$).

both initial and terminal moments, but the absolute value of the remaining joint torque values do not exceed 2 [N m], and the maximum value of the whole process does not exceed 6 [N m], which demonstrates the appropriateness of the introduced method to small gait locomotion in practice.

After the optimal control is obtained, we can calculate the value of the cost function according to (4) in the time range of $[t_0, t]$. Figure 6 gives the trajectory of the energy related function value in a gait cycle. As can be seen from the figure, the value of the function monotonically increases, and the growth rate after $t > 0.6$ [s] is faster than the time period of $0 < t < 0.6$ [s], which shows that the energy change rate during the robots leg retracting phase is greater than the starting and raising legs the rate at which the energy changes in stages.

## 4.2 Locomotion via CPG Based Method

We plan a gait cycle using the same initial state of the robot as the previous subsection for the CPG based algorithm. For convenience, we set the key parameters $\mu_1 = \mu_2$, $\omega_1 = \omega_2 = 2\pi$. Based on this, we give the trajectories of joint angles in Fig. 7, where we take ten cycles of simulation time. The blue and red curves in the two figures represent the changes in the hip and knee joint angles of the robot left front (LF) and right front (RF) legs. Because the robot is in trot gait, the amplitude and phase of the diagonal legs are exactly the same.

In order to compare with the double GFs based method, we need to calculate the energy consumed by the CPG based method here. We observe from the original dynamic model (1) that the $M$, $N$, and $G$ on the left hand side are matrices containing $\theta$ and its first and second derivatives. You need to substitute the variables $\theta$, $\dot{\theta}$, and $\ddot{\theta}$ to get the trajectory of the joint torque u. However, the bionic CPG based method only obtains the joint angle value. We get the value of $\dot{\theta}$ and $\ddot{\theta}$ in the corresponding time by the difference algorithm, which is listed below.

$$f_1(m-1) = f(m+1) - f(m-1)/2h$$

$$f_2(m-1) = f(m+1) + f(m-1) - 2f(m)$$

where *f (m)* is the θ angle curve obtained by CPG planning, m is the number of discrete sampling points, h is the step size, and $f_1(m)$ and $f_2(m)$ are the first and second derivatives of *f(m)*, which is shown in Fig. 7.

Because the CPG method provides the relative angle, we need to add the corresponding initial joint angle $x_0$ of the robot in the process of calculating the torque. The trajectory of the joint torque can be obtained by substituting into the dynamic equation, which is shown in Fig. 8(a). It shows that the quadruped robot walks exactly one step (one gait cycle). The calculated joint torques of the robot during walking (the diagonal leg torque values remain the same). It also shows that during one step process of the quadruped robot, the hip and knee joint torques of the supporting leg (RF leg) are very small in amplitude, while the torque of the swinging leg (LF leg) is relatively large, and the knee joint amplitude is about 3 [N m], the amplitude of the hip joint is about 4 [N m].

After obtaining the input torque u, we still calculate the value of the cost function in the time range of $[t_0, t]$ according to (4). The current value of the energy cost function $J(t)$ over a gait period T is shown in Fig. 8(b). It can be seen from the figure that when the robot planned by the CPG bionic method walks one step under the same conditions, the cumulative value of the energy related cost function monotonically increases with the increase of the time t, and the total amount is significantly greater than that of the double GFs based method (almost six times).

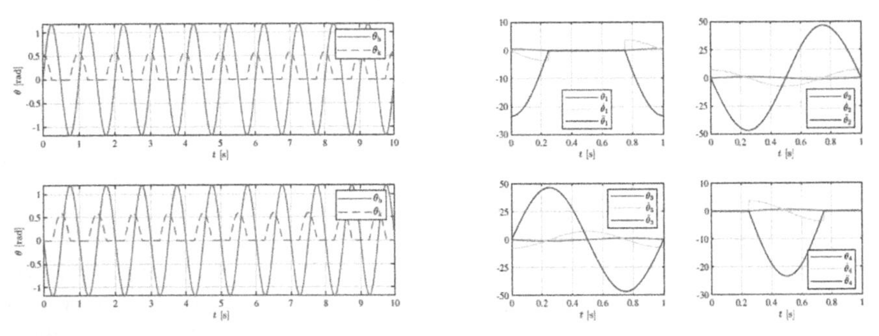

**Fig. 7.** Trajectories of hip and knee joint angles of the LF and RF legs and trajectories of angle, angular velocity, and angular acceleration of each joint.

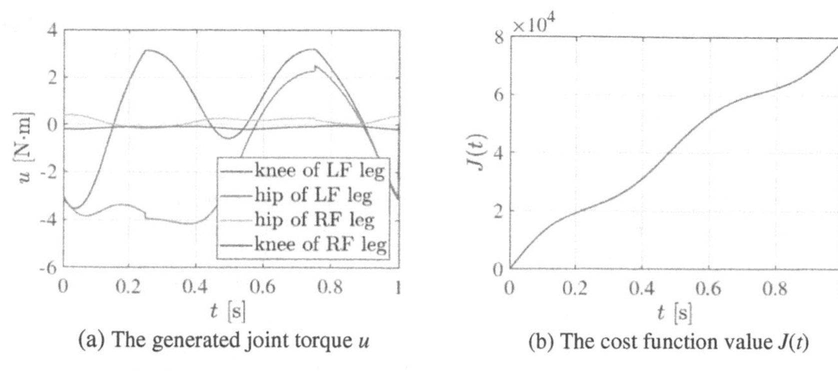


(a) The generated joint torque $u$

(b) The cost function value $J(t)$

**Fig. 8.** The generated joint torque and cost function trajectories.

## 5 Conclusions

In this paper, the double GFs method is introduced to the trot gait locomotion of the quadruped robot. This method can put a lot of tedious differential equation calculation process offline, and has the advantages of high efficiency of online planning and the ability to plan for different boundary conditions at the same time. Though the double GFs method belongs to the model based method, such a particular algorithm structure makes the method as efficient as the bionic based method, hence is more efficiency than almost all the other model based locomotion approaches.

In addition, the robot dynamics model needs to be linearized during application, because the method is planned for a linear system, and the modeling errors caused by linearization can be adjusted and controlled by the PD tracking control technique involved. In the simulation, we compare the double GFs based method with the CPG based method in the domain of energy consumption in the planning step by using the equivalent phase synchronization length $d = 0.28$ [m] and the same gait period $T = 1$ [s] with the same boundary conditions. The energy consumption obtained from the introduced method is about 17.5% of the CPG based method in this paper.

**Acknowledgments.** This work is supported by the National Natural Science Foundation of China (Grant No. 62003321), and Natural Science Foundation of Zhejiang Province, China (Grant No. LQ19F030006 and LQ20E050017).

## References

1. Gehring, S., Coros, M., Hutler, M., et al.: Practice makes perfect: an optimization-based approach to controlling agile motions for a quadruped robot. J. IEEE Robot. Autom. Mag. **23**(1), 34–43 (2016)
2. Pouya, S., Khodabakhsh, M., Spröwitz, A., Ijspeert, A.: Spinal joint compliance and actuation in a simulated bounding quadruped robot. Auton. Robot. **41**(2), 437–452 (2015). https://doi.org/10.1007/s10514-015-9540-2

3. Sprowitz, A., Tuleu, M., Vespignani, M., et al.: Towards dynamic trot gait locomotion: design, control, and experiments with Cheetah-cub, a compliant quadruped robot. J. Int. J. Robot. Res. **32**(8), 932–950 (2013)
4. Raibert, M., Chepponis, H., Brown, H.B.J.R.: Running on four legs as though they were one. J. IEEE J. Roboti. Autom. **2**(2), 70–82 (1986)
5. Raibert, M.H.: Legged Robots That Balance. MIT Press, Cambrige (1986)
6. Gong, P., Wang, S., Zhao, S., et al.: Bionic quadruped robot dynamic gait control strategy based on twenty degrees of freedom. J. IEEE/CAA J. Automatica Sinica. **5**(1), 382–388 (2018)
7. Winkler, W., Bellicoso, C.D., Hutter, M., et al.: Gait and trajectory optimization for legged systems through phase-based end-effector parameterization. J. IEEE Robot. Autom. Lett. **3**(3), 1560–1567 (2018)
8. Chen, X., Gao, F., Qi, C., et al.: Gait planning for a quadruped robot with one faulty actuator. J. Chin. J. Mech. Eng. **28**, 11–19 (2015)
9. Li, H., Zhang, H., et al.: Development of adaptive locomotion of a caterpillar-like robot based on a sensory feedback CPG model. J. Adv. Robot. **28**(6), 389–401 (2014)
10. Kimura, H., Akiyama, S., Sakurama, K.: Realization of dynamic walking and running of the quadruped using neural oscillator. J. Autonom. Robot. **7**, 247–258 (1999)
11. Fukuoka, Y., Kimura, H.: Dynamic locomotion of a biomorphic quadruped Tekken robot using various gaits: walk, trot, free-gait and bound. J. Appl. Bion. Biomech. **6**(1), 63–71 (2009)
12. Zheng, X., Zhang, X., Guan, X., et al.: Quadruped robot based on biological central pattern generator. J. J. Tsinghua Univ. (Sci. Technol.) **2**, 166–169 (2004)
13. Hu, J., Liang, T., Wang, T.: Parameter synthesis of coupled nonlinear oscillators for CPG-based robotic locomotion. J. IEEE Trans. Ind. Electron. **61**(11), 6183–6191 (2014)
14. Bramburger, B., Dionne, V., LeBlanc, G.: Zero-Hopf bifurcation in the Van der Pol oscillator with delayed position and velocity feedback. J. Nonlinear Dyn. **78**, 2959–2973 (2014)
15. Geering, P.: Optimal Control with Engineering Applications. 1st Edition. Springer-Verlag, Berlin (2007)
16. Rostami, M., Bessonnet, G.: Sagittal gait of a biped robot during the single support phase Part 2: optimal motion. J. Robotica. **19**(3), 241–253 (2001)
17. Saidouni, T., Bessonnet, G.: Generating globally optimised sagittal gait cycles of a biped robot. J. Robotica **21**(2), 199–210 (2003)
18. Park, D., Scheeres, J.: Determination of optimal feedback terminal controllers for general boundary conditions using generating functions. J. Automatica **42**(5), 869–875 (2006)
19. Hao, Z., Fujimoto, K., Hayakawa, Y.: On-demand optimal gait generation for a compass biped robot based on the double generating function method. In: Proceedings of 2013 IEEE/RSJ International Conference on Intelligent Robots and Systems, Tokyo, pp. 3108–3113 (2013)
20. Cherouvim, A., Papadopoulos, E.: Speed and height control for a special class of running quadruped robots. In: Proceedings of 2008 IEEE International Conference on Robotics and Automation, Pasadena, CA, pp. 825–830 (2008)

# An End-to-End Deep Learning Model to Predict Surface Roughness

Jin-tao Lv[1], Xi-ning Huang[1], JunJiang Zhu[1(✉)], and Zhao-jie Zhang[2]

[1] School of Mechanical and Electrical Engineering, China Jiliang University, Hangzhou, People's Republic of China
zjj602@yeah.net

[2] Wuhan Heavy Duty Machine Tool Group Corporation, Wuhan, People's Republic of China

**Abstract.** This paper proposes an end-to-end deep learning prediction model to improve predicting the workpiece's surface roughness from the milling process's vibration signal in the intelligent production process. First, use the CNN model to automatically extract the characteristics of the vibration signal and train the data; secondly, use the LSTM model suitable for time series sensitive signal training to take on the CNN training data and continue training; finally, the FC classification accepts the data output prediction model to perform Predictive analysis. The experimental results show that through the CNN-LSTM-FC model, the automatic extraction of vibration signal features can be achieved efficiently and with an average error of 0.0066 μm and an average deviation of 3.45%. It has a high surface roughness prediction accuracy and eliminates the need for processing personnel reliance on experience.

**Keywords:** Surface roughness · Vibration signals · Deep learning predict

## 1 Introduction

In mechanical processing, improving the quality of processed workpieces has an essential impact on reducing the failure rate of assembled mechanical products. Surface roughness is one of the criteria for evaluating processing quality [1]. In the milling process, the cutter head of the milling cutter will periodically touch the metal surface during the rotation, thereby forming a corresponding contour on the workpiece's surface to accurately reflect the processing state through the surface roughness. Therefore, the research direction of measuring and predicting surface roughness has been widely discussed [2]. The traditional method of improving the surface roughness is generally offline detection after the workpiece is processed, and then the cutting parameters of the machine tool are adjusted according to the experience of the processing workers. This method is lagging, very time-consuming, and relies on the processing personnel's experience, which is not easy to improve production efficiency. Therefore, the research on real-time online surface roughness prediction has received extensive attention.

Currently, there are two mainstream online surface roughness prediction methods, and one is based on physical or deterministic methods [3–6]: with the help of tool


Q. Han et al. (Eds.): LSMS 2021/ICSEE 2021, CCIS 1469, pp. 607–616, 2021.
https://doi.org/10.1007/978-981-16-7213-2_58

performance, machine tool characteristics, dynamic characteristics, thermal effects, and cutting attributes [7], the surface roughness has an impact on Big attribute, surface roughness prediction model is established. Nevertheless, it is also because these six attributes interact and influence each other to make surface roughness prediction difficult. The other is a research method based on big data [8–16] install the corresponding sensor, use efficient signal processing technology to extract the signal information related to the monitored parameter, establish the corresponding mapping relationship between the signal data and the workpiece quality parameter, and realize the signal data to predict the cutting process Zhang, N. [8] predicts the surface roughness based on the LS-SVM model. Huang, L. [9] pointed out the possibility of accelerometers used in vibration signal detection and analyzed the great value of vibration signals as input signals in surface roughness prediction. Upadhyay, V. [10] applies the vibration signal and cutting parameters to the ANN model to predict the surface degree. Plaza, E.G. [11–13] uses SSA, G-WPT, E-WPT, SE-WPT, and other models to analyze vibration signals and cutting parameters to predict surface roughness. Li, Z. [14] predicts the surface roughness based on machine learning based on the temperature plus vibration signal. Based on the Bayesian model, Kong, D. [15, 16] developed BLR, Bayesian plus RVM, and other surface roughness prediction and tool wear assessment models.

The methods in these studies all have high experience requirements for the performer, and there is no quantitative analysis of the selected features. Convolutional layer, the core of learning depends on a specific data set, without the need to manually extract features [17]. Convolutional Neural Network (CNN) [19, 20, 26] and has excellent performance in the detection and classification of rotating machinery fault diagnosis. LeCun Y [18] proposed that CNN is not completely suitable for learning time series. In the face of time series sensitive problems and tasks, LSTM is usually more suitable. Both convolutional neural networks and long- and short-term memory networks have distinct features. The convolution kernel inside the convolutional neural network can automatically extract features of the original input signal. The LSTM network has a better predictive function for time series signals, but it cannot perform feature extraction operating. Therefore, this article will try to combine the neural network model of CNN and LSTM to automatically extract the characteristics of vibration signals, reduce the sensitivity to time series, and establish a connection between the input vibration signal and the prediction, and greatly improve the accuracy of surface roughness prediction.

## 2 Research Methodology

This experiment is to apply the emerging artificial intelligence technology to the traditional surface roughness prediction of processed workpieces. The original vibration signal is collected through the acceleration sensor installed on the spindle. Furthermore, preprocess the signal. Since the vibration signal is a time series signal, to automatically extract the characteristics of the vibration signal and improve the efficiency of data processing, this experiment uses a convolutional neural network (CNN) and a long and short-term memory network (LSTM) as a combined model for training and finally connects to a fully connected network Surface roughness prediction model.

## 2.1 Vibration Signal Preprocessing

Because the vibration signal is sensitive to background noise, and the low-pass filter allows low-frequency or DC components in the signal to pass and suppresses high-frequency components or interference and noise [21], the low-pass filter can smooth the signal. The calculation formula for the critical frequency of low-pass filtering is as follows:

$$f_c = 1/2\pi RC. \tag{1}$$

## 2.2 Convolutional Neural Network (CNN)

Compared with the ordinary fully connected neural network, the convolutional neural network's most significant advantage is that it automatically extracts signal features and directly processes the original vibration signal [17]. The data processed in this experiment is a vibration signal, which is a signal that periodically changes with time. In the actual experiment, the signal is collected and stored in the computer at a set sampling frequency, so it belongs to a one-dimensional data type. Because 1D-CNN can better process time series data [22], 1D-CNN is selected as the convolutional neural network part of the prediction model.

The convolutional neural network has five parts: input layer, convolution layer, pooling layer, fully connected layer, and output layer. As the first layer, the convolutional layer can directly process the original input vibration signal. The internal convolution kernel traverses the vibration data in chronological order, filters useless signals, and extracts corresponding features to generate feature maps. These features are related to the predicted surface roughness. Next, the top pooling layer selects the maximum value in a specific range of the feature map, and these selected maximum values will be combined into a new feature map to reduce the size of the feature map. The data processed in the pooling layer is finally compared with the surface roughness value measured by the instrument after the fully connected layer and the output layer, and the weight of the parameters in the neural network is adjusted through the corresponding algorithm. After a certain number of iterations, the output value is close to the measured value, thus completing the training. The calculation formulas of the convolutional layer and the pooling layer are as follows:

$$Y_{mk} = f\left(\Sigma_{i=1}^{n} X_i * w + b\right). \tag{2}$$

$$Z_{ml} = \max(Y_{mk}). \tag{3}$$

$Y_{mk}$ is the output of the convolutional layer, $X_i$ is the number of samples, $w$ is the size of the convolution kernel, $f$ is the activation function, and $Z_{ml}$ is the output of the maximum pooling layer.

## 2.3 Long-Short-Term Memory Network (LSTM)

Because the LSTM network model can avoid gradient disappearance in the recurrent neural network (RNN) [23], this article uses the LSTM network model. LSTM can recall the time series data at the last moment and has a significant advantage in determining which are essential features. Three main gates control the unit's state, namely the forget gate, the input gate, and the output gate. The input gate controls the new information that can be stored in the unit. The forget gate controls how much information from the previous moment is retained in the current unit state. The output gate controls the unit's information.

## 2.4 1D-CNN + LSTM + FCN Combined Prediction Model

Both convolutional neural networks and extended- and short-term memory networks have distinct features. The convolution kernel inside the convolutional neural network can automatically extract features of the original input signal. The LSTM network has an excellent predictive function for time-series signals, but it cannot perform feature extraction. Therefore, combining the two artificial neural networks can give full play to each other's advantages and make up for the existing shortcomings. Therefore, this experiment will establish a combined prediction model of CNN + LSTM. The model consists of a convolutional neural network, an LSTM network, and a fully connected network layer, and the output value is the predicted surface roughness value. The process framework of the entire experiment is shown in Fig. 1.

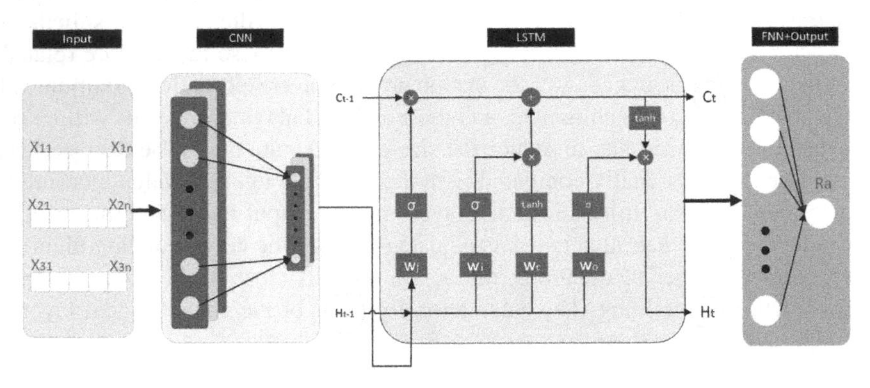


**Fig. 1.** 1D-CNN + LSTM + FCN combined prediction model framework

First, the original vibration signal is input to the one-dimensional convolutional neural network layer. The data size after the convolutional layer's output and the pooling layer will be reduced due to the convolution kernel and the pooling operation. After the pooling layer, connect the LSTM network to process time-series data. In the first section of this chapter, the output of CNN is $Z_{ml}$, which will be used as the LSTM network's input. Then the gate, input gate, and output gate are forgotten in the calculation process of the LSTM network Part of the formula is (4)–(8), $ht^{-1}$ represents the output of the LSTM at the last moment. After the LSTM network, a fully connected layer and an output layer are added. The output value of the model is the predicted surface roughness of the workpiece. During the model training process, the output predicted value would be compared with the experimental measurement value to establish MSE loss function. In the model training iteration, the weights in the training model are continuously adjusted through the algorithm. Reduce the value of the loss function so that the predicted value reaches the optimal solution as much as possible,

$$f_t = \sigma\left(W_f\left[h_{t-1}, Z_t\right] + b_f\right). \tag{4}$$

$$i_t = \sigma\left(W_i\left[h_{t-1}, Z_t\right] + b_i\right). \tag{5}$$

$$\hat{C}_t = \tanh\left(W_c\left[h_{t-1}, Z_t\right] + b_c\right). \tag{6}$$

$$o_t = \sigma\left(W_o\left[h_{t-1}, Z_t\right] + b_o\right). \tag{7}$$

$$h_t = o_t \tanh(C_t). \tag{8}$$

Using the loss function, a set number of iterations in the reverse propagation algorithm to constantly adjust weight values of the model's parameters to minimize the error value, the completion of training a predictive model. After the model training process is over, a part of the data set is selected as the test set and input into the trained prediction model to test the model's accuracy.

## 3 Experiments

This experiment aims to collect the vibration signal data generated by CNC machine tools in milling, preprocess the data, and train the prediction model so that the model can accurately predict the surface roughness of the workpiece during milling. In this chapter, we will introduce the experiment's relevant settings, including describing the vibration signal generated in the milling process, the relevant introduction of the sensor and its collection process, and the specific process of the experiment.

### 3.1 Data Description

The vibration signal generated in the milling process is a waveform that continuously changes within a specific amplitude over time. In this experiment, the vibration signal is measured by an acceleration sensor installed on the workpiece. The equipment model and processing parameters used in the experiment are as follows.

Choose the Dytran 3263A2 three-axis acceleration sensor, and install them on the spindle, fixture, and metal block, respectively. The installation positions are shown in Fig. 2. Figure 3 shows a schematic diagram of milling processing and collecting vibration signals. The INV3062T0 24-bit 4-channel vibration collector is used to collect the vibration sensor signals and then transmit them to the computer for storage. The sampling frequency is 20480 Hz. The metal block selected for processing is a 6061 aluminum alloy block with 12 mm*12 mm*24 mm. The overall processing of the milling machine is shown in Fig. 4. After processing, use the SJ-210SJ-210 roughness meter to measure the surface roughness value. The roughness value is recorded as Ra. Formula (9) is the calculation formula for Ra. L represents the length of the workpiece surface during measurement, h(x) Indicates the distance of each point on the surface from the reference line. To reflect the overall surface roughness of the processing area, the metal milling surface is divided into four central areas for measurement during measurement.

**Fig. 2.** The left side shows the Spindle vibration sensor; the right side shows Fixture and metal block vibration sensor.

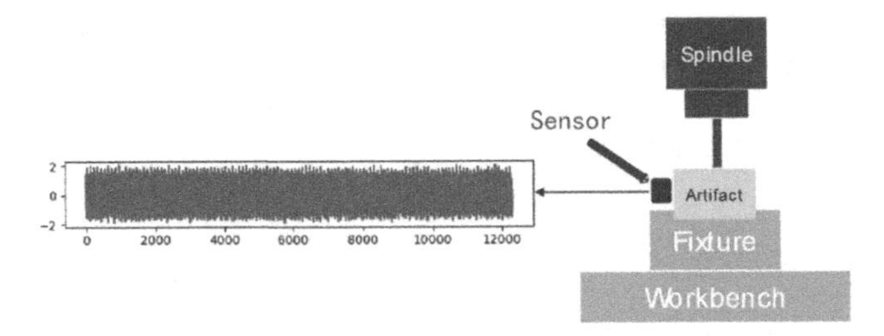


**Fig. 3.** Schematic diagram of milling processing and collecting vibration signals

**Fig. 4.** Physical image of experimental processing

$$Ra = \frac{1}{L}\int_{0}^{L} |h(x)|dx. \quad (9)$$

During the experimental processing, the machine tool's machining state is directly controlled by the input cutting parameters. The cutting parameters set in this experiment are shown in Table 1.

**Table 1.** Milling processing parameters

| Parameter | Data setting | | | | |
|---|---|---|---|---|---|
| Cutting speed (r/min) | 6k | 9k | 12k | 15k | 18k |
| Feed speed (mm/min) | 1k | 1.65k | 2.3k | 3k | |
| Measure the amount of knife (mm) | 1 | 0.7 | 0.5 | 0.3 | |

## 3.2 Experimental Data Preparation

Eighty metal blocks were processed and tested in this process, and 80 data sets were obtained. Low-pass filtering is performed on each group of data sets. Considering that the number of data samples collected during processing is not large, more data is obtained by sliding window. The window length is 2048, and the sliding step length is 10, and finally, each group is used. The number of data samples analyzed is 2048, and a total of 80,000 data sets are planned. Among them, 64,000 groups are used as training sets, and the remaining 12,800 groups are used as test sets. The MAPE and RMSE used to evaluate the predicted value of surface roughness are defined in formulas (10) and (11):

$$\text{MAPE} = \frac{1}{n}\sum\nolimits_{i=1}^{n}\left|\frac{y_i - x_i}{y_i}\right| * 100\%. \tag{10}$$

$$\text{RMSE} = \sqrt{\frac{\sum_{i=1}^{n}(y_i - x_i)^2}{n}}. \tag{11}$$

## 3.3 Experimental Model Setting

To compare the prediction performance of the models proposed in this experiment, a control group is set to compare the prediction results. The control group uses several different types and neural networks as prediction models. These models optimize the best accuracy by increasing or decreasing the convolutional layer and adjusting the parameters of each level of CNN and LSTM. These two models are: (1) CNN + FC, (2) CNN + LSTM + FC. At the same time, compare the experimental results of other articles.

## 4 Experimental Results and Discussion

Select the vibration data in the Z-axis direction of the three-axis acceleration sensor in the data selection. The model structure combination and its quantity are input layer, one-dimensional convolution layer*2, maximum pooling layer*1, LSTM layer*1, fully connected layer*1, and output layer. MSE as loss function. The number of nodes in the input layer is 2048.

**Table 2.** Surface roughness prediction results

| Rotating speed rpm | Feed mm/min | Side knife Mm | Measured Ra value/μm | Average predicted Ra value/μm | Average forecast error/μm | MAPE | RMSE |
|---|---|---|---|---|---|---|---|
| 9000 | 3000 | 0.3 | 0.66175 | 0.673296 | 0.01276 | 1.93% | 0.01463 |
| 15000 | 1000 | 0.5 | 0.08775 | 0.084386 | 0.00356 | 4.06% | 0.00411 |
| 6000 | 1650 | 1 | 0.52225 | 0.52522 | 0.00398 | 0.76% | 0.00526 |
| 9000 | 2300 | 1 | 0.365 | 0.357857 | 0.00778 | 2.13% | 0.00947 |
| 6000 | 1650 | 0.5 | 0.468 | 0.468384 | 0.00388 | 0.83% | 0.00510 |
| 18000 | 1650 | 1 | 0.0725 | 0.07368 | 0.00188 | 2.59% | 0.00252 |
| 18000 | 1000 | 0.5 | 0.07175 | 0.072635 | 0.00133 | 1.86% | 0.00172 |
| 6000 | 2300 | 0.7 | 0.83375 | 0.835675 | 0.00586 | 0.70% | 0.00756 |

The test set detection time is 2.537 s. The surface roughness prediction results of some vibration signals are shown in Table 2. The error range is between 0.001331–0.01679 μm, the average error is only 0.00605 μm, the average deviation is 3.36%, and the root means the square error is $0.76 \times 10^{-2}$. This paper's prediction results are very close to the actual results, reflecting the accuracy and reliability of the prediction model store. Simultaneously, when the CNN + FC model is used, and the same data is input, the average error is only 0.0102 μm, the average deviation is 5.08%, and the root means the square error is $1.41 \times 10^{-2}$.

**Table 3.** Comparison of research method results

| Method | MAPE | RMSE |
|---|---|---|
| CNN + LSTM + FC | 3.36% | $0.76 \times 10^{-2}$ |
| CNN + FC | 5.08% | $1.41 \times 10^{-2}$ |
| ANFIS [24] | – | 1.8533 |
| GRANN [25] | – | 0.1550 |
| FFT + LSTM [26] | 6.57% | – |
| Standard_SBLR[16] | – | $3.17 \times 10^{-2}$ |

Table 3 shows the average deviation and root mean square error of the method proposed in the literature. It can be seen from Table 3 that for the vibration signal data generated by CNC machine tools in milling processing, the combined neural network model method proposed in this paper predicts an average deviation of surface roughness of 3.36% and an RMSE of $0.76 \times 10^{-2}$, which is higher than that of LSTM, the average deviation is 3.21% lower, and the RMSE is 23% lower than Standard_SBLR. The method proposed in this paper significantly improves the accuracy of surface roughness prediction. Simultaneously, the model can extract insensitive feature information, which is helpful for prediction in the high-precision milling process.

## 5 Conclusions

In this paper, the vibration signal is applied to the deep learning network to predict, evaluate and predict the surface roughness. According to the CNN-FC and CNN-LSTM-FC models, their training and testing are evaluated. The experimental results show that the average deviation of surface roughness predicted by the CNN + LPOSTM + FC combined neural network model method is 3.36%, and the RMSE is $0.76 \times 10^{-2}$. The CNN-LSTM-FC model can automatically extract the vibration signal and reduce the sensitivity to time series and realize the accurate prediction of surface roughness. In terms of surface roughness prediction of vibration signals, the CNN-LSTM-FC model significantly improves the accuracy of predicting surface roughness, saves the time and cost of online detection system analysis and feature extraction, and greatly reduces the experience of processing personnel. To reduce the labor cost of inspection and improve the operating efficiency of the real-time inspection system for online processing. There is still a lot of room for optimization of computational efficiency, which is intelligent in the future Surface roughness online detection method can be researched.

**Acknowledgments.** This work was supported by the Key research and development plan of Zhejiang Province (2019C03114).

## References

1. Wang, H., To, S., Chan, C.: Investigation on the influence of tool-tip vibration on surface roughness and its representative measurement in ultra-precision diamond turning. Int. J. Mach. Tools Manuf. **69**, 20–29 (2013)
2. Chang, H.-K., Kim, J.-H., Kim, I.H., Jang, D.Y., Han, D.C.: In-process surface roughness prediction using displacement signals from spindle motion. Int. J. Mach. Tools Manuf. **47**, 1021–1026 (2007)
3. Liu, C., Lu, R., Chen, L., Wang, J.: Progress of surface roughness measurement based on optical method. Semicond. Optoelectr. **31**, 495–500 (2010)
4. Urbikain, G., de Lacalle, L.L.: Modelling of surface roughness in inclined milling operations with circle-segment end mills. Simulat. Modell. Pract. Theory **84**, 161–176 (2018)
5. Wojciechowski, S., Wiackiewicz, M., Krolczyk, G.: Study on metrological relations between instant tool displacements and surface roughness during precise ball end milling. J. Measur. **129**, 686–694 (2018)

6. Nguyen, T.-T.: Prediction and optimization of machining energy, surface roughness, and production rate in SKD61 milling. J. Measur. **136**, 525–544 (2019)
7. Khorasani, A.M., Yazdi, M.R.S., Safizadeh, M.S.: Analysis of machining parameters effects on surface roughness: a review. Int. J. Comput. Mater. Sci. Surf. Eng. **5**, 68–84 (2012)
8. Zhang, N., Shetty, D.: An effective LS-SVM-based approach for surface roughness prediction in machined surfaces. J. Neurocomput. **198**, 35–39 (2016)
9. Huang, L., Chen, J.C.: A multiple regression model to predict in-process surface roughness in turning operation via accelerometer. J. Indust. Technol. **17**, 1–8 (2001)
10. Upadhyay, V., Jain, P., Mehta, N.: In-process prediction of surface roughness in turning of Ti–6Al–4V alloy using cutting parameters and vibration signals. J. Measur. **46**, 154–160 (2013)
11. Plaza, E.G., López, P.N.: Surface roughness monitoring by singular spectrum analysis of vibration signals. J. Mech. Syst. Signal Process. **84**, 516–530 (2017)
12. Plaza, E.G., López, P.N.: Application of the wavelet packet transform to vibration signals for surface roughness monitoring in CNC turning operations. J. Mech. Syst. Signal Process. **98**, 902–919 (2018)
13. Plaza, E.G., López, P.N.: Analysis of cutting force signals by wavelet packet transform for surface roughness monitoring in CNC turning. J. Mech. Syst. Signal Process. **98**, 634–651 (2018)
14. Li, Z., Zhang, Z., Shi, J., Wu, D.: Prediction of surface roughness in extrusion-based additive manufacturing with machine learning. J. Robot. Comput. Integr. Manuf. **57**, 488–495 (2019)
15. Kong, D., Chen, Y., Li, N.: Gaussian process regression for tool wear prediction. J. Mech. Syst. Signal Process. **104**, 556–574 (2018)
16. Kong, D., Zhu, J., Duan, C., Lu, L., Chen, D.: Bayesian linear regression for surface roughness prediction. J. Mech. Syst. Signal Process. **142**, 106770 (2020)
17. Qu, S., Li, J., Dai, W., Das, S.: Understanding audio pattern using convolutional neural network from raw waveforms (2016). https://arxiv.org/abs/1611.09524
18. LeCun, Y., Bengio, Y.: Convolutional networks for images, speech, and time series. Handbook Brain Theory Neural Netw. **3361**, 1995 (1995)
19. Pan, H., He, X., Tang, S., Meng, F.: An improved bearing fault diagnosis method using one-dimensional CNN and LSTM. J. Mech. Eng. **64**, 443–452 (2018)
20. Janssens, O., et al.: Convolutional neural network based fault detection for rotating machinery. J. Sound Vib. **377**, 331–345 (2016)
21. Christov, I., Neycheva, T., Schmid, R., Stoyanov, T., Abächerli, R.: Pseudo-real-time low-pass filter in ECG, self-adjustable to the frequency spectra of the waves. Med. Biol. Eng. Compu. **55**(9), 1579–1588 (2017)
22. Ha, S., Choi, S.: Convolutional neural networks for human activity recognition using multiple accelerometer and gyroscope sensors. In: 2016 International Joint Conference on Neural Networks (IJCNN), pp. 381–388. IEEE, (2016)
23. Jozefowicz, R., Zaremba, W., Sutskever, I.: An empirical exploration of recurrent network architectures. In: International Conference on Machine Learning, pp. 2342–2350. PMLR (2015)
24. Kumar, R., Hynes, N.R.J.: Prediction and optimization of surface roughness in thermal drilling using integrated ANFIS and GA approach. Eng. Sci. Technol. Int. J. **23**(1), 30–41 (2020)
25. Yuan, L., Zeng, S.: Prediction of milling surface roughness based on grey relational neural network. Mach. Des. Manuf. **3**, 293-296+300 (2021)
26. Lin, W.J., Lo, S.H., Young, H.T., et al.: Evaluation of deep learning neural networks for surface roughness prediction using vibration signal analysis. J. Appl. Sci. **9**(7), 1462 (2019)

# Finite Element Analysis of Ultrasonic Guided Wave Propagation and Damage Detection in Complex Pipeline

Cheng Bao, Lei Gu, Sainan Zhang, and Qiang Wang(✉)

College of Automation and College of Artificial Intelligence, Nanjing University of Posts and Telecommunications, Nanjing 210023, China
wangqiang@njupt.edu.cn

**Abstract.** During the service of complex pipeline structures, such as nuclear power plant pipe (a kind of variable diameter pipe), different defects such as weld, corrosion and so on, often happen due to poor service conditions. For the concern of safety, it's important to monitor and detect the proper structural defects. In this paper, guided wave based damage detection method is adopted to solve this problem for variable diameter pipelines. Finite element analysis is carried out firstly to investigate the propagation of the guided wave in complex pipeline. ABAQUS, one of the finite element analysis tools, is adopted to build the simulation model of pipe joints. Thus, the responses of ultrasonic guided wave caused by defects in structure are analyzed by means of signal amplitudes and wave packet area, etc., and the damage-related information is extracted, so as to realize the defcect identification of variable diameter pipelines. The results of finite element simulation show that guided wave based damage monitoring method is effective in online damage monitoring for variable diameter pipelines.

**Keywords:** Pipeline · Finite element analysis · Defects · Ultrasonic guided waves

## 1 Introduction

With the development of industrial society, nuclear power has gradually come into people's vision, and plays an increasingly important role. Variable diameter pipeline structures are widely used in nuclear power plant. Due to the particularity of nuclear power, the pipeline carries high temperature and high pressure liquid for a long time, resulting in corrosion and leakage. Because of the complexity of its structure, the pipeline joint often becomes a vulnerable area, resulting in major safety accidents. It is of great significance to study the defect detection of in-service nuclear power plant pipeline joints.

Ultrasonic guided wave is often used to detect the damage of pipeline structures because of its characteristics such as large detection range, small attenuation and sensitivity to small damage. It is a frontier direction of nondestructive testing technology. Wang [1] developed a sparse sensor network to actively excite and acquire cylindrical waves for damage identification and health monitoring of pipeline structures.


Q. Han et al. (Eds.): LSMS 2021/ICSEE 2021, CCIS 1469, pp. 617–626, 2021.
https://doi.org/10.1007/978-981-16-7213-2_59

Finite element analysis is one of the most commonly used detection methods for pipeline defects, such as ABAQUS. At present, a large number of scholars have used finite element software for pipeline health detection, and have done a lot of numerical simulation and experimental research for in-service pipeline. In the traditional ultrasonic guided wave pipeline detection, the signal transmission will be disturbed by many factors, such as the inability to place the sensor in the complex structure, the coupling between two signal lines, and the crosstalk caused by mutual inductance and mutual capacitance between the signal lines. In the finite element analysis software, these negative factors can be well avoided, and ideal signals can be obtained.

## 2 Finite Element Modeling

ABAQUS is a powerful finite element software for engineering simulation. It can solve the problems from relatively simple linear analysis to many complex nonlinear ones. ABAQUS has a library of units that can simulate any geometry, and this library of material models includes almost every type. So we can analyze the complex mechanical system of solid mechanical structure. In this paper, ABAQUS CAE is used to model the pipe head of nuclear power plant, the physical and simulation model parameters are consistent.

### 2.1 Establish 3D Model

The three-dimensional model and physical objective are shown in the Fig. 1. The pipe head of nuclear power plant is made up of two parts: the inner part and the outer part. The inner pipe is nested into the outer joint. The joint is connected by welding. At the same time, the joint is also the place where damage often occurs. The cross section sketch of the inner and outer parts is shown in the Fig. 2, on which are marked the welds where there are cavities formed by the inner and outer pipe walls.

**Fig. 1.** Pipeline model in ABAQUS and physical picture

It can be clearly seen from the above four drawings that the pipe joint of nuclear power plant is an irregularly reduced pipe with the largest pipe diameter at the head,

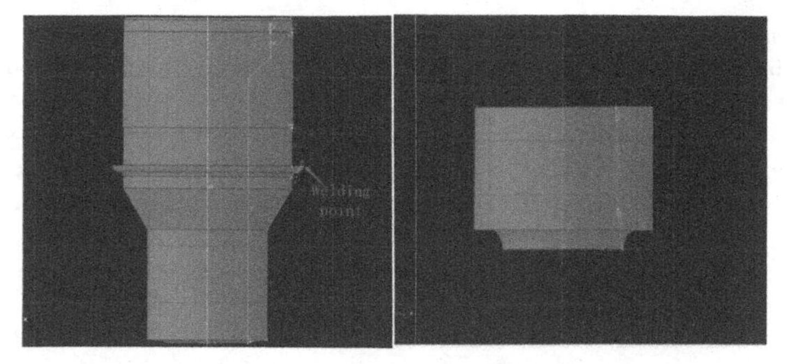

**Fig. 2.** Internal specimen (left) and external specimen (right)

then the internal pipe diameter becomes smaller while the external pipe diameter remains unchanged, and then the external diameter below the welding site gradually decreases and finally becomes a regular pipe.

### 2.2 Defect Modeling

Two models of health and damage are established in this experiment for carrying out comparative study. In order to simulate the actual pipeline damage, the defect of the damage model is established at the welding site, and the location is shown in the Fig. 3. Defects are modeled directly by ABAQUS, and the type is penetration. Through the cutting and stretching function of ABAQUS CAE, the original health model is processed and the defect model is obtained. In this experiment, only one defect is established, which can better obtain the signal of the damage.

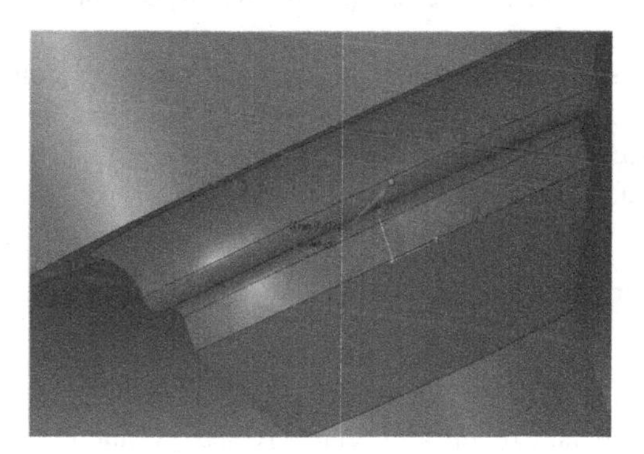

**Fig. 3.** Damage model

### 2.3 Section and Subsection Headings

For the study of the propagation characteristics of ultrasonic guided wave in the pipe head of nuclear power plant, the geometric parameters of the pipe head are as follows:

the total length of the joint is 290 mm, the inner diameter of the inner pipe is 70 mm, the outer diameter of the outer joint is 240 mm, the defect is a rectangular damage of 2 mm long and 1.9 mm wide, the outer diameter of the damage is 82.3 mm, and the wall thickness is 1.9 mm. The density, Young's modulus and Poisson's ratio of the specimens are 7850 $kg/m^3$, 206 gpa and 0.3 respectively. Because of the irregularity of the specimen with defects, the hexahedral mesh can not divide the whole part, so the tetrahedral mesh can only be selected.

To ensure the accuracy of the model simulation and avoid the error caused by the mesh size, the maximum mesh size should meet [3]:

$$\frac{\lambda_{min}}{\max(\Delta x, \Delta y, \Delta z)} > 8 \tag{1}$$

In formula (1): $\lambda_{min}$ is the minimum wavelength of ultrasonic guided wave; $\Delta x$, $\Delta y$, $\Delta z$ is the distance between two adjacent nodes, namely, is the size of the grid. The speed of guided wave propagation in the pipeline is 5000 m/s, the frequency used in simulation is 200 kHz, and the wavelength is 25 mm, so the grid size should be less than 3.125 mm. Considering that reducing the grid size will greatly increase the amount of calculation and the configuration of the laboratory computer, the approximate global size of the grid selected in this experiment is 3 mm.

## 3 Experimental Research

In order to obtain the influence of component damage on ultrasonic guided wave propagation, the experimental procedures are as follows:

1) Import ultrasonic guided wave signal into ABAQUS and select the appropriate frequency.
2) Select the appropriate signal excitation point and receiving point.
3) Under the same conditions, the healthy specimen and the damaged specimen are stimulated respectively to obtain the excitation waveform.
4) Analyze the data.

### 3.1 Excitation Waveform

In the experiments, sinusoidal modulation signal with five wave peaks is used as the excitation signal in the form of narrow-band excitation. The signal frequency band generated by narrow-band excitation is relatively narrow, and the position of structural damage can be reflected by the arrival time of the signal. Moreover, the ultrasonic guided wave signal generated by narrowband excitation is relatively simple, which can effectively suppress the dispersion phenomenon and reduce the complexity of signal processing in the later stage. As shown in the Fig. 4, the guided wave narrow-band excitation signal used in this experiment is:

$$I(t) = A[H(t) - H(t - N/f_c)] \times (1 - \cos(2\pi f_c/N)) \sin 2\pi f_c t. \tag{2}$$

In the expression (2), $A$ is Lamb wave signal amplitude modulation, $H(t)$ is Heaviside step function, it is the center frequency of the excitation signal, and $N$ is the number of wave peaks of the signal. Usually, the center frequency is 60 kHz–200 kHz. Through multi-group comparison experiments, 200 kHz is selected as the center frequency of the excitation signal in this experiment.

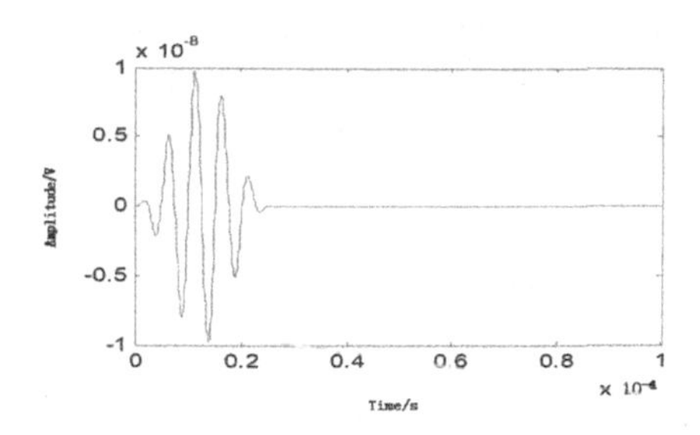


**Fig. 4.** Damage time domain diagram of excitation signal with center frequency of 200 kHz

## 3.2 Experimental Results Analysis

As shown in Fig. 5, the grid of the pipe head event model is a tetrahedral model, which is stronger than the hexahedral grid in the generality of generating complex geometry. The signal excitation point of the tetrahedral mesh model is a node (the node of the triangle box in Fig. 5 (right)), and the signal receiving point is a tetrahedral unit (the triangle box in Fig. 5 (right)). The red dots are excitation points and receiving units. Sensor placement (i.e., the positions of excitation points and sensing points in ABAQUS) usually takes many forms. The most common form is to place the excitation receiving points around the pipeline. In this experiment, a variety of arrangement methods are tried. Finally, the excitation receiving points are placed adjacent to each other (in the damaged specimen, two points are set below the damage). When the excitation receiving points are close to each other, the reflected wave can be effectively suppressed, thus simplifying the waveform and facilitating the analysis of damage data.

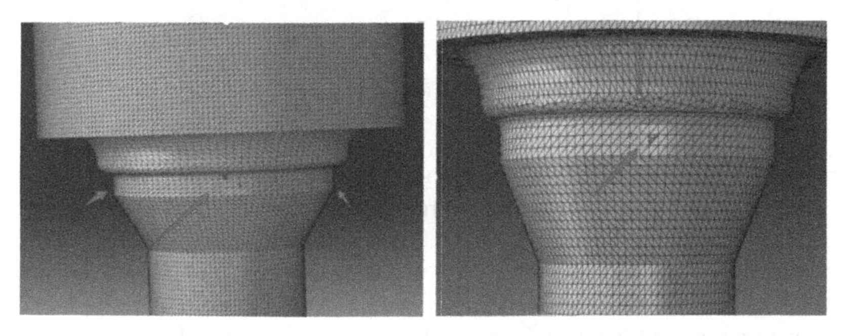

**Fig. 5.** Set around the pipe (left) and set next to each other (right)

The waveforms of the two placing modes are shown in Fig. 6 and Fig. 7. Typical features are circled. As can be seen from Fig. 6, when the sensor is placed relative to the pipeline, several large wave packets can be clearly seen, but the wave packets are chaotic and the source of each wave packet cannot be analyzed. The shape, amplitude and other information of the first several wave packets can be clearly seen in Fig. 7, which is conducive to the identification of damage. Therefore, the arrangement of excitation sensing points as shown in Fig. 5 (right) is selected in this experiment. In the damaged pipeline, the excitation point and the receiving point are placed below the damaged place, and the distance between the two points is two grid cells.

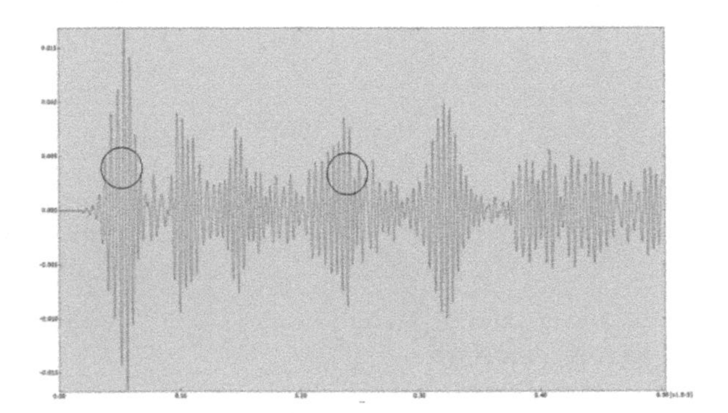

**Fig. 6.** Waveform of sensing points placed around pipes

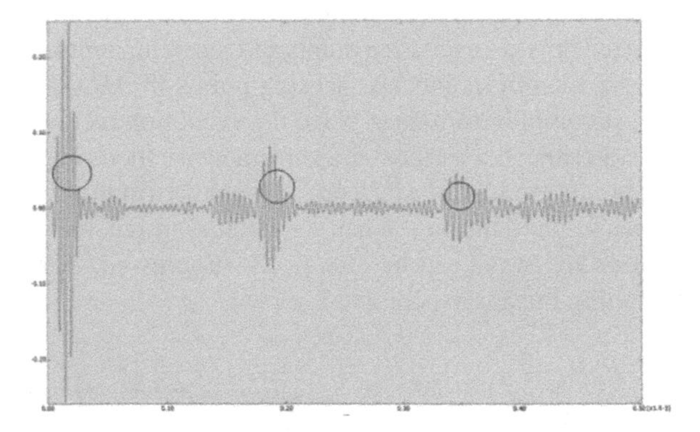

**Fig. 7.** Waveform of adjacent sensor points

The experimental data are shown in Fig. 8 and 9. The time step is 0.5 ms and the center frequency is 200 kHz. In the waveform diagram of the healthy specimen, the first wave packet is the direct wave of the excitation point, with an amplitude of about 0.17. The second clear wave packet appears at about 0.2 ms with an amplitude of about 0.8. After 0.3 ms, several chaotic wave packets appear. At this time, a variety of reflected waves have arrived, and the waveform is relatively complex, which is not conducive to the extraction of characteristic information.

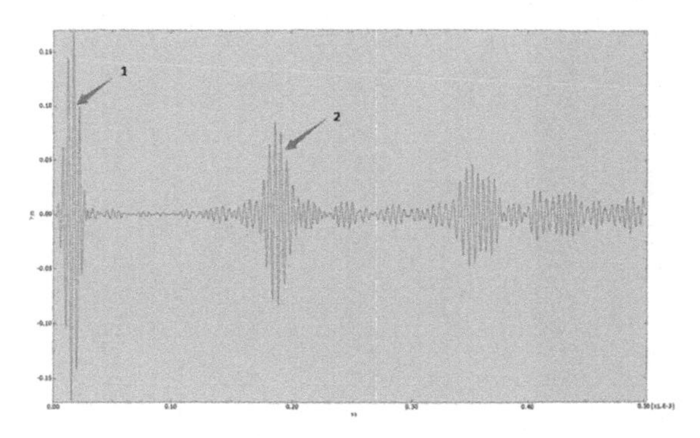


**Fig. 8.** Health data

In the waveform diagram of the damaged specimen, as shown in Fig. 9, the amplitude of the first direct wave is about 0.17, which is about the same as that of the healthy specimen. In the second wave packet, the difference can be seen, and the amplitude is about 0.05, which is obviously lower than that of the healthy specimen. The third wave packet after this shows a significant difference, with a amplitude of around 0.025. It can be seen from the damage waveform that the defect will affect the amplitude of the waveform and the size of the wave packet.

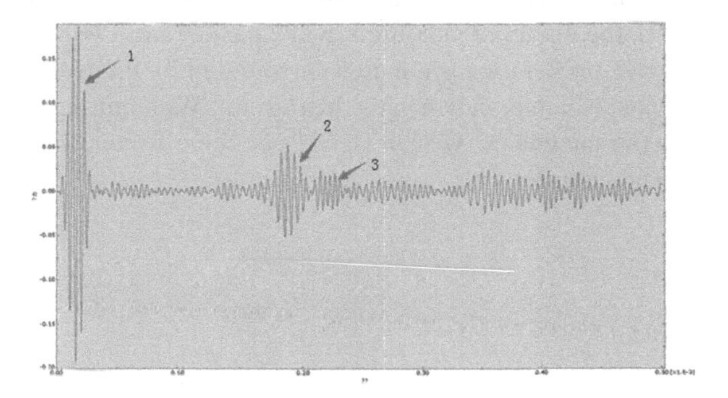


**Fig. 9.** Damage data

The comparison diagram of health data and damage data is shown in Fig. 10. It can be seen that there are obvious differences between health data and damage data at the three wave packets pointed by the arrow. The amplitude of wave packets 1 and 3 of health data is larger, while there is an extra 2 wave packet in damage data.

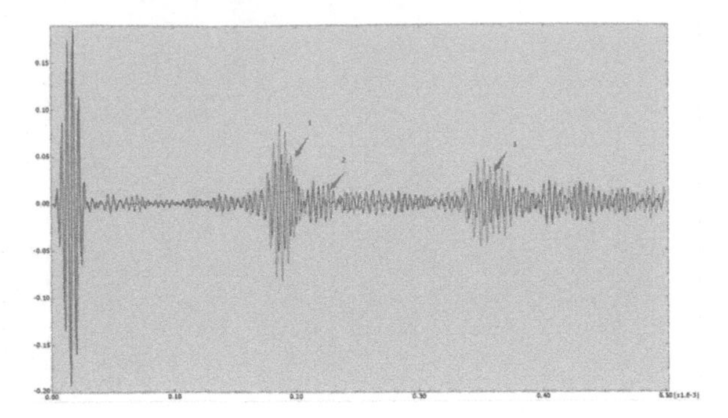

**Fig. 10.** Comparison of health data and damage data

### 3.3 Monitoring Results

The sensitivity of ultrasonic guided wave to damage will cause reflection when the signal is transmitted to the damage, thus changing the shape of the output waveform. This paper analyzes the characteristics of damage waveform by using this feature, and compares the damage data with health data from the wave packet area of ABAQUS output waveform. Simulation data is processed in MATLAB to verify the monitoring effect. In the simulation waveform, the maximum points of the data are firstly obtained, and then the maximum points are connected into lines (i.e. the upper half envelope of the waveform). Then, the top three maximum wave packet areas within the first 0.3 ms are obtained. The wave packet area is the area surrounded by the five maximum points and the coordinate axis, which is shown in the histogram. With putting the two groups of comparative data in the same table, we can clearly see the differences between different groups of data. As shown in Fig. 11:

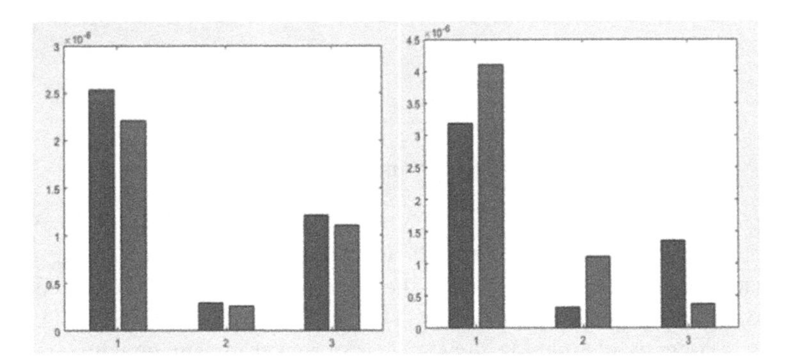

**Fig. 11.** Data comparison chart

The diagram on the left of Fig. 11 shows the comparison diagram of health data. The blue and red data are health data, and there is little difference in wave packet area in the same case. On the right side of the figure is the comparison graph of wave packet area between health data and damage data. Blue is health data and red is damage data.

It can be seen that the wave packet area of damage data of the first wave packet and the second wave packet is larger when the amplitude of the first wave packet and the second wave packet are the same, while the area of the third wave packet is healthier. In this paper, based on a group of health data, the area difference of the total wave packet is calculated. Through several groups of reference experiments, the boundary line of the damage health data is found and expressed in a bar chart:

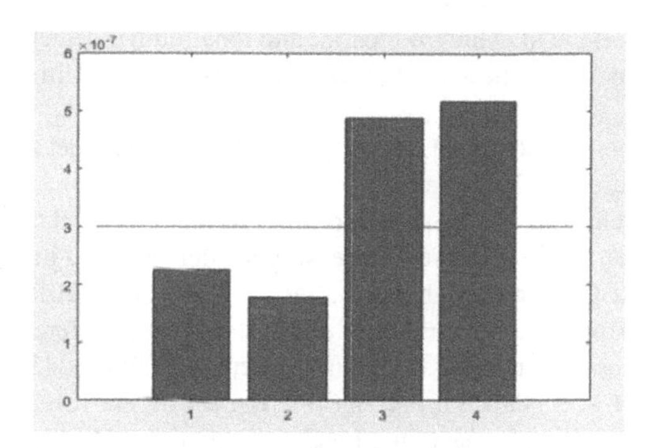


**Fig. 12.** Threshold graph

As shown in Fig. 12, the two groups of bar graphs on the right show the difference between the wave packet area of damage data and the wave packet area of health data, while the two groups on the left show the difference between the health data. You can see the obvious difference from the graph. In this paper, through multiple groups of experiments, set $3 \times 10^{-7}$ as the threshold value. When the area difference of wave packet is less than the threshold value, this data can be determined as health data; when the area difference of wave packet is greater than the threshold value, it is damage data.

## 4 Conclusion

In this paper, finite element analysis is adopted to validate the effective of guided wave based damage monitoring for the variable diameter pipe, such as nuclear power plant pipe head. From the experimental results, it is found that pipeline's damage would have an impact on the waveform amplitude and wave packet area. Then the damage can be analyzed by means of the wave packet area, and it is also found that the damage can be determined when the difference between the wave packet area of the output data and the health data is greater than the threshold.

## References

1. Wang, Q., Hong, M., Su, Z.: A sparse sensor network topologized for cylindrical wave-based identification of damage in pipeline structures. Smart Mater. Struct. **25**(7), 075015 (2016)
2. Keshwani, R.T.: Analysis of magnetic flux leakage signals of instrumented pipeline inspection gauge using finite element method. IETE J. Res. **55**(2), 73–82 (2014)

3. Alleyne, D.N., Lowe, M.J.S., Cawley, P.: The reflection of guided waves from circumferential notches in pipes. J. Appl. Mech. **65**(3), 635–641 (1998)
4. Liu, Z., Fan, J., Hu, Y., et al.: Torsional mode magnetostrictive patch transducer array employing a modified planar solenoid array coil for pipe inspection. NDT E Int. **69**, 9–15 (2015)
5. Li, Z., He, C., Liu, Z., et al.: Quantitative detection of lamination defect in thin-walled metallic pipe by using circumferential Lamb waves based on wavenumber analysis method. NDT E Int. **102**, 56–67 (2019)
6. Liu, Z., He, C., Wu, B., et al.: Circumferential and longitudinal defect detection using T(0,1) mode excited by thickness shear mode piezoelectric elements. Ultrasonics **44**, 1135–1138 (2006)
7. He, J., Huo, H., Guan, X., et al.: A Lamb wave quantification model for inclined cracks with experimental validation. Chin. J. Aeronaut. **34**(2), 601–611 (2020)
8. García-Gómez, J., Gil-Pita, R., Rosa-Zurera, M., Romero-Camacho, A., Jiménez-Garrido, J., García-Benavides, V.: Smart sound processing for defect sizing in pipelines using EMAT actuator based multi-frequency lamb waves. Sensors **18**(3), 802 (2018)
9. Sukatskas, V., Volkovas, V.: Investigation of the lamb wave inference in a pipeline with sediments on the inner surface. Russ. J. Nondestruct. Test. **39**(6), 445–452 (2003)
10. Thien, A.B., Puckett, A.D., Park, G., Farrar, C.R.: The use of time reversal methods with Lamb waves to identify structural damage in a pipeline system. In: Health Monitoring and Smart Nondestructive Evaluation of Structural and Biological Systems V, vol. 6177, p. 617708. International Society for Optics and Photonics (2006)

# Research on the Automatic Detection Device for Working Glass Container

Jie Chen, Zhenqi Shen(✉), Bin Li, Sicheng Wang, and Haoran Zhuo

Shanghai University, Shanghai 200444, China

**Abstract.** Working glass containers are usually calibrated manually, which is not only inefficient, but also has large error. In this paper, a set of automatic verification device for working glass container is developed, which integrates automatic filling, automatic reading, data acquisition and processing, certificate preparation and printing. Firstly, the basic requirements and design scheme of the hardware part of the automatic verification system are analyzed and determined. Secondly, the software system of automatic verification system is analyzed and designed. By using the method of image processing, the software solves the problems for the camera positioning scale line and the problems for determining the position of concave liquid level. It also gives the scheme of automatic liquid adding. Finally, the measuring cylinder of different specifications is tested with the device. The test results show that the device meets the requirements of technical indicators and improves the detection efficiency and accuracy.

**Keywords:** Glass measuring instrument verification · Liquid level detection · Image processing

## 1 Introduction

Working glass containers include single-line volumetric flask, single-line pipette, indexing pipette, burette, measuring cylinder, measuring cup, etc. [1]. The accuracy of Working glass container is the basis and guarantee for the development of teaching, scientific research and testing. Domestic and foreign scientific and technical personnel engaged in liquid level testing have studied a variety of liquid level testing programs, and the technical indicators, characteristics and economic benefits of the liquid level testing programs have been analyzed and studied, in the joint efforts of scientific and technical personnel, effectively promoted the application and development of liquid level testing.

In 2018, Harbin engineering university Qiao Renjie [2] implements the design of the liquid level automatic tracking system based on vision, fuzzy PID control and liquid level automatic tracking algorithm to realize automatic tracking, based on the level of the image preprocessing, get a clear liquid level image, and use the methods such as template matching, realize the scale value and calibration of recognition. A dynamic water line recognition algorithm based on deep learning was designed. Finally, the detection accuracy reached ±1 mm, and the real-time detection reached 50 frame/s.


Q. Han et al. (Eds.): LSMS 2021/ICSEE 2021, CCIS 1469, pp. 627–636, 2021.
https://doi.org/10.1007/978-981-16-7213-2_60

In November 2019, Li Chen and Shi Qufei et al. [3] from Beijing Institute of Metrology and Testing designed a set of automatic capacity verification device for small capacity containers based on image processing in the patent "An Automatic Capacity Verification Device for Small Capacity Containers and Its Usage Method". The device using portable water injection needle to add fluid, use of a removable camera device to obtain the bottom of the container and the scale line position, and then judge algorithm through images concave liquid level height, whether to reach the scale line according to the height of the decided to add the fluid to be achieved. The design is granted a patent but ignored eyelevel shooting request, and subsequent detailed study was not found, the feasibility of the unknown.

According to "JJG 196-2006 Verification Regulation of Working Glass Container" [4], glass measuring apparatuses have the characteristics of large quantity, miscellaneous types and different models. The verification process steps are tedious and repetitive, requiring the verifier to manually load the verification medium (distilled water or deionized Water), the human eye observes the reading, not only is easy to cause visual fatigue and measurement error, but also due to the difference of the tester, the difference in the measurement results is also large. These problems are particularly prominent in the measurement of large quantities of samples. At the same time, the verifier needs to consult a large number of parameter tables and tolerance tables when processing the verification result data because there are many types and specifications of glass measuring devices, which is time-consuming and laborious. This leads to high labor intensity of the inspectors, low inspection efficiency and high human error rate.

In order to better adapt to the rapid development of the measurement business of commonly used glass measuring devices and to solve the problems existing in the verification work, this paper designs an automatic verification and measurement device for commonly used glass measuring devices and proposes an automatic technology for capacity indication verification based on the image processing method and feedback control technology. The device's automatic liquid filling program and computer vision-based automatic reading program greatly reduce human reading errors. Finally, this paper designs a set of software system integrating data collection and processing, report preparation and printing, which greatly reduces the repetitive work of operators, and improves efficiency and measurement accuracy.

## 2 Technical Index and Hardware System Design

### 2.1 Technical Index

There are many kinds of commonly used glass measuring devices. This device is mainly for the automatic verification of single-line volumetric flask (25–200 mL), measuring cylinder (5–2000 mL) and measuring cup (5–2000 mL). The calibration accuracy requirements for different gauges are also different. According to the accuracy requirements of common glass gauges in Verification Regulation of Common Glass Gauges JJG 196-2006, the technical indicators of device design are as follows:

(1) Applicable environment: room temperature (20 ± 5) °C, and the change of room temperature shall not be more than 1 °C/h.

(2) The difference between water temperature and room temperature shall not be more than 2 °C.
(3) The verification medium is pure water (distilled water or deionized water), which should meet the requirements of GB6682-1992.
(4) The measurement range, allowable error and extended uncertainty requirements are shown in Table 1.
(5) Stability and repeatability meet the assessment requirements of corresponding measurement standards.

**Table 1.** Experimental data.

| Measurement object | Measuring range | Allowable error (mL) | Extended uncertainty U(k = 2) (mL) |
|---|---|---|---|
| One-mark volumetric flasks | 25 mL–2000 mL | ±0.03–±0.60 | 0.017–0.26 |
| Measuring cylinder | 5 mL–2000 mL | ±0.05–±20 | 0.01–2.2 |
| Measuring cup | 5 mL–2000 ml | ±0.2–±20 | 0.05–2.2 |

## 2.2 Hardware System Design

According to the technical assessment indicators and system requirements, the measurement method is determined to verify the gauges, and the automation solutions of different units are designed. The distribution of the scale line and the liquid level in the gauge are obtained by using a camera, and are used as the basis of image processing. Lifting unit of device, the image processing unit, filling unit were analyzed, and put forward the design scheme: the device should be fixed with liquid mouth position, moving with the saddle again to add liquid mouth position, then move the lifting platform to waiting for fixed position, carrying a camera through the process of liquid level and liquid, until the liquid level reaches the target location, so as to realize automatic calibration. Figure 1 is the overall structure diagram of the device.

The device is mainly divided into photographing unit, moving unit, automatic filling unit, lighting and light source unit, data acquisition unit. Wherein, the camera in the photography unit is installed on the slide rail 1; The moving unit is composed of two servo motors and two slide rails. Slide rail 1 is used to drive the camera to move, and slide rail 2 is used to drive the measuring device to be detected to move. The automatic filling unit is composed of a peristaltic pump and a micro pump. In the automatic filling unit, the filling speed of the micro pump and a peristaltic pump is controlled by the upper computer program to add liquid to the measuring device. First, a peristaltic pump is used for adding liquid at a fast flow rate to reach the scale to be measured in a relatively short time and reduce the verification time. When the volume of added water accounts for 90% of the nominal volume of the scale to be detected, the micro pump is controlled to add liquid at a slow flow rate until the liquid reaches the scale to be detected. The lighting and light source unit adopts the medical viewing lamp, which is placed behind the measuring device and serves as the light source plate of the device. The data acquisition unit is composed of industrial computer, temperature sensor, liquid level sensor, motion control card, data acquisition card, etc.

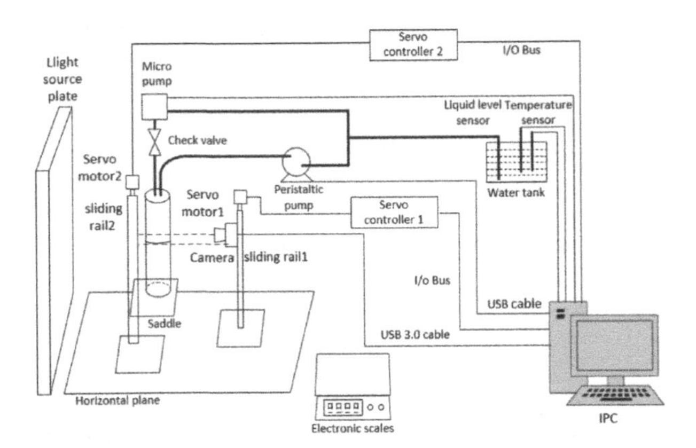


**Fig. 1.** Schematic diagram of automatic detection device for Working glass container.

The upper computer is developed by the industrial control host of Advantech Corporation and LabWindows CVI upper computer of NI Corporation. The IPC external equipment includes: temperature sensor, used for real-time monitoring of water temperature in the water tank; Liquid level sensor for real-time monitoring of the liquid level height in the water tank; ADAM4117 data acquisition board card, collect the temperature sensor, liquid level sensor information in the water tank; PCI-1245 motion control card and its corresponding cable and terminal board, used for communication with two servo controllers, so as to control the movement of the servo motor; Digital camera for obtaining the image information of the gauge, such as the position of the scale line, the position of the liquid level, etc. Electronic balance for weighing the mass of the measuring apparatus; Peristaltic pump is used to control fast flow rat so that the liquid level quickly reached the scale near the line; The micro pump is used to control slow flow rate and accurately add liquid to target position.

Figure 2 is the physical diagram of the device after the completion of machining and the physical diagram of the electric cabinet, Fig. 3 is the connection diagram of the electrical equipment of the device.

**Fig. 2.** Physical diagram of the automatic detection device for Working glass container and electric cabinet.

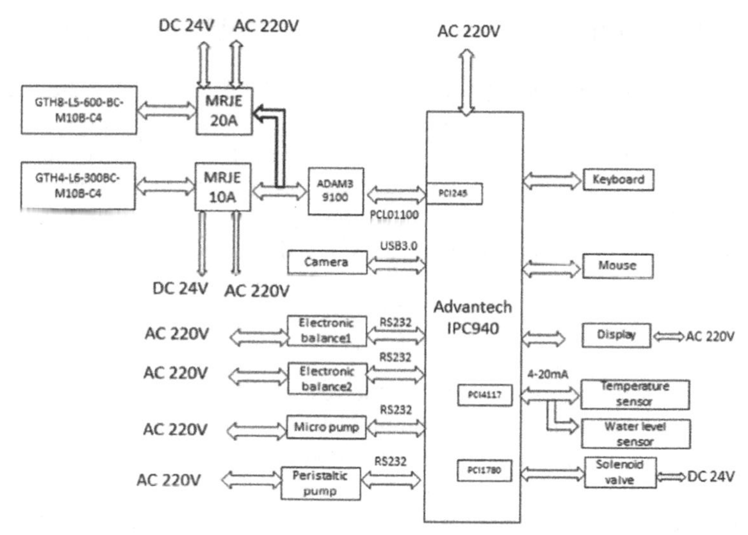


**Fig. 3.** Schematic diagram of electrical equipment connection of automatic detection device of working glass container.

## 3 Software System Design

The software system designed in this paper refers to the principle of using the measuring method to verify glassware in the "JJG196-2006 Commonly Used Glassware Verification Regulations": At a specified temperature, the nature of pure water is fixed, that is, the density, mass, and volume are constant. Generally, the capacity of the measuring device at 20 °C is regarded as the nominal capacity. In order to verify the capacity of the container, it is necessary to take a specific quality of pure water, and then convert the mass to volume according to the density of pure water at the current temperature, and then convert the mass to volume of the container at 20 °C. Finally, the actual volume of pure water measured by the container to be tested is compared with the nominal capacity of the container at 20 °C. If the error between the two is within the allowable error range, the container to be tested is qualified; otherwise, it is unqualified.

### 3.1 Overall Process Design

According to Verification Regulation of Common Glass Gauge JJG196-2006 and device design requirements and technical indicators, the verification process of automatic verification device for common glass gauges is determined as follows:

(1) Select the type, capacity and detection point of the calibrator;
(2) Weighing empty containers;
(3) System initialization;
(4) Put the meter on the bracket, and the camera will automatically position the scale line of the meter;
(5) the peristaltic pump adds liquid at a fast flow rate to 90% of the fixed point capacity to be detected;
(6) micro-pump accurate adding liquid so that concave liquid level at the target scale line;
(7) Weigh the weight of the measuring device, calculate the net weight of adding liquid, calculate the actual volume of the liquid according to the verification regulations. For example, when the standard temperature is 20 °C, the actual capacity of the glass measuring device at this temperature has the following calculation formula:

$$V_{20} = \frac{m(\rho_B - \rho_A)}{\rho_B(\rho_W - \rho_A)}[1 + \beta_1(20 - t)]. \tag{1}$$

Where $V_{20}$ is the actual capacity of the tester at 20 °C, mL; $\rho_B$ is the weight density, take 8.00 g/cm$^3$; $\rho_A$ is the ambient air density at the time of measurement, take 0.0012 g/cm$^3$; $\rho_W$ is the density of distilled water at $t$°C, g/cm$^3$; $\beta_1$ is the volume expansion coefficient of the measured device, °C$^{-1}$; $t$ is the temperature of distilled water at the time of verification, °C$^{-1}$; $m$ is the apparent mass of water that can be contained in the measuring device, g.
(8) The actual capacity is compared with the nominal capacity, and the allowable error of the measuring device is combined to draw a conclusion.

### 3.2 Calibration Line Positioning

For the measuring cylinder, the scale lines are printed neatly at equal intervals. Based on this, we draw the process of finding the graduation line of the graduated cylinder as shown in Fig. 4. Since the measuring cup is tapered, the scale lines are not evenly spaced, we use a method different from the measuring cylinder to find graduation lines. The single-marked volumetric flask has only one graduation line, so the position of the graduation line can be easily found by taking pictures and image processing.

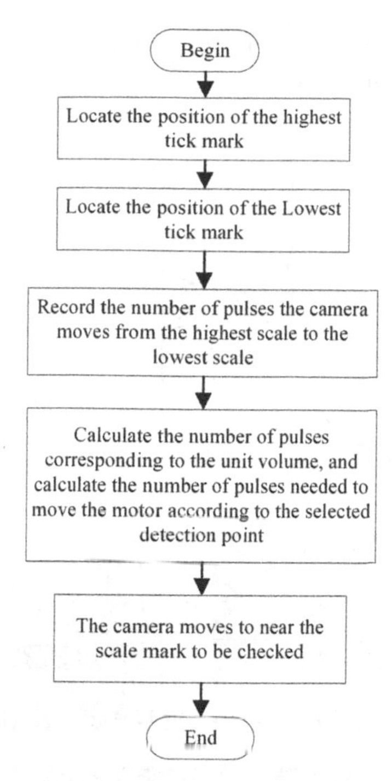


**Fig. 4.** Flow chart of measuring cylinder looking for calibration line.

## 3.3 Liquid Level Recognition

After adding liquid into the glassware, an arc-shaped concave liquid surface will be formed. The meniscus is caused by the capillary phenomenon [5], which is the result of the combined effect of liquid surface tension and wetting. We should take the position of the lowest point of the meniscus as the actual position of the liquid level.

The traditional method for human eyes to observe the liquid level and read is: the measuring cylinder must be placed flat when measuring, and the line of sight must be level with the lowest part of the liquid concave surface in the measuring instrument. If you look down or look up at the reading, errors will be caused [6]. To observe the liquid level, you need to look up at the liquid level and judge whether the liquid level reaches the scale line. This step determines the weight of the filling liquid and plays a decisive role in determining whether the gauge is qualified. Therefore, it is very important to correctly obtain the liquid level regardless of manual verification or automatic verification and plays a decisive role in determining whether the gauge is qualified. Therefore, it is very important to correctly obtain the liquid level regardless of manual verification or automatic verification.

As shown in Fig. 5, we should first keep our sight level with the lowest point of the meniscus, and then read the volume of the liquid taken, that is: the line of sight, the graduation line, and the lowest point of the meniscus are on the same horizontal line [7]. The traditional human eye reading method is to find the lowest point C of the meniscus first, use the human eye to judge that points A, C, and B are on the same straight line, and then read the volume of the liquid level. The data read by this method is accurate and conforms to the specification [8].

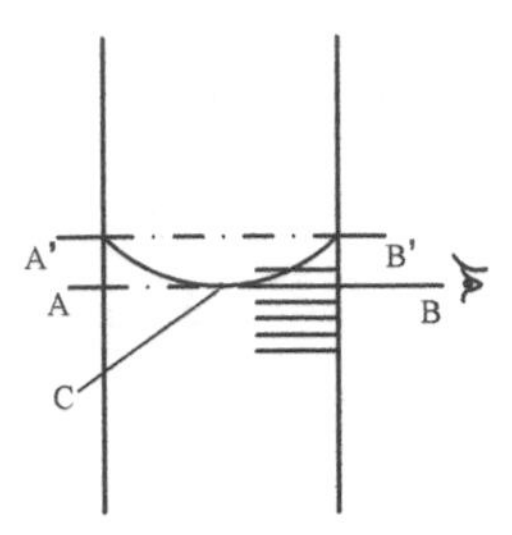


**Fig. 5.** Traditional human eye readings.

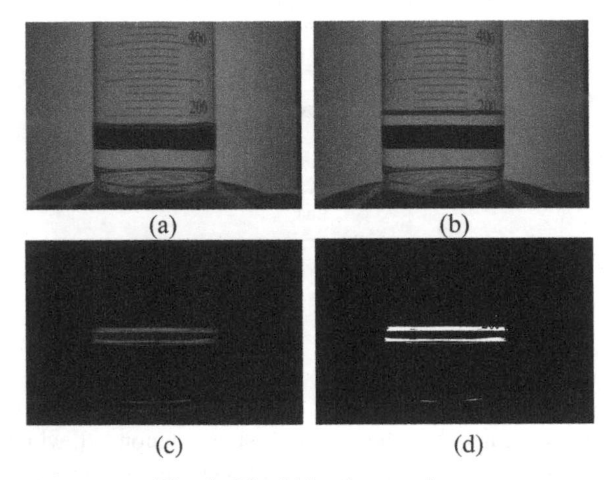


**Fig. 6.** Liquid level extraction.

In the automatic verification device, a camera is used to take pictures to find and determine the position of the meniscus instead of human eye reading. This process is called automatic liquid level recognition. When locating the position of the meniscus, in order to eliminate the influence of the scale line on the liquid level recognition, we take the method of image subtraction to extract the liquid level. For example, we subtract Fig. 6(a) form Fig. 6(b), the resulting picture is Fig. 6(c). Figure 6(d) is the result of the binary process of Fig. 6(c). It can be seen that in Fig. 6(d), except for the two liquid levels in Fig. 6(a) and Fig. 6(b), the information on the picture has been eliminated.

### 3.4 Automatic Filling Scheme

The solution we adopted is: (1) The host computer sends a command to start the peristaltic pump, and the volume of liquid added to the glassware is 90% of the capacity of the point to be verified. (2) The camera detects whether the meniscus reaches the target scale line, if not, the micro pump starts to add liquid, whose volume is 90% of the remaining volume. (3) Repeat the procedure of (2), and stop adding liquid when our device judges that the meniscus reaches the scale line.

## 4 Experiment

We used the processed device to verify the 100 mL graduated cylinder and the 2000 mL graduated cylinder. Each verification point of each graduated cylinder has been verified twice. The experimental data is shown in Table 2, where the volume is equal to the mass multiplied by $K(t)$. The value of $K(t)$ is 1.00285.

According to the requirements in the verification regulations, the difference between the two verification data should not exceed 1/4 of the allowable error of the glassware being verified. The difference between the two verifications for each verification point in Table 2 is obviously less than 1/4 of the allowable error, which proves that the repeatability of our device is good.

**Table 2.** Experimental data.

| Type | Verification point(mL) | Mass (g) | Volume (mL) | Difference between the two verifications (mL) | Allowable error (mL) |
|---|---|---|---|---|---|
| 100 ml graduated cylinder | 10 | 9.25 | 9.28 | 0.01 | ±1 |
| | 10 | 9.26 | 9.29 | | |
| | 50 | 49.21 | 49.35 | 0.03 | |
| | 50 | 49.24 | 49.38 | | |
| | 100 | 99.41 | 99.69 | 0.06 | |
| | 100 | 99.47 | 99.75 | | |
| 2000 ml graduated cylinder | 200 | 200.99 | 201.56 | 1.47 | ±20 |
| | 200 | 199.52 | 200.09 | | |
| | 1000 | 997.32 | 1000.16 | 0.97 | |
| | 1000 | 998.28 | 1001.13 | | |
| | 2000 | 1993.93 | 1999.61 | 0.88 | |
| | 2000 | 1994.80 | 2000.49 | | |

## 5 Conclusion

This paper has developed an automatic detection device for commonly used glassware, completed the mechanical structure system design and electrical system design of the device. It solves the problems of Camera head-up shooting, calibration line positioning, automatic liquid level recognition and automatic liquid filling. Experiments on 100 ml and 2000 ml graduated cylinders suggest that the automatic detection device fully meets the requirements of the verification regulations. The device can replace manual measuring instrument detection, therefore people's repetitive labor when verifying glassware can be greatly reduced.

## References

1. Xu, L.: Several problems in the metrological verification of common glass gauge. Technol. Market. **25**(04), 123–125 (2018)
2. Qiao, R.J.: Automatic Recognition and Tracking Design of Liquid Level Based on Vision. D. Harbin Engineering University (2018)
3. Li, C., Shi, Q.F., Jiang, Z.R., Chen, H.T.: An Automatic Capacity Verification Device for Small Capacity Container and Its Application Method. P. CN110715712A (2020)
4. JJG 196-2006 Verification Regulation of Working Glass Container. S. General Administration of Quality Supervision, Inspection and Quarantine of the People's Republic of China (2007)
5. Lin, C.S.: Design of Real-time Liquid Level Measurement System Based on CIS. Hefei University of Technology, Anhui (2018)
6. Liu, Y.: Analysis of reading problem of common chemical gauge. J. Baoding Univ. **02**, 24–25 (2008). (in Chinese)
7. Compilation group of book title of Beijing Normal University. M. Chemical experiment standard (1987). (in Chinese)
8. Liu,H.L.: Correction of "taking liquid volume with measuring cylinders". Teach. Instrum. Exp. **5**, 33–33 (2011)

# Fault Feature Transfer and Recognition of Bearings Under Multi-condition

Xiangwei Wu, Mingxuan Liang(✉), Ying Wang, and Heng Ma

China Jiliang University, Hangzhou 310018, China

**Abstract.** Although increasingly being applied, the deep learning fault diagnosis method is hard to deliver high-precision accuracy due to sample size limit under practical working conditions. To solve this problem, a new bearing fault diagnosis method is proposed. Firstly, converted the original vibration signals of 10 types of states into two-dimensional image data and proportionally divided into three different datasets. Then, the structure of ResNet34 network is improved to extract the weight parameters pre-trained on ImageNet dataset, and the feature parameters are transferred to the Case Western Reserve University (CWRU) dataset by transfer learning to process the small sample fault diagnosis. Finally, the fault diagnosis model is obtained by repeatedly adjusting the hyperparameter training. Compared with the experimental results of several other methods, the final accuracy of this improved method can reach 99.21%. The test results under different working conditions also demonstrate that this transfer learning method has higher identification accuracy than the existing methods and can meet the requirements of fault diagnosis in actual industrial production.

**Keywords:** Fault diagnosis · ResNet · Transfer learning · Multi-condition

## 1 Introduction

The operation of rolling bearings is related to the safe operation of the entire rotating machinery system, because of it is a key part of rotating machinery. Serious accidents and enormous economic losses will be caused by failure to identify the fault accurately and timely. Therefore, the research of rolling bearing fault identification has become an important direction for researchers [1]. Traditional manual fault feature extraction is hard to extract useful features, the emergence of deep learning successfully has successfully improved this problem. Deep learning has been widely applied in many fields in recent years, such as image processing [2], audio classification [3], which fully proves that deep learning has broad application prospects. Especially the CNN has been widely applied in original vibration signals due to its powerful nonlinear mapping capacity, and it is also often used in rolling bearing fault diagnosis [4, 5]. However, deep learning also exposed some issues in actual working conditions, a lot of vibration data have been collected, but fault labeled sample data is relatively few to collect, collecting these labeled data is usually time-consuming. Thus, training a high precision fault diagnosis model by using small amount of marked vibration signal data is hard.


Q. Han et al. (Eds.): LSMS 2021/ICSEE 2021, CCIS 1469, pp. 637–643, 2021.
https://doi.org/10.1007/978-981-16-7213-2_61

The emergence of transfer learning has solved these problems. Literature [6] proposed an intelligent bearing fault diagnosis system by using time-frequency images and AlexNet, which has good performance. Literature [7] carried out experiments on three fault datasets by migrating the parameters of VGG-16 model, and verified it on the CWRU dataset. Inspired by these examples, we proposed a bearing fault diagnosis method based on fault feature transfer under multi-condition, the weight parameters of the ResNet34 network pre-trained on the large dataset were directly applied to bearing fault diagnosis, and verified on the CWRU dataset. It solves the problem of low diagnostic accuracy caused by small amount of labeled fault sample data in actual industrial application. By comparing the results of cross training and testing of dataset under multi-conditions, final fault identification results of this model are better than the experimental results of other networks, and it has good diagnostic performance.

## 2 Data Preprocessing

The data preprocessing method proposed transforms all types of original signals into two-dimensional images. Intercept N signal slices of length M from the continuous signal to form a two-dimensional matrix K, after normalizing the matrix from 0 to 255, output N gray-scale images of M × M pixels. The transformation theoretical formula can be expressed as:

$$L = round\left(\frac{T - T_{(Min)}}{T_{(Max)} - T_{(Min)}} \times 255\right). \quad (1)$$

Where round () represents the rounding function, which normalizes all pixel values to the pixel value range of the grayscale image: 0–255, T is the original signal. Considering the input format of the ResNet is 224 × 224 × 3, we convert the grayscale image into a 3-channel format after copying and superimposing. The converted image is an ideal input data format for CNN.

## 3 Improved ResNet Network

As one kind of CNN, ResNet can ideally avoid a series of problems such as gradient disappearance which often occur in the training process of traditional deep plain network. The basic structure of the residual block is shown in Fig. 1, which is different from the traditional plain network. Identity mapping is added to make the output of one layer directly cross several layers as the input of the next layer, which effectively solved the problem of network degradation to some extent.

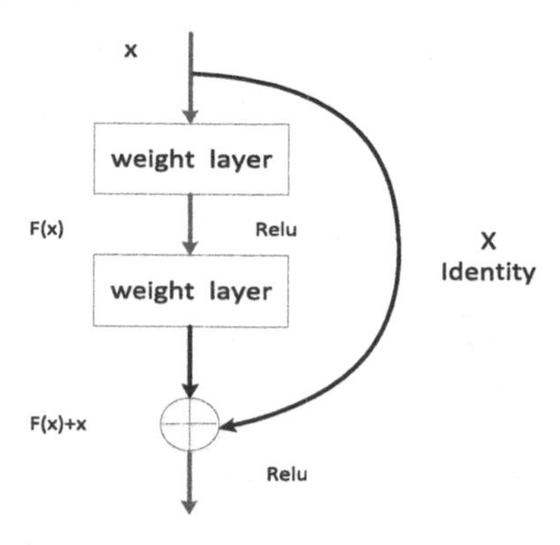


**Fig. 1.** The basic structure of the residual block.

When the residual blocks are stacked repeatedly, a network structure of different depths can be formed. Based on the comprehensive consideration of experimental results and calculation, the ResNet34 network is selected. The first layer of the ResNet network is a 7 × 7 convolution layer with a large receptive field, which is sufficient to extract features for ImageNet dataset, but it is not fully applicable to complex bearing fault data. For the purpose of obtaining more important features from complex fault signal data, the concatenation of two convolution layers is used to replace the 7 × 7 convolutional layer, which effectively reduced the number of parameters and calculations. Supposed that the size of input and output feature map of the convolution layer is A, the parameter of a 7 × 7 convolution layer is only $(7 \times 7 \times A) \times A = 49 \times A^2$, while the parameters of the concatenation of the 1 × 7 layer and the 7 × 1 layer are $2 \times (1 \times 7 \times A) \times A = 14 \times A^2$, only $(1 \times 7 + 7 \times 1)/(7 \times 7) = 28.6\%$ of the calculation is used, which can bring better performance to the model. The improved ResNet34 network structure is shown in Fig. 2.

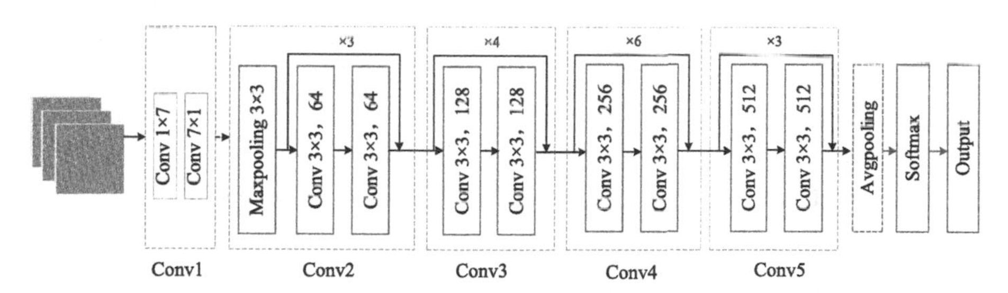


**Fig. 2.** The improved network structure.

# 4 Transfer Learning

Transfer learning means that knowledge is transferred from the source large dataset to the target small dataset. Knowledge and large amounts of data labels are in the source domain, which is the object to be transferred, while the target domain is the object to which we ultimately assign knowledge and labels. Transfer learning transfer parameters from related tasks that have already been trained, thus improving the tasks that need to be trained without the need to learn from nothing. The process of transfer is shown in Fig. 3.

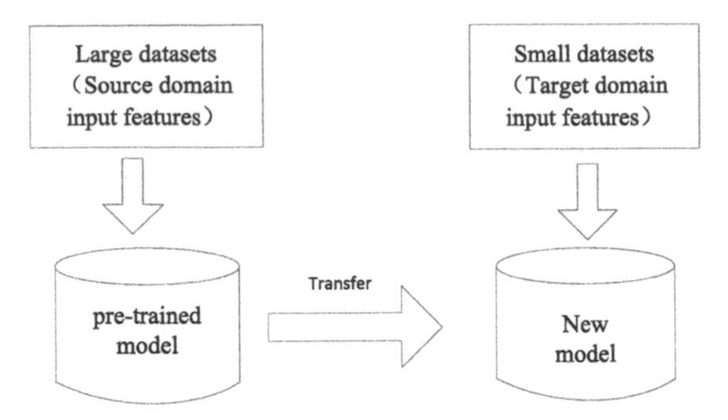


**Fig. 3.** The method of transfer learning.

# 5 Experiments and Results

## 5.1 Experiments Environment

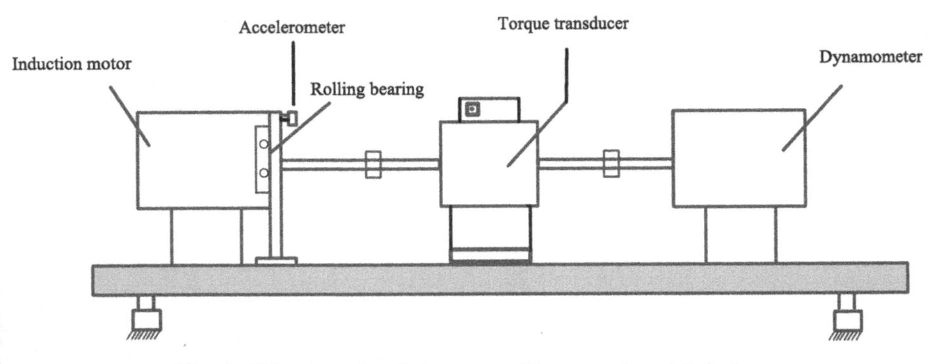


**Fig. 4.** Vibration signal data acquisition experimental device.

To test the performance of this method, we testify it on Bearing Data Center of Case Western Reserve University (CWRU). Vibration signal data acquisition equipment is shown in Fig. 4. The drive end bearing with a sampling frequency of 12 kHz has been studied, which has one healthy condition and three single point fault conditions with three different diameters respectively. The signal of normal state is labeled as label 0, remaining signals of the bearing are marked as 9 different labels from 1 to 9, with a total of 10 different labels. The size of one sample is 400, and 300 such non overlapping samples are intercepted from the signal data of each condition. Finally, there are totally 3000 samples applied to the experiment.

### 5.2 Fault Diagnosis Method Based on Transfer Learning

In order to avoid the consequence of low diagnosis accuracy caused by less fault sample data in actual production, this paper uses transfer learning to make up for it. The high-accuracy network parameters trained on the ImageNet dataset are used as the initial parameters of the new network that this paper is modified, then the top convolution network is trained in the fault dataset to make it better applied to bearing fault diagnosis tasks. Frozen the convolutional layer of the network and taken the parameters prepared by pre-training as the initial parameters of the small sample dataset of rolling bearings. The full connection layer is reconstructed and trained based on the number of labels for fault diagnosis to adapt the classification tasks. The process of network training and transfer learning is shown in Fig. 5.

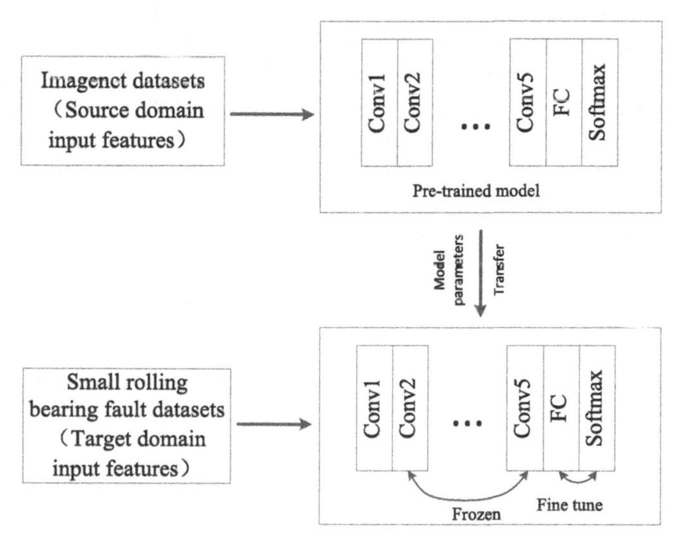


**Fig. 5.** The process of parameter transfer and network training.

### 5.3 Compare and Analyze Experimental Results

We have compared and analyzed the fault identification accuracy results of various methods to demonstrate the advantage of this method, data under the 1hp, 2hp and 3hp motor load are defined as dataset A, B and C, respectively. Each data set was proportionally divided into 8:1:1 for training, validation and test data. The results of identification accuracy obtained from the experiment on dataset A are shown in Table 1. Five experiments were performed for each type of network to eliminate the influence of randomness. The average identification accuracy with the proposed method is 99.21%, it is higher than the general CNN, PCA-SVM, and the general transfer learning, which are 85.67%, 88.67%, and 96.95%, respectively.

**Table 1.** Comparison of identification results under dataset A.

| Method | CNN | PCA-SVM | General transfer learning | Proposed |
|---|---|---|---|---|
| Identification accuracy (%) (3000 samples) | 90.74 | 88.67 | 97.38 | 99.17 |
| | 87.17 | 88.67 | 97.26 | 99.34 |
| | 85.89 | 85.83 | 97.32 | 99.26 |
| | 83.69 | 88.83 | 96.54 | 98.89 |
| | 80.86 | 88.00 | 96.23 | 99.39 |
| Average: | 85.67 | 88.67 | 96.95 | 99.21 |

We plot a diagnostic accuracy comparison trend chart under variable working loads to further prove the advantages of the proposed method. All precision comparison under different experimental combinations is shown in Fig. 6. Nine different representation methods of dataset combination are used to list these nine different combination experiments. For instance, C $\rightarrow$ A notes training and validating in dataset C and testing in dataset A. When the dataset from same distribution is trained and tested in the network, those four methods all show better performance, and all accuracy is markedly higher than the experimental results of the training dataset and test dataset come from different distribution. It can be remarked that the average diagnosis accuracy of this method in different combination operations are much higher than other methods.

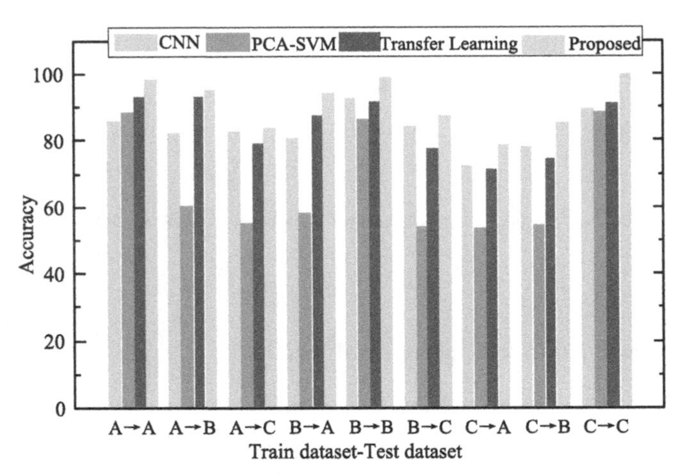


**Fig. 6.** Identification accuracy under different combinations.

## 6 Conclusion

In this paper, aiming at the low accuracy of traditional fault diagnosis methods due to the small dataset, we proposed a diagnosis method based on transfer learning. The weight and the shallow structure in the pre-trained neural network model are retained, and fine-tuned the weight in the higher layer of the model by using the target datasets, then the optimized model is applied to the classification of bearing faults. The proposed model can maintain superior algorithm performance under different loads by validation results on public dataset show.

## References


1. Yang, L., Chen, H.: Fault diagnosis of gearbox based on RBF-PF and particle swarm optimization wavelet neural network. Neural Comput. Appl. **31**(9), 4463–4478 (2018). https://doi.org/10.1007/s00521-018-3525-y
2. Chen, G., Zhang, X., Wang, Q., Dai, F., Gong Y., Zhu, K.: Symmetrical dense-shortcut deep fully convolutional networks for semantic segmentation of very-high-resolution remote sensing images. IEEE J. Sel. Top. Appl. Earth Obs. Remote Sens. **11**(5), 1633–1644 (2018). https://doi.org/10.1109/JSTARS.2018.2810320
3. Abdul Qayyum, A.B., Arefeen, A., Shahnaz, C.: Convolutional neural network (CNN) based speech emotion recognition. In: 2019 IEEE International Conference on Signal Processing, Information, Communication & Systems (SPICSCON), pp. 28–30 (2019)
4. Qu, J., Yu, L., Yuan, T., Tian, Y., Gao, F.: Adaptive fault diagnosis algorithm for rolling bearings based on one-dimensional convolutional neural network. Chin. J. Sci. Instrum. **39**(07), 134–143 (2018). (in Chinese)
5. Li, J., Liu, Y., Yu, Y.: Application of convolutional neural network and kurtosis in fault diagnosis of rolling bearing. J. Aerosp. Power **34**(11), 2423–2431 (2019). (in Chinese)
6. Ma, P., Zhang, H., Fan, W., Wang, C., Wen, G., Zhang, X.: A novel bearing fault diagnosis method based on 2D image representation and transfer learning-convolutional neural network. Meas. Sci. Technol. **30**(5), 055402 (2019)
7. Shao, S., McAleer, S., Yan, R., Baldi, P.: Highly accurate machine fault diagnosis using deep transfer learning. IEEE Trans. Ind. Inform. **15**(4), 2446–2455 (2019)

# Impedance Based Damage Monitoring Method Under Time-Varying Conditions

Yifei Lu, Sainan Zhang, and Qiang Wang(✉)

College of Automation and College of Artificial Intelligence, Nanjing University of Posts and Telecommunications, Nanjing 210023, China
wangqiang@njupt.edu.cn

**Abstract.** Due to the sensitivity of electro-mechanical impedance and guided wave to microstructural changes, the detection of local structural damage has been widely studied. In this paper, a damage monitoring system based on electro-mechanical impedance technology is proposed to improve the detection ability of crack damage types at different temperatures. Firstly, the piezoelectric chip attached to the target structure is used to collect the health admittance signal in a wide range of temperature. Then, it is decomposed into real part and imaginary part of admittance. The real part is used for damage diagnosis and the imaginary part is used for temperature compensation. In order to verify the effectiveness of the system, an experimental study on crack damage detection of composite plates was carried out in the temperature range of 30–50 °C.

**Keywords:** Structural health monitoring · Electro-mechanical impedance · Time-varying conditions · Crack damage

## 1 Introduction

Impedance technology has been widely used in structural damage detection because of its high sensitivity to the changes of micro structural damage. By monitoring the change of measured impedance signal, the potential defects in the main structure can be simply identified. However, many works have proved that impedance signal is also affected by other environmental changes, such as temperature, which makes it vulnerable to false alarms caused by these non-damage related changes [1].

Thus, the main problems faced by impedance technology in structural health monitoring is how to minimize the false alarm caused by temperature change. In order to solve this problem, researchers have put forward many methods. For example, Fabricio G. Baptista [2] studied the influence of temperature on the impedance of piezoelectric sensor through experiments, and proved that the temperature effect has a strong frequency dependence. T. Wandowski [3] adopted the compensation algorithm based on signal cross-correlation to eliminate the influence of temperature effect. Park [4] simulated the damage by using the loosening and fasting state of the bolt. Improved impedance based damage monitoring method is studied in this paper to eliminate the problems caused by time-varying conditions. In the monitoring by using the new method, the admittance


Q. Han et al. (Eds.): LSMS 2021/ICSEE 2021, CCIS 1469, pp. 644–652, 2021.
https://doi.org/10.1007/978-981-16-7213-2_62

signal will be decomposed to obtain the real part and imaginary part, and these two parts are adopted to evaluate the conditions and damages respectively [5]. The results show that the method can minimizes false alarms due to temperature variation and has a certain engineering application prospect.

## 2 Theoretical Development

### 2.1 Analysis of Impedance Signal

As a new type of structural health monitoring technology, the basic idea of electromechanical impedance technology is to fix the low-power piezoelectric sensor as a driver and sensor on the complex structure, and use the electromechanical coupling characteristics of piezoelectric elements and the interaction between piezoelectric elements and structure, the electromechanical impedance of the structure is reflected by the electrical impedance of the piezoelectric sensor and structure after coupling. Piezoelectric sensors act as driving and sensing elements to excite and obtain the dynamic response of the structure. In short, the working principle of piezoelectric impedance technology is to reflect the change of structural state caused by defects or other physical changes through the change of mechanical impedance of structural parts.

Figure 1 shows the coupling principle of piezoelectric plate and structure [6]. Keeping the parameters and properties of the piezoelectric sheet constant, the only way to determine the mechanical impedance of the structure is the electrical impedance, so any change of the electrical impedance reflects the defects, damage or other physical changes in the structure.

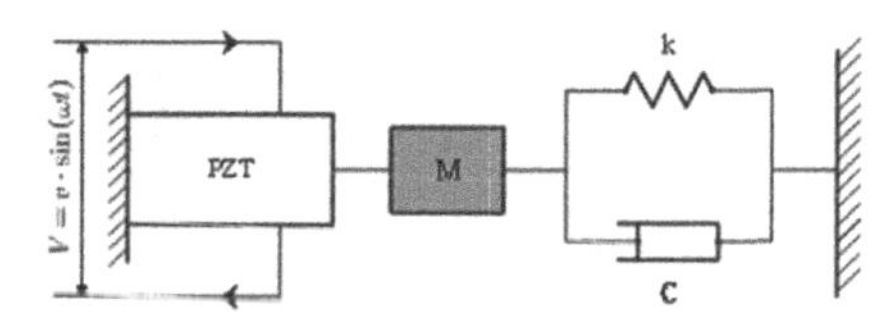


**Fig. 1.** Coupling model diagram of piezoelectric and structure.

### 2.2 Analysis of Voltage Current Method

The principle of measuring impedance by voltage current method is shown in Fig. 2. In this experiment, because of its positive and negative piezoelectric effect, the piezoelectric excites and senses simultaneously. An experimental circuit is built on the board, and the sliding resistance and the piezoelectric ceramic chip constitute a series circuit to realize voltage division. According to the derivation of the following series of mathematical formulas, the electrical admittance of the piezoelectric sheet is calculated, and the impedance information is obtained.

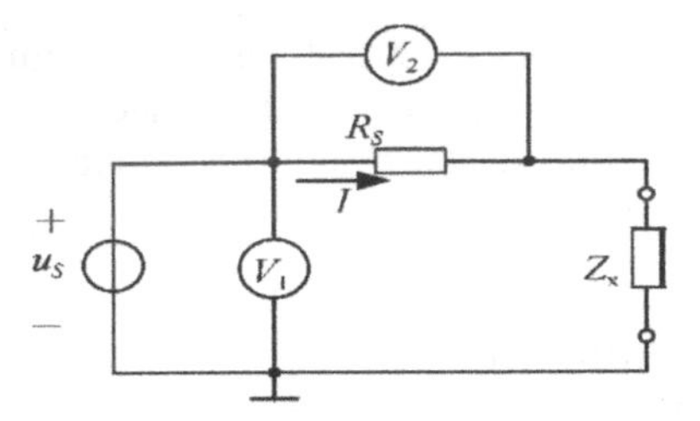


**Fig. 2.** Circuit diagram of voltage current method.

Where $u_s$ is the AC voltage source, the specific circuit is the function signal generator; $V_1$ is the voltage value of the function signal generator; $R_s$ is the partial resistance; $V_2$ is the partial voltage value of the partial resistance $R_s$, $Z_x$ is the effective impedance value.

$Z_x$ can be obtained from Eq. (1):

$$Z_x = \frac{V_1 - V_2}{I} = \frac{V_1 - V_2}{V_2} R_s. \tag{1}$$

According to the existing research results, the real part of admittance of piezoelectric sensor is sensitive to structural damage and temperature change, but the imaginary part of admittance is only sensitive to temperature change, so we can analyze the real part and imaginary part of admittance. Admittance signal is the reciprocal of impedance signal. We decompose admittance signal.

$$Y_x = \frac{V_2}{(V_1 - V_2)R_s}. \tag{2}$$

Where $Y_x$ is the admittance of PZT. Because of the capacitance characteristic of piezoelectric sensor, the voltage signal has a change of amplitude and phase. Assuming that the phase difference between the two signals is θ, then:

$$Y_x = \frac{\left(\frac{U_{1m}}{U_{2m}}\right)^{e^{j\theta}-1}}{R_s} = G + jB. \tag{3}$$

Where $U_{1m}$ and $U_{2m}$ are the amplitudes of $U_1$ and $U_2$ respectively, $G$ is the real part of electrical admittance and $B$ is the imaginary part of electrical admittance.

$$\mathrm{G} = \frac{U_{1m}}{R_s U_{2m}} cos\theta - \frac{1}{R_s}. \tag{4}$$

$$\mathrm{B} = \frac{U_{1m}}{R_s U_{2m}} sin\theta. \tag{5}$$

The admittance decomposition flow chart is shown in the following Fig. 3.

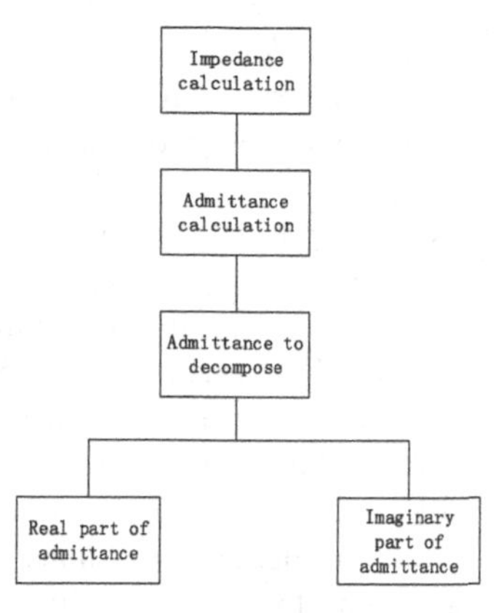


**Fig. 3.** Admittance decomposition flow chart.

## 3 Design of New Methodology Scheme

The overall steps are shown in Fig. 4. Firstly, the admittance signal is collected from the baseline and test conditions of the structure. Then, as described in Eq. (4) and Eq. (5), the measured admittance signal is decomposed into a real part of admittance and an imaginary part of admittance. The imaginary part of admittance is used for temperature compensation and the real part of admittance is used for damage detection. Then, the damage index and damage diagnosis threshold are calculated according to the real part of admittance.

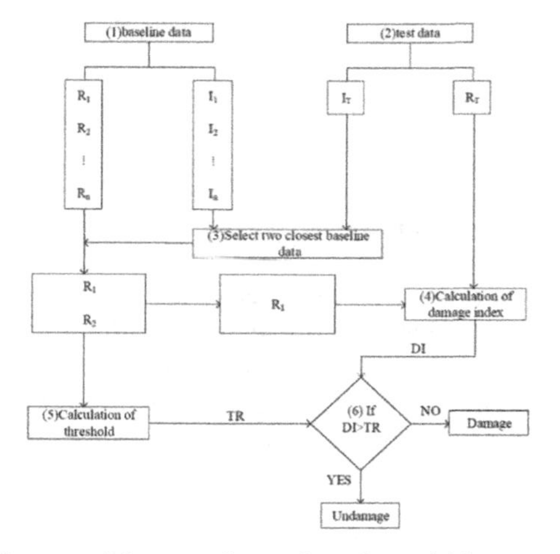


**Fig. 4.** Schematic diagram of damage diagnosis under variable temperature environment.

The experiment consists of six steps. *R* and *I* is the real part and imaginary part of admittance signal, subscripts *1*, *2*, and *n* are baseline data, and subscript *t* is test data. *DI* and *TR* are the damage index and damage diagnosis threshold calculated according to Eq. (6).

Firstly, we collect multiple sets of admittance signals from healthy structures under various temperature conditions. In order to reduce false alarms, we should cover most of the temperature conditions as much as possible and decompose the admittance signal into real part and imaginary part according to Eq. (4) and (5).

Secondly, when there may be damage in the structure, the real part and imaginary part of the test admittance are collected in the same way as in the first step.

Thirdly, the imaginary part of the test admittance is compared with all the imaginary parts of the baseline data, and the two nearest imaginary parts of the admittance are selected according to the variance, and the corresponding real parts of the admittance are selected to calculate the damage index and the damage diagnosis threshold.

Fourth, when the closest baseline data is selected, *DI* and *TR* are calculated according to the cross-correlation coefficient formula.

$$\rho_{XY} = \frac{E([X(u) - \mu(u)][Y(u - \Delta u) - \mu_Y])}{\sigma_X \sigma_Y}. \tag{6}$$

Where $X(u)$ and $Y(u)$ are two admittance signals in the frequency domain.$\mu$, $\sigma$ and $E[]$ represent the mean, standard deviation and expectation. $\Delta u$ is the frequency shift of admittance signal.

Fifthly, the damage diagnosis threshold is calculated as the same as the damage index. Two admittance signals are selected from the baseline data, and the threshold is obtained according to Eq. (6).

Finally, if $DI < TR$, a damage alarm will be issued, otherwise the alarm will not be triggered.

# 4 Experimental Research

## 4.1 Experimental System and Process

The experimental object used in this experiment is composite plate, its size is 450 mm × 450 mm × 3 mm, the density of the material is 1960 kg/m$^3$, Young's modulus of elasticity is 20 GPa, Poisson's coefficient ratio is 0.17. The frequency range of the excitation signal in the experiment is determined to be 20 kHz–120 kHz, the interval of each frequency scanning is 1 kHz, and the interval of the excitation signal of two adjacent frequencies is set to be 20 ms.The voltage value of function signal generator is 9 V, the selected partial resistance is 8100 Ω, and the partial voltage value is about 4.5 V. The experimental equipment is shown in Fig. 5. According to the previous layout design, the crack damage exists between the two sensors, as shown in Fig. 6.

**Fig. 5.** Experimental equipment.

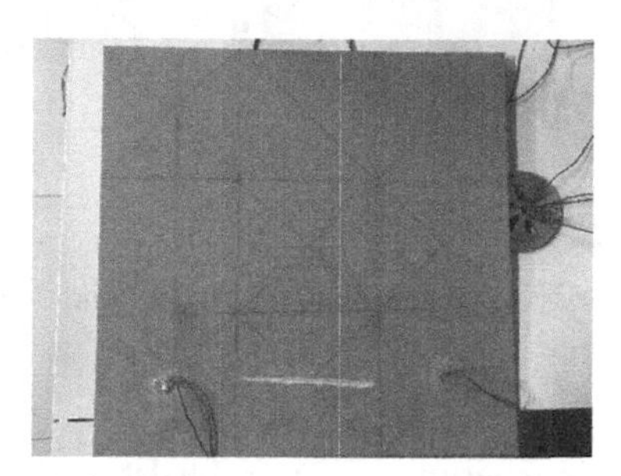

**Fig. 6.** Physical picture of crack damage.

## 4.2 Analysis of Experimental Results

Figure 7 show the real part and imaginary part of the admittance of the composite plate under different temperature conditions, respectively. On the left is the real part and the figure on the right is the imaginary part. It can be seen that the real part and imaginary part of the admittance will cause the signal frequency shift with the change of temperature. As the temperature increases, the real part and imaginary part of the admittance in the healthy condition shifts to the left.

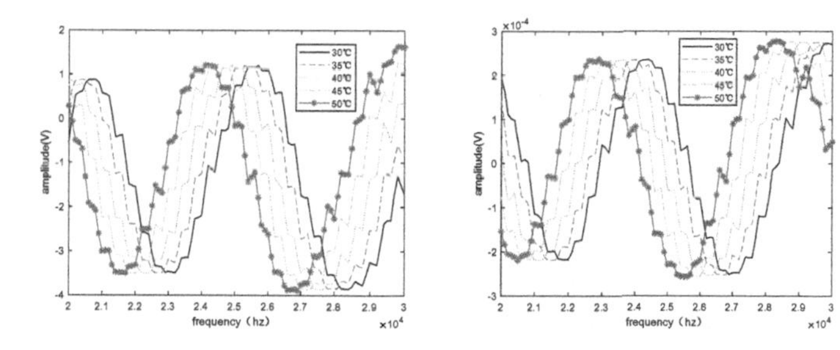


**Fig. 7.** Real part and imaginary part of admittance in 30–50° health.

Figure 8 respectively show the comparison of the real and imaginary parts of the admittance of the composite plate under the condition of health and crack damage at different temperatures. On the left is the real part and the figure on the right is the imaginary part. It can be seen that the frequency of the real part of the admittance is obviously shifted to the left, and the amplitude also has obvious changes, but the amplitude change of the imaginary part of the admittance can be ignored. So we can choose the real part of admittance for damage monitoring and the imaginary part of admittance for temperature compensation.

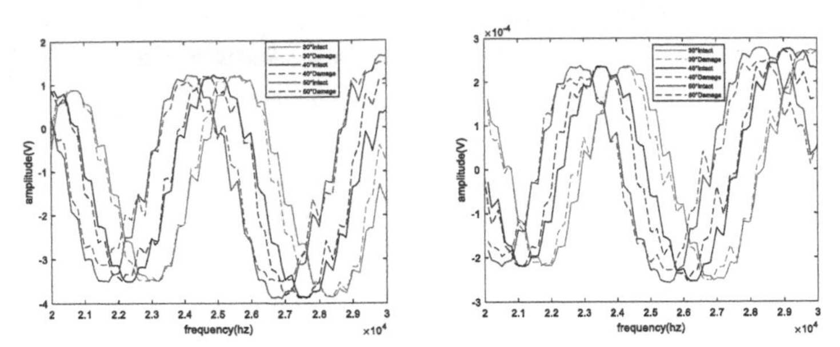


**Fig. 8.** Real part and imaginary part of admittance to health and damage conditions at different temperatures.

As shown in Fig. 9, in this experiment, 35° health data and damage data were selected as test data. The near-baseline data sets were 30° and 40°, respectively. According to the root mean square calculation, the closest baseline data set is the 30 °C baseline data set. According to Eq. (6), the number of cross-relations between the two groups of signals can be calculated, where the maximum value can be regarded as *TR* and *DI* and the *TR* value is 481.

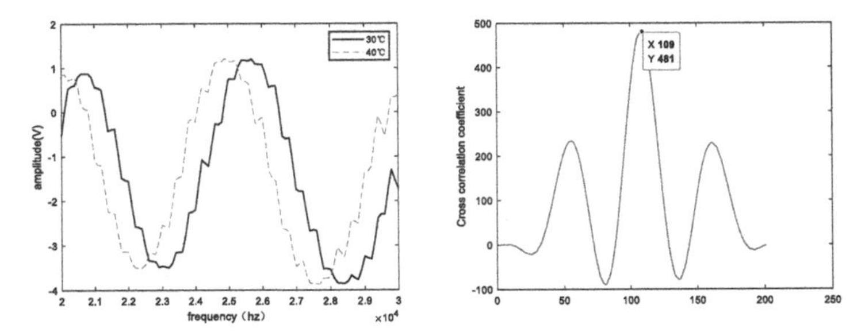


**Fig. 9.** The real part of baseline signal in health and the TR value.

As can be seen from Fig. 10, the damage index in the healthy state is 488.2, and the damage index is greater than *TR*, which verifies that the composite material plate has no crack damage. As can be seen from Fig. 11, the damage index in the damage state is 438.4, and the damage index is less than *TR*, which verifies the existence of crack damage in the composite material plate.

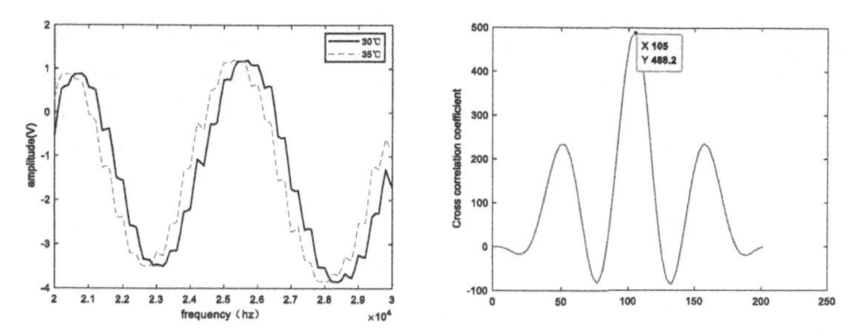


**Fig. 10.** The real part at 30 and 35° in health and the DI(H) value.

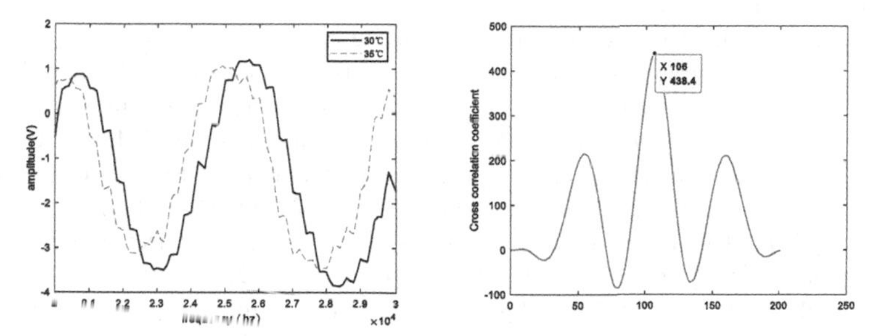


**Fig. 11.** The healthy real part at 30 and damage real part at 35° and the DI(D) value.

Then according to the above method again measured 40° and 45° of the composite material plate damage.

The results of crack damage diagnosis are summarized in Table 1.

**Table 1.** Monitoring of cracks in composite plates at 35°.

| CASE | Temperature (°C) | TR | DI | Damage diagnosis |
|---|---|---|---|---|
| INTACT | 35 | 481 | 488.2 | Intact |
| INTACT | 40 | 472.7 | 481.1 | Intact |
| INTACT | 45 | 463.6 | 470.1 | Intact |
| CRACK | 35 | 481 | 438.4 | Damage |
| CRACK | 40 | 472.7 | 467.9 | Damage |
| CRACK | 45 | 468.6 | 459.7 | Damage |

## 5 Conclusion

In this paper, the electromechanical impedance technology is used to obtain the admittance signal by using the piezoelectric plate attached to the target structure. At the same time, enough baseline data are collected at different temperatures. The admittance signal is decomposed into real part and imaginary part. The real part is used for damage diagnosis, and the imaginary part is used for temperature compensation, so as to effectively diagnose whether there is crack damage in the composite at different temperatures. In addition, we can continue to study the influence of external cyclic load on structural health monitoring.

## References

1. Campos, F.D.S., Castro, B.A.D., Budoya, D.E.: Feature extraction approach insensitive to temperature variations for impedance-based structural health monitoring. J. IET Sci. Measure. Technol. **13**(4), 536–543 (2019)
2. Baptista, F.G., et al.: An experimental study on the effect of temperature on piezoelectric sensors for impedance-based structural health monitoring. J. Sens. **14**(1), 1208–1227 (2014)
3. Wandowski, T., Malinowski, P.H., Ostachowicz, W.M.: Delamination detection in CFRP panels using EMI method with temperature compensation. J. Compos. Struct. **151**, 99–107 (2016)
4. Park, S., Yun, C.B., Inman, D.J.: Structural health monitoring using electro-mechanical impedance sensors. J. Fatigue Fract. Eng. Mater. Struct. **52**, 714–724 (2008)
5. Liang, Y., Li, D., Parvasi, S.M, et al.: Load monitoring of pin-connected structures using piezoelectric impedance measurement. J. Smart Mater. Struct. **25**(10), 105011 (2016)
6. Liang, C., Sun, F.P., Rogers, C.A.: An impedance method for dynamic analysis of active material systems. J. Intell. Mater. Syst. Struct. **8**(4), 323–334 (1997)

# Comparative Analysis Between New Combined Profile Scroll and Circular Involute Profile Scroll

Zhenghui Liu[1], Lijuan Qian[1,2](✉), and Xianyue Shao[3]

[1] College of Mechanical and Electrical Engineering, China Jiliang University, Hangzhou 310018, Zhejiang, China
qianlj@cjlu.edu.cn

[2] Key Laboratory of Intelligent Manufacturing Quality Big Data Tracing and Analysis of Zhejiang Province, China Jiliang University, Hangzhou 310018, Zhejiang, China

[3] Zhejiang Ruifeng Wufu Pneumatic Tools Co., Ltd., Taizhou 317500, Zhejiang, China

**Abstract.** Scroll compressor is widely used in machines for gases compression, and scroll profile is the basic of a scroll compressor. To further study the combined profile of scroll compressor, a new type of combined profile scroll composed of base circle involutes, variable radius and base circle involutes is introduced. Firstly, it is modeled according to the parametric equation of the vortex disk generatrix and the isometric method. Then, compression ratio is analyzed, deformation of the scroll and pressure distribution in internal scroll are simulated using ANSYS software. The results are compared with circular involute profile scroll. The results show that, under the condition of a certain number of scrolls, the new profile scroll compressor has larger compression ratio, less deformation, and better pressure condition than the circular involute profile scroll.

**Keywords:** Combined profile · Modeling · Compression ratio · Deformation · Pressure

## 1 Introduction

Scroll compressor is a new kind of displacement fluid machinery with the characteristics of high efficiency and low noise. Compared with other types of displacement compressors, it has a smaller volume, lighter weight, higher volumetric efficiency and lower noise [1, 2]. The compressor is the core component of the air conditioner. In the air conditioner field, in order to improve the utilization of the refrigerant, it is necessary to improve the compression ratio. By increasing the number of turns to improve the compression ratio, the leakage length will increase and the internal heat transfer performance of the compressor will deteriorate [3]. The scroll compressor using the combined profile can use a few turns to achieve a higher compression ratio. There are many types for combined profiles. The specific research on the composite profile includes: JW Bush et al. proposed a kind of combined profile scroll composed of involute, high power curves, and arc. Under the same compression ratio, this profile has fewer turns than the circular involute profile, and the length of the leakage line has been reduced [4]; Li Xueqin et al. proposed a new


Q. Han et al. (Eds.): LSMS 2021/ICSEE 2021, CCIS 1469, pp. 653–661, 2021.
https://doi.org/10.1007/978-981-16-7213-2_63

type of equal-wall-thickness arc-line segment combined profile scroll. Compared with the circular involute profile, this profile has a larger internal volume ratio and volume utilization, and the compression ratio has been improved, and the size of the scroll has also been reduced [5]; Wu Zaixin et al. established the geometric theory and related parameter calculation equations for the involute, high power curve of combined profile. It is proved that the deformation and stress of the combined profile scroll are smaller than that of a single involute profile [6]; Wang Jun et al. established the construction theory of circular involute and circular arc combined profile [7, 8]. In addition to above mentioned researches, some researches on combined profile scroll have not been considered.

The combined profile is very important to improve the overall performance of the scroll compressor. For this reason, this paper analyzes the performance of a new type of combined profile scroll composed of base circle involute, variable radius and base circle involute. Mainly considering the function of gas pressure, the performance analysis is carried out from three aspects of compression ratio, deformation of static structural analysis and pressure distribution of the internal flow field in the scroll. The comparative analysis is made with circular involute profile scroll.

## 2 The Establishment of Physical Model

### 2.1 The Establishment of the Generatrix

The generatrix of the scroll profile is the basis for the formation of the inner and outer wall profile of the scroll compressor. This paper focuses on a new combined profile. The concrete expression of the combined profile is: the base circle involute is used in the expansion angle which ranges from 0 to $2\prod$, variable radius involute is used in the expansion angle which ranges from $2\prod$ to $4\prod$, the base circle involute is used in the expansion angle which ranges from $4\prod$ to $5.5\prod$. The generatrix of the combined profile scroll is generated on the basic of Eqs. (1), (2) and (3). The generatrix of the circular involute profile scroll is generated on the basic of Eq. (4). The main geometry parameters of the scroll profile can be seen as Table 1. The values of r, k, $\alpha$, $\beta$, and $\gamma$ are set. The values of $c_1$ and $c_2$ are acquired by constraints that there are the same normal line and tangent line in the points of junction. The two kinds of scrolls have the same rolls. Figure 1 displays the generatrix of the combined profile scroll. Figure 2 displays the generatrix of the combined profile scroll.

$$\begin{cases} x = r_g cos\varphi + r_s sin\varphi \\ y = r_g sin\varphi - r_s cos\varphi \end{cases}. \tag{1}$$

$$r_g = \begin{cases} \mathrm{r} & [0, \alpha] \\ \mathrm{r} + \mathrm{k}\varphi + \mathrm{c}_1 & [\alpha, \beta] \\ \mathrm{r} + \mathrm{k}(\beta - \alpha) & [\beta, \gamma] \end{cases}. \tag{2}$$

$$r_s = \begin{cases} \mathrm{r}\varphi & [0, \alpha] \\ 0.5\mathrm{k}\varphi^2 + \mathrm{r}\varphi + \mathrm{c}_1\varphi + \mathrm{c}_2 & [\alpha, \beta] \\ [\mathrm{r} + \mathrm{k}(\beta - \alpha)]\varphi - \mathrm{k}\beta - \alpha[\beta - 0.5(\beta - \alpha)] & [\beta, \gamma] \end{cases}. \tag{3}$$

**Table 1.** The main geometry parameters of the scroll profile.

| Parameter | Value |
|---|---|
| r | 4 |
| k | 0.4 |
| $c_1$ | $-\mathrm{k}\,\alpha$ |
| $c_2$ | $0.5\mathrm{k}\,\alpha^2$ |
| $\alpha$ | $2\,\pi$ |
| $\beta$ | $4\,\pi$ |
| $\gamma$ | $5.5\,\pi$ |

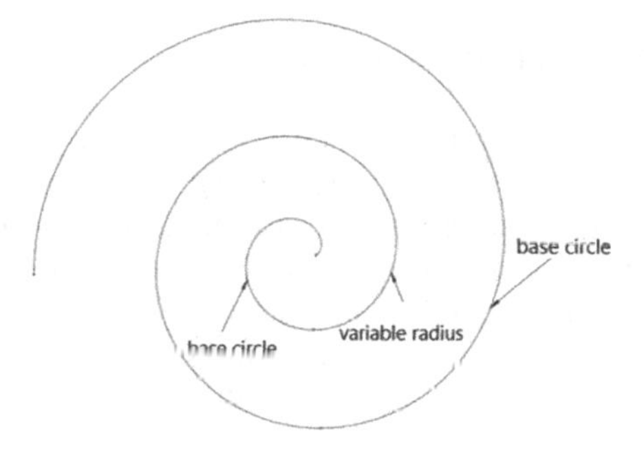


**Fig. 1.** The generatrix of the combined profile scroll involute profile scroll.

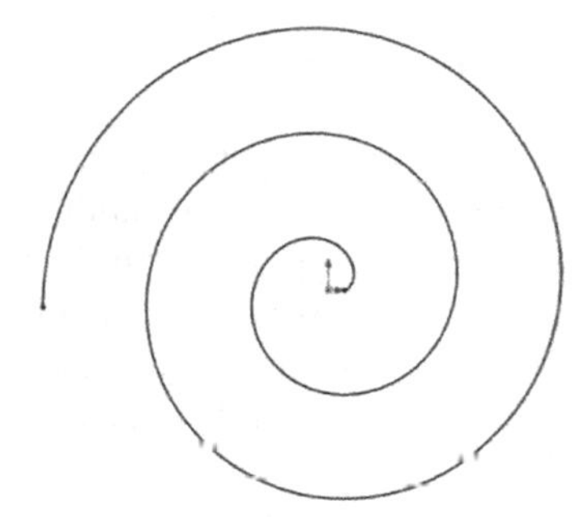

**Fig. 2.** The generatrix of the circular.

$$\begin{cases} x = r(cos\theta + \theta sin\theta) \\ y = r(sin\theta - \theta cos\theta) \end{cases} [0, \gamma]. \tag{4}$$

## 2.2 The Formation of the Scroll

The two kind of scrolls are generated based on the generatrix using normal-equidistant-curve method [9]. Firstly, the generatrix is drawn according to the parametric equation. Then it is mirrored about X axis and next mirrored about Y axis to obtain another generatrix. Finally, the inner and outer wall profile of the orbiting scroll and static scroll are acquired by deviating the two generatrices in two normal directions respectively. The deviation distance is $R_{or}$. $R_{or}$ represents the radius of gyration of the orbiting scroll around the static scroll. The two kinds of scrolls are shown in Fig. 3 and Fig. 4.

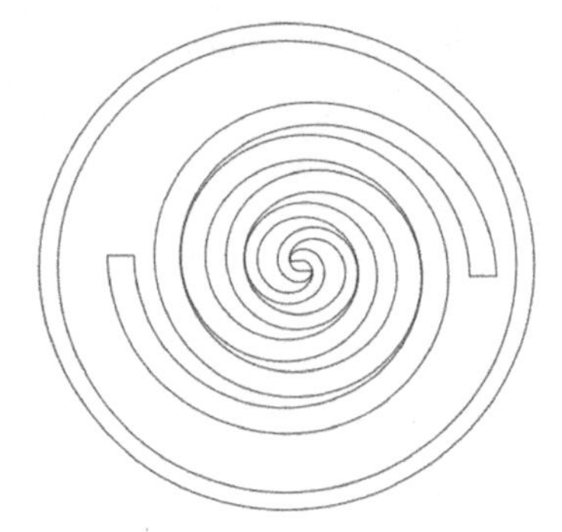

**Fig. 3.** The combined profile scroll.

**Fig. 4.** The circular involute profile scroll.

# 3 Comparative Analysis of Performance Parameters

The circular involute profile scroll has the characteristics of high volume ratio, short leakage line, high area utilization rate, and stable spindle torque change. It is the most widely used single profile. In order to understand the performance parameters of the new combined profile scroll more intuitively, the circular involute profile scroll is taken as the comparison item. The circular involute profile scroll and the new combined profile have the same base circle radius 'r' and the same number of vortex turns. The height of scroll-'H' is set as 30 mm. The radius of gyration of the orbiting scroll around the static scroll-$R_{or}$ is set as 6 mm.

## 3.1 The Comparative Analysis of Compression Ratio

When the scroll compressor starts to discharge, the volumes of the suction chamber, compression chamber, and discharge chamber are calculated respectively. The calculation formula for the volume is: $V = 2H\ R_{or}\ l$. 'l' represents the generatrix length of each working chamber. This paper uses SolidWorks to draw the generatrix of the scroll according to the parameter equation, and then uses the measuring tool in the software to measure the generatrix length of the corresponding working chamber. The results are shown in Table 2, and the corresponding calculated working chamber volumes are shown in Table 3. The compression ratio of the combined profile scroll compressor $\varepsilon = V_1/V_3 = 13.16$, and the compression ratio of the base circle involute scroll compressor $\varepsilon = V_1/V_3 = 10.18$.

**Table 2.** Generatrix length of each working chamber (mm).

| Working chamber | The combined profile scroll | The circular involute profile scroll |
|---|---|---|
| Suction chamber | 438.94 | 339.51 |
| Compression chamber | 186.53 | 181.59 |
| Discharge chamber | 33.36 | 33.36 |

**Table 3.** Volume of each working chamber ($mm^3$).

| Working chamber | The combined profile scroll | The circular involute profile scroll |
|---|---|---|
| Suction chamber/$V_1$ | 158018.4 | 122223.6 |
| Compression chamber/$V_2$ | 67150.8 | 65372.4 |
| Discharge chamber/$V_3$ | 12009.6 | 12009.6 |

## 3.2 The Comparative Analysis of Deformation Simulation

### 3.2.1 The Establishment of Three-Dimensional Model

Since the orbiting and static scrolls are symmetrically distributed, it is feasible to select the orbiting scroll for simulation analysis. SolidWorks software is used to build a scroll compressor orbiting scroll model, and save the established 3D model as a ".x_t" file. Then the file is imported into ANSYS Workbench. The materials of scrolls are designed as aluminum alloy. The density is 2770 kg/mm$^3$ the Poisson's ratio is 0.33, and the elastic modulus is $7.1 \times 10^{11}$ Pa.

### 3.2.2 The Generation of Mesh

ANSYS Meshing is used to divide the meshes of the two types of profile scrolls separately. "Automatic Method" is used to generation grids. The element size is 5mm. Having divided the meshes, the number of mesh elements for the combined profile scroll is 93689, and the number of mesh elements for the circular involute profile scroll is 92861. The average mesh quality for the combined profile scroll is 0.84, and the average mesh quality for the combined profile scroll is 0.84. The two kinds of meshes can satisfy the demands. The meshes are shown in Fig. 5 and Fig. 6.

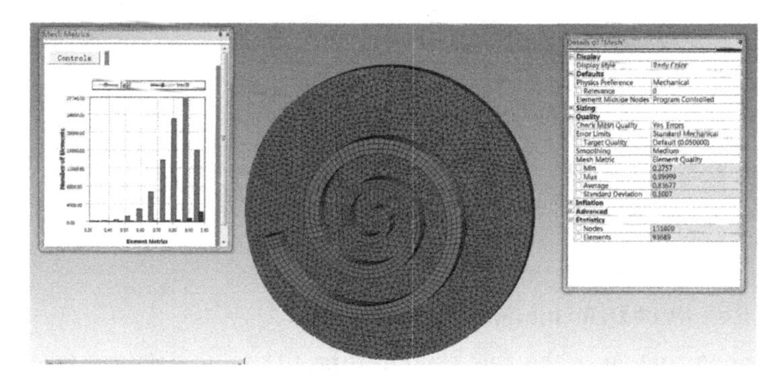


**Fig. 5.** Mesh of the combined profile scroll.

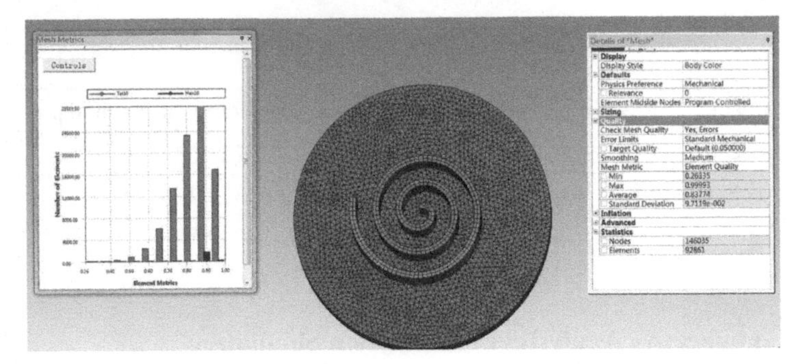


**Fig. 6.** Mesh of the circular involute profile scroll.

### 3.2.3 Apply Constraints and Loads

After the grid is divided, constraints are imposed, and full constraints are imposed on the side of the main bearing hole of the orbiting scroll [10, 11]. The pressure loading surfaces that need to be loaded are the scroll surface and the bottom surface corresponding to each enclosed working chamber. The concrete loading surfaces are the curved surface between the two meshing lines, and the crescent bottom surface. Assuming that the gas is the state of adiabatic compression in the working chamber of the compressor, the ideal gas equation of state can be used to obtain the pressure calculation formula of each working chamber: $p_i = (\frac{V}{V_i})^k$ p. 'V' represents the suction volume, '$V_i$' represents the volume of the chamber 'i', 'p' represents the suction pressure whose value is set as 0.6 MPa, k is the isentropic coefficient of the gas which is set as 0.3. The volume of each working chamber is based on the data in Table 3. The calculated pressure of each working chamber is shown in Table 4.

**Table 4.** Pressure of each working chamber (MPa)

| Working chamber | The combined profile scroll | The circular involute profile scroll |
|---|---|---|
| Suction chamber | 0.600 | 0.600 |
| Compression chamber | 1.005 | 0.997 |
| Discharge chamber | 1.300 | 1.204 |

### 3.2.4 The Results of Deformation Simulation

The deformation contour of the new combined profile scroll obtained from the static structure analysis are shown in Fig. 7. Figure 8 displays the deformation contour of the circular involute profile scroll. It is found that the maximum deformation for the two kinds of profile scrolls is in the upper part of the tooth head. The maximum deformation of the new combined profile scroll is $8.3637 \times 10^{-4}$ mm, and the maximum deformation of the circular involute profile scroll is $1.6832 \times 10^{-3}$ mm. Obviously, the deformation of the combined profile scroll is less than that of the circular involute profile scroll.

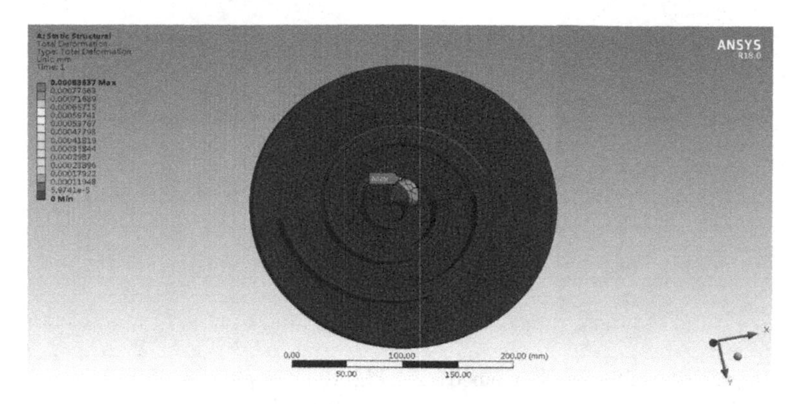


**Fig. 7.** The deformation contour of the new combined profile scroll.

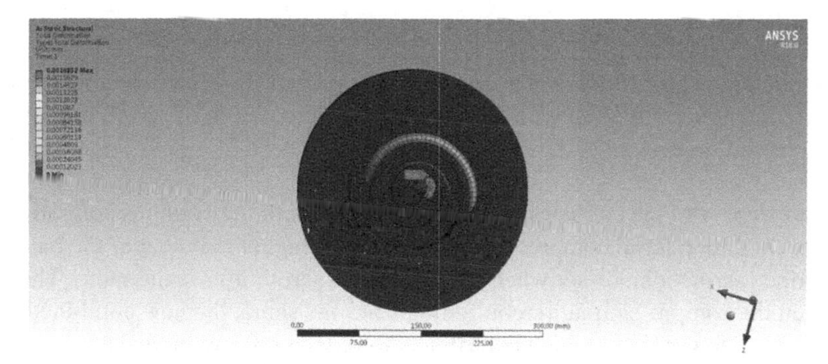


**Fig. 8.** The deformation contour of the circular involute profile scroll.

### 3.3 The Comparative Analysis of the Pressure in the Internal Flow Field

The pressure is the larger when the the orbiting scroll is closer to the center of scroll [12]. So selecting the state that the orbiting scroll is near the center of scroll to analyse is suitable. The pressure distribution of the combined profile of scroll and the circular involute profile of scroll are respectively shown in Fig. 9 and Fig. 10. The maximum pressure of the combined profile of scroll is $1.13 \times 10^6$ Pa, and the maximum pressure of the circular involute profile of scroll is $1.15 \times 10^6$ Pa. So in this situation, the combined profile of scroll has favorable conditions for bearing the pressure.

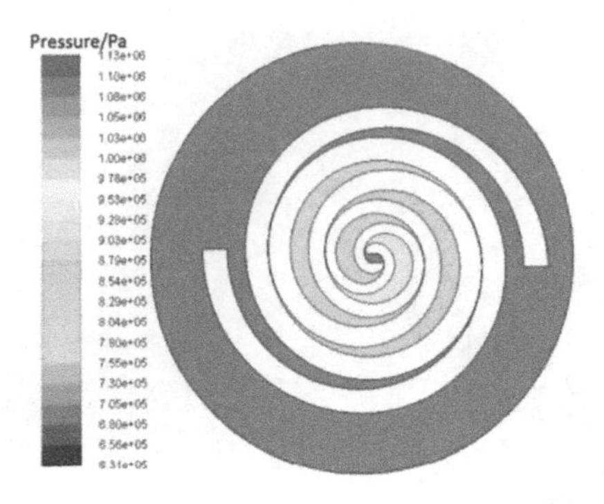


**Fig. 9.** The pressure of the new combined profile scroll.

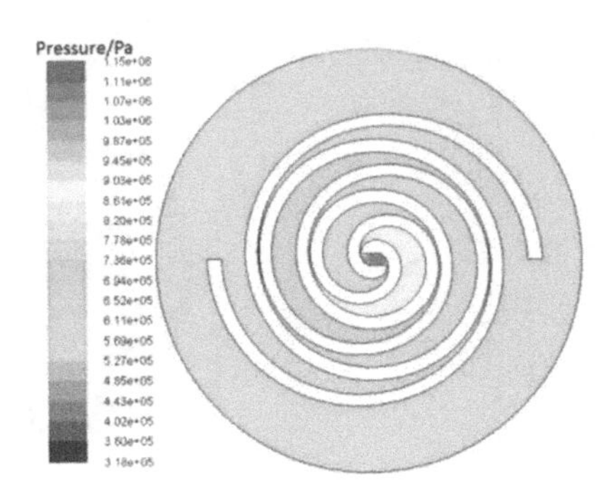


**Fig. 10.** The pressure of the circular involute profile scroll.

# 4 Conclusions

The new combined profile of scroll composed of base circle involute, variable radius and base circle involute comparing with the circular involute profile of scroll, the following conclusions can be drawn:

(1). Compared with the circular involute profile of scroll, the compression ratio of the new combined scroll compressor is significantly higher than that of the base circle involute scroll compressor when the number of scroll turns is the same. Therefore, when the compression ratio requirements are the same, the new combined profile scroll has fewer turns, which contributes to reduce refrigerant leakage.

(2) For the combined profile scroll and the circular involute profile scroll, In the case of the same number of scroll turns, the maximum deformation of the new scroll compressor is less than that of the circular involute profile of scroll. By comparison, it is found that the maximum pressure of the circular involute profile scroll is higher than that of the combined profile scroll. Therefore, the new combined profile scroll is more favorable to bear the pressure than the circular involute profile of scroll.

**Acknowledgments.** The research was supported by Key Study Program of Zhejiang Province (Grant No. 2020C01054).

# References

1. Li, L., Jiang, Y.: Scroll Compressor (in Chinese). China Machine Press, Beijing (1998)
2. Wu, Y., Li, H., Zhang, H., et al.: Refrigeration Compressor (in Chinese). China Machine Press, Beijing (2010)
3. Ling, F.A.N., Zongchang, Q.U., Chunmei, Q.I., et al.: Present situation and progress of scroll Compressor general profile design (in Chinese). J. Fluid Machin. **28**(1), 27–30 (2000)
4. Bush, J.W., Beagle, W.P., Housmen, M.E.: Maximizing scroll compressor displacement using generalized wrap geometry. In: Proceeding of the International Compressor Engineering Conference at Purdue, pp. 205–210 (1994)

5. Li, X., Wang, J.: A novel combined profile of arc and line segment for scroll compressor (in Chinese). J. Fluid Mach. **38**(8), 27–30 (2010)
6. Zaixin, W.U., Wenwu, D.U., Tao, L.I.U., et al.: Design of involute higher curve combined profile for scroll compressor and finite element analysis (in Chinese). J. Compressor Technol. **2**, 5–9 (2011)
7. Wang, J., Guan, H., Dai, K.: Construction of involute and arc combination profile for multi-wrap scroll compressor (in Chinese). J. Fluid Mach. **36**(5), 14–17 (2008)
8. Wang, J., Liu, Z., Li, C.: Design calculation of involute and circular arc combined profile for scroll compressor (in Chinese). J. Fluid Mach. **32**(10), 10–13 (2004)
9. Wang, J., Liu, Q., Cao, C.Y., et al.: Design methodology and geometric modeling of complete meshing profiles for scroll compressors. J. Int. Refrig. **91**, 199–210 (2018)
10. Liu, G., Zhang, G., Li, Y., et al.: Deformation and stress simulated analysis of orbiting scroll under the coupling function (in Chinese). J. Chinese Hydraulics Pneumatics (5), 60–64 (2016)
11. Li, C., Xie, W., Zhao, M.: Deformation and stress analysis of orbiting scroll under the multi-field coupling (in Chinese). J. Fluid Mach. **41**(8), 16–20 (2013)
12. Wu, Z., Feng, Z.G., Su, Y.F.: Simulation and analysis of internal flow field in scroll compressor with circular involute profile. J. Mech. Sci. Technol. Aerosp. Eng. **38**(12), 1840–1846 (2019)

# Flow Characteristic of Scroll Compressor with Combined Profile Scroll

Zhenghui Liu[1], Zhongli Chen[1](✉), Lijuan Qian[1,2], Jian Lin[3], and Yongqing Liu[3]

[1] College of Mechanical and Electrical Engineering, China Jiliang University, Hangzhou 310018, Zhejiang, China
chenzhongli@cjlu.edu.cn

[2] Key Laboratory of Intelligent Manufacturing Quality Big Data Tracing and Analysis of Zhejiang Province, China Jiliang University, Hangzhou 310018, Zhejiang, China

[3] Zhejiang Santian Automotive Air Conditioning Compressor Co., Ltd., Lishui 323000, Zhejiang, China

**Abstract.** Scroll compressor is widely used in machines for gases compression, and scroll profile is the basic of a scroll compressor. The combined profile of scroll has advantages of several single scroll profiles. For improving the performance of a scroll compressor, it is necessary to investigate characteristics of the internal flow of combined profile of scroll compressors. In this paper, the combined profile is composed of the base circle involute-variable radius and base circle involute. A two-dimensional transient flow model of the combined profile of scroll refrigeration compressor operating with ammonia is established. The pressure, temperature, and velocity distribution of the working chamber are collected using the computational fluid dynamic method. Based on comparison between the combined profile and the circular involute profile, it is found that the combined profile is more favorable to bear the pressure, decrease the leakage, and decrease the fluctuation of mass flow rate than the circular involute profile.

**Keywords:** Scroll compressor · Combined profile of scroll · Circular involute profile of scroll · Transient flow · Computational fluid dynamic

## 1 Introduction

Scroll compressor is mainly used in small refrigeration system with its advantages of small volume, low noise, high efficiency and energy-saving. Therefore, it is necessary to study the internal flow characteristic of working fluid for improving the performance in terms of energy conservation. For improving the performance of a scroll compressor, many methods have been used. The theoretical method is an effective means to study the scroll compressor. Winandy et al. and Cuevas et al. proposed a simplified mathematical model of scroll compressors considering heat transference in the suction process [1, 2]. However, the model did not consider the impact of leakage and heat transference during the compression process. Wang et al. established a mathematical model that could predict inner compression process of scroll compressor [3, 4]. The model is validated with experimental results. Chen et al. put forward a model that predicted the variation of


Q. Han et al. (Eds.): LSMS 2021/ICSEE 2021, CCIS 1469, pp. 662–674, 2021.
https://doi.org/10.1007/978-981-16-7213-2_64

temperature and pressure during the process of compression [5, 6]. However, the relative parameter distribution in the working chamber cannot be obtained by these mathematical models. Therefore, it is necessary to find a more complicated model to predict the internal field of the working chamber. Ooi et al. established a two-dimensional steady flow model to investigate the internal fluid flow in the working chamber, but the clearance was neglected in the model [7]. Lin et al. exploited the finite method to simulate the steady temperature distribution on scrolls based on measured data [8]. The distribution of the thermal deformation on scroll was obtained. Nevertheless, the steady model was not enough to simulate the transient flow in a scroll compressor accurately.

Computational fluid dynamics (CFD) has been used widely in recent years, becoming a practical means to study the transient flow field of scroll machinery on the basis of dynamic mesh technology. Cui et al. firstly carried out three-dimensional transient flow simulation on scroll compressor for air conditioning [9]. The temperature, pressure and velocity distribution were obtained. Xiao et al. investigated the flow distribution of the circular involute profile of scroll using dynamic mesh with a two-dimensional model, getting the variation of pressure, temperature, and velocity [10]. Morini et al. made a two-dimensional CFD simulation on scroll expander and introduced the wave of the inlet mass flow rate [11]. Song et al. and Wei et al. set up a three-dimensional CFD model of a scroll expander and study the unsteady flow field and the suction process of the expander [12, 13]. The flow mechanisms of vortexes existing in the suction chamber were revealed. Zha et al. made three-dimensional simulation on the circular involute profile of scroll with double arc correction [14]. Wu et al. studied the flow distribution of the circular involute profile of scroll using dynamic mesh with a three-dimensional model, making a comparison with the results of the trial [15]. Rak and Pietrowicz established a two-dimensional transient flow numerical model including leakages to investigate the heat transfer based on CFD [16]. Sun et al. established a three -dimensional unsteady CFD model to investigate the modification of the suction flow passage in a scroll refrigeration compressor [17]. Zheng et al. studied the transient flow characteristic of trans-critical $CO_2$ in the scroll compressor using a three-dimensional unsteady CFD approach [18]. Cavazzini et al. investigated the impact of the axial gap on the resulting performance of the inner fluid dynamics of a scroll machine [19].

Given existing literatures above, the CFD method is suitable for investigating the internal flow field. Although the mechanisms of internal flow phenomena have been discussed and analyzed in detail, the study is mainly on the single profile of scroll, and combined profile of scroll is not enough. In order to optimize the performance of combined profile of scroll, it is necessary to study the internal flow field of combined profile of scroll.

In the present paper, the combined profile of scroll composed of base circle involute-variable radius - base circle involute is established. Then the transient flow simulation based on CFD method is carried out, and the mesh quality is checked. The results are compared with existing literature, proving the CFD model. The internal flow parameter distributions are obtained. Lastly, the comparison analysis between the combined profile of scroll and the circular involute profile of scroll is made.

## 2 Numerical Method

### 2.1 Physical Model

The scroll compressor is applied in refrigeration system. The generatrix of the combined scroll profile is generated on the basic of Eqs. (1), (2) and (3). The circular involute scroll profile is generated on the basic of Eq. (4). The two kind of scroll profiles are generated based on the generatrix using normal-equidistant-curve method [20]. The main geometry parameters of the scroll profile can be seen as Table 1. The values of r, k, $\alpha$, $\beta$, and $\gamma$ are set. The values of $c_1$ and $c_2$ are acquired by constraints that there are the same normal line and tangent line in the points of junction. The two kinds of scrolls have the same rolls.

$$\begin{cases} x = R_g cos\varphi + R_s sin\varphi \\ y = R_g sin\varphi - R_s cos\varphi \end{cases} \tag{1}$$

$$R_g = \begin{cases} r \quad [0, \alpha] \\ r + k\varphi + c_1[\alpha, \beta] \\ r + k(\beta - \alpha)[\beta, \gamma] \end{cases} \tag{2}$$

$$R_s = \begin{cases} r\varphi \quad [0, \alpha] \\ 0.5k\varphi^2 + r\varphi + c_1\varphi + c_2[\alpha, \beta] \\ [\mathrm{r} + \mathrm{k}(\beta - \alpha)]\varphi - k\beta - \alpha \\ [\beta - 0.5(\beta - \alpha)][\beta, \gamma] \end{cases} \tag{3}$$

$$\begin{cases} x = rcos\varphi + r\varphi sin\varphi \\ y = rsin\varphi - r\varphi cos\varphi \end{cases} \tag{4}$$

**Table 1.** The main geometry parameters of the scroll profile.

| Parameter | Value |
|---|---|
| r | 4 |
| k | 0.4 |
| $c_1$ | $-\mathrm{k}\alpha$ |
| $c_2$ | $0.5\mathrm{k}\alpha^2$ |
| $\alpha$ | $2\,\pi$ |
| $\beta$ | $4\,\pi$ |
| $\gamma$ | $5.5\,\pi$ |

## 2.2 CFD Model and Boundary Condition

In order to simplify the physical model, the computational model is designed as a two-dimensional model. The radial clearance is 0.3 mm. Figure 1 and Fig. 2 display the model of the combined profile of scroll and the circular involute profile of scroll for simulation respectively. The suction port is designed at the sidewall. The discharge port is designed at static scroll. The rotational speed of orbiting scroll is 2500 r/min. The time step is $10^{-5}$s. When every time step gets over, the orbiting scroll rolls 0.15°. The RNG k-$\varepsilon$ turbulence model is applied because of considering the turbulence of vortex. Standard wall function is conducted. Ammonia is used as the working fluid. The boundary conditions are set before the simulation. The relational operation parameters used during the numerical simulation are as follows: the pressure of suction port Ps = 0.6 MPa, the pressure of discharge port Pd = 1.1 MPa, the temperature of inlet is set to 300K, the temperature of outlet is set to 400K. The conditions of sidewall are set to adiabatic.

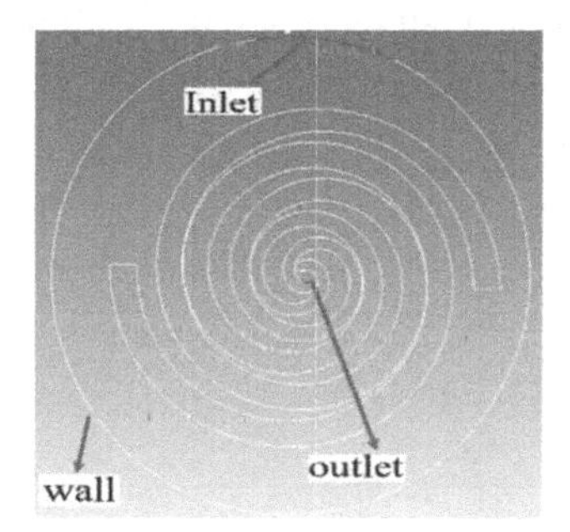


**Fig. 1.** The model for the combined profile of scroll.

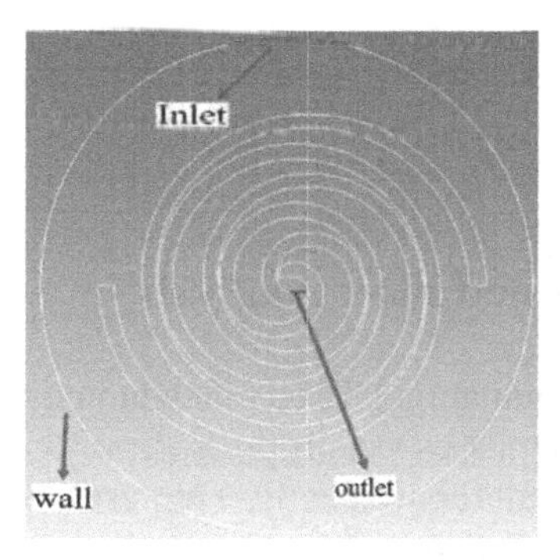


**Fig. 2.** The model for the circular involute profile of scroll.

### 2.3 The Numerical Grid and Grid Quality Check

The computational numerical grid of scroll models are produced by ANSYS Meshing. The mesh of the whole computational domain is created using triangle grid. Owing to the movement of orbiting scroll, the grid in the computational domain will be deformed with time. The mesh of fluid domain is created using triangle grids which is suitable for dynamic mesh. Smoothing and remeshing techniques are used in the dynamic grid control for fluid domains to obtain high quality mesh. In the smoothing method, Spring/Laplace/Boundary Layer method is selected. In the remeshing method, Local Cell method is applied [21]. The movement of the orbiting scroll is defined by user defined function (UDF). The mesh is consistently remeshed during the whole working process to cooperate with the moving deforming boundary.

For the combined profile of scroll, the computational grids include 332,637 elements. By grid quality check shown in Fig. 3, all mesh quality is all beyond 0.6. For the circular involute scroll profile, the computational grids include 236,021 elements. By grid quality check shown in Fig. 4, all mesh quality is all beyond 0.55. So the two kind of computational grids can meet the demand.

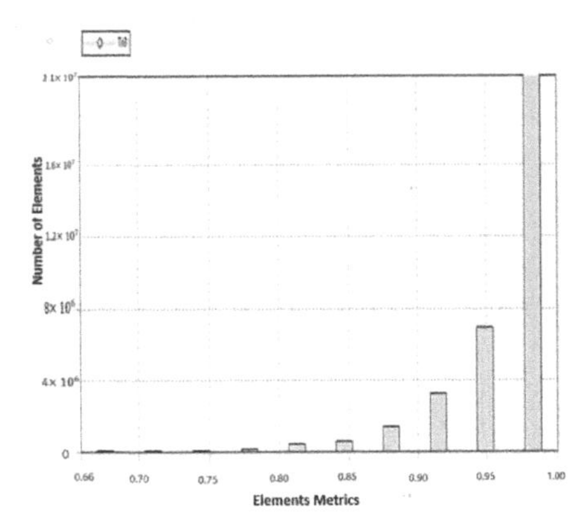


**Fig. 3.** Grid quality check of the combined profile of scroll.

## 3 Simulation Results and Flow Analysis

The output from the CFD simulation gives an insight into the evolution of the internal flow while the orbiting scroll rolls, especially pressure, velocity, temperature distribution. To analyze the simulation results conveniently, $\theta$ representing the crank angle, is defined as 0° at the beginning of suction process.

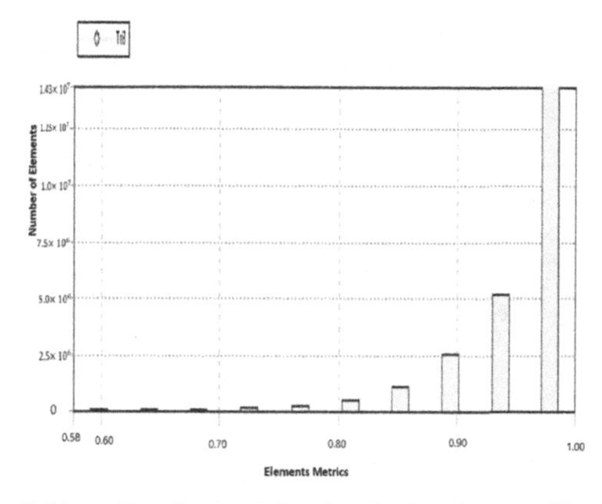


**Fig. 4.** Grid quality check of the circular involute profile of scroll.

### 3.1 Pressure Distribution

The pressure distribution of the combined profile of scroll and the circular involute profile of scroll are respectively shown in Fig. 5 and Fig. 6 which include $\theta$ of 0°, 90°, 180°, 270°. The pressure increases gradually from the suction chamber to the discharge chamber. As the suction port and discharge port are not absolutely symmetrical, the

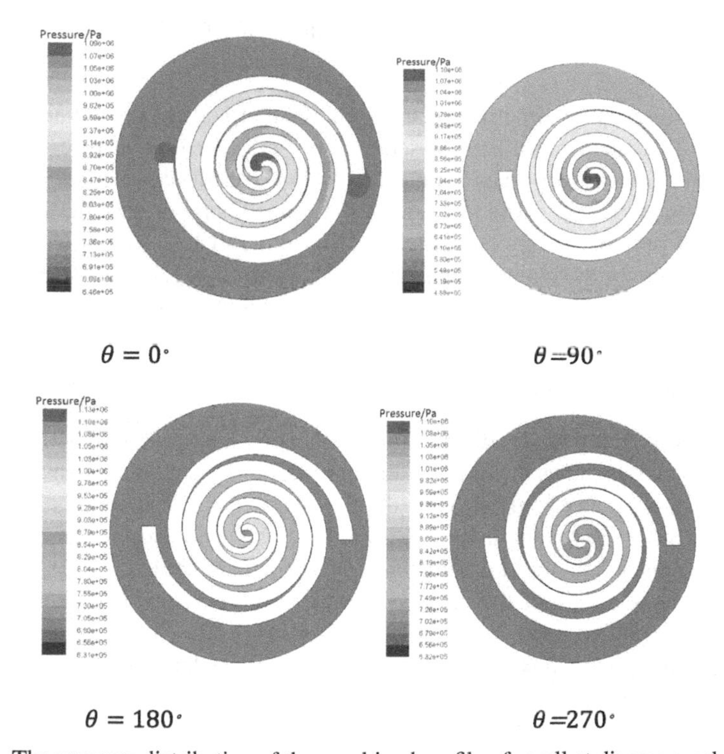


**Fig. 5.** The pressure distribution of the combined profile of scroll at diverse crank angles.

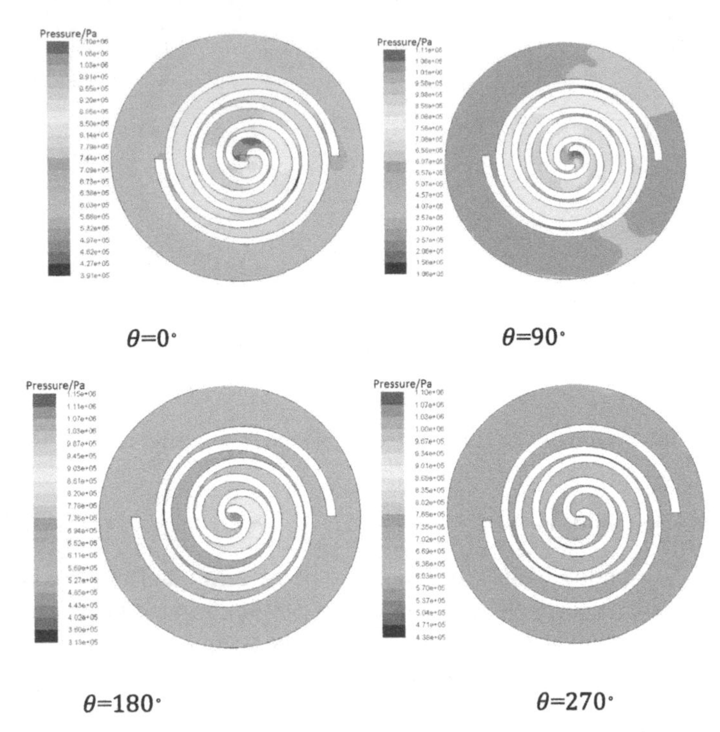


**Fig. 6.** The pressure distribution of the circular involute profile of scroll at diverse crank angles.

pressure distribution is asymmetrical in a pair of symmetrical crescent-shaped chamber. Excluding fields adjacent to leakage gaps, the pressure distribution inside every crescent-shaped chamber can be deemed as homogeneous. Such results correspond with existing researches on the topic [22]. The pressure of discharge chamber rises firstly, then falls. The maximum pressure of the combined profile of scroll is $1.13° \times 10^6$Pa, and the maximum pressure of the circular involute profile of scroll is $1.15° \times 10^6$Pa. So in this situation, the combined profile of scroll has favorable conditions for bearing the pressure.

## 3.2 Temperature Distribution

Figure 7 shows the temperature distribution of the combined profile of scroll at diverse crank angle. Figure 8 shows the temperature distribution of the circular involute profile of scroll at diverse crank angle. Obviously, the temperature is inhomogeneous in the fluid domain. Such results are analogous to ones which are described in the literature [22]. From the given contour, it is seen that when the crank angle is 0°, 180° and 270°, the minimum temperature of the circular involute profile of scroll is higher than the combined profile of scroll. However, the maximum temperature of the combined profile of scroll is 400K, and the maximum temperature of the circular involute profile of scroll

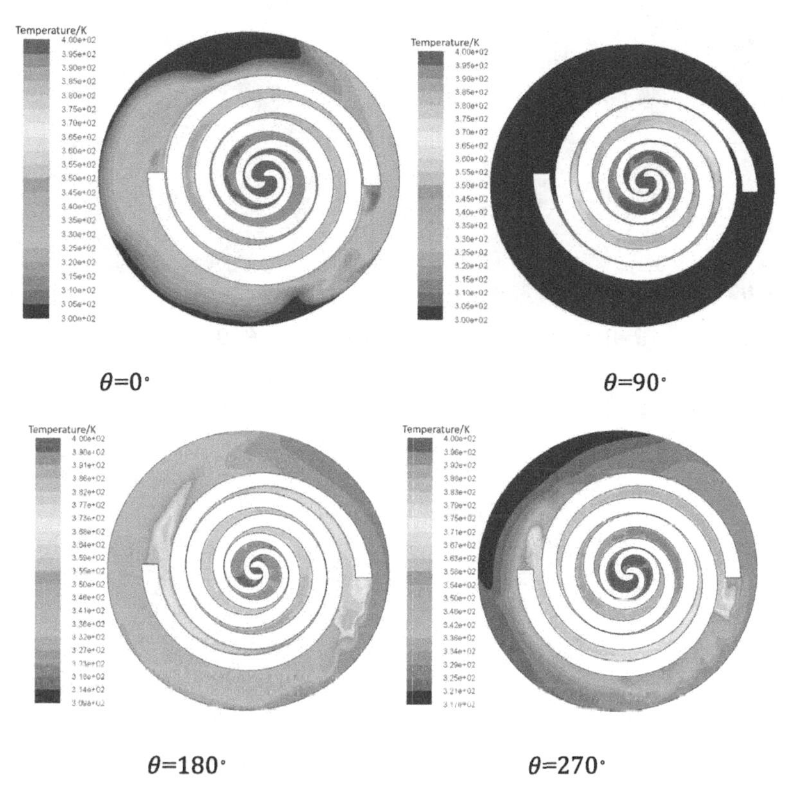


**Fig. 7.** The temperature distribution of the combined profile of scroll at diverse crank angles.

is 401K. The difference of the maximum temperature between the two kind of profiles of scrolls is small. So the influence of the temperature on the two kind of profile of scrolls is similar.

### 3.3 Velocity Distribution

Figure 9 shows the velocity distribution of the combined profile of scroll at diverse crank angle. Figure 10 shows the velocity distribution of the circular involute profile of scroll at diverse crank angle. The velocity distribution is inhomogeneous in the two kind of profiles of scrolls. The maximum velocity is in the leakage gap. On the basis of Bernoulli equation: $p + \frac{1}{2}\rho v^2 + \rho gh = C$, $p$ represents pressure of the fluid, $\rho$ represents density of the fluid, v represents velocity of the fluid, $h$ represents vertical height, $g$ represents acceleration of gravity, $C$ represents constant quantity. It is known that the smaller the pressure, the greater the flow velocity when other conditions are the same. Owing to the

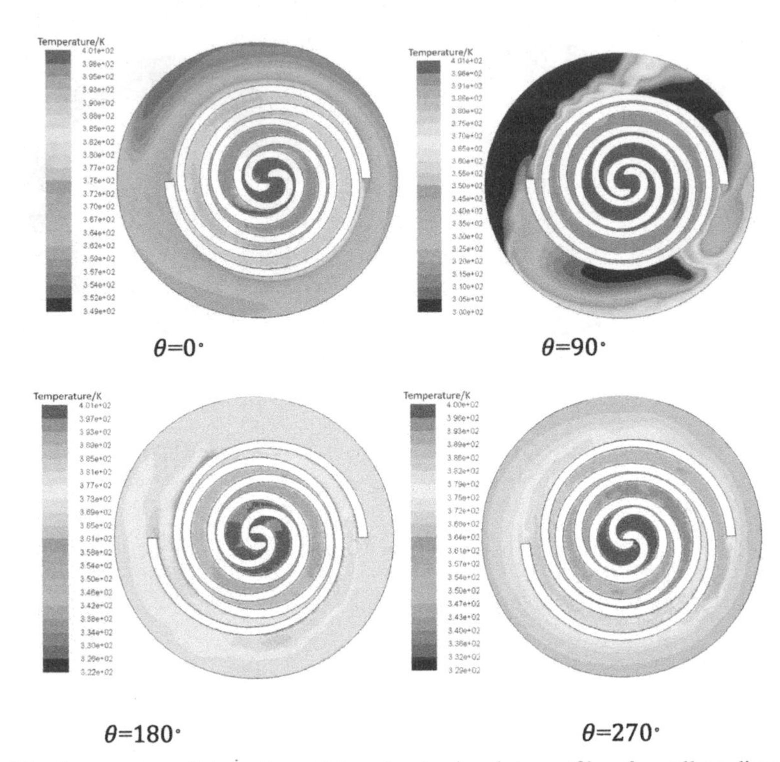


**Fig. 8.** The temperature distribution of the circular involute profile of scroll at diverse crank angles.

smaller pressure of the leakage gap shown in the Fig. 6 and Fig. 7, the flow velocity is greater. From the given contour, it is seen that the maximum velocity of the combined profile of scroll is 108 m/s, and the maximum velocity of the circular involute profile of scroll is 137 m/s. So the combined profile of scroll is favorable to decrease the leakage which is from high pressure chamber to low pressure chamber.

### 3.4 Mass Flow Rate

Figure 11 shows the inlet mass flow rate of the combined profile of scroll and the circular involute profile of scroll at different time. Figure 12 shows the outlet mass flow rate of the combined profile of scroll and the circular involute profile of scroll at different time. It concludes that mass is to keep conservation during the simulation. The inlet mass flow rate and outlet mass flow changes with the working time. It is shown that the inlet mass flow rate and outlet mass flow rate of the circular involute profile of scroll is higher than that of the combined profile of scroll in the whole period, owing to the bigger chamber

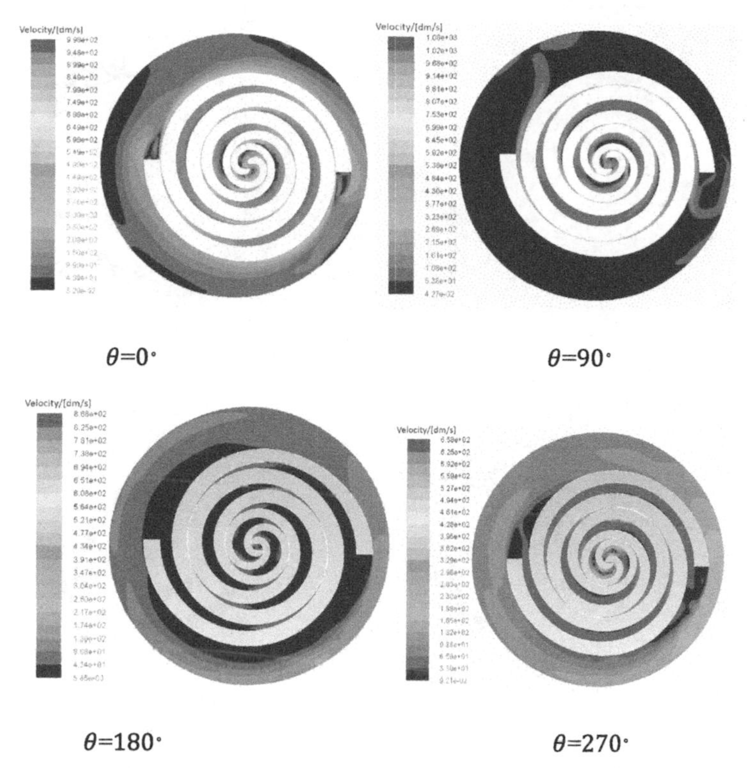


**Fig. 9.** The velocity distribution of the combined profile of scroll at diverse crank angles.

volume of the circular involute profile of scroll. It is found that the absolute fluctuation of the mass flow rate of the combined profile of scroll is: 1.8–0.2 = 1.6, the absolute fluctuation of the mass flow rate of the circular involute profile of scroll is: 3.35–0.5 = 2.85. So the absolute fluctuation of the mass flow rate of the circular involute profile of scroll is larger than that of the combined profile of scroll. The relative fluctuation of the mass flow rate is defined as "maximum value subtracts minimum value, then divides average value". The mass flow rate fluctuation of the circular involute profile of scroll is: (3.35–0.5)/2.05 = 1.39. The relative fluctuation of mass flow rate of the combined profile of scroll is: (1.8–0.2)/1.02 = 1.57. So the difference of relative fluctuation of mass flow rate for the two kind of profiles is not obvious.

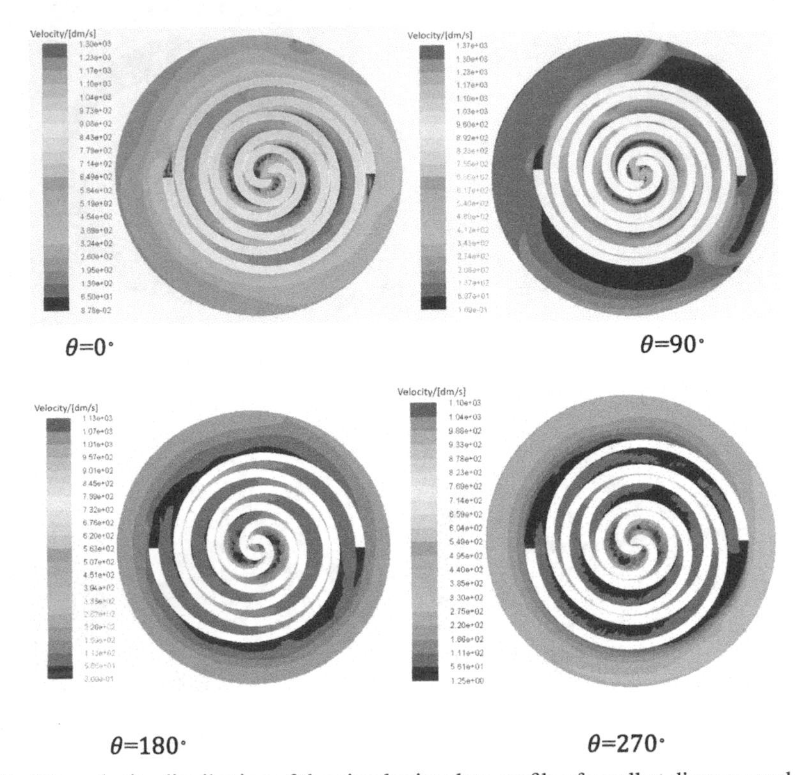


**Fig. 10.** The velocity distribution of the circular involute profile of scroll at diverse crank angles.

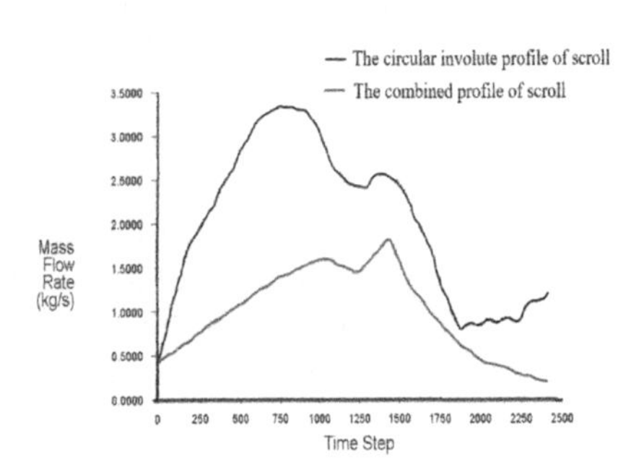


**Fig. 11.** The inlet mass flow rate at different time.

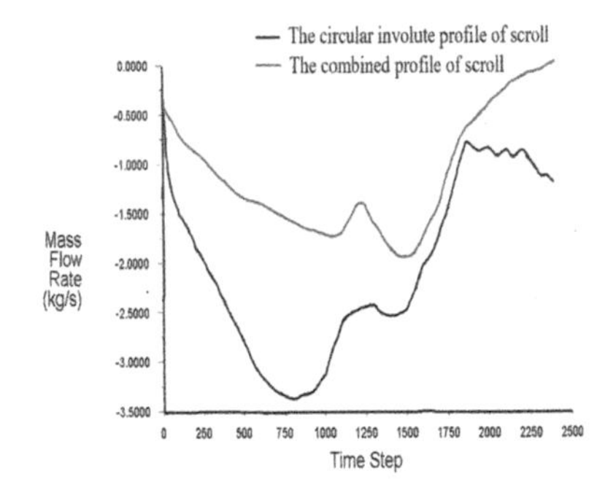


**Fig. 12.** The outlet mass flow rate at different time.

## 4 Conclusion

Scroll compressor is a key component of the refrigeration system. In this paper, a two-dimensional transient computational fluid dynamic model of the combined profile of scroll operating with ammonia is established to study the internal fluid characteristic, and the comparison is carried out with the circular involute profile of scroll. Some conclusions are summarized as follows.

(1) For the combined profile of scroll and the circular involute profile of scroll, the pressure distribution is almost homogeneous inside every crescent-shaped chamber. The temperature and velocity distribution are inhomogeneous and asymmetrical. It is demonstrated that the influence of leakage flow on the temperature distribution is stronger than that on the pressure distribution. The maximum velocity happens in the leakage gap.
(2) By comparison, it is found that the maximum pressure and velocity of the circular involute profile of scroll is higher than that of the combined profile of scroll, and the structures of profiles exert influence on the absolute fluctuation of mass flow rate when the number of rolls and the boundary condition are the same. The combined profile of scroll is more favorable to bear the pressure, decrease the leakage and decrease the absolute fluctuation of mass flow rate than the circular involute profile of scroll.

**Acknowledgment.** The research was supported by Key Study Program of Zhejiang Province (Grant No. 2020C01054).

## References

1. Winandy, E., Claudio, S.O., Lebrun, J.: Experimental analysis and simplified modelling of a hermetic scroll refrigeration compressor. J. Appl. Therm. Eng. **22**, 107–120 (2002)
2. Cuevas, C., Lebrun, J.: Testing and modelling of a variable speed scroll compressor. J. Appl. Therm. Eng. **29**, 469–478 (2009)
3. Wang, B., Shi, W., Li, X., et al.: Numerical research on the scroll compressor with refrigeration injection. J. Appl. Therm. Eng. **28**, 440–419 (2008)
4. Wang, B., Shi, W., Li, X.: Numerical analysis on the effects of refrigerant injection on the scroll compressor. J. Appl. Therm. Eng. **29**, 37–46 (2009)
5. Chen, Y., Halm, N.P., Groll, E.A., et al.: Mathematical modeling of scroll compressors-Part I: compression process modeling. Int. J. Refrig. **25**, 731–750 (2002)
6. Chen, Y., Halm, N.P., Groll, E.A., et al.: Mathematical modeling of scroll compressors-Part II: overall scroll compressor modeling. Int. J. Refrig. **25**, 751–764 (2002)
7. Ooi, K.T., Zhu, J.: Convective heat transfer in a scroll compressor chamber: a 2-D simulation. Int. J. Therm. Sci. **43**, 677–688 (2004)
8. Lin, C., Chang, Y., Liang, K., et al.: Temperature and thermal deformation analysis on scroll of scroll compressor. J. Appl. Therm. Eng. **25**, 1924–1939 (2005)
9. Cui, M.M.: Numerical study of unsteady flows in a scroll compressor. J. Fluids Eng. **2006**(128), 947–955 (2002)
10. Xiao, G.F., Liu, G.P., Wang, J.T., et al.: Numerical simulation for transient flow in a scroll compressor using dynamic mesh technique. Mach. Tool Hydraulics **41**(1), 146–149 (2013)
11. Morini, M., Pavan, C., Pinelli, M., et al.: Analysis of a scroll machine for micro ORC applications by means of a RE/CFD methodology. J. Appl. Therm. Eng. **80**, 132–140 (2015)
12. Song, P., Wei, M., Liu, Z., et al.: Effects of suction port arrangements on a scroll expander for a small scale ORC system based on CFD approach. J. Appl. Energy **150**, 274–285 (2015)

13. Wei, M., Song, P., Zhao, B., et al.: Unsteady flow in the suction process of a scroll expander for an ORC waste heat recovery system. J. Appl. Therm. Eng. **78**, 460–470 (2015)
14. Zha, H.B., Zhang, X.H., Wang, J.: Numerical simulation of three-dimensional unsteady flow in the scroll compressor. J. Fluid Mach. **44**(2), 17–23 (2016)
15. Wu, Z., Feng, Z.G., Su, Y.F.: Simulation and analysis of internal flow field in scroll compressor with circular involute profile. Mech. Sci. Technol. Aerosp. Eng. **38**(12), 1840–1846 (2019)
16. Rak, J., Pietrowicz, S.: Internal flow field and heat transfer investigation inside the working chamber of a scroll compressor. Energy **202**, 117700 (2020)
17. Sun, S., Wang, X., Guo, P., et al.: Investigation on the modifications of the suction flow passage in a scroll refrigeration compressor. Appl. Therm. Eng. **170**, 115031 (2020)
18. Zheng, S.Y., Wei, M.S., Song, P.P., et al.: Thermodynamics and flow unsteadiness analysis of trans-critical in a scroll compressor for mobile heat pump air-conditioning system. Appl. Therm. Eng. **175**, 115368 (2020)
19. Cavazzini, G., Giacomel, F., Benato, A.: Analysis of the inner fluid-dynamics of scroll compressors and comparison between CFD numerical and modelling approaches. Energies **14**, 1158 (2021)
20. Wang, J., Liu, Q., Cao, C.Y., et al.: Design methodology and geometric modeling of complete meshing profiles for scroll compressors. Int. J. Refrig. **91**, 199–210 (2018)
21. Zhao, R.C., Li, W.H., Zhuge, W.L.: Unsteady characteristic and flow mechanism of a scroll compressors with novel discharge port for electric vehicle air conditioning. Int. J. Refrig. **118**, 403–414 (2020)
22. Sun, S.H., Wu, K., Guo, P.C.: Analysis of the three-dimensional transient flow in a scroll refrigeration compressor. Appl. Therm. Eng. **127**, 1086–1094 (2017)

# Design of a New Pneumatic Atomizer with V-shaped Slits and Numerical Study on the Spray Flow Field

Wentong Qiao[1], Lijuan Qian[1,2(✉)], Chenlin Zhu[1], Xianyue Shao[3], and Peng Wang[4]

[1] College of Mechanical and Electrical Engineering, China Jiliang University, Hangzhou 310018, China
[2] Key Laboratory of Intelligent Manufacturing Quality Big Data Tracing and Analysis of Zhejiang Province, China Jiliang University, Hangzhou 310018, China
[3] Zhejiang Ruifeng Wufu Pneumatic Tools Co., Ltd., Taizhou 317500, China
[4] Zhejiang Santian Car Air Conditioning Compressor Co., Ltd., Lishui 323000, China

**Abstract.** In order to better atomize the water-based paint, a new pneumatic atomizer with V-shaped slits is proposed in this paper. The atomizing air hole is improved by changing its configuration from the ordinary annular structure to a V-shaped groove structure. A 3D model of this atomizer is established for numerical simulation, where the turbulent transport is modeled using the realizable $k - \varepsilon$ model. By means of computational fluid dynamics (CFD), the velocity and pressure distributions in the interference spray flow field are numerically investigated by varying the shaping air pressure when different numbers of V-shaped slits are employed. The results show that the higher the shaping air pressure is, the more even the airflow is distributed on the target plate. With increasing the number of the slits, the local high-speed airflow near the nozzle in the spray flow field will be enhanced, which is conducive to atomizing water-based paint.

**Keywords:** Pneumatic atomizer · V-shaped slit · Spray flow field · Numerical simulation

## 1 Introduction

Water-based paint is attracting more and more attention in numerous industrial coating applications, such as automotive painting, interior decoration, furniture manufacturing [1–3]. This type of paint mainly uses water as a diluent, which is harmless to human health. In contrast, the solvent-based paint will cause a series of environmental pollution problems that result from the emissions of volatile organic compounds (VOCs) [4–6]. Traditional spray equipment is usually developed for atomizing solvent-based paint, but not suitable for the water-based paints that have different physical properties. For example, the pneumatic atomizers are widely used for atomizing solvent-based paint as they can provide ultrafine coatings with the requested high optical qualities, especially in car body painting [7–9]. Nevertheless, the annular structure of atomizing air hole in


Q. Han et al. (Eds.): LSMS 2021/ICSEE 2021, CCIS 1469, pp. 675–684, 2021.
https://doi.org/10.1007/978-981-16-7213-2_65

pneumatic atomizer is designed for solvent-based paint. The surface tension of water-based paint is relatively large, and its viscosity value is related to the water content in it. When this atomizer is used to spray water-based paints, it cannot get a sufficient atomization performance due to larger surface tension and possibly higher viscosity in paint liquid [1, 5]. Therefore, in this study, the original structure of atomizing air hole in pneumatic atomizer needs to be optimized to make it suitable for atomizing water-based paints.

So far, scholars have mainly devoted to studying the spray flow field and atomization performance of traditional pneumatic atomizer, as well as the amelioration of atomizer structure for the solvent-based paint [10–13]. However, there are few detailed reports concerning the improvement of a pneumatic atomizer for better atomizing the water-based paints. In particular, it remains unclear how the shaping air pressure and the V-shaped groove structure affect the velocity and pressure distributions in the interference spray flow field, which motivates the present work.

## 2 Design of a New Atomizer with V-shaped Slits

The basic structure of a conventional pneumatic atomizer consists of paint hole, atomizing air hole, auxiliary air holes and shaping air holes. The atomizing air hole is an annular channel for the gas flow. The high-speed gas flow shaped as a ringlike column surrounds the paint hole and breaks the liquid paint into small droplets, as shown in Fig. 1(a). However, this annular atomizing air hole is only suitable for atomizing solvent-based paint, but not referring to the water-based paint since too much surface tension in it. If several V-shaped slit structures are designed on the atomizing air hole, things will be going differently. A depicted in Fig. 1(b), the gas flow ejected from the V-shaped slit cuts the paint column like a sharp blade. The water-based paint will be broken evenly, and the degree of atomization of the paint can be controlled effectively. The waterborne paint coatings obtained on the target plate are dense and more uniform. This process only requires lower air pressure and lower air consumption to complete.

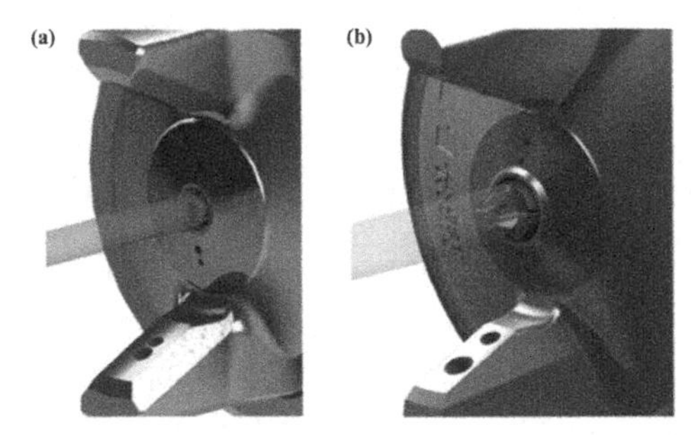


**Fig. 1.** Comparison of the ordinary and special air cap of the pneumatic atomizer: (a) Ordinary air cap; (b) Special air cap with slits.

A new pneumatic atomizer with V-shaped slits was proposed, as presented in Fig. 2. The atomizing air hole has been improved from the ordinary annular structure to a

V-shaped groove structure. The V-shaped slits are designed on the inner surface of atomizing air hole and evenly distributed around the paint hole. The number of the V-shaped slits was chosen as three or six in this study. The opening angle of each V-shaped slit is 37°. Table 1 shows the main structure parameters of the new atomizer.

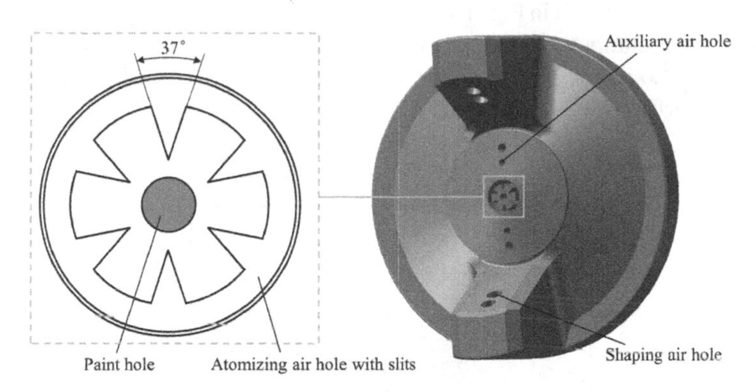


**Fig. 2.** Design of the new atomizer with V-shaped Slits.

**Table 1.** Structure parameters of the pneumatic atomizer with V-shaped slits.

| Structure parameters | Dimension |
|---|---|
| Paint hole diameter | 0.8 mm |
| Atomizing air hole outer diameter | 2.8 mm |
| Auxiliary air hole diameter | 0.5, 0.7 mm |
| Shaping air hole diameter | 1.3 mm |
| Angle of V-shaped slit | 37° |

# 3 Numerical Setup

## 3.1 Mathematical Model

The gas flow between the atomizer and the target plate is turbulent, which is governed by the conservation equations of mass, momentum and energy [14, 15]. Since the gas flow near the multi-holes is under a supersonic condition with the complexity of movement mechanism, the realizable $k - \varepsilon$ turbulence model was adopted to describe the turbulent gas phase [7, 16].

## 3.2 Computational Domain and Numerical Methods

As schematically shown in Fig. 3, a cuboid block of 400 mm × 200 mm × 200 mm is used as the computational domain of the numerical model. The pneumatic atomizer is located in the center of plane ABCD. The distance from the atomizer to the target plate EFGH is 194 mm. As shown in Fig. 4, unstructured meshes were adopted to discretize the three-dimensional computational domain, which could better fit the complex geometry of the atomizer. Local mesh refinements around the nozzle axis, elliptical spray cone and the target surface were carried out.

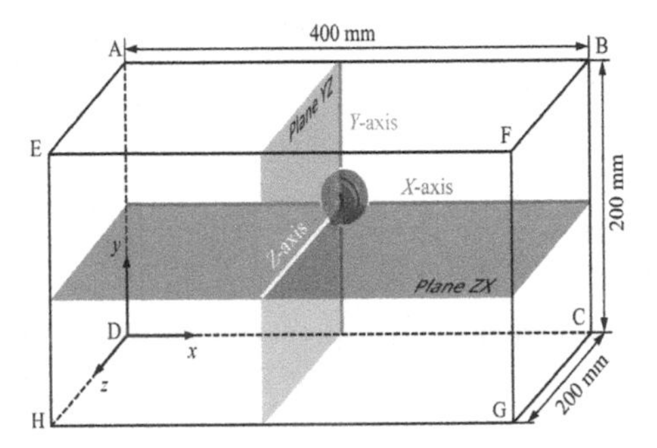


**Fig. 3.** Schematic diagram of computational domain.

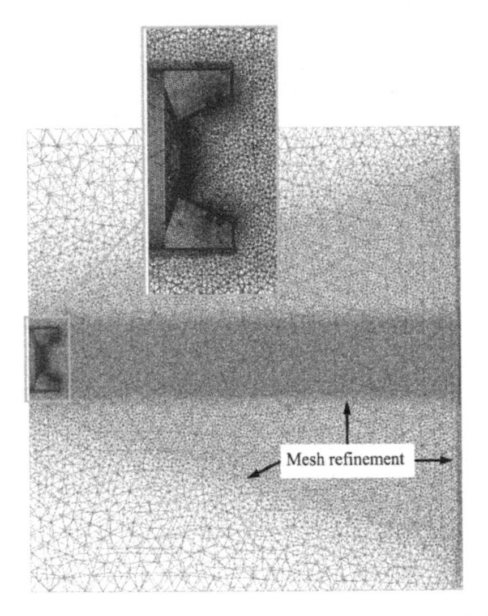


**Fig. 4.** Mesh in the plane *YZ* of the computational domain.

The pressure-based solver and the semi-implicit method for pressure linked equations (SIMPLE) algorithm were employed in ANSYS Fluent. The turbulent kinetic energy and turbulent dissipation rate were both set as first-order upwind, and the other variables used second-order upwind. Pressure-inlet boundary conditions were set at atomizing air hole, auxiliary air holes and shaping air holes. The target plate EFGH and the nozzle surface were set as no-slip wall boundary conditions. The other boundaries (planes ABFE, CDHG, ADHE, BCGF and ABCD) were set as pressure-outlet conditions. The inlet pressure of the atomizing air hole and auxiliary air holes was set as 150 kPa. The inlet pressure of the shaping air holes was respectively set as 0, 20, 50, 80 kPa in the next section. The outlet pressure was set to 1 atm.

## 4 Results and Discussion

### 4.1 The Influence of Shaping Air Pressure

In current section, four different shaping air pressures of 0 kPa, 20 kPa, 50 kPa and 80 kPa were chosen to study the influence on the velocity and pressure distributions in the spray flow field. Figure 5 show the velocity contours in the plane *ZX* for four shaping air pressures. The gas velocity near the nozzle is very high. When there is no shaping air pressure, the shape of the gas flow field is approximately a cylinder. It can be found that the gas flow diffuses in the space under the action of the shaping air pressure, and then flows out along the target plate. With increasing the shaping air pressure gradually, the gas flow field becomes an elliptical cone progressively. That means that the extension along *X*-axis is wider while narrow along the *Y*-axis. The greater the shaping air pressure is, the wider the range of airflow on the target surface can be obtained.

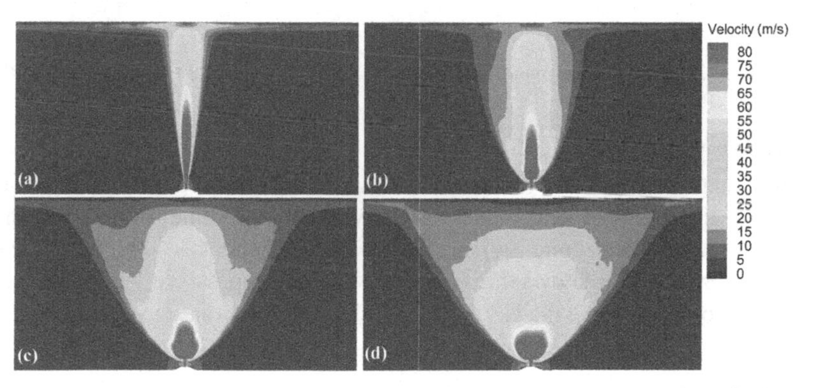


**Fig. 5.** Velocity contours on the plane *ZX* for four shaping air pressure: (a) 0 kPa; (b) 20 kPa; (c) 50 kPa; (d) 80 kPa.

Figure 6 shows the pressure contours on the plane *YZ* for different shaping air pressures. There are two high-pressure regions following the negative pressure region. It can be seen from Fig. 6 that the distribution range of the second high-pressure region is wider than that of the first high-pressure region. When the shaping air pressure increases, the

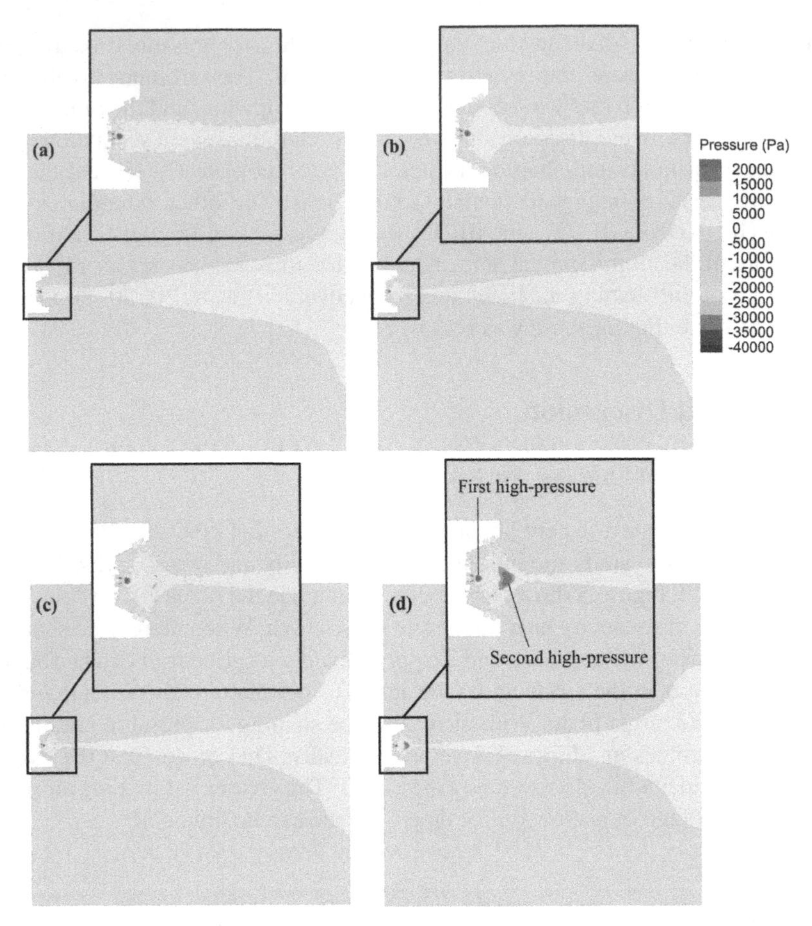


**Fig. 6.** Pressure contours on the plane *YZ* for four shaping air pressure: (a) 0 kPa; (b) 20 kPa; (c) 50 kPa; (d) 80 kPa.

distribution range of the second high-pressure region becomes larger. It can be seen from Fig. 7 that the first pressure peak values basically remain the same for different shaping air pressures. While the second pressure peak value will increase with the increase of shaping air pressure. This shows that elevating shaping air pressure will again increase the relative velocity between the gas and the liquid, which is conducive to the secondary atomization of the liquid.

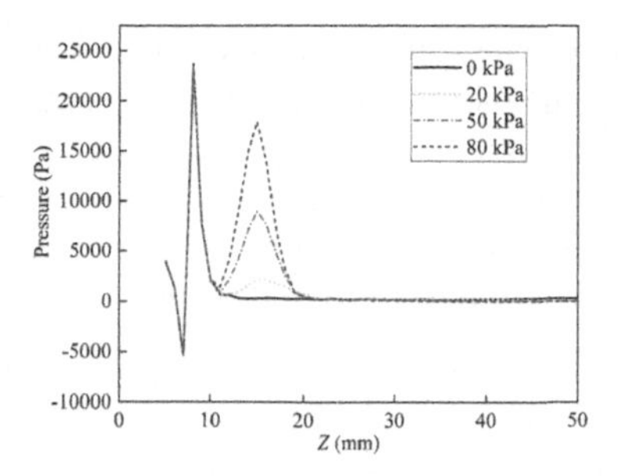


**Fig. 7.** Pressure distributions near the nozzle along the $Z$-axis for different shaping air pressure.

## 4.2 The Influence of the Number of V-shaped Slits

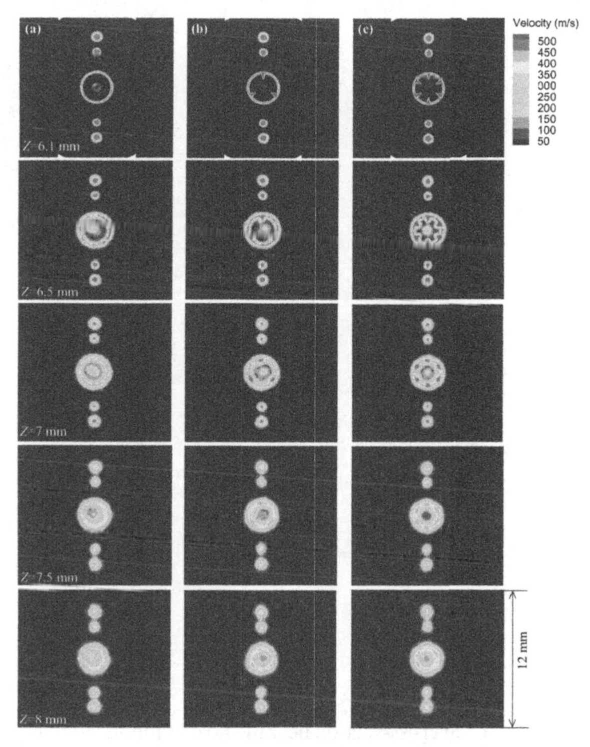


**Fig. 8.** Velocity contours on each cross-section near the nozzle for different numbers of V-shaped slits: (a) 0 slit; (b) 3 slits; (c) 6 slits.

Figure 8 shows the velocity contours on the cross-sections of $Z = 6.1, 6.5, 7, 7.5, 8$ mm, which illustrates the velocity evolution process for the atomizers that have different numbers of V-shaped slits. As shown in Fig. 8, as the position is away from the nozzle, the velocity in the ring-shaped or sawtooth ring-shaped area around the paint hole gradually decreases. This is different from the change trend of the velocity at the center. The velocity at the center will increase and then may decrease due to the existence of negative pressure region. In addition, when the number of slits is increased from 0 to 6, the velocity distribution around the paint liquid is more even and symmetrical, which helps the liquid column to break up stably. For example, when 6 slits are selected, the gas flow is like several sharp blades, cutting the liquid column from multiple positions around the liquid simultaneously. This can produce finer and uniform droplets, which improves the atomization performance of the water-based paint.

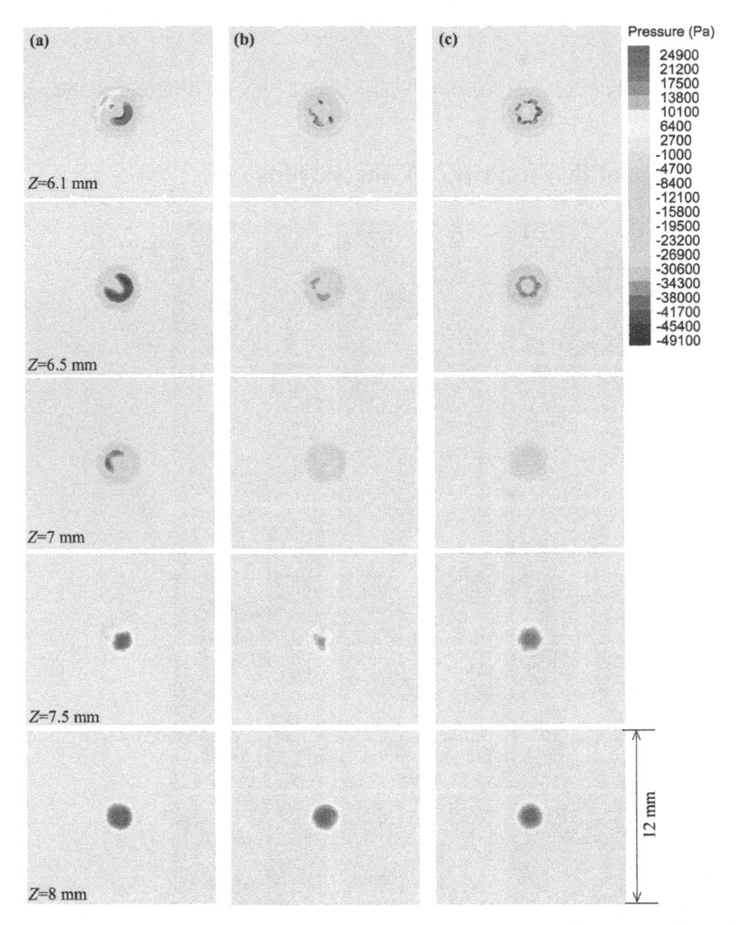


**Fig. 9.** Pressure contours on each cross-section near the nozzle for different numbers of V-shaped slits: (a) 0 slit; (b) 3 slits; (c) 6 slits.

Figure 9 shows the pressure contours on the cross-sections of $Z = 6.1, 6.5, 7, 7.5, 8$ mm, which reveals the pressure evolution process for the atomizers with different numbers of V-shaped slits. As shown in Fig. 9, ring-shaped or plum-like regions with high negative pressure are formed around the paint hole. When the cross-section is closed to the nozzle, the pressure in the center is much larger than that around the paint liquid. It can be also seen that if no V-shaped slit is designed in the atomizer, the negative pressure distribution around the paint hole is uneven and asymmetrical. Once the slits are designed in the atomizing air hole, this issue can be significantly improved, which contributes to making the water-based paint evenly broken and atomized.

Figure 10 shows the pressure distribution near the nozzle along the $Z$-axis. When different numbers of V-shaped slits are chosen, the deviation of pressure distribution is mainly reflected in the negative pressure region. When there is no slit in atomizing air hole, a larger negative pressure can be obtained, which helps to effectively suck the liquid from the paint hole. However, with increasing the number of slits, the pressure in the negative pressure region will rise. This means that when using an atomizer with V-shaped slits, a diaphragm pump may have to be used to help the paint liquid transport to the outlet of paint hole.

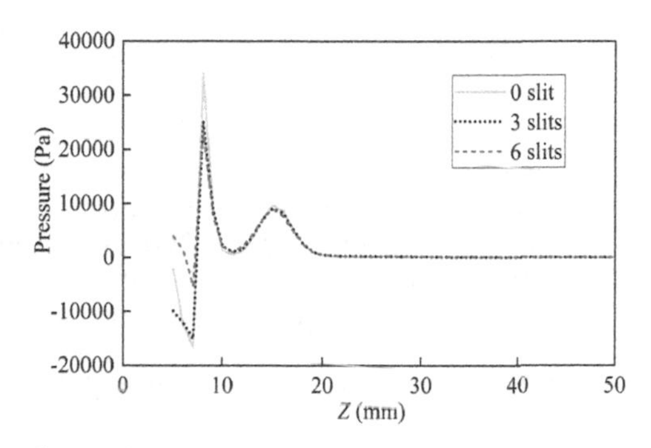


**Fig. 10.** Pressure distributions along the $Z$-axis for different numbers of V-shaped slits.

## 5 Conclusions

In this study, a new pneumatic atomizer with V-shaped slits has been proposed and designed for better atomizing the water-based paint. By the means of CFD, the spray flow field between the atomizer and the target plate has been numerically studied, especially in the influence of the shaping air pressure and the number of V-shaped slits. The evolution processes of velocity and pressure distribution were discussed in detail. The main conclusions are drawn as follows.

1. With increasing the shaping air pressure, the gas flow field becomes an elliptical cone progressively, which means the extension along $X$-axis is wider but narrow along $Y$-axis.
2. There are two high-pressure regions following the negative pressure region. As the shaping air pressure increases, the distribution range of the second high-pressure region becomes larger, the first pressure peak values basically remain the same, and the second pressure peak value will increase.
3. As the position of the cross-section near the nozzle is away from the nozzle, the velocity in the ring-shaped or sawtooth ring-shaped area around the paint hole gradually decreases. When the number of V-shaped slits increases, the velocity and negative pressure distributions on the cross-sections are more even and symmetrical, which helps the water-based paint evenly break and atomize.

4. When fewer V-shaped slits were designed in the atomizer, a larger negative pressure can be obtained, which may help to effectively suck the liquid from paint hole.

In future work, the water-based paint needs to be added to the paint hole for spray simulation to study the influence of the shaping air pressure and the number of V-shaped slits on the atomization performance, such as droplet diameter, droplet velocity and coating thickness.

**Acknowledgments.** This work was supported by the National Natural Science Foundation of China (11872352; 11632016), the Natural Science Foundation of Zhejiang Province (LR21E060001) and the Key Research and Development Program of Zhejiang Province (2020C01054).

## References

1. Poozesh, S., Akafuah, N., Saito, K.: Effects of automotive paint spray technology on the paint transfer efficiency – a review. P. I. Mech. Eng. D-J. Aut. **232**(2), 282–301 (2017)
2. Chen, W., Chen, Y., Zhang, W., He, S., Li, B., Jiang, J.: Paint thickness simulation for robotic painting of curved surfaces based on Euler–Euler approach. Braz. Soc. Mech. Sci. **41**(4), 199 (2019)
3. Streitberger, H.J., Dossel, K.F.: Automotive Paints and Coatings. Wiley-VCH Verlag GmbH & Co. KGaA, Weinheim, Germany (2008)
4. Papasavva, S., Kia, S., Claya, J., Gunther, R.: Characterization of automotive paints: an environmental impact analysis. Prog. Org. Coat. **43**, 193–206 (2001)
5. Akafuah, N., Poozesh, S., Salaimeh, A., Patrick, G., Lawler, K., Saito, K.: Evolution of the automotive body coating process – a review. Coatings **6**(2), 24 (2016)
6. Honarkar, H.: Waterborne polyurethanes: a review. Disper. Sci. Technol. **39**(4), 507–516 (2018)
7. Ye, Q., Domnick, J.: Analysis of droplet impingement of different atomizers used in spray coating processes. J. Coat. Technol. Res. **14**(2), 467–476 (2017)
8. Xie, X., Wang, Y.: Research on distribution properties of coating film thickness from air spraying gun-based on numerical simulation. Coatings **9**(11), 721 (2019)
9. Ye, Q., Shen, B., Tiedje, O.: Breakup simulation of a viscous liquid using a coaxial high-speed gas jet. In: ILASS–Europe 2019, 29th Conference on Liquid Atomization and Spray Systems, Paris, France, 2–4 September 2019 (2019)
10. Fogliati, M., Fontana, D., Garbero, M., Vanni, M., Baldi, G., Dondè, R.: CFD simulation of paint deposition in an air spray process. J. Coat. Technol. Res. **3**, 117–125 (2006)
11. Li, W., Qian, L., Song, S., Zhong, X.: Numerical study on the influence of shaping air holes on atomization performance in pneumatic atomizers. Coatings **9**(7), 410 (2019)
12. Wang, Y., Xie, X., Lu, X.: Design of a double-nozzle air spray gun and numerical research in the interference spray flow field. Coatings **10**(5), 475 (2020)
13. Niblett, D., et al.: Development and evaluation of a dimensionless mechanistic pan coating model for the prediction of coated tablet appearance. Int. J. Pharmaceut. **528**, 180–201 (2017)
14. Lin, J., Qian, L., Xiong, H., Chan, T.L.: Effects of operating conditions on droplet deposition onto surface of atomization impinging spray. Surf. Coat. Technol. **203**(12), 1733–1740 (2009)
15. Aydin, O., Unal, R.: Experimental and numerical modeling of the gas atomization nozzle for gas flow behavior. Comput. Fluids. **42**(1), 37–43 (2011)
16. Qian, L., Lin, J., Bao, F.: Numerical models for viscoelastic liquid. Energies **9**(12), 1079 (2016)

# Design of Scratching System for Scattered Stacking Workpieces Based on Machine Vision

Jun Xiang(✉), Jian Sun, and Hongwei Xu

College of Mechanical and Electrical Engineering, China Jiliang University, Hangzhou 310018, Zhejiang, China

**Abstract.** The traditional industrial production line uses vibration plate to solve the loading and sorting work of randomly stacked workpiece. However, for soft magnetic, ceramic sheet, and other materials with weak impact toughness, the use of vibration plate feeding will lead to the workpiece broken, crack, and other consequences. Therefore, in order to improve the efficiency and reduce the damage rate, a structured optical camera is used to collect the 3D data of the scattered stacked workpiece and obtain the 3D point cloud data of the workpiece. Then, the collected 3D point cloud data is segmented to obtain the point cloud data of a single workpiece. Finally, the pre-established workpiece template is matched with the segmented point cloud data to obtain the 3D pose of the scattered stacked workpiece, which is convenient for the manipulator to grasp the workpiece with different postures.

**Keywords:** Industrial robot · Machine vision · Pose recognition · Calculate disorderly stack · Workpiece sorting

## 1 Introduction

Mechanical vibration device is widely used in industrial production of workpiece sorting and feeding links, has the advantages of low cost, simple technology and easy to achieve. But some materials are fragile and vulnerable to external force damage of the workpiece, the use of mechanical vibration device for sorting work, not only the sorting efficiency is low, but also lead to large-scale damage to the workpiece. For this type of workpiece, only manual sorting and feeding can be used, which greatly improves the cost of production and sorting.

Therefore, in order to improve the sorting efficiency [1], reduce production and sorting costs, the use of machine vision adaptability, no contact and industrial robot technology high efficiency, high precision advantages, the combination of machine vision technology and industrial robot technology to establish automatic sorting system of workpiece will become a development trend [2].

Aiming at the difficulties of robot sorting technology [3–5] based on machine vision for scattered stacked workpieces, a beneficial exploration is made, and a feasible solution is proposed.


Q. Han et al. (Eds.): LSMS 2021/ICSEE 2021, CCIS 1469, pp. 685–693, 2021.
https://doi.org/10.1007/978-981-16-7213-2_66

## 2 System Scheme Design

In order to obtain detailed and accurate 3D point cloud data of scattered and stacked workpieces, a structured light 3D camera is used to obtain 3D point cloud data. The principle of the structured light camera is shown in Fig. 1.

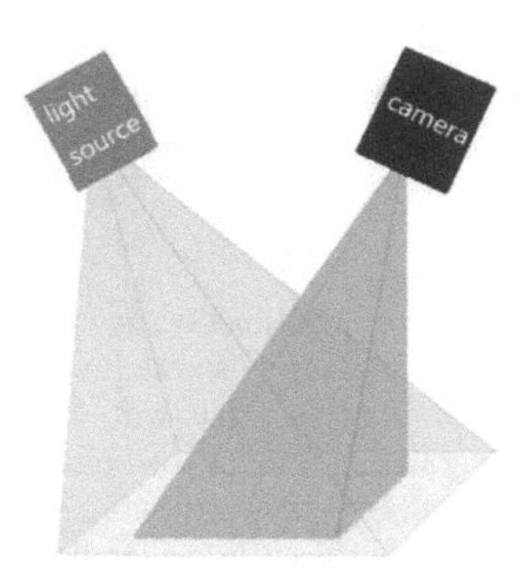


**Fig. 1.** Schematic diagram of structured light camera.

This system mainly consists of 4 parts.

They are (1) industrial robot (2) 3D structured light camera (3) adjustable camera bracket (4) working platform. The specific composition is shown in Fig. 2.

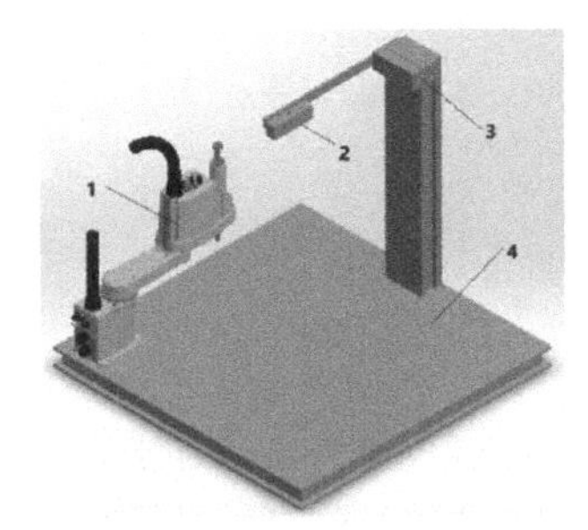


**Fig. 2.** Overall structure of the system.

A physical scene in which artifacts are placed on top of each other. This is shown in Fig. 3.

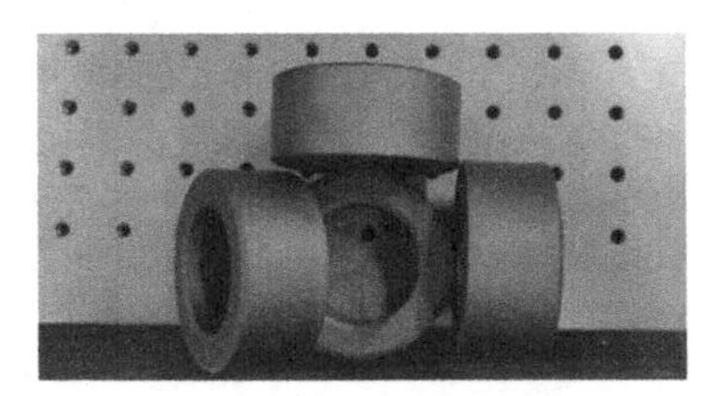

**Fig. 3.** Artifact SceneDiagram.

3D point cloud effect of scene reconstructed by 3D structured light camera. This is shown in Fig. 4.

**Fig. 4.** 3D reconstruction diagram of workpiece physical scene.

## 3 Object Segmentation, Pose Recognition and Matching

### 3.1 Workpiece Template Point Cloud Generation

Before determining the pose of the scattered stacked workpieces, it is necessary to obtain the template point cloud for three-dimensional matching. The template point cloud is matched with the field scenic spot cloud of the stacked workpieces obtained by actual shooting through the algorithm, and finally the position and pose information of the target workpieces in the field scenic spot cloud is obtained.

Because the workpiece is a circular cylinder, there are mainly three positions of the workpiece, namely horizontal posture, vertical posture and oblique posture, and the original point cloud effect of the reconstruction of the three positions of the workpiece. As shown in Fig. 5.

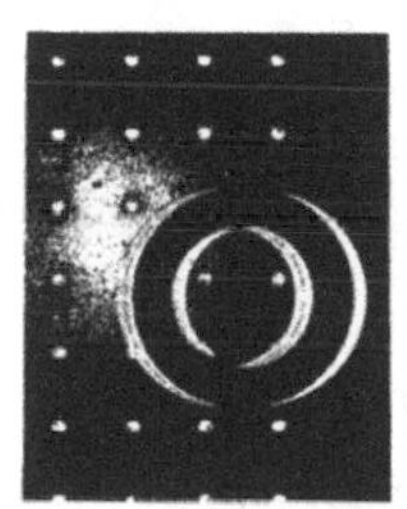 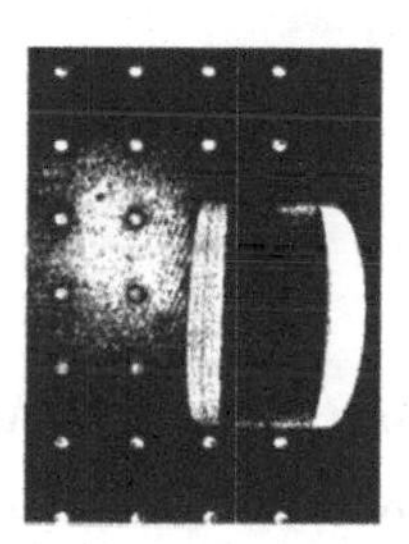 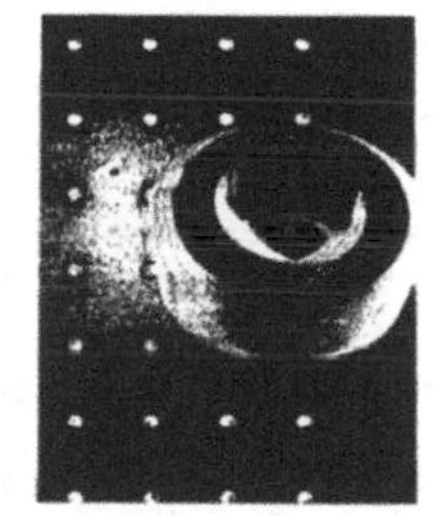

**Fig. 5.** Original point cloud map of horizontal attitude, Original point cloud map of vertical attitude and Original point cloud map of oblique attitude.

Since the original point cloud data of the workpiece template contains a lot of useless point data, it is filtered to achieve the point cloud effect of the workpiece template after filtering. As shown in Fig. 6.

**Fig. 6.** Horizontal attitude filtering point cloud image, Vertical attitude filtering point cloud image and Oblique attitude filtering point cloud image.

### 3.2 Target Workpiece Segmentation Algorithm

In the collected field cloud with scattered and stacked artifacts, the artifacts are not completely separated, so it is necessary to use the region growth algorithm based on normal vector and curvature to segment the originally connected point cloud images, to achieve the extraction of a single target artifact, and to achieve the 3D point cloud effect after segmentation by the algorithm. As shown in Fig. 7.

**Fig. 7.** Image segmentation diagram of artifact scene.

### 3.3 Workpiece Matching and Pose Recognition Algorithm

The matching between 3D point clouds has been realized in many ways, such as feature point matching, ICP matching [6–8], etc.

The feature point [9] method determines the feature sub of the sub-region mainly through the local point cloud distribution of the 3D point cloud. By comparing the features [10] of the target and the template, the closest combination is found, and the spatial rigid body transformation matrix between the target and the template point cloud is obtained by further solving. The matching speed of feature points is fast, but if the feature points are few or distributed in irrelevant areas (such as the workpiece boundary), the matching failure is easy to occur.

ICP matching is an algorithm that establishes matching point pairs through the nearest distance principle and iteratively calculates the three-dimensional rigid body space transformation matrix between the target point cloud and the template point cloud.

But because the workpiece to be identified is a cylindrical ring, so between the three positions of the workpiece, some characteristics are the same. Therefore, if ICP algorithm is directly used for matching, local optimal matching will appear, with large computation and long matching time.

Therefore, two-way KDTree is first used to process the point cloud of the workpiece template and the scenic spot cloud of the workpiece physical field.

In fact, KdTree is a high-dimensional binary tree, which can quickly search target data according to certain rules through continuous division of space. The establishment of KDTree firstly selects the segmentation axis, which needs to calculate the maximum variance of the data points in k spatial dimensions. The two sides of the dividing point are left and right subtrees of the node respectively, in which the left and right subtrees are divided according to the following rules until all the points are divided. The structure of two-dimensional KDTree is shown in Figs. 8 and 9.

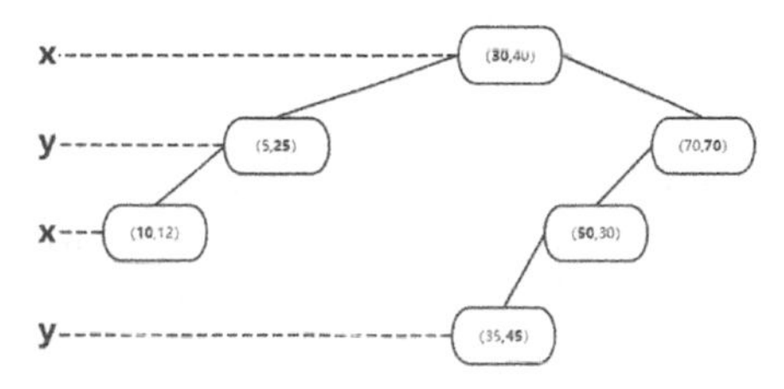


**Fig. 8.** Two-dimensional KdTree data structure diagram.

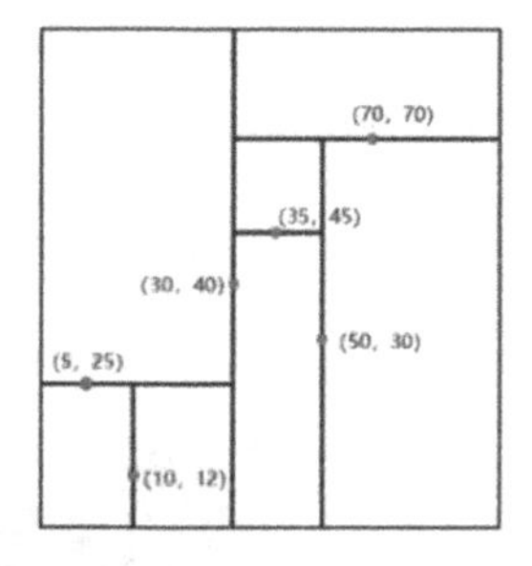


**Fig. 9.** Patial structure diagram of two-dimensional KDTree.

Bidirectional KDTree is used to optimize the target point cloud $P$ and source point cloud $Q$ to be matched respectively. In the optimized target point cloud $P$ and source point cloud $Q$, according to certain constraints, the nearest point $(p_i, q_i)$ is found, and then the optimal matching parameters $R$ and $t$ are calculated to minimize the error function $E(R, t)$.The error function is $E(R, t)$:

$$E(R, t) = \frac{1}{n} \sum\nolimits_{i=1}^{n} ||q_i - (Rp_i + t)||^2 \tag{1}$$

Where n is the number of nearest point pairs, $p_i$ is a point in the target point cloud $P$, $q_i$ is the nearest point corresponding to $p_i$ in the source point cloud $Q$, $R$ is the rotation matrix, and $t$ is the translation vector. The point set $p_i \in P$ is taken from the target point cloud $P$.

The corresponding point set $q_i \in Q$ in the source point cloud $Q$ is found, to make

$$||q_i - p_i|| = min. \tag{2}$$

Calculate the rotation matrix $R$ and the shift vector $t$ to minimize the error function $E(R, t)$.$p_i$ is transformed by the rotation matrix $R$ and the transformation vector $t$ obtained in the previous step, and a new set of corresponding points is obtained

$$p_i^{'} = \left\{p_i^{'} = Rp_i + t, p_i \in P\right\}. \tag{3}$$

Calculate the average distance $d$ between $p_i^{'}$ prime and the corresponding point set $q_i$.

$$d = \frac{1}{n}\sum_{i=1}^{n}\left|\left|p_i^{'} - q_i\right|\right|^2. \tag{4}$$

If $d$ is less than a given threshold or greater than the preset maximum number of iterations, the iterative calculation is stopped. Otherwise, return (2) to continue to find out the corresponding point set $q_i \in Q$ in the source point cloud $Q$, to make

$$||q_i - p_i|| = min. \tag{5}$$

Until the convergence condition is satisfied.

## 4 Experimental Results and Analysis

The grasping object and the original image information collected by the camera in this experiment are shown in the Fig. 10.

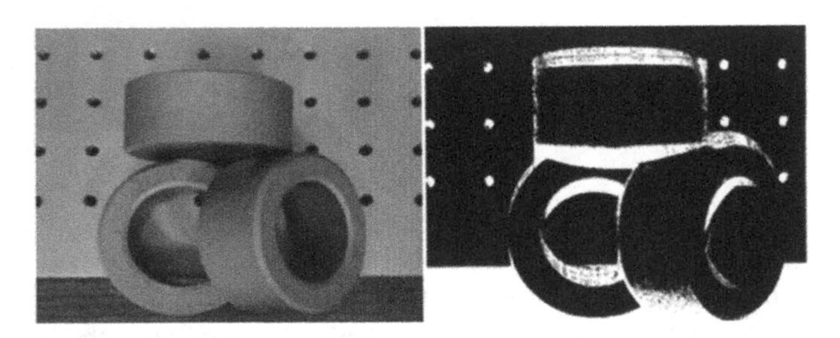

**Fig. 10.** Grab the object and the original image collected by the camera.

Since the point cloud data collected by the camera contains the metal platform on which the workpiece is placed, the point cloud data is too large, which is not conducive to the pose recognition and matching of the workpiece in the later stage. Therefore, straight-through filtering is adopted to obtain the $X, Y, Z$ axis coordinate data of the lowest point cloud data, and filter the point cloud data in the $Z$ axis direction of the point cloud data to filter out the point cloud data of the metal platform. In the point cloud data after the metal platform is removed, there are still some discrete points and useless points far away from the main body of the workpiece, in order to reduce the calculation amount of attitude recognition in the later stage. Statistical filtering was used to filter outliers, as shown in Fig. 11.

**Fig. 11.** Statistical filtering point cloud map.

The point cloud data processed by filtering only contains the point cloud data of the workpiece to be identified, but these point cloud data are not completely separate and independent. In order to achieve the extraction of a single target workpiece, it is necessary to segment the point cloud data. The region growing algorithm based on normal vector and curvature is used to segment the data, which is convenient to distinguish and identify different postures of the workpiece. Figure 12 shows the effect drawing after segmentation.

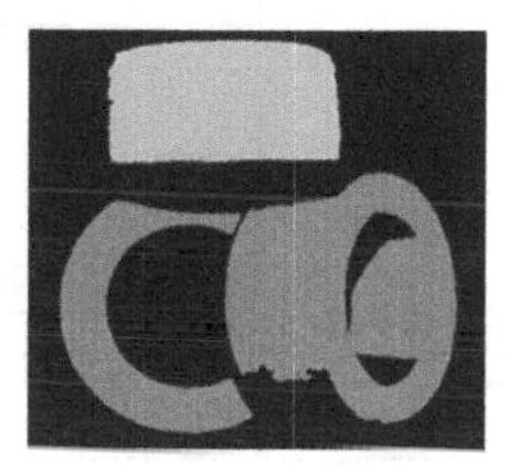

**Fig. 12.** Workpiece segmentation diagram.

The target point cloud $P$ and source point cloud $Q$ are processed by two-way KDTree. A large number of points with low registration accuracy are eliminated and those with high registration quality are retained. Then the ICP algorithm is used to calculate the target point cloud $P$ and source point cloud $Q$, and the matching results are obtained, as shown in the Fig. 13.

**Fig. 13.** Horizontal attitude filtering point cloud image, Vertical attitude filtering point cloud image and Oblique attitude filtering point cloud image.

## 5 Conclusion

The straight-through filter is used to remove the background of the target workpiece to be captured and the template of the workpiece, and statistical filter is used to remove the outliers in the point cloud data. The corresponding point pairs with low matching degree are removed by KDTree to reduce the amount of matching calculation. Then ICP algorithm is used to match and identify the target object, and its three-dimensional pose information is obtained. The calculation process is shown in Fig. 14.

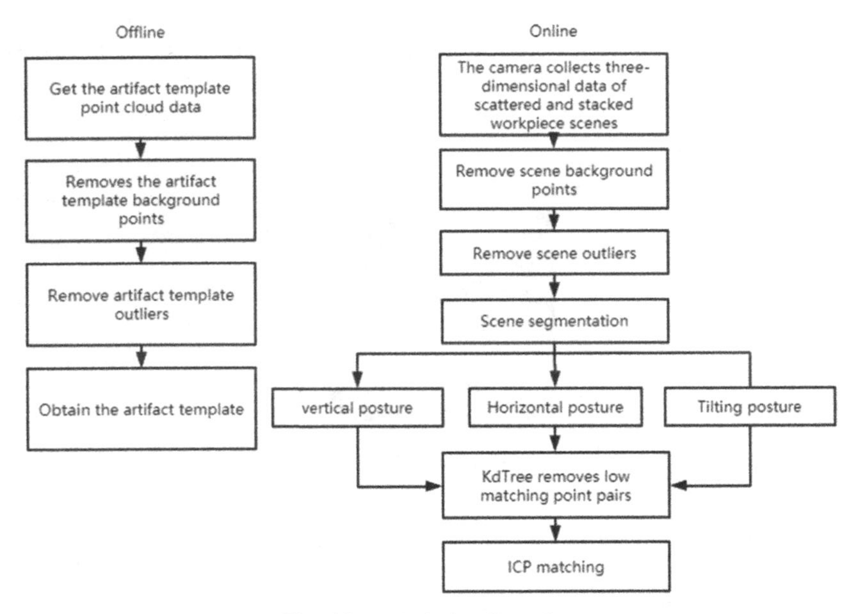


**Fig. 14.** Calculation flow chart.

Experimental results show that this method can identify the target workpiece and obtain its three-dimensional pose information.

## References

1. Cui, X.D., Tan, H., Wang, P.J., Yu, L.B., Chen, J.H.: Research on robot grabbing technology of spreadly stacked workpieces based on 3D vision. Manufact. Technol. Mach. Tool **2021**(02), 36–41 (2021)
2. Jia, K.: Design of automatic sorting system for automobile parts based on machine vision. Automobile Digest **2021**(01), 48–52 (2021)
3. Li, S.C., Zhang, J., Zhang, H., Liu, M.L., Yang, H.Y., Liu, L.X.: Target pose estimation method in the process of fetching for robot. J. Sens. Micro Syst. **38**(7), 32–34 (2019)
4. Long, Y.R., Xu, Y.L., Fei, X.Y.: Object grabbling of JACO manipulator based on PCL. Ind. Control Comput. **32**(04), 65–67 (2019)
5. Zou, Y., Xiong, H.G., Tao, Y., Ren, F., Chen, C.Y., Jiang, S.: Robot grasping control method based on 3D point cloud depth information and centroid distance. High Technol. Commun. **30**(05), 508–517 (2020)
6. Chen, T.J., Qin, W., Zou, D.W.: Object detection and pose estimation based on semantic segmentation and point cloud registration. Electron. Technol. **49**(01), 36–40 (2020)
7. Tian, Q.H., Bai, R.L., Li, D.: Point cloud registration algorithm for scattered workpieces based on SHOT feature fusion. J. Small Microcomput. Syst. **40**(02), 275–279 (2019)
8. Amin, N., Amin, T.G., Zhang, Y.D.: Image-based deep learning automated sorting of date fruit. Postharvest Biol. Technol. **153**, 133–141 (2019)
9. Dąbrowski, P.S., Zienkiewicz, M.H.: 3D point-cloud spatial expansion by total least-squares line fitting. Photogramm. Rec. **35**(172), 509–527 (2020)
10. Kumar, R.P., RamirezSerrano, A., Kumar, P.D.: Real time moving object detection and removal from 3D pointcloud data for humanoid navigation in dense GPS-denied environments. Eng. Rep. **2**(12), 34–38 (2020)

# A Discrete Sparrow Search Algorithm for Robot Routing Problem

Yang Han and Zhen Zhang[(✉)]

School of Mechatronic Engineering and Automation,
Shanghai University, Shanghai 200444, China
zhangzhen_ta@shu.edu.cn

**Abstract.** Arranging a reasonable route satisfying constraints for robots to serve is crucial for its application in hospital logistics. The problem was modeled as a single depot delivery problem with capacitated constraint. For solving the problem, a discrete sparrow search algorithm was proposed in this study with multiple enhancement strategies. Based on the original sparrow search algorithm, local search and genetic operators were combined to make it feasible for the aforementioned problem. In addition, the solution representation had also been ameliorated to fit the application. These defined operations hold a good balance between exploration and exploitation. Experiment results showed that our proposed approach gives a promising solution when compared to other novel heuristic methods.

**Keywords:** Robot routing · Sparrow search algorithm · Local search

## 1 Introduction

COVID-19, the global pandemic, has drawn public attention to contactless delivery services and robotic automation and logistics [1, 2]. Especially in hospitals, medicine delivery and medical consumables by robots give the possibility to prevent infection of medical staff and patients under the conditions of increased epidemiological danger. Efficient path planning is the cornerstone of designing a powerful autonomous mobile robot delivery system.

Various approaches have been used by scholars to solve this complicated optimization problem over the past few years. They contain three types: (1) classical approaches; (2) nature-inspired algorithms; (3) Hybrid algorithms. Some classical methods, for instance, potential field method [3], A* algorithm [4] and rapidly exploring random tree (RRT) [5] have been proven that they have a stable performance on the problem. Gaining experience from the swarm intelligence of animals or other objects, researchers created lots of algorithms to solve the optimization problem. These algorithms can be named by the source of their inspiration. For example, the artificial bee colony (ABC) algorithm [6] proposed by Karaboga, is inspired by the foraging behaviors of honey bees in a colony. The existence of multiple types of bees makes a balance between global optimal and local search. Annand [7] improve basic ABC's performance by the concept of the Arrhenius equation and apply it to the robot path planning. The same way could be found


Q. Han et al. (Eds.): LSMS 2021/ICSEE 2021, CCIS 1469, pp. 694–703, 2021.
https://doi.org/10.1007/978-981-16-7213-2_67

in Genetic algorithm (GA) [8], Ant Colony Optimization (ACO) [9], Bat Algorithm (BA) [10], Cuckoo Search Algorithm (CS) [11], etc. However, just like "no free lunch" (NFL) goes, there is no such algorithm that fits all conditions. That is where Hybrid algorithms come in. They were also obtained better performance than any of the methods. Similar research could be found in [12–14] and so on.

Sparrow search algorithm (SSA) [15] is recently contributed to the community swarm intelligence algorithm for numerical optimization problems. The new bionic optimization algorithm is inspired by the sparrow population's foraging and anti-predator performance. Based on its global optimal capacity and preferable convergence, many researchers used it as an optimization method for engineering problems, such as stochastic configuration networks [16], fault diagnosis of wheelset-bearing [17], and renewable energy systems [18].

To design reasonable routes for robots to deliver supplies in hospitals, this paper presents an improved SSA. It is named the Discrete Sparrow Search Algorithm (DSSA). In a nutshell, the main contribution of our research is listed as:

a. To address the optimization problem for robot delivery medical supplies under limited load, we build a mathematical model for it.
b. A new discrete meta-heuristic algorithm DSSA, which enhances with the genetic operator and local search is proposed for the sake of solving the defined model.
c. Sets of the experiment have been adopted to evaluate our algorithm's effectiveness and rate of convergence under different environments.

## 2 Problem Statement and Model

Mathematically speaking, the problem can be considered as a capacitated vehicle routing problem (CVRP).

In the hospital, each ward has a corresponding drug demand, and these drugs are delivered by a drug delivery robot. We set the number of wards as $n$, $V' = V \cup \{0\} = \{0, 1, \ldots n\}$, where $V$ is the collection of wards, and 0 represents the location where the pharmacy and the delivery robot are located. A graph $G = (V, E)$ is given, where $E$ is $G$ set of edges. For each edge, we define $d_{ij}$ to represent the distance between ward $i$ and ward $j$. In the pharmacy, there are $k$ delivery robots, and the drug-carrying capacity of the delivery robots is $Q$. Each ward $i(i = 1, 2, \ldots n)$ has a corresponding drug demand $d_i$, and the drug demand set of the ward is defined as $D_i$. $g_{ij}$ represents the loading capacity of the drug delivery robot on the side $i, j$.

When the drug delivery robot $\pi$ passes through the edge $i, j$, $x_{ij}^{\pi} = 1$; otherwise, $x_{ij}^{\pi} = 0$. According to the goal of minimizing the total distance to complete a distribution task, the following model is established in this paper:

$$Min \sum_{i=0}^{N} \sum_{j=0}^{N} \sum_{\pi=0}^{K} C_{ij} x_{ij}^{\pi}. \tag{1}$$

$s.t.$

$$\sum_{j \in V'} x_{ij}^{\pi} = 1 \qquad \forall i \in V, i \neq j. \tag{2}$$

$$\sum_{j\in V'} x_{ij}^{\pi} = \sum_{j\in V'} x_{ji}^{\pi} \qquad \forall i \in V', i \neq j. \tag{3}$$

$$\sum_{j\in V'} x_{0j} \leq K. \tag{4}$$

$$\sum_{j\in V'} g_{ji} - \sum_{j\in V'} g_{ij} = d_i \qquad \forall i \in V, i \neq j. \tag{5}$$

$$0 \leq g_{ij} \leq Q \times x_{ij}^{\pi} \forall i \in V, j \in V', i \neq j. \tag{6}$$

$$x_{ij}^{\pi} = \{0, 1\} \qquad \forall i \in V, j \in V', i \neq j. \tag{7}$$

Equation (1) is the objective function, minimizing the total path of the robot to deliver drugs; Constraint (2) indicates that there is a drug delivery robot in each ward to deliver medicine, and it only delivers once to meet the drug demand of the ward. Constraint (3) indicates that the delivery robot will not stay in the hospital room, and each delivery robot will eventually return to the pharmacy. Constraint (4) indicates that the number of robots for delivery cannot exceed the total number of robots in the hospital. Constraint (5) shows the quantity of drug change of the drug delivery robot during the delivery process, which is also the drug demand of Ward $i$. Constraint (6) indicates that the drug-carrying of robots in the delivery process cannot exceed the maximum capacity $Q$. Constraint (7) shows the value range of the variable.

# 3 Materials and Method

## 3.1 Sparrow Search Algorithm

SSA is proposed by imitating the sparrow group's foraging and anti-predation behavior. It has the advantages of less adjustable parameters, stronger search capacity, and faster efficiency. The algorithm and variants have remarkable performance in continuous optimization problems compared with other meta-heuristic algorithms [16]. The main procedures of basic SSA can be described as below:

**Step 1:** Initialization and produce the initial solution. Population size, the maximum number of iterations, proportion of producers, and sparrows who take charge of strongly vigilant are set in this step. The initial position of the sparrow population is displayed as follows. They are produced randomly.

$$X = \begin{bmatrix} x_{1,1} & \cdots & x_{1,d} \\ \vdots & \ddots & \vdots \\ x_{n,1} & \cdots & x_{n,d} \end{bmatrix}. \tag{8}$$

In the matrix, $n$ is the number of population and $d$ indicates the dimension of the decision parameters. The fitness is calculated for later operation. It will be stored as a $n \times 1$ matrix.

**Step 2:** Based on fitness, the whole group can be divided into producers and scroungers. Producers update their position by Eq. (9) and others use Eq. (10).

$$x_i^{g+1} = \begin{cases} x_i^g \cdot \exp\left(\frac{-i}{\alpha \cdot g_{max}}\right) & R < ST \\ x_i^g + Q \cdot L & R \geq ST \end{cases}. \tag{9}$$

$$x_i^{g+1} = \begin{cases} Q \cdot \exp\left(\frac{G_{worst} - x_i^g}{i^2}\right) & Q \cdot \exp\left(\frac{G_{worst} - x_i^g}{i^2}\right) \\ S_{best} + \left|x_i^g - S_{best}\right| \cdot A^+ \cdot L \; else \end{cases}. \tag{10}$$

Where $i = 1, 2, 3, \ldots, n$, $g$ is the current iteration. $\alpha$ is a value in the range of [0, 1]. $Q$ is a random number that follows a normal distribution. $L$ represents a $1 \times$ D identity matrix. $R$ and $ST$ are shown as alarm value and safety threshold respectively. The worst location of sparrows is $G_{worst}$ and the current best location of producers is $S_{best}$. $A$ is a matrix ($1 \times$ d) whose elements are randomly assigned 1 or −1. Then, $A^+ = A^T (A^+ A)^{-1}$.

**Step 3:** After the population updated position, some sparrows (10–20%) are selected as scouters who are responsible for detection and warning. Updating their location by Eq. (11).

$$x_i^{g+1} = \begin{cases} G_{best} + \eta \cdot \left|x_i^g - G_{best}\right| & f\left(x_i^g\right) > f(G_{best}) \\ x_i^g + K \cdot \left(\frac{\left|x_i^g - G_{worst}\right|}{\left(f\left(x_i^g\right) - f(G_{worst})\right) + \sigma}\right) & f\left(x_i^g\right) > f(G_{best}) \end{cases}. \tag{11}$$

Where $G_{best}$ is the current global optimal location, and $\eta$ represents a random number that obeys normal distribution $\left(E = 0,\ \sigma^2 = 1\right)$. $K$ is a random number between −1 and 1. $\sigma$ is the smallest constant to avoid the zero-division error. The fitness value of the present sparrow can be calculated by $f\left(x_i^g\right)$.

**Step 4:** Comparing every individual current position with the last iteration. If the new location is better than before, update it and save the best position.

**Step 5:** If the current iterations < maximum number of iterations, move on to Step 2; Otherwise, stop the algorithm and output the best solution.

### 3.2 Discrete Sparrow Search Algorithm (DSSA)

As introduced in the previous content, the basic SSA is intended to solve the continuous optimization problem and is unsuited for settling discrete optimization problems. The SSA has two main steps: (1) producers and scroungers update their position by corresponding rule; (2) randomly select some sparrows to update location in scouters' behavior during the iteration process. To improve the capacity of this algorithm in discrete optimization, we perform different improvements in the related process.

A. Solution representation and fitness
For convenience, we utilized the so-called path encoding for addressing the problem in this paper. A vector $X$ is used to represent the position of a sparrow. Each element $x_j$ in $X$ represents a node that the robot needs to visit. The rounding function can make them acceptable if emerge decimal during later iteration process. And the position is discretized using rounding.

The DSSA randomly generates a group of sparrows to establish an initial solution. The construction of every route is employing random numbers. Last, the fitness or objective function which we called is the total cost of all complete routes given in Eq. (1).

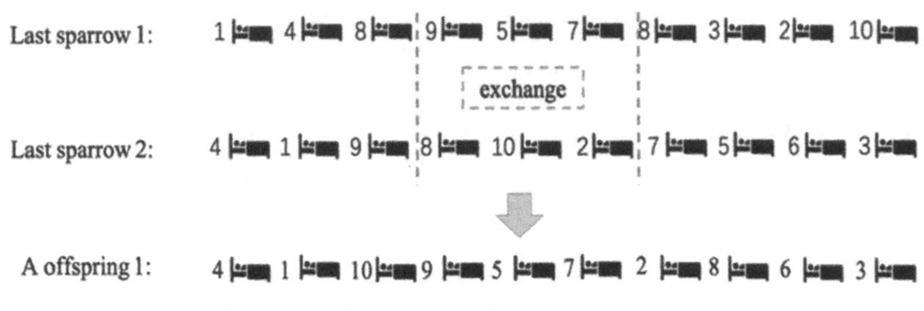


**Fig. 1.** The illustration of a crossover operator.

B. Producers and scroungers update their position
   During updating individual location, some policies should be adopted to ensure that every solution can be used for decoding. Meanwhile, the balance between the original principle and new items' features is also should be considered. Based on it, the genetic operator that is widely used in hybrid algorithms is associated with DSSA. The new solution location may gain meaningful experience from the last generation. Figure 1 shows a simple illustration of crossover operation.
C. *Scouters' Behavior*

Different from previous behavior, the scouters' actions will improve the global optimization power of the population. A local enhancement technique named local search is integrated into the scouters' behavior in DSSA. The first swap operator randomly selects two elements in all nodes and exchanges their position. The insertion selects and extracts one randomly chosen node from the entire route and re-insert again in a position by randomly selected. As its name implies, the existence of a reverse operation is to make a reversal of the position that has been chosen.

All the above and other not discussed details, the DSSA has been accomplished. Figure 2 shows the flowchart of DSSA and our reinforcement.

## 4 Experiment and Results Analysis

In this study, we used a PC with Intel i5-9600k at 3.7 GHz and 16 GB RAM on 64-bit Windows 10. The DSSA is implemented in MATLAB R2018b. Five standard instances have been adopted to test the performance of the DSSA.

All of the used instances can be downloaded in CVRPLIB (http://vrp.atd-lab.inf.puc-rio.br/index.php/en/). Each CVRP instance contains several nodes with X and Y

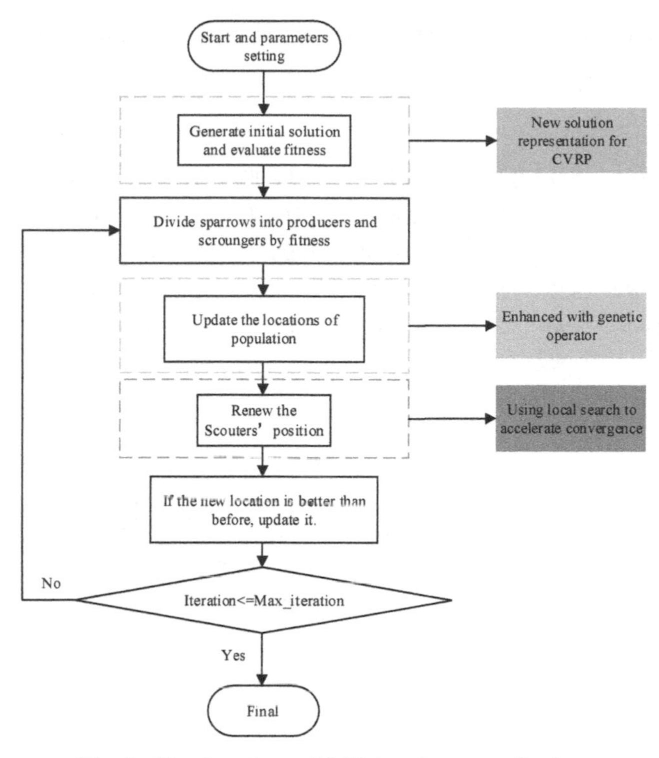


**Fig. 2.** The flowchart of DSSA and our contribution.

coordinate and theirs demand good is also included. All results obtained by the DSSA are from 10 independent trials.

The size of the population in the sparrow algorithm is 200. The maximum number of iterations is set to 100 and SD = 20%, respectively, and ST = 20%. The comparison is conducted based on the following publications:(1) Hybrid firefly [19], (2) PSO [20], (3) Adaptive ABC [21], (4) CS [22], (5) IDE [23].

Table 1 shows the results of DSSA compared with those methods. The first column indicates the instances' name from CVRPLIB, the second column stands for the best-known solution (BKS) from the library. The percentage deviation of the best solution length over -BKS is calculated as shown in Eq. (12).

$$PDBest(\%) = \frac{(Best - BKS)}{BKS} \times 100\%. \tag{12}$$

From the result, we can find that the DSSA has achieved superior results and perform certain advantages compared with matched groups. In total, $PDBest(\%)$ is no more than 3% for all these instances. The route of instance A-32n-k5 with distance 785 is shown in Fig. 3. With the increase in the problem scale, our approach still maintains the ability to find an optimal solution. Figure 4 shows the convergence curve when the DSSA algorithm solves the CVRP instance A-n33-k6, A-n36-k5. It's noticed that the proposed method has a high convergence speed during the initial stage.

**Table 1.** The results of DSSA in comparison with other algorithms for instances.

| Instance | *BKS* | [21] | PSO | [23] | CS | IDE | DSSA | |
|---|---|---|---|---|---|---|---|---|
| | | | | | | | *Best* | *PDBest*(%) |
| A-n32-k5 | *784* | 831 | 829 | 818.11 | 1065.40 | 876.25 | **785** | 0.1 |
| A-n33-k5 | *661* | 711 | 705 | 698.19 | 914.80 | 724.15 | **671** | 1.5 |
| A-n33-k6 | *742* | 783 | – | 751.94 | 1005.33 | 814.40 | **752** | 1.3 |
| A-n34-k5 | *778* | 827 | 832 | 814.55 | 1083.91 | 842.46 | **789** | 1.4 |
| A-n36-k5 | *799* | 870 | – | 823.04 | 1092.47 | 883.56 | **816** | 2.1 |

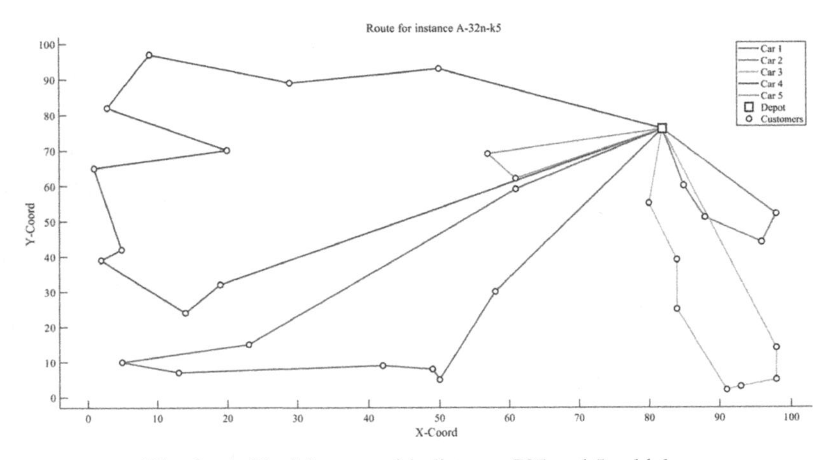


**Fig. 3.** A-32n-k5 route with distance 785 and 5 vehicles.

Furthermore, two classical meta-heuristic algorithms were also been used for evaluating the proposed method's convergence. The parameters of related algorithms are shown in Table 2. Np represents the population size. The maximum generation number and number of runs are chosen to be the same as the compared techniques. The maximum iterations of those algorithms are 100.

**Table 2.** Parameter settings for each related algorithm.

| Algorithm | Parameter |
|---|---|
| DSSA | Np = 200, SD = 20%, ST = 20% |
| GA | Np = 200, P-crossover = 0.9, P-mutation = 0.05 |
| ACO | Np = 200, Alpha = 1, Beta = 3, rho = 0.85, Q = 5 |

Figure 4 demonstrates the convergence curve from DSSA deal with instance A-n33-k6, A-n36-k5 compared with GA and ACO. The figure shows how quickly DSSA finds the optimal solution. The combination of basic SSA and genetic operator has a better performance than GA, ACO in the same iterations. It means the proposed algorithm can get a better solution with less time.

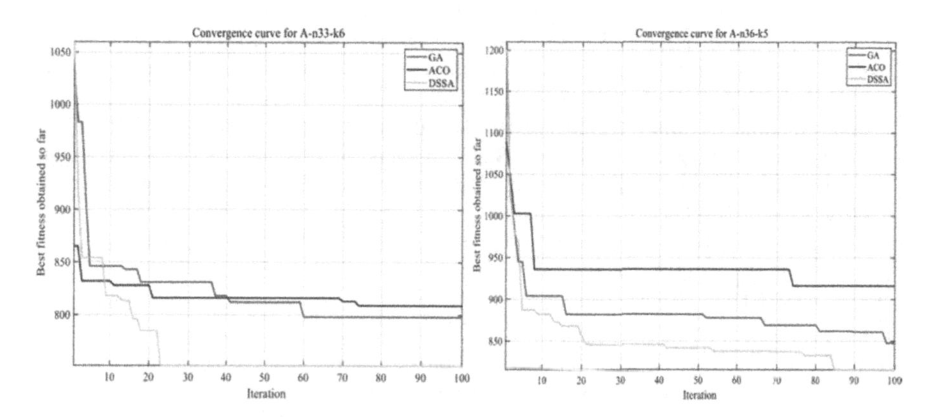


**Fig. 4.** Convergence curve of DSSA with two classic algorithms.

## 5 Conclusion

In this paper, a robot delivery routing problem for minimizing travel distance is considered. The task is to find a routing plan for robots to transport medical supplies in hospitals by minimizing the length of all routes.

We present a DSSA method that extends and improves SSA by discrete tactics and local search. We also evaluated the proposed method in comparison with some existing algorithms under various scenarios. Results between the DSSA proposed in this paper and other methods (Hybrid firefly, PSO, Adaptive ABC, etc.) indicates that our improved algorithm has an outstanding performance on the tested instances. In future work, the algorithm can be tested in a real hospital logistics scene. The DSSA could be improved with complex strategies to solve routing problems with other constraints, such as dynamic demand, time windows, and so on.

## References

1. Gao, A., Murphy, R.R., Chen, W., Dagnino, G., Yang, G.Z.: Progress in robotics for combating infectious diseases. J. Sci. Rob. **6**, 52 (2021)
2. Wang, X.V., Wang, L.: A literature survey of the robotic technologies during the COVID-19 pandemic. J. Manuf. Syst. **60**, 823–836 (2021)
3. Chiang, H., Malone, N., Lesser, K., Oishi, M., Tapia, L.: Path-guided artificial potential fields with stochastic reachable sets for motion planning in highly dynamic environments. In: 2015 IEEE International Conference on Robotics and Automation (ICRA), pp. 2347–2354. Seattle (2015)

4. Duchoň, F., et al.: Path planning with modified a star algorithm for a mobile robot. J. Procedia Eng. **96**, 59–69 (2014)
5. Jiang, C., Hu, Z., Mourelatos, Z.P., Gorsich, D., Majcher, M.: R2-RRT*: reliability-based robust mission planning of off-road autonomous ground vehicle under uncertain terrain environment. IEEE Trans. Autom. Sci. Eng. **99**, 1–17 (2021)
6. Karaboga, D.: An idea based on honey bee swarm for numerical optimization. Erciyes University, Engineering Faculty, Computer Engineering Department, Technical report-tr06 (2015)
7. Jain, S., Sharma, V., Kumar, S.: Robot path planning using differential evolution. In: Sharma, H., Govindan, K., Poonia, R.C., Kumar, S., El-Medany, W.M. (eds.) Advances in computing and intelligent systems. AIS, pp. 531–537. Springer, Singapore (2020). https://doi.org/10.1007/978-981-15-0222-4_50
8. Davoodi, M., Panahi, F., Mohades, A., Hashemi, S.N.: Clear and smooth path planning. J. Appl. Soft Comput. **32**, 568–579 (2015)
9. Wang, X., Shi, H., Zhang, C.: Path planning for intelligent parking system based on improved ant colony optimization. IEEE Access. **8**, 65267–65273 (2020)
10. Yang, X.-S.: A new metaheuristic bat-inspired algorithm. In: Gonz, J.R., Pelta, D.A., Cruz, C., Terrazas, G., Krasnogor, N. (eds.) Nature Inspired Cooperative Strategies for Optimization (NICSO 2010), pp. 65–74. Springer, Heidelberg (2010). https://doi.org/10.1007/978-3-642-12538-6_6
11. Gandomi, A.H., Yang, X.-S., Alavi, A.H.: Cuckoo search algorithm: a metaheuristic approach to solve structural optimization problems. J. Eng. Comput. **29** (1), 17–35 (2013)
12. Liang, Y., Xu, L.: Global path planning for mobile robot based genetic algorithm and modified simulated annealing algorithm. In: Proceedings of the first ACM/SIGEVO Summit on Genetic and Evolutionary Computation, pp. 303–308. ACM, Shanghai (2009)
13. Kala, R., Shukla, A., Tiwari, R.: Robotic path planning in static environment using hierarchical multi-neuron heuristic search and probability based fitness. J. Neurocomput. **74**(14–15), 2314–2335 (2011)
14. Ajeil, F.H., Ibraheem, I.K., Sahib, M.A., Humaidi, A.J.: Multi-objective path planning of an autonomous mobile robot using hybrid PSO-MFB optimization algorithm. J. Appl. Soft Comput. **89**, 106076 (2020)
15. Xue, J., Shen, B.: A novel swarm intelligence optimization approach: sparrow search algorithm. Syst. Sci. Control Eng. **8**(1), 22–34 (2020)
16. Zhang, C., Ding, S..: A stochastic configuration network based on chaotic sparrow search algorithm. Knowl. Based Syst. **220**, 106924 (2021)
17. Xing, Z., Yi, C., Lin, J., Zhou, Q.: Multi-component fault diagnosis of wheelset-bearing using shift-invariant impulsive dictionary matching pursuit and sparrow search algorithm. J. Measure. **109375** (2021)
18. Liu, B., Rodriguez, D.: Renewable energy systems optimization by a new multi-objective optimization technique: a residential building. J. Build. Eng. **35**, 102094 (2021)
19. Altabeeb, A.M., Mohsen, A.M., Ghallab, A.: An improved hybrid firefly algorithm for capacitated vehicle routing problem. Appl. Soft Comput. **84**, 1568–4946 (2019)
20. Bin, W., Wanliang, W., Yanwei, Z., Xinli, X., Fengyu, Y.: A novel real number encoding method of particle swarm optimization for vehicle routing problem. In: Sixth World Congress on Intelligent Control and Automation, pp. 3271–3275. IEEE, Dalian (2006)
21. Mingprasert, S., Masuchun, R.: Adaptive artificial bee colony algorithm for solving the capacitated vehicle routing problem. In: 9th International Conference on Knowledge and Smart Technology (KST), pp. 23–27. IEEE (2017)

22. Santillan, J.H., Tapucar, S., Manliguez, C., Calag, V.: Cuckoo search via Lévy flights for the capacitated vehicle routing problem. J. Ind. Eng. Int. **14**(2), 293–304 (2017). https://doi.org/10.1007/s40092-017-0227-5
23. Song, L., Dong, Y.: An improved differential evolution algorithm with local search for capacitated vehicle routing problem. In: Tenth International Conference on Advanced Computational Intelligence (ICACI), pp. 801–806. IEEE, Xiamen (2018)

# Trajectory Tracking Control for An Underactuated Unmanned Surface Vehicle Subject to External Disturbance and Error Constraints

Bo Kou[1], Yu-Long Wang[1(✉)], Zhao-Qing Liu[1], and Xian-Ming Zhang[2(✉)]

[1] School of Mechatronic Engineering and Automation, Shanghai University, Shanghai 200444, China
yulongwang@shu.edu.cn

[2] School of Software and Electrical Engineering, Swinburne University of Technology, Melbourne, VIC 3122, Australia
xianmingzhang@swin.edu.au

**Abstract.** This paper is concerned with trajectory tracking control for an underactuated unmanned surface vehicle (USV) subject to external disturbance and error constraints. Firstly, a disturbance observer is constructed to observe the unknown external disturbance induced by wind, waves, and currents. Secondly, in order to avoid the "explosion of complexity" problem caused by virtual control variable differentiation, a first order filter and a time-varying tan-type barrier Lyapunov function (BLF) are developed. Then, by introducing the backstepping and dynamic surface control techniques, feedback control laws ensuring trajectory tracking are designed such that the errors of position and heading angle between the unmanned surface vehicle and a virtual leader converge to a bounded region. Finally, the effectiveness of the designed control scheme is verified by a simulation example.

**Keywords:** Time-varying tan-type BLF · Disturbance observer · Underactuated USV · Trajectory tracking control

## 1 Introduction

As an effective equipment for the exploration and development of ocean resources, USVs are widely used in military and civil fields such as maritime battlefield reconnaissance, scientific experiments, salvage and lifesaving [1]. In recent years trajectory tracking control of USVs has attracted growing attention. However, for the trajectory tracking problem of USVs, there exists some significant challenges, which include

This work was supported in part by the National Key R & D Program of China (Grant No. 2018AAA0102804); the National Science Foundation of China (Grant Nos. 61873335, 61833011); the Natural Science Foundation of Shanghai Municipality, China under Grant No. 20ZR1420200; and the 111 Project, China under Grant No. D18003.


Q. Han et al. (Eds.): LSMS 2021/ICSEE 2021, CCIS 1469, pp. 704–713, 2021.
https://doi.org/10.1007/978-981-16-7213-2_68

1) in order to reduce manufacturing costs, most of the deployed and developing USVs are underactuated, while most of the existing methods about trajectory tracking of USVs are designed for fullactuated USVs, which means these methods cannot be applied directly to underactuated USVs;
2) due to the existence of external disturbance induced by wind, waves, and ocean currents, the tracking performance of USVs is inevitably affected;
3) when carrying out tasks such as target tracking or stalking, USVs are usually needed to track the target as soon as possible, which results in a high requirement to the tracking speed and accurace of USVs.

Since the unknown external disturbance can severely affect the trajectory tracking performance of USVs, thus how to compensate the negative influences induced by disturbance is practically valuable and attractive. In order to solve this problem, the disturbance must be accurately estimated first. In [2], a disturbance observer was developed to estimate the external disturbance. However, this developed disturbance observer could only estimate slow time-varying disturbance. By taking into account the coexistence of slow time-varying disturbance and non-slow time-varying disturbance, a more capable disturbance observer was presented in [3]. Moreover, based on the presented disturbance observer in [3], an efficient trajectory tracking control strategy and a high-precision dynamic positioning control strategy were proposed in [4] and [5], respectively. Therefore, constructing a disturbance observer is an effective way to estimate the unknown external disturbance, which gives the first motivation of this paper.

As mentioned above, the trajectory tracking control of USVs has attracted growing attention. Accordingly, a number of notable results have been proposed in the literature. To name a few, under backstepping control methods, a trajectory tracking control strategy was proposed in [6] with double closed-loop structure. However, the introduction of backstepping technique will increase computation and decrease convergence speed. In order to solve these shortcomings, an adaptive dynamic surface control (DSC) method was presented in [7]. Moreover, this DSC method can effectively used to reduce the complexity of the controllers designed. Based on DSC technique, an adaptive updating law, which could effectively improve the computational efficiency of trajectory tracking control algorithms, was proposed in [8]. On the basis of [8], a adaptive control method was proposed in [9] by taking "explosion of complexity" problem into consideration. Consider the advantages of backstepping and DSC techniques synthetically, how to propose a rapid and effective trajectory tracking control scheme for USVs is important, which gives the second and main motivation of this paper.

In addition, by constraining state errors during the process of USV trajectory tracking controller design, the transient and steady-state performance of control system can be significantly improved. An effective way to constrain these errors is to construct a barrier Lyapunov function (BLF). For example, by maintaining the boundedness of a BLF in the closed loop system, the state constraint was implemented in [10]. Motivated by [10], an asymmetric BLF was proposed in [11]. To solve the problem of multiple output constraints in the tracking control, a symmetric BLF was established in [12]. Compared with traditional USV control schemes, constructing a time-varying tan-type BLF to constrain the position errors is more practically valuable and attractive. Thus, how to develop an appropriate error constrain function is the third motivation of this paper.

This paper aims to solve the trajectory tracking problem of underactuated USVs. The main contributions of this paper are highlighted as follows:

1) In order to avoid the "explosion of complexity" problem induced by virtual control variable differentiation, a first order filter and a time-varying tan-type BLF are constructed to reduce the complexity of calculation.
2) An appropriate trajectory tracking control scheme, which can guarantee tracking errors exponentially converge to a bounded region, are proposed for underactuated USV systems.

## 2 Problem Formulation and Preliminaries

### 2.1 Modeling for Underactuated USVs

Motivated by [13], the three degrees of freedom (surge-sway-yaw) dynamic model of an underactuated USV can be expressed as

$$\begin{cases} \dot{x} = u\cos\varphi - v\sin\varphi \\ \dot{y} = u\sin\varphi + v\cos\varphi \\ \dot{\varphi} = r \end{cases} \quad \begin{cases} \dot{u} = f_u(\nu) + \frac{1}{m_u}\tau_u + \frac{1}{m_u}\tau_{wu} \\ \dot{v} = f_v(\nu) + \frac{1}{m_v}\tau_{wv} \\ \dot{r} = f_r(\nu) + \frac{1}{m_r}\tau_r + \frac{1}{m_r}\tau_{wr} \end{cases} \tag{1}$$

with

$$\begin{cases} f_u(\nu) = \frac{m_v}{m_u}vr - \frac{d_u}{m_u}u - \frac{d_{u2}}{m_u}|u|\,u - \frac{d_{u3}}{m_u}u^3 \\ f_v(\nu) = -\frac{m_u}{m_v}ur - \frac{d_v}{m_v}v - \frac{d_{v2}}{m_v}|v|\,v - \frac{d_{v3}}{m_v}v^3 \\ f_r(\nu) = \frac{(m_u - m_v)}{m_r}uv - \frac{d_r}{m_r}r - \frac{d_{r2}}{m_r}|r|\,r - \frac{d_{r3}}{m_r}r^3 \end{cases} \tag{2}$$

All the parameters in (1) and (2) can be found in [13].

The reference trajectory of the USV is generated by a virtual leader with the dynamic form as $\eta_d = [x_d,\ y_d,\ \varphi_d]^T$. Without loss of generality, the following assumption is introduced. It is assumed that the reference trajectory $\eta_d$ and the external disturbance $\tau_w$ are bounded.

### 2.2 Error Constraint Function

By improving the constant BLF proposed in [10], a time-varying tan-type BLF is developed in this paper, which can be expressed as $V_b = \frac{k_b^2}{\pi}\tan\left(\frac{\pi z_1^T z_1}{2k_b^2}\right)$, where $k_b$ is a time-varying upper bound function of $|z_1|$ with $|z_1(0)| \le k_b(0)$.

*Remark 1.* Note that the initial value of $\|z_1\|$ satisfies $\|z_1(0)\| \le k_b(0)$ and $V_b$ is bounded. If there exist $\lim_{\|z_1\| \to k_b} \frac{k_b^2}{\pi}\tan\left(\frac{\pi z_1^T z_1}{2k_b^2}\right) = \infty$, one can conclude that $|z_1|$ will not exceed $k_b$. Moreover, If $\|z_1\|$ is not required to be constrained. Then the following equation $\lim_{k_b \to \infty} \frac{k_b^2}{\pi}\tan\left(\frac{\pi z_1^T z_1}{2k_b^2}\right) = \frac{1}{2}z_1^T z_1$ is obtained. It means that when there is no constraint function or the constraint function is infinite, the time-varying BLF will degrade to the commonly form, i.e., $V_b = \frac{1}{2}z_1^T z_1$.

## 3 Design of Control Scheme

In order to solve the trajectory tracking problem of the underactuated USV, an appropriate trajectory tracking control scheme is proposed in this section by taking into account error constraints.

The position and heading angle errors in the earth-fixed frame are defined as $x_e$, $y_e$, and $\varphi_e$. Meanwhile, these errors in the body-fixed frame are defined as $e_x$, $e_y$, and $e_\varphi$. The conversion relationship between the errors in two different frames is:

$$\begin{bmatrix} e_x \\ e_y \\ e_\varphi \end{bmatrix} = \begin{bmatrix} \cos\varphi & \sin\varphi & 0 \\ -\sin\varphi & \cos\varphi & 0 \\ 0 & 0 & 1 \end{bmatrix} \begin{bmatrix} x_e \\ y_e \\ \varphi_e \end{bmatrix} \tag{3}$$

Taking the time derivative of (3), one has

$$\begin{bmatrix} \dot{e}_x \\ \dot{e}_y \\ \dot{e}_\varphi \end{bmatrix} = \begin{bmatrix} u + re_y - v_p \cos e_\varphi \\ v - re_x + v_p \sin e_\varphi \\ r - \dot{\varphi}_d \end{bmatrix} \tag{4}$$

where $v_p = \sqrt{\dot{x}_d^2 + \dot{y}_d^2}$ and $\varphi_d = \arctan(\frac{\dot{y}_d}{\dot{x}_d})$.

To guarantee the convergence of position errors $e_x$ and $e_y$, a time-varying tan-type BLF function is constructed as follows:

$$V_1 = \frac{k_{b1}^2}{\pi} \tan\left(\frac{\pi e_x^2}{2k_{b1}^2}\right) + \frac{k_{b2}^2}{\pi} \tan\left(\frac{\pi e_y^2}{2k_{b2}^2}\right) \tag{5}$$

where $k_{b1}$ and $k_{b2}$ are time-varying parameters to be designed later.

Taking the time derivative of $V_1$ in (5), we have

$$\begin{aligned} \dot{V}_1 = & \frac{2k_{b1}\dot{k}_{b1}}{\pi} \tan\left(\frac{\pi e_x^2}{2k_{b1}^2}\right) + \frac{2k_{b2}\dot{k}_{b2}}{\pi} \tan\left(\frac{\pi e_y^2}{2k_{b2}^2}\right) - \frac{\dot{k}_{b1}}{k_{b1}} \frac{e_x^2}{\cos^2\left(\frac{\pi e_x^2}{2k_{b1}^2}\right)} \\ & - \frac{\dot{k}_{b2}}{k_{b2}} \frac{e_y^2}{\cos^2\left(\frac{\pi e_y^2}{2k_{b2}^2}\right)} + \frac{e_x \dot{e}_x}{\cos^2\left(\frac{\pi e_x^2}{2k_{b1}^2}\right)} + \frac{e_y \dot{e}_y}{\cos^2\left(\frac{\pi e_y^2}{2k_{b2}^2}\right)} \end{aligned} \tag{6}$$

Let $l_1 = \frac{1}{\cos^2\left(\frac{\pi e_x^2}{2k_{b1}^2}\right)}$, $l_2 = \frac{1}{\cos^2\left(\frac{\pi e_y^2}{2k_{b2}^2}\right)}$, $k_{11} = \sup_t \sqrt{(\frac{\dot{k}_{b1}}{k_{b1}})^2 + \sigma_1}, k_{22} = \sup_t \sqrt{(\frac{\dot{k}_{b2}}{k_{b2}})^2 + \sigma_2}$, where $\sigma_1 > 0$ and $\sigma_2 > 0$ are two predefined scalar. Substituting (4) into (6) yields

$$\begin{aligned} \dot{V}_1 < & 2k_{11} \frac{k_{b1}^2}{\pi} \tan\left(\frac{\pi e_x^2}{2k_{b1}^2}\right) + 2k_{22} \frac{k_{b2}^2}{\pi} \tan\left(\frac{\pi e_y^2}{2k_{b2}^2}\right) + k_{11} l_1 e_x^2 + k_{22} l_2 e_y^2 \\ & + l_1 e_x (u + re_y - v_p \cos e_\varphi) + l_2 e_y (v - re_x + v_p \sin e_\varphi) \end{aligned} \tag{7}$$

To guarantee the convergence of yaw velocity error, an auxiliary variable $h = v_p \sin\varphi_e$ is introduced. Meanwhile, a virtual control laws are designed as:

$$\begin{aligned} u_\kappa &= -(2k_{11}+k_1)\tfrac{k_{b1}^2}{\pi}\tfrac{1}{e_x}\sin\left(\tfrac{\pi e_x^2}{2k_{b1}^2}\right)\cos\left(\tfrac{\pi e_x^2}{2k_{b1}^2}\right) - k_{11}e_x - re_y + v_p\cos e_\varphi \\ h_\kappa &= -(2k_{22}+k_2)\tfrac{k_{b2}^2}{\pi}\tfrac{1}{e_y}\sin\left(\tfrac{\pi e_y^2}{2k_{b2}^2}\right)\cos\left(\tfrac{\pi e_y^2}{2k_{b2}^2}\right) - k_{22}e_y - v + re_y \end{aligned} \tag{8}$$

where $u_k$ and $h_k$ are the virtual control signals of $u$ and $h$, respectively; $k_1$, $k_2$, $k_{11}$, and $k_{22}$ are positive parameters to be designed later; $k_{b1}$ and $k_{b2}$ are the time-varying bound values of $\|e_x\|$ and $\|e_y\|$, respectively. To avoid the derivative of the virtual control signals $u_k$ and $h_k$, the DSC technique is adopted in the following analysis.

Let $u_k$ and $h_k$ pass through two first-order low-pass filters with the filter time constants $\lambda_1$ and $\lambda_2$ to obtain $u_f$ and $h_f$, respectively. There are

$$\begin{cases} \lambda_1 u_f + u_f = u_\kappa \\ u_f(0) = u_\kappa(0) \end{cases} \quad \begin{cases} \lambda_2 h_f + h_f = h_\kappa \\ h_f(0) = h_\kappa(0) \end{cases} \tag{9}$$

Then, define the following new error functions:

$$z_u = u_f - u_\kappa,\ e_u = u - u_f,\ z_h = h_f - h_\kappa,\ e_h = h - h_f \tag{10}$$

Taking the time derivative of $e_h$ in (10), one has

$$\dot{e}_h = \dot{v}_p \sin e_\varphi + v_p \cos e_\varphi (r - \dot{\varphi}_d) + \frac{1}{\lambda_2} z_h \tag{11}$$

Obviously, (7) is converted to

$$\dot{V}_1 < -k_1\frac{k_{b1}^2}{\pi}\tan\left(\frac{\pi e_x^2}{2k_{b1}^2}\right) - k_2\frac{k_{b2}^2}{\pi}\tan\left(\frac{\pi e_y^2}{2k_{b2}^2}\right) + l_1 e_x(e_u+z_u) + l_2 e_y(e_h+z_h) \tag{12}$$

Let $T = -k_1\frac{k_{b1}^2}{\pi}\tan\left(\frac{\pi e_x^2}{2k_{b1}^2}\right) - k_2\frac{k_{b2}^2}{\pi}\tan\left(\frac{\pi e_y^2}{2k_{b2}^2}\right)$, then (12) can be rewritten as

$$\dot{V}_1 < T + l_1 e_x(e_u + z_u) + l_2 e_y(e_h + z_h) \tag{13}$$

Consider the following Lyapunov function candidate:

$$V_2 = V_1 + \frac{1}{2}e_h^2 + \frac{1}{2}e_\varphi^2 \tag{14}$$

Taking the time derivative of (14), one obtains

$$\dot{V}_2 = \dot{V}_1 + e_h(\dot{v}_p \sin e_\varphi + v_p\cos e_\varphi(r - \dot{\varphi}_d) + \tfrac{1}{\lambda_2}z_h) + e_\varphi(r - \dot{\varphi}_d) \tag{15}$$

Let $r_k$ be the virtual control signal of yaw velocity $r$. Then, design the following virtual control law:

$$r_\kappa = k_3(e_\varphi + e_h v_p \cos e_\varphi) + \dot{\varphi}_d \tag{16}$$

Similar to (9), let $r_k$ pass through a first-order low-pass filter with the filter time constant $\lambda_3$ to obtain $r_f$, that is

$$\lambda_3 r_f + r_f = r_\kappa,\ r_f(0) = r_\kappa(0) \tag{17}$$

Similarly, construct the following new error functions:

$$z_r = r_f - r_\kappa,\ e_r = r - r_f \tag{18}$$

Substituting (16), (17), and (18) into (15) yields

$$\begin{aligned}\dot{V}_2 < T - k_3(e_\varphi + e_h v_p \cos e_\varphi)^2 + l_1 e_x(e_u + z_u) + l_2 e_y(e_h + z_h) \\ + e_h(\dot{v}_p \sin e_\varphi + \tfrac{1}{\lambda_2} z_h) + (e_r + z_r)(e_\varphi + e_h v_p \cos e_\varphi)\end{aligned} \tag{19}$$

According to (10) and (18), one has

$$\dot{e}_u = f_u(\nu) + \frac{1}{m_u}\tau_u + \frac{1}{m_u}\tau_{wu} + \frac{1}{\lambda_1} z_u,\ \dot{e}_r = f_r(\nu) + \frac{1}{m_r}\tau_r + \frac{1}{m_r}\tau_{wr} + \frac{1}{\lambda_3} z_r \tag{20}$$

Consider the following Lyapunov function candidate:

$$V_3 = V_2 + \frac{1}{2} m_u e_u^2 + \frac{1}{2} m_r e_r^2 \tag{21}$$

Taking the time derivative of (21), we have

$$\dot{V}_3 = \dot{V}_2 + e_u(m_u f_u(\nu) + \tau_u + \tau_{wu} + \tfrac{m}{u}\lambda_1 z_u) + e_r(m_r f_r(\nu) + \tau_r + \tau_{wr} + \tfrac{m_r}{\lambda_3} z_r) \tag{22}$$

Substituting (20) into (22) yields

$$\begin{aligned}\dot{V}_3 < T - k_3(e_\varphi + e_h v_p \cos e_\varphi)^2 + e_u(l_1 e_x + m_u f_u(\nu) + \tau_u + \tau_{wu} + \tfrac{m_u}{\lambda_1} z_u) \\ + e_r(e_\varphi + m_r f_r(\nu) + \tau_r + \tau_{wr} + \tfrac{m_r}{\lambda_3} z_r) + l_1 e_x z_u + l_2 e_y z_h \\ + e_h(l_2 e_y + \dot{v}_p \sin e_\varphi + \tfrac{1}{\lambda_2} z_h) + z_r(e_\varphi + e_h v_p \cos e_\varphi)\end{aligned} \tag{23}$$

Then, the actual feedback control laws $\tau_u$ and $\tau_r$ can be designed as

$$\tau_u = -k_4 e_u - l_1 e_x - m_u f_u(\nu) - \tau_{wu} - \frac{m_u}{\lambda_1} z_u \tag{24}$$

$$\tau_r = -k_5 e_r - e_\varphi - m_r f_r(\nu) - \tau_{wr} - \frac{m_r}{\lambda_3} z_r \tag{25}$$

Since the unmeasurable disturbance $\tau_{wu}$ and $\tau_{wr}$ in (24) and (25) are unavailable for the control law design. To compensate the negative influences induced by external disturbance, the following disturbance observers are established

$$\hat{\tau}_{wu} = \beta_1 + k_{wu} m_u u,\ \dot{\beta}_1 = -k_{wu}\beta_1 - k_{wu}(m_u f_v(\nu) + \tau_u + k_{wu} m_u u) \tag{26}$$

$$\hat{\tau}_{wr} = \beta_2 + k_{wr} m_r r,\ \dot{\beta}_2 = -k_{wr}\beta_2 - k_{wr}(m_r f_r(\nu) + \tau_r + k_{wr} m_r r) \tag{27}$$

where $\beta_1$ and $\beta_2$ are the observers state; $k_{wu}$ and $k_{wr}$ are positive parameters to be designed later; $\hat{\tau}_{wu}$ and $\hat{\tau}_{wr}$ are the estimated values of $\tau_{wu}$ and $\tau_{wr}$, respectively; Define the observer errors as $\tilde{\tau}_{wu} = \hat{\tau}_{wu} - \tau_{wu}$ and $\tilde{\tau}_{wr} = \hat{\tau}_{wr} - \tau_{wr}$. Then, (24) and (25) can be rewritten as

$$\tau_u = -k_4 e_u - l_1 e_x - m_u f_u(\nu) - \hat{\tau}_{wu} - \frac{m_u}{\lambda_1} z_u \tag{28}$$

$$\tau_r = -k_5 e_r - e_\varphi - m_r f_r(\nu) - \hat{\tau}_{wr} - \frac{m_r}{\lambda_3} z_r \tag{29}$$

Consider the following Lyapunov function candidate:

$$V_4 = V_3 + \frac{1}{2}\tilde{\tau}_{wu}^2 + \frac{1}{2}\tilde{\tau}_{wr}^2 \tag{30}$$

whose derivative is

$$\dot{V}_4 = \dot{V}_3 + \tilde{\tau}_{wu}(\dot{\tau}_{wu} - k_{wu}\tilde{\tau}_{wu}) + \tilde{\tau}_{wr}(\dot{\tau}_{wr} - k_{wr}\tilde{\tau}_{wr}) \tag{31}$$

Substituting (28) and (29) into (30) yields

$$\dot{V}_4 < T - k_3(e_\varphi + e_h v_p \cos e_\varphi)^2 - k_4 e_u^2 - k_5 e_r^2 - k_{wu}\tilde{\tau}_{wu}^2 - k_{wr}\tilde{\tau}_{wr}^2 + \delta \tag{32}$$

where $\delta = l_1 e_x z_u + l_2 e_y z_h + e_h(l_2 e_y + \dot{v}_p \sin e_\varphi + \frac{1}{\lambda_2} z_h) + z_r(e_\varphi + e_h v_p \cos e_\varphi) + e_u \tilde{\tau}_{wu} + e_r \tilde{\tau}_{wr} - \tilde{\tau}_{wu}\dot{\tau}_{wu} - \tilde{\tau}_{wu}\dot{\tau}_{wu}$.

## 4 Stability Analysis

In this section, we analyze the stability of the designed trajectory tracking control system. Consider a Lyapunov function candidate as

$$V_5 = V_4 + \frac{1}{2}(z_u^2 + z_h^2 + z_r^2) \tag{33}$$

whose derivative is

$$\dot{V}_5 = \dot{V}_4 - \tfrac{1}{\lambda_1} z_u^2 - \tfrac{1}{\lambda_2} z_h^2 - \tfrac{1}{\lambda_3} z_r^2 - z_u \Psi_1 - z_h \Psi_2 - z_r \Psi_3 \tag{34}$$

where $\dot{z}_u = \dot{u}_f - \dot{u}_\kappa = -\frac{1}{\gamma_1} z_u - \Psi_1$, $\dot{z}_h = \dot{h}_f - \dot{h}_\kappa = -\frac{1}{\lambda_2} z_h - \Psi_2$, $\dot{z}_r = \dot{r}_f - \dot{r}_\kappa = -\frac{1}{\gamma_3} z_r - \Psi_3$.

It is assumed that $\Psi_1$, $\Psi_2$, and $\Psi_3$ are continuous functions with maximum $\bar{\Psi}_1$, $\bar{\Psi}_2$, and $\bar{\Psi}_3$. Then, substituting (32) into (34) yields

$$\begin{aligned}\dot{V}_5 < T &- k_3 e_\varphi^2 - k_3(v_p \cos e_\varphi)^2 e_h^2 - k_4 e_u^2 - k_5 e_r^2 - 2k_3 e_h e_\varphi v_p \cos e_\varphi \\ &- k_{wu}\tilde{\tau}_{wu}^2 - k_{wr}\tilde{\tau}_{wr}^2 - \tfrac{1}{\lambda_1} z_u^2 - \tfrac{1}{\lambda_2} z_h^2 - \tfrac{1}{\lambda_3} z_r^2 - z_u \Psi_1 - z_h \Psi_2 - z_r \Psi_3 + \delta\end{aligned} \tag{35}$$

It is assumed that $\bar{\gamma}$, $\bar{\tau}_{wu}$, $\bar{\tau}_{wr}$, $\bar{k}_{b1}$ and $\bar{k}_{b2}$ are continuous functions with maximum $\gamma$, $\dot{\tau}_{wu}$, $\dot{\tau}_{wr}$, $e_x$ and $e_y$, where $\gamma = \dot{v}_p \sin e_\varphi + \frac{z_h}{\lambda_2}$. According to Young's inequality, by expanding $\delta$, we have

$$\begin{aligned}\delta \le\ & \tfrac{l_1}{2} z_u^2 + \tfrac{l_1}{2}\bar{k}_{b1}^2 + \tfrac{l_2}{2} z_h^2 + \tfrac{l_2}{2}\bar{k}_{b2}^2 + \tfrac{l_2}{2} e_h^2 + \tfrac{l_2}{2}\bar{k}_{b2}^2 + \tfrac{v_p}{2} e_h^2 + \tfrac{v_p}{2} z_r^2 + \tfrac{1}{2} e_\varphi^2 + \tfrac{1}{2} z_r^2 \\ &+ \tfrac{1}{2} e_h^2 + \tfrac{1}{2}\bar{\gamma}^2 + \tfrac{1}{2} e_u^2 + \tfrac{1}{2}\tilde{\tau}_{wu}^2 + \tfrac{1}{2} e_r^2 + \tfrac{1}{2}\tilde{\tau}_{wr}^2 + \tfrac{1}{2}\tilde{\tau}_{wu}^2 + \tfrac{1}{2}\bar{\tau}_{wu}^2 + \tfrac{1}{2}\tilde{\tau}_{wr}^2 + \tfrac{1}{2}\bar{\tau}_{wr}^2\end{aligned} \tag{36}$$

Then, by sorting out the above equation, one can have

$$\begin{aligned} \dot{V}_5 < & -a_1 \frac{k_{b1}^2}{\pi} \tan\left(\frac{\pi e_x^2}{2k_{b1}^2}\right) - a_2 \frac{k_{b2}^2}{\pi} \tan\left(\frac{\pi e_y^2}{2k_{b2}^2}\right) - a_3 e_\varphi^2 \\ & - a_4 e_u^2 - a_5 e_r^2 - a_6 e_h^2 - a_7 z_u^2 - a_8 z_h^2 - a_9 z_r^2 - a_{10} \tilde{\tau}_{wu}^2 - a_{11} \tilde{\tau}_{wr}^2 + \Delta \end{aligned} \tag{37}$$

where $a_1 = k_1$, $a_2 = k_2$, $a_3 = k_3 - k_3 v_p - \frac{1}{2}$, $a_4 = k_4 - \frac{1}{2}$, $a_5 = k_5 - \frac{1}{2}$, $a_6 = k_3 v_p - \frac{l_2}{2} - \frac{1}{2}$, $a_7 = \frac{1}{\lambda_1} - \frac{l_1}{2} - \frac{1}{2}$, $a_8 = \frac{1}{\lambda_2} - \frac{l_2}{2} - \frac{1}{2}$, $a_9 = \frac{1}{\lambda_3} - \frac{v_p}{2} - 1$, $a_{10} = k_{wu} - 1$, $a_{11} = k_{wr} - 1$, and $\Delta = \frac{1}{2}\bar{\Psi}_1^2 + \frac{1}{2}\bar{\Psi}_2^2 + \frac{1}{2}\bar{\Psi}_3^2 + \frac{1}{2}\bar{\gamma}^2 + \frac{l_1}{2}\bar{k}_{b1}^2 + l_2 \bar{k}_{b2}^2 + \frac{1}{2}\bar{\tau}_{wu}^2 + \frac{1}{2}\bar{\tau}_{wr}^2$.

**Theorem 1.** *Consider underactuated USV (1) satisfying Assumption 1, control laws $\tau_u$ in (28) and $\tau_r$ in (29), disturbance observers (26) and (27). If given any $\varsigma$ for all initial conditions satisfying $V(0) \leq \varsigma$ with the preselected error constrains satisfying $\|e_x\| \leq k_{b1}$ and $\|e_y\| \leq k_{b2}$, and the design parameters satisfying $a_i > 0$. Then we have the observer errors $\tilde{\tau}_{wu}$ and $\tilde{\tau}_{wu}$ converge to an adjustable neighborhood of the origin by appropriately choosing the design parameters.*

*Proof.* Let $a = \min\{a_1, a_2 ... a_{11}\}$, then (37) can be simplified to $\dot{V}_5 \leq -aV_5 + \Delta$. Solving the above inequality, we have

$$V_5 \leq \left(V_5(0) - \frac{\Delta}{a}\right) e^{-at} + \frac{\Delta}{a}, \forall t > 0 \tag{38}$$

Based on the above analysis, by adjusting parameters $k_{11}$, $k_{22}$, $k_{b1}$, $k_{b2}$, $k_{wu}$, $k_{wr}$ and $k_j (j = 1, 2, ...5)$, we can obtain a larger $a$. When satisfied $a > \frac{\Delta}{V_5}$, we have $\dot{V}_5 < 0$. Hence, it is observed from (38) that $V_4$ is eventually bounded by $\frac{\Delta}{a}$. $e_x$, $e_y$, $e_\varphi$, $e_u$, $e_h$ and $e_r$ are globally uniformly bounded. Theorem 1 is proved.

## 5 Simulation Results and Discussion

In this section, the effectiveness of the proposed tracking control scheme is verified by a simulation example.

The model parameters in (1) are given as $m_u = 1.956$, $m_v = 2.045$, $m_r = 0.043$, $d_u = 0.0358$, $d_{u2} = 0.0179$, $d_{u3} = 0.0089$, $d_v = 0.1183$, $d_{v2} = 0.0591$, $d_{v3} = 0.0295$, $d_r = 0.0308$, $d_{r2} = 0.0154$, and $d_{r3} = 0.0077$. The design parameters are selected as $k_1 = k_2 = 1$, $k_3 = 3$, $k_4 = 5$, $k_5 = 6$, $k_{wu} = 2$, $k_{wr} = 2$ and $\lambda_1 = \lambda_2 = \lambda_3 = 0.1$. $k_{b1}$ and $k_{b2}$ are constructed as $k_{b1} = 0.43 + 6.35e^{-t}$, $k_{b2} = 0.43 + 3.55e^{-t}$. In addition, the external disturbance are given by $\tau_{wu} = 1 + 0.1sin(0.2t) + 0.2cos(0.5t)$, $\tau_{wr} = 1 + 0.2sin(0.1t) + 0.1cos(0.2t)$. The initial states of the USV are $[x, y, \varphi]^T = [0m, 5m, 0rad]^T$. The initial states of the virtual leader are $[x_d, y_d, \varphi_d]^T = [0m, 0m, 0rad]^T$. The surge reference velocity $u_d$ is chosen as $u_d = 2m/s$, while the yaw reference velocity $r_d$ is designed as $r_d = \frac{\pi}{160} rad/s$ in the first $40s$, $r_d = \frac{\pi}{165} rad/s$ for $40s \leq t < 80s$, $r_d = \frac{\pi}{170} rad/s$ for $80s \leq t < 130s$, and $r_d = 0 rad/s$ for $130s \leq t \leq 200s$. The observed and actual external disturbance are shown in Fig. 1.

The real-time tracking performance is shown in Fig. 2. The tracking errors of position and heading angle are depicted in Fig. 3. Meanwhile, the tracking errors

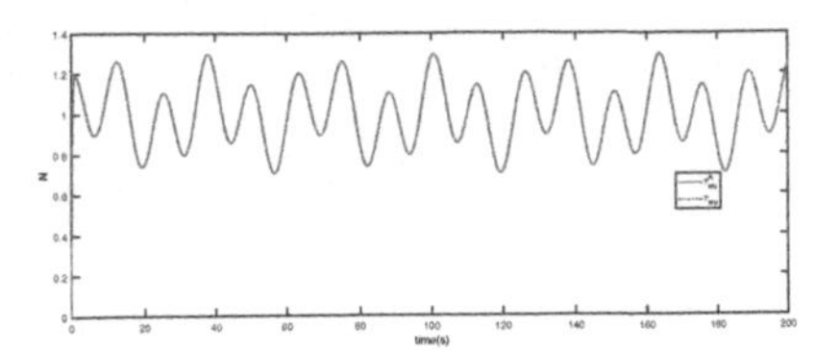

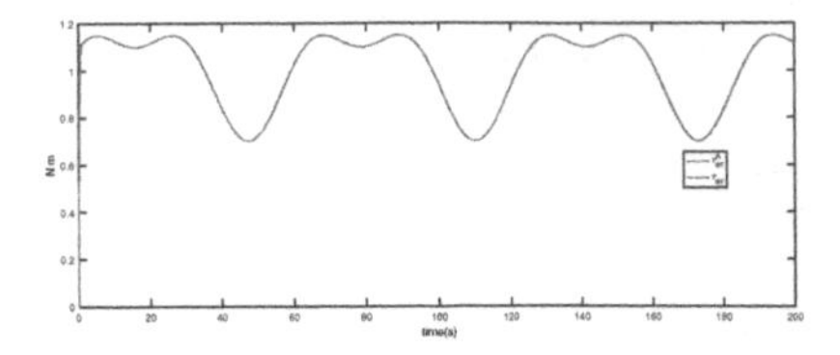

**Fig. 1.** Estimated and actual disturbance.

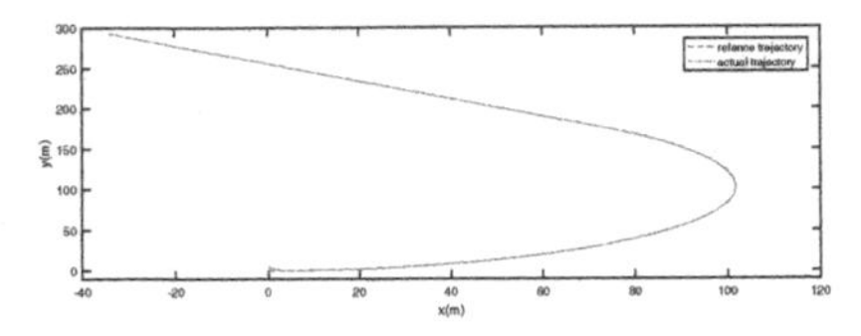

**Fig. 2.** Trajectory tracking performance.

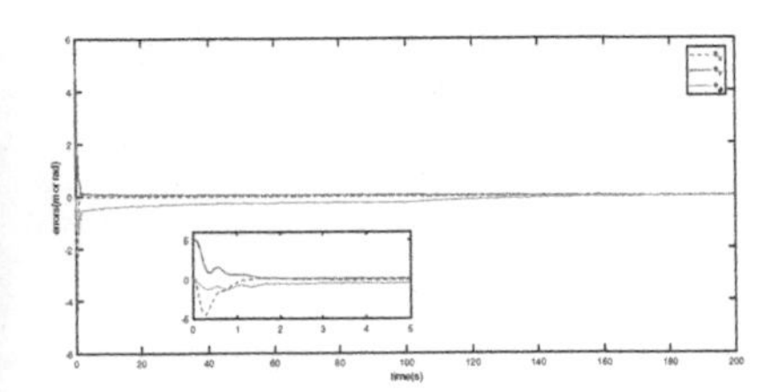

**Fig. 3.** Tracking errors of position and heading angle.

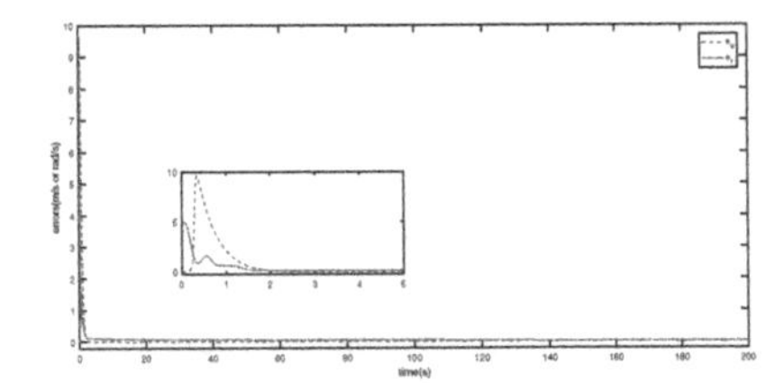

**Fig. 4.** Tracking errors of velocities.

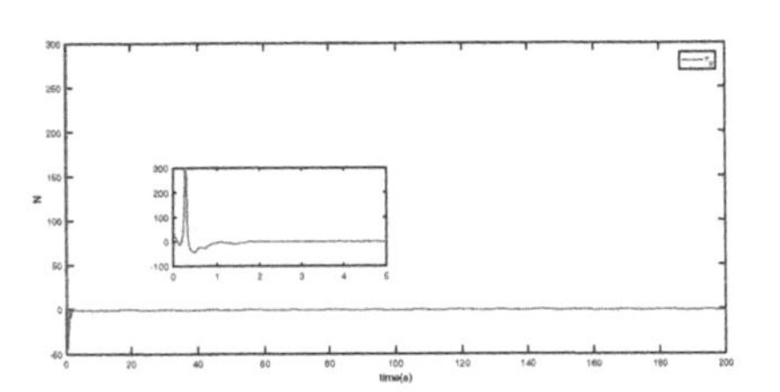

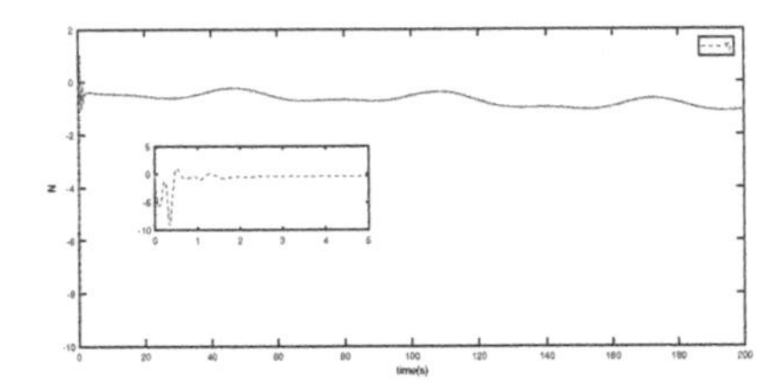

**Fig. 5.** Control inputs of the USV.

of velocities are shown in Fig. 4. The control inputs of the USV are depicted in Fig. 5. From Fig. 3 and Fig. 4, it can be seen that the proposed trajectory tracking control scheme can guarantee satisfying performance for the underactuated USV. Since the heading angle changes constantly during the process of trajectory tracking, thus the control input $\tau_r$ fluctuates in a bounded region. The above simulation results prove the effectiveness of the proposed scheme.

## 6 Conclusion

In this paper, a trajectory tracking control scheme for underactuated USVs has been proposed. By constructing a disturbance observer, the external disturbance has been estimated effectively. To avoid the problem of "explosion of complexity" induced by the derivation of the virtual control variables, a first order filter and a time-varying tan-type BLF have been established. By introducing the backstepping and dynamic surface control techniques, the trajectory tracking control laws have been designed. Under the designed control laws, the trajectory tracking of underactuated USVs has been achieved. Finally, a simulation example has been presented to verify the applicability of the results derived in this paper.

## References


1. Pei, Z.Y., Dai, Y.S., Li, Z.G., Jin, J.C., Shao, F.: Overview of unmanned surface vehicle motion control methods. Mar. Sci. **44**(03), 153–162 (2020)
2. Zhang, Y., Hua, C.C., Li, K.: Disturbance observer-based fixed-time prescribed performance tracking control for robotic manipulator. Int. J. Syst. Sci. **50**(13), 2437–2448 (2019)
3. Van, M.: Adaptive neural integral sliding-mode control for tracking control of fully actuated uncertain surface vessels. Int. J. Robust Nonlinear Control **29**(5), 1537–1557 (2019)
4. Mu, D.D., Wang, G.F., Fan, Y.S., Zhao, Y.: Modeling and identification of podded propulsion unmanned surface vehicle and its course control research. Math. Probl. Eng. 1–13 (2017)
5. Zhao, Y., Mu, D.D., Wang, G.F., Fan, Y.S.: Trajectory tracking control for unmanned surface vehicle subject to unmeasurable disturbance and input saturation. IEEE Access **8**, 191278–191285 (2020)
6. Li, J., Guo, H., Zhang, H.: Double-Loop structure integral sliding mode control for UUV trajectory tracking. Proc. IEEE **7**, 101620–101632 (2019)
7. Swaroop, D., Hedrick, J.K., Yip, P.P., Gerdes, J.C.: Dynamic surface control for a class of nonlinear system. IEEE Trans. Autom. Control **45**(10), 1893–1899 (2000)
8. Pan, C., Zhou, L., Xiong, P., Xiao, X.: Robust adaptive dynamic surface tracking control of an underactuated surface vessel with unknown dynamics. In: 2018 37th Chinese Control Conference, pp. 592–597 (2018)
9. Zhang, C.J., Wang, C., Wei, Y., Wang, J.Q.: Robust trajectory tracking control for underactuated autonomous surface vessels with uncertainty dynamics and unavailable velocities. Ocean Eng. **218**, 108099 (2020)
10. He, W., Yin, Z., Sun, C.Y.: Adaptive neural network control of a marine vessel with constraints using the asymmetric barrier lyapunov function. IEEE Trans. Cybern. **47**(7), 1641–1651 (2017)
11. Yin, Z., He, W., Yang, C.G.: Tracking control of a marine surface vessel with full-state constraints. Int. J. Syst. Sci. **48**(3), 535–546 (2017)
12. Zheng, Z.W., Sun, L., Xie, L.H.: Error-constrained LOS path following of a surface vessel with actuator saturation and faults. IEEE Trans. Syst. Man Cybern. **48**(10), 1794–1805 (2018)
13. Fossen, T.I.: Handbook of marine craft hydrodynamics and motion control (2011)

# Lane Line Detection Based on DeepLabv3+ and OpenVINO

Risheng Yang and Xiai Chen(✉)

College of Mechanical and Electrical Engineering, China Jiliang University, Hangzhou 310018, Zhejiang, People's Republic of China
xachen@cjlu.edu.cn

**Abstract.** With the deepening of technology, unmanned driving has developed rapidly, and road detection is one of the key points. At present, the recognition accuracy and reasoning speed of lane line detection technology are mature, but there is a problem that large computing resources are required for reasoning, and it is not easy to deploy to low-cost mobile devices. Therefore, this paper proposes to use the DeepLabV3+ network model of semantic segmentation to detect lane lines, train the model with its own data set, and then export it as a pb file, and then convert it to the intermediate representation IR file of OpenVINO through Intel's Model Optimizer, using Intel's CPU Perform model reasoning. The final result shows that there is no difference in the accuracy of reasoning with Intel CPU after optimization, and the reasoning time is greatly shortened. The cost can be greatly reduced while getting better recognition results.

**Keywords:** Unmanned · DeepLabv3+ · Lane line · Semantic segmentation · OpenVINO

## 1 The Background and Significance of the Research

With the development of technology and society, the application of smart cars has become more and more extensive, and the accompanying assisted driving and unmanned driving are particularly important. The most important thing in this technology is the detection part of the lane line, which uses algorithms to identify the lane line so that the control system can adjust the state of the car. Early algorithms mostly used traditional algorithms for detection [1]. In recent years, deep learning methods have been used to solve lane line detection, and the results obtained are very significant, but the requirements for computing power are relatively high. In this paper, the trained pb file is converted to the OpenVINO IR file, and the CPU is used to infer the lane line detection. It can greatly alleviate the problem of requiring higher resources. Improve inference accuracy without losing too much speed [2].


Q. Han et al. (Eds.): LSMS 2021/ICSEE 2021, CCIS 1469, pp. 714–723, 2021.
https://doi.org/10.1007/978-981-16-7213-2_69

## 2 DeepLabv3+ and OpenVINO

### 2.1 DeepLabv3+ Network

Compared with V3, DeepLabv3+ makes an article on the model architecture. In order to fuse multi-scale information, it introduces the commonly used encoder-decoder form of semantic segmentation. The DeepLabv3 part is actually the encoder part of V3+, and a decoder part is added on this basis. Compared with other semantic segmentation models, its biggest feature is the use of hollow convolution, which enlarges the receptive field, and each convolution can contain a larger range of information, which can extract more effective features. Combining the hole convolution and the depth separable convolution, it is applied to the hole space convolution pooling pyramid module, which greatly reduces the complexity of the model. The use of the decoding module allows it to optimize the subdivision results, especially the area along the target boundary [3]. The backbone uses the improved Xception_65 structure. In the decoding area, the feature map is up-sampled and merged with the shallow features. The shallow features are used to optimize and supplement the pixel information that is not restored by the up-sampling, and the final semantic segmentation result is obtained. The model structure is shown in Fig. 1.

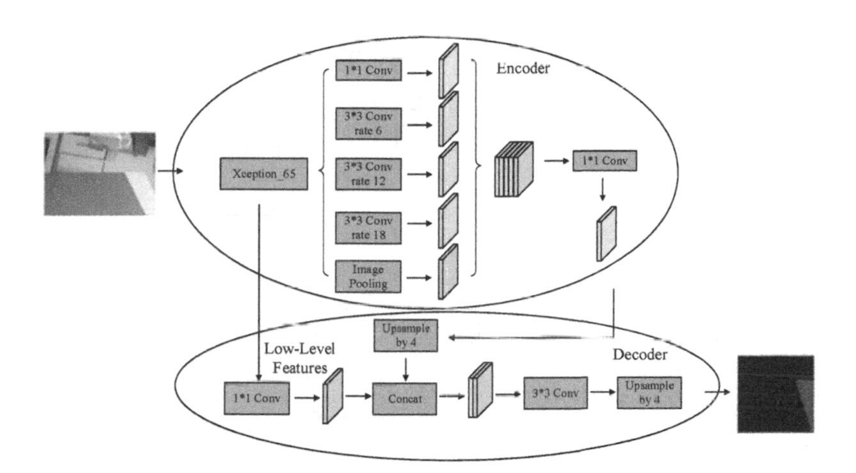


**Fig. 1.** DeepLabv3+ model structure diagram

Hole convolution can increase the receptive field while keeping the size of the feature map unchanged. The schematic diagram of the two-dimensional image cavity convolution is shown in Fig. 4. A 3*3 convolution kernel with an expansion rate of 2, the receptive field is the same as the 5*5 convolution kernel, and only requires 9 parameters [4–6]. A larger receptive field is required at the network layer and computing resources are limited. It is impossible to increase the volume. When the number or size of the product cores, the hole convolution becomes very effective [7] (Fig. 2).

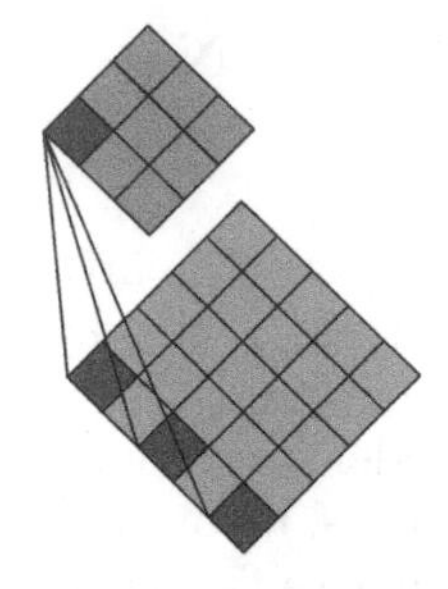

**Fig. 2.** Two-dimensional hole convolution

Spatial pyramid pooling, namely ASPP, is a method of fusing multi-scale information at the pooling layer. DeepLabV2 proposes ASPP. In DeepLabv3, ASPP with different void rates can effectively obtain multi-scale information [8]. ASPP is shown in Fig. 3 below.

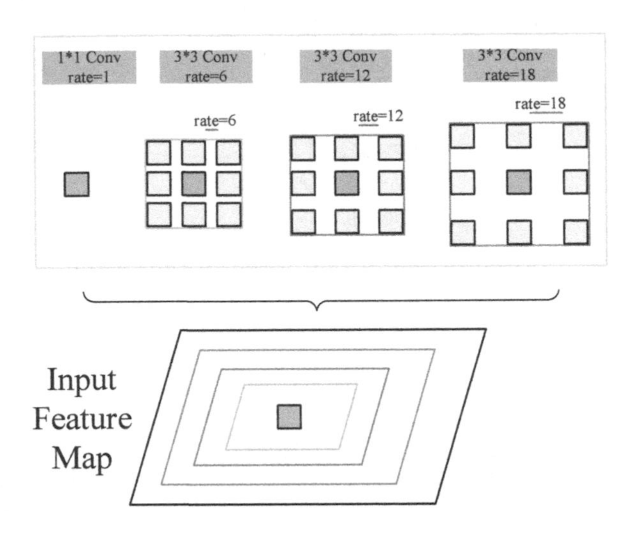


**Fig. 3.** ASPP structure diagram

## 2.2 OpenVINO Reasoning Optimization Tool

OpenVINO is a deployment toolkit launched by Intel Corporation for the rapid development of high-performance computer vision. Contains a series of functions related to deep learning model deployment, such as inference library and model optimization. The OpenVINO toolkit can be used for rapid application development, and can accelerate the inference process on Intel's hardware, thereby improving performance to a greater extent [9,10]. OpenVINO is currently a very mature and rapidly developing inference engine library, providing most of the examples for everyone to get started. And open source a lot of already trained IR model files. The model file can be used directly for some projects. OpenVINO has optimized various graphics processing algorithms, which greatly expands the high performance of Intel hardware. And delete unnecessary network structure in the process of model optimization, optimize the model structure [11]. It mainly includes a tool for optimizing neural network models, namely, a model optimizer and an inference engine for inference acceleration calculation. The OpenVINO deployment workflow is shown in Fig. 4.

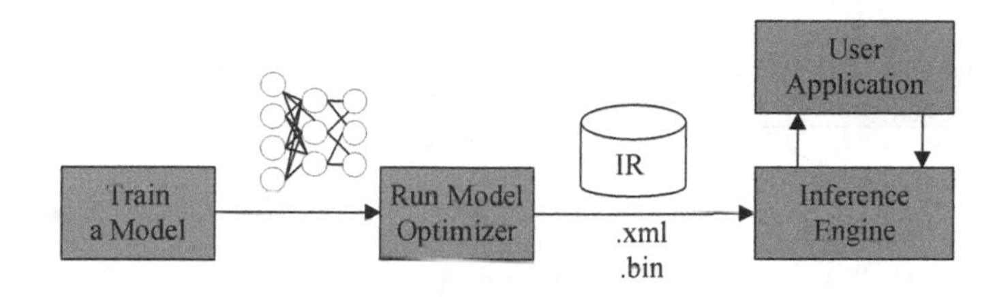


**Fig. 4.** OpenVINO deployment workflow

The model optimizer is a python script tool used to convert models trained by other open source frameworks into intermediate expression files that can be recognized by the inference engine. One is the bin file containing the weights and deviations of the network, and the other is the xml file containing the description of the network structure [12]. The model optimizer will delete some useless operations during inference, Such as the integration of the Dropout layer and the optimization of memory. The reasoning engine is an API interface that supports C++ and Python, which can independently realize the development of the reasoning process.

### 2.3 Process

First use the camera to collect the data set. The data is processed, and then the data set is marked with labelme. The labels are divided into two types: road and background. Then use labelme's built-in json conversion dataset function to convert the data into labels and mask images, and then perform the specified pixel conversion. Due to the small hardware memory, the image is specified to the size of 321*321. Use the initial DeepLabV3+ model code provided by Tensorflow. Next, divide the data set, and divide the training set: validation set: test set at a ratio of 8:1:1. And convert the data to Tfrecord format, which can read the data more efficiently, which can reduce resource consumption and training time [13,14]. Next, train the model. After training, export the model as a pb file. Convert the model file into a bin file and an xml file, and then put it into the Intel NUC to call the Inference Engine to complete the inference calculation, and get the final recognition result. The flow chart is shown in Fig. 5.

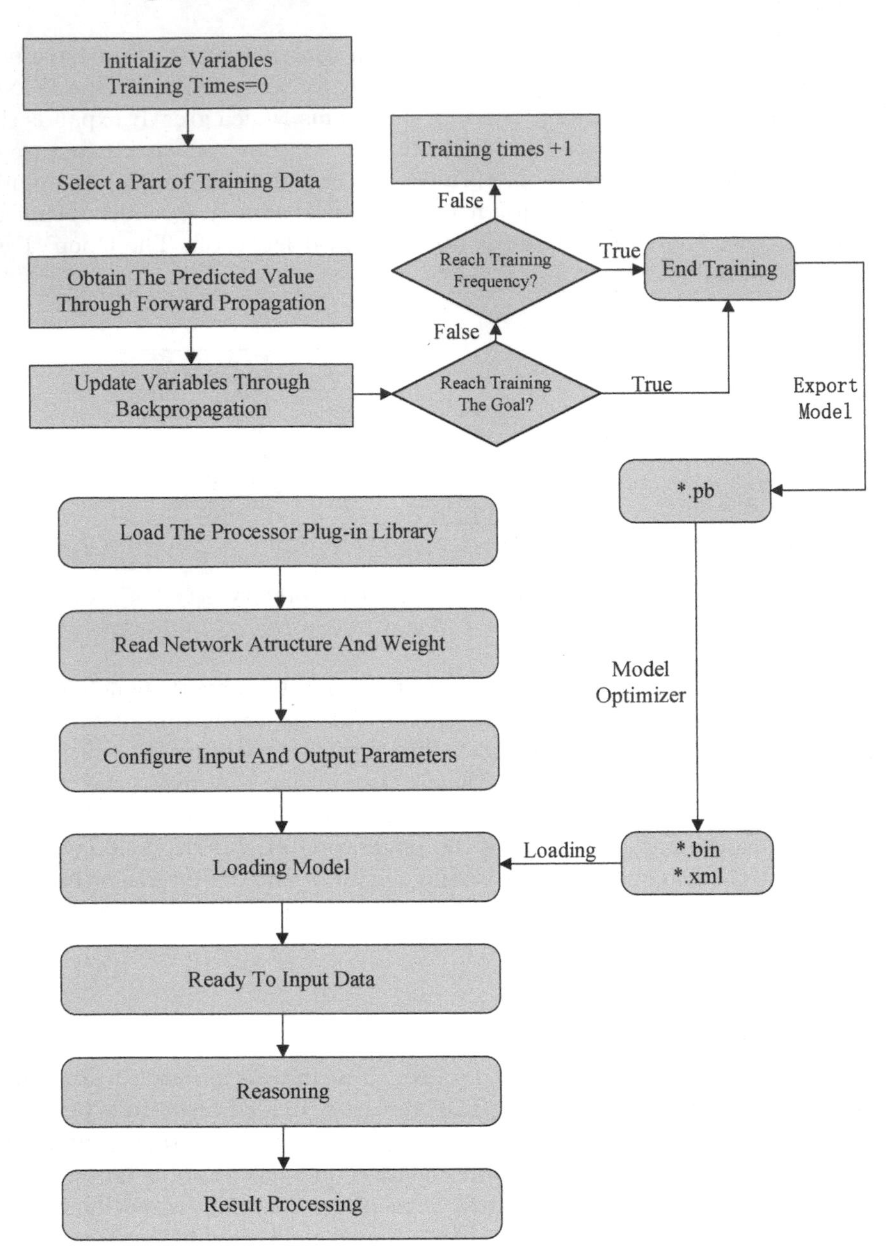


**Fig. 5.** Flow chart

# 3 Application and Analysis

## 3.1 Experimental Environment

The data training operating environment is Ubuntu 18.04. The CPU is 3600X, the GPU model is NVIDIA GeForce GTX 1070Ti, the programming language is Python, and the deep learning framework is Tensorflow 1.14. This article uses the official source code of Tensorflow of DeepLabv3+. The configuration is shown in Table 1.

**Table 1.** Training environment configuration diagram

| | |
|---|---|
| CPU | AMDR5 3600X |
| GPU | NVIDIA GeForce GTX 1070Ti(8GB) |
| Memory | 32 GB |
| Operation system | Ubuntu 18.04.5 LTS |
| Programming language | Python |
| Deep learning framework | Tensorflow 1.14 |

The final reasoning uses the hardware device as Intel NUC. The OpenVINO version is 2020.4. The configuration is shown in Table 2.

**Table 2.** Training environment configuration diagram

| | |
|---|---|
| CPU | Intel(R) Core(TM) i5-8265U CPU @ 1.60 GHz 1.80 GHz |
| Memory | 8.00 GB |
| Operation System | Windows 10 1809 |
| OpenVINO Version | OpenVINO 2020.4 |

## 3.2 Data Collection and Enhancement

The experiment uses self-collected data sets because it only detects lane lines, so two classifications are used. In this experiment, the picture resolution is reduced to 321 × 321 pixels [15,16], and convert the data set into eight Tfrecord files for efficient reading. Label with labelme, the result is shown in Fig. 8.

The experiment uses online and offline data set enhancement methods to enhance the lane line image. Offline enhancement adopts the method of brightness enhancement. The enhanced picture is shown in Fig. 9. Then use online enhancement methods during training, including contrast and cropping (Fig. 6 and 7).

## 3.3 Model Training and Results

This paper does not use the given pre-training weights. The batch training size is 8 and the number of model iterations is set to 15000; the learning rate is

**Fig. 6.** Marking diagram

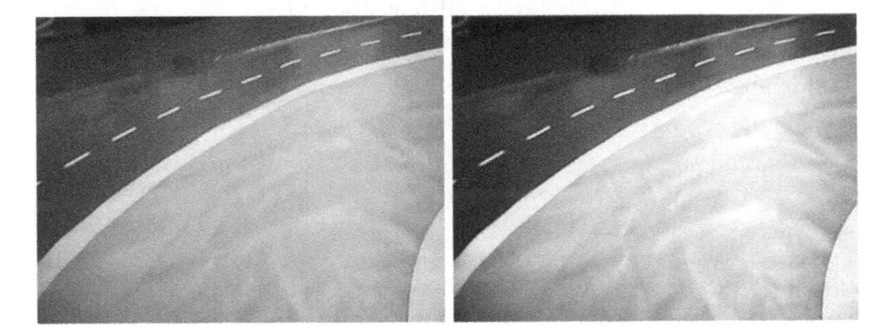

**Fig. 7.** Data enhancement map

dynamically adjusted, and the initial setting is 0.0001. After 1000 iterations, the iterative learning rate is one-tenth of the original. The initial training loss value is 0.8812, finally stabilized at about 0.2. The training loss value curve is shown in Fig. 8.

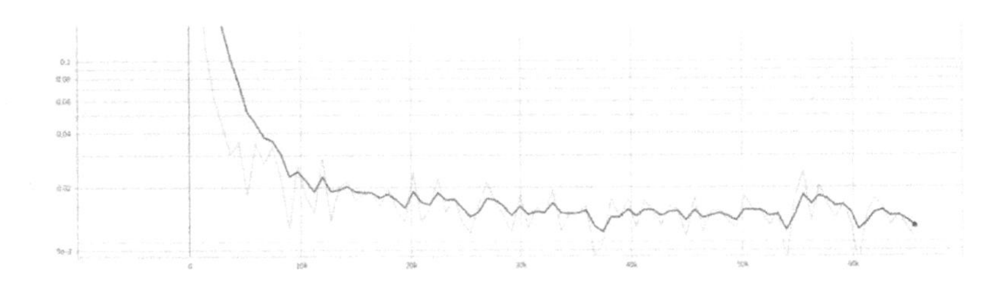

**Fig. 8.** Loss value change curve

After the model training is completed, the segmentation results on the test set are shown in Fig. 9.

**Fig. 9.** Model prediction results

# 4 Export Model and Inference Output

Use Tensorflow's own model export program to export the model to a pb file, then put the model in Intel NUC, use Model Optimizer to convert the model to an intermediate representation (IR) file, and specify the input size as 321*321. Other default output. Then load the IR file into the program for inference output. The final result is shown in Fig. 10.

**Fig. 10.** Model inference output results

The relative performance comparison results of different inference methods are shown in Table 3.

**Table 3.** Comparison of relevant performance of different inference methods

| Reasoning method | Average reasoning time/s | Accuracy |
|---|---|---|
| Tensorflow+*.pb(GPU) | 3.4944 | 0.9801 |
| OpenVINO(CPU) | 0.9776 | 0.9764 |

From the above, it can be concluded that for the same model, after the OpenVINO tool suite model optimizer is optimized, there is no difference in the inference accuracy of the model after optimization. The optimized model has a significant increase in the inference time, and the accuracy rate is basically not reduced.

## 5 Summary and Outlook

Research based on deep learning has become a key and popular direction in the field of artificial intelligence today, and more and more researchers are researching and applying deep learning. We used the DeepLabv3+ model of Xception's deep learning semantic segmentation with backbone to train the lane line data. From the results, it can be seen that this experiment can detect the lane line well. Next, optimize for longer reasoning time. This paper proposes to optimize the model based on the Intel model optimization tool suite OpenVINO. The model can be inferred and output results by borrowing a low-cost Intel CPU. Effectively solve the problem of inference speed and cost of model deployment. Later, consider using Intel NCS Neural Compute Stick to optimize model detection speed and further optimize model performance. In addition, it will try to simplify the architecture to improve its inference speed. In the future, this model will be considered to be deployed on a trolley with Arm as the lower machine for warehouse inspections, factory inspections and other applications.

## References

1. Xu, H., Li, L., Fang, M., Hu, L.: A method of real time and fast lane line detection. In: 2018 Eighth International Conference on Instrumentation & Measurement, Computer, Communication and Control (IMCCC), pp. 1665–1668 (2018)
2. Feng, J., Wu, X., Zhang, Y.: Lane detection base on deep learning. In: 2018 11th International Symposium on Computational Intelligence and Design (ISCID), pp. 315–318 (2018)
3. Sun, Z.: Vision based lane detection for self-driving car. In: 2020 IEEE International Conference on Advances in Electrical Engineering and Computer Applications (AEECA), pp. 635–638 (2020)
4. Demidovskij, A., et al.: OpenVINO deep learning workbench: a platform for model optimization, analysis and deployment. In: 2020 IEEE 32nd International Conference on Tools with Artificial Intelligence (ICTAI), Baltimore, MD, pp. 661–668 (2020)

5. Mathew, G., Sindhu Ramachandran, S., Suchithra, V.S.: Lung nodule detection from low dose CT scan using ptimization on Intel Xeon and core processors with Intel distribution of OpenVINO Toolkit. In: TENCON 2019 IEEE Region 10 Conference (TENCON), pp. 1783–1788 (2019)
6. Jin, Z., Finkel, H.: Analyzing deep learning model inferences for image classification using OpenVINO. In: 2020 IEEE International Parallel and Distributed Processing Symposium WoDeepLabV3rkshops (IPDPSW), pp. 908–911 (2020)
7. Liu, M., et al.: Comparison of multi-source satellite images for classifying marsh vegetation using DeepLabV3 Plus deep learning algorithm. Ecol. Indicators **125**, 107562 (2021)
8. Shelhamer, E., et al.: Fully convolutional networks for semantic segmentation. IEEE Trans. Pattern Anal. Mach. Intell. **39**, 640–651 (2017)
9. Gu, D., Wang, N., Li, W.-C., Chen, L.: Method of lane line detection in low illumination environment based on model fusion. J. Northeastern Univ. (Nat. Sci.) **42**(3), 305–309 (2021)
10. Lo, S., Hang, H., Chan, S., Lin, J.: Multi-class lane semantic segmentation using efficient convolutional networks. In: 2019 IEEE 21st International Workshop on Multimedia Signal Processing (MMSP), pp. 1–6 (2019)
11. Mottaghi, R., Chen, X., Liu, X., et al.: The role of context for object detection and semantic segmentation in the wild. In: IEEE Conference on Computer Vision and Pattern Recognition, Columbus, United States of America, pp. 891–898 (2014)
12. Kustikova, V., et al.: Intel distribution of OpenVINO toolkit: a case study of semantic segmentation. In: van der Alast, W.M.P., et al. (eds.) AIST 2019. LNCS, vol. 11832, pp. 11–23. Springer, Cham (2019). https://doi.org/10.1007/978-3-030-37334-4_2
13. Chen, L.-C., et al.: DeepLab: semantic image segmentation with deep convolutional nets, atrous convolution, and fully connected CRFs. IEEE Trans. Pattern Anal. Mach. Intell. **40**, 834–848 (2018)
14. Scimeca, D.: Industrial computer includes toolkit for deep learning models. Vis. Sys Des.t. **24**(8) (2019)
15. Castro-Zunti, R.D., et al.: License plate segmentation and recognition system using deep learning and OpenVINO. IET Intell. Transp. Syst. **14**, 119–126 (2020)
16. He, K., et al.: Spatial pyramid pooling in deep convolutional networks for visual recognition. IEEE Trans. Pattern Anal. Mach. Intell. **37**, 1904–1916 (2015)

# Design of a Smart Community Inspection Robot

Longhai Zhu, Shenglong Xie(✉), and Jingwen Wang

School of Mechanical and Electrical Engineering, China Jiliang University, Hangzhou 310018, Zhejiang, China

**Abstract.** Aiming at the security requirements of smart community, an intelligent inspection robot was present in this paper. The environmental characteristics of the smart community were described in detail, and the characteristics of community inspection robot were summarized. According to the idea of modularization, the mechanical structure of the inspection robot was designed, which makes the inspection robot has the advantages of good off-road performance, strong climbing ability, and good ability of adapting to the complex terrain environment. The control system of this robot was designed based on the distributed structure design concept, the multi-sensor information acquisition system of this robot was constructed, and the motion control scheme of this robot was designed. Finally, the prototype of this inspection robot was designed, and the functional test and demonstration application were carried out in indoor and outdoor environment. The experimental results indicated that the proposed robot has a good motion performance, and meets the requirements of intelligent community inspection task.

**Keywords:** Smart community · Inspection robot · Control system · Motion control · Modularization

## 1 Introduction

Patrol inspection is a regular or random inspection for product manufacture, machine operation or jurisdiction safety. It aims to find product quality problems, equipment operation faults and ensure the safety of people's lives and property. However, manual inspection usually consumes large number of resources and is inefficient. Intelligent inspection robot can ensure the efficient and reliable implementation of inspection tasks [1, 2], thus freeing inspectors from monotonous and repetitive inspection tasks. Therefore, the inspection robot has been widely used in bridge and tunnel maintenance, agriculture, cable abnormal investigation and substation maintenance [3, 4].

A farmland crop detection robot based on the fusion of machine vision and GPS is developed in refer [5]. It can track crop lines through image processing, and use fuzzy control to realize navigation, which can realize the recognition of low and short crops effectively. A police patrol robot is introduced in refer [6], which can be used in outdoor environment such as shopping malls, residential areas and indoor environment such as offices, so as to protect community security effectively. Salvucci [7] introduced an automatic detection robot for crop information in greenhouse environment. The robot


Q. Han et al. (Eds.): LSMS 2021/ICSEE 2021, CCIS 1469, pp. 724–732, 2021.
https://doi.org/10.1007/978-981-16-7213-2_70

navigation line recognition and crop information collection were realized by camera, and PID algorithm was used to realize the navigation control of this robot. Wang [8] designed a substation inspection robot, which can collect substation equipment information in real time. It can realize the detection of substation equipment fault through image processing, pattern recognition and other technologies. Combined with equipment image expert database, it can realize online intelligent detection, intelligent diagnosis, alarm and analysis of equipment. A modular pipeline robot is designed based on the modular design method, which can complete a variety of pipeline internal operations, and possess the functions of pipeline internal detection, obstacle removal, grinding and so on [9]. In Ref. [10], a floating inspection robot for transformer interior is proposed. This robot possesses the characteristics of zero turning radius and flexible movement. However, most of the existing research focuses on substation inspection robot, transmission line inspection robot and pipe gallery inspection robot, while few literatures focus on community inspection robot.

In order to promote the development of community security inspection towards automation and intelligence, this paper presents a smart community inspection robot, which can replace manual inspection based on the combination of the modular design idea, wireless sensor technology and distributed structure design concept. Firstly, the mechanical structure of the inspection robot is designed by using the modular design idea after the demand analysis of community inspection; Secondly, the control system of the inspection robot is designed based on the theory of distributed structure, and the robot control system is developed; Finally, the performance verification experiment of this robot is carried out in the indoor and outdoor environment, and the relevant conclusions are obtained.

## 2 Structural Design

### 2.1 Requirement Analysis

As the mobility of urban population is more obvious and the population structure is always changing, the community environment is more complex. Therefore, the community inspection robot needs to meet the following requirements:

(1) Uninterrupted Patrol: In order to realize unmanned duty, it is necessary to supply the whole day on-line service. Therefore, it is necessary to carry out lightweight design to reduce the energy consumption of the robot at the same time of low power auto-charging;
(2) Intelligent path planning: The robot can locate independently, navigate intelligently, avoid obstacles, and plan the optimal path;
(3) Environment perception: The robot can be equipped with a variety of sensors to monitor the identification and warning of dangerous and suspicious objects, and transmit the information to the background in real time;
(4) Terminal cooperation: The robot should possess the functions of data acquisition, upload, alarm with the help of the intelligent analysis and recognition system of the robot. Meanwhile, the monitoring video can be sent back to the terminal to realize the synchronization of decision-making and execution;

(5) Multi inspection mode: In order to ensure the safety of the community, the inspection robot should have the ability of setting a variety of inspection modes, such as automatic timing and fixed-point inspection, remote control inspection and designated path inspection.

## 2.2 Mechanical Structure

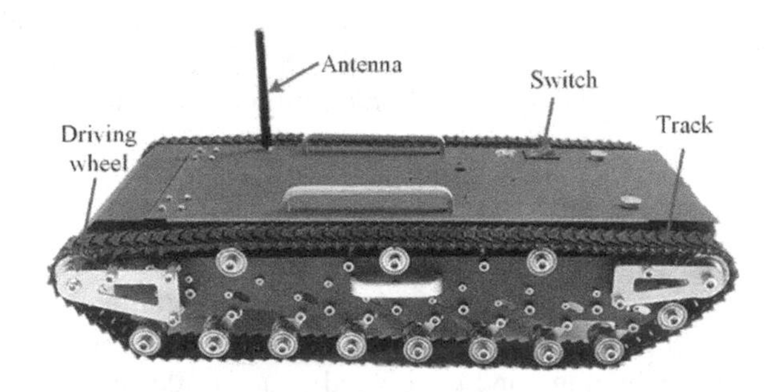


(a) Mobile chassis

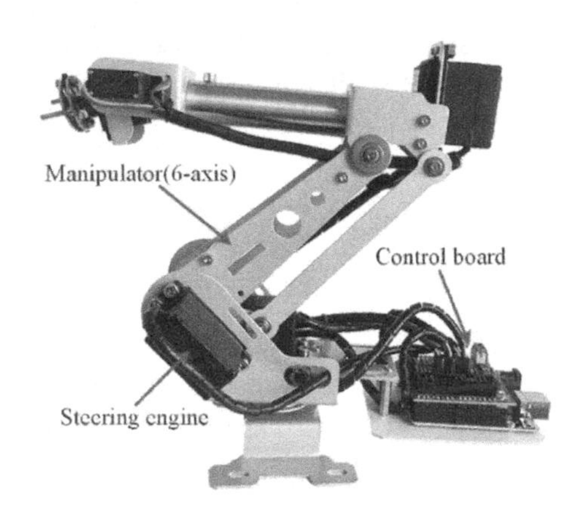


(b) Manipulator

**Fig. 1.** Mechanical structure of inspection robot

According to the idea of modular design [11], the mechanical structure of the inspection robot can be divided into two parts: the mobile platform and the manipulator. Figure 1 illustrates the structure of this robot. The mobile platform is the main body of this inspection robot, which provides power for the movement of the whole robot. At the same time, it can install depth cameras, lidar, infrared sensors, etc., which constitute the perception system of the robot. The manipulator is the executive part of this inspection robot, which carries various tools to complete the community inspection and foreign body cleaning management and other tasks.

The mobile platform of inspection robot adopts twin pedrail type structural, so it has the advantages of large traction, low grounding ratio, good stability, good off-road performance and strong climbing ability. It can suit to various geological conditions and climate environment in the community, so it has the advantage of reaching all areas in the community. The robot arm is a 6-DOF serial robot, six joints are driven by the steering gear, and the hollow structure design is adopted in the design of 6-axis. Therefore, it has the advantages of simple structure, light weight, low cost, simple control, large movement space and so on. It can easily realize the tasks of community inspection and foreign body cleaning in the process of inspection.

## 3 Control System Design

### 3.1 Overall Scheme of Control System

According to the idea of distributed structure design, the robot control system adopts the two-level distributed structure of communication between upper computer and lower computer [12]. The upper computer is mainly used for data analysis and processing, including control signal processing, communication with the lower computer and providing the window for human-computer interaction. The lower computer is mainly responsible for the robot's action execution, environmental perception and monitoring, including the acquisition of various sensor signals and motor drive. The upper computer communicates with the lower computer through WiFi, supporting real-time control and distributed control. With the help of effective fusion of the upper computer, the lower computer, the motion controller and the sensor, the hardware block diagram of the detection and control system is established as shown in Fig. 2.

The control architecture of this robot is divided into robot autonomous control system and terminal operation control system. In order to ensure the flexibility and security of robot movement in the process of inspection, wireless communication is adopted for data transmission of the control system. The communication module adopts DDL2350 module, which includes two RS232 serial ports, one Ethernet interface and one HDMI interface. It can transmit high-definition video, lidar data and control command data at the same time, and possess the advantages of ultra-low delay, long transmission distance and data encryption. The control system of this robot mainly includes lidar, wireless communication module, HD camera, core control system, ultrasonic sensor and the attitude and heading reference system (AHRS), et al. The control unit of this control system is designed by STM32 single chip microcomputer, which possess the advantages of low power consumption and rich interface. The TTL level of MCU is converted to RS232 level by using serial port chip MAX232, and connected with wireless transmission module and AHRS. The STM32 microcontroller integrated with AD interface, and the analog voltage of sensor is converted by LM358 chip into the microcontroller. The operation and control terminal includes a micro industrial computer, a display, a joystick, a communication module and a video acquisition card. The operation control terminal receives, displays and stores lidar data, high-definition video, robot running posture and other data in real time, and collects the data of the joystick to control the speed and direction of this robot.

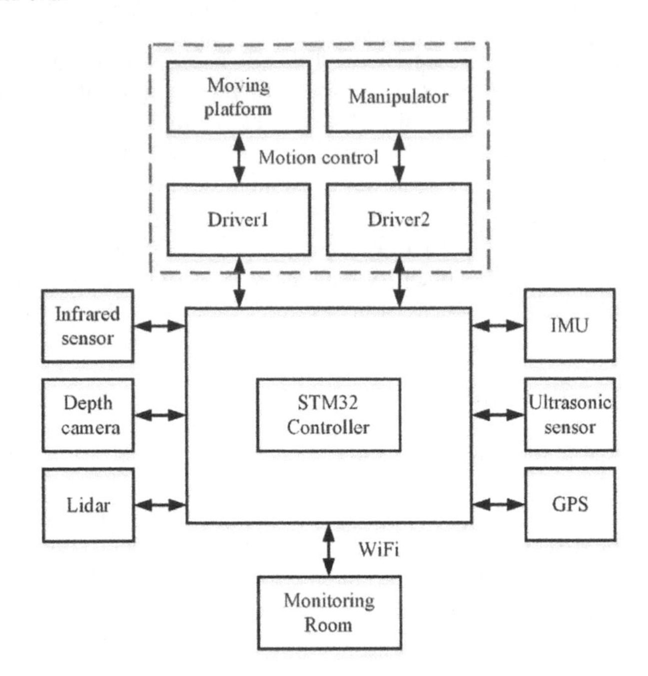


**Fig. 2.** Hardware framework of inspection robot

### 3.2 Sensor System of Robot

The robot sensor system is composed of camera, depth meter, lidar, ultrasonic sensor and AHRS, which is used to sense the external environment and measure the robot's running posture and position. The main sensors used in this robot is shown in Fig. 3. The AHRS adopts 9-axis inertial navigation module gy953, which integrates 3-axis acceleration, 3-axis gyroscope and 3-axis magnetometer sensors. The output of GY953 has the ability to resist instantaneous electromagnetic interference with the help of Kalman filter data fusion algorithm. The ultrasonic sensor adopts US-015 ultrasonic module, which possess the advantages of high precision, corrosion resistance and long service life. The laser radar adopts M2M1 of Slamtec Mapper company, which can realize high-precision map construction and real-time positioning. This makes it is competent for various complex mapping and positioning application scenarios. The camera adopts the ALIENTEK ATK-OV2640 camera module, which has the characteristics of high sensitivity, high flexibility and JPEG output.

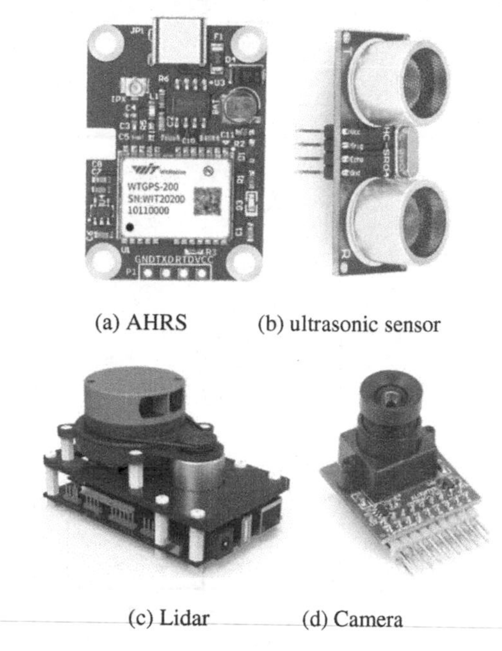


(a) AHRS (b) ultrasonic sensor

(c) Lidar (d) Camera

**Fig. 3.** Sensor system of inspection robot

### 3.3 Software Design of Robot

The software of robot control system integrates data receiving, sensor data acquisition, data sending, motion control and other functions. In addition, the software uses the timer to detect the communication link with the operation control end in real time. If the communication is abnormal, the program will turn off the motor, and the robot will stop moving, waiting for the monitoring personnel to repair. The software flow chart is shown in Fig. 4. It receives the command from the operation control end in the way of interruption, controls the robot movement, and reads the data of each sensor in turn in the way of rotation training. After reading the data, the data sending program is called to send the robot motion posture to the operation control end. When receiving the autonomous inspection instruction, the robot calls the autonomous inspection program to start the inspection work.

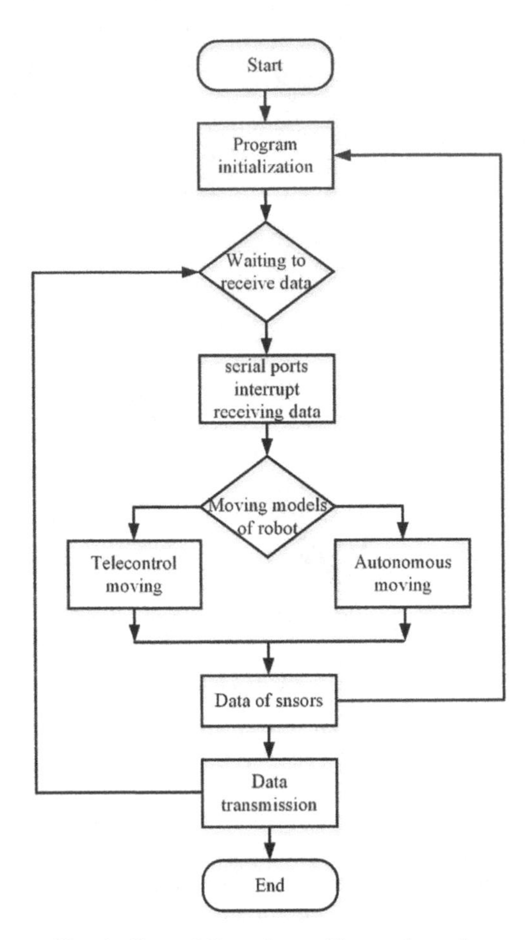


**Fig. 4.** Control flow chart of inspection robot

## 4 Experimental Verification

In order to verify the effectiveness of the robot body and its control system, the indoor and outdoor simulated patrol environment is established to carry out relevant verification experiments, and the outdoor environment is used to verify the robot climbing, complex terrain driving and obstacle avoidance experiments, which is shown in Fig. 5. The experimental results show that the designed robot system has high real-time performance, and the robot can well adapt to various complex terrain environment.

Figure 6 illustrates the velocity control of this robot, the horizontal coordinate-axis represents time, and the vertical coordinate-axis represents the speed of the motor. It is obvious that the motor can track the desired velocity very well.

(a) Complex terrain (b) Climbing slope

(c) The lawn (d) Obstacle avoidance

**Fig. 5.** The performance verification experiment of robot

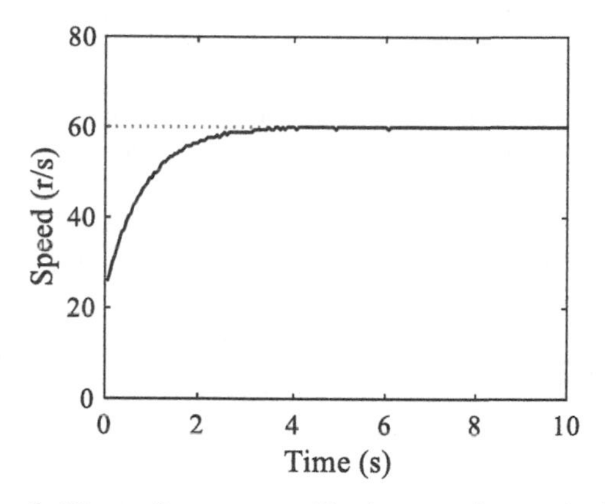


**Fig. 6.** The performance verification experiment of robot

# 5 Conclusion

This paper introduces a smart community inspection robot. The characteristics of inspection robot are summarized based on the detailed analysis of the environment characteristics of smart community. Subsequently, the mechanical structure and control system of this inspection robot is designed, which has the advantages of good stability, good off-road performance and strong climbing ability. Finally, the prototype of this inspection robot is designed, and the functional test and demonstration application are carried out in indoor and outdoor environment. The experimental results indicate that this robot possess good motion performance, and meets the requirements of intelligent community inspection task. In the future, the research on human-computer interaction control of this inspection robot will be carried out to realize the intelligent inspection of this robot.

**Acknowledgments.** This work was supported in part by the Postdoctoral Science Foundation of Zhejiang Province under Grant ZJ2020007, in part by the Natural Science Foundation of Zhejiang Province under Grant LQ20E050017, and in part by the National College Students' Innovation and Entrepreneurship Training Program under Grant 00927-200084.

## References

1. Shukla, A., Karki, H.: Application of robotics in onshore oil and gas industry-a review (part I). Robot. Auton. Syst. **75**, 490–507 (2016)
2. Shukla, A., Karki, H.: Application of robotics in offshore oil and gas industry-a review (part II). Robot. Auton. Syst. **75**, 508–524 (2016)
3. Montero, R., Victores, J.G., Martinez, S., et al.: Past, present and future of robotic tunnel inspection. Autom. Constr. **59**, 99–112 (2015)
4. Khadhraoui, A., Beji, L., Otmane, S., et al.: Stabilizing control and human scale simulation of a submarine ROV navigation. Ocean Eng. **114**, 66–78 (2016)
5. José, B.G., Jesus, C.M., Dionisio, A., et al.: Merge fuzzy visual servoing and GPS-based planning to obtain a proper navigation behavior for a small crop-inspection robot. Sensors **16**(3), 276 (2016)
6. Mahmud, M., Abidin, M., Mohamed, Z.: Development of an autonomous crop inspection mobile robot system. In: 2015 IEEE Student Conference on Research and Development (SCOReD), Kuala Lumpur, pp. 105–110. IEEE Press (2015)
7. Salvucci, V., Baratcart, T., Koseki, T.: Increasing isotropy of intrinsic compliance in robot arms through biarticular structure. IFAC Proc. Volumes **47**(3), 332–337 (2014)
8. Wang, T.B., Wang, H.P., Qi, H., et al.: The humanoid substation inspection robot modelling design. Appl. Mech. Mater. **365–366**, 771–774 (2013)
9. Li, Q., Xie, T.Y., Yang, H.J., et al.: Design of modular pipeline operation robot. J. Mech. Eng. (2021). https://kns.cnki.net/kcms/detail/detail.aspx?dbcode=CAPJ&dbname=CAPJLAST&filename=JXXB2021021901Y&uniplatform=NZKPT&v=70CtNECuetsdFlt2KdG1Dz5Gh7xo6Sg%25mmd2FiGZaOSWXVgiDVEcJjOgRwhQ7c%25mmd2B%25mmd2F3p5Up (Published online)
10. Feng, Y.B., Yu, Y., Gao, H.W.: Floating robot design method for inspecting the inside of transformer. J. Mech. Eng. **56**(7), 52–59 (2021)
11. Shuai, L.G., Su, H.Z., Zheng, L.Y., et al.: Study on steering movement of track - wheel mobile robot. J. Harbin Eng. Univ. **38**(10), 1630–1634 (2017)
12. Zhu, H., You, S.Z.: Research and experiment of a new type of coal mine rescue robot. J. China Coal Soc. **45**(6), 2170–2181 (2020)

# Cartographer Algorithm and System Implementation Based on Enhanced Pedestrian Filtering of Inspection Robot

Ganghui Hu[1], Jianbo Sun[2], Xiao Chen[3], and Binrui Wang[1](✉)

[1] College of Mechanical and Electrical Engineering, China Jiliang University, Hangzhou 310018, Zhejiang, China
[2] Hangzhou Ai Rui Technology Co., Ltd., Zhejiang 310018, China
[3] Zhejiang Qiantang Robot and Intelligent Equipment Research Co., Ltd., Hangzhou 310018, Zhejiang, China

**Abstract.** Aiming at the problem that the pedestrian filtering function is unstable in the subway station hall with heavy traffic in the cartographer and the generated map has related noise that affects later path planning and causes other issues, an enhanced pedestrian filtering based on obstacle detection is designed. The method realizes an autonomous positioning and navigation system combined with the enhanced pedestrian filtering cartographer algorithm. Laboratory tests and field tests in subway stations and halls have verified the effectiveness of the enhanced pedestrian filtering cartographer algorithm and the usability of the autonomous positioning and navigation system, providing convenience for subsequent environmental monitoring of subway station hall and real-time data updates.

**Keywords:** Inspection robot · Cartographer · Pedestrian filtering · Autonomous positioning · Navigation

## 1 Introduction

The subway station hall is a densely populated public place, which requires regular inspections of infrastructure operations and environmental parameters to prevent disasters. The traditional monitoring method is manual periodic inspection, but the efficiency is low, the cost is high and the data update is slow, so the use of robot inspection has become the general trend [1] instead of manual inspection. The design of autonomous positioning and navigation system is the key technology to realize automation of inspection robots. The positioning accuracy and navigation efficiency determine the extent to which the robot is popularized and applied. At present, the indoor environment generally has problems such as complex object layout, many dynamic factors, and lack of texture in some environments, which limit the application of mobile robots in subway station halls. In response to the above problems, a variety of robot positioning and navigation technologies such as visual SLAM (simultaneous localization and


Q. Han et al. (Eds.): LSMS 2021/ICSEE 2021, CCIS 1469, pp. 733–742, 2021.
https://doi.org/10.1007/978-981-16-7213-2_71

mapping) and laser SLAM have emerged in recent years. Based on the image feature extraction method, the visual SLAM positioning and navigation technology is divided into: feature point method and direct method. The feature point method [2,3] has high stability and is not sensitive to lighting and dynamic objects, but it needs to extract image feature points, which are easy to lose in an environment that lack texture and it takes a lot of time to extract. The direct method [4,5] can save time of extracting feature points, but it requires the image to have a pixel gradient, and the direct method has no discrimination to a single pixel, and it is easily affected by illumination, which leads to positioning failure [6]. Laser SLAM-based positioning and navigation technology uses lidar to perceive environmental information, which is less affected by light, not affected by texture loss, and high stability. It performs better than visual SLAM in complex indoor environments.

Mobile robot path planning is mainly divided into: global path planning and local path planning, where local path planning reflects the robot's ability to handle dynamic obstacles during the movement. Xian et al. [7] and Lu et al. [8] improved the dynamic window approach (DWA, dynamic window approach) algorithm to plan local paths. The algorithm improved efficiency by limiting the feasible set of effective trajectories to a subset with a constant velocity segment, but the robot moving time may not be optimal [9]. At present, the proposed autonomous positioning and navigation system is widely used in dynamic environment inspection, which has problems such as positioning inaccurate and failure to avoid dynamic obstacles in time [10].

The innovation points of this paper are as follows:

The obstacle detection algorithm was used to improve the pedestrian filtering function of Cartoans algorithm on widely used patrol robots based on the ROS (Robot Operating System) platform. An autonomous positioning and navigation system was implemented combining the Cartographer algorithm with enhanced pedestrian filtering to sense the environment and construct maps based on laser-point cloud and odometer information, and was validated in laboratory and subway station Settings.

# 2 Cartographer Algorithm Based on Enhanced Pedestrian Filtering

## 2.1 Cartographer's Original Pedestrian Filtering Function

The submap in Cartographer is a map representation model using probability grid. When new scan data is inserted into the probability grid, the state of the grid will be calculated. Each grid has two states, Hit and miss. The grid that is hit will insert its adjacent grid into the hit set. In the missing set, add all the relevant points on the connecting ray between the scan center and the scan point. For each grid that has not been observed before, Set a probability value, and the observed grid will be probabilistically updated according to the following formula. The schematic diagram is shown in Fig. 1. The shaded and crossed grid in the figure indicates a hit, and only the shaded grid indicates loss [11].

$$odds(p) = \frac{p}{1-p} \tag{1}$$

$$M_{new}(x) = clamp(odds^{-1}(odds(M_{old}(x))odds(p_{hit}))) \tag{2}$$

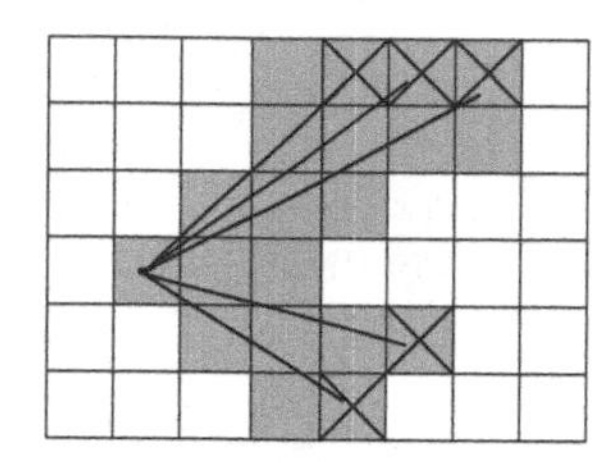

**Fig. 1.** Submap grid status update diagram

When Cartographer constructs the submap and the overall map, it essentially updates the probability of the grid map. The accumulation of the probability of each node on the grid can remove the noise generated by general moving objects during map construction. For example, during the map building process, after a few pedestrians, as long as they do not stay within the lidar scanning range for a long time, the probability of the grid map will be updated in time, and the resulting map will not contain pedestrians. However, there are often many people around the car body in the actual construction of the subway station hall, resulting in pedestrian noise in part of the generated submap. If this path is not repeated in the later stage, the final generated map will have related noise points, and the later path planning. It will be regarded as an obstacle, which will affect the navigation effect.

## 2.2 Principle of Obstacle Detection Algorithm

The obstacle detection algorithm is a local geometric obstacles detection and tracking method based on two-dimensional laser scanning proposed by Mateusz Przybya [12]. The detected obstacles are represented by linear or circular models. The detection method of the linear model is:

(1) The point cloud is classified into different combinations through a specific distance threshold judgment method.

$$d < d_{group} + R_i d_p \tag{3}$$

Where $d$ represents the Euclidean distance between points $P1$ and $P2$, $R_i$ represents the range of point $P1$ (Fig. 2), $d_{group}$ is the user-defined threshold for grouping and $d_p$ is the user-defined distance proportion which loosens the criterion for distant points (Fig. 3).

(2) Use the least squares method to fit the line segment to construct the segmented point cloud data $R_i$.

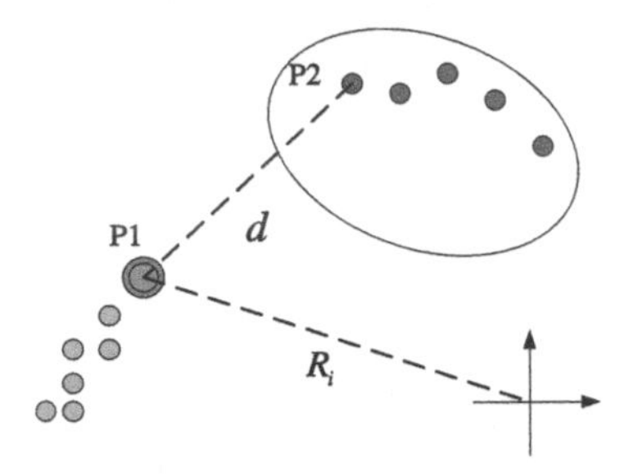


**Fig. 2.** Graphic representation of variables involved in synthetic points grouping

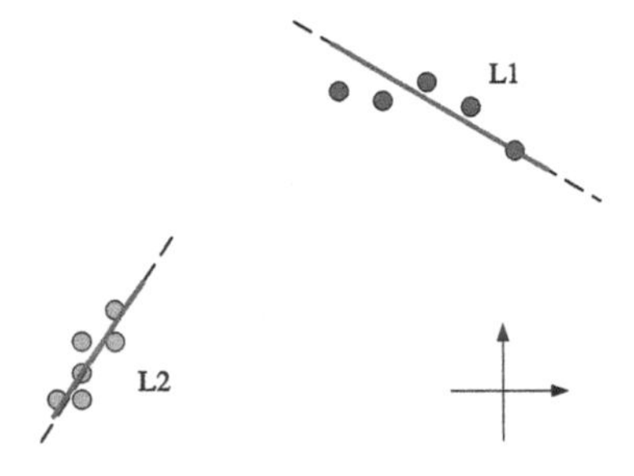


**Fig. 3.** Graphic representation of segments obtained from synthetic data

## 2.3 Cartographer Algorithm Based on Obstacle Detection to Enhance Pedestrian Filtering

The original pedestrian filtering in the above-mentioned cartographer algorithm, under certain circumstances, is likely to have pedestrian noise in the generated submap, and it cannot be removed through its own probability update. For this reason, this paper designs a method based on obstacle detection to enhance pedestrian filtering. Figure 4 gives a detailed description of the method.

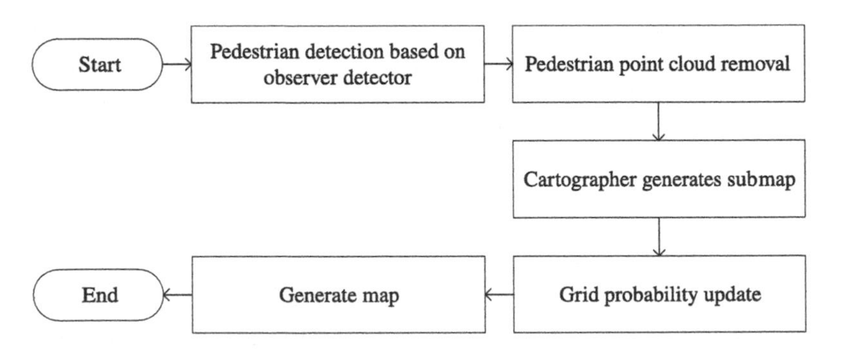


**Fig. 4.** Enhanced pedestrian filtering

This method first passes the radar point cloud data through the obstacle detection algorithm for pedestrian point cloud detection and removal. The processed point cloud data is released in real-time as the input of the cartographer algorithm, and then the laser point cloud is updated by the cartographer's probability. Noise filtering. Through two noise filtering, the cartographer's ability to filter pedestrians is greatly enhanced. As shown in Fig. 5, the left image is the original point cloud, and the right image is the point cloud after the pedestrian point cloud is removed by the obstacle detection.

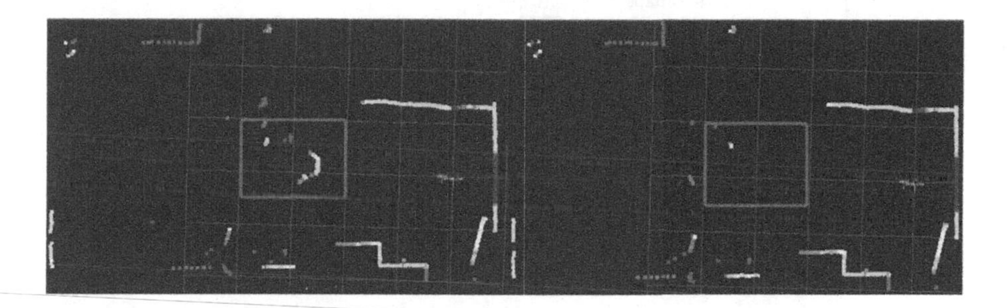

**Fig. 5.** Original point cloud (left), pedestrian filtered point cloud (right)

## 3 Design and Implementation of Autonomous Positioning and Navigation System

### 3.1 Realization of Positioning Function

Only when the inspection robot achieves autonomous positioning can it achieve autonomous navigation. It provides a robot environment map, collects itself and environment information through sensors such as lidar, wheel odometer, and processes the algorithm to match the known environment map. The final output of the robot is The pose in the map [13].

The positioning uses the global SLAM of the cartographer algorithm to realize the robot repositioning, and find a large number of mutual constraints between the frozen trajectory of the environment map and the current trajectory to realize the robot repositioning. However, when the map is large and there are a large number of frozen trajectories, searching for constraints is very time-consuming. In order to speed up the positioning speed, the closed-loop constraint generation rate is reduced to reduce the closed-loop detection time, so as to ensure the real-time nature of the relocation process. The positioning process is shown in Fig. 6.

### 3.2 Navigation Function Realization

The path planning of the global map and the path planning of the local map are necessary to realize the autonomous navigation of the inspection robot. The

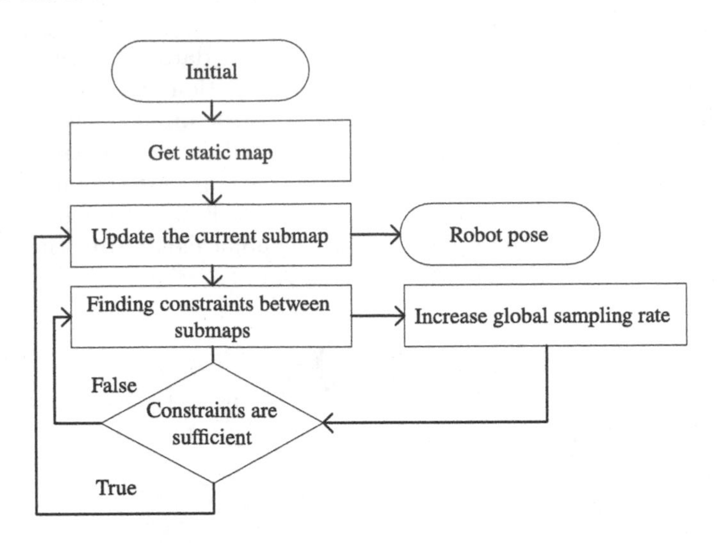


**Fig. 6.** Relocation process of cartographer algorithm

global path planning search obtains the shortest path without collision, which is obtained through the existing global cost map and the given target point location information. This shortest path is used as the input of the path planning of the local map to control the movement of the car body. Through local path planning, the car can avoid dynamic obstacles, and calculate the optimal path and optimal speed of the current position to avoid the collision. The global path planning uses the A* algorithm [14]. For the 2D grid map, the mathematical description of the algorithm is

$$g(n) = x(n) + y(n) \tag{4}$$

Among them: $g(n)$ is the score estimation function of the car from the starting point to the ending point through position $n$; $x(n)$ is the actual scoring function from the starting point to the position $n$; $y(n)$ represents the distance score between the current position n and the ending point function. As shown in Fig. 7, given a global cost map and a target point, the global path to the target point is calculated through the $A^*$ algorithm. The local path planning uses the TEB algorithm [15], which establishes the trajectory execution time target equation, optimizes the actual trajectory of the robot according to the safety distance between the robot and the obstacle and dynamic constraints, and finally outputs the trajectory and wheel speed information of the robot.

# 4 Experimental Result

## 4.1 Cartographer Algorithm Experiment to Enhance Pedestrian Filtering

The system used by the patrol robot platform is Ubuntu 16.04, and its kernel is 4.4.0-139-generic x86 64. It collects and sends sensor data based on the ROS

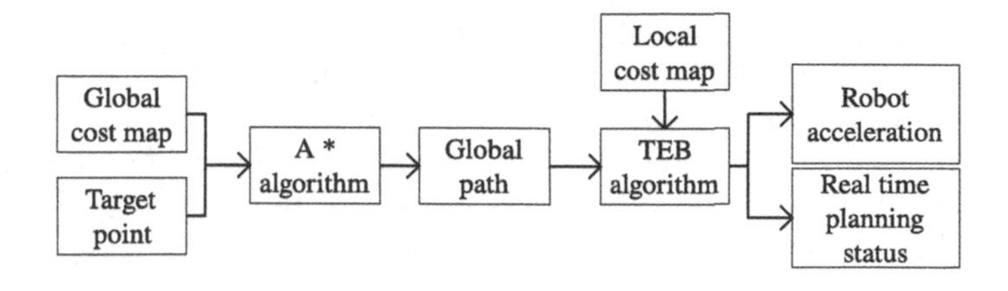


**Fig. 7.** Navigation data flow

platform. Equipped with VLP-16 lidar, the maximum detection distance can reach 100 m, the release frequency is 10 HZ, and it is equipped with IMU and odometer sensors used by SLAM.

Based on the inspection robot platform, the cartographer algorithm for enhanced pedestrian filtering is tested in the laboratory corridor to verify its effect. During the test, two students were arranged to walk back and forth at the designated location. The robot mapping process passed through these two places, recorded the data packet, and used the two algorithms for offline mapping experiment comparison. The results are shown in Fig. 8. Compared with the original algorithm, the map created by the cartographer algorithm with enhanced pedestrian filtering is cleaner, and there is no noise, which is convenient for later path planning and navigation. It verifies the effectiveness of pedestrian filtering when a small number of people are walking around.

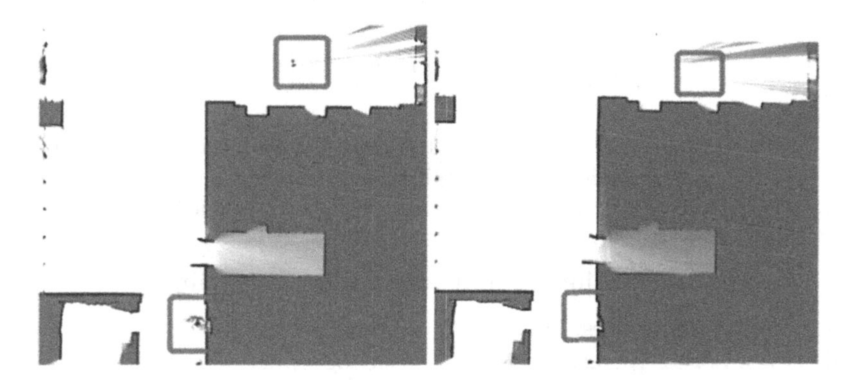

**Fig. 8.** Map constructed by original algorithm (left), map constructed by improved algorithm (right)

## 4.2 Navigation and Positioning Accuracy Experiment

Experimental conditions: test in the public indoor field of the school laboratory, the inspection robot walks on the map to lay tiles, and the slope is 0.

The test method is to specify a target point and obtain the target point coordinates through the laser tracker. The robot starts from any position about 10 m away from the target point and reaches the target point. After the robot stops normally, the laser tracker is used to record the position of each stop point. Stops are recorded repeatedly 30 times, and the position accuracy is calculated using the following formula:

$$AP_p = \sqrt{(\bar{x} - x_c)^2 + (\bar{y} - y_c)^2} \tag{5}$$

Where $\bar{x} = \frac{1}{k}\sum_{i=1}^{k} x_i, \bar{y} = \frac{1}{k}\sum_{i=1}^{k} y_i, x_c$ represents the x-axis coordinates, $y_c$ represents the x-axis coordinates. The results are shown in Fig. 9.

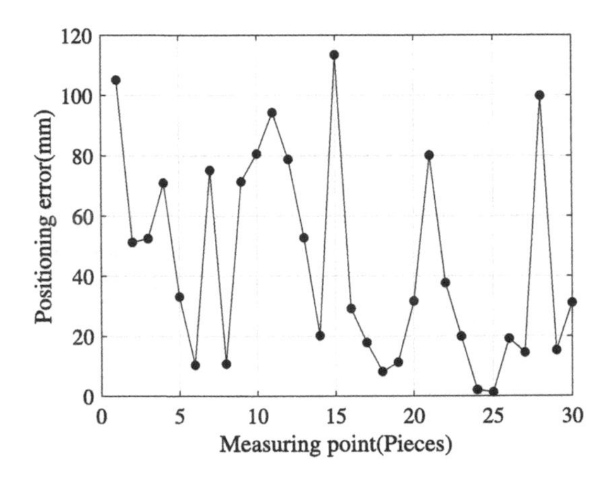


**Fig. 9.** Navigation positioning error

It can be seen that the maximum navigation and positioning error of the system is 113.5 mm, and the average navigation and positioning error is 44.6 mm, which meets the accuracy requirements for practical applications.

### 4.3 Actual Site Positioning and Navigation Experiment

Because the laboratory test environment is relatively small and the flow of people is small, in order to verify whether the cartographer algorithm for enhanced pedestrian filtering can build a clean map and the designed autonomous positioning and navigation system can complete the inspection task. Then the actual site test was carried out, and the test site was selected as the Shanghai Century Avenue Metro Station Hall, which is the largest subway transfer station in Shanghai. For this test, Zone A of Century Avenue Metro Station was selected, with an inspection area of approximately 6,000 $m^2$.

The established environment map is shown in Fig. 10, and the resulting map has almost no pedestrian noise, which verifies the effectiveness of the cartographer algorithm for enhancing pedestrian filtering. After obtaining the environment map, plan the robot's inspection route. The robot's inspection start

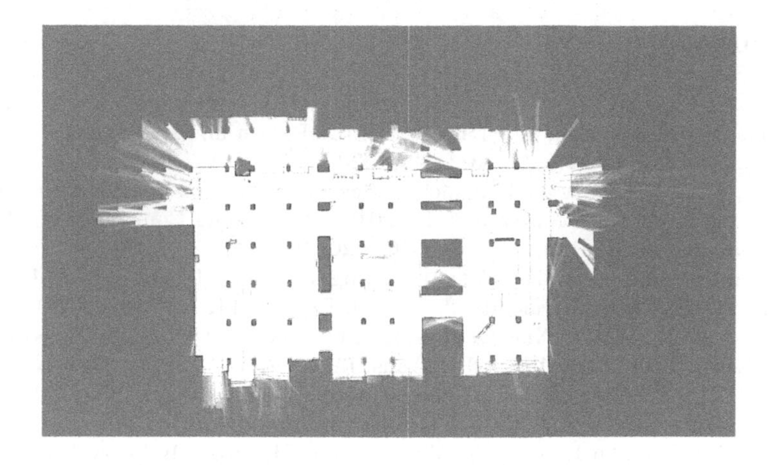

**Fig. 10.** 2D grid map of subway station hall

point, position, and endpoint are all in the upper right corner. In order to effectively collect environmental information, the inspection speed is about 0.3m/s. Finally, the inspection robot starts from the starting point. Along the designated inspection route, back to the endpoint, and successfully completed the inspection task, which verified the feasibility of the designed autonomous positioning and navigation system.

# 5 Conclusion

This article mainly analyzes and researches the current laser SLAM algorithm. Based on the obstacle detection, a cartographer algorithm that enhances pedestrian filtering is designed to achieve the purpose of constructing a relatively clean map in the subway station hall environment with many pedestrians. It also analyzes the existing navigation and positioning algorithms, uses cartographer's relocation to replace the commonly used filter positioning, and uses A* algorithm and TEB algorithm to realize the robot's navigation function. Designed and implemented a mapping system and an autonomous positioning and navigation system based on the enhanced cartographer algorithm on the ROS platform. Finally, through laboratory experiments and actual field experiments, the effectiveness of the enhanced pedestrian filtering cartographer algorithm and the feasibility of the autonomous positioning and navigation system were verified Sex.

**Acknowledgment.** The authors would like to thank the National Key R&D Program of China under Grant (2018YFB2101004).

## References

1. Hara, S., Shimizu, T., Konishi, M., Yamamura, R., Ikemoto, S.: Autonomous mobile robot for outdoor slope using 2D LiDAR with uniaxial gimbal mechanism. J. Rob. Mechatron. **32**(6), 1173–1182 (2020)
2. Zheng, F., Liu, Y.H.: Visual-odometric localization and mapping for ground vehicles using SE(2)-XYZ constraints. In: 2019 International Conference on Robotics and Automation (ICRA), pp. 3556–3562 (2019)
3. Fang, B., Mei, G., Yuan, X., Wang, L., Wang, Z., Wang, J.: Visual SLAM for robot navigation in healthcare facility. J. Pattern Recogn. **113**, 107822 (2021)
4. Li, G., Gan, Y., Wu, H., Xiao, N., Lin, L.: Cross-modal attentional context learning for RGB-D object detection. IEEE Trans. Image Process. **28**(4), 1591–1601 (2019)
5. Qin, T., Li, P., Shen, S.: VINS-mono: a robust and versatile monocular visual-inertial state estimator. IEEE Trans. Rob. **34**(4), 1004–1020 (2018)
6. Xiang, G.: Fourteen Lectures on Visual Slam. Electronic Industry Press, Beijing (2017)
7. Ji, X., Feng, S., Han, Q., Yu, S.: Improvement and fusion of A* algorithm and dynamic window approach considering complex environmental information. Arab. J. Sci. Eng. **46**, 7445–7459 (2021)
8. Chang, L., Shan, L., Jiang, C., Dai, Y.: Reinforcement based mobile robot path planning with improved dynamic window approach in unknown environment. J. Auton Robot. **45**, 51–76 (2021)
9. Li, X., Liu, F., Liu, J., Liang, S.: Obstacle avoidance for mobile robot based on improved dynamic window approach. J. Electric. Eng. Computer Sci. **25**(2), 666–676 (2017)
10. Azizi, M.R., Rastegarpanah, A., Stolkin, R.: Motion planning and control of an omnidirectional mobile robot in dynamic environments. J. Rob. **10**, 48 (2021)
11. Hess, W., Kohler, D., Rapp, H., Andor, D.: Real-time loop closure in 2D LIDAR SLAM. In: 2016 IEEE International Conference on Robotics and Automation (ICRA), pp. 1271–1278 (2016)
12. Przybyła, M.: Detection and tracking of 2D geometric obstacles from LRF data. In: 2017 11th International Workshop on Robot Motion and Control (RoMoCo), pp. 135–141 (2017)
13. Paulo Moreira, A., Costa, P., Lima, J.: New approach for beacons based mobile robot localization using Kalman filters. Procedia Manuf. **51**, 512–519 (2020)
14. Chaari, I., Koubaa, A., Bennaceur, H., Ammar, A., Alajlan, M., Youssef, H.: Design and performance analysis of global path planning techniques for autonomous mobile robots in grid environments. Adv. Rob. Syst. **14** (2017)
15. Rösmann, C., Hoffmann, F., Bertram, T.: Integrated online trajectory planning and optimization in distinctive topologies. Rob. Auton. Syst. **88**, 142–153 (2017)

# Tire Body Defect Detection: From the Perspective of Industrial Applications

Xin Yi[1], Chen Peng[1(✉)], Mingjin Yang[1], and Suhaib Masroor[2(✉)]

[1] School of Mechatronic Engineering and Automation, Shanghai University, Shanghai 200444, China
c.peng@shu.edu.cn

[2] Biomedical Engineering Department, Sir Syed University of Engineering and Technology, Karachi 75300, Pakistan
smasroor@ssuet.edu.pk

**Abstract.** Radial tires have a large market share in the tire market due to their better wear resistance and puncture resistance. However, the complexity of the production process makes impurities and irregular cord spacing defects often appear in the body cord area. In this paper, a method relying on X-ray images to detect impurities and irregular cord spacing defects is proposed. The detection problems of these two types of defects are transformed into calculating cord pixel spacing and background connected domains. Firstly, the tire crown, tire ring, and tire body regions are segmented by a novel semantic segmentation network (SSN). Then the background and the cord are separated by an adaptive binarization method. Finally, the irregular cord spacing defects are detected through the refinement and column statistics. The impurities of tire body are located by the marks of the connected domains. The experimental results of the X-ray images show that this method can meet the positioning requirements of irregular shaped impurities. A idea of setting the threshold column effectively improves the detection speed of irregular cord spacing. In addition, the detection accuracy rates for both types of defects are higher than 90%, which is helpful for further research on various types of tire defects and the design of an automatic tire defect identification system.

**Keywords:** Tire X-ray image · Impurity detection · Irregular cord spacing measurement · Semantic segmentation · Morphology

## 1 Introduction

As the process of industrialization moves towards a mature stage, the scale of the automobile industry is becoming larger and larger. Obviously the tire industry is one of the important supporting industries of the automobile industry. The production process of tires, especially radial tires, contains many processes [1], including a large number of processes involving humans. In addition, the production process of rubber raw materials, belt processing, vulcanization molding


Q. Han et al. (Eds.): LSMS 2021/ICSEE 2021, CCIS 1469, pp. 743–752, 2021.
https://doi.org/10.1007/978-981-16-7213-2_72

and other processing processes can easily cause the final tire product to show various types of defects. The quality defects of tires generally reduce their service life, and may cause traffic accidents in severe cases, which not only cause safety hazards to vehicle drivers, but also increase the risk of claims settlement for tire manufacturers. It can be seen that the best treatment plan is to join the quality inspection link before leaving the factory. At present, X-ray image technology is widely used in the detection of tire defects [2]. This technology can detect various typical defects distributed in tire body, tire crown, and tire ring. In the critical defect determination link, most tire production plants still rely on workers to observe images to achieve defect detection. This method is not only inefficient, but also prone to missed inspection due to fatigue. The study of automatic detection methods based on computer vision is an effective solution to this problem.

So far, many defect detection algorithms have emerged in the application scenarios of industrial production. In recent years, X-ray defect detection based on computer vision has attracted the research interest of many scholars. Some progress in theory and application has also been made in the cooperative research with the industry. Zhao *et al.* Proposed to use the local inverse difference moment (LIDM) to characterize the texture of the tire X-ray image [3]. It is worth mentioning it effectively improves the robustness of defect detection through the background suppression method proposed in the paper, but the calculation speed of LIDM feature distribution needs to be improved. A dissimilarity of the weighted texture in the neighborhood window of each pixel coordinate was calculated to locate the impurities [4]. This method is effective in detecting defects such as irregular cord spacing measurement and impurities. But unfortunately proposed method requires a preset binarization threshold. Based on the principle of wavelet multi-scale analysis, Zhang *et al.*. Statistically analyzed the optimal scale and threshold parameters corresponding to the defect edge and the normal texture [5]. During the experiment, this method had a good description of the impurity defect edge. The Curvelet operator was used to enhance the image texture. Then the improved Canny algorithm to locate the coordinates of impurities [6]. Obviously this method is susceptible to interference from original noise. The author also proposed a method to measure the total variation, that is, to separate the original image into a texture image and a cartoon image, and eliminate the impurities according to the background [7]. It can be seen from the existing research that only relying on traditional image processing methods cannot achieve the detection of complex images.

In recent years, the use of deep learning methods has made great progress in the field of image processing. With regard to tire inspection, the method of taking X-rays one by one has led tire manufacturers to have a large number of normal and defective images [8], which are enough to prepare training samples that meet the needs of neural networks. It cannot be ignored that the requirements for the accuracy and real time of defect detection in industrial production sites are far greater than the standards of life products. A mixture of traditional and deep learning methods is a wise choice.

In this paper, the design of defect detection algorithm for field application is fully considered. A method of segmenting the overall tire structure is covered. Various morphometric treatment solutions are also applied to identify the defects and measure the cord distance. The main innovations of this paper are divided into the following three parts:

1. This paper proposes a lightweight semantic segmentation network, which combines statistical methods to reduce the error of network output boundary. The process completes the separation of tire body, tire ring and crown area.
2. Based on the tire body image extracted in stage 1, an adaptive binarized image is generated. The cord skeleton is obtained by using the thinning algorithm. At this time, statistical analysis showed that the certain cords areas are uneven.
3. This paper designs a connected domain labeling method, which is not only suitable for the detection of strip and block impurities, but also has advantages in detection speed.

## 2 X-Ray Image Area Segmentation

Region segmentation of the original tire X-ray image is a feasible method to achieve rapid defect detection. There are two main methods for texture image analysis: statistical method [11] and texture model construction method [12]. Gabor transform was introduced into the image texture feature extraction as a method of texture model construction. Unfortunately, The tire pattern has a great influence on image segmentation. To improve the effect of region segmentation, this paper introduces a semantic segmentation scheme.

### 2.1 Overview of Semantic Segmentation

Semantic segmentation technology is an important branch of computer vision. Since 2015, Full Convolutional Network (FCN), the DeepLab, the SegNet and the RefineNet networks are proposed one after another. The structure of the SegNet network is expressed in the form of a combination of encoder-decoder modules, which effectively reduces the training parameters and storage and calculation costs of the network. It has been widely used for its excellent understanding of real world scenarios [13]. The lightweight SegNet network is constructed in this paper and applied to the task of tire defect detection to complete the pixel level online tire region segmentation.

### 2.2 Construction of Lightweight SegNet Network

X image is divided into five parts using lightweight Segnet network: left tire ring, left tire body, crown, right tire body, and right ring, as shown in Fig. 1 below.

Taking into account the actual application of real-time detection of tire defects, a encoder structure uses an improved MobileNet network to replace

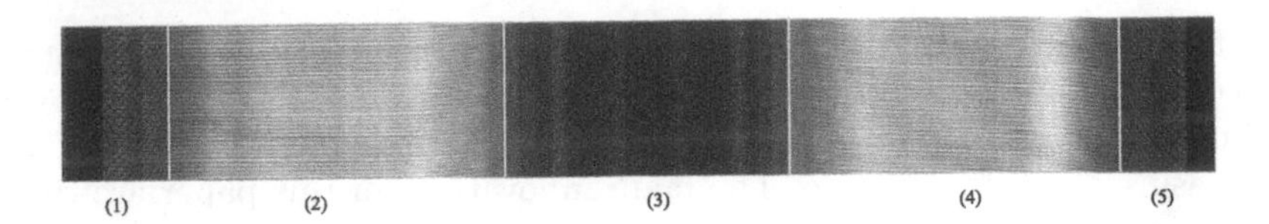


**Fig. 1.** The rules for the division of X-ray images of tires (1) Left tire ring (2) Left tire body (3) Tire crown (4) Right tire body (5) Right tire ring.

the original VGG16 network. Through the use of depth separable convolution, the calculation parameters of the convolution kernel are significantly reduced at a small loss cost. After the feature extraction of the encoder, the salient features of the tire crown, tire body and tire ring boundary are stored in the low-resolution image. The decoder realizes the decoding and mapping of the boundary position from the low-resolution image to the size of the input end. By averaging the boundaries of multiple sub-images, the partition coordinates are completely recorded. A depthwise separable convolution and pointwise convolution are used together to improve the efficiency of defect classification. The specific segmentation process is shown in Fig. 2.

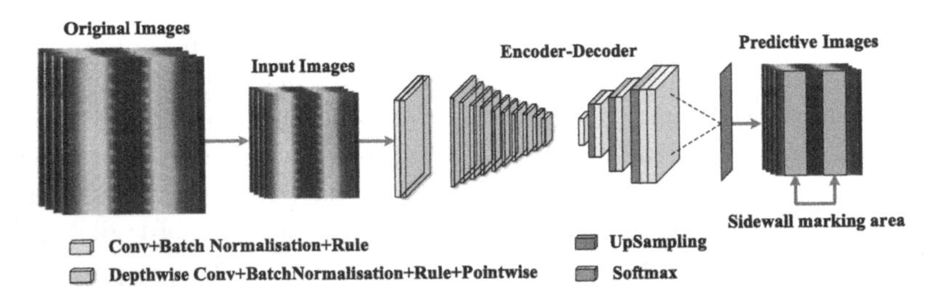


**Fig. 2.** Lightweight semantic segmentation network.

After down sampling, the original image input image with a size of 2464 × 2464 becomes a size of 416 × 416. The decoding process sequentially uses upsampling and convolution kernel operations. This process complete the mapping of boundary features to the input image boundary features. As shown in Fig. 3, it can be found that some of the segmentation boundary results make the tire body area exceed the actual range, and some are less than the real space range. Therefore, the structure still needs to be improved.

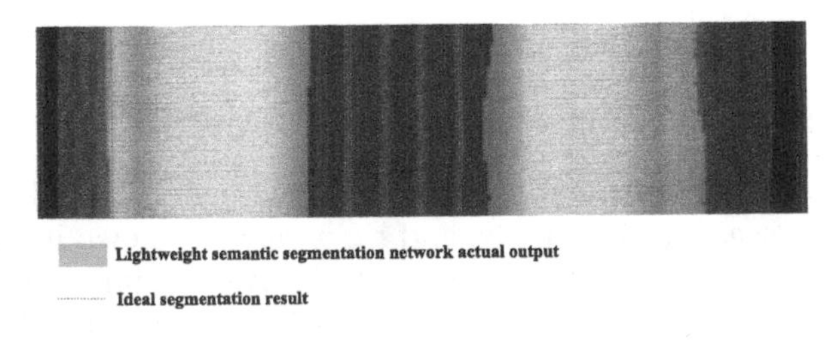


**Fig. 3.** The result of semantic segmentation.

In the preprocessing stage, the idea of dividing an image into multiple images in the horizontal direction is considered. As a result, the specific design idea of tire body segmentation is in the following Algorithm 1.

**Algorithm 1.** Improved segmentation algorithm for tire regions

**Input:** original image, size: $h(height) \times w(width)$, network input image size: $h1(height) \times w1(width)$, initialize count the number of subgraphs: $h/h_1$;

**Output:** sidewall boundary coordinates;

1: Semantic segmentation network operation;
2: Recording boundary coordinates of subgraphs $\Omega$;
3: Computing the sidewall border of the $k$ subgraph $C_k = \{c_{1k}, c_{2k}, c_{3k}, c_{4k}\}$, $c_{mk} = 1/h_1 \sum_{i=1}^{h_1} d_{mi}$, $d \epsilon \Omega$;
4: $O = \{\max(c_{1k}), \min(c_{2k}), \max(c_{3k}), \min(c_{4k})\}$, $k = 1, 2, ...N \leftarrow$ Finally computing the sidewall of original image.

The design idea considers each sub-image as an independent part to predict the boundaries of the tire ring, tire body and crown. First of all, the four boundary columns of each sub-image are averaged pixel by pixel to obtain the four-column boundary of the sub-image. Secondly, under the condition that the predicted boundary is close to the true boundary. The best value of the corresponding coordinates in all the sub-images is selected. The final expected goal is that the predicted tire body area cannot overlap the real tire crown and tire ring area.

## 3 X-Ray Image Defect Detection

The temperature and viscosity of the rubber material used in the casting process will affect the distribution distance of cord. Similarly, various impurities have a wide range of sources, which may not only come from the produced rubber raw materials and production equipment. Irregular cord spacing is shown in Fig. 4(a) and tire body impurties is shown in Fig. 4(b).

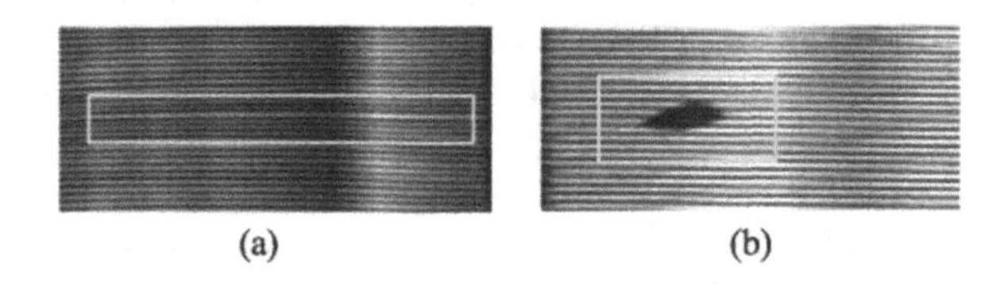


**Fig. 4.** Irregular cord spacing and tire body impurity.

### 3.1 Identification Algorithm of Cord Spacing Irregularity

After the lightweight semantic segmentation network is applied, the tire body image can be obtained. Identification of cord spacing irregularity is mainly completed by three steps: binarization, refinement and distance measurement.

Binarization uses an $11 \times 11$ sliding window and takes the average value in the window as the threshold to divide it into 0 or 255. As shown in Fig. 5 left, the wire in the binarization image has a certain width. In this paper, a thinning algorithm is introduced to extract the central axis of the steel cord to represent the position of the steel wire without destroying the connectivity of the cord. A improved Zhang-Suen thinning algorithm is used in the paper. The image extracted in this way does not guarantee that the width of the wire cord on the steel drawing is a single pixel. Some areas of the wire cord are multiple pixels wide on the image, as shown on the right of Fig. 5.

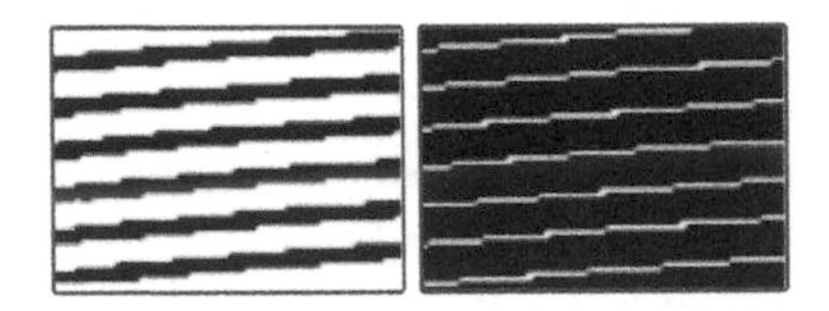

**Fig. 5.** Binarized image of tire body and refined image of tire body.

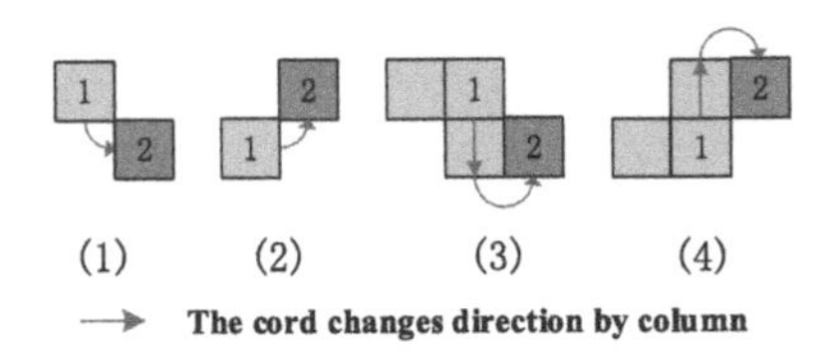


**Fig. 6.** Inter-row jump mode of cord.

It should be noted that a true cord is not completely horizontal but that each cord has a jump in the vertical direction. Pixel position 1 indicates the position of the pixel of a certain cord in the previous column. Position 2 indicates the position of the pixel of this cord in the next column. Due to the existence of incomplete refinement, the pixel point distribution of the cord in two adjacent columns is shown in Fig. 6. There are 4 possible distribution modes. This paper introduces a method of counting the spacing by column. Rn represents the two types of spacing types with the most occurrences in the nth column, which is abbreviated as $Top_2n$ frequency. Algorithm 2 shows how to search for tire body cord spacing irregularities. The cord coordinate matrix is recorded through a custom search method to generate the cord spacing matrix.

Through this method, the marking of the approximate area of the irregular cord spacing can be realized. The column proportion statistical scheme proposed in this paper can effectively reduce the influence of local mean binarization error on the calculation of distance.

**Algorithm 2.** Irregular identification of cord spacing

**Input:** refined image: img, the minimum proportion of columns exceeding the $Top_2$ threshold: $\alpha = 0.85$;
**Output:** Coordinate profile of irregular cord spacing;
1: $P \leftarrow$ Marking the cord coordinates of the first column of refined image (img);
2: $C \leftarrow$ Taking the $P$ as the starting position and generating a coordinate matrix from the refined image(img) representing each cord;
3: $D \leftarrow$ Generating cord distance matrix from $C$;
4: **for** $i = 1;i < D.column;i++$ **do** Calculate $Top_2 i$
5: **end for**
6: **for** $j = 1;j < D.row;j++$ **do** Initialize the $N1$=0
7: **for** $k = 1;k < D.column;k++$ **do**
8: **if** $D[j,k] > Top_{2k}$ **then** $N1++$
9: **end if**
10: **if** $N1/D.column > \alpha$ **then**
11: return the position of cord
12: **end if**
13: **end for**
14: **end for**

### 3.2 Location of Impurity in Tire Body

The detection scheme of tire body impurities is divided into four steps: pre-process, local binarization, marking connected demain, locating impurity area. Histogram equalization was used in the pretreatment stage. The size of the processing window set in this paper is 300 × 4. Due to the pattern on the sidewall, if the original image is directly binarized, the cord is likely to be accidentally disconnected at the pattern, which is not conducive to the subsequent connected domain marking process. Due to the obvious changes of gray values in the tire body area of different tire types, the image can not be divided into three parts by the double threshold method: background, cord and impurities. By using local binarization, the background is divided into pure white, while the cord and impurities are divided into pure black, as shown in Fig. 7. The areas where the width of the white connected domain is not consistent with the width of the original image can be considered to be caused by impurities. According to this idea, the following Algorithm 3 is proposed.

Through the step search of Algorithm 3, the approximate area of impurities is determined. It should be pointed out that this method is mainly for the detection of impurities in the vertical direction.

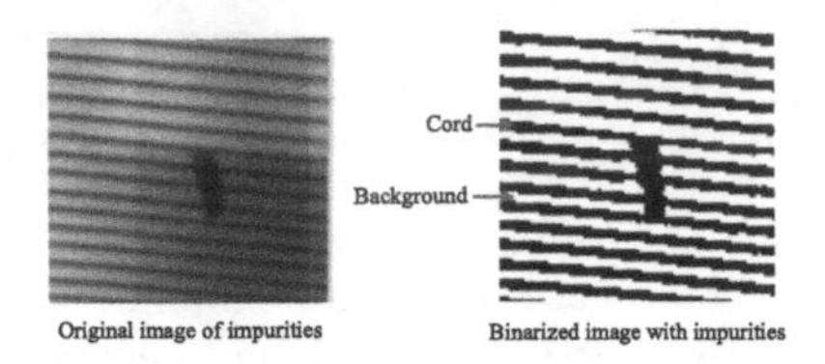


**Fig. 7.** Principle of tire body impurities search.

---

**Algorithm 3.** Sidewall impurities recognition method

---

**Input:** Tire body original image: img1, number of continuous incomplete contours:$num = 0$, minimum positioning accuracy of impurities: $\delta = 4$;
**Output:** Impurity coordinate area
1: eq $\leftarrow$ Histogram equalization $\leftarrow$ img1;
2: The number of connected domains corresponding to the white area: $n \leftarrow$ bin $\leftarrow$ Local binarization $\leftarrow$ eq;
3: $\Phi_i \leftarrow$ Record the $i$th contour coordinates of the white connection field to mark the background $\leftarrow$ bin;
4: $\phi_{ik} \epsilon \Phi_i (k = 1, 2, ..., q)$, $q$ represents the number of contour coordinates;
5: **for** $j = 1, ..., n$ **do**
6:     **if** max ($\phi_{jr}.x$) <img1.column **then**
7:         $num++$
8:         **if** $num >= \delta$ **then**
9:             return impurity coordinate
10:         **else**
11:             continue
12:         **end if**
13:     **else**
14:         $num = 0$
15:     **end if**
16: **end for**

---

## 4 Experimental Analysis and Verification

In the experiment, the computer processor Intel i7-6700U runs 16 GB of memory. Firstly, a semantic segmentation network proposed is compared with the method of Gabor transform. It has been verified that the proposed scheme is superior in the accuracy of dividing the crown, tire body and tire ring. The parameters of the Gabor transformation selected in the experiment include the wavelength of the factor $\lambda = 2\pi$, the normal direction of the parallel stripes of the Gabor function $\theta = \pi/2$, the standard deviation of the Gaussian envelope $\sigma = 1$ and the gamma space aspect ratio g = 0.5. This method has a certain filtering effect on the horizontal cord, but it is easily affected by the change of image brightness. Thus, this paper proposes to use a semantic segmentation network to divide the tire ring, tire body and tire crown. In order to improve the accuracy of the final segmentation, Algorithm 1 is proposed on this basis, which relies on multiple sub-images to calculate the average segmentation line. In comparison, the accuracy of the region segmentation is significantly improved.

The data set to verify the uneven line spacing and impurity defect detection algorithm was provided by a tire factory. The resolution of all images is 2469 × 10000, including 80 images with irregular line spacing, 80 images with impurity and 80 images without defects. As shown in Fig. 8(a), the images from top to bottom represent the original input image, the binarized image and the refined image. Figures 8(b) and 8(c) are the detection effects of tire body impurities, showing the original defect map, binarized image, and location mark image from top to bottom. It can be seen that the tire body impurities location method proposed in this paper not only locates the massive impurities, but also meets the requirements of vertical impurity defect position detection.

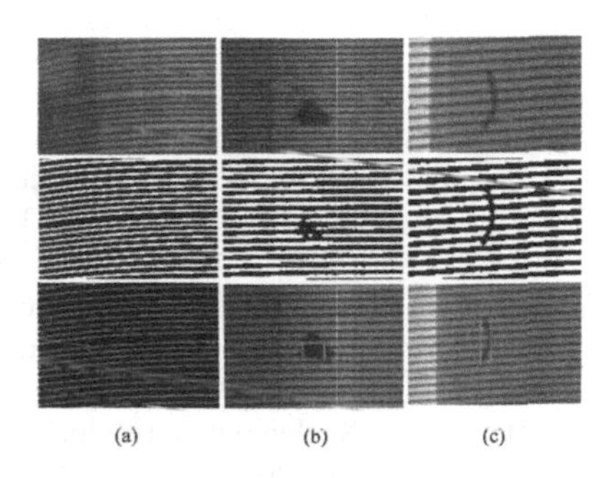


**Fig. 8.** Tire body cord spacing irregularities and impurities identification results

In the test of the performance of the algorithm proposed in this paper, the average detection speed of tire body impurities is 3s. The identification of the irregular cord spacing is also completed within 5s. When measuring the performance of the algorithm, the misjudgment rate (MR) of the defect-free images and the accuracy rate (AR) of the image defect types are designed. Through the verification of the data set composed of 240 images, the performance of the algorithm is shown in Table 1.

**Table 1.** Defect detection performance results

| Type of defect | MR | AR |
|---|---|---|
| Sidewall impurities | 0.018 | 0.91 |
| Irregular cord spacing | 0.014 | 0.93 |

## 5 Conclusion

An improved semantic segmentation network is used to extract the tire body. In the following irregular cord spacing defect detection, based on the refined cord pitch, the presence of irregular cord spacing is determined by the column specific gravity threshold method. At the same time, considering that the tire body is prone to impurity defects, this paper proposes to use the connected domain labeling method to locate the area of the incomplete connected domain. In the end, the experimental results have showed that the proposed detection method for tire body impurities and irregular cord spacing not only meets the low time complexity, but also meets the requirements of industrial production in the accuracy of defect identification. Furthermore, the effect of impurity defect size measurement is not ideal. It is planned to focus on the outline extraction of impurity defects in the next research.

## References


1. Zhou, Y.: Design of manufacturing execution system in tire enterprises. Trans. Tech. Publ. Ltd. **411**, 2343–2346 (2013)
2. Zhang, Y., Cui, X., Liu, Y., Yu, B.: Tire defects classification using convolution architecture for fast feature embedding. Int. J. Comput. Intell. Syst. **11**(1), 1056–1066 (2018)
3. Zhao, G., Qin, S.: High-precision detection of defects of tire texture through X-ray imaging based on local inverse difference moment features. Sensors **18**(8), 2524 (2018)
4. Guo, Q., Zhang, C., Liu, H., Zhang, X.: Defect detection in tire X-ray images using weighted texture dissimilarity. J. Sens. (2016)
5. Zhang, Y., Lefebvre, D., Li, Q.: Automatic detection of defects in tire radiographic images. IEEE Trans. Autom. Sci. Eng. **14**(3), 1378–1386 (2015)
6. Zhang, Y., Li, T., Li, Q.: Defect detection for tire laser shearography image using curvelet transform based edge detector. Opt. Laser Technol. **47**, 64–71 (2013)
7. Zhang, Y., Li, T., Li, Q.: Detection of foreign bodies and bubble defects in tire radiography images based on total variation and edge detection. Chin. Phys. Lett. **30**(8), 084205 (2013)
8. Deng, L., Yu, D.: Deep learning: methods and applications. Found. Trends Sig. Process. **7**(3), 197–387 (2014)
9. Chen, J., Li, Y., Zhao, J.: X-ray of tire defects detection via modified faster R-CNN. In: 2019 2nd International Conference on Safety Produce Informatization (IICSPI), pp. 257–260 (2019)
10. Cui, X., Liu, Y., Zhang, Y., Wang, C.: Tire defects classification with multi-contrast convolutional neural networks. Int. J. Pattern Recogn. Artif. Intell. **32**(4), 1850011 (2018)
11. Zheng, G., et al.: Development of a gray-level co-occurrence matrix-based texture orientation estimation method and its application in sea surface wind direction retrieval from SAR imagery. IEEE Trans. Geosci. Remote Sens. **56**(9), 5244–5260 (2018)
12. Jacob, N., Kordi, B., Sherif, S.: Assessment of power transformer paper ageing using wavelet texture analysis of microscopy images. IEEE Trans. Dielectr. Electr. Insul. **27**(6), 1898–1905 (2020)
13. Lin, C.M., Tsai, C.Y., Lai, Y.C., Li, S.A., Wong, C.: Visual object recognition and pose estimation based on a deep semantic segmentation network. IEEE Sens. J. **18**(22), 9370–9381 (2018)

# Computational Intelligence and Applications

# TPE-Lasso-GBDT Method for BV-2 Cell Toxicity Classifier

Qing Liu[1], Dakuo He[1,2](✉), Jinpeng Wang[1], and Yue Hou[3](✉)

[1] College of Information Science and Engineering, Northeastern University, Shenyang 110819, People's Republic of China

[2] State Key Laboratory of Synthetical Automation for Process Industries and the College of Information Science and Engineering, Northeastern University, Shenyang 110819, People's Republic of China

hedakuo@mail.neu.edu.cn

[3] College of Life and Health Sciences, Northeastern University, Shenyang 110169, People's Republic of China

houyue@mail.neu.edu.cn

**Abstract.** Neuroinflammation mediated by microglia cells plays an important role in the pathogenesis of Alzheimer's disease. Classification of the BV-2 cell toxicity is a significant stage in early drug discovery process. The prediction models based on computer aided drug design with rapid and high accuracy characteristics were widely applied in drug toxicity domain. In this study, BV-2 cell toxicity data set containing 191 compounds was collected from actual experimental results and relevant literature, which contains 63 nontoxicity and 128 toxicity compounds. This study faces the problems of data imbalance, low sample size and high dimension features. In order to solve these problems, this study proposes a TPE-Lasso-GBDT method for BV-2 cell toxicity classifier, of which, GBDT as BV-2 cell toxicity classifier based on boosting ensemble method, Tree Parzen Estimator (TPE) was used to optimize the model hyper-parameter so as to realize better performance of the model, Lasso feature selection method was used to eliminate redundant features. The validity of the proposed method was verified by comparative experiments.

**Keywords:** BV-2 cell toxicity classifier · Computer aided drug design · Data imbalance · TPE-Lasso-GBDT

## 1 Introduction

Drug toxicity is defined as the side effect on cells and organs caused by the metabolism or action of a compound [1]. Drug toxicity measurement is one of the most significant and challenging step involved in the drug design and discovery,

Thanks are due to the National Natural Science Foundation of China under grant nos. 61773105, 61533007, 61873049, 61873053, 61703085 and 61374147 and the Fundamental Research Funds for the Central Universities under grant no. N182008004 for supporting this research work.



Q. Han et al. (Eds.): LSMS 2021/ICSEE 2021, CCIS 1469, pp. 755–764, 2021.
https://doi.org/10.1007/978-981-16-7213-2_73

which has a direct effect on the drug development cycle [1]. High-throughput screening for toxicity that is capable of assessing and prioritizing the cell toxicity of compound for further experimental verification, but there are time-consuming and expensive problems [2]. Therefore, an in silico calculation model has been widely applied for drug toxicity that is the advantage of relatively cheap, rapid and it is a credible substitution to massive scale in vivo and in vitro bioassays [1,3].

In recent years, many domestic and foreign researchers have discussing the potential roles of machine learning methods in toxicity prediction field. Most of the publication results can be divided into classical machine learning [4–6] and deep learning algorithms [7–10]. In which, the two key factors are the selection of molecular representations that are most relevant to the task and an appropriate modeling algorithm was selected to build a toxicity classification model. For ML-based approaches, it is important to select suitable ML algorithm to build the toxicity prediction model for a specific data set. At present, various ML algorithms that contains SVM, RVM, kNN, RF, multilayer perceptron ensemble (MLPE), XGBoost, CT, NB were widely applied for QSAR modeling. Although well-performing toxicity predictive models have been developed using the above ML methods, which inputs are in the form of specific molecular representations that cannot be automatically extracted task-specific chemical features [1]. With the rise of DL technologies [7–10], it has been possible to generate new molecular representations with abstract and task-specific chemical features. Deep learning methods strongly depend on encoding functions for mapping molecular structural information into a fixed-length vector as the molecular representation, and it is closely related to the quantity and quality of the data. Several groups have proposed DL networks that predict toxicity by automatically learning feature vectors from molecular structure [9,10]. Due to the limitation of data quantity, it is not applicable to establish BV-2 cell toxicity prediction model based on DL algorithm. In this study, six molecular representations, three machine learning methods and three feature selection methods were assessed and compared by the predictive accuracy and F1-score of the BV-2 cell toxicity classifier.

## 2 Methods

### 2.1 Molecular Representation

In the field of computer aided drug design, it is extremely important to characterize drug molecules in machine-readable form. In recent years, molecular descriptors has been widely used in QSAR modeling, which mainly focuses on chemical structure information which contains fragment information, molecular topology, connectivity, pharmacophore and shape patterns. A brief description of six molecular descriptors as follows:

**Avalon Fingerprint** enumerates feature types of the molecule graphics and certain paths by using molecular generator, which contain the information of atom, bond, ring and feature pairs [11]. By using a hash function, all molecules were coding to fingerprint bit implicitly when they are enumerated.

**MACCS** pre-defined 166 substructure fragments known as MDL Public keys [12], which consisting of a string of binary digits that describe the structural characteristics of molecules, for the detail information of substructure fragments in the literature [12].

**BRICS** was automatic fragment molecules based on seventeen types synthetic bonds, which obtained from vendor catalogue sources and biologically active compounds [13]. It exploits the medicinal chemistry information contained in existing compounds and integrates it with domain knowledge of certain chemical elements [13].

**Rdkit topological fingerprint** was inspired by Daylight Fingerprint, which calculate the molecular subgraphs that contain the information of an atomic types, aromaticity and bond types between minPath and maxPath, which was encoding as numeric identifiers using a hash function [14].

**ECFP** is formed by setting a radius from a particular atom and counting the number of parts of the molecular structure within that radius [15]. It is not pre-defined, and represent a large number of different molecular characteristics, including stereochemical information, and has the advantages of fast calculation speed and flexible scene usage [15].

**WHALES** [16] is a holistic molecular representation incorporating pharmacophore and shape pattern, which capture simultaneously molecular shape, the partial charge distribution and geometric interatomic distance. The above six molecular descriptors were calculated using open-source software RDKit.

## 2.2 Machine Learning Models

**Support Vector Machine** was known for its ability to deal with small samples [17]. The basic theory of SVM classifier is find a hyperplane in a multidimensional space that maximizes geometric margin area and minimize empirical error [17]. In this paper, the value of C within its range $[0.001, 1000)$, kernel function that contains[linear, rbf, sigmoid, poly], the range of Gamma is $[0.0001, 8)$, which will be optimized by TPE, note that there is no need to optimize the gamma parameter for linear kernel function.

**Random Forest** is an ensemble algorithm that integrates multiple decision trees based on bagging algorithm, which aim is to improve the generalization ability of the model by reducing the variance of individual decision trees [18]. RF introduced the randomness of samples and variables to minimize correlation between different decision trees [18], this procedure is repeated N times to obtain N trees. The main hyper-parameter in RF are $N\_estimators$,$Max\_depth$, $Max\_features$, the range of which are $[50, 100]$, $[5, 20]$, $[5, 30]$, respectively.

**Gradient Boosting Decision Tree** is an ensemble algorithm that integrates multiple CART regression trees based on boosting algorithm, which combined the additive model with forward distribution algorithm [19]. The regression tree as the base learner was trained under the condition of initial value, then the prediction value can be gained in the leaf node and obtained the residual of the base learner [19]. The subsequent CART regression tree will learn based on the gradient direction of residual reduction until the termination condition is met [19]. The main hyper-parameter in GBDT are *N_estimators*, *Subsample*, *Learning_rate*, the range of which are [50, 100], [0.5, 1), [0.05, 0.06), respectively.

## 2.3 Feature Selection Methods

Molecular descriptors have high dimensional features that contain various chemical and biological information, it is necessary that we select important features to QSAR modeling so as to improve the generalization ability and prevent overfitting. Mutual information, Lasso and Extremely randomized trees are adopted to select important features in this study.

**Mutual information** is used to measure the correlation between a feature variable and a particular category. If the amount of feature information is greater, the correlation between the feature and the category is greater and vice versa [20,21].

**Lasso.** The basic idea of which is that the sum of the absolute values of regression coefficients is less than a constant under the constraint condition [22], the penalty function is minimized. The optimization objective function is:

$$\min_{W} \frac{1}{2} \left\| X^T w - Y \right\|_2^2 + \lambda \|w\|_1 \tag{1}$$

Of which,$w$ represents the regression coefficient of the eigenvector, $\|w\|_1$is the regularization term uses the $L_1$ norm to generate a sparse solution in the eigenvector space, the coefficients corresponding to the irrelevant and redundant features will be set to 0, while others will be retained for subsequent classification task [22].

**Extremely randomized trees** (ET) was proposed by Pierregeurts et al. [23] in 2006, it has been widely applied in supervised classification and regression problems. ET assigns a score to each feature in the data as its importance, the higher the score, the more important the feature is. By using the element importance attribute of the model, the element importance of each element in the feature set can be obtained. Therefore, features are screened to delete redundant features and retain important features.

## 2.4 Tree Parzen Estimator (TPE)

TPE was used as hyper-parameter optimizetion method that converts the configuration space into a nonparameteric density distribution. The configuation space is described by using uniform, log-uniform, q-uniform, and categorical variables [20]. TPE can produce different densities over the configuration space, using different observations $\{x^{(1)}, x^{(2)}, \cdots, x^{(k)}\}$, result in different non-parametric densities:

$$p(x|y) = \begin{cases} \ell(x) \text{ if } y < y^* \\ g(x) \text{ if } y \geq y^* \end{cases} \tag{2}$$

where $\ell(x)$ formed for those observations $\{x^{(i)}\}$ that satisfy corresponding condition(the corresponding loss $f(x^{(i)})$ was less than $y^*$ and $g(x)$ formed for the remaining observations. TPE was used to obtain the optimal model hyper-parameter to improve the prediction performance. In order to evaluate the model reasonably and improve the stability of the prediction model, 5-fold cross validation was used for the above modeling methods.

## 3 Simulation

### 3.1 Dataset Description and Model Performance Evaluation

One hundred ninety-one samples of BV-2 cell toxicity were provided by actual experiments [24] and relevant literature [25–27], among which one hundred twenty-eight were toxicity and sixty-three were nontoxicity. BV-2 cell viability was measured using MTT reduction method as an evaluation index for cell toxicity. At the highest measured concentration, the ratio of BV-2 cell viability more than 80% are considered nontoxicity, and vice versa. For detailed experimental procedures please refer to the literature [24].

In order to comprehensively evaluate the model performance, prediction accuracy($Acc$), F1-score were applied for evaluate BV-2 cell toxicity classifier, $Acc \in [0, 1]$ reflects the ratio of accurate prediction samples, F1-Score $\in [0, 1]$ as the harmonic mean of precision and recall. The bigger the above metrics are, the better the model performance will be.

### 3.2 Experiment Results

Figures 1, 2 and 3 compare of the model performance evaluation that are the prediction accuracy and the F1-score of training set and test set for six molecule representations based on GBDT, SVM, RF modeling methods, among which hyper-parameter were optimized by TPE. As shown in Fig. 1, compare with six molecular descriptors based on GBDT modeling method, the best molecule representation is Rdkit molecular descriptor, the ECFP4 and Avalon molecular descriptor came in second. This is because the Rdkit fingerprint can describe preferably the information of atomic types, aromaticity and bond types between minPath and maxPath. As shown in Fig. 2, the best result is Rdkit molecular descriptor, the Avalon and ECFP4 molecular descriptor came in second for RF modeling method. As shown in Fig. 3, take SVM as the modeling method, the overall results were inferior to GBDT and RF modeling method, it is verified that the ensemble learning model had better performance for the BV-2 cell toxicity classifier in this paper. Of which, the F1-Score performance index is particularly low. For the above three modeling methods, the optimal model performance correspond to molecular representation is Rdkit molecular descriptor, individually. Therefore, the feature selection methods were discussed subsequently based on Rdkit molecular descripitor and three different modeling methods.

Figures 4, 5 and 6 compare the prediction accuracy and F1-score of training set and test set of three feature selection methods with three modeling methods

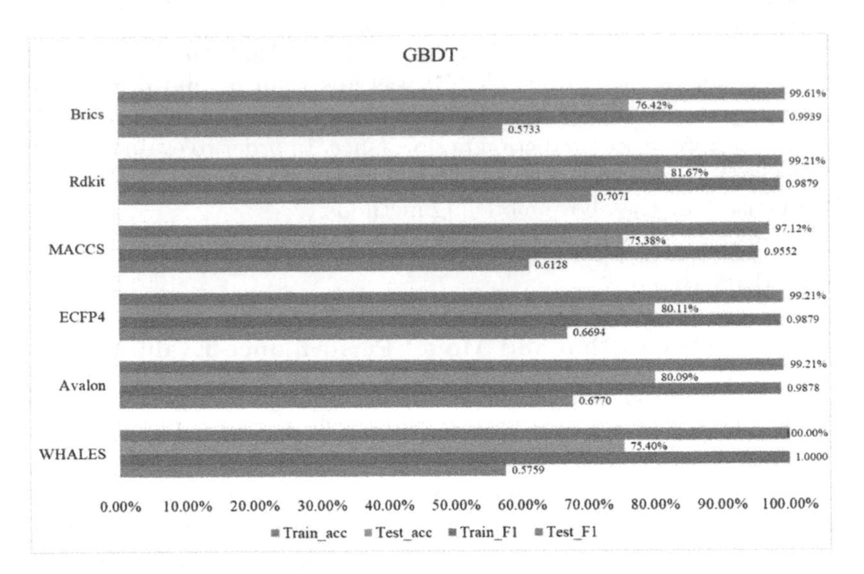


**Fig. 1.** Comparsion of the performance indexes for different molecule representations based on GBDT.

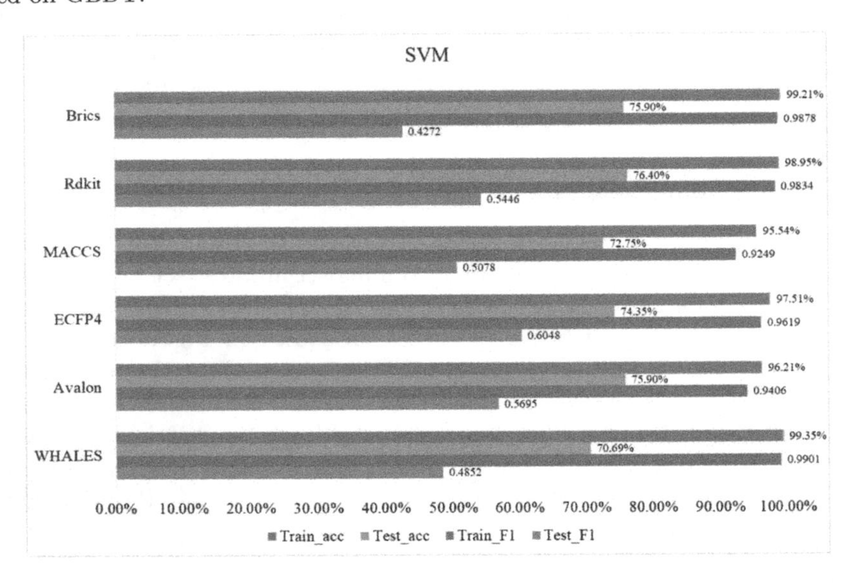


**Fig. 2.** Comparsion of the performance indexes for different molecule representations based on SVM.

based on Rdkit molecular descriptor. Compare to Figs. 1, 2 and 3, the results shown that three different feature selection methods can improve the prediction performance of the BV-2 cell toxicity classifier. Among the three feature selection methods, Lasso feature selection method had the best performance in GBDT, SVM, RF modeling methods. Compared to the basic three modeling methods,

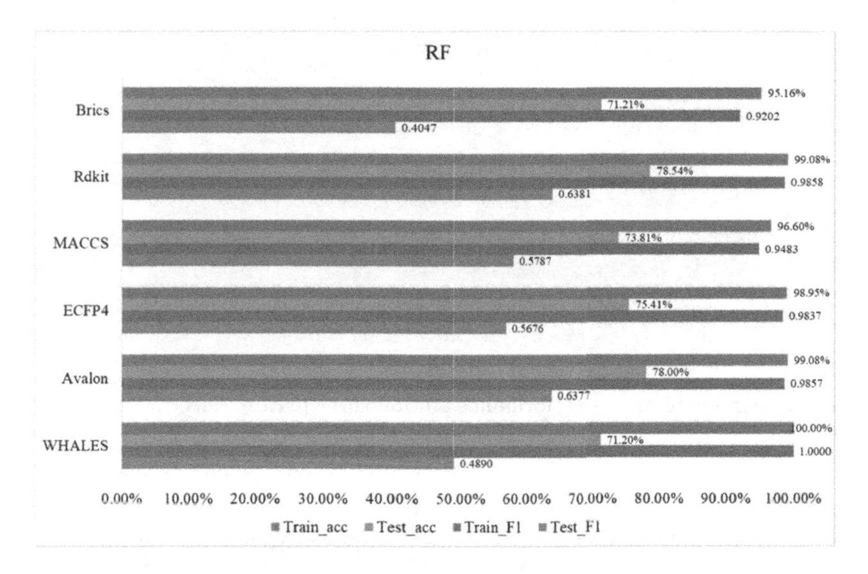


**Fig. 3.** Comparsion of the performance indexes for different molecule representations based on RF.

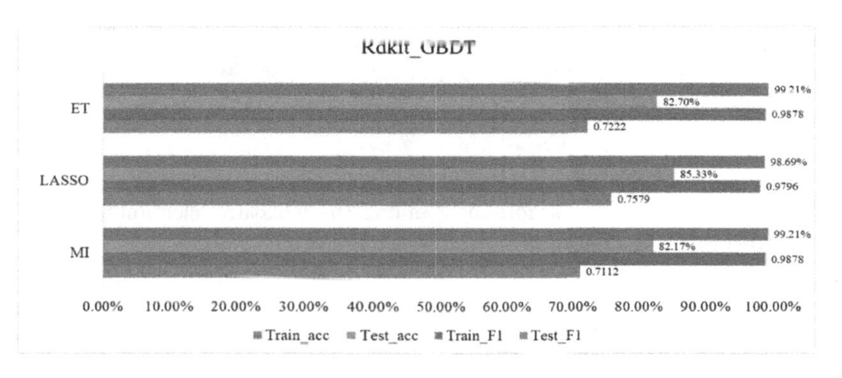


**Fig. 4.** Comparsion of model performance among three feature selecting methods based on Rdkit_GBDT.

the results showed that the prediction accuracy and F1-Score of test set based on Lasso-GBDT, Lasso-SVM and Lasso-RF method have improved 3.66% and 0.0508, 7.9% and 0.2030, 5.75% and 0.0984, respectively. The effectiveness of the Lasso feature selection method can be verified by the above experimental results, it used to gain important features that relevant to BV-2 cell toxicity. Of which, Lasso-GBDT based on Rdkit molecular descriptor for BV-2 cell toxicity classifier has the best predcition performance.

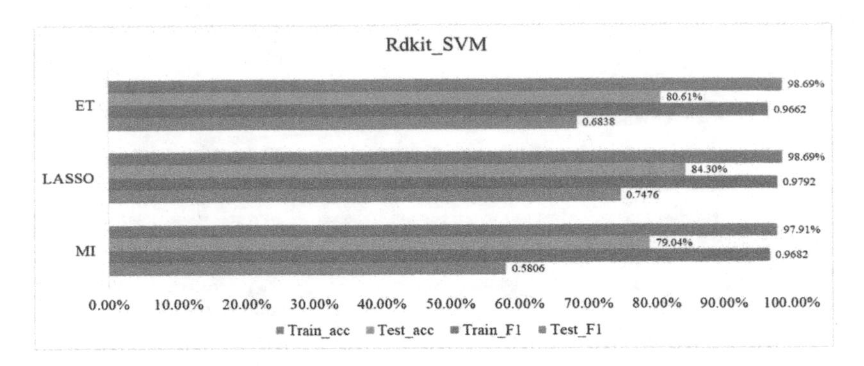


**Fig. 5.** Comparsion of model performance among three feature selection methods based on Rdkit_SVM.

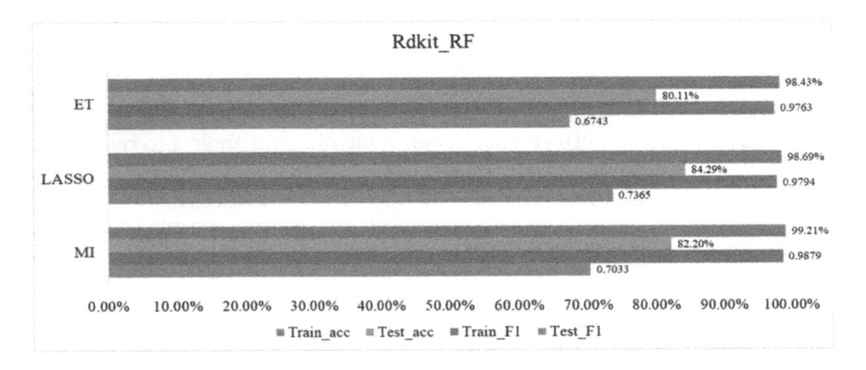


**Fig. 6.** Comparsion of model performance among three feature selection methods based on Rdkit_RF.

## 4 Conclusion

The aim of this study is to compass BV-2 cell toxicity classifier based on computer aided drug design method. In order to solve the problem of small size samples and data imbalance, six different molecular descriptors combined with three modeling methods were used to eatablish BV-2 cell toxicity classifier and the corresponding hyper-parameter can be optimized by TPE, the results shown Rdkit molecular descriptor was the best molecular representation and the ensemble modeling methods (RF, GBDT) was better than SVM. Among which, Rdkit_GBDT is the best one in this study. The Lasso feature selection method was introduced to address the problem of high feature dimension. TPE-Lasso-GBDT was obtained by couple TPE hyper-parameter optimization method with GBDT modeling method and Lasso feature selection method. The results show that the accuracy and F1-Score were 98.69% and 0.9796 for the training set and 85.33% and 0.7579 for the test set, respectively. Due to the small amount of data available at present, DL-based algorithms cannot be applied in this study. We will further collect relevant data and adopt DL-based algorithms to improve the prediction accuracy of BV-2 cell toxicity classifier in the future.

## References


1. Yang, X., Wang, Y., Byrne, R., et al.: Concepts of artificial intelligence for computer-assisted drug discovery. Chem. Rev. **119**(18), 10520–10594 (2019)
2. Zhu, X.W., Xin, Y.J., Chen, Q.H.: Chemical and in vitro biological information to predict mouse liver toxicity using recursive random forests. SAR QSAR Environ. Res. **27**(7), 559–572 (2016)
3. Zhang, C., Cheng, F., Li, W., et al.: In silico prediction of drug induced liver toxicity using substructure pattern recognition method. Mol. Inf. **35**(3–4), 136–144 (2016)
4. Lei, T., Li, Y., Song, Y., et al.: ADMET evaluation in drug discovery: 15. Accurate prediction of rat oral acute toxicity using relevance vector machine and consensus modeling. J. Cheminf. **8**(1), 1–19 (2016)
5. Lei, T., Chen, F., Liu, H., et al: ADMET evaluation in drug discovery. Part 17: development of quantitative and qualitative prediction models for chemical-induced respiratory toxicity. Mol. Pharm. **14**(7), 2407–2421 (2017)
6. Lei, T., Sun, H., Kang, Y., et al.: ADMET evaluation in drug discovery. 18. Reliable prediction of chemical-induced urinary tract toxicity by boosting machine learning approaches. Mol. Pharm. **14**(11), 3935–3953 (2017)
7. Mayr, A., Klambauer, G., Unterthiner, T., et al.: DeepTox: toxicity prediction using deep learning. Front. Environ. Sci. **3**, 80 (2016)
8. Gawehn, E., Hiss, J.A., Schneider, G.: Deep learning in drug discovery. Mol. Inf. **35**(1), 3–14 (2016)
9. Xu, Y., Dai, Z., Chen, F., et al.: Deep learning for drug-induced liver injury. J. Chem. Inf. Model. **55**(10), 2085–2093 (2015)
10. Xu, Y., Pei, J., Lai, L.: Deep learning based regression and multiclass models for acute oral toxicity prediction with automatic chemical feature extraction. J. Chem. Inf. Model. **57**(11), 2672–2685 (2017)
11. Gedeck, P., Rohde, B., Bartels, C.: QSAR-how good is it in practice? comparison of descriptor sets on an unbiased cross section of corporate data sets. J. Chem. Inf. Model. **46**(5), 1924–1936 (2006)
12. Durant, J.L., Leland, B.A., Henry, D.R., et al.: Reoptimization of MDL keys for use in drug discovery. J. Chem. Inf. Comput. Sci. **42**(6), 1273–1280 (2002)
13. Degen, J., Wegscheid-Gerlach, C., Zaliani, A., et al.: On the art of compiling and using'drug-like'chemical fragment spaces. ChemMedChem Chem. Enabling Drug Discov. **3**(10), 1503–1507 (2008)
14. Nilakantan, R., Bauman, N., Dixon, J.S., et al.: Topological torsion: a new molecular descriptor for SAR applications. Comparison with other descriptors. J. Chem. Inf. Comput. Sci. **27**(2), 82–85 (1987)
15. Rogers, D., Hahn, M.: Extended-connectivity fingerprints. J. Chemi. Inf. Model. **50**(5), 742–754 (2010)
16. Grisoni, F., Merk, D., Consonni, V., et al.: Scaffold hopping from natural products to synthetic mimetics by holistic molecular similarity. Commun. Chem. **1**(1), 1–9 (2018)
17. Cortes, C., Vapnik, V.: Support-vector networks. Mach. Learn. **20**(3), 273–297 (1995)
18. Ho, T.K.: The random subspace method for constructing decision forests. IEEE Trans. Pattern Anal. Mach. Intell. **20**(8), 832–844 (1998)
19. Friedman, J.H.: Greedy function approximation: a gradient boosting machine. Ann. Stat. 1189–1232 (2001)

20. Dong, H., He, D., Wang, F.: SMOTE-XGBoost using tree parzen estimator optimization for copper flotation method classification. Powder Technol. **375**, 174–181 (2020)
21. Church, K., Hanks, P.: Word association norms, mutual information, and lexicography. Comput. Linguist. **16**(1), 22–29 (1990)
22. Tibshirani, R.: Regression shrinkage and selection via the lasso: a retrospective. J. Roy. Stat. Soc. Ser. B (Stat. Methodol.) **73**(3), 273–282 (2011)
23. Geurts, P., Ernst, D., Wehenkel, L.: Extremely randomized trees. Mach. Learn. **63**(1), 3–42 (2006)
24. Tang, Y., Su, G., Li, N., et al.: Preventive agents for neurodegenerative diseases from resin of Dracaena cochinchinensis attenuate LPS-induced microglia overactivation. J. Nat. Med. **73**(1), 318–330 (2019)
25. Li, N., Ma, Z., Li, M., et al.: Natural potential therapeutic agents of neurodegenerative diseases from the traditional herbal medicine Chinese dragon' s blood. J. Ethnopharmacol. **152**(3), 508–521 (2014)
26. Zhou, D., Li, N., Zhang, Y., et al.: Biotransformation of neuro-inflammation inhibitor kellerin using Angelica sinensis (Oliv.) Diels callus. RSC Adv. **6**(99), 97302–97312 (2016)
27. Li, J., Jiang, Z., Li, X., et al.: Natural therapeutic agents for neurodegenerative diseases from a traditional herbal medicine Pongamia pinnata (L.) Pierre. Bioorg. Med. Chem. Lett. **25**(1), 53–58 (2015)

# High-Resolution Digital Mapping of Soil Total Nitrogen in Hilly Region Using Multi-variables Based on Landform Element Classification

Yuchen Wei[1], Changda Zhu[1], Xiuxiu Zhang[1], Xin Shen[2], and Jianjun Pan[1(✉)]

[1] Nanjing Agricultural University, College of Resources and Environmental Sciences, Anhui Polytechnic University, Weigang 1, Nanjing 210095, China
jpan@njau.edu.cn

[2] Anhui Key Laboratory of Electric Drive and Control, Anhui Polytechnic University, Weigang 1, Nanjing 210095, China

**Abstract.** In this study, we simulated the complex and nonlinear relationship between STN and environmental variables, and evaluated the importance of each variable for accurate STN mapping. Machine learning models were trained to map STN contents in a small watershed (1: 25000) of hilling area using high-resolution landform elements classification maps, DEM derivatives, and optical and SAR remote sensing data. The performance was evaluated using all GM variables under different hyperparameter settings. Three machine-learners were applied with different combinations of variables. The results showed that more accurate predictions of STN content were achieved with the introduction of GM variables compared to only using individual DEM derivatives. In addition, the GM map with 20 cells searching radius ($L$) and 5° flatness threshold ($t$) showed the highest relative importance within GM variables in three models.

**Keywords:** Soil total nitrogen · Landform elements · Combination of variables · Spatial distribution · Machine learning

## 1 Introduction

As an important soil attribute reflecting the level of soil quality and fertility, the correct assessment of the spatial variability of soil total nitrogen (STN) at the small watershed scale is of great significance for guiding the promotion of soil management measures [1]. However, it is not only time-consuming and laborious but also costly to obtain relevant information through traditional soil surveys. Therefore, digital soil mapping (DSM) that predicts soil properties and categories through discrete samples within a certain area has been widely recognized and applied due to the reduction of the cost of sampling and analysis to a certain extent [2].

In DSM research, terrain factors always play an important role. The resolution of digital elevation models (DEM) directly affect the accuracy of the generated terrain model and derived variables, and high precision DEM data could better serve the prediction of soil properties [3]. Jasiewicz and Stepinski [4] proposed a new pixel-based


Q. Han et al. (Eds.): LSMS 2021/ICSEE 2021, CCIS 1469, pp. 765–774, 2021.
https://doi.org/10.1007/978-981-16-7213-2_74

terrain classification method called Geomorphons (GM) in 2013, that is, to determine the terrain element type corresponding to its spatial location according to the relative difference of elevation value in the neighborhood window with a specified size.

In this paper, the various GM terrain classification results obtained under different super parameter settings were used as environmental covariates to participate in STN content mapping. The prediction results had been compared with different combinations of traditional DEM derivatives and remote sensing variables in three different machine learning algorithms to evaluate the final effectiveness of the simulation.

# 2 Methods and Materials

## 2.1 Site Description

The study was conducted in the north of the Changjiang Delta Plain in east central part of China between latitude 32°3′20″–32°5′50″ N and longitude 119°10′20″–119°12′30″ E (Fig. 1a). The average height is about 82 m, while there is a 225 m ascent on sloping hills northwards. This region, of about 5.37 km$^2$, is characterized by a north subtropical warm humid climate with a mean annual temperature of 15.2 °C, and an average rainfall of 1060 mm. The maximum and average slope in this region is around 69° and 12°, respectively. The hilly terrain is prominent in the study area, especially in the northern region, while dryland and irrigation agricultural (mainly with wheat, rice and other cash crops) are located in the southern region on the flatter area (Fig. 1b).

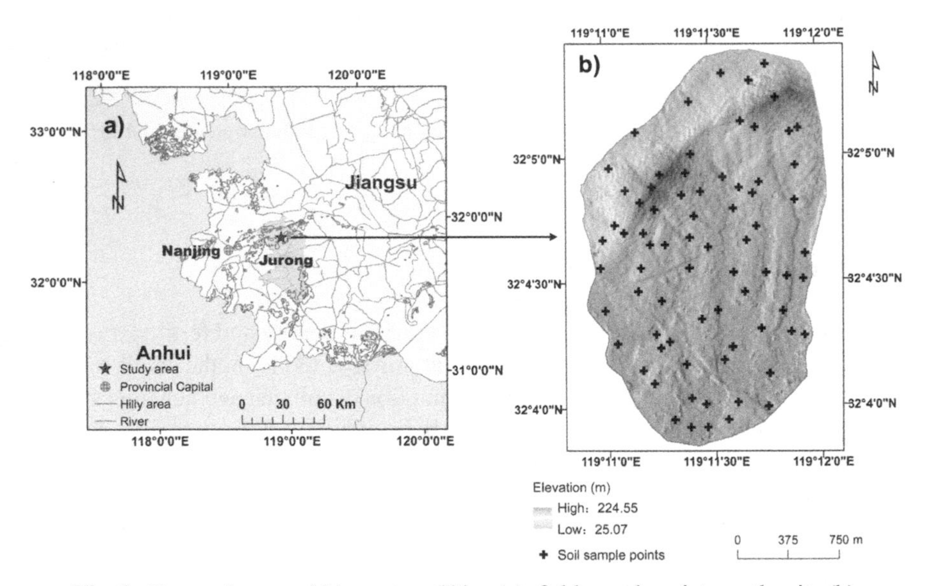


**Fig. 1.** Research area within eastern China (a); field sample points on the site (b).

## 2.2 Soil Sampling and Analysis

In order to represent landscape heterogeneity as fine as possible, 74 soil samples were collected in the past five years. Sampling design was based on landscape classification, and integrated environmental information closely related to the soil formation process. A total of 25 soil profiles and 49 soil boring data were collected. Surface (0–20 cm) soil samples were air dried and the content of STN was then analyzed in an ISO-certified laboratory.

## 2.3 Remote Sensing Variables

The remote sensing data were extracted from GF-2 and GF-3 images downloaded from China Resources Satellite Application Center (http://www.cresda.com/CN/). The multi-spectral sensor on GF-2 has four bands with 4 m spatial resolution. The imaging time of GF-2 was September 19, 2019 and there was no cloud coverage. The synthetic aperture radar (SAR) image data is GF-3 standard strip image, with dual polarization, 130 km width and 25 m spatial resolution. The imaging time was January 19, 2020 and also no cloud coverage in the data. The ENVI 5.3 and PIE SAR 6.0 software were used to pre-process GF-2 and GF-3 images, respectively.

## 2.4 GM Variables and DEM Derivatives

The GM terrain classification maps were derived from the *r.geomorphon* extension tool in GRASS GIS (v 7.8) [4]. The two most important parameters in the process of terrain feature recognition are search radius ($L$) and flatness threshold ($t$). The former represents the maximum distance for the calculation of zenith angle and nadir angle, while the latter is the maximum slope gradient for horizontal area, which is determined by reference direction and search distance. We used two settings on both $L$ (5 and 20 cells) and $t$ values (1° and 5°) in order to find the most fitting outcome. In addition, the GM terrain classification map was aggregated to generate GM (group) variable. The original 10 terrain elements were clustered into 4 types according to the adjacent borders and similar geomorphic types: "*crest zone*" (composed of "peak", "ridge", and "shoulder"), "*footslope zone*" ("flat" and "footslope"), "*slope zone*" ("spur", "slope", and "hollow"), and "*valley zone*" ("valley" and "pit"). The standard GM terrain classification map and GM aggregation map are shown in Fig. 2b.

The 5 m resolution DEM in the study area was extracted from the stereo pairs of ZY3-02 images through photogrammetry technology (ENVI 5.3.1 software DEM extraction module), and 46 ground control points (GCPs) file was used as calibration. Eight DEM derivatives were calculated by SAGA GIS software for comparison with GM variables to participate in STN prediction.

After collinearity analysis, a total of 18 environmental variables were selected (Table 1). Among them, four representative GM variables with large difference in hyperparameter settings were selected.

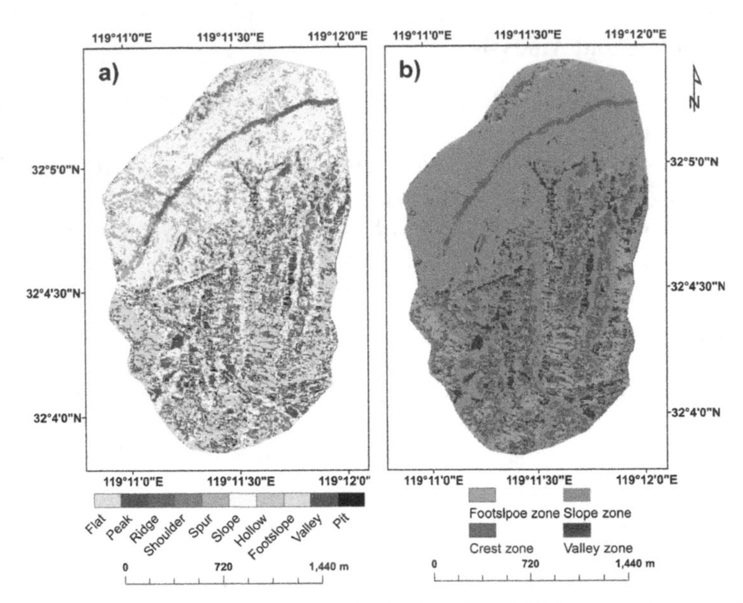


**Fig. 2.** Images of GM standard map (20 *L*, 5 *t*) (a) and GM aggregation map (b).

**Table 1.** The composition of environmental variables.

| Categories of variables | Indexes | Abbreviation | Native resolution | Type |
|---|---|---|---|---|
| GM terrain classification | GM with 5 cells *L* and 5° *t* | GM (5 *L*, 5 *t*) | 5 m | Category |
| | GM with 20 cells *L* and 5° *t* | GM (20 *L*, 5 *t*) | 5 m | |
| | GM with 20 cells *L* and 1° *t* | GM (20 *L*, 1 *t*) | 5 m | |
| | Aggregated GM | GM (group) | 5 m | |
| DEM derivates | Elevation | DEM | 5 m | Quantitative |
| | Convergence index | CI | 5 m | |
| | Length-slope factor | LSF | 5 m | |
| | Terrain surface convexity | TSC | 5 m | |
| | Multi-resolution index of valley bottom flatness | MRVBF | 5 m | |
| | Total catchment area | TCA | 5 m | |
| | Valley depth | VD | 5 m | |
| | Topographic wetness index | TWI | 5 m | |
| Optical and SAR imagery | GF-2 spectral band 2 | B2 | 3.2 m | Quantitative |
| | GF-2 spectral band 5 | B5 | 3.2 m | |
| | Normalized difference vegetation index | NDVI | 3.2 m | |

(*continued*)

**Table 1.** (*continued*)

| Categories of variables | Indexes | Abbreviation | Native resolution | Type |
|---|---|---|---|---|
| | Anthocyanin reflectance index 2 | ARI2 | 3.2 m | |
| | The backscatter coefficients of the VH polarization from the GF 3 image | VH | 25 m | |
| | The backscatter coefficients of the VV polarization from the GF 3 image | VV | 25 m | |

## 2.5 Modelling Techniques

This study compared three machine learning techniques for STN content prediction: bagged classification and regression trees (Bagged CART), RF and Cubist. The "caret" package in R software was used for parameter regulation [5]. Then the performance of different combinations of predictor variable types were evaluated to fit the best model for the final STN mapping. CART algorithm is a nonparametric data mining technology for regression and classification problems, which has been greatly improved in recent years and was widely available for the prediction of soil attributes [6]. RF is also a tree-based ensemble learning method. Multiple bootstrap samples are used to obtain random samples, and corresponding decision trees are established through these samples to form a random forest. Cubist is another advanced non-parametric regression tree algorithm, which is suitable for processing the nonlinear relationship between STN and predictor variables [7].

## 2.6 Statistical Analyses

SPSS 25 software was used for descriptive statistical analysis of STN. Regression analysis was performed to detect the collinearity among environmental variables and Pearson correlation analysis was also processed. The environmental variables with higher variance inflation factor (VIF $\geq$ 10) were deleted from the model [8].

Based on three machine learning techniques, the prediction models of STN content in the study area were constructed using five different combinations of environmental variables. The 10-fold cross-validation method was performed to evaluate the model performance. The performance of different models is verified by three indexes: consistency index (C-index), root mean square error (RMSE), and coefficient of determination ($R^2$). For the sake of evaluating prediction uncertainty, model with the best performance among was selected from the prediction outcomes and 100 STN prediction maps were generated. The mean value, 90% limit and standard deviation (SD) of each pixel were calculated as the final STN mapping and prediction uncertainty.

# 3 Results and Discussion

## 3.1 Descriptive Statistics of Sampled STN Content

The statistical characteristics of the STN content in the study area were calculated. The mean value of STN was 1.09 g/kg, and the maximum and minimum value were 2.48 and 0.48 g/kg, respectively. The standard deviation (SD) was 0.43 g/kg and the skewness coefficient of 1.17 performed a skewed distribution of the STN.

## 3.2 Evaluation of Model Predictions

To assess the ability of GM terrain classification variables for predicting STN content, three machine learning methods included Bagged CART, RF and Cubist were used to fit models: Model I, Model II, and Model IV included GM variables, DEM derivatives and remote sensing variables, respectively. Model III and Model V represented terrain environment variables (GM variables and DEM derivatives) and total environmental variables (Table 2). The performance of these models in predicting STN content were calculated (Table 3).

Compared with the one-dimensional estimators, adding the GM variables to the DEM derivatives implying increased performance in predicting $R^2$ of STN content, for example in Cubist (from 0.30 to 0.48). This result validated the capacity of proposed GM data that led to a significant improvement in predictive performance.

As expected, the RMSE decreased with an increasing prediction horizon for all three predictor types in all setups. Compared with the application of terrain environment variables, the $R^2$ (from 0.48 to 0.55) obtained by using the Cubist model was improved by 14.6%, and similar advancement could be observed for the other algorithms. All three verification indicators of model V had the highest accuracy among all machine learning techniques, of which the Cubist method performed best (C-index = 0.57, RMSE = 0.27, $R^2$ = 0.55). The $R^2$ value indicated that this model was sufficient to describe the variation of STN (about 55%) across a given area.

**Table 2.** Different combinations of predictor variables.

| NO. | Model | Environmental variables |
|---|---|---|
| a | Model I | GM |
| b | Model II | DEM derivatives |
| c | Model III | GM + DEM derivatives |
| d | Model IV | Remote sensing data |
| e | Model V | GM + DEM derivatives + Remote sensing data |

**Table 3.** Comparison of prediction accuracies based on three modelling techniques using different combinations of predictors.

| Modelling techniques | Model | C-index | RMSE | $R^2$ |
|---|---|---|---|---|
| Bagged CART | I | 0.38 | 0.56 | 0.31 |
| | II | 0.46 | 0.52 | 0.38 |
| | III | 0.49 | 0.47 | 0.41 |
| | IV | 0.42 | 0.50 | 0.33 |
| | V | 0.50 | 0.45 | 0.48 |
| RF | I | 0.37 | 0.39 | 0.31 |
| | II | 0.45 | 0.33 | 0.37 |
| | III | 0.48 | 0.32 | 0.45 |
| | IV | 0.45 | 0.35 | 0.36 |
| | V | 0.55 | 0.3 | 0.52 |
| Cubist | I | 0.41 | 0.35 | 0.34 |
| | II | 0.44 | 0.33 | 0.40 |
| | III | 0.49 | 0.30 | 0.45 |
| | IV | 0.46 | 0.34 | 0.37 |
| | V | **0.57** | **0.27** | **0.55** |

Note: The most accurate results are shown in bold.

### 3.3 The Relative Importance of Environmental Data

For the STN prediction using Model V, the feature importance rankings were shown in Fig. 3 (predictor variables were normalized to 100%). The importance of all variables in the two prediction techniques were slightly different, indicating that the environmental characteristics dominated in these models were also discrepant. DEM derivatives were the main explanatory variables in both prediction techniques, followed by remote sensing data, and GM variables had the lowest percentage of relative importance. Although the arrangement characteristics of variables were different after subdivision, the Elevation and MRVBF variable were both located in the top two most important environmental variables. In addition, the GM variables explained 17% and 19% of STN spatial variation in the RF and Cubist models, indicating that the GM terrain classification maps had the potential to be used for predicting and mapping the STN content in the hilling area.

In this study, four GM variables performed different importance contribution of model fitting and could explain 19% of STN spatial variation by using Cubist method. Among them, the GM terrain classification map generated with 20 cells *L* and 5° *t* showed the highest relative importance using two machine learning methods. This indicated that the setting of 20 cells L was more proper for the small watershed scale to collect more terrain heterogeneity. For the same search radii, the hilly terrain was obviously more adaptable to a larger flatness threshold (5° t) than the default value (1° t) due to the hypsography in this area, as there were much of farmland in this area cultivated on gradual ascent instead of flat terrain. In addition, the aggregated GM map (Fig. 2b) also performed more than 5% contribution in both learning techniques. Although part of the feature information was lost, the GM (group) map with lower heterogeneity could play a better role in DSM research on a larger scale.

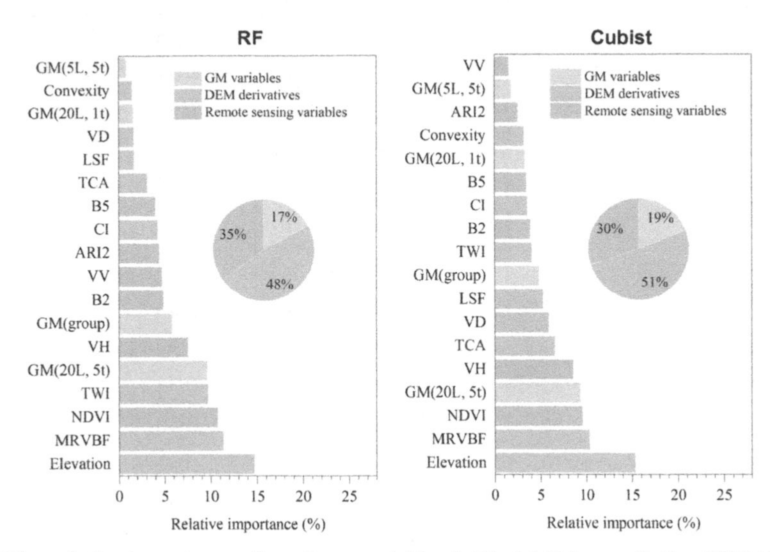


**Fig. 3.** The relative importance of predictor variables in Model V for predicting STN (Model V: GM variables + DEM derivatives + Remote sensing variables).

## 3.4 The Spatial Prediction of STN Content

Based on the evaluation of modeling accuracy level, the Cubist model was selected to run 100 times in model V to assess the prediction uncertainty [9], and the mean value (final outcome), 90% limit and SD (prediction uncertainty) maps after 100 runs were generated (Fig. 4). The final average STN content was 1.04 g/kg, and the prediction model also performed lower uncertainty. For 100 predicted STN distribution maps, the average SD value was 0.34 g/kg, which also showed Cubist model had a relatively stable forecasting ability.

Although the overall similar distribution characteristics of three models could be observed, the remarkable differences existed among the levels of STN content. The content of the STN map generated by the Bagged CART model was low as a whole. The maximum content predicted was 1.89 g/kg, which was the lowest among three models and had the narrowest interval of prediction. The RF model was slightly higher than the former, with a maximum value of 2.20 g/kg, while the Cubist model had the highest predicted STN content of 2.33 g/kg and the widest prediction interval.

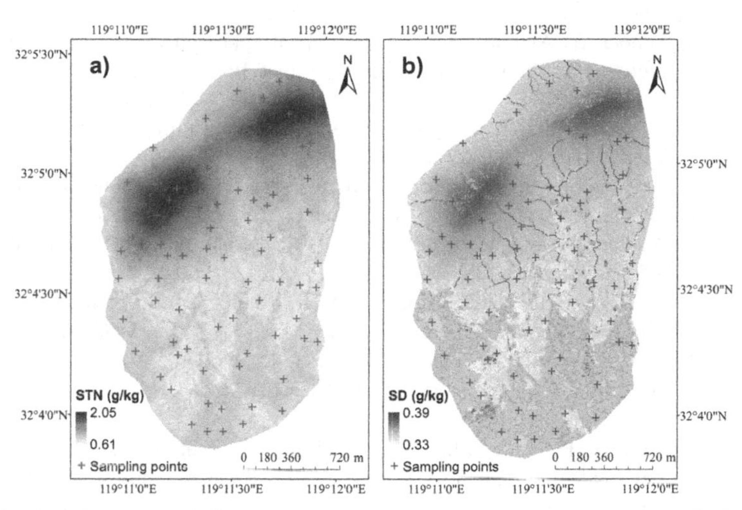


**Fig. 4.** The 90% lower prediction map (a), mean STN map (b), 90% upper prediction map (c) and standard deviation map generated from Model V after 100 runs of the Cubist method (Model V: GM variables + DEM derivatives + Remote sensing variables).

## 4 Conclusion

The main conclusions of this work can be summarized as follows: (i) The Cubist model had better performance than the Bagged CART and RF model in predicting STN content in the hilly area at the small watershed scale. (ii) The application of GM terrain classification data could enhance the accuracy level of STN prediction: the GM (20 L, 5 t) had the highest model contribution, indicating that hilly terrain was more adaptable to larger search radii and flatness threshold; and the GM (group) showed the potential to be predictor variable in large scale DSM research. (iii) The combination of GM variables, DEM derivatives and remote sensing variables had the highest prediction accuracy. The addition of the GM data improved the predictive performance, with C-index, RMSE and R2 enhancing by 11.4%, 9.1% and 12.5%, respectively.

**Acknowledgments.** This work was supported by two of the General Program of National Natural Science Foundation of China (NSFC) [grant number 41971057 and 41771247] and the Open Research Fund of Anhui Key Laboratory of Detection Technology and Energy Saving Devices under grant DTESD2020A02. We would like to thank Dr. Lisheng Wei, Dr. Xiang Gao, and Dr. Xiaosan Jiang for their beneficial reviews of the manuscript.

## References

1. Xu, S., Wang, M., Shi, X., Yu, Q., Zhang, Z.: Integrating hyperspectral imaging with machine learning techniques for the high-resolution mapping of soil nitrogen fractions in soil profiles. J. Sci. Total Environ. **754**, 142135 (2021)
2. McBratney, A., Mendonça Santos, M., Minasny, B.: On digital soil mapping. Geoderma **117**, 3–52 (2003)

3. Aguilar, F.J., Agüera, F., Aguilar, M.A.: A theoretical approach to modeling the accuracy assessment of digital elevation models. J. Photogrammet. Eng. Remote Sens. **73**, 1367–1380 (2007)
4. Jasiewicz, J., Stepinski, T.F.: Geomorphons-a pattern recognition approach to classification and mapping of landforms. J. Geomorphol. **182**, 147–156 (2013)
5. Team, R.: R: a language and environment for statistical computing. J. MSOR Connect. **1** (2014)
6. Heung, B., Ho, H.C., Zhang, J., Knudby, A., Bulmer, C.E., Schmidt, M.G.: An overview and comparison of machine-learning techniques for classification purposes in digital soil mapping. J. Geoderma. **265**, 62–77 (2016)
7. Pouladi, N., Møller, A.B., Tabatabai, S., Greve, M.H.: Mapping soil organic matter contents at field level with Cubist Random Forest and kriging. J. Geoderma **342**, 85–92 (2019)
8. Lombardo, L., Saia, S., Schillaci, C., Mai, P.M., Huser, R.: Modeling soil organic carbon with quantile regression: dissecting predictors' effects on carbon stocks. J. Geoderma **318**, 148–159 (2018)
9. Jeong, G., Oeverdieck, H., Park, S.J., Huwe, B., Ließ, M.: Spatial soil nutrients prediction using three supervised learning methods for assessment of land potentials in complex terrain. J. Catena **154**, 73–84 (2017)

# Research on Adulteration Detection of Rattan Pepper Oil Based on BAS_WOA_SVR

Yong Liu(✉), Ce Wang, and Suan Xu

College of Mechanical and Electrical Engineering, China Jiliang University, Hangzhou, China

**Abstract.** In this paper, the adulteration detection of rattan pepper oil was studied by near-infrared spectroscopy. Standard Normal Transformation (SNV) pretreatment method was used to process the near-infrared spectra data of rattan pepper oil. Use cars algorithm to extract spectral feature points. Establish BAS_WOA_SVR model and WOA_SVR model based on characteristic light data. The experimental results showed that BAS_WOA_SVR model has the best performance, which is 23% higher than the traditional SVR model. The accuracy of model BAS_WOA_SVR is 9% higher than that of model WOA_SVR. The prediction accuracy of model BAS_WOA_SVR is as high as 89%. BAS_WOA_SVR model has a good function of quantitative prediction of rattan pepper oil adulteration.

**Keywords:** Near-infrared spectroscopy · Rattan pepper oil · BAS · WOA · SVR

## 1 Introduction

Rattan pepper is rich in iron and vitamins, which is a good raw material for refining medicine, nutrition and health care products. The price of rattan pepper oil is higher than other kinds of vegetable oil. In order to make profits, some businesses add low price vegetable oil into rattan pepper oil, which infringes the interests and health of consumers. It is of great practical value to detect the adulteration of rattan pepper oil by near-infrared spectroscopy [1, 2]. Because of the low absorption resolution and serious spectral overlap in traditional near infrared spectroscopy, data preprocessing before modeling is usually carried out. Then, the characteristic wavelength optimization method is selected to optimize the near infrared spectrum data, and finally, the quantitative prediction model of rattan pepper oil is established. The modeling method is support vector machine [3–5] regression (SVR). Particle swarm optimization (PSO) and genetic algorithm (GA) are usually used to optimize the C and G parameters of SVR [5–7] model, but the model is easy to fall into local optimum. In this paper, the combination of BAS [8–10] and WOA [11–13] is used to optimize the SVR model. The model is no longer easy to fall into local optimum, thus improving the accuracy of the quantitative prediction model.

### 1.1 Detection Principle of Near Infrared Spectroscopy

Near-infrared spectroscopy (NIRS) is essentially an analytical method to determine the molecular structure and identify compounds based on the information of relative


Q. Han et al. (Eds.): LSMS 2021/ICSEE 2021, CCIS 1469, pp. 775–782, 2021.
https://doi.org/10.1007/978-981-16-7213-2_75

vibration among atoms in molecules. The absorption of infrared light was recorded by a spectrometer. Wavelength (λ) or wave number (σ) is usually used as the abscissa to represent the position of absorption peak, and transmittance (T%) or absorbance (A) is used as the ordinate to represent absorption intensity.

The HL-2000 halogen tungsten lamp light source sends out the laser through the optical fiber transmission to reach the colorimeter dish, the colorimeter dish contains the rattan pepper oil adulteration sample, the light beam enters the near infrared spectrometer through the optical fiber after the colorimeter dish, through the computer and the spectrometer connection, display the spectra of the adulterated samples on the computer.

The original data is preprocessed by SNV, then the spectral data is extracted by CARS, the data is divided into training set and test set, the BAS model is established, and the quantitative prediction results are obtained.

### 1.2 Principles of SVR Model

SVR is full name of support vector regression machine, it is a Support vector machine of regression problems to upgrade the use. As shown in Fig. 1, the difference between SVR and SVM is that SVR only has one class of sample points, and the optimal hyperplane it seeks is not the "Most open" of two or more class of sample points, as SVM does, it minimizes the total deviation of all sample points from the hyperplane. The process of finding the optimal solution by using the characteristic attributes of SVR is the process of finding the best description by using hyperplane technique. SVR itself has strict mathematical theory characteristics, intuitive geometric interpretation and good generalization ability, can approximate any function with any precision, and it can effectively avoid the gradient explosion caused by the lack of learning and over-learning, especially in dealing with small sample training learning problems has unique advantages.

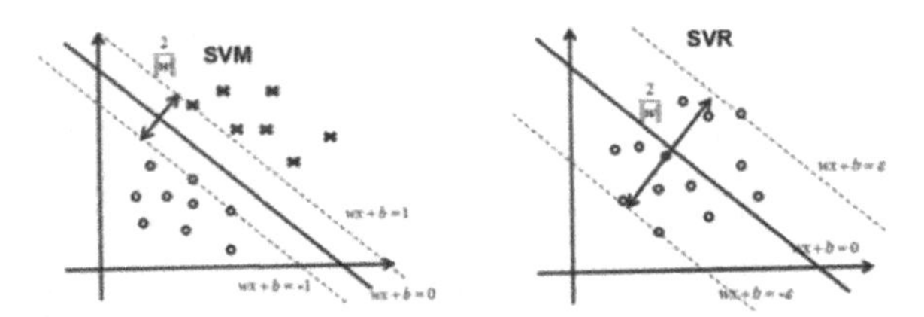


**Fig. 1.** SVR and SVR comparison chart

### 1.3 Principle of the Beetle Antennae Search Algorithm (BAS)

The Beetle Antennae search algorithm (BAS) is a new technique for multi-objective function optimization based on the beetle foraging principle, which was proposed in 2017. Its biological principle is: when the longicorn foraging, it does not know where is the food, but according to the strength of the smell of food to look for food. The longhorn beetle has two long horns. If the left antenna picks up a stronger smell than the right, the next step is for the longhorn beetle to fly to the left, and vice versa. Similar to genetic algorithm and particle swarm optimization, BAS can automatically realize the optimization process without knowing the specific form of function and gradient information. The algorithm works like this:

(1) create the random vector toward which the longicorn beetles must be oriented and normalize it

$$\vec{b} = \frac{rand(k, 1)}{\|rands(k, 1)\|}. \tag{1}$$

In the formula: rand() is a random function; $\|rands()\|$ Represents a spatial dimension.

(2) Create the spatial coordinates of the left and right whiskers of longhorn beetle

$$\begin{cases} x_{lt} = x_t - d_0 * \vec{b}/2 \\ x_{rt} = x_t + d_0 * \vec{b}/2 \end{cases} (t = 0, 1, \cdots, n). \tag{2}$$

(3) Judging the odor intensity of the left and right whiskers according to the fitness function, namely the intensity of $f(x\_l)$ and $f(x\_r)$, Function $f()$ is the fitness function.

(4) Iterative updating of the positions of the Antennae beetles

$$x_{t+1} = x_t - \delta_t * \vec{b} * sign(f(x_{rt}) - f(x_{lt})). \tag{3}$$

In the formula: $\delta_t$ represents the step factor at the i-th iteration, sign() is a symbolic function.

## 1.4 Principle of the Whale Optimization Algorith (WOA)

### 1.4.1 Encircling Prey

The algorithm is a mathematical model based on the predation behavior of humpback whales. Because the position of the prey is uncertain, the WOA algorithm first assumes that the current best candidate solution is the position of the target prey or the position closest to the prey, and then continuously updates its position. This behavior uses the following (4) and (5):

$$D = |CX^*(t) - X(t)|. \tag{4}$$

$$X(t+1) = X^*(t) - AD. \tag{5}$$

Among them, D is the encircling step length, X(t + 1) is the position vector of the solution after the next iteration, X*(t) is the position vector of the current optimal solution. X(t) is the position vector of the current solution; t is the current iteration number; A and C are random coefficient vectors; X*(t) needs to be updated every time a better solution appears during each iteration.

A and C are obtained by the following formula:

$$\begin{cases} A = 2\mathrm{ar} - \mathrm{a} \\ \mathrm{a} = 2 - \dfrac{2\mathrm{t}}{\mathrm{t_{max}}} \end{cases}. \tag{6}$$

$$\mathrm{C} = 2\mathrm{r}. \tag{7}$$

Where a is the control parameter linearly decreases from 2 to 0 in the iterative process,$t_{max}$ is the maximum number of iterations, r is a random vector of [0,1].

### 1.4.2 Bubble-Net Attacking Method

1) Shrink the enveloping mechanism. The new individual position can be defined at any position between the current individual whale and the best individual.
2) Spiral update location. Calculate the distance D' between the whale and the prey, and construct the equation:

$$X(t+1) = D' e^{bl} \cos(2\pi l) + X^*(t). \tag{8}$$

Among them, $D' = |X^*(\mathrm{t}) - X(t)|$ defines the shape of the logarithmic spiral, b is constant; l is a random number of [−1,1].

### 1.4.3 Search for Prey

When |A|>1, it enters the random search stage, and randomly selects whales to update the position, rather than according to the best known whale individuals. The algorithm terminates when the maximum number of iterations is reached. The mathematical model of random search is as follows:

$$D = |CX_{\mathrm{rand}} - X(t)|. \tag{9}$$

$$X(t+1) = X_{rand} - AD. \tag{10}$$

$X_{rand}$ is a random position vector in. A determines whether the humpback whale enters the random search state.

## 1.5 BAS Combined with WOA

Take the optimal whale of each whale school as the starting position of the current baleen, explore the left and the right baleen separately, then move forward, and then calculate the objective function after moving forward, if the objective function because of the current optimal whale value, the whale position is replaced by the forward and backward positions, and the improvement of the whale algorithm is realized (Fig. 2).

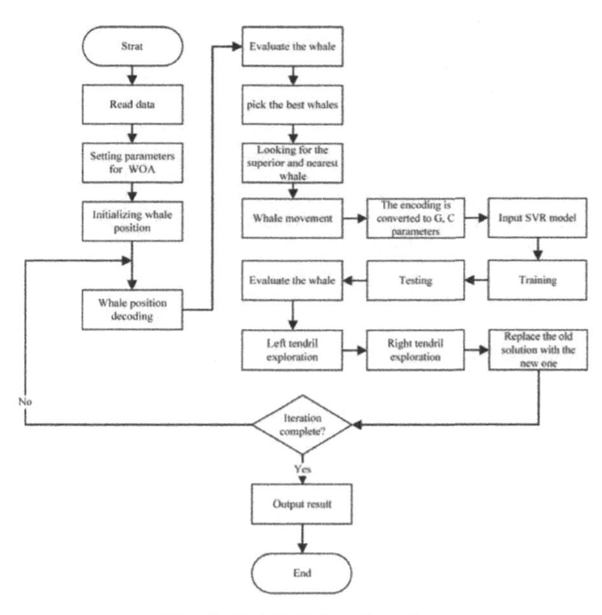


**Fig. 2.** BAS_WOA flowchart

### 1.6 Model Evaluation Criteria

The root mean square error (RMSE) is used to measure the error between the predicted value and the true value. The smaller the value of RMSE, the better the prediction effect of the model.

$$RMSE = \sqrt{\frac{1}{N}\sum_{I=1}^{N}\left[y_{pre,i} - f(x_i)_{true}\right]}. \quad (11)$$

## 2 The Experimental Part

### 2.1 Sample Preparation

Method for preparing adulterated samples of rattan pepper oil:Pure rattan pepper oil is used as the base oil, and soybean oil is mixed into the base oil according to the percentage of the total oil sample weight to prepare a binary adulterated oil sample. The blending ratio column is based on a 0–50 concentration gradient with 2% concentration intervals, and a 50–100 concentration gradient with 5% concentration intervals.

### 2.2 Near-Infrared Spectroscopy Collection and Pre-processing

Use a cuvette with a light path of 1 mm and an ambient temperature of 20 °C. Try to ensure the indoor temperature, humidity, light, noise, etc. when collecting spectra Consistency of experimental conditions. The effect of SNV-preprocessed [14] is to eliminate the influence of particle size, surface scattering and optical path transformation on the diffuse reflectance spectra. The characteristic spectrum data is extracted by CARS algorithm [15] (Figs. 3, 4, and 5).

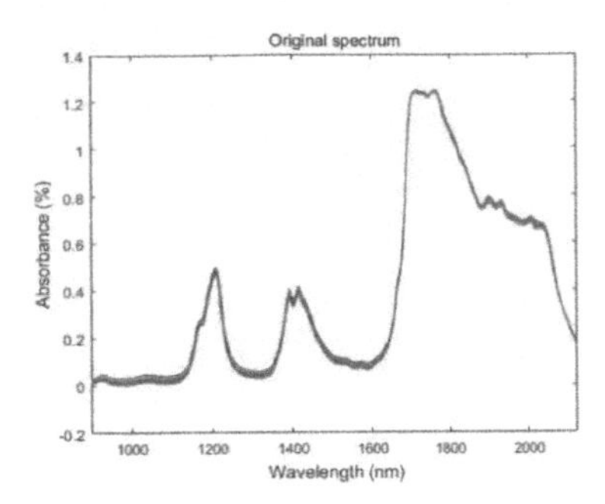


Fig. 3. Raw spectral data

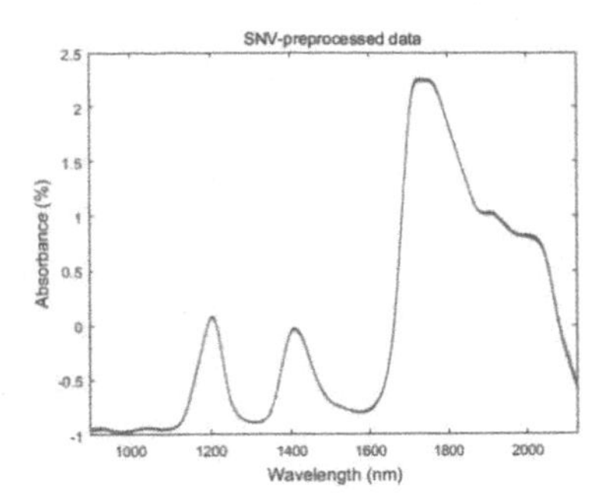


Fig. 4. SNV-preprocessed data

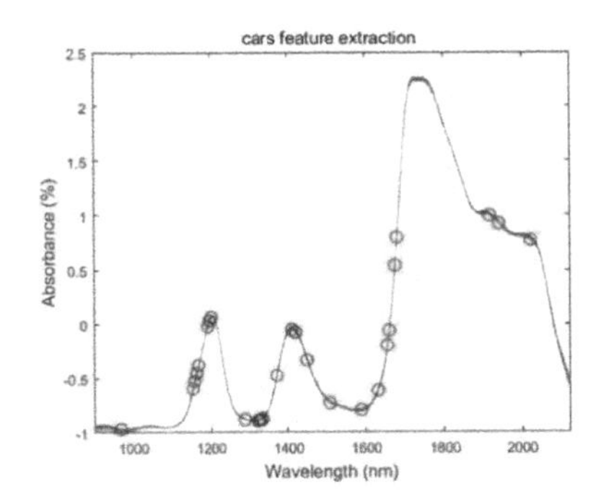


Fig. 5. Screening of CARS spectral characteristic data

## 2.3 Model Accuracy Test

Table 1. Model accuracy test

| | SVR | | WOA_SVR | | BAS_W0A_SVR | |
|---|---|---|---|---|---|---|
| | $R^2$ | RMSE | $R^2$ | RMSE | $R^2$ | RMSE |
| Training set | 0.732 | 21.261 | 0.836 | 8.951 | 0.921 | 3.617 |
| Prediction set | 0.665 | 17.251 | 0.800 | 10.314 | 0.897 | 4.351 |

The spectral data after SNV-preprocessing and CARS characteristic wavelength extraction were used as the data basis of the SVR model, and the SVR quantitative detection model was established. Optimize the SVR model with WOA and BAS_WOA respectively.

It can be seen from Table 1 that the BAS_WOA_SVR model has high model accuracy and generalization ability, and the algorithm can effectively optimize the (c, g) parameters The prediction set has the highest coefficient of determination ($R^2 = 0.897$) and the smallest mean square error (RMSE = 4.351). Figure 6 is the iterative graph of fitness curve of WOA_SVR model and BAS_WOA_SVR model. Figure 7 shows the prediction results of the adulteration samples of rattan pepper oil.

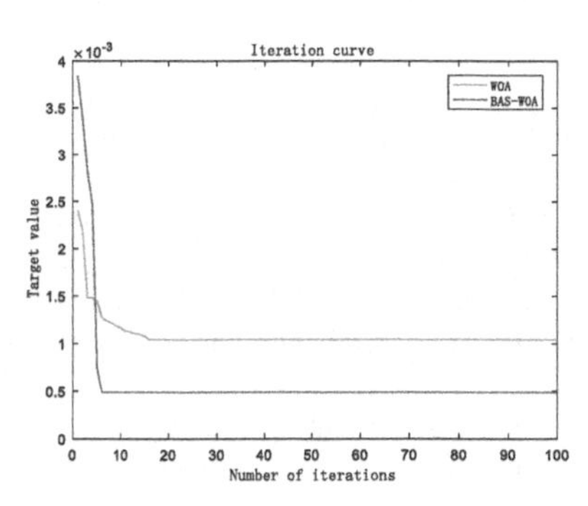


**Fig. 6.** Iterative graph

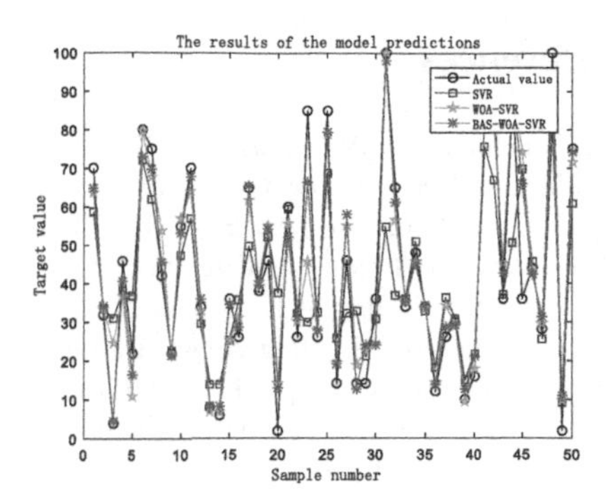


**Fig. 7.** Prediction set result

## 2.4 Analysis of Results

The beard whisker algorithm (BAS) was used to optimize the whale algorithm (WOA) to optimize the kernel function parameters and penalty factors of the support vector regression machine (SVR), and the BAS_WOA_SVR rattan pepper oil adulteration prediction model was established. The accuracy of the model to distinguish the adulteration content of tengjiao oil is as high as 89%.

It can be seen from the fitness curve in Fig. 6, that compared with the traditional whale algorithm, the whale algorithm optimized by the beetle antennae search algorithm faster and has a small root mean square error (RMSE), which greatly improves the accuracy of the model. The forecast is better. It shows that the effect of improving the whale algorithm is more obvious.

# 3 Conclusion

This paper proposes a set of optimized support vector regression (SVR) models based on BAS combined with WOA. Through the comparison and analysis of the prediction results of the adulteration of vine pepper oil, the following conclusions are obtained:

(1) In the rattan pepper oil adulteration quantitative prediction model, WOA optimized by BAS is used to optimize the parameters of the SVR model, and the best parameter modeling is obtained. Compared with the traditional whale algorithm (WOA) Parameter optimization modeling, which greatly improves the accuracy of the model, and improves the prediction effect of the rattan pepper oil adulteration model.
(2) The BAS_WOA_SVR rattan pepper oil adulteration prediction model was established. The test results showed that compared with the traditional SVR quantitative prediction model, the average accuracy of the prediction set was greatly improved, and the accuracy was increased by 9.7% compared with the traditional WOA_SVR model, which proved this article The accuracy and effectiveness of the proposed method for detecting adulteration of rattan pepper oil.

## References


1. Hutchinson, M.R., Rojas, R.F.: Bilateral connectivity in the somatosensory region using near-infrared spectroscopy (NIRS) by wavelet coherence. SPIE BioPhoton. Austr. (2016)
2. Ha, U., Yoo, H.J.: An EEG-NIRS ear-module SoC for wearable drowsiness monitoring system. In: Solid-State Circuits Conference. IEEE (2017)
3. Sahu, S.K., Pujari, A.K., Kagita, V.R., Kumar, V., Padmanabhan, V.: GP-SVM: tree structured multiclass SVM with greedy partitioning. In: 2015 International Conference on Information Technology (ICIT), pp. 142–147(2015)
4. Parveen, S.A.: Detection of brain tumor in MRI images, using combination of fuzzy c-means and SVM. In: International Conference on Signal Processing and Integrated Networks (SPIN). IEEE (2015)
5. Yao, Y., Cheng, D., Peng, G., Huang, X.: Fault prognosis method of industrial process based on PSO-SVR. In: Ju, Z., Yang, L., Yang, C., Gegov, A., Zhou, D. (eds.) UKCI 2019. AISC, vol. 1043, pp. 331–341. Springer, Cham (2020). https://doi.org/10.1007/978-3-030-29933-0_28
6. Zhang, Z., Gao, G., Tian, Y., Yue, J.: Two-phase multi-kernel LP-SVR for feature sparsification and forecasting. Neurocomputing **214**, 594–606 (2016)
7. Zhu, Z., Peng, J., Zhou, Z., Zhang, X., Huang, Z.: PSO-SVR-based resource demand prediction in cloud computing. J. Adv. Comput. Intell. Intell. Inf. **20**(2), 324–331 (2016)
8. Jile, L., Yeping, R., Luqi, Q.: Optimization of extraction process of fengyin decoction based on Bas-ga-bp neural network and entropy weight method. J. Chin. J. Tradit. Chin. Med. 116–123 (2020)
9. Yu, J., Yuan, H., Yang, Y., Zhang, S., Fei, Q.: Optimization of aeroengine pipeline layout based on adaptive tennessee algorithm. J. Chin. J. Mech. Eng. 188–198(2020)
10. Wang, Z., Zeng, Y., Wangjunling, H.: Intrusion detection based on BP neural network optimized by improved algorithm of Tianniu Group. J. Sci. Technol. Eng. 170–178 (2020)
11. Srivastava, V., Srivastava, S.: Whale optimization algorithm (WOA) based control of nonlinear systems. In: International Conference on Power Energy, Environment and Intelligent Control (2019)
12. Shao, L., Zhan, X.: Identification of mine water inrush source based on Iwoa-hkelm. J. Chin. J. Saf. Sci. 117–122 (2019)
13. Kong, D., Chen, Y., Li, N.: Tool wear estimation in end-milling of titanium alloy using NPE and a novel WOA-SVM model. J. IEEE Trans. Instrum. Measur. 1–5 (2019)
14. Li, J., Li, S., Jiang, L.: Nondestructive identification of green tea based on near infrared spectroscopy and chemometrics. J. Acta Analyt. Sin. 47–53 (2020)
15. Yuan, N., Wei, L., Zhang, Y.: Hyperspectral inversion analysis of arsenic content in farmland soil based on optimized cars and PSO-SVM algorithm. J. Spectrosc. Spect. Anal. 241–247 (2020)

# Feature Selection Based on a Modified Adaptive Human Learning Optimization Algorithm

Siyue Yu[1], Yihao Jia[1], Xuelian Hu[1], Haoqi Ni[2], and Ling Wang[1](✉)

[1] Shanghai Key Laboratory of Power Station Automation Technology, School of Mechatronics Engineering and Automation, Shanghai University, Shanghai 200072, China
{april07,jiayh,huxuelian,wangling}@shu.edu.cn

[2] Department of Electrical and Computer Engineering, North Carolina State University, Raleigh, NC 27606, USA
hni2@ncsu.edu

**Abstract.** This paper proposes a new wrapper feature selection method based on the modified adaptive Human Learning Optimization (MAHLO) algorithm and Support Vector Machine (SVM). To achieve better results, the initialization and random learning strategies are modified in MAHLO to solve the feature selection problem more efficiently and the adaptive strategies are used to enhance the performance and relieve the effort of parameter setting. Besides, a two-stage evaluation function is adopted to eliminate the useless and redundant features, which is easier to operate in applications. The simulation results indicate that MAHLO can solve high-dimensional feature selection problems more efficiently, and the classification results on the UCI benchmark problems further demonstrate the efficiency and the advantage of the proposed MAHLO-based wrapper feature selection method.

**Keywords:** Human Learning Optimization · Feature selection · Meta-heuristic · Support vector machines

## 1 Introduction

In data mining and machine learning [1], feature selection (FS) is one of the most important pre-processing steps [2] because high-dimensional data sets will greatly affect the performance of problems such as classification and clustering. Feature selection methods can extract useful information from complex features, eliminate useless and redundant or even misleading parts [3], thereby reducing the difficulty of searching and improving the accuracy of evaluation.

According to the evaluation criteria, the feature selection can be divided into filter approaches and wrapper approaches [4]. The wrapper-based feature selection can optimize the parameters of the classifier at the same time, which typically results in a wrapper that achieves better performance. However, feature selection belongs to the class of NP-hard problems [5] because there are solutions for n-dimensional space [6]. Therefore, meta-heuristics have the advantages over traditional methods, and a large number of meta-heuristic algorithms have been applied to feature selection, such as Genetic Algorithms [7] and Particle Swarm Optimization [8].


Q. Han et al. (Eds.): LSMS 2021/ICSEE 2021, CCIS 1469, pp. 783–794, 2021.
https://doi.org/10.1007/978-981-16-7213-2_76

The results in the previous study [9] show that: 1) the performance of the optimization algorithms has an important impact on the result of the high-dimensional feature selection; 2) for the ultra-high-dimensional feature selection problems, it is hard to effectively eliminate redundant features. Feature selection is naturally represented as a binary-coding problem, but almost all the meta-heuristics except Genetic Algorithms are originally proposed for continuous or discrete problems, and therefore the binary variants of these meta-heuristics need be developed for feature selection, which usually unavoidably causes a performance loss since the mechanisms and operators of the algorithms need be re-defined for binary space. This motivates to propose a new wrapper feature selection method based on Human Learning Optimization (HLO), which is a novel inborn binary-coding meta-heuristic algorithm and has its unique advantage.

Human Learning Optimization [10] is a novel metaheuristic inspired by a simplified human learning model. To enhance the performance for FS, a Modified Adaptive Human Learning Optimization (MAHLO) algorithm is developed, in which the learning rates of the learning operators will adaptively change according to the problem dimension as well as the number of iterations, and therefore it is very suitable for feature selection, especially for high-dimensional cases. Then this paper further improves MAHLO based on the characteristics and difficulties of FS and proposes a wrapper feature selection based on MAHLO.

The rest of the paper is organized as follows. Section 2 introduces MAHLO in details. The proposed feature selection methods based on MAHLO is described in Sect. 3. Section 4 verifies the proposed method on the benchmark problems. At last, the conclusions are given in Sect. 5.

## 2 Modified Adaptive Human Learning Optimization Algorithm for Feature Selection

Like the standard HLO and previous adaptive HLO [11], MAHLO uses the random learning operator, the individual learning operator and the social learning operator to generate new candidates to search for the optima. However, the modified initiation and search strategies have been developed in MAHLO to enhance the performance for feature selection.

### 2.1 Initialization

The binary-coding framework is adopted in MAHLO and a solution can be represented as Eq. (1), in which each bit stands for a basic component of knowledge for solving problems.

$$x_i = \left[ x_{i1}\ x_{i2} \cdots x_{ij} \cdots x_{iM} \right],\ \ x_{ij} \in \{0, 1\}, 1 \le i \le N, 1 \le j \le M. \tag{1}$$

where $x_i$ denotes the $i$-th individual, N is the size of the population, and M is the dimension of the solution. However, for feature selection, especially for complex high-dimensional cases, the bit need be set to ‘1’ with more chance to guarantee that the important features are chosen during initialization. In this paper, the probability of being ‘1’ is set to 0.7

based on experimental experience. After generating N individuals, an initial population is obtained as Eq. (2)

$$X = \begin{bmatrix} x_1 \\ x_2 \\ \vdots \\ x_i \\ \vdots \\ x_N \end{bmatrix} = \begin{bmatrix} x_{11} & x_{12} & \cdots & x_{1j} & \cdots & x_{1M} \\ x_{21} & x_{22} & \cdots & x_{2j} & \cdots & x_{2M} \\ \vdots & \vdots & & \vdots & & \vdots \\ x_{i1} & x_{i2} & \cdots & x_{ij} & \cdots & x_{iM} \\ \vdots & \vdots & & \vdots & & \vdots \\ x_{N1} & x_{N2} & \cdots & x_{Nj} & \cdots & x_{NM} \end{bmatrix}. \tag{2}$$

## 2.2 Learning Operators

### 2.2.1 Random Learning Operator

To simulate the random learning behavior, the random learning operator, defined as Eq. (3), is used in MAHLO. Note that the probability of random learning taking '0' is lowered to 0.3 in MAHLO, instead of 0.5 in the HLO and the other variants, to enhance the capability of picking up features,

$$x_{ij} = RE(0, 1) = \begin{cases} 0, \ 0 \leq r \leq 0.3 \\ 1, \ else \end{cases}. \tag{3}$$

where $r$ represents a random number between 0 and 1.

### 2.2.2 Individual Learning Operator

Individual learning is defined as the ability to build knowledge through personal reflections on external stimuli and sources [12]. To simulate this learning behavior in MAHLO, the individual best solution is stored in the Individual Knowledge Database (IKD), as in Eqs. (4) and (5)

$$IKD = \begin{bmatrix} ikd_1 \\ ikd_2 \\ \vdots \\ ikd_i \\ \vdots \\ ikd_N \end{bmatrix}, 1 \leq i \leq N. \tag{4}$$

$$IKD_i = \begin{bmatrix} ikd_{i1} \\ ikd_{i2} \\ \vdots \\ ikd_{ip} \\ \vdots \\ ikd_{iL} \end{bmatrix} = \begin{bmatrix} ik_{i11} & ik_{i12} & \cdots & ik_{i1j} & \cdots & ik_{i1M} \\ ik_{i21} & ik_{i22} & \cdots & ik_{i2j} & \cdots & ik_{i2M} \\ \vdots & \vdots & & \vdots & & \vdots \\ ik_{ip1} & ik_{ip2} & \ldots & ik_{ipj} & \ldots & ik_{ipM} \\ \vdots & \vdots & & \vdots & & \vdots \\ ik_{iL1} & ik_{iL2} & \cdots & ik_{iLj} & \cdots & ik_{iLM} \end{bmatrix}, 1 \leq i \leq N, 1 \leq p \leq L, 1 \leq j \leq M \quad (5)$$

where $ikd_i$ denotes the IKD of individual $i$, $L$ indicates the pre-defined number of solutions saved in the IKD, and $ikd_{ip}$ is known as the $p$-th best solution of individual $i$.

When MAHLO performs the individual learning operator, it generates new solutions by copying the corresponding knowledge stored in the IKD as Eq. (6).

$$x_{ij} = ik_{ipj}. \quad (6)$$

### 2.2.3 Social Learning Operator

Everyone can gain self-improvement by learning the experiences of others. In MAHLO, the Social Knowledge Data (SKD) is used to reserve the knowledge of the population as Eq. (7),

$$SKD = \begin{bmatrix} skd_1 \\ skd_2 \\ \vdots \\ skd_q \\ \vdots \\ skd_H \end{bmatrix} = \begin{bmatrix} sk_{11} & sk_{12} & \cdots & sk_{1j} & \cdots & sk_{1M} \\ sk_{21} & sk_{22} & \cdots & sk_{2j} & \cdots & sk_{2M} \\ \vdots & \vdots & & \vdots & & \vdots \\ sk_{q1} & sk_{q2} & \ldots & sk_{qj} & \ldots & sk_{qM} \\ \vdots & \vdots & & \vdots & & \vdots \\ sk_{H1} & sk_{H2} & \cdots & sk_{Hj} & \cdots & sk_{HM} \end{bmatrix}, 1 \leq q \leq H, 1 \leq j \leq M. \quad (7)$$

where $skd_q$ denotes the $q$-th solution in the SKD and $H$ is the size of the SKD. Learning from the best solutions of the population, the social learning operator is operated in MAHLO as Eq. (8).

$$x_{ij} = sk_{qj}. \quad (8)$$

### 2.2.4 Re-learning Operator

When an individual does not improve the fitness value for certain iterations successively, it is considered that it encounters a bottleneck and is trapped in the local optima. Then the re-learning strategy is activated, which discards the experience stored in the IKD, i.e., re-initializing the current IKD, to help the individual jump out of the local optimum.

In summary, MAHLO generates a new solution by implementing individual learning, social learning, and random learning with a specific probability as Eq. (9) and performs a re-learning operator when an individual encounters a bottleneck.

$$x_{ij} = \begin{cases} Rand(0, 1), & 0 \leq rand \leq pr \\ ik_{ipj}, & pr < rand \leq pi \\ sk_{qj}, & else \end{cases}. \tag{9}$$

### 2.3 Adaptive Strategies

As can be seen from the above, the parameters *pi* and *pr* have a great impact on the performance of the algorithm, because they directly determine the balance between exploration and exploitation. To strengthen the search efficiency and relieve the effort of the parameter setting, the adaptive *pr* and *pi* are used in MAHLO as Eqs. (10) and (11),

$$pr = pr_{\max} - \frac{pr_{\max} - pr_{\min}}{Ite_{\max}} \times Ite. \tag{10}$$

$$pi = pi_{\min} + \frac{pi_{\max} - pi_{\min}}{Ite_{\max}} \times Ite. \tag{11}$$

where $pr_{\min}/pi_{\min}$ and $pr_{\max}/pi_{\max}$ are the minimum and maximum values of $pr/pi$, respectively; $Ite$ and $Ite_{\max}$ are the current iteration number and maximum iteration number of the search, respectively.

## 3 Feature Selection Based on MAHLO

### 3.1 Wrapper Feature Selection Based on MAHLO and SVM

In the proposed wrapper feature selection based on MAHLO (FSMAHLO), the Support Vector Machine (SVM) is adopted as the classifier. The solution in FSMAHLO is divided into two parts as Eq. (12); one part is used for feature selection problems, and the other part is used to optimize the parameters of SVM.

$$x_i = \left[ x_{i1}\ x_{i2} \cdots x_{ij} \cdots x_{im} \cdots x_{iM} \right],\ \ x_{ij} \in \{0, 1\}, 1 \leq i \leq N, 1 \leq j \leq M. \tag{12}$$

In the above binary solution, the first part includes the bits from $x_{i1}$ to $x_{im}$, and each bit represents a feature of problems, of which the value, i.e., '1' or '0', is used to indicate that the feature is selected or not. The second part consists in the bits from $x_{im}$ to $x_{iM}$, which is used to find the best parameter values used in SVM.

In this work, the radial base function (RBF) is used as the kernel function, as Eq. (13), and the MAHLO algorithm is used to search the best values of the penalty parameter C and the RBF parameter $\sigma$, denoted by the bits from $x_{im}$ to $x_{iM}$ in an individual of MAHLO. The penalty parameter C represents the degree of punishment for classification errors in the case of inseparable linearity.

$$K_{RBF}(x, y) = \exp\left\{ -\frac{\|x - y\|^2}{2\sigma^2} \right\}. \tag{13}$$

In the processing of the data sets, the original data are divided by 7:3, i.e. the training dataset and test dataset. 70% of the training data is used for the training of the classifier and the remaining part is used as the validation set to calculate the fitness in MAHLO. For the data training of the support vector machine, the k-fold cross-validation method is adopted to ensure the stability of the model.

The fitness function of MAHLO-SVM is defined as Eq. (14)

$$Fitness\left(I_i^t\right) = \frac{1}{K}\sum_{k=1}^{K}\frac{1}{N}\sum_{j=1}^{N}\delta\left[c\left(x_j\right), y_j\right]. \tag{14}$$

where $t$ represents the $t$-th iteration of an individual, $K$ represents the fold number of cross-validation, and $N$ represents the sample number of the validation data set. $c(x_j)$ and $y_i$ represent the classification result of the $j$-th sample and the true category of that. The relationship between $c(x_j)$ and $y_i$ is defined as Eq. (15).

$$\delta\left[c\left(x_j\right), y_j\right]=\begin{cases} 1 \ if \ c\left(x_j\right) = y_j \\ 0 \quad otherwise \end{cases}. \tag{15}$$

When the classification result of the $j$-th sample by the SVM classifier is consistent with the actual category, that is, when the SVM classifies a sample correctly, the value of $\delta$ is 1, otherwise it is 0.

To efficiently eliminate the redundant features and useless features, a two-stage strategy is adapted in the updating of the IKD and SKD, which first compares the fitness value of the candidate with those of the solutions in the IKD and SKD; if the fitness values are equal, it furthers compares the numbers of features selected by those solutions, and the one with fewer features is considered as a better solution and saved/remained in the IKD and SKD.

The optimal solution obtained after training is the selected features and the model of the optimized SVM, which is tested with the test dataset to verify the FSMAHLO method.

### 3.2 Procedure

The implementation of feature selection based on MAHLO can be described as follows:

Step 1: Divide the data into the training dataset, validation dataset and test dataset.

Step 2: Set the parameters of the algorithms, including the size of the population (SP), the maximum generation (MG) and the control parameters of algorithms, such as $pr_{max}$, $pr_{min}$, $pi_{max}$ and $pi_{min}$, and the search value range of the parameters of SVM.

Step 3: Initialize the population, IKDs and SKD.

Step 4: Use the MAHLO's learning operators to generate new candidate solutions as Eq. (9).

Step 5: Use the training data to train the SVM model and calculate the fitness value with the validation dataset as Eq. (14) according to the selected features and kernel parameter values given by the new candidates of MAHLO.

Step 6: Update the IKD and SKD according to the two-stage strategy.

Step 7: Stop the search and output the optimal solution if the preset MG is met, otherwise go to Step 4.

Step 8: Output the selection features and optimal SVM model given by the best solution of MAHLO.

# 4 Experiments and Analysis

## 4.1 Simulative Benchmark Problems

Usually, the theoretical optimal classification results are not known. Therefore, to clearly and exactly check the performance of the proposed method, the simulative benchmark is used as suggested in [9]. Seven standard problems are generated and all of them include relevant features (RFs), major relevant features (MRFs), redundant features (ReFs), misleading features (MFs), useless features, and correlation features (CFs). Table 1 lists the number of these features in each case. The features are randomly assigned contribution values according to their types to indicate their impact on the classification.

**Table 1.** The simulative benchmark problems.

| Features | RFs | MRFs | ReFs | CFs | MFs |
|---|---|---|---|---|---|
| 200 | 10 | 3 | 2 | 2 | 15 |
| 1000 | 20 | 5 | 5 | 4 | 100 |
| 2000 | 25 | 5 | 5 | 4 | 300 |
| 5000 | 30 | 10 | 8 | 4 | 600 |
| 10000 | 50 | 12 | 15 | 6 | 1600 |
| 15000 | 60 | 15 | 20 | 8 | 4000 |
| 20000 | 80 | 15 | 30 | 10 | 6000 |

The theoretical optimal value of these seven all standard test problems is 100. As mentioned above, if the fitness values of the two individuals are equal, and the individual with fewer features is superior.

For comparison, the standard Human Learning Optimization (HLO) [10] and two other state-of-art metaheuristics, i.e. Modified Binary Differential Evolution (MBDE) [9] and Time-varying Transfer Binary Particle Swarm Optimization (TVTBPSO) [13], are also used for feature selection. The parameter settings of all algorithms are listed in Table 2. Each algorithm ran on each case 10 times independently. The results of feature selection are shown in Table 3.

**Table 2.** Parameter settings of all the meta-heuristics.

| Algorithm | Parameter settings |
|---|---|
| MAHLO | $pr_{max} = 10/M$ , $pr_{min} = 1/M$ , $pi_{max} = 0.9 + 1/M$ , $pi_{min} = 0.8 + 1/M$ |
| HLO | $pr = 5/M$ , $pi = 0.85 + 2/M$ |
| MBDE | $\mathrm{CR} = 0.2, b = 20, F = F_{\max} - (F_{\max} - F_{\min}) * t/G, F_{\max} = 0.8, F_{\min} = 0.005$ |
| TVTBPSO | $w = 1, \varphi_{\max} = 5, \varphi_{\min} = 1, c_1 = c_2 = 2, v_{\max} = 2.6655 \ln(M) - 4.10$ |

**Table 3.** The results of feature selection on the simulative benchmark problems

| Dimension | Algorithm | Fitness | RFs | MRFs | ReFs | CFs | MFs |
|---|---|---|---|---|---|---|---|
| 200 | Optima | 100 | 7 | 3 | 2 | 15 | 0 |
| | MAHLO | 100 | 7 | 3 | 2 | 15 | 0 |
| | HLO | 100 | 7 | 3 | 6 | 15 | 0 |
| | MBDE | 100 | 7 | 3 | 2 | 15 | 0 |
| | TVTBPSO | 100 | 7 | 3 | 6 | 15 | 0 |
| 1000 | Optima | 100 | 15 | 5 | 5 | 23 | 0 |
| | MAHLO | 100 | 15 | 5 | 5 | 23 | 0 |
| | HLO | 100 | 15 | 5 | 15 | 23 | 0 |
| | MBDE | 100 | 15 | 5 | 15 | 23 | 0 |
| | TVTBPSO | 100 | 15 | 5 | 16 | 23 | 0 |
| 2000 | Optima | 100 | 20 | 5 | 5 | 20 | 0 |
| | MAHLO | 100 | 20 | 5 | 5 | 20 | 0 |
| | HLO | 100 | 20 | 5 | 10 | 20 | 0 |
| | MBDE | 100 | 20 | 5 | 12 | 20 | 0 |
| | TVTBPSO | 100 | 20 | 5 | 11 | 20 | 0 |
| 5000 | Optima | 100 | 20 | 10 | 8 | 25 | 0 |
| | MAHLO | 100 | 20 | 10 | 8 | 25 | 0 |
| | HLO | 100 | 20 | 10 | 22 | 25 | 0 |
| | MBDE | 100 | 20 | 10 | 26 | 25 | 0 |
| | TVTBPSO | 100 | 20 | 10 | 24 | 25 | 0 |
| 10000 | Optima | 100 | 38 | 12 | 15 | 42 | 0 |
| | MAHLO | 100 | 38 | 12 | 15 | 42 | 0 |
| | HLO | 94.8498 | 38 | 12 | 22 | 42 | 0 |
| | MBDE | 96.4711 | 38 | 12 | 36 | 42 | 55 |

*(continued)*

**Table 3.** (*continued*)

| Dimension | Algorithm | Fitness | RFs | MRFs | ReFs | CFs | MFs |
|---|---|---|---|---|---|---|---|
| | TVTBPSO | 98.9986 | 38 | 12 | 40 | 42 | 17 |
| 15000 | Optima | 100 | 45 | 15 | 20 | 52 | 0 |
| | MAHLO | 100 | 45 | 15 | 20 | 52 | 0 |
| | HLO | 95.7063 | 45 | 15 | 32 | 31 | 0 |
| | MBDE | 72.3681 | 44 | 15 | 64 | 46 | 1472 |
| | TVTBPSO | 83.5844 | 43 | 15 | 64 | 49 | 371 |
| 20000 | Optima | 100 | 65 | 15 | 30 | 62 | 0 |
| | MAHLO | 100 | 65 | 15 | 30 | 62 | 0 |
| | HLO | 92.2476 | 65 | 15 | 40 | 29 | 0 |
| | MBDE | 21.1238 | 63 | 15 | 82 | 38 | 2379 |
| | TVTBPSO | 67.9778 | 60 | 15 | 84 | 41 | 762 |

Table 3 shows that all algorithms can achieve the excellent results in the low-dimensional cases (dimensions less than 10000). However, when the dimension of the problems is more than 10000, only the proposed MAHLO can efficiently find the optimal feature sets. HLO, MBDE, and TVTBPSO can't eliminate the misleading MFs completely, and the fitness values are inferior to those of MAHLO, which demonstrates that MAHLO has a better global search ability.

### 4.2 UCI Benchmark Problems

To further evaluate the proposed FSMAHLO, the UCI benchmark problems, including Australian, Diabetes, German, Heart, Ionosphere, Liver, and Sonar, are used [14], of which the details are given in Table 4. MAHLO, as well as the standard Human Learning Optimization (HLO) [10], Modified Binary Differential Evolution (MBDE) [9], Time-varying Transfer Binary Particle Swarm Optimization (TVTBPSO) [13], and Improved Swarming and Elimination-dispersal Bacterial Foraging Optimization (ISEDBFO) [15], is used to select features and optimize the parameters of SVM. For a fair comparison, the experimental parameter settings follow the instructions given in [15]. All data sets were randomly divided proportionally using 10-fold cross validation technique [16]. The average classification results are shown in Tables 5 and 6, and the bold values denote the best results.

**Table 4.** Data sets information.

| Data set | No. of instances | No. of features |
|---|---|---|
| Australian | 690 | 14 |
| Diabetes | 768 | 8 |
| German | 1000 | 24 |
| Heart | 303 | 13 |
| Ionosphere | 351 | 34 |
| Liver | 345 | 6 |
| Sonar | 208 | 60 |

**Table 5.** Number of selected features via five algorithms on each data set.

| Data set | Original features | TVTBPSO | MBDE | ISEDBFO | HLO | MAHLO |
|---|---|---|---|---|---|---|
| Australian | 14 | 12 | 13 | 8 | 5 | **3** |
| Diabetes | 8 | 5 | 6 | 5 | **4** | **4** |
| German | 24 | 20 | 18 | **12** | 13 | **12** |
| Heart | 13 | 5 | 7 | 7 | **4** | 6 |
| Ionosphere | 34 | 25 | 23 | 16 | 17 | **14** |
| Liver | 6 | 4 | 6 | 5 | **3** | **3** |
| Sonar | 60 | 34 | 30 | 25 | 32 | **24** |

**Table 6.** Average classification accuracy on each data set.

| Data set | SVM | TVTBPSO | MBDE | ISEDBFO | HLO | MAHLO |
|---|---|---|---|---|---|---|
| Australian | 75.3 | 83.7 | 84.3 | 87.3 | 84.4 | **88.4** |
| Diabetes | 61.8 | 73.5 | 71.3 | 77.6 | 78.7 | **79.1** |
| German | 60.4 | 69.8 | 72.7 | 77.4 | 77.3 | **78.0** |
| Heart | 79.0 | 85.3 | 84.2 | 86.1 | 91.4 | **91.5** |
| Ionosphere | 88.7 | 92.8 | 95.6 | 96.6 | **97.1** | **97.1** |
| Liver | 67.8 | 72.9 | 71.7 | 74.8 | 75.8 | **76.0** |
| Sonar | 81.2 | 86.7 | 89.2 | 92.8 | 98.4 | **100** |

The number of selected features and the classification accuracy rate are the two indicators to measure the performance in this work. The accuracy rate represents the fundamental meaning of the classification problem. The feature number is the embodiment of the algorithm's optimization ability, and it also affects the classification accuracy.

When the optimization results have fewer features and higher accuracy, the performance of the algorithm is better.

Tables 5 and 6 show that MAHLO can find the least representative features on the six problems and get the highest classification accuracy on the all seven problems. For the Diabetes, German and Liver data sets, although MAHLO has the same numbers of features captured as HLO or ISEDBFO, its accuracy rates are better. For Ionosphere, MAHLO and HLO have the same classification accuracy, but MAHLO selects fewer features. For Heart, HLO selects fewer features, but its accuracy rate decreases, which shows that the performance of the algorithm is insufficient. For Sonar, MAHLO can get a classification accuracy of 100%.

The above results show that the proposed MAHLO algorithm can adapt to different sizes of search space and has a higher classification accuracy.

## 5 Conclusions

In this work, we propose a new MAHLO algorithm for feature selection problems in which the initialization and random learning strategies are modified to solve the feature selection problem more efficiently and the adaptive strategies are used to enhance the performance and relieve the effort of parameter setting. A two-stage evaluation function is adopted to reduce the number of selected features. The simulation results demonstrate that MAHLO can efficiently handle complex high-dimensional problems, remove redundant, useless and misleading features. The proposed MAHLO-FS can obtain the perfect fitness value on the 20000-dimensional case and its convergence speed is significantly faster than the other algorithms, which indicates that the proposed method is very powerful for high-dimensional problems. In the classification experiments, MAHLO-FS obtains the best classification accuracy rates with fewer features, which further demonstrate the efficiency and the advantage of MAHLO-FS.

**Acknowledgments.** This work is supported by National Key Research and Development Program of China (No. 2019YFB1405500), National Natural Science Foundation of China (Grant No. 92067105 & 61833011), Key Project of Science and Technology Commission of Shanghai Municipality under Grant No. 19510750300 & 19500712300, and 111 Project under Grant No. D18003.

## References

1. Han, J., Kamber, M., Pei, J.: Data mining concepts and techniques third edition. J. Morgan Kaufmann Ser. Data Manag. Syst. **5**(4), 83–124 (2011)
2. Arora, S., Anand, P.: Binary butterfly optimization approaches for feature selection. J. Exp. Syst. Appl. **116**, 147–160 (2019)
3. Li, Y., Li, T., Liu, H.: Recent advances in feature selection and its applications. Knowl. Inf. Syst. **53**(3), 551–577 (2017)
4. Liu, H., Yu, L.: Toward integrating feature selection algorithms for classification and clustering. J. IEEE Trans. Knowl. Data Eng. **17**(4), 491–502 (2005)

5. Yusta, S.C.: Different metaheuristic strategies to solve the feature selection problem. J. Pattern Recogn. Lett. **30**(5), 525–534 (2009)
6. Guyon, I., Elisseeff, A.: An introduction to variable and feature selection. J. Mach. Learn. Res. **3**, 1157–1182 (2003)
7. Yang, J., Honavar, V.: Feature subset selection using a genetic algorithm. In: Feature Extraction, Construction and Selection, pp. 117–136. Springer, Boston (1998) https://doi.org/10.1007/978-1-4615-5725-8_8
8. Rostami, M., Forouzandeh, S., Berahmand, K., Soltani, M.: Integration of multi-objective PSO based feature selection and node centrality for medical datasets. J. Genom. **112**(6), 4370–4384 (2020)
9. Wang, L., Ni, H., Yang, R., Pappu, V., Fenn, M.B., Pardalos, P.M.: Feature selection based on meta-heuristics for biomedicine. J. Optimiz. Method. Softw. **29**(4), 703–719 (2014)
10. Wang, L., Ni, H., Yang, R., Fei, M., Ye, W.: A simple human learning optimization algorithm. In: Fei, M., Peng, C., Su, Z., Song, Y., Han, Q. (eds.) LSMS/ICSEE 2014. CCIS, vol. 462, pp. 56–65. Springer, Heidelberg (2014). https://doi.org/10.1007/978-3-662-45261-5_7
11. Wang, L., Ni, H., Yang, R., Pardalos, P.M., Du, X., Fei, M.: An adaptive simplified human learning optimization algorithm. J. Inf. Sci. **320**, 126–139 (2015)
12. Forcheri, P., Molfino, M.T., Quarati, A.: ICT driven individual learning: new opportunities and perspectives. J. Educ. Technol. Soc. **3**(1), 51–61 (2000)
13. Islam, M.J., Li, X., Mei, Y.: A time-varying transfer function for balancing the exploration and exploitation ability of a binary PSO. J. Appl. Soft Comput. **59**, 182–196 (2017)
14. Blake, C.: UCI repository of machine learning databases (1998). http://www.ics.uci.edu/~mlearn/MLRepository.html
15. Chen, Y.P., et al.: A novel bacterial foraging optimization algorithm for feature selection. J. Exp. Syst. Appl. **83**, 1–17 (2017)
16. Dietterich, T.G.: Approximate statistical tests for comparing supervised classification learning algorithms. J. Neural Comput. **10**(7), 1895–1923 (1998)

# New BBO Algorithm with Mixed Migration and Gaussian Mutation

Benben Yang[1](✉), Lisheng Wei[2], Yiqing Huang[2], and Xu Sheng[3]

[1] Anhui Polytechnic University, Wuhu, An-hui 241000, China
[2] Anhui Key Laboratory of Electric Drive and Control, Wuhu, An-hui 241002, China
[3] Nantong Vocational University, Nantong, Jiang-su 226007, China

**Abstract.** This paper proposes a new type of biogeography-based optimization with mixed migration and Gaussian mutation. First, a hyperbolic tangent migration model is adopted, which is more in line with natural science; secondly, for the core step of the algorithm-migration operation, a hybrid migration mode with the addition of a perturbation factor $\gamma$ is used to enhance the algorithm's global search capabilities, At the same time, Gaussian mutation is added to the original BBO mutation operation. These two mutation methods work together on the mutation operation, and the final results show that the proposed improvement strategy can on the one hand improve searching accuracy, on the other hand accelerate convergence speed of the algorithm.

**Keywords:** Biogeography-based optimization · Migration model · Mixed migration · Gaussian mutation

## 1 Introduction

Dan Simon first proposed the Biogeography-based Optimization (BBO) algorithm in 2008, The migration and mutation of organisms between habitats are the core steps of the algorithm. Through the research of scholars, people have concluded that the algorithm has problems such as low solution accuracy and slow convergence speed in the search process, so it has important academic and engineering value for the theory and application of BBO algorithm.

In recent years, there are mainly three forms of improvement strategies for the BBO algorithm. The first type is an in-depth theoretical type. For example, Simon used Markov chain model to analyze the convergence of the BBO algorithm [1]. According to the distribution of species in biogeography, Ma et al. introduced four different mobility models [2]. Wang et al. applied three new nonlinear mobility models to the BBO algorithm [3]; The second type is hybrid algorithm type, such as the GBBO algorithm based on GA algorithm proposed by Wang et al. [4], and the biogeographic optimization algorithm of dual-mode migration strategy proposed by Li et al. [5]; The third type is the algorithm expansion application field type, such as Liang et al. introduced the secondary mutation operation of adjacent dimensions in the BBO algorithm, and applied it to the optimization setting of the current limiter of a 500 kV regional power grid [6]. Luo et al.


Q. Han et al. (Eds.): LSMS 2021/ICSEE 2021, CCIS 1469, pp. 795–804, 2021.
https://doi.org/10.1007/978-981-16-7213-2_77

combined BBO algorithm adds a three-dimensional mutation operation, and the results show that the improved BBO algorithm can provide an effective method for the target assignment of maritime joint strikes [7]. Although these improvements have a certain effect on improving the algorithm, there is still room for improvement in convergence accuracy, speed and stability [8].

The structure of this paper is as follows: The first part is the introduction, which briefly introduces the research background and development status of this paper; The second chapter introduces the algorithm from two aspects: basic idea and core steps. The third part studies the algorithm model, migration operation and mutation operation, and then proposes the improved algorithm in this paper. The experimental simulation is arranged in the fourth chapter, and the algorithm performance is compared through the experimental data, and then the conclusion is given in the fifth part.

## 2 Basic BBO Algorithm

### 2.1 The Basic Idea of BBO Algorithm

The basic idea of the algorithm is roughly as follows: In the vegetation distribution, if light intensity, humidity and other factors of a habitat $H_i$ meet the survival requirements of most biological populations, it can be said that the habitat suitability index ($HSI$) is higher. It is conceivable that Animals move back and forth between habitats to improve their chances of survival.

The BBO algorithm is such an evolutionary algorithm based on the migration and mutation of biological populations. Each habitat in the algorithm has two natural attributes, namely the immigration rate $\lambda$ and the emigration rate $\mu$. Their size is proportional to the $HSI$ of the habitat, and $HSI$ is determined by the suitable index vector ($SIV$). The vegetation distribution, temperature and other factors mentioned above are all $SIV_s$.

### 2.2 Migration Operation

The migration operation completes the exchange and sharing of information by simulating the mutual migration of organisms in the habitat. The migration process can be shown by Eq. (1), $H_i(SIV_g)$ represents the type g $SIV$ of the habitat $H_i$.

$$H_i(SIV_g) \leftarrow H_j(SIV_g). \tag{1}$$

The immigration rate $\lambda_k$ and the emigration rate $\mu_k$ of a certain habitat $H_k$ can be defined as follows, where $k$ is the number of species contained in the habitat $H_k$, $S_{\max} = n$.

$$\lambda_k = I\left(1 - \frac{k}{n}\right). \tag{2}$$

$$\mu_k = E\frac{k}{n}. \tag{3}$$

### 2.3 Mutation Operation

A large number of facts show that the habitat with a lower probability $P_i$ is more prone to mutation, in other words the mutation rate $M_i$ is inversely proportional to the species probability $P_i$, the specific expression is as follows:

$$M_i = M_{\max} \times \left(1 - \frac{P_i}{P_{\max}}\right). \tag{4}$$

where the maximum mutation rate $M_{\max}$ is defined by scholars, and $P_{\max} = \max\{P_i\}$.

## 3 New BBO Algorithm

### 3.1 Select the Hyperbolic Tangent Transfer Model

The basic BBO algorithm is a linear migration model. Literature [3] introduces three non-linear mobility models suitable for the BBO algorithm. After comparison, this paper selects a hyperbolic tangent migration model that is more in line with natural science. The migration model is shown in Fig. 1.

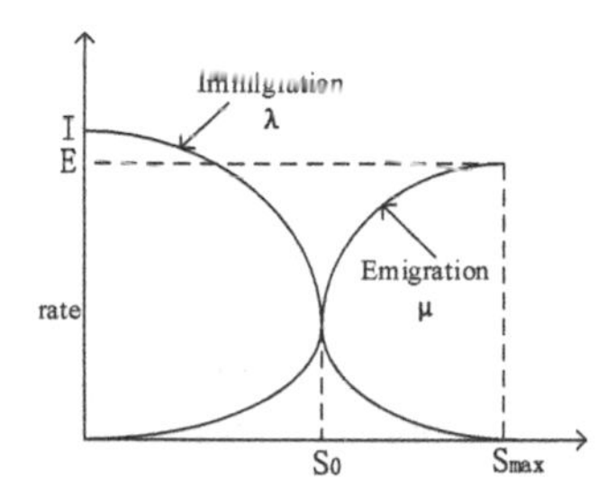


**Fig. 1.** Hyperbolic tangent migration model of species.

It can be seen from Fig. 1 that when the number of species is small, the migration rate λ is greater than the migration rate $\mu$, and the number of species in the entire habitat increases; when the number of species is large, the migration rate $\mu$ is greater than the migration rate λ, and the number of species in the entire habitat decreases; when the number of species is around $S_0$, the immigration rate is roughly the same as the emigration rate, and the species is in dynamic equilibrium at this time. The specific expressions of the immigration rate $\lambda_k$ and the emigration rate $\mu_k$ of a certain habitat under this model are as follows, where $\alpha$ is the impact factor, and the literature takes it as 1.1, $k$ is the number of biological populations in the habitat, and $n$ is the habitat's maximum number of biological populations.

$$\lambda_k = \frac{I}{2}\left(-\frac{\alpha^{k-n/2} - \alpha^{-k+n/2}}{\alpha^{k-n/2} + \alpha^{-k+n/2}} + 1\right) \tag{5}$$

$$\mu_k = \frac{E}{2}\left(\frac{\alpha^{k-n/2} - \alpha^{-k+n/2}}{\alpha^{k-n/2} + \alpha^{-k+n/2}} + 1\right) \tag{6}$$

### 3.2 Mixed Migration Mode with Perturbation Factor

The Eq. (1) shows that the migration operator of basic BBO is just a simple replacement of *SIV* between habitats. This single migration mode makes the newly generated solution obtain less information, so the weakness of the BBO algorithm is often manifested as low development ability and falling into a local optimum. Most scholars advocate the improvement of the migration operator. This article does not follow the rules, but adopts the mixed migration mode with the addition of the perturbation factor $\gamma$ to step up the global convergence speed and avoid the algorithm from falling into the local optimum. The expression of the mixed migration and perturbation factor $\gamma$ is as follows:

$$\begin{cases} H_i(SIV_g) = \gamma \times H_j(SIV_g) + (1-\gamma) \times H_i(SIV_g) \\ \gamma = \left(1 - \sin\left(\frac{\pi}{2} \times \frac{G_{index}}{G_{\max}}\right)\right) \times (1 - rand(0,1)) \end{cases}. \tag{7}$$

where $\gamma$ is the perturbation factor, $G_{index}$ is the current iteration number, and $G_{\max}$ is the maximum evolutionary algebra of the improved algorithm, $rand(0, 1)$ represents a random number from 0 to 1. We can know that $\gamma \in (0, 1)$, this strategy makes the value of $\gamma$ more flexible by introducing a sine function.

### 3.3 Improved Mutation Operation

This paper introduces the Gaussian mutation of the differential evolution algorithm, and combines the mutation operation of the basic BBO algorithm with the Gaussian mutation to improve the mutation operation. When realizing Gaussian mutation, assuming that there are 12 uniformly distributed random numbers $q_i(i = 1, 2, ...12)$ in the range of [0, 1], so a random number $L$ can be obtained by formula (8):

$$L = \mu + \delta\left(\sum_{i=1}^{12} q_i - 6\right). \tag{8}$$

In the improved algorithm, Gaussian mutation is to add a random disturbance item that obeys its random distribution to a certain component of the habitat. In other words, it is in progress by $H_k(SIV) = (SIV_1, SIV_2, ..., SIV_s, SIV_{s+1}, ...SIV_d)$ to $H_k(SIV) = \left(SIV_1, SIV_2, ..., SIV_{s'}, SIV_{s+1}, ...SIV_d\right)$, the expression of $SIV_{s'}$ can be obtained by Eq. (9):

$$H_k\left(SIV_{s'}\right) = \frac{G_{index}}{G_{\max}}(H_k(SIV_s) \times (1 + L)). \tag{9}$$

In order not to destroy the high-quality solution, the population suitability *HSI* is sorted from high to low. The first half of the habitat still uses the basic BBO mutation due to its own suitability, while the Gaussian mutation is for t the poorer second half of the population's habitat. In the later stage of the iteration, $G_{index}$ is getting closer to $G_{\max}$, and the outstanding local search ability of Gaussian mutation promotes the accurate search of candidate solutions in a local range, which improves the optimization accuracy of the algorithm.

### 3.4 Improved Algorithm Flow Chart

The process of the improved algorithm is shown in Fig. 2.

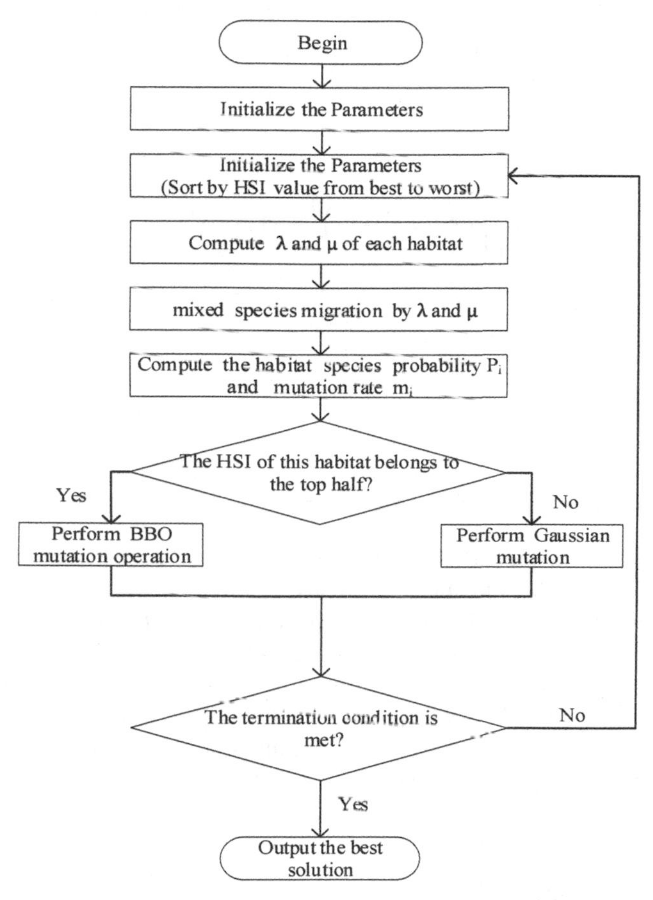


**Fig. 2.** Improved algorithm flow chart.

## 4 Research on Simulation Examples

### 4.1 Select Test Functions

In order to verify the performance of the improved algorithm, the 6 test functions selected in this paper are shown in Table 1.

**Table 1.** Benchmark test functions.

| Function | Expression | Domain |
|---|---|---|
| Ackley | $f_1(x) = -20\exp\left(-0.2\sqrt{\frac{\sum_{i=1}^{n} x_i^2}{n}}\right) - \exp\left(\frac{\sum_{i=1}^{n}\cos(2\pi x_i)}{n}\right) + 20 + e$ | $-32 \le x_i \le 32$ |
| Griewank | $f_2(x) = \sum_{i=1}^{n} \frac{x_i^2}{4000} - \prod_{i=1}^{n}\left(\cos\left(\frac{x_i}{\sqrt{i}}\right)\right) + 1$ | $-600 \le x_i \le 600$ |
| Rastrigin | $f_3(x) = 10n + \sum_{i=1}^{n}\left(x_i^2 - 10\cos(2\pi x_i)\right)$ | $-10 \le x_i \le 10$ |
| Rosenbrock | $f_4(x) = \sum_{i=1}^{n-1}\left[100\left(x_{i+1} - x_i^2\right)^2 + (1 - x_i)^2\right]$ | $-2 \le x_i \le 2$ |
| Sphere | $f_5(x) = \sum_{i=1}^{n} x_i^2$ | $-5.12 \le x_i \le 5.12$ |
| Step | $f_6(x) = \sum_{i=1}^{n} ([x_i + 0.5])^2$ | $-100 \le x_i \le 100$ |

### 4.2 Choose Various Comparison Algorithms

The experimental environment of the improved algorithm in this paper is: a notebook with a CPU of 2.10 GHz, a memory of 16 GB, and a Window10 system, and the programming language uses MATLAB R2016a.Because the improved algorithm contains mixed migration and Gaussian mutation, the algorithm in this paper is named MGBBO. In order to verify the effect of the improved algorithm, this article chooses to compare the algorithms ES, DE, and BBO. In addition, since the GBBO based on GA improvement proposed by Wang N also involves Gaussian mutation [4], this article also compares with it. In order to reduce the interference of random factors, each experiment was run independently 20 times.

### 4.3 Parameter Settings

The parameters set in this paper based on experimental experience are as follows: population size $N = 50$, feature dimension $D = 20$, global migration rate $P_{\text{mod}} = 1$, $I$ and $E$ of the habitat are both 1, and the maximum mutation rate $M_{\max} = 0.05$, the mutation rate of Gaussian mutation is also set to 0.05, and the maximum number of iterations $G_{\max} = 100$.

### 4.4 Analysis of Results

The following Table 2 is the simulation results of each test function. In this paper, two test indicators, the average value (Mean) and the optimal value (Optimal), are selected to evaluate the performance of the improved algorithm. Among them, the average value is the average solution of all values of the algorithm, which can show the optimization accuracy of the algorithm; the optimal value is the optimal solution found by the algorithm, which reflects the optimization ability of the algorithm.

**Table 2.** Test results of each algorithm on the benchmark function.

| Function | Index | ES | DE | BBO | GBBO | MGBBO |
|---|---|---|---|---|---|---|
| Ackley | Mean | 19.9668 | 7.19963 | 8.24865 | 7.27635 | 6.75396 |
| | Optimal | 19.8829 | 6.11326 | 4.72426 | 3.97185 | 3.74225 |
| Griewank | Mean | 1734.52 | 2.75174 | 16.6506 | 8.99743 | 7.59375 |
| | Optimal | 159.873 | 1.27446 | 3.17738 | 1.88562 | 1.44461 |
| Rastrigin | Mean | 556.604 | 150.534 | 32.4234 | 11.7768 | 9.56821 |
| | Optimal | 243.505 | 113.156 | 16.3229 | 3.03265 | 0 |
| Rosenbrock | Mean | 32702.6 | 129.023 | 487.932 | 427.825 | 331.998 |
| | Optimal | 2667.67 | 78.4821 | 140.353 | 108.773 | 19.6358 |
| Sphere | Mean | 506.446 | 15.2307 | 13.8301 | 10.2537 | 6.1442 |
| | Optimal | 90.3152 | 0.31765 | 1.12774 | 1.00526 | 0 |
| Step | Mean | 95706.6 | 289.567 | 1374.98 | 2855.23 | 1671.28 |
| | Optimal | 18364.7 | 95.6874 | 196.368 | 222.987 | 135.984 |

It can be seen from the data in the table that for functions $f_2$ and $f_6$, the average and optimal results of the improved algorithm in this paper are not as good as DE, but are better than other algorithms; for function $f_4$, only the average value is worse than DE, but other The data is fully dominant; for the functions $f_1, f_3$, and $f_5$, the results obtained by the improved algorithm in this paper are the best regardless of the average value or the optimal value, especially the optimal values of $f_3$ and $f_5$ are all 0. That is to say, in the case of the same parameter settings, whether it is facing a unimodal function or a multimodal function, compared with other classic intelligent algorithms, most of the improved algorithms in this paper can converge to a smaller value, and the result is closer to the most merit. In addition, the convergence curves of five different algorithms on the six benchmark functions are shown in Figs. 3, 4, 5, 6, 7 and 8.

The abscissa of the Figs. 3, 4, 5, 6, 7 and 8 is Iterations, and the ordinate is the Benchmark function value. There are two points that can be clearly observed from Figs. 3, 4, 5, 6, 7 and 8.

1) The optimal value of the improved algorithm in this article is less than other intelligent algorithms at the initial stage of the iteration. The specific performance is that the improved algorithm in this article is 18 for function $f_1$, and the others are more

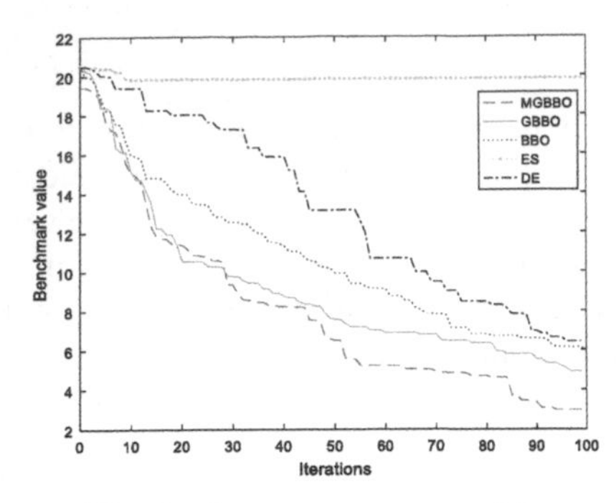


Fig. 3. Convergence curve of $f_1$.

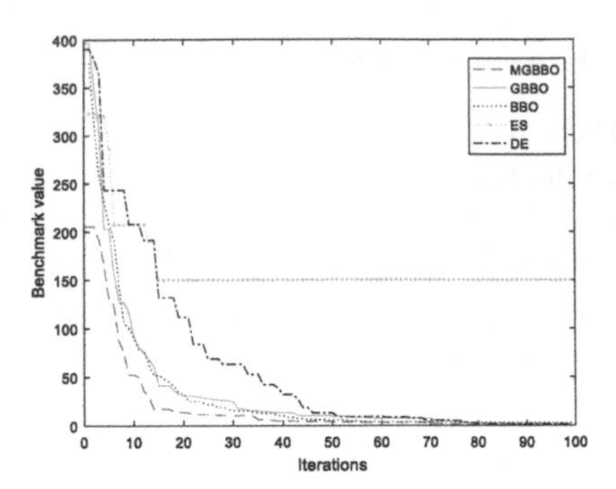


Fig. 4. Convergence curve of $f_2$.

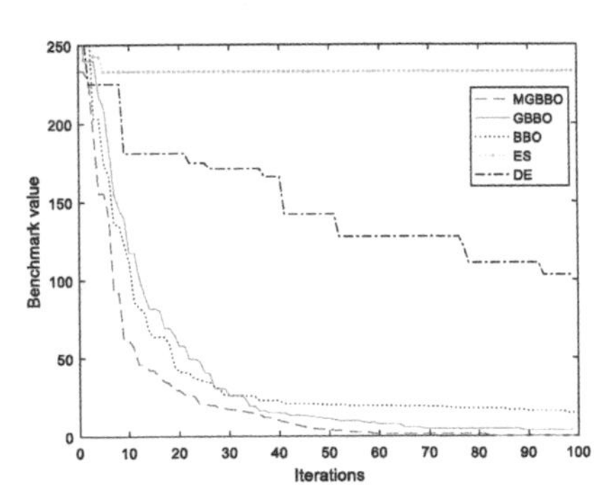


Fig. 5. Convergence curve of $f_3$.

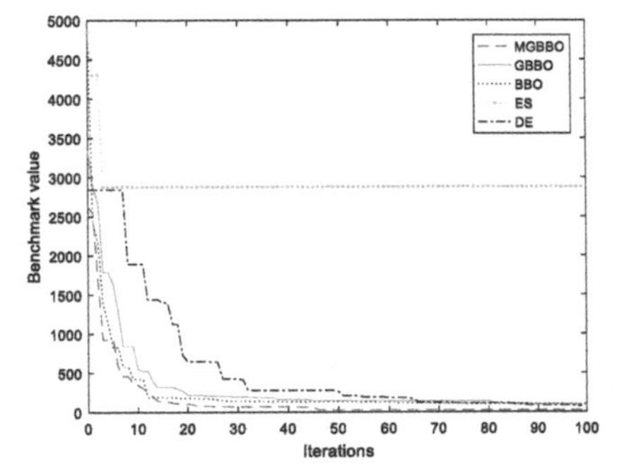


Fig. 6. Convergence curve of $f_4$.

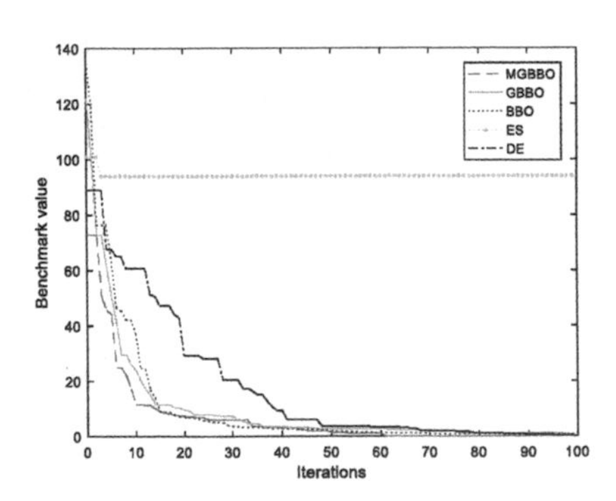


Fig. 7. Convergence curve of $f_5$.

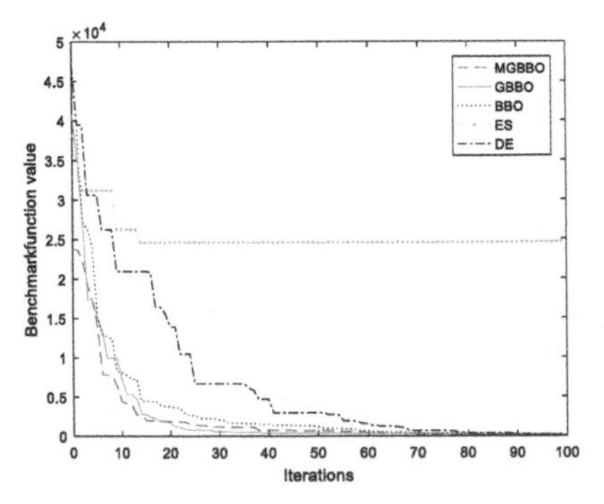


Fig. 8. Convergence curve of $f_6$.

than 20. For the function $f_2$ in this paper, it is around 210. As a comparison with other algorithms, it is greater than 300. The comparison of the initial solutions of $f_3$, $f_4$, and $f_5$ can also be seen. The performance is more obvious in function $f_6$. The initial solution of the improved algorithm in this paper is $2.4 \times 10^4$, while other intelligent algorithms are around $4 \times 10^4$.

2) In terms of convergence speed and optimization accuracy, it can be seen from $f_1$–$f_4$ that the improved algorithm in this paper is better than other intelligent optimization algorithms. On $f_5$ and $f_6$, the improved algorithm in this paper is comparable to GBBO and is still better than other algorithms. The improved algorithm in this paper has a better fitness value at the beginning of the iteration, which shows that the algorithm in this paper has better stability. At the beginning of the iteration, the improved algorithm in this paper showed a better evolution curve, which shows its stronger exploration ability. Except for the function $f_1$, all other functions almost reached the local optimum when the number of iterations was about 60 times, so it demonstrates their excellent development capabilities.

## 5 Conclusions

In this paper, a new BBO algorithm with mixed migration and Gaussian mutation is proposed to solve the problems of slow convergence speed and easy to fall into local optimum. The experimental results show that the improved algorithm in this paper has a good effect on improving the convergence accuracy and speed, and can be used in some fields such as multi-objective programming and complex system optimization. The following work focuses on the application of the algorithm in the optimization of energy dispatching of microgrid, which also has a certain practical significance.

**Acknowledgments.** This work was supported by the Natural Science Research Programme of Colleges and Universities of Anhui Province under grant KJ2020ZD39, and the Open Research Fund of Anhui Key Laboratory of Detection Technology and Energy Saving Devices under grant DTESD2020A02, the Scientific Research Project of "333 project" in Jiangsu Province under grant BRA2018218 and the Postdoctoral Research Foundation of Jiangsu Province under grant 2020Z389, and Qing Lan Project of colleges and universities in Jiangsu province.

## References

1. Simo, D.: Biogeography-based optimization. J. IEEE Trans. Evol. Comput. **12**(6), 702–713 (2009)
2. Ma, H.P., Li, X., Lin, S.D.: Analysis of migration rate models for biogeography-based optimization. J. Southeast Univ. **39**(S1), 16–21 (2009)
3. Wang, Y.P., Zhang, Z.J., Yan, Z.H., Jin, Y.Z.: Biogeography-based optimization algorithms based on improved migration rate models. J. Comput. Appl. **39**(09), 2511–2516 (2019)
4. Wang, N., Wei, L.S.: A Novel biogeography-based optimization algorithm research based on GA. J. Syst. Simul. 1717–1723 (2020)
5. Li, C.X., Zhang, Y.: Biogeography-based optimization algorithm based on the dual-mode migration strategy. J. Xi'an Univ. Posts Telecommun. **24**(01), 73–78+84 (2019)

6. Liang, Y.S., Chen, L.X., Li, H.F., Wang, G., Zeng, D.H., Huang, Z.J.: Configuration method for fault current limiter based on improved biogeography-based optimization algorithm with second mutation. J. Autom. Electr. Power Syst. **44**(01), 183–191 (2020)
7. Luo, R.H., Li, S.M.: Optimization of firepower allocation based on improved BBO algorithm. J. Nanjing Univ. Aeronaut. Astronaut. **52**(06), 897–902 (2020)
8. Wang, C.R., Wang, N.N., Duan, X.D., Zhang, Q.L.: Survey of Biogeography-based Optimization. J. Comput. Sci. **37**(07), 34–38 (2010)

# Clustering Ensemble Algorithm Based on an Improved Co-association Matrix

Xiaowei Zhou, Fumin Ma(✉), and Mengtao Zhang

College of Information Engineering, Nanjing University of Finance and Economics, Nanjing 210023, China

**Abstract.** The co-association matrix plays a critical role in constructing the consistency function for ensemble algorithms. However, the differences between basic clustering results can be ignored in the traditional co-association matrix method, hurting the division ability. Thus, this paper proposes an improvement strategy based on the granular computing theory for dealing with uncertain problems. First, the granularity distance is utilized to filter and refine the initial clustering results. Then a new method for computing the co-association matrix is developed, by incorporating the normalized average entropy in matrix elements. Finally, a clustering ensemble algorithm based on the improved co-association matrix is proposed, which can achieve more reasonable divisions. Experiment results show that the proposed algorithm can significantly improve clustering accuracy and NMI index.

**Keywords:** Clustering ensemble · Co-association matrix · Base clustering · Information entropy

## 1 Introduction

The concept of clustering ensemble was first proposed by Strehl et al. [1], which combined the clustering analysis with the ensemble learning. Several partitions are primarily obtained by clustering, and the final division can be eventually completed by some combination methods. The core tasks of clustering ensemble can be mainly divided into the design of base clustering and the construction of consistency function.

The construction of consistency function plays a critical role in clustering ensemble algorithm. The current research about the construction of consistency function could be sorted as the following: (1) Voting Based Methods. Each base clustering was regarded as a vote that the samples belong to a certain cluster. (2) Hypergraph Based Methods. Each sample was usually treated as a vertex of the graph in these methods. (3) Hybrid Construction Methods. Several methods ware usually mixed to construct a consistency function. (4) Co-Association Matrix (CAM) Methods. CAM methods ware proposed based on the idea of statistics in literature [2], which used the average weight of all relation matrices to construct a co-association matrix. CAM was translated into the similarity matrix based on path by Zhong et al. [3], and the spectral clustering algorithm was adopted at last. The generation method of the elements to CAM was improved by


Q. Han et al. (Eds.): LSMS 2021/ICSEE 2021, CCIS 1469, pp. 805–815, 2021.
https://doi.org/10.1007/978-981-16-7213-2_78

the idea of adding knowledge granularity, and the clustering ensemble algorithm based on granular computing (CEAGC) was proposed in literature [4].

Voting based methods and hypergraph based methods are heuristic algorithms, which complexity is too high. Hybrid construction methods are well suitable for all kinds of base clustering results, which could be easily fall into the local optimal situation. The CAM method is widely used due to the calculation simplicity without any initial parameters setting, at the same time, the number of clusters is related to the base clustering, the shapes of clusters have no effect on the result. However, the quality of base clustering results in most of CAM methods was ignored, which will affect the calculation of CAM directly. It is unreasonable to regard the probability of two samples belonging to the same cluster in each base clustering result as equal.

The selection of base clustering algorithms could be sorted as the following: (1) Selection based on Heterogeneous Algorithm. Using the inherent randomness of the algorithm, such as the initial point selection of classic K-means algorithm; Besides, selecting different clustering algorithms will also obtain the different base clustering results [5]. (2) Selection based on Heterogeneous Data. Include selecting different data subsets selecting different feature subsets [6] and feature projection [7].

The base clustering selection methods based on Heterogeneous data have become a hot research topic in recent years because of its obvious advantages in handling high-dimensional and large samples data sets. However, the operation of these methods are complicated. So, it is favorable to combine the heterogeneous algorithm selection with the heterogeneous data selection based on the characteristics of the data sets.

Considering the simple tractability and wide applicability, clustering ensemble algorithm based on an improved co-association matrix (CEICAM) is proposed in this paper. The design of base clustering is combining different feature subsets and well applicable clustering algorithm based on the characteristics of the data sets. After that, granular computing theory is innovatively used to filter the initial base clustering and get the refined base clustering results. Finally, the co-association matrix is improved by incorporating the normalized average entropy into the calculation of the matrix elements.

The structure of this paper is as follows: Sect. 2 introduces the knowledge of clustering ensemble and CAM. In Sect. 3, the design of base clustering design and proposed CEICAM algorithm are comprehensively discussed. Several groups of comparative experiments are designed respectively in Sect. 4. Section 5 offers some conclusions.

# 2 Clustering Ensemble Algorithm

## 2.1 Introduction to Clustering Ensemble

Clustering ensemble is an algorithm to improve the accuracy, stability and robustness of clustering results. A better result can be produced by combining multiple base clustering results. Therefore, clustering ensemble algorithm is mainly divided into base clustering and clustering ensemble.

### 2.2 Introduction to CAM

The CAM method plays an important role in clustering ensemble algorithm due to the simple process and high accuracy. A matrix that reflects overall division of samples in different base clustering results is constructed [8].

The definition of co-association matrix is $A = (A_{ij})_{n \times n}$, where $A_{ij} = \sum_{h=1}^{H} \delta_{ij}^{h} \Big/ H$, $\delta_{ij}^{h} = \begin{cases} 1, \ if \ C^h(x_i) = C^h(x_j) \\ 0, \ if \ C^h(x_i) \neq C^h(x_j) \end{cases}$, $H$ represents the total number of base clustering results, $\delta_{ij}^{h}$ represents whether the sample $x_i$ and $x_j$ are divided into the same cluster. Therefore, $A_{ij}$ means the similarity between $x_i$ and $x_j$, which represents the frequency of two samples being clustered in the same cluster. Moreover, the co-association matrix is also a real symmetric matrix, and the main diagonal elements are 1, $A_{ii} = 1$, $A_{ij} \in [0, \ 1]$, $\forall i = 1, \ 2, \ \ldots, \ n$. The probability of samples belonging to same cluster can be inferred from the value of $A_{ij}$. In literature [2], the threshold value is 0.5.

For example, in Fig. 1, $H_1$ and $H_2$ are the results of two base clustering, and thc final co-association matrix is shown in Fig. 2.

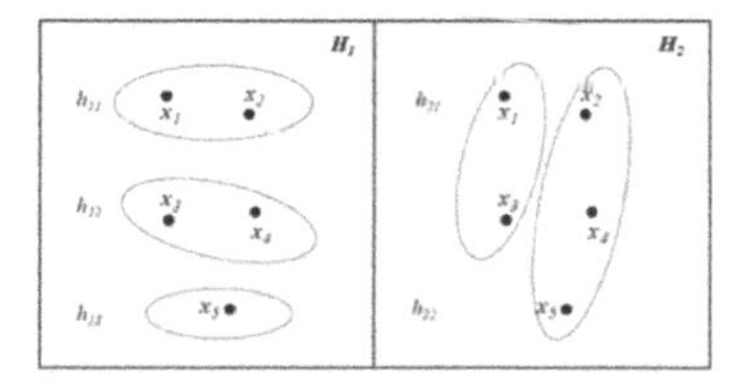

**Fig. 1.** Result of base clustering.

| | $x_1$ | $x_2$ | $x_3$ | $x_4$ | $x_5$ |
|---|---|---|---|---|---|
| $x_1$ | 1 | 0.5 | 0.5 | 0 | 0 |
| $x_2$ | 0.5 | 1 | 0 | 0.5 | 0.5 |
| $x_3$ | 0.5 | 0 | 1 | 0.5 | 0 |
| $x_4$ | 0 | 0.5 | 0.5 | 1 | 0.5 |
| $x_5$ | 0 | 0.5 | 0 | 0.5 | 1 |

**Fig. 2.** Co-Association Matrix corresponding to the base clustering in Fig. 1.

## 3 The Clustering Ensemble Algorithm Based on An Improved Co-Association Matrix (CEICAM)

### 3.1 Design of Base Clustering

The design of base clustering is the first step of clustering ensemble. In order to get better ensemble result, there should be some differences between individual learners. The characteristics of a data set will be revealed from different base clustering results from different perspectives. The ensemble of different base clustering members will take a positive impact on the results of the algorithm.

The design of base clustering algorithm is based on the characteristics of a data set itself in this paper. For the data set with four or more dimensions, the selection strategy is combining different feature subsets and heterogeneous clustering algorithm.

After the selection of basc clustering algorithm, several base clustering results are obtained, which can be called initial base clustering $C$. However, not all individuals in the initial base clustering contribute well to the final clustering ensemble. Diversity and individual precision are two critical factors to improve generalization ability that was revealed in literature [9, 10]. It is a crucial step to select the high-precision individuals from the initial base clustering results with diversity. Considering the advantages of granular computing theory in dealing with uncertain problems, granularity distance could be introduced to measure the difference between the results of initial base clustering. The individual precision of ensemble learning can be improved after filtering by granularity distance. The optimized results are called the refined base clustering.

The main advantage of granular computing to deal uncertainty and vague information is that it does not require any additional information about data, which was proposed by Zadeh in 1979 [11]. Let $K = (U, \mathrm{R})$ is a knowledge base, where $U$ represents a nonempty finite set that be called a domain and R represents a cluster of equivalent relationships [12]. $R \in \mathrm{R}$ represents an equivalent relationship (knowledge), where $R \subseteq U \times U$, the knowledge granularity is defined as: $GD(R) = |R|/\left|U^2\right| = |R|/|U|^2$, where $|R|$ represents the cardinal number of $R \subseteq U \times U$, the minimum granular of $R$ is $|U|/|U|^2 = 1/|U|$, If $R$ equals the whole domain, the maximum granular is 1. In general, the smaller $GD(R)$ is, the stronger the resolution.

$K = (U, \mathrm{R})$ is a knowledge base, $P,\ Q \in \mathrm{R}$ represent equivalent relationship (knowledge) separately, $K(P),\ K(Q)$ represent two partitions derived from the equivalent relationship $P$, $Q$, where $K(P) = \{[x_i]_P|x_i \in U\}$,$K(Q) = \{[x_i]_Q|x_i \in U\}$. The granularity distance is defined as:

$$dis(K(P), K(Q)) = \left(\sum_{i=1}^{|U|} \left|[x_i]_P \oplus [x_i]_Q\right|/|U|\right) \Big/ |U|. \tag{1}$$

Where $\left|[x_i]_P \oplus [x_i]_Q\right| = \left|[x_i]_P \cup [x_i]_Q\right| - \left|[x_i]_P \cap [x_i]_Q\right|$. The granularity distance is applied to filter out the initial base clustering $C^i$ with the smallest distance from the rest of base clustering in $C$, where the difference between $C^i$ and the whole initial base clustering $C$ is $Dis(C^i,\ C) = \sum_{j=1}^{M} dis(p_i,\ p_j)$. $C^i$ is considered to be the closest clustering result to the real division and the highest individual accuracy since it shares the most information in $C$. The process is repeated until the specified number of results being filtered out in set $C$ that have removed $C^i$. And refined base clustering result $H$ could be obtained eventually.

The specific steps to design the base clustering are as follows:

Input: original data set $D$.
Output: refined base clustering result $H$.
Step1: the appropriate algorithm $f$ will be selected according to the characteristics of the data set $D$.
Step2: the initial base clustering $C = \left\{C^1,\ C^2,\ \ldots,\ C^{\mathrm{N}}\right\}$ will be obtained.
Step3: $Dis(C^i,\ C)$ will be calculated in turn, where $i = 1,\ 2,\ \ldots,\ N$.
Step4: $C^i$ will be added to set $H$, where $C^i = \min\left\{Dis(C^k,\ C),\ k = 1,\ 2,\ \ldots,\ N\right\}$.
Step5: Step4 will be repeated in set $C$-$H$ until the specified number of results being filtered out, $H = \left\{C^1,\ C^2,\ \ldots,\ C^M\right\}$.

### 3.2 Design of CEICAM

In CEICAM, the normalized base clustering average entropy $\phi(C^m)$ will be introduced to the elements of co-association matrix. Information entropy is a mathematically abstract concept, which was defined as the probability of occurrence of discrete random events. Related uncertainty of random variables could be measured by information entropy. The degree to which one base clustering differs from the others could be measured by using the normalized base clustering average entropy $\phi(C^m)$. The advantage is that the final co-association matrix will be calculated on the basis of the comprehensive measurement of the base clustering quality.

The information entropy is defined as $H(x) = -\sum_{i=1}^{n} p(x_i)\log(p(x_i))$, where $p(x_i)$ represents the probability of random event $x_i$. Information entropy is the expected value in essence, which measures the information brought by a specific event, while entropy is the expectation of the amount of information that may be generated before the result comes out.

For any $c_i^m \in C^M$, $c_j^n \in C^N$ where $C^M$, $C^N \in H$, and the intersection of $c_i^m$, $c_j^n$ is not empty in the base clustering $H$, then the uncertainty measure of $c_i^m$ relative to $c_j^n$ is $H(c_i^m, c_j^n) = -p(c_i^m, c_j^n)\log p(c_i^m, c_j^n)$, $p(c_i^m, c_j^n) = \left|c_i^m \cap c_j^n\right| / \left|c_i^m\right|$. Let $C^n = \left\{c_1^n, c_2^n, \ldots, c_{N_n}^n\right\}$ is a base clustering in $H$, then the uncertainty of any cluster $c_i^n \in C^n$ relative to other base clustering $C^m$ is the sum of the uncertainty measures, so the uncertainty measure of cluster $c_i^n \in C^n$ relative to base clustering $C^m$ is as follows:

$$H^m(c_i^n) = \sum_{j=1}^{N_m} p(c_i^n, c_j^m)\log p(c_i^n, c_j^m). \quad (2)$$

Taking all base clustering results $H$ into account, the uncertainty measure of cluster $c_i^n \in C^n$ relative to the total base clustering $H$ should be $H(c_i^n) = \sum_{m=1}^{M} H^m(c_i^n)$, Therefore, for the base clustering $C^m \in H$, the average base clustering entropy is:

$$H_\mu(c^m) = \sum_{i=1}^{n^m} H(c_i^m) \Big/ n^m. \quad (3)$$

Where the value range of function $H_\mu(C^m)$ is $[0, +\infty)$, and the normalization results are $\varphi(C^m) = e^{-H_\mu(C^m)}$. The value range of $\varphi(C^m)$ is changed to $(0, 1]$ after normalization due to the monotonicity of information entropy. If the value of $\varphi(C^m)$ is closer to 1, it means that the difference between the results of base clustering $C^m \in H$ and the others is smaller.

Based on the idea of granularity, each base clustering is regarded as a new attribute of data, and the clustering result is regarded as a partition of data domain by attributes. The elements of the new co-association matrix are defined:

$$A_{ij} = \sum_{\lambda=1}^{H} \delta(x_i, x_j) \Big/ \sum_{\lambda=1}^{H} \left|V_{p\lambda}\right|. \quad (4)$$

$\delta(x_i, x_j) = \begin{cases} \varphi(C^{p_i})|V_{p\lambda}|, & if \quad C^{p\lambda}(x_i) = C^{p\lambda}(x_j) \\ 0, & if \quad C^{p\lambda}(x_i) \neq C^{p\lambda}(x_j) \end{cases}$, where $|V_{p_\lambda}|$ represents the number of clusters contained in the division result $p_\lambda$. $C^{p_\lambda}(x_i)$ represents the cluster label of sample $x_i$ in $p_\lambda$, and $\varphi(C^{p_\lambda})$ is the function of normalized average entropy.

The advantage of the new definition is that the quality of base clustering will affect the value of matrix elements accordingly. Finally, the threshold is set to calculate the final clustering result according to the value of the elements in the co-association matrix.

There are specific steps to design of CEICAM:

Input: refined base clustering result $H$.
Output: final clustering result.
Step1: the average entropy of base clustering $H_\mu(C^m)$ and the normalized result $\varphi(C^m)$ will be calculated.
Step2: the elements $A_{ij}$ of new matrix and the co-association matrix $A$ will be calculated.
Step3: the final clustering result will be calculated through $A$.

## 4 Experiment Analysis

The common indicators to measure the results of clustering ensemble algorithm are *Accuracy* and *NMI* (normalized mutual information):

$$Accuracy = Max\left(\sum_{C_k, L_m} T(C_k, L_m)\right) \Big/ N. \tag{5}$$

Where $C_k$ represents the $k$-th cluster in final clustering result, $L_m$ represents the $m$-th cluster in real division of the data set. $T(C_k,\ L_m)$ represents the number that belongs to the $m$-th cluster in real division of the data set while be divided into the $k$-th cluster in the final clustering ensemble algorithm.

$$NMI(F, L) = -2\sum_{i=1}^{k_a}\sum_{j=1}^{k_b}\frac{n_{ij}}{n}\log(\frac{n_{ij}\cdot n}{n_{ai}n_{bj}}) \Big/ \left(\sum_{i=1}^{k_a}\frac{n_{ai}}{n}\log(\frac{n_{ai}}{n}) + \sum_{i=1}^{k_b}\frac{n_{bj}}{n}\log(\frac{n_{bj}}{n})\right). \tag{6}$$

Where $k_a$, $k_b$ represent the number of clusters in final clustering ensemble result $F$ and real division $L$ respectively. $n_{ai}$ represents the number of samples that be divided the $i$-th cluster in the final clustering ensemble. $n_{bj}$ represents the number of samples that be divided the $j$-th cluster in the real division. $n_{ij}$ represents the number of samples that be divided into the $i$-th cluster in clustering ensemble algorithm and the $j$-th cluster in the real division at the same time. And $NMI \in (0,\ 1]$.

The data sets used in experiments are a group of artificial data set and four standard data sets in UCI database (as shown in Table 1): iris, glass identification, page blocks, red wine quality. The experimental environment is windows 10 operating system, AMD ryzen7 processor, 16 GB memory, python language programming, python version 3.8.

**Table 1.** The information of data sets.

| Name | Samples Number | Dimension | Category |
|---|---|---|---|
| Iris | 150 | 4 | 3 |
| Glass Identification | 214 | 9 | 7 |
| Page Blocks | 5473 | 10 | 5 |
| Red Wine Quality | 1599 | 11 | 6 |

### 4.1 Experiment Results on Artificial and Standard Data Sets

In order to verify the performance of the algorithm proposed in this paper, several groups of comparative experiments are designed respectively. In the experiments, according to literature [2], the threshold value of the matrix elements is 0.5. The relationship between initial base clustering results $C$ and refined base clustering results $H$ are satisfied $|H| = |C| \times 80\%$.

Firstly, comparative experiments of CEAGC and CEICAM are designed by using a group of artificial data set. The results are shown in Figs. 3, 4, 5 and 6.

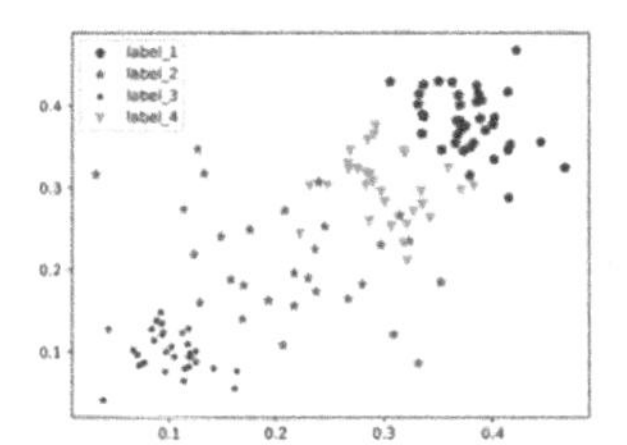

**Fig. 3.** Original artificial data set.

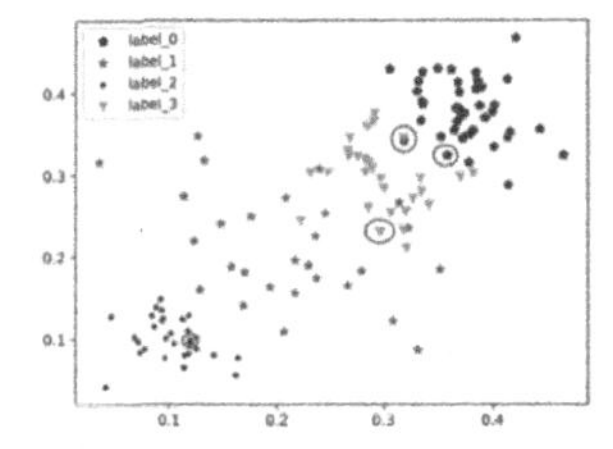

**Fig. 4.** The clustering ensemble result of CEICAM.

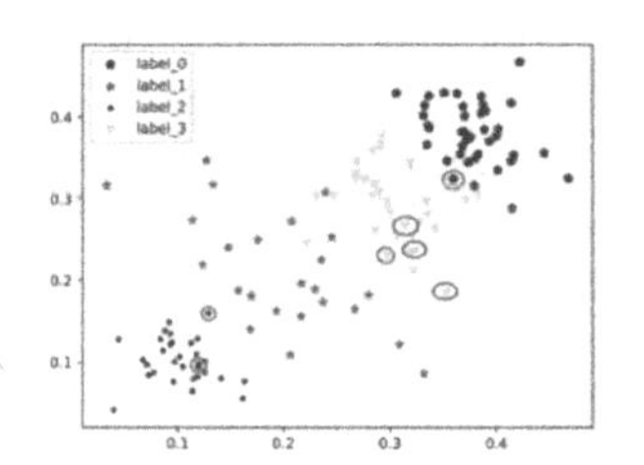

**Fig. 5.** The clustering ensemble result of CEAGC.

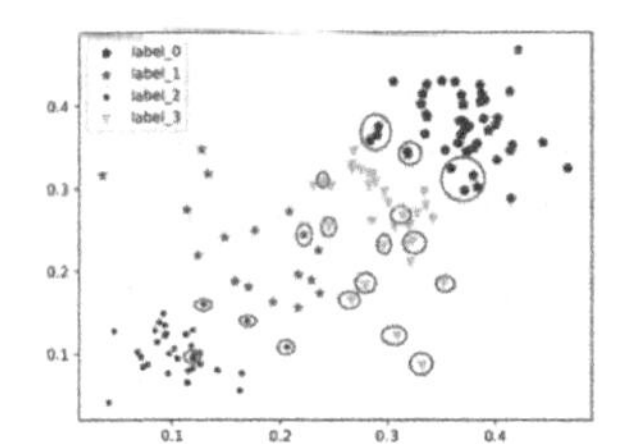

**Fig. 6.** The result of a base clustering.

From the experiment results on artificial data set, it can be seen that the algorithm of CEICAM, CEAGC and the base clustering are all incorrectly divided some samples at the junction of the real clusters. There are 4 wrong division points in CEICAM algorithm, 7 wrong division points in CEAGC algorithm and totally 22 wrong division points in base clustering algorithm. The misjudgment rate of CEICAM algorithm is lower than other algorithms, and in the cluster boundary areas, CEICAM algorithm has better resolving capacity. So, the accuracy of CEICAM algorithm is significantly improved in artificial data set.

Several groups of comparative experiments are designed by using standard data sets as shown in Table1. The final results of experiments are shown in Tables 2 and 3.

In experiment group 1, the K-means clustering algorithm and different feature subsets are selected to generate the base clustering results. The final results of the experiments are shown in Table 2 third and fourth columns. In experiment group 2, the different feature subsets and the heterogeneous algorithms which include hierarchical clustering algorithm and K-means clustering algorithm are selected to generate the base clustering results. The base clustering results are increased accordingly. The final results of the experiments are shown in Table 2 fifth and sixth columns.

**Table 2.** The comparative results between base clustering and CEICAM by using different feature subsets and heterogeneous algorithms.

| Name | Indicators | Base clustering | CEICAM | Base clustering | CEICAM |
|---|---|---|---|---|---|
| Iris | *Accuracy* | 90.67% | **94%** | 89.33% | **91.33%** |
| | *NMI* | 0.8586 | **0.9107** | 0.8373 | **0.8592** |
| Glass Identification | *Accuracy* | 85.47% | **87.85%** | 85.05% | **89.25%** |
| | *NMI* | 0.7563 | **0.8131** | 0.7663 | **0.8159** |
| Page Blocks | *Accuracy* | 65.83% | **66.82%** | 77.01% | **79.84%** |
| | *NMI* | 0.5047 | **0.5321** | 0.5832 | **0.6253** |
| Red Wine Quality | *Accuracy* | 78.74% | **80.55%** | 80.74% | **83.24%** |
| | *NMI* | 0.6351 | **0.6879** | 0.6913 | **0.7354** |

In experiment group 3, the ensemble clustering algorithms of CEAGC and CEICAM will be compared based on the different UCI datasets. The same base clustering results in experiment group 1 are used in both of clustering ensemble algorithms. The final results of comparative experiments are shown in Table 3 third and fourth columns. In experiment group 4, the comparative algorithms of ensemble clustering are still CEAGC and CEICAM, but the base clustering adopts the results in experiment group 2. The final results of experiment group 4 are shown in Table 3 fifth and sixth columns.

**Table 3.** The comparative results between CEAGC and CEICAM by using the same base clustering results in experiment group 1 and experiment group 2.

| Name | Indicators | CEAGC | CEICAM | CEAGC | CEICAM |
|---|---|---|---|---|---|
| Iris | *Accuracy* | 92% | **94%** | **91.33%** | **91.33%** |
| | *NMI* | 0.8797 | **0.9107** | **0.8592** | **0.8592** |
| Glass Identification | *Accuracy* | 86.92% | **87.85%** | 86.45% | **89.25%** |
| | *NMI* | 0.7977 | **0.8131** | 0.7899 | **0.8159** |
| Page Blocks | *Accuracy* | **67.42%** | 66.82% | 78.40% | **79.84%** |
| | *NMI* | **0.5513** | 0.5321 | 0.6089 | **0.6253** |
| Red Wine Quality | *Accuracy* | 79.42% | **80.55%** | 82.23% | **83.24%** |
| | *NMI* | 0.6676 | **0.6879** | 0.7178 | **0.7354** |

## 4.2 Experimental Analysis

From Table 2, comparing the proposed CEICAM with the base clustering algorithm, it can be seen that the clustering *Accuracy* and *NMI* index of CEICAM are improved on standard data sets. There is a positive correlation between the quality of base clustering and the effect of the improved algorithm, and the higher accuracy results of the base clustering algorithm, the more obvious improvement effect of the improved algorithm is.

From Table 3, comparing the proposed CEICAM with the ensemble clustering algorithm CEAGC. If the base clustering results adopt K-means clustering algorithm, the *Accuracy* and *NMI* of CEICAM are improved on the data sets of iris, glass identification and red wine quality, but decreased on the data set of page blocks. If the base clustering results adopt heterogeneous clustering algorithms, the *Accuracy* and *NMI* of CEICAM on these four data sets are also improved. For iris data set, the results of the two algorithms are the same. The *Accuracy* and *NMI* of the other three data sets is improved. Since each element of co-association matrix in the proposed CEICAM contains the normalized average entropy, the influence of some low-quality base clustering members is reduced to a certain extent compared with other algorithms. But, when the quality of all the base clustering results are not ideal, each corresponding normalized average entropy will be relatively improved which results in that the improvement of the CEICAM on the data set of the page blocks is not obvious in experiment group 3. In general, the positive impact of high-quality base clustering on the clustering ensemble algorithm CEICAM is improved, which enhancing the stability and accuracy of clustering. From the above experiment results, the feasibility and effectiveness of CEICAM could be verified.

The running time of CEICAM and CEAGC are shown in Table 4.

**Table 4.** Running time of the experiments in CEICAM and CEAGC.

| Name | Algorithms | Running time in group 3(s) | Running time in group 4(s) |
|---|---|---|---|
| Iris | CEICAM | 0.673 | 1.312 |
| | CEAGC | 0.509 | 1.127 |
| Glass Identification | CEICAM | 1.423 | 2.014 |
| | CEAGC | 1.230 | 1.793 |
| Page Blocks | CEICAM | 33.159 | 55.701 |
| | CEAGC | 27.601 | 48.507 |
| Red Wine Quality | CEICAM | 5.329 | 7.662 |
| | CEAGC | 4.481 | 6.905 |

In experiment group 3, the time complexity of CEICAM is $O(n \log n + n + n^2)$, while time complexity of CEAGC is $O(n \log n + n^2)$. The increase of the time complexity in CEICAM algorithm mainly depends on the calculation of normalized average entropy. From the experimental results and running time, the time will be prolonged when the size of data set increases. In experiment group 4, the time complexity has increased by $O(n^2 \log n)$ due to the hierarchical clustering algorithm added. However, from the *Accuracy* and *NMI*, it can be seen that although the running time of CEICAM algorithm is longer than CEAGC algorithm, the improvement of CEICAM is quite obvious. It is worth sacrificing some time complexity for higher accuracy.

## 5 Conclusion

For the clustering ensemble algorithm based on co-association matrix, the problem that ignoring base clustering quality will result in unreasonable division. In the proposed ensemble clustering algorithm CEICAM, the quality of the base clustering results is incorporated into each element of co-association matrix through the normalized average entropy. The positive impact of high-quality base clustering is improved, and the influence of some low-quality base clustering members is reduced. The comparative experiments on artificial data set and UCI data sets show that CEICAM algorithm can effectively improve the accuracy of clustering ensemble. But the improvement of *Accuracy* and *NMI* index brings more time consumption. In order to improve the clustering efficiency, the parallel algorithm of clustering ensemble based on an improved co-association matrix will be researched in the near future.

**Acknowledgments.** This work is supported by National Natural Science Foundation of China (No.61973151), the Natural Science Foundation of Jiangsu Province (BK20191406), Postgraduate Research and Practice Innovation Program of Jiangsu Province (KYCX20_1328).

## References


1. Strehl, A., Ghosh, J.: Cluster ensembles-a knowledge reuse framework for combining multiple partitions. J. Mach. Learn. Res. **3**(1), 583–617 (2003)
2. Fred, A.: Finding consistent clusters in data partitions. In: Kittler, J., Roli, F. (eds.) MCS 2001. LNCS, vol. 2096, pp. 309–318. Springer, Heidelberg (2001). https://doi.org/10.1007/3-540-48219-9_31
3. Zhong, C., Yue, X., Zhang, Z., et al.: A clustering ensemble: two-level-refined co-association matrix with path-based transformation. J. Pattern Recog. **48**(8), 2699–2709 (2015)
4. Xu, L.: Research on clustering ensemble algorithm based on granular computing. D. China Univ. Mining Technol. (2018)
5. Wang, Y., Wang, C.Y., Shen, M.: Improved bee colony optimization clustering ensemble joint similarity recommendation algorithm. J. Comput. Eng. **46**(10), 88–94+102 (2020)
6. Yang, Y., Jiang, J.: Hybrid sampling-based clustering ensemble with global and local constitutions. IEEE Trans. Neural Netw. Learn. Syst. **27**(5), 952–965 (2016)
7. Huang, H., Wang, C., Xiong, Y., et al.: A weighted K-means clustering method integrating the distance between clusters and within clusters. J. Compu. Appl. **42**(12), 2836–2848 (2019)
8. Wang, T.: Research on Clustering Ensemble Algorithm Based on Co-Association Matrix. D. ShanXi University (2020)
9. Galar, M., Fernandez, A., Barrenechea, E., Bustince, H., Herrera, F.: A review on ensembles for the class imbalance problem: bagging-, boosting-, and hybrid-based approaches. IEEE Trans. Syst. Man, Cybern., Part C (Applications and Reviews) **42**(4), 463–484 (2012)
10. Gomes, H.M., Barddal, J.P., Enembreck, F., Bifet, A.: A survey on ensemble learning for data stream classification. ACM Comput. Surv. **50**(2), 1–36 (2017). https://doi.org/10.1145/3054925
11. Akram, M., Luqman, A., Kenani A.N.: Certain models of granular computing based on rough fuzzy approximations. J. Intell. Fuzzy Syst. **39**(3), 2797–2816 (2020)
12. Wang, Y., Zhang, H., Ma, X., et al.: Granular computing model for knowledge uncertainty. J. Software. **22**(04), 676–694 (2011)

# Optimal Design of Closed-Loop Supply Chain Network Considering Supply Disruption

Xiulei Liu(✉) and Yuxiang Yang

China Jiliang University, Hangzhou 310018, China

**Abstract.** Considering the supply interruption in supply chain network, a closed-loop supply chain network model with dual objectives including total profit and total carbon emissions based on recycling mechanism was proposed. In this paper, we considered not only the elasticity of supply chain network but also the recycling prices. The model investigated the impact of the recycling mechanism on the supply interruption. The numerical results revealed the relationship between the degree of supply disruption and the recovery price, as well as the influence of the loss of suppliers' supply capacity and the tolerance coefficient of shortage on the planning results.

**Keywords:** Closed-loop supply chain network · Supply disruption · Recovery mechanism

## 1 Introduction

In recent years, closed-loop supply chain network (CLSCN) has been increasingly researched in order to achieve the dual development goals of low-carbon and sustainability [1]. Compared with the traditional supply chain network, CLSCN further includes the recycling and remanufacturing of used products, so its design process is more complicated [2]. Due to the complexity of the market environment, enterprises need to consider various uncertainty factors while designing supply chain networks, and the supply disruption problem is one of them [3].

To address the supply disruption problem in the supply chain operation, Li et al. [4] proposed to force suppliers to restore supply capacity quickly by signing penalty clauses and providing financial assistance; In order to reduce the risks associated with supply chain network disruptions, Ruiz-Torres et al. [5] used decision tree approach to analyze all possible supply disruption scenarios. Xiao et al. [6] constructed a three-level supply chain network optimization model with contingency capability for the uncertainty of node supply and demand. Di and Wang [7] considered the case of multi-level node disruptions, they built an inventory decision model with the objective of minimizing cost. Rohaninejad et al. [8] developed a mixed-integer nonlinear programming model to optimize a supply chain network where disruptions may occur at all levels of facilities. Aiming at the problem that both the market demand and the supply of remanufactured products are disturbed, Huang and Wang [9] established a model to study how manufacturers should make pricing and production decisions. When facing the issue of interrupt,


Q. Han et al. (Eds.): LSMS 2021/ICSEE 2021, CCIS 1469, pp. 816–825, 2021.
https://doi.org/10.1007/978-981-16-7213-2_79

in order to cope with random demand, Giri and Sharma [10] established models to solve the best production quantity and inventory decisions; Ghomi-Avili et al. [11] considered both pricing decisions and possible disruptions in their CLSCN design. It can be seen that the methods proposed in these studies mainly involve production decisions, inventory decisions, and market pricing decisions for used products [12]. However, few scholars have considered the impact of the recycling scale of used products on supply disruptions.

When the forward supply chain is interrupted, the recycling of used products is an important part of CLSCN design, whether it can effectively bridge the gap of forward supply disruption is well worth studying. In this paper, a closed-loop supply chain network model based on recycling mechanism is designed. Considering that the recycling price is the most important factor affecting the scale of recycling, so the recycling price of used products is taken as a decision variable. And under the dual objectives including total carbon emission and total profit optimization, the relationship between the degree of supply disruption and recycling price is studied, and the impact of the recovery mechanism on the design of the closed-loop supply chain network is analyzed. Because the tolerance coefficient of shortage can reflect the elasticity of supply chain network, it is used as a constraint to analyze the impact of elasticity constraints on the planning results.

## 2 Model Description and Assumptions

This paper studies the facility location of the five-level supply chain network composed of $I$ suppliers, $J$ manufacturers, $M$ distributors, $N$ customer demand points, $K$ recycling centers, and the product transportation among nodes. In this supply chain network, the facilities of manufacturers, recycling centers need to be located, and the locations of these candidate facility nodes which are known. As shown in Fig. 1, the suppliers supply the raw materials to the manufacturers, the manufacturers sell the finished products to the consumers through the distributors. The recycling centers recycle the used products from the consumers and decompose them, and then resupply the available parts to the manufacturers for remanufacturing.

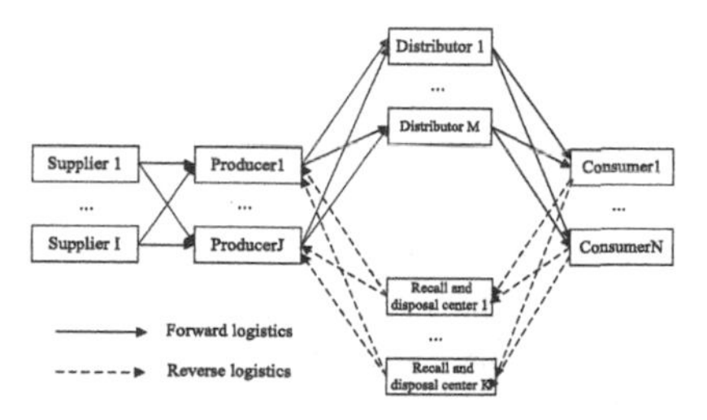


**Fig. 1.** Closed-loop supply chain network

Suppliers are vulnerable to supply interruption caused by external factors. This paper wants to solve this problem by establishing a closed-loop supply chain network with recovery mechanism. The design idea of recycling mechanism is shown in Fig. 2. Firstly, a closed-loop supply chain network optimization model with recycling mechanism is established, and then the recycling price of used products is taken as the decision variable. By analyzing the relationship between the degree of supplier supply capacity loss and recovery price, a reasonable pricing strategy is formulated to stimulate the recovery scale. Finally, according to the loss degree of supply capacity, the reasonable network facility layout and recovery scale are determined to reduce the interruption risk caused by the supply shortage of upstream suppliers. The model needs to follow the following assumptions [13].

(1) The remanufactured products and new products produced by the manufacturer have the same performance and price.
(2) The spare parts produced by the recycling center are meeting the production standards.
(3) The quantity of used products returned is mainly affected by the recycling price, and there is a positive linear correlation between them.
(4) The productivity of the candidate manufacturers is limited, the capacity of distributors and recycling centers is limited, and the supply capacity of suppliers is limited.

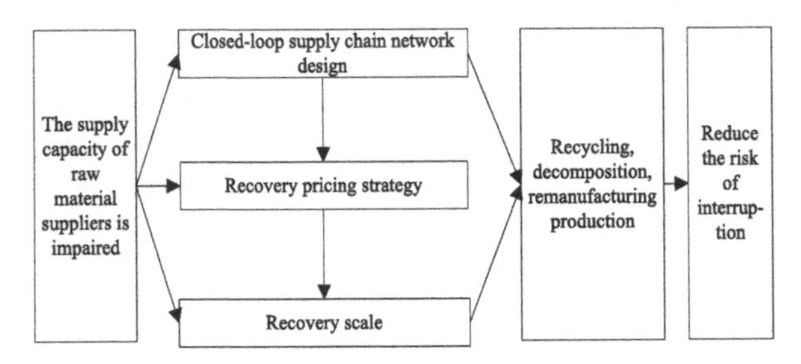


**Fig. 2.** Design idea diagram of recovery mechanism

## 3 Symbol Description

(1) Node set:
$I(i = 1, 2...\hat{i})$: set of locations for the suppliers; $J(j = 1, 2...\hat{j})$: set of candidate locations for manufacturers; $M(m = 1, 2...\hat{m})$: set of locations for distributors; $N = (n = 1, 2...\hat{n})$: set of locations for consumer groups; $K(k = 1, 2...\hat{k})$: set of candidate locations for recycling centers.
(2) Decision variables:
$X_j$: 0–1 integer variable, 1 if a factory is constructed in candidate location $j$, 0 otherwise; ($Y_k$ is the same). $q_{ij}$: quantity of raw materials shipped from supplier $i$ to manufacturer $j$; ($q_{jm}$, $q_{mn}$, $q_{rk}$, $q_{kj}$ are the same). $c$: recycling price of unit product by the recycling disposal center.

(3) Other parameters:
$Q$: total demand, obey the mean distribution of expectations. $R$: total profit. $p$: price of the product. $F_j$: fixed cost for building manufacturer $j$; ($F_k$ is the same). $h_m$: the cost of storing a product by distributor $m$. $c_i$: unit cost of raw materials supplied by supplier $i$. $u$: the cost of producing a product. $r$: the cost of recovering and disposing a product. $o$: unit out-of-stock cost. $t_{ij}$: unit transportation cost from supplier $i$ to manufacturer $j$; ($T_{jm}$, $T_{mn}$, $T_{nk}$, $T_{kj}$ are the same). $T_{ij}$: distance between supplier $i$ and manufacturer $j$; ($T_{\mathrm{jm}}$, $T_{\mathrm{mn}}$, $T_{\mathrm{nk}}$, $T_{\mathrm{kj}}$ are the same). $\varepsilon$: disposal efficiency of used products. $\alpha$, $\beta$: represents the correlation coefficient of the linear relationship between the recovery quantity and the recovery price. $\delta$: the loss ratio of supplier's supply capacity. $\rho$: tolerance coefficient for shortage. $S_i$: the maximum supply capacity of supplier $i$; ($S_j$, $S_{\mathrm{m}}$, $S_k$ are the same).

(4) Carbon emission related parameter:
$co$: carbon emissions required for one product transportation; $co_j$: carbon emissions for building manufacturer $j$; $co_j$: carbon emissions for producing one new product by manufacturer $j$; $co_m$: carbon emissions for storing one product by distributor $m$; $co_k$: carbon emissions for building recycling center $k$; $co^k$: carbon emission required for unit product recovered and treated by the recycling center $k$; *Tax*: carbon tax rate.

## 4 Mathematical Modelling

(1) Basic Model

The objective function, total profit to be maximized is thus as given in (1):

$$\max z_1 = \sum_M\sum_N q_{mn}p - [\sum_J F_jX_j + \sum_K F_kY_k + \sum_J\sum_M (u+\mathrm{h}_m)q_{jm} + \sum_N\sum_K (r+c)q_{nk} + \sum_I\sum_J q_{ij}c_i + (Q - \sum_M\sum_N q_{mn})o + \sum_I\sum_J t_{ij}T_{ij}q_{ij} + \sum_J\sum_M t_{jm}T_{jm}q_{jm} + \sum_M\sum_N t_{mn}T_{mn}q_{mn} + \sum_N\sum_K t_{nk}T_{nk}q_{nk} + \sum_K\sum_J t_{kj}T_{kj}q_{kj}]. \tag{1}$$

In the objective function, the first item is the total sales revenue, the second is the fixed construction cost of manufacturers, the third is the fixed construction cost of the recycling center, the fourth is the production cost of manufacturers and the inventory cost of distributors, the fifth is the recovery cost in the process of recycling and disposal cost, the sixth is supply cost, the seventh is shortage cost, items 8 to 12 are transportation cost between nodes.

Constraint conditions:

$$\sum_J q_{ij} \le (1-\delta)\ \ S_i \quad \forall i \in I. \tag{2}$$

$$\sum_I q_{ij} + \sum_K q_{kj} \ge \sum_M q_{jm},\ \forall j \in J. \tag{3}$$

$$\sum_{J} q_{jm} \geq \sum_{N} q_{mn}, \quad \forall m \in M. \tag{4}$$

$$\varepsilon \sum_{N} \sum_{K} q_{nk} \geq \sum_{K} \sum_{J} q_{kj}. \tag{5}$$

$$\alpha c + \beta \geq \sum_{N} \sum_{K} q_{nk}. \tag{6}$$

$$p \geq c + r. \tag{7}$$

$$\sum_{M} q_{jm} \leq S_j X_j, \forall j \in J. \tag{8}$$

$$\sum_{N} q_{nk} \leq S_k Y_k, \forall k \in K. \tag{9}$$

$$\sum_{J} q_{jm} \leq S_m, \forall m \in M. \tag{10}$$

$$1 - \frac{\sum_{M} \sum_{N} q_{mn}}{Q} \leq \rho. \tag{11}$$

$$\begin{aligned} & q_{ij}, q_{jm}, q_{mn}, q_{nk}, q_{kj} \geq 0, \forall i \in I, j \in J, \\ & m \in M, n \in N, k \in K; c \geq 0; X_j, Y_k \in \{0, 1\}, \quad \forall j \in J, k \in K. \end{aligned} \tag{12}$$

Constraint (2) indicates the limit of the quantity supplied by the supplier suffering from the loss of supply capacity; constraint (3) indicates the limit of the manufacturer's production quantity; constraint (4) indicates the limit of the sales volume of the distributor; constraint (5) indicates that only part of the recycled old products can be remanufactured; constraint (6) indicates the linear constraint between the recovered price and the recycled quantity; constraint (7) indicates that the total cost of recovery and disposal of unit product should be less than the sales price of unit product; constraint (8), constraint (9) and constraint (10) are the maximum capacity constraints of each facility; constraint (11) is a supply chain elastic constraint; constraint (12) is a decision variable constraint.

(2) Objective Function of Carbon Emission

The carbon emission objective function is as follows:

$$\begin{aligned} \min z_2 = & \sum_{J} co_j X_j + \sum_{J} \sum_{M} q_{jm} co^j + \sum_{J} \sum_{M} q_{jm} co_m + \sum_{K} Y_k co_k \\ & + \sum_{N} \sum_{K} q_{nk} co^k + co[\sum_{I} \sum_{J} T_{ij} q_{ij} + \sum_{J} \sum_{M} T_{jm} q_{jm} + \sum_{M} \sum_{N} T_{mn} q_{mn} + \\ & \sum_{N} \sum_{K} T_{nk} q_{nk} + \sum_{K} \sum_{J} T_{kj} q_{kj}]. \end{aligned} \tag{13}$$

In the formula, the first is the carbon emissions during the construction of the manufacturer, the second is the carbon emissions during the production of the products, the third is the carbon emissions of the finished products in the storage process, the fourth is the carbon emissions during the construction of the recycling center, the fifth is the carbon emissions generated by the recycling of waste products, and the sixth to the tenth are the carbon emissions generated by the transportation process between the nodes.

(3) Total Objective Function

By introducing the carbon tax rate into the model, according to formula (1) and formula (13), the total objective function can be obtained as the following formula.

$$\max z = z_1 - z_2 Tax. \tag{14}$$

## 5 Numerical Examples

An electrical appliance manufacturer wants to recover and remanufacture used products by constructing a closed-loop supply chain. There are three suppliers, six distributors and four consumer groups, with fixed and known locations; in addition, the location of the manufacturer and the recycling center needs to be determined, there are four known manufacturer candidate locations and six recycling center candidate locations, other parameters are shown in Table 1 [2, 14, 15].

**Table 1.** Values of model parameters

| Parameter | Value | Parameter | Value |
|---|---|---|---|
| $Q$ | U (36500,92400) | $t_{ij}$ | 0.021 |
| $P$ | 1160 | $T_{ij}$ | U (32,150) |
| $F_j$ | 5000000 | $\varepsilon$ | U (0.3,0.6) |
| $F_k$ | 2500000 | $\delta$ | U (0.1,0.6) |
| $h_m$ | 1.2 | $Tax$ | 0.39 |
| $c_i$ | 410 | $S$ | U (15000, 40000) |
| $u$ | 75 | $co$ | 120 |
| $r$ | 25 | $co_j \backslash co_k$ | 54000000 |
| $o$ | 600 | $co^j$ | 3000 |
| $\alpha$ | 212 | $co_m$ | 1200 |
| $\beta$ | 3080 | $co^k$ | 1800 |

### 5.1 Comparative Analysis of Recovery Mechanism Model and Non-recovery Mechanism Model

The tolerance coefficient of shortage $\rho$ is set to 0.3. The maximum profit with and without the recovery mechanism are calculated when facing different supply capacity loss ratios. The results are shown in Fig. 3. It can be seen that the total profit of the supply chain in both states tends to decrease as the supply capacity loss ratio $\delta$ increases. However, the profit after establishing the recovery mechanism is significantly larger than the profit without the recovery mechanism, and the gap between the total profits in the two states also increases with the increase of $\delta$, this indicates that when suppliers' supply capacity is impaired, the method of establishing the recycling center to recycle used products can reduce the risk of supply disruption.

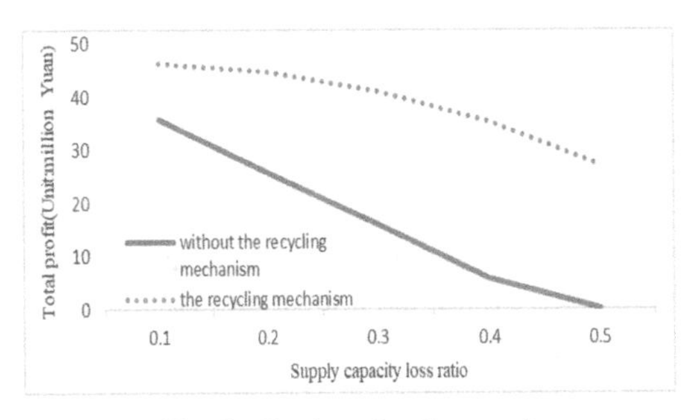


**Fig. 3.** Total profit of network

### 5.2 Impact of Supply Capacity Loss Ratio on Planning Results

The $\rho$ is set to 0.3. The planning results corresponding to different supply capacity loss ratios are calculated as shown in Table 2. It can be seen that the change of the supply capacity loss ratio $\delta$ has a great impact on the site selection results. As $\delta$ increases, the manufacturer's site selection results increase from 1, 4 to 2, 3, 4 and then decrease to 2, 3. The reason is when the supplier's supply capacity loss is small, recycling used products ensures the supply of raw materials, and the manufacturers can build plants to expand production to meet market demand. However, as suppliers' supply capacity continued to decline, the scale of recycling reached a certain limit again. When the supply of raw materials is no longer able to meet the production needs, the only way to reduce the loss of total profit is to reduce the construction of unnecessary manufacturers, thus reducing the total cost.

On the other hand, as $\delta$ increases, the result of recycling center site selection changes from 4 to 1, 2, 3, 4, 6 and reaches a steady state. This is because in order to make up for the shortage of raw material supply, the only way is to establish more recycling centers to recycle more used products for remanufacturing. However, when $\delta$ is greater than 0.5, the result of recycling center location will stabilize. That is because the utility of the recycling mechanism is maximized. The marginal utility of increasing the scale of

**Table 2.** Optimal results under different supply capacity loss ratios

| Supply capacity loss ratio $\rho$ | Total profit/Yuan | Location decision | |
|---|---|---|---|
| | | J | K |
| 0.1 | 46192000 | 1 4 | 4 |
| 0.2 | 44686000 | 1 3 4 | 3 4 |
| 0.3 | 41018000 | 1 3 4 | 3 4 6 |
| 0.4 | 35197000 | 2 3 4 | 3 4 6 |
| 0.5 | 27235000 | 2 3 4 | 1 2 3 4 6 |
| 0.6 | 17515000 | 2 3 | 1 2 3 4 6 |

recycling by increasing the recycling price, and thus reducing the economic profit loss caused by stockouts, has tended to zero.

In order to investigate the relationship between the supplier's supply capacity loss ratio $\delta$ and the recycling price of used products, the maximum profit of the supply chain under different supply capacity loss ratios is calculated (set $\rho = 0.3$) and the corresponding recycling price of used products is found. The results are shown in Fig. 4. It can be seen that when $\delta$ is less than 0.5, there is a general linear relationship between $\delta$ and the recovery price. This has great reference value for enterprises. For example, when the supply of raw materials in the forward supply chain is disturbed, the recycling price of used products can be flexibly adjusted according to the degree of supply capacity loss of upstream suppliers, so that the recycling quantity can reach a desirable scale, and then the parts obtained from the decomposition process are supplied to the manufacturer for reproduction.

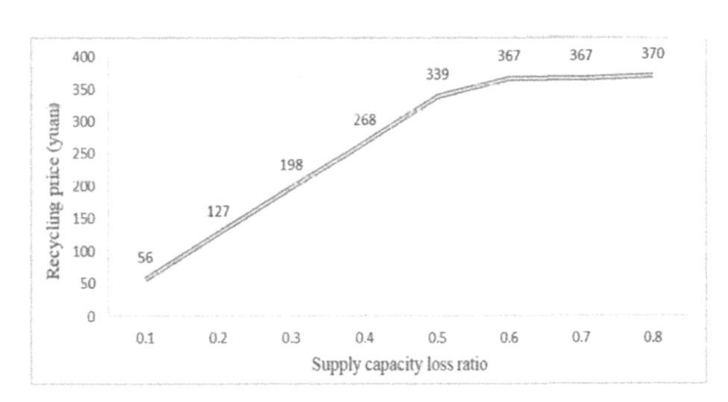


**Fig. 4.** Relationship between supply capacity loss ratio and recovery price

However, the recycling mechanism is not always effective. From the data in Fig. 4, it can be seen that when the supply capacity loss ratio is too large, that is, when $\delta > 0.5$, the recovery price will tend to be stable. In contrast, it can be seen from Table 2 that the decline of the total profit increases after $\delta > 0.5$. This suggests that there is a limit to the utility of establishing a recycling mechanism to mitigate supply disruptions. On the one hand, it is due to the limitation of the processing capacity of the recycling center itself, and on the other hand, it is due to the increase of the production cost caused by the increase of the recycling price. When the production cost increases to a certain extent, it will inevitably erode part of the profit of the supply chain.

### 5.3 The Influence of Tolerance Coefficient of Shortage on Planning Results

When the supply capacity loss ratio $\delta$ is the same, the change of $\rho$ has no effect on the final location result. In order to further study the influence of $\rho$ on supply chain network planning, we set the supply capacity loss ratio $\delta$ to 0.3 and calculate the total profit of the supply chain for different $\rho$. The results are shown in Fig. 5. It can be seen that the lower the tolerance coefficient of shortage is not the better, $\rho$ is changed from 0.1 to 0.2, which is equivalent to give up some customers actively, but the total profit of the whole supply chain has increased; the higher tolerance coefficient of shortage is not the better, for example, when $\rho$ is higher than 0.3, the amount of shortage allowed by enterprises is too large, profits will be reduced, and at the same time, it will cause a large number of customers loss.

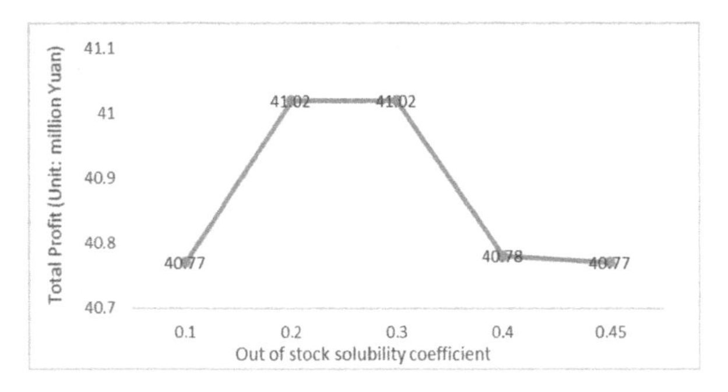


**Fig. 5.** Total profit of supply chain under different $\rho$

## 6 Conclusion

In this paper, to address the supply disruption problem in the supply chain network design process, recycling price is introduced as a decision variable and a closed-loop supply chain network design model with recycling mechanism is established. The relationship between the supply capacity loss ratio of upstream suppliers and the recycling price is explored, as well as the superiority of establishing a recycling mechanism model. The tolerance coefficient of shortage, which can reflect the elasticity of the supply chain network, is introduced into the model, and its impact on the planning results is analyzed. The results of the analysis have certain practical significance.

Of course, this paper is too idealistic about the assumption of a linear relationship between the recovery price and the recovery size. In reality, the relationship between them may be more complicated. Therefore, future research can consider to introduce other factors into the model to make the model optimization results more reliable.

**Acknowledgements.** The authors acknowledge the support from National Natural Science Foundation of China (no. 71972172), Humanity and Social Science Planning Foundation of Ministry of Education of China (no.19YJA630101), Philosophy and Social Sciences Planning Foundation of Zhejiang Province (no. 20NDJC114YB).

## References

1. Ma, Z.J., Hu, S., Dai, Y., Ye, Y.-S.: Pay-as-you-throw versus recycling fund system in closed-loop supply chains with alliance recycling. J. Int. Trans. Oper. Res. **25**, 1811–1829 (2016)
2. Li, J.: Multi-objective robust fuzzy optimization problem for closed-loop supply chain network design under low-carbon environment. J. Control Decis. **33**, 29–300 (2018)
3. Liu, Y., Ma, L., Liu, Y.: A novel robust fuzzy mean-upm model for green closed-loop supply chain network design under distribution ambiguity. J. Appl. Math. Model. **92**, 99–135 (2020)
4. Li, Y., Zhen, X., Qi, X., Cai, G.: Penalty and financial assistance in a supply chain with supply disruption. J. Omega **61**, 167–181 (2016)
5. Ruiz-Torres, A.J., Mahmoodi, F., Zeng, A.Z.: Supplier selection model with contingency planning for supplier failures. J. Ind. Eng. **66**(2), 374–382 (2013)
6. Xiao, J.H., Liu, X., Shang, S., Chen, P.: Optimization model and algorithm of elastic supply chain network based on node failure and demand uncertainty. J. Stat. Decis. **34**, 50–53 (2018)
7. Di, W.M., Wang, R.: Supply chain location-inventory decision model and its optimization algorithm with multi-echelon facility disruptions. J. Comput. Integr. Manuf. Syst. **27**(01), 270–283 (2021)
8. Rohaninejad, M., Sahraeian, R., Tavakkoli-Moghaddam, R.: An accelerated benders decomposition algorithm for reliable facility location problems in multi-echelon networks. J. Ind. Eng. **124**, 523–534 (2018)
9. Huang, Y., Wang, Z.: Demand disruptions, pricing and production decisions in a closed-loop supply chain with technology licensing. J. Cleaner Prod. **191**, 248–260 (2018)
10. Giri, B.C., Sharma, S.: Optimal production policy for a closed-loop hybrid system with uncertain demand and return under supply disruption. J. Cleaner Prod. **112**, 2015–2028 (2016)
11. Ghomi-Avili, M., Jalali-Naeini, S.G., Tavakkoli-Moghaddam, R., Jabbarzadeh, A.: A fuzzy pricing model for a green competitive closed-loop supply chain network design in the presence of disruptions. J. Cleaner Prod. **188**, 425–442 (2018)
12. YuChi, Q.L., He, Z.W., Wang, N.M.: Integration optimization research on location-inventory-routing problem considering shortage strategy in closed-loop supply chain. J. Oper. Res. Manage. Sci. **30**, 53–60 (2021)
13. Yu, H., Solvang, W.D.: A fuzzy-stochastic multi-objective model for sustainable planning of a closed-loop supply chain considering mixed uncertainty and network flexibility. J. Cleaner Prod. **266**, 1–20 (2020)
14. Hu, H.T., Bian, Y.Y., Guo, S.Y., Wang, S.A., Yan, W.: Research on supply chain network optimization with the relationship of pricing and demand. J. Chin. J. Manage. Sci. **28**, 165–171 (2020)
15. Azaron, A., Venkatadri, U., Doost, A.F.: A stochastic approach for designing supply chain networks under uncertainty. J. IFAC-PapersOnLine. **51**, 1465–1469 (2018)

# Author Index

Printed in the United States
by Baker & Taylor Publisher Services